Control Systems Engineering Principles and Practice

This textbook reimagines control systems instruction by integrating theoretical precision with practical understanding through carefully crafted real-world analogies, extensive worked examples, and hands-on reinforcement. It offers a transformative approach to control systems education that unites mathematical rigor with intuitive clarity.

Control Systems Engineering: Principles and Practice stands out for its innovative approach and practical focus. It offers intuitive explanations through extensive real-life analogies, making complex concepts such as poles, zeros, characteristic modes, and advanced topics like internal stability and nonminimum-phase systems accessible and relatable. Practical reinforcement is provided through hundreds of worked examples and numerical problems, along with end-of-chapter multiple-choice questions that enable self-assessment and knowledge verification. Ready-to-use simulation codes allow readers to experiment immediately, bridging the gap between theory and practice effortlessly. The book balances mathematical depth with clear, engaging prose, presenting rigorous content in a way that respects both the complexity of the subject and the intelligence of the reader. Additionally, its experience-driven pedagogy systematically addresses common learning obstacles, drawing on years of classroom-tested teaching insights to enhance understanding.

This textbook is ideal for students learning control theory for the first time and professionals seeking to deepen their expertise. It covers foundational topics like open-loop and closed-loop architectures and progresses to advanced methods such as state-space techniques and optimal control strategies.

Control Systems Engineering Principles and Practice

Classical and Modern Control

Akshya Swain

CRC Press
Taylor & Francis Group
Boca Raton London New York

CRC Press is an imprint of the
Taylor & Francis Group, an **informa** business

Designed cover image: Akshya Swain

First edition published 2026
by CRC Press
2385 NW Executive Center Drive, Suite 320, Boca Raton, FL 33431

and by CRC Press
4 Park Square, Milton Park, Abingdon, Oxon, OX14 4RN

CRC Press is an imprint of Taylor & Francis Group, LLC

ISBN: 978-1-032-94484-5 (hbk)
ISBN: 978-1-032-94538-5 (pbk)
ISBN: 978-1-003-57130-8 (ebk)

DOI: 10.1201/9781003571308

Typeset in CMR10 font
by KnowledgeWorks Global Ltd.

You are the sandalwood,
I am the water that carries its fragrance.

You are the pearl,
I am the thread; in your association, I find my value.

You are the brilliant lamp,
I am the wick that glows in your flame.

You are the inspired writer,
I am the pen that gives life to your words.

To the Supreme Lord and my Spiritual Masters,
whose wisdom and subtle guidance brought this book to completion

To my parents,

Annapurna *and* ***Alekh Chandra****,*

and my parents-in-law,

Suprava *and* ***Bairagi Charan****,*

with love and gratitude.

Contents

Preface

The rapid advancement of Artificial Intelligence (AI), Machine Learning (ML), and data-centric technologies is transforming the landscape of engineering practice. While these developments offer unprecedented opportunities, they also underscore the continuing need for engineers to master the foundational tools required to model, analyze, and control complex physical systems in a mathematically rigorous and structured manner. This book was written in response to that need and to the exciting challenges that lie at the intersection of traditional control theory and modern technological innovation.

In our interconnected world, control systems are the invisible engines driving everything from the precision robotics manufacturing your smartphone to the autonomous systems navigating our skies and roads. The same fundamental principles that govern a simple DC motor also orchestrate the complex dance of thousands of variables in a smart power grid or enable a surgical robot to perform delicate operations with superhuman precision. Understanding these principles isn't just academically satisfying; it's the key to shaping the future of technology.

Why Master the Fundamentals in a Digital Age?

Learning basic control concepts without relying solely on computer tools like MATLAB® and Simulink is essential for developing a strong foundation in control systems engineering. While these tools are invaluable for solving complex problems quickly and accurately, an over-reliance on them can hinder a deep understanding of fundamental principles. By working through problems manually, students and engineers gain insight into the underlying mathematics, physical interpretations, and practical limitations of control systems.

This hands-on approach fosters intuition about system behavior, understanding viscerally how poles and zeros affect system response, why certain feedback configurations lead to instability, and how performance trade-offs manifest in real designs. Moreover, this conceptual foundation ensures that when you do use advanced computational tools, you're wielding them with expertise rather than blind faith. In critical moments, whether troubleshooting a malfunctioning system, designing under constraints, or innovating beyond existing paradigms, this deep understanding becomes your most valuable asset.

A Bridge Between Worlds

Control Systems Engineering: Principles and Practice[1] seeks to offer a balanced and integrated treatment of both classical and modern control theory. It is grounded in the belief that a deep understanding of the physics of dynamic systems alongside the formal methodologies of control engineering remains essential, especially in an era increasingly dominated by data-driven approaches. The ability to synthesize classical insight with modern computational tools is becoming a defining feature of competent engineers and system designers.

This textbook recognizes that today's control engineer must be bilingual: fluent in both the time-tested methods of classical control and the sophisticated state-space techniques that enable modern multi-variable system design. Whether you're stabilizing an aircraft, optimizing a chemical process, or designing the next generation of autonomous vehicles, you'll need both the intuitive insights of frequency-domain analysis and the analytical power of state-space methods.

The Journey Through Control Theory

This book takes you on a carefully structured journey that mirrors how control theory itself evolved. We begin with a review of fundamental concepts in signals and systems, introducing essential ideas such as system dynamics, modes, poles, and eigenvalues, the building blocks that will appear throughout your engineering career.

The exploration continues through the modeling of physical systems using transfer functions and differential equations, followed by detailed discussions on time-domain response, stability analysis, and steady-state error. Here you'll develop the intuition that distinguishes excellent control engineers: the ability to predict system behavior and design specifications before running a single simulation.

Root locus and frequency-domain techniques follow, providing you with the graphical insights and design tools that have guided control engineers for decades. These classical methods remain relevant because they offer unparalleled insight into the fundamental trade-offs in control design;the eternal tension between performance and robustness, speed and stability. The final chapters present a structured approach to modern control design, utilizing state-space representations, encompassing modeling, analysis, and linear state-feedback design. Here, the diverse practical systems you encountered earlier, from DC motors to biological processes, unite under a common mathematical framework, revealing the elegant universality of control theory.

[1] A companion volume focusing on digital control and nonlinear systems is under development. It extends the foundational material presented in this book and provides advanced tools for modern control design and analysis.

Bridging Theory and Practice

To facilitate comprehension, each chapter is supported by numerous illustrative examples, exercises, and conceptual highlights. MATLAB codes and solutions to selected problems are made available on the companion website, enhancing both learning and practical application. Importantly, control theory is presented not as a purely mathematical discipline, but as an engineering framework rooted in physical intuition, feedback principles, and real-world performance trade-offs.

Key Features

- Unified treatment of classical and modern control techniques
- Emphasis on modeling physical systems using both transfer function and state-space approaches
- Structured coverage of time-domain, frequency-domain, and state-space analysis tools
- Analogies from real-life examples to explain abstract control concepts
- Conceptual clarity reinforced through examples, MCQs, and practice problems
- Practical orientation linking theory to engineering design and control implementation
- Balanced approach combining manual problem-solving with computational tools (MATLAB/Simulink)

Supplementary Materials

A companion website provides comprehensive supplementary resources to support learning and teaching. Solutions to all concept reinforcement problems are available, along with MATLAB files used in generating figures and implementing controller design procedures presented in the book. Solutions to selected knowledge assessment problems are also provided. An Instructor's Manual with complete solutions to all knowledge assessment problems is available to qualified instructors and can be obtained by contacting the publisher.

Feedback from readers is most welcome, and any errors or clarifications will be addressed via the book's errata page hosted on its companion website.

Looking Forward

As we advance into a future defined by cyber-physical systems, distributed intelligence, and collaborative robotics, the foundations in this book become our essential toolkit for navigating an increasingly complex technological landscape. The state-space representations that describe dynamic systems are the same mathematical frameworks that will govern smart cities, renewable energy grids, and precision medicine. The control strategies you develop here will empower you to architect tomorrow's autonomous fleets, space exploration missions, and adaptive manufacturing ecosystems.

I hope this work will serve as a valuable resource for students beginning their journey into control systems and for practicing engineers seeking a practical reference in an evolving field. More than that, I hope it inspires you to see control theory not just as a collection of mathematical techniques, but as a powerful lens for understanding and shaping the dynamic world around us.

Akshya Swain
The University of Auckland,
New Zealand

Gratitude

A book is never the product of an author's effort alone. It takes a collective contribution from many individuals who inspire, guide, and support the journey from concept to completion. It is a pleasure to acknowledge the invaluable support I have received from them throughout this process.

The idea of writing this book had always been at the back of my mind. However, the credit must go to my brother-in-law, Nilasundar, who lives in New Jersey, USA, with his wife, Sandhya, and their children, Sumedha and Tuhin. His persistent encouragement was the catalyst for this book. Without his continuous prodding, this idea might still be just an idea.

My interest in control theory stems from Professor A. K. Mohanty, who was instrumental in sparking my passion for the field. He, along with his wife, Mrs. Saroj Mohanty, provided parental guidance and care during the early stages of my academic career, for which I am deeply grateful.

I would like to express my gratitude to one of my former PhD scholars, Faizal Hafiz, for his valuable assistance during various stages of the manuscript preparation. Whenever technical challenges related to software or formatting issues arose, his expertise and willingness to help were invaluable. I am thankful to Gunawan Dewantoro and Chathura Wanigasekara for taking the time to review the initial version of the manuscript.

I would also like to extend my gratitude to Partha Roop, who is not only my colleague but also like a younger brother to me. I deeply appreciate the invaluable support he, along with his wife Bindu, has extended to my family since I joined the University of Auckland.

I wish to acknowledge the unwavering support of my academic colleague Dr. Nitish Patel, with whom I have had the privilege of sharing teaching responsibilities in Control Engineering courses at The University of Auckland for over two decades. I also extend my sincere appreciation to my other colleagues, both past and present, whose support and indirect contributions deserve special mention, including Zoran Salcic, John Boys, Kevin Sowerby, Karl Stol, Abhisek Ukil, Jagdish Patra, and B. Gangadhar Prusty.

During this period, my regular visits to cafes at Auckland Airport and Mission Bay with Samaya Muduli, CEO, CredenceSoft, provided the required strength at the right time. While it is impossible to name everyone, I would like to mention a few, including Avishek Mishra, Derick Singh, and Sailaja Rajaretnam. I would like to express my appreciation to all who have shown continuous interest and enthusiasm in the progress of this book.

I extend my gratitude to the esteemed reviewers of this book, whose constructive feedback has greatly enhanced its quality and clarity: James Peyton Jones (USA), Eduardo Mendes (Brazil), Subhas Mukhopadhyaya (Australia), N. Premchand Mahalik (USA), and Bidyadhar Subudhi (India).

Special thanks to one of my former students, Laxmidhar Behera, Director, IIT Mandi, and Anjan Swain for suggesting to publish this book with CRC Press and for introducing me to the right team. I am thankful to Marc Gutierrez and Jennifer Petruzziello of CRC Press for their support and guidance in bringing this book to fruition. Their professionalism, expertise, and commitment to excellence have been instrumental in shaping this work.

To my son, Ayaskanta, my hero, who fills my life with joy and inspires me profoundly. Through him, I have gained invaluable insights into controlling even the most uncontrollable systems. As a firmware engineer, his expertise and mentorship have guided me in many technical aspects of my work. His enduring support during the preparation of this manuscript has been nothing short of priceless. To my daughter-in-law, Thussheeta, whose boundless curiosity and infectious joy have been a source of inspiration and a driving force in the completion of this book. The recent arrival of our beloved grandson, Aashay, has brought immense happiness and a renewed sense of purpose to our lives. His presence is a precious reminder of the legacy we pass on through love, learning, and the pursuit of meaningful work.

In my tradition, husband and wife are regarded as inseparable partners. This belief holds especially true in my life, as this book is as much a creation of my wife, Kabyashree, as it is of me. Her love, patience, and steadfast support have been the cornerstone of this work. I am profoundly grateful for her enduring belief in me and her remarkable tolerance of my idiosyncrasies over the years. This book is a testament to the love and support she and all my cherished loved ones have so generously bestowed upon me.

Lastly, to my readers, thank you for your interest in this book. I hope you find the principles and practices discussed within both informative and inspiring.

Akshya Swain
The University of Auckland,
New Zealand

About the author

Akshya Swain is a leading expert in control systems and nonlinear system identification, with over thirty years of experience in teaching and research. For over two decades, he has been affiliated with the University of Auckland, New Zealand, where he leads the Control Systems Research Group. His research spans both theoretical foundations and real-world applications of control engineering, and his work has been widely published in leading international journals and conferences.

Chapter 1

Signals, Systems, and Control: Introductory Concepts

A signal without noise is like a joke without a punchline: technically perfect but completely unrealistic. A system without feedback is like texting without read receipts—you never know if it landed.

DESIRED LEARNING OUTCOMES

Upon completion of this chapter, readers should be able to:

- Classify different types of signals, and distinguish between power and energy signals, odd and even signals, and periodic and aperiodic signals.
- Explain the importance of the impulse and exponential functions in control and signal processing.
- Understand the concept of complex frequency and the complex frequency plane (s-plane).
- Differentiate between various types of systems.
- Compute the time response of linear time-invariant (LTI) systems.
- Derive the transfer function of LTI systems from their differential equation models.
- Interpret the physical significance of poles and zeros.
- Analyze the effects of feedback on various system properties, including sensitivity.

DOI: 10.1201/9781003571308-1

1.1 Introduction

The study of signals and systems is fundamental in engineering and science. Signals and systems serve as the backbone of engineering, providing the theoretical foundation and analytical tools necessary for understanding, designing, and implementing a wide range of systems, including electrical and electronic systems. Application of signals and systems theory is not limited to any specific field of engineering; it has wide-ranging applications in various domains. Many physical phenomena can be represented and analyzed using signals and systems concepts, making them indispensable in diverse areas of science and engineering. Control systems utilize signals and systems theory to regulate the behavior of dynamic systems. A thorough understanding of signals and systems is therefore essential for students and practitioners in the field of control engineering. This chapter is intended to provide a basic treatment of signals and systems at an introductory level, which serves as a basis for understanding other topics covered in subsequent chapters.

1.2 What Is a Signal?

It is difficult to find a unique definition of *signal*. However, a workable definition that suits most of our purposes is:

Definition 1.1. *A signal conveys information about a physical phenomenon which evolves in time or space.*

Signals can be found everywhere in our day-to-day life. A signal is a physical representation of a phenomenon that evolves in time (temporal phenomena) or space (spatial phenomena). Some examples of temporal signals include audio signals (such as speech or music), video signals (such as moving images), and sensor data (such as temperature or pressure measurements over time). In these cases, the signal's amplitude or other characteristics vary with time, allowing us to capture and analyze dynamic behaviors. On the other hand, signals which convey information about spatial attributes, such as location, distance, shape, and arrangement of objects, are spatial signals. Some examples of spatial signals include those used in remote sensing (e.g., electromagnetic waves for imaging), global positioning systems (e.g., GPS signals for location), and spatial data collected through techniques such as lidar or sonar.

1.2.1 Size of a Signal

The "size" of a signal is a multifaceted concept that can encompass various aspects of the signal's characteristics, such as amplitude, energy, power, and duration, depending on the specific context and the desired measure. Each aspect provides valuable information about the signal's characteristics and behavior, and understanding these aspects is essential for signal analysis, processing, and interpretation in various applications. Common measures of signal size include energy and power.

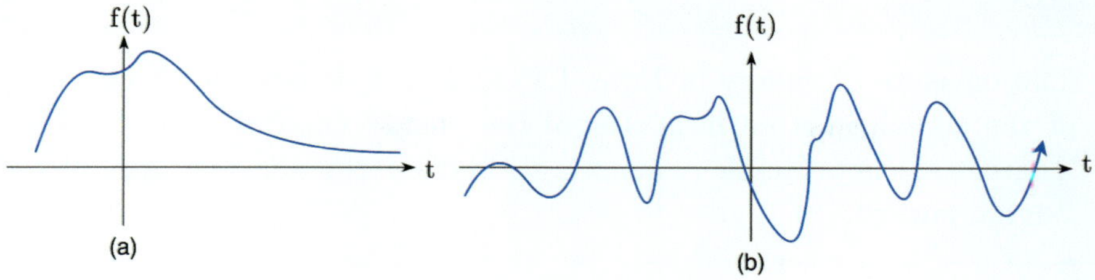

Figure 1.1: (a) Energy signal, (b) Power signal

Signal Energy:

Definition 1.2. The energy of a real-valued signal x(t) is defined as

$$E = \int_{-\infty}^{\infty} x^2(t)dt \tag{1.1}$$

For complex-valued signal,

$$E = \int_{-\infty}^{\infty} |x(t)|^2 dt \tag{1.2}$$

Note that if signal energy is to be considered as a meaningful measure of the signal size, it must be finite. A necessary condition for the energy to be finite is that the amplitude must tend to zero as time approaches infinity (see Figure 1.1a). However, in certain cases, when the amplitude of x(t) does not tend to zero as $|t| \to \infty$, the signal energy becomes infinite (see Figure 1.1b). In such a case, a more meaningful measure of signal size is the time average of the energy, provided it exists. This measure is referred to as the signal's power.

Signal Power:

Definition 1.3. The power of a real-valued signal x(t) is defined as

$$P = \lim_{T\to\infty} \frac{1}{T} \int_{-T/2}^{T/2} x^2(t)dt \tag{1.3}$$

For complex-valued signal,

$$P = \lim_{T\to\infty} \frac{1}{T} \int_{-T/2}^{T/2} |x(t)|^2 dt \tag{1.4}$$

Remark

The measure of energy in Eqns(1.1) and (1.2) is not the actual energy of the signal; it is an indicator of the energy capability of the signal. Therefore, the principle of conservation of energy does not apply to this "signal energy."

1.3 Classification of Signals

Although the signals can be classified into several classes, we classify them into the following five classes based on different features.

1. Continuous-time and discrete-time signals
2. Analog and digital signals
3. Periodic and aperiodic signals
4. Energy and power signals
5. Deterministic and random signals

Continuous-Time and Discrete-Time Signals

Definition 1.4.

A signal x(t) is said to be a *continuous-time* signal if it is defined for all time t.

A signal x(t) is said to be a *discrete-time* signal if it is defined only at discrete instants of time t.

Figure 1.2(a) shows a continuous-time signal whose amplitude varies continuously with time. The discrete-time signal is obtained by sampling a continuous-time signal. In practice, the sampling is carried out at a uniform rate. If T denotes the sampling period and k denotes an integer, which may assume both positive and negative values, then sampling the continuous-time signal x(t) at time $t = kT$ gives sample of value x(kT). The discrete time signal is usually represented by dropping the sampling time T as:

$$x(k) = x(kT), \quad k = 0, \pm 1, \pm 2, \ldots$$

The relationship between a continuous-time signal x(t) and the discrete-time signal $x(k) = x(kT)$ is shown in Figure 1.2b.

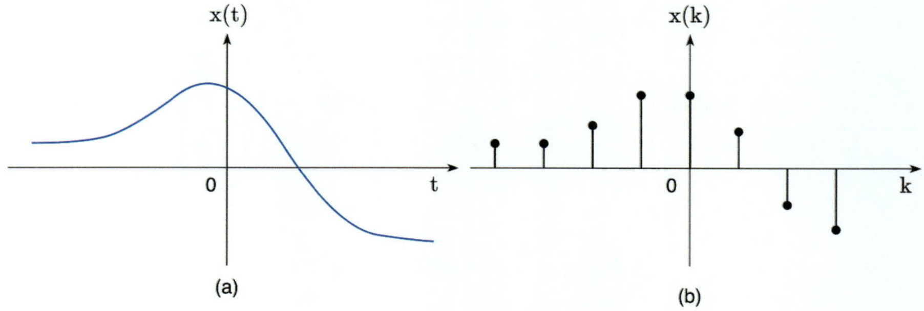

Figure 1.2: (a) Continuous-time signal x(t) and (b) Discrete equivalent of x(t) as x(k)

Even and Odd Signals

Definition 1.5.

A continuous-time signal x(t) is said to be an even signal if it satisfies the condition

$$x(-t) = x(t) \quad \text{for all t} \tag{1.5}$$

Definition 1.6.

A continuous-time signal x(t) is said to be an odd signal if it satisfies the condition

$$x(-t) = -x(t) \quad \text{for all t} \tag{1.6}$$

Examples of even signals are shown in Figure 1.3, and odd signals are shown in Figure 1.4.

Properties of Even Signal

1. The even signals are symmetrical about the vertical axis.
2. The value of an even signal at time (t) is the same as at time (-t).

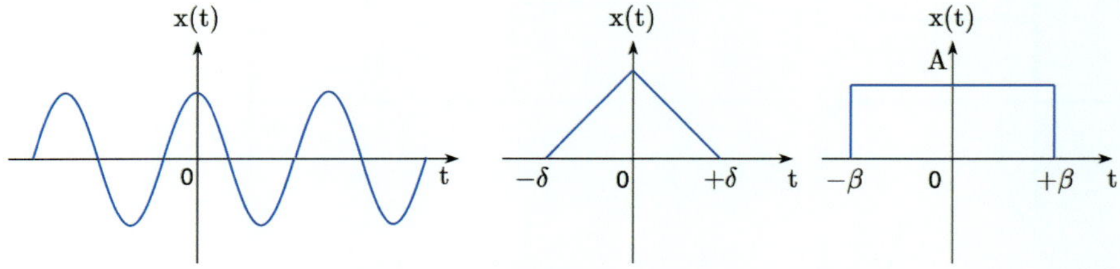

Figure 1.3: Even signal

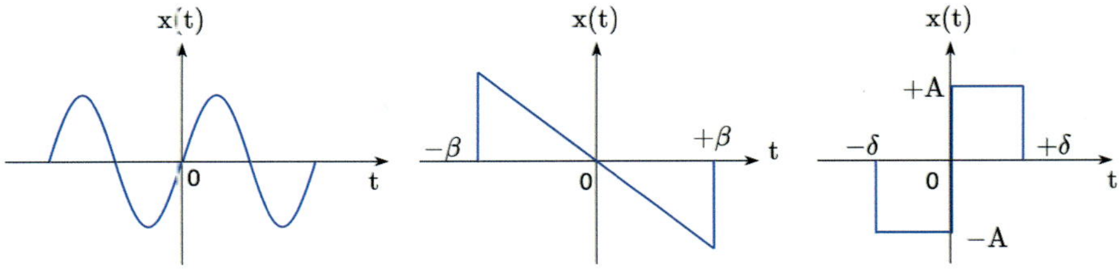

Figure 1.4: Odd signal

3. The even signal is identical to its reflection about the origin.
4. The area under the even signal is twice its one-sided area.

Properties of Odd Signal

1. The odd signal is anti-symmetric about the origin.
2. The value of odd signal at time (t) is negative of its value at time (−t) for all t, i.e., $-\infty < \mathrm{t} < \infty$.
3. The odd signal must necessarily be zero at time $\mathrm{t} = 0$ to hold $\mathrm{x}(0) = -\mathrm{x}(0)$.
4. Area under the odd signal is always zero.

Note: A continuous-time signal is said to be neither even nor odd if it does not satisfy the condition of the even signal or that of the odd signal. Some examples of such signals (neither even nor odd) are shown in Figure 1.5.

Consider an example of the exponentially damped signal $\mathrm{x(t)} = \mathrm{e}^{-\mathrm{at}}$. Now

$$\mathrm{x(t)}= \mathrm{e}^{-\mathrm{at}}, \quad \mathrm{x(-t)}= \mathrm{e}^{\mathrm{at}} \quad \text{and} \quad -\mathrm{x(t)}= -\mathrm{e}^{-\mathrm{at}}$$

Thus $\mathrm{x(t)} \neq \mathrm{x(-t)}$ and $\mathrm{x(-t)} \neq -\mathrm{x(t)}$. Hence, it is neither an even nor an odd signal.

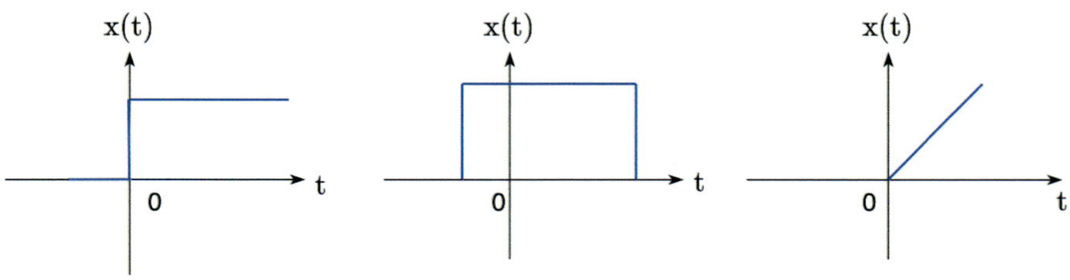

Figure 1.5: Neither even nor odd signal

Theorem 1.1.

Any signal x(t) can be decomposed into even and odd parts as:

$$x(t) = x_e(t) + x_o(t) \tag{1.7}$$

where

$$x_e(t) = \frac{x(t) + x(-t)}{2} \quad \text{and} \quad x_o(t) = \frac{x(t) - x(-t)}{2} \tag{1.8}$$

EXAMPLE 1.1

Problem

Decompose the following signals into odd and even parts.

a. $x(t) = \left(a + t^3\right)\sin^3(bt)$
b. $x(t) = a\cos t + b\sin t + c\sin t\cos t$

SOLUTION:

a. Now $x(t) = \left(a + t^3\right)\sin^3(bt)$. Hence
$x(-t) = \left(-a + t^3\right)\sin^3(bt)$.
It can be decomposed into even and odd parts as:

$$x_e(t) = \frac{x(t) + x(-t)}{2} = t^3\sin^3(bt)$$
$$x_o(t) = \frac{x(t) - x(-t)}{2} = a\sin^3(bt)$$

b. Now $x(t) = a\cos t + b\sin t + c\sin t\cos t$.
Hence $x(-t) = a\cos t - b\sin t - c\sin t\cos t$. The even and odd components are

$$x_e(t) = \frac{x(t) + x(-t)}{2} = a\cos t$$
$$x_o(t) = \frac{x(t) - x(-t)}{2} = b\sin t + c\sin t\cos t$$

CONCEPT REINFORCEMENT PROBLEM 1.1

Problem

Decompose the following signals into odd and even parts.

a. $x(t) = e^{-at}\sin t$

b. $x(t) = a_0 + a_1 t + a_2 t^2 + a_3 t^3 + a_4 t^4$

Periodic and Aperiodic Signals

Definition 1.7.

A signal x(t) is said to be *periodic signal* if for some positive constant T,

$$x(t) = x(t+T) \quad \text{for all } t \tag{1.9}$$

where T is a positive constant.

Definition 1.8.

A signal x(t) is said to be an *aperiodic signal* if there exists no value of T which satisfies Eqn(1.9).

The smallest value of T which satisfies Eqn(1.9) is called the *fundamental period* of x(t). The reciprocal of the fundamental period T is called the *fundamental frequency* and is denoted as f, where

$$f = \frac{1}{T} \tag{1.10}$$

EXAMPLE 1.2

Problem

For each of the following signals, determine whether it is periodic or aperiodic. If periodic, find the period.

a. $x(t) = 5 + \cos(10t)$

b. $x(t) = 10\sin^2(at)$

c. $x(t) = e^{-4t}$

d. $x(t) = e^{j\left(8t+\frac{\pi}{2}\right)}$

e. $x(t) = e^{-4t}\cos(20t)$

SOLUTION:

a. It is periodic with period $T = \frac{2\pi}{10} = \frac{\pi}{5}$ sec.

b. It is periodic with period $T = \frac{\pi}{a}$ sec.

$\left(\text{Hint: } \sin^2(x) = \frac{1 - \cos 2x}{2}\right)$.

c. This is an exponentially damped signal and is therefore aperiodic.

d. This is a complex exponential function. It is periodic with period $T = \frac{2\pi}{8} = \frac{\pi}{4}$ sec.

e. This is aperiodic, because the signal $\cos(20t)$ is multiplied by a decaying exponential.

Let the periods of continuous time signals $x_1(t)$ and $x_2(t)$ be equal to T_1 and T_2, respectively. Then, the signal $x(t)$, which is a linear combination of $x_1(t)$ and $x_2(t)$, will be periodic if and only if there exist integers m and k such that

$$mT_1 = kT_2 \quad \text{and} \quad \frac{T_1}{T_2} = \frac{k}{m} = \text{rational number} \tag{1.11}$$

The fundamental period of $x(t)$ is $mT_1 = kT_2$ provided that the values of m and k are chosen such that the greatest common divisor (gcd) between m and k is 1.

EXAMPLE 1.3

Problem

For each of the following signals, determine whether it is periodic or aperiodic. If periodic, find the period.

a. $x(t) = 7\sin(6\pi t) + 10\sin(4\pi t)$

b. $x(t) = 5\cos(10\pi t) + 11\cos(12t)$

SOLUTION:

a. Let $x_1(t) = 7\sin(6\pi t)$ and $x_2(t) = 10\sin(4\pi t)$. The fundamental period T_1 of $x_1(t)$ and T_2 of $x_2(t)$ are

$$T_1 = \frac{2\pi}{6\pi} = \frac{1}{3}, \quad T_2 = \frac{2\pi}{4\pi} = \frac{1}{2}$$

The ratio $\frac{T_1}{T_2} = \frac{2}{3}$ is a rational number. Hence $x(t)$ is a periodic signal. The fundamental period of the signal $x(t)$ is obtained from the relation $3T_1 = 2T_2$, which gives 1sec.

b. Let $x_1(t) = 5\cos(10\pi t)$ and $x_2(t) = 11\cos(12t)$. The fundamental periods T_1 of $x_1(t)$ and T_2 of $x_2(t)$ are

$$T_1 = \frac{2\pi}{10\pi} = \frac{1}{5}, \quad T_2 = \frac{2\pi}{12} = \frac{\pi}{6}$$

The ratio $\frac{T_1}{T_2} = \frac{6}{5\pi}$ is not a rational number. Hence x(t) is an aperiodic signal.

1.3.1 Energy and Power Signal

Definition 1.9.

A signal x(t) is referred to as an *energy signal* if and only if the total energy of the signal satisfies the condition

$$0 < E < \infty$$

A signal x(t) is referred to as a *power signal* if and only if the average power of the signal satisfies the condition

$$0 < P < \infty$$

Key Points

a. All practical signals have finite energy; thus, they are energy signals.

b. In practice, the physical generation of a power signal is impossible since it requires infinite duration and infinite energy.

c. All finite-duration signals of finite amplitude are energy signals.

d. Both energy and power signals are mutually exclusive, i.e., no signal can be both a power signal and an energy signal.

e. A signal is neither an energy nor a power signal if both the energy and power of the signal are equal to infinity.

f. The sum of an energy signal and a power signal is a power signal.

g. A signal whose amplitude is constant over an infinite duration is a power signal.

h. The energy of a signal is not affected by time shifting and time inversion. It is only affected by the time scaling.

EXAMPLE 1.4

Problem

For each of the following signals, determine whether it is an energy signal, a power signal, or neither.

a. $x(t) = A\cos\omega t$
b. $x(t) = e^{-at}u(t)$, where $u(t)$ is unit step signal $a > 0$
c. $x(t) = tu(t)$

SOLUTION:

a. The energy of the signal is

$$E = \int_{-\infty}^{\infty} |x(t)|^2 dt = \int_{-\infty}^{\infty} A^2\cos^2(\omega t)dt = \infty$$

The signal $x(t)$ is periodic with $T = 2\pi/\omega$. Hence, the average power of the signal is

$$P = \frac{1}{T}\int_0^T |x(t)|^2 dt = \frac{\omega}{2\pi}\int_0^{2\pi/\omega} A^2\cos^2(\omega t)dt$$

$$= \frac{A^2\omega}{2\pi}\int_0^{2\pi/\omega}\frac{1}{2}[1+\cos 2\omega t]\,dt$$

$$= \frac{A^2\omega}{2\pi}\left[\int_0^{2\pi/\omega}\frac{1}{2}dt + \int_0^{2\pi/\omega}\frac{1}{2}\cos 2\omega t dt\right]$$

$$= \frac{A^2}{2} < \infty$$

Thus, $x(t)$ is a power signal.

b. Now, let us calculate the energy of the signal

$$E = \int_{-\infty}^{\infty}[x(t)]^2\,dt = \int_0^{\infty}\left[e^{-at}\right]^2 dt = \int_0^{\infty} e^{-2at}dt = \frac{1}{2a} < \infty$$

Thus, $x(t) = e^{-at}u(t)$ is an energy signal.

c. Here $x(t) = tu(t)$ is a ramp signal. Its energy and power are

$$E = \lim_{T\to\infty}\int_{-T/2}^{T/2}|x(t)|^2 dt = \lim_{T\to\infty}\int_0^{T/2} t^2 dt = \lim_{T\to\infty}\frac{1}{3}\left(\frac{T}{2}\right)^3 = \infty$$

$$P = \lim_{T\to\infty}\frac{1}{T}\int_{-T/2}^{T/2}|x(t)|^2 dt = \lim_{T\to\infty}\frac{1}{T}\int_0^{T/2} t^2 dt$$

$$= \lim_{T\to\infty}\frac{1}{T}\frac{1}{3}\left(\frac{T}{2}\right)^3 = \lim_{T\to\infty}\frac{T^2}{24} = \infty$$

Since both the energy and power are infinite, the ramp signal is neither an energy nor a power signal.

CONCEPT REINFORCEMENT PROBLEM 1.2

Problem

For each of the following signals, determine whether it is an energy signal, a power signal, or neither.

a. $x(t) = u(t)$, the unit step signal
b. $x(t) = e^{-at}$, considering following two cases
   i. a is real
   ii. a is imaginary

Deterministic and Random Signals

Definition 1.10.

A signal x(t) is said to be a *deterministic signal* if its value at any point in time can be defined precisely either by a mathematical equation or by a graph.

Definition 1.11.

A signal x(t) is said to be a *random signal* if its value at any point in time cannot be defined precisely but are known only in terms of probabilistic description, such as mean value, mean squared value, and so on.

1.4 Elementary Functions: Impulse and Exponential Functions

In the following, we will discuss the two most common candidates for elementary signals, which are used for linear systems analysis: the unit impulse function and the exponential function.

1.4.1 The Unit Impulse Function $\delta(t)$

The *unit impulse* function $\delta(t)$, also called as *delta function* or the *Dirac delta function* (after the name of P.A.M. Dirac who first defined it), is defined as

$$\begin{aligned} &\delta(t) = 0, \quad t \neq 0 \\ &\int_{-\epsilon}^{\epsilon} \delta(\tau)\, d\tau = 1, \quad \text{for any real number } \epsilon > 0 \end{aligned} \tag{1.12}$$

The first condition implies that the unit impulse function is zero for all values of t except for $t = 0$, where it is infinitely high. The second condition states that the area under the impulse is 1, so $\delta(t)$ has unit area. Thus, an ideal impulse signal is of zero width and infinite amplitude. While this signal cannot be

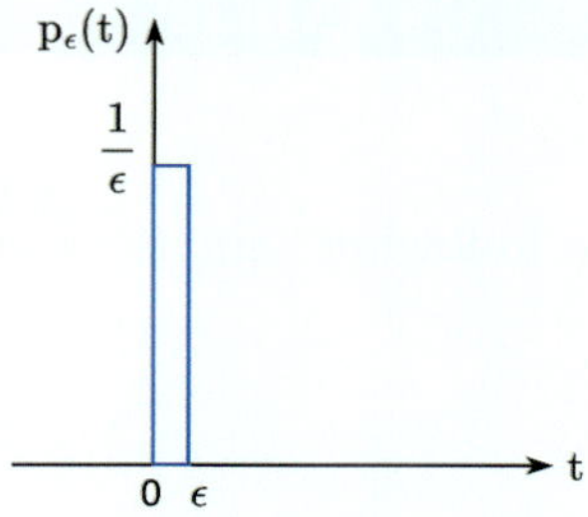

Figure 1.6: An impulse signal

physically generated, it is an essential tool in system analysis for determining the impulse response, which completely characterizes a linear time-invariant system. Usually, it is approximated by a pulse of large amplitude and narrow width and is defined as:

$$p_\epsilon(t) = \begin{cases} 0, & \text{if } t < 0. \\ \dfrac{1}{\epsilon} & 0 \le t < \epsilon \\ 0 & t \ge \epsilon \end{cases} \tag{1.13}$$

This is a pulse of duration ϵ and amplitude $\dfrac{1}{\epsilon}$, as shown in Figure 1.6.

The unit impulse function $\delta(t)$ can be defined to be the limit of the signal in Eqn(1.14) as $\epsilon \to 0$:

$$\delta(t) = \lim_{\epsilon \to 0} p_\epsilon(t) \tag{1.14}$$

Multiplication of a Function by an Impulse

Let us see what happens when we multiply the unit impulse $\delta(t)$ by a function $f(t)$ which is assumed to be continuous at $t = 0$. Since the impulse exists only at $t = 0$, and the value of $f(t)$ at $t = 0$ is $f(0)$, we get the following relation

$$\boxed{f(t)\delta(t) = f(0)\delta(t)} \tag{1.15}$$

Similarly, if $f(t)$ is multiplied by an impulse $\delta(t - T)$, then

$$\boxed{f(t)\delta(t - T) = f(T)\delta(t - T)} \tag{1.16}$$

where we assume that $f(t)$ is continuous at $t = T$.

EXAMPLE 1.5

Problem

Simplify each of the following signals using the relation in Eqns(1.15) and (1.16).

a. $\left(\frac{\cos t}{t^2+5}\right)\delta(t)$

b. $\left(\frac{a+j\omega}{\omega^2+b}\right)\delta(\omega)$

c. $\left[e^{-2t}\cos(5t-60^0)\right]\delta(t)$

d. $\left(\frac{a}{c+j\omega}\right)\delta(\omega+5)$

SOLUTION:

a. Now $\left(\frac{\cos t}{t^2+5}\right)\delta(t) = f(t)\delta(t) \quad \Longrightarrow \quad f(t) = \frac{\cos t}{t^2+5}$

Using Eqn(1.15), this is equal to

$$f(0)\delta(t) = \frac{\cos(0)}{0+5}\delta(t) = \frac{1}{5}\delta(t)$$

b. $\left(\frac{a+j\omega}{\omega^2+b}\right)\delta(\omega) = f(\omega)\delta(\omega) \quad \Longrightarrow \quad f(\omega) = \frac{a+j\omega}{\omega^2+b}$

This is simplified to $f(0)\delta(\omega) = \frac{a+j.0}{0+b} = \frac{a}{b}\delta(\omega)$

c. $\left[e^{-2t}\cos(5t-60^0)\right]\delta(t) = f(t)\delta(t) \quad \Longrightarrow \quad f(t) = e^{-2t}\cos(5t-60^0)$.

It is equal to $f(0)\delta(t) = \cos(-60^0)\delta(t) = \frac{1}{2}\delta(t)$

d. Now $\left(\frac{a}{c+j\omega}\right)\delta(\omega+5) = f(\omega)\delta(\omega-T)$.

This implies $f(\omega) = \frac{a}{c+j\omega}$ and $T=-5$.

Using Eqn(1.16), this is simplified to

$$f(T)\delta(\omega-T) = f(-5)\delta(\omega+5) = \frac{a}{c-j5}\delta(\omega+5)$$

1.4.2 Representation of Signals in Terms of Impulses

Let us define p(t) to be a pulse of unit length and width $\Delta\tau$, starting at $t = 0$ as shown in Figure 1.7(a).

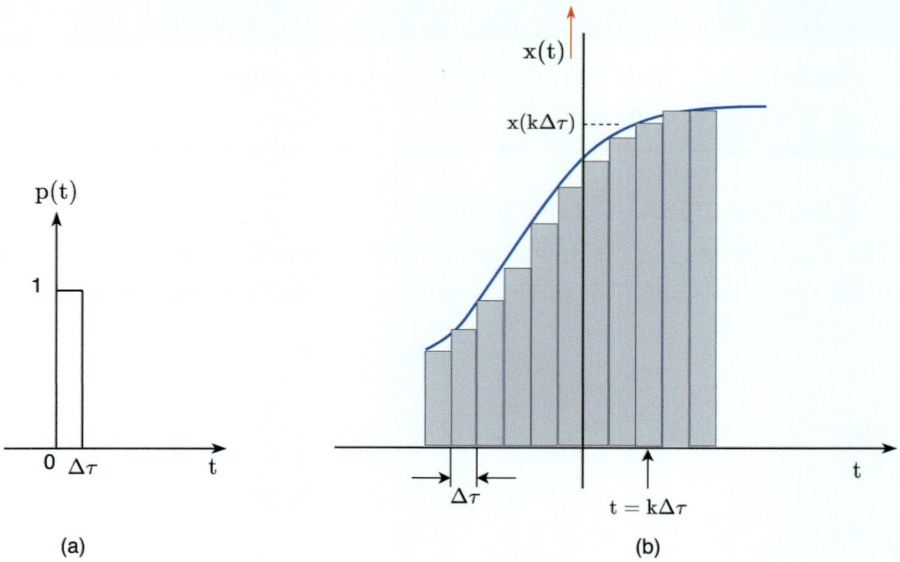

Figure 1.7: Representation of an arbitrary input as a weighted sum of impulses

Consider an arbitrary signal x(t) which can be approximated by a sum of rectangular pulses of width $\Delta\tau$ and of varying heights as the one shown in Figure 1.7(b). The pulse starting at $t = k\Delta\tau$ has a height $x(k\Delta\tau)$, and is expressed as:

$$\begin{aligned} x(t) &= \lim_{\Delta\tau \to 0} \sum_{\tau} x(k\Delta\tau)p(t - k\Delta\tau) \\ &= \lim_{\Delta\tau \to 0} \sum_{\tau} \left[\frac{x(k\Delta\tau)}{\Delta\tau}\right] p(t - k\Delta\tau)\Delta\tau \end{aligned} \tag{1.17}$$

Note that the term $[x(k\Delta\tau)/\Delta\tau]p(t - k\Delta\tau)$ represents a pulse $p(t - k\Delta\tau)$ with height $[x(k\Delta\tau/\Delta\tau]$. As $\Delta\tau \to 0$, the height of this strip $\to \infty$; however, its area remains $x(k\Delta\tau)$. Hence, this strip approaches an impulse $x(k\Delta\tau)\delta(t - k\Delta\tau)$ as $\Delta\tau \to 0$. Hence,

$$x(t) = \lim_{\Delta\tau \to 0} \sum_{\tau} x(k\Delta\tau)\delta(t - k\Delta\tau)\Delta\tau \tag{1.18}$$

Further, as $\Delta\tau \to 0$, the summation in Eqn(1.18) becomes an integral and hence

$$x(t) = \int_{-\infty}^{\infty} x(\tau)\delta(t - \tau)d\tau \tag{1.19}$$

Note

An arbitrary signal can be expressed as a weighted superposition (sum) of time-shifted impulses

1.4.3 The Exponential Function e^{st}

The role of exponential functions in control systems and signal processing is significant. This signal is of the form e^{st} and has complex frequency s where

$$s = \sigma + j\omega \tag{1.20}$$

Thus

$$e^{st} = e^{(\sigma + j\omega)t} = e^{\sigma t}e^{j\omega t} = e^{\sigma t}\left(\cos\omega t + j\sin\omega t\right) \tag{1.21}$$

If $s^* = \sigma - j\omega$ denotes the complex conjugate of s, then

$$e^{s^* t} = e^{(\sigma - j\omega)t} = e^{\sigma t}e^{-j\omega t} = e^{\sigma t}\left(\cos\omega t - j\sin\omega t\right) \tag{1.22}$$

Using Euler's formulae,

$$e^{\sigma t}\cos\omega t = \frac{1}{2}\left(e^{st} + e^{s^* t}\right) \quad \text{and} \quad e^{\sigma t}\sin\omega t = \frac{1}{2j}\left(e^{st} - e^{s^* t}\right) \tag{1.23}$$

Key Takeaway

The exponential function e^{st} of complex frequency $s = \sigma + j\omega$ describes an exponentially varying sinusoid

This function encompasses a large class of functions. For example, the following functions can be considered as special cases of the exponential signal e^{st}.

1. A constant $K = Ke^{0t}$, $(s = 0)$
2. A monotonically decreasing exponential $e^{-\sigma t}$, $(\omega = 0, s = -\sigma)$
3. A monotonically increasing exponential $e^{\sigma t}$, $(\omega = 0, s = +\sigma)$
4. A pure sinusoid $\cos\omega t$ $(\sigma = 0, s = \pm j\omega)$
5. An exponentially varying sinusoid $e^{\sigma t}\cos\omega t$ $(s = \sigma \pm j\omega)$

These signals are illustrated in Figure 1.8.

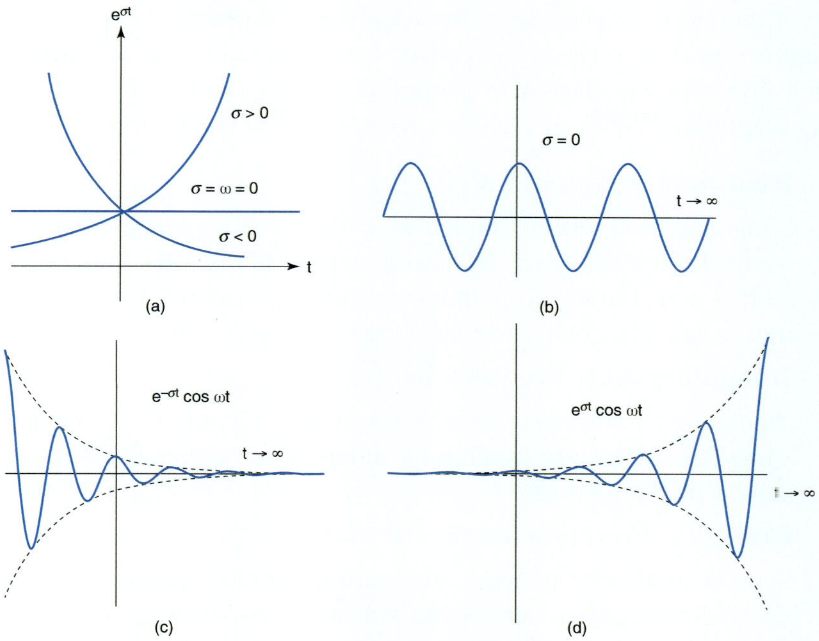

Figure 1.8: Complex frequency signals

1.4.4 Complex Frequency Plane

The complex frequency s is represented on a *Complex frequency plane*, called s-plane as shown in Figure 1.9.

The horizontal axis is the real axis and is denoted as σ-axis, and the vertical axis is the imaginary axis, denoted as $j\omega$-axis. The real part σ of the complex frequency s is called the **neper frequency** and provides information about the

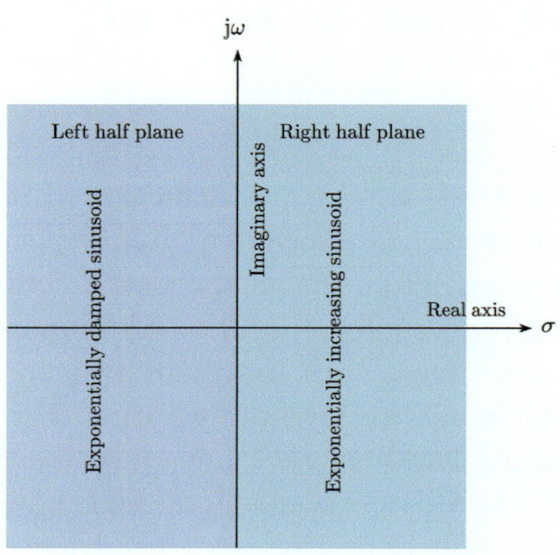

Figure 1.9: Complex frequency plane

rate of increase or decrease of the amplitude of complex exponential e^{st}. The imaginary part ω of the complex frequency s is called the **radian frequency** which gives the oscillation frequency of e^{st}. As discussed before, signals can be characterized by their complex frequencies in the s-plane:

a. **Real-axis Frequencies** ($\omega = 0$)**:**

Signals with complex frequencies on the real-axis represent either monotonically increasing or decreasing exponentials, depending on the sign of σ. If $\sigma > 0$, the signal is monotonically increasing; If $\sigma < 0$, the signal is monotonically decreasing, as shown in Figure 1.8a.

b. **Imaginary-axis Frequencies** ($\sigma = 0$)**:**

Signals with complex frequencies on the imaginary axis represent pure sinusoids. These are oscillatory signals with a frequency of ω rad/sec, as shown in Figure 1.8b.

c. **Complex Frequencies** ($\sigma \neq 0$ **and** $\omega \neq 0$)**:**

Signals with both non-zero real and imaginary parts represent exponentially decreasing or increasing sinusoids, depending on the sign of σ. If $\sigma < 0$, the signal is an exponentially decreasing sinusoid (see Figure 1.8c); if $\sigma > 0$, the signal is an exponentially increasing sinusoid (see Figure 1.8d).

1.5 What Is a System?

The term *system* is derived from the Greek word *systema*, which means an organized relationship among functioning units or components. It is often used to describe any orderly arrangement of ideas or constructs. According to the International Council on Systems Engineering (INCOSE)[89]

> **Definition 1.12.** *A system is an arrangement of parts or elements that together exhibit behavior or meaning that the individual constituents do not.* The elements, or parts, can include people, hardware, software, facilities, policies, and documents; that is, all things required to produce systems-level results.

It isn't easy to give a single and precise definition of the term *system*, which will satisfy all perspectives, as its meaning often depends on the context and objectives of a given study. From the perspective of control engineering, a system can be understood as an interconnection of components working together to achieve specific objectives. Such a system is defined by its inputs, outputs, and the governing rules or laws of operation. For instance, in electrical systems, the operational laws include Ohm's law, which defines the voltage-current relationships for resistors, capacitors, and inductors, and Kirchhoff's laws, which dictate the interconnection rules for various electrical components. Similarly, in mechanical systems, Newton's laws of motion govern the relationships between forces, masses, and motion. These foundational laws form the

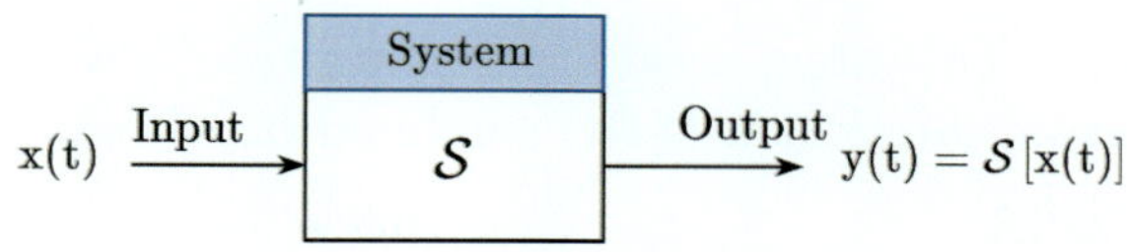

Figure 1.10: System as an operator

basis for deriving mathematical models of the respective systems, enabling their analysis and control.

System as an Operator

Definition 1.13.

A system is defined mathematically as a transformation which maps an input signal $x(t)$ into an output signal $y(t)$ as shown in Figure 1.10. For a continuous-time system, the input-output mapping is expressed as:

$$y(t) = \mathcal{S}[x(t)] \tag{1.24}$$

where $\mathcal{S}$ is an operator.

1.6 Classification of Systems

The basis of classifying systems is many. For example, they can be classified according to the **time frame** (*discrete, continuous, or hybrid*); **system complexity** (*physical, conceptual, and esoteric*); **uncertainties** (*deterministic and stochastic*); **nature and type of components** (*static or dynamic, linear or nonlinear, time-invariant or time variant, lumped or distributed, etc.*). In this book, we will broadly classify the systems into the following categories and discuss their characteristics.

1. Linear and nonlinear systems;
2. Time-invariant and time-varying systems;
3. Static (memory-less) and dynamic (with memory) systems;
4. Causal and Non-causal systems;
5. Lumped and distributed parameter systems;
6. Deterministic and stochastic systems;
7. Continuous and discrete systems

1.6.1 Linear and Nonlinear Systems

Consider a system which is excited by an input $x(t)$. Let the corresponding output be $y(t)$.

Definition 1.14. The system is said to be linear provided it satisfies *principle of superposition*, which is the combination of *additive* and *homogeneity* properties.

Additive Property: Assume that a system is initially at rest. Suppose an input $x_1(t)$ to this system produces an output $y_1(t)$ and an input $x_2(t)$ produces an output $y_2(t)$. If the system is linear, then application of the input $x_1(t) + x_2(t)$ will produce an output $y_1(t) + y_2(t)$. Thus if

$$x_1(t) \rightarrow y_1(t) \quad \text{and} \quad x_2(t) \rightarrow y_2(t), \quad \text{then}$$
$$x_1(t) + x_2(t) \rightarrow y_1(t) + y_2(t)$$

Alternatively we can express this property as follows. If

$$y_1(t) = \mathcal{S}\left[x_1(t)\right], \quad \text{and}$$
$$y_2(t) = \mathcal{S}\left[x_2(t)\right]$$

then due to *additive* property for linear systems,

$$\boxed{\mathcal{S}\left[x_1(t) + x_2(t)\right] = \mathcal{S}\left[x_1(t)\right] + \mathcal{S}\left[x_2(t)\right] = y_1(t) + y_2(t)} \tag{1.25}$$

Principle of Homogeneity or Scaling: Let an input (*cause*) x(t) produce an output (*effect*) y(t). If the system is linear, then scaling the input by a factor a will scale the output by the same factor. In other words, if x(t) produces y(t), then the scaled input ax(t) produces ay(t), where a can be any real or complex number. Thus, for a linear system

$$x(t) \rightarrow y(t) \quad \Longrightarrow \quad ax(t) \rightarrow ay(t)$$

In terms of the system operator, this is expressed as:

$$\boxed{\mathcal{S}\left[x(t)\right] \rightarrow y(t) \quad \Longrightarrow \quad \mathcal{S}\left[ax(t)\right] \rightarrow a\mathcal{S}\left[x(t)\right]} \tag{1.26}$$

Principle of Superposition: This combines both the additive and homogeneity properties of a linear system. It states that if the zero state response of a linear system to N-number of inputs $x_1(t)$, $x_2(t)$,..., $x_N(t)$ is $y_1(t)$, $y_2(t)$,..., $y_N(t)$, respectively, then the response of the system to any linear combination of these inputs

$$a_1x_1(t) + a_2x_2(t) + a_3x_3(t) + \cdots + a_Nx_N(t)$$

is the corresponding linear combination of the individual responses:

$$a_1y_1(t) + a_2y_2(t) + a_3y_3(t) + \cdots + a_Ny_N(t)$$

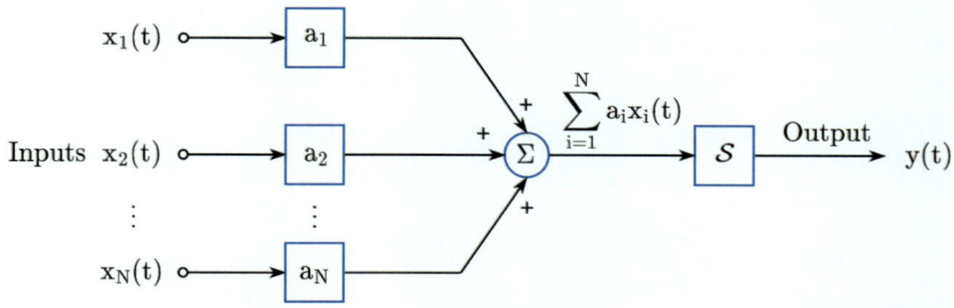

Figure 1.11: Linearity property: The summation precedes the system operator $\mathcal{S}$

where $a_1, a_2, \ldots, a_N$ are arbitrary constants (either real or imaginary). Thus, if

$$x_1(t) \rightarrow y_1(t), \;\; x_2(t) \rightarrow y_2(t), \ldots \; x_N(t) \rightarrow y_N(t),$$

then, for all $a_1, a_2, \ldots, a_N$

$$a_1x_1(t) + a_2x_2(t) + \cdots a_Nx_N(t) \rightarrow a_1y_1(t) + a_2y_2(t) + \cdots + a_Ny_N(t)$$

We will now alternately express using the system operator.

Consider a continuous-time system represented as an operator $\mathcal{S}[.]$. Let us apply a signal $x(t)$ to the input of this system, where

$$x(t) = a_1x_1(t) + a_2x_2(t) + \cdots + a_Nx_N(t) = \sum_{i=1}^{N} a_ix_i(t)$$

where $x_1(t), x_2(t), \ldots, x_N(t)$ are a set of input signals which are weighted by the constants $a_1, a_2, \ldots, a_N$, respectively. This gives the output as

$$\boxed{y(t) = \mathcal{S}\left[x(t)\right] = \mathcal{S}\left[a_1x_1(t) + \cdots + a_Nx_N(t)\right] = \mathcal{S}\left[\sum_{i=1}^{N} a_ix_i(t)\right]} \tag{1.27}$$

Equation (1.27) is schematically represented in Figure 1.11.

If the system is linear, then according to the principle of superposition

$$y(t) = a_1y_1(t) + a_2y_2(t) + \cdots + a_Ny_N(t) = \sum_{i=1}^{N} a_iy_i(t) \tag{1.28}$$

where $y_i(t) = \mathcal{S}\left[x_i(t)\right]$ is the output of the system due to input $x_i(t)$. Thus Eqn(1.28) can be expressed as:

$$\boxed{y(t) = \sum_{i=1}^{N} a_iy_i(t) = \sum_{i=1}^{N} a_i\mathcal{S}\left[x_i(t)\right]} \tag{1.29}$$

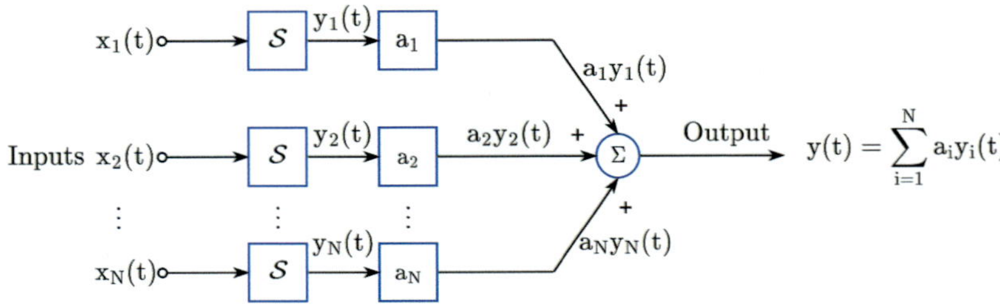

Figure 1.12: Linearity property: The system operator $\mathcal{S}$ precedes amplitude scaling of each input

Equation (1.29) is schematically represented in Figure 1.12.

Note that in a linear system, the operator $\mathcal{S}$ *commute*; therefore, interchanging the position of the system operator does not change the output.

The system which does not satisfy the superposition property is said to be nonlinear(more on features of nonlinear systems will be discussed in the companion book [1]).

EXAMPLE 1.6

Problem

Determine if the systems described below are linear or nonlinear.

a. $y(t) = tu(t)$
b. $y(t) = x^p(t), \quad p \text{ is an integer and } p \neq 0, 1$
c. $\dfrac{dy(t)}{dt} + t^2 y(t) = (5t + 2)\, u(t)$

SOLUTION:

a. Let us first apply two distinct inputs $u(t) = u_1(t)$ and $u(t) = u_2(t)$ to the system. The outputs corresponding to these two inputs $y_1(t)$ and $y_2(t)$ are

$$y_1(t) = tu_1(t), \qquad \text{and} \quad y_2(t) = tu_2(t)$$

Let us apply the input $u(t) = au_1(t) + bu_2(t)$. This gives the output

$$y(t) = t\,[au_1(t) + bu_2(t)] = ay_1(t) + by_2(t)$$

This shows

$$y(t) = ay_1(t) + by_2(t)$$

[1]A companion book on Digital and Nonlinear Control is in preparation

Hence, this system is linear.

b. Given

$$y(t) = x^p(t)$$

Let $y_1(t)$ and $y_2(t)$ be the outputs of the system due to the inputs $x_1(t)$ and $x_2(t)$, respectively, that is $y_1(t) = x_1^p(t)$ and $y_2(t) = x_2^p(t)$. Let $x(t) = x_1(t) + x_2(t)$ and $y(t)$ is the corresponding output. Note that

$$y(t) = [x_1(t) + x_2(t)]^p \neq y_1(t) + y_2(t)$$

Hence, the system is nonlinear.

c. Now the system equation is

$$\frac{dy(t)}{dt} + t^2 y(t) = (5t + 2)\, u(t) \tag{1.30}$$

Let $y_1(t)$ and $y_2(t)$ be the outputs of the system due to the inputs $u_1(t)$ and $u_2(t)$, respectively. Then

$$\frac{dy_1(t)}{dt} + t^2 y_1(t) = (5t + 2)\, u_1(t) \tag{1.31}$$

and

$$\frac{dy_2(t)}{dt} + t^2 y_2(t) = (5t + 2)\, u_2(t) \tag{1.32}$$

Multiplying Eqn(1.31) by k_1 and Eqn(1.32) by k_2 and adding them gives

$$\begin{aligned} &\frac{d}{dt}\left[k_1 y_1(t) + k_2 y_2(t)\right] + t^2 \left[k_1 y_1(t) + k_2 y_2(t)\right] \\ &= (5t + 2)\left[k_1 u_1(t) + k_2 u_2(t)\right] \end{aligned} \tag{1.33}$$

Note that Eqn(1.33) is the system of Eqn(1.30) with

$$\begin{aligned} u(t) &= k_1 u_1(t) + k_2 u_2(t) \quad \text{and} \\ y(t) &= k_1 y_1(t) + k_2 y_2(t) \end{aligned}$$

Therefore, when the input is $k_1 u_1(t) + k_2 u_2(t)$, the system response is $k_1 y_1(t) + k_2 y_2(t)$. Hence, the system is linear.

EXAMPLE 1.7

Problem

The output $y(t)$ of a system is related to its input $u(t)$ as:

$$y(t) = \frac{u^2(t)}{\dot{u}(t)}$$

where $\dot{u}(t) = \frac{du(t)}{dt}$. Show that this system satisfies the homogeneity property, but not the additive property.

SOLUTION:

Testing Homogeneity Property:

Let $y_1(t)$ denotes the output for an input $u(t)$ and $y_2(t)$ denotes the output for an input $ku(t)$, where k is a constant. Now, from the input-output relation, we have

$$y_1(t) = \frac{u^2(t)}{\dot{u}(t)}$$

and

$$y_2(t) = \frac{k^2u^2(t)}{k\dot{u}(t)} = k\frac{u^2(t)}{\dot{u}(t)} = ky_1(t)$$

Thus, when the input is multiplied by a constant k, the output is multiplied by the same constant. Thus, this system satisfies the homogeneity property.

Testing Additive Property:

Now application of $u_1(t)$ and $u_2(t)$ gives respectively the outputs $y_1(t)$ and $y_2(t)$, that is,

$$u_1(t) \rightarrow \quad y_1(t) = \frac{u_1^2(t)}{\dot{u}_1(t)}, \quad u_2(t) \rightarrow \quad y_2(t) = \frac{u_2^2(t)}{\dot{u}_2(t)}$$

If we apply the input $u_1(t) + u_2(t)$, the output becomes

$$y(t) = \frac{[u_1(t) + u_2(t)]^2}{\dot{u}_1(t) + \dot{u}_2(t)} \neq y_1(t) + y_2(t)$$

Hence, the system does not satisfy the additive property.

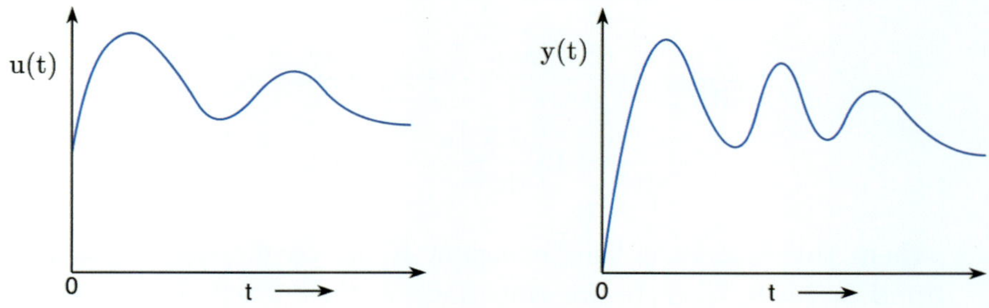

Figure 1.13: Time invariance property

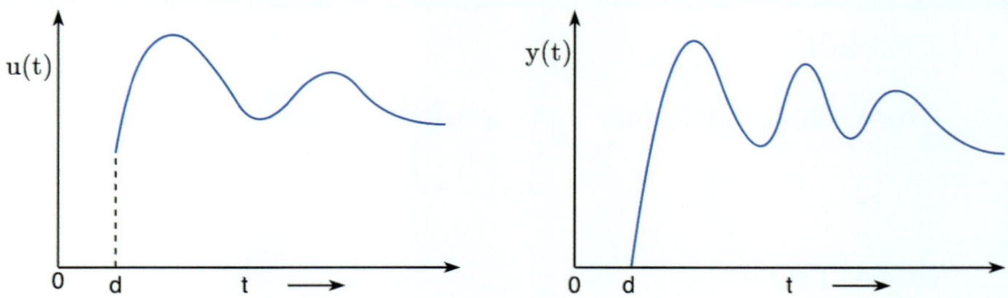

Figure 1.14: Time invariance property

1.6.2 Time-invariant and Time-varying Systems

Definition 1.15. A system is *time-invariant* if a time shift in the input signal results in an identical time shift in the output signal.

For example, in a time-invariant system, if an input u(t) produces an output y(t), then application of a delayed input u(t − d) results in y(t − d).

Mathematically, if

$$\mathrm{y(t)} = \mathcal{S}[\mathrm{u(t)}], \quad \text{then} \quad \mathrm{y(t-d)} = \mathcal{S}[\mathrm{u(t-d)}] \tag{1.34}$$

This is depicted in Figures 1.13 and 1.14.

Remark

If the input-output relation of a system is described by a linear differential equation of the form

$$\frac{d^n y}{dt^n} + a_1\frac{d^{n-1}y}{dt^{n-1}} + \cdots + a_{n-1}\frac{dy}{dt} + a_n y$$
$$= b_0\frac{d^m u}{dt^m} + b_1\frac{d^{m-1}u}{dt^{m-1}} + \cdots + b_{m-1}\frac{du}{dt} + b_m u$$

then, this system is time-invariant if the coefficients $a_i, i = 1, 2, \ldots, n$ and $b_i, i = 0, 1, \ldots, m$ are constants. If these coefficients are functions of time, then the system is a **linear time-varying** system.

EXAMPLE 1.8

Problem

Check the linearity and time-invariance properties of the system

$$y(t) = \cos[x(t)]$$

SOLUTION:

Testing Linearity:

Let us first apply two distinct inputs $x(t) = ax_1(t)$ and $x(t) = bx_2(t)$ to the system. The outputs corresponding to these two inputs $y_1(t)$ and $y_2(t)$ are

$$y_1(t) = \cos[ax_1(t)], \quad \text{and} \quad y_2(t) = \cos[bx_2(t)]$$

Let us apply the sum of the inputs i.e., $x(t) = ax_1(t) + bx_2(t)$. This gives the output

$$y(t) = \cos[ax_1(t) + bx_2(t)]$$

This shows

$$y(t) = \cos[ax_1(t) + bx_2(t)] \neq ay_1(t) + by_2(t)$$

Hence, this system is nonlinear.

Testing Time Invariance:

The output at $y(t)$ at delayed time d is given by

$$y(t - d) = \cos[x(t - d)]$$

Let us delay the input by d which gives $x_1(t) = x(t-d)$. Now the output for this input

$$y_1(t) = [\cos(x(t-d))]$$

Thus,

$$y_1(t) = y(t-d))$$

Hence, the system is time-invariant.

CONCEPT REINFORCEMENT PROBLEM 1.3

Problem

Determine if the systems described below are time-invariant or time-varying.

a. $y(t) = tu(t-3)$
b. $y(t) = \left(\frac{du(t)}{dt}\right)^2$
c. $y(t = u(at)$
d. $y(t) = \int_{-10}^{10} u(\tau)d\tau$

EXAMPLE 1.9

Consider the system, a resistive circuit, as shown in Figure 1.15.

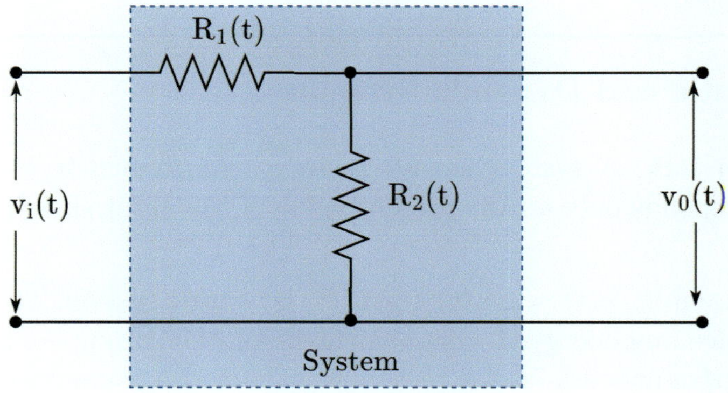

Figure 1.15: Resistive circuit

The output of the system for the input $v_i(t)$ is

$$y_1(t) = \frac{R_2(t)}{R_1(t) + R_2(t)} v_i(t) \tag{1.35}$$

For the input $v_i(t + \tau)$, the output is

$$y_2(t) = \frac{R_2(t)}{R_1(t) + R_2(t)} v_i(t + \tau) \tag{1.36}$$

The system is time-invariant if $y_2(t) = y_1(t + \tau)$ for any $v_i(t)$. That is

$$y_2(t) = \frac{R_2(t + \tau)}{R_1(t + \tau) + R_2(t + \tau)} v_i(t + \tau) \tag{1.37}$$

From Eqns(1.36) and (1.37), we find that the system is time-invariant if

$$\frac{R_2(t)}{R_1(t) + R_2(t)} = \frac{R_2(t + \tau)}{R_1(t + \tau) + R_2(t + \tau)} \tag{1.38}$$

It is obvious that the system is time-invariant if the resistances do not vary with time. However, Eqn(1.38) is also satisfied for all τ, if

$$R_2(t) = kR_1(t)$$

where k is a constant. Hence, the system can be time-invariant even if some of the elements vary with time. [a]

[a] A system is time-invariant if its operator $\mathcal{S}[.]$ does not vary with time, even though the elements of the system do vary with time. Thus, a time-invariant system does not preclude the possibility that the elements of the system might vary with time.

1.6.3 Static and Dynamic Systems

Definition 1.16. A system is said to be *static* or memoryless if its present output depends only on the present value of the input signal.

For example, $y(t) = a_1u(t) + a_2u^2(t)$ is a static system. The static systems do not have any energy-storing elements. Another example of a static system is a resistive network; as the resistance cannot store energy. The voltage-current relationship in a resistor is described as:

$$i(t) = \frac{1}{R} v(t)$$

where R is the resistance of the resistor. The current i(t) flowing through the resistor at time t is due to the voltage v(t) at time t.

Definition 1.17. A system is said to be *dynamic* (or a system with memory) if its output depends on the past or future values of the input.

The dynamic systems can further be categorized depending on how far the memory of the system extends into the past or future, as:

i. *Finite memory system:* If the response at time "t" is completely determined by the input signals over the past T seconds, then the system is called a finite-memory system with memory of T seconds. For example,

$$y(t) = u(t) + b_1 u(t-4) + b_2 u^2(t-5)$$

is a finite memory system.

ii. *Infinite memory system:* If the response at time "t" is determined by the input signals over the entire past $(-\infty, t)$, then the system is called an infinite memory system. Unlike the resistor, the inductor and capacitor have memory. The current flowing through the inductor $i_L(t)$ and the voltage across a capacitor $v_c(t)$ can be expressed as:

$$i_L(t) = \frac{1}{L}\int_{-\infty}^{t} v(\tau)d\tau$$

$$v_c(t) = \frac{1}{C}\int_{-\infty}^{t} i(\tau)d\tau$$

These are energy-storing elements where the inductor stores energy in its magnetic field and the capacitor stores energy in its electric field. Therefore, systems with capacitance or inductance are infinite memory systems.

1.6.4 Causal and Non-Causal Systems

Definition 1.18. A system is said to be *causal* if its output at present time depends only on the present and past values of input signal, i.e., its output at time t_0 depends only on the value of the input u(t) for $t \leq t_0$.

This means, in a causal system, the output is not dependent on the future values of the input. It does not anticipate future inputs and hence is also called a **non-anticipative** system. Note that any practical system which operates in *real time* must be causal.

Definition 1.19.

A system is said to be *non-causal* if its output at the present time depends on future values of the input signal; that is, the output at time t_0 depends on the value of the input $u(t)$ for $t \geq t_0$.

Note on Non-Causal Systems

At first glance, it might seem that non-causal systems do not exist in reality. However, this is not the case; this argument is valid if we only deal with *temporal systems*, i.e., the systems where the independent variable is time. Time is a privileged parameter in science and engineering and is embedded in our consciousness, leading to the assumption that all systems are temporal. As a result, we conclude that non-causal temporal systems cannot be realized in practice. However, if the independent variable is something other than time (e.g., **space**), non-causal systems can indeed be physically realized. Examples of non-causal systems can be found in spatial filtering in image processing, prediction algorithms in finance, some encoding and decoding algorithms in signal processing, temperature control in a rod, or fluid flow control in a pipe, and so on.

EXAMPLE 1.10

Problem

Determine if the following systems are static or dynamic, causal or non-causal.

a. **S1:** $y(t) = u(t) + u^3(t)$
b. **S2:**$y(t) = \cos[u(t)] + \cos[u(t-2)]$
c. **S3:**$y(t) = 10\sin[u(t+1)] + \cos^2[u(t-2)]$
d. **S4:**$y(t) = u(t-1) + u^2(t)$
e. **S5:**$y(t) = u(t+2) + u(t-1) - u^3(t-3)$

SOLUTION:

a. **S1:** For the system $y(t) = u(t) + u^3(t)$, the output at time t depends only on the current input. Hence, this system is causal and static (without memory).
b. **S2:** For the system $y(t) = \cos[u(t)] + \cos[u(t-2)]$, the output at time t depends both on the current input as well as on past input. Hence, this system is causal and dynamic (with memory).
c. **S3:** For the system $y(t) = 10\sin[u(t+1)] + \cos^2[u(t-2)]$, the output at time t depends both on past and future inputs. This is an example of a non-causal system with memory (dynamic).

d. **S4:** For the system, $y(t) = u(t-1) + u^2(t)$, the output at time t depends both on the current input as well as on past input. Hence, this is a causal and dynamic system (with memory).

e. **S5:** For the system, $y(t) = u(t+2) + u(t-1) - u^3(t-3)$, the output at time t depends both on past and future inputs. This is an example of a non-causal system with memory (dynamic).

1.6.5 Lumped and Distributed Parameter Systems

Definition 1.20. A *lumped system* is one where the parameters are assumed to be concentrated at a point.

Some of the key characteristics of such systems include:

i. **Concentration of Parameters:** The parameters of these systems are concentrated at specific points or elements within the system. For example, the mass of a pendulum in simple harmonic motion can be assumed to be concentrated at a single point in space. In an inverted pendulum, the mass of the rod may be considered to be negligible, and the entire mass of the system can be lumped at a point, and the dependent variables of interest are a function of time alone.

ii. **Instantaneous Interaction:** The interaction between elements in a lumped parameter system is instantaneous. This means that changes in the state or behavior of one element immediately affect the states of connected elements. As an example, for a physical electric circuit to be considered lumped, its physical dimension must be small enough so that electromagnetic waves propagate across the circuit virtually instantaneously.

iii. **Ordinary Differential Equations:** The behavior of lumped parameter systems can be described using ordinary differential equations (ODEs), which relate the system's input, output, and state variables as functions of time only (not of space).

Definition 1.21. A *distributed system* is one whose parameters vary continuously over space rather than being concentrated at specific points.

Some of the key characteristics of such systems include:

i. **Continuous Variation:** The parameters of the system, such as mass, capacitance, and resistance, are distributed continuously over space or along a length. This continuous variation distinguishes distributed parameter systems from lumped parameter systems, where parameters are concentrated at discrete points.

ii. **Spatial Dynamics:** The behavior of distributed parameter systems is influenced by the spatial distribution of their parameters. Changes in parameter values at different spatial locations affect the system's overall behavior.

iii. **Partial Differential Equations:** These systems are often described using partial differential equations (PDEs) that relate the system's input, output, and state variables as functions of both time and space.

Some of the examples of distributed parameter systems are:

i. **Electromagnetic systems,** such as transmission lines or wave-guides, where electric and magnetic fields vary continuously over space.

ii. **Wave propagation phenomena**, such as acoustic waves in air or water, where pressure or displacement varies continuously in both space and time.

iii. **Heat conduction** in a material, where temperature varies continuously over space and time according to Fourier's law of heat conduction.

1.7 Time Response of Linear Dynamic Systems

As discussed, the dynamic systems have memory and have at least one energy-storing element. Therefore, even in the absence of any input excitation, the output will be excited exclusively due to the energy storage (initial conditions, say $\mathbf{x}(t_0)$). This output is called the *zero-input response* and will be denoted by $y_{zi}(t)$ (see Figure 1.16(a)). This response is due to internal system conditions, such as energy storage.

If the initial conditions $x(t_0)$ are zero, then the output is excited due to the input alone. This output is called the *zero-state response* and will be denoted by $y_{zs}(t)$ (see Figure 1.16(b)).

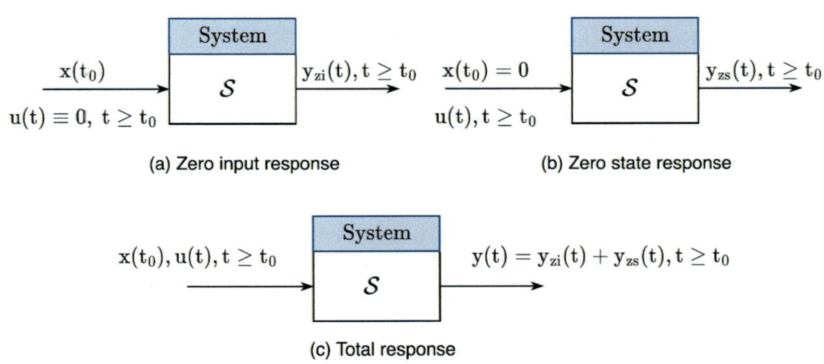

Figure 1.16: (a) Zero-input response, (b) zero-state response, and (c) total response

Due to additive property of linear system, the total response of linear system is the sum of *zero-input* and the *zero-state response*, that is

$$\text{Output due to}\begin{cases}\mathbf{x}(\mathrm{t_0})\\ \mathbf{u}(t),\ \mathrm{t}\geq \mathrm{t_0}\end{cases} = \text{Output due to}\begin{cases}\mathbf{x}(\mathrm{t_0})\\ \mathbf{u}(t)\equiv 0,\ \ \mathrm{t}\geq \mathrm{t_0}\end{cases}$$
$$+\,\text{Output due to}\begin{cases}\mathbf{x}(\mathrm{t_0})=0\\ \mathbf{u}(t),\ \ \mathrm{t}\geq \mathrm{t_0}\end{cases} \tag{1.39}$$

Key Takeawy

Total response = zero-input response (natural response)+zero-state response (forced response)

The response of any linear system can be split into two components: the *zero-input component*, arising from initial conditions, and the *zero-state component*, arising from the input. These components can be studied separately, and their sum gives the complete response. This is known as the decomposition property. Being linear, the system also obeys the principle of superposition: if the initial condition is scaled by k, the zero-input component scales by k; similarly, doubling the input u(t) doubles the zero-state component. Nonlinear systems do not follow this property, as their total response may differ from the sum of zero-input and zero-state responses. Further details are provided in Section 1.10.

1.8 Impulse Response and Transfer Functions of Linear Systems

Most commonly, the mathematical relationship between the input and output variables in the linear time-invariant systems is characterized by transfer functions. These functions can be defined either using the impulse response function of the system or by taking the Laplace transform of inputs and outputs.

1.8.1 Impulse Response of a System

The *impulse response* of a system is defined as the output response when the input is a unit impulse function $\delta(t)$, **with all initial conditions equal to zero**. It is denoted by $g(t)$. Mathematically,

$$\mathrm{g(t)} = \mathcal{S}\left[\delta(\mathrm{t})\right] \tag{1.40}$$

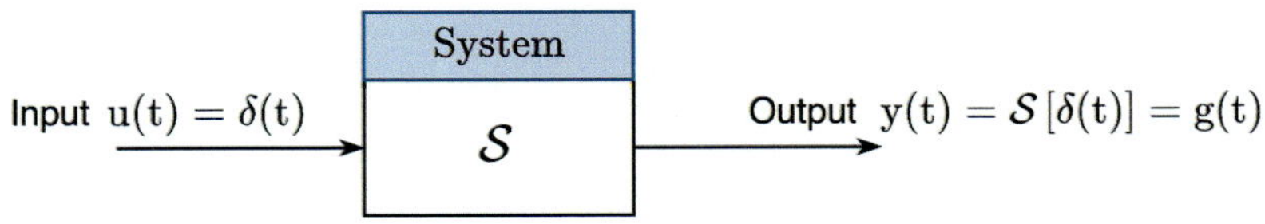

Figure 1.17: Impulse response of a system

We know that a continuous time signal u(t) can be represented/approximated as a weighted superposition of shifted impulses as:

$$\hat{u}(t) = \int_{-\infty}^{\infty} u(\tau)\delta(t-\tau)d\tau \tag{1.41}$$

where $u(\tau)dt$ on the delayed impulse $\delta(t-\tau)$ is determined from the value of input signal u(t) at the occurrence of each impulse. Hence, the output of a system to an arbitrary input u(t), which is expressed as a weighted superposition of shifted impulses (as in Eqn(1.19), is given by

$$y(t) = \mathcal{S}[u(t)] = \mathcal{S}\left[\int_{-\infty}^{\infty} u(\tau)\delta(t-\tau)d\tau\right] \tag{1.42}$$

Applying linearity property to Eqn(1.42) gives

$$y(t) = \int_{-\infty}^{\infty} u(\tau)\mathcal{S}[\delta(t-\tau)]\,d\tau \tag{1.43}$$

Using the time-invariance property, we have

$$\mathcal{S}[\delta(t-\tau)] = g(t-\tau)$$

Thus, the output of a linear time-invariant system to an arbitrary input u(t) is

$$y(t) = \int_{-\infty}^{\infty} u(\tau)g(t-\tau)d\tau \tag{1.44}$$

Eqn(1.44) is referred to as the **convolution integral** and is denoted by the symbol "*" as:

$$\boxed{y(t) = u(t) * g(t) = \int_{-\infty}^{\infty} u(\tau)g(t-\tau)d\tau} \tag{1.45}$$

Key Takeaway

If the impulse response function of a system, that is, its response to a unit impulse input is known, then the response of the system to any arbitrary input can be computed by convolving the input with the impulse response (see Eqn(1.45)).

1.8.2 Transfer Function of a Linear Time-Invariant System

Definition 1.22. The transfer function of a linear time-invariant system can be defined as the Laplace transform of its impulse response function, that is

$$G(s) = \mathcal{L}\left[g(t)\right] = \int_{-\infty}^{\infty} g(\tau)e^{-s\tau}d\tau \tag{1.46}$$

We know that convolution in the time domain is equivalent to multiplication in the frequency domain. Hence, in frequency domain, the convolution in Eqn(1.45) is expressed as:

$$Y(s) = G(s)U(s) \tag{1.47}$$

The transfer function can alternately be defined by applying the convolution property.

Definition 1.23. The transfer function G(s) of a linear time-invariant system is defined as the ratio of the Laplace transform of output to the Laplace transform of the input with all initial conditions zero. That is

$$G(s) = \left.\frac{Y(s)}{U(s)}\right|_{\text{Initial Conditions Zero}} \tag{1.48}$$

EXAMPLE 1.11

Problem

Compute the transfer function of the following systems from their impulse response g(t):

a. $g(t) = 5 + 2e^{-4t}$
b. $g(t) = 10e^{-t} + 4e^{-2t}$
c. $g(t) = 5e^{-2t} + \cos(2t)$

d. $g(t) = 1 + e^{-t}\sin(t)$

SOLUTION:

a. Now $g(t) = 5 + 2e^{-4t}$. Hence

$$G(s) = \mathcal{L}[g(t)] = \frac{5}{s} + \frac{2}{s+4} = \frac{7s+20}{s(s+4)}$$

b. For $g(t) = 10e^{-t} + 4e^{-2t}$

$$G(s) = \mathcal{L}[g(t)] = \frac{10}{s+1} + \frac{4}{s+2} = \frac{14s+24}{(s+1)(s+2)}$$

c. When $g(t) = 5e^{-2t} + \cos(2t)$,

$$G(s) = \mathcal{L}[g(t)] = \frac{5}{s+2} + \frac{s}{s^2+4} = \frac{6s^2+2s+20}{(s+2)(s^2+4)}$$

d. For $g(t) = 1 + e^{-t}\sin(t)$

$$G(s) = \mathcal{L}[g(t)] = \frac{1}{s} + \frac{1}{(s+1)^2+1} = \frac{s^2+3s+2}{s(s^2+2s+2)}$$

EXAMPLE 1.12

Problem

The output of an LTI system excited by the input $u(t) = 3\sin(3t)$ is given as

$$y(t) = 9e^{-t} + 3\sin(3t) - 9\cos(3t)$$

Compute the transfer function of the system

SOLUTION:

Given

$$y(t) = 9e^{-t} + 3\sin(3t) - 9\cos(3t)$$

Taking the Laplace transform of the above gives

$$Y(s) = \frac{9}{s+1} + \frac{9}{s^2+9} - \frac{9s}{s^2+9} = \frac{90}{(s+1)(s^2+9)}$$

For an input $u(t) = 3\sin(3t)$, $U(s) = \dfrac{9}{s^2+9}$.

Therefore, the transfer function is

$$G(s) = \frac{Y(s)}{U(s)} = \frac{10}{s+1}$$

1.9 Transfer Function Computation from Differential Equation Model

Although we can compute the transfer function of a system using various methods, the most convenient method is to derive it from the differential equation model of the system as described below.

Consider a single-input single-output (SISO) linear time-invariant system which is described by an n-th order linear differential equation

$$\begin{aligned} &\frac{d^n y(t)}{dt^n} + a_1 \frac{d^{n-1} y(t)}{dt^{n-1}} + \cdots + a_{n-1} \frac{dy(t)}{dt} + a_n y(t) \\ &= b_0 \frac{d^m u(t)}{dt^m} + b_1 \frac{d^{m-1} u(t)}{dt^{m-1}} + \cdots + b_{m-1} \frac{du(t)}{dt} + b_m u(t) \end{aligned} \tag{1.49}$$

where y(t) is the output, u(t) is the input, and a_i's and b_i's are real coefficients of differential equation model.

Taking the Laplace transform of both sides of Eqn(1.49) with zero initial conditions gives

$$\begin{aligned} &\left[s^n + a_1 s^{n-1} + a_2 s^{n-2} + \cdots + a_{n-1} s + a_n\right] Y(s) \\ &= \left[b_0 s^m + b_1 s^{m-1} + \cdots + b_{m-1} s + b_m\right] U(s) \end{aligned} \tag{1.50}$$

The transfer function between the output y(t) and input u(t) is given by

$$G(s) = \frac{Y(s)}{U(s)} = \frac{b_0 s^m + b_1 s^{m-1} + \cdots + b_{m-1} s + b_m}{s^n + a_1 s^{n-1} + a_2 s^{n-2} + \cdots + a_{n-1} s + a_n} \tag{1.51}$$

The transfer function of a system can be represented by a block diagram, as shown in Figure 1.18.

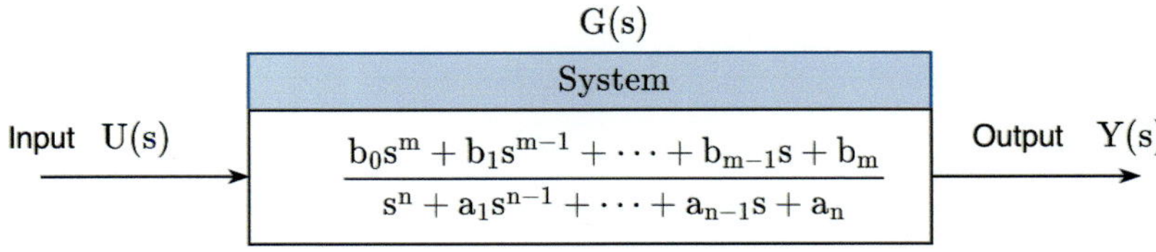

Figure 1.18: Block diagram of a transfer function

From Figure 1.18, the relationships between the output and input in the frequency domain can be expressed as:

$$\boxed{Y(s) = G(s)U(s)} \tag{1.52}$$

Let us illustrate the transfer function computation by a few examples. Some of the common Laplace transform pairs to be frequently used during the transformation are

$$\begin{aligned} x(t) &\longleftrightarrow X(s) \\ \frac{dx(t)}{dt} &\longleftrightarrow sX(s) \\ \frac{d^n x(t)}{dt^n} &\longleftrightarrow s^n X(s) \\ x(t-T) &\longleftrightarrow e^{-sT} X(s) \\ \int x(t) &\longleftrightarrow \frac{1}{s} X(s) \end{aligned}$$

EXAMPLE 1.13

Problem

Determine the transfer function models of the systems which are represented by the following differential equation models.

a. $\frac{d^4y}{dt^4} + 5\frac{d^3y}{dt^3} + 7\frac{d^2y}{dt^2} + 11\frac{dy}{dt} + 9y = u$

b. $\frac{d^3y}{dt^3} + 2\frac{d^2y}{dt^2} + 5\frac{dy}{dt} + 7y = 3\frac{du}{dt} - 5u$

c. $\frac{d^2y}{dt^2} + 3\frac{dy}{dt} + 6y = 5u + 3u(t-2)$

d. $\frac{d^3y}{dt^3} + 6\frac{d^2y}{dt^2} + 2\frac{dy}{dt} + 7y + 3\int_0^t y(t)\, dt = 5\frac{du}{dt} + 9u$

SOLUTION:

a. Taking the Laplace transform of both sides of

$$\frac{d^4y}{dt^4} + 5\frac{d^3y}{dt^3} + 7\frac{d^2y}{dt^2} + 11\frac{dy}{dt} + 9y = u$$

gives

$$\left[s^4 + 5s^3 + 7s^2 + 11s + 9\right] Y(s) = U(s)$$

Hence, the transfer function is

$$G(s) = \frac{Y(s)}{U(s)} = \frac{1}{s^4 + 5s^3 + 7s^2 + 11s + 9}$$

b. Taking the Laplace transform of both sides of

$$\frac{d^3y}{dt^3} + 2\frac{d^2y}{dt^2} + 5\frac{dy}{dt} + 7y = 3\frac{du}{dt} - 5u \quad \text{gives}$$

$$\left[s^3 + 2s^2 + 5s + 7\right] Y(s) = [3s - 5]\, U(s)$$

Hence, the transfer function is

$$G(s) = \frac{Y(s)}{U(s)} = \frac{3s - 5}{s^3 + 2s^2 + 5s + 7}$$

c. Similar to above, taking the Laplace transform of

$$\frac{d^2y}{dt^2} + 3\frac{dy}{dt} + 6y = 5u + 3u(t-2) \quad \text{gives}$$

$$\left[s^2 + 3s + 6\right] Y(s) = \left[5 + 3e^{-2s}\right] U(s)$$

Hence, the transfer function is

$$G(s) = \frac{Y(s)}{U(s)} = \frac{5 + 3e^{-2s}}{s^2 + 3s + 6}$$

d. Let us take the Laplace transform of

$$\frac{d^3y}{dt^3} + 6\frac{d^2y}{dt^2} + 2\frac{dy}{dt} + 7y + 3\int_0^t y(t)\, dt = 5\frac{du}{dt} + 9u$$

This results

$$\left[s^3 + 6s^2 + 2s + 7 + \frac{3}{s}\right] Y(s) = [5s + 9]\, U(s)$$

Hence, the transfer function is

$$G(s) = \frac{Y(s)}{U(s)} = \frac{s(5s + 9)}{s^4 + 6s^3 + 2s^2 + 7s + 3}$$

EXAMPLE 1.14

Problem

Determine the differential equation model of the system whose transfer function model is

$$G(s) = \frac{Y(s)}{U(s)} = \frac{2s+1}{s^3+10s^2+5s+11}$$

SOLUTION:

Now

$$G(s) = \frac{Y(s)}{U(s)} = \frac{2s+1}{s^3+10s^2+5s+11}$$

or,

$$s^3Y(s) + 10s^2Y(s) + 5sY(s) + 11Y(s) = 2sU(s) + U(s)$$

Taking Laplace inverse of the above gives the differential equation as:

$$\frac{d^3y}{dt^3} + 10\frac{d^2y}{dt^2} + 5\frac{dy}{dt} + 11y = 2\frac{du}{dt} + u$$

1.9.1 Proper, Strictly Proper, and Improper Transfer Functions

Consider a system with a transfer function

$$\begin{aligned} G(s) &= \frac{Y(s)}{U(s)} = \frac{N(s)}{D(s)}, \quad \text{where} \qquad (1.53)\\ N(s) &= b_0s^m + b_1s^{m-1} + \cdots + b_{m-1}s + b_m, \quad \text{and}\\ D(s) &= s^n + a_1s^{n-1} + a_2s^{n-2} + \cdots + a_{n-1}s + a_n \end{aligned}$$

Definition 1.24. The transfer function in Eqn(1.53) is said to be

1. **Strictly proper if** $n > m$; that is, if the order of the denominator polynomial is greater than that of the numerator polynomial.
2. **Proper if** $n = m$; that is, if the order of the denominator polynomial is equal to that of the numerator polynomial.
3. **Improper if** $m > n$; that is, if the order of the numerator polynomial is greater than that of the denominator polynomial.

EXAMPLE 1.15

Examples of Different types of transfer functions

System-1: $\frac{10}{s(s+2)}$. This is a *strictly proper* transfer function.

System-2: $\frac{s^3+3s^2+2s+1}{s^2+5s+9}$. This is an *improper transfer function.*

System-3: $\frac{s^2+3s+2}{s^2+7s+12}$. This is a *proper* transfer function.

System-4: $\frac{5(s+1)}{s^3+2s^2+9s+20}$. This is a *strictly proper* transfer function.

1.9.2 Poles and Zeros

Before discussing the physical significance of poles and zeros of a system, we first introduce the concept of poles and zeros of a system from its transfer function model.

Consider a system represented by a proper transfer function

$$G(s) = \frac{N(s)}{D(s)}$$

where N(s) and D(s) are polynomials with real coefficients as defined in Eqn(1.53) with $m \leq n$.

Definition 1.25. A finite real or complex number λ is a pole of G(s) if $|G(\lambda)| = \infty$. It is a zero of G(s) if $G(\lambda) = 0$.

Consider a system whose transfer function model is given by

$$G(s) = \frac{N(s)}{D(s)} = \frac{10(s^2-s-2)}{(s-2)(s^2+5s+6)} \tag{1.54}$$

To examine whether or not $s = 2$ is a pole of G(s), compute G(2):

$$G(2) = \frac{N(2)}{D(2)} = \frac{0}{0}$$

This is not defined. Applying l'Hôpital's rule gives

$$G(s) = \left.\frac{N(s)}{D(s)}\right|_{s=2} = \left.\frac{N'(s)}{D'(s)}\right|_{s=2} = \left.\frac{10(2s-1)}{3s^2+6s-4}\right|_{s=2} = \frac{30}{20} \neq \infty$$

Thus, s = 2 is not a pole of G(s). This is because the numerator and denominator polynomial of G(s) has a common factor, i.e.,

$$G(s) = \frac{10(s^2 - s - 2)}{(s-2)(s^2+5s+6)} = \frac{10(s-2)(s+1)}{(s-2)(s+2)(s+3)} = \frac{10(s+1)}{(s+2)(s+3)} \quad (1.55)$$

The poles of the transfer function are at −2 and −3, and the zeros are at −1. The observations from this example are summarized as:

Key Takeaways

1. If the polynomials N(s) and D(s) of transfer function

$$G(s) = \frac{N(s)}{D(s)}$$

have no common factors, then the roots of N(s) are the **zeros** of the system, and the roots of the D(s) are the poles of the system (transfer function model of the system).

2. More generally, the poles of G(s) are the roots of D(s) **after canceling any common factors of** N(s).

1.10 Zero-Input Response and Zero-State Response

Consider a linear time-invariant system described by Eqn(1.53). The total response y(t) of the system can be split into:

1. **Zero-Input Response (ZIR):** This is the response due to the initial conditions alone when the external input u(t) = 0 (see Figure 1.16a).
2. **Zero-State Response (ZSR):** This is the response due to the external input u(t), assuming the initial conditions are zero (see Figure 1.16b).

Before introducing some of the common terminologies associated with the total response, let us illustrate the concept using a simple system.

Consider a linear time-invariant (LTI) system described by the differential equation:

$$\frac{d^2y(t)}{dt^2} + 5\frac{dy(t)}{dt} + 6y(t) = 2\frac{du(t)}{dt} + u(t) \quad (1.56)$$

Applying Laplace transform to Eqn(1.56) gives

$$s^2Y(s) - sy(0^-) - \dot{y}(0^-) + 5\left[sY(s) - y(0^-)\right] + 6Y(s) = 2\left[sU(s) - u(0^-)\right] + U(s) \quad (1.57)$$

Note that the differential equation (1.56) is transformed into an algebraic equation (1.57) using the Laplace transform. This is an important advantage as it is easier to manipulate algebraic equations compared to differential equations. Now Eqn(1.57) can be reorganized as:

$$\left[s^2 + 5s + 6\right] Y(s) = sy(0^-) + \dot{y}(0^-) + 5y(0^-) - 2u(0^-) + [2s + 1]\, U(s) \tag{1.58}$$

or,

$$Y(s) = \underbrace{\frac{(s+5)\, y(0^-) + \dot{y}(0^-) - 2u(0^-)}{s^2 + 5s + 6}}_{\text{Zero Input Response}} + \underbrace{\frac{2s+1}{s^2 + 5s + 6} U(s)}_{\text{Zero State Response}} \tag{1.59}$$

Eqn(1.59) shows that the response consists of two parts. The first part is excited exclusively by the initial conditions $y(0^-), \dot{y}(0^-)$ and is called the *zero-input response.* **These initial conditions, called as initial state, encodes the effects of the past input** $u(t), t < 0$**, on the future output** $y(t)$**, for** $t \geq 0$. The second part is excited exclusively by the input and is called the *zero-state response.* Since the system is linear, we can study the *zero-input response* and *zero-state response* separately. Let us first study the *zero-input response* followed by *zero-state response.*

1.10.1 Zero-Input Response and the Characteristic Equation

Consider again the differential equation in Eqn(1.56). If the input $u(t) = 0$, for $t \geq 0$, then Eqn(1.56) becomes

$$\frac{d^2y(t)}{dt^2} + 5\frac{dy(t)}{dt} + 6y(t) = 0 \tag{1.60}$$

This is a *homogeneous equation.* Applying Laplace transform to Eqn(1.60) gives

$$s^2Y(s) - sy(0^-) - \dot{y}(0^-) + 5\left[sY(s) - y(0^-)\right] + 6Y(s) = 0 \tag{1.61}$$

or,

$$Y(s) = \frac{(s+5)\, y(0^-) + \dot{y}(0^-)}{s^2 + 5s + 6} = \frac{(s+5)\, y(0^-) + \dot{y}(0^-)}{(s+2)(s+3)} \tag{1.62}$$

Eqn(1.62) can alternately be expressed as:

$$Y(s) = \frac{k_1}{s+2} + \frac{k_2}{s+3} \tag{1.63}$$

where

$$k_1 = \left.\frac{(s+5)y(0^-) + \dot{y}(0^-)}{s+3}\right|_{s=-2} = 3y(0^-) + \dot{y}(0^-)$$

and

$$k_2 = \left.\frac{(s+5)y(0^-) + \dot{y}(0^-)}{s+2}\right|_{s=-3} = -\left[2y(0^-) + \dot{y}(0^-)\right]$$

Thus, the *zero-input response* is

$$\boxed{y(t) = k_1 e^{-2t} + k_2 e^{-3t}} \tag{1.64}$$

Note that irrespective of what the initial conditions are, the zero-input response is always a linear combination of the two functions e^{-2t} and e^{-3t}, where -2 and -3 are two roots of the denominator of Eqn(1.62). The denominator of Eqn(1.62) is called the *characteristic equation* of the system, and its roots are called *modes* (*natural modes*) of the system. These are also the *internal modes* or *eigenvalues of the system* (to be discussed more later).

Key Point

Modes(characteristic modes) of a system essentially govern the zero-input response of a system.

Let us extend the method to a more general case of an nth order linear system described by

$$\frac{d^n y(t)}{dt^n} + a_1 \frac{d^{n-1} y(t)}{dt^{n-1}} + \cdots + a_{n-1} \frac{dy(t)}{dt} + a_n y(t)$$
$$= b_0 \frac{d^m u(t)}{dt^m} + b_1 \frac{d^{m-1} u(t)}{dt^{m-1}} + \cdots + b_{m-1} \frac{du(t)}{dt} + b_m u(t) \tag{1.65}$$

The zero-input response is computed by making the input $u(t) = 0$. Thus, Eqn(1.65) becomes an homogeneous equation:

$$\frac{d^n y(t)}{dt^n} + a_1 \frac{d^{n-1} y(t)}{dt^{n-1}} + \cdots + a_{n-1} \frac{dy(t)}{dt} + a_n y(t) = 0 \tag{1.66}$$

To determine the zero-input solution, define

$$Q(\lambda) = \lambda^n + a_1 \lambda^{n-1} + a_2 \lambda^{n-2} + \cdots + a_{n-1}\lambda + a_n \tag{1.67}$$

The equation

$$Q(\lambda) = 0 \tag{1.68}$$

is called the *characteristic equation* of the system of Eqn(1.66) because it governs the *free, unforced, or natural* response of (1.66). The roots of the *characteristic equation* are termed either as **eigenvalues, poles, or natural frequencies**. The relationship and differences between eigenvalues and poles will be discussed in Section 10.8.

Case-1: Distinct Roots: If the roots λ_1, $\lambda_2, \ldots$, λ_n of the characteristic equation are distinct, then the zero input solution is given by

$$y(t) = c_1 e^{\lambda_1 t} + c_2 e^{\lambda_2 t} + \cdots + c_n e^{\lambda_n t} \tag{1.69}$$

where $c_1, c_n, \ldots, c_n$ are arbitrary constants determined from n constraints (auxiliary conditions). The exponentials $e^{\lambda_i t}$, $i = 1, 2, \ldots, n$ are called the *characteristic modes* or *natural modes* of the system.

Case-2: Repeated Roots: If some of the roots of the characteristic equation are repeated, the solution is slightly modified. As an example, consider a system with the characteristic polynomial

$$Q(\lambda) = (\lambda - \lambda_1)^r (\lambda - \lambda_{r+1}) \ldots (\lambda - \lambda_n) \tag{1.70}$$

The characteristic modes of this system are $e^{\lambda_1 t}, te^{\lambda_1 t}, \ldots, t^{r-1}e^{\lambda_1 t}, e^{\lambda_{r+1} t}$, $\ldots, e^{\lambda_n t}$. The zero-input response is

$$y(t) = \left(c_1 + c_2 t + \cdots + c_r t^{r-1}\right) e^{\lambda_1 t} + c_{r+1} e^{\lambda_{r+1} t} + \cdots + c_n e^{\lambda_n t} \tag{1.71}$$

1.10.2 Zero-State Response and Its Relation to Transfer Function

Consider the system described in Eqn(1.60), that is

$$\frac{d^2y(t)}{dt^2} + 5\frac{dy(t)}{dt} + 6y(t) = 2\frac{du(t)}{dt} + u(t) \tag{1.72}$$

Note that the *zero-state* response disregards all initial conditions and is exclusively due to the input. Taking the Laplace transform of Eqn(1.72), with zero initial conditions, gives

$$Y(s) = \frac{2s+1}{s^2+5s+6}U(s) = G(s)U(s) \tag{1.73}$$

where $G(s) = \dfrac{2s+1}{s^2+5s+6}$ is the *transfer function* of this system and Eqn(1.73) is the *zero-state* response of the system.

Thus, the transfer function of a system describes only the *zero-state* response which completely disregards the initial conditions. Note that the analysis and design in classical control are carried out using a transfer function, which disregards the zero-input response completely. The question arises: can we disregard the zero-input response? The answer depends on the context and requirements of the system. For example, if the system is stable, the zero-input response decays to zero, and the steady-state behavior is dominated by the zero-state response. In such cases, ignoring the zero-input response is reasonable, particularly for tasks such as steady-state error analysis, frequency response design, and stability assessment. However, if the system is unstable or marginally stable (e.g., oscillatory systems), the zero-input response does not decay and can significantly impact the overall system behavior. In such cases, ignoring the zero-input response is not valid, and a more comprehensive state-space analysis that includes both ZIR and ZSR is necessary. Further, in applications where the transient response or the system's response to initial conditions is critical (e.g., systems involving energy storage or systems with safety-critical requirements), the zero-input response must be considered. For nonlinear systems, the response cannot always be decomposed into ZIR and ZSR as in LTI systems, so the initial conditions inherently affect the system behavior. Disregarding the zero-input response in such cases can lead to inaccurate or incomplete analysis.

Note that the zero-input response is determined by the roots of the characteristic polynomial $D(s) = 0$, which are called the modes of the system. The zero-state response, on the other hand, is governed by the poles of $G(s)$, i.e., the roots of $D(s)$ after canceling any common factors with $N(s)$. Thus, the poles of a system are not always identical to its characteristic modes. The poles can also be referred to as *external modes.*

Key Takeaways

1. If $N(s)$ and $D(s)$ have **no common factors**:

    a. The **poles** of the system are equal to the **characteristic modes** of the system; that is, the internal and external modes are identical.

    b. The system is completely characterized by its transfer function.

2. If $N(s)$ and $D(s)$ have **common factors (say** $R(s)$**):**

    a. The poles of the system are not equal to the characteristic modes; the internal and external modes differ. Some missing modes in the output correspond to the roots of $R(s)$.

    b. The system is **not completely characterized** by its transfer function. [a]

[a]The concept of characteristic modes and poles is fundamental to the behavior of dynamic systems and is independent of any specific model or representation. They are determined by the intrinsic properties of the system's dynamics. However, having a mathematical model such as a state-space representation, a differential equation, or a transfer function makes it easier to identify and calculate these characteristic modes and poles.

EXAMPLE 1.16

Problem:

Consider the systems that are represented by their differential equation models. Compute the transfer functions and determine the characteristic modes and poles of the system, and identify which of these systems are not completely characterized by their transfer functions.

a. $\ddot{y} + 8\dot{y} + 15y = \dot{u} + 5u$

b. $\dddot{y} + 3\ddot{y} + 2\dot{y} = \dot{u} + 4u$

c. $\ddddot{y} + 2\dddot{y} + 5\ddot{y} = \ddot{u} + 2\dot{u} + 5u$

d. $\dddot{y} - \ddot{y} - 2\dot{y} = \dot{u} + u$

SOLUTION:

a. $G(s) = \dfrac{N(s)}{D(s)} = \dfrac{s+5}{s^2+8s+15} = \dfrac{s+5}{(s+3)(s+5)} = \dfrac{1}{s+3}.$

There are common factors between N(s) and D(s). The characteristic modes are at -3 and -5. The system has a simple pole at -3. Hence, the system is not characterized by its transfer function. The missing pole is at $s = -5$.

b. $G(s) = \dfrac{N(s)}{D(s)} = \dfrac{s+4}{s^3+3s^2+2s} = \dfrac{s+4}{s(s+1)(s+2)}.$

There are no common factors between N(s) and D(s). The characteristic modes are equal to the poles and are located at 0, -1, and -2. Hence, the system is completely characterized by its transfer function.

c. $G(s) = \dfrac{N(s)}{D(s)} = \dfrac{s^2+2s+5}{s^2(s^2+2s+5)} = \dfrac{1}{s^2} = \dfrac{1}{s+3}.$

There are common factors between N(s) and D(s). Hence, the system is not characterized by its transfer function. The missing pole is at $s = -1 \pm j2$.

d. $G(s) = \dfrac{N(s)}{D(s)} = \dfrac{s+1}{s^3-s^2-2s} = \dfrac{s+1}{s(s+1)(s-2)} = \dfrac{1}{s(s-2)}.$

There are common factors between N(s) and D(s). Hence, the system is not characterized by its transfer function. The missing pole is at $s = -1$.

Justification of Transfer Function Based Analysis

Justification of Transfer Function Based Analysis

1. The transfer function determines completely the *zero-state response* of linear time-invariant system.

2. When a system is completely characterized by its transfer function, with no pole-zero cancellation, the internal and external modes of the system are the same. This means that the modes of the system (which are determined by the poles) will appear both in the zero-input response and the zero-state response. Therefore, the transfer function alone suffices to describe the system's behavior, and there is no need to separately account for non-zero initial conditions when designing or analyzing the system.

3. If there is pole-zero cancellation and the system is not completely characterized by a transfer function, then transfer function-based analysis and design may be carried out with caution.

Note

Another type of response, commonly used in control engineering, is the **forced response**; it is the response of the system due to the input signal but may include the effects of **initial conditions**. Although it is theoretically distinct from the **zero state response (ZSR)**, which assumes zero initial conditions, the forced response is often used interchangeably with ZSR in practice. This is because, in most practical control systems, the effects of initial conditions decay quickly (in stable systems), leaving the input-driven behavior dominant. Thus, in subsequent discussion, the total system response c(t) will be decomposed as

$$c(t) = c_f(t) + c_n(t) \tag{1.74}$$

where the first component $c_f(t)$, called as the **forced response**, typically approximated as the zero state response when initial condition effects are negligible and the second component $c_n(t)$, called as **natural response** arises from the system's initial energy storage and corresponds to the zero input response.

EXAMPLE 1.17

Problem

a. Determine the zero-state response of the system described in Eqn(1.56) for a unit step input $u(t) = 1$.

b. Find an input $u(t)$, so that the pole -2 will not be excited.

c. Find an input $u(t)$, so that the pole -3 will not be excited.

SOLUTION:

a. For the step input $u(t) = 1$, $U(s) = \frac{1}{s}$. Hence, the zero-state response is computed as:

$$\begin{aligned} Y(s) &= \frac{2s+1}{s^2+5s+6}U(s) = \frac{2s+1}{s(s+2)(s+3)} \\ &= \frac{k_1}{s+2} + \frac{k_2}{s+3} + \frac{k_3}{s} \end{aligned}$$

where $k_1 = \lim_{s\to-2} \frac{2s+1}{s(s+3)} = \frac{3}{2}$, $k_2 = \lim_{s\to-3} \frac{2s+1}{s(s+2)} = -\frac{5}{3}$ and

$$k_3 = \lim_{s\to 0} \frac{2s+1}{(s+2)(s+3)} = \frac{1}{6}$$

Thus, the *zero-state* response is

$$y(t) = \underbrace{\frac{3}{2}e^{-2t} - \frac{5}{3}e^{-3t}}_{\text{Due to the poles of } G(s)} + \underbrace{\frac{1}{6}}_{\text{Due to the poles of Input } U(s)} \tag{1.75}$$

b. Let us select an arbitrary input with a strictly proper function

$$U(s) = \frac{s+2}{s(s+5)} \implies u(t) = 0.4 + 0.6e^{-5t}$$

This input is bounded. The output due to this input becomes

$$\begin{aligned} Y(s) &= \frac{2s+1}{(s+2)(s+3)} \times \frac{s+2}{s(s+5)} = \frac{2s+1}{s(s+5)(s+3)} \\ &= \frac{1}{15}\frac{1}{s} + \frac{5}{6}\cdot\frac{1}{s+3} - \frac{9}{10}\cdot\frac{1}{s+5} \end{aligned}$$

Hence the output

$$y(t) = \frac{5}{6}e^{-3t} - \frac{9}{10}e^{-5t} + \frac{1}{15}$$

The second and the third terms are due to the input, the first term is due to the pole -3. The term e^{-2t} does not appear in $y(t)$, thus the pole -2 is not excited by the input.

c. Let us select an arbitrary input with a strictly proper function

$$U(s) = \frac{s+3}{s(s+4)} \implies u(t) = 0.75 + 0.25e^{-4t}$$

This input is bounded. The output due to this input becomes

$$\begin{aligned} Y(s) &= \frac{2s+1}{(s+2)(s+3)} \times \frac{s+3}{s(s+4)} = \frac{2s+1}{s(s+2)(s+4)} \\ &= \frac{1}{8}\frac{1}{s} + \frac{3}{4} \cdot \frac{1}{s+2} + \frac{7}{8} \cdot \frac{1}{s+4} \end{aligned}$$

Hence the output

$$y(t) = \frac{3}{4}e^{-2t} + \frac{7}{8}e^{-4t} + \frac{1}{8}$$

The second and the third terms are due to the input, the first term is due to the pole -2. The term e^{-3t} does not appear in $y(t)$, thus the pole -3 is not excited by the input.

In the last example, we found that the effects of two poles located at -2 and -3 appear in the output response due to the step input, that is, the two poles are excited by the step input. However, this is not the case when the system is excited by other types of inputs, as was demonstrated in the example. Our observations are summarized as follows

Summary

1. If the input signal $U(s)$ has no zeros, it can excite all the poles of the system. For example, a step input with

$$U(s) = \frac{1}{s}$$

or a sinusoidal input, with

$$U(s) = \frac{\omega}{s^2 + \omega^2}$$

can excite all the poles of a linear time-invariant system, because these functions have no zeros.

2. Conversely, if an input $u(t)$ or $U(s)$ has a zero that coincides with a system pole, that particular pole will not be excited.

1.10.3 The Role of Zeros

We saw that the general form of zero-state response is governed by the poles of the system. For example, for a system

$$Y(s) = G(s)U(s) = \frac{(s+5)(s+3)(s+1)^2}{s^3(s+2)^2(s+4+j6)(s+4-j6)}U(s)$$

the zero state response due to input U(s) is of the form

$$y(t) = k_1 + k_2 t + k_3 t^2 + k_4 e^{-2t} + k_5 te^{-2t} + k_6 e^{-(4-j6)t} + k_7 e^{-(4+j6)t} \\ + \text{Terms due to poles of } U(s) \tag{1.76}$$

The question arises: what are the roles of zeros of a system? Do they affect the zero-state response? The answer is yes. Certainly, the zeros affect the system response because they affect the values of the constants k_is, that is, the residues. These constants depend on both poles and zeros of G(s) as well as on U(s). The following example illustrates the effects of zeros on the nature of the system response.

EXAMPLE 1.18

Consider four systems whose transfer function models are:

$$G_1(s) = \frac{10}{(s+5)(s^2+6s+18)} = \frac{10}{s^3+11s^2+48s+90}$$

$$G_2(s) = \frac{5(s+2)}{(s+5)(s^2+6s+18)} = \frac{5s+10}{s^3+11s^2+48s+90}$$

$$G_3(s) = \frac{-5(s-2)}{(s+5)(s^2+6s+18)} = \frac{-5s+10}{s^3+11s^2+48s+90}$$

$$G_4(s) = \frac{(5/3)\,(s+1)(s+6)}{(s+5)(s^2+6s+18)} = \frac{1.667s^2+11.67s+10}{s^3+11s^2+48s+90}$$

The transfer function $G_1(s)$ does not have any zeros, $G_2(s)$ has a single zero located at -2. The zero of $G_3(s)$ is in the right half of the s-plane (located at $s = +2$). It is a non-minimum phase transfer function (to be discussed in Chapter 7) and $G_4(s)$ has two zeros located at -1 and -6. All the transfer functions have the same set of poles. The unit step responses of all these systems are of the form

$$y(t) = k_0 + k_1 e^{-5t} + k_2 e^{(-3+j3)t} + k_3 e^{(-3-j3)t} \tag{1.77}$$

where the constant k_3 is the complex conjugate of k_2.

The unit step responses of all these systems are shown in Figure 1.19.

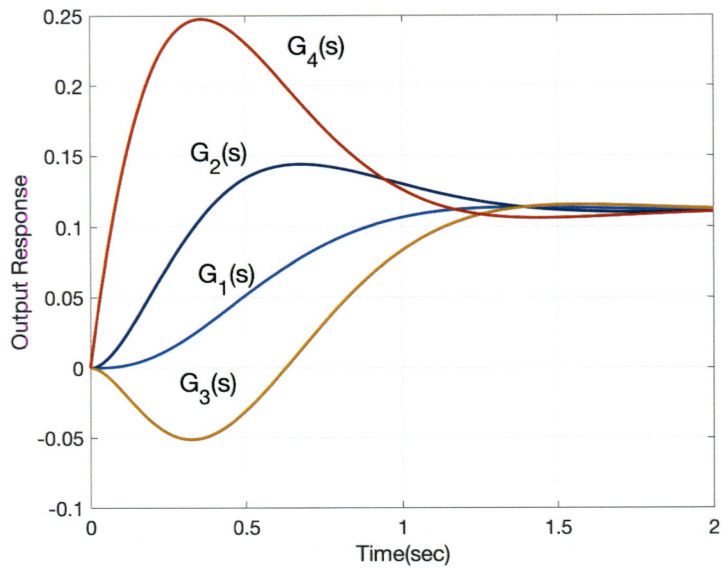

Figure 1.19: Effects of zero on system response

We observe that the responses are significantly different from each other, although all of them have the same poles.

Key Point

Even though the basic form of the response is determined from the poles, the exact response is determined by both the poles and the zeros, and the input. Hence, we cannot completely ignore the zeros of a system during analysis and design.

1.10.4 Physical Significance of Zeros

As shown in Example 1.18, the contribution of zeros in the system response cannot be ignored. Consider a linear time-invariant system that is stable. Let the transfer function G(s) of this system have a zero at $j\omega_z$ on the $j\omega$ axis. If this system is excited by a sinusoid of frequency ω_z rad/sec, the steady state output of the system will be zero. For example, the system with the transfer function

$$G(s) = \frac{s^2+4}{(s+1)(s+5)(s+10)} \tag{1.78}$$

has a pair of complex conjugate zeros at $\pm j2$. If it is excited by a sinusoid of frequency 2 rad/sec (say $\sin 2t$), the output will be zero.

This concept of system zeros can be further reinforced using relatable analogies:

Understanding Zeros

i. **The Selective Hearing Analogy:**
   - A teenager hears everything clearly
   - But when parents say "clean your room" → zero response
   - Certain inputs get completely blocked

ii. **The Noise-Canceling Headphones Analogy:**
   - Headphones work normally for most sounds
   - But at specific frequencies → complete cancellation (zero output)
   - The system "refuses" to respond at those frequencies

iii. **The Personality Resonance Analogy:**
   - When someone has characteristics that completely clash with yours → zero connection
   - They talk about things you have no interest in → you "tune out" completely
   - No matter how hard they try to engage you on that topic → zero response
   - This is like a zero; certain inputs get completely blocked

Key Points

1. The zeros of a system are the frequencies that are blocked by the system. These are the frequencies to which the system does not respond.
2. The zeros create a "notch" in the frequency response. The gain at those frequencies is attenuated.

1.11 The Unit Impulse Response g(t) of Linear Differential Systems

The impulse response $g(t)$ of a system equals the response of the system to an impulse input $\delta(t)$ applied at $t = 0$ with all the initial conditions zero at $t = 0^-$. Examples of impulse input are many. For example

i. **Hammer Striking a Bell:** When a hammer strikes a bell, the impact can be considered as an impulse input, and the resulting sound that is produced is the impulse response of the bell system.

ii. **Cricket Batsman Hitting the Ball:** When a cricket batsman hits the ball, the force is applied to the ball over a very short duration, which resembles an impulse input. Subsequently, the trajectory and speed of the ball following the hit are the system's response to this impulse.

iii. **Lightning Strike:** In an electrical circuit, a sudden voltage spike (like a lightning strike or a very brief pulse from a signal generator) can be modeled as an impulse input. The circuit's voltage and current response to this spike characterize its impulse response.

iv. **Martial Art:** When a martial artist intends to break hard objects such as bricks or wooden planks, etc., using his/her hands, he/she moves his/her hand swiftly and hits the object at a faster rate. By increasing the speed, the reaction time gets significantly reduced, and a huge magnitude of impulsive force is developed. This impulse force is strong enough to break the hard objects into pieces.

v. **Pestle and Mortar:** When a pestle strikes the mortar, a short-time force gets exerted on the spices contained in it. This force is very strong and has a small response time. Thus, the impulse force causes the objects present in the mortar to be crushed and break into smaller pieces.

An impulse input $\delta(t)$ acts on the system momentarily at $t = 0$, and then vanishes. However, during that brief moment, it stores energy in the system; that is, it creates non-zero initial conditions instantaneously within the system at $t = 0^+$. Although the impulse input $\delta(t)$ becomes zero for $t > 0$, and there is no input after this, the system exhibits a response at $t > 0$ due to these newly generated initial conditions. The impulse input, thus, excites the natural behavior (characteristic modes) of the system for $t \geq 0^+$. Hence

$$g(t) = \text{Terms consisting of characteristic modes}, \qquad t \geq 0^+$$

This response is valid for $t > 0$. However, at $t = 0$, there can at most be an impulse. So the general form of the complete impulse response $g(t)$ is

$$g(t) = K_0\delta(t) + \text{Terms consisting of characteristic modes}, \qquad t \geq 0^+$$

where K_0 is a constant.

Let us first illustrate the impulse response computation, considering the following two examples.

The dynamics of a system are described by

$$\frac{d^2y(t)}{dt^2} + 3\frac{dy(t)}{dt} + 2y(t) = 3\frac{du(t)}{dt} + u(t) \tag{1.79}$$

Applying Laplace transform to Eqn(1.79) gives

$$s^2Y(s) - sy(0^-) - \dot{y}(0^-) + 3\left[sY(s) - y(0^-)\right] + 2Y(s) = 3\left[sU(s) - u(0^-)\right] + U(s) \tag{1.80}$$

Since the response is computed assuming all initial conditions are zero, Eqn(1.80) becomes

$$Y(s) = \frac{3s+1}{s^2+3s+2}U(s) = \frac{3s+1}{s^2+3s+2}, \quad \text{since} \quad U(s) = 1 \tag{1.81}$$

or,

$$Y(s) = \frac{5}{s+2} - \frac{2}{s+1} \tag{1.82}$$

Thus, the impulse response is

$$\boxed{g(t) = y(t) = 5e^{-2t} - 2e^{-t}} \tag{1.83}$$

Note that -2 and -1 are the *characteristic modes* of this system.

Consider another example where the dynamics are described by

$$\frac{d^2y(t)}{dt^2} + 3\frac{dy(t)}{dt} + 2y(t) = \frac{d^2u(t)}{dt^2} + 8\frac{du(t)}{dt} + 15u(t) \tag{1.84}$$

Eqn(1.84) in the Laplace domain with zero initial conditions is expressed as

$$Y(s) = \frac{s^2+8s+15}{s^2+3s+2}U(s) = \frac{s^2+8s+15}{s^2+3s+2}, \quad \text{since} \quad U(s) = 1 \tag{1.85}$$

or,

$$Y(s) = 1 + \frac{8}{s+1} - \frac{3}{s+2} \tag{1.86}$$

Thus, the impulse response is

$$\boxed{g(t) = y(t) = \delta(t) + 8e^{-t} - 3e^{-2t}} \tag{1.87}$$

Let us extend the method to a more general case of an nth order linear system described by

$$\begin{aligned} &\frac{d^ny(t)}{dt^n} + a_1\frac{d^{n-1}y(t)}{dt^{n-1}} + \cdots + a_{n-1}\frac{dy(t)}{dt} + a_ny(t) \\ &= b_0\frac{d^mu(t)}{dt^m} + b_1\frac{d^{m-1}u(t)}{dt^{m-1}} + \cdots + b_{m-1}\frac{du(t)}{dt} + b_mu(t) \end{aligned} \tag{1.88}$$

Note that, for all practical systems, the order $m \leq n$. Consider the following two possible cases:

Case-1: m=n: In this case $b_0 \neq 0$, and the impulse response will therefore be of the form

$$g(t) = b_0\delta(t) + \text{Terms consisting of characteristic modes}, \qquad t \geq 0^+ \quad (1.89)$$

If the roots of the characteristic equation $\lambda_1, \lambda_2, \ldots, \lambda_n$ are distinct, then

$$\boxed{g(t) = b_0\delta(t) + c_1e^{\lambda_1 t} + c_2e^{\lambda_2 t} + \cdots + c_ne^{\lambda_n t}} \quad (1.90)$$

where c_1, c_2, ..., c_n are constants which can be determined using various methods such as impulse matching(not discussed here).

Case-2: m < n: In this case $b_0 = 0$ and the impulse response will be of the form

$$g(t) = \text{Terms consisting of characteristic modes}, \qquad t \geq 0^+ \quad (1.91)$$

1.12 Significance of Characteristic Modes and Their Relation to System Poles

The characteristic modes of a system play a pivotal role in determining its entire behavior. As we have established, the zero-input response of a system consists of these characteristic modes. For stable systems, these modes decay exponentially to zero over time. However, this decay does not diminish their profound influence on system behavior. On the contrary, their impact permeates every aspect of the system's dynamics.

Important Note

The characteristic modes of a system are integral to its behavior and dynamics, and leave their imprint on every aspect of system behavior.

1.12.1 Philosophical Perspective of Characteristic Modes

Everything in the cosmos is fundamentally energy, with every object vibrating at specific frequencies. When interpreted through this lens, we discover a profound truth about the interconnected nature of existence: at the most fundamental level, everything can be understood in terms of energy and vibrations. This concept bridges both physics and metaphysics.

Einstein's renowned equation, $E = mc^2$, demonstrates the fundamental equivalence of energy and matter, revealing their intrinsic interconnectedness. This scientific principle resonates with metaphysical interpretations, suggesting that a universal energy or consciousness connects all things in the universe. Such perspectives align beautifully with Eastern philosophical traditions, which emphasize the harmonious and vibrational nature of existence.

From this perspective, a system's characteristic modes can be viewed as its unique **"vibrational fingerprint"**; the distinctive energy patterns that define its behavioral essence. Just as every entity in the cosmos vibrates in harmony with universal frequencies, a system's modes determine its natural frequencies and resonances, fundamentally shaping its dynamic responses. These modes reflect the intricate interplay between inherent properties and external influences, symbolizing the profound interconnectedness that underlies all existence.

Natural Oscillation and System Response

When an impulse disturbance is applied to an object or a system and left to respond freely, it tends to vibrate preferentially at its characteristic modes. These modes represent the natural frequencies at which the system oscillates most efficiently. That is, every physical object or system has a unique set of characteristic modes, which are the frequencies at which it naturally oscillates when disturbed. These frequencies are determined by the physical properties of the object, such as its mass, stiffness, and shape.

Extending this concept to the cosmic scale, celestial bodies such as planets, stars, and galaxies all exhibit their own natural vibrational frequencies. These cosmic vibrations arise from various sources: internal processes like planetary seismic activity or stellar oscillations, and external influences such as gravitational interactions between celestial bodies. This cosmic perspective illuminates the universal presence of natural frequencies, creating a bridge between philosophical understanding and physical reality.

The DNA Analogy: Understanding Persistent Influence

For stable systems, characteristic modes decay exponentially and eventually vanish as time approaches infinity. Yet this temporal decay does not diminish their profound impact on system behavior. A compelling analogy can be drawn with human DNA, which encodes the intrinsic blueprint of life.

The Genetic Blueprint Parallel

1. **Fundamental Definition** Just as DNA contains the genetic code that defines a human being's fundamental traits such as physical features, growth patterns, and predispositions to specific abilities, the characteristic modes of a system represent its intrinsic behavioral blueprint, specifying natural frequencies, resonances, and damping patterns.

2. **Development Through Interaction**

   Human development arises from the interplay between the genetic code and the environment. Likewise, a system's response is shaped by the dynamic interaction of disturbances with its characteristic modes, determined by mass, stiffness, and damping.

In the same way that a person's life is the product of both genetic inheritance and environment, a system's behavior emerges from the interplay between its inherent modal characteristics and external perturbations.

Resonance Phenomenon and Characteristic Modes in Dynamic Systems

There is a clear relationship between resonance and characteristic modes, which governs how a system reacts to inputs aligned with its natural frequencies. As discussed, characteristic modes represent the intrinsic frequencies at which a system naturally oscillates when disturbed, determined by its physical properties. Resonance occurs when the frequency of an external input matches one of these modes. At resonance, the system absorbs energy most efficiently, leading to significant amplification of its oscillatory response. In essence, resonance is the system's intensified reaction to inputs perfectly aligned with its characteristic modes, much like a swing gaining momentum when pushed in sync with its motion. This alignment maximizes energy transfer, producing pronounced oscillations.

The Personality Resonance Analogy

The concept of poles and zeros in system response can be understood through an everyday human analogy.

Understanding Resonance

- **Poles:** Meeting someone whose personality perfectly matches yours creates an instant and amplified emotional response. Shared humor, values, and interests turn a small initial interaction into an overwhelming response, similar to how a system exhibits a large output when the input frequency coincides with its natural frequency.

- **Zeros:** In contrast, meeting someone with completely mismatched traits results in no connection at all. No matter how much effort they make, the response is effectively zero; just as certain inputs are fully blocked in a system due to the presence of zeros.

The following numerical example illustrates this resonance-mode relationship in greater detail.

EXAMPLE 1.19

Consider a simple first-order system with a transfer function.

$$G(s) = \frac{K}{s+a} \tag{1.92}$$

where K is a constant. This system has a single characteristic mode e^{-at}. Let us excite this system with an input e^{-bt} and compute its output, considering two cases.

Case-1: Input is similar (not identical) to the characteristic mode Now

$$\begin{aligned} Y(s) = G(s)U(s) &= \frac{K}{s+a} \times \frac{1}{s+b} \\ &= \frac{K}{a-b}\left[\frac{1}{s+b} - \frac{1}{s+a}\right] \end{aligned} \tag{1.93}$$

Hence

$$y(t) = \frac{K}{a-b}\left[e^{-bt} - e^{-at}\right] \tag{1.94}$$

From Eqn(1.94), it is obvious that if the input e^{-bt} is similar to e^{-at}, $(a-b)$ would be small and the system response becomes large. On the contrary, if the input is significantly different from the characteristic mode, $(a-b)$ becomes large and the system response becomes poor(small). Thus,

1. The closer the input u(t) to the characteristic mode, the stronger (larger) the system response.
2. The farther the input u(t) from the characteristic mode, the poorer (smaller) the system response.

The remarks made for a first-order system with a single mode can be generalized for an n-th order system with n-number of characteristic modes. Since the impulse response g(t) of an n-th order system is a linear combination of its n-modes, if the input signal is similar to any of the modes, the corresponding response will be high. However, if it is not similar to any of the modes, the response will be small.

Case-2: Input is identical to the characteristic mode

Intuitively, if a system is excited by a signal whose frequency is equal to the natural frequency of the system, the system will resonate. Let us examine whether this is the case. As before, consider a first-order system with transfer function

$$G(s) = \frac{K}{s + a} \tag{1.95}$$

For an input $u(t) = e^{-at}$, the output becomes

$$Y(s) = G(s)U(s) = \frac{K}{(s + a)^2} \tag{1.96}$$

Hence

$$y(t) = Kte^{-at} \tag{1.97}$$

Assume that $a > 0$. Hence, the factor e^{-at} decays faster than t and $y(t) \to 0$ as $t \to \infty$. Note that the resonance phenomenon is present. However, it could not be manifested due to the signal's own exponential decay. The response of the system of Eqn(1.95) to an input e^{-at} for $a = 2$ is shown in Figure 1.20.

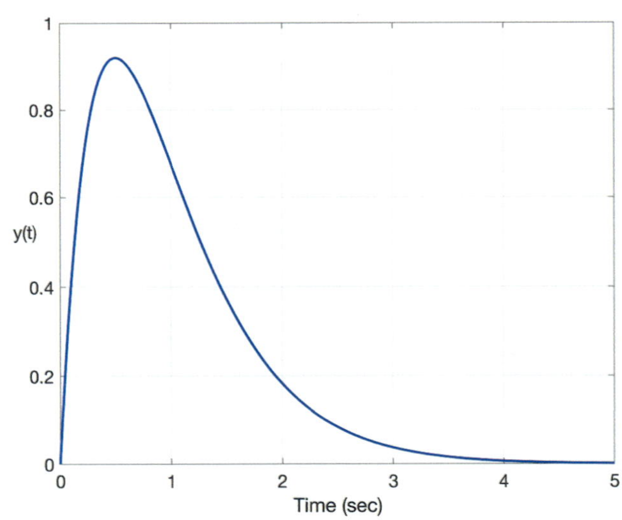

Figure 1.20: System response build up during resonance for system of Eqn(1.95) for K = 5 and a = 2

This shows that the resonance phenomenon is not instantaneous; it is cumulative. In this example, the response increases linearly with t. However, if the characteristic root repeats r times, the resonance effect increases as t^{r-1}. However, $t^{r-1}e^{-at} \to 0$ as $t \to \infty$ for any value of r provided $a < 0$ or in a more general case if the real part of the characteristic mode is negative. Thus, if the mode decays exponentially, the signal decays too fast to zero before the resonance effects build up.

The resonance effects can be seen in practice if the mode decays at a rate less than 1/t. For a system with a mode $e^{\lambda t}$, this is possible if $\mathrm{Real}(\lambda) \geq 0$. For example, consider a system whose natural modes lie purely imaginary, that is

$$\lambda = \pm j\omega$$

The transfer function of the system is

$$G(s) = \frac{K}{(s + j\omega)(s - j\omega)} = \frac{K}{s^2 + \omega^2} \tag{1.98}$$

If we apply an input $u(t) = \cos\omega t$, the output becomes

$$Y(s) = G(s)U(s) = \frac{Ks}{(s^2 + \omega^2)^2} \tag{1.99}$$

Thus, the output is

$$y(t) = \frac{K}{2\omega} t \sin\omega t \tag{1.100}$$

From Eqn(1.100), it is clear that the system response to its characteristic mode increases linearly with time, eventually reaching ∞, as shown in Figure 1.21 for $K = 5$ and $\omega = 10$ rad/sec.

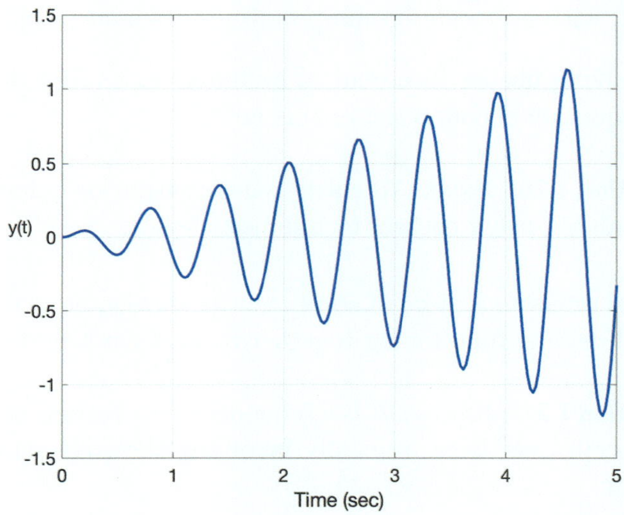

Figure 1.21: System response build up during resonance for system of Eqn(1.98) for $K = 5$ and $\omega = 10$ rad/sec

From the discussion about the characteristic modes and definition of the poles, we summarize the following:

Key Points

1. The roots of the characteristic equation (characteristic polynomial) determine the natural (or characteristic) modes of a system. These are referred to as the *internal modes* of the system.

2. The poles of a system form a subset of these characteristic modes and describe the *external modes*. In other words,

$$\text{Number of poles} \leq \text{Number of characteristic modes.}$$

1.13 What Is a Control System?

A **control system** is a system composed of an interconnection of various subsystems and processes designed to achieve a desired performance. For example, consider a thermostat controlling the temperature of a room. The thermostat (the controller) switches the heater on or off to minimize the difference (or error) between the actual temperature and the desired temperature. Control systems are developed to fulfil diverse objectives. For instance, they may optimize the operation of a plant to achieve maximum efficiency at minimal cost, enhance the performance of systems by accelerating their dynamic response with zero steady-state error, or stabilize a passenger cruise system to mitigate turbulence and improve passenger comfort. Regardless of the specific purpose of a control system, system stability remains the foremost priority. Stability is critical, with no room for compromise in its assurance.

Before delving deeper into control systems, let us first define some common terms frequently encountered in this field.

Definition 1.26. *Plant:* An entity whose output or behavior is controlled or regulated using control techniques.

This often refers to a piece of equipment or device, or a combination of equipment and devices functioning together to perform a particular operation.

Definition 1.27. *Process:* It is a sequence of operations or transformations that occur over time, typically involving the conversion of inputs into outputs.

Processes in control systems often exhibit gradual changes over time as inputs are transformed into outputs. They typically follow a series of steps or stages that occur in a relatively fixed or predictable manner. These steps may be interconnected, with the output of one step serving as the input to the next. Some of the examples include chemical manufacturing processes where raw materials undergo a series of chemical reactions and separations to produce desired products; biological processes such as fermentation, where microorganisms convert sugars into alcohol and carbon dioxide.

Definition 1.28. *System:* A "system" refers to an interconnected set of components or elements that work together to achieve a specific objective.

Systems can range from simple to highly complex, and they can encompass physical, biological, or abstract entities. Systems can also include the plants or processes being controlled, along with any additional components or elements involved in the control process, such as sensors, actuators, and controllers.

While there may be subtle distinctions between the terms **plants, processes, and systems** in specific contexts, they often overlap in meaning and will be used interchangeably in this book.

Definition 1.29. *Disturbances:*. It refers to external influences or changes that impact the behavior of the controlled process.

Disturbances can arise from various sources and can have significant impacts on the performance and stability of the controlled process. In a control system, they can originate internally as well as externally. While external disturbances come from factors outside the control system, internal disturbances arise from within the system itself or from closely related subsystems. Examples of internal disturbances include variations in component performance due to aging or wear, changes in system dynamics caused by temperature fluctuations, or coupling effects between control loops in a multi-loop system. Some examples of external disturbances include temperature variations, changes in humidity, fluctuations in supply voltages, variations in load conditions, and disturbances from adjacent machinery or processes, and so on.

Definition 1.30. *Sensor:* It is a device or component that detects and responds to a physical stimulus or environmental change, converting it into an electrical, optical, or other measurable signal.

Sensors come in a wide variety of types and designs, each tailored to specific applications and operating environments. They can be categorized based on their sensing principles, such as resistive, capacitive, inductive, optical, piezoelectric, magnetic, or biochemical sensing, and so on. Common types of sensors include temperature sensors, pressure sensors, proximity sensors, motion sensors, light sensors (photo-detectors), accelerometers, gyroscopes, humidity sensors, gas sensors, biosensors, and more. They are indispensable tools for capturing data, monitoring processes, and control systems.

> **Definition 1.31.** *Controller:* It is a device or component that generates control signals based on feedback from the system to regulate its behavior or operation

Controllers play a central role in control systems by continuously monitoring the system's output, comparing it to a desired reference or setpoint, and adjusting the system's inputs or control variables to maintain or achieve the desired performance or behavior.

> **Definition 1.32.** *Actuator:* It is a device that translates control signals from the controller into physical action or movement, exerting control over the system's output or manipulated variables.

Upon receiving the control signal, the actuator translates it into physical action or movement to achieve the desired control objective. They may produce linear motion (such as extending or retracting a piston), rotary motion (such as rotating a shaft or valve), or other types of motion depending on the application and design of the actuator. Actuators come in various types and designs, each tailored to specific applications and operating conditions. Common types of actuators include electric actuators (such as motors and solenoids), pneumatic actuators (such as cylinders and valves powered by compressed air), hydraulic actuators (such as hydraulic cylinders and pumps), and electromechanical actuators (such as piezoelectric actuators and stepper motors).

1.14 Classification of Control Systems

Control systems can broadly be classified as:

1. Open-loop control system
2. Closed-loop control system

Before we differentiate between open and closed-loop systems and formally define these systems, let us understand from a few relatable analogies.

Understanding Control Systems

1. **The Cooking with Mom Analogy:** Think about cooking with your mother watching.
   - An open-loop system is like cooking alone; you add salt, cross your fingers, and hope for the best.

- A closed-loop system is like cooking with mom hovering over your shoulder, constantly tasting and saying things like "Add more salt," "Too much salt!" "What did you do to my recipe?!" Overwhelming? Perhaps. Effective? Absolutely. Your dish turns out perfect.

2. **The Backseat Driver Analogy:** Let us understand from the universal driving experience.

   - An open-loop system is like driving solo to a new place, relying solely on your memory of someone's verbal directions: "Just head toward the city center, then look for the big sports field, turn left somewhere around there." You drive with complete confidence, making your best guesses at every junction, never checking if you're actually on the right path, and inevitably end up hopelessly lost.
   - A closed-loop system is like driving with your spouse as a backseat driver: "Slow down! Speed up! You're too close! There's a parking spot! No wait, that one! Actually, go around the roundabout again!" They provide constant "feedback" to optimize your driving, whether you want it or not. Overwhelming? Perhaps. But you'll actually reach your destination instead of ending up at a gas station in the middle of nowhere, asking directions from someone who's lived in Fulwood for 30 years and still does not know where Endcliffe Avenue is.

Having grasped the concept through these relatable scenarios, we can now formally define these two fundamental system configurations.

1.14.1 Open-Loop Control System

Definition 1.33. A control system where the control input is not dependent on or affected by the output is called an open-loop control system. It is a system without feedback.

The schematic of an open-loop system is shown in Figure 1.22 which consists of a controller, an actuator, and the plant.

Observe that the output is neither measured nor given feedback for comparison with the input.

Many commonly used home appliances serve as excellent examples of open-loop control systems, where operation is based solely on predefined settings

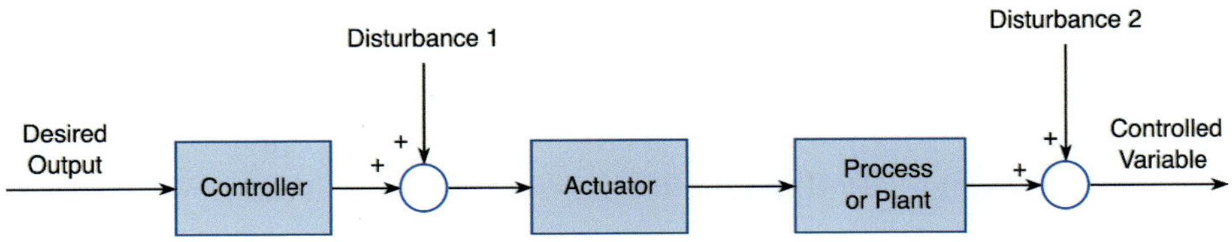

Figure 1.22: Open-loop control system

without feedback to adjust or correct performance. Here are a few examples of open-loop control systems commonly found in home appliances:

Electric Dryer

The block diagram of an electric dryer is shown in Figure 1.23.

It is an open-loop system where the user manually sets the operating duration by adjusting a timer based on their estimated drying needs. The dryer operates for the specified time without monitoring or adjusting for factors such as the actual moisture level of the clothes or ambient conditions, making it a simple and user-dependent system.

Some other examples include **toasters** which operates for a preset time based on user settings, producing toast regardless of the actual browning level achieved; **microwave ovens (manual setting)** where we heat food for a specified duration and power level without sensing the food's temperature; **electric kettles (without thermostat)** where we heat water until manually turned off, without monitoring the water temperature; **hair dryers** which function at a fixed heat and airflow setting, without detecting the hair's moisture level or temperature; **basic washing machine** which executes wash cycles according to preset programs, without checking the cleanliness of the clothes or the water condition. Some other examples of open-loop control systems are traffic light control systems, which operate on a pre-programmed timing sequence regardless of the actual traffic conditions at the intersection; an **irrigation sprinkler system** that operates on a timer, watering plants at predetermined intervals. These systems are straightforward, reliable, and

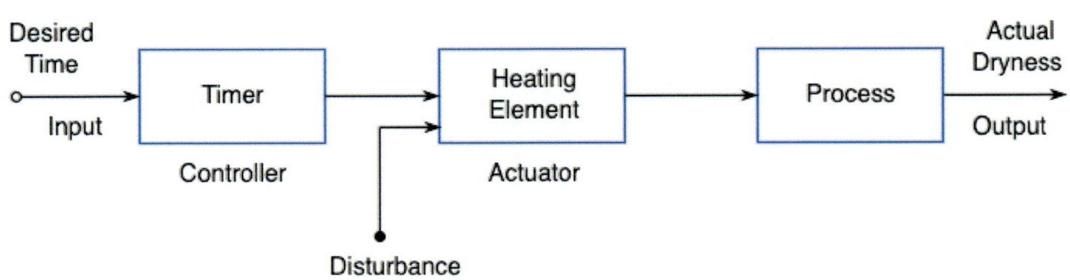

Figure 1.23: Open-loop electric-dryer control system

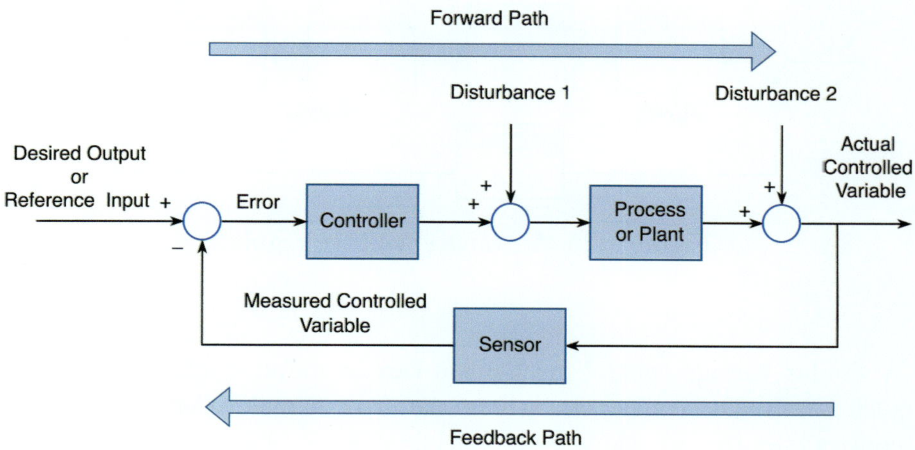

Figure 1.24: Closed-loop control system

cost-effective, but can lack efficiency and precision in scenarios where feedback-based adjustments are needed.

1.14.2 Closed-Loop Control System

Definition 1.34. A control system where the control input is dependent on or influenced by the output is called a closed-loop control system. It is a system with feedback.

The schematic of a closed-loop control system is shown in Figure 1.24, which consists of a controller, an actuator, a plant, and a sensor.

Observe that, in a closed-loop control system, the output is continuously monitored and the input to the system is adjusted to give the desired output or response. Thus, a closed-loop control system is characterized by its reliance on feedback from the system's output to determine the control input. This feedback mechanism enables the system to regulate its behavior and maintain desired performance, making it a versatile and widely used approach in various engineering and control applications.

In the following, we discuss the operating principles of some simple yet commonly encountered closed-loop control systems to provide a clear and relatable understanding of their fundamental concepts.

1.14.3 Examples of Closed-Loop Control Systems

1. **Automatic Voltage Regulator**

Figure 1.25 shows the schematic of an automatic voltage regulator (AVR) in a power system. It compares the reference voltage with the actual generator voltage, and the resulting error signal is used to adjust the excitation current to ensure a stable output voltage.

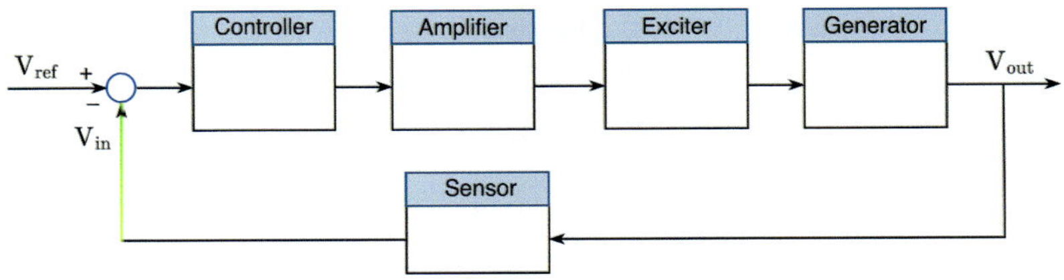

Figure 1.25: Automatic voltage regulator

A similar voltage control mechanism can be found in other control systems, such as DC-DC converters, where the error signal is fed to the converter's control circuitry to regulate the output voltage. Another example is wireless power transfer systems, where voltage control is achieved using power converters, such as buck (step-down) or boost (step-up) converters, which maintain the appropriate voltage level regardless of fluctuations in input power or load conditions.

2. **Cruise Control System**

   This can either be automatic or driver-based. In an automatic cruise control system, commonly found in automobiles, the speed is maintained at a speed set by the driver without the need for continuous manual input on the accelerator pedal. This is achieved by controlling the throttle (actuator) input based on feedback from vehicle speed sensors (see Figure 1.26).

   While automatic cruise control systems offer automation and convenience for highway driving, driver-based speed control provides drivers with the flexibility and control necessary to navigate diverse driving environments safely and efficiently. The block diagram of a driver-based speed control system is shown in Figure 1.27.

3. **Automatic Wiper in Automobiles**

   The schematic block diagram of a smart windshield wiper is shown in Figure 1.28.

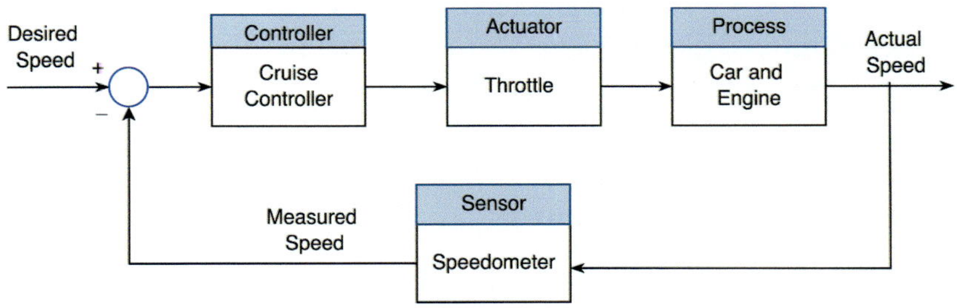

Figure 1.26: Automatic cruise control system

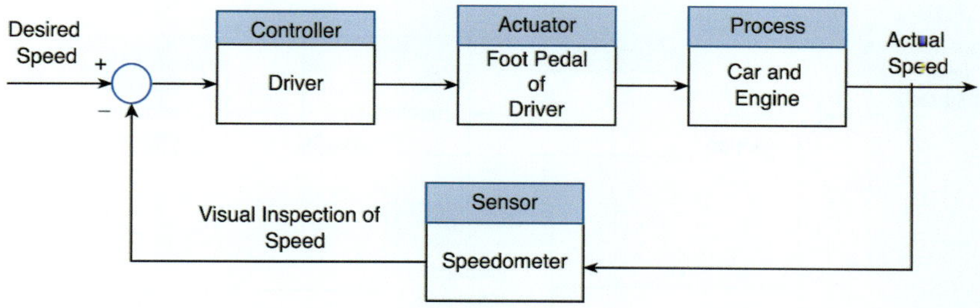

Figure 1.27: Driver-based cruise control system

The smart wiper automatically turns itself on or off and adjusts its wiping cycle (speed) according to the intensity of rain. There are several components working together to automate the wiper operation based on the intensity of rain. The rain intensity sensor detects the amount of rain falling on the wind shield and provides this information to the control system. The control system uses the rain intensity information to look up the desired wiper speed from the lookup table, which stores predefined mappings between rain intensity levels and desired wiper speeds. Based on the desired wiper speed obtained from the lookup table, the control system generates control signals to adjust the wiper speed accordingly.

4. **Autonomous Navigation System: A Robotic Vacuum Cleaner**

An autonomous navigation system is a technology framework that enables machines, robots, or vehicles to navigate their environment without human intervention. It leverages a combination of sensors, algorithms, and decision-making processes to perceive surroundings, determine position, plan a path, and execute movements. Autonomous navigation plays a crucial role in applications such as self-driving cars, drones, robotic systems, and underwater or space exploration vehicles.

A simple and practical example of this system is a robotic vacuum cleaner, as shown in Figure 1.29.

This device uses onboard sensors and control algorithms to autonomously navigate its environment, detect obstacles, and efficiently clean surfaces without manual guidance.

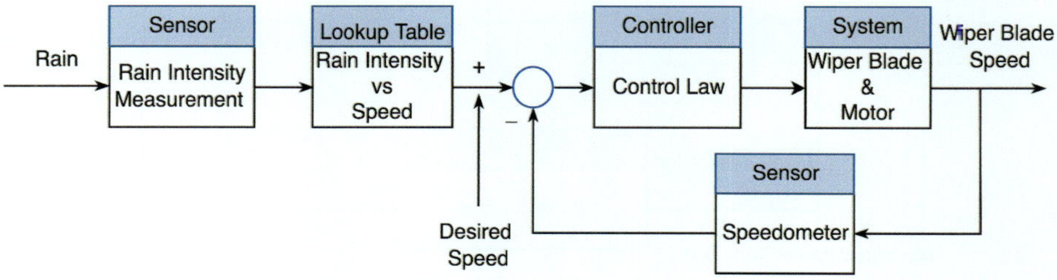

Figure 1.28: Smart wiper control system

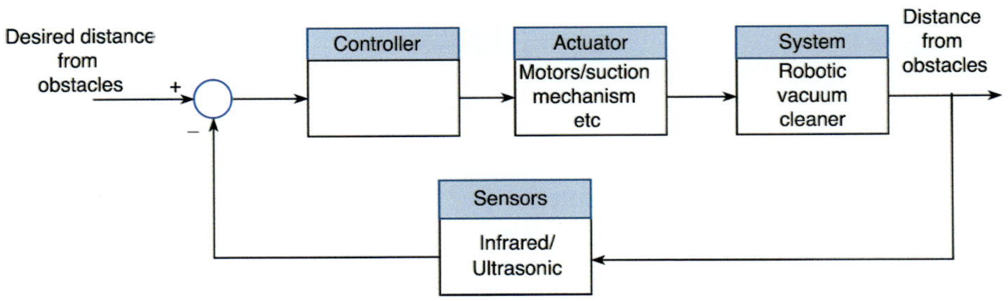

Figure 1.29: Robotic vacuum cleaner: an autonomous navigation system

5. **Home Shower System**

The schematic block diagram of a typical home shower control system is shown in Figure 1.30.

This system is designed to deliver water at a desired temperature and flow rate. It is modeled as a two-input system, where the inputs are the flow contributions from the hot water supply and the cold water supply. The user adjusts the mixing valve to regulate these two inputs. The outputs of the system are the resulting water temperature and the overall flow rate. Thus, the shower system serves as a practical example of a multi-variable control problem: achieving a comfortable shower requires balancing the hot and cold water inputs.

6. **Control of HVAC Systems**

An HVAC (Heating, Ventilation, and Air Conditioning) control system is designed to regulate indoor environmental conditions, including temperature,

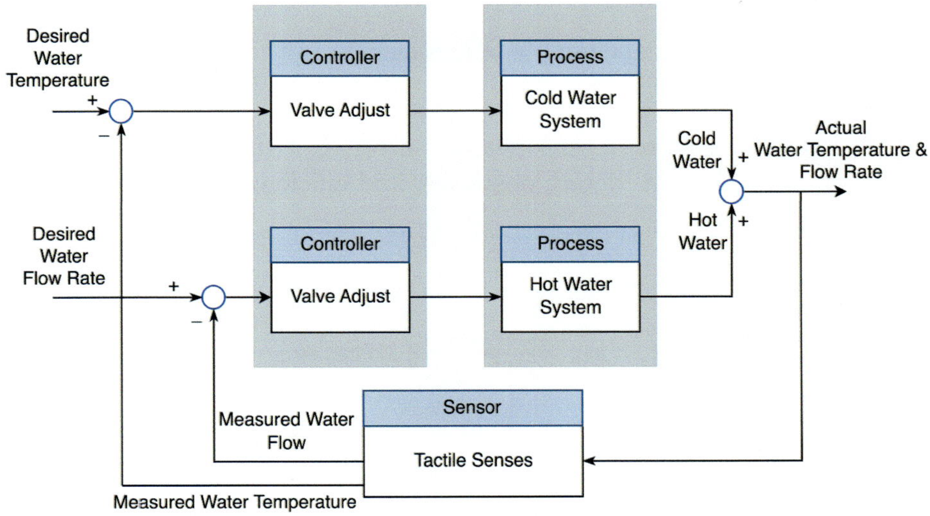

Figure 1.30: Home shower control system

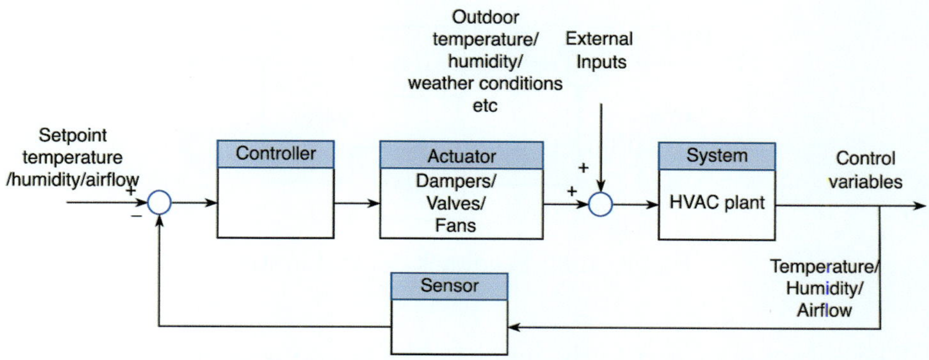

Figure 1.31: Heating, ventilation, and air conditioning system

humidity, air quality, and airflow. Its primary objective is to ensure a comfortable and healthy indoor environment while maximizing energy efficiency. A simplified block diagram of an HVAC system is shown in Figure 1.31.

The specific control variables depend on the system's design and purpose, but commonly controlled parameters include temperature, humidity, and air quality. The user sets predefined values, or set points, for these variables, which are continuously compared with the actual values to calculate error signals. The controller processes these error signals and generates appropriate control commands to achieve the desired conditions. These commands are executed by actuators, such as dampers, valves, or fans, which adjust the operation of the physical HVAC system. The physical system consists of components for heating, cooling, ventilation, and air circulation, working together to maintain the specified indoor environment.

1.15 Characteristics of Feedback Systems

Feedback systems can indeed be both advantageous and challenging, depending on their design and application. Understanding these effects is essential for creating systems that meet performance objectives while mitigating potential drawbacks. In the following sections, we briefly discuss the characteristics and implications of feedback systems to provide a comprehensive understanding of their role in control applications.

1.15.1 Effects of Feedback on System Gain

The block diagram of an open-loop system with a feed-forward path gain G(s) is shown in Figure 1.32.

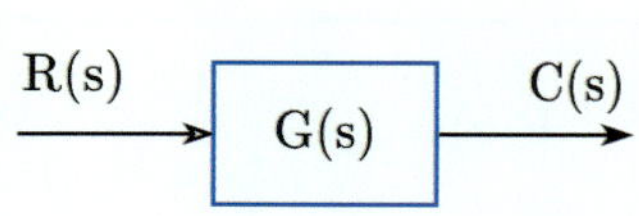

Figure 1.32: Open-loop control system

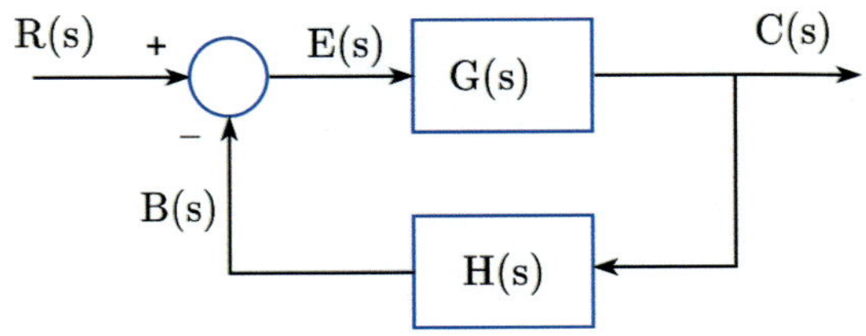

Figure 1.33: Feedback control system

The output is related to the input by the following relation:

$$C(s) = G(s)R(s), \text{ or } \frac{C(s)}{R(s)} = G(s) \tag{1.101}$$

The block diagram of a feedback system is shown in Figure 1.33.

From the schematic, the transform of the error is given as:

$$E(s) = R(s) - B(s) = R(s) - C(s)H(s) \tag{1.102}$$

The transform of the output is given as:

$$C(s) = G(s)E(s) \tag{1.103}$$

Substituting the value of E(s) from Eqn(1.102) in Eqn(1.103) gives

$$\begin{aligned} C(s) &= G(s)E(s) = G(s)\left[R(s) - C(s)H(s)\right] \\ &= G(s)R(s) - G(s)C(s)H(s) \end{aligned}$$

or,

$$C(s)\left[1 + G(s)H(s)\right] = G(s)R(s)$$

The output is expressed as

$$\boxed{C(s) = \frac{G(s)}{1 + G(s)H(s)}R(s)} \tag{1.104}$$

The actuating error is given by

$$\boxed{E(s) = \frac{1}{1 + G(s)H(s)}R(s)} \tag{1.105}$$

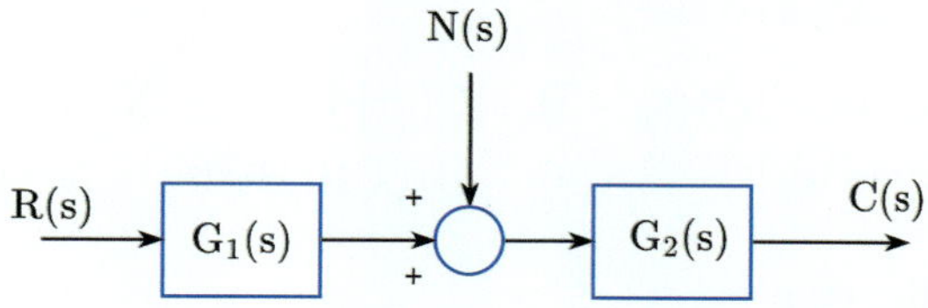

Figure 1.34: Open-loop control system with noise

Key Point

The feedback affects the gain of the open-loop system by a factor $1 + G(s)H(s)$. The error can be reduced by keeping the loop gain $G(s)H(s)$ large in the frequency range of interest.

1.15.2 Effect of Feedback on Noise

To know the effect of feedback on noise, let us compare the transfer function relations with and without feedback due to the noise signal alone. Consider an open-loop control system with the noise signal as shown in Figure 1.34.

From Figure 1.34, the output is expressed as:

$$\begin{aligned} C(s) &= [G_1 R(s) + N(s)]\, G_2 \\ &= G_1 G_2 R(s) + G_2 N(s) \end{aligned} \tag{1.106}$$

The open-loop transfer function, due to noise alone, can be obtained by making $R(s) = 0$, which gives

$$\boxed{\frac{C(s)}{N(s)} = G_2(s)} \tag{1.107}$$

Next, consider the closed-loop control system with noise signal as shown in Figure 1.35.

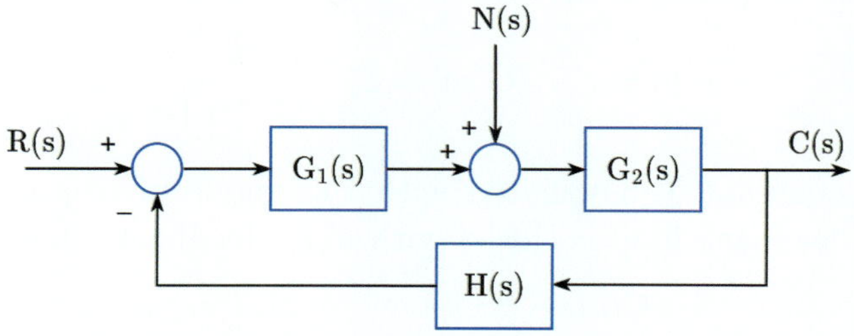

Figure 1.35: Feedback control system

The output is expressed as:

$$\begin{aligned} C(s) &= [\{R(s) - H(s)C(s)\}\, G_1(s) + N(s)]\, G_2(s) \\ &= [G_1(s)R(s) - G_1(s)H(s)C(s) + N(s)]\, G_2(s) \end{aligned} \tag{1.108}$$

Further simplification gives

$$C(s) + G_1G_2HC(s) = G_1(s)G_2(s)R(s) + G_2(s)N(s)$$

or,

$$C(s)\,[1 + G_1(s)G_2(s)H(s)] = G_1(s)G_2(s)R(S) + G_2(s)N(S) \tag{1.109}$$

Thus

$$C(s) = \frac{G_1(s)G_2(s)}{1 + G_1(s)G_2(s)H(s)}R(s) + \frac{G_2(s)}{1 + G_1(s)G_2(s)H(s)}N(s) \tag{1.110}$$

The closed-loop transfer function due to noise alone can be obtained by making $R(s) = 0$, which gives

$$\boxed{\frac{C(s)}{N(s)} = \frac{G_2(s)}{1 + G_1(s)G_2(s)H(s)}N(s)} \tag{1.111}$$

From Eqns(1.107) and (1.111), we conclude that

Key Point

In the closed-loop control system, the gain due to noise signal is decreased by a factor of $(1 + G_1(s)G_2(s)H(s))$. Note that, in most practical control systems, $(1 + G_1(s)G_2(s)H(s))$ is greater than one.

1.15.3 Effect of Feedback on Speed of Response

Consider an open-loop system

$$G(s) = \frac{C(s)}{R(s)} = \frac{K}{s + a} \tag{1.112}$$

The impulse response of this system is given by

$$c(t) = \mathcal{L}^{-1}[C(s)] = \mathcal{L}^{-1}\left[\frac{K}{s + a}\right] = Ke^{-at} \tag{1.113}$$

The time constant T associated with this mode of response equals to $1/a$. When the feedback loop is closed with unity feedback, then

$$\begin{aligned} \frac{C(s)}{R(s)} &= \frac{G(s)}{1 + G(s)} = \frac{K/(s + a)}{1 + K/(s + a)} \\ &= \frac{K}{s + (a + K)} \end{aligned} \tag{1.114}$$

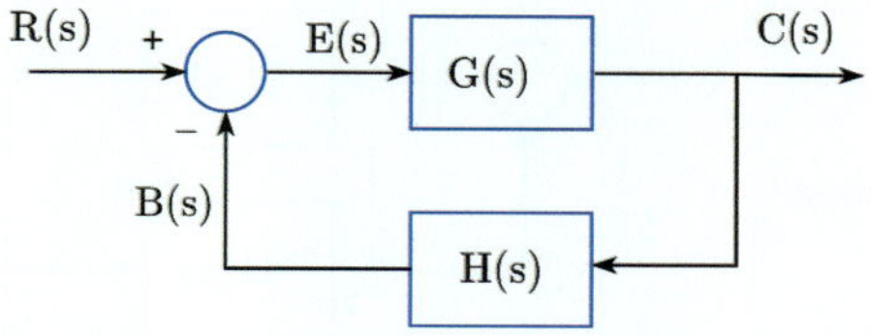

Figure 1.36: Feedback control system

The impulse response of this system is given by

$$\begin{aligned} c(t) &= \mathcal{L}^{-1}\left[C(s)\right] = \mathcal{L}^{-1}\left[\frac{K}{s+a+K}\right] \\ &= Ke^{-(a+K)t} \end{aligned} \tag{1.115}$$

The time constant T, associated with this mode of response, equals to $1/(a+K)$. Thus, the speed of response can be changed by selecting different values of gain K.

Key Takeaway

The feedback can increase the speed of response

1.15.4 Effect of Feedback on System Stability

The block diagram of a feedback system is shown in Figure 1.36.

The relation between the input and output is expressed as:

$$\boxed{\frac{C(s)}{R(s)} = \frac{G(s)}{1+G(s)H(s)}} \tag{1.116}$$

From Eqn(1.116), it is obvious that when $G(s)H(s) = -1$, the output of the system will be infinite for any finite input and the system will be unstable. Thus

Now, we will demonstrate that feedback can stabilize an unstable system. Let us introduce another feedback loop through a negative feedback gain F(s) as shown in Figure 1.37.

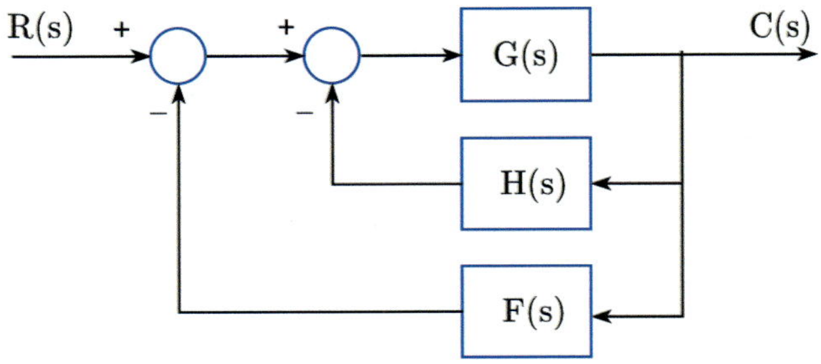

Figure 1.37: Two-loop feedback control system

The relationship between the output and input is given as:

$$\frac{C(s)}{R(s)} = \frac{G(s)}{1 + G(s)H(s) + G(s)F(s)} \tag{1.117}$$

Suppose the gains of the inner feedback loop are such that $G(s)H(s) = -1$. This will make the inner loop unstable. However, from Eqn(1.117), it is obvious that the overall system can be made stable by proper selection of outer loop gain $F(s)$.

Remark

Feedback can cause a system that is originally stable to become unstable and vice versa.

1.15.5 Reduction of Effects of Parameter Variations

The various parameters of a control system often change with environment and age, and several other unknown factors that affect the system. In an open-loop system, the variations in the parameters change the model of the system and thereby give inaccurate output. However, a closed-loop system senses the change in the output and can correct these errors. So, intuitively, the feedback can reduce the effects of parameter variations.

Consider the case of an open-loop control system where

$$C(s) = G(s)R(s) \tag{1.118}$$

Suppose, due to parameter variations $G(s)$ changes to $[G(s) + \Delta G(s)]$. The output of the open-loop system, therefore, changes to

$$C(s) + \Delta C(s) = [G(s) + \Delta G(s)]\, R(s) \tag{1.119}$$

Thus

$$\Delta C(s) = \Delta G(s)R(s) \tag{1.120}$$

Similarly, in the closed-loop case, the output is given by

$$C(s) = \frac{G(s)}{1 + G(s)H(s)} R(s) \tag{1.121}$$

Due to variation $\Delta G(s)$ in the forward path transfer function, this changes to

$$\begin{aligned} C(s) + \Delta C(s) &= \frac{G(s) + \Delta G(s)}{1 + [G(s) + \Delta G(s)]H(s)} R(s) \\ &= \frac{G(s) + \Delta G(s)}{1 + G(s)H(s) + \Delta G(s)H(s)} R(s) \end{aligned} \tag{1.122}$$

Since $|G(s)| \gg |\Delta G(s)|$, the variation in the output can be expressed as:

$$\boxed{\Delta C(s) = \frac{\Delta G(s)}{1 + G(s)H(s)} R(s)} \tag{1.123}$$

Key Takeaway

Thus, compared to the open-loop system, the change in the output of the closed-loop system due to variations in $G(s)$ is reduced by a factor of $[1 + G(s)H(s)]$ which is often much greater than unity.

1.15.6 Effect of Feedback on Sensitivity

One of the most important objectives of a feedback control system is to reduce the sensitivity of the closed-loop system due to variations in the process transfer function. In fact, this was the principal reason for the development of feedback amplifiers. A good control system should be very insensitive to parameter variations, but sensitive to input commands.

To understand what sensitivity means in practical terms, consider the analogy of **driving a car on different road surfaces**:

Understanding Sensitivity

- **High Sensitivity (like driving on ice):** Small steering inputs cause large changes in direction. The car "overreacts" to every little adjustment you make, requiring extremely gentle and precise control. Small disturbances (like a gust of wind) can send you sliding off course, and recovery from mistakes is difficult and can lead to over-correction.
- **Low Sensitivity (like driving in thick mud):** You have to turn the wheel significantly to get any response. The car feels "sluggish"

and unresponsive, requiring substantial input to change course. While external disturbances barely affect you, you also can't react quickly when needed, and you might not notice you're off course until it's too late to correct easily.

- **Optimal Sensitivity (like driving on dry pavement):** Steering inputs produce predictable, proportional responses. You can make smooth corrections without overcompensating, the car responds well to your intentions but isn't jumpy, and you can handle reasonable disturbances while maintaining control.

Similarly, in control systems, we want optimal sensitivity; the system should be responsive enough to follow commands accurately, but not so sensitive that small parameter variations or disturbances cause erratic behavior.

In the following, we will define several sensitive functions.

1. **System Sensitivity due to Variations in Process Transfer Function**

This describes the relative variation in the closed-loop transfer function $T(s) = C(s)/R(s)$ due to variation in process transfer function $G(s)$ and is defined as:

$$\text{Sensitivity} = \frac{\text{Percentage change in } T(s)}{\text{Percentage change in } G(s)} \tag{1.124}$$

For small incremental variation in $G(s)$, the sensitivity of $T(s)$ with respect to $G(s)$ is expressed quantitatively as:

$$S_G^T = \frac{\partial T/T}{\partial G/G} = \frac{\partial \, \mathrm{Ln}T}{\partial \, \mathrm{Ln}G} \tag{1.125}$$

The sensitivity of the closed-loop system is

$$\begin{aligned} S_G^T &= \frac{\partial T}{\partial G} \times \frac{G}{T} = \frac{(1+GH) - GH}{(1+GH)^2} \times \frac{G}{G/(1+GH)} \\ &= \frac{1}{1+GH} \end{aligned} \tag{1.126}$$

Thus

$$\boxed{S_G^T = \frac{1}{1+GH}} \tag{1.127}$$

2. **Sensitivity of Open-Loop System**

   The sensitivity of the open-loop system to changes in G(s) is given by

$$S_G^T = \frac{\partial T}{\partial G} \times \frac{G}{T} = 1, \tag{1.128}$$

   Thus,

Key Point

Due to variations in G, the sensitivity of the closed-loop system is reduced by a factor of $(1 + GH)$ compared to the open-loop system.

3. **System Sensitivity to Variations in Feedback Sensor**

   The sensitivity of T(s) with respect to the feedback sensor H(s) is given as:

$$S_H^T = \frac{\partial T}{\partial H} \times \frac{H}{T} = G\left[\frac{-G}{(1+GH)^2}\right] \times \frac{H}{G/(1+GH)}$$
$$= \frac{-GH}{1+GH} \tag{1.129}$$

   Thus

$$S_H^T = \frac{-GH}{1+GH} \tag{1.130}$$

   This shows that for large values of G(s)H(s), the sensitivity of feedback system with respect to H(s) is unity. This implies that any changes in H(s) directly affect the system output. Therefore, it is important to use feedback elements that remain substantially constant and do not vary with environmental changes.

4. **Parameter Sensitivity**

   Very often, we are interested in finding the sensitivity of the process as well as the closed-loop system with respect to variation in a particular parameter or parameters. We will first derive the open-loop sensitivity, followed by the closed-loop sensitivity.

   a. **Open-Loop Sensitivity Due to Process Parameter Variation:** Let α be a parameter of the process transfer function G(s), i.e., $G(s) = G(s, \alpha)$. Then the sensitivity of G(s) with respect to the parameter α is defined as:

$$S_\alpha^G = \frac{d \ln G(s)}{d \ln \alpha} = \frac{dG/G}{d\alpha/\alpha}$$
$$= \frac{\alpha}{G} \times \frac{dG}{d\alpha} \tag{1.131}$$

This is called the open-loop sensitivity to variations in parameter α. Thus

$$\boxed{S_{\alpha}^{G} = \frac{\alpha}{G} \times \frac{dG}{d\alpha}} \tag{1.132}$$

b. **Closed-Loop Sensitivity Due to Process Parameter Variation:** The closed-loop sensitivity of the system to parameter variations can be computed as follows: Since

$$T(s) = \frac{C(s)}{R(s)} = \frac{G(s)}{1 + G(s)H(s)}$$

The sensitivity of T with respect to parameter α is computed as:

$$\begin{aligned} S_{\alpha}^{T} &= \frac{d \ln T}{d \ln \alpha} = \frac{\alpha}{T}\frac{dT}{d\alpha} \\ &= \frac{\alpha(1+GH)}{G}.\frac{dT}{dG}.\frac{dG}{d\alpha} \\ &= \frac{\alpha(1+GH)}{G}\frac{dG}{d\alpha}.\frac{1}{(1+GH)^2} \end{aligned} \tag{1.133}$$

Hence

$$\boxed{S_{\alpha}^{T} = \frac{1}{(1+GH)}.S_{\alpha}^{G}} \tag{1.134}$$

It can be seen that the feedback has reduced the sensitivity to variations in α by a factor of $1/[(1+GH)]$. Usually, $GH >> 1$ over the frequency range of interest.

c. **Closed-Loop Sensitivity Due to Parameter Variation in Feedback Transfer Function:**

Let α be a parameter in the feedback transfer function H(s). The sensitivity of T with respect to parameter α is computed as:

$$\begin{aligned} S_{\alpha}^{T} &= \frac{\alpha}{T}\frac{dT}{d\alpha} = \frac{\alpha}{T}.\frac{dT}{dH}.\frac{dH}{d\alpha} = \frac{\alpha}{H}.\frac{dH}{d\alpha}.\frac{H}{T}\frac{dT}{dH} \\ &= S_{\alpha}^{H}\frac{H}{T}.\frac{(-G).G}{(1+GH)^2} = -S_{\alpha}^{H}.\frac{H(1+GH)}{G}.\frac{G^2}{(1+GH)^2} \\ &= -S_{\alpha}^{H}.\frac{GH}{(1+GH)} \end{aligned} \tag{1.135}$$

Hence

$$\boxed{S_{\alpha}^{T} = -S_{\alpha}^{H}.\frac{GH}{(1+GH)}} \tag{1.136}$$

Note that when G(s)H(s) is much greater than one, the magnitudes of S_{α}^{T} and S_{α}^{H} are nearly equal. This implies that there is no reduction of sensitivity due to parameter variations in the feedback path.

d. **Closed-Loop Sensitivity Due to Parameter Variation in System Transfer Function:**

Let the transfer function of the system be expressed as

$$T(s) = \frac{N(s,\alpha)}{D(s,\alpha)}; \quad \alpha = \text{parameter under consideration}$$

The sensitivity of T with respect to parameter α is given by

$$\begin{aligned} S_{\alpha}^{T} &= \left.\frac{\partial \text{LnN}}{\partial \text{Ln}\alpha}\right|_{\alpha_0} - \left.\frac{\partial \text{LnD}}{\partial \text{Ln}\alpha}\right|_{\alpha_0} \\ &= S_{\alpha}^{N} - S_{\alpha}^{D} \end{aligned} \tag{1.137}$$

where α_0 is the nominal value of the parameter around which the variation occurs.

EXAMPLE 1.20

Problem

Consider the system shown in Figure 1.38. This consists of an amplifier with gain K_1 and the system is represented by a simple transfer function K_2

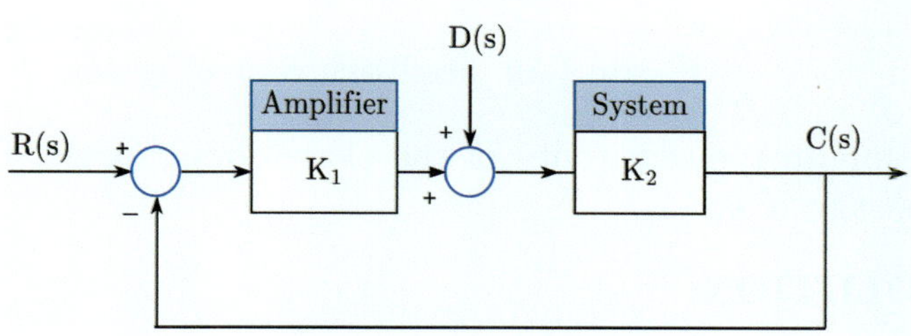

Figure 1.38: Feedback control system

Compute the sensitivity of the closed-loop transfer function with respect to K_2. Assume $D(s) = 0$

SOLUTION:

This is unity feedback system with feedback path transfer function $H(s) = 1$. The forward path transfer function is given by

$$G(s) = K_1 K_2$$

The closed-loop transfer function of the system is given by

$$T(s) = \frac{G(s)}{1 + G(s)} = \frac{K_1K_2}{1 + K_1K_2}$$

Let us compute S_G^T, and $S_{K_2}^G$. Now

$$S_G^T = \frac{1}{1 + GH} = \frac{1}{1 + K_1K_2} \quad \text{and}$$

$$S_{K_2}^G = \frac{K_2}{G}\frac{dG}{dK_2} = \frac{K_2}{K_1K_2} \times K_1 = 1$$

$$\text{Hence} \quad S_{K_2}^T = S_G^T \times S_{K_2}^G = \frac{1}{1 + K_1K_2}$$

EXAMPLE 1.21

Problem

Consider the system shown in Figure 1.39.

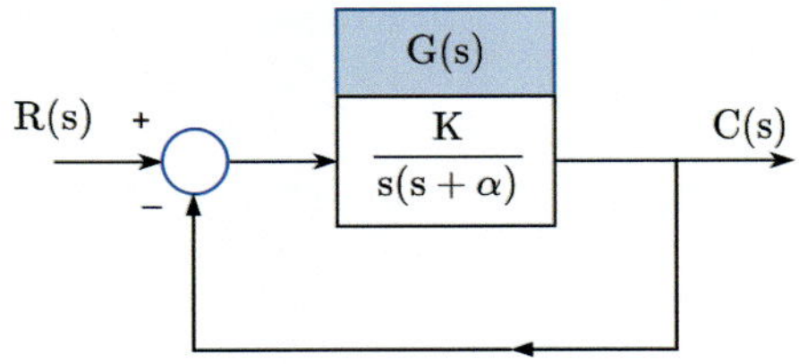

Figure 1.39: Feedback control system

Compute the sensitivity of the closed-loop transfer function with respect to K and α.

SOLUTION:

This is a unity feedback system with feedback path transfer function $H(s) = 1$. The forward path transfer function is given by

$$G(s) = \frac{K}{s(s + \alpha)}$$

The closed-loop transfer function of the system is given by

$$T(s) = \frac{G(s)}{1 + G(s)} = \frac{K}{s^2 + \alpha s + K}$$

Let us compute S_G^T, S_K^G, and S_α^G. Now

$$S_G^T = \frac{1}{1+GH} = \frac{s(s+\alpha)}{s^2+\alpha s+K}$$

$$S_K^G = \frac{K}{G}\frac{dG}{dK} = \frac{Ks(s+\alpha)}{K}\left[\frac{s(s+\alpha)}{(s(s+\alpha))^2}\right] = 1$$

$$S_\alpha^G = \frac{\alpha}{G}\frac{dG}{d\alpha} = \frac{\alpha s(s+\alpha)}{K}\left[\frac{-Ks}{(s(s+\alpha))^2}\right] = -\frac{\alpha}{s+\alpha}$$

Let us compute S_K^T and S_α^T. Now

$$S_K^T = S_G^T \times S_K^G = \frac{s(s+\alpha)}{s(s+\alpha)+K} \times 1 = \frac{s(s+\alpha)}{s(s+\alpha)+K}$$

$$S_\alpha^T = S_G^T \times S_\alpha^G = \frac{s(s+\alpha)}{s(s+\alpha)+K} \times -\frac{\alpha}{s+\alpha} = \frac{-\alpha s}{s^2+\alpha s+K}$$

EXAMPLE 1.22

Problem

Consider the system shown in Figure 1.40.

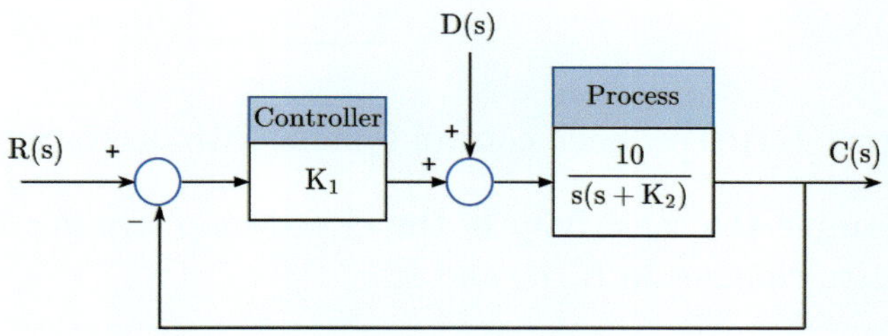

Figure 1.40: Feedback control system

Compute the sensitivity of the closed-loop transfer function with respect to changes in K_2. Assume $D(s) = 0$.

SOLUTION:

This is a unity feedback system with feedback path transfer function $H(s) = 1$. The forward path transfer function is given by

$$G(s) = \frac{10K_1}{s(s+K_2)}$$

The closed-loop transfer function of the system is given by

$$T(s) = \frac{G(s)}{1+G(s)} = \frac{10K_1}{s^2+K_2 s+10K_1}$$

Now

$$S_{K_2}^{T} = \frac{K_2}{T}\frac{dT}{dK_2} = \frac{K_2}{T}\left[\frac{-10K_1 s}{(s^2 + K_2 s + 10K_1)^2}\right]$$
$$= \frac{K_2(s^2 + K_2 s + 10K_1)}{10K_1} \times \frac{-10K_1 s}{(s^2 + K_2 s + 10K_1)^2}$$
$$= \frac{-K_2 s}{s^2 + K_2 s + 10K_1}$$

CONCEPT REINFORCEMENT PROBLEM 1.4

Problem

Consider the system with derivative feedback shown in Figure 1.41.

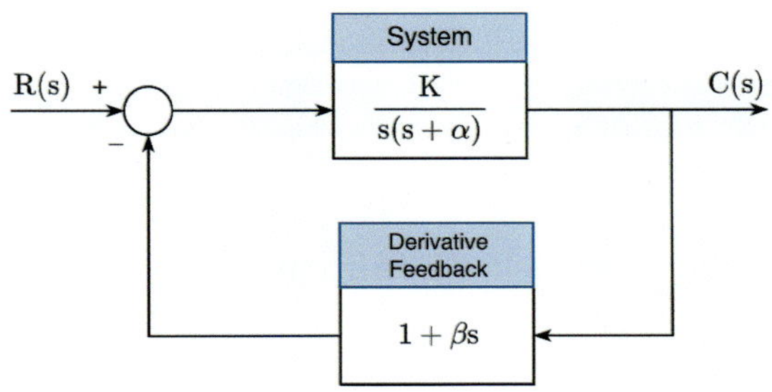

Figure 1.41: Feedback control system with derivative feedback

Determine the sensitivity of the closed-loop transfer function with respect to changes in K, α, and β.

1.16 Multiple Choice Skills Test Questions

M 1.1. Which of the following signals are periodic signals?

$$x_1(t) = 1 + \sin(10t + 1),\ \ x_2(t) = e^{-3t}$$
$$x_3(t) = e^{j10t + \frac{\pi}{2}} \cos(100t),\ \ x_4(t) = e^{-4t}\cos(100t)$$

A. $x_2(t)$ and $x_4(t)$ **C.** $x_1(t)$ and $x_3(t)$

B. $x_1(t)$ and $x_2(t)$ **D.** $x_1(t)$ and $x_3(t)$

M 1.2. Consider the following statements related to the two signals $x_1(t)$ and $x_2(t)$ where

$$x_1(t) = 10\sin(8\pi t) \ \text{ and } \ x_2(t) = 20\cos(6\pi t)$$

I. Both $x_1(t)$ and $x_2(t)$ are periodic signals

II. $x_1(t)$ is periodic; but $x_2(t)$ is aperiodic

III. Signal $x(t) = x_1(t) + x_2(t)$ is a periodic signal

IV. Signal $x(t) = x_1(t) + x_2(t)$ is an aperiodic signal

A. Statement I is only correct

B. Statements I and III are correct

C. Statement IV is correct

D. Statements II and IV are correct

M 1.3. Let $x_1(t)$ and $x_2(t)$ be two periodic signals with fundamental periods T_1 and T_2, respectively. The signal $x(t) = x_1(t) + x_2(t)$ is a periodic signal if

A. $\frac{T_1}{T_2} = \frac{1}{\sqrt{2}}$

B. $\frac{T_1}{T_2} = \frac{1}{\sqrt{3}}$

C. $\frac{T_1}{T_2} = \frac{2}{\sqrt{5}}$

D. $\frac{T_1}{T_2} = \frac{2}{3}$

M 1.4. Consider the following statements

1. The power of the energy signal is zero and its energy is finite

2. The power of energy signal is infinity

3. The energy of power signal is infinity; but its average power is finite

4. The energy of power signal is zero

Which of the following statements is true?

A. 1 & 3

B. 2 & 3

C. 3 & 4

D. 2 & 4

M 1.5. The power and energy of a unit step sequence are respectively

A. 2 and 0

B. $\frac{1}{2}$ and 0

C. $\frac{1}{2}$ and ∞

D. 2 and ∞

M 1.6. Consider the example of four systems whose dynamics are as follows:

S1: $\dot{y}(t) + 5ty(t) = t^2u(t)$ **S2:** $y(t) = 7u(t) + 3$

S3: $\dot{y}(t) + 6y(t) = 15u(t)\dot{u}(t)$ **S4:** $y(t) = \int_{-\infty}^{t} u(\tau)d\tau$

Which of the following statements is correct?

A. S1 and **S3** are linear systems

B. S1 and **S4** are linear systems

C. S2 and **S3** are linear systems

D. S2 is the only linear system

M 1.7. The impulse and state response of a system are respectively denoted by h(t) and s(t). The output response y(t) for any input u(t) is given by

A. $\frac{d}{dt}\int_0^t s(t-\tau)u(\tau)d\tau$

B. $\frac{d}{dt}\int_0^t h(t-\tau)u(\tau)d\tau$

C. $\int_0^t s(t-\tau)u(\tau)d\tau$

D. $\int_0^t \int_0^{\tau} s(t-\tau_1)u(\tau_1)d\tau_1 d\tau$

M 1.8. The impulse response of an initially relaxed linear system is $e^{-at}u(t)$. To produce a response of $te^{-at}u(t)$, the input must be equal to

A. $\frac{1}{a}e^{-at}u(t)$

B. $ae^{-at}u(t)$

C. $e^{-at}u(t)$

D. $2e^{-t}u(t)$

M 1.9. The output of a linear time-invariant system for an input r(t) equals to c(t). If the input r(t) passes through a block with transfer function e^{-2s} and then applied to the system, the modified output of the system would be

A. $\frac{c(t)}{1+e^{2t}}$

B. $\frac{c(t)}{1+e^{-2t}}$

C. $c(t-2)u(t-2)$

D. $c(t)u(t-2)$

M 1.10. The response y(t) of a linear time-invariant system (initially at rest) to a unit step input is $te^{-t},\ \ t > 0$. The transfer function of the system is

A. $\frac{1}{(s+1)^2}$ **C.** $\frac{1}{s(s+1)^2}$

B. $\frac{s}{(s+1)^2}$ **D.** $\frac{1}{s(s+1)}$

M 1.11. The system $y(t) = x(3t) + 5$ is

A. Linear and time-invariant **C.** Linear and memoryless (static)

B. Nonlinear and time-invariant **D.** Nonlinear and time variant

M 1.12. The dynamics of a system is described by the differential equation model

$$\ddot{y}(t) + 5\dot{y}(t) + 4y(t) = 2\dot{u}(t) + u(t)$$

The modes and the general form of the zero-input response of the system are respectively

A. $-1, -4$ and $k_1e^{-t} + k_2e^{-4t}$ **C.** $-1, -0.5$ and $k_1e^{-t} + k_2e^{-0.5t}$

B. $1, 4$ and $k_1e^{t} + k_2e^{4t}$ **D.** $-4, -0.5, -1$ and $k_1e^{-t} + k_2e^{-4t} + k_3e^{-0.5t}$

M 1.13. The differential equation model of a system is described by

$$\frac{d^4y}{dt^4} - 4\frac{d^3y}{dt^3} + 4\frac{d^2y}{dt^2} = 3\frac{d^2u}{dt^2} - 10 + u(t)$$

The modes and the general form of the zero-input response of the system are respectively

A. 0,0,-2,-2 and $k_1 + k_2t + k_3e^{-2t} + k_4te^{-2t}$ **C.** 0,+2 and $k_1 + k_2e^{2t}$

B. 0,0,+2,+2 and $k_1 + k_2t + k_3e^{2t} + k_4te^{2t}$ **D.** 0,-2,-2 and $k_1 + k_2e^{-2t} + k_3te^{-2t}$

M 1.14. Consider the following rational functions

S1: $s^2 + 1$ **S2:** $\frac{10}{s+4}$ **S3:** $\frac{s^2+3s+2}{s^2+8s+15}$ **S4:** $\frac{s^2-1}{s+1}$

Which of the following are correct classifications of these functions such as proper, strictly proper, biproper, improper?

A. **S1:** improper, **S2:** proper, **S3:** strictly proper, and **S4:** proper

B. **S1:** proper, **S2:** improper, **S3:** strictly proper, and **S4:** strictly proper

C. **S1:** improper, **S2:** strictly proper, **S3:** proper, and **S4:** improper

D. **S1:** strictly proper, **S2:** improper, **S3:** strictly proper, and **S4:** proper

M 1.15. A system with zero initial conditions has the closed-loop transfer function

$$T(s) = \frac{s^2 + 16}{(s+5)(s+1)}$$

The system output becomes zero at the frequency

A. 1 rad/sec

B. 4 rad/sec

C. 0.5 rad/sec

D. 2 rad/sec

M 1.16. The unit impulse response of a linear time-invariant system is the unit step function u(t). For $t > 0$, the response of the system to an excitation $e^{-at}u(t),\ \ a > 0$ is

A. ae^{-at}

B. $a(1 - e^{-at})$

C. $\frac{1}{a}(1 - e^{-at})$

D. $1 - e^{-at})$

M 1.17. Given that

$$\mathcal{L}[f(t)] = \frac{s+4}{s^2+2}, \quad \text{and} \quad \mathcal{L}[g(t)] = \frac{s^2+2}{(s+1)(s+4)}$$

If $h(t) = \int_0^t f(\tau)g(t-\tau)d\tau$, then $\mathcal{L}[h(t)]$ is

A. $\frac{s^2+2}{s+1}$

B. $\frac{1}{s+1}$

C. $\frac{s^2+2}{(s+1)(s+4)}$

D. $\frac{1}{(s+1)(s+4)}$

M 1.18. A system with an input x(t) and output y(t) is described by the relation y(t) = tx(t). The system is

A. nonlinear and time varying
B. linear and time-invariant
C. nonlinear time-invariant
D. linear and time varying

M 1.19. A linear time-invariant system has an impulse response e^{-2t}, $t > 0$. If the initial conditions are zero and the input is e^{-t}, then the output for $t > 0$ is

A. $e^{-t} + e^{-2t}$
B. $e^{t} - e^{2t}$
C. $e^{t} + e^{2t}$
D. $e^{-t} - e^{-2t}$

M 1.20. A system with zero initial conditions has the closed-loop transfer function

$$T(s) = \frac{Y(s)}{U(s)} = \frac{s+1}{(s+4)(s+6)(s^2+4)}$$

The system output becomes infinity if

A. $u(t) = \sin 4t$
B. $u(t) = \sin 2t$
C. $u(t) = \sin t$
D. $u(t) = \sin\sqrt{3}t$

Answers

M 1.1. C **M 1.2**. B **M 1.3**. D **M 1.4**. A **M 1.5**. C **M 1.6**. B **M 1.7**. A **M 1.8**. A **M 1.9**. D **M 1.10**. B **M 1.11**. D **M 1.12**. A **M 1.13**. A **M 1.14**. C **M 1.15**. B **M 1.16**. C **M 1.17**. B **M 1.18**. D **M 1.19**. D **M 1.20**. B

1.17 Knowledge Assessment Problems

KaP 1.1 Prove the following

a. Multiplication of an even signal by an odd signal produces an odd signal.

b. Multiplication of an even signal by an even signal produces an even signal.

c. Multiplication of an odd signal by an odd signal produces an even signal.

KaP 1.2 Decompose the following signals into odd and even parts.

a. $x(t) = e^{j5t}$

b. $x(t) = e^{-at}\cos t$

c. $x(t) = \left(a + t^3\right)\cos^3(bt)$

d. $x(t) = a_0 + a_1 t\cos t + a_2 t^2 \sin t + a_3 t^3 \sin t \cos t$

KaP 1.3 For each of the following signals, determine whether it is an energy signal, a power signal, or neither.

a. $x(t) = 2e^{-2|t|}$

b. $x(t) = e^{j(2t+\pi/4)}$

c. $x(t) = \sin t$

d. $x(t) = e^t u(t)$ (a growing exponential)

KaP 1.4 Simplify each of the following signals

a. $\left(\dfrac{\sin t - \cos t}{t^3 - 4}\right)\delta(t)$

b. $\left(\dfrac{3 + j\omega}{\omega^2 + 18}\right)\delta(\omega)$

c. $\left[e^{-2t}\cos(2t - 30^0) + e^{-5t}\sin(4t + 60^0)\right]\delta(t)$

d. $\left(\dfrac{\sin\left[\dfrac{\pi}{2}(t-3)\right]}{t^2 + 4}\right)\delta(t-4)$

KaP 1.5 Determine if the systems described below are linear or nonlinear.

a. $y(t) = u(t)u(t-1)$

b. $y(t) = \int_{-\infty}^{t} u(\tau)d\tau$

c. $y(t) = \sqrt{u(t)}$

d. $\dfrac{dy(t)}{dt} + y^2(t) = u(t)$

KaP 1.6 Consider a continuous-time system whose input and output are related by

$$y(t) = \begin{cases} \dfrac{u^2(t)}{u(t-1)} & \text{if } u(t-1) \neq 0 \\ 0 & \text{if } u(t-1) = 0 \end{cases}$$

Show that the system satisfies the homogeneity property but not the additive property.

KaP 1.7 Determine if the systems described below are time-invariant or time-varying.

a. $y(t) = u(-t)$

b. $y(t) = u(t-5)$

c. $y(t) = [a + \sin t]\, u(t)$

d. $y(t) = \dfrac{du(t)}{dt}$

KaP 1.8 Determine if the following systems are static or dynamic, causal or non-causal.

a. **S1:** $y(t) = 10u(t) + u^3(t-1)$

b. **S2:**$y(t) = \cos[u(t)] + u^2(t+1)$

c. **S3:**$y(t) = 5\sin^3[u(t-1)] + \cos[u(t-2)]$

d. **S4:**$y(t) = 10u(t) + 0.5u^2(t)$

e. **S5:**$y(t) = 0.7u(t) + 2u(t-1) + u^2(t-5)$

KaP 1.9 Compute the transfer function of the following systems from their impulse response $g(t)$:

a. $g(t) = 10 + e^{-t} + e^{-5t}$

b. $g(t) = 2e^{-4t} + \sin(5t)$

c. $g(t) = 4e^{-t} + 3\sin(2t) - 4\cos(2t)$

d. $g(t) = \dfrac{7}{3}e^{-2t} - 2e^{-t} + \dfrac{2}{3}e^{t}$

KaP 1.10 Compute the transfer function and the impulse response of the following systems from their differential equation models.

a. $\dot{y}(t) + 5y(t) = 10u(t)$

b. $\ddot{y}(t) + 4\dot{y}(t) - 5y(t) = \dot{u}(t) + 2u(t)$

c. $\ddot{y}(t) + 2\dot{y}(t) + y(t) = 2\dot{u}(t) - u(t)$

KaP 1.11 The step response of an LTI system is given by

$$y(t) = 5 + e^{-t}\sin(3t)$$

What is the transfer function of the system?

KaP 1.12 Consider the systems that are represented by their differential equation models. Determine which of these are completely characterized by their transfer functions? For the systems that are not completely characterized by transfer functions, find the missing poles.

a. $\ddot{y} + 4\dot{y} + 3y = \dot{u} + 3u$

b. $\ddot{y} + 2\dot{y} - 8y = \dot{u} - 2u$

c. $\ddot{y} + 2\dot{y} + 2y = 10u$

d. $\dddot{y} + 3\ddot{y} + 2\dot{y} = \dot{u} + u$

KaP 1.13 Consider the system shown in Figure 1.42.

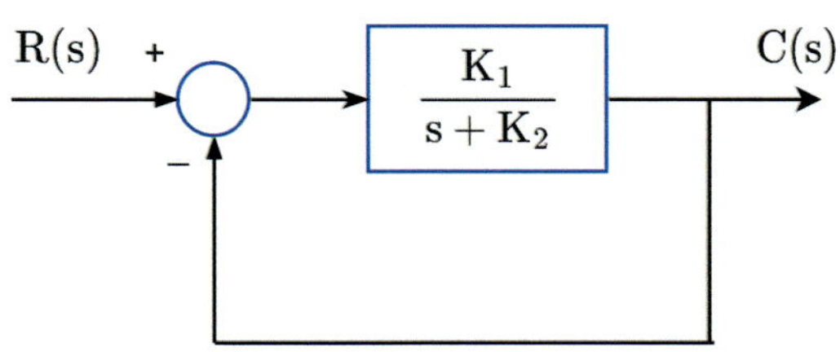

Figure 1.42: Feedback control system

Compute the sensitivity of the closed-loop transfer function with respect to K_1 as well as K_2.

KaP 1.14 The block diagram of an amplifier with feedback is shown in Figure 1.43.

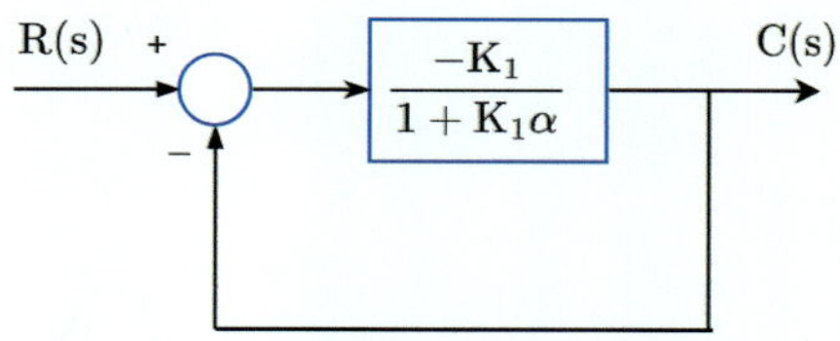

Figure 1.43: Feedback control system

Determine the sensitivity of the closed-loop transfer function with respect to the amplifier gain K_1.

KaP 1.15 For the system shown in Figure 1.44, determine the sensitivity of the closed-loop transfer function with respect to the amplifier gain K_1, K_2 and the feedback gain K_3.

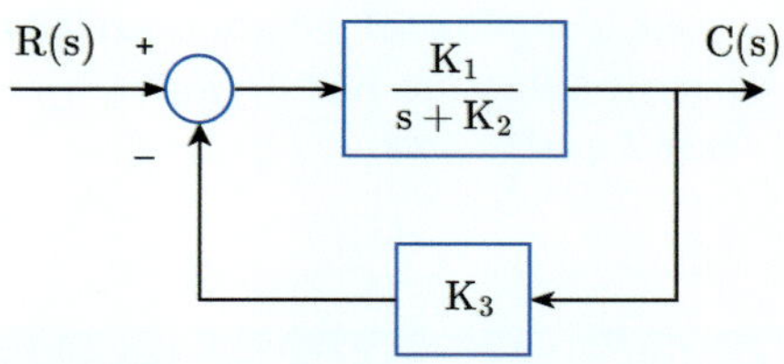

Figure 1.44: Feedback control system

KaP 1.16 Consider the system shown in Figure 1.45.

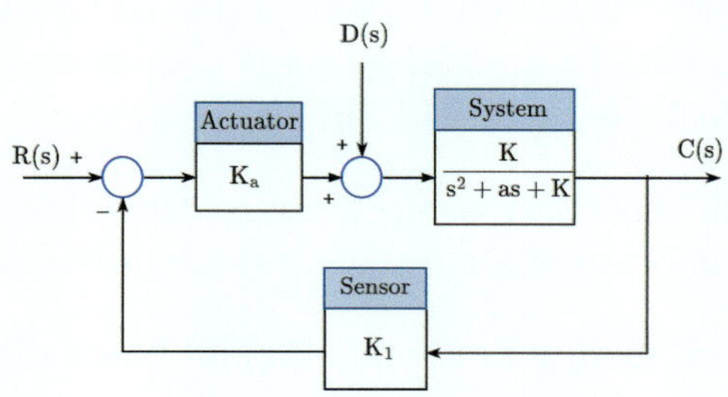

Figure 1.45: Feedback control system

Compute the sensitivity of the closed-loop transfer function with respect to changes in actuator gain K_a and sensor gain K_1.

Chapter 2

System Modeling, Block Diagrams, and Signal Flow Graphs

The art of modeling: turning a messy, complicated reality into a neat, complicated equation-pretending we've simplified things, then spending the rest of your career explaining why reality doesn't quite agree.

DESIRED LEARNING OUTCOMES

Upon completion of this chapter, readers should be able to:

- Understand the fundamental principles of modeling dynamic systems across multiple physical domains
- Derive differential equation models from first principles using Newton's laws, Kirchhoff's laws, and energy conservation principles for electrical, mechanical, thermal, and fluid systems
- Establish the relationship between time-domain differential equations and frequency-domain transfer functions through the Laplace transform
- Construct, manipulate, and simplify block diagrams to represent complex interconnected systems
- Convert system representations into signal flow graphs and compute overall transfer functions using Mason's gain formula

DOI: 10.1201/9781003571308-2

2.1 Introduction

Determining a mathematical model of a real-world system is a difficult task in almost all fields of engineering and science. The difficulties arise for various reasons. Some of them include diverse nature and complexities of dynamic systems ranging from linear to nonlinear; time-invariant to time variant; lumped parameter to distributed parameter, and so on. A mathematical model should capture the underlying physics of the system and define the cause-and-effect relationships between various variables in the system in a quantitative manner. One of the most important requirements of modeling is that it must be suitable for the intended purpose. This suitability ensures that the model not only captures the essential dynamics of the system but also meets the specific needs of the application. For example, as will be discussed in later chapters, the state variable model provides a structured framework that facilitates the application of modern control techniques to design pole placement, optimal, and robust controllers. On the contrary, the discrete parametric model is particularly useful for adaptive control applications, where the controller parameters are adjusted in real-time based on observed system behavior.

Desired Characteristics of a Model

A model[1] should strike a balance between simplicity and complexity to be both useful and accurate for analysis and controller design. Here are two important considerations when developing a mathematical model:

1. **Avoid Over-Complication:** Do not make the model overly complex by including every microscopic detail, rendering it impractical for analysis and controller design. Include only the relevant dynamics and phenomena that significantly impact the system's behavior.

2. **Avoid Over-Simplification:** Do not over-simplify the model to the extent that it loses the dynamic resemblance to the original system and therefore provides incorrect predictions. A model must retain enough detail to capture the key behaviors and interactions of the system to ensure that the controller designed based on this model will perform adequately

[1]It is essential to recognize that everything in the cosmos is fundamentally energy vibrating at different frequencies. The discrete objects and boundaries we draw—"system" versus "environment," "input" versus "output"—are pragmatic conventions imposed on a continuous, interconnected field. The various models and mathematical equations used in science and engineering do not exist in physical reality; they are mathematical abstractions imposed upon this continuous energetic flow. A physical phenomenon can exist independently of its mathematical model, whereas a model cannot exist independently of the phenomenon it represents. As Korzybski [41] remarked, "the map is not the territory"—our models are not literal replicas of ultimate reality but conceptual maps of an underlying reality that may be far more fluid and unified than our discrete representations suggest. While these models serve as powerful tools for understanding, predicting, and controlling physical systems, they remain human constructs designed to navigate our perceptual experience rather than capture ultimate reality. Use them, but don't mistake them for the territory.

in real-world scenarios.

Modeling Approaches

Basically, there are two approaches that are followed by the system designer to find a model.

1. Break down the complex system into smaller components/elements/modules and establish a mathematical model for each of these elements based on basic laws of physics. This is known as a *first-principles model.*
2. Collect a set of input-output data by conducting experiments and process these data using various techniques to get a mathematical model. This approach comes under the purview of *system identification.*

A separate area of control engineering, called *system identification*, is dedicated to the second approach. This chapter will follow the first principle approach to model dynamic systems whose behavior evolves. The dynamic systems to be considered here are either electrical, mechanical, thermal, or pneumatic components of a control system. Although these components are distributed (in space) and their characteristics are often nonlinear, we make the following assumptions to get linear ordinary differential equation models of these components.

Assumptions

1. The elements of the system, such as resistors, capacitors, inductors, mass, inertia, etc., are concentrated at singular points in space, i.e., these are assumed to be *lumped* instead of *distributed.*
2. The system is assumed to be ideal, where each of its elements has a single property. For example, an inductor is assumed to have pure inductance with no resistance or capacitance.
3. The system/device/component operates around a small region of its nominal operating point and can be described by linear differential equations.

In this chapter, we explore the modeling of fundamental physical systems, including electrical, mechanical, thermal, and fluid systems. While these systems originate from diverse fields and involve different physical quantities and components, their underlying dynamic behavior often exhibits striking similarities. This similarity stems from the common mathematical structures governing their responses, typically described by linear or nonlinear differential equations. By recognizing these parallels, we can apply unified modeling techniques—such as the use of analogous variables and energy-based methods —to simplify the analysis and design of complex systems across various engineering disciplines.

2.2 Classification of Variables and Elements in System Modeling

The dynamic systems across diverse domains share significant similarities in their components/elements, behavior, and the type of variables. Recognizing these similarities and classifying them into various categories allows for a unified approach to modeling, analysis, and control, leveraging analogous techniques and tools across different fields.

Classification of Variables

The variables in a dynamic system can be categorized as **across variables** and **through variables**.

> **Definition 2.1.** The variable, which is measured across an element, is called an **across variable**.

Examples of across variables include velocity, and displacement in mechanical systems, voltage in electrical systems, pressure difference in fluid systems, and temperature difference in thermal systems.

> **Definition 2.2.** The variable, which is transmitted or flows through an element unaltered, is called a **through variable**.

Examples of through variables include force in mechanical systems, current in electrical systems, fluid flow rate in fluid systems, heat transfer rate, mass flow rate, entropy flow rate in thermal systems.

Classification of Elements

From the perspective of modeling, the elements/components in a dynamic system can be classified into three groups based on their behavior and the type of variable they store, transform, or dissipate, regardless of whether the component belongs to mechanical, electrical, fluid, thermal, or another type of system. These classifications often include **A-type**, **T-type,** and **D-type** elements, each corresponding to a specific characteristic in the system and helping in the systematic construction and analysis of dynamic models, ensuring a consistent representation of energy flow and transformations across various physical domains. Here's a detailed look at each type:

> **Definition 2.3. A-Type Element (Accumulation/Storage Element):** A-type elements are those that store energy or material in the system. They accumulate the quantity over time.

These elements store potential energy, kinetic energy, electrical charge, fluid volume, or thermal energy. Some of the examples of this element are: the mass (stores kinetic energy) and spring (stores potential energy) in mechanical systems; the capacitor (stores electrical energy) in electrical systems, the tank (stores fluid volume) in fluid systems, and thermal mass (stores thermal energy) in thermal systems.

Definition 2.4. T-Type Element (Through/Transmission Element): T-type elements are those that transmit the through variable without storing or dissipating it.

These elements allow the flow of energy or material without significant storage or dissipation. The through variable is directly transmitted through this element. Some of the examples of this element are: ideal rod (transmits force without deformation), gears(converts mechanical energy between different rotational speeds and torques) in mechanical systems; ideal wire (transmits current without resistance), transformers (convert electrical energy between different voltage/current levels) in electrical systems, ideal pipe (transmits fluid flow without pressure drop) in fluid systems and ideal thermal conductor (transmits heat without temperature drop) in thermal systems.

Definition 2.5. D-Type Element (Dissipation/Damping Element): D-type elements provide resistance to the flow of the through variable and dissipate energy within the system, often converting it to non-recoverable forms such as heat.

These elements convert the energy of the through variable into heat or another form of dissipation. In these elements, the across variable is related to the through variable via a resistive coefficient. Some of the examples of this element are: damper (dissipates kinetic energy) in mechanical systems, resistor (dissipates electrical energy) in electrical systems, valve or orifice (dissipates fluid energy) in fluid systems, and thermal resistance (dissipates thermal energy) in thermal systems.

We can integrate these concepts to develop better models that capture the essential dynamics of real-world systems, leading to better performance prediction, control, and optimization.

2.3 Differential Equation Modeling of Passive Electrical Circuits

The basic elements of electrical systems are the **resistor, inductor and capacitor**, and the variables of interest are **voltages** and **currents**. Differential equation models of electrical systems are derived by applying Ohm's law and Kirchhoff's laws, which enforce the balance of voltages and currents in a circuit. For a linear resistor, with resistance R, the voltage drop, v(t), is

Table 2.1: Electrical elements and voltage current relations

Electrical element	**Symbol**	**Voltage-current relation**
Resistor	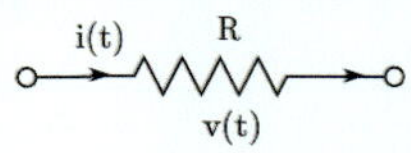	$v(t) = Ri(t)$
Inductor	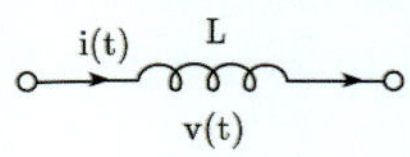	$v = L\frac{di(t)}{dt},\ i = \frac{1}{L}\int v dt$
Capacitor	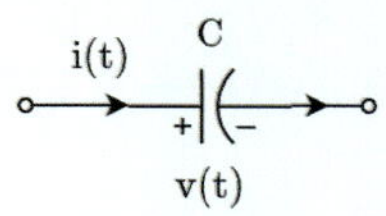	$v = \frac{1}{C}\int idt,\ i = C\frac{dv(t)}{dt}$

proportional to the current i(t) flowing through it. In the case of an inductor with inductance L, the voltage drop is proportional to the rate of change of current. For a capacitor with capacitance C, the voltage drop is proportional to the integral of the current flowing through it.

These relationships are summarized in Table 2.1.

Consider the electrical circuit shown in Figure 2.1a consisting of a resistor and an inductor. Applying Kirchhoff's voltage law to this circuit, we have

$$Ri(t) + L\frac{di(t)}{dt} = v(t) \tag{2.1}$$

Next, let us consider the electrical circuit consisting of a resistor and a capacitor shown in Figure 2.1b. Applying Kirchhoff's voltage law to this circuit, we

(a) R-L circuit (b) R-C circuit

Figure 2.1: Simple R-L and R-C circuit

have

$$\mathrm{Ri(t)} + \frac{1}{\mathrm{C}}\int \mathrm{i(t)dt} = \mathrm{v(t)} \implies \mathrm{R}\frac{\mathrm{di(t)}}{\mathrm{dt}} + \frac{1}{\mathrm{C}}\mathrm{i(t)} = \frac{\mathrm{dv(t)}}{\mathrm{dt}} \tag{2.2}$$

Let us model a circuit where all three passive elements, a resistor, an inductor, and a capacitor, are connected in series (see Figure 2.2)

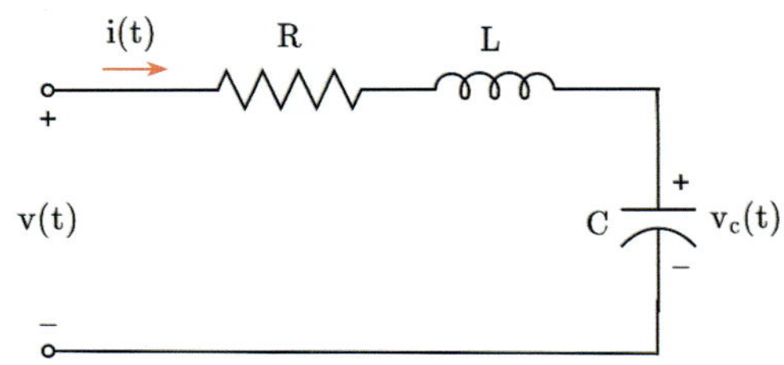

Figure 2.2: R-L-C Circuit

Applying Kirchhoff's voltage law to this circuit, we have

$$\mathrm{Ri(t)} + \mathrm{L}\frac{\mathrm{di(t)}}{\mathrm{dt}} + \frac{1}{\mathrm{C}}\int \mathrm{i(t)dt} = \mathrm{v(t)} \implies \mathrm{L}\frac{\mathrm{d^2i(t)}}{\mathrm{dt^2}} + \mathrm{R}\frac{\mathrm{di(t)}}{\mathrm{dt}} + \frac{1}{\mathrm{C}}\mathrm{i(t)} = \frac{\mathrm{dv(t)}}{\mathrm{dt}} \tag{2.3}$$

If q denotes the charge, then the dynamics of this circuit can alternatively be expressed as:

$$\mathrm{L}\frac{\mathrm{d^2q(t)}}{\mathrm{dt^2}} + \mathrm{R}\frac{\mathrm{dq(t)}}{\mathrm{dt}} + \frac{1}{\mathrm{C}}\mathrm{q(t)} = \mathrm{v(t)} \tag{2.4}$$

Note

1. Each of the R-L and R-C circuits, shown in Figure 2.1a and 2.1b, has one energy storing element, i.e., an inductor and a capacitor. The dynamics of this circuit is described by a first-order differential equation.

2. The R-L-C circuit of Figure 2.2 contains two energy-storing elements: one inductor and one capacitor. The dynamics of this circuit are described by a second-order differential equation.

3. The **order** of a system equals the number of independent energy-storing elements present in the system.

2.4 Transfer Function of Modeling of Electrical Networks by Transform Approach

The transform approach, specifically using the Laplace transform, is a powerful method for modeling electrical networks by transfer functions. This approach converts differential equations governing the behavior of electrical circuits into algebraic equations, making it easier to analyze and design systems.

Steps in Transfer Function Modeling by Transform Approach

1. Identify the components and write the governing differential equations using Kirchhoff's laws (KVL and KCL).
2. Transform the time-domain differential equations into the s-domain (Laplace domain), assuming zero initial conditions. The key transforms: $\frac{di(t)}{dt} \longleftrightarrow sI(s)$, $\frac{dv(t)}{dt} \longleftrightarrow sV(s)$ and $\int idt \longleftrightarrow \frac{I(s)}{s}$
3. Solve for the ratio of the output Laplace transform to the input Laplace transform by algebraic manipulation.

Let's go through a few examples that illustrate the procedure of transfer function modeling by the transform approach for different types of electrical networks.

EXAMPLE 2.1

Problem

For the R-L-C circuit shown in Figure 2.2, determine the transfer function considering

a. The current is the output, and the applied voltage is the input.
b. Voltage across the capacitor is the output, and the applied voltage is the input.

SOLUTION:

a. The differential equation model of this circuit is given in Eqn(2.3) which is reproduced here

$$Ri(t) + L\frac{di(t)}{dt} + \frac{1}{C}\int i(t)dt = v(t) \tag{2.5}$$

Taking the Laplace transform of Eqn(2.5) gives

$$\left[R + Ls + \frac{1}{Cs}\right] I(s) = V(s)$$

Hence

$$G(s) = \frac{I(s)}{V(s)} = \frac{Cs}{LCs^2 + RCs + 1}$$

b. The voltage across the capacitor is

$$V_C(s) = \frac{1}{Cs} I(s) = \frac{1}{Cs} \times \frac{Cs}{LCs^2 + RCs + 1} V(s)$$

Hence

$$G(s) = \frac{V_C(s)}{V(s)} = \frac{1}{LCs^2 + RCs + 1} = \frac{\frac{1}{LC}}{s^2 + \frac{R}{L}s + \frac{1}{LC}}$$

EXAMPLE 2.2

Problem

Consider the circuit shown in Figure 2.3 where v_i is the input voltage to the circuit and v_0 denotes the output.

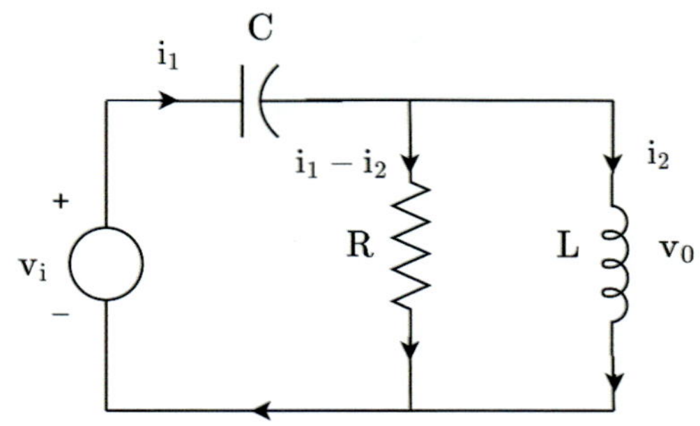

Figure 2.3: Two loop electrical network

Determine the transfer function between the voltage across the inductor and the input voltage, i.e., $\frac{V_0(s)}{V_{in}(s)}$.

SOLUTION:

Applying KVL to the two loops gives

$$\frac{1}{C}\int i_1 \, dt + R(i_1 - i_2) = v_{in} \tag{2.6a}$$

$$L\frac{di_2}{dt} - R(i_1 - i_2) = 0 \tag{2.6b}$$

Taking Laplace transform of Eqns(2.6a), and (2.6b) gives

$$\frac{1}{Cs}I_1(s) + R\left[I_1(s) - I_2(s)\right] = V_{in}(s) \tag{2.7a}$$

$$LsI_2(s) - R\left[I_1(s) - I_2(s)\right] = 0 \tag{2.7b}$$

The output voltage is given by $v_0 = L\frac{di_2}{dt}$ whose Laplace transform is

$$V_0(s) = LsI_2(s) \tag{2.8}$$

From Eqn(2.8) it is seen that the output depends on $I_2(s)$. Therefore, let us find the relation between $I_2(s)$ and $V_{in}(s)$ as:

From Eqn(2.7b) we get

$$LsI_2(s) - R\left[I_1(s) - I_2(s)\right] = 0 \ \text{ or, } I_1(s) = \frac{(R + Ls)}{R}I_2(s) \tag{2.9}$$

Substituting the value of $I_1(s)$ from Eqn(2.9) in Eqn(2.7a) gives

$$\left[\frac{(1 + RCs)}{Cs}\right]\left[\frac{(R + Ls)}{R}\right] I_2(s) - RI_2(s) = V_{in}(s) \tag{2.10}$$

or

$$I_2(s) = \frac{RCs}{(R + Ls)(RCs + 1) - R^2Cs}V_{in} = \frac{RCs}{RLCs^2 + Ls + R}V_{in} \tag{2.11}$$

Substituting the value of $I_2(s)$ from Eqn(2.11) into Eqn(2.8) gives

$$V_0(s) = Ls \times \frac{RCs}{(RLCs^2 + Ls + R}V_{in}(s)$$

Thus, the transfer function is given by

$$\frac{V_0(s)}{V_{in}(s)} = \frac{s^2}{s^2 + \frac{1}{RC}s + \frac{1}{LC}} \tag{2.12}$$

CONCEPT REINFORCEMENT PROBLEM 2.1

Problem

Consider the circuit shown in Figure 2.4 where v_i is the input voltage to the circuit and v_0 denotes the output.

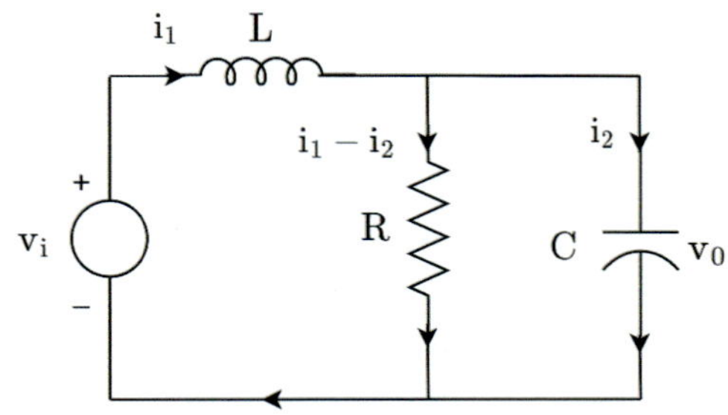

Figure 2.4: Two loop RLC network

Determine the transfer function between the voltage across the capacitor and the input voltage, i.e., $\dfrac{V_0(s)}{V_i(s)}$.

2.5 Transfer Function Modeling of Electrical Circuits by Impedance Approach

The impedance approach is a powerful and convenient technique for deriving transfer function models of electrical networks. It utilizes the concept of impedance to streamline the analysis of circuits in the frequency (Laplace) domain, especially for linear time-invariant (LTI) systems. In this approach, the voltage-current relationship for each circuit element can be expressed as:

$$\boxed{\frac{V(s)}{I(s)} = Z(s)} \tag{2.13}$$

The impedance expressions for common elements such as resistors, inductors, and capacitors are summarized in Table 2.2.

One of the advantages of this method is that we can algebraically manipulate the impedances, similar to the manipulation of resistances. The only difference is that the impedances are rational functions, rather than real numbers, as in the case of resistances. The procedure to compute the transfer function using the transformed method is summarized as follows:

Table 2.2: Voltage current relations in transformed (frequency) domain

Resistor	**Inductor**	**Capacitor**
$V(s) = RI(s)$ $Z(s) = R$	$V(s) = LsI(s)$ $Z(s) = Ls$	$V(s) = \dfrac{1}{Cs}I(s)$ $Z(s) = \dfrac{1}{Cs}$

1. Redraw the original electrical network replacing all the time variables such as v(t), i(t), $v_C(t)$, etc. by their Laplace transforms V(s), I(s), and $V_C(s)$, respectively.
2. Replace the component values by their respective impedance values; that is replace R by R, L by Ls, C by $\frac{1}{Cs}$.
3. Analyze the network considering it as a resistive network.

Let us now consider a few examples to illustrate how the impedance approach can be used to systematically derive the transfer functions of electrical networks.

EXAMPLE 2.3

Problem

Obtain the transfer function $\frac{V_0(s)}{V_i(s)}$ of the electrical network shown in Figure 2.3 using transform method

SOLUTION:

The circuit of Figure 2.3 is redrawn in the transformed (Laplace) domain and is shown in Figure 2.5.

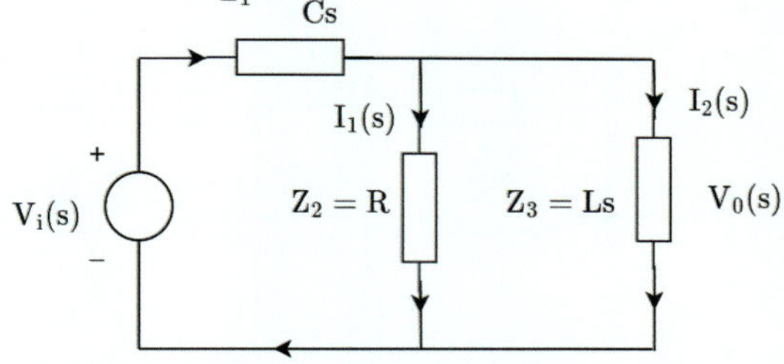

Figure 2.5: Two loop electrical network

The impedance of parallel connection of Z_2 and Z_3 is

$$Z_p = \frac{Z_2 Z_3}{Z_2 + Z_3}$$

From the circuit, the total current from the source is

$$I(s) = \frac{V_i(s)}{Z_1 + Z_p}$$

Using the standard rule of current division, we have

$$I_2(s) = I(s)\frac{Z_2}{Z_2 + Z_3} = \frac{V_i(s)Z_2}{(Z_1 + Z_p)(Z_2 + Z_3)}$$

The output voltage is $V_0(s) = I_2(s) * Z_3 = \dfrac{V_i(s)Z_2Z_3}{(Z_1 + Z_p)(Z_2 + Z_3)}$

Hence, the transfer function

$$\frac{V_0(s)}{V_i(s)} = \frac{Z_2Z_3}{(Z_1 + Z_p)(Z_2 + Z_3)} = \frac{Z_p}{Z_1 + Z_p} \tag{2.14}$$

Substituting the values of $Z_1 = \dfrac{1}{Cs}$, $Z_2 = R$, and $Z_3 = Ls$ in Eqn(2.14) gives

$$\frac{V_0(s)}{V_i(s)} = \frac{s^2}{s^2 + \dfrac{1}{RC}s + \dfrac{1}{LC}}$$

Compare this with Eqn(2.12).

CONCEPT REINFORCEMENT PROBLEM 2.2

Problem

Determine the transfer function $\dfrac{V_o(s)}{V_i(s)}$ of the bridge T circuit shown in Figure 2.6.

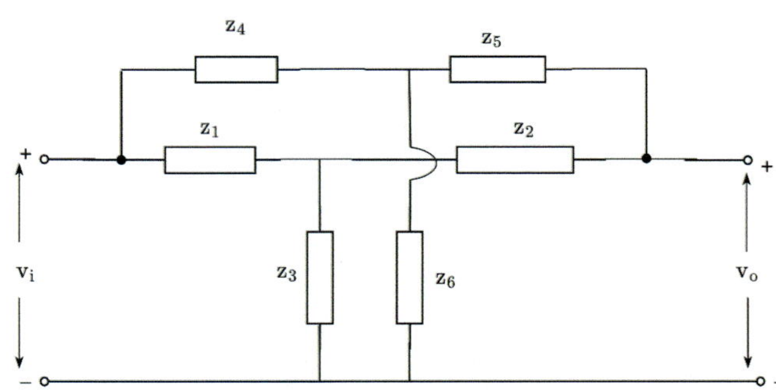

Figure 2.6: Bridge-T circuit

EXAMPLE 2.4

Problem

Write down the mesh equations for the 3-loop network shown in Figure 2.7a and compute the transfer function of the system $G(s) = \dfrac{V_o(s)}{V_i(s)}$.

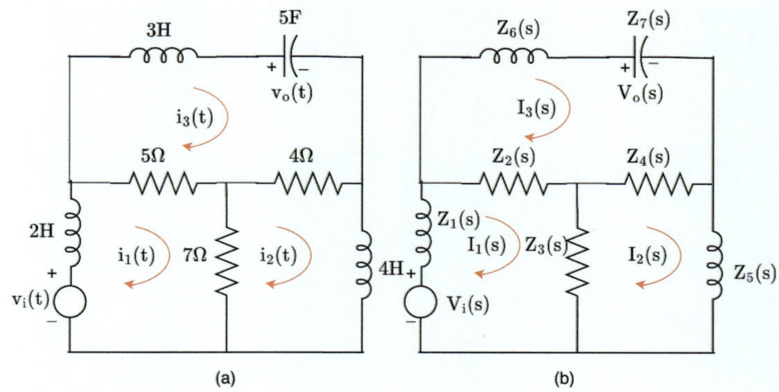

Figure 2.7: 3-Loop network

SOLUTION:

The network in s-domain is shown in Figure 2.7b. Comparing this figure with Figure 2.7a, we have

$$Z_1(s)= 2s,\ Z_2(s) = 5,\ Z_3(s) = 7,\ Z_4(s) = 4,\ Z_5(s) = 4s$$
$$Z_6(s)= 3s, \text{ and } Z_7(s) = \frac{1}{5s}$$

$$\text{Thus,}\quad Z_{11}(s)= 2s + 12,\ Z_{12}(s) = -7,\ Z_{13}(s) = -5$$
$$Z_{22}(s)= 4s + 11,\ Z_{23}(s) = -4, \text{ and } Z_{33}(s) = 9 + 3s + \frac{1}{5s}$$

Note that Z_{ii} denotes the sum of all impedances in loop-i and Z_{ij} is the sum of all impedances common between loop i and loop j (with a negative sign, indicating a shared impedance).

The mesh equations are expressed as:

$$\begin{aligned} Z_{11}I_1 + Z_{12}I_2 + Z_{13}I_3 &= V_i(s) \\ Z_{12}I_1 + Z_{22}I_2 + Z_{23}I_3 &= 0 \\ Z_{13}I_1 + Z_{23}I_2 + Z_{33}I_3 &= 0 \end{aligned} \tag{2.15}$$

Let

$$\Delta = \begin{vmatrix} Z_{11} & Z_{12} & Z_{13} \\ Z_{12} & Z_{22} & Z_{23} \\ Z_{13} & Z_{23} & Z_{33} \end{vmatrix}$$

The output voltage $V_o(s) = I_3(s)Z_7(s)$ where the current $I_3(s)$ is computed from

$$I_3(s) = \frac{1}{\Delta}\begin{vmatrix} Z_{11} & Z_{13} & V_i \\ Z_{12} & Z_{23} & 0 \\ Z_{13} & Z_{33} & 0 \end{vmatrix} = \frac{5s(83+20s)V_i}{120s^4+1410s^3+3743s^2+70s+83} \tag{2.16}$$

This implies

$$V_o(s) = I_3(s) \times \frac{1}{5s} = \frac{(83+20s)V_i}{120s^4+1410s^3+3743s^2+70s+83}$$

Hence, the transfer function is

$$G(s) = \frac{V_o(s)}{V_i(s)} = \frac{0.1667s+0.6917}{s^4+11.75s^3+31.19s^2+0.5833s+0.6917}$$

CONCEPT REINFORCEMENT PROBLEM 2.3

Problem

Write down the mesh equations for the 3-loop network shown in Figure 2.8 and compute the transfer of the network $G(s) = \dfrac{V_0(s)}{V_i(s)}$ shown in Figure 2.9 for $R = 1\Omega$, $L_1 = L_2 = L_3 = 1H$ and $C = 1F$.

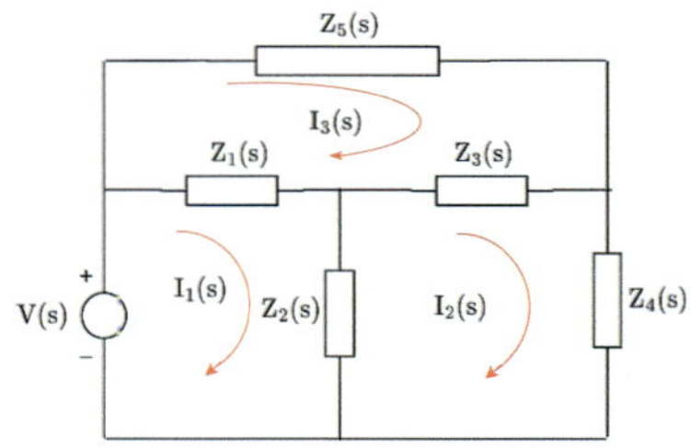

Figure 2.8: 3-Loop network

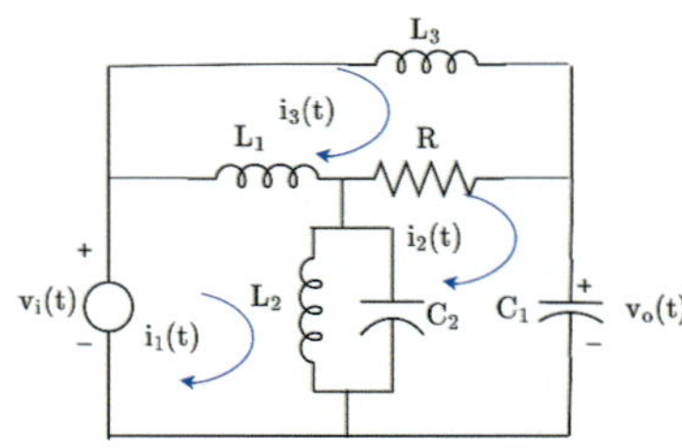

Figure 2.9: 3-Loop network

CONCEPT REINFORCEMENT PROBLEM 2.4

Problem

Compute the transfer function $\dfrac{V_c(s)}{V_1(s)}$ of the electrical network shown in Figure 2.10.

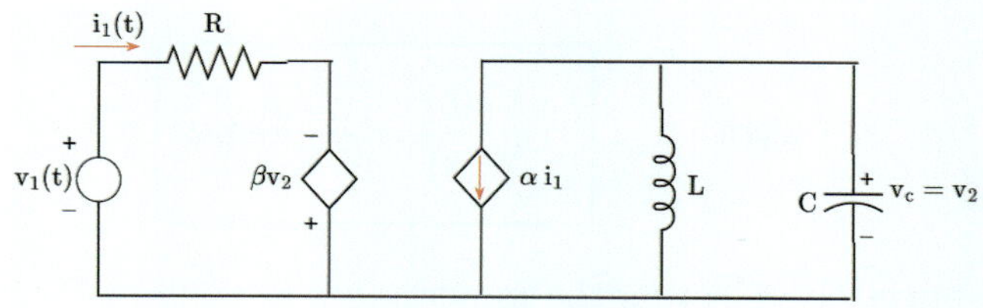

Figure 2.10: Electrical network with dependent source

2.6 Modeling of Active Electrical Networks

The operational amplifiers belong to a special class of analog integrated circuits and are extensively used in implementing various control functions. Some of the characteristics of an ideal operational amplifier include:

a. High input impedance, i.e., $Z_i = \infty$

b. Low output impedance, $Z_0 = 0$

c. High amplification gain, $A = \infty$

The schematic of an inverting OP-AMP is shown in Figure 2.11.

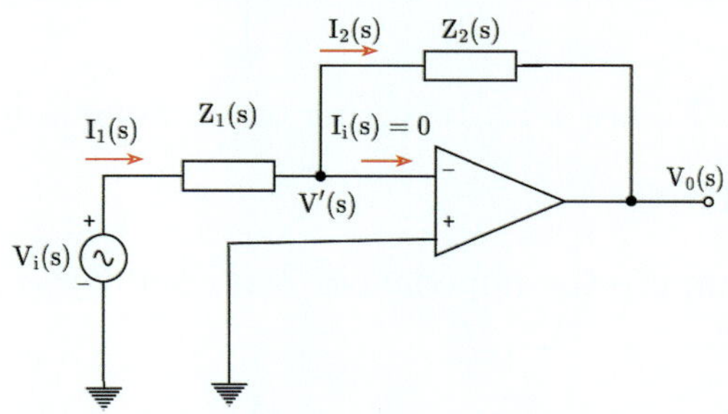

Figure 2.11: Inverting operational amplifier

Note that in an inverting operational amplifier, one of the inputs is grounded. Further, since the input impedance of OP-AMP is very high, $I_i(s) = 0$.

Let us derive the transfer function $V_0(s)/V_i(s)$ following impedance approach using nodal analysis. From the circuit shown in Figure 2.11, we have

$$\frac{V_i(s) - V'(s)}{Z_1(s)} = \frac{V'(s) - V_0(s)}{Z_2(s)} \tag{2.17}$$

Since the gain **A** of OP-AMP is very very high, the voltage $V'(s) = 0$. Thus

$$\frac{V_i(s)}{Z_1(s)} = -\frac{V_0(s)}{Z_2(s)} \tag{2.18}$$

Rearranging Eqn(2.18) gives

$$\frac{V_O(s)}{V_i(s)} = -\frac{Z_2(s)}{Z_1(s)} \tag{2.19}$$

By appropriately selecting the impedances impedance $Z_1(s)$ and $Z_2(s)$, various controllers such as integral, derivative, and compensating networks can be synthesized (see Table 8.1).

EXAMPLE 2.5

Problem

Obtain the transfer function $\frac{V_o(s)}{V_i(s)}$ of the operational amplifier circuit shown in Figure 2.12 for $R_1 = 1k\Omega$, $R_2 = 8k\Omega$, $R_3 = 50k\Omega$, and $C = 12.5\mu F$.

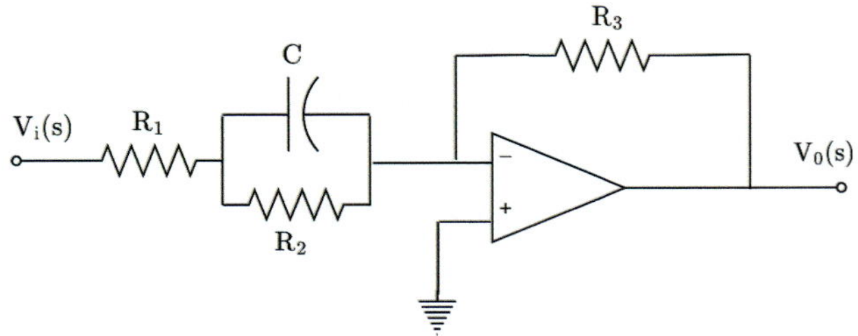

Figure 2.12: Inverting operational amplifier

SOLUTION:

Let us compute the impedances $Z_1(s)$ and $Z_2(s)$ (compare with Figure 2.11).

$$Z_1(s) = R_1 + R_2 \parallel \frac{1}{C_2 s} = R_1 + \frac{R_2}{R_2Cs + 1}$$
$$= \frac{R_1R_2Cs + R_1 + R_2}{R_2Cs + 1}$$

and $Z_2(s) = R_3$. Hence, the transfer function is

$$\frac{V_o(s)}{V_i(s)} = -\frac{Z_2(s)}{Z_1(s)} = -\frac{R_3(R_2Cs + 1)}{R_1R_2Cs + R_1 + R_2} \tag{2.20}$$

The above can alternatively be expressed as:

$$\frac{V_o(s)}{V_i(s)} = -\frac{K(s + a)}{(s + b)} \quad \text{where} \tag{2.21}$$

$$K = \frac{R_3}{R_1}, \quad a = \frac{1}{R_2C}, \quad \text{and } b = \frac{R_1 + R_2}{R_1R_2C}$$

For $R_1 = 1k\Omega$, $R_2 = 8k\Omega$, $R_3 = 50k\Omega$, and $C = 12.5\mu F$, the transfer function is

$$\frac{V_o(s)}{V_i(s)} = -\frac{50(s+10)}{s+90}$$

2.6.1 Non-inverting Operational Amplifier

Let us derive the transfer function of the noninverting operational amplifier shown in Figure 2.13.

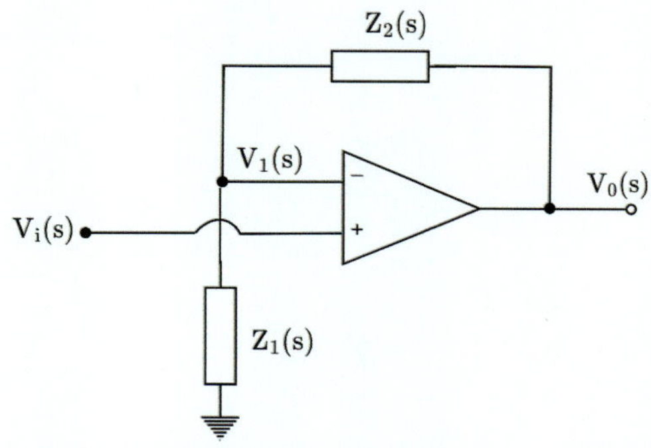

Figure 2.13: Noninverting operational amplifier circuit

From the figure, we observe that

$$V_o(s) = A\left[V_i(s) - V_1(s)\right] \tag{2.22}$$

But

$$V_1(s) = V_o(s) \times \frac{Z_1(s)}{Z_1(s) + Z_2(s)} \tag{2.23}$$

Substituting Eqn(2.23) into Eqn(2.22) and simplifying we get

$$\frac{V_o(s)}{V_i(s)} = \frac{A}{1 + \dfrac{AZ_1(s)}{Z_1(s) + Z_2(s)}} \tag{2.24}$$

Since $A >> 1$, we can disregard 1 in the denominator of Eqn(2.24) which gives

$$\boxed{\frac{V_o(s)}{V_i(s)} = \frac{Z_1(s) + Z_2(s)}{Z_1(s)}} \tag{2.25}$$

EXAMPLE 2.6

Problem

Obtain the transfer function $\frac{V_o(s)}{V_i(s)}$ of the noninverting operational amplifier circuit shown in in Figure 2.14.

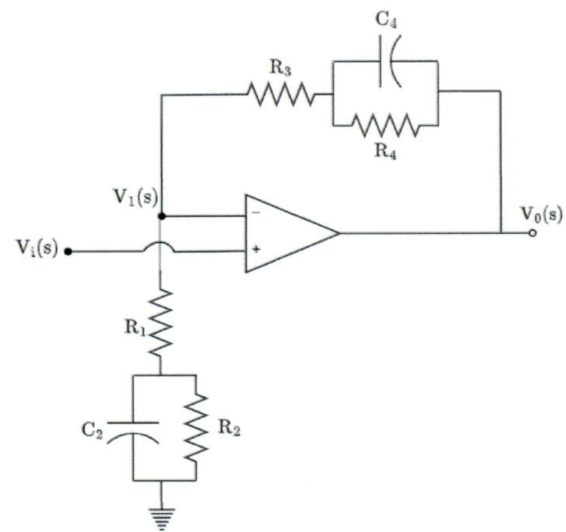

Figure 2.14: Noninverting operational amplifier

SOLUTION:

Now let us compute the impedances $Z_1(s)$ and $Z_2(s)$ (compare with Figure 2.13).

$$
\begin{aligned}
Z_1(s) &= R_1 + \frac{R_2}{R_2C_2s+1} = \frac{R_1R_2C_2s+(R_1+R_2)}{R_2C_2s+1} \\
&= \frac{R_1R_2C_2\left[s+\dfrac{(R_1+R_2)}{R_1R_2C_2}\right]}{R_2C_2\left[s+\dfrac{1}{R_2C_2}\right]} = \frac{R_1\left[s+\dfrac{(R_1+R_2)}{R_1R_2C_2}\right]}{\left[s+\dfrac{1}{R_2C_2}\right]} \\
&= K_1\left[\frac{s+a_1}{s+b_1}\right] \qquad (2.26)
\end{aligned}
$$

where

$$K_1 = R_1,\ a_1 = \frac{R_1+R_2}{R_1R_2C_2},\ \text{and}\ b_1 = \frac{1}{R_2C_2} \qquad (2.27)$$

Similarly, the impedance in the feedback path $Z_2(s)$ is

$$Z_2(s) = K_2\left[\frac{s+a_2}{s+b_2}\right] \qquad (2.28)$$

where

$$K_2 = R_3, \ a_2 = \frac{R_3 + R_4}{R_3R_4C_4}, \text{ and } b_2 = \frac{1}{R_4C_4} \tag{2.29}$$

The transfer function is

$$\begin{aligned}\frac{V_0(s)}{V_i(s)} &= 1 + \frac{Z_2(s)}{Z_1(s)}\\ &= 1 + K\frac{(s+a_2)(s+b_1)}{(s+a_1)(s+b_2)}, \text{ where } K = \frac{K_2}{K_1}\end{aligned} \tag{2.30}$$

Eqn(2.30) can be simplified as:

$$\frac{V_0(s)}{V_i(s)} = \frac{\alpha_0 s^2 + \alpha_1 s + \alpha_2}{(s+a_1)(s+b_2)} \tag{2.31}$$

where

$$\alpha_0 = K + 1, \quad \alpha_1 = a_1 + b_2 + K(a_2 + b_1), \text{ and } \alpha_2 = a_1b_2 + Ka_2b_1$$

2.7 Modeling of Mechanical Systems

Mechanical systems can be classified as **translational, rotational**, or a combination of both. Their differential equation models are derived by applying Newton's laws of motion and D'Alembert's principle, either directly or indirectly, through the balance of forces, energy, and mass.

In translational systems, the fundamental components are mass, spring, and damper, with key variables including displacement, velocity, acceleration, and force. For rotational systems, the analogous elements are moment of inertia, torsional springs, and rotational dampers, with the corresponding variables being angular displacement, angular velocity, angular acceleration, and torque.

2.7.1 Translational System

When an object or a body moves along a straight or curved path, the resulting motion is termed translational motion. Three variables characterize this motion: **displacement**, **velocity**, and **acceleration**. The three basic components of translational mechanical systems are *mass*, *spring*, and *dash pot (damper)*.

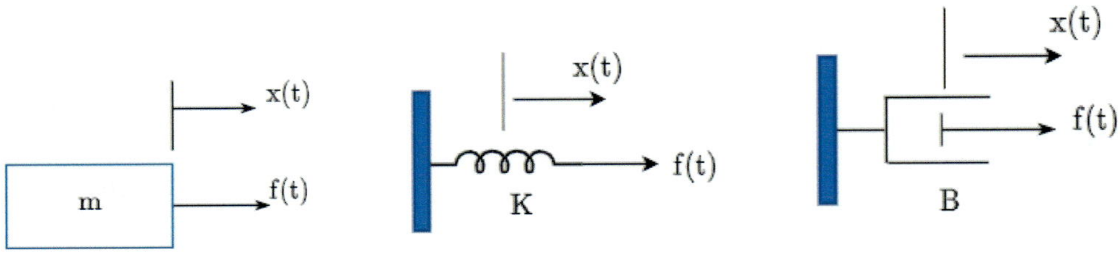

(a) Force-Mass System (b) Force-Spring System (c) Viscous Damper

Figure 2.15: Mechanical Elements

1. **Mass:** This is a property of an element/body/object which stores the kinetic energy of translational motion. It is the mechanical inertia of the object and is analogous to the inductance of electric networks, which is called the electrical inertia. For a rigid body with mass m, the relationship between the applied force f(t) and its acceleration a(t) is given by Newton's second law which states that the algebraic sum of external forces acting on a rigid body in a particular direction equals to the product of the mass of the body and its acceleration in the same direction. Thus,

$$f(t) = ma(t) = m\frac{d^2x(t)}{dt^2} = m\frac{dv(t)}{dt} \tag{2.32}$$

where v(t) denotes linear velocity and x(t) is the displacement. This is shown in Figure 2.15a.

2. **Spring:** A spring is an elastic object that stores mechanical (potential) energy. A linear spring element is shown in Figure 2.15b. For a linear spring, the force f(t) is directly proportional to the displacement (deformation) x(t), that is,

$$f(t) = Kx(t) = K\int v(t)dt \tag{2.33}$$

where K is the spring constant termed as spring stiffness (Nm^{-1}), and v is the velocity.

3. **Damper:** Friction is one of the most significant phenomena in the physical world. It arises whenever one body moves or tries to move over another. It is a force between two surfaces, which opposes movement between those surfaces and is usually nonlinear. Three most commonly observed frictions, in practice, are **viscous friction**, **static friction**, and **Coulomb friction**. In this chapter, we will model the viscous friction as a **viscous damper** or **dash pot**, where the friction force is linearly proportional to the velocity of the mass and

is given by

$$f(t) = B\frac{dx(t)}{dt} = Bv(t) \tag{2.34}$$

where B is the viscous frictional coefficient. The functional relationship between the viscous frictional force and the velocity is shown in Figure 2.15c.

The development of mathematical models for mechanical systems is presented next, progressing from basic to more complex systems.

EXAMPLE 2.7

Consider a mass-spring-damper system shown in Figure 2.16

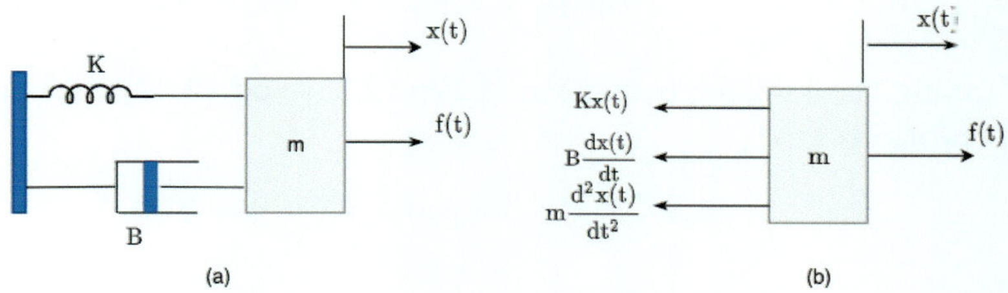

Figure 2.16: (a) Mass spring and damper system (b) Free body diagram

The body moves in the horizontal direction. From the free body diagram shown in Figure 2.16(b), we apply Newton's law to get the force equation

$$m\frac{d^2x(t)}{dt^2} + B\frac{dx(t)}{dt} + Kx(t) = f(t) \tag{2.35}$$

where x(t) is the displacement and $\frac{dx(t)}{dt}$ and $\frac{d^2x(t)}{dt^2}$ denote respectively the velocity and acceleration. We will compute other types of models, such as the transfer function and state variable model of this system, later.

EXAMPLE 2.8

Problem

Obtain the transfer function $\frac{X_2(s)}{F(s)}$ of the system shown in Figure 2.17.

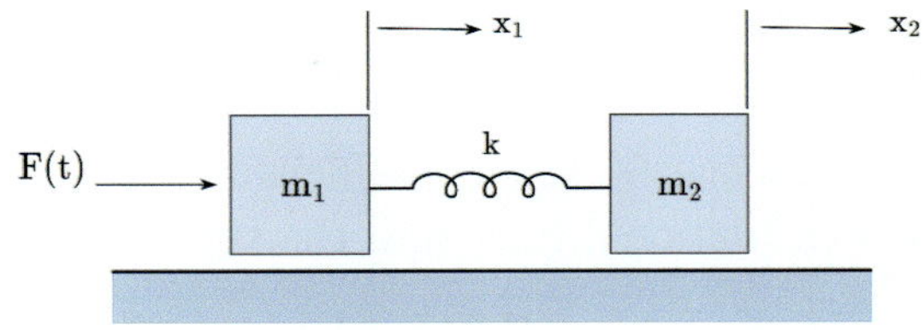

Figure 2.17: Two masses in friction-less surface

SOLUTION:

The motion dynamics can be expressed as:

$$m_1\ddot{x}_1 + k(x_1 - x_2) = F \tag{2.36}$$

$$m_2\ddot{x}_2 + k(x_2 - x_1) = 0 \tag{2.37}$$

Taking the Laplace transform of Eqn(2.36) and (2.37) with zero-initial conditions give

$$m_1s^2X_1(s) + kX_1(s) - kX_2(s) = F(s) \tag{2.38}$$

or,

$$\left[m_1s^2 + k\right]X_1(s) - kX_2(s) = F(s) \tag{2.39}$$

and

$$m_2s^2X_2(s) + kX_2(s) - kX_1(s) = 0 \tag{2.40}$$

or,

$$X_1(s) = \frac{(m_2s^2 + k)}{k}X_2(s) \tag{2.41}$$

Substituting the value of $X_1(s)$ from Eqn(2.41) in Eqn(2.39) gives

$$\frac{\left(m_1s^2 + k\right)\left(m_2s^2 + k\right)}{k}X_2(s) - kX_2(s) = F(s) \tag{2.42}$$

or,

$$\left[\left(m_1s^2 + k\right)\left(m_2s^2 + k\right) - k^2\right]X_2(s) = kF(s) \tag{2.43}$$

Hence, the transfer function between $X_2(s)$ and $F(s)$ is given by

$$\frac{X_2(s)}{F(s)} = \frac{k}{(m_1s^2 + k)(m_2s^2 + k) - k^2} = \frac{k}{s^2\left[m_1m_2s^2 + k(m_1 + m_2)\right]} \tag{2.44}$$

EXAMPLE 2.9

Problem

Obtain the transfer function $\frac{Y_1(s)}{F(s)}$ of the system shown in Figure 2.18.

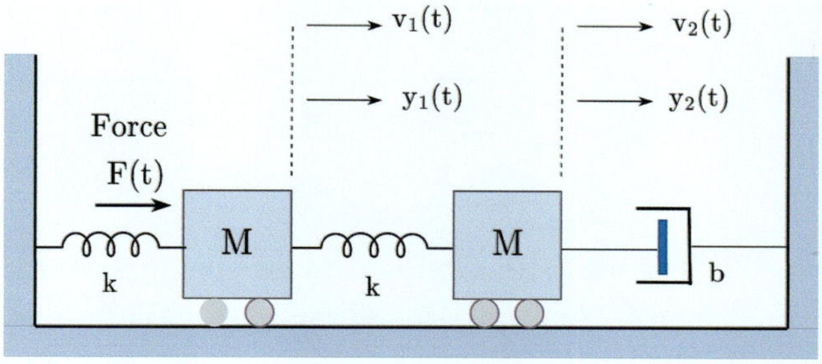

Figure 2.18: Coupled mass spring system

SOLUTION:

The motion dynamics can be expressed as:

$$\begin{aligned} M\ddot{y}_1 + ky_1 + k(y_1 - y_2) &= F \\ M\ddot{y}_2 + b\dot{y}_2 + k(y_2 - y_1) &= 0 \end{aligned} \tag{2.45}$$

Taking the Laplace transform of Eqn(2.45) with zero-initial conditions give

$$\begin{aligned} \left(Ms^2 + 2k\right) Y_1(s) - kY_2(s) &= F(s) \\ -kY_1(s) + \left(Ms^2 + bs + k\right) Y_2(s) &= 0 \end{aligned} \tag{2.46}$$

Expressing Eqn(2.46) in matrix form gives

$$\begin{bmatrix} Ms^2 + 2k & -k \\ -k & Ms^2 + bs + k \end{bmatrix} \begin{bmatrix} Y_1(s) \\ Y_2(s) \end{bmatrix} = \begin{bmatrix} F(s) \\ 0 \end{bmatrix}, \quad \text{Thus} \tag{2.47}$$

$$\begin{aligned} \begin{bmatrix} Y_1(s) \\ Y_2(s) \end{bmatrix} &= \begin{bmatrix} Ms^2 + 2k & -k \\ -k & Ms^2 + bs + k \end{bmatrix}^{-1} \begin{bmatrix} F(s) \\ 0 \end{bmatrix} \\ &= \frac{1}{\Delta} \begin{bmatrix} Ms^2 + bs + k & k \\ k & Ms^2 + 2k \end{bmatrix} \begin{bmatrix} F(s) \\ 0 \end{bmatrix} \end{aligned} \tag{2.48}$$

where $\Delta = \left(Ms^2 + bs + k\right)\left(Ms^2 + 2k\right) - k^2$

Hence the transfer function between $Y_1(s)$ and $F(s)$ is given by

$$\frac{Y_1(s)}{F(s)} = \frac{Ms^2 + bs + k}{\Delta} = \frac{Ms^2 + bs + k}{(Ms^2 + bs + k)(Ms^2 + 2k) - k^2} \tag{2.49}$$

CONCEPT REINFORCEMENT PROBLEM 2.5

Problem

Obtain the transfer function $\dfrac{Y(s)}{F(s)}$ of the system shown in Figure 2.19.

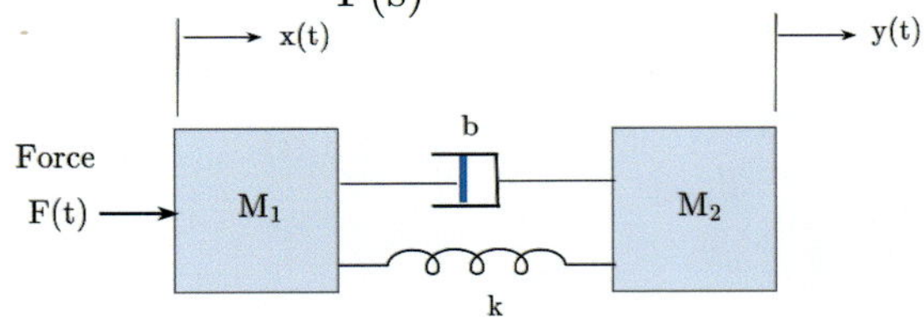

Figure 2.19: Coupled mass spring system

2.8 Modeling of Mechanical (Rotational) Systems

When an object or a body moves about a fixed axis, the resulting motion is called rotational motion. The three basic components of rotational systems are *inertia*, *torsional spring*, and *rotational dash pot (damper)* (see Figure 2.20).

1. **Inertia:** The inertia of a rotational system is represented by the moment of inertia of a mass. The moment of inertia plays a role analogous to that of mass in the case of rectilinear and uniform motion. It is a property of a body that stores the kinetic energy of rotational motion. The inertia of a body depends on several factors, which include its geometric composition about its axis of rotation, density, and the forces involved in the movement. For example, the inertia of a solid cylinder or disc of radius R and mass m about its geometric

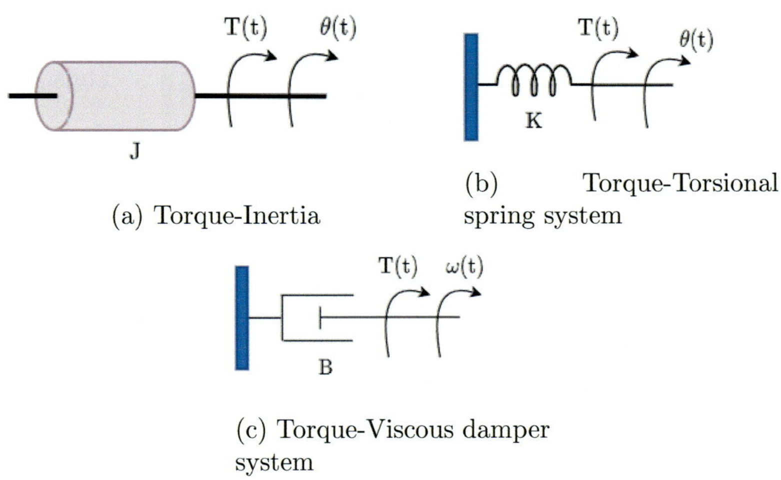

(a) Torque-Inertia

(b) Torque-Torsional spring system

(c) Torque-Viscous damper system

Figure 2.20: Elements of rotational system

axis is given by

$$J = \frac{1}{2}mR^2$$

The inertia of a solid sphere of radius R and mass m about its geometric axis is

$$J = \frac{2}{5}mR^2$$

whereas the inertia of a thin rod of mass m, length L with its rotational axis at the end of the rod is given by

$$J = \frac{1}{3}mL^2$$

For a body with inertia J, the relationship between the applied torque T(t) and the resulting angular acceleration $\alpha(t)$ is obtained using Newton's second law for rotational motion which states that the algebraic sum of moments or torque about a fixed axis is equal to the product of the inertia and the angular acceleration about the axis. Thus

$$\boxed{T(t) = J\alpha(t) = J\frac{d\omega(t)}{dt} = J\frac{d^2\theta(t)}{dt^2}} \tag{2.50}$$

where $\theta(t)$ denotes the angular displacement, and $\omega(t)$ is the angular velocity.

2. **Torsional Spring:**. The torsional spring can store and release rotational energy, operating with a twisting force. For a torsional spring, the torque T(t) is directly proportional to the angular displacement $\theta(t)$, that is

$$\boxed{T(t) = \mathbf{K}\theta(t) = \mathbf{K}\int \omega(t)dt} \tag{2.51}$$

where **K**, in torque-per-unit-angular displacement, is the **torsional spring constant**.

3. **Rotational Dash Pot/Damper:** Similar to translational motion, the commonly observed friction in a rotational system is **viscous friction**, **static friction**, and **Coulomb friction**. The rotary dampers/dash pot can absorb and slow down rotary motion, helping to reduce vibration, noise, and wear on machine components. In this chapter, we will model the viscous friction as a **viscous damper** or **dash pot**, where the friction force is linearly proportional to the angular velocity and is given by

$$\boxed{T(t) = B\frac{\theta(t)}{dt} = B\omega(t)} \tag{2.52}$$

where B is the rotary viscous frictional coefficient.

The following examples illustrate the modeling of simple rotational systems.

EXAMPLE 2.10

Consider the rotational system shown in Figure 2.21.

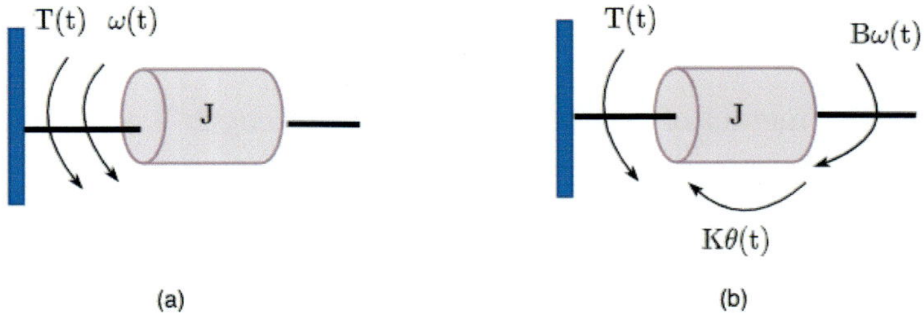

Figure 2.21: (a) Rotational system (b) Free body diagram

The body rotates at a speed $\omega(t)$. From the free body diagram shown in Figure 2.21(b), we apply Newton's law to get the torque equation

$$\boxed{J\frac{d^2\theta(t)}{dt^2} + B\frac{d\theta(t)}{dt} + K\theta(t) = T(t)} \tag{2.53}$$

where $\theta(t)$ is the angular displacement and $\frac{d\theta(t)}{dt}$ and $\frac{d^2\theta(t)}{dt^2}$ denote respectively the angular velocity and angular acceleration.

EXAMPLE 2.11

Determine the transfer function $G(s) = \frac{\theta_1(s)}{T(s)}$ of the rotational system shown in Figure 2.22

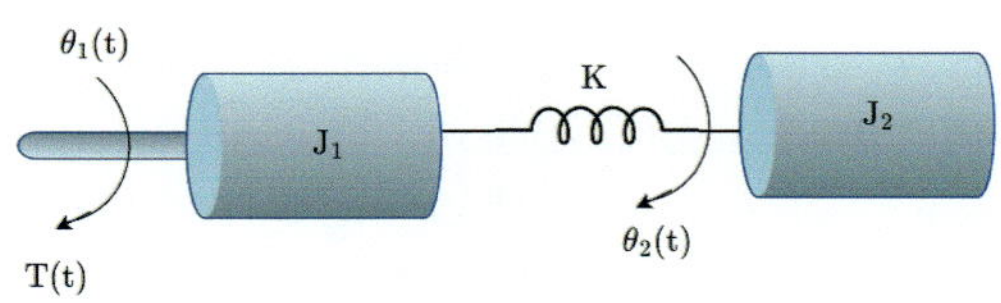

Figure 2.22: Rotational system

SOLUTION: The free body diagram of the system shown in Figure 2.22 is shown in Figure 2.23.

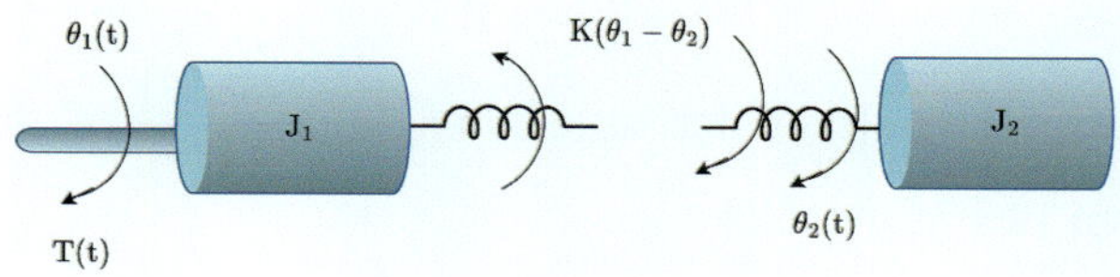

Figure 2.23: Rotational system free body diagram

The differential equations governing the dynamics of this system, from the free body diagram, can be expressed as:

$$J_1 \frac{d^2\theta_1}{dt^2} + K(\theta_1 - \theta_2) = T(t) \tag{2.54}$$

and

$$J_2 \frac{d^2\theta_2}{dt^2} + K(\theta_2 - \theta_1) = 0 \tag{2.55}$$

Taking Laplace transform of Eqn(2.54) gives

$$J_1 s^2 \theta_1(s) + K[\theta_1(s) - \theta_2(s)] = T(s)$$

or

$$\left(J_1 s^2 + K\right)\theta_1(s) = K\theta_2(s) + T(s) \tag{2.56}$$

Taking Laplace transform of Eqn(2.55) gives

$$\left[J_2 s^2 + K\right]\theta_2(s) = K\theta_1(s)$$

or

$$\theta_2(s) = \frac{K\theta_1(s)}{J_2 s^2 + K} \tag{2.57}$$

Substituting the value of $\theta_2(s)$ from Eqn(2.57) into Eqn(2.56) gives

$$\left(J_1 s^2 + K\right)\theta_1(s) = K \times \left[\frac{K\theta_1(s)}{J_2 s^2 + K}\right] + T(s)$$

or

$$\left[\left(J_1 s^2 + K\right) - \frac{K^2}{J_2 s^2 + K}\right]\theta_1(s) = T(s)$$

or

$$G(s) = \frac{\theta_1(s)}{T(s)} = \frac{J_2 s^2 + K}{(J_1 s^2 + K)(J_2 s^2 + K) - K^2} \tag{2.58}$$

CONCEPT REINFORCEMENT PROBLEM 2.6

Determine the transfer function $G(s) = \dfrac{\theta_2(s)}{T(s)}$ of the rotational system shown in Figure 2.24

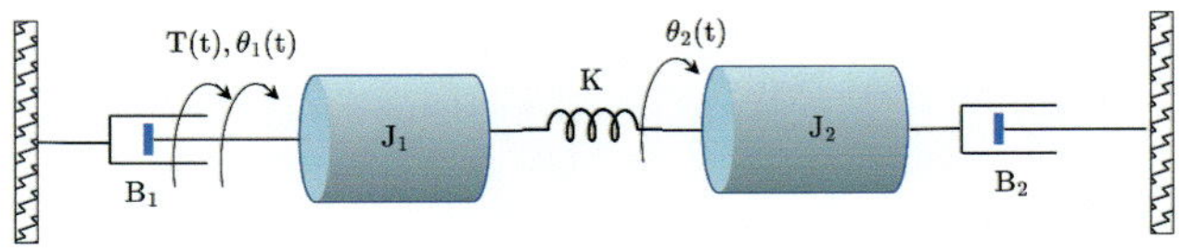

Figure 2.24: Rotational system

EXAMPLE 2.12

Determine the transfer function $G(s) = \dfrac{\theta_2(s)}{T(s)}$ of the rotational system shown in Figure 2.25

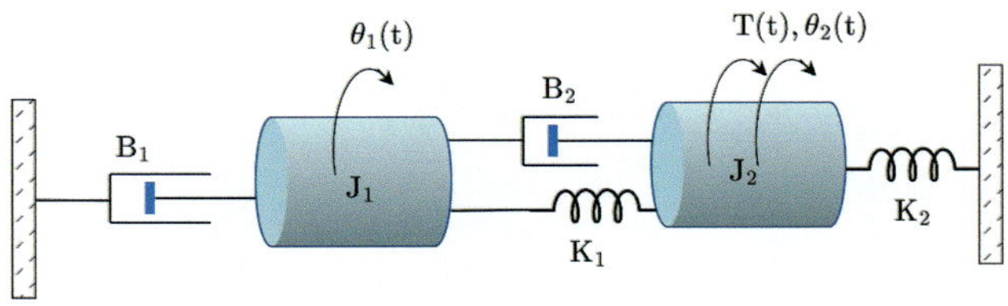

Figure 2.25: Rotational system

SOLUTION:

The dynamics of the system at angular displacement θ_1 is

$$J_1\ddot{\theta}_1 + B_1\dot{\theta}_1 + B_2\left(\dot{\theta}_1 - \dot{\theta}_2\right) + K_1\left(\theta_1 - \theta_2\right) = 0 \tag{2.59}$$

and at angular displacement θ_2 is

$$J_2\ddot{\theta}_2 + B_2\left(\dot{\theta}_2 - \dot{\theta}_1\right) + K_2\theta_2 + K_1\left(\theta_2 - \theta_1\right) = T(t) \tag{2.60}$$

Taking Laplace transform of Eqn(2.59) gives

$$\left[J_1s^2 + (B_1 + B_2)s + K_1\right]\theta_1(s) - \left[B_2s + K_1\right]\theta_2(s) = 0 \tag{2.61}$$

Similarly, the Laplace transform of Eqn(2.60) gives

$$\left[-B_2s - K_1\right]\theta_1(s) + \left[J_2s^2 + B_2s + (K_1 + K_2)\right]\theta_2(s) = T(s) \tag{2.62}$$

For $J_1 = J_2 = 2\text{kg} - \text{m}^2$, $B_1 = 8\text{N} - \text{m} - \text{s/rad}$, $B_2 = 2\text{N} - \text{m} - \text{s/rad}$, $K_1 = 10\text{N} - \text{m/rad}$ and $K_2 = 5\text{N} - \text{m/rad}$, Eqns(2.61) and (2.62) become

$$\begin{aligned} \left(2s^2 + 10s + 10\right)\theta_1(s) - (2s + 10)\,\theta_2(s) &= 0 \\ -(2s + 10)\,\theta_1(s) + \left(2s^2 + 2s + 15\right)\theta_2(s) &= T(s) \end{aligned} \tag{2.63}$$

Now

$$\theta_2(s) = \frac{\begin{vmatrix} 2s^2 + 10s + 10 & 0 \\ -(2s + 10) & T \end{vmatrix}}{\begin{vmatrix} 2s^2 + 10s + 10 & -(2s + 10) \\ -(2s + 10) & 2s^2 + 2s + 15 \end{vmatrix}}$$
$$= \frac{s^2 + 5s + 5}{2s^4 + 12s^3 + 33s^2 + 65s + 25} T(s)$$

Hence the transfer function

$$\frac{\theta_2(s)}{T(s)} = \frac{s^2 + 5s + 5}{2s^4 + 12s^3 + 33s^2 + 65s + 25}$$

2.9 Transfer Function for Systems with Gears

In Section 2.8, we derived the transfer function of rational systems. The systems we considered have no gear trains; a rare thing in a rotational system. Most of the rotational systems have gears (gear trains) that transmit energy from one part of the system to another. In control systems, gear trains are often used to match the speed and torque between different parts of a mechanical system, to ensure that power is transferred efficiently and effectively, optimizing the performance of the system. For example, in a servo control system, the servo motors operate at high speed with low torque. If such motors are to be used to drive a load with high torque and low speed, we require magnification of torque and reduction in speed. This is achieved using gear trains. Thus, in mechanical systems, gear trains act as a matching device similar to transformers in electrical systems.

Consider the gear train, shown in Figure 2.26, where two gears are coupled together.

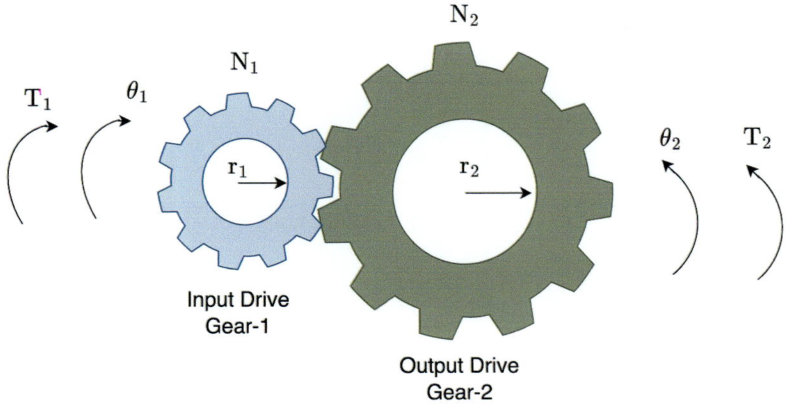

Figure 2.26: A gear system

Note that gears often exhibit *backlash* because of the lose fit between two meshed gears. The drive gear rotates through a small angle before making contact with the meshed gear. Therefore, the output gear does not rotate until the input gear rotates by a small angle. In this section, we derive the relationships between the torques T_1, and T_2, angular displacements θ_1, θ_2, the number of teeth N_1, and N_2, angular velocities ω_1 and ω_2 and the radii r_1 and r_2 of input (primary) and output(secondary) gears, respectively. It is assumed that the gears are ideal with no inertia, friction, or backlash.

When the input gear with radius r_1 and teeth N_1 rotates through an angle θ_1 due to a torque T_1, the output gear with radius r_2 and N_2 number of teeth rotates by an angle θ_2 and delivers a torque T_2. The following principles hold for gear trains shown in Figure 2.26.

1. The distance traveled along the circumference of each gear is the same. Thus

$$\boxed{r_1\theta_1 = r_2\theta_2} \tag{2.64}$$

2. The number of teeth on the surface of gears is proportional to the radii r_1 and r_2 of the gears; that is

$$\boxed{r_1N_2 = r_2N_1} \tag{2.65}$$

3. Assuming that the gears do not absorb or store energy, that is, there is no loss, the work done by each gear is the same. Further, since the translational energy of force times displacement becomes rotational energy of torque times angular displacement, this gives the relation

$$\boxed{T_1\theta_1 = T_2\theta_2} \tag{2.66}$$

Eqns(2.64), (2.65), and (2.66) give the relationship

$$\frac{T_1}{T_2} = \frac{\theta_2}{\theta_1} = \frac{N_1}{N_2} = \frac{r_1}{r_2} \tag{2.67}$$

Mechanical Impedance

When different parts of a rotational dynamic system are connected by gears, it is crucial to understand the concept of reflected impedance (inertia, viscous friction coefficient, spring constant). The reflected impedance refers to how a variable (e.g., inertia) of one part of a system, when observed through a gear train, appears at another part of the system.

Consider a rotational system shown in Figure 2.27.

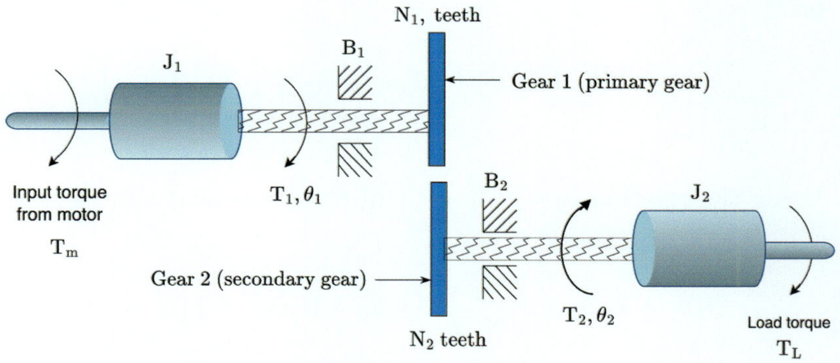

Figure 2.27: A gear system

The rotational dynamics of this system on the side of gear 1 is

$$J_1\ddot{\theta}_1 + B_1\dot{\theta}_1 + T_1 = T_m \tag{2.68}$$

where T_m is the torque developed by the motor and T_1 is the load torque due to the rest of the gear train.

The shaft dynamics of the output gear 2 are

$$J_2\ddot{\theta}_2 + B_2\dot{\theta}_2 + T_L = T_2 \tag{2.69}$$

where T_2 is the torque transmitted to gear 2 and T_L is the load torque.

Objective: To describe the rotational dynamics of the whole system in terms of gear 1 variables.

Now from Eqn(2.67), we have

$$\frac{T_1}{T_2} = \frac{N_1}{N_2} \implies T_1 = T_2\left(\frac{N_1}{N_2}\right) \quad \text{and}$$

$$\frac{\theta_2}{\theta_1} = \frac{N_1}{N_2} \implies \theta_2 = \theta_1\left(\frac{N_1}{N_2}\right) \tag{2.70}$$

Substituting the value of T_2 from Eqn(2.69) into Eqn(2.68) and using the relations in Eqn(2.70) gives

$$J_1\ddot{\theta}_1 + B_1\dot{\theta}_1 + T_1 = T_m$$

or

$$J_1\ddot{\theta}_1 + B_1\dot{\theta}_1 + T_2\left(\frac{N_1}{N_2}\right) = T_m$$

or

$$J_1\ddot{\theta}_1 + B_1\dot{\theta}_1 + \left[J_2\ddot{\theta}_2 + B_2\dot{\theta}_2 + T_L\right]\left(\frac{N_1}{N_2}\right) = T_m \tag{2.71}$$

Using the relation $\theta_2 = \theta_1\left(\frac{N_1}{N_2}\right)$ in the above and eliminating θ_2 yields

$$J_1\ddot{\theta}_1 + B_1\dot{\theta}_1 + \left(\frac{N_1}{N_2}\right)^2\left[J_2\ddot{\theta}_1 + B_2\dot{\theta}_1\right] + T_L\left(\frac{N_1}{N_2}\right) = T_m \tag{2.72}$$

or

$$J_{eq1}\ddot{\theta}_1 + B_{eq1}\dot{\theta}_1 + T_L\left(\frac{N_1}{N_2}\right) = T_m \tag{2.73}$$

where J_{eq1} and B_{eq1} are respectively the equivalent inertia and friction referred to the primary and are given by

$$\boxed{J_{eq1} = J_1 + J_2\left(\frac{N_1}{N_2}\right)^2} \tag{2.74}$$

and

$$\boxed{B_{eq1} = B_1 + B_2\left(\frac{N_1}{N_2}\right)^2} \tag{2.75}$$

Similarly, gear parameters and variables can be reflected from gear 1 to gear 2, and the corresponding equivalent inertia and viscous friction referred to the output is

$$\boxed{J_{eq2} = J_2 + J_1 \left(\frac{N_2}{N_1}\right)^2} \tag{2.76}$$

and

$$\boxed{B_{eq2} = B_2 + B_1 \left(\frac{N_2}{N_1}\right)^2} \tag{2.77}$$

The dynamics with respect to the output (load shaft) is

$$J_{eq2}\ddot{\theta}_2 + B_{eq2}\dot{\theta}_2 + T_L = T_m \left(\frac{N_2}{N_1}\right) \tag{2.78}$$

Before generalizing the above relations, consider a multistage gear train with three gears as follows:

1. Gear-1 (driving gear) with N_1=20 teeth.
2. Gear-2 (intermediate gear) with N_2=50 teeth.
3. Gear-3 (driven gear) with N_3=100 teeth.

The rotational inertia of Gear-3 (J_3) is given as 4kg.m^2. Now reflected inertia of Gear-3 to Gear-2 is

$$J_{3,2} = J_3 \left(\frac{N_2}{N_3}\right)^2 = 4\left(\frac{50}{100}\right)^2 = 4 \times \frac{1}{4} = 1 \text{ kg.m}^2$$

Reflected inertia of $J_{3,2}$ to Gear-1 is

$$J_{3,2,1} = J_{3,2} \left(\frac{N_1}{N_2}\right)^2 = 1\left(\frac{20}{50}\right)^2 = 1 \times \frac{4}{25} = 0.16\text{kg.m}^2$$

The total reflected inertia from Gear-3 to Gear-1 is $J_{reflected} = J_{3,2,1} = 0.16\text{kg.m}^2$

For n-stage gear trains, the reflected inertia from the n-th gear to the first gear is:

$$\boxed{J_{n,1} = J_n \prod_{i=2}^{n} \left(\frac{N_{i-1}}{N_i}\right)^2} \tag{2.79}$$

which reduces to

$$J_{n,1} = J_n \left(\frac{N_1}{N_n}\right)^2 \tag{2.80}$$

For n number of gears with gear **i** having N_i teeth, the reflected inertia of gear n back to gear m is

$$J_{n,m} = J_n \left(\frac{N_m}{N_n}\right)^2$$

EXAMPLE 2.13

Problem

For the rotational system, shown in Figure 2.28, determine the transfer function $G(s) = \dfrac{\theta_2(s)}{T_m(s)}$.

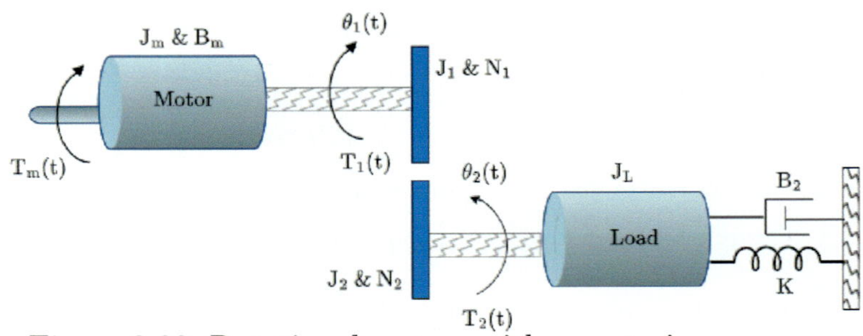

Figure 2.28: Rotational system with gear trains

SOLUTION:

The transfer function will be derived following two approaches as outlined below.

Method-1:

Dynamics at node θ_1 (i.e., gear-1)

$$(J_m + J_1)\ddot{\theta}_1 + B_m\dot{\theta}_1 + T_1(t) = T_m(t)$$

or

$$\left[(J_m + J_1)s^2 + B_m s\right]\theta_1(s) + T_1(s) = T_m(s) \tag{2.81}$$

Dynamics at node θ_2 (i.e., gear-2)

$$(J_L + J_2)\ddot{\theta}_2 + B_2\dot{\theta}_2 + K\theta_2(t) = T_2(t)$$

or

$$\left[(J_L + J_2)\, s^2 + B_2 s + K\right] \theta_2(s) = T_2(s) \tag{2.82}$$

Further, the system satisfies the following relations:

$$\frac{T_1}{T_2} = \frac{\theta_2}{\theta_1} = \frac{N_1}{N_2} = n \tag{2.83}$$

The block diagram of the system satisfying the relations described in Eqns(2.81), (2.82), and (2.83) is shown in Figure 2.29.

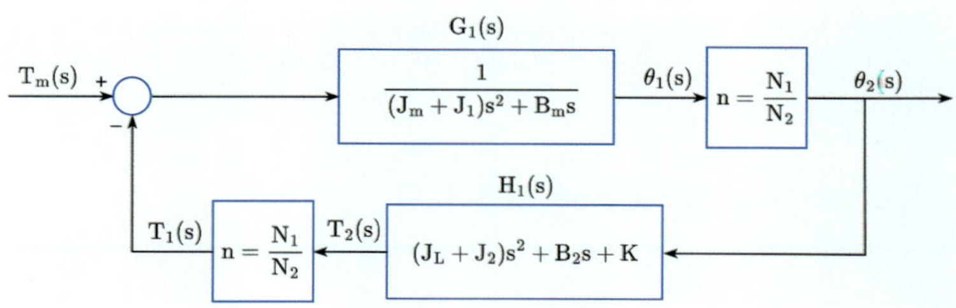

Figure 2.29: Transfer function block diagram of rotational system

From the block diagram, the transfer function

$$\begin{aligned} G(s) = \frac{\theta_2(s)}{T_m(s)} &= \frac{nG_1(s)}{1 + n^2 G_1(s) H_1(s)} \\ &= \frac{n}{\left[(J_1 + J_m) + n^2 (J_2 + J_L)\right] s^2 + (B_m + n^2 B_2)\, s + n^2 K} \end{aligned} \tag{2.84}$$

Method-2:

Let us denote

$$J_1' = J_1 + J_m, \quad J_2' = J_2 + J_L$$

Let us reflect all the variables and parameters on to load side. Now reflected inertia J_{eq2} and friction coefficient B_{eq2} at the load side is

$$J_{eq2} = J_2' + \frac{J_1'}{n^2} = (J_2 + J_L) + \frac{(J_1 + J_m)}{n^2}, \quad \text{and} \quad B_{eq2} = B_2 + \frac{B_m}{n^2}$$

The reflected torque at output frame is $T_m \dfrac{N_2}{N_1}$.

The dynamics of the complete system are

$$J_{eq2}\ddot{\theta}_2 + B_{eq2}\dot{\theta}_2 + K\theta_2 = T_m \frac{N_2}{N_1} \tag{2.85}$$

The transfer function is

$$G(s) = \frac{\theta_2(s)}{T_m(s)} = \frac{\frac{1}{n}}{J_{eq2}s^2 + B_{eq2}s + K_2}$$

$$= \frac{\frac{1}{n}}{\left(J_2' + \frac{J_1'}{n^2}\right)s^2 + \left(B_2 + \frac{B_m}{n^2}\right)s + K}$$

$$= \frac{n}{(n^2 J_2' + J_1')\,s^2 + (n^2 B_2 + B_m) + n^2 K]}$$

$$= \frac{n}{[(J_1 + J_m) + n^2 (J_2 + J_L)]\,s^2 + (B_m + n^2 B_2)\,s + n^2 K} \tag{2.86}$$

Compare Eqn(2.86) and (2.84).

CONCEPT REINFORCEMENT PROBLEM 2.7

Problem

For the rotational system, shown in Figure 2.30, determine the transfer function $G(s) = \dfrac{\theta_2(s)}{T(s)}$.

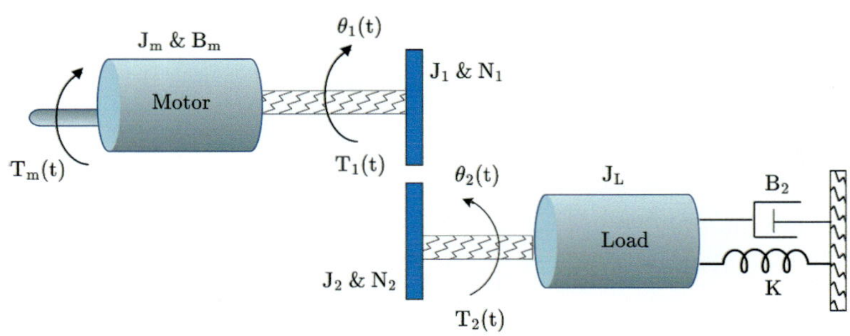

Figure 2.30: Rotational system

EXAMPLE 2.14

Problem

A motor-load system coupled through a gear train with gear ratio $n = \dfrac{N_1}{N_2}$ is shown in Figure 2.31. The motor develops a torque $T_m(t)$ and the $T_L(t)$ is the load torque.

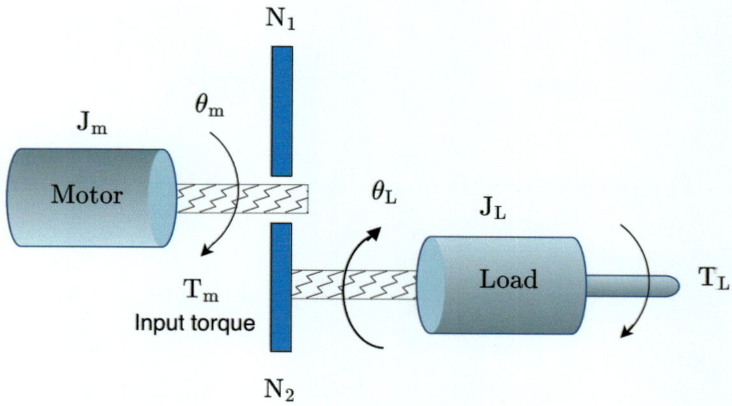

Figure 2.31: Rotational System

a. Determine the optimum gear ratio n∗ such that the load acceleration $\ddot{\theta}_L$ is maximum.

b. Repeat part(a) when the load torque is zero.

SOLUTION:

a. Let us reflect on all impedances to the load frame. The equivalent inertia in the output frame is

$$J_{eq} = J_L + J_m\left(\frac{N_2}{N_1}\right)^2 = J_L + \frac{J_m}{n^2}$$

The reflected motor torque in the load frame is

$$T_{m2} = T_m\frac{N_2}{N_1} = \frac{T_m}{n}$$

The rotational dynamics of the system are

$$J_{eq}\frac{d^2\theta_L}{dt^2} + T_{m2} = T_L$$

or

$$\left[J_L + \frac{J_m}{n^2}\right]\frac{d^2\theta_L}{dt^2} + T_L = \frac{T_m}{n}$$

$$\Longrightarrow \quad \left[J_L + \frac{J_m}{n^2}\right]n\frac{d^2\theta_L}{dt^2} + nT_L = T_m$$

or

$$\left[nJ_L + \frac{J_m}{n}\right]\frac{d^2\theta_L}{dt^2} + nT_L = T_m$$

This gives the acceleration as:

$$\ddot{\theta}_L = \frac{nT_m - n^2T_L}{J_m + n^2J_L} \tag{2.87}$$

The optimum value of n is determined by setting $\frac{d\ddot{\theta}_L}{dn} = 0$ which gives

$$(T_m - 2nT_L)(J_m + n^2J_L) - 2nJ_L(nT_m - n^2T_L) = 0$$

or

$$n^2J_LT_m + 2nJ_mT_L - J_mT_m = 0 \tag{2.88}$$

Selecting a positive sign, the optimum gear ratio is

$$n^* = -\frac{J_mT_L}{J_LT_m} + \frac{\sqrt{J_m^2T_L^2 + J_mJ_LT_m^2}}{J_LT_m} \tag{2.89}$$

b. When $T_L = 0$, the optimum gear ratio is obtained by replacing $T_L = 0$ in the last expression. This yields

$$n^* = \sqrt{\frac{J_m}{J_L}} \tag{2.90}$$

CONCEPT REINFORCEMENT PROBLEM 2.8

Problem

For the rotational system with gear trains, shown in Figure 2.32, find the transfer function $G(s) = \frac{\theta_4(s)}{T(s)}$. The inertia, bearing friction, and number of teeth of gears are shown in the figure.

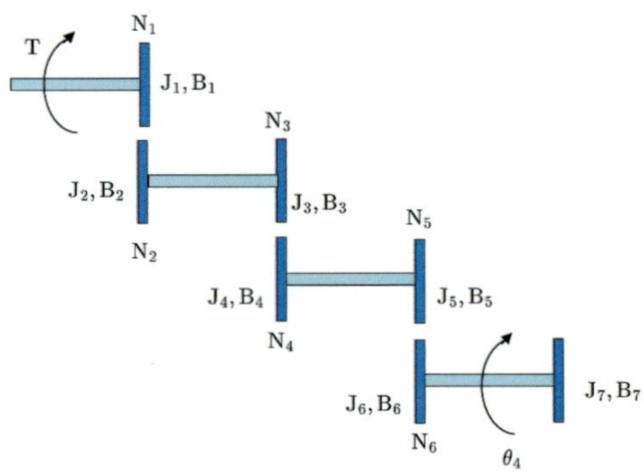

Figure 2.32: Rotational System

2.10 Transfer Function Modeling of Electromechanical Systems

We have discussed how to obtain transfer function models of electrical and mechanical (both translational and rotational) systems. But many practical systems are *electromechanical systems* which are constructed by combining electrical and mechanical elements. Electromechanical systems can be found in a wide range of applications, including **home appliances** (e.g., washing machines, refrigerators, dishwashers), **automotive systems** (e.g. anti-lock braking systems, electric and hybrid vehicles), **robotics** (e.g., robotic vacuum cleaners, lawn mowers), **medical devices** (e.g., prosthetic limbs), **building systems**(e.g., elevators, HVAC systems) and many more.

The dynamic models of such hybrid systems can be derived by integrating the dynamics of electrical and mechanical components using fundamental electrical and mechanical principles such as Kirchhoff's laws, Newton's laws, etc. This typically involves the following steps:

1. Identify the electrical and mechanical elements involved in the system
2. Write governing differential equations applying related electrical and mechanical laws.
3. Integrate the electrical and mechanical equations to get a complete system model
4. Use Laplace transforms to convert the time-domain differential equations into algebraic equations in the s-domain and solve for the transfer function.

In the following, we derive transfer function models of DC servomechanisms that use DC servo motors. Servo motors in industrial applications are different from regular motors because of their capability to precisely rotate the motor shaft. This is achieved by special electronics built into the motors. A motor is an actuator, converting electrical energy into rotational mechanical energy. A motor requiring a DC power supply for operation is termed a DC motor. DC motors are widely used in control applications like robotics, tape drives, machines, and many more. Separately excited DC motors are suitable for control applications because of a separate field and armature circuit. Two ways to control DC separately excited motors are: Armature Control and Field Control.

2.10.1 Transfer Function of Armature Controlled DC Servo Motor

Consider an armature-controlled DC servo motor shown in Figure 2.33

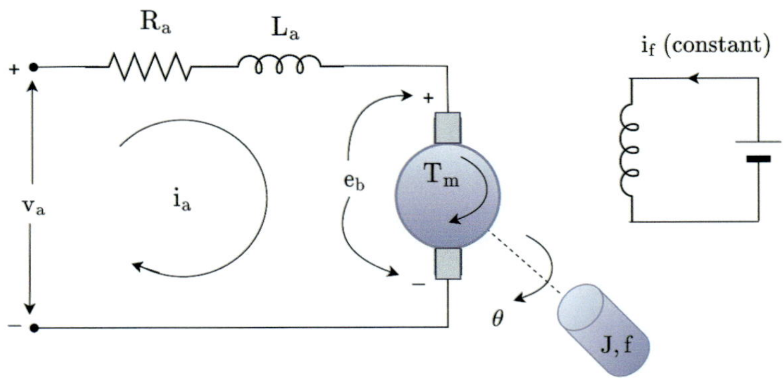

Figure 2.33: Armature-controlled DC motor

where

R_a = Armature resistance in ohm (Ω)

L_a = Armature inductance in Henry (H)

i_a = Armature current in ampere (A)

i_f = Field current in ampere(A); held constant

v_a = Applied armature voltage in volts (V)

e_b = Back emf in volts(V)

T_m = Torque developed by the motor in Newton-Meter (NM)

θ = Angular displacement of motor shaft in radian (rad)

ω = Angular velocity of motor shaft in radian per second (rad/sec)

J = Equivalent moment of inertia of motor and load referred to motor shaft in $kg - m^2$

f = Equivalent viscous friction coefficient of motor and load referred to motor shaft in $\left(\frac{Nm}{rad/sec}\right)$

In an armature-controlled DC motor, the desired speed is obtained by varying the armature, where the field current is assumed to be constant. When we apply a voltage $v_a(t)$ to the armature, it will cause the armature to rotate. The speed of the motor is directly proportional to the armature voltage and inversely proportional to the field flux. Let us derive the relationship between $v_a(t)$ and $\theta(t)$ where θ is the position of the DC servo motor. This is an electromechanical system having both electrical and mechanical dynamics.

A DC motor consists of two parts: a rotor and a stator. The stator consists of field windings, while the rotor (also called the armature) consists of an armature winding. When both the armature and the field windings are excited by a DC supply, current flows through the windings, and a magnetic flux proportional to the current is produced. When the flux from the field interacts with the flux from the armature, it results in motion of the rotor. Armature control is the most common control technique for DC motors. In order to implement this control, the stator flux must be kept constant. To achieve this, either the stator voltage is kept constant or the stator coils are replaced by a permanent magnet. In the latter case, the motor is said to be a permanent magnet DC motor and is driven by the armature coils only.

When a DC motor is used in control applications as an actuator, it operates in the linear range of the magnetization curve. Thus the air gap flux ϕ is proportional to the field current $\mathrm{i_f}$ i.e.,

$$\phi \propto \mathrm{i_f} \quad \text{or,} \quad \phi = \mathrm{K_f i_f} \tag{2.91}$$

where $\mathrm{K_f}$ is a constant.

The torque $\mathrm{T_m}$ developed by the motor is proportional to the product of the flux and the armature current. Thus,

$$\mathrm{T_m} \propto \phi \mathrm{i_a} \quad \text{or,} \quad \mathrm{T_m} \propto \mathrm{K_f i_f i_a} \quad \text{or,} \quad \mathrm{T_m} = \mathrm{K_a K_f i_f i_a} \tag{2.92}$$

where $\mathrm{K_a}$ is a constant.

Since in the armature-controlled DC motor, the field current is kept constant

$$\boxed{\mathrm{T_m} = \mathrm{K_t i_a}} \tag{2.93}$$

where $\mathrm{K_t} = \mathrm{K_a K_f i_f}$ is called the motor torque constant.

When the motor rotates, an emf is generated in the armature, called back emf $\mathrm{e_b}$. This is proportional to the speed and is expressed as:

$$\mathrm{e_b} \propto \frac{\mathrm{d}\theta}{\mathrm{dt}} \quad \text{or} \quad \mathrm{e_b} = \mathrm{K_b}\frac{\mathrm{d}\theta}{\mathrm{dt}} = \mathrm{K_b}\omega \tag{2.94}$$

or

$$\boxed{\mathrm{e_b} = \mathrm{K_b}\frac{\mathrm{d}\theta}{\mathrm{dt}} = \mathrm{K_b}\omega} \tag{2.95}$$

where $\mathrm{K_b}$ is the back emf constant.

Electrical Dynamics of DC Servo Motor

Applying Kirchhoff's voltage law gives

$$\mathrm{R_a i_a} + \mathrm{L_a}\frac{\mathrm{di_a}}{\mathrm{dt}} + \mathrm{e_b} = \mathrm{v_a} \quad \Longrightarrow \quad \mathrm{R_a i_a} + \mathrm{L_a}\frac{\mathrm{di_a}}{\mathrm{dt}} = \mathrm{v_a} - \mathrm{e_b} \tag{2.96}$$

Taking the Laplace transform of Eqn(2.96) gives

$$(\mathrm{R_a} + \mathrm{L_a s})\,\mathrm{I_a(s)} = \mathrm{V_a(s)} - \mathrm{E_b(s)} \implies \mathrm{I_a(s)} = \frac{\mathrm{V_a(s)} - \mathrm{E_b(s)}}{\mathrm{L_a s} + \mathrm{R_a}} \tag{2.97}$$

Since

$$\mathrm{E_b(s)} = \mathrm{K_b s}\theta(\mathrm{s}) = \mathrm{K_b}\omega(\mathrm{s}) \tag{2.98}$$

Eqn(2.97) is expressed as

$$\boxed{\mathrm{I_a(s)} = \frac{\mathrm{V_a(s)} - \mathrm{K_b s}\theta(\mathrm{s})}{\mathrm{L_a s} + \mathrm{R_a}}} \tag{2.99}$$

The torque developed by the armature

$$\boxed{\mathrm{T_m(s)} = \mathrm{K_t I_a(s)}} \tag{2.100}$$

Mechanical Dynamics of DC Servo Motor

The rotational dynamics of the DC servo motor are described as:

$$\mathrm{J}\frac{\mathrm{d}^2\theta}{\mathrm{dt}^2} + \mathrm{f}\frac{\mathrm{d}\theta}{\mathrm{dt}} = \mathrm{T_m} - \mathrm{T_d} \tag{2.101}$$

Assume that the losses in the motor are negligible. Then, the electrical power developed in the motor equals the available mechanical power. Thus,

$$\mathrm{E_b i_a} = \mathrm{T_m}\omega \quad \text{or,} \quad \mathrm{K_b}\omega\mathrm{i_a} = \mathrm{K_t i_a}\omega, \quad \text{or,} \quad \mathrm{K_b} = \mathrm{K_t} \tag{2.102}$$

Integration of Electrical and Mechanical Dynamics

Now Eqn(2.101) becomes

$$\mathrm{J}\frac{\mathrm{d}^2\theta}{\mathrm{dt}^2} + \mathrm{f}\frac{\mathrm{d}\theta}{\mathrm{dt}} = \mathrm{K_t i_a} - \mathrm{T_d} \tag{2.103}$$

Taking the Laplace transform of Eqn(2.103) gives

$$\left(\mathrm{Js}^2 + \mathrm{fs}\right)\theta(\mathrm{s}) = \mathrm{K_t I_a(s)} - \mathrm{T_d(s)} \tag{2.104}$$

Substituting the value of $I_a(s)$ from Eqn(2.99) into Eqn(2.104) gives

$$\left[Js^2 + fs\right]\theta(s) = K_t\left[\frac{V_a - K_b s\theta(s)}{R_a + L_a s}\right] - T_d(s)$$

or

$$\left[Js^2 + fs + \frac{K_t K_b s}{R_a + sL_a}\right]\theta(s) = \frac{K_t V_a(s)}{R_a + L_a s} - T_d(s)$$

or

$$\left[\left(Js^2 + fs\right)\left(R_a + sL_a\right) + K_t K_b s\right]\theta(s) = K_t V_a(s) - \left(R_a + sL_a\right)T_d(s)$$

This gives

$$\theta(s) = \frac{K_t}{s\left[(Js + f)\left(R_a + sL_a\right) + K_t K_b\right]}V_a(s) - \frac{R_a + sL_a}{s\left[(Js + f)\left(R_a + sL_a\right) + K_t K_b\right]}T_d(s) \qquad (2.105)$$

The transfer function between the position $\theta(s)$ and armature voltage $V_a(s)$, with $(T_d = 0)$ is

$$\frac{\theta(s)}{V_a(s)} = \frac{K_t}{s\left[(Js + f)\left(R_a + sL_a\right) + K_t K_b\right]} \qquad (2.106)$$

The transfer function between the position $\theta(s)$ and $T_d(s)$, with $(V_a(s) = 0)$ is

$$\frac{\theta(s)}{T_d(s)} = -\frac{R_a + sL_a}{s\left[(Js + f)\left(R_a + sL_a\right) + K_t K_b\right]} \qquad (2.107)$$

Eqn(2.106) can alternately be expressed as:

$$\frac{\theta(s)}{V_a(s)} = \frac{K_t/R_a f}{s\left(1 + \tau_m s\right)\left(1 + \tau_a s\right) + sK_t K_b} \qquad (2.108)$$

where $\tau_a = \dfrac{L_a}{R_a}$ is the electrical time constant of armature circuit and $\tau_m = \dfrac{J}{f}$ is the motor time constant.

CONCEPT REINFORCEMENT PROBLEM 2.9

Problem

The schematic of a field-controlled DC motor is shown in Figure 2.34

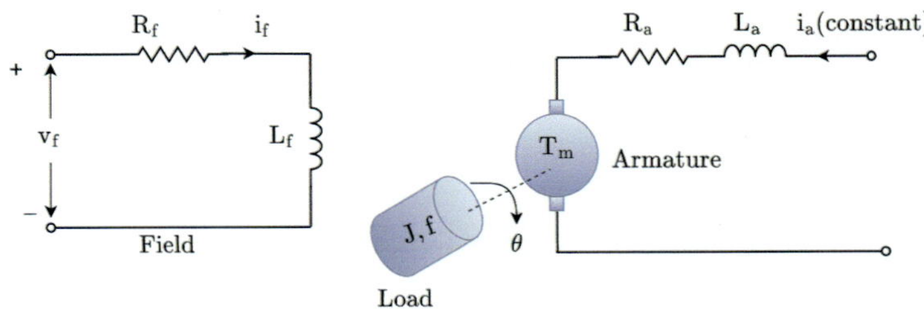

Figure 2.34: Field-controlled DC motor

Compute the transfer function of the system between the angular position and field voltage and the angular position and disturbance torque.

2.10.2 Modeling of AC Servo Motor

AC servomotors play a critical role in modern control systems, offering precision, reliability, and efficiency across a wide range of applications. These motors have no brush contacts and are preferred for low-power applications because of their light weight. They are primarily two-phase induction motors with several distinctive features that make them suitable for precision control applications. The following are some of the distinctive features of AC servo motors:

a. **High Resistance Rotor**

The rotor of the AC servomotor is designed with high resistance to keep the X/R ratio (the ratio of rotor reactance X to rotor resistance R) small.

b. **Extended Stability Region**

A small X/R ratio extends the stability region of the motor. This characteristic is critical for applications requiring stable and precise control over a wide range of operating conditions. The extended stability region helps maintain consistent performance and prevents oscillations or instability in the control system.

c. **Torque Characteristics**

A high X/R ratio results in achieving maximum torque close to the operating region (within 5% slip). This means that the motor can deliver high torque at low slip values, which is beneficial for applications requiring quick response and high torque at low speeds. The ability to produce maximum torque near the operating speed enhances the motor's efficiency and performance in dynamic applications.

d. **Small Rotor Diameter**

The diameter of the rotor is kept small, which reduces the inertia and thereby enables fast acceleration. This design feature is crucial for applications that demand rapid changes in speed and direction, such as robotics and CNC machines. A smaller rotor diameter minimizes the rotational inertia, allowing the motor to accelerate and decelerate quickly, thus improving the overall responsiveness of the system.

Operation of Two Phase AC Servo Motor

The schematic diagram of a two-phase AC servo motor (induction motor) is shown in Figure 2.35.

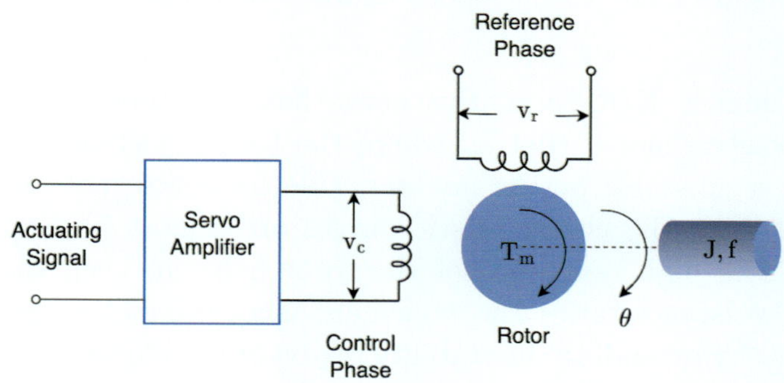

Figure 2.35: Two-phase AC servo motor

The stator has two sets of windings, known as **control winding** and **reference winding** which are displaced from each other by 90^0 electrical. The rotor is usually a squirrel-cage type with high resistance to achieve desirable control characteristics. When the stator windings are excited by two-phase AC voltages that are 90 degrees out of phase with each other, a rotating magnetic field is produced in the stator, rotating at synchronous speed concerning the stator windings. The rotating magnetic field induces a voltage in the rotor bars. Since the rotor is short-circuited, a current flows in the rotor, which produces a magnetic field that interacts with the rotating magnetic field of the stator. The interaction between the rotor's induced magnetic field and the stator's rotating magnetic field generates a torque. The direction and magnitude of this torque depend on the phase difference and magnitudes of the currents in the stator windings.

Torque-Speed Characteristics

The torque-speed characteristic of the motor is shown in Figure 2.36 whose shape is dependent on the ratio of rotor reactance X to rotor resistance R.

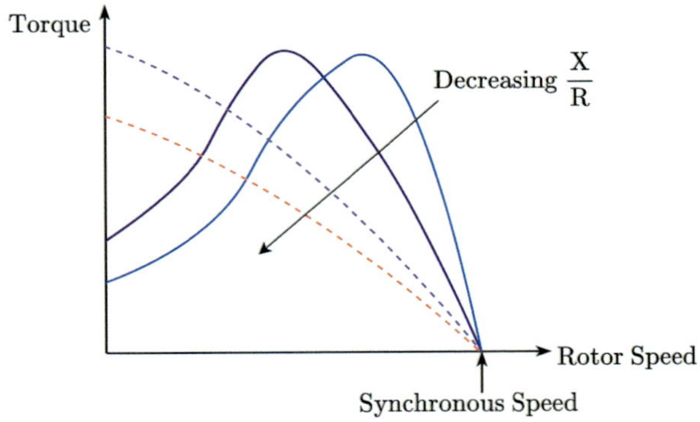

Figure 2.36: Torque-speed characteristics of AC servo motor

For a higher X/R ratio, this curve has both positive and negative slope. The positive slope, that is, where the torque increases with an increase in speed, is unstable and is not desirable for servo motors. As seen from the Figure 2.36, the characteristic curves are almost linear for a smaller X/R ratio. The high resistance of the rotor helps to keep the X/R ratio (reactance/resistance ratio) low, which enhances the motor's stability and responsiveness by extending the stability region of the motor. It also ensures that the maximum torque is achieved at low slip values, which is desirable for precise control.

In servo applications, the voltages applied to two stator windings are seldom balanced. The reference winding is excited by a constant voltage, but the control winding is excited by an amplified error signal which varies in magnitude and polarity ($\pm 90^0$ w.r.t. reference phase). The direction of the motor reverses with the reversal of the error signal polarity. The output torque is proportional to the magnitude of the control voltage, and the direction of the torque is determined from the polarity of the error signal.

The torque generated by the motor depends both on the rms control voltage V_c and speed $\omega(=\frac{d\theta}{dt})$. In servo applications, the motor normally operates at low speeds. Therefore, the torque-speed curves are approximated by straight lines as shown in Figure 2.37

Transfer Function of AC Servo Motor

Consider Figure 2.37 which shows ideal torque-speed characteristic of an AC servomotor.

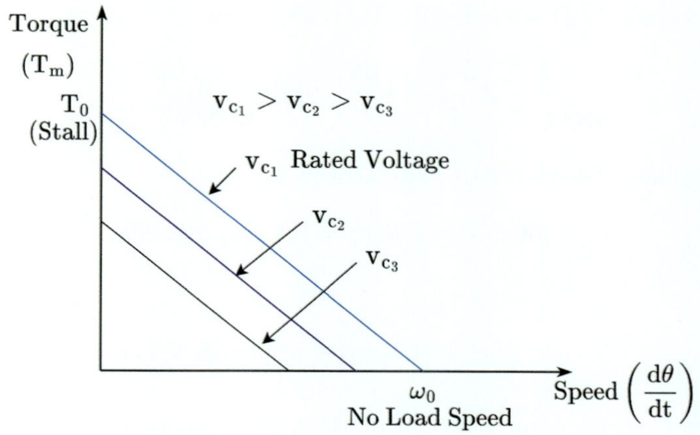

Figure 2.37: Ideal Torque-speed characteristics of AC servo motor(straight line approximation)

Let $\mathrm{T_0}$ denote the stalling torque, that is, the torque at zero rotor speed or when the rotor is held stationary. This is proportional to the voltage applied to the control phase. Let K denote a constant which is defined as stalled torque (blocked rotor torque) per unit rated control voltage, that is

$$\boxed{\mathrm{K} = \frac{\mathrm{T_0}}{\mathrm{V_{c_1}}}} \tag{2.109}$$

Since the output torque decreases linearly with speed (see Figure 2.37), it is equal to the stalled torque minus the torque absorbed in damping. The damping coefficient is proportional to the slope of the torque-speed curve. For rated voltage $\mathrm{V_{c_1}}$, the slope m equals to

$$\boxed{\mathrm{m} = -\frac{\mathrm{T_0}}{\omega_0}} \tag{2.110}$$

where, in Eqn(2.110), ω_0 is the no load speed at rated voltage and $\mathrm{T_0}$ is the stalled torque at rated voltage.

The family of straight lines shown in Figure 2.37, can be expressed as:

$$\mathrm{T_m} = \mathrm{KV_c} + \mathrm{m}\frac{\mathrm{d}\theta}{\mathrm{dt}} \tag{2.111}$$

Let J=moment of inertia of motor plus load referred to motor and f=viscous friction coefficient of motor plus load referred to motor. Then, the torque can be expressed as:

$$\mathrm{J}\frac{\mathrm{d}^2\theta}{\mathrm{dt}^2} + \mathrm{f}\frac{\mathrm{d}\theta}{\mathrm{dt}} = \mathrm{T_m} \tag{2.112}$$

Equating Eqn(2.112) with Eqn(2.111), we have

$$J\frac{d^2\theta}{dt^2} + f\frac{d\theta}{dt} = KV_c + m\frac{d\theta}{dt} \tag{2.113}$$

Taking Laplace transform of Eqn(2.113)

$$Js^2\theta(s) + fs\theta(s) = KV_c(s) + ms\theta(s) \tag{2.114}$$

or

$$\left[Js^2 + (f-m)\,s\right]\theta(s) = KV_c(s) \tag{2.115}$$

Hence, the transfer function of the AC servo motor is

$$G(s) = \frac{\theta(s)}{V_c(s)} = \frac{K}{Js^2 + s(f-m)} = \frac{K}{s(f-m)\left[1 + s\dfrac{J}{f-m}\right]} \tag{2.116}$$

or

$$\boxed{G(s) = \frac{\theta(s)}{V_c(s)} = \frac{K_m}{s(1+s\tau_m)}} \tag{2.117}$$

where $K_m = \dfrac{K}{f-m}$ is the **motor gain constant** and $\tau_m = \dfrac{J}{f-m}$ is the **motor time constant**.

2.11 Analogous Systems

Analogous systems allow engineers and scientists to draw parallels between different physical domains (mechanical, electrical, fluidic, thermal) using similar mathematical models. By understanding the analogies, one can apply known solutions from one domain to another, simplifying the analysis and design of complex systems. This cross-domain thinking can reveal insights that might not be obvious when considering a single domain in isolation.

Force-Voltage Analogy

One of the most common analogies in dynamic systems is the force-voltage analogy, which maps mechanical systems to electrical systems. This analogy allows the application of electrical engineering techniques to solve mechanical problems and vice versa. The through variables in mechanical translational and rotational systems are, respectively, force and torque, and they correspond to the current (through variable) in an electrical system. Similarly, in mechanical systems, the across variables are linear and angular velocity, which correspond to current in an electrical system. The analogous quantities in both systems can be found by comparing Eqn(2.4) with Eqns(2.35) and (2.53) and are tabulated in Table 2.3.

Table 2.3: Analogous quantities in force–voltage analogy

Translational System	**Rotational System**	**Electrical System**
Force f	Torque T	Voltage v
Mass (m)	Moment of inertia (J)	Inductance (L)
Damping coefficient (B)	Damping coefficient (B)	Resistance (R)
Spring constant (K)	Spring constant (K)	Reciprocal of capacitance (1/C)
Linear displacement (x)	Angular displacement (θ)	Charge (q)
Linear velocity ($\dot{x}$)	Angular velocity ($\dot{\theta}$)	Current (i)

Force-Current Analogy:

Consider a RLC-circuit shown in Figure 2.38

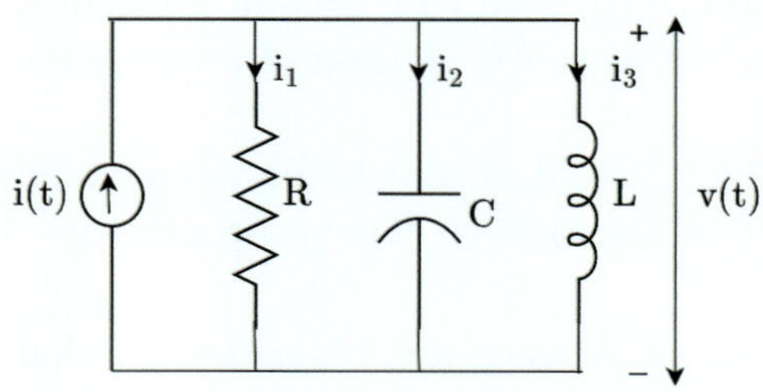

Figure 2.38: R-L-C Circuit

Applying KCL gives

$$i = i_1 + i_2 + i_3 \tag{2.118}$$

which can alternately be described by

$$\frac{1}{R}v + \frac{1}{L}\int v(t)dt + C\frac{dv(t)}{dt} \tag{2.119}$$

Since the voltage v(t) and the flux ϕ are related as

$$\frac{d\phi}{dt} = v(t) \tag{2.120}$$

Eqn(2.119) can be expressed as:

$$\boxed{C\frac{d^2\phi}{dt^2} + \frac{1}{R}\frac{d\phi}{dt} + \frac{1}{L}\phi = i} \tag{2.121}$$

Similar to force-voltage analogy discussed before, the force-current analogies are found by comparing Eqn(2.121) with Eqn(2.35) and (2.53) and are tabulated in Table 2.4

Table 2.4: Analogous quantities in force–current analogy

Translational System	Rotational System	Electrical System
Force f	Torque T	Current i
Mass (m)	Moment of inertia (J)	Capacitance (C)
Damping coefficient (B)	Damping coefficient (B)	Conductance (G)
Spring constant (K)	Spring constant (K)	Reciprocal of inductance (1/L)
Linear displacement (x)	Angular displacement (θ)	Magnetic flux (ϕ)
Linear velocity ($\dot{x}$)	Angular velocity ($\dot{\theta}$)	Voltage (v)

2.12 Modeling of Fluid Systems

Fluid is a general term used to represent a gas or a liquid. It can be classified into two types: *compressible fluids* and *incompressible fluids*.

> **Definition 2.6.** A fluid is said to be *compressible* if its density changes significantly in response to changes in pressure and temperature.

These variations in density are often due to changes in the fluid's volume when subjected to pressure or temperature variations. All gases are considered compressible. A common example is air, because in the atmosphere, the density of air decreases with altitude due to a decrease in pressure and temperature. When air is compressed, like in an air compressor, its density increases significantly.

> **Definition 2.7.** A fluid is said to be *incompressible* if its density remains nearly constant regardless of changes in pressure and temperature.

All liquids can be considered incompressible, although they are slightly compressible at high pressure. For most practical engineering applications, water can be considered *incompressible* because its density changes very little with changes in pressure and temperature, especially under moderate conditions.

A fluid system uses one or more fluids to achieve its purpose. Dampers and shock absorbers are examples of fluid systems because they depend on the viscous nature of a fluid to provide damping. In addition to providing damping, other applications of fluid systems include actuators and processes that involve mixing, heating, and cooling of fluids. Fluid systems can be divided into two categories: *hydraulic*, where the fluid is a liquid and is assumed to be

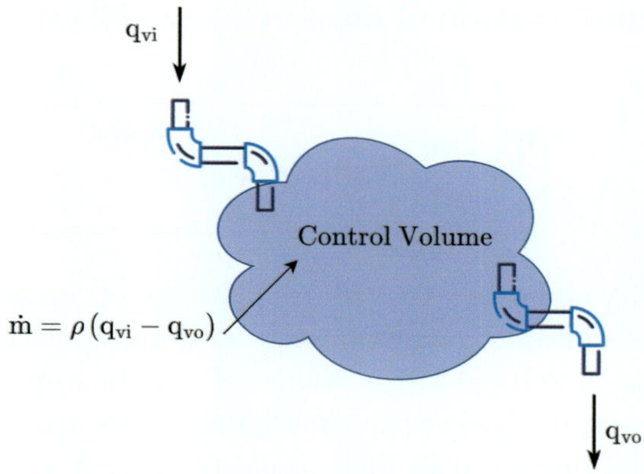

Figure 2.39: Control volume and net mass flow rate

incompressible, and *pneumatic*, where it is a gas that can be compressed and consequently shows a density change.

The key parameters to be considered while modeling fluid systems are *pressure, flow rate, temperature, density,* and *flow volume.*

The operating principle of fluid systems is essentially based on the law of conservation of mass known otherwise as the equation **of fluid continuity**.

Consider the control volume shown in Figure 2.39.

Let q_{vi} represent the amount of fluid entering the control volume per unit time, q_{vo} represent the amount of fluid leaving the control volume per unit time, and q be the net flow rate of the fluid. Hence,

$$\boxed{q = q_{vi} - q_{v0}} \tag{2.122}$$

If ρ denotes the density of the fluid and m is the net mass flow, then

$$m = \int \rho q dt = \int \rho\,(q_{vi} - q_{vo})\,dt = \int (q_{mi} - q_{mo})dt \tag{2.123}$$

where q_{mi} and q_{mo} denote mass flow rate into and out of the container, respectively.

Let V denotes the volume of the container and $m_{cv} = \rho V$ is the mass of the control volume. The **conservation of mass** states

$$\frac{dm}{dt} = \rho q = \frac{d}{dt}\,(m_{cv}) = \frac{d}{dt}\,(\rho V) \tag{2.124}$$

or

$$\boxed{\frac{dm}{dt} = \rho\dot{V} + V\dot{\rho}} \tag{2.125}$$

The law of **conservation of mass** in Eqn(2.123) can alternately be expressed as:

$$\frac{dm}{dt} = \rho(q_{vi} - q_{vo}) = q_{mi} - q_{mo} \tag{2.126}$$

The continuity equation, derived from the law of conservation of mass, ensures that the mass of fluid remains constant within a control volume, accounting for any mass flow in and out of the volume. It can be applied to both compressible and incompressible fluids, with simplifications for the incompressible case. The equation is fundamental in fluid dynamics and is widely used in engineering applications such as pipe flow, aerodynamics, and hydraulic systems.

Note that for incompressible fluids, the **conservation of volume** states that the volume of an incompressible fluid remains constant as it flows through a system. This means that the fluid cannot be compressed or expanded, so the volume flow rate (also known as volumetric flow rate) must be conserved. In practical terms, for incompressible fluids, this principle translates to the fact that the amount of fluid entering a control volume must equal the amount of fluid exiting the control volume, since the fluid's volume cannot change and can be expressed as:

$$\frac{dV}{dt} = q_{vi} - q_{vo} \tag{2.127}$$

which represents the rate of change of volume V within a control volume in terms of the inflow rate q_{vi} and the outflow rate q_{vo}. The above equation is a simple yet powerful expression of the conservation of volume for incompressible fluids, describing how the volume of fluid within a control volume changes over time based on the balance of inflow and outflow rates.

Since for an incompressible fluid ρ is constant,

$$\frac{dm}{dt} = \dot{m} = \rho\dot{V} \tag{2.128}$$

Building Blocks of Fluid Flow Systems

Fluid flow systems, like electrical circuits, can be analyzed using analogous components to electrical resistance, capacitance, and inductance. These three fundamental building blocks help in understanding and modeling the behavior of fluid systems and can be defined either using volumetric flow rate or mass flow rate.

Key Point

In fluid system modeling, it's crucial to maintain consistency in the units and parameters we use throughout the analysis. Mixing volumetric flow rate with mass flow rate parameters can lead to errors and confusion.

Definitions and Parameters

Definition 2.8. **Volumetric Flow Rate** q_v**:** It is the volume of fluid flowing per unit time, typically measured in cubic meters per second (m^3/sec)

Definition 2.9. **Mass Flow Rate** q_m**:** It is the mass of fluid flowing per unit time, typically measured in kilograms per second (kg/sec)

For such systems, the input is the volumetric rate of flow q_i, which is equivalent to current in electrical circuits, and the output is the pressure difference $p_1 - p_2$, which is equivalent to electrical potential difference (see Figure 2.40).

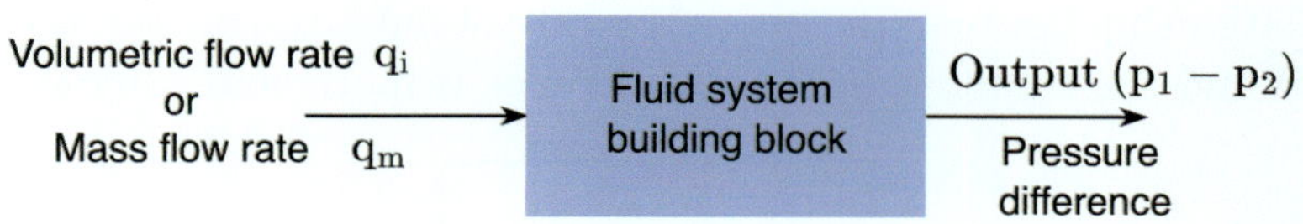

Figure 2.40: Fluid system block

Resistance of Incompressible Fluids

When a fluid flows through a pipe, a valve, or an orifice, it encounters resistance that leads to a pressure drop. This resistance is a result of frictional forces and changes in flow path geometry (e.g., change in a pipe diameter) (see Figure 2.41).

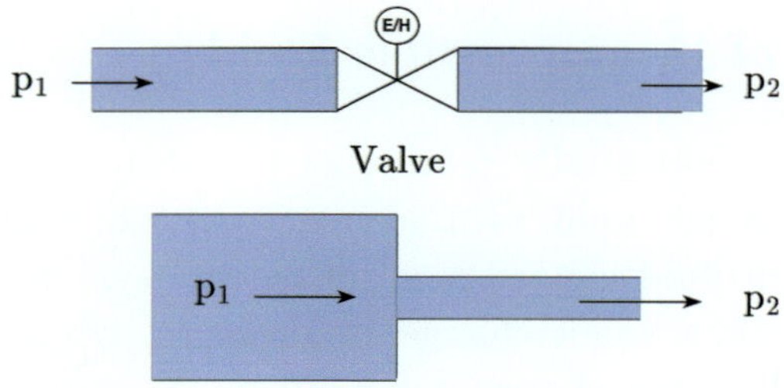

Figure 2.41: Fluid resistance

Definition 2.10. Fluid resistance is a measure of the opposition to the flow of fluid through a conduit or system.

It is analogous to electrical resistance in circuits, where it quantifies how much the flow rate is reduced for a given pressure difference across the fluid system. The fluid resistance R can be defined using either the volumetric flow rate or the mass flow rate, but not a mix of both:

Using Volumetric Flow Rate:

The relationship between the volume rate of flow of fluid q_v through the resistance element and the resulting pressure difference is mathematically expressed as:

$$R_v = \frac{\Delta p}{q_v} \tag{2.129}$$

where Δp is the pressure drop and q_v is the volumetric flow rate (m^3/sec).

Using Mass Flow Rate:

The relationship between the mass flow rate of fluid q_m through the resistance element and the resulting pressure difference is mathematically expressed as:

$$R_m = \frac{\Delta p}{q_m} \tag{2.130}$$

where Δp is the pressure drop and q_m is the mass flow rate (kg/sec). Since $q_m = \rho q_v$, from Eqns(2.129) and (2.130), we have

$$R_m = \frac{R_v}{\rho} \tag{2.131}$$

Capacitance of Incompressible Fluids

Fluid capacitance essentially represents the storage capacity of the system. It reflects how much fluid volume a system can store for a given change in pressure. High fluid capacitance means the system can store a large volume of fluid with a small change in pressure, while low fluid capacitance means a small volume change for a given pressure change. Similar to resistance, it can be defined either using volumetric flow rate or mass flow rate:

Using Mass Flow Rate:

It gives the relation between the stored fluid mass and the resulting pressure caused by the stored mass. As the fluid flows into the control volume, the fluid mass will change, as well as the pressure (see Figure 2.39). Mathematically, it is defined as the ratio of the change in stored mass to the change in fluid pressure:

$$\boxed{\mathrm{C_f} = \mathrm{C_m} = \frac{(\mathrm{dm}/\mathrm{dt})}{(\mathrm{dp}/\mathrm{dt})} = \frac{\mathrm{dm}}{\mathrm{dp}}} \tag{2.132}$$

Using Volumetric Flow Rate:

For an incompressible fluid, the density (ρ) remains constant regardless of pressure changes; the concept of fluid capacitance primarily depends on the elasticity of the container or system housing the fluid. This is because an incompressible fluid itself does not change volume under pressure, but the container might expand or contract slightly, leading to a volume change. Therefore, the fluid capacitance $\mathrm{C_f}$ for incompressible fluids is defined as the change in volume of the container per unit change in pressure and is expressed as:

$$\boxed{\mathrm{C_f} = \mathrm{C_v} = \frac{(\mathrm{dV}/\mathrm{dt})}{(\mathrm{dp}/\mathrm{dt})} = \frac{\mathrm{dV}}{\mathrm{dp}}} \tag{2.133}$$

Note that the resistance and capacitance, defined in terms of volumetric and mass flow rate, are summarized as:

$$\begin{aligned} \mathrm{R_v} &= \frac{\Delta \mathrm{p}}{\mathrm{q_v}}, \quad \mathrm{R_m} = \frac{\Delta \mathrm{p}}{\mathrm{q_m}}, \quad \Longrightarrow \quad \mathrm{R_m} = \frac{\mathrm{R_v}}{\rho} \\ \mathrm{C_v} &= \frac{\mathrm{dV}}{\mathrm{dp}}, \quad \mathrm{C_m} = \frac{\mathrm{dm}}{\mathrm{dp}}, \quad \Longrightarrow \quad \mathrm{C_m} = \rho \mathrm{C_v} \end{aligned} \tag{2.134}$$

Thus

$$\boxed{\mathrm{R_v C_v} = \mathrm{R_m C_m}} \tag{2.135}$$

EXAMPLE 2.15

Consider a liquid container shown in Figure 2.42.

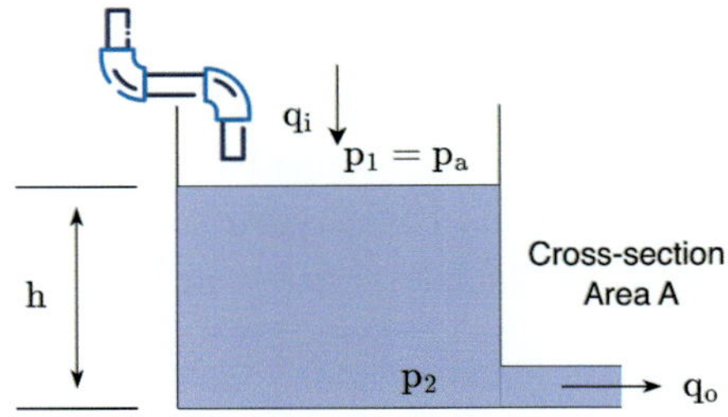

Figure 2.42: Fluid capacitance

The rate of change of volume V in the container is equal to the difference between the volumetric rate at which liquid enters the container q_i and the rate at which it leaves q_o, that is

$$q_i - q_o = \frac{dV}{dt} \tag{2.136}$$

But V = Ah, where A is the cross-sectional area of the container and h is the height of liquid in it. Hence

$$q_i - q_o = \frac{d(Ah)}{dt} = A\frac{dh}{dt} \tag{2.137}$$

Now let the pressure difference between the input and the output be denoted as p where

$$p = \rho g h$$

where ρ being the fluid density and g the acceleration due to gravity. If the liquid is assumed to be incompressible, Eqn(2.137) can be expressed as:

$$q_i - q_o = A\frac{d\,(p/\rho g)}{dt} = \frac{A}{\rho g}\frac{dp}{dt} \tag{2.138}$$

Hence the fluid capacitance is

$$\boxed{C_v = \frac{A}{\rho g}} \tag{2.139}$$

Deriving Capacitance in Terms of Mass

Assume that the liquid of density ρ is stored in a container whose cross-sectional area varies with the liquid height. If A(h) denotes the cross-sectional area, where h is the height of the liquid level above the bottom of the vessel, then the mass stored in the container can be determined by integrating $\rho A(h)$ from the base of the container to the top of the liquid,

$$m = \int_0^h \rho A(\lambda) d\lambda \tag{2.140}$$

where λ is a dummy variable of integration. Now

$$C = C_m = \frac{dm}{dp} = \frac{dm}{dh}\frac{dh}{dp} \tag{2.141}$$

From Eqn(2.140)

$$\frac{dm}{dh} = \rho A(h) \tag{2.142}$$

Further, for a liquid of density ρ, the absolute pressure p and the liquid height h are related by

$$p = \rho g h + p_a \tag{2.143}$$

where p_a is the atmospheric pressure, g is the gravitational acceleration. Hence

$$\frac{dp}{dh} = \frac{d}{dh}(\rho g h + p_a) = \rho g \tag{2.144}$$

Thus, the fluid capacitance, in terms of mass, is given by

$$C = C_m = \rho A(h) \frac{1}{\rho g} = \frac{A(h)}{g} \tag{2.145}$$

If the cross-sectional area of the container is constant, then we have

$$C_m = \frac{A}{g} = \rho \left[\frac{A}{\rho g}\right] = \rho C_v \tag{2.146}$$

Inertance (Inductance) of Incompressible Fluids

Fluid inertance, also known as fluid inductance, represents the inertia of the fluid in motion within a system. It is analogous to inductance in electrical circuits or a spring in mechanical systems. As in electrical systems, the inductance resists changes in current flow, and in mechanical systems, the spring resists changes in displacement; similarly, in fluid systems, the fluid inertance resists changes in fluid flow rate. This accounts for the kinetic energy of the moving fluid stream and is usually negligible.

Definition 2.11. Fluid inertance (L_f) is defined as opposition to the change in flow rate, which arises due to the mass of the fluid in the system.

Consider a section of a pipe with cross-sectional area A and length L, filled with fluid of mass m whose density is ρ shown in Figure 2.43.

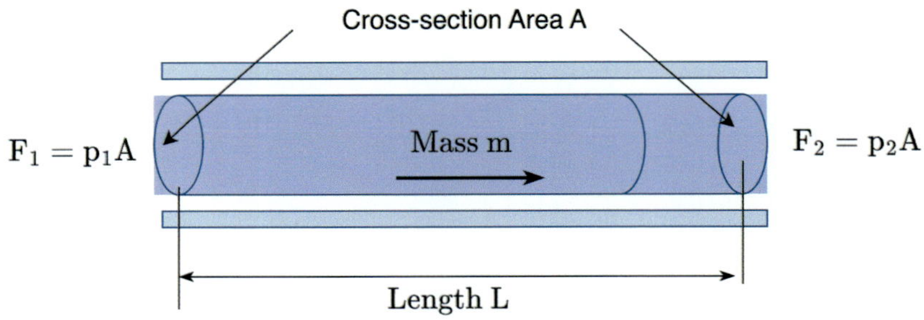

Figure 2.43: Fluid Inertance

To accelerate the fluid and increase its velocity, a force is required. Now, the net force acting on the liquid is

$$F_1 - F_2 = p_1A - p_2A = (p_1 - p_2)\,A$$

where $(p_1 - p_2)$ is the pressure difference and A is the cross-sectional area. This force causes the mass to accelerate with an acceleration a, and therefore

$$(p_1 - p_2)\,A = ma$$

Since a is the rate of change of velocity dv/dt,

$$(p_1 - p_2)\,A = m\frac{dv}{dt}$$

Note that the mass of the liquid considered here is in the length L of the pipe with cross-sectional area A. Hence, the volume of the mass of the fluid is equal to AL. If the liquid has a density ρ, then $m = AL\rho$ and so

$$(p_1 - p_2)\,A = AL\rho\frac{dv}{dt} \tag{2.147}$$

But the volume rate of flow $Q = Av$. Thus,

$$\frac{dv}{dt} = \frac{1}{A}\frac{dQ}{dt} \tag{2.148}$$

Substituting value of (dv/dt) from Eqn(2.148) in Eqn(2.147) gives

$$(p_1 - p_2)\,A = L\rho\frac{dQ}{dt}$$

or

$$(p_1 - p_2) = I\frac{dQ}{dt}$$

where

$$\boxed{I = \frac{L\rho}{A}}$$

is the *fluid inertance*. Note that in the context of compressible fluids and gases, inductance is less frequently discussed because other properties like compressibility, capacitance, and resistance are typically more significant. While fluid inductance can be relevant in specific high-speed or high-frequency applications, it is generally not a primary consideration in the analysis of compressible fluid systems. Thus, in many practical discussions and analyses, fluid inductance is not emphasized.

EXAMPLE 2.16

Problem

Show that the pressure at the bottom of a tank filled with fluid to a height h, shown in Figure 2.44, is given by the hydro-static pressure

$$P = \rho g h$$

where P is the pressure at the bottom of the tank, ρ is the density of the fluid, g is the acceleration due to gravity, h is the height of the fluid column above the point where the pressure is measured.

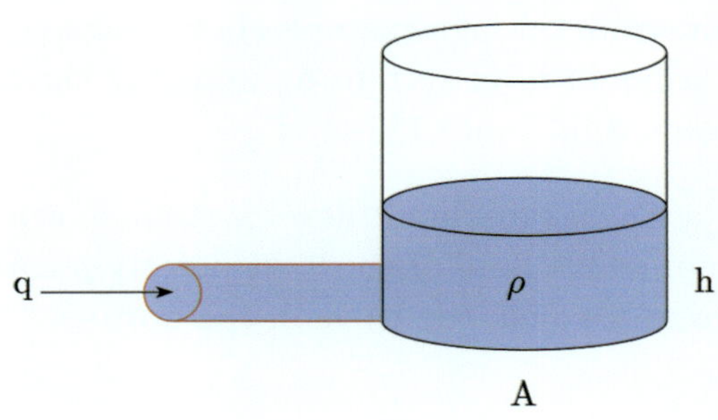

Figure 2.44: Incompressible fluid flow into an open-top cylindrical container

SOLUTION:

The pressure at a depth h in a fluid at rest is due to the weight of the fluid column above that depth. The weight W of the fluid column can be expressed as:

$$\mathrm{W} = \mathrm{m} \times \mathrm{g}, \quad \mathrm{m} = \rho \times \mathrm{volume}$$

where m is the mass, ρ is the density of the fluid, and g is the acceleration due to gravity.

If the cross-sectional area of the tank is A, the volume V of the fluid column is

$$\mathrm{V} = \mathrm{Ah}$$

Thus, the mass of the fluid column is:

$$\mathrm{m} = \rho \times \mathrm{A} \times \mathrm{h}$$

Therefore, the weight

$$\mathrm{W} = \rho \times \mathrm{A} \times \mathrm{h} \times \mathrm{g}$$

The pressure is defined as the force per unit area, the force here being the weight of the fluid column:

$$\mathrm{P} = \frac{\mathrm{W}}{\mathrm{A}} = \rho \mathrm{gh}$$

EXAMPLE 2.17

We will illustrate through examples how to derive a differential equation model of a liquid level system by applying the law of conservation of mass presented in Eqn(2.132).

Consider a single-tank liquid level system shown in Figure 2.45 where the incompressible fluid of constant density ρ enters the tank from the top and exits through the valve in the bottom.

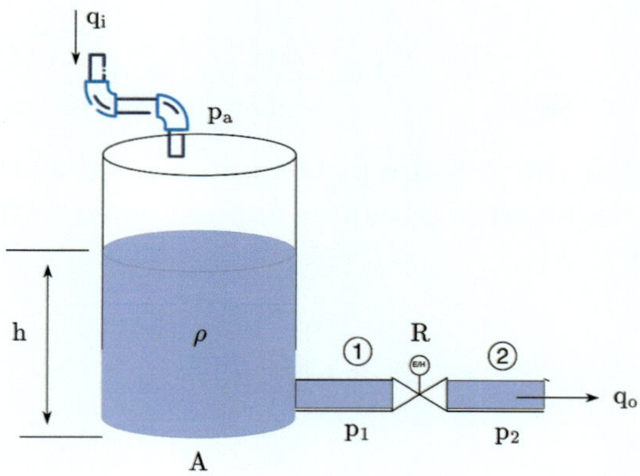

Figure 2.45: A single tank liquid level system with a valve

The volume flow rate at the inlet and at the outlet are q_i and q_o, respectively. The cross-sectional area of the tank is A and the fluid height in the tank is h. The resistance of the valve is R and the atmospheric pressure is p_a.

Let us derive the differential equation model of the system considering the input q_i and output, h.

Applying the law of conservation of mass to the system gives

$$\frac{dm}{dt} = \frac{d(\rho V)}{dt} = \frac{d(\rho Ah)}{dt} = \rho q_i - \rho q_o = q_{mi} - q_{mo} \tag{2.149}$$

where ρq_i and ρq_o are the mass flow rate in and out of the valve, respectively. Since the fluid density ρ is assumed to be constant, we can apply the law of conservation of volume of Eqn(2.137), which gives

$$\frac{dV}{dt} = \frac{d(Ah)}{dt} = q_i - q_o \tag{2.150}$$

Recall Eqn(2.132)

$$C = \frac{dm/dt}{dp/dt} \tag{2.151}$$

or

$$\frac{dm}{dt} = C\frac{dp}{dt} \tag{2.152}$$

Using Eqn(2.149), we

$$C\frac{dp}{dt} = \rho q_i - \rho q_o \tag{2.153}$$

Assuming laminar flow, the fluid resistance R, of the valve is

$$R = \frac{\Delta p}{\Delta q_m} = \frac{p_1 - p_2}{\rho q_o} \tag{2.154}$$

Assume that the pressure p_1 is approximated as the hydro-static pressure and the pressure p_2 can be approximated as the atmospheric pressure, that is

$$\begin{aligned} p_1 &= p_a + \rho g h \\ p_2 &= p_a \end{aligned} \tag{2.155}$$

Hence

$$R = \frac{\rho g h}{\rho q_o} \implies \rho q_o = \frac{\rho g h}{R} \tag{2.156}$$

Substituting this in Eqn(2.149), we get

$$\rho A \frac{dh}{dt} = \rho q_i - \frac{\rho g h}{R} \tag{2.157}$$

Rearranging the above gives

$$\boxed{\frac{RA}{g}\frac{dh}{dt} + h = \frac{R}{g} q_i} \tag{2.158}$$

Since the capacitance of the fluid $C = \frac{A}{g}$, Eqn(2.158) can be expressed as

$$\boxed{RC\frac{dh}{dt} + h = \frac{R}{g} q_i} \tag{2.159}$$

Now, let us consider the single tank fluid system shown in Figure 2.46.

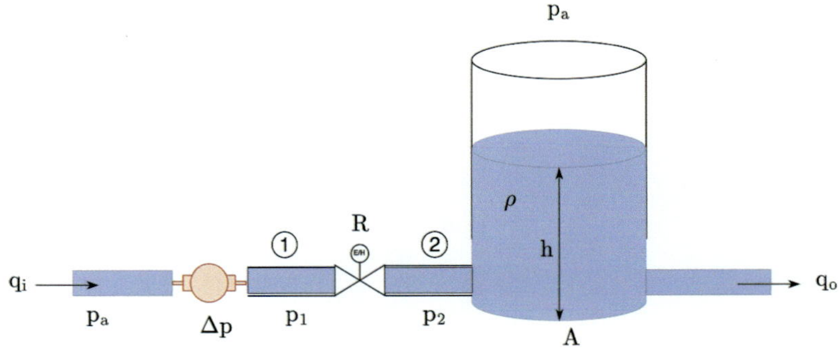

Figure 2.46: A single tank liquid level system with a pump

This is different from the system shown in Figure 2.45, because here a pump is connected to the bottom of the tank through a valve of linear resistance R. The inlet to the pump is open to the atmosphere, and the pressure of the fluid increases by Δp when crossing the pump. We derive the differential equation model of the system relating the fluid height h with the volume flow rate q_o. As before, the fluid density ρ as well as the cross-sectional area A are constant.

Let us begin by applying the law of conservation of mass to the tank

$$\frac{dm}{dt} = q_{mi} - q_{mo} \tag{2.160}$$

The fluid mass inside the tank is ρAh and for constant fluid density ρ and constant cross-sectional area

$$\frac{dm}{dt} = \rho A \frac{dh}{dt} \tag{2.161}$$

Now the mass flow rate into the tank

$$q_{mi} = \frac{p_1 - p_2}{R} \tag{2.162}$$

where

$$p_1 = p_a + \Delta p \quad \text{and} \quad p_2 = p_a + \rho g h$$

Thus

$$\boxed{q_{mi} = \frac{\Delta p - \rho g h}{R}} \tag{2.163}$$

The mass flow rate out of the tank can be expressed in terms of the volume flow rate q_o as

$$q_{mo} = \rho q_o$$

Substituting these expressions into law of conservation of mass in Eqn(2.160) gives

$$\rho A \frac{dh}{dt} = \frac{\Delta p - \rho g h}{R} - \rho q_0 \tag{2.164}$$

or

$$\boxed{\rho A \frac{dh}{dt} + \frac{\rho g}{R} h - \frac{\Delta p}{R} = -\rho q_0} \tag{2.165}$$

CONCEPT REINFORCEMENT PROBLEM 2.10

Problem

A two-tank liquid level system is shown in Figure 2.47.

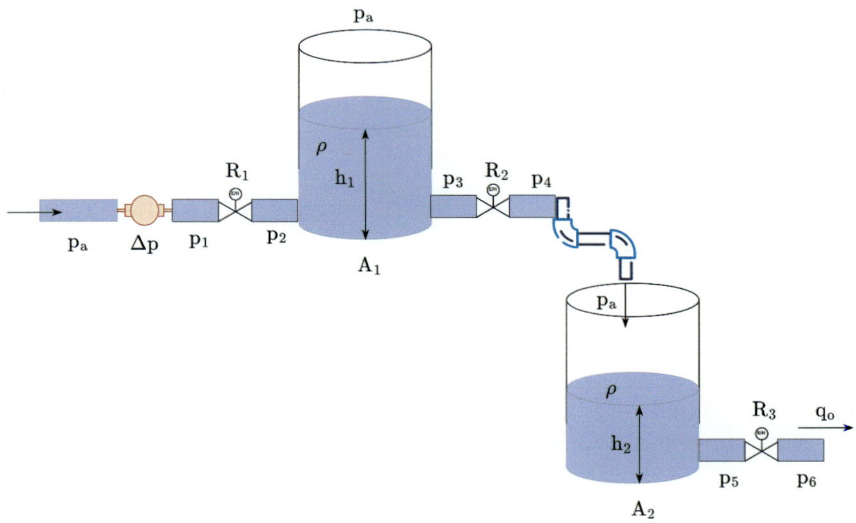

Figure 2.47: A two tank liquid level system

The cross-sectional areas of two tanks are A_1 and A_2, respectively. A pump is connected to the bottom of tank 1 through a valve of linear resistance R_1. The liquid flows from tank 1 to tank 2 through a valve of linear resistance R_2, and leaves tank 2 through a valve of linear resistance R_3. The density ρ of the liquid is constant. Derive a differential equation model of the system. Assume the pump pressure Δp as the input and liquid heights h_1 and h_2 as outputs.

2.13 Modeling of Thermal Systems

A thermal system is a system where thermal energy (heat) is transferred, stored, and converted. These systems can be found in a wide range of applications, from industrial processes and heating, ventilation, and air conditioning (HVAC) systems to electronic device cooling and natural environmental processes. These systems obey the fundamental principles of thermodynamics, particularly the conservation of energy. When an object is at a different temperature from its surroundings or another object, heat will be transferred from the object with a higher temperature to the one with a lower temperature, ensuring that the total energy is conserved.

Similar to fluid systems, which are modeled using fluid capacitance and fluid resistance based on the law of conservation of mass, thermal systems can be modeled using **two basic building blocks**: *thermal resistance* and *thermal capacitance* based on the **principle of conservation of energy**.

First Law of Thermodynamics

The first law of thermodynamics is a more general form of the law of conservation of energy. It states that energy can neither be created nor destroyed, but only can be transferred or converted from one form to another. For a system with well-defined boundaries, this law is expressed as:

$$\Delta \mathrm{E} = \mathrm{Q} - \mathrm{W} \tag{2.166}$$

where ΔE is the change in the internal energy of the system, Q is the heat added to the system (positive if heat is added, negative if heat is removed), and W is the work done by the system (positive if work is done by the system on its surroundings, negative if work is done on the system). Eqn(2.166) can alternately be expressed as:

$$\Delta \mathrm{E} = (\mathrm{Q}_{\mathrm{in}} - \mathrm{Qout}) - (\mathrm{W}_{\mathrm{in}} - \mathrm{W}_{\mathrm{out}}) \tag{2.167}$$

where Q_{in} and Q_{out} respectively denote the heat flow into the system and out of the system and W_{in} and W_{out} denote the work done by the system or it is done to the system.

The first law of thermodynamics can be precisely formulated to account for all forms of energy in a system. The net amount of energy added to the system is equal to the net increase in internal energy plus any change in mechanical energy of the system's center of mass. Mathematically, this can be expressed as:

$$\Delta \mathrm{E} = \Delta \mathrm{U} + \Delta \mathrm{ME}_{\mathrm{C}} \tag{2.168}$$

where ΔE is the total energy change in the system, ΔU is the change in internal energy (i.e., the energy stored at the molecular level, including kinetic and potential energy of molecules). ΔME_C is the change in mechanical energy of the system's center of mass (kinetic and potential energy associated with the mass center). For systems where the change in mechanical energy is negligible, the equation simplifies to

$$\Delta \mathrm{U} = \Delta \mathrm{E} = \mathrm{Q} - \mathrm{W} = (\mathrm{Q}_{\mathrm{in}} - \mathrm{Qout}) - (\mathrm{W}_{\mathrm{in}} - \mathrm{W}_{\mathrm{out}}) \tag{2.169}$$

This expression states that the change in internal energy is equal to the amount of energy gained by heating minus the amount of energy lost by doing work on the environment. Here Q is the heat transfer at the molecular level and W is the energy(work) transfer capable of producing macroscopic mechanical motion. For thermal systems with pure heat transfer and no work involved,

($W_{in} = W_{out} = 0$), the law of conservation of energy of Eqn(2.169) simplifies to

$$\Delta U = Q = (Q_{in} - Q_{out}) \tag{2.170}$$

This means when $Q = \Delta U$, the total heat energy transferred to the system (Q), results in a corresponding change in the internal energy of the system (ΔU).

In terms of **rate of heat flow**,(say Q_r), Eqn(2.170) can be written as:

$$Q_r = \frac{dQ}{dt} = \frac{d(\Delta U)}{dt} = \frac{dQ_{in}}{dt} - \frac{dQ_{out}}{dt} = q_{in} - q_{out} \tag{2.171}$$

where

q_{in}= rate of flow of heat into a system and

q_{out} = rate of flow of heat out of a system

Thermal Capacitance

Definition 2.12. Thermal capacitance, denoted as C, is a measure of an object's ability to store thermal energy. It is defined as the ratio of the change in heat flow to the change in temperature of the object.

$$C = \frac{dQ}{dT} \tag{2.172}$$

where dQ is the change in heat flow and dT is the change in temperature.

Thermal capacitance can be expressed in different units, depending on the system of measurement. The most common units are: Joules per Kelvin (J/K) (SI Unit); Joules per Celsius ($J/^0C$) and Foot-pounds per Fahrenheit $ft.lb/^0F$. For consistency, we will use the unit Joules/Kelvin (J/K).

Thermal capacitance quantifies how much heat energy an object can store per unit change in temperature. A higher thermal capacitance means that the object can store more heat energy for a given temperature change, indicating a higher thermal inertia. This concept is analogous to electrical capacitance in electrical systems, where a capacitor stores electrical energy.

The thermal capacitance of a substance depends on the thermodynamic process it undergoes. For a **constant-volume** process, where no work is done by the system (since volume does not change), all the heat added to the system goes into internal energy. The relationship is given by

$$Q = \Delta U = mc_v \Delta T \tag{2.173}$$

where Q is the heat added to the system, ΔU is the change in internal energy, m is the mass of the substance, ΔT is the change in temperature of the substance, and c_v is the specific heat capacity at **constant-volume** (units of $J - K/kg$) (amount of heat required to raise the temperature of 1 kg of the material by 1 Kelvin).

For a **constant-pressure** process, the heat added to the system changes the enthalpy H of the system. The relationship for a constant-pressure process is given by:

$$\boxed{Q = \Delta H = mc_p \Delta T} \tag{2.174}$$

where Q is the heat added to the system, ΔH is the change in enthalpy, and c_p is the specific heat capacity at **constant-pressure** (units of J/Kg.K). Combining Eqn(2.172) with Eqns(2.173) and (2.174) gives

$$C = mc_v \tag{2.175}$$

or

$$C = mc_p \tag{2.176}$$

For incompressible liquids and solids, the volume remains constant under pressure. This implies that the heat flow in a constant-pressure process is equivalent to the change in internal energy ($Q = \Delta U$), and the specific heat capacities at constant pressure (c_p) and at constant volume (c_v) are equal. Assuming that ρ the density of the substance, V is the the volume of the substance, and c is the specific heat capacity (since $c_p = c_v$), the thermal capacitance C can be expressed as

$$\boxed{C = mc} \tag{2.177}$$

where m the mass of the substance. For a substance with density (ρ) and volume (V), the thermal capacitance becomes:

$$\boxed{C = mc = (\rho V)\,c} \tag{2.178}$$

As an example, consider a block of copper with mass $m = 10$ kg and specific heat capacity $c = 385J/kg - K$. The thermal capacitance is

$$C = mc = 10 \times 385 = 3850J/K$$

This implies 3850 Joules of heat are required to raise the temperature of the copper block by 1 Kelvin.

Now from Eqn(2.172) we have

$$C = \frac{dQ}{dT}$$

For a time-varying process, where the temperature changes with time, we can write:

$$CdT = dQ \tag{2.179}$$

Dividing both sides of above by dt, the infinitesimal time interval, gives us:

$$C\frac{dT}{dt} = \frac{dQ}{dt} = \frac{d(\Delta U)}{dt} \tag{2.180}$$

Thus,

$$\text{rate of change change of internal energy} = C \times \text{ rate of change in temperature}$$

Eqn(2.180) can be expressed as

$$\boxed{C\frac{dT}{dt} = Q_r = q_{in} - q_{out}} \tag{2.181}$$

where Q_r denotes the rate of net heat flow. This equation is a fundamental relationship in thermal dynamics that links the rate of heat flow to the rate of change of temperature in a system with a given thermal capacitance.

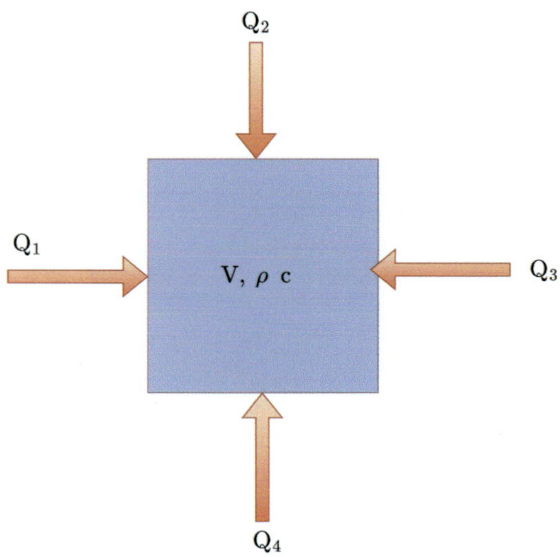

Figure 2.48: A control volume of thermal system

Thermal Resistance

Heat can flow from one substance to another through three primary mechanisms: *conduction*, *convection*, and *radiation*. In conduction, heat is transferred between substances that are in direct contact with each other and flows from the hotter part of the substance to the cooler part due to the temperature gradient. In convection, heat is transferred due to a flowing fluid, which can either be a gas or a liquid. It occurs either due to density differences caused by temperature variations within the fluid (natural convection) or when the fluid is forced to flow over the surface by external means such as fans or pumps (forced convection). In radiation, heat transfer takes place through empty space by electromagnetic waves.

A thermal resistor provides resistance to heat transfer in a body or a medium.

Definition 2.13. The thermal resistor R is defined as the ratio of the change in temperature difference to the change in heat flow rate.

$$R = \frac{\Delta T}{Q_r}, \quad \text{where } Q_r = \frac{dQ}{dt} \tag{2.182}$$

In the following, we derive the expressions for computing resistances for each of these heat transfer mechanisms.

Thermal Resistance in Conduction:

The principle of heat flow via conduction is based on Fourier's law of heat conduction, which states that the rate of heat conduction is proportional to the negative temperature gradient. For simple one-dimensional conduction, shown in Figure 2.49a, it is expressed as:

$$Q_r = -kA\frac{\partial T}{\partial x} \tag{2.183}$$

where x is the direction of heat transfer; A is the cross-sectional area of the element along which heat transfer takes place, and k is the thermal conductivity. The negative sign in Eqn(2.183) indicates that heat flows from regions of higher temperature to regions of lower temperature. It ensures that the direction of the heat flow rate Q aligns with the physical reality of heat transfer.

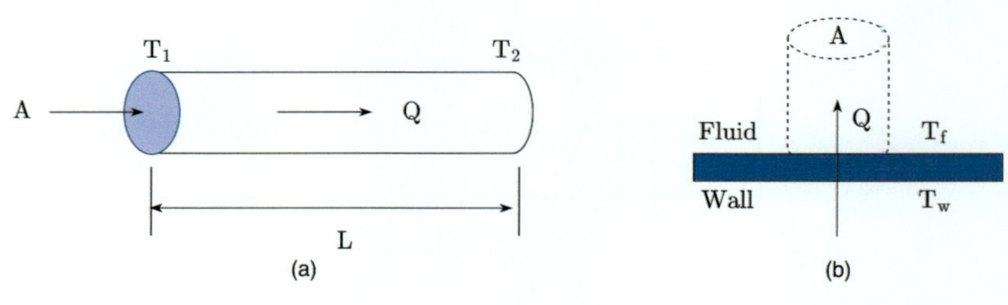

Figure 2.49: (a) An element of 1-D heat conduction (b).A control volume of heat conduction by convection

If we consider a finite object of length L and cross-section A with temperatures T_1 and T_2 at the two ends, as shown in Figure 2.49a, the one-dimensional heat transfer rate Q can be written as:

$$Q_r = kA\frac{(T_1 - T_2)}{L}, \quad T_1 > T_2 \tag{2.184}$$

Eqn(2.184) can alternately be expressed as:

$$Q_r = \frac{1}{R_k}(T_1 - T_2) \tag{2.185}$$

where

$$R_k = \frac{L}{kA} = \text{conductive thermal resistance} \tag{2.186}$$

Thermal Resistance in Convection

The principle of heat flow via convection is based on Newton's law of cooling, which states that the rate of change of temperature of an object is proportional to the difference between its temperature and the ambient temperature (the temperature of its surroundings). Mathematically, it is expressed as:

$$Q_r = h_c A \Delta T = h_c A (T_w - T_f) \tag{2.187}$$

where A is the cross-sectional area normal to the direction of heat flow, T_w is the temperature of the wall against which the fluid is flowing; T_f is the fluid temperature at the wall interface and h_c is the convection heat transfer coefficient (see Figure 2.49b).

In practice, h_c is a nonlinear function of temperature, and hence Eqn(2.187) is nonlinear. However, by making a linear approximation, we can combine Eqn(2.182) with (2.187), which gives

$$Q_r = \frac{1}{R_c}(T_w - T_f) \tag{2.188}$$

where

$$R_c = \frac{1}{h_c A} = \text{convective thermal resistance} \tag{2.189}$$

EXAMPLE 2.18

Problem

Consider a thermal system shown in Figure 2.50 consisting of a room with an electric heater. The heater emits heat at the rate q_1 and the room loses heat at the rate q_2. Assume that the room is at uniform temperature T and the walls do not store any heat. The temperature outside the room is T_o. Derive a dynamic model of the system.

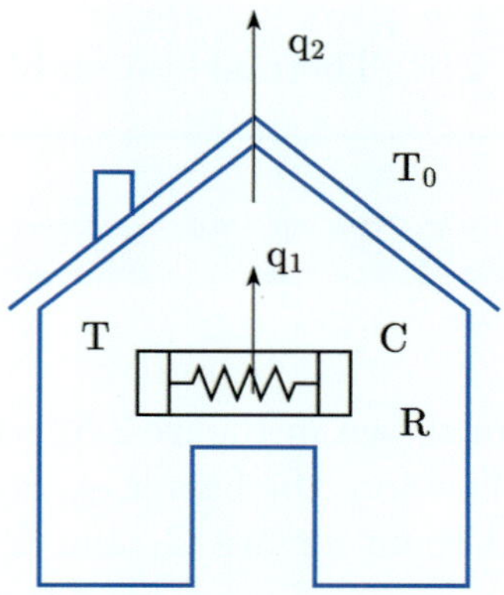

Figure 2.50: Thermal system

SOLUTION:

Let the thermal capacitance of the room is C. Hence

$$q_1 - q_2 = C\frac{dT}{dt} \quad \text{(see Eqn(2.181))} \tag{2.190}$$

If R is the resistance of the walls, then

$$q_2 = \frac{T - T_o}{R}$$

Substituting the value of q_2 from above in Eqn(2.190) gives

$$q_1 - \frac{T - T_o}{R} = C\frac{dT}{dt}$$

or

$$RC\frac{dT}{dt} + T = Rq_1 + T_0$$

or

$$[Rcs + 1]\,T(s) = Rq_1(s) + T_o(s)$$

The block diagram of the system is shown in Figure 2.51.

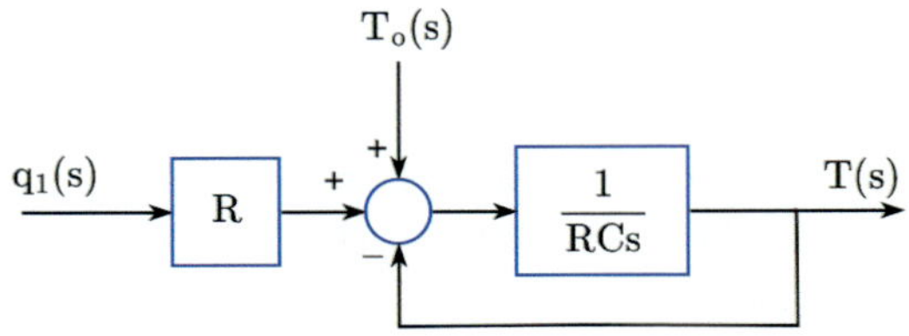

Figure 2.51: Thermal system block diagram

CONCEPT REINFORCEMENT PROBLEM 2.11

Problem

Consider the room shown in Figure 2.52 where it is heated by a heater. The rate of heat flow into the heater q_{hi} is q_0. The thermal capacitance of the heater and room air are C_1 and C_2, respectively. The thermal resistance of the heater and room air interface is R_1, and that between the room-wall and ambient-air interface is R_2. The temperatures of the heater and room are T_1 and T_2, respectively. The outside temperature is T_0, which is assumed to be constant.

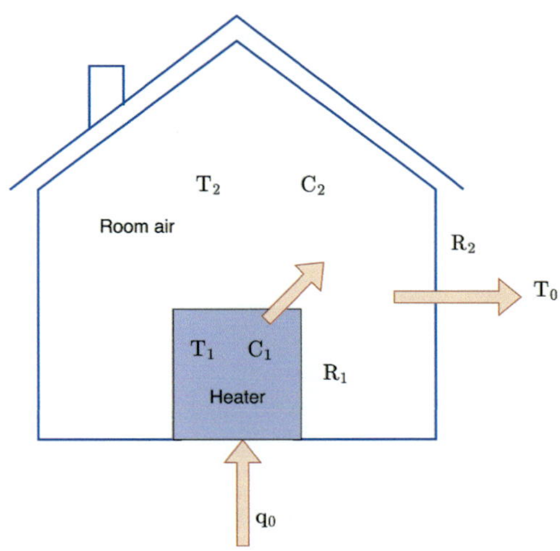

Figure 2.52: A room with a heater

a. Determine the differential equation models of the system which relate the temperatures T_1 and T_2 to the inputs q_o and the output temperature T_0.

b. From the differential equation models derived above, determine the state space model of the system with T_1 and T_2 as state variables and outputs.

2.14 Linearization of Nonlinear Systems

Many real-world systems exhibit nonlinear behavior due to physical characteristics, component interactions, or operating conditions. However, analyzing and designing controllers for nonlinear systems can be mathematically complex and often intractable using standard techniques. To simplify the problem, it is common practice to approximate the nonlinear system by a linear model in the vicinity of a chosen operating point, typically an equilibrium point. This process, known as linearization, allows the use of powerful linear control tools while still capturing the system's essential behavior near that point.

The linearization method presented here relies on the Taylor series expansion of the nonlinear functions about a given operating point, retaining only the first-order (linear) terms. Since higher-order terms are discarded, the accuracy of the approximation improves when the system operates within a small neighborhood of the chosen operating point,that is, when the deviations from the selected operating point are minimal. The general procedure to obtain a linearized model is outlined below:

Step-1: Choose the operating point (often the equilibrium point) around which the system is to be linearized.

Step-2: Expand the nonlinear function using the Taylor series around the selected operating point. The expansion expresses the system dynamics as an infinite series in terms of deviations from the steady state, involving linear, quadratic, cubic, and higher-order terms.

Step-3: Retain only the first-order (linear) terms of the expansion. Neglect all higher-order terms to obtain a linear approximation that accurately represents the system behavior near the operating point.

To illustrate the linearization process, we consider a few generic examples where the system output is a nonlinear function of one or more variables.

1. **Single-variable case:**

Let the system be represented by a nonlinear function of one variable x:

$$y = f(x) \tag{2.191}$$

Let us choose an operating point x_0, and approximate f(x) using the first-order Taylor series expansion about x_0 (retaining only the first-order term and neglecting higher-order terms):

$$y \approx f(x_0) + \left.\frac{df(x)}{dx}\right|_{x=x_0} (x - x_0) \tag{2.192}$$

Eqn(2.192) can alternately be expressed as

$$y = y_0 + K(x - x_0), \quad \text{where } K = \left.\frac{df(x)}{dx}\right|_{x=x_0} \tag{2.193}$$

Eqn(2.193) may be expressed as:

$$y - y_0 = K(x - x_0) \tag{2.194}$$

2. **Two-variable case:**

Consider a system described by

$$y = f(x_1, x_2) \tag{2.195}$$

To linearize around an operating point $x_0 = (x_{10}, x_{20})$, we use a multivariable Taylor series expansion

$$y \approx f(x_{10}, x_{20}) + \left.\frac{\partial f}{\partial x_1}\right|_{x_0} (x_1 - x_{10}) + \left.\frac{\partial f}{\partial x_2}\right|_{x_0} (x_2 - x_{20}) \tag{2.196}$$

The linear approximation of this system around this operating point is

$$y - y_0 = K_1(x_1 - x_{10}) + K_2(x_2 - x_{20}) \tag{2.197}$$

where

$$y_0 = f(x_{10}, x_{20}), \quad K_1 = \left.\frac{\partial f}{\partial x_1}\right|_{x_0} \text{ and } K_2 = \left.\frac{\partial f}{\partial x_2}\right|_{x_0}$$

Multivariable case:

For a general nonlinear function of n variables:

$$y = f(x_1, x_2, \ldots, x_n) \tag{2.198}$$

Let the operating point be $\mathbf{x_0} = (x_{10}, x_{20}, \ldots, x_{n0})^T$, the linearized approximation is

$$y \approx f(x_{10}, x_{20}, \ldots, x_{n0}) + \left.\frac{\partial f}{\partial x_1}\right|_{x=x_0} (x_1 - x_{10}) + \left.\frac{\partial f}{\partial x_2}\right|_{x=x_0} (x_2 - x_{20})$$
$$\cdots + \left.\frac{\partial f}{\partial x_n}\right|_{x=x_0} (x_n - x_{n0}) \tag{2.199}$$

This can be expressed in compact form as:

$$y \approx f(\mathbf{x_0}) + \sum_{i=1}^{n} \left.\frac{\partial f}{\partial x_i}\right|_{x=x_0} (\mathbf{x} - \mathbf{x_0}) \tag{2.200}$$

EXAMPLE 2.19

Problem

Linearize the nonlinear system

$$y = x_1x_2x_3$$

with the variable ranges $x_1 \in [2, 6]$, $x_2 \in [3, 7]$, and $x_1 \in [4, 8]$.

SOLUTION:

Step-1: Choose an operating point

We select the midpoint of each interval as the operating point:

$$x_{10} = \frac{2+6}{2} = 4,\ x_{20} = \frac{3+7}{2} = 5,\ x_{30} = \frac{4+8}{2} = 6,$$

Step-2: Evaluate the function at the operating point

$$y_0 = f(x_{10}, x_{20}, x_{30}) = 4 \times 5 \times 6 = 120$$

Step-3: Compute partial derivatives at the operating point

$$K_1 = \left.\frac{\partial y}{\partial x_1}\right|_{x_0} = x_{20}x_{30} = 5 \times 6 = 30$$

$$K_2 = \left.\frac{\partial y}{\partial x_2}\right|_{x_0} = x_{10}x_{30} = 4 \times 6 = 24$$

$$K_3 = \left.\frac{\partial y}{\partial x_2}\right|_{x_0} = x_{10}x_{20} = 4 \times 5 = 20$$

Step-4: Linearized model

Using first-order Taylor expansion:

$$y \approx 120 + 30(x_1 - 4) + 24(x_2 - 5) + 20(x_3 - 6)$$

This can be simplified as:

$$y \approx 30x_1 + 24x_2 + 20x_3 - 240$$

CONCEPT REINFORCEMENT PROBLEM 2.12

Problem

The flow through a valve $q(m^3s^{-1})$ is related to the pressure head across the valve $h(m)$ by the relation

$$q = C_d\sqrt{2gh}$$

where C_d and g are constants. Find a linear relationship between q and h around the steady state value of h, namely h_0.

2.15 Block Diagram and Signal Flow Graphs

A typical control system is often complex and consists of the interconnection of several components/elements, which are practically nonlinear. Although their input-output behavior can be approximated by linear differential equations, it is difficult to derive the complete behavior of a complex control system by a single equation/function. Therefore, it is always desirable to represent the dynamics of each of the components separately and connect them together to represent the whole system. This is the motivation for the block diagram representation of control systems, where each component of the control system is represented with a block.

Some Features of Block Diagram Representation

Block Diagram: Merits and Demerits

Advantages:

1. Possible to represent a complex system by merely connecting the blocks of the components according to the signal flow.
2. The contribution of each component to the overall system performance can be evaluated.
3. It becomes easier to visualize the functional operation of the system by examining the block diagram.
4. The overall transfer function can easily be computed.

Disadvantages:

1. No information about the physical construction of the system can be obtained. Therefore, a block diagram may represent many dissimilar and unrelated systems
2. The main source of energy is not explicitly shown.
3. The same system can be represented by different block diagrams.

2.16 Basic Elements of Block Diagram

A block diagram essentially has three basic elements such as a block, the summing point, and a junction or pick-off/take-off point. In the following, we will discuss each of these elements.

1. **A Block:** A block is used to represent the proportional relationship between two Laplace transformed signals, which are the input and output of the block. The proportionality function is the transfer function or transmittance that relates the incoming and outgoing signals of the block. Thus, each of the blocks is the symbolic representation of the transfer function of a particular component of the control system. The complete control system is represented by the interconnection of a required number of blocks. The steps for block diagram representation are as follows:

> 1. Express the dynamics of each component by linear differential equations.
> 2. Take the Laplace transform of these differential equations, which convert these equations into algebraic equations in complex frequency variables s.
> 3. Select the input and output variables of each component.
> 4. Compute the transfer function of each component, which is defined as the ratio of the Laplace transform of the output and the input under zero initial conditions.
> 5. Represent this transfer function in block form as shown in Figure 2.53.

Note that for each block, the input may be regarded as the "cause" and the output as the "effect." Since the "effect" cannot produce the "cause," the **block diagram is "unidirectional**."

The transfer function G(s) relates the Laplace transform Y(s) of the output y(t) to the Laplace transform X(s) of the input x(t) through the relationship

$$Y(s) = G(s)X(s)$$

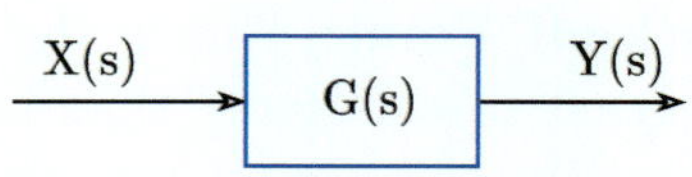

Figure 2.53: Block diagram of a linear system

Main Block Diagram Rule: *The output signal of a block is the product of the transfer function of the given block and the input signal of the block.*

EXAMPLE 2.20

Consider a system represented by the differential equation model:

$$\frac{d^3y(t)}{dt^3} + a_1\frac{d^2y(t)}{dt^2} + a_2\frac{dy(t)}{dt} + a_3 = b_0\frac{d^2u(t)}{dt^2} + b_1\frac{du(t)}{dt} + b_2$$

where y is the output and u is the input of the system.

Taking the Laplace transform of this system with zero initial conditions gives

$$s^3Y(s) + a_1s^2Y(s) + a_2sY(s) + a_3Y(s) = b_0s^2U(s) + b_1sU(s) + b_3U(s)$$

The transfer function of this system is given by

$$G(s) = \frac{Y(s)}{U(s)} = \frac{b_0s^2 + b_1s + b_2}{s^3 + a_1s^2 + a_2s + a_3}$$

This is represented in block diagram in Figure 2.54.

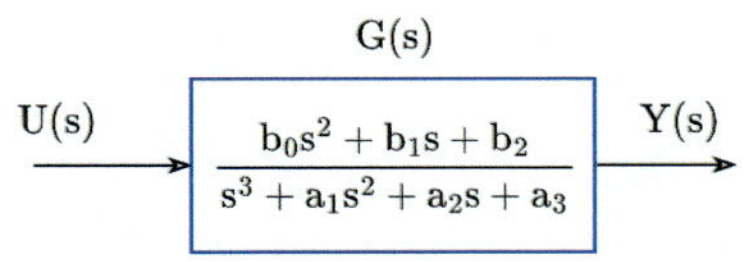

Figure 2.54: Block diagram of a linear system

2. **Summing Point:** This block is used to represent addition and subtraction of signals. It can have any number of incoming signals. However, it has only one outgoing signal. The algebraic sign of each of the incoming signals is shown next to the arrowhead of this signal. Figure 2.55(a) shows the summing point of three inputs $(X_1(s), X_2(s), X_3(s))$ and one output Y(s) where $X_1(s), X_2(s)$ have positive signs and $X_3(s)$ has negative sign. Figure 2.55(b) shows the summing point of four inputs $(X_1(s), X_2(s), X_3(s), X_4(s))$ and one output Y(s) where $X_1(s), X_3(s)$ have positive signs and $X_2(s), X_4(s)$ have negative sign.

3. **Junction or Pick-off Point:** This is the point in the block diagram where the same signal goes to more than one different place, as shown in Figure 2.56 where the same input goes to the inputs of three different blocks.

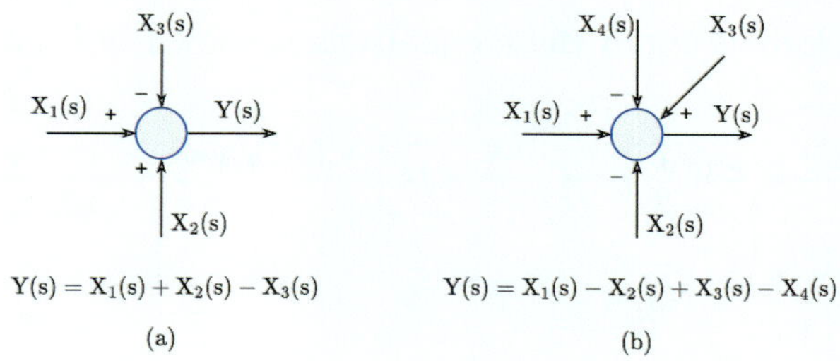

Figure 2.55: Summing point of block diagram

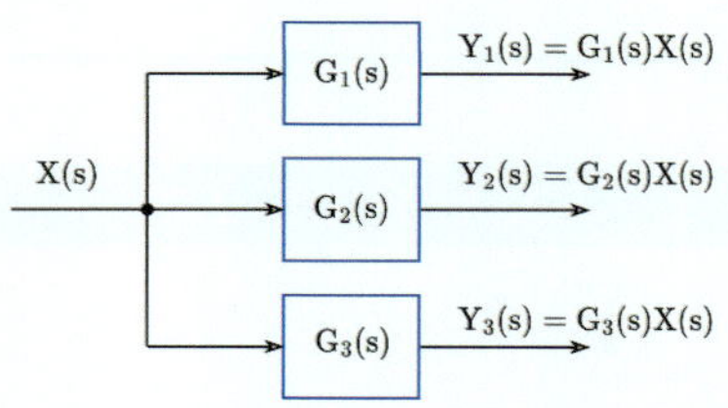

Figure 2.56: Pick-off point of block diagram

We illustrate this concept through the following examples.

EXAMPLE 2.21

Problem

Depict the dynamic behavior of a simple R–L circuit shown in Figure 2.57 using a block diagram.

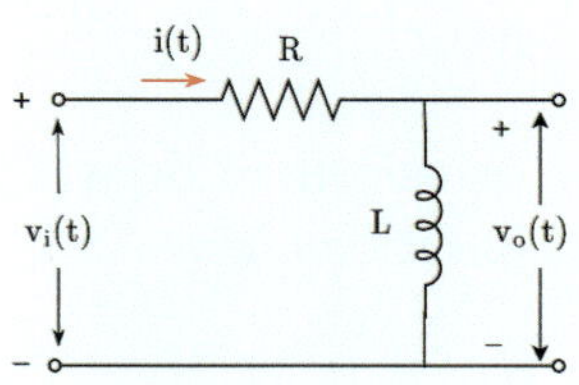

Figure 2.57: Simple R-L circuit

The equations of current and output voltage are

$$i(t) = \frac{v_i(t) - v_o(t)}{R} \quad \text{and} \quad v_o(t) = L\frac{di(t)}{dt}dt \tag{2.201}$$

In the Laplace domain, the above equations are expressed as

$$I(s) = \frac{V_i(s) - V_o(s)}{R}, \quad \text{and} \quad V_O(s) = LsI(s) \tag{2.202}$$

The block diagram of these equations is shown in Figure 2.58

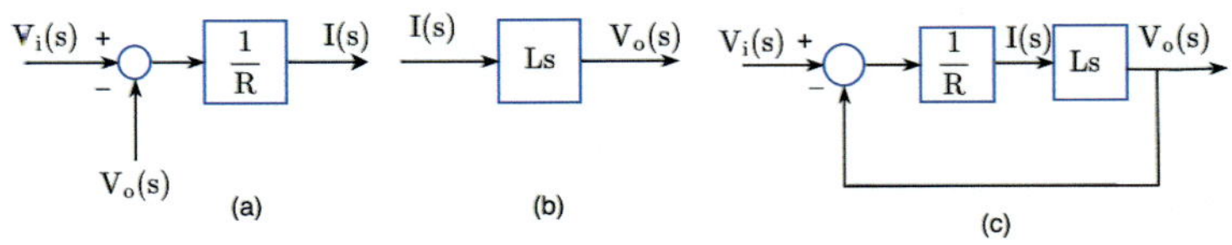

Figure 2.58: Block diagram representation of the dynamics of simple R-L circuit

EXAMPLE 2.22

Problem

Draw the block diagram representing the dynamics of the armature-controlled DC motor as described in Section 2.10.1.

SOLUTION:

The dynamics of DC motor, discussed in section 2.10.1 are reproduced here for quick reference. The electrical dynamics is

$$R_a i_a + L_a \frac{di_a}{dt} + e_b = v_a \quad \text{where} \quad e_b = K_b \frac{d\theta}{dt} \tag{2.203}$$

The mechanical dynamics of d.c motor is given by

$$J\frac{d^2\theta}{dt^2} + f\frac{d\theta}{dt} + T_d = T_m \quad \text{where} \quad T_m = K_T i_a \tag{2.204}$$

Taking the Laplace transform of Eqns(2.203) and (2.204) with assumption of zero initial conditions, gives

$$(L_a s + R_a)\, I_a(s) = V_a(s) - E_b(s) \tag{2.205a}$$

$$E_b(s) = K_b s\theta(s) \tag{2.205b}$$

$$T_m(s) = K_T I_a(s) \tag{2.205c}$$

$$\left(Js^2 + fs\right)\theta(s) = T_m(s) - T_d(s) \tag{2.205d}$$

Let us represent each of the equations in (2.205) in block diagram. Figure 2.59 shows the block diagram of Eqns(2.205a) and (2.205b).

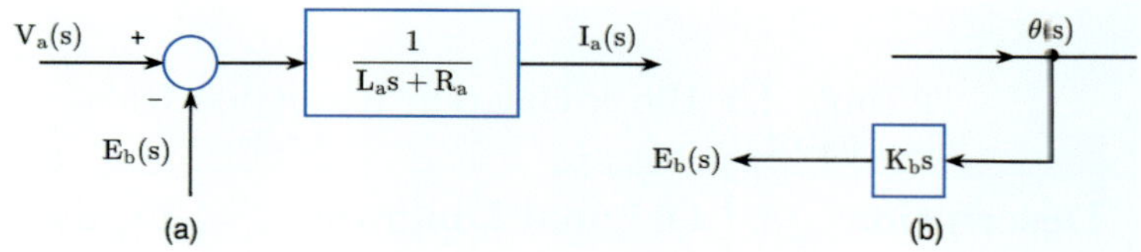

Figure 2.59: Block diagram of Eqns(2.205a) and (2.205b)

The block diagram of mechanical system corresponding to Eqns(2.205c) and (2.205d) is shown in Figure 2.60.

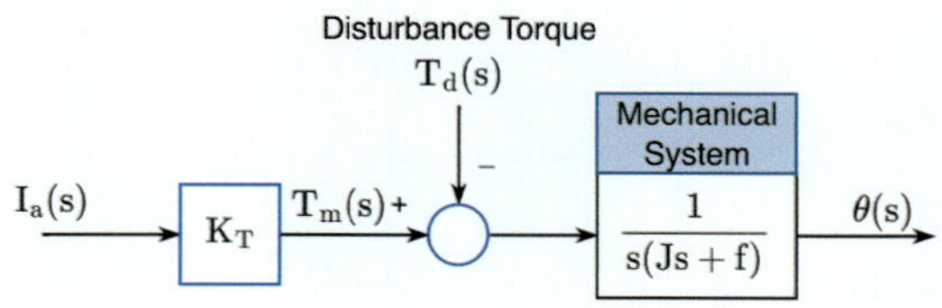

Figure 2.60: Block diagram of Eqns(2.205c) and (2.205d)

The complete system is shown in the block diagram of Figure 2.61

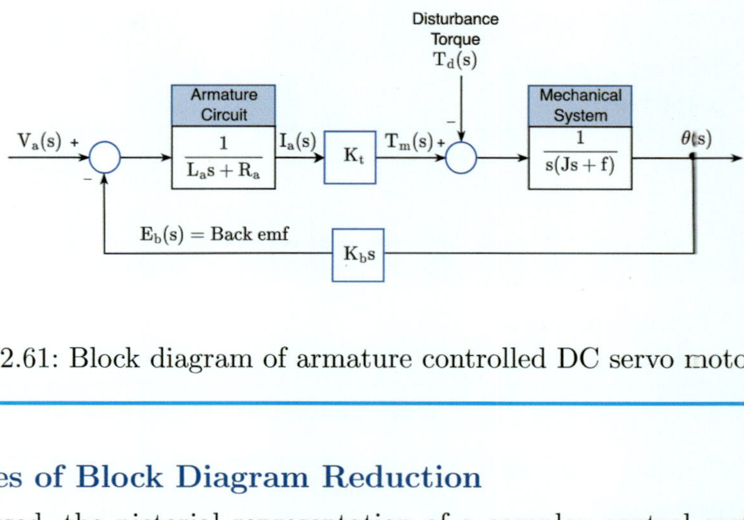

Figure 2.61: Block diagram of armature controlled DC servo motor

2.16.1 Rules of Block Diagram Reduction

As discussed, the pictorial representation of a complex control system with several components interconnected by a block diagram can provide better physical information on the structure of the control system. Suppose various components of the system are noninteracting. In that case, i.e., there is no loading effect of one component on another. It is possible to obtain the overall transfer function of the system by algebraically manipulating the block diagram using some simple rules of block diagram transformations shown in Table-2.5.

Table 2.5: Block Diagram Reduction Rules

<table>
<tr><th>Rule No.</th><th>Description</th><th>Original Diagram</th><th>Equivalent Diagram</th></tr>
<tr><td>1</td><td>Combining blocks in cascade</td><td>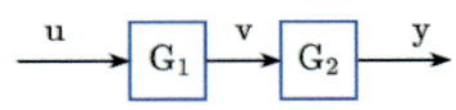
</td><td>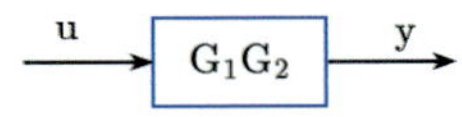
</td></tr>
<tr><td>2</td><td>Blocks in parallel</td><td>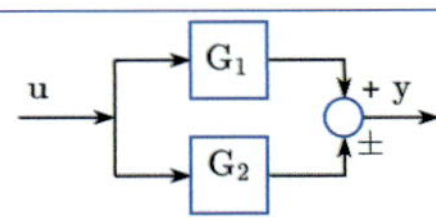
</td><td>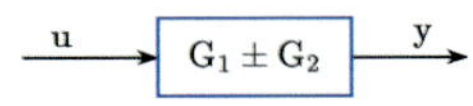
</td></tr>
<tr><td>3</td><td>Eliminating a feedback loop</td><td>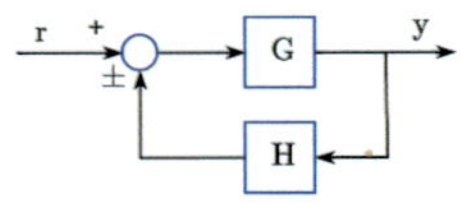
</td><td>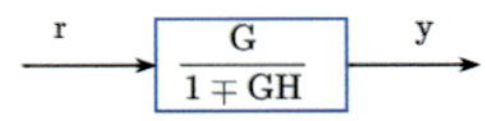
</td></tr>
<tr><td>4</td><td>Shifting summing point back</td><td>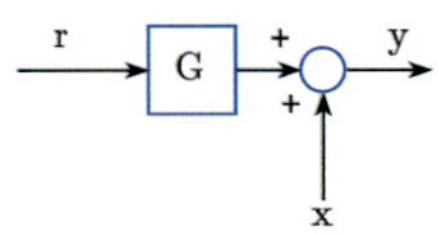
</td><td>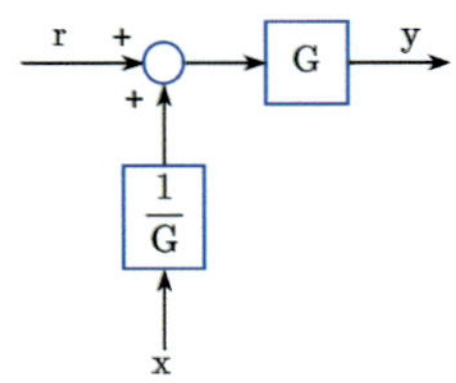
</td></tr>
<tr><td>5</td><td>Shifting summing point forward</td><td>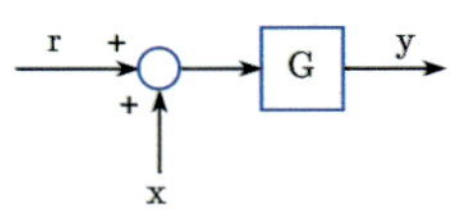
</td><td>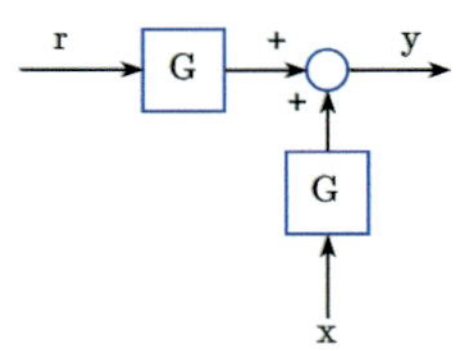
</td></tr>
<tr><td>6</td><td>Shifting take off point back</td><td>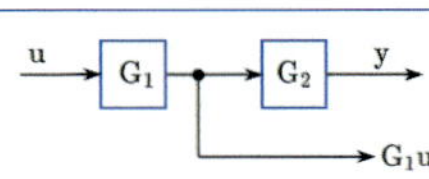
</td><td>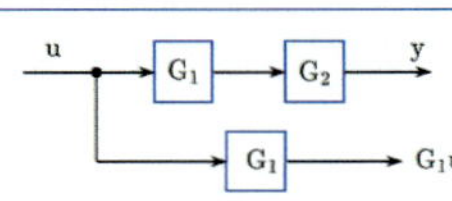
</td></tr>
<tr><td>7</td><td>Shifting take off point forward</td><td>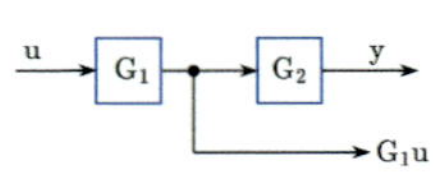
</td><td>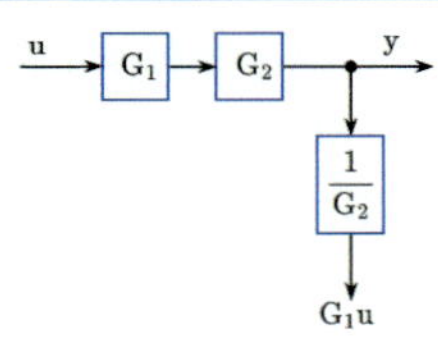
</td></tr>
</table>

2.17 Guidelines of Block Diagram Reduction

Using the block diagram algebra, we can reduce a complex block diagram following some simple guidelines (GL):

GL-1: Check for the blocks which are connected in series and simplify.

GL-2: Check for the blocks which are connected in parallel and simplify.

GL-3: Check for the blocks which are connected in a feedback loop and simplify.

GL-4: If there is any difficulty with the take-off point, shift it to the right.

GL-5: If there is any difficulty with the summing point, shift it toward the left.

GL-6: Repeat the above steps until a simplified form is obtained.

This method is best illustrated by several examples.

EXAMPLE 2.23

Problem

Determine $\frac{C(s)}{R(s)}$ for the system shown in Figure 2.62 using block diagram reduction techniques.

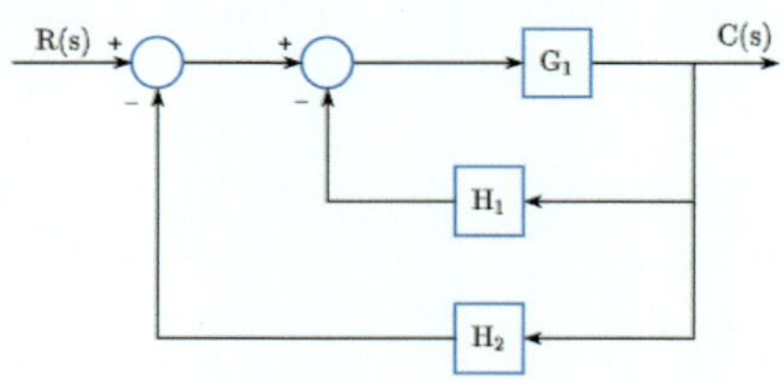

Figure 2.62: Multi loop system

SOLUTION:

Step-1: Now this system has two loops; the inner loop I and the outer loop II, as shown in Figure 2.63 (I and II).

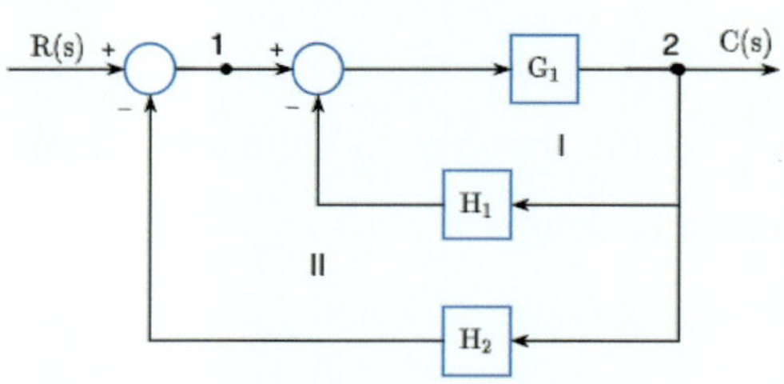

Figure 2.63: Multi loop system

Let us eliminate the inner feedback-loop I between the points 1 and 2. We can simplify this loop by the transfer function $G = \dfrac{G_1}{1+G_1H_1}$. Thus, Figure 2.63 becomes

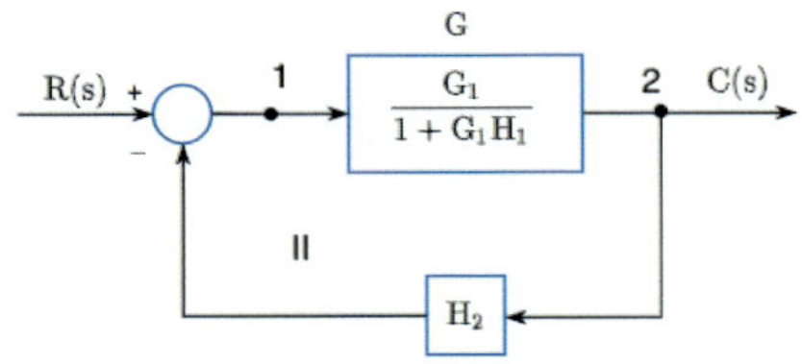

Figure 2.64: Multi loop system

Step-2: Let us eliminate the feedback loop II. This gives

$$\frac{C(s)}{R(s)} = \frac{G}{1+GH_2} = \frac{\dfrac{G_1}{1+G_1H_1}}{1+\dfrac{G_1H_2}{1+G_1H_1}}$$

$$= \frac{G_1}{1+G_1H_1+G_1H_2}$$

The reduced block diagram is shown in Figure 2.65.

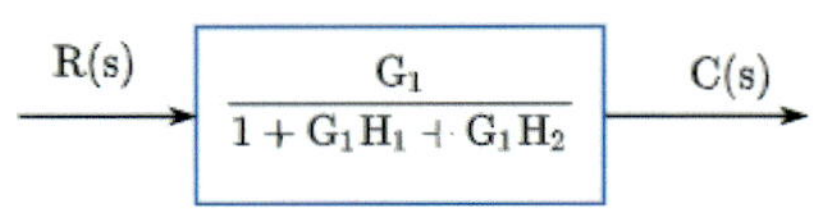

Figure 2.65: Multi loop system

EXAMPLE 2.24

Problem

Determine $\dfrac{C(s)}{R(s)}$ for the system shown in Figure 2.66 using block diagram reduction techniques.

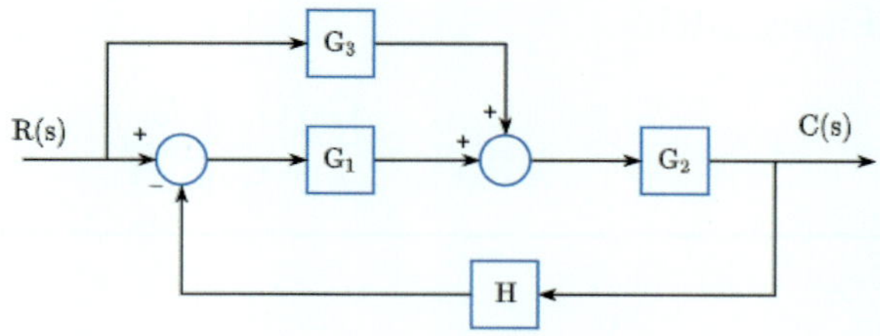

Figure 2.66: Multi loop system

SOLUTION:

Step-1: Now this system has two summing points; as shown in Figure 2.66. Let us shift the summing point to the right and apply Rule 4. This gives

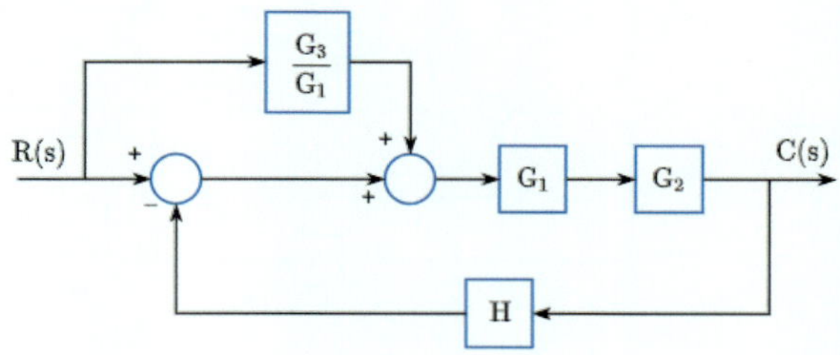

Figure 2.67: Multi loop system

Step-2: Let us interchange the summing points and combine the series blocks G_1 and G_2. This gives

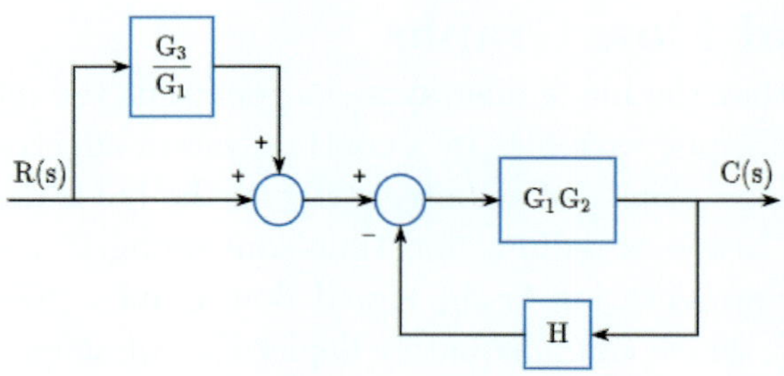

Figure 2.68: Multi loop system

Step-3: Now applying the parallel connection and feedback rules, the block diagram gets reduced to Figure 2.69.

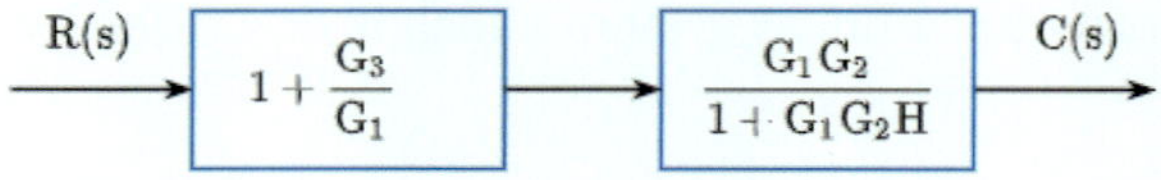

Figure 2.69: Multi loop system

From Figure 2.69,

$$\frac{C(s)}{R(s)} = \frac{G_2(G_3 + G_1)}{1 + G_1G_2H}$$

CONCEPT REINFORCEMENT PROBLEM 2.13

Problem

Determine $\frac{C(s)}{R(s)}$ for the system shown in Figure 2.70 using block diagram reduction techniques.

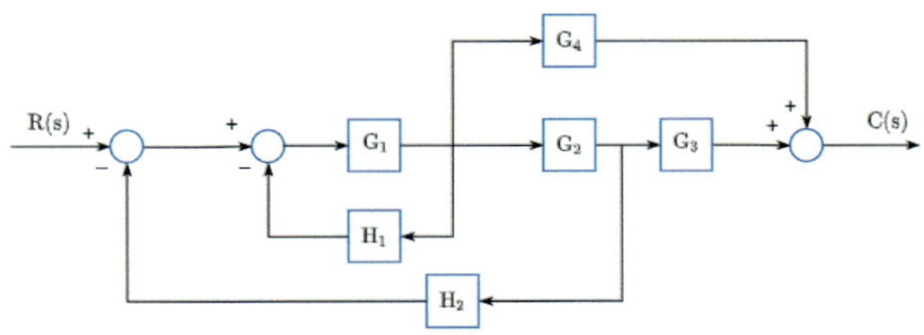

Figure 2.70: Multi loop system

2.18 Signal Flow Graphs

We know that the block diagrams can represent the interrelationships of controlled and input variables in a control system effectively. However, with the increasing complexity of a control system, the block diagram reduction procedure often becomes tedious and time-consuming. An alternative to the block diagram representation is the signal flow graphs (SFG), introduced by S.J. Mason [55], which can adequately depict the topological structure of a control system. The relation between the variables can easily be determined using a flow graph gain formula without the need for any reduction procedure or manipulation of the flow graph.

The SFG is essentially a pictorial method of portraying a set of linear algebraic equations that describes the inter-dependencies among system variables. For example, consider a linear system which is described by N number of algebraic equations:

$$\begin{aligned} x_1 &= a_{11}x_1 + a_{21}x_2 + a_{31}x_3 + \cdots + a_{N1}x_N \\ x_2 &= a_{12}x_1 + a_{22}x_2 + a_{32}x_3 + \cdots + a_{N2}x_N \\ &\vdots \qquad \vdots \qquad \vdots \\ x_N &= a_{1N}x_1 + a_{2N}x_2 + a_{3N}x_3 + \cdots + a_{NN}x_N \end{aligned}$$

or,

$$x_j = \sum_{k=1}^{N} a_{kj}x_k, \qquad j = 1, 2, \ldots, N$$

Note that these algebraic equations are in the form of cause-effect relations where a_{kj} is the gain from k to j. The j-th effect can be expressed as:

$$\text{jth effect} = \sum (\text{gain from k to j}) \times (\text{kth cause})$$

If we want to represent a set of integro-differential equations using a signal flow graph, then, at first, these equations have to be transformed into algebraic equations using Laplace transforms as:

$$X_j(s) = \sum_{k=1}^{N} G_{kj}(s)X_k(s), \qquad j = 1, 2, \ldots, N$$

2.18.1 Basic Elements of a Signal Flow Graph

There are essentially two basic elements in a signal flow graph *node* and *branch*.

Definition 2.14. *Node:* This is a point in the graph that represents a system variable. The basic rule for nodes is that *a signal at a node is equal to the sum of signals coming into the node from branches.*

Definition 2.15. *Branch:* This is a line segment which joins two nodes according to cause-effect equations. It relates the dependency of an input and output variable, analogous to a block in a block diagram. A branch is always associated with gains and directions. It consists of an input node, an output node, and an arrow showing the direction of signal flow.

Note

A signal can transmit through a branch only in the direction of the arrow.

Consider a linear system which is represented by the algebraic equation

$$x_2 = a_{12}x_1 \tag{2.206}$$

where x_1 is the input, x_2 is the output, and a_{12} is the gain, or transmittance between the two variables. This can be represented by the signal flow graph shown in Figure 2.71

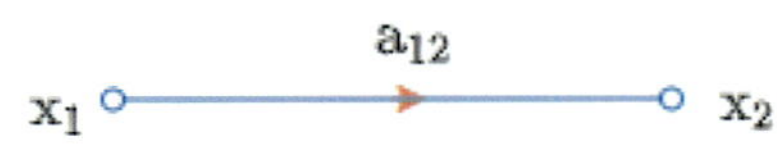

Figure 2.71: Signal flow graph of $x_2 = a_{12}x_1$

It is worth emphasizing that the signal flows in the SFG is *unidirectional.* Therefore, the branch directed from the node x_1 to the node x_2 implies that the variable x_2 depends on the variable x_1, but not the reverse.

Although, algebraically, we can express Eqn(2.206) as:

$$x_1 = \frac{1}{a_{12}}x_2 \tag{2.207}$$

The signal flow graph of Figure 2.71 does not correspond to this relationship. If Eqn(2.207) represents a valid *cause-effect* relation, its signal flow graph can be drawn, as in Figure 2.72, by taking x_2 as the input and x_1 as the output.

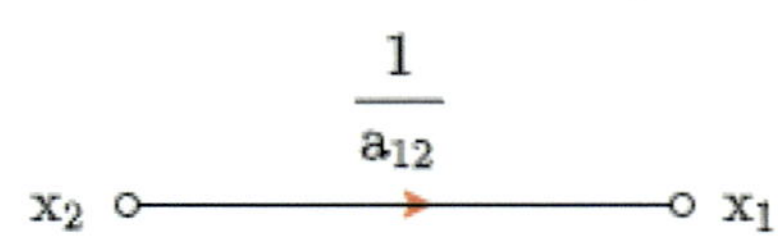

Figure 2.72: Signal flow graph of $x_1 = \frac{1}{a_{12}}x_2$

In the following, we illustrate with an example of constructing a signal flow graph from algebraic system equations

Example: Construction of Signal Flow Graph

Consider a system described by the following set of equations

$$\begin{aligned} x_2 &= a_{12}x_1 + a_{42}x_4 \\ x_3 &= a_{23}x_2 + a_{43}x_4 \\ x_4 &= a_{24}x_2 + a_{34}x_3 + a_{44}x_4 \\ x_5 &= a_{35}x_3 + a_{45}x_4 \end{aligned} \tag{2.208}$$

Show the step-by-step procedure for constructing the signal flow graph. The signal flow graph for these equations is constructed, step by step, in Figure 2.73

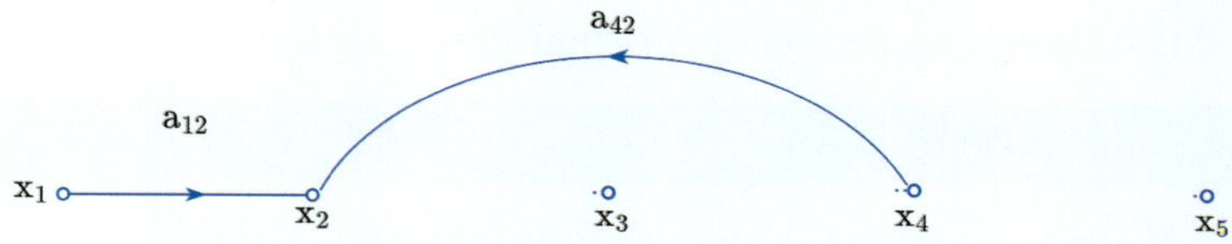

(a) $x_2 = a_{12}x_1 + a_{42}x_4$

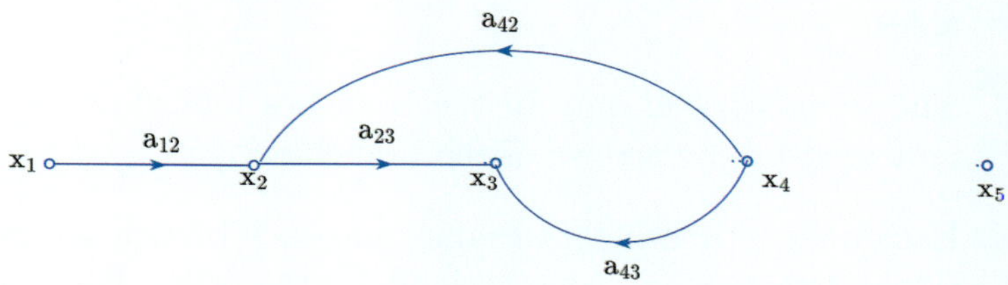

(b) $x_2 = a_{12}x_1 + a_{42}x_4, \;\; x_3 = a_{23}x_2 + a_{43}x_4$

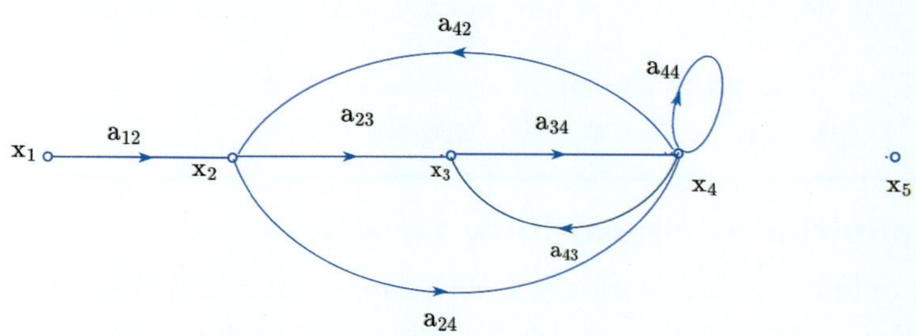

(c) $x_2 = a_{12}x_1 + a_{42}x_4, \;\; x_3 = a_{23}x_2 + a_{43}x_4, \;\; x_4 = a_{24}x_2 + a_{34}x_3 + a_{44}x_4$

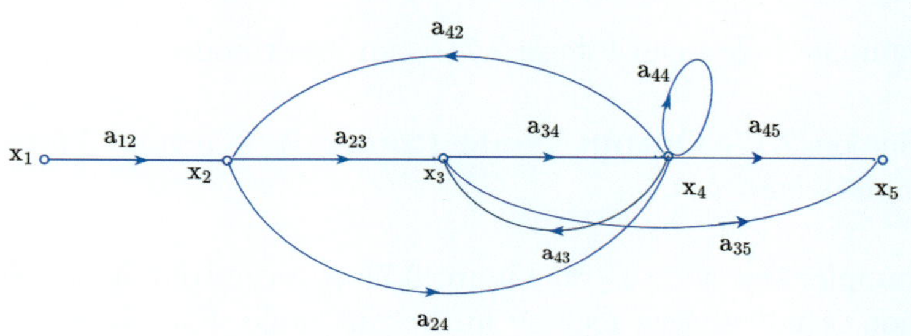

Figure 2.73: Complete Signal Flow Graph

2.18.2 Properties of Signal Flow Graph

Some of the basic properties of a signal flow graph are:

Key Points

1. It applies only to linear time-invariant systems.
2. The signal flow graph can represent the systems that are described by algebraic equations.
3. The nodes are arranged from left to right, from the input to the output, following the cause-and-effect relations in the system variables.
4. The signals travel only in the branches following the directions indicated by the arrows of the branches.
5. Each node represents a variable, and each branch shows the functional dependence of one variable on the other. For example, the branch directing from node x_k to x_j implies that x_j depends on x_k but not the reverse.
6. Each branch is associated with a gain or transmittance. When signal x_k travels along a branch between x_k to x_j, then a signal $a_{kj}x_k$ is delivered at x_j where a_{kj} is the transmittance of the branch.
7. A node adds the signals of all incoming branches and transmits this sum to all outgoing branches.

2.18.3 Definition of Signal Flow Graph Terms

Let us define a few commonly used terms in a signal flow graph, which will be used for identification and execution of signal flow algebra.

Definition 2.16. Input Node (Source): It is *a node which has only outgoing branches*

For example, node x_1 in Figure 2.71 is an input node.

Definition 2.17. Output Node (Sink): It is *a node which has only incoming branches.*

For example, the node x_2 in Figure 2.71 is an output node. Note that this condition is not always met by an output node. For example, consider the signal flow graph shown in Figure 2.74(a). Here, none of the nodes satisfy the condition of an output node. The question is, can we convert any non-input node into an output node?

Key Point

Any non-input node in a signal flow graph can be converted into an output node by adding an extra branch of unity gain to the existing node.

Thus, if we want to treat x_2 as an output node, then an additional branch of unity gain need to be connected to the existing node x_2 to the new node which is also designated as x_2 as shown in Figure 2.74(Similarly, the node x_4 can be treated as an output node by connecting an additional branch with unity gain. Thus, any non-input node can be converted to an output node by the procedure described above

The next obvious question: Is it possible to convert a non-input node into an input node?

Key Point

A non-input node cannot be converted into an input node

Consider for example the signal flow graph in Figure 2.75.

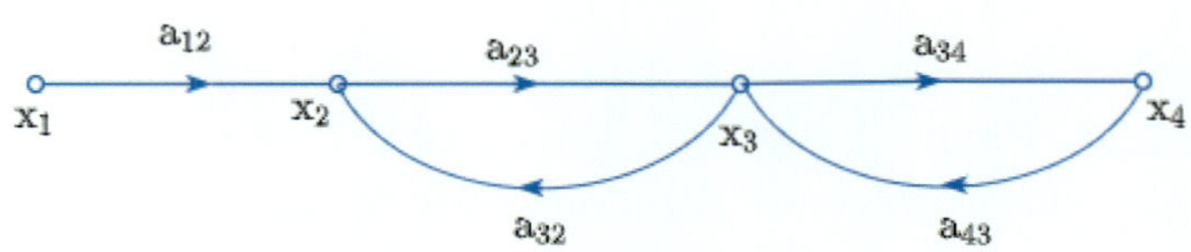

(a). Original signal-flow graph

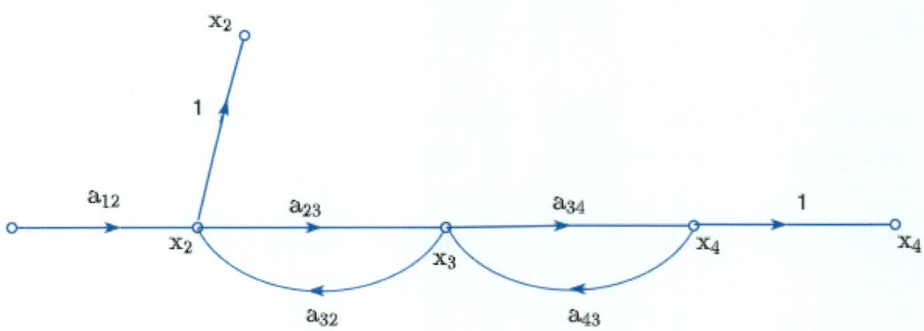

(b). Modified signal-flow graph

Figure 2.74: Modification of a signal-flow graph such that x_2 and x_4 satisfy the condition of output nodes

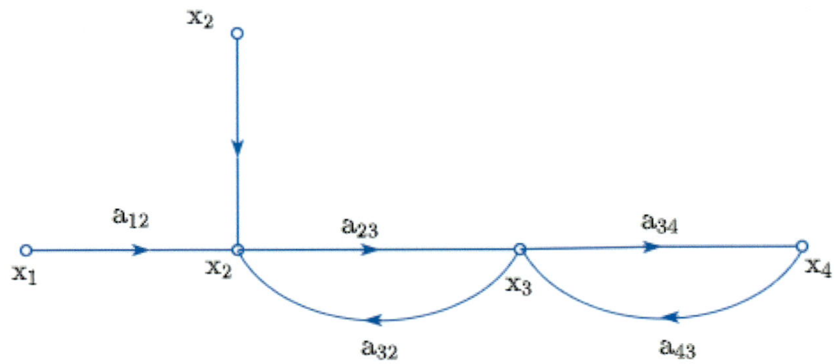

Figure 2.75: Erroneous way to make x_2 an input node

The node x_2 is not an input node. If we follow a similar procedure described for output nodes and connect an additional incoming branch of unity gain from another identical node x_2, this will create the signal flow graph shown in Figure 2.75. The algebraic equation at node x_2 is described as:

$$x_2 = x_2 + a_{12}x_1 + a_{32}x_3$$

which is different from the original relation at node x_2, i.e.,

$$x_2 = a_{12}x_1 + a_{32}x_3$$

Definition 2.18. **Path:** *A path is a branch or a continuous sequence of branches that can be traversed from one node to another node.*

Definition 2.19. **Forward Path:** *It is a path which starts at an input node and ends at an output node, and along which no node is traversed more than once.*

Definition 2.20. **Path Gain:** *It is the product of the branch gains encountered while traversing a path.*

For example, the path-gain for the path $x_1 - x_2 - x_3 - x_4$ in Figure 2.73c is $a_{12}a_{23}a_{34}$.

Definition 2.21. **Loop:**It is a *path which originates and terminates at the same node and along which no other node is encountered more than once.*

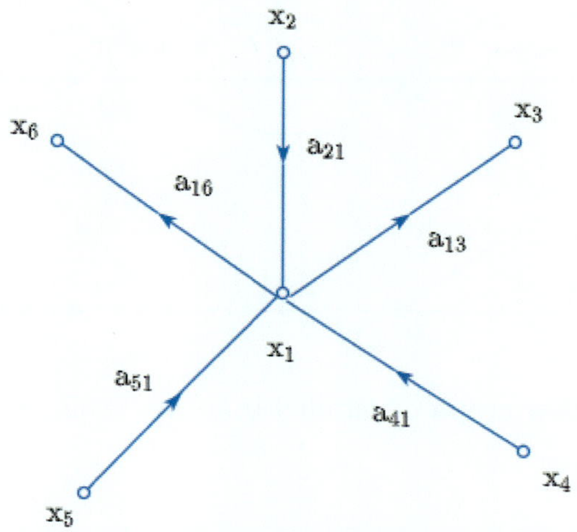

Figure 2.76: Node as a summing point and as a transmitting point

Definition 2.22. **Forward Path Gain:** It is the *product of the branch gains encountered while traversing a forward path.*

Definition 2.23. **Loop Gain:** It is the *product of the branch gains encountered while traversing the loop.*

Definition 2.24. **Nontouching Loops:** *Loops are said to be nontouching if they do not share a common node.*

More about these will be illustrated through examples in later sections.

2.18.4 Signal Flow Algebra

In the following, we discuss some basic rules used to manipulate signal flow graphs.

Rule-1: *Node as a summing point:* The value of a variable at a node is equal to the sum of all the signals entering that node.

As an example, for the signal flow graph of Figure 2.76, the value of variable x_1 is expressed as:

$$x_1 = a_{21}x_2 + a_{41}x_4 + a_{51}x_5$$

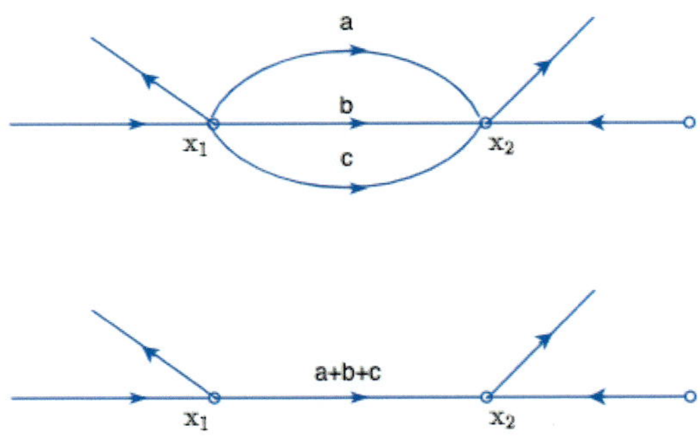

Figure 2.77: Signal flow graph of parallel branches replaced with a single branch

> **Rule-2:** *Node as a transmitting point:* The variable at the node is transmitted through all branches leaving the node.

From the signal flow graph of Figure 2.76, we have

$$x_3 = a_{13}x_1$$
$$x_6 = a_{16}x_1$$

> **Rule-3:** *Simplifying Parallel Branches:* The parallel branches connecting the two nodes in the same direction can be replaced by a single branch whose gain equals to the sum of the gains of the parallel branches.

In Figure 2.77 we have shown how three parallel branches with different gains can be replaced by a single gain.

> **Rule-4:** *Simplification of Series Connections:* A series connection of unidirectional branches can be replaced by a single branch whose gain equals to the product of the gains of the unidirectional branches.

This is illustrated in Figure 2.78.

2.18.5 Relation Between Block Diagrams and Signal Flow Graphs

Table-2.6 gives the signal flow graphs of three simple block diagrams.

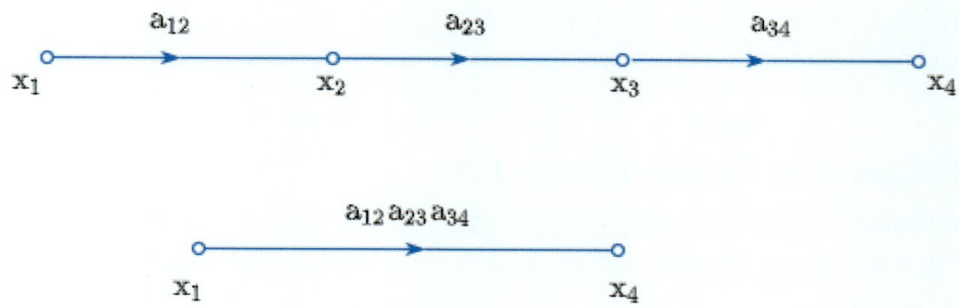

Figure 2.78: Signal flow graph with cascade unidirectional branches replaced by a single branch

Table 2.6: Block diagram and its equivalent signal flow graphs

Transfer Function	**Block Diagram**	**Signal Flow Diagram**
$\frac{Y(s)}{U(s)} = G(s)$	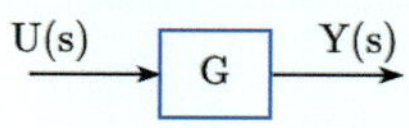	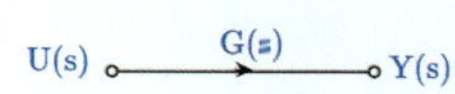
$\frac{Y(s)}{U(s)} = G_1(s) + G_2(s)$	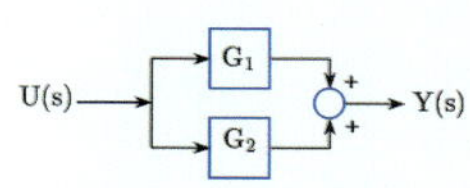	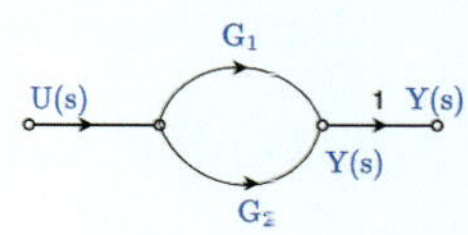
$\frac{Y(s)}{U(s)} = \frac{G(s)}{1 + G(s)H(s)}$	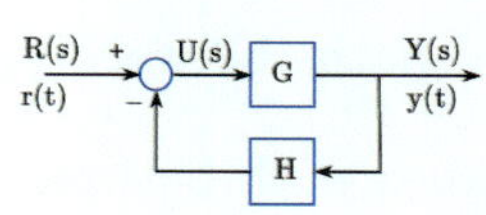	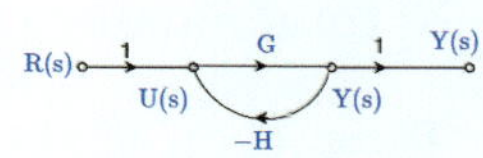

2.18.6 Construction of Signal Flow Graph from Block Diagram

The conversion of a block diagram can be carried out using the following structured procedure:

Construction Procedure of Signal Flow Graph

i. Identify each summing and take-off point. If the summing and take-off points are near each other, represent them as separate nodes.

ii. Represent the blocks in the block diagram as branches in a signal flow graph.

iii. Represent the transfer functions inside the blocks of the block diagram as gains or transmittance of the branches in the signal flow graph.

iv. Identify the input and output nodes and connect the nodes by branches, and show explicitly the gain of each branch.

2.19 Mason's Gain Formula

The transfer function of a system can be directly determined from its signal flow graph using Mason's rule, which is stated as follows:

Assume the signal flow graph contains N forward paths and K individual loops. Then, the gain between the input node y_{in} and the output node y_{out} is given by

$$G = \frac{y_{out}}{y_{in}} = \frac{1}{\Delta}\sum_{k=1}^{N} T_k \Delta_k$$

where

y_{in}:Input-node variable

y_{out}: Output-node variable

G: Overall gain between y_{in} and y_{out}

N:Total number of forward paths between y_{in} and y_{out}

T_k =: Gain of the kth forward path between y_{in} y_{out}

Δ: Determinant of the signal flow graph, defined as:

$$\begin{aligned}\Delta = 1 - \Sigma\,&(\text{individual loop gains})\\ &+\Sigma\,(\text{gain products of all possible combinationsof two non} - \text{touching loops})\\ &-\Sigma\,(\text{gain products of all possible combinationsof three non} - \text{touching loops})\\ &+\Sigma\,(\text{gain products of all possible combinationsof four non} - \text{touching loops})\\ &-\cdots\cdots+\cdots\cdots-\cdots\cdots\end{aligned}$$

Δ_k: Co-factor associated with the k^{th}forward path, defined as:

$\Delta_k = \Delta - \Sigma$ loop gain terms in Δ that touch k^{th} forward path. Note that Δ_k can easily be found from Δ by eliminating those loop gains which touch the kth forward path.

This can alternatively be expressed as:

$$\Delta = 1 - \sum_i L_{i1} + \sum_i L_{j2} - \sum_i L_{k3} + \cdots$$

L_{mr}= gain product of mth $(m = i, j, k, \ldots)$ possible combination of r non-touching loops $(1 \leq r \leq K)$.

Let us now illustrate the procedure with a series of examples.

EXAMPLE 2.25

Problem

Convert the block diagram of Figure 2.79 into signal-flow graph and determine the transfer function $\frac{C(s)}{R(s)}$ using Mason's rule.

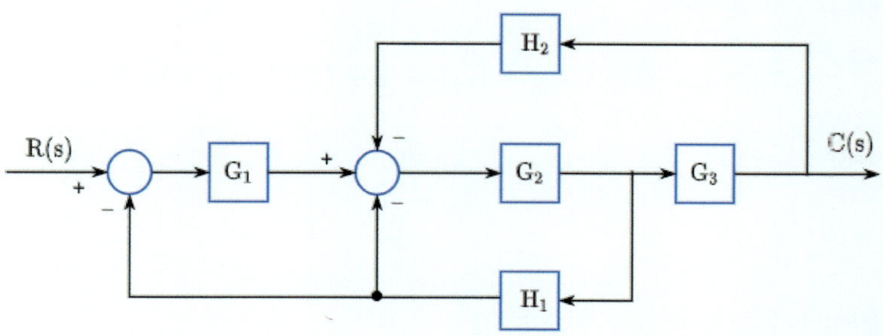

Figure 2.79: Block diagram of Example-2.25

SOLUTION:

The signal flow graph of Figure 2.79 is shown in Figure 2.80.

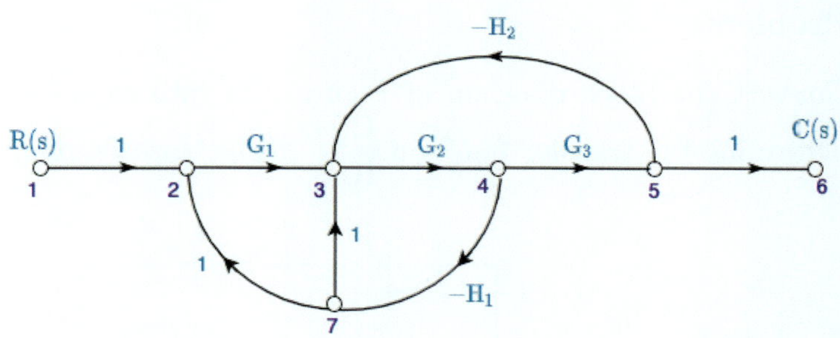

Figure 2.80: Signal flow graph of Figure 2.79

There is only one forward path connecting the nodes 1-2-3-4-5-6. The forward path gain

$$T_1 = G_1 G_2 G_3$$

There are three loops:

Loop-1: Connected Nodes 2-3-4-7-2

Loop-2: Connected Nodes 3-4-5-3 and

Loop-3: Connected Nodes 3-4-7-3.

The loop gains are

$$L_1 = -G_1G_2H_1$$
$$L_2 = -G_2G_3H_2$$
$$L_3 = -G_2H_1$$

All the loops are touching; therefore the number of non-touching loops=0. Hence

$$\Delta = 1 - (L_1 + L_2 + L_3) = 1 + G_1G_2H_1 + G_2G_3H_2 + G_2H_1$$

and

$$\Delta_1 = 1$$

Hence the transfer function

$$\frac{C(s)}{R(s)} = \frac{T_1\Delta_1}{\Delta} = \frac{G_1G_2G_3}{1 + G_1G_2H_1 + G_2G_3H_2 + G_2H_1}$$

EXAMPLE 2.26

Problem

Convert the block diagram of Figure 2.81 into signal-flow graph and determine the transfer function $\frac{C(s)}{R(s)}$ using Mason's rule.

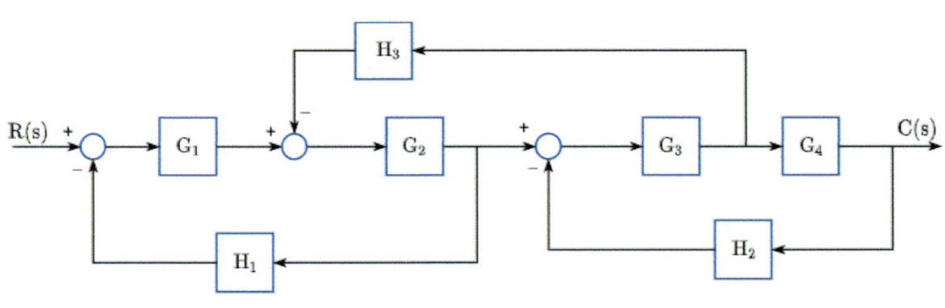

Figure 2.81: Block diagram of Example-2.26

SOLUTION:

The signal flow graph of Figure 2.81 is shown in Figure 2.82.

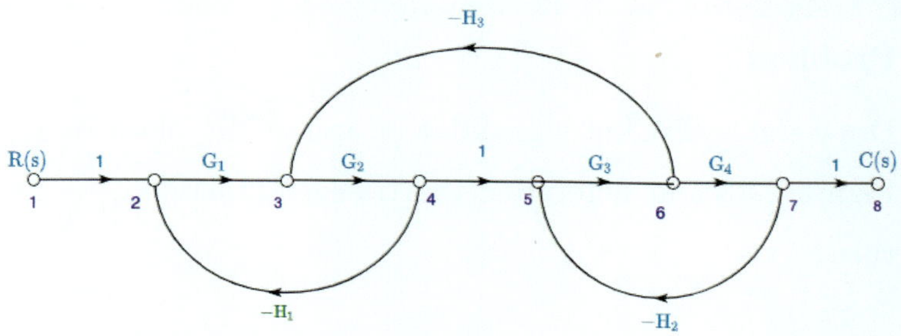

Figure 2.82: Signal flow graph of Figure 2.81

There is one forward path connecting the nodes 1-2-3-4-5-6-7-8. The forward path gain is

$$T_1 = G_1G_2G_3G_4$$

There are three loops:

Loop-1: Connected Nodes 2-3-4-2

Loop-2: Connected Nodes 5-6-7-5 and

Loop-3: Connected Nodes 6-3-4-5-6.

The loop gains are

$$\begin{aligned} L_1 &= -G_1G_2H_1 \\ L_2 &= -G_3G_4H_2 \\ L_3 &= -G_2G_3H_3 \end{aligned}$$

The loops L_1 and L_2 are non-touching; therefore the number of non-touching loops=2. Hence

$$\begin{aligned} \Delta &= 1 - (L_1 + L_2 + L_3) + L_1L_2 \\ &= 1 + G_1G_2H_1 + G_3G_4H_2 + G_2G_3H_3 + G_1G_2G_3G_4H_1H_2 \\ &= G_3G_4H_2\left(1 + G_1G_2H_1\right) + \left(1 + G_1G_2H_1\right) + G_2G_3H_3 \\ &= \left(1 + G_1G_2H_1\right)\left(1 + G_3G_4H_2\right) + G_2G_3H_3 \end{aligned}$$

Since the forward path touches all the loops

$$\Delta_1 = 1$$

Hence the transfer function

$$\frac{C(s)}{R(s)} = \frac{T_1\Delta_1}{\Delta} = \frac{G_1G_2G_3G_4}{\left(1 + G_1G_2H_1\right)\left(1 + G_3G_4H_2\right) + G_2G_3H_3}$$

CONCEPT REINFORCEMENT PROBLEM 2.14

Problem

Draw the equivalent signal-flow graph of the block diagram shown in Figure 2.83 and determine the transfer function $\frac{C(s)}{R(s)}$ using Mason's rule.

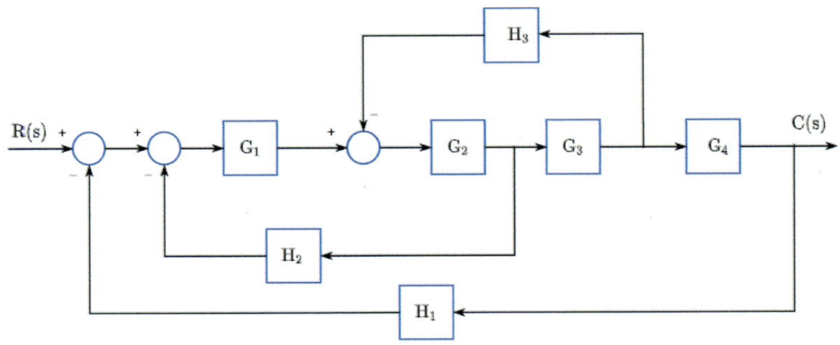

Figure 2.83: Block Diagram of CRP 2.14

EXAMPLE 2.27

Problem

Convert the block diagram of Figure 2.84 into signal-flow graph and determine the transfer function $\frac{C(s)}{R(s)}$ using Mason's rule.

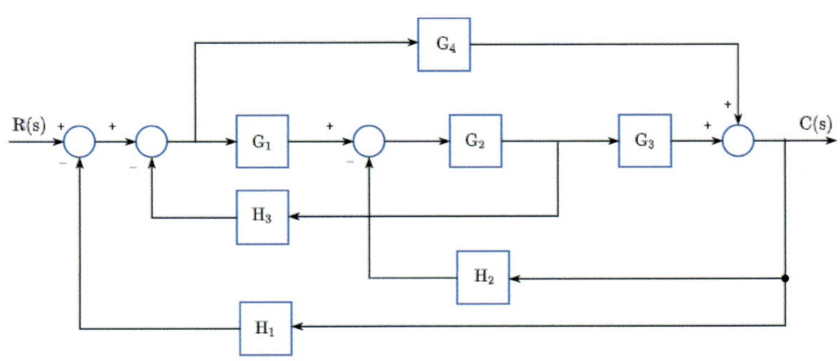

Figure 2.84: Block diagram of Example-2.27

SOLUTION:

The signal flow graph of Figure 2.84 is shown in Figure 2.85.

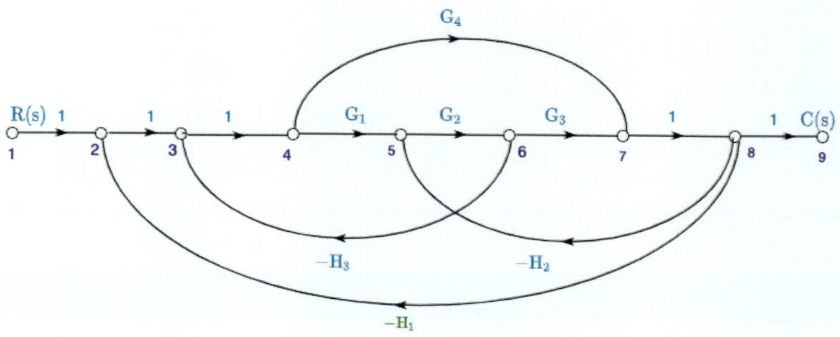

Figure 2.85: Signal flow graph of Figure 2.84

There are two forward paths; the first one connecting the nodes 1-2-3-4-5-6-7-8-9 and the second path connects the nodes 1-2-3-4-7-8-9. The forward path gains i.e the transmittances are

$$T_1 = G_1G_2G_3$$
$$T_2 = G_4$$

There are five loops:

Loop-1: Connected Nodes 3-4-5-6-3

Loop-2: Connected Nodes 5-6-7-8-5 and

Loop-3: Connected Nodes 2-3-4-5-6-7-8-2.

Loop-4: Connected Nodes 2-3-4-7-8-2.

Loop-5: Connected Nodes 4-7-8-5-6-3-4.

The loop gains are

$$L_1 = -G_1G_2H_3$$
$$L_2 = -G_2G_3H_2$$
$$L_3 = -G_1G_2G_3H_1$$
$$L_4 = -G_4H_1$$
$$L_5 = G_4H_2G_2H_3$$

All the loops are touching; therefore the number of non-touching loops = 0. Hence

$$\Delta = 1 - (L_1 + L_2 + L_3 + L_4 + L_5)$$
$$= 1 + G_1G_2H_3 + G_2G_3H_2 + G_1G_2G_3H_1 + G_4H_1 - G_4H_2G_2H_3$$

and $\Delta_1 = 1$, and $\Delta_2 = 1$. Hence, the transfer function

$$\frac{C(s)}{R(s)} = \frac{T_1\Delta_1 + T_2\Delta_2}{\Delta}$$

$$= \frac{G_1G_2G_3 + G_4}{1 + G_1G_2H_3 + G_2G_3H_2 + G_1G_2G_3H_1 + G_4H_1 - G_4H_2G_2H_3}$$

EXAMPLE 2.28

Problem

Compute the transfer function $\frac{C(s)}{R(s)}$ of the system whose signal flow graph is shown in Figure 2.86.

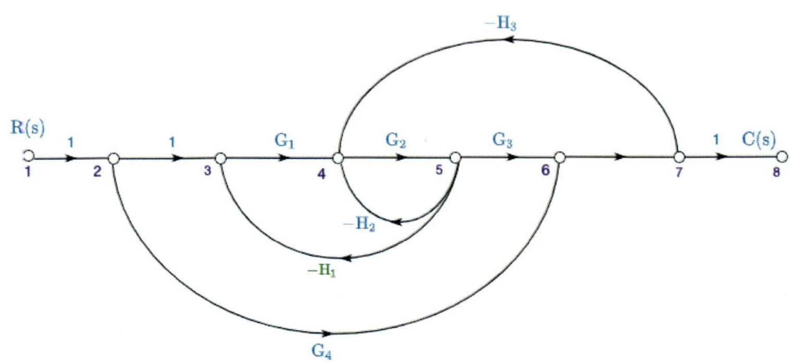

Figure 2.86: Signal flow graph of Example-2.28

SOLUTION:

The signal flow graph has two forward paths: Path-1 connects the nodes 1−2−3−4−5−6−7−8 and Path-2 connects the nodes 1−2−6−7−8. The transmittance of these two forward paths is:

$$T_1 = G_1G_2G_3, \quad \text{and} \quad T_2 = G_4$$

There are three loops:

Loop-1: (3-4-5-3), Gain $L_1 = -G_1G_2H_1$

Loop-2: (4-5-4), Gain $L_2 = -G_2H_2$

Loop-3: (4-5-6-7-4), Gain $L_3 = -G_2G_3H_3$

All the loops are touching. Hence

$$\Delta = 1 - (L_1 + L_2 + L_3) = 1 + G_1G_2H_1 + G_2H_2 + G_2G_3H_3$$

The forward path-1 touches all the loops. However, path-2 touches only loop-3. Hence

$$\Delta_1 = 1, \quad \text{and} \quad \Delta_2 = 1 + G_1G_2H_1 + G_2H_2$$

The transfer function

$$\frac{C(s)}{R(s)} = \frac{T_1\Delta_1 + T_2\Delta_2}{\Delta}$$
$$= \frac{G_1G_2G_3 + G_4\left(1 + G_1G_2H_1 + G_2H_2\right)}{1 + G_1G_2H_1 + G_2H_2 + G_2G_3H_3}$$

EXAMPLE 2.29

Problem

Compute the transfer function $\frac{C(s)}{R(s)}$ of the system whose signal flow graph is shown in Figure 2.87.

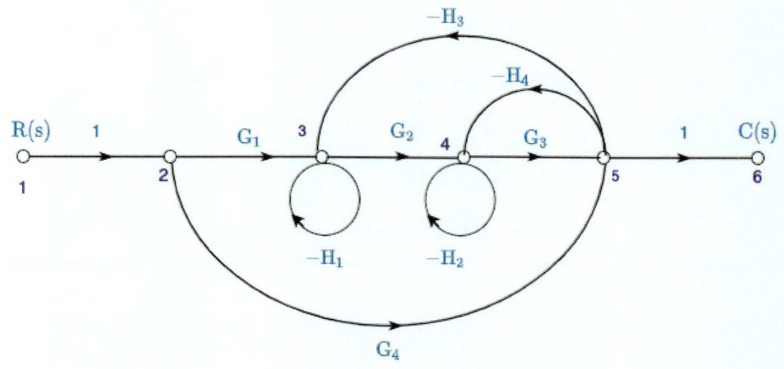

Figure 2.87: Signal flow graph of Example-2.29

SOLUTION:

The signal flow graph has two forward paths: Path-1 connects the nodes $1-2-3-4-5-6$ and Path-2 connects the nodes $1-2-5-6$. The transmittance of these two forward paths is:

$$T_1 = G_1G_2G_3, \quad \text{and} \quad T_2 = G_4$$

There are four loops:

Loop-1: (3-3), Gain $L_1 = -H_1$

Loop-2: (4-4), Gain $L_2 = -H_2$

Loop-3: (4-5-4), Gain $L_3 = -G_3H_4$

Loop-4: (3-4-5-3), Gain $L_4 = -G_2G_3H_3$

Non-touching Loops:

1. The loops L_1 and L_2 are non-touching
2. The loops L_1 and L_3 are non-touching.

Hence

$$\begin{aligned}\Delta &= 1 - (L_1 + L_2 + L_3) + L_1L_2 + L_1L_3\\ &= 1 + H_1 + H_2 + G_3H_4 + G_2G_3H_3 + H_1H_2 + G_3H_1H_4\end{aligned}$$

The forward path-1 touches all the loops. However, path 2 does not touch loop 1 and loop 2. Hence

$$\Delta_1 = 1, \quad \text{and} \quad \Delta_2 = 1 + H_1 + H_2 + H_1H_2$$

The transfer function

$$\begin{aligned}\frac{C(s)}{R(s)} &= \frac{T_1\Delta_1 + T_2\Delta_2}{\Delta}\\ &= \frac{G_1G_2G_3 + G_4(1 + H_1 + H_2 + H_1H_2)}{1 + H_1 + H_2 + G_3H_4 + G_2G_3H_3 + H_1H_2 + G_3H_1H_4}\end{aligned}$$

CONCEPT REINFORCEMENT PROBLEM 2.15

Problem

Convert the block diagram of Figure 2.88 into signal-flow graph and determine the transfer function $\frac{C(s)}{R(s)}$ using Mason's rule.

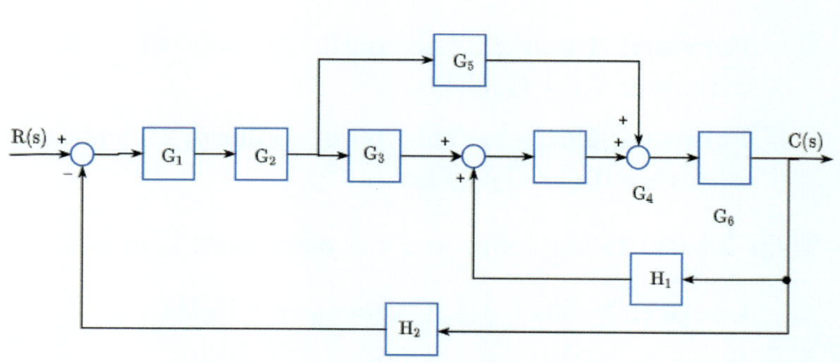

Figure 2.88: Block Diagram

EXAMPLE 2.30

Problem

Determine the transfer function $\dfrac{C(s)}{R(s)}$, for the system which is represented by the signal-flow graph in Figure 2.89.

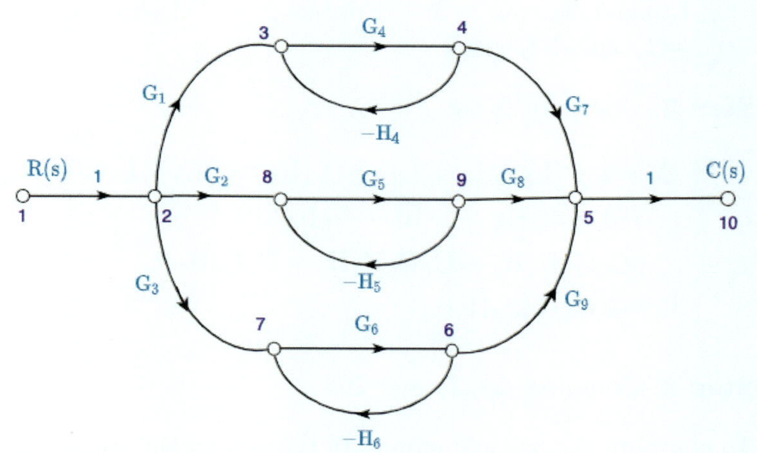

Figure 2.89: Signal flow graph for Example-2.30

SOLUTION:

In Figure 2.89, the nodes are numbered from 1 to 10.

Step-1: Identify the forward paths and determine the gain of each forward path.

Forward Path-1: This path connects the nodes- 1-2-3-4-5-10 with gain $T_1 = G_1G_4G_7$.

Forward Path-2: This path connects the nodes- 1-2-8-9-5-10 with gain $T_2 = G_2G_5G_8$.

Forward Path-3: This path connects the nodes- 1-2-7-6-5-10 with gain $T_3 = G_3G_6G_9$.

Step-2: Identify the loops and the associated loop gain.

Loop-1: Nodes 3-4-3, Gain=$L_1 = -G_4H_4$

Loop-2: Nodes 8-9-8, Gain=$L_2 = -G_5H_5$

Loop-3: Nodes 7-6-7, Gain=$L_3 = -G_6H_6$

Step-3: Identify the non-touching loops taken two at a time and compute the product of these loop gains.

Loop-1 & 2: Product of Loop Gains=$L_1L_2 = G_4G_5H_4H_5$

Loop-2 & 3: Product of Loop Gains=$L_2L_3 = G_5G_6H_5H_6$

Loop-3 & 1: Product of Loop Gains=$L_3L_1 = G_4G_6H_4H_6$

Step-4: Identify the non-touching loops taken three at a time and compute the product of these loop gains.

Loop-1,2 & 3: Product of Loop Gains=$L_1L_2L_3 = -G_4G_5G_6H_4H_5H_6$

Step-5: Compute Δ as:

$$\begin{aligned}\Delta &= 1 - (L_1 + L_2 + L_3) + (L_1L_2 + L_2L_3 + L_3L_1) - L_1L_2L_3 \\ &= 1 + G_4H_4 + G_5H_5 + G_6H_6 \\ &+G_4G_5H_4H_5 + G_5G_6H_5H_6 + G_4G_6H_4H_6 \\ &+G_4G_5G_6H_4H_5H_6\end{aligned}$$

Step-6: Compute Δ_1,Δ_2 and Δ_3:

To compute Δ_k, we will eliminate the term with coefficient L_k from Δ. Thus

$$\begin{aligned}\Delta_1 &= 1 - (L_2 + L_3) + L_2L_3 \\ \Delta_2 &= 1 - (L_1 + L_3) + L_3L_1 \\ \Delta_3 &= 1 - (L_1 + L_2) + L_1L_2\end{aligned}$$

Step-7: Compute the overall transfer function as:

$$\frac{C(s)}{R(s)} = \frac{T_1\Delta_1 + T_2\Delta_2 + T_3\Delta_3}{\Delta}$$

CONCEPT REINFORCEMENT PROBLEM 2.16

Problem

Compute the transfer function $\dfrac{C(s)}{R(s)}$ of the system whose signal flow graph is shown in Figure 2.90.

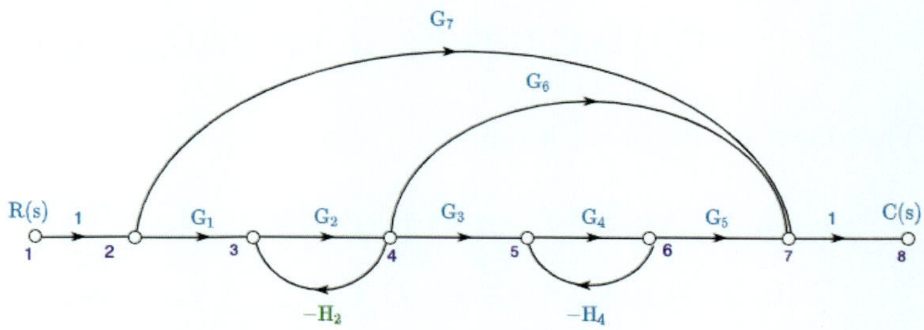

Figure 2.90: Signal flow graph of CRP 2.16

EXAMPLE 2.31

Problem

Compute the transfer function $\dfrac{C(s)}{R(s)}$ of the system whose signal flow graph is shown in Figure 2.91.

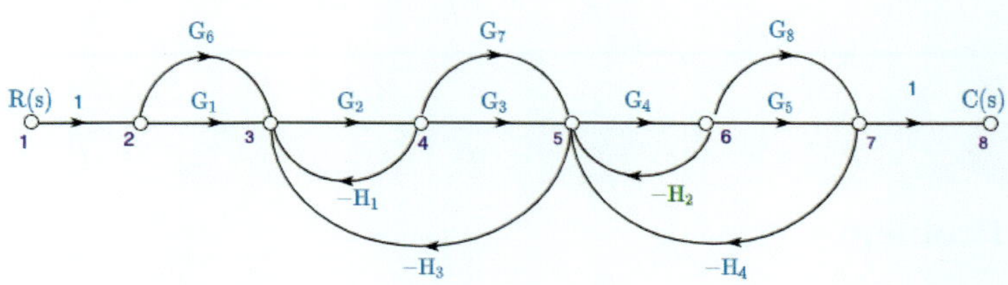

Figure 2.91: Signal flow graph of Example-2.31

SOLUTION:

The signal flow graph has eight forward paths. The transmittance of these forward paths is:

$$T_1 = G_1G_2G_3G_4G_5, \quad T_2 = G_1G_2G_3G_4G_8$$

$$T_3 = G_1G_2G_7G_4G_5, \quad T_4 = G_1G_2G_7G_4G_8$$

$$T_5 = G_6G_2G_3G_4G_5, \quad T_6 = G_6G_2G_3G_4G_8$$

$$T_7 = G_6G_2G_7G_4G_8, \quad T_8 = G_6G_2G_7G_4G_5$$

There are six loops with gains:

$$L_1 = -G_2H_1, \quad L_2 = -G_2G_3H_3, \quad L_3 = -G_2G_7H_3$$

$$L_4 = -G_4H_2, \quad L_5 = -G_4G_5H_4, \quad L_6 = -G_4G_8H_4$$

The loop with gain L_1 does not touch the loops with gain L_4,L_5 and L_6. Hence

$$\Delta = 1 - (L_1 + L_2 + L_3 + L_4 + L_5 + L_6) + (L_1L_4 + L_1L_5 + L_1L_6)$$

All the forward paths touch all the loops. Hence

$$\Delta_i, \ i = 1, 2, \ldots, 8 = 1$$

The transfer function

$$\frac{C(s)}{R(s)} = \frac{\sum_{k=1}^{8} T_k}{\Delta}$$

CONCEPT REINFORCEMENT PROBLEM 2.17

Problem

Compute the transfer function $\frac{C(s)}{R(s)}$ of the system whose signal flow graph is shown in Figure 2.92.

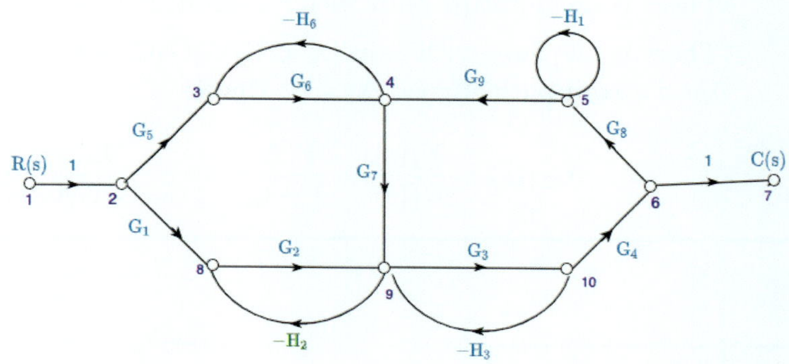

Figure 2.92: Signal flow graph of CRP 2.17

EXAMPLE 2.32

Problem

Find the expressions for $Y_1(s)$ and $Y_2(s)$ for the system whose signal flow graph is shown in Figure 2.93.

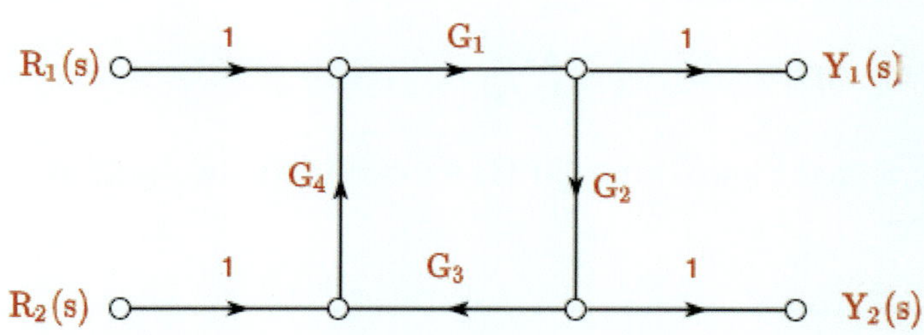

Figure 2.93: Signal flow graph of Example-2.32 of MIMO system

SOLUTION:

Since this is a multiple-input, multiple-output (MIMO) system, we have to apply Mason's rule repeatedly for each different combination of output and input. The signal flow graph in Figure 2.93 has two inputs and two outputs. The four associated transfer functions between different combinations can be computed by constructing the single-input, single-output signal flow graphs as follows:

a. **Computations of $G_{11}(s)$:** This is computed from the signal flow graph shown in Figure 2.94(a), which is obtained from Figure 2.93 by making $R_2(s) = 0$.

There is one forward path with transmittance $T_1 = G_1$.

There is one loop with gain $L_1 = G_1G_2G_3G_4$ which is touched by the forward path. Hence $\Delta_1 = 1$. Hence

$$G_{11}(s) = \frac{Y_1(s)}{R_1(s)} = \frac{T_1\Delta_1}{1 - L_1} = \frac{G_1}{1 - G_1G_2G_3G_4}$$

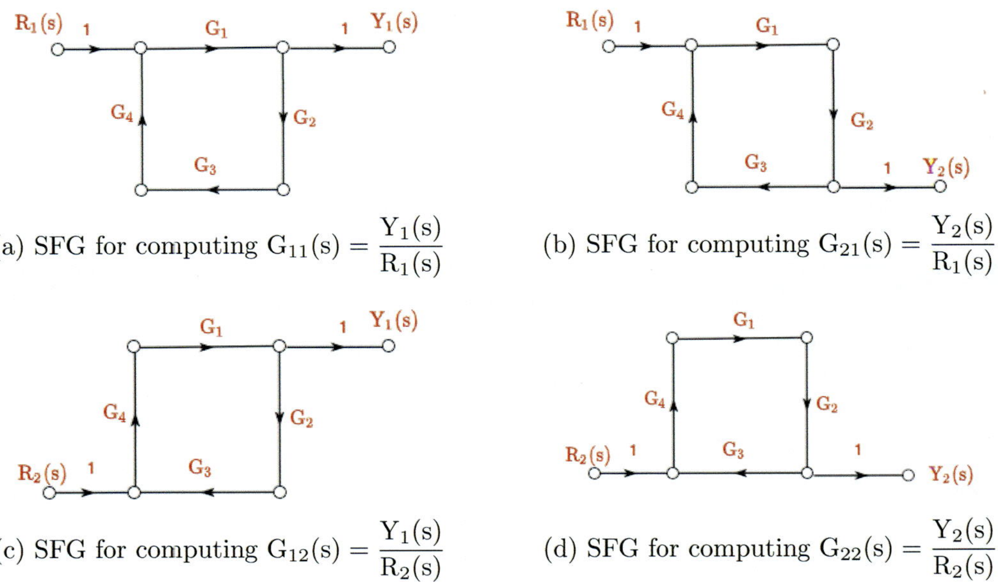

(a) SFG for computing $G_{11}(s) = \dfrac{Y_1(s)}{R_1(s)}$

(b) SFG for computing $G_{21}(s) = \dfrac{Y_2(s)}{R_1(s)}$

(c) SFG for computing $G_{12}(s) = \dfrac{Y_1(s)}{R_2(s)}$

(d) SFG for computing $G_{22}(s) = \dfrac{Y_2(s)}{R_2(s)}$

Figure 2.94: Signal Flow Graph (SFG) for two-input-two-output system of Example-2.32

b. **Computations of $G_{21}(s)$:** This is computed from the signal flow graph shown in Figure 2.94(b), which is obtained from Figure 2.93 by making $R_2(s) = 0$.

There is one forward path with transmittance $T_1 = G_1G_2$.

There is one loop with gain $L_1 = G_1G_2G_3G_4$ which is touched by the forward path. Hence $\Delta_1 = 1$. Hence

$$G_{21}(s) = \frac{Y_2(s)}{R_1(s)} = \frac{T_1\Delta_1}{1 - L_1} = \frac{G_1G_2}{1 - G_1G_2G_3G_4}$$

c. **Computations of $G_{12}(s)$:** This is computed from the signal flow graph shown in Figure 2.94(c), which is obtained from Figure 2.93 by making $R_1(s) = 0$.

There is one forward path with transmittance $T_1 = G_1G_4$.

There is one loop with gain $L_1 = G_1G_2G_3G_4$ which is touched by the forward path. Hence $\Delta_1 = 1$. Hence

$$G_{12}(s) = \frac{Y_1(s)}{R_2(s)} = \frac{T_1\Delta_1}{1 - L_1} = \frac{G_1G_4}{1 - G_1G_2G_3G_4}$$

d. **Computations of** $G_{22}(s)$**:** This is computed from the signal flow graph shown in Figure 2.94(d), which is obtained from Figure 2.93 by making $R_1(s) = 0$.

There is one forward path with transmittance $T_1 = G_1G_2G_4$.

There is one loop with gain $L_1 = G_1G_2G_3G_4$ which is touched by the forward path. Hence $\Delta_1 = 1$. Hence

$$G_{22}(s) = \frac{Y_2(s)}{R_2(s)} = \frac{T_1\Delta_1}{1 - L_1} = \frac{G_1G_2G_4}{1 - G_1G_2G_3G_4}$$

Hence

$$Y_1(s) = G_{11}(s)R_1(s) + G_{12}(s)R_2(s), \quad \text{and}$$

$$Y_2(s) = G_{21}(s)R_1(s) + G_{22}(s)R_2(s)$$

2.20 Multiple Choice Skills Test Questions

M 2.1. The order of a dynamic system equals

A. total number of independent energy dissipating components

B. total number of both independent energy dissipating and storing components

C. total number of independent energy storing components

D. total number of dependent energy storing components

M 2.2. Block diagrams can be used to represent

A. both linear and nonlinear systems

B. only linear systems

C. only nonlinear systems

D. only time-invariant systems

M 2.3. Four blocks with gains 2,5, 10, 0.01 are connected in cascade. The total gain of the configuration is

A. 100 **B.** 10 **C.** 0.1 **D.** 1.0

M 2.4. Three blocks with gains of 5, 10, and 25 are connected in parallel. The overall gain of the system is

A. 1250 **B.** 40 **C.** 625 **D.** 80

M 2.5. The transfer function $\frac{C}{R}$ of the system shown in Figure 2.95 is

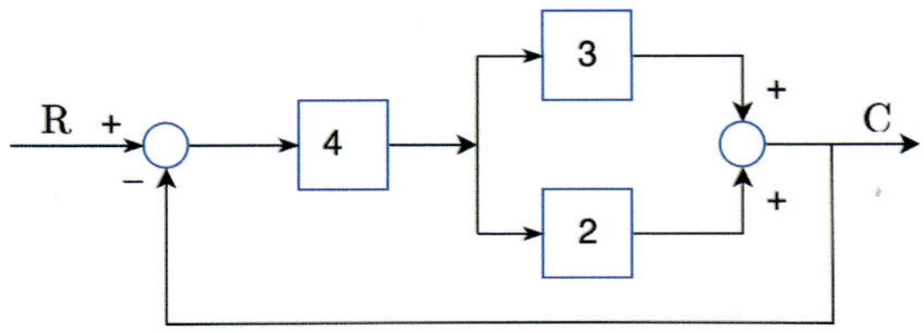

Figure 2.95

A. $\frac{20}{21}$ **B.** $\frac{24}{25}$ **C.** $\frac{9}{21}$ **D.** $\frac{20}{25}$

M 2.6. The transfer function $\frac{C}{R}$ of the system shown in Figure 2.96 is

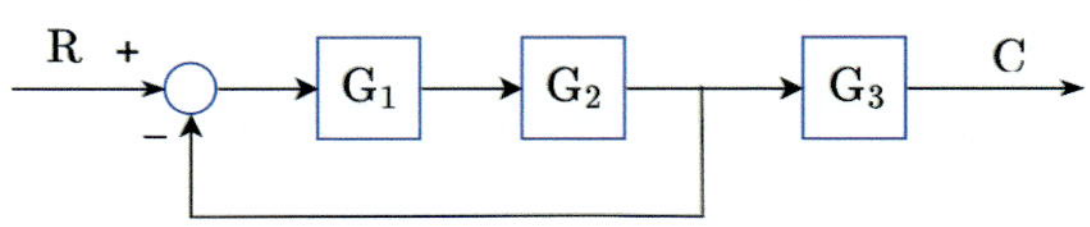

Figure 2.96

A. $\frac{G_1G_2}{1+G_1G_2G_3}$ **C.** $\frac{G_1G_2G_3}{G_3(1+G_1G_2)}$

B. $\frac{G_1G_2}{G_3(1+G_1G_2)}$ **D.** $\frac{G_1G_2G_3}{1+G_1G_2}$

M 2.7. The transfer function $\frac{C}{R}$ of the system shown in Figure 2.97 is

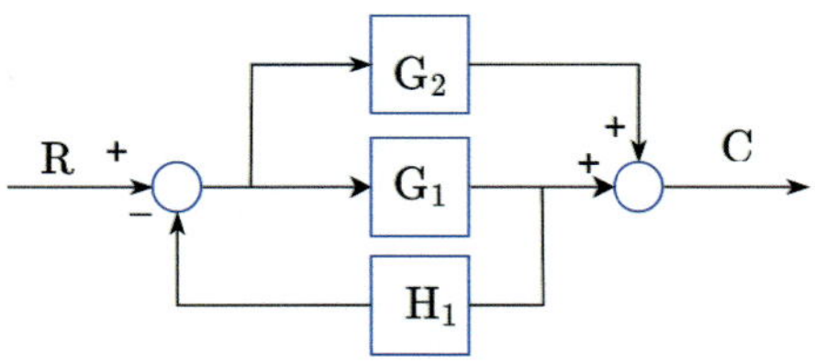

Figure 2.97

A. $\dfrac{G_1}{G_2(1+G_1H_1)}$

B. $\dfrac{G_1-G_2}{1-G_1H_1}$

C. $\dfrac{G_1+G_2}{1+G_1H_1}$

D. $\dfrac{G_1G_2}{1+G_1H_1}$

M 2.8. By performing cascading and/or summing/differencing operations using transfer function blocks $G_1(s)$ and $G_2(s)$, the form of the transfer function which cannot be realized is

A. $G_1(s)G_2(s)$

B. $G_1(s)\left[\dfrac{1}{G_1(s)}+G_2(s)\right]$

C. $G_2(s)\left[\dfrac{1}{G_2(s)}-G_1(s)\right]$

D. $\dfrac{G_1(s)}{G_2(s)}$

M 2.9. The transfer function $\dfrac{C}{R}$ of the system shown in Figure 2.98 is

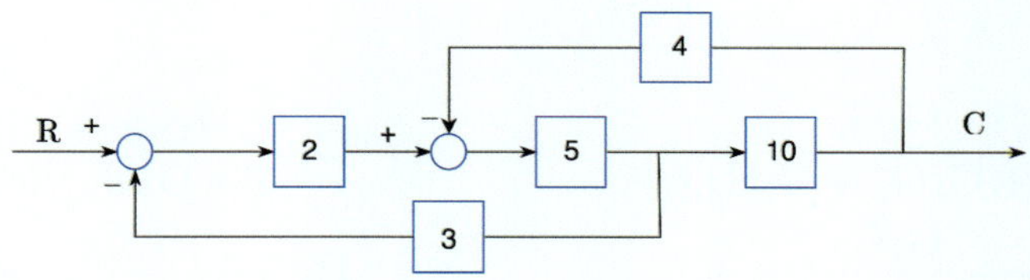

Figure 2.98

A. $\dfrac{100}{231}$ **B.** $\dfrac{100}{1201}$ **C.** $\dfrac{100}{1101}$ **D.** $\dfrac{231}{100}$

M 2.10. Consider the block diagram of a system shown in Figure 2.99.

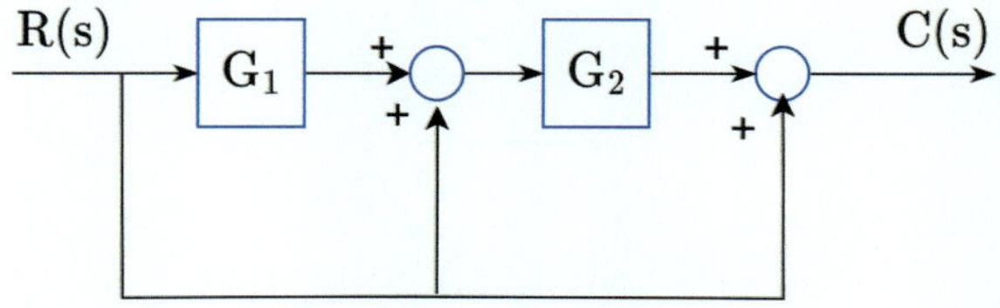

Figure 2.99

The transfer function $\frac{C(s)}{R(s)}$ is

A. $\frac{G_1G_2}{1+G_1G_2}$

B. $\frac{G_1}{1+G_1G_2}$

C. $G_1G_2+G_1+1$

D. $G_1G_2+G_2+1$

M 2.11. Consider the block diagram of a system shown in Figure 2.100.

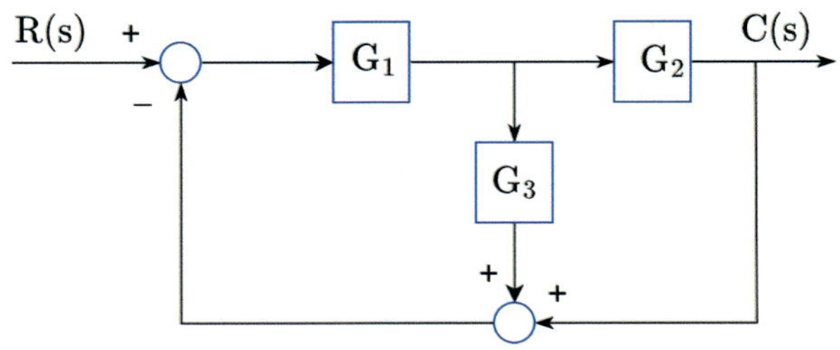

Figure 2.100

The transfer function $\frac{C(s)}{R(s)}$ is

A. $\frac{G_1G_2}{1+G_1G_2+G_1G_3}$

B. $\frac{G_1G_2}{1+G_1G_2-G_1G_3}$

C. $\frac{G_1G_2}{1+G_1G_2}$

D. $\frac{G_1G_2}{G_1G_2+G_1G_3}$

M 2.12. Consider the block diagram of a system shown in Figure 2.101.

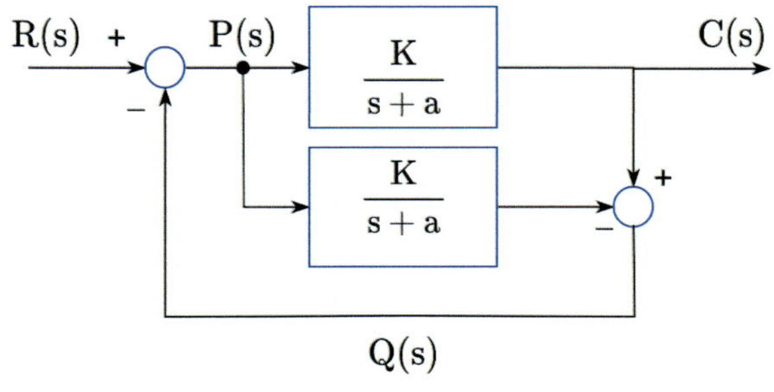

Figure 2.101

The transfer function $\dfrac{C(s)}{R(s)}$ is

A. $\dfrac{K}{(s+a)^2}$ **B.** 0 **C.** $\dfrac{K}{s+a}$ **D.** $\dfrac{2K}{s+a}$

M 2.13. For the signal flow graph shown in Figure 2.102, the transfer function $\dfrac{x_3}{x_1}$ is

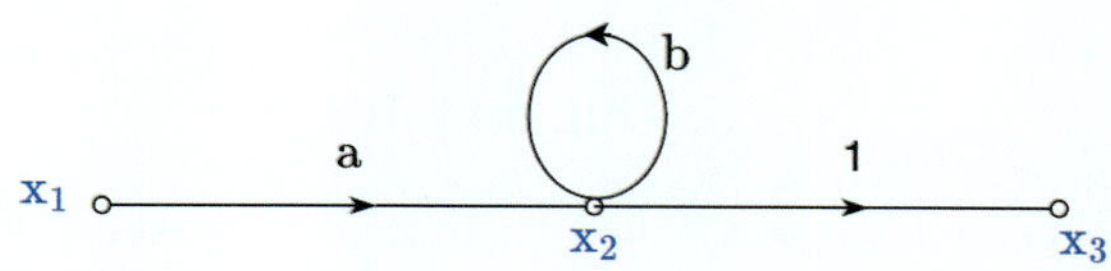

Figure 2.102

A. $\dfrac{a}{1+b}$ **B.** $\dfrac{b}{1-a}$ **C.** $\dfrac{a}{1-b}$ **D.** $\dfrac{a}{b}$

M 2.14. For the signal flow graph shown in Figure 2.103, the total number of loops equal to

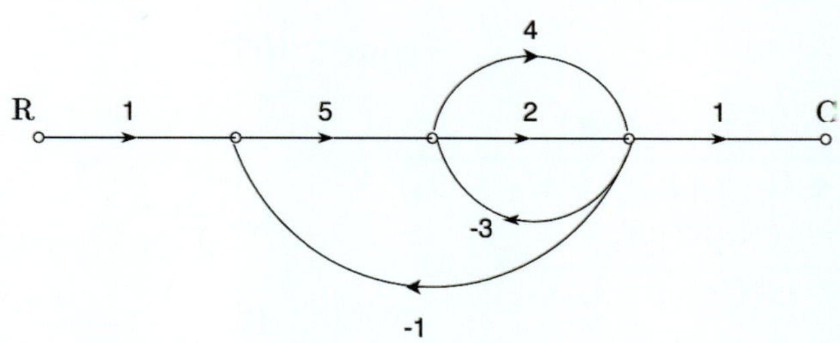

Figure 2.103

A. 2 **B.** 4 **C.** 3 **D.** 1

M 2.15. For the signal flow graph shown in Figure 2.103, the ratio $\dfrac{C}{R}$ is

A. $\dfrac{30}{49}$ **B.** $\dfrac{10}{49}$ **C.** $\dfrac{20}{49}$ **D.** $\dfrac{15}{49}$

M 2.16. For the signal flow graph shown in Figure 2.104, the ratio $\frac{C}{R}$ is

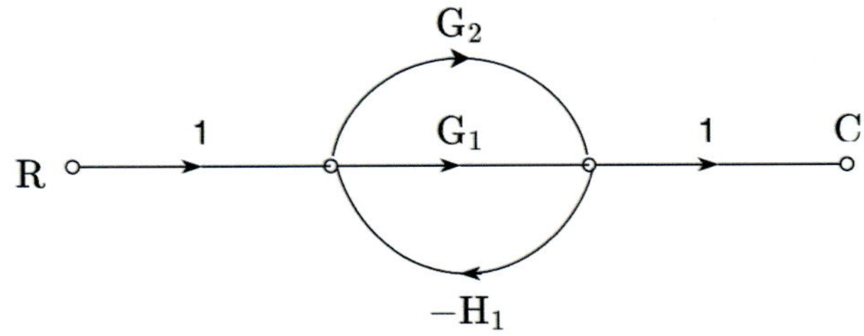

Figure 2.104

A. $\dfrac{G_1G_2}{1+H_1(G_1+G_2)}$

B. $\dfrac{G_1+G_2}{1-H_1(G_1+G_2)}$

C. $\dfrac{G_1+G_2}{1+H_1(G_1+G_2)}$

D. $\dfrac{G_1-G_2}{1-H_1(G_1+G_2)}$

M 2.17. The ratio $\frac{C}{R}$ for the signal flow graph shown in Figure 2.105 is

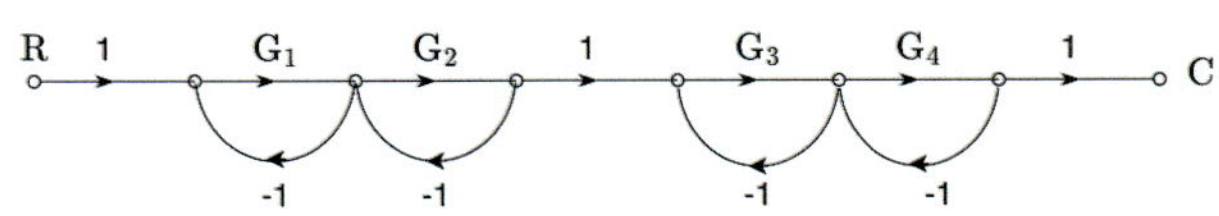

Figure 2.105

A. $\dfrac{G_1G_2G3G_4}{(1+G_1+G_2)(1+G_3+G_4)}$

B. $\dfrac{G_1G_2G_3G_4}{(1+G_1G_2)(1+G_3G_4)}$

C. $\dfrac{G_1G_2G_3G_4}{(1+G_1+G_2+G_3+G_4)}$

D. $\dfrac{G_1G_2G_3G_4}{(1-G_1-G_2)(1-G_3-G_4)}$

M 2.18. For the signal flow graph shown in Figure 2.106, the transfer function $\frac{C}{R}$ is

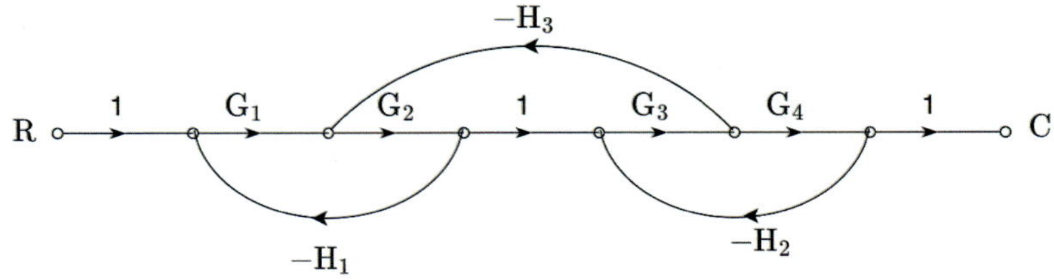

Figure 2.106

A. $\dfrac{1}{1 - G_1G_2H_1 - G_3G_4H_2 - G_2G_3H_3 - G_1G_2G_3G_4H_1H_2}$

B. $\dfrac{1}{1 + G_1G_2H_1 + G_3G_4H_2 + G_2G_3H_3 + G_1G_2G_3G_4H_1H_2}$

C. $\dfrac{G_1G_2G_3G_4}{1 - G_1G_2H_1 - G_3G_4H_2 - G_2G_3H_3 - G_1G_2G_3G_4H_1H_2}$

D. $\dfrac{G_1G_2G_3G_4}{1 + G_1G_2H_1 + G_3G_4H_2 + G_2G_3H_3 + G_1G_2G_3G_4H_1H_2}$

M 2.19. For the signal flow graph shown in Figure 2.107, the transfer function $\dfrac{C}{R}$ is

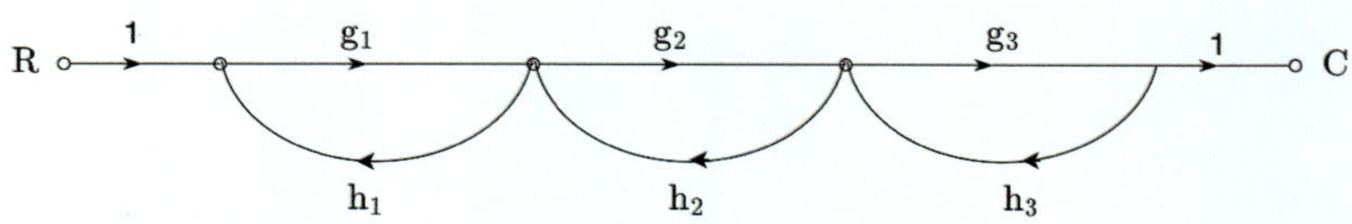

Figure 2.107

A. $\dfrac{1 - (g_1h_1 + g_2h_2 + g_3h_3)}{g_1g_2g_3}$

B. $\dfrac{g_1g_2g_3}{1 - (g_1h_1 + g_2h_2 + g_3h_3)}$

C. $\dfrac{g_1g_2g_3}{1 - (g_1h_1 + g_2h_2 + g_3h_3) + g_1g_3h_1h_3}$

D. $\dfrac{g_1g_2g_3}{1 + g_1h_1 + g_2h_2 + g_3h_3 + g_1g_3h_1h_3}$

M 2.20. The signal flow graph of a system is shown in Figure 2.108.

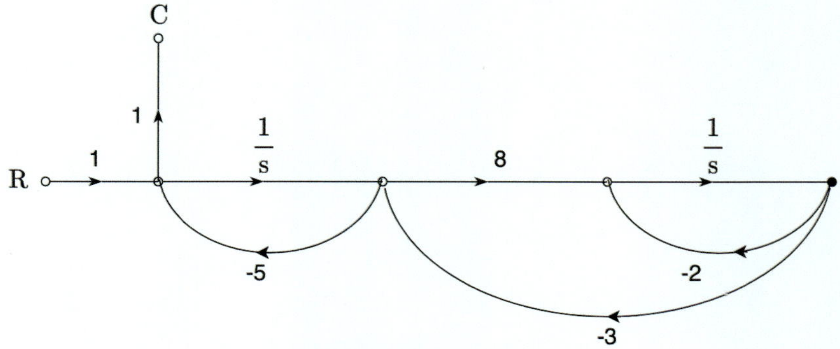

Figure 2.108

The transfer function $\dfrac{C}{R}$ of the system is

A. $\dfrac{8s}{s^2 + 31s + 10}$

B. $\dfrac{8}{s^2 + 31s + 10}$

C. $\dfrac{s(s + 8)}{s^2 + 31s + 10}$

D. $\dfrac{s(s + 26)}{s^2 + 31s + 10}$

Answers

M 2.1. C **M 2.2**. A **M 2.3**. D **M 2.4**. B **M 2.5**. A **M 2.6**. D **M 2.7**. C **M 2.8**. D **M 2.9**. A **M 2.10**. D **M 2.11**. A **M 2.12**. C **M 2.13**. C **M 2.14**. B **M 2.15**. A **M 2.16**. C **M 2.17**. A **M 2.18**. D **M 2.19**. C **M 2.20**. D

2.21 Knowledge Assessment Problems

KaP 2.1 Find the transfer function $G(s) = \dfrac{V_L(s)}{V_i(s)}$, for the network shown in Figure 2.109.

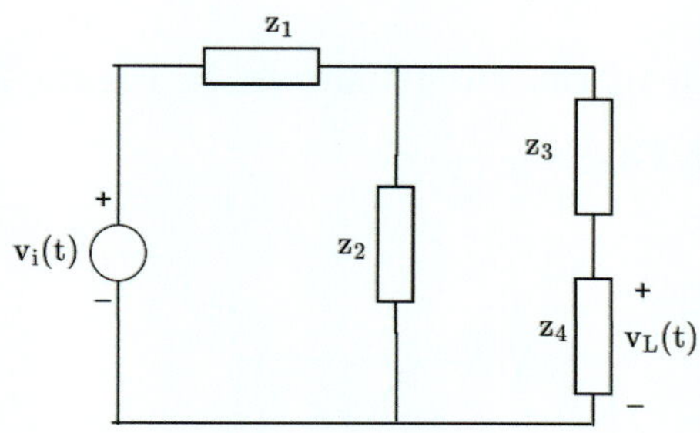

Figure 2.109: Passive R-L-C circuit

KaP 2.2 Consider the circuit shown in Figure 2.110.

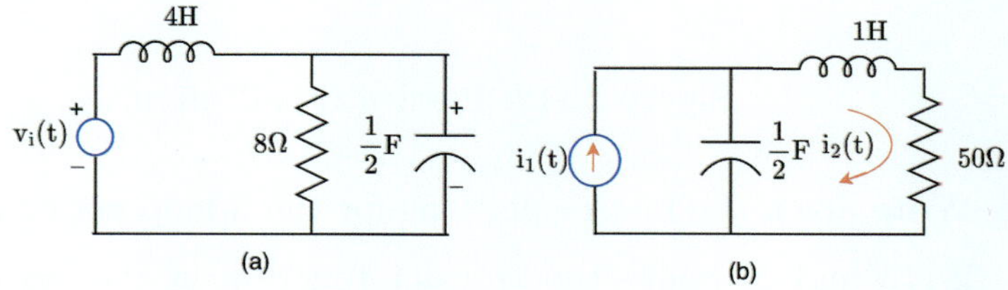

Figure 2.110: Passive R-L-C circuit

a. For the circuit shown in Figure 2.110a, determine the transfer function $G(s) = \dfrac{V_L(s)}{V_i(s)}$.

b. For the circuit shown in Figure 2.110b, compute the transfer function $G(s) = \dfrac{I_2(s)}{I_1(s)}$.

KaP 2.3 Consider the network shown in Figure 2.111.

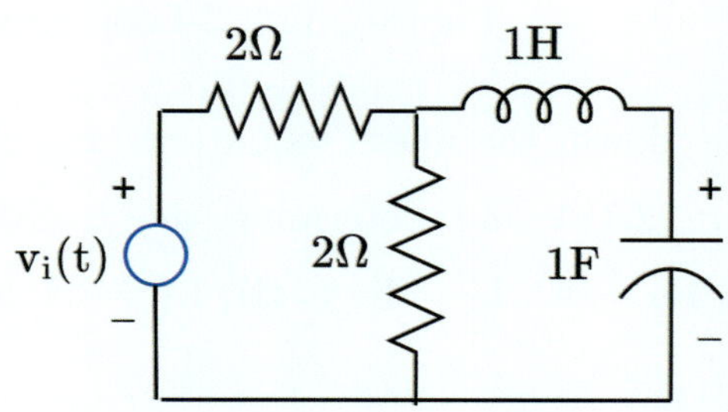

Figure 2.111: Passive R-L-C circuit

Find the transfer function $G(s) = \dfrac{V_L(s)}{V_i(s)}$ when

a. $v_L(t)$ is the voltage across the capacitor.

b. $v_L(t)$ is the voltage across the inductor.

KaP 2.4 For each of the circuit shown in Figure 2.112, determine the transfer function $G(s) = \dfrac{V_L(s)}{V_i(s)}$.

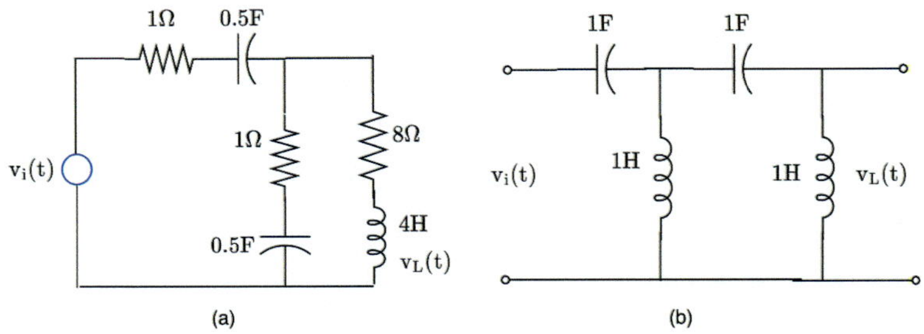

Figure 2.112: Passive R-L-C circuit

KaP 2.5 Write down the mesh equations for the 3-loop network shown in Figure 2.113 and compute the transfer function of the system $G(s) = \dfrac{V_o(s)}{V_i(s)}$. Assume the parameter values are $R_1 = R_2 = 10\Omega$, $L_1 = L_2 = 2H$, and $C = 1F$.

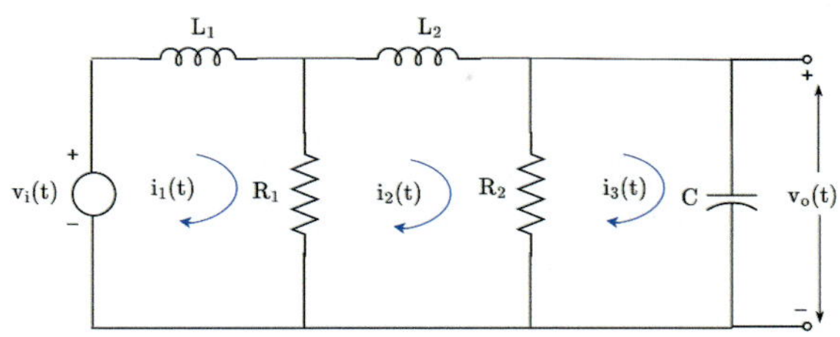

Figure 2.113: 3-Loop network

KaP 2.6 Write down the mesh equations for the 3-loop network shown in Figure 2.114 and determine the transfer function $G(s) = \dfrac{V_o(s)}{V_i(s)}$ for $R_1 = R_2 = 2\Omega$, $L_2 = L_3 = 4H$, $L_4 = 2H$, and $C = 1F$.

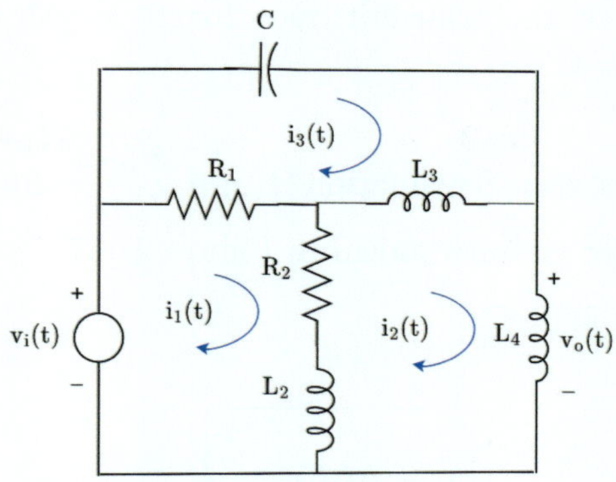

Figure 2.114: 3-Loop network

KaP 2.7 Write down the mesh equations for the 3-loop network shown in Figure 2.115 and determine the transfer function $G(s) = \dfrac{V_o(s)}{V_i(s)}$ for $R_1 = 2\Omega$, $R_2 = R_3 = 1\Omega, R_4 = 5\Omega, L_1 = 4H, L_2 = 3H$, $R_4 = 5$ and $C = 1F$.

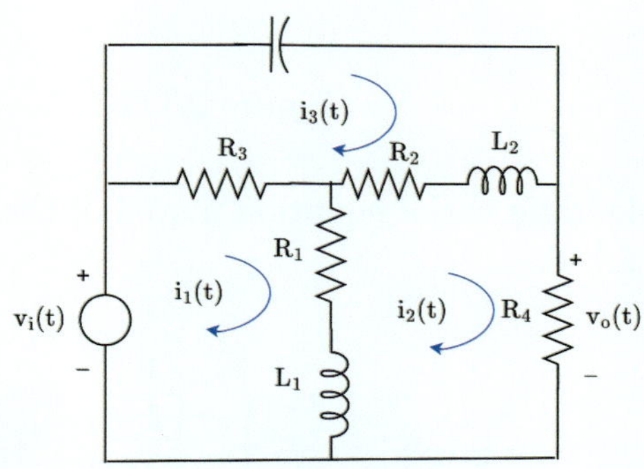

Figure 2.115: 3-Loop network

KaP 2.8 Find the transmittance $\dfrac{I_1(s)}{V_1(s)}$ of the network shown in Figure 2.116.

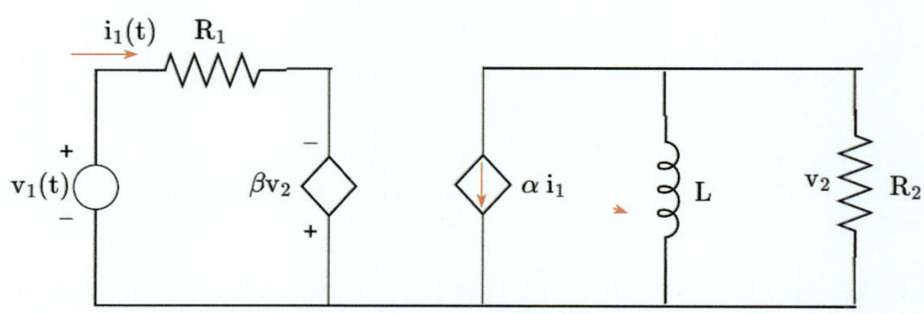

Figure 2.116

Compute the transmittance for $R_1 = 2\Omega$, $R_2 = 1\Omega$, $L = 0.5H$, $\alpha = 10$, and $\beta = 0.1$.

KaP 2.9 Find the transfer function $G(s) = \dfrac{V_o(s)}{V_i(s)}$, for the noninverting operational amplifier circuits shown in Figure 2.117.

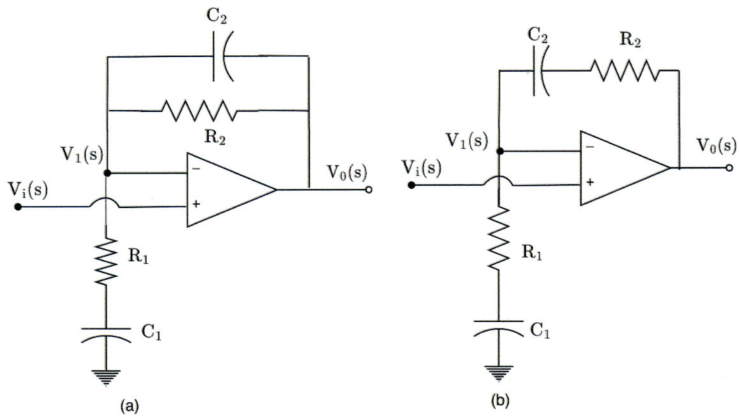

Figure 2.117

KaP 2.10 Write the differential equation model of the rotational systems shown in Figure 2.118.

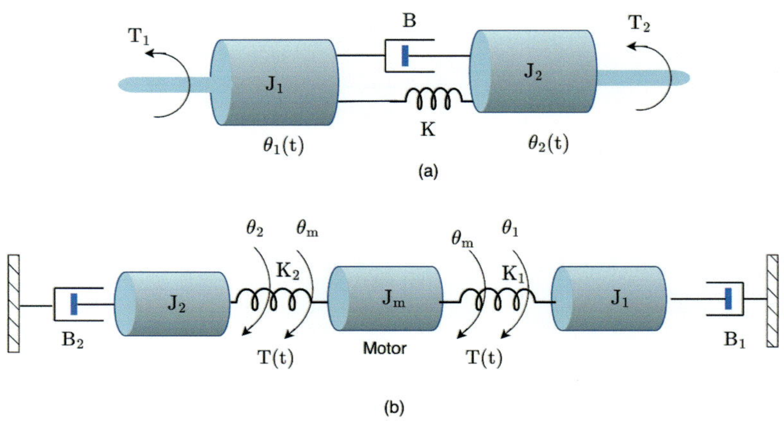

Figure 2.118

KaP 2.11 Write the dynamics of the rotational systems shown in Figure 2.119 both in time and the Laplace domain.

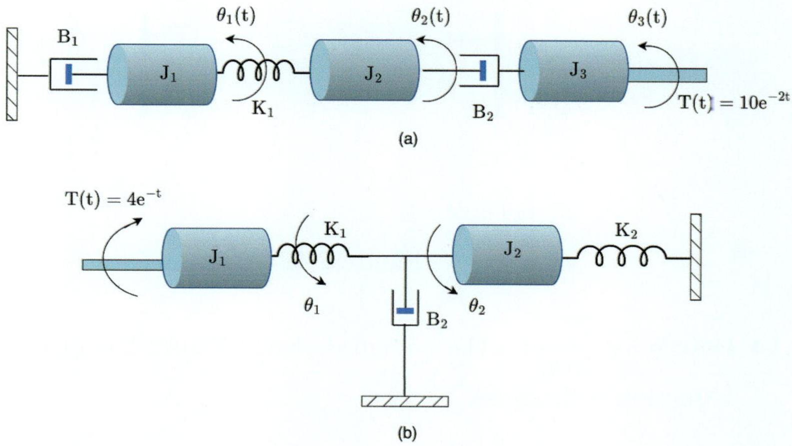

Figure 2.119

KaP 2.12 Write the dynamics of the rotational systems shown in Figure 2.120 both in time and Laplace domain and determine the general expression of transfer function $G(s) = \frac{\theta_2(s)}{T(s)}$. Compute its value for $J_1 = 2\ \mathrm{kg-m^2}$, $J_2 = 100\ \mathrm{kg-m^2}$, $J_3 = 100\ \mathrm{kg-m^2}$. The spring constants $K_1 = 2\ \mathrm{N-m/rad}$ and $K_2 = 200\ \mathrm{N-m/rad}$. The damping is $B_3 = 1000\ \mathrm{N-m-s/rad}$. Assume $N_1 = 2, N_2 = 20$ and $N_3 = 10$.

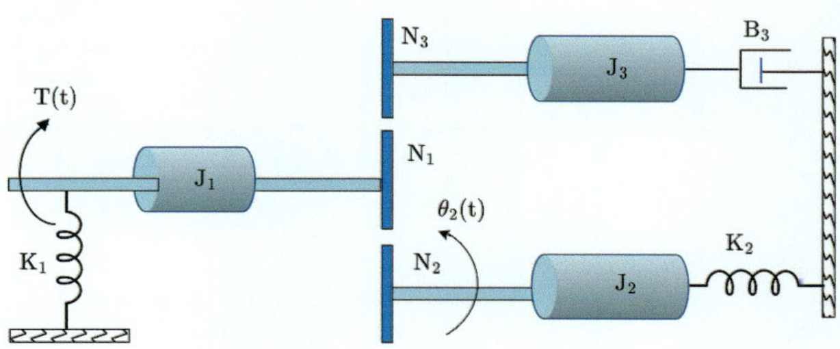

Figure 2.120

KaP 2.13 Find the transfer function $G(s) = \frac{\theta_L(s)}{T(s)}$ for the rotational system shown in Figure 2.121. Assume $J_2 = 10\mathrm{kg-m^2}$, $B_2 = 2\mathrm{N-m-s/rad}$, $B_3 = 0.08\mathrm{N-m-s/rad}$.

The spring constant $K_2 = 1\mathrm{N-m/rad}$. Take $N_1 = 5, N_2 = 50, N_3 = 25$ and $N_4 = 5$.

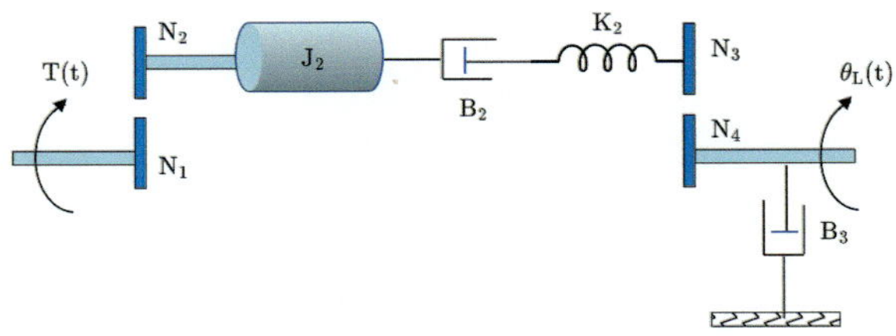

Figure 2.121

KaP 2.14 Determine $\dfrac{C(s)}{R(s)}$ for the system shown in Figure 2.122 using block diagram reduction techniques.

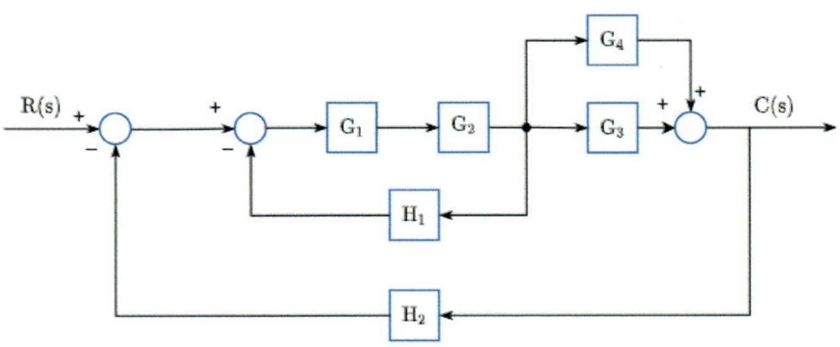

Figure 2.122: Multi loop system

KaP 2.15 Determine the transfer function of the system whose block diagram is shown in Figure 2.123.

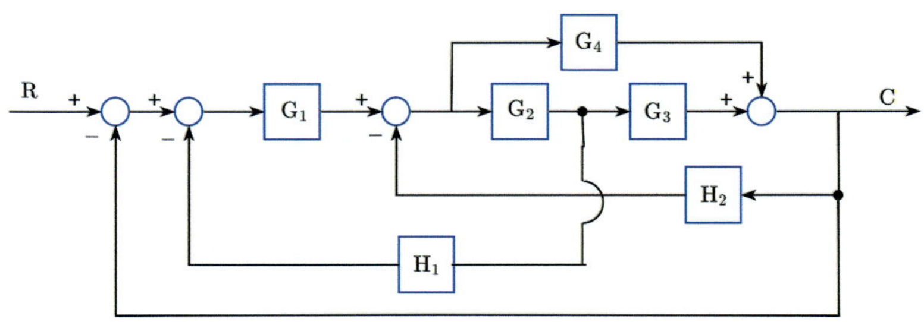

Figure 2.123

KaP 2.16 Determine the transfer function of the system whose block diagram is shown in Figure 2.124.

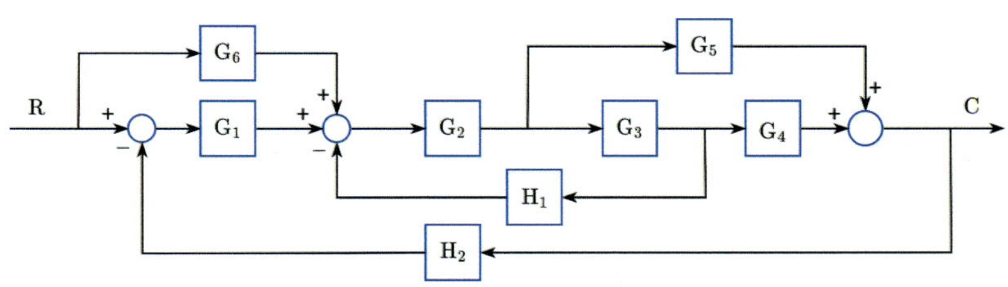

Figure 2.124

KaP 2.17 Using Mason's gain formulae, determine the transfer function of the system whose block diagram is shown in Figure 2.125.

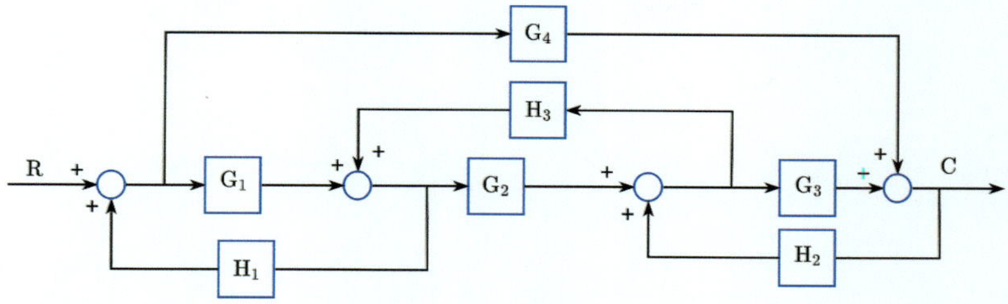

Figure 2.125

KaP 2.18 Using Mason's gain formulae, determine the transfer function of the system whose block diagram is shown in Figure 2.126.

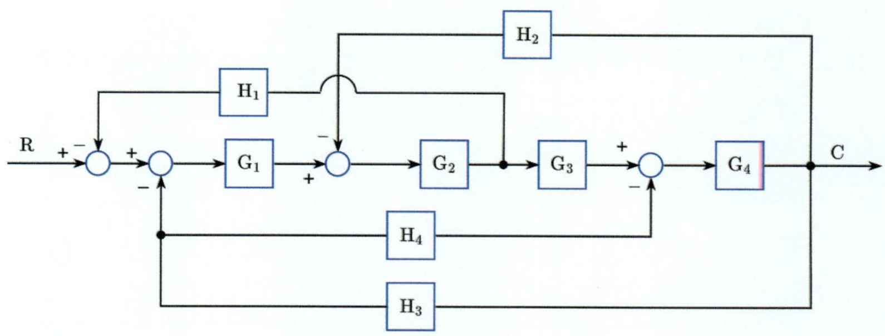

Figure 2.126

KaP 2.19 Determine the transfer function $\frac{C}{R}$ from the signal flow graph shown in Figure 2.127.

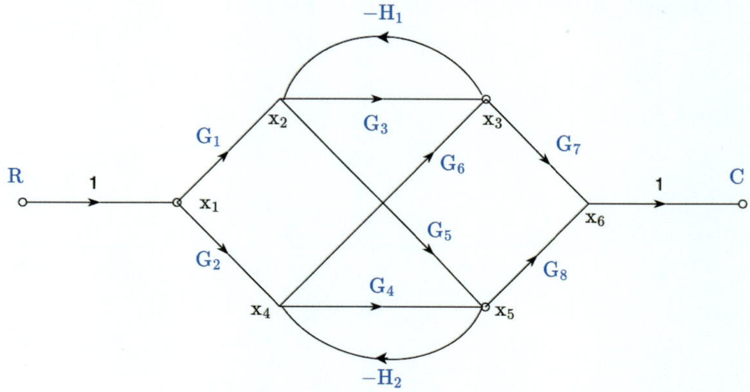

Figure 2.127

KaP 2.20 Determine the transfer function $\frac{C}{R}$ from the signal flow graph shown in Figure 2.128.

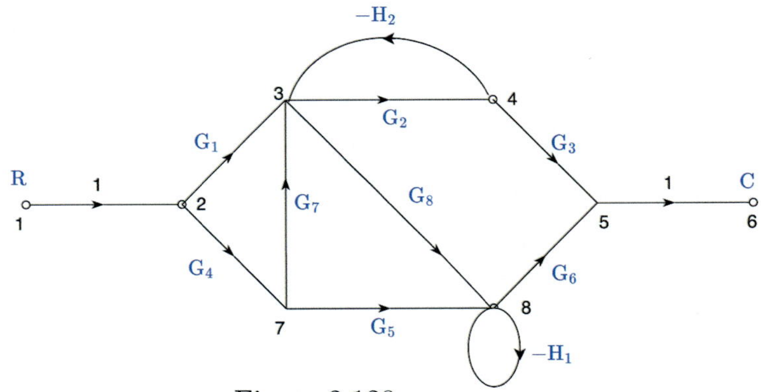

Figure 2.128

KaP 2.21 Determine the transfer function $\frac{C}{R}$ from the signal flow graph shown in Figure 2.129.

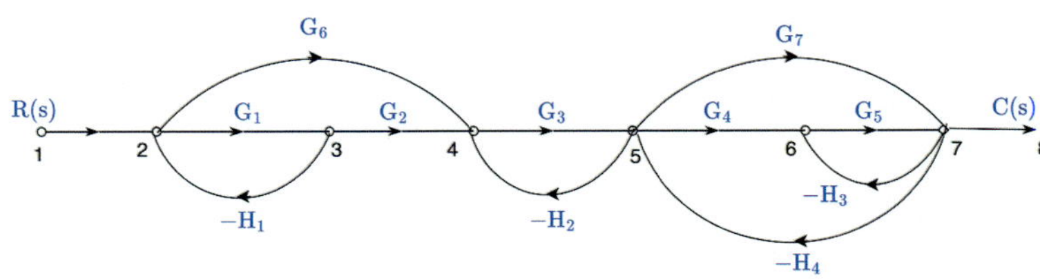

Figure 2.129

KaP 2.22 Determine the transfer function of the system from the signal flow graph shown in Figure 2.130. Also determine $\frac{x_2}{x_1}, \frac{x_4}{x_1}, \frac{x_7}{x_2}, \frac{x_4}{x_2}$, and $\frac{x_7}{x_4}$

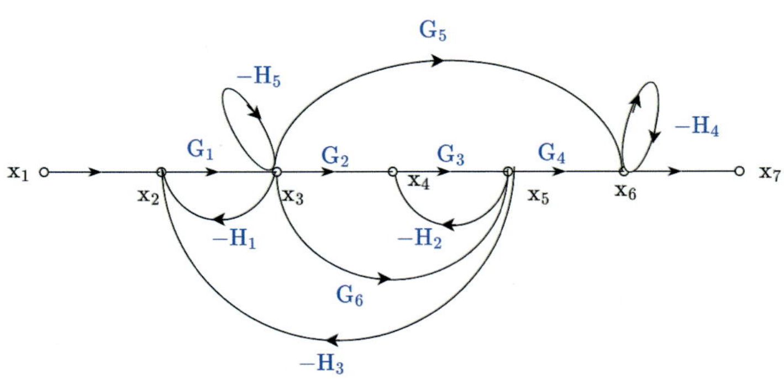

Figure 2.130

KaP 2.23 Determine the gains $\frac{x_6}{x_1}$, $\frac{x_2}{x_1}$, and $\frac{x_6}{x_2}$ from the signal flow graph shown in Figure 2.131.

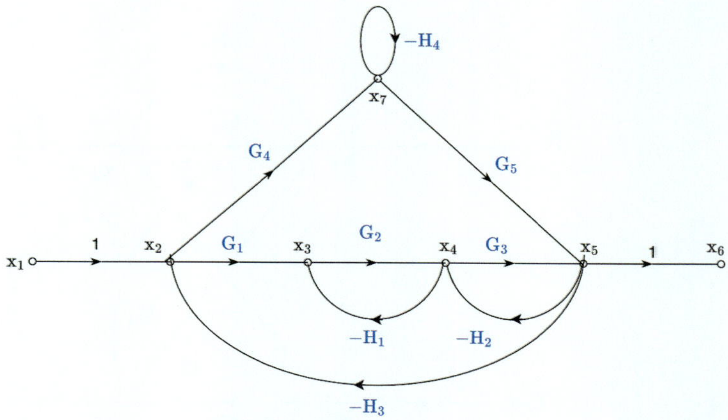

Figure 2.131

KaP 2.24 Determine the following transfer functions from the signal flow graph shown in Figure 2.132.

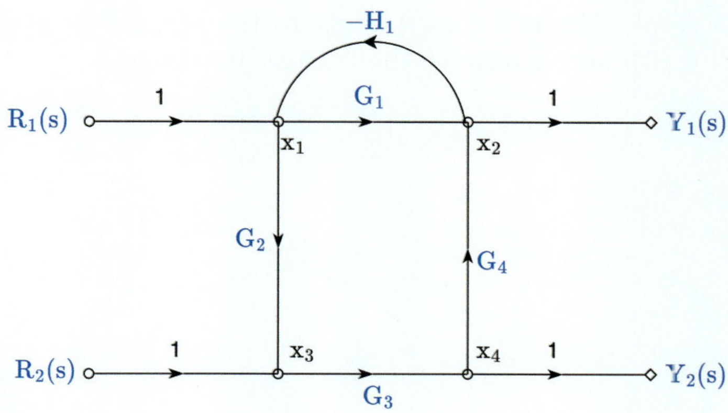

Figure 2.132

a. $T_{11}(s) = \frac{Y_1(s)}{R_1(s)}$

b. $T_{21}(s) = \frac{Y_2(s)}{R_1(s)}$

c. $T_{12}(s) = \frac{Y_1(s)}{R_2(s)}$

d. $T_{22}(s) = \frac{Y_2(s)}{R_2(s)}$

KaP 2.25 Determine the transfer function $\frac{C}{R}$ from the signal flow graph shown in Figure 2.133.

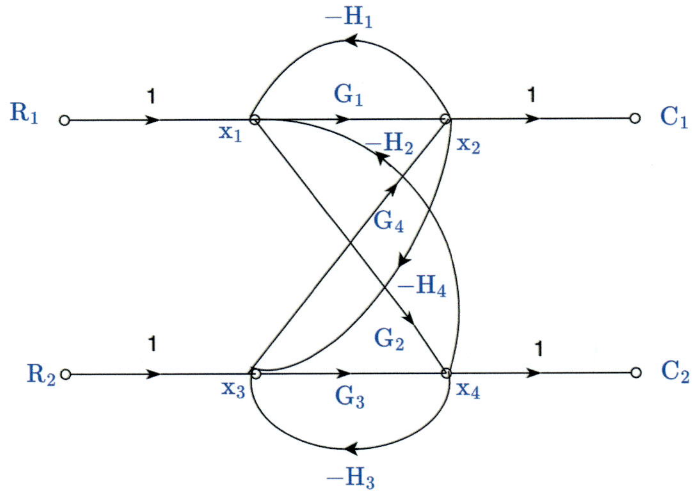

Figure 2.133

KaP 2.26 Apply Mason's gain formula to the signal flow graph shown in Figure 2.134. and determine the transfer functions:

$$\frac{x_5}{x_1}, \frac{x_4}{x_1}, \frac{x_2}{x_1}, \frac{x_5}{x_2}, \text{ and } \frac{x_4}{x_2}$$

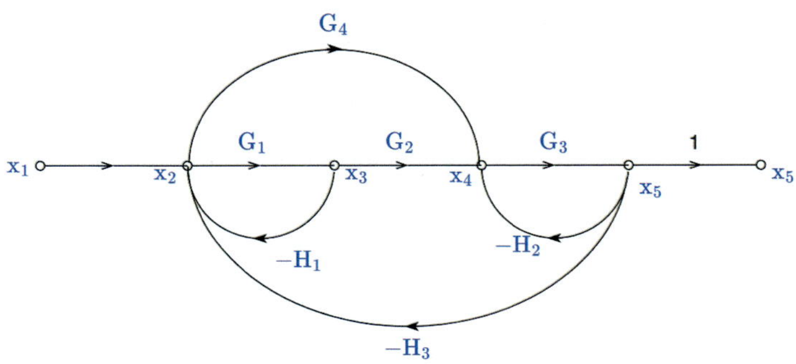

Figure 2.134

KaP 2.27 Apply Mason's gain formula to the signal flow graph shown in Figure 2.135 and determine the transfer functions:

$$\frac{x_7}{x_1} \text{ and } \frac{x_2}{x_1}$$

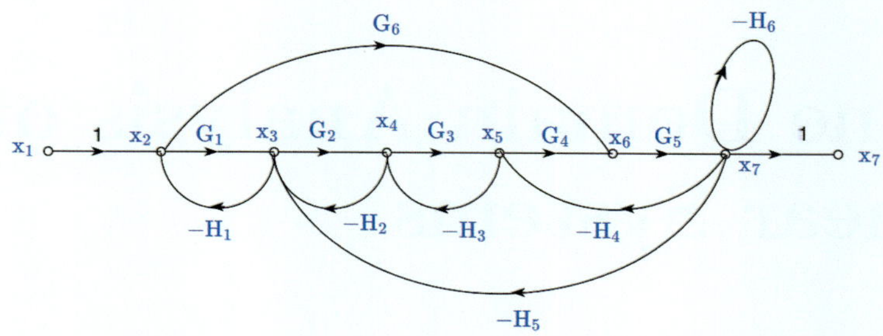

Figure 2.135

Chapter 3

Time Domain Analysis of Linear Systems

Studying systems over time is like dealing with teenagers: you ask them to do something, they eventually respond, but not without some drama first.

DESIRED LEARNING OUTCOMES

Upon completion of this chapter, readers should be able to:

- Explain the philosophy behind selecting test signals for analyzing the transient response of linear systems.
- Characterize both transient and steady-state responses of first-order systems.
- Understand the concept of DC gain in a system.
- Define and interpret time-domain specifications of a system.
- Compute commonly used time-domain specifications for a prototype second-order system.
- Explain how the location of poles affects the nature of the transient response.
- Approximate higher-order systems by lower-order models.

3.1 Introduction

In a feedback control system, it is possible to get the desired transient and steady state behavior by designing a suitable controller. However, before

DOI: 10.1201/9781003571308-3

designing (synthesizing) any controller, we must analyze the performance of the system, i.e., *analysis must precede synthesis* either in the time domain or in the frequency domain. The two domains are complementary to each other; each domain offers unique insights into system dynamics, stability, and performance, and together they provide a comprehensive understanding of control system behavior. In this chapter, we focus on time domain analysis. Details of frequency domain analysis will be covered later in Chapter 7.

Due to the presence of energy storing elements such as inductor, capacitor, mass, spring, etc., the control systems are inherently dynamic. Therefore, the time response of a control system consists of two parts: the *transient response* and the *steady-state response*. The steady-state response is the part of the total response that remains after the transient has died out. It refers to the condition where the behavior of the system becomes time-invariant or settles into a repeating pattern, but this pattern itself can vary depending on the system's dynamics and the nature of the input signal. **Note that, during steady state conditions, the system might exhibit a fluctuating or oscillating response, especially when it is subjected to periodic inputs or disturbances.**

Note

a. Steady-state response doesn't necessarily imply a constant output over time.

b. The steady-state response can vary in a fixed pattern, such as a ramp function or a parabolic function that increases with time.

The first step in time domain analysis of control systems is to excite the system with various types of input signals and study its behavior during transient and steady state conditions. In practice, the designer is provided with a set of desired transient and steady state performance characteristics such as *rise time, peak time, settling time, peak overshoot, steady state error*, etc., and the controllers are designed to achieve these specifications.

In the following section, we discuss typical input signals that are often used in time domain analysis.

3.2 Typical Test Signals for Time Domain Analysis

Selecting appropriate input signals for system analysis is fundamental in control engineering, as the choice of test signals directly influences both system response characteristics and the quality of insights obtained from behavioral studies. In practice, actual input signals are typically unknown *a priori* and exhibit unpredictable variations, presenting a significant design challenge. Control engineers must develop systems that maintain satisfactory performance across diverse input scenarios, making the selection of representative test signals a critical design consideration.

This raises fundamental questions: Which input signals should be employed to comprehensively characterize system behavior, and what criteria should guide the selection of these test signals? The key considerations and criteria for input signal selection are outlined below.

1. **Correlation:** There should be a reasonable correlation between the responses due to a standard test signal and the actual reference (input) signal, which the system will be subjected to most frequently under normal operation.

2. **Frequency content:** The input signals should contain a range of frequencies relevant to the system's dynamics, which would enable analyzing the system's behavior across different frequency bands. For instance, low-frequency signals may be used to analyze the system's steady-state response, while high-frequency signals may reveal details about transient behavior.

3. **Amplitude characteristics:** Test signals should have appropriate amplitude characteristics to test the system across its expected operating range. For example, to assess the linearity/nonlinearity of a system, we should apply signals with varying magnitudes.

4. **Randomness:** Random signals, such as white noise or pseudo-random binary sequences (PRBS), can be used to analyze the system's response to stochastic disturbances or to perform system identification.

5. **Harmonic signals:** Sinusoidal or harmonic signals can be used to analyze frequency response and resonance behavior of the system.

6. **Transient signals:** Impulse or step inputs can be used to study transient response characteristics such as rise time, settling time, and overshoot. Here are some guidelines for selecting transient signals.

    i. If the system is subjected to shock inputs, an impulse function is the best choice. Some common examples of impulse inputs include striking a bell with a hammer, where the bell experiences a very high force over a very short period. The hammer injects a finite amount of energy into the bell, which swings back and forth. Another example is the injection of a drug into the body; dumping a bucket of fluid or a bag of solids into a chemical reactor; the suspension system of a car experiences an impulse-type input when the car suddenly encounters a bump in the road at high speed.

ii. If the system is often subjected to sudden, instantaneous changes in the reference input and thereafter held at a constant level, a step signal is a good choice. Some of the examples include closing of a switch in an electric circuit where a voltage is applied; opening of a valve in the pipe, which causes the flow rate to rapidly change from one value to another.

iii. If our objective is to test the ability of the system to follow a linearly increasing input (e.g., to track a constant velocity target, a tracking control system), then a ramp signal may be appropriate. For example, to track the motion of a satellite that moves across the sky at a constant angular velocity

iv. If our interest is to know how the system responds to acceleration input, then we use a parabolic signal for testing. Consider the example of driving a car. If we are in first gear and raise the accelerator, its speed can be increased to a certain extent, but not fully. This is because the gear system of the car determines the amount of power available from the engine. Hence, if we want to investigate the performance of the acceleration mechanism of the car, a parabolic input is useful.

7. **Practical considerations:** The choice of input signals should also consider practical constraints such as signal generation capabilities, experimental setup limitations, and safety concerns.

Ultimately, the selection of input signals should be tailored to the specific characteristics and requirements of the system under study, aiming to provide comprehensive insights into its behavior across different operating conditions. Following these guidelines, the designer chooses some standard test signals to study the performance of the system.

The standard test input signals used for time domain analysis are **step, ramp, parabolic, and impulse signals;** as many control systems experience similar input signals during normal operation. One added advantage is that these signals are easy to handle mathematically.

Note: In practice, the system should not be tested by a signal which is faster than a parabolic signal. This will be clear in our later discussion, where we show that in order to accurately track a higher-order input, we require a larger number of integrators in the system. This may lead to stability problems, which are not acceptable (zero tolerance on stability).

Step Function Input:

This input changes from one level (usually 0) to another level A in zero time as shown in Figure 3.1(a). This signal can be expressed mathematically as:

$$\begin{aligned} r(t) &= A; \quad t \geq 0 \\ &= 0; \quad t < 0 \end{aligned} \tag{3.1}$$

where A is a real constant. This can alternatively be expressed as:

$$r(t) = Au_s(t) \tag{3.2}$$

where $u_s(t)$ is the unit step function.

Ramp Function Input:

This is a signal which increases linearly with time as shown in Figure 3.1(b). This is expressed mathematically as:

$$\begin{aligned} r(t) &= At; \quad t \geq 0 \\ &= 0; \quad t < 0 \end{aligned} \tag{3.3}$$

where A is a real constant. This can alternatively be expressed as:

$$r(t) = Atu_s(t) \tag{3.4}$$

Parabolic Function Input:

This is a signal which is one order faster than the ramp function as shown in Figure 3.1(c). This is expressed mathematically as:

$$\begin{aligned} r(t) &= \frac{At^2}{2}; \quad t \geq 0 \\ &= 0; \quad t < 0 \end{aligned} \tag{3.5}$$

where A is a real constant. This can alternatively be expressed as

$$r(t) = \frac{At^2}{2}u_s(t) \tag{3.6}$$

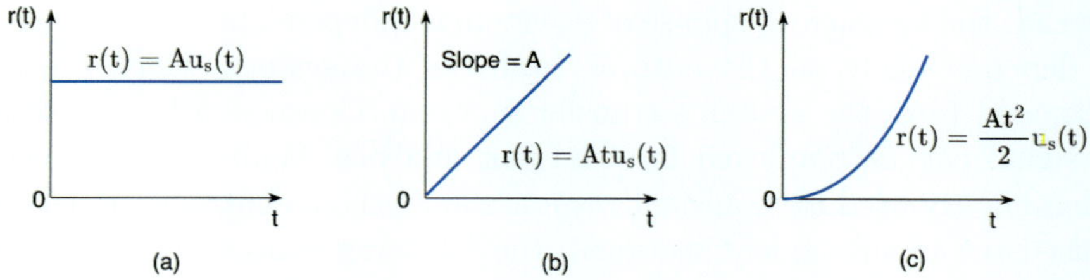

Figure 3.1: Common time-domain test signals for control systems (a) Step function (b) Ramp function (c) Parabolic function

Unit Impulse Function Input:

A unit impulse signal $\delta(t)$ is defined as:

$$\begin{aligned} \delta(t) &= 0; \quad t \neq 0 \\ &= \infty; \quad t = 0 \end{aligned} \tag{3.7}$$

and has the following properties

$$\int_{-\infty}^{\infty} \delta(t)dt = 1 \quad \text{and} \quad \int_{-\infty}^{\infty} \delta(t-\tau)g(t)dt = g(\tau) \tag{3.8}$$

Note that an impulse signal having infinite magnitude and zero duration is not physically realizable. In practice, it is generated by a rectangular function

$$f_a(t) = \begin{cases} \frac{1}{a}, & -\frac{a}{2} \leq t \leq \frac{a}{2}, \\ 0 & \text{otherwise} \end{cases} \tag{3.9}$$

and is shown in Figure 3.2.

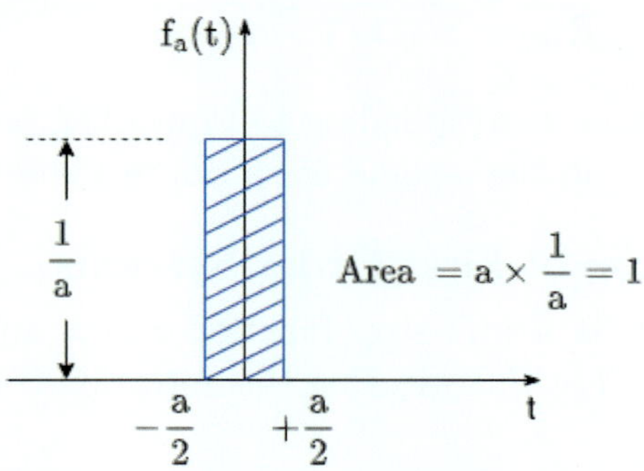

Figure 3.2: Practical impulse function

This is a pulse signal with a very short duration compared to the significant time constant of the system. Thus, as "a" becomes very small (almost zero), 1/a becomes infinitely tall. This means that the width of the above-shown rectangle becomes essentially 0, and the height becomes infinite; yet their area remains equal to 1.

Note that the transient response of a system is independent of the input signal and depends solely on the natural modes of the system, which are readily determined from the system's transfer function. Consequently, any standard test signal can be employed for transient analysis. While the step signal is predominantly used for transient response analysis in control systems (as will be discussed in subsequent sections), the following analysis initially presents the response of a first-order system to step, ramp, and impulse inputs, followed by second-order system analysis, to provide comprehensive insight into system behavior.

3.3 Time Domain Analysis of First Order Systems

Consider a first-order system without zero shown in Figure 3.3

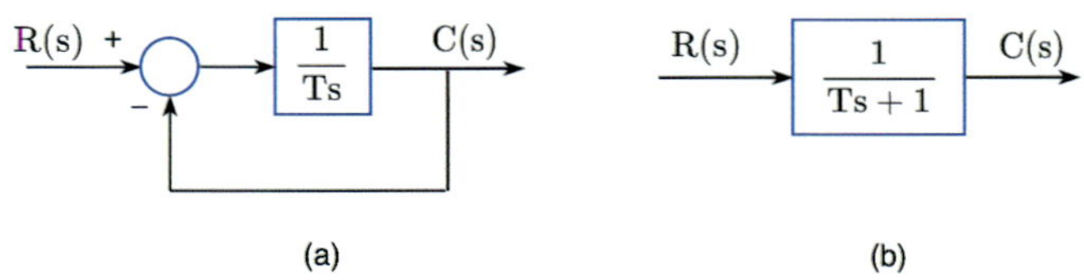

Figure 3.3: (a) Block diagram of a first-order system (b) Simplified block diagram

Although this is a generic first-order system, it may represent an R-L circuit, R-C circuit, a simple thermal system, and so on. The relationship between the input and output is given by

$$\frac{C(s)}{R(s)} = \frac{1}{Ts + 1} \tag{3.10}$$

Eqn(3.10) can alternately be expressed as:

$$\frac{C(s)}{R(s)} = \frac{1/T}{s + 1/T} = \frac{a}{s + a}, \quad \text{where} \quad a = \frac{1}{T} \tag{3.11}$$

The block diagram, corresponding to Eqn(3.11), is shown in Figure 3.4a and the pole location, in the s-plane, is shown in Figure 3.4b.

Unit-Step Response of First Order Systems

When the input is a unit-step function $r(t) = u_s(t)$, its Laplace transform equals to $R(s) = 1/s$. Substituting this into Eqn(3.10) gives

$$C(s) = \frac{1}{Ts + 1}\frac{1}{s} = \frac{1/T}{s(s + 1/T)} \tag{3.12}$$

Let us expand Eqn(3.12) into partial fraction as:

$$C(s) = \frac{1/T}{s(s + 1/T)} = \frac{k_1}{s} + \frac{k_2}{s + 1/T} \quad \text{where} \tag{3.13}$$

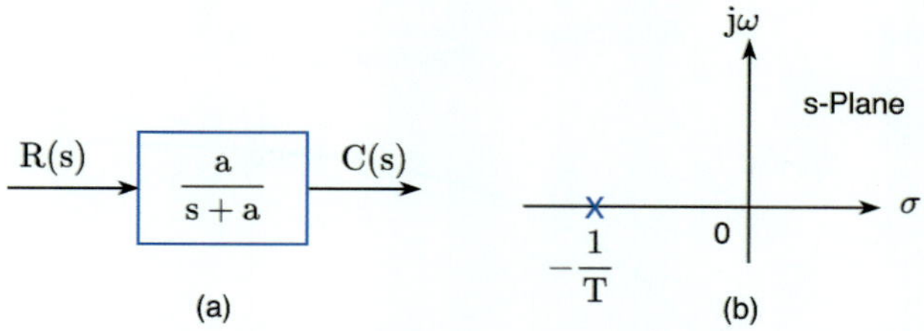

Figure 3.4: (a) Block diagram of first-order system corresponding to Eqn(3.11) (b) Pole plot

$$k_1 = \lim_{s\to 0} s\frac{1/T}{s(s+1/T)} = 1 \quad \text{and } k_2 = \lim_{s\to -1/T}(s+1/T)\frac{1/T}{s(s+1/T)} = -1. \text{ Hence}$$

$$C(s) = \frac{1}{s} - \frac{1}{s+1/T} \tag{3.14}$$

Taking the inverse Laplace transform of Eqn(3.14) gives

$$\boxed{c(t) = c_f(t) + c_n(t) = 1 - e^{-t/T}} \tag{3.15}$$

where the forced response $c_f(t) = 1$ is generated due to pole of the input, which is located at the origin and, the natural response $c_n(t) = e^{-t/T}$ is generated due to the system pole at $-1/T$. From Eqn(3.15), we observe that

$$c(0) = 1 - e^0 = 1 - 1 = 0 \quad \text{and} \quad c(\infty) = 1 - e^{-\infty/T} = 1 - 0 = 1$$

Thus, the response rises exponentially from zero to one as time increases from zero to infinity, as shown in Figure 3.5.

The slope of the response is $\frac{dc(t)}{dt} = \frac{1}{T}e^{-t/T}$. Hence, the initial slope $\left.\frac{dc(t)}{dt}\right|_{t=0} = \frac{1}{T}$ and the final slope $\left.\frac{dc(t)}{dt}\right|_{t=\infty} = 0$. Thus, the slope of the response decreases monotonically from $1/T$ at $t = 0$ to zero at $t = \infty$. From Eqn(3.15), it is seen that the transient response is dependent only on the parameter T. When $t = T$ we have

$$\left.e^{-t/T}\right|_{t=T} = e^{-1} = 0.37 \quad \text{and } \left.c(t)\right|_{t=T} = 1 - \left.e^{-t/T}\right|_{t=T} = 1 - e^{-1} = 0.63 \tag{3.16}$$

In the following, we define three transient response specifications which can be used to judge the performance of the system in the time domain.

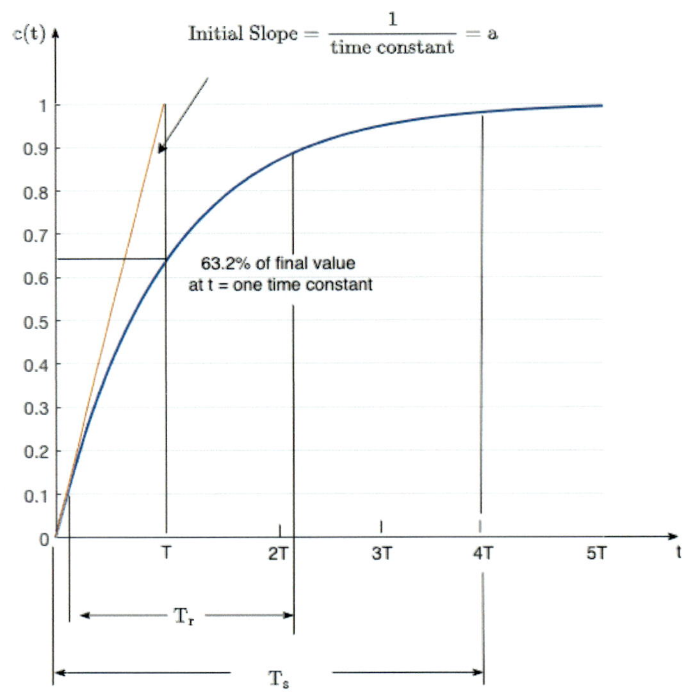

Figure 3.5: First order system response to a unit step input

Time Constant

Definition 3.1. This is defined as the time required for the step response to rise from 0 to 63% of its final value. This is equal to T (see Eqn(3.15)) and indicates how fast the system tends to reach the final value.

The smaller the time constant T, the faster is the system response; that is, the closer a pole is to the imaginary axis (jω-axis), the slower is the system's response, and the farther it is, the faster the response. As an example, consider three systems S_1, S_2 and S_3 with different pole locations depicted in Figure 3.6.

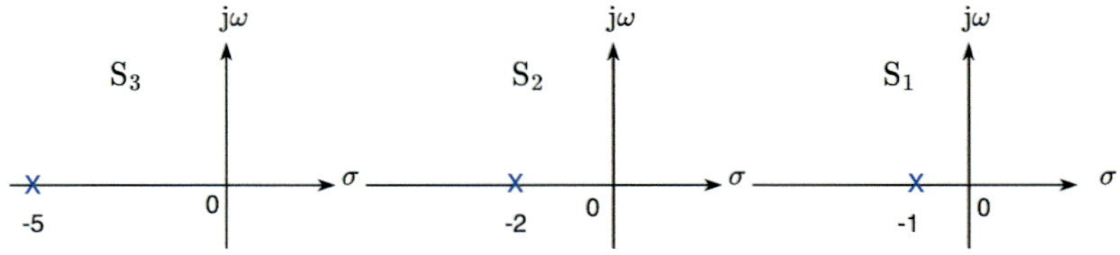

Figure 3.6: Pole location of 3 first order systems

The pole of system S_1 is closest to the jω-axis. This indicates that it has the slowest response among the three systems. The proximity of the pole to the imaginary axis implies that the system's dynamics are dominated by low-frequency behavior, leading to a sluggish response to input changes. The pole of system S_2 is located at an intermediate distance from the $j\omega$-axis. This suggests that system S_2 has a moderate response speed compared to S_1 and S_3. It will exhibit faster dynamics than S_1, but slower than S_3. The pole of system S_3 is farthest from the $j\omega$-axis. As a result, it exhibits the fastest response among the three systems. The distance of the pole from the imaginary axis indicates that the system's dynamics are dominated by high-frequency behavior, allowing for rapid response to input changes.

Rise Time T_r

Definition 3.2. The *rise time* is defined as the time it takes for the response to go from 10% to 90% of its final value. This can be computed from Eqn(3.15) by determining the difference in time at $c(t) = 0.9$ and $c(t) = 0.1$ and is given by

$$\boxed{T_r = 2.31T - 0.11T = 2.2T} \tag{3.17}$$

Settling Time T_s

Definition 3.3. The *settling time* is defined as the time it takes for the response to reach and stay within 2% of its final value. This can be computed from Eqn(3.15) by determining the time when $c(t) = 0.98$. This gives

$$\boxed{T_s = 4T} \tag{3.18}$$

Unit-Ramp Response of First Order Systems.

When the input is a unit-ramp function $r(t) = tu_s(t)$, its Laplace transform equals to $R(s) = 1/s^2$. Substituting this into Eqn(3.10) and resolving it into partial fraction gives

$$C(s) = \frac{1}{s^2(Ts+1)} = \frac{1}{s^2} - \frac{T}{s} + \frac{T^2}{Ts+1} \tag{3.19}$$

Taking the inverse Laplace transform of Eqn(3.19) gives

$$\boxed{c(t) = t - T + Te^{-t/T}, \quad \text{for} \quad t \geq 0} \tag{3.20}$$

The error signal is given by

$$e(t) = r(t) - c(t) = T\left(1 - e^{-t/T}\right) \tag{3.21}$$

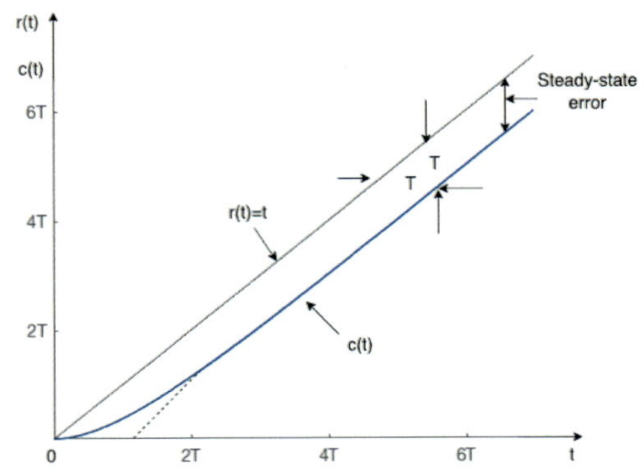

Figure 3.7: First order system response to a unit ramp input

The initial error $e(0) = 0$ and the final error $e(\infty) = T$. The response to the unit ramp input is shown in Figure 3.7.

From Figure 3.7, we observe that T affects both the transient behavior as well as the steady state error. The steady state error decreases with a decrease in the time constant T.

Unit-Impulse Response of First Order Systems.

When the input is a unit-impulse function $r(t) = \delta(t)$, its Laplace transform equals to $R(s) = 1$. Substituting this into Eqn(3.10) gives

$$C(s) = \frac{1}{Ts+1} \implies c(t) = \frac{1}{T}e^{-t/T}, \quad \text{for } t \geq 0 \tag{3.22}$$

Thus, $c(t)\Big|_{t=0} = \frac{1}{T}$ and $c(t)\Big|_{t=\infty} = 0$. The response to the unit impulse input is shown in Figure 3.8.

From the foregoing analysis, we observe the following for a first-order linear time-invariant system

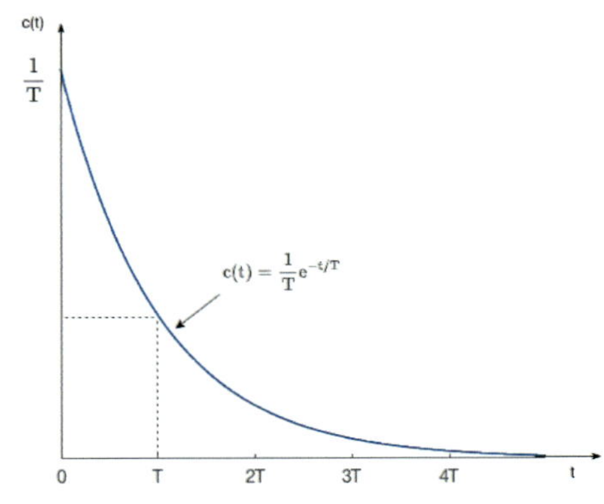

Figure 3.8: First order system response to a unit impulse input

a. The response c(t) to a unit ramp input is given by

$$c(t) = t - T + Te^{-t/T}, \quad \text{for } t \geq 0 \qquad \text{[see Eqn(3.20)]}$$

b. The response c(t) to a unit-step input (which is the derivative of the unit-ramp input) is given by

$$c(t) = 1 - e^{-t/T}, \quad \text{for } t \geq 0 \qquad \text{[see Eqn(3.15)]}$$

c. The response c(t) to a unit-impulse input (which is the derivative of the unit-step input) is given by

$$c(t) = \frac{1}{T}e^{-t/T}, \quad \text{for } t \geq 0 \qquad \text{[see Eqn(3.22)]}$$

Thus, for linear time-invariant systems

1. The response to the derivative of an input can be obtained by differentiating the response of the system to the original input.
2. The response to the integral of an input signal can be obtained by integrating the response of the system to the original input and by determining the integration constants from the initial condition.

The following example illustrates the calculation of time domain specifications of first-order systems.

EXAMPLE 3.1

Problem

Determine the output step response, time constant, rise time, and settling time for the first-order system with transfer function

$$G(s) = \frac{C(s)}{R(s)} = \frac{K}{s + a}$$

SOLUTION:

Now $G(s) = \dfrac{K}{s + a}$ and $R(s) = \dfrac{1}{s}$ (since the input is a step function). Hence

$$C(s) = G(s)R(s) = \frac{K}{s(s + a)} = \frac{K}{a}\left[\frac{1}{s} - \frac{1}{s + a}\right]$$

$$\Longrightarrow \quad c(t) = \underbrace{\frac{K}{a}}_{\substack{\text{Forced} \\ \text{Response}}} - \underbrace{\frac{K}{a}e^{-at}}_{\substack{\text{Natural} \\ \text{Response}}}$$

The time constant $T = \frac{1}{a}$sec, rise time $T_r = 2.2 \times T = \frac{2.2}{a}$ sec and settling time $T_s = 4 \times T = \frac{4}{a}$ sec.

CONCEPT REINFORCEMENT PROBLEM 3.1

Problem

Determine the output step response, time constant, rise time, and settling time for each of the first-order systems whose transfer functions $G(s) = \frac{C(s)}{R(s)}$ are given below.

a. $G(s) = \frac{10}{s+2}$ b. $G(s) = \frac{10}{s+5}$ c. $G(s) = \frac{40}{s+10}$ d. $G(s) = \frac{40}{s+20}$

3.3.1 DC Gain of a System

This is the gain of the system when the input frequency is zero. Therefore, it is also known as the zero frequency gain of the system. It is the ratio of the magnitude of the steady-state output to that of the constant input, provided the system output is finite. This can be computed as follows:

Consider the system with the transfer function

$$G(s) = \frac{C(s)}{R(s)} \tag{3.23}$$

If we apply a unit step input $R(s) = 1/s$, the steady state value of the output becomes

$$\lim_{t\to\infty} c(t) = \lim_{s\to 0} sG(s)R(s) = \lim_{s\to 0} sG(s)\frac{1}{s} = \lim_{s\to 0} G(s) \tag{3.24}$$

Since the input is a constant of magnitude unity, the steady state output is also the gain in the steady state; that is $G(0)$. Thus

$$\boxed{\text{DC gain} = \lim_{s\to 0} G(s) = G(0)} \tag{3.25}$$

When computing the DC gain of a system, it's essential to ensure that the system response has settled into steady-state. Further, we must note that to compute the DC gain, it is not necessary to apply an input that remains constant throughout. Because the settling time of a system is typically considered to be around four to five times the system's time constant. Therefore, to accurately compute the DC gain, the input signal $r(t)$ should remain constant for a duration longer than four to five times the system's time constant. This ensures that the system has reached its steady-state response, and any transient effects have sufficiently decayed. This is demonstrated in Figure 3.9.

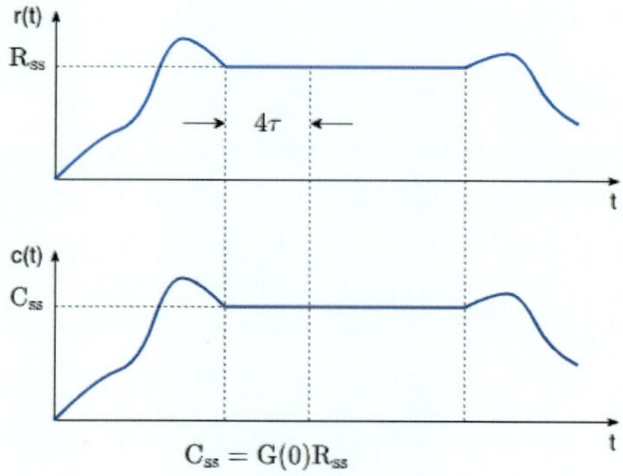

Figure 3.9: Input and output of a system for computing DC gain

Although the input, shown in Figure 3.9, changes over time, it remains constant at a value R_{ss} for a period longer than the five time constants of the system. The output settles at a value C_{ss}. The DC gain is then calculated from

$$\text{DC gain} = \frac{C_{ss}}{R_{ss}} \tag{3.26}$$

EXAMPLE 3.2

Problem: Find the DC gain of a system whose transfer function is

$$G(s) = \frac{10(s+2)}{s^2+4s+5}$$

SOLUTION: Now

$$\text{DC gain} = \lim_{s \to 0} G(s) = \lim_{s \to 0} \frac{10(s+2)}{s^2+4s+5} = \frac{10 \times 2}{5} = 4$$

3.4 Time Domain Analysis of Second Order Systems

We will now extend the method of time domain analysis of first-order systems to second-order systems. Although the order of most practical systems is more than two, there are several compelling reasons why we restrict our analysis to second-order systems. Some of them are:

1. This is the lowest-order system that is capable of exhibiting an oscillatory response.

    - Note that for oscillations to occur, the system must have at least two independent energy storage elements, such as the inductor and the capacitor in RLC circuits, spring and mass in a mass-spring-damper system, and so on.

2. The second-order system can exhibit a wide range of behavior depending on the values of the poles (the roots of its characteristic equation), which can be in one of the four configurations, such as

   i. Both the poles are real and distinct and lie on the negative real axis of the complex s-plane; the system is *overdamped.*

   ii. The poles are complex conjugate with a negative real part; the system is *underdamped.*

   iii. The poles are purely imaginary with zero real part; the system is *undamped* and

   iv. There are double poles on the negative real axis; the system is *critically damped.*

3. Most of the higher-order systems can be approximated by second-order systems within practically acceptable levels of accuracy.

3.4.1 Response Classification of General Second Order Systems

Before we derive the time response of a prototype second-order system, let us first analyze how the impulse response of a general second-order system will change with variations of the parameters "K" and "a" of the general model shown in Figure 3.10. Note that the parameters "K" and "a" can easily be related to physical parameters of second-order systems such as RLC circuits, mass-spring-damper systems, rotational systems, and so on (see Sections 2.3 and 2.8 of Chapter 2).

The poles of this system are the roots of the characteristic equation $s^2 + as + K = 0$ and are given by

$$\lambda_1 = \frac{-a}{2} + \frac{\sqrt{a^2 - 4K}}{2}, \quad \text{and} \quad \lambda_2 = \frac{-a}{2} - \frac{\sqrt{a^2 - 4K}}{2}$$

The nature of these roots (poles) depends on the parameters "K" and "a" and significantly influences their transient response characteristics, such as *overdamped, underdamped, critically damped* and *undamped.*

Understanding damping is like understanding human maturity. Some of us bounce through life endlessly, some overreact and wobble, some respond per-

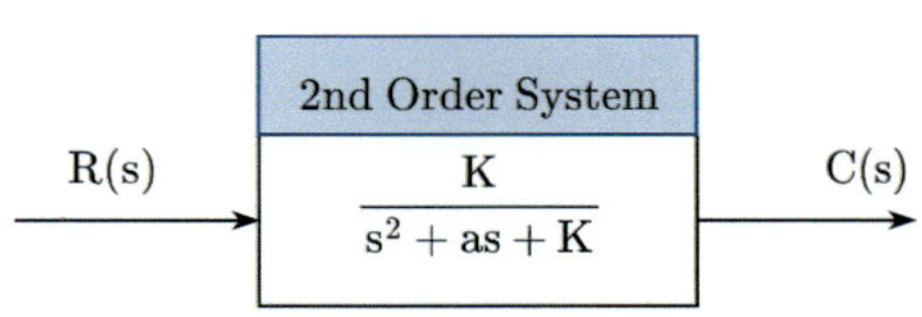

Figure 3.10: General second-order linear system

fectly (rare!), and some take forever to do anything. To build intuition for these different response types, consider this analogy:

Understanding Damping

The Maturity Analogy for Damping

- **Undamped ($\xi = 0$):** The toddler on a sugar rush who bounces off the walls forever, never settles down, pure chaos with no end in sight.
- **Underdamped ($0 < \xi < 1$):** The dramatic teenager who overreacts to everything, swings between extremes, but eventually calms down after making a scene.
- **Critically damped ($\xi = 1$):** The wise adult handles challenges smoothly, responds perfectly with no overshoot and no wasted energy.
- **Overdamped ($\xi > 1$):** The overly cautious elder who takes forever to make any decision, avoids all risks, eventually gets there, but very slowly.

Let us examine how these parameters affect the transient behavior through impulse response analysis.

1. **Overdamped responses:** *Roots are real and distinct;* $\lambda_1 = -\sigma_1$ *and* $\lambda_2 = -\sigma_2$

   The impulse response consists of two exponential terms with time constants equal to the reciprocal of the pole locations and is given by

   $$\boxed{\mathrm{c(t) = A_1 e^{-\sigma_1 t} + A_2 e^{-\sigma_2 t}}} \quad (3.27)$$

2. **Underdamped responses:** *Roots are complex conjugate with real part;* $\lambda_1, \lambda_2 = -\sigma_\mathrm{d} \pm \mathrm{j}\omega_\mathrm{d}$

   The impulse response consists of a damped sinusoid with an exponential envelope whose time constant is equal to the reciprocal of the real part of the pole (σ_d) and is expressed as:

   $$\boxed{\mathrm{c(t) = A e^{-\sigma_d t} \cos(\omega_d t - \phi)}} \quad (3.28)$$

3. **Critically damped responses:** *Roots are real and equal;* $\lambda_1, \lambda_2 = -\sigma_1$

The response due to an impulse input consists of two exponential terms with time constants equal to the reciprocal of the pole location, with one multiplied by time, t, and is given by

$$c(t) = A_1 e^{-\sigma_1 t} + A_2 t e^{-\sigma_1 t}$$

4. **Undamped responses:** *Roots are purely imaginary* $\lambda_1, \lambda_2 = \pm j\omega_1$

The impulse response is an undamped sinusoid with radian frequency equal to the imaginary part of the pole, and is

$$c(t) = A\cos(\omega_1 t - \phi) \tag{3.29}$$

CONCEPT REINFORCEMENT PROBLEM 3.2

Problem

Write the general form of the step response for each of the systems whose closed-loop poles and zeros are shown in Figure 3.11.

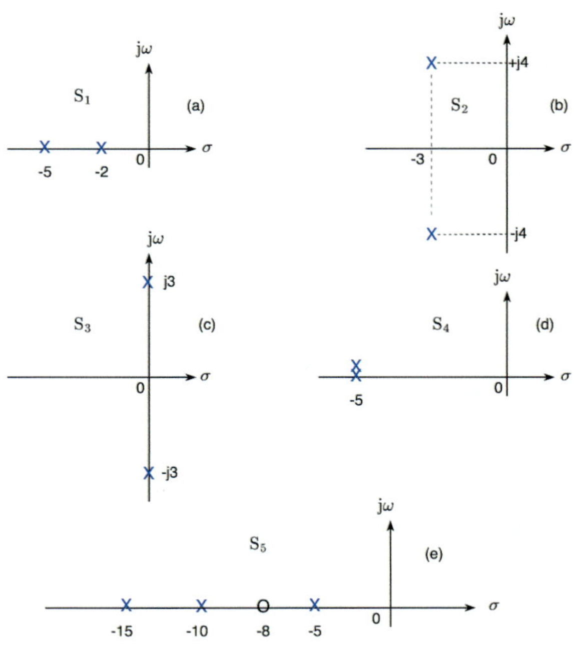

Figure 3.11: Pole-zero plots of closed-loop system

3.5 Time Domain Analysis of Prototype Second Order Linear Systems

In the previous section, we studied the time response of different second-order systems to impulse inputs. It should be noted that if the impulse response of a linear system is known, the response to any arbitrary input can be determined using convolution. However, in practice, step response analysis is more commonly employed because step inputs are easier to implement experimentally and provide direct insight into key performance metrics. Furthermore, the impulse response can be obtained by differentiating the step response, establishing a direct mathematical relationship between these two fundamental system characterizations. Therefore, analyzing the step response of a standard second-order system offers fundamental insights into system stability, transient behavior, and damping characteristics, which are crucial for understanding overall system performance. Thus, in the following analysis, we will examine the transient behavior of a second-order system subjected to a step input.

Consider a prototype second order system shown in Figure 3.12.

The closed-loop transfer function of this system is given by

$$\mathrm{T(s)} = \frac{\mathrm{C(s)}}{\mathrm{R(s)}} = \frac{\omega_\mathrm{n}^2}{\mathrm{s}^2 + 2\xi\omega_\mathrm{n}\mathrm{s} + \omega_\mathrm{n}^2} = \frac{\mathrm{N(s)}}{\mathrm{D(s)}} \tag{3.30}$$

where ω_n is the *natural frequency*, ξ is the *damping ratio*. In many cases, the closed-loop transfer function of a second-order system is expressed as:

$$\mathrm{T(s)} = \frac{\mathrm{C(s)}}{\mathrm{R(s)}} = \frac{\mathrm{K}\omega_\mathrm{n}^2}{\mathrm{s}^2 + 2\xi\omega_\mathrm{n}\mathrm{s} + \omega_\mathrm{n}^2} = \frac{\mathrm{N(s)}}{\mathrm{D(s)}} \tag{3.31}$$

where "K" is the system gain. Note that although the term ω_n^2 appears in the numerator of both Eqns(3.30) and (3.31), the **natural frequency** is determined **exclusively from the denominator** as it represents the inherent dynamics of the system (poles of the transfer function). The numerator $\mathrm{K}\omega_\mathrm{n}^2$

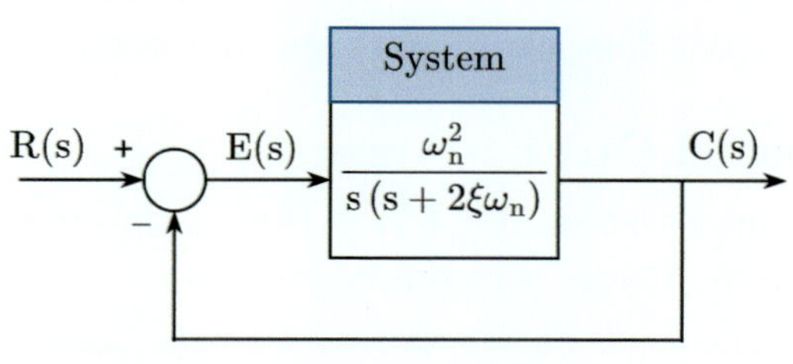

Figure 3.12: Standard (prototype) second-order system

Table 3.1: Response classification according to damping

Damping Ratio	Poles	Nature of Response
$\xi = 0$	Purely imaginary	Undamped
$0 < \xi < 1$	Complex conjugate	Underdamped
$\xi = 1$	Repeated real poles	Critically damped
$\xi > 1$	Two distinct real poles	Overdamped

does not influence the natural frequency; it affects only the system's gain or amplitude response. Without loss of generality, we assume $\mathrm{K} = 1$ when deriving transient response specifications.

The time response of any system is characterized by the poles of the system, which can be determined from the roots of the denominator polynomial D(s) of its transfer function model. The denominator polynomial D(s) is often called the *characteristic equation.* The characteristic equation of the general second-order system (shown in Figure 3.12) is

$$\boxed{\mathrm{s}^2 + 2\xi\omega_\mathrm{n}\mathrm{s} + \omega_\mathrm{n}^2 = 0} \tag{3.32}$$

The roots λ_1 and λ_2 of this characteristic equation are given by

$$\mathrm{s}^2 + 2\xi\omega_\mathrm{n}\mathrm{s} + \omega_\mathrm{n}^2 = (\mathrm{s} - \lambda_1)(\mathrm{s} - \lambda_2) \tag{3.33}$$

This system will exhibit *underdamped, overdamped, critically damped* and *undamped* behavior depending on the values of its roots, which alternately depend on the value of the damping ratio ξ. For example, if $0 < \xi < 1$, the system is *underdamped*; if $\xi > 1$, the system is *overdamped.* For $\xi = 1$ and $\xi = 0$, the system exhibits a critically damped response and an undamped response, respectively. These are summarized in Table 3.1.

In the following we will compute the step response of a second order system considering four different cases: the *underdamped* $(0 < \xi < 1)$, the *undamped* $(\xi = 0)$, the *critically damped* $(\xi = 1)$ and the *overdamped* $(\xi > 1)$.

Underdamped Second Order Systems

For an underdamped system, $0 < \xi < 1$. Hence, the characteristic roots (poles) become complex conjugate, which are given by

$$\boxed{\lambda_1, \lambda_2 = -\xi\omega_\mathrm{n} \pm \mathrm{j}\omega_\mathrm{n}\sqrt{1 - \xi^2} = -\sigma \pm \mathrm{j}\omega_\mathrm{d}} \tag{3.34}$$

where $\omega_d = \omega_n\sqrt{1-\xi^2}$, is called the *damped natural frequency*, $\sigma = \xi\omega_n$ is the *damping factor*. The closed-loop system is given by

$$\frac{C(s)}{R(s)} = \frac{\omega_n^2}{(s+\xi\omega_n+j\omega_d)\,(s+\xi\omega_n-j\omega_d)} \tag{3.35}$$

For a unit step input,

$$C(s) = \frac{\omega_n^2}{s(s^2+2\xi\omega_n s+\omega_n^2)} = \frac{1}{s} - \left[\frac{s+\xi\omega_n}{(s+\xi\omega_n)^2+\omega_d^2} + \frac{\xi}{\sqrt{1-\xi^2}}\frac{\omega_d}{(s+\xi\omega_n)^2+\omega_d^2}\right] \tag{3.36}$$

Hence, the inverse transform of Equation(3.36) gives

$$c(t) = 1 - \frac{e^{-\xi\omega_n t}}{\sqrt{1-\xi^2}}\left[\sqrt{1-\xi^2}\cos\omega_d t + \xi\sin\omega_d t\right] \tag{3.37}$$

Consider the trigonometric identity

$$a\sin x + b\cos x = A\sin(x+\phi), \quad \text{where} \quad A = \sqrt{a^2+b^2}, \quad \text{and} \quad \phi = \tan^{-1}(b/a)$$

Using the above identity, Eqn(3.37) is expressed as:

$$c(t) = 1 - \frac{e^{-\xi\omega_n t}}{\sqrt{1-\xi^2}}\sin(\omega_d t+\phi), \text{ for } t \geq 0. \text{ where } \phi = \tan^{-1}\left[\frac{\sqrt{1-\xi^2}}{\xi}\right] \tag{3.38}$$

The error of the system is given by

$$e(t) = r(t) - c(t) = \frac{e^{-\xi\omega_n t}}{\sqrt{1-\xi^2}}\sin(\omega_d t+\phi) \tag{3.39}$$

The error signal exhibits a damped sinusoidal oscillation whose steady state value is equal to zero, and therefore the steady state value of the output is one. That is

$$\lim_{t\to\infty} e(t) = 0 \quad \Longrightarrow \quad c_{ss} = \lim_{t\to\infty} c(t) = 1.0$$

Undamped Second Order Systems

For this case, the damping ratio $\xi = 0$ and the output oscillates indefinitely. This response can be obtained from Equation (3.38) by substituting $\xi = 0$ which gives

$$c(t) = 1 - \sin(\omega_n t + 90^0) = 1 - \cos\omega_n t, \quad \text{for } t \geq 0 \tag{3.40}$$

Critically Damped Second Order Systems

For this case, the damping ratio $\xi = 1$ and the two poles of the system are equal and lie on the real axis of the complex s-plane. For a unit step input, the output becomes

$$C(s) = \frac{\omega_n^2}{s(s+\omega_n)^2} \tag{3.41}$$

The inverse Laplace transform of Equation (3.41) gives

$$\boxed{c(t) = 1 - e^{-\omega_n t}\left(1+\omega_n t\right), \quad \text{for } t \geq 0} \tag{3.42}$$

Overdamped Second Order Systems

For this case, the damping ratio $\xi > 1$ and the two poles of the system are real and distinct. For a unit step input, the output becomes

$$C(s) = \frac{\omega_n^2}{s\left(s+\xi\omega_n+\omega_n\sqrt{\xi^2-1}\right)\left(s+\xi\omega_n-\omega_n\sqrt{\xi^2-1}\right)} \tag{3.43}$$

The inverse Laplace transform of Eqn(3.43) gives

$$\boxed{c(t) = 1 + \frac{\omega_n}{2\sqrt{\xi^2-1}}\left[\frac{e^{-s_1 t}}{s_1} - \frac{e^{-s_2 t}}{s_2}\right], \quad s_1, s_2 = \left(\xi \pm \sqrt{\xi^2-1}\right)\omega_n} \tag{3.44}$$

Thus, the response c(t) includes two decaying exponential terms.

The step responses of a prototype second-order system due to various damping factors are shown in Figure 3.13.

CONCEPT REINFORCEMENT PROBLEM 3.3

Problem

Consider the closed-loop transfer functions of systems given below

a. $T(s) = \dfrac{2}{(s+1)(s+5)}$

b. $T(s) = \dfrac{100}{s^2 + 10s + 100}$

c. $T(s) = \dfrac{s+1}{(s^2 + 25)}$

d. $T(s) = \dfrac{s+2}{(s+3)^2}$

For each of the transfer functions

i. Find the damping ratio and natural frequency

ii. Express the step response in general form without solving using the inverse Laplace transform. State the nature of the response, such as overdamped, underdamped, undamped, etc., and show that the damping value conforms with the response characteristics.

3.6 Transient Response Specifications

Many natural and engineered systems exhibit underdamped behavior, which includes mechanical systems such as vehicles, structures, and machinery, as well as electrical circuits, biological systems, and numerous other applications.

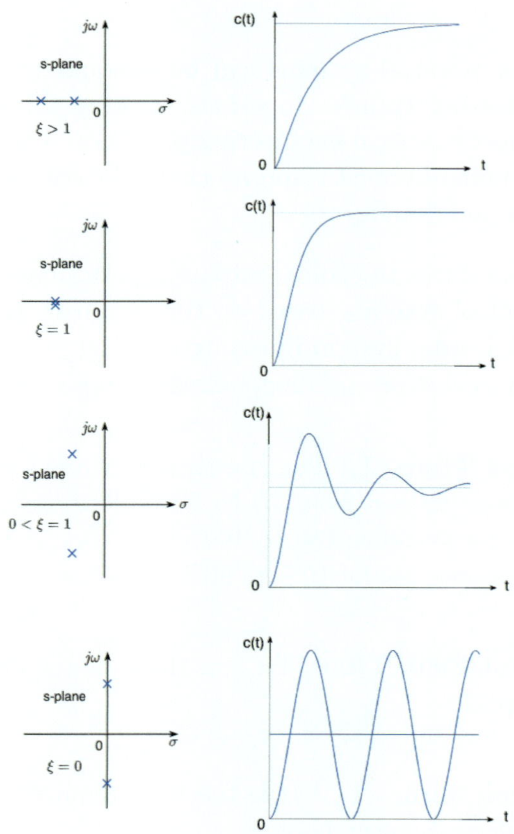

Figure 3.13: Step response of second-order system with variations of damping

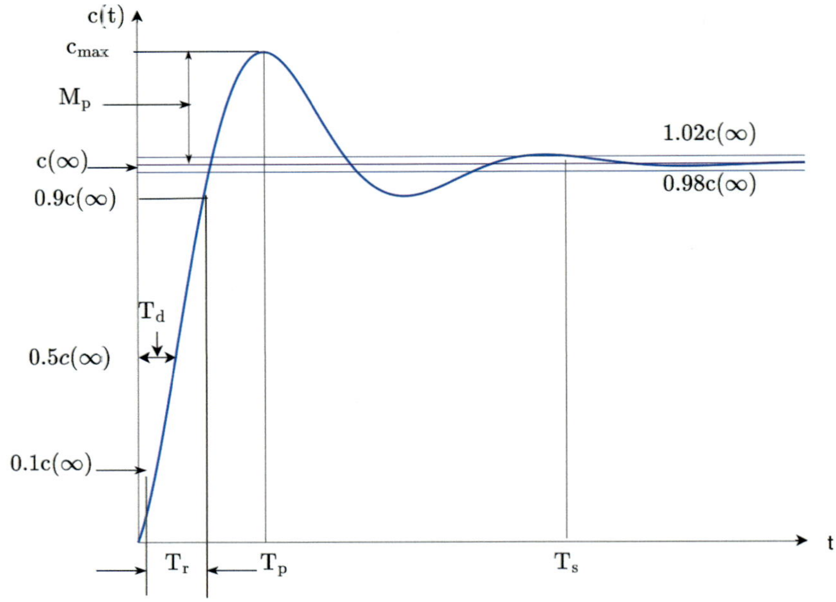

Figure 3.14: Step response of a prototype second-order underdamped system

Moreover, many physical systems can be adequately modeled using underdamped second-order transfer functions, making this a fundamental representation for most practical engineering problems. Therefore, a thorough understanding of underdamped response characteristics is essential for effective system analysis and design.

This section introduces the commonly used performance indices that characterize linear control systems, based on the step response of an underdamped prototype second-order system illustrated in Figure 3.12. These performance metrics are defined below and demonstrated graphically in Figure 3.14.

Definition 3.4. **Rise Time, T_r:** The rise time is defined as the time required for the step response to rise from 10 to 90% of its final value. For underdamped second-order systems, 0% to 100% rise time is normally used, and for overdamped systems, use of 10% to 90% rise time is very common.

Definition 3.5. **First Peak Time, T_f:** It is the time required for the response to reach the first peak.

Definition 3.6. **Peak Time, T_p:** It is the time required for the response to reach the first peak of the overshoot.

Remark

a. For second-order underdamped systems, the first peak consistently represents the maximum peak, i.e., $T_p = T_f$, with all subsequent peaks exhibiting progressively smaller magnitudes until the system reaches steady state. This characteristic behavior stems from the presence of a single dominant oscillatory mode that governs the transient response.

b. Conversely, higher-order systems, particularly those with multiple underdamped modes, may exhibit markedly different behavior. The step response can display multiple peaks with varying magnitudes, where the first peak does not necessarily correspond to the maximum peak of the overall response. This phenomenon arises from the coexistence of multiple oscillatory modes, each contributing distinct frequency and damping characteristics to the total response. The complex interaction between these modes can generate intricate transient behavior characterized by multiple peaks of different amplitudes occurring at different time instants.

To illustrate this concept, consider the step response of a fourth-order system with the transfer function:

$$G(s) = \frac{50}{(s^2 + 0.4s + 4)(s^2 + 0.1s + 3.24)}$$

The response, shown below, demonstrates this multi-peak behavior

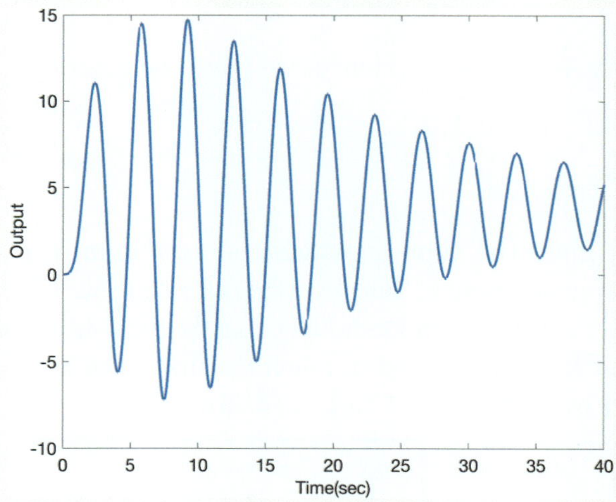

Step response of a higher-order system

Definition 3.7. Maximum (Percent) Overshoot, M_p: The maximum overshoot is the maximum peak value of the response due to unity step input measured from unity. However, if the steady state value of the response differs from unity, then we define maximum percent overshoot as:

$$\text{Maximum percent overshoot} = \frac{c(T_p) - c(\infty)}{c(\infty)} \times 100\%$$

where $c(T_p)$ is the peak value of the response at peak time and $c(\infty)$ is the steady state value of the response.

Definition 3.8. Settling Time, T_s: It is the time required for the response to reach and stay within a range of its final value (usually 2% or 5%).

In the following, we will derive expressions for *rise time*, *peak time*, *maximum overshoot*, and the *settling time* of an underdamped second order control system.

3.6.1 Computation of Peak time T_p

This is obtained by differentiating c(t) in Equation (3.38) with respect to time and making this derivative equal to zero. Now

$$\frac{dc(t)}{dt} = 0 - \frac{e^{-\xi\omega_n t}}{\sqrt{1-\xi^2}}\left[\cos(\omega_d t + \phi)\right]\omega_d + \frac{\xi\omega_n}{\sqrt{1-\xi^2}} e^{-\xi\omega_n t}\sin(\omega_d t + \phi) \quad (3.45)$$

This can be simplified as:

$$\tan(\omega_d t + \phi) = \frac{\omega_d}{\xi\omega_n} = \frac{\sqrt{1-\xi^2}}{\xi} = \tan\phi \quad \left(\because \phi = \tan^{-1}\left[\frac{\sqrt{1-\xi^2}}{\xi}\right]\right)$$

But $\tan(n\pi + \phi) = \tan\phi$. Hence, we have the solution

$$\omega_d t = n\pi, \quad n = 0, 1, 2, \ldots \implies t = \frac{n\pi}{\omega_d} = \frac{n\pi}{\omega_n\sqrt{1-\xi^2}}, \quad n = 0, 1, 2, \ldots \quad (3.46)$$

Eqn(3.46) gives the time at which the extremum (i.e., maxima or minima) of the response occurs. The overshoots occur at odd values of n, that is, $n = 1, 3, 5, \ldots$ and the undershoots occur at even values of n, that is, $n = 2, 4, 6, \ldots$. Since the time required to reach the first peak is defined as the peak time, it is found by letting $n = 1$ in Eqn(3.46):

$$\boxed{T_p = \frac{\pi}{\omega_d} = \frac{\pi}{\omega_n\sqrt{1-\xi^2}}} \quad (3.47)$$

Note that irrespective of the nature of the extremum, that is, overshoots or undershoots, they occur at a time given by Eqn(3.46). Further, although the unit step response for $\xi \neq 0$ is not periodic, the overshoots and undershoots occur at periodic intervals, as shown in Figure 3.15.

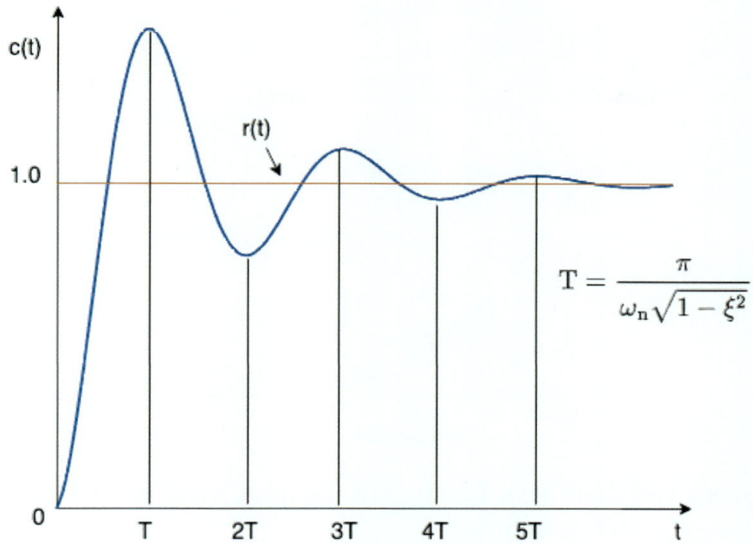

Figure 3.15: Periodic overshoots and undershoots of a prototype second-order underdamped system

3.6.2 Computation of Maximum Overshoot M_p

Assume that the final output of the prototype second-order system $c(\infty)$ is unity. Then, the magnitudes of overshoots or undershoots can be obtained by substituting Eqn(3.47) into Eqn(3.38). This gives

$$c(t)\Big|_{\text{max or min}} = 1 - \frac{e^{-n\pi\xi/\sqrt{1-\xi^2}}}{\sqrt{1-\xi^2}}\sin(n\pi + \phi), \quad n = 1, 2, \ldots \tag{3.48}$$

or

$$c(t)\Big|_{\text{max or min}} = 1 + (-1)^{n-1}e^{-n\pi\xi/\sqrt{1-\xi^2}}, \qquad n = 1, 2, \ldots \tag{3.49}$$

The maximum overshoot is obtained by letting $n = 1$ in Eqn(3.49) which gives

$$M_p = e^{-\pi\xi/\sqrt{1-\xi^2}}$$

The maximum percent overshoot is

$$\boxed{\%M_p = e^{-\pi\xi/\sqrt{1-\xi^2}} \times 100} \tag{3.50}$$

Thus, given the damping ratio ξ, we can determine the maximum percent overshoot using Eqn(3.50). We can also determine ξ for a given maximum

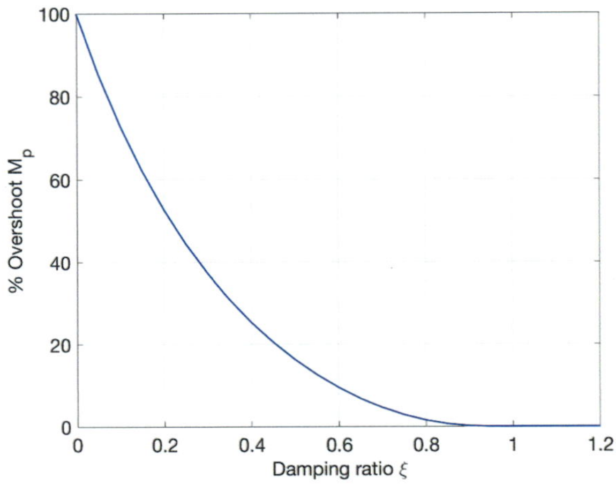

Figure 3.16: Percent overshoot as a function of damping ratio for the step response of a prototype second-order system

percent overshoot by manipulating Eqn(3.50) which gives

$$\xi = \frac{\gamma}{\sqrt{\pi^2 + \gamma^2}}, \quad \text{where} \quad \gamma = -\ln\left(\%\mathrm{M_p}/100\right) \tag{3.51}$$

If the final value $c(\infty)$ is not unity, then the maximum overshoot is to be calculated from the following

$$\mathrm{M_p} = \frac{c(\mathrm{T_p}) - c(\infty)}{c(\infty)} \tag{3.52}$$

The variation of percentage peak overshoot with damping ratio ξ is illustrated in Figure 3.16.

The following observations are noteworthy:

1. Although the peak time $\mathrm{T_p}$ (see (Eqn(3.47)) is a function of both the undamped natural frequency ω_n and the damping ratio ξ, the peak overshoot $\mathrm{M_p}$ is independent of ω_n and is a function of damping ratio ξ only.
2. $\mathrm{M_p}$ is a monotonically decreasing function of ξ, as shown in Figure 3.16.

3.6.3 Computation of Rise Time $\mathrm{T_r}$ and Delay Time $\mathrm{T_d}$

It is often difficult to find a precise mathematical expression for the rise time $\mathrm{T_r}$, the delay time $\mathrm{T_d}$. However, we derive an expression for $\mathrm{T_r}$ where it is defined as the time required to reach from 0% to 100% of its final value. Thus, at $t = T_r$, the output $c(T_r) = 1$. Thus from Eqn(3.38)

$$c(T_r) = 1 = 1 - e^{-\xi\omega_n T_r}\left[\cos\omega_d T_r + \frac{\xi}{\sqrt{1-\xi^2}}\sin\omega_d T_r\right] \tag{3.53}$$

Since $e^{-\xi\omega_n T_r} \neq 0$, we have from Eqn(3.53)

$$\cos\omega_d T_r + \frac{\xi}{\sqrt{1-\xi^2}}\sin\omega_d T_r = 0 \tag{3.54}$$

Simplification of Eqn(3.54) gives

$$\tan\omega_d T_r = -\frac{\sqrt{1-\xi^2}}{\xi} = -\frac{\omega_d}{\sigma} \tag{3.55}$$

where $\omega_d = \omega_n\sqrt{1-\xi^2}$, and $\sigma = \xi\omega_n$.

Thus, the rise time T_r is given by

$$\boxed{T_r = \frac{1}{\omega_d}\tan^{-1}\left(\frac{\omega_d}{-\sigma}\right) = \frac{\pi-\beta}{\omega_d}, \quad \text{where } \beta = \tan^{-1}\frac{\omega_d}{\sigma}} \tag{3.56}$$

An alternate expression of the rise time expression is obtained by approximating it by a first-order equation given by

$$\boxed{T_r = \frac{0.8 + 2.5\xi}{\omega_n} \qquad 0 < \xi < 1.0} \tag{3.57}$$

A better approximation is obtained using a second-order equation

$$\boxed{T_r = \frac{1 - 0.4167\xi + 2.917\xi^2}{\omega_n} \qquad 0 < \xi < 1.0} \tag{3.58}$$

Delay Time T_d

Although we defined delay time for a second-order system, in practice, it is commonly associated with first-order systems or systems with transport delays. It is often difficult to precisely determine the delay time T_d even for a simple prototype second-order system. Additionally, in many cases, the delay time may not be a significant factor in characterizing the behavior of a second-order system, as its response is mainly defined by other parameters such as *damping ratio, natural frequency,* and *settling time.* Nonetheless, approximating the delay time can provide some insight into the system's initial response dynamics. One commonly used approximation of delay time is

$$T_d \cong \frac{1 + 0.7\xi}{\omega_n} \qquad 0 < \xi < 1.0 \tag{3.59}$$

A better second-order approximation of delay time is

$$T_d \cong \frac{1.1 + 0.125\xi + 0.469\xi^2}{\omega_n} \qquad 0 < \xi < 1.0 \tag{3.60}$$

3.6.4 Computation of Settling Time T_s

It is the time when the response reaches and stays within a specified tolerance band (usually ±2% to ±5% of its steady state value). Although it is difficult to obtain a precise mathematical expression for T_s, we can obtain an approximate expression from the envelope of the step response curve, which is shown in Figure 3.17.

The curves $1 \pm e^{-\xi\omega_n t}/\sqrt{1-\xi^2}$ are the envelope of the unit step response. The time constant of this envelope equals $1/\xi\omega_n$. We can obtain an approximate value of T_s considering only exponential decay $(0 < \xi < 0.7)$.

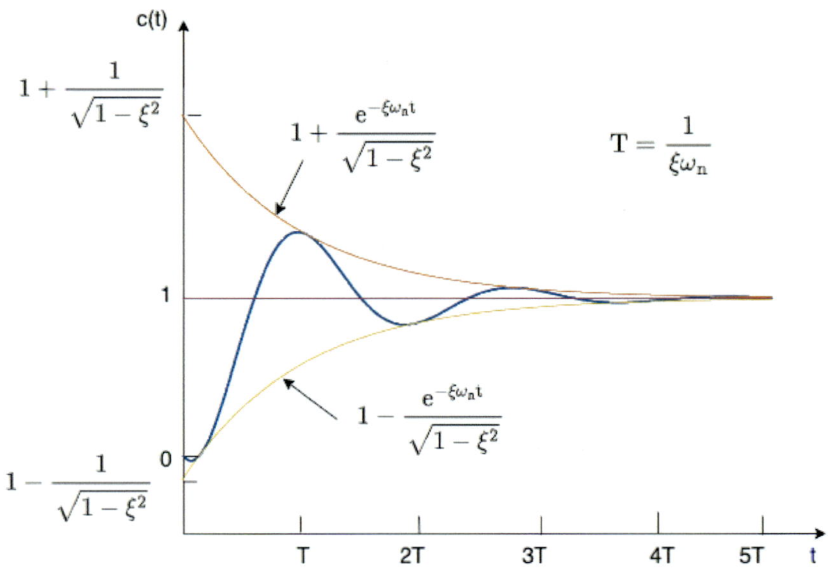

Figure 3.17: Periodic overshoots and undershoots of a prototype second-order underdamped system

For 2% tolerance band $e^{-\xi\omega_n T_s} = 0.02$ which gives

$$T_s = \frac{3.912}{\xi\omega_n} \approx \frac{4}{\xi\omega_n} \tag{3.61}$$

For 5% tolerance band $e^{-\xi\omega_n t_s} = 0.05$ which gives

$$T_s = \frac{2.9957}{\xi\omega_n} \approx \frac{3}{\xi\omega_n} \tag{3.62}$$

Key Takeaways

Key Points:

1. The settling time is inversely proportional to the product of the damping ratio ξ and undamped natural frequency ω_n.
2. Since the damping ratio ξ is selected according to the permissible overshoot M_p, for a desired overshoot, the settling time becomes a function of ω_n.
3. The transient duration can be varied by varying ω_n; the larger the value of ω_n, the smaller the settling time.

In the following, we illustrate computing time-domain specifications through a few examples.

EXAMPLE 3.3

Problem

Determine the damping ratio ξ, natural frequency ω_n, setting time T_s, peak time T_p, rise time T_r and percentage overshoot $\%M_p$ for the system with closed-loop transfer function

$$T(s) = \frac{C(s)}{R(s)} = \frac{4}{s^2 + 0.4s + 4}$$

SOLUTION:

Given, the closed-loop transfer function

$$T(s) = \frac{4}{s^2 + 0.4s + 4}$$

Compare this with the closed-loop transfer function of the prototype second order system of Eqn(3.30) as

$$\frac{\omega_n^2}{s^2 + 2\xi\omega_n s + \omega_n^2} = \frac{4}{s^2 + 0.4s + 4}. \qquad \text{This gives}$$

Natural Frequency and Damping Ratio:

$$\omega_n = \sqrt{4} = 2 \text{ rad/sec}, \quad \xi = \frac{0.4}{2\omega_n} = \frac{0.4}{4} = 0.1$$

Damped Frequency of Oscillation ω_d:

$$\omega_d = \omega_n\sqrt{1 - \xi^2} = 2\sqrt{1 - 0.1^2} = 1.99 \text{ rad/sec}$$

Rise Time T_r:

$$T_r = \frac{\pi - \beta}{\omega_d} = \frac{3.14 - \beta}{\omega_d}, \quad \text{where}$$

$$\beta = \tan^{-1}\frac{\omega_d}{\sigma} = \tan^{-1}\frac{\omega_d}{\xi\omega_n} = \tan^{-1}\frac{\sqrt{1-\xi^2}}{\xi} \quad \text{(Note: } \sigma = \xi\omega_n\text{). Thus}$$

$$T_r = \frac{\pi - \beta}{\omega_d} = \frac{3.14 - 1.4706}{1.99} = 0.8397 \text{ sec}$$

Peak Time T_p, Maximum Overshoot M_p

$$T_p = \frac{\pi}{\omega_d} = \frac{\pi}{1.99} = 1.5787 \text{ sec}$$

$$M_p = e^{\frac{-\pi\xi}{\sqrt{1-\xi^2}}} = 0.7292 \ (72.92\%)$$

Settling Time T_s:

For 2% criterion, the setting time is

$$T_s = \frac{4}{\xi\omega_n} = 20 \text{ sec}$$

For 5% criterion, the setting time is

$$T_s = \frac{3}{\sigma} = 15 \text{ sec}$$

CONCEPT REINFORCEMENT PROBLEM 3.4

Problem

Determine the damping ratio ξ, natural frequency ω_n, setting time T_s, peak time T_p, rise time T_r and percentage overshoot $\%M_p$ for the system with closed-loop transfer function

$$T(s) = \frac{C(s)}{R(s)} = \frac{100}{s^2 + 2s + 400}$$

EXAMPLE 3.4

Problem

The closed-loop transfer function of a second-order system is given by

$$T(s) = \frac{10s}{7s^2 + 14s + \beta + 1}$$

Find the value of β such that the system has a damping ratio $\xi = 0.4$.

SOLUTION:

Let us express the transfer function as:

$$T(s) = \frac{(10/7)s}{s^2 + 2s + \dfrac{\beta + 1}{7}}$$

Comparing this with the transfer function of the standard second-order system gives

$$\omega_n^2 = \frac{\beta + 1}{7} \quad \Longrightarrow \quad \omega_n = \sqrt{\frac{\beta + 1}{7}} \quad \text{and}$$

$$2\xi\omega_n = 2 \times 0.4\sqrt{\frac{\beta + 1}{7}} = 2 \quad \Longrightarrow \quad \sqrt{\frac{\beta + 1}{7}} = \frac{10}{4}$$

or,

$$\frac{\beta + 1}{7} = \frac{100}{16} \quad \Longrightarrow \quad \beta + 1 = \frac{700}{16} = 43.75$$

Hence $\beta = 42.75$.

EXAMPLE 3.5

Consider a unity feedback system with forward path transfer function

$$G(s) = \frac{K}{s(s+a)}$$

Find the minimum value of gain K such that the step response of the closed-loop system will exhibit an underdamped response, that is, it will have an overshoot.

SOLUTION:

Now the closed-loop transfer function is

$$T(s) = \frac{K}{s^2 + as + K}$$

Comparing with the prototype second-order system, we have

$$\omega_n^2 = K \quad \text{and} \quad 2\xi\omega_n = a \quad \Longrightarrow \quad \xi = \frac{a}{2\omega_n} = \frac{a}{2\sqrt{K}}$$

For the system to be underdamped $0 < \xi < 1$. This is satisfied when

$$\xi = \frac{a}{2\sqrt{K}} < 1 \quad \text{or,} \quad K > \frac{a^2}{4}$$

EXAMPLE 3.6

Problem

The ratio between the settling time (2% criterion) T_s and the peak time T_p of the step response of a second-order unity feedback system equals 6.2376

a. Compute the damping ratio.
b. Compute the percentage peak overshoot.
c. If the natural frequency of the system $\omega_n = 10$ rad/sec, compute the settling time, peak time, and the closed-loop transfer function.

SOLUTION:

a. Now $T_s = \dfrac{4}{\xi\omega_n}$ and $T_p = \dfrac{\pi}{\omega_d} = \dfrac{\pi}{\omega_n\sqrt{1-\xi^2}}$.

   Hence $\dfrac{T_s}{T_p} = \dfrac{4\sqrt{1-\xi^2}}{\pi\xi}$

Manipulating the above gives

$$\xi = \frac{4}{\sqrt{16+\alpha^2\pi^2}}, \quad \text{where} \quad \alpha = \frac{T_s}{T_p}$$

Substituting the value of $\alpha = 6.2376$ in the above gives $\xi = 0.2$

b. **Maximum Overshoot** M_p

$$M_p = e^{\frac{-\pi\xi}{\sqrt{1-\xi^2}}} = 0.5266 = 52.66\%$$

c. For $\omega_n = 10$ rad/sec

$$T_s = \frac{4}{\xi\omega_n} = \frac{4}{2} = 2\text{sec}$$

$$T_p = \frac{\pi}{\omega_d} = \frac{\pi}{\omega_n\sqrt{1-\xi^2}} = \frac{\pi}{9.798} = 0.3206\text{sec}$$

$$\frac{C(s)}{R(s)} = \frac{\omega_n^2}{s^2+2\zeta\omega_n s+\omega_n^2} = \frac{100}{s^2+4s+100}$$

CONCEPT REINFORCEMENT PROBLEM 3.5

Problem

The block diagram of a position control system with velocity feedback is shown in Figure 3.18

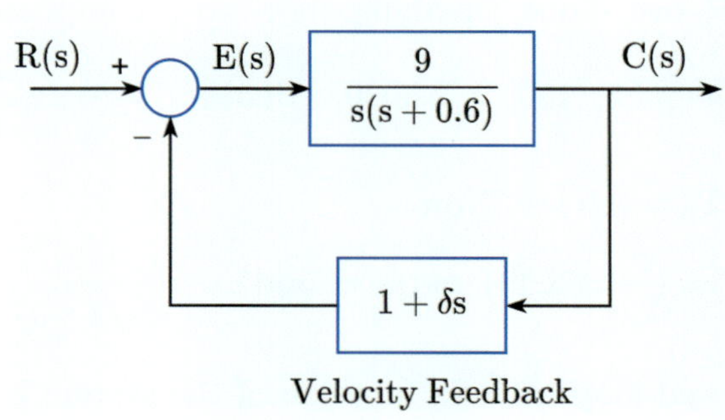

Figure 3.18: Position control system with velocity feedback

Determine the damping ratio, peak overshoot, peak time, and settling time (2% criterion) for:

a. $\delta = 0$ b. For $\delta = 0.1$ c. For $\delta = 0.3$

Comment on your observations.

EXAMPLE 3.7

Problem

The block diagram of a position control system with velocity feedback is shown in Figure 3.19

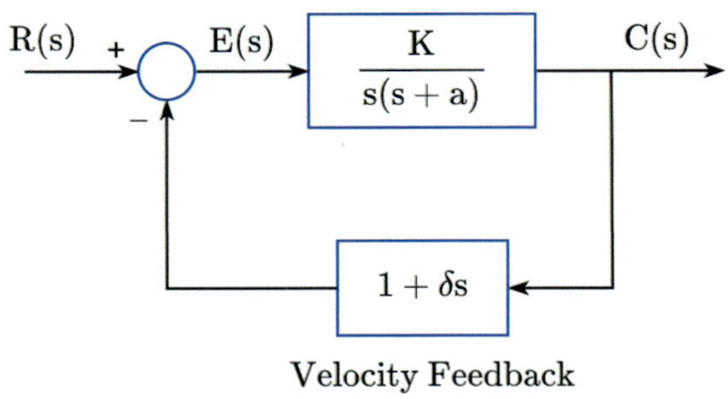

Figure 3.19: Position control system with velocity feedback

Determine the value of K and δ which ensures that the maximum overshoot in the step response is 10% and peak time $T_p = 0.9$ sec. Assume $a = 1$.

SOLUTION:

Now

$$\xi = \frac{\gamma}{\sqrt{\pi^2 + \gamma^2}}, \quad \text{where} \quad \gamma = -\ln(\%M_p/100)$$

For 10% overshoot, the damping ratio is computed as:

$$\gamma = -\ln(\%M_p/100) = 2.3026, \quad \text{Hence} \quad \xi = \frac{2.3026}{\sqrt{\pi^2 + (2.3026)^2}} = 0.5912$$

Given $T_p = 0.9$ sec. Now

$$\omega_d = \frac{\pi}{T_p} = 3.491 \text{ rad/sec} \quad \text{and} \; \omega_n = \frac{\omega_d}{\sqrt{1 - \xi^2}} = 4.328 \text{ rad/sec}$$

The closed-loop transfer function of the system is given by

$$T(s) = \frac{K}{s^2 + (K\delta + a)s + K}$$

Comparing this with the standard closed-loop transfer function of a standard second-order system:

$$\frac{\omega_n^2}{s^2 + 2\xi\omega_n s + \omega_n^2} \quad \text{gives}$$

$$\omega_n = \sqrt{K} \implies K = \omega_n^2 = 18.730$$

Derivative Feedback δ

$$\delta = \frac{2\xi\omega_n - a}{K} = \frac{2 * 0.5912 * 4.328 - 1}{18.730} = 0.2198$$

Damped Frequency of Oscillation ω_d

$$\omega_d = \omega_n\sqrt{1-\xi^2} = 4.328\sqrt{1-0.5912^2} = 3.491 \text{ rad/sec}$$

Rise Time T_r

$$T_r = \frac{\pi - \beta}{\omega_d} = \frac{3.14 - \beta}{\omega_d}, \quad \beta = \tan^{-1}\frac{\omega_d}{\sigma}, \quad \sigma = \xi\omega_n$$
$$= 0.6312\text{sec}$$

Settling Time T_s

For the 2% criterion, the setting time is

$$T_s = \frac{4}{\xi\omega_n} = 1.5635\text{sec}$$

CONCEPT REINFORCEMENT PROBLEM 3.6

Problem:

Determine the closed-loop transfer function of the second-order system whose step response is shown in Figure 3.20.

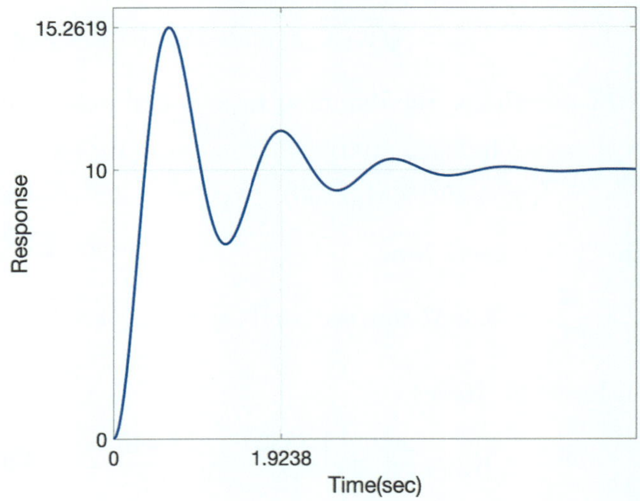

Figure 3.20: Step response of a second order system

EXAMPLE 3.8

Problem

The block diagram of a position control system with velocity feedback is shown in Figure. 3.21

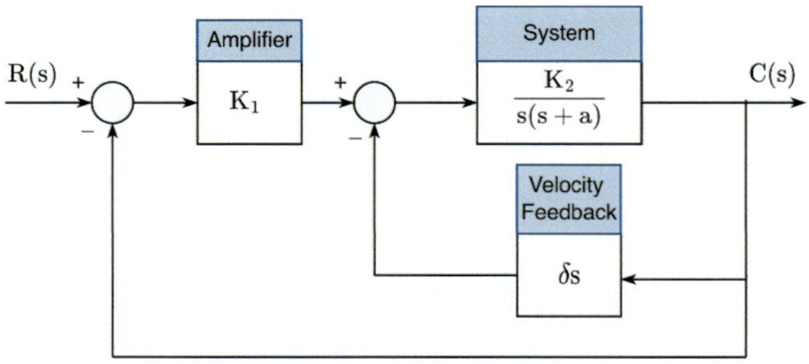

Figure 3.21: Position control system with velocity feedback

Determine the value of "K_2" and "a" which ensures that the maximum overshoot in the step response is 10% and settling time $T_s = 2.0$ sec. Assume $\delta = 0.1$ and $K_1 = 10$.

SOLUTION:

The closed-loop transfer function of the system is

$$T(s) = \frac{C(s)}{R(s)} = \frac{K_1K_2}{s^2 + (a + K_2\delta)\,s + K_1K_2}$$

Comparing this with the closed-loop transfer function of the prototype second-order system gives

$$\omega_n = \sqrt{K_1K_2}, \quad \text{and} \quad 2\xi\omega_n = a + K_2\delta$$

For 10% overshoot, the damping ratio is computed as:

$$\xi = \frac{-\ln(\%M_p/100)}{\sqrt{(\pi^2 + \ln^2(\%M_p/100)}} = \frac{-\ln(10/100)}{\sqrt{(\pi^2 + \ln^2(10/100)}} = 0.5912$$

Given $T_s = 2.0$ sec. Now

$$\omega_n = \frac{4}{\xi T_s} = 3.3832 \text{ rad/sec} \quad \text{and } \omega_d = \omega_n\sqrt{1-\xi^2} = 2.729 \text{ rad/sec}$$

Given $K_1 = 10$. Hence

$$K_2 = \frac{\omega_n^2}{K_1} = \frac{3.3832^2}{10} = \frac{11.446}{10} = 1.1446$$

Further

$$a + K_2\delta = 2\xi\omega_n \quad \Longrightarrow \quad a = 2\xi\omega_n - K_2\delta = 4 - 0.11446 = 3.8855$$

EXAMPLE 3.9

Problem

The block diagram of a missile attitude control system is shown in Figure. 3.22

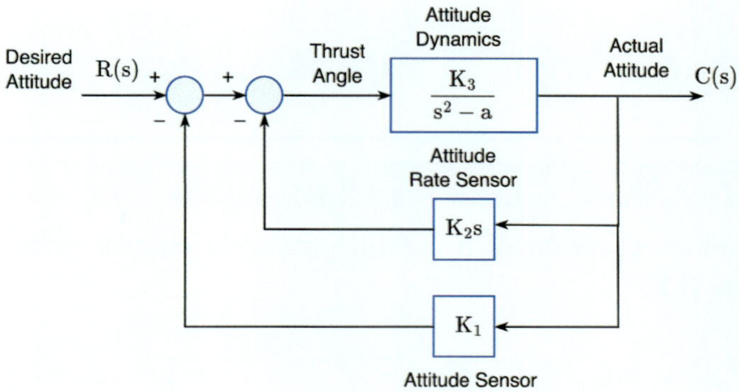

Figure 3.22: Attitude control system of a missile

a. Determine the values of K_1 and K_2 such that the closed-loop system will have an undamped natural frequency $\omega_n = 12$ rad/sec and damping ratio $\xi = 0.5$.
b. Compute the maximum overshoot and settling time for a unit step input.

The parameters $a = 6$ and $K_3 = 15$.

SOLUTION:

The closed-loop transfer function is

$$\frac{C(s)}{R(s)} = \frac{K_3}{s^2 + K_2K_3s + (K_1K_3 - a)}$$

a. Comparing with the transfer function of a prototype second-order system, we have

$$K_1K_3 - a = \omega_n^2 \quad \Longrightarrow \quad K_1 = \frac{\omega_n^2 + a}{K_3} = \frac{150}{15} = 10$$

and

$$2\xi\omega_n = K_2K_3 \quad \Longrightarrow \quad K_2 = \frac{2\xi\omega_n}{K_3} = \frac{12}{15} = 0.8$$

b The maximum percentage overshoot

$$\%M_p = e^{-\frac{\pi\xi}{\sqrt{1-\xi^2}}} \times 100 = 16.3\%$$

and the settling time

$$T_s = \frac{4}{\xi\omega_n} = 0.67\text{sec}$$

CONCEPT REINFORCEMENT PROBLEM 3.7

The block diagram of a satellite attitude control system is shown in Figure 3.23.

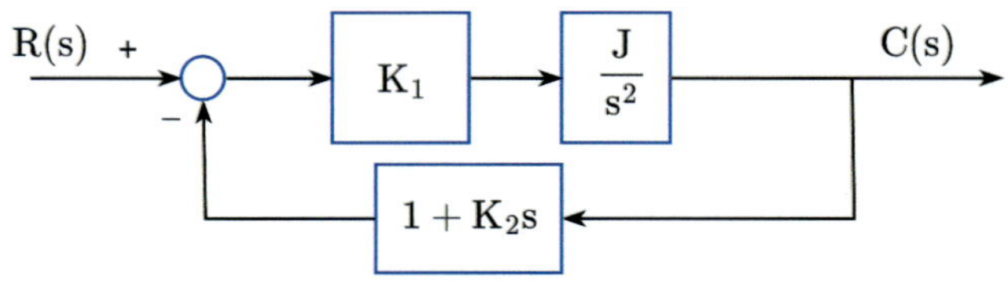

Figure 3.23: Satellite control system

Determine K_1 and K_2 of the closed-loop system which results in maximum overshoot of 10% at peak time 1.5 sec. Assume $J = 1$ kg-m^2.

EXAMPLE 3.10

Problem

The open-loop transfer function of a unity feedback control system is given by

$$G(s) = \frac{K}{s(1 + sT)}$$

a. By what factor should the amplifier gain K be multiplied such that the damping ratio is increased from 0.15 to 0.60?

b. By what factor should the time constant T be multiplied so that the damping ratio is reduced from 0.8 to 0.4?

SOLUTION:

The closed-loop transfer function of the system is

$$T(s) = \frac{G(s)}{1 + G(s)} = \frac{K}{s^2T + s + K} = \frac{K/T}{s^2 + \frac{1}{T}s + \frac{K}{T}} \tag{3.63}$$

Comparing Eqn(3.63) with Eqn(3.30)

$$\omega_n = \sqrt{\frac{K}{T}} \text{ and } \xi = \frac{1}{2T\omega_n} = \frac{1}{2\sqrt{TK}} \tag{3.64}$$

a. For the same time constant, the damping ratio is inversely proportional to the square root of the gain. If $K = K_1$ and K_2 denote the amplifier gains corresponding to the damping factor ξ_1 and ξ_2, then

$$\xi_1 = \frac{1}{2\sqrt{TK_1}} \quad \xi_2 = \frac{1}{2\sqrt{TK_2}}$$

Thus $\dfrac{\xi_1}{\xi_2} = \sqrt{\dfrac{K_2}{K_1}}$.
If $\xi_1 = 0.15$ is increased to $\xi_2 = 0.6$, then

$$\frac{\xi_1}{\xi_2} = \sqrt{\frac{K_2}{K_1}} = \frac{0.15}{0.6} = \frac{1}{4} \quad \Longrightarrow \quad K_2 = \frac{K_1}{16}$$

Thus, the original gain should be decreased 16 times for the damping to be increased 4 times.

b. For the same gain K, $\xi_1 = \dfrac{1}{2\sqrt{T_1K}}$ and $\xi_2 = \dfrac{1}{2\sqrt{T_2K}}$
where T_1 and T_2 are the time constants corresponding to ξ_1 and ξ_2. Hence

$$\frac{\xi_1}{\xi_2} = \sqrt{\frac{T_2}{T_1}}$$

If $\xi_1 = 0.8$ is decreased to $\xi_2 = 0.4$, then

$$\frac{\xi_1}{\xi_2} = \sqrt{\frac{T_2}{T_1}} = \frac{0.8}{0.4} = 2 \quad \Longrightarrow \quad T_2 = 4T_1$$

Thus, the original time constant should be increased 4 times for the damping to be decreased 2 times.

3.7 Effects of Pole Motion on Transient Response Specifications of Second Order Underdamped System

The location of poles in the transfer function of a control system directly influences its response characteristics, including response speed, overshoot, settling time, and oscillatory behavior. In this section, we will discuss how the response of an underdamped second-order system changes as the poles move

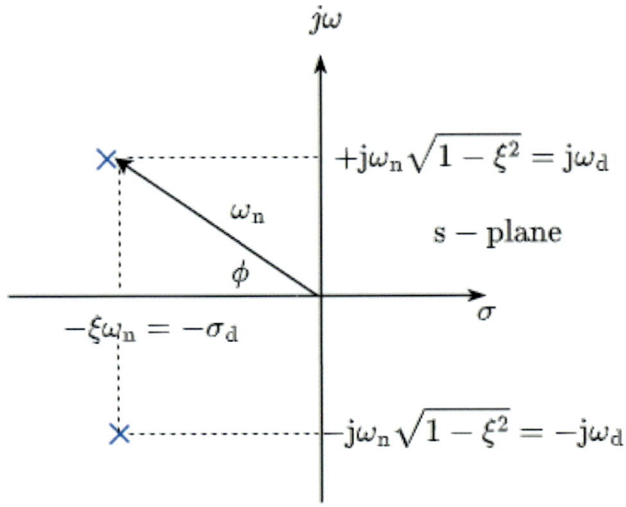

Figure 3.24: Pole plot of a prototype second-order underdamped system

in the s-plane in some defined manner.

The pole plot of the prototype underdamped second-order system is shown in Figure 3.24.

From the plot, it is seen that the radial distance from the origin to the pole location equals the natural frequency, ω_n and $\cos\phi = \xi$.

Impact of Pole Location on Peak Time T_p:

Since the peak time is given by $T_p = \frac{\pi}{\omega_d}$, it has an inverse relationship with the damped natural frequency ω_d, which represents the imaginary component of the pole. In the s-plane, horizontal lines represent constant imaginary values, making them equivalent to lines of constant peak time, as illustrated in Figure 3.25.

Impact of Pole Location on Settling Time T_s:

The settling time $T_s = \frac{4}{\xi\omega_n} = \frac{4}{\sigma_d}$ relationship shows that it varies inversely with σ_d, the real component of the pole. In the s-plane, vertical lines maintain constant real values, therefore representing lines of constant settling time, as depicted in Figure 3.25.

Impact of Pole Location on Peak Overshoot $\%M_p$:

The peak overshoot expression derived earlier is reproduced below.

$$\%M_p = e^{-\pi\xi/\sqrt{1-\xi^2}} \times 100$$

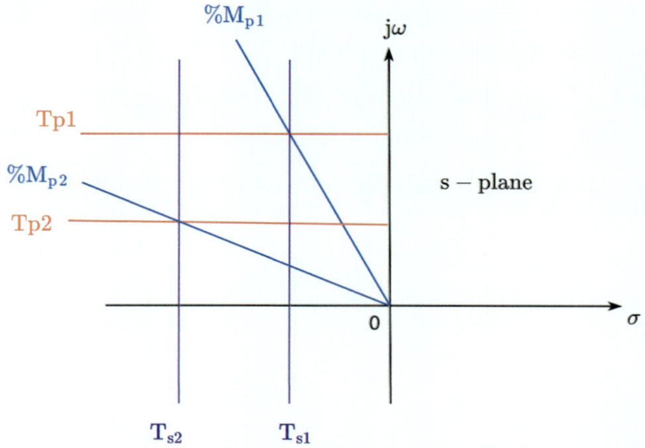

Figure 3.25: Lines of constant peak time T_p, settling time T_s and percent overshoot $\%M_p$ of second order underdamped system: Note $T_{s2} < T_{s1}$; $T_{p2} < T_{p1}$;$\%M_{p2} < \%M_{p1}$

This is a function soley of ξ. Since $\xi = \cos\theta$, radial lines of constant ξ in the s-plane, correspond to lines of constant percent overshoot, as shown in Figure 3.25.

Impact of Pole Location on Step Response:

Let us analyze how step response characteristics vary with changes in pole location.

Case-1: Vertical pole movement with constant real part:

When the real component $\sigma = \xi\omega_n$ remains fixed while the imaginary part ω_d varies, the following effects occur:

a. The response envelope remains unchanged, decaying with a time constant of $\frac{1}{\xi\omega_n}$.
b. The settling time T_s remains constant due to constant $\xi\omega_n$.
c. The peak time T_p decreases as the imaginary part of the pole ω_d increases, resulting in more output cycles $n = \frac{T_s}{T_p}$.
d. The effective damping ratio decreases as ω_d increases, leading to greater overshoot and reduced rise time.

Case-2:Horizontal pole movement with constant imaginary part:

When the imaginary component ω_d remains fixed while the real part $\sigma = \xi\omega_n$ varies, the following effects occur:

a. As the real part increases, settling time decreases, meaning the response dampens more quickly.
b. Since ω_d remains constant, the frequency of damped oscillations in the step response remains unchanged.
c. Peak time also stays constant as ω_d is unchanged.

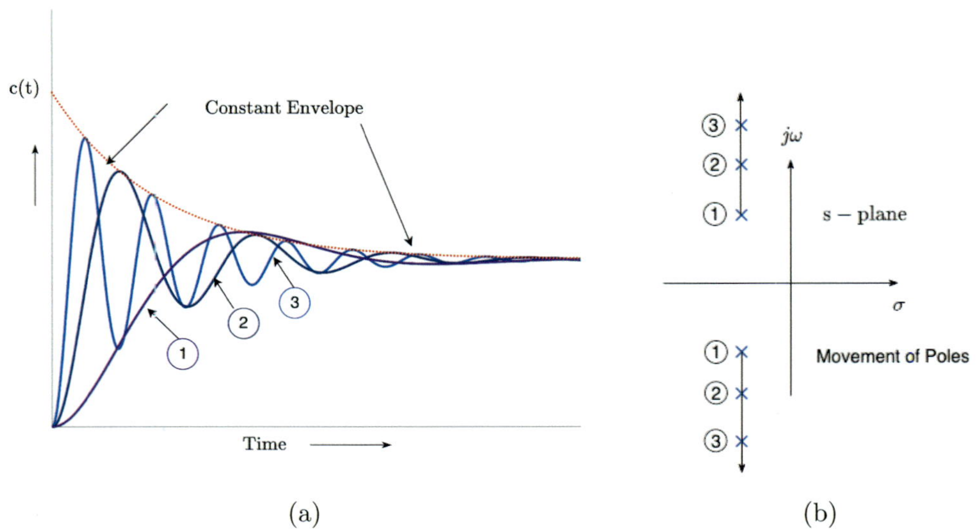

Figure 3.26: (a) Response variations when the real part σ is constant (b) Pole motion keeping the real part constant

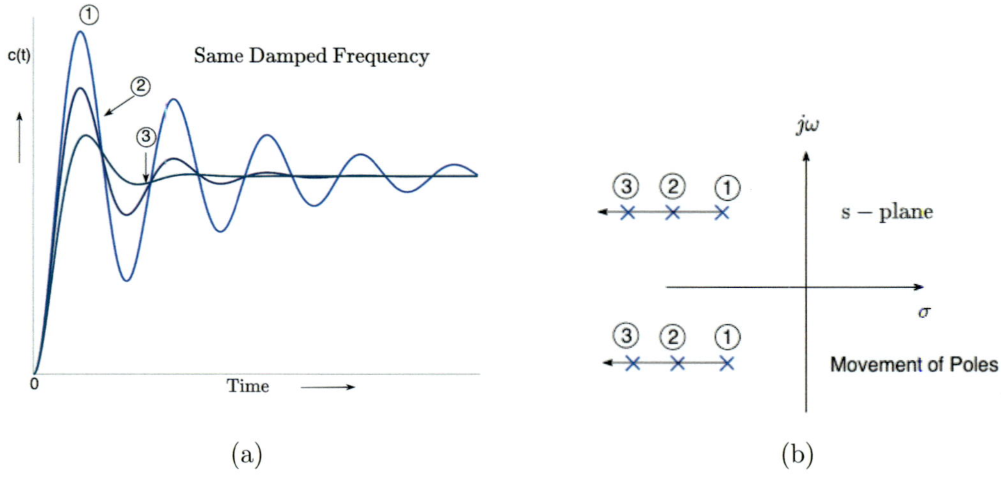

Figure 3.27: (a) Response variations when imaginary part ω_d is constant (b) Pole motion keeping imaginary part constant

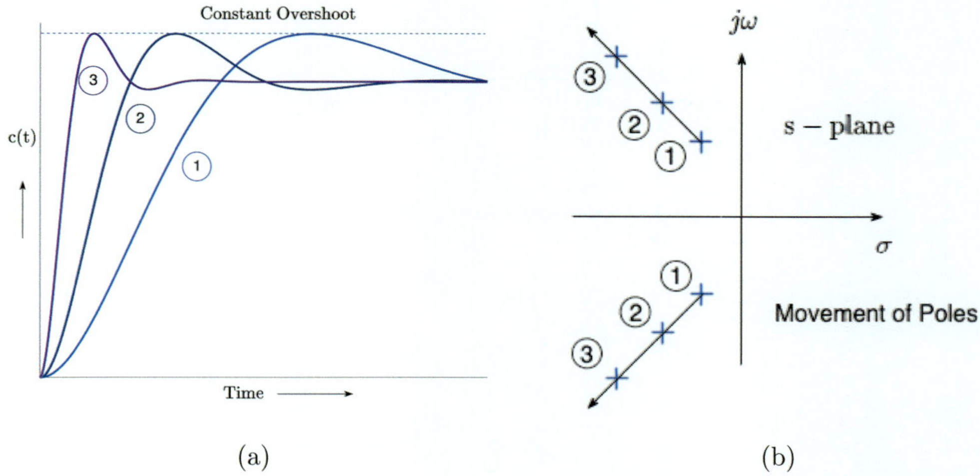

Figure 3.28: Effects of pole movement on step response of a second order under-damped system

Case-3: Pole movement along constant radial lines:

a. Percent overshoot remains unchanged.
b. The speed of response increases as the pole moves further from the imaginary axis.

3.8 Effects of Adding Poles and Zeros to Transfer Functions

We have analyzed the transient behavior of a prototype second-order system and derived formulas for various time domain performance specifications, including percent overshoot, peak time, rise time, and settling time. However, practical systems are typically higher-order with transfer function models containing additional poles and zeros. Therefore, it is important to understand how adding poles and zeros to the prototype second-order transfer function modifies the transient response.

3.8.1 Addition of a Pole to the Forward Path Transfer Function

Consider a second-order system with an open-loop transfer function

$$G(s) = \frac{K}{s(s+a)} \tag{3.65}$$

When compared with the prototype second order system of Eqn(3.30), we have $a = 2\xi\omega_n$ and $K = \omega_n^2$. When a pole at $s = -1/\beta$ is added to the open-loop transfer function, the modified open-loop transfer function becomes

$$G(s) = \frac{K}{s(s+a)(1+\beta s)} \tag{3.66}$$

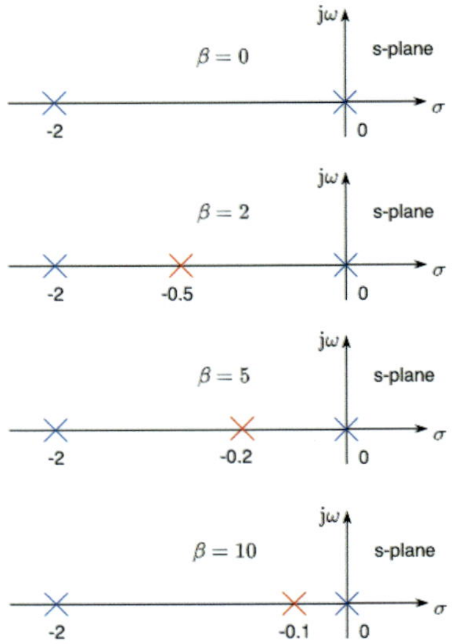

Figure 3.29: Pole plots of open-loop system of Eqn(3.66) for a $= 2$, K $= 1$, $\beta = 0, 2, 5,$ and 10

The closed-loop transfer function becomes

$$T(s) = \frac{G(s)}{1 + G(s)} = \frac{C(s)}{R(s)} = \frac{K}{\beta s^3 + (1 + a\beta)\, s^2 + as + K} \tag{3.67}$$

A single pole is added to the open-loop transfer function by varying the parameter β. The pole-zero plots of the open-loop system for a $= 2$, and $\beta = 0, 2, 5,$ and 10 are shown in Figure 3.29,

The unit step responses of the closed-loop system are plotted in Figure 3.30 for various locations of the additional pole, assuming $\xi = 1.0$, $\omega_n = 1.0$, i.e., a $= 2.0$.

From the plot of Figure 3.30, we found the following:

1. Adding a pole to the forward path transfer function increases the peak overshoot. As the pole moves closer to the imaginary axis in the s-plane, the peak overshoot increases; meaning the peak overshoot increases with higher values of β.

2. The rise time of the system also increases with an increasing value of β. Therefore, both overshoot and rise time increase when the pole moves nearer to the imaginary axis.

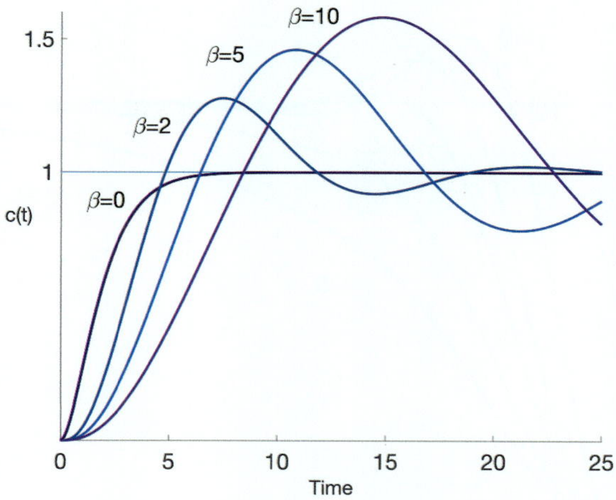

Figure 3.30: Unit step response from Eq.(3.67) showing effects of adding a pole to forward path loop transfer function: $a = 2$, $K = 1$, $\beta = 0, 2, 5,$ and 10

3.8.2 Addition of a Pole to the Closed-Loop Transfer Function

The closed-loop transfer function of the prototype second-order system of Eq. (3.65) is given by

$$T(s) = \frac{C(s)}{R(s)} = \frac{K}{s^2 + as + K} \tag{3.68}$$

If we add a pole at $s = -\frac{1}{\beta}$ to Eqn(3.68), then the modified closed-loop transfer function becomes

$$\begin{aligned} T(s) = \frac{C(s)}{R(s)} &= \frac{K}{(s^2 + as + K)(1 + \beta s)} \\ &= \frac{K}{\beta s^3 + (1 + a\beta)s^2 + (a + K\beta)s + K} \end{aligned} \tag{3.69}$$

The following observations are drawn from Figure 3.31.

1. As the pole moves closer to the imaginary axis, meaning as β increases, the rise time increases.

2. Nevertheless, the overshoot reduces as β increases. Thus, from the perspective of overshoot, adding a pole to the closed-loop transfer function produces the opposite effect compared to adding a pole to the forward-path transfer function.

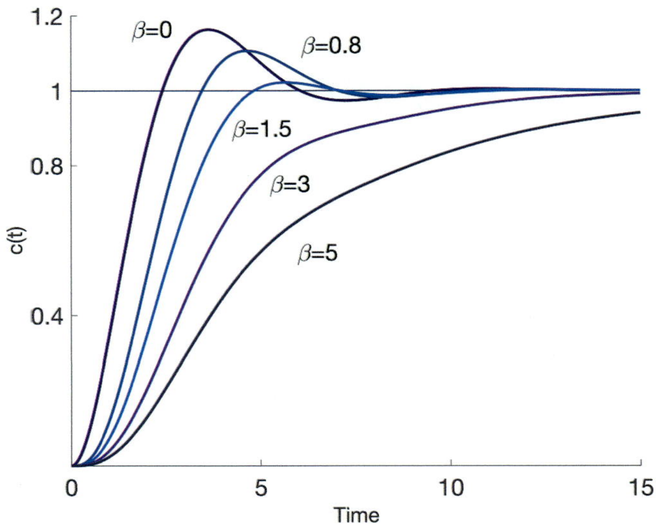

Figure 3.31: Unit step response from Eqn(3.69) showing effects of adding a pole to closed-loop transfer function: $a = 1$, $K = 1$, $\beta = 0, 0.8, 1.5, 3,$ and 5

3.8.3 Addition of a Zero to the Closed-Loop Transfer Function

If we add a zero at $s = -1/\beta$ to the closed-loop transfer function of Eqn(3.68), this modifies to

$$\begin{aligned} T(s) = \frac{C(s)}{R(s)} &= \frac{K(1+\beta s)}{s^2 + as + K} \\ &= \frac{K}{s^2 + as + K} + \beta s \frac{K}{s^2 + as + K} \end{aligned} \tag{3.70}$$

Figure 3.32 shows the unit step response of the system of Eqn(3.70) where $a = K = 1$; and $\beta = 0, 1, 3, 5$ and 10. Due to the addition of the zero, we observe the following from the step response of the closed-loop system:

a. The rise time decreases and the peak overshoot increases as the value of β increases, meaning as the zero is positioned closer to the imaginary axis.

b. While the peak overshoot will tend toward infinity as β approaches infinity, the system will still exhibit a stable response provided $\xi > 0$.

Eqn(3.70) can be expressed as:

$$C(s) = \underbrace{\frac{K}{s^2 + as + K}R(s)}_{C_1(s)} + \underbrace{\beta s \frac{K}{s^2 + as + K}R(s)}_{\beta s C_1(s)} \tag{3.71}$$

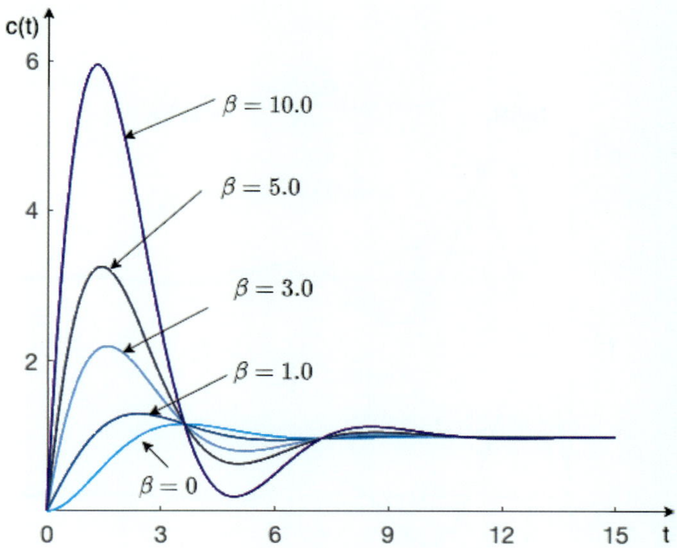

Figure 3.32: Unit step responses showing effects of adding a zero to the closed-loop transfer function in Eqn(3.70): a=1, K=1, $\beta = 0, 1, 3, 5$, and 10

Therefore, the total response c(t) of the system can be written as

$$c(t) = c_1(t) + \beta \frac{dc_1(t)}{dt} \tag{3.72}$$

which shows that the total response is the sum of $c_1(t)$ plus the scaled version of its first derivative. The unit step response c(t) of a system, along with the constituent responses $c_1(t)$ and $\beta\dot{c}_1(t)$ is shown in Figure 3.33

3.8.4 Addition of a Zero to the Forward Path Transfer Function

Let us investigate what happens if a zero at $s = -1/\beta$ is added to the feedforward path transfer function. Thus, the feedforward path transfer function becomes

$$G(s) = \frac{K(1 + \beta s)}{s(s + a)} \tag{3.73}$$

This gives the closed-loop transfer function

$$T(s) = \frac{C(s)}{R(s)} = \frac{K(1 + \beta s)}{s^2 + s(a + K\beta) + K} \tag{3.74}$$

From Eqn(3.74), we notice that the parameter β, which decides the location of the additional zero, is present both in the numerator as well as in the denominator. From our analysis of a standard second-order system, we conclude the following:

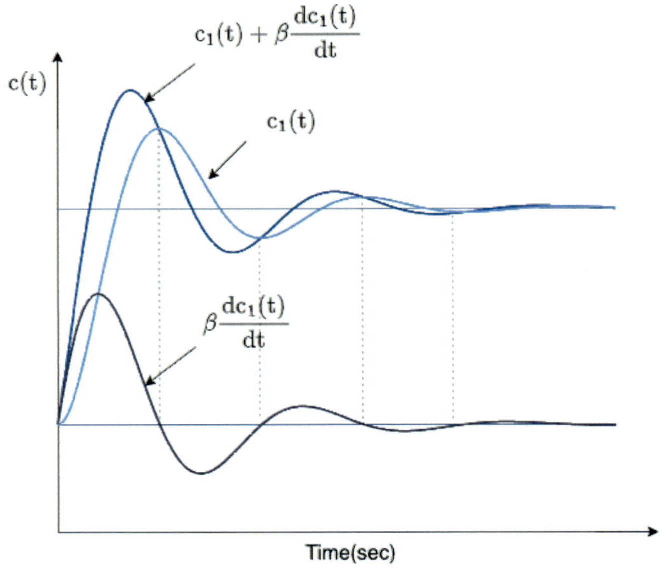

Figure 3.33: Unit step responses showing effects of adding a zero to the closed-loop transfer function

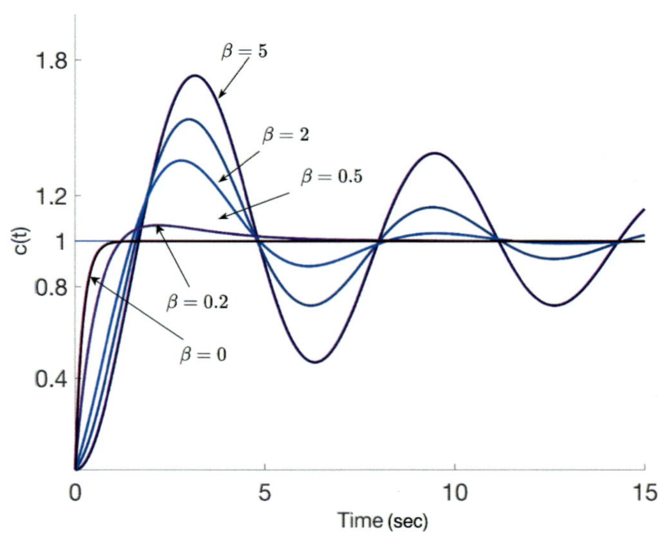

Figure 3.34: Unit step responses showing effects of adding a zero to the forward path transfer function

a. Since the coefficient of **s** in the denominator of Eqn(3.74) changes from "a" to "$a + K\beta$," the damping of the closed-loop system is enhanced. This decreases the peak overshoot.

b. However, the presence of the term $(1 + \beta s)$ in the numerator of Eqn(3.74) will increase the overshoot.

c. From the observations (a) and (b), we note that the presence of the term β in the numerator increases the overshoot, while its presence in the denominator decreases the overshoot.

3.9 Approximating Higher Order Systems

In the preceding sections, we have precisely derived different time domain performance measures that characterize the transient response of a prototype second-order system. These performance measures are shown to depend on the undamped natural frequency ω_n and the damping ratio ξ. However, most practical control systems have an order greater than two. Consequently, the formulas derived for second-order system performance measures are no longer applicable to higher-order systems. To apply the concepts and formulas for second-order systems in analyzing higher-order systems, we must first approximate the higher-order system to a lower-order one (e.g., order 2).

Approximating higher-order systems to lower order based on the concept of dominant poles and pole-zero cancellation is a common technique employed in control system analysis and design. In the following, we provide guidelines for approximating higher-order systems to lower orders using these concepts.

3.9.1 Dominant Poles and Zeros of Transfer Functions

Consider an n-th order system with transfer function

$$G(s) = \frac{C(s)}{R(s)} = \frac{N(s)}{D(s)}$$

Assume that this system has n real poles $-p_1, -p_2, \ldots, -p_n$ which are distinct and lie in the left half of the s-plane. Hence, the impulse response of the system can be expressed as:

$$C(s) = \frac{k_1}{s + p_1} + \frac{k_2}{s + p_2} + \cdots + \frac{k_n}{s + p_n}$$

where $k_1, k_2, \ldots, k_n$ are real constants. The time response of the system is

$$c(t) = k_1 e^{-p_1 t} + k_2 e^{-p_2 t} + \cdots + k_n e^{-p_n t}$$

Assume that p_1 is closer to $j\omega$-axis while $p_2, \ldots, p_n$ are farther away. In this case, p_1 will dominate the transient response of the system due to its slower

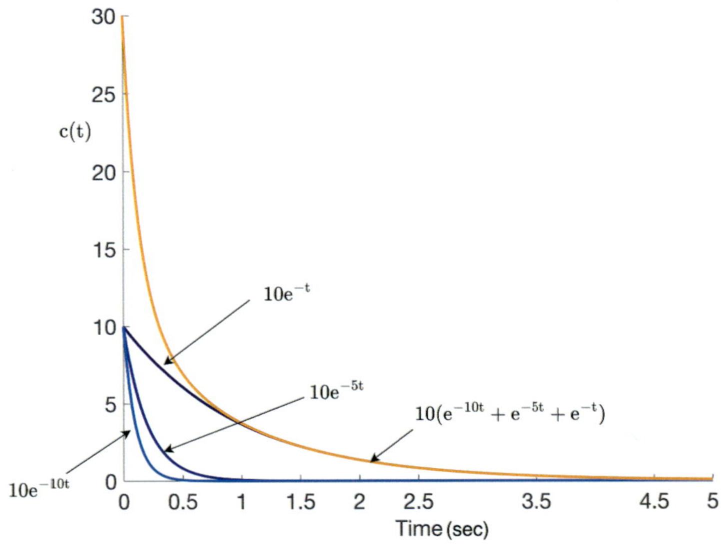

Figure 3.35: Impulse response of third-order system

decay rate. The poles $p_2, \ldots, p_n$ will have faster decay rates, contributing less to the overall transient response. As an example, consider a third-order system

$$G(s) = \frac{C(s)}{R(s)} = \frac{10}{s+1} + \frac{10}{s+5} + \frac{10}{s+10}$$

The impulse response is given by

$$c(t) = 10e^{-t} + 10e^{-5t} + 10e^{-10t}$$

Figure 3.35 shows the contribution of each pole to the impulse response of a third-order system. The contribution of the term $10e^{-10t}$ becomes zero very quickly (after 0.5sec) and of the term $10e^{-5t}$ becomes negligible after 1sec. A major part of the complete response is due to the pole at -1 ($10e^{-t}$).

From these observations, we conclude that although all the poles contribute to the transient response, the poles that are close to the imaginary axis contribute significantly, whereas the contribution of poles far away from the imaginary axis is negligible, except for the initial part of the transient response. Thus,

Key Point

Poles that are closer to the $j\omega$-axis are termed dominant poles, while those located far from the $j\omega$-axis are called non-dominant or insignificant poles.

The distance D between the dominant region and the non-dominant (insignificant) region is shown in Figure 3.36. Now comes the important question: how far from the imaginary axis the poles should be located so that they can be designated as insignificant or non-dominant. We do not have any clear-cut rules to determine this. However, as a rule of thumb,

> If the magnitude of the real part of a pole is at least 5 to 10 times greater than that of a dominant pole or a pair of complex dominant poles, then the pole may be considered non-dominant (insignificant) for the purpose of preserving transient response with acceptable error.

Similar to the poles, the zeros which are close to the jω-axis in the left-half s-plane affect the transient response significantly compared to the zeros which are located far away from the jω-axis (relative to dominant poles).

Although, the regions shown in Figure 3.36 give us a rough idea about the dominant and non-dominant regions in the s-plane, while designing the controllers (e.g., the pole placement controller), we often desire the dominant poles to be located near the regions shown in Figure 3.37 which will ensure $\xi = 0.707$. (more about this to be discussed in Chapter 8).

3.9.2 Correct Procedure of Neglecting Insignificant Poles

While approximating a higher order system to a lower order one, we must ensure that both the transient and the steady-state responses of the original system are preserved as closely as possible (with practically acceptable error). This is essential for maintaining the accuracy of the approximation while simplifying the system for analysis or design. The procedure to achieve this is as follows:

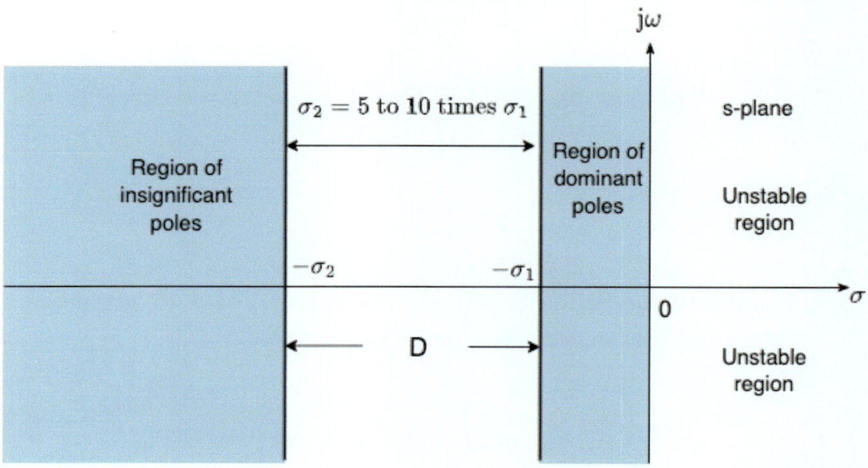

Figure 3.36: Regions of dominant and insignificant poles in the s-plane

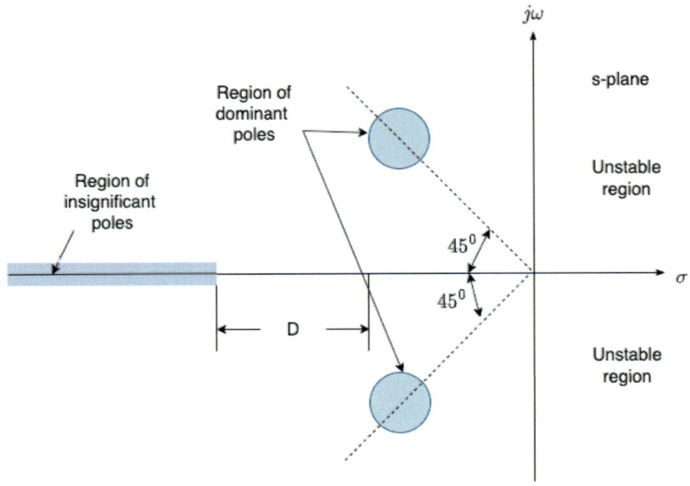

Figure 3.37: Regions of dominant and insignificant poles in the s-plane for the purpose of design

Procedure for Approximating Higher Order Systems

a. Suppose the system is represented as:

$$G(s) = \frac{K(s+z_1)(s+z_2)\ldots(s+z_m)}{s^N(s+p_1)(s+p_2)\ldots(s+p_\alpha)(s+p_{\alpha+1})\ldots(s+p_n)} \tag{3.75}$$

b. Assume that there are α-number of insignificant poles located at $s = -p_1, -p_2, \ldots, -p_\alpha$.

c. Express the terms associated with these poles in time-constant form. The transfer function of Eqn(3.75) becomes

$$G(s)=\left(\frac{K}{p_1p_2\ldots p_\alpha}\right)\frac{(s+z_1)(s+z_2)\ldots(s+z_m)}{s^N(1+T_1s)\ldots(1+T_\alpha s)(s+p_{\alpha+1})\ldots(s+p_n)} \tag{3.76}$$

where $T_i = \frac{1}{p_i}, \quad i = 1, 2, \ldots, \alpha.$

d. Eliminate the terms $(1 + T_i s)$, $i = 1, 2, \ldots, \alpha$. This gives the approximated transfer function as:

$$G(s) = \left(\frac{K}{p_1p_2\ldots p_\alpha}\right)\frac{(s+z_1)(s+z_2)\ldots(s+z_m)}{s^N(s+p_{\alpha+1})\ldots(s+p_n)} \tag{3.77}$$

We will illustrate this procedure with an example.

EXAMPLE 3.11

Consider a system with a closed-loop transfer function

$$T(s) = \frac{16000(s+1)}{(s+2)(s+8)(s+40)(s+50)} \tag{3.78}$$

The poles, located at s = −40 and s = −50, are far to the left of $j\omega$-axis. Their contribution to the transient response is negligible (they decay very quickly to zero as e^{-40t} and e^{-50t}). These poles can be neglected without appreciable error as follows:

Let us first represent the terms associated with poles in time-constant form as:

$$T(s) = \frac{16000(s+1)}{40 \times 50\cancel{(1+\frac{1}{40}s)}\cancel{(1+\frac{1}{50}s)}(s+2)(s+8)} \tag{3.79}$$

The approximated transfer function equals

$$T(s) = \frac{8(s+1)}{(s+2)(s+8)} \tag{3.80}$$

Note

a. Following the above method, the steady state value of the original system is equal to the steady state value of the approximated system. For a unit step input, it is equal to 0.5.

b. However, if we simply remove the terms $(s+40)$ and $(s+50)$ in Eqn(3.78), the approximated second order system will have a different steady state value, i.e 1000 to a step input.

The unit step response of the original transfer function in Eqn(3.78) and reduced order (approximated) transfer function in Eqn(3.80) is compared and shown in Figure 3.38. It is observed that the step responses of the original and approximated reduced-order systems are almost identical with the same steady-state value.

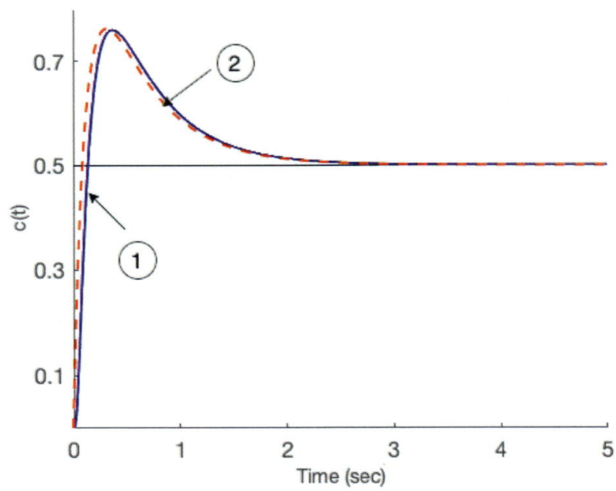

Figure 3.38: Unit step response: (1) Original system (2) Approximated system

CONCEPT REINFORCEMENT PROBLEM 3.8

Problem

Obtain the second-order approximate model of the system with closed-loop transfer function

$$T(s) = \frac{10(s+4)}{(s^2+2s+2)(s^2+20s+101)}$$

and compare the step response of the original system with the approximated system. Use MATLAB to get the response.

3.9.3 Pole-Zero Cancellation

If some of the poles of a transfer function are either exactly equal to or very near to the zeros of the transfer function, it may be possible to simplify or reduce the order of the system by cancelling those poles and zeros. This process is known as pole-zero cancellation. For example, consider a third-order system with a zero given by

$$T(s) = \frac{K(s+z_1)}{(s+p_3)(s^2+\alpha s+\beta)} \tag{3.81}$$

a. If the pole at $s = -p_3$ is exactly equal to the zero at $s = -z_1$, then the system is reduced to lower order as

$$T(s) = \frac{K\cancel{(s+z_1)}}{\cancel{(s+p_3)}(s^2+\alpha s+\beta)} = \frac{K}{s^2+\alpha s+\beta} \tag{3.82}$$

b. However, if the pole at $s = -p_3$ is very close to the zero at $s = -z_1$, then express the transfer function into partial fractions and determine the residues. If the residue of the pole $-p_3$ is very small, then we can cancel the pole-zero pair.

We will illustrate this by an example.

EXAMPLE 3.12

Problem

The output response function of two systems is given by

$$C_1(s) = \frac{130(s+5)}{s(s+4.5)(s+7)(s+9)}$$
$$C_2(s) = \frac{130(s+5)}{s(s+4.9)(s+7)(s+9)}$$

Determine for which of these functions the pole-zero cancellation is valid and find its response.

SOLUTION:

The partial fraction expansion of $C_1(s)$ is given by

$$C_1(s) = \frac{2.0635}{s} + \frac{3.9394}{s+10} - \frac{4.9524}{s+7} - \frac{1.0505}{s+4.5} \tag{3.83}$$

The pole at -4.5 is closest to the zero at -5. The residue of this pole equals 1.0505. Compared to other residues, this is not negligible. Thus, a lower order approximation of $C_1(s)$ should not be made.

The partial-fraction expansion of $C_2(s)$ is

$$C_2(s) = \frac{1.8950}{s} + \frac{4.2484}{s+10} - \frac{5.8957}{s+7} - \frac{0.2477}{s+4.9} \tag{3.84}$$

The residue of the pole at -4.9, which is closest to the zero at -5, is equal to 0.2477. This is much less compared to the residues at other poles. Hence, we can neglect the contribution of the pole at -4.9. The approximated response is given by

$$c_2(t) = 1.8950 + 4.2484e^{-10t} - 5.8957e^{-7t} \tag{3.85}$$

The response of the approximated system is compared to the original system and is shown in Figure 3.39, which shows an acceptable approximation error.

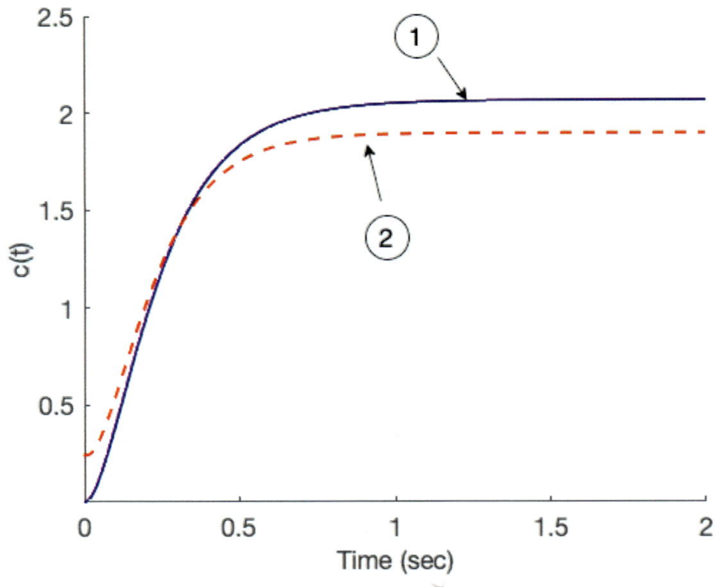

Figure 3.39: Unit step response: (1) Original system (2) Approximated systems with pole-zero cancellation

3.10 Multiple Choice Skills Test Questions

M 3.1. A system with a large time constant usually corresponds to

A. an underdamped system

B. an overdamped system

C. faster system

D. sluggish system

M 3.2. The unit step response of a system starting from rest is given by

$$c(t) = 1 - e^{-5t}$$

The transfer function of the system is

A. $\frac{1}{s+5}$ **B.** $\frac{s}{s+5}$ **C.** $\frac{5}{s+5}$ **D.** $\frac{5s}{s+5}$

M 3.3. The closed-loop transfer function of a system is given by

$$T(s) = \frac{C(s)}{R(s)} = \frac{100}{s^2 + 8s + 100}$$

The damping ratio ξ, natural frequency ω_n and the nature of time response are respectively

A. $\xi = 0.8$, $\omega_n = 10$ rad/sec, underdamped

B. $\xi = 0.4$, $\omega_n = 10$ rad/sec, underdamped

C. $\xi = 1.0$, $\omega_n = 8$ rad/sec, critically damped

D. $\xi = 2.0$, $\omega_n = 2$ rad/sec, overdamped

M 3.4. The closed-loop transfer function of a system is given by

$$T(s) = \frac{C(s)}{R(s)} = \frac{100}{s^2 + 10s + 25}$$

The damping ratio ξ, natural frequency ω_n and the nature of time response are respectively

A. $\xi = 1.0$, $\omega_n = 5$ rad/sec, critically damped

B. $\xi = 0.5$, $\omega_n = 10$ rad/sec, underdamped

C. $\xi = 2.0$, $\omega_n = 5$ rad/sec, overdamped

D. $\xi = 0.0$, $\omega_n = 10$ rad/sec, undamped

M 3.5. Control systems are usually designed with a damping factor

A. $\xi = 0$ **B.** $\xi > 1$ **C.** $\xi = 1$ **D.** $\xi < 1$

M 3.6. For a second-order critically-damped system, the poles are

A. complex conjugate pair with negative real part

B. purely imaginary

C. real and equal

D. real and unequal

M 3.7. For a second-order undamped system, the poles are

A. complex conjugate pair with negative real part

B. purely imaginary

C. real and equal

D. real and unequal

M 3.8. For a second-order overdamped system, the poles are

A. complex conjugate pair with negative real part

B. purely imaginary

C. real and equal

D. real and unequal

M 3.9. For underdamped systems, the damping ratio

A. $\xi > 1$ **B.** $\xi < 1$ **C.** $\xi = 0$ **D.** $\xi = \infty$

M 3.10. For undamped systems, the damping ratio

A. $\xi > 1$ **B.** $\xi < 1$ **C.** $\xi = 0$ **D.** $\xi = \infty$

M 3.11. For overdamped systems, the damping ratio

A. $\xi = \infty$ **B.** $\xi < 1$ **C.** $\xi = 0$ **D.** $\xi > 1$

M 3.12. The open-loop transfer function of a unity feedback system is given by

$$G(s) = \frac{K}{s(s+5)}$$

If the gain K is increased to infinity, then the damping ratio will tend to become

A. $\xi = \infty$ **B.** $\xi = 1$ **C.** $\xi = 0$ **D.** $\xi = 0.5$

M 3.13. The unit impulse response function of a linear time-invariant second-order system is $g(t) = e^{-6t} \sin 8t$. The natural frequency and damping factor of the system are respectively

A. 6 rad/sec and 1.0

B. 8 rad/sec and 0.75

C. 12 rad/sec and 0.5

D. 10 rad/sec and 0.6

M 3.14. For a second-order system, if the damping ratio ξ is increased from zero, the response becomes

A. progressively less oscillatory

B. progressively more oscillatory

C. critically damped

D. zero

M 3.15. The output of a prototype second-order system for a unit step input is given as

$$c(t) = 1 - \frac{2}{\sqrt{3}} e^{-2t} \sin(3.4641 + \theta)$$

The closed-loop transfer function is

A. $\dfrac{16}{s^2 + 4s + 16}$

B. $\dfrac{1}{s^2 + 4s + 16}$

C. $\dfrac{2}{(s+2)(s+\sqrt{3}}$

D. $\dfrac{4}{s^2 + 2s + 8}$

M 3.16. If the closed-loop poles of a prototype second order system are located at $-a \pm jb$, then the damping coefficient ξ and natural frequency ω_n of the system are

A. $\xi = \dfrac{a}{\sqrt{a+b}}$ and $\omega_n = \sqrt{a+b}$

B. $\xi = \dfrac{a}{\sqrt{a^2+b^2}}$ and $\omega_n = a^2 + b^2$

C. $\xi = \dfrac{b}{\sqrt{a^2+b^2}}$ and $\omega_n = \sqrt{a^2+b^2}$

D. $\xi = \dfrac{a}{\sqrt{a^2+b^2}}$ and $\omega_n = \sqrt{a^2+b^2}$

M 3.17. The roots of the characteristic equation of a prototype second-order system are located at $-a \pm jb$. The step response of the system will settle (2% bound) at

A. $\dfrac{4}{a}$ sec

B. $\dfrac{8}{a}$ sec

C. $\dfrac{a}{4}$ sec

D. $\dfrac{4}{b}$ sec

M 3.18. Consider the unity feedback control system shown in Figure 3.40

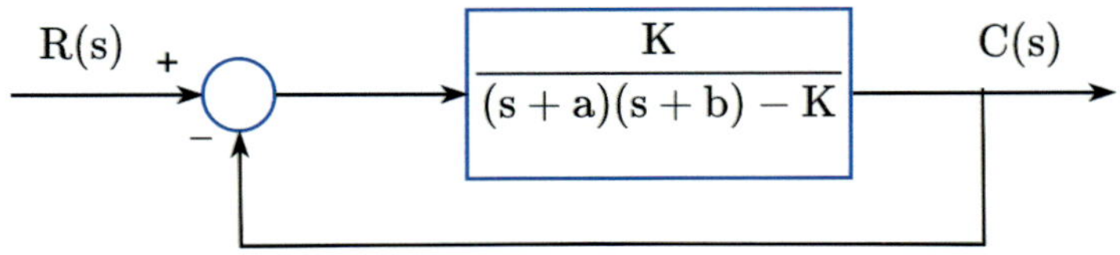

Figure 3.40: Closed-loop control system

The natural frequency ω_n of the closed-loop system is

A. $\omega_n = \sqrt{K}$

B. $\omega_n = \sqrt{ab}$

C. $\omega_n = ab$

D. $\omega_n = \dfrac{\sqrt{K}}{ab}$

M 3.19. For the system shown in Figure 3.40 the settling time t_s (considering 2% band) is

A. $t_s = \dfrac{8}{a+b}$ **B.** $t_s = \dfrac{a+b}{8}$ **C.** $t_s = \dfrac{4}{a+b}$ **D.** $t_s = \dfrac{8}{ab}$

M 3.20. If the parameters a and b of the system shown in Figure 3.40 are doubled, the settling time

A. is halved

B. remains unchanged

C. is doubled

D. is quadrupled

M 3.21. The unity feedback control system shown in Figure 3.41 has a damping ratio 0.6 and a damped frequency of oscillation of 10 rad/sec. The values of K and a are respectively

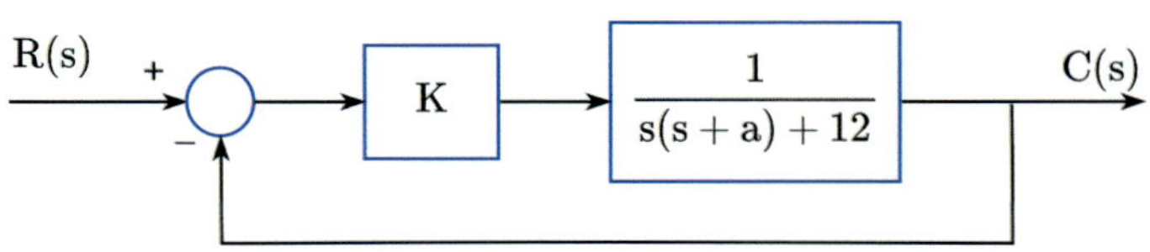

Figure 3.41: Closed-loop control system

A. $K = 3.5355$ and $a = -7.75$

B. $K = 12.5$ and $a = 3$

C. $K = 156.25$ and $a = 3$

D. $K = 156.25$ and $a = 1$

M 3.22. The roots of the characteristic equation of a prototype second-order system are located at $-a \pm jb$. The peak time t_p of the system is

A. $\frac{2}{\pi b}$ sec **B.** $\frac{\pi}{b}$ sec **C.** $\frac{\pi}{4b}$ sec **D.** $\frac{1}{b}$ sec

M 3.23. The closed-loop transfer function of a system is given by

$$T(s) = \frac{10K}{s^2 + 10s + K}$$

For this system to be critically damped, the value of K is

A. $K = 10$ **B.** $K = 25$ **C.** $K = 5$ **D.** $K = 2.5$

M 3.24. A second-order control system exhibits 100% overshoot. Its damping ratio ξ is

A. $\xi = 1$ **B.** $\xi > 1$ **C.** $\xi < 1$ **D.** $\xi = 0$

M 3.25. The roots of the characteristic equation of a prototype second-order system are located at $-a \pm jb$. The first undershoot occurs at

A. $\frac{2\pi}{b}$ sec **B.** $\frac{\pi}{b}$ sec **C.** $\frac{\pi}{4b}$ sec **D.** $\frac{1}{b}$ sec

M 3.26. For a typical mass-spring-damper system with mass $m = 1$, damping constant $f = 2$, and spring constant $k = 1$, the transient response of displacement due to a force would be

A. overdamped

B. underdamped

C. critically damped

D. the nature of the transient response depends on the magnitude of the applied force

M 3.27. Consider the following unity feedback systems with open-loop transfer functions:

S1: $\frac{16}{s(s+1.6)}$ **S2:** $\frac{100}{s(s+16)}$ **S3:** $\frac{25}{s(s+4)}$ **S4:** $\frac{64}{s(s+9.6)}$

What is the correct sequence of these systems in increasing order of settling time?

A. S4,S2,S3,S1

B. S2,S4,S1,S3

C. S2,S3,S1,S4

D. S2,S4,S3,S1

M 3.28. Consider the system shown in Figure 3.42

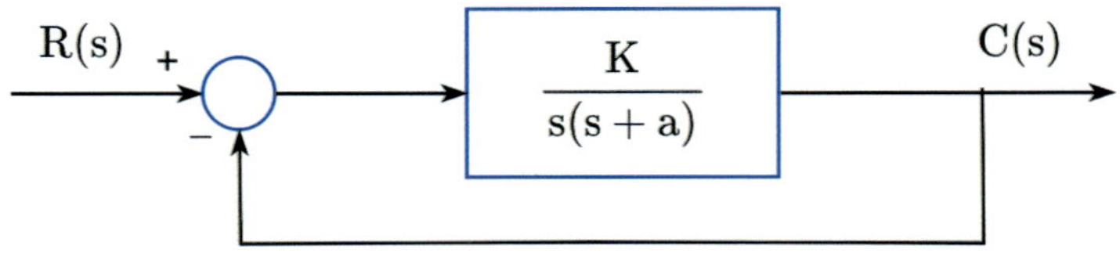

Figure 3.42: Closed-loop control system

What is the value of K such that the step response does not have any overshoot with minimum settling time.

A. $K = \frac{4}{a^2}$ **B.** $K = \frac{a^2}{2}$ **C.** $K = \frac{a^2}{4}$ **D.** $K = \frac{a}{4}$

M 3.29. Match the transfer functions of the second-order systems with the nature of the system given below

Transfer functions	Nature of system
S1: $\frac{25}{s^2 + 10s + 25}$	I. Underdamped
S2: $\frac{100}{s^2 + 2s + 100}$	II. Overdamped
S3: $\frac{81}{s^2 + 81}$	III. Critically damped
S4: $\frac{16}{s^2 + 12s + 16}$	IV. Undamped

A. S1-II,S2-I,S3-IV, S4-III

B. S1-III,S2-IV,S3-I, S4-II

C. S1-III,S2-I,S3-IV, S4-II

D. S1-I,S2-III,S3-IV, S4-II

M 3.30. Which of the following systems exhibits maximum overshoot for a step input? The closed-loop transfer function of these systems is:

A. S1: $\frac{100}{s^2 + 10s + 100}$

B. S2: $\frac{100}{s^2 + 2s + 100}$

C. S3: $\frac{100}{s^2 + 16s + 100}$

D. S4: $\frac{100}{s^2 + 25s + 100}$

M 3.31. The dynamics of a second-order system is represented by the differential equation

$$\frac{d^2y}{dt^2} + 12\frac{dy}{dt} + 36y = u$$

The response of the system to a unit impulse input is of the form

A. $k_1 e^{-t} + k_2 e^{-6t}$

B. $k e^{-6t} \sin(6t + \phi)$

C. $(k_1 + k_2 t)\, e^{-6t}$

D. $(k_1 + k_2 t)\, e^{6t}$

M 3.32. If two identical first-order stable low-pass filters are cascaded non-interactively, then the unit step response of the composite filter will be

A. overdamped

B. undamped

C. underdamped

D. critically damped

M 3.33. Match the pole locations with the corresponding step responses shown in Figure 3.43

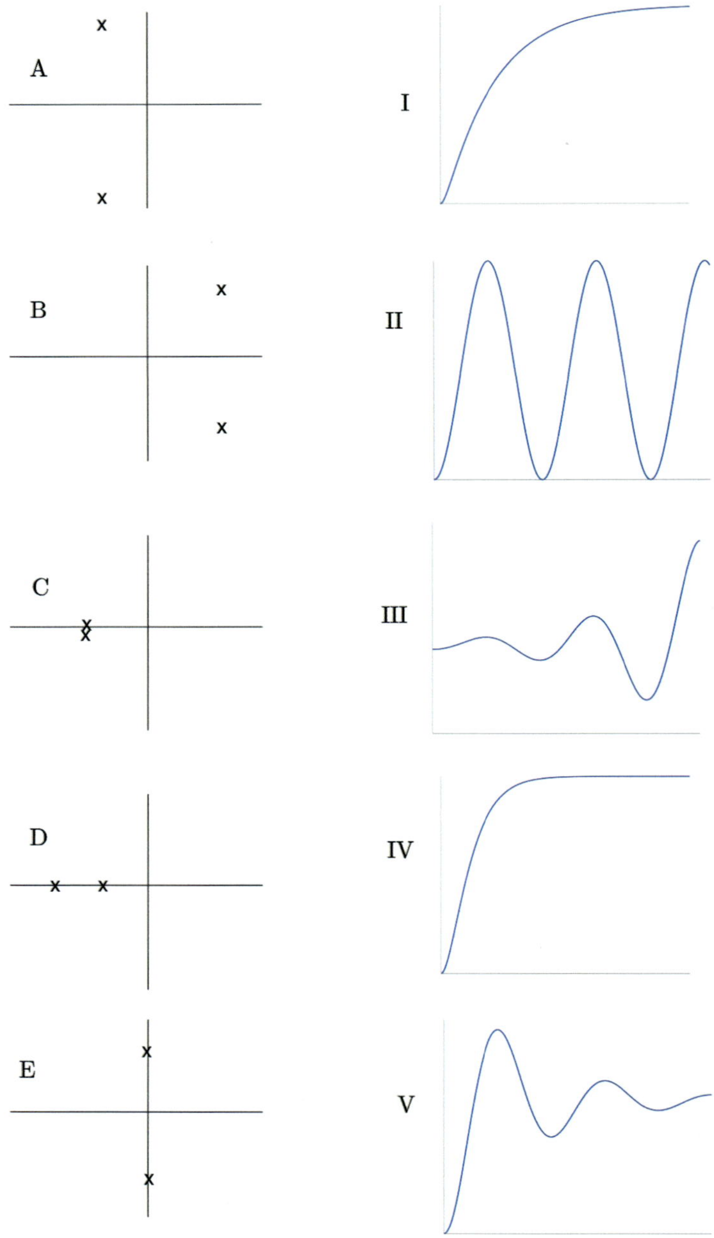

Figure 3.43: Location of poles and system step responses

A. A – V, B – III, C – I, D – IV, E – II

B. A – V, B – III, C – IV, D – I, E – II

C. A – II, B – III, C – I, D – IV, E – V

D. A – V, B – II, C – IV, D – I, E – III

M 3.34. The closed-loop poles of a second-order system are shown in Figure 3.44

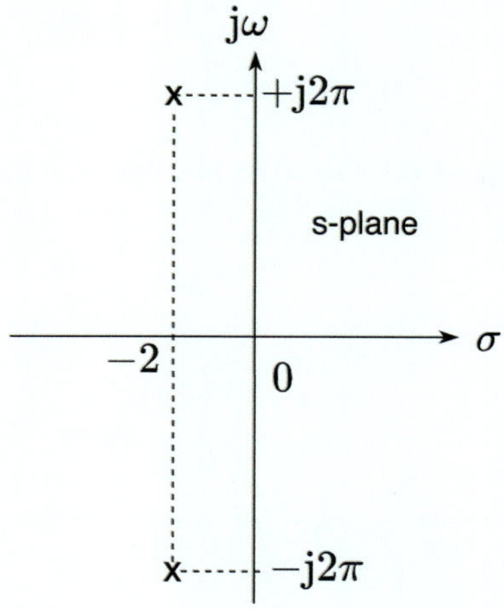

Figure 3.44: Location of poles of a second order system

The settling time t_s (2% criterion) and peak time t_p of the system are

A. $t_s = 3$ sec and $t_p = 1.0$ sec
B. $t_s = 1.5$ sec and $t_p = 0.5$ sec
C. $t_s = 2$ sec and $t_p = 0.5$ sec
D. $t_s = 4$ sec and $t_p = 2$ sec

M 3.35. The transfer function of a system is given by

$$G(s) = \frac{b_0 + b_1 s + b_2 s^2 + \cdots + b_m s^m}{a_0 + a_1 s + a_2 s^2 + \cdots + a_n s^n}, \quad m \leq n$$

The DC gain of the system is

A. $\frac{a_0}{b_0}$ **B.** $\frac{b_m}{a_n}$ **C.** $\frac{a_n}{b_m}$ **D.** $\frac{b_0}{a_0}$

M 3.36. Consider a system with a closed-loop transfer function

$$T(s) = \frac{6000(s+5)}{(s+2)(s+3)(s+20)(s+30)}$$

Its lower order approximation is

A. $T(s) = \dfrac{6000(s+5)}{(s+2)(s+3)}$

B. $T(s) = \dfrac{10(s+5)}{(s+2)(s+3)}$

C. $T(s) = \dfrac{1000(s+5)}{(s+20)(s+30)}$

D. $T(s) = \dfrac{1200(s+5)}{(s+2)(s+3)}$

M 3.37. The pole-zero map of a system is shown in Figure 3.45

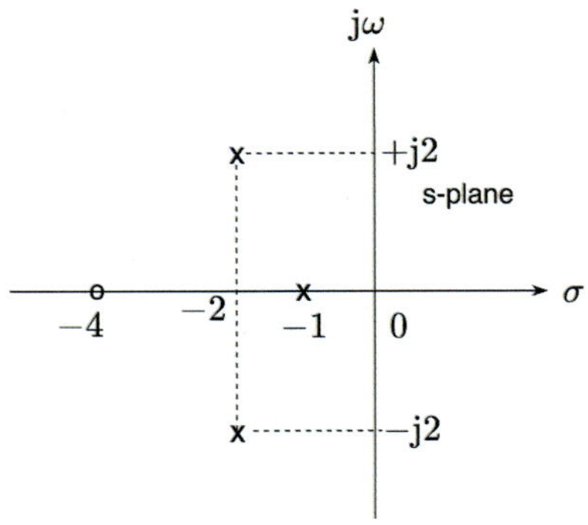

Figure 3.45: Pole-Zero map of a control system

The value of K which would yield a DC gain of 40 is

A. 80 **B.** 20 **C.** 40 **D.** 100

M 3.38. The closed-loop transfer function of a unity feedback control system is

$$T(s) = \frac{s+2}{s^2+7s+27}$$

The open-loop DC gain is

A. $\dfrac{2}{27}$ **B.** $\dfrac{2}{25}$ **C.** $\dfrac{1}{25}$ **D.** 2

M 3.39. By adding a pole to the forward path transfer function

A. decreases the peak overshoot whose magnitude increases with the decrease of the distance of the added pole from jω-axis

B. increases the peak overshoot whose magnitude increases with the increase of the distance of the added pole from jω-axis

C. increases the peak overshoot whose magnitude increases with the decrease of the distance of the added pole from jω-axis

D. decreases the peak overshoot whose magnitude increases with the increase of the distance of the added pole from jω-axis

M 3.40. Addition of a zero to the closed-loop transfer function

A. decreases the rise time and increases the peak overshoot as the zero is placed closer to the jω-axis

B. increases both the rise time and the peak overshoot as the zero is placed closer to the jω-axis

C. decreases both the rise time and the peak overshoot as the zero is placed closer to the jω-axis

D. increases the rise time and decreases the peak overshoot as the zero is placed closer to the jω-axis

Answers

M 3.1. D **M 3.2**. C **M 3.3**. B **M 3.4**. A **M 3.5**. D **M 3.6**. C **M 3.7**. B **M 3.8**. D **M 3.9**. B **M 3.10**. C **M 3.11**. D **M 3.12**. C **M 3.13**. D **M 3.14**. A **M 3.15**. A **M 3.16**. D **M 3.17**. A **M 3.18**. B **M 3.19**. A **M 3.20**. A **M 3.21**. C **M 3.22**. B **M 3.23**. B **M 3.24**. D **M 3.25**. A **M 3.26**. C **M 3.27**. D **M 3.28**. C **M 3.29**. C **M 3.30**. B **M 3.31**. C **M 3.32**. D **M 3.33**. A **M 3.34**. C **M 3.35**. D **M 3.36**. B **M 3.37**. A **M 3.38**. B **M 3.39**. B **M 3.39**. C **M 3.40**. A

3.11 Knowledge Assessment Problems

KaP 3.1 Determine the output step response, time constant, rise time, and settling time for each of the first-order systems whose transfer functions $G(s) = \frac{C(s)}{R(s)}$ are given below.

a. $G(s) = \frac{15}{s+3}$

b. $G(s) = \frac{15}{s+10}$

c. $G(s) = \frac{50}{s+5}$

d. $G(s) = \frac{50}{s+100}$

KaP 3.2 For each of the transfer functions $T(s) = \frac{C(s)}{R(s}$, express the natural response in general form without solving using inverse Laplace transform.

a. $T(s) = \frac{20}{(s^2+6s+9)(s^2+2s+17)}$

b. $T(s) = \frac{10s+15}{s(s^2+4s+29)}$

c. $T(s) = \frac{5s^2}{(s^2+4s+4)(s^2+2s+26)}$

d. $T(s) = \frac{2s^3-5s^2+3s+50}{(s^2+2s+2)(s^2+4s+8)}$

e. $T(s) = \frac{20s^2}{(s+3)^2(s^2+2s+3)(s^2+6s+25)}$

KaP 3.3 For each of the second-order systems with the following closed-loop transfer functions, determine the undamped natural frequency, the damping ratio, and the damped oscillatory frequency.

a. $T(s) = \frac{36}{(s^2+6s+36)}$

b. $T(s) = \frac{100}{s^2+5s+16}$

c. $T(s) = \frac{s^2+10}{s^2+8s+25}$

d. $T(s) = \frac{-5s+2}{(s^2+0.45s+20)}$

KaP 3.4 Determine the damping ratio ξ, natural frequency ω_n, setting time T_s, peak time T_p, rise time T_r and percentage overshoot $\%M_p$ for each of the systems whose closed-loop transfer functions are given below.

a. $T(s) = \frac{25}{s^2+4s+25}$

b. $T(s) = \frac{100}{s^2+12s+100}$

c. $T(s) = \frac{10}{s^2+1.6s+1}$

KaP 3.5 Consider a unity feedback system with forward path transfer function

$$G(s) = \frac{K}{s(s+a)}$$

Compute the rise time T_r, peak time T_p, maximum overshoot M_p, and settling time T_s when the system is subjected to step input.

KaP 3.6 Consider the system shown in the Figure 3.46 where $\xi = 0.6$ and $\omega_n = 5\ rad/sec$. Compute the rise time T_r, peak time T_p, maximum overshoot M_p, and settling time T_s when the system is subjected to step input.

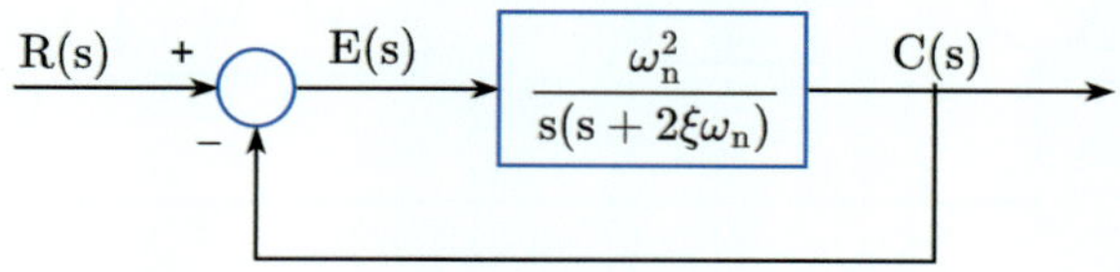

Figure 3.46: Second Order Unity Feedback System

KaP 3.7 Determine the value of β for the following closed-loop transfer functions, which exhibit the given second order specification.

a. $T(s) = \dfrac{50}{s^2 + 12s + \beta}$, Desired $\xi = 0.6$.

b. $T(s) = \dfrac{s^2 + 10}{\beta s^2 + 2s + 10}$, Desired $\omega_n = 5$.

c. $T(s) = \dfrac{15s}{5s^2 + 15s + \beta + 3}$, Desired $\omega_n = 4$.

KaP 3.8 For the system shown in Figure 3.47, determine the value of gain K, such that the damping ratio of the closed-loop system is 0.75.

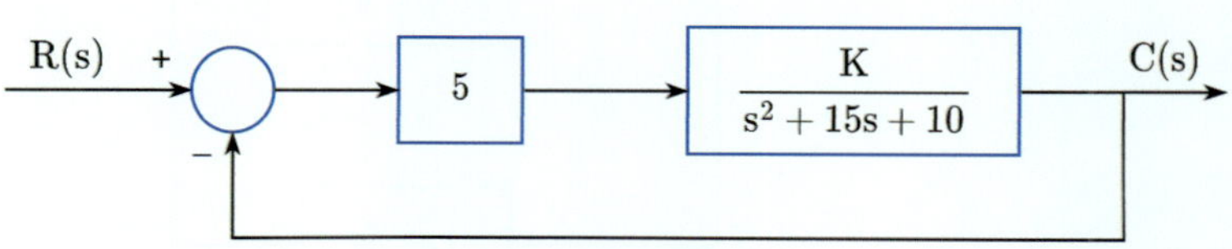

Figure 3.47: Closed-loop control system

KaP 3.9 For the system shown in Figure 3.48, find the value of K_1, in terms of other parameters, for which the natural component of the output decays at least as rapidly as $\exp(-\beta t)$. Assume that the closed-loop system exhibits an underdamped response. Determine the value of K_1 for $K_2 = 5$, $a = 4$ and $\beta = 15$.

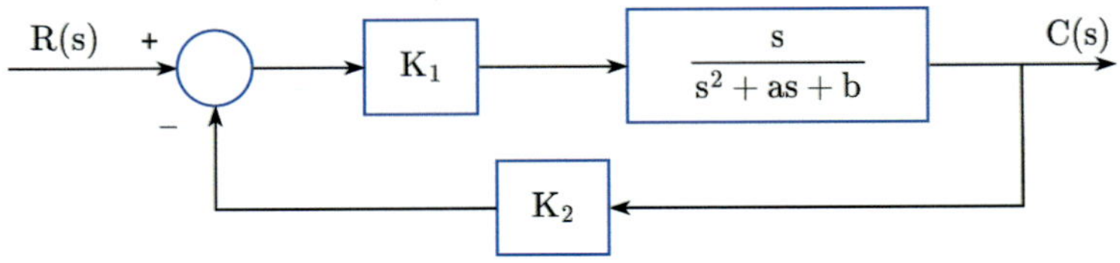

Figure 3.48: Closed-loop control system

KaP 3.10 Consider the system shown in Figure 3.49.

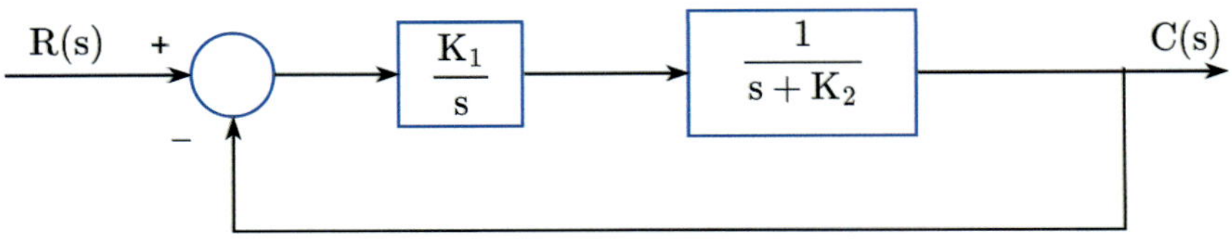

Figure 3.49: Control system

a. Determine the values of K_1 and K_2 so that the damping ratio of the system $\xi = 0.7$ and the damped frequency of oscillation $\omega_d = 12$ rad/sec.
b. The percentage peak overshoot.
c. The time required for the step response to stabilize is 2% of the final value.
d. Number of cycles completed before the output is stabilized.

KaP 3.11 The block diagram of a position control system with velocity feedback is shown in Figure 3.50

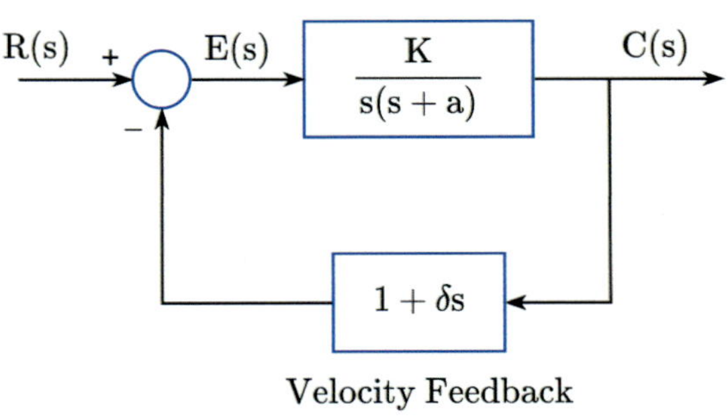

Figure 3.50: Position Control System with Velocity Feedback

Determine the value of δ which ensures that the maximum overshoot in the step response is 5%.Assume $K = 16$ and $a = 3$.

KaP 3.12 The block diagram of a position control system with velocity feedback is shown in Figure 3.51

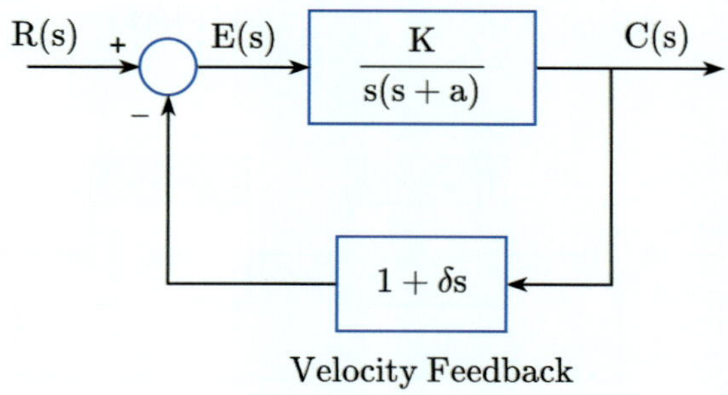

Figure 3.51: Position Control System with Velocity Feedback

Determine the value of K and a which ensures that the maximum overshoot in the step response is 10% and settling time $T_s = 2.0$ sec. Assume $\delta = 0.1$

KaP 3.13 The block diagram of a feedback control system is shown in Figure 3.52

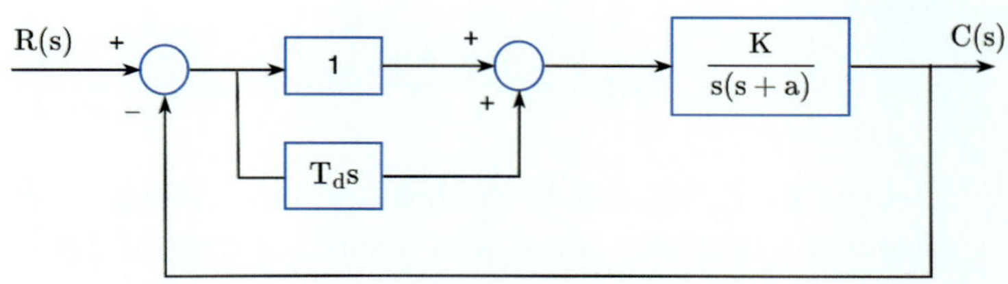

Figure 3.52: Closed-loop control system

Determine the value of T_d which would make the system critically damped. Assume $K = 16$ and $a = 4$.

KaP 3.14 For the system shown in Figure 3.52, determine the value of '**a**' such that the system is

i. Critically damped

ii. Underdamped with damping ratio $\xi = 0.8$.

Assume $K = 9$ and $T_d = 0.5$.

KaP 3.15 The controller and plant transfer function of a unity feedback control system shown in Figure 3.53 are

$$G_c(s) = \frac{s+\alpha}{s+\beta} \quad \text{and} \quad G(s) = \frac{K}{s+a}$$

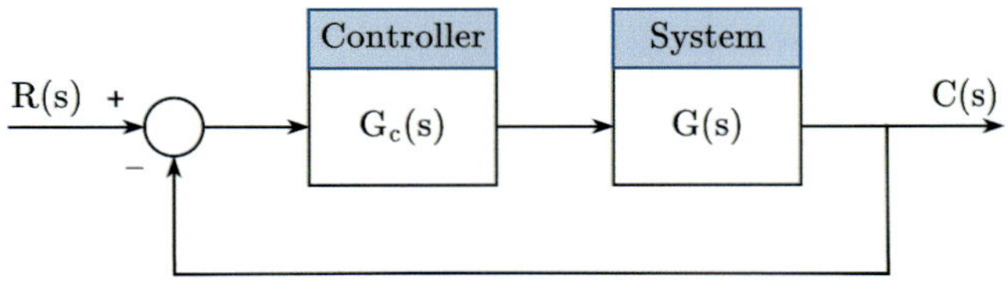

Figure 3.53: Feedback control system cascaded in series with a controller

Determine the value of α and β so that the overshoot in the closed-loop system does not exceed 20% and the settling time is 0.8sec (2% criterion). Assume K = 1 and a = 4.

KaP 3.16 Consider the feedback control system shown in Figure 3.53

where

$$G_c(s) = \frac{K(s+a)}{s+b} \quad \text{and} \quad G(s) = \frac{1}{s(s+5)}$$

Determine K, a, and b so that the closed-loop system exhibits a 10% overshoot to a step input and a settling time of 2.0 sec (2% criterion).

KaP 3.17 The closed-loop transfer function of a unity feedback control system is

$$T(s) = \frac{80}{s^2 + 1.6s + 16}$$

Determine

i. Natural frequency, damped natural frequency, and damping ratio.
ii. Peak time, settling time, and the time at which 1st undershoot occurs
iii. Number of cycles the output undergoes before settling down

KaP 3.18 The open-loop transfer function of a unity feedback system is given by

$$G(s) = \frac{K}{s(1+Ts)}$$

By what factor the gain $K_1 = K$ should be multiplied such that the peak overshoot is reduced from 60% to 20%. Assume T remains constant.

KaP 3.19 Figure 3.54 shows the block diagram of a DC servo mechanism.

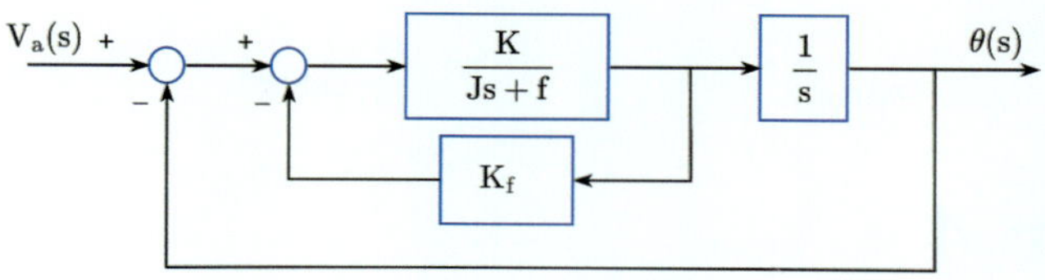

Figure 3.54: DC servo control system

Assume J=1 kg-m^2 and f=1 N-m-/rad/sec. The maximum overshoot of the unit step response is 0.15, and the peak time is 1sec.

i. Determine the damping ratio and natural frequency.

ii. Find the gain K and the velocity feedback gain K_f.

iii. Compute the rise and settling time

KaP 3.20 Find the percent overshoot, peak time, rise time, and settling time for those systems shown below that can be approximated as second-order systems. Represent the approximated system and compare step responses with the original system using MATLAB.

$$\textbf{S1:}\quad G(s) = \frac{100}{(s+12)(s^2+4s+25)}$$

$$\textbf{S1:}\quad G(s) = \frac{20}{(s+1)(s^2+s+10)}$$

$$\textbf{S3:}\quad G(s) = \frac{50}{(s+5)(s+10)(s^2+2s+9)}$$

Chapter 4

Stability Analysis of Linear Systems

A stable system is like a happy marriage—arguments happen, but you make up over dessert; an unstable system throws tantrums like a toddler on a sugar rush, and a marginally stable system is like that friend who's been 'leaving in five minutes' for the past three hours.

DESIRED LEARNING OUTCOMES

Upon completion of this chapter, readers should be able to:

- Understand the concept of stability of dynamic systems
- Differentiate between stable, unstable, and marginally stable systems
- Understand the concept of bounded-input-bounded-output(BIBO) stability
- Identify the conditions for a system to be BIBO stable
- Investigate stability using the Hurwitz criterion
- Construct the Routh array and investigate stability using the Routh-Hurwitz criterion
- Determine the range of a given parameter that ensures stability
- Investigate the relative stability of a system using the Routh-Hurwitz criterion
- Understand the limitations of the Ruth-Hurwitz criterion

DOI: 10.1201/9781003571308-4

4.1 Introduction

In Chapter 3, we have discussed certain performance specifications such as *rise time, peak time, percentage overshoot, settling time*, etc., which are used in the design of controllers. While these performance specifications are crucial for achieving desired system behavior, stability is the foundation upon which all other aspects of control system design are built. It is often possible to make compromises on other performance specifications in control system design; however, stability is non-negotiable. Stability must always be ensured as the top priority in control system design and operation. Any compromises on stability can indeed have serious consequences for safety, reliability, and regulatory compliance

It is not possible to give a general definition of stability that will be applicable for all types of systems: linear, nonlinear, time invariant, time varying, and so on. While stability may present itself differently across various types of systems and mathematical formulations, the fundamental idea remains consistent: a stable system maintains its desired behavior and returns to equilibrium following disturbances or external excitations. Roughly, we can say that a system is stable if small changes in system inputs or initial conditions, or system parameters, do not result in large changes in system behavior. Intuitively, a system is stable if it remains at rest unless excited by an external source and returns to rest if all excitations are removed. Let us understand this concept intuitively through some simple examples.

Consider an example of a cone shown in Figure 4.1.

Suppose it is placed on its bottom side on the flat surface as shown in Figure 4.1 a. If we apply a small force, the cone swings on its bottom for some time, but returns to its initial position thereafter. This is an example of **stable dynamics**. Suppose the cone is placed on its side as shown in Figure 4.1b. If it is displaced from its previous position to a new position, it remains in its new position. It neither returns to its previous position nor goes farther from the new position. It is called a neutral equilibrium and the cone is **neutrally stable**. Finally, suppose the cone stands on its top as shown in Figure 4.1c.

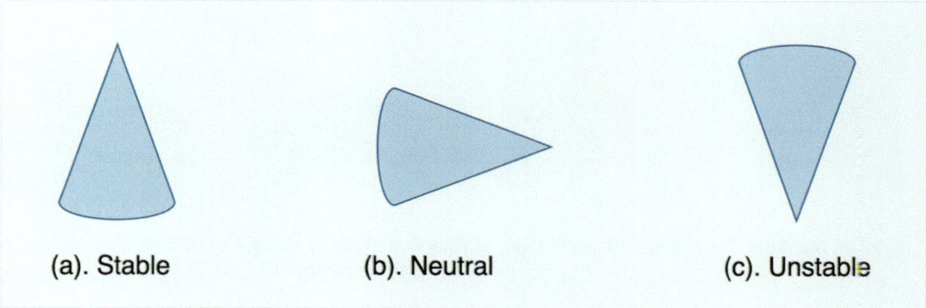

Figure 4.1: Stable, neutrally stable, and unstable positions of a cone

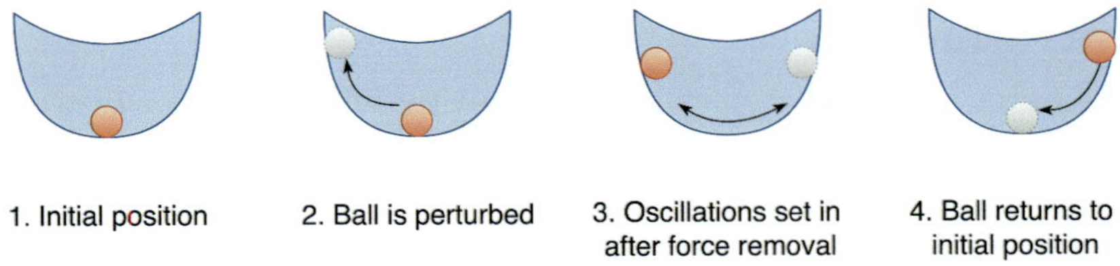

Figure 4.2: Absolute stability: ball and bowl example

It is obvious that even a small disturbance or unbalance will cause the cone to fall off. This is an example of **unstable dynamics**.

Let us discuss this concept further through an example of a ball and bowl where the ball is resting at different positions of the bowl, which is depicted in Figure 4.2. Suppose the ball is initially inside the bowl in position 1.

If it is perturbed from its initial position by a small force, it will cause the ball to move. When the force is removed, the ball will oscillate and eventually return to its initial position. This is an example of an absolutely stable dynamics.

Next, consider the situation where the bowl is turned upside down and the ball is placed on the top of the bowl. When the ball is slightly perturbed by the application of a force, it begins to move on its own without any additional force applied. It will never return to its original position (see Figure 4.3). This is an example of unstable dynamics or unstable equilibrium.

If the ball is placed on a flat surface, then after the application of a small force, the ball will move. But when the force is withdrawn, the ball stops and remains in its new position as shown in Figure 4.4. It is an example of neutral stability/neutral equilibrium.

From foregoing discussions, we find that, in all three cases, at the initial position, the ball is in equilibrium. However, only in one case, the ball absolutely stable. We must therefore distinguish between a system being in equilibrium and a system being stable. These two things are quite different.

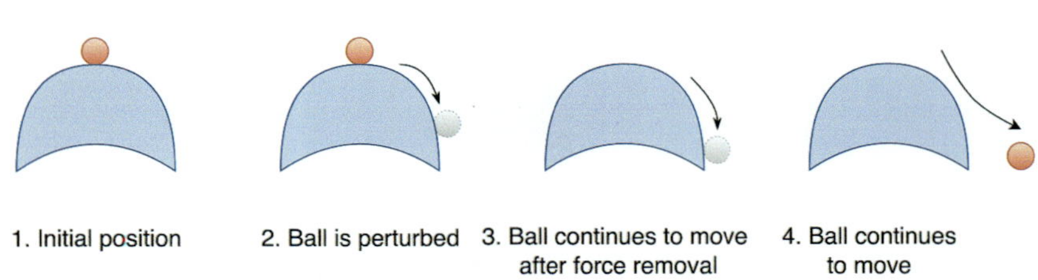

Figure 4.3: Absolute instability: ball and bowl example

Figure 4.4: Neutral stability: ball and bowl example

Let us understand the concept of stability from a broader perspective. Suppose a system has an equilibrium point at $x = x_e$. If the system is perturbed from its equilibrium position, this begins to move. In stability studies, we generally address the following questions:

1. If the system with zero input is perturbed from its equilibrium point x_e at $t = t_0$, will the system output $y(t)$
   a. Return to x_e?
   b. Remain close to x_e? or
   c. Diverge from x_e?
2. If the system is relaxed, will a **bounded input** produce **a bounded output**?

4.2 Definitions of Stability, Instability, and Marginal Stability

Before formally defining the stability of LTI systems, let us review two types of responses discussed in Chapter 1.

1. **Zero-state response (ZSR)** This is the part of the response due to input excitation only, with all initial conditions of the system being zero.
2. **Zero-input response (ZIR)** This is the part of the response due to the initial conditions only; with all inputs being zero. This is also called as **natural response or transient response**.

The total response of a system c(t) of the dynamic system is expressed as:

$$c(t) = c_{ZSR}(t) + c_{ZIR}(t)$$

The natural response (ZIR) of a dynamic system is solely due to initial conditions. This response arises from the energy stored in the energy-storing elements of the system, such as capacitors and inductors in electrical circuits or springs and masses in mechanical systems. In contrast, a static system does not have a natural response. Its response is solely determined by the input excitation. Note that to distinguish between natural response and forced response from the total response is challenging.

Let us define various types of stability based on the natural response and forced response. Based on the **natural response**, we define stability, instability, and marginal stability as follows:

Definition 4.1. **Stable:** A linear time-invariant system is stable provided its natural response converges to zero as time approaches infinity.

Definition 4.2. **Unstable:** A linear time-invariant system is unstable if its natural response grows without bound as time approaches infinity.

Definition 4.3. **Marginally Stable:** A linear time-invariant system is marginally stable if its natural response neither decays nor grows unbounded but remains constant or oscillates as time approaches infinity.

The above definitions of stability are determined by the natural response, which is the component of the total response solely due to the initial conditions of the system. Let us examine the stability considering the total response. If the system is excited by an unbounded input and produces an unbounded response, nothing can be said about its stability. However, if it is subjected to a bounded input, but results in an unbounded response, then it is difficult to say whether the natural response is unbounded or the zero-state response is unbounded. This ambiguity is removed by an alternate definition of stability, to be called **BIBO stability**. This stability is based on total response and implies zero input stability.

Definition 4.4.

BIBO Stability: A system is said to be BIBO (bounded-input, bounded-output) stable if for every bounded input, the output is bounded.

In summary, a linear time-invariant system is stable if the following two notions of system stability are satisfied.

a. When the system is excited by a bounded input, the output is bounded. (**BIBO stability**)

b. With zero input and arbitrary initial conditions, the output tends toward zero—the equilibrium state of the system (**Asymptotic stability**).

These two notions of stability are equivalent for a linear time-invariant system.

Stability conditions in control systems can be described based on either impulse response or transfer function. Both approaches are valid and commonly used in stability analysis, and they provide complementary insights into the stability behavior of control systems. The choice of approach often depends on the specific characteristics of the system and the analysis being performed. In the following, we derive the conditions of BIBO stability in terms of the impulse response function.

4.3 Conditions of BIBO Stability of LTI System Based on Impulse Response Function

Consider a single-input, single-output system with transfer function

$$\frac{C(s)}{R(s)} = G(s) = \frac{b_0 s^m + b_1 s^{m-1} + \cdots + b_{m-1}s + b_m}{a_0 s^n + a_1 s^{n-1} + \cdots + a_{n-1}s + a_n}; \quad (m \leq n) \tag{4.1}$$

Assuming that the initial conditions are zero, the output of the system is given by

$$c(t) = \mathcal{L}^{-1}\left[G(s)R(s)\right]$$

The output c(t) of the system due to an input r(t) is computed by convolution of the impulse response function with the input and is expressed as:

$$c(t) = \int_0^{\infty} g(\tau)r(t-\tau).d\tau \tag{4.2}$$

where $g(t) = \mathcal{L}^{-1}\left[G(s)\right]$ is the *impulse response* of the system of Eqn(4.1).

The absolute value of c(t) is

$$\left|c(t)\right| = \left|\int_0^{\infty} g(\tau)\ r(t-\tau)\ d\tau\right|$$

Since the absolute value of an integral is not greater than the integral of the absolute value of the integrand,

$$\begin{aligned}\left|c(t)\right| &\leq \int_0^{\infty} \left|g(\tau)r(t-\tau)\right|\ d\tau \\ &\leq \int_0^{\infty} \left|g(\tau)\right| \cdot \left|r(t-\tau)\right|\ d\tau \end{aligned} \tag{4.3}$$

Since the input r(t) is bounded,

$$\left|r(t)\right| \leq M \tag{4.4}$$

where M is a finite positive number. Then

$$\left|c(t)\right| \leq M \int_0^{\infty} \left|g(\tau)\right| \, d\tau \tag{4.5}$$

To ensure that the system is BIBO stable, the output c(t) has to be bounded, i.e.,

$$\left|c(t)\right| \leq N < \infty$$

or,

$$M \int_0^{\infty} \left|g(\tau)\right| \, d\tau \leq N < \infty \tag{4.6}$$

where N is a finite positive number. Thus

Key Point

The output is bounded provided the impulse response g(t) is absolutely integrable i.e., $\int_0^{\infty} \left|g(\tau)\right| d\tau$ is finite (the area under $|g(\tau)|$ versus τ curve is finite).

$$\int_0^{\infty} \left|g(\tau)\right| d\tau < \infty \tag{4.7}$$

This guarantees that the system is BIBO stable.

4.4 Conditions of Stability of Linear Time-Invariant Systems Based on Transfer Function

Stability conditions based on transfer functions often involve analyzing the poles of the transfer function in the complex plane. For a system with a closed-loop transfer function T(s), the stability (closed-loop) can be determined by examining the location of poles (closed-loop poles) in the Laplace domain. Let us first illustrate the concept of stability through simple examples before delving into general conditions.

EXAMPLE 4.1

Consider a system with a closed-loop transfer function

$$T(s) = \frac{C(s)}{R(s)} = \frac{15}{(s+3)(s+5)}$$

The output of this system for unit step input $R(s) = 1/s$ is

$$C(s) = \frac{15}{s(s+3)(s+5)} = \frac{1}{s} - \frac{2.5}{s+3} + \frac{1.5}{s+5}$$

This gives

$$\boxed{c(t) = 1 - 2.5e^{-3t} + 1.5e^{-5t}} \tag{4.8}$$

Since the closed-loop poles are real and located in the left-half s-plane, the output response contains exponential terms with negative indices, i.e., e^{-3t} and e^{-5t}. As time $t \to \infty$, both exponential terms will approach zero and the output will reach its steady state value of 1, i.e., $c(\infty) = 1$.

Such types of systems where the poles are in the left-half of the s-plane are **absolutely stable systems**.

Key Takeaway

- If the closed-loop poles are located in the left half of the s-plane, it introduces exponential terms with negative exponents in the time-domain response of the system.
- These negative exponents indicate exponential decay in the system's response over time, leading to system stability.

EXAMPLE 4.2

Consider a system with one closed-loop pole located in the right half s-plane, with the closed-loop transfer function is

$$T(s) = \frac{C(s)}{R(s)} = \frac{10}{(s-2)(s+3)}$$

The output of this system for unit step input $R(s) = 1/s$ is

$$C(s) = \frac{10}{s(s-2)(s+3)} = \frac{-1.666}{s} + \frac{1}{s-2} + \frac{0.666}{s+3}$$

This gives

$$\boxed{c(t) = -1.666 + e^{2t} + 0.666e^{-3t}} \qquad (4.9)$$

Due to a pole located in the right half, there is one exponential term with positive index (term e^{2t}) which goes on increasing in amplitude as time $t \to \infty$.

Systems where any of the poles are in the right-half of the s-plane are **unstable** systems.

Key Takeaway

- The presence of any closed-loop poles in the right half s-plane introduces exponential terms with positive exponents in the time-domain response of the system.
- This positive exponent indicates exponential growth in the system's response over time, causing instability.

EXAMPLE 4.3

Consider a system with a closed-loop transfer function

$$T(s) = \frac{C(s)}{R(s)} = \frac{25}{s^2 + 25}$$

The closed-loop poles are purely imaginary and are located on the $j\omega$-axis.

The output of this system for unit step input $R(s) = 1/s$ is

$$C(s) = \frac{25}{s(s^2+25)} = \frac{1}{s} - \frac{1}{2(s-j5)} - \frac{1}{2(s+j5)}$$

This gives

$$\boxed{c(t) = 1 - \left[\frac{e^{-j5t}}{2} + \frac{e^{j5t}}{2}\right] = 1 - \cos 5t} \qquad (4.10)$$

Due to the presence of purely imaginary poles, the response is oscillatory. Systems where the poles are purely imaginary are **marginally stable** systems.

Key Takeaway

- If the closed-loop poles are purely imaginary, it exhibits sustained oscillations without growing or decaying over time, and the system is **marginally stable.**

Let us generalize this concept by considering an n-th order single-input,single-output linear time-invariant system. The closed-loop transfer function of this system is expressed as:

$$T(s) = \frac{C(s)}{R(s)} = \frac{b_0 s^m + b_1 s^{m-1} + \cdots + b_{m-1}s + b_m}{s^n + a_1 s^{n-1} + \cdots + a_{n-1}s + a_n} = \frac{P(s)}{Q(s)}; \quad (m \le n) \tag{4.11}$$

The closed-loop *characteristic equation* of the system is

$$Q(s) = s^n + a_1 s^{n-1} + a_2 s^{n-2} + \cdots + a_{n-1}s + a_n = 0 \tag{4.12}$$

The roots of $Q(s) = 0$ gives the closed-loop poles. Let us find the solution to the differential equation corresponding to the characteristic equation (4.12), considering the following cases.

1. **Case-1: All the roots of the characteristic equation (closed-loop poles), $\lambda_1, \lambda_2, \ldots, \lambda_n$ are distinct**. They can either be real or complex. Then, the output is

$$\boxed{c(t) = \sum_{i=1}^{n} A_i e^{\lambda_i t}} \tag{4.13}$$

where the constant A_i depends on initial conditions and locations of zeros.

1.1 If all the roots $\lambda_1, \lambda_2, \ldots, \lambda_n$ are real, their contribution to the output is of the form

$$\boxed{c(t) = A_1 e^{\lambda_1 t} + A_2 e^{\lambda_2 t} + \cdots + A_n e^{\lambda_n t}} \tag{4.14}$$

1.2 The contribution of a complex root pair $\lambda = \sigma_i \pm j\omega_i$ to the output is of the form

$$\boxed{A e^{\sigma_i t} \sin(\omega_i t + \phi_i)} \tag{4.15}$$

Note that, since the coefficients of the characteristic equation are assumed to be real, each complex root is accompanied by its complex conjugate.

The necessary and sufficient condition for the system to be stable is that every term in Eqn(4.13) converges to zero as $t \to \infty$, that is

$$e^{\lambda_i t} \to 0 \quad \text{for all} \quad \lambda_i$$

This is satisfied if all the roots of the characteristic equation (poles of the system) have a negative real part, i.e., they are strictly in the left half plane, where

$$\boxed{\mathrm{Re}\left[\lambda_i\right] < 0}$$

2. **Case-2: If some of the roots of the characteristic equation are repeated.**

   2.1 Without loss of generality, assume that $\lambda_1 = \lambda_2$ and they are real. Then the contribution of this to the output is of the form $A_1 e^{\lambda_1 t} + A_2 t e^{\lambda_1 t}$. Similarly, if there is a repeated root of multiplicity k at

   λ_1, i.e., $\lambda_1 = \lambda_2 = \cdots = \lambda_k$, then their contribution to the output is of the form

$$\boxed{\left[A_1 + A_2 t + A_3 t^2 + \cdots + A_k t^{k-1}\right] e^{\lambda_1 t}} \tag{4.16}$$

   2.2 If there are complex conjugate root pairs $\lambda = \sigma \pm j\omega$ **of multiplicity** k, their contribution to the output is of the form

$$\boxed{\left[A_1 \sin(\omega t + \phi_1) + A_2 t \sin(\omega t + \phi_2) + \cdots + A_k t^{k-1} \sin(\omega t + \phi_k)\right] e^{\sigma t}} \tag{4.17}$$

Remark

- When a system has repeated poles of k-th order in its transfer function, the output response can include terms like $t, t^2, \ldots, t^{k-1}$ multiplied by an exponential term.
- Unless the effects of these polynomial terms are counteracted by decaying exponential terms, stability cannot be ensured. Note that **an exponential decay is stronger than a polynomial growth of any order.**

3. **Case-3: If some of the roots are purely imaginary**

   3.1 If the roots are pure complex conjugate pair located at $\pm j\omega$, their contribution to the output is of the form

   $$\boxed{A\sin(\omega t + \phi)} \tag{4.18}$$

   Such purely imaginary pole pairs produce an oscillatory (sinusoidal) natural response.

   3.2 If there are purely **complex conjugate root pairs** $\pm j\omega$ **of multiplicity** k, then their contribution to the output is

   $$\boxed{A_1\sin(\omega t + \phi_1) + A_2 t\sin(\omega t + \phi_2) + \cdots + A_k t^{k-1}\sin(\omega t + \phi_k)} \tag{4.19}$$

   This gives rise to an unbounded response, and the system is unstable.

Thus, the response curve of a stable higher-order system can often be represented as the sum of several exponentially damped and exponentially damped sinusoidal curves. This behavior arises due to the presence of multiple poles in the left-half plane (LHP) in the system's transfer function. As discussed above, each pole in the left-half plane contributes to an exponentially damped term in the time-domain response. If the pole is not purely real, but instead has an imaginary component, it introduces an oscillatory component to the response as well, resulting in a damped sinusoidal term. The specific form of the response depends on the location and multiplicity of the poles.

The stability analysis discussed above can be summarized as follows:

1. If all the roots of the characteristic equation lie in the left half s-plane, the system is stable. In this case, the impulse response is bounded and eventually converges to zero, and therefore $\int_0^\infty |g(\tau)|d\tau$ is finite and the system is BIBO stable.

2. If any root of the characteristic equation lies in the right-half s-plane or if there is a repeated root on the $j\omega$-axis, the system is*unstable*. In this case, the impulse response is unbounded and $\int_0^\infty |g(\tau)|d\tau$ is infinite, leading to instability.

3. If the characteristic equation has one or more non-repeated roots on the $j\omega$-axis but no right half plane roots, then the system is *marginally stable* or limitedly stable. In this case, the impulse response is finite, but $\int_0^\infty |g(\tau)|d\tau$ is infinite.

Exceptions

There are a few exceptions to the foregoing discussion and definitions of stability.

- When a unit step input ($R(s) = 1/s$) is applied to a perfect integrator ($G(s) = 1/s$), the output response will be a ramp function, starting from zero and increasing linearly with time (unbounded).
- Similarly, a system with roots on $j\omega$ axis at $s = \pm j\omega_0$ when excited by sinusoidal input $\sin\omega_0 t$ results infinite response of the form $t\sin\omega_0 t$. However, the same system with roots at $s = \pm j\omega_0$ when excited by an impulse input results in a response of the form $\sin\omega_0 t$. This response is finite but not asymptotic, i.e., $\lim_{t\to\infty} c(t) \neq 0$.

There is no need to be misled by the complexity of occasional special cases. The stability conditions discussed above are often useful in a vast majority of cases.

The region of stability in the s-plane are illustrated in Figure 4.5

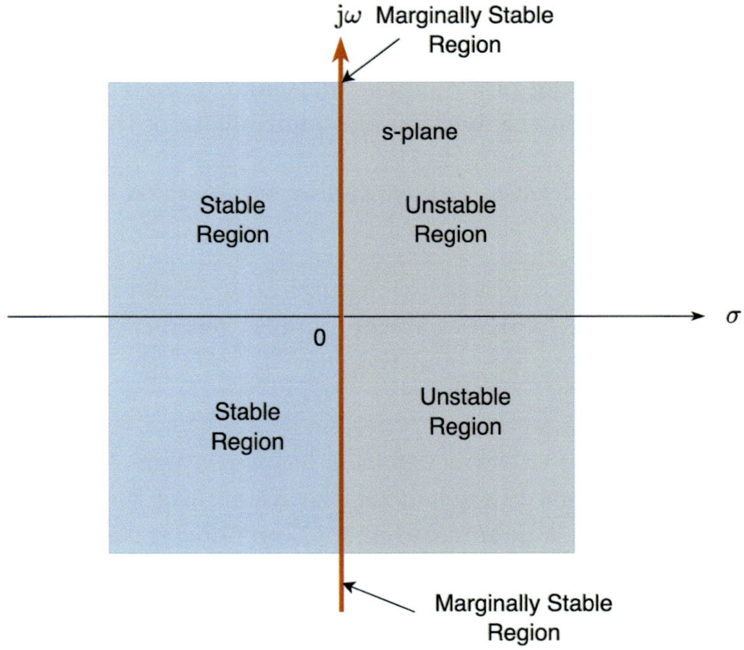

Figure 4.5: Stable, unstable, and marginally stable region in the s-plane

4.5 Necessary Conditions of Stability

In many cases, determining the stability of a system does not require computing the complete response. Instead, for linear time-invariant (LTI) single-input, single-output (SISO) systems, the stability can be inferred by examining the location of the roots of the characteristic equation of the system.

Consider the closed-loop characteristic equation of a linear time-invariant system, which is of the form

$$Q(s) = a_0 s^n + a_1 s^{n-1} + \cdots + a_{n-1} s + a_n = 0 \tag{4.20}$$

To ensure that there are no roots of the characteristic equation with positive real parts, it is **necessary but not sufficient** that

1. All the coefficients of the polynomial have the same sign.
2. None of the coefficients vanishes.

Note that while these conditions are necessary to avoid having roots with positive real parts, it does not guarantee stability by themselves, especially for higher order systems. Additional analysis, such as examining the locations of the roots or performing stability tests, is required to ensure system stability. These are further discussed in the following propositions.

Theorem 4.1. The necessary and sufficient conditions for the first and second order systems to be stable are that all the coefficients of the characteristic equation must have the same sign.

EXAMPLE 4.4

The characteristic equation of a first-order system is

$$a_0 s + a_1 = 0$$

which has a single root

$$s_1 = -\frac{a_1}{a_0} \tag{4.21}$$

It is obvious from Eqn(4.21) that if both a_0 and a_1 have the same sign, the characteristic roots are negative and the system is stable.

Next, consider the characteristic equation of a second-order system

$$a_0 s^2 + a_1 s + a_2 = 0$$

whose roots are

$$s_1, s_2 = \frac{-a_1 \pm \sqrt{a_1^2 - 4a_0a_2}}{2a_0} \tag{4.22}$$

From Eqn(4.22) it is found that if the coefficients a_0, a_1 and a_2 have same sign, then both the roots are either **negative and real** or they are **complex conjugate with negative real parts**.

Note

- If all the coefficients of the characteristic equation of a system have the same sign, this ensures that the real roots of the system are negative.
- However, this does not ensure the negativeness of the real parts of the complex roots for third and higher order systems. Thus, it cannot be a **sufficient condition of stability** for third and higher-order systems.

Example:

Consider a third-order system with characteristic equation

$$s^3 + 2s^2 + 2s + 40 = 0 \tag{4.23}$$

Eqn(4.23) can be expressed as

$$(s + 4)(s - 1 + j3)(s - 1 - j3) = 0$$

Notice that the real part of the complex roots is positive, indicating system instability, although all the coefficients of the characteristic equation have the same sign (positive). Thus, when the order of the characteristic equation is higher than two, it becomes insufficient to infer the stability of the system solely by examining the signs of all the coefficients. This is because higher-order systems can exhibit more complex behavior, and the relationship between the coefficients and the stability of the system becomes more intricate. The stability analysis of higher order (> 2) systems should therefore be carried out by examining their characteristic equation as follows:

Step-1: If the signs of all the coefficients of the characteristic equation are not the same and/or if some of the coefficients are zero, then it is indicative of potential instability in the system.

Step-2: If all the coefficients of the characteristic equation have the same sign, the possibility of instability cannot be excluded, because this is a

necessary condition of stability. We, therefore, do further analysis to determine sufficient conditions for stability.

A Hurwitz and E. J. Routh independently developed sufficient conditions for the stability of linear time-invariant systems. While they address the same fundamental question of stability, they do so using different mathematical approaches. The Hurwitz criterion is based on determinants, and Routh's criterion is in terms of array formulation. In the following, we discuss about Hurwitz stability criterion; the Routh criterion, which can be derived from the Hurwitz criterion, is presented next.

4.6 Hurwitz's Stability Criterion

Hurwitz's criterion, named after Adolf Hurwitz, who developed it in the late 19th century, is a mathematical method used to determine the stability of a linear time-invariant (LTI) system by examining the signs of certain determinants constructed from the coefficients of the characteristic equation. It does not require explicitly computing the roots of the characteristic polynomial. Here's a step-by-step procedure for conducting Hurwitz's test on stability:

Step-1: Formulate the characteristic polynomial

$$\mathrm{Q(s) = a_0 s^n + a_1 s^{n-1} + \cdots + a_{n-1} s + a_n = 0} \tag{4.24}$$

Step-2: Construct Hurwitz's determinant H as follows:

$$\mathrm{H} = \begin{vmatrix} a_1 & a_0 & 0 & 0 & 0 & 0 & \dots & 0 & 0 \\ a_3 & a_2 & a_1 & a_0 & 0 & 0 & \dots & 0 & 0 \\ a_5 & a_4 & a_3 & a_2 & a_1 & a_0 & \dots & 0 & 0 \\ \vdots & \vdots & \vdots & \vdots & \vdots & \vdots & \vdots & \vdots & \vdots \\ \vdots & \vdots & \vdots & \vdots & \vdots & \vdots & \vdots & \vdots & \vdots \\ a_{2n-1} & a_{2n-2} & a_{2n-3} & a_{2n-4} & a_{2n-5} & \dots & \dots & a_{n-1} & a_n \end{vmatrix} \tag{4.25}$$

While formulating this determinant, the coefficients with indices larger than n or with negative indices are replaced by zeros.

Step-3: Construct sub-determinants $\Delta_1, \Delta_2, \dots \Delta_n$ of the Hurwitz determinant H as follows:

$$\Delta_1 = \begin{vmatrix} a_1 \end{vmatrix} \quad \Delta_2 = \begin{vmatrix} a_1 & a_0 \\ a_3 & a_2 \end{vmatrix}, \quad \Delta_3 = \begin{vmatrix} a_1 & a_0 & 0 \\ a_3 & a_2 & a_1 \\ a_5 & a_4 & a_3 \end{vmatrix}, \dots, \Delta_n = \begin{vmatrix} \mathrm{H} \end{vmatrix} \tag{4.26}$$

Step-4: Examine the Signs of the above Sub-determinants:

1. If all the sub-determinants are positive, the system is stable. That is, for the system to be stable, $\Delta_1 > 0,\ \Delta_2 > 0, \ldots, \Delta_n > 0$.
2. If any of the sub-determinants are zero or negative, the system is unstable.
3. If the sub-determinant Δ_{n-1}, that is, the determinant of the sub-matrix obtained by removing the last row and last column of the Hurwitz matrix, equals zero, the system is marginally stable.

EXAMPLE 4.5

Problem

Using Hurwitz's criterion, determine the stability of the system with the characteristic equation

$$s^4 + 4s^3 + 6s^2 + 4s + 1 = 0$$

SOLUTION:

Let us compare the given characteristic equation with the standard form of the characteristic equation, that is,

$$s^4 + 4s^3 + 6s^2 + 4s + 1 = a_0 s^4 + a_1 s^3 + a_2 s^2 + a_3 s + a_4 = 0$$

This gives $a_0 = 1$,$a_1 = 4$, $a_2 = 6$, $a_3 = 4$ and $a_4 = 1$. The Hurwitz determinant is

$$H = \begin{vmatrix} a_1 & a_0 & 0 & 0 \\ a_3 & a_2 & a_1 & a_0 \\ a_5 & a_4 & a_3 & a_2 \\ a_7 & a_6 & a_5 & a_4 \end{vmatrix} = \begin{vmatrix} 4 & 1 & 0 & 0 \\ 4 & 6 & 4 & 1 \\ 0 & 1 & 4 & 6 \\ 0 & 0 & 0 & 1 \end{vmatrix}$$

The sub-determinants are

$$\Delta_1 = |4| > 0, \quad \Delta_2 = \begin{vmatrix} 4 & 1 \\ 4 & 6 \end{vmatrix} = 20 > 0$$

$$\Delta_3 = \begin{vmatrix} 4 & 1 & 0 \\ 4 & 6 & 4 \\ 0 & 1 & 4 \end{vmatrix} = 64 > 0, \quad \text{and} \quad \Delta_4 = \begin{vmatrix} 4 & 1 & 0 & 0 \\ 4 & 6 & 4 & 1 \\ 0 & 1 & 4 & 6 \\ 0 & 0 & 0 & 1 \end{vmatrix} = 64 > 0$$

Since all the sub-determinants are positive, the system is stable.

CONCEPT REINFORCEMENT PROBLEM 4.1

Problem Using Hurwitz's criterion, determine the stability of systems with the following characteristic equations

a. **S1:** $s^3 + 2s^2 + 2s + 40 = 0$

b. **S2:** $s^3 + 2s^2 + 9s + 18 = 0$

Limitations of Hurwitz's Method

While Hurwitz's criterion is a powerful method for analyzing the stability of linear time-invariant systems, it does have some limitations and disadvantages: Some of the limitations are:

1. For higher-order systems, computing all the determinants required by Hurwitz's criterion can be computationally intensive and time-consuming.
2. Cannot determine the number of roots located in the right-half s-plane.

These limitations can be overcome by the method proposed by E. J. Routh.

4.7 Routh's (Routh-Hurwitz) Stability Criteria

The Routh's method, also known as Routh-Hurwitz method, provides a straightforward and computationally efficient approach to stability analysis, particularly for higher-order systems, where the Hurwitz criterion may be more cumbersome to apply. This method offers a numerical procedure for determining the stability of a system without explicitly solving for the roots of the characteristic polynomial. It provides a systematic approach to assess the location of the roots, particularly whether any roots lie in the right-half plane (RHP) or on the imaginary axis based on the properties of a tabular representation called Routh's array. This method essentially requires two steps:

Step-1: Construct a data table or array called *Routh table* or *Routh array* using the coefficients of the characteristic polynomial.

Step-2: Analyze the first column of the Routh array, to determine how many roots of the characteristic equation (closed-loop poles) are in the left-half plane, right-half plane, and on the $j\omega$-axis.

Question arises: Is it necessary to study the Routh-Hurwitz criterion when modern computational tools can precisely determine the exact location of poles? The answer is affirmative; there are several reasons why studying the Routh-Hurwitz criterion remains valuable. First, the power of the method lies in the design rather than the analysis. If our interest is to know how the stability of the system is affected due to parametric uncertainties and variations, the Routh-Hurwitz criterion can provide a closed-form expression for the range of parameters and regions in parameter space where the system remains stable. Therefore, by understanding how stability is affected by parameter variations, we can make informed design decisions to achieve desired system behavior while ensuring stability under uncertain operating conditions. In summary, the Routh-Hurwitz criterion remains valuable for its theoretical insight, robustness, ease of use, and historical significance in control theory. It provides a fundamental understanding of stability analysis principles and serves as a useful tool for engineers and researchers in the field.

4.7.1 Construction of a Basic Routh Table

Consider the closed-loop characteristic equation of a linear time-invariant system, which is of the form

$$a_0 s^n + a_1 s^{n-1} + \cdots + a_{n-1} s + a_n = 0 \tag{4.27}$$

where all the coefficients are real numbers. It is assumed that $a_n \neq 0$, that is, any zero root has been removed. The procedure for forming the Routh array is given below

Procedure to Construct Routh Table

Step-1: Represent the characteristic equation (polynomial) in descending powers of s.

Step-2: Label the rows with powers of s from the highest power of the characteristic equation to s^0.

Step-3: Arrange the coefficients of polynomials into two rows. The first row consists of the **first, third, fifth**,...coefficients, and the second row consists of the **second, fourth, sixth**,...coefficients.

Step-4: Entries in Remaining Rows: Each entry is the negative determinant of entries in the previous two rows divided by the entry in the first column directly above the calculated row. The left-hand column of the determinant is always the first column of the previous two rows, and the right-hand column is the elements of the column above and to the right.

Step-5: The table is complete when all of the rows are completed down to s^0.

Let's illustrate the process of constructing Routh's array with an example of a 6th-order characteristic polynomial given by

$$a_0 s^6 + a_1 s^5 + a_2 s^4 + a_3 s^3 + a_4 s^2 + a_5 s + a_6 \tag{4.28}$$

Label the rows from s^6 to s^0 and arrange the coefficients in 2 rows. This gives the initial layout of the Routh table with all possible entries as shown in Table 4.1.

Table 4.1: Initial layout of Routh Table

s^6	a_0	a_2	a_4	a_6
s^5	a_1	a_3	a_5	0
s^4				
s^3				
s^2				
s^1				
s^0				

The complete Routh table will look as shown in Table 4.2.

Table 4.2: Complete Routh Table

s^6	a_0	a_2	a_4	a_6
s^5	a_1	a_3	a_5	0
s^4	b_1	b_2	b_3	0
s^3	c_1	c_2	c_3	0
s^2	d_1	d_2	d_3	0
s^1	e_1	e_2	e_3	0
s^0	f_1	0	0	0

The various entries in Table 4.2 are computed as follows:

Computation of Elements in s^4 row:

$$b_1 = \frac{-\begin{vmatrix} a_0 & a_2 \\ a_1 & a_3 \end{vmatrix}}{a_1} = \frac{a_1a_2 - a_0a_3}{a_1}, \quad b_2 = \frac{-\begin{vmatrix} a_0 & a_4 \\ a_1 & a_5 \end{vmatrix}}{a_1} = \frac{a_1a_4 - a_0a_5}{a_1},$$

$$b_3 = \frac{-\begin{vmatrix} a_0 & a_6 \\ a_1 & 0 \end{vmatrix}}{a_1} = \frac{a_1a_6 - a_0 \times 0}{a_1} = a_6$$

Computation of Elements in s^3 row:

$$c_1 = \frac{-\begin{vmatrix} a_1 & a_3 \\ b_1 & b_2 \end{vmatrix}}{b_1} = \frac{b_1a_3 - a_1b_2}{b_1}, \quad c_2 = \frac{-\begin{vmatrix} a_1 & a_5 \\ b_1 & b_3 \end{vmatrix}}{b_1} = \frac{b_1a_5 - a_1b_3}{b_1},$$

$$c_3 = \frac{-\begin{vmatrix} a_1 & 0 \\ b_1 & 0 \end{vmatrix}}{b_1} = 0$$

Computation of Elements in s^2 row:

$$d_1 = \frac{-\begin{vmatrix} b_1 & b_2 \\ c_1 & c_2 \end{vmatrix}}{c_1} = \frac{b_2c_1 - b_1c_2}{c_1},\ d_2 = \frac{-\begin{vmatrix} b_1 & b_3 \\ c_1 & c_3 \end{vmatrix}}{c_1} = \frac{b_3c_1 - b_1c_3}{c_1} = b_3, d_3 = 0$$

Computation of Elements in s^1 row:

$$e_1 = \frac{-\begin{vmatrix} c_1 & c_2 \\ d_1 & d_2 \end{vmatrix}}{d_1} = \frac{c_2d_1 - c_1d_2}{d_1},\ e_2 = \frac{-\begin{vmatrix} c_1 & 0 \\ d_1 & 0 \end{vmatrix}}{d_1} = 0$$

Computation of Elements in s^0 row:

$$f_1 = \frac{-\begin{vmatrix} d_1 & d_2 \\ e_1 & e_2 \end{vmatrix}}{e_1} = \frac{e_1d_2 - d_1e_2}{e_1} = \frac{e_1d_2 - d_1 \times 0}{e_1} = d_2$$

Key Observations

1. The complete array of coefficients is triangular
2. The missing terms in Routh's array are regarded as zero
3. An entire row may be multiplied or divided by a positive number to simplify subsequent calculation without altering the stability conclusion.

4.7.2 Routh-Hurwitz Stability Criteria for Linear Feedback Control System

After constructing the Routh table, the next and final step is to interpret it. As discussed, the primary goal is to investigate the stability of a system. Therefore, in the stability study, the polynomial used to construct Routh's table is the closed-loop characteristic equation of a control system whose roots correspond to the poles of the closed-loop system.

Routh's Stability Criterion:

Theorem 4.2. The necessary and sufficient condition for the stability of a linear time-invariant (LTI) system is that all elements in the first column of Routh's array must have the same sign. If there are no sign changes in this column, all roots of the characteristic polynomial lie in the left-half of the s-plane (LHP), and the system is therefore stable.

i. If there are sign changes in the first column of Routh's table, then some roots of the characteristic polynomial (i.e., the closed-loop poles) lie in the right half of the s-plane(RHP), indicating system instability.

ii. The number of sign changes in the first column of Routh's table equals the number of roots of the characteristic polynomial in the right-half of s-plane (RHP).

4.8 Different Configurations of Routh's Table

While formulating the Routh's table, we must consider the following four cases, where each case requires a different procedure to complete the Routh's table.

Case-1: No element in the first column is zero. This is the normal case, and we have to follow the procedure detailed above.

Case-2: The first element in a row of Routh's table is zero, but all other elements in the row are non-zero, or there are no remaining terms.

Case-3: All the elements of a row are zero.

Case-4: All the elements of a row are zero, but with repeated roots on the jω-axis.

In the following, we illustrate through examples for each case, considering systems with different characteristic equations.

4.8.1 Case-1: No Element in the First Column is Zero

EXAMPLE 4.6

Problem:

Consider a system with a closed-loop characteristics equation:

$$s^4 + 10s^3 + 35s^2 + 50s + 24 = 0$$

Construct the Routh array and determine the stability of the system

SOLUTION:

The complete Routh table for this system is shown in Table 4.3.

Table 4.3: Completed Routh Table

s^4	1	35	24
s^3	10	50	
s^2	$\dfrac{(10 \times 35) - 50}{10} = 30$	24	
s^1	$\dfrac{(30 \times 50) - (10 \times 24)}{30} = 42$	0	
s^0	24		

Since there are no sign changes in the first column of the Routh array, all the closed poles lie in the left half of the s-plane and the system is stable.

CONCEPT REINFORCEMENT PROBLEM 4.2

Problem:

Using Routh's table, investigate the stability of the systems with the following closed-loop characteristic equations.

a. $P(s) = s^3 + 13s^2 + 20s - 100$
b. $P(s) = s^5 - 5s^4 - 15s^3 + 65s^2 + 74s - 120$

The following two examples consider both a second and a third order system and find the condition of stability from the coefficients.

EXAMPLE 4.7

Problem:

Using Routh's table, determine if the system with the following closed-loop characteristic equation is stable.

$$P(s) = a_0s^2 + a_1s + a_2$$

SOLUTION: The Routh's table is constructed as follows:

Table 4.4: Completed Routh Table

s^2	a_0	a_2
s^1	a_1	0
s^0	a_2	0

From the first column of the Routh's array, we can conclude that

A second-order system is stable provided all the coefficients of the characteristic equation are either positive or negative.

EXAMPLE 4.8

Problem:

Using Routh's table, determine if the system with the following closed-loop characteristic equation is stable.

$$P(s) = a_0 s^3 + a_1 s^2 + a_2 s + a_3$$

SOLUTION: The Routh's table is constructed as follows:

Table 4.5: Completed Routh Table

s^3	a_0	a_2
s^2	a_1	a_3
s^1	$\dfrac{a_1 a_2 - a_0 a_3}{a_1}$	0
s^0	a_3	0

From the Routh's table we can conclude that

a. A third-order system is stable provided $a_1 a_2 > a_0 a_3$.

b. This system is marginally stable if $a_1 a_2 = a_0 a_3$.

The Routh-Hurwitz criterion can be used to study the effect of increasing gain on stability, particularly in the context of linear time-invariant (LTI) systems. The following example illustrates how we can analyze the effect of increasing gain on the stability of a system using the Routh-Hurwitz criterion:

EXAMPLE 4.9

Problem:

Consider the closed-loop system shown in Figure 4.6

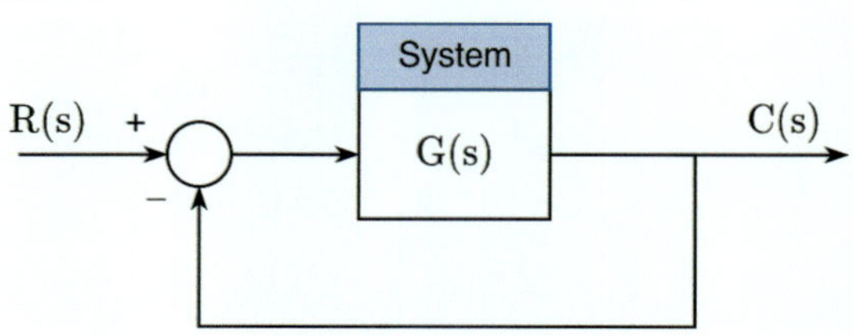

Figure 4.6: Unity feedback control System

where

$$G(s) = \frac{K}{(s+1)(s+2)(s+3)(s+4)}$$

Investigate the stability of the closed-loop system for

a. $K = 100$
b. $K = 130$

SOLUTION: The closed-loop transfer of the system is given by

$$T(s) = \frac{G(s)}{1+G(s)} = \frac{K}{s^4 + 10s^3 + 35s^2 + 50s + 24 + K}$$

a. For $K = 100$, the closed-loop characteristic equation becomes

$$s^4 + 10s^3 + 35s^2 + 50s + 124 = 0$$

The Routh's table is given as:

Table 4.6: Completed Routh Table

s^4	1	35	124
s^3	10	50	0
s^2	30	124	0
s^1	$\frac{26}{3}$	0	0
s^0	124	0	0

Since there are no sign changes in the first column of Routh's array, the system is stable for $K = 100$.

b. The closed-loop characteristic equation for $K = 130$ is given by

$$s^4 + 10s^3 + 35s^2 + 50s + 154 = 0$$

The Routh's table is given as:

Table 4.7: Completed Routh Table

s^4	1	35	154
s^3	10	50	0
s^2	30	154	0
s^1	$-\frac{4}{3}$	0	0
s^0	154	0	0

Since there are sign changes in the first column of Routh's array, the system is unstable for K = 130. Further, since there are two sign changes, this implies that the closed-loop system has two roots in the right-half of the s-plane.

4.8.2 Case-2: The First Element in a Row is Zero; Other Elements are Non-zero

If the first element of a row is zero, then we cannot proceed further in forming the Routh's table, as the next row formation requires the division by the first element of the previous row. In this case:

1. Replace the zero element by a very small positive number ϵ and continue forming Routh's table.
2. Examine the sign of the first column of the table.

EXAMPLE 4.10

Problem:

Investigate the stability of a system with the characteristic equation

$$P(s) = s^3 + 3s^2 + 2s + 6$$

SOLUTION:

Let us construct Routh's Table as follows

Table 4.8: Completed Routh Table

s^3	1	2
s^2	3	6
s^1	$\not{0} \approx \epsilon$	
s^0	6	

From the first column, it is seen that there are no sign changes, because ϵ is a small positive number. So the system is stable.

EXAMPLE 4.11

Problem:

Investigate the stability of the system with the characteristic polynomial

$$P(s) = s^4 + 2s^3 + s^2 + 2s + 10$$

SOLUTION:

Let us construct Routh's Table as follows

Table 4.9: Completed Routh Table

s^4	1	1	10
s^3	2	2	
s^2	$0 \approx \epsilon$	10	
s^1	$\frac{2\epsilon - 20}{\epsilon} = 2 - \frac{20}{\epsilon}$		
s^0	10		

Since ϵ is a very small positive number, $\frac{20}{\epsilon}$ is a very large number. Hence, $2 - \frac{20}{\epsilon}$ is negative. Therefore, in the first column, there are two sign changes; first sign change occurs from s^2 row to s^1 row and the next sign change occurs from s^1 row to s^0 row; because ϵ is a small positive number. So the system is unstable and there are two poles in the right half s-plane. You can verify this in MATLAB with

```
roots([1 2 1 2 10])
```

This gives that the poles are located at $-1.7817 \pm j1.0013$ and $+0.7817 \pm j1.3352$.

CONCEPT REINFORCEMENT PROBLEM 4.3

Problem

Investigate the stability of the unity feedback closed-loop system with forward path transfer function

$$G(s) = \frac{s + 8}{s^5 - s^4 + 4s^3 - 4s^2 + 2s - 10}$$

4.8.3 Case-3: All the Elements in an Entire Row are Zero

In some cases, we find that all the terms in a row of Routh's table are zero. This occurs if some roots of the characteristic polynomial are of equal magnitude and lie radially opposite in the s-plane. This can happen if

1. Pairs of real roots with opposite signs (see Figure 4.7a).
2. Pairs of imaginary roots (see Figure 4.7b).
3. Pairs of complex-conjugate roots forming symmetry about the origin of the s-plane (see Figure 4.7c).

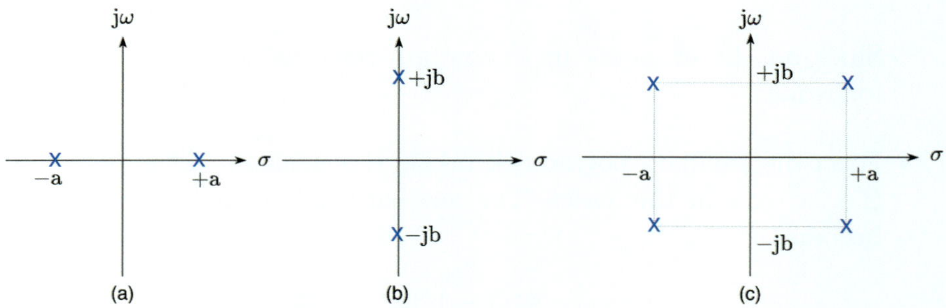

Figure 4.7: Possible conditions of premature termination of Routh's array

In such cases, we will use the following procedure:

1. Form an auxiliary polynomial A(s) using the coefficients of the row just preceding the row containing all zeros.
2. Compute the derivative of the auxiliary polynomial with respect to s.
3. Replace the row of all zeros with the coefficients of the resultant equation $\dfrac{dA(s)}{ds}$.
4. Construct the rest of the Routh array and examine the sign of the elements in the first column.

EXAMPLE 4.12

Problem

Investigate the stability of the system with the characteristic polynomial

$$P(s) = s^6 + 14s^5 + 70s^4 + 140s^3 + 49s^2 - 154s - 120$$

SOLUTION The Routh's table is constructed as follows:

Table 4.10: Completed Routh Table

s^6	1	70	49	−120	
s^5	14	140	−154	0	
s^4	60	60	−120	0	
s^3	126	−126	0		
s^2	120	−120	0		
s^1	~~0~~	~~0~~	~~0~~	⟵	zero row
s^0	STOP	STOP			

Since all the elements in s^1 row are zero, we stop here and do the following:

- Form the auxiliary polynomial taking the elements of the previous row (i.e., s^2 row in this case). The auxiliary polynomial for this example becomes

$$A(s) = 120s^2 - 120$$

- Differentiate this polynomial with respect to s. This gives

$$\frac{dA(s)}{ds} = 240s$$

- Replace the row of all zeros with these coefficients and continue forming the table.

The complete Routh's table is given as:

Table 4.11: Completed Routh Table

s^6	1	70	49	−120
s^5	14	140	−154	0
s^4	60	60	−120	0
s^3	126	−126	0	
s^2	120	−120	0	
s^1	~~0~~ (240)	0	0	
s^0	−120	0		

Since there is one sign change in the first column of the Routh-Hurwitz array, the system is unstable and one pole is in the right half of the s-plane. You can verify this by determining the roots of the characteristic equation in MATLAB using the command poles=roots[1 14 70 140 49 -154 -120]

The poles are located at $p_1 = -1$, $p_2 = -2$, $p_3 = -3$, $p_4 = -5$, $p_5 = -5$ and $p_6 = +1.0$

CONCEPT REINFORCEMENT PROBLEM 4.4

Problem:

Using Routh's table, investigate the stability of a system with a closed-loop characteristic polynomial

$$P(s) = s^5 + 3s^4 + 0.5s^3 + 14s^2 - 14s + 8$$

Determine how many roots of the polynomial are on the $j\omega$-axis.

4.8.4 Case-4: Repeated Roots of the characteristic equation on the $j\omega$-axis

We know that if there are simple roots on the $j\omega$-axis, the system is marginally stable. However, if there are repeated roots on the $j\omega$-axis, the system response will be of the form **$Bt\sin(\omega t + \phi)$** (B being a constant). **This is an exceptional case where Routh's criteria will fail to provide information about the stability**. Therefore, if all the elements in a row of Routh's table are zero, we must determine the location of the roots on the $j\omega$-axis. If there are any repeated roots, then the system is unstable. This will be illustrated by the following example.

EXAMPLE 4.13

Problem:

The closed-loop characteristic equation of a system is given

$$P(s) = s^5 + 2s^4 + 2s^3 + 4s^2 + s + 2$$

Using Routh's table, determine if there are any repeated roots on the $j\omega$-axis.

SOLUTION:

The Routh's table is constructed as follows:

Table 4.12: Completed Routh Table

s^5	1	2	1
s^4	2	4	2
s^3	STOP	STOP	STOP

Since all the elements in s^3 row are zero, the Routh's test breaks down, and we cannot proceed further to complete the Routh's table. We stop here and do the following:

1. Form the auxiliary polynomial taking the elements of the previous row (i.e., s^4 row in this case). The auxiliary polynomial for this example becomes

$$A(s) = 2s^4 + 4s^2 + 2$$

2. Differentiate this polynomial with respect to s, which gives

$$\frac{dA(s)}{ds} = 8s^3 + 8s$$

3. Replace the zeros of s^3 row with these coefficients and continue forming the table.

The Routh's table formulation proceeds as:

Table 4.13: Completed Routh Table

s^5	1	2	1
s^4	2	4	2
s^3	8	8	0
s^2	2	2	0
s^1	STOP	STOP	

Since all the elements in s^1 row are zero, the Routh's test breaks down and we cannot proceed further to complete the Routh's table. We stop here and do the following:

1. Form the auxiliary polynomial taking the elements of the previous row (i.e., s^2 row in this case). The auxiliary polynomial for this example becomes

$$A(s) = 2s^2 + 2$$

2. Differentiate this polynomial with respect to s which gives

$$\frac{dA(s)}{ds} = 4s$$

3. Replace the zeros of s^1 row with these coefficients and complete the table.

The Routh's table formulation proceeds as:

Table 4.14: Completed Routh Table

s^5	1	2	1
s^4	2	4	2
s^3	8	8	0
s^2	2	2	0
s^1	4	0	
s^0	2		

The roots on the $j\omega$-axis are computed solving the auxiliary equation

$$2s^4 + 4s^2 + 2 = 0$$

The roots are at $+j, +j$ and $-j, -j$. This means there are repeated roots on the $j\omega$-axis. The impulse response of the system will be of the form $Bt\sin(t + \phi)$ and will increase over time. However, if we examine the sign of the first column of the Routh's array, there are no sign changes. So applying the conventional Routh-Hurwitz criteria will falsely indicate that the system is stable.

CONCEPT REINFORCEMENT PROBLEM 4.5

Problem:

The closed-loop characteristic equation of a system is given

$$P(s) = s^6 + 3s^5 + 10s^4 + 24s^3 + 32s^2 + 48s + 32$$

Using Routh's table, determine if there are any repeated roots on the $j\omega$ axis.

4.9 Application of Routh's Stability Test to Control System Design

The Routh criterion provides valuable insights into the stability of a system and can be used to determine the range of the open-loop gain for which the feedback system remains stable. Although this method gives no guidelines to improve the relative stability of the system, it can be used to determine the ultimate gain/ critical gain and corresponding frequency of oscillations, which are defined below.

Definition 4.5.

Ultimate Gain: The gain at which the system is marginally stable is called the ultimate gain. It represents the maximum gain that can be applied to the system without causing instability.

Definition 4.6.

Ultimate Frequency: The frequency of sustained oscillations at the ultimate gain is referred to as the ultimate frequency.

The ultimate gain, as well as the ultimate frequency, are important parameters used in controller design, especially in PID (Proportional-Integral-Derivative) controllers. Various tuning methods, such as the Ziegler-Nichols method, use these parameters as a starting point to adjust the proportional, integral, and derivative gains of the PID controller to achieve desired system performance, such as stability, overshoot, and settling time (to be discussed in Chapter 8).

Let's go through several examples to illustrate how to determine these parameters using the Routh stability criterion.

EXAMPLE 4.14

Problem:

Determine the range of K to ensure the stability of a unity feedback control system with feed-forward transfer function

$$G(s) = \frac{K(s+2)}{s(s-1)(s+4)}$$

SOLUTION:

The closed-loop transfer of the system is given by

$$T(s) = \frac{G(s)}{1+G(s)} = \frac{\dfrac{K(s+2)}{s(s-1)(s+4)}}{1+\dfrac{K(s+2)}{s(s-1)(s+4)}}$$

$$= \frac{K(s+2)}{s^3+3s^2+(K-4)s+2K}$$

The closed-loop characteristic equation is given by

$$s^3+3s^2+(K-4)s+2K=0$$

The Routh's table is given as:

Table 4.15: Completed Routh Table

s^3	1	$K-4$
s^2	3	$2K$
s^1	$\frac{3K-12-2K}{3}$	
s^0	$2K$	

For the system to be stable,

$$2K>0, \implies K>0$$
$$\frac{K-12}{3}>0 \implies K-12>0 \implies K>12$$

Thus, for the system to be stable

$$K>12$$

The critical gain of the system, which makes the system marginally stable, is given by

$$K_{cu}=12$$

The frequency of oscillation when the system is marginally stable is obtained from

$$3s^2+2K_{cu}=3s^2+24=0 \implies \omega=\sqrt{8}\ \text{rad/sec}$$

CONCEPT REINFORCEMENT PROBLEM 4.6

Problem:

Determine the range of K, for the system shown in Figure 4.8, which will cause the system to be stable, unstable, and marginally stable.

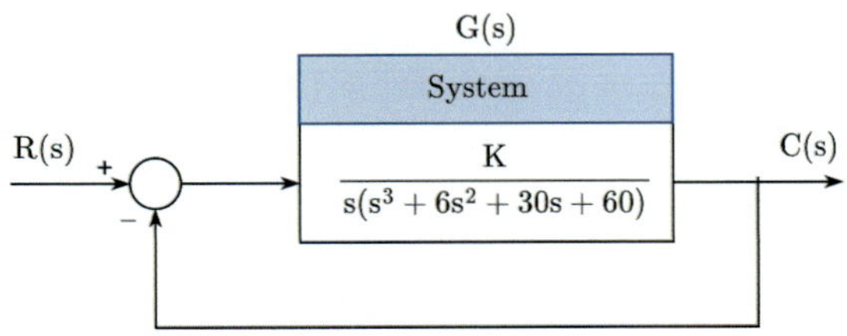

Figure 4.8: Feedback control system

EXAMPLE 4.15

Problem:

Determine the range of K to ensure the stability of a unity feedback control system with feed-forward transfer function

$$G(s) = \frac{K}{s(s+a)(s+b)}$$

SOLUTION:

The closed-loop transfer of the system is given by

$$T(s) = \frac{G(s)}{1+G(s)} = \frac{\frac{K}{s(s+a)(s+b)}}{1+\frac{K}{s(s+a)(s+b)}} = \frac{K}{s^3+s^2(a+b)+abs+K}$$

The closed-loop characteristic equation is given by

$$s^3 + s^2(a+b) + abs + K = 0$$

The Routh's table is given as:

Table 4.16: Completed Routh Table

s^3	1	ab
s^2	$a+b$	K
s^1	$\frac{ab(a+b)-K}{a+b}$	
s^0	K	

For the system to be stable

$$K > 0, \implies K > 0$$
$$\frac{ab(a+b) - K}{a+b} > 0 \implies ab(a+b) - K > 0 \implies K < ab(a+b)$$

Thus, for the system to be stable

$$0 < K < ab(a+b)$$

The frequency of oscillation when the system is marginally stable is obtained from

$$(a+b)s^2 + ab(a+b) = 0 \implies \omega = \sqrt{ab}$$

CONCEPT REINFORCEMENT PROBLEM 4.7

Problem:

a. Consider a unity feedback system with a forward path transfer function

$$G(s) = \frac{K}{(s+a)(s+b)(s+c)}$$

Using Routh's criteria, find the value of K which will cause sustained oscillation.

b. A unity feedback system with a forward path transfer function

$$G(s) = \frac{K(s+1)}{s^3 + \alpha s^2 + 4s + 1}$$

oscillates with frequency 3 rad/sec. Determine the value of α and the ultimate gain.

EXAMPLE 4.16

Problem:

Find the range of gain K which will ensure stability for the unity feedback system with forward path transfer function

$$G(s) = \frac{K(s+3)}{(s^2+1)(s+2)(s-1)}$$

SOLUTION:

The closed-loop transfer of the system is given by

$$T(s) = \frac{G(s)}{1+G(s)} = \frac{\frac{K(s+3)}{(s^2+1)(s+2)(s-1)}}{1+\frac{K(s+3)}{(s^2+1)(s+2)(s-1)}}$$

$$= \frac{K(s+3)}{s^4+s^3-s^2+(K+1)s+(3K-2)}$$

The closed-loop characteristic equation is given by

$$s^4+s^3-s^2+(K+1)s+(3K-2)=0$$

The Routh's table is given as:

Table 4.17: Completed Routh Table

s^4	1	-1	$3K-2$
s^3	1	$K+1$	
s^2	$-(K+2)$	$3K-2$	
s^1	$\frac{K(K+6)}{K+2}$	0	
s^0	$3K-2$		

For the system to be stable

$$-(K+2)>0, \implies K<-2$$
$$K(K+6)>0 \implies 0<K<-6$$
$$3K-2>0, \implies K>\frac{2}{3}$$

These conditions cannot be satisfied simultaneously. Hence, the system is not stable for any value of K

EXAMPLE 4.17

Problem:

For the closed-loop system shown in Figure 4.6 where

$$G(s) = \frac{K}{(s+a)(s^2+bs+b)}, \quad a, b > 0$$

find the range of gain K which ensures stability of the closed-loop system.

SOLUTION: The closed-loop transfer of the system is given by

$$T(s) = \frac{G(s)}{1+G(s)} = \frac{\dfrac{K}{(s+a)(s^2+bs+b)}}{1+\dfrac{K}{(s+a)(s^2+bs+b)}}$$

$$= \frac{K}{s^3+(a+b)s^2+b(a+1)s+(ab+K)}$$

The closed-loop characteristic equation is given by

$$s^3+(a+b)s^2+b(a+1)s+(ab+K) = 0$$

The Routh's table is given as:

Table 4.18: Completed Routh Table

s^3	1	$b(a+1)$
s^2	$a+b$	$K+ab$
s^1	$\dfrac{b(a+b)(a+1)-(K+ab)}{a+b} = \dfrac{ab(a+b)+b^2-K}{a+b}$	
s^0	$K+ab$	

For the system to be stable

$$\frac{ab(a+b)+b^2-K}{a+b} > 0, \implies K < ab(a+b)+b^2$$

$$\text{and } K+ab > 0, \implies K > -ab$$

Since both $a, b > 0$, this condition implies $K > 0$.

Thus,for the system to be stable $0 < K < ab(a+b)+b^2$

The ultimate gain $K_{cu} = ab(a+b)+b^2$ and the frequency of oscillation at ultimate gain is obtained from

$$(a+b)s^2+K_{cu}+ab = 0 \text{ which gives}$$

$$\omega = \sqrt{b(a+1)} \text{ rad/sec}$$

CONCEPT REINFORCEMENT PROBLEM 4.8

Problem:

The forward path transfer function of a unity feedback system is

$$G(s) = \frac{K}{as^2(s^2+b)}$$

i. Find the range of gain K which ensures stability of the closed-loop system.
ii. Determine the frequency where the system is marginally stable

EXAMPLE 4.18

Problem:

Consider a unity feedback system with forward path transfer function

$$G_c(s)G(s) = \frac{K(s+a)}{s(1+Ts)(1+bs)}$$

where a, b and T are positive. Determine the conditions of absolute stability.

SOLUTION:

The closed-loop characteristic equation of a system is given by

$$1 + G_c(s)G(s) = bTs^3 + (T+b)s^2 + (K+1)s + Ka = 0$$

The Routh's table is given as:

Table 4.19: Completed Routh Table

s^3	bT	$K+1$
s^2	$b+T$	Ka
s^1	$\frac{(T+b)(K+1)-KTab}{b+T}$	
s^0	Ka	

Thus, for the system to be stable, we require

i. $bT > 0 \implies T > 0$
ii. $T+b > 0$. This is always true as T and b is assumed positive

iii. $Ka > 0 \implies K > 0$

iv.

$$\frac{(T+b)(K+1) - KTab}{b+T} > 0 \implies (T+b)(K+1) - KTab > 0$$
$$\implies T[1 - K(ab-1)] + b(K+1) > 0$$

Let us denote

$$N(T,K) = T[1 - K(ab-1)] + b(K+1)$$

and solve $N(T,K) > 0$.

- If $0 < K < \frac{1}{ab-1}$, then $1 - K(ab-1) > 0$. Because $T > 0$ and $b(K+1) > 0$, $N > 0$ automatically.
- If $K = \frac{1}{ab-1}$, then $N = b(K+1) > 0$. Thus, the system is stable for all $T > 0$.
- If $K > \frac{1}{ab-1}$, then $1 - K(ab-1) < 0$. Rearranging gives

$$T < \frac{b(K+1)}{K(ab-1)-1}$$

Stability Region in $T - K$ plane

The marginal-stability boundary is the curve

$$T = \frac{b(K+1)}{K(ab-1)-1}$$

Further

$$\lim_{K\to\infty} T = \lim_{K\to\infty} \frac{b(K+1)}{K(ab-1)-1} = \frac{b}{ab-1}$$

In summary

$$\begin{cases} K \le \dfrac{1}{ab-1} & \implies \text{Stable if} \quad T > 0 \\ K > \dfrac{1}{ab-1} & \implies \text{Stable if} \quad 0 < T < \dfrac{b(K+1)}{K(ab-1)-1} \end{cases}$$

CONCEPT REINFORCEMENT PROBLEM 4.9

Problem:

The closed-loop characteristic equation of a system is given by

$$s^4 + as^3 + K_1s^2 + bs + K_2 = 0$$

where a, b, K_1 and K_2 are positive. Determine the conditions of absolute stability and plot the region of stability in the $K_2 - K_1$ plane.

4.10 Relative Stability Analysis

Routh's stability criterion primarily provides information about the absolute stability of a system, meaning it answers the question of whether the system is stable or unstable without considering the degree of stability. While determining absolute stability is essential, understanding the relative stability of a system is equally important in practical control system design and analysis. Note that relative stability refers to the system's ability to maintain stability in the face of parameter variations, disturbances, or changes in operating conditions. It encompasses not only whether a system is stable but also the degree of stability and its robustness to perturbations. It tells us if the system is stable, what is the degree of its stability, or when the system is unstable, it offers insights into actions that can be taken to stabilize unstable systems.

Although this criterion is not the preferred approach to investigate the relative stability of a system, it is possible to use it to gain insights into relative stability, particularly in determining how many roots of the characteristic equation lie to the right of a specified vertical line in the complex plane, represented by $s = -\sigma$.

Before we proceed with the mathematical steps, let us understand the fundamental concept underlying this analysis: the idea of shifting our reference frame. This technique of axis translation or variable transformation appears throughout engineering, from linearizing around equilibrium points to analyzing perturbations. To make this abstract concept concrete, consider the following analogy:

The Relativity of Roots: A Control Theory Parable

Understanding Relativity of Roots

"You exist because I'm watching. I exist because you're here."

The professor's opening line made everyone wonder if they'd wandered into the philosophy department by mistake. But then he drew a timeline on the board.

Three Observers, One Life:

Assume **you live from 1990 to 2100. Are you alive?**

Observer 1 (Year 2025):

Yes. You're here, observable, real.

Their origin: t = 0 at t = 2025. You exist at t = −35 to t = +75

Observer 2 (Year 2110):

"No. You're already gone, a historical footnote."

Their origin: t = 0 at 2110. You exist at t = −120 to t = −10. Pure past.

Observer 3 (Year 1970):

"Not yet. You're still science fiction."

Their origin: t = 0 at 1970. You exist at t = +20 to t = +130. Pure future.

The Lesson

Your lifespan never changed. Only the reference frame did. Your existence is invariant. Your classification is not. Whether you're "alive," "dead," or "not yet born" depends entirely on who's measuring and from where. **Reality is absolute. Interpretation requires a reference frame.**

Connection to Relative Stability Analysis:

When we analyze a characteristic polynomial P(s) using the standard Routh criterion, we count roots relative to the imaginary axis ($\text{Re}(s) = 0$), as if measuring from "now." But what if we want to know how many roots have $\text{Re(s)} > -\sigma$ for some $\sigma > 0$? This is equivalent to asking: "From the perspective of someone standing at a different year, how many of your life years are still to come?"

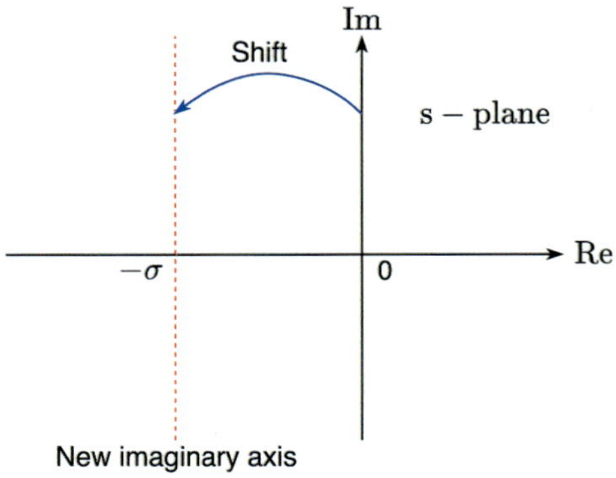

Figure 4.9: Relative stability analysis by Routh Criterion

We accomplish this by shifting the axis through the substitution $s = \hat{s} - \sigma$, creating a new polynomial $\hat{P}(\hat{s})$ with a new reference point. The imaginary axis in the $\hat{s}$-plane corresponds to the line $Re(s) = -\sigma$ in the original s-plane. Roots with $Re(\hat{s}) > 0$ (right of the new origin) are roots with $Re(s) > -\sigma$ (right of the shifted line). The number of sign changes in the Routh array for $\hat{P}(\hat{s})$ tells us how many roots lie to the right of $s = -\sigma$.

Let us apply this concept of axis translation to investigate the relative stability of a system using Routh's stability criterion. The procedure to do this is given below.

4.10.1 Procedure for Relative Stability Analysis

Using the following steps, we can determine the number of roots lying to the right of the vertical line $s = -\sigma$ using Routh's criterion:

1. Shift the imaginary axis σ units to the left as shown in Figure 4.9.
2. Substitute $s = \hat{s} - \sigma$ in the characteristic polynomial of the system.
3. Apply Routh's test to the new polynomial in $\hat{s}$.
4. The number of sign changes in the first column of the array gives the number of roots that are located to the right of the vertical line $s = -\sigma$.

EXAMPLE 4.19

Problem:

The characteristic equation of a unity feedback system is given by

$$P(s) = s^2 + 5s + 4$$

a. Determine how many roots of the characteristic equation lie to the right of the vertical line s = −2
b. Determine how many roots of the characteristic equation lie to the right of the vertical line s = −5

SOLUTION:

Let us first find the closed-loop characteristic equation.

a. Substitute s in P(s) by $\hat{s} - \sigma = \hat{s} - 2$ and compute the characteristic equation in the $\hat{s}$ plane. This gives

$$\begin{aligned} P(\hat{s}) &= (\hat{s} - 2)^2 + 5(\hat{s} - 2) + 4 \\ &= \hat{s}^2 + \hat{s} - 2 \end{aligned}$$

The Routh's table is given as:

Table 4.20: Completed Routh Table

$\hat{s}^2$	1	−2
$\hat{s}^1$	1	
$\hat{s}^0$	−2	

Since there is one sign change in the first column of Routh's array, there is one root that is greater than −2.

b. Substitute s in P(s) by $\hat{s} - \sigma = \hat{s} - 5$ and compute the characteristic equation in the $\hat{s}$ plane. This gives

$$\begin{aligned} P(\hat{s}) &= (\hat{s} - 5)^2 + 5(\hat{s} - 5) + 4 \\ &= \hat{s}^2 - 5\hat{s} + 4 \end{aligned}$$

The Routh's table is given as:

Table 4.21: Completed Routh Table

$\hat{s}^2$	1	4
$\hat{s}^1$	−5	
$\hat{s}^0$	4	

Since there are two sign changes in the first column of Routh's array, there are two roots that are greater than −5.

CONCEPT REINFORCEMENT PROBLEM 4.10

Problem:

The characteristic equation of a unity feedback system is given by

$$P(s) = s^3 + 12s^2 + 39s + 28$$

a. Determine how many roots of the characteristic equation lie to the right of the vertical line $s = -2$

b. Determine how many roots of the characteristic equation lie to the right of the vertical line $s = -6$

c. Determine how many roots of the characteristic equation lie to the right of the vertical line $s = -8$

4.11 Limitations of Routh-Hurwitz Criterion

While the Routh-Hurwitz criterion is a valuable tool for evaluating the stability of linear time-invariant systems, it has several limitations, as outlined below.

1. **Applicability Constraints:**

   i. The criterion applies only to systems with an algebraic characteristic equation where all coefficients are real.

   ii. It cannot be used if any coefficient of the characteristic equation is complex or time-varying.

   iii. This criterion cannot be applied to systems with time delays whose characteristic equation contains exponential terms like e^{-Ts}.

2. **Restriction to Linear Time-Invariant Systems:** The criterion is limited to linear time-invariant systems and cannot be directly applied to nonlinear or time-varying systems.

3. **Lack of Relative Stability Analysis:** It offers information only on the absolute stability of the system and does not indicate how close the roots of the characteristic equation are to the imaginary axis in the s-plane, which is crucial for assessing relative stability.

Despite these limitations, the Routh-Hurwitz criterion remains a valuable and widely used tool for initial stability analysis due to its simplicity and ease of application. However, we should be aware of its limitations and consider complementary stability analysis methods for a more comprehensive assessment of system stability (to be discussed in later chapters).

4.12 Multiple Choice Skills Test Questions

M 4.1. Routh-Hurwitz criterion is used to determine

A. time response of a system
B. absolute stability of a system
C. relative stability of the system
D. frequency response of a system

M 4.2. The signs of the first element of each of the rows of the Routh-Hurwitz array of a system are shown below.

Rows	1	2	3	4	5	6	7
Signs	+ve	+ve	-ve	+ve	+ve	-ve	+ve

The number of roots of the system lying on the right half s-plane is

A. 2 **B.** 0 **C.** 3 **D.** 4

M 4.3. The signs of the first element of each of the rows of the Routh-Hurwitz array of a system are shown below.

Rows	1	2	3	4	5	6
Signs	+ve	+ve	+ve	+ve	-ve	+ve

The number of roots of the system lying on the right half s-plane is

A. 2 **B.** 3 **C.** 4 **D.** 0

M 4.4. The terms in the first column of the Routh array of the characteristic equation of a system are 4, 3, 8,-6,2,4. The number of roots of the characteristic equation in the right half of the s-plane is equal to

A. 3 **B.** 2 **C.** 1 **D.** 0

M 4.5. The terms in the first column of the Routh array of the characteristic equation of a system are 5, 12, 10,-6,-11,-4. The number of roots of the characteristic equation in the right half of the s-plane is equal to

A. 3 **B.** 2 **C.** 1 **D.** 0

M 4.6. Consider the characteristic polynomial of four third-order systems:

S1: $s^3 + 4s^2 + s - 6$ **S2:** $s^3 + 6s^2 + 11s + 6$

S3: $s^3 + 12s^2 + 44s + 48$ **S4:** $s^3 + s^2 - 14s - 24$

By inspection, determine which of the following statements is true

A. **S1** and **S4** are stable

B. **S1**, **S3** and **S4** are stable

C. **S2** and **S4** are stable

D. **S2**, **S3** and **S4** are stable

M 4.7. The closed-loop characteristic polynomials of several systems are given below.

S1: $4s^2 + 2s + 6$ **S2:** $s^2 - 6s + 10$, **S3:** $-2s^2 - 5s - 20$

S4: $15s^2 + 25s + 90$, **S5:** $3s^2 + 8s - 10$

Consider the following statements

I: Systems **S1**, **S3**, and **S4** are stable

II: Systems **S1** is stable; but **S3** is unstable

III: Systems **S2**, and **S5** are unstable

IV: Systems **S2**, **S3**, and **S5** are unstable

Which of the above is true

A. I and III are true

B. II and IV are true

C. IV is only true

D. III is only true

M 4.8. The closed-loop characteristic equation of a system is

$$q(s) = a_0s^3 + a_1s^2 + a_2s + a_3$$

The system is stable if

A. $a_1a_0 > a_2a_3$ **B.** $a_1a_2 < a_0a_3$ **C.** $a_1a_2 = a_0a_3$ **D.** $a_1a_2 > a_0a_3$

M 4.9. The closed-loop characteristic equation of a system is

$$q(s) = s^3 + a_1s^2 + a_2s + a_3$$

The condition when the system will be marginally stable and the corresponding oscillation frequency ω_u are respectively

A. $a_2 = a_1a_3,\ \omega_u = \frac{a_1}{a_3}$ rad/sec

B. $a_3 = a_1a_2,\ \omega_u = \frac{a_3}{a_1}$ Hz

C. $a_3 = a_1a_2,\ \omega_u = \frac{a_3}{a_1}$ rad/sec

D. $a_1 = a_2a_3,\ \omega_u = \frac{a_3}{a_1}$ rad/sec

M 4.10. The closed-loop characteristic equation of a system is given by

$$q(s) = s^3 + 20s^2 + 4s + 80 = 0$$

This system is

A. unstable

B. marginally stable

C. stable

D. exhibits overdamped response

M 4.11. The closed-loop characteristic equation of a system is

$$q(s) = s^3 + Ks^2 + 2Ks + K$$

The system is stable if

A. $0 < K < 0.5$

B. $0 < K < 1$

C. $K > 0.5$

D. $K = 0.5$

M 4.12. The open-loop transfer function of a unity feedback control system is

$$G(s) = \frac{20}{s(s+1)(s+T)}$$

where T is a variable parameter. The closed-loop system is stable if

A. $T > 0$ **B.** $T > 4$ **C.** $T < 4$ **D.** $0 < T < 4$

M 4.13. The open-loop transfer function of a unity feedback control system is given by

$$G(s) = \frac{10(s+1)}{s^2(s+2)}$$

The stability characteristics of open-loop and closed-loop configuration are respectively

A. stable, stable

B. unstable, unstable

C. stable, unstable

D. unstable, stable

M 4.14. Consider a system with closed-loop characteristic equation

$$s^3 + Ts^2 + (K + a)s + (K + b) = 0$$

The system is marginally stable if

A. $T = \dfrac{K+b}{K+a}$ **B.** $T = \dfrac{K+a}{K+b}$ **C.** $T = \dfrac{K-a}{K-b}$ **D.** $T = \dfrac{K-b}{K-a}$

M 4.15. The open-loop transfer function of a unity feedback system is

$$G(s) = \frac{K}{s(s+2)(s+3)}$$

The range of value of K which ensures closed-loop stability is

A. $0 < K < 30$

B. $0 < K < 15$

C. $0 < K < 20$

D. $0 < K < 25$

M 4.16. Consider the unity feedback system shown in Figure 4.10

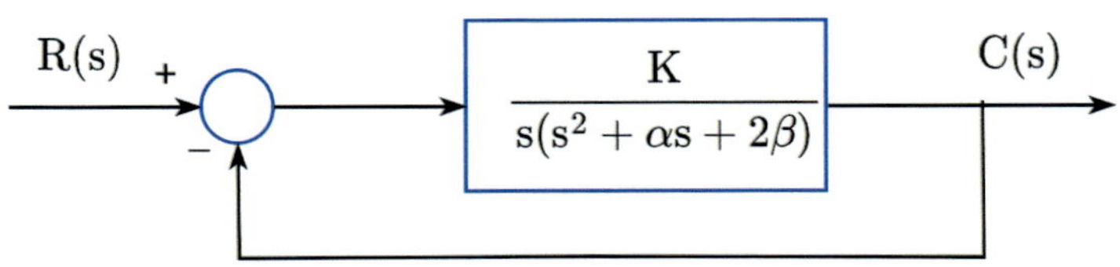

Figure 4.10: Closed-loop control system

The system will be marginally stable if

A. $K = 2\beta$ **B.** $K = 2\alpha\beta$ **C.** $K = \alpha\beta$ **D.** $K = \dfrac{2}{\alpha\beta}$

M 4.17. For the system shown in Figure 4.10, the value of gain $K = 25$. If the system oscillates with frequency 5rad/sec, the vales of α and β are

A. $\alpha = 12.5$ and $\beta = 1.0$

B. $\alpha = 6.25$ and $\beta = 2.0$

C. $\alpha = 1$ and $\beta = 12.5$

D. $\alpha = 2.0$ and $\beta = 6.25$

M 4.18. The parameters of the unity feedback system, shown in Figure 4.11, have been selected such that it acts as an oscillator.

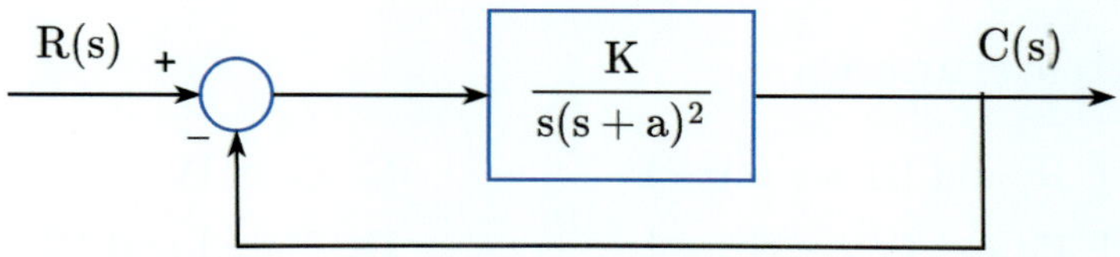

Figure 4.11: Closed-loop control system

What is the value of K and the corresponding frequency of oscillation ω_u

A. $K = a^3$ and $\omega_u = \sqrt{a}$ rad/sec

B. $K = a^2$ and $\omega_u = a$ Hz

C. $K = \frac{2}{a^3}$ and $\omega_u = a$ rad/sec

D. $K = 2a^3$ and $\omega_u = a$ rad/sec

M 4.19. The characteristic equation of a system is

$$s^3 + 3s^2 + (K + 1)s + 9 = 0$$

The system is stable if

A. $K > 6$ **B.** $K < 2$ **C.** $K > 2$ **D.** $K = 2$

M 4.20. Consider the unity feedback system with time delay shown in Figure 4.12

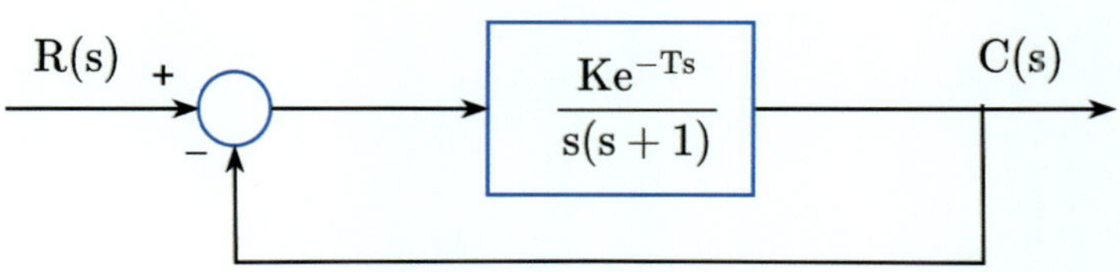

Figure 4.12: Closed-loop control system

The closed-loop characteristic equation of the system is

$$s^2 + s + Ke^{-Ts} = 0, \quad \text{Assume } K > 0$$

Consider the following statements

I The characteristic equation has an infinite number of roots

II The term e^{-Ts} can be approximated by $1 - Ts$ for stability analysis

III The system is stable if $T < 1$

IV The system is stable if $T < \frac{1}{K}$

Of these statements

A. I, II, and III are correct

B. I, II, and IV are correct

C. Only IV is correct

D. Only I and IV are correct

Answers

M 4.1. B **M 4.2.** D **M 4.3.** A **M 4.4.** B **M 4.5.** C **M 4.6.** C **M 4.7.** A **M 4.8.** D **M 4.9.** C **M 4.10.** B **M 4.11.** C **M 4.12.** B **M 4.13.** D **M 4.14.** A **M 4.15.** A **M 4.16.** B **M 4.17.** C **M 4.18.** D **M 4.19.** C **M 4.20.** B

4.13 Knowledge Assessment Problems

KaP 4.1 Using Routh's table, determine if the system with the following closed-loop characteristic equation is stable.

$$P(s) = s^3 + s^2 - 17s + 15$$

KaP 4.2 Using Routh's table, determine if the system with the following closed-loop characteristic equation is stable.

$$P(s) = s^4 - 5s^3 - 7s^2 + 29s + 30$$

KaP 4.3 Using Routh's table, determine if the system with the following closed-loop characteristic equation is stable.

$$P(s) = s^5 + 9.5s^4 + 30s^3 + 32.5s^2 - s - 12$$

KaP 4.4 Using Routh's table, determine if the system with the following closed-loop characteristic equation is stable.

$$P(s) = s^6 - 0.5s^5 - 11.25s^4 - 4.375s^3 + 32.75s^2 + 38.625s + 11.25$$

KaP 4.5 Using Routh's table, determine how many roots of the polynomial

$$P(s) = s^5 + 3s^4 + 0.5s^3 + 14s^2 - 14s + 8$$

are on the $j\omega$-axis.

KaP 4.6 Investigate the stability of the system with the characteristic polynomial

$$P(s) = s^3 + 4s^2 - 3s - 12$$

KaP 4.7 Using Routh's table, determine how many roots of the polynomial

$$P(s) = s^5 + 2s^4 - 3s^3 - 6s^2 + 5s + 1$$

are in the right-half of the s-plane.

KaP 4.8 The closed-loop characteristic equation of a system is given by

$$P(s) = s^3 + Kas^2 + (K + b)s + c$$

Determine the conditions of stability using the Routh-Hurwitz method. Assume that $K, b, c > 0$.

KaP 4.9 Find the range of gain K which will ensure stability for the unity feedback system with forward path transfer function

$$G(s) = \frac{K(s-1)(s+2)(s+3)}{s^2+5}$$

KaP 4.10 Investigate the stability of the closed-loop system shown in Figure 4.6 where

$$G(s) = \frac{K}{(s+1)(s+2)(s+3)(s-4)}$$

KaP 4.11 Consider the closed-loop system shown in Figure 4.6 where

$$G(s) = \frac{K(s+\alpha)}{s(s+2)(s+4)}$$

Investigate the stability of the closed-loop system for

a. $\alpha = 10$

b. $\alpha = 7$

c. $\alpha = 3$

KaP 4.12 The block diagram of a Type 2 system, stabilized by a lead compensator, is shown in Figure 4.13. Determine the range of values of K which will ensure stability of the system. Consider $\alpha = 2$, $\beta = 6$, and $a = 4$.

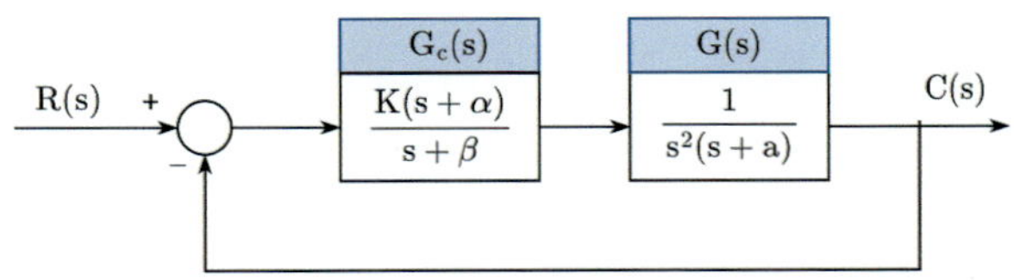

Figure 4.13: Type 2 feedback control system

KaP 4.13 Consider the closed-loop system shown in Figure 4.6 where

$$G(s) = \frac{K(s+a)}{s(s+b)(s+c)}$$

Find the range of gain K which ensures stability of the closed-loop system.

KaP 4.14 Determine the range of K to ensure the stability of a unity feedback control system with feed-forward transfer function

$$G(s) = \frac{K(s+a)}{s(s-b)(s+c)}$$

KaP 4.15 Determine the range of K to ensure the stability of a unity feedback control system with feed-forward transfer function. Find the ultimate gain and ultimate frequency of oscillation.

$$G(s) = \frac{K(s+2)}{s(s-1)(s+4)}$$

KaP 4.16 Given the unity feedback system of Figure 4.6 with

$$G(s) = \frac{K(s-2)(s-3)}{(s+3)(s^2+4s+4)}$$

Find the following

a. The range of K that ensures stability of the system.

b. The ultimate gain K_{cu} which makes the system oscillate and

c. The frequency of oscillation is when the system is marginally stable.

KaP 4.17 Using Routh's criteria, find the value of K which will cause sustained oscillation for the unity feedback system shown in Figure 4.6 with

$$G(s) = \frac{K}{(s+5)(s+10)(s+15)}$$

KaP 4.18 Using Routh's criteria, find the value of K which will cause sustained oscillation for the unity feedback system shown in Figure 4.6 with

$$G(s) = \frac{K}{s(s+a)(s+b)(s+c)}$$

KaP 4.19 Given the unity feedback system of Figure 4.6 with

$$G(s) = \frac{Ks(s+1)}{(s+2)(s^2-6s+18)}$$

Find the following

a. The range of K that ensures stability of the system.

b. The ultimate gain K_{cu} which makes the system oscillate and

c. The frequency of oscillation is when the system is marginally stable.

KaP 4.20 Determine the range of gain K which ensures stability of the unity feedback closed-loop system with forward path transfer function

$$G(s) = \frac{K(s+2)(s+5)}{(s-3)(s-7)}$$

KaP 4.21 For the closed-loop system shown in Figure 4.6, where

$$G(s) = \frac{K(s^2-a^2)}{s^2+b}, \quad a, b > 0$$

find the range of gain K which ensures stability of the closed-loop system.

KaP 4.22 For the closed-loop system shown in Figure 4.6, where

$$G(s) = \frac{K}{(s+a)^3}, \quad a > 0$$

Find the range of gain K which ensures stability of the closed-loop system.

KaP 4.23 For the closed-loop system shown in Figure 4.6, where

$$G(s) = \frac{K}{(s+5)(s^2+2s+2)}$$

Determine the range of gain K which ensures stability of the closed-loop system.

KaP 4.24 Consider a unity feedback closed-loop system with forward path transfer function

$$G(s) = \frac{K(s+a)(s+b)}{(s-c)(s-d)}, \quad a, b, c, d > 0$$

i. Determine the range of gain K which ensures stability
ii. Find the frequency of oscillation when the system is marginally stable.

KaP 4.25 Consider a unity feedback system with forward path transfer function

$$G_c(s)G(s) = \frac{(s+m)(s+a)}{s}\frac{K}{s^2-b}$$

where a, b, and K are positive. Determine the conditions of absolute stability and plot the region of stability in the K-m plane.

Chapter 5

Steady State Error of Linear Systems

Steady-state error is your GPS confidently announcing 'You have arrived' while you're still three houses away.

DESIRED LEARNING OUTCOMES

Upon completion of this chapter, the readers should have learned the following:

- Understand the concept of system type and its physical interpretation
- Calculate static position, velocity, and acceleration error constants for a unity feedback system
- Determine the steady state error of a unity feedback system due to different inputs
- Relate system type to steady-state error performance and system robustness
- Design and adjust system gain to achieve desired steady-state error specifications
- Analyze steady-state error due to disturbances in a unity feedback system
- Compute steady-state error from the closed-loop transfer function of a unity feedback system
- Compute steady-state error for non-unity feedback systems
- Evaluate the sensitivity of steady-state error to various parameters of the system

DOI: 10.1201/9781003571308-5

5.1 Introduction

The term steady-state error is used for the error that occurs between the reference input to a control system and its output after it has settled down to its steady value. It serves as a metric for assessing the system's ability to accurately track the desired reference signal over time. Before exploring the mathematical foundations, let's visualize this concept through a practical example that illustrates why steady-state error matters.

Understanding Steady State Error

Imagine a robotic arm tasked with feeding a baby. The reference input is the baby's mouth position, and the system output is the position of the spoon.

- **If the system tracks perfectly** → the spoon aligns with the mouth, and the food goes in.
- **If there is a slight steady-state error** → the spoon consistently stops just a little short, or slightly off to the side. As a result, despite the effort, the food never quite reaches the mouth.
- **In control terms** → even a small steady-state error can have significant consequences, because the system output fails to achieve the desired goal in practice.

Although one of the primary objectives of most control systems is for the system output to accurately follow the reference signal in the steady state, a discrepancy often arises between the reference and the output due to various reasons, including imperfect modeling, disturbances, noise, and limitations in the control algorithm. Further, in many control systems, steady-state errors arise from nonlinear characteristics of control system elements, such as nonlinear friction or a dead zone. Many practical control systems often have limitations such as actuator saturation, sensor noise, or bandwidth constraints. These limitations can introduce steady-state errors, particularly if they are not adequately accounted for in the control system design. Since most of the practical systems are nonlinear to a certain degree, there exists no error-free control system in the real world.

One of the major objectives of a control system designer is to eliminate or reduce the steady state error by careful design, analysis, and tuning of the control parameters such that the system is both stable and exhibits acceptable transient response. Various techniques can mitigate these errors and improve the overall performance of the control system; some of these will be discussed later in the book. In this chapter, the focus is on analysing the steady state error of linear systems that arise from the configuration of the control system.

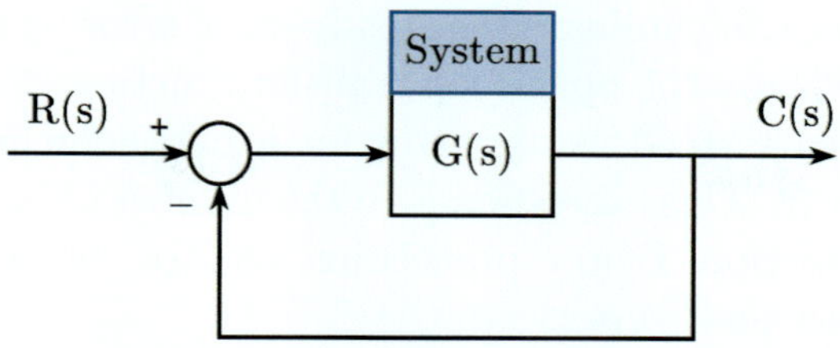

Figure 5.1: Unity Feedback Control System

5.2 Steady State Error in Unity Feedback Systems

Consider a unity feedback control system shown in Figure 5.1

The closed-loop transfer function is given by

$$\frac{C(s)}{R(s)} = \frac{G(s)}{1 + G(s)} \tag{5.1}$$

The error E(s) between the input R(s) (reference signal), and the output C(s) is given by

$$E(s) = R(s) - C(s) = R(s) - G(s)E(s) \qquad (\because\ C(s) = G(s)E(s)) \tag{5.2}$$

Solving for E(s) gives

$$E(s) = \frac{R(s)}{1 + G(s)} \tag{5.3}$$

The steady state error e_{ss} of a feedback control system is defined as the error when the time reaches infinity. Thus,

$$\boxed{e_{ss} = e(\infty) = \lim_{t\to\infty} e(t) = \lim_{s\to 0} sE(s)} \tag{5.4}$$

Assume that the closed-loop system is stable. Substituting Eqn(5.3) into Eqn(5.4) gives

$$\boxed{e_{ss} = \lim_{s\to 0} \frac{sR(s)}{1 + G(s)}} \tag{5.5}$$

From Eqn(5.5), we find that the steady state error depends on both the forward path transfer function G(s) and the input R(s). We will demonstrate later that a system may exhibit zero steady-state error for a step input, yet the same system may show non-zero steady-state error for a ramp input. As will be shown, we can eliminate the steady-state error by modifying the system structure, specifically the open-loop transfer function. Note that the system's ability to eliminate steady state error for a particular input depends on the *Type* of the system. This corresponds to the number of integrators in the open-loop transfer function. Before proceeding further, let us first classify control systems based on their types.

5.2.1 Classification of Control Systems: System *Types*

Consider the unity feedback system with the open-loop transfer function, represented in time constant form, as:

$$G(s) = \frac{K(1+T_a s)(1+T_b s)\dots(1+T_m s)(1+T_{z_1}s+T_{z_2}s^2)}{s^N(1+T_1 s)(1+T_2 s)\dots(1+T_n s)(1+T_{p_1}s+T_{p_2}s^2)}e^{-T_d s} \tag{5.6}$$

where K and T's are real constants and T_d is the delay. The s^N term in the denominator of Eqn(5.6) represents a pole of multiplicity N at the origin of the s-plane. Since a division by "s" is equivalent to the time-domain operation of integration, the term N indicates that there are N integrators in the forward path of the system. We will classify the systems based on the number of integrations in the forward path.

Definition 5.1. System Type: A closed-loop system with forward-path transfer function of Eqn(5.6) is of *Type*-N if there are N number of integrators in the open-loop transfer function.

Therefore, a system is designated as *Type* 0,*Type* 1,*Type* 2, $\dots$, when $N = 0$, $N = 1$, $N = 2, \dots$ respectively. This classification affects the system's capability to track various types of input signals accurately in a steady state. Note that this is distinct from the order of a system and is independent of the number of terms and their coefficients in the numerator and denominator of G(s).

EXAMPLE 5.1

This example determines the type of few system by comparing it with the standard transfer function of Eqn(5.6).

a. **S1:** $G(s) = \dfrac{K(s+2)}{(s+3)(s+5)(s+7)}$, Type 0

b. **S2:** $\dfrac{K(s+1)}{s(s+2)(s^2+2s+8)}$, Type 1

c. **S3:** $\dfrac{Ke^{-0.4s}}{(1+0.5s)(1+5s)(1+10s)}$, Type 0

d. **S4:** $\dfrac{K(s+1)(s^2+s+4)}{s^2(s+5)(s+9)(s^2+3s+10)}$, Type 2

e. **S5:** $\dfrac{20}{s^3(1+0.5s)(1+2s)(1+s)}$, Type 3

f. **S6:** $\dfrac{10(s+1)}{s^3(1+0.5s)}$, Type 3

The **Type number** is a critical parameter of a control system as it directly indicates the accuracy of the steady-state response of a stable closed-loop system to various forcing inputs. Subsequently, we will demonstrate that a system's ability to track different inputs improves as the type-number increases. However, this improvement comes with a trade-off: an increase in the type-number reduces the stability margin of the system. More about the stability margin will be discussed in Chapter 7.

Let us now investigate how the steady state error is influenced by the system type and nature of the inputs, focusing on step, ramp, and parabolic inputs. We will define steady state error performance specifications using *static error constants* and derive formulae to compute these constants.

5.2.2 Steady State Error for Step Function Input and Static Position Error Constant

If a step input of magnitude A, that is, $r(t) = Au_s(t)$, $R(s) = \dfrac{A}{s}$ is applied to the system of Figure 5.1, then the steady state error (from Eqn(5.5)) is given by

$$e_{ss} = \lim_{s\to 0} \frac{sR(s)}{1+G(s)} = \lim_{s\to 0} \frac{A}{1+G(s)} = \frac{A}{1+\lim\limits_{s\to 0} G(s)} \tag{5.7}$$

Define the *static position error constant* K_p as:

$$\boxed{K_p = \lim_{s\to 0} G(s) = G(0)} \tag{5.8}$$

Therefore, the steady-state error, expressed in terms of static position error constant K_p, is given by

$$\boxed{e_{ss} = \frac{A}{1+K_p}} \tag{5.9}$$

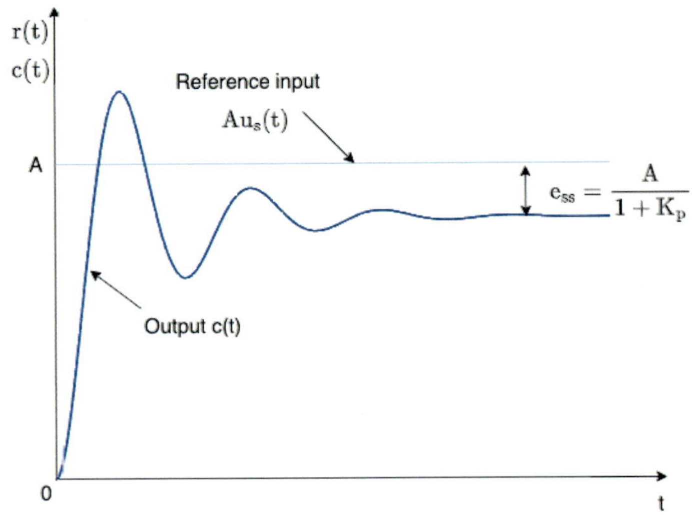

Figure 5.2: Typical steady state error due to step input

The static position error constant for systems of different types is:

$$K_p = \lim_{s\to 0} \frac{K(1+T_a s)\ldots\ldots}{(1+T_1 s)\ldots\ldots} e^{-T_d s} = K \quad \text{(For N = 0; Type 0 System)}$$

$$K_p = \lim_{s\to 0} \frac{K(1+T_a s)\ldots\ldots}{s^N(1+T_1 s)\ldots\ldots} e^{-T_d s} = \infty, \quad \text{for } N \geq 1$$

The output of a typical *Type 0* system, which exhibits finite steady state error due to step-input, is shown in Figure 5.2.

From Eqn(5.9), we observe that if the constant K_p is finite, then the steady-state error will be non-zero. Higher values of gain K can reduce this error, but cannot eliminate it. Furthermore, higher values of K will reduce the stability margin of the system and may compromise its stability. If we desire to get zero steady-state error, K_p must be infinite. This is possible if the *Type* of the system is one or higher; that is, to get zero steady-state error due to a step input, there must be at least one integrator in the forward path.

The steady state error e_{ss} due to step input are summarized as:

$$\boxed{\begin{aligned} e_{ss} &= \frac{A}{1+K_p}, \quad &&\text{for Type 0 system} \\ e_{ss} &= \frac{A}{1+K_p} = 0, \quad &&\text{for Type 1 or higher order system} \end{aligned}} \tag{5.10}$$

5.2.3 Steady State Error for Ramp Function Input and Static Velocity Error Constant

When the input to the control system is ramp function with a slope A, that is, $r(t) = Atu_s(t)$, and $R(s) = \frac{A}{s^2}$, then the steady state error (from Eqn(5.5)) is given by

$$e_{ss} = \lim_{s\to 0} \frac{s}{1+G(s)} \frac{A}{s^2} = \lim_{s\to 0} \frac{A}{s\,[1+G(s)]} = \frac{A}{\lim\limits_{s\to 0} sG(s)} \tag{5.11}$$

Define the *static velocity error constant* K_v as:

$$\boxed{K_v = \lim_{s\to 0} sG(s)} \tag{5.12}$$

Thus the steady-state error, in terms of velocity error constant K_v is given by

$$\boxed{e_{ss} = \frac{A}{K_v}} \tag{5.13}$$

The static velocity error constant for systems of different Type is:

$$\boxed{\begin{aligned} K_v &= \lim_{s\to 0} \frac{sK(1+T_a s)\ldots\ldots}{(1+T_1 s)\ldots\ldots} e^{-T_d s} = 0 \quad (\text{For } N=0,\ \text{Type 0 system}) \\ K_v &= \lim_{s\to 0} \frac{sK(1+T_a s)\ldots\ldots}{s(1+T_1 s)\ldots\ldots} e^{-T_d s} = K \quad (\text{For } N=1,\ \text{Type 1 system}) \\ K_v &= \lim_{s\to 0} \frac{sK(1+T_a s)\ldots\ldots}{s^2(1+T_1 s)\ldots\ldots} e^{-T_d s} = \infty, \quad (\text{For } N\geq 2,\ \text{Type 2 \& higher}) \end{aligned}}$$

The output of a typical *Type 1* system, which exhibits finite steady state error due to ramp-input, is shown in Figure 5.3.

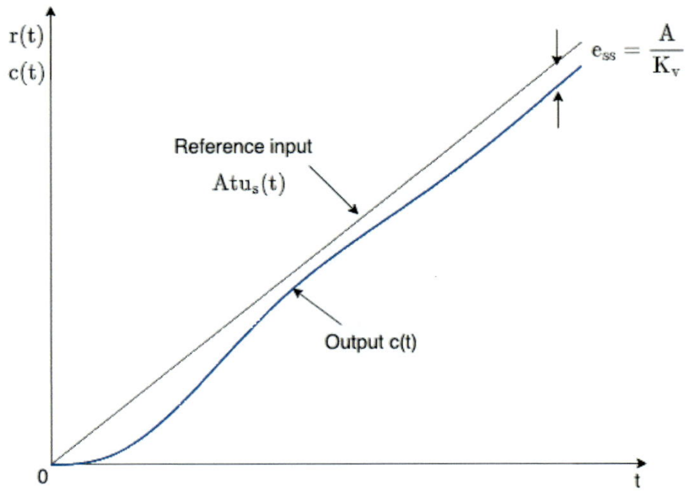

Figure 5.3: Typical steady state error due to ramp input

In order to make the this error zero, K_v must be infinite. From Eqns(5.6) and (5.11),

$$K_v = \lim_{s \to 0} sG(s) = \lim_{s \to 0} \frac{K}{s^{N-1}}, \quad N = 0, 1, 2, \ldots \tag{5.14}$$

For K_v to be infinite, N must be at least equal to 2, or there must be at least two integrators in the forward path, that is, the system must be of *Type 2* or higher.

The steady state error e_{ss} due to ramp input are summarized as:

$$\begin{aligned} e_{ss} &= \frac{A}{K_v} = \infty, && \text{for Type 0 system} \\ e_{ss} &= \frac{A}{K_v} = \frac{A}{K}, && \text{for Type 1 system} \\ e_{ss} &= \frac{A}{K_v} = 0, && \text{for Type 2 and higher} \end{aligned} \tag{5.15}$$

5.2.4 Steady State Error for Parabolic Function Input and Static Acceleration Error Constant

When the input to the control system is a standard parabolic function, that is, $r(t) = \frac{At^2}{2}u_s(t)$, $R(s) = \frac{A}{s^3}$, then the steady state error (from Eqn(5.5) is given by

$$e_{ss} = \lim_{s\to 0} \frac{s}{1+G(s)} \frac{A}{s^3} = \lim_{s\to 0} \frac{A}{s^2\left[1+G(s)\right]} = \frac{A}{\lim\limits_{s\to 0} s^2 G(s)} \tag{5.16}$$

Define the *static acceleration error constant* K_a as:

$$\boxed{K_a = \lim_{s\to 0} s^2 G(s)} \tag{5.17}$$

Thus, the steady-state error, in terms of acceleration error constant K_a, is given by

$$\boxed{e_{ss} = \frac{A}{K_a}} \tag{5.18}$$

The static acceleration error constant of systems of different Type is:

$$K_a = \lim_{s\to 0} \frac{s^2 K(1+T_a s)\ldots\ldots}{(1+T_1 s)\ldots\ldots} e^{-T_d s} = 0 \text{ (N = 0, Type-0 system)}$$

$$K_a = \lim_{s\to 0} \frac{s^2 K(1+T_a s)\ldots\ldots}{s(1+T_1 s)\ldots\ldots} e^{-T_d s} = 0 \text{ (N = 1, Type-1 system)}$$

$$K_a = \lim_{s\to 0} \frac{s^2 K(1+T_a s)\ldots\ldots}{s^2(1+T_1 s)\ldots\ldots} e^{-T_d s} = K \text{ (N = 2, Type-2 system)}$$

$$K_a = \lim_{s\to 0} \frac{s^2 K(1+T_a s)\ldots}{s^N(1+T_1 s)\ldots} e^{-T_d s} = \infty, \text{ (}N \geq 3\text{ Type-3 \& higher)}$$

The output of a typical *Type 2* system, which exhibits finite steady state error due to parabolic input, is shown in Figure 5.4.

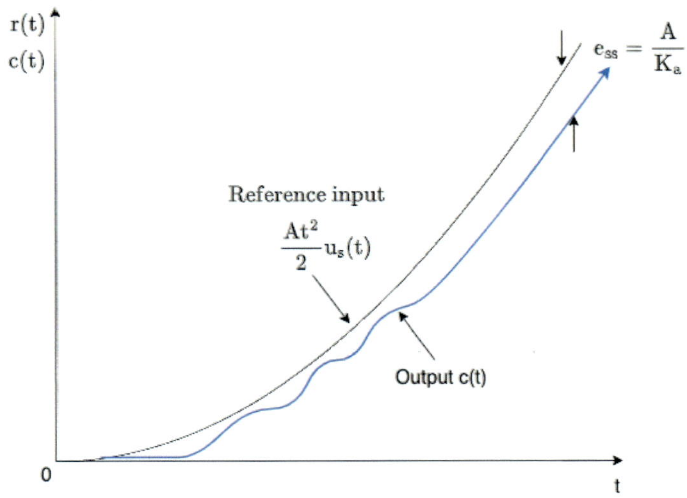

Figure 5.4: Typical steady state error due to parabolic input

Following similar arguments for step and ramp inputs, the steady state error due to parabolic input will be zero if the system is of *Type 3* or higher.

The steady state error e_{ss} due to parabolic input are summarized as:

$$\begin{aligned} e_{ss} &= \frac{A}{K_a} = \infty, && \text{for Type 0 and Type 1 system} \\ e_{ss} &= \frac{A}{K_a}, && \text{for Type 2 system} \\ e_{ss} &= \frac{A}{K_a} = 0, && \text{for Type 3 or higher} \end{aligned} \tag{5.19}$$

The observations from the preceding analysis can be summarized as follows. The steady-state error in a control system depends on both the number of integrators in the forward path and the characteristics of different input signals. A Type 0 system produces an error with a step input and infinite error with ramp or parabolic inputs. A Type 1 system produces no error with a step input, constant error with a ramp input, and infinite error with a parabolic input. A Type 2 system produces zero error for step or ramp inputs, but finite error with a parabolic input. These observations are fundamental to understanding control system behavior regarding steady-state error. They arise from the relationship between the system's type and the characteristics of various input signals.

Summary of Steady State Error

The steady state errors for *Type 0, Type 1, and Type 2* systems are summarized in Table 5.1 when the system is subjected to the test inputs.

	Step Input $r(t) = 1$	Ramp Input $r(t) = t$	Acceleration Input $r(t) = \frac{1}{2}t^2$
Type 0 system	$\frac{1}{1 + K_p}$	∞	∞
Type 1 system	0	$\frac{1}{K_v}$	∞
Type 2 system	0	0	$\frac{1}{K_a}$

Table 5.1: Relationships between input, system type, error constants, and steady state errors

5.2.5 System Type as a Robustness Property

As seen from Table 5.1, a *Type 0* system produces a finite steady-state error due to a step input. For a system of *Type 1* or higher, the steady-state error to a step input remains zero regardless of variations in the system parameters. Similarly, for a system of *Type 2* or higher, the steady-state error due to both ramp and step inputs remains zero regardless of parametric variations. This implies that System Type is one of the robustness measures of a system from the perspective of steady-state error. This robustness is particularly valuable in practical applications where system parameters may drift due to aging, temperature variations, or manufacturing tolerances. However, this improved robustness comes at the expense of reduced stability margins, as higher-type systems are inherently more difficult to stabilize.

Remark

As the system type increases, the robustness of the system with respect to steady-state error to a polynomial input (step, ramp, parabolic, etc.) improves.

EXAMPLE 5.2

Problem

Compute the static error constants and corresponding steady state errors for three basic types of inputs for unity feedback systems whose open-loop transfer functions $G(s)$ are as follows:

a. $G(s) = \dfrac{10(s+5)}{(s+1)(s+2)(s+3)(s+4)}$

b. $G(s) = \dfrac{K}{s(s+2)(s+5)(s+10)}$

c. $G(s) = \dfrac{K(s+3)}{s^2(s+6)}$

SOLUTION:

a. Now $G(s) = \dfrac{10(s+5)}{(s+1)(s+2)(s+3)(s+4)}$. This is a Type-0 system.

Step:	Position error constant $K_p = \dfrac{50}{24}$	$e_{ss} = \dfrac{1}{1+K_p} = \dfrac{24}{74}$
Ramp:	Velocity error constant $K_v = 0$	$e_{ss} = \dfrac{1}{K_v} = \infty$
Parabolic:	Acceleration error constant $K_a = 0$	$e_{ss} = \dfrac{1}{K_a} = \infty$

b. $G(s) = \dfrac{K}{s(s+2)(s+5)(s+10)}$ is a Type-1 system.

Step:	Position error constant $K_p = \infty$	$e_{ss} = \dfrac{1}{1+K_p} = 0$
Ramp:	Velocity error constant $K_v = \dfrac{K}{100}$	$e_{ss} = \dfrac{1}{K_v} = \dfrac{100}{K}$
Parabolic:	Acceleration error constant $K_a = 0$	$e_{ss} = \dfrac{1}{K_a} = \infty$

Note that these are valid results provided the value of K is in the range that ensures closed-loop stability. In this case $0 < K < 435.98$.

c. $G(s) = \dfrac{K(s+3)}{s^2(s+6)}$ is a Type-2 system.

Step:	Position error constant $K_p = \infty$	$e_{ss} = \dfrac{1}{1+K_p} = 0$
Ramp:	Velocity error constant $K_v = \infty$	$e_{ss} = \dfrac{1}{K_v} = 0$
Parabolic:	Acceleration error constant $K_a = \dfrac{K}{2}$	$e_{ss} = \dfrac{1}{K_a} = \dfrac{2}{K}$

CONCEPT REINFORCEMENT PROBLEM 5.1

Problem

Determine the static error constants and corresponding steady state errors steady for following test inputs: $30u(t)$, $16tu(t)$, $12.5t^2u(t)$ for unity feedback systems with following open-loop transfer function G(s).

a. $G(s) = \dfrac{5000(s+1)(s+3)}{(s+10)(s+15)(s+20)}$

b. $G(s) = \dfrac{1000(s+4)(s+6)(s+8)}{s(s+10)(s+15)(s+20)}$

c. $G(s) = \dfrac{250(s+1)(s+2)(s+4)(s+6)(s+8)}{s^2(s+10)(s+12)(s+16)}$

EXAMPLE 5.3

Problem

Find the system type for the system shown in Figure 5.5, and determine the appropriate static error constant.

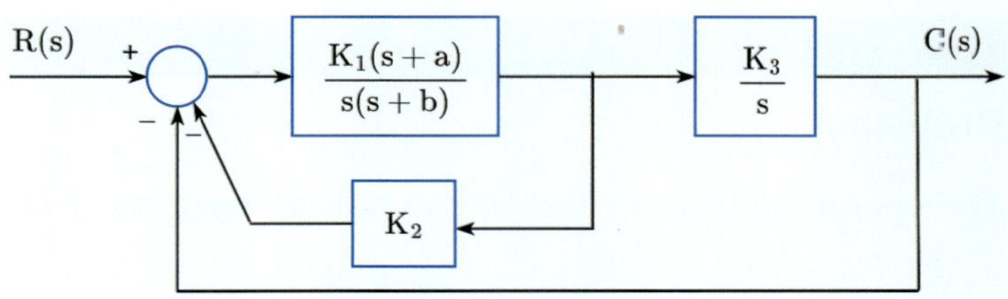

Figure 5.5: Feedback control system

SOLUTION:

Let us denote

$$G_1(s) = \frac{K_1(s+a)}{s(s+b)}, \quad G_2(s) = K_2, \quad \text{and} \quad G_3(s) = \frac{K_3}{s}$$

Simplifying first the inner loop consisting of $G_1(s)$ and $G_2(s)$ by a single function $G_4(s)$ gives

$$G_4(s) = \frac{G_1(s)}{1+G_1(s)G_2(s)} = \frac{K_1(s+a)}{s^2+(K_1K_2+b)s+K_1K_2a}$$

Combining above with $G_3(s) = \dfrac{K_3}{s}$ gives the equivalent forward path transfer function

$$G_e(s) = G_4(s)G_3(s) = \frac{K_1K_3(s+a)}{s\,(s^2+(K_1K_2+b)s+K_1K_2a)}$$

Hence, the system is a Type-1 system, and the appropriate static error constant is the velocity error constant, which is given by

$$K_v = \lim_{s\to 0} sG_e(s) = \frac{K_1K_3a}{K_1K_2a} = \frac{K_3}{K_2}$$

CONCEPT REINFORCEMENT PROBLEM 5.2

Problem

The open-loop transfer function of a unity feedback system is given by

$$G(s) = \frac{K(s+4)}{(s+1)(s+5)(s+6)}$$

Determine

i. The static position error constant and the position error for a unit step input

ii. How does the steady state error change with an increase in the gain K? Using MATLAB, plot the error variation with the gain and comment on your observation.

EXAMPLE 5.4

Problem

The open-loop transfer function of a unity feedback system is given by

$$G(s) = \frac{K}{(s+a)(s+b)(s+c)} \tag{5.20}$$

Determine

i. The static position error constant and the position error for a unit step input

ii. Compute the minimum possible steady state error due to a unit step input.

SOLUTION:

i. Now the static position error constant is

$$K_p = \lim_{s \to 0} G(s) = \lim_{s \to 0} \frac{K}{(s+a)(s+b)(s+c)} = \frac{K}{abc}$$

The steady state error due to a unit step input is

$$e_{ss} = \frac{1}{1+K_p} = \frac{abc}{K+abc} \tag{5.21}$$

ii. The ultimate gain of the system can be determined from the Routh-Hurwitz test (See CRP 4.7 in Chapter 4), which gives

$$K_{cu} = (a+b+c)(ab+bc+ca) - abc \tag{5.22}$$

Hence, the minimum steady state error becomes

$$e_{ss} = \frac{abc}{K_{cu} + abc} = \frac{abc}{(a+b+c)(ab+bc+ca)}$$

EXAMPLE 5.5

Problem

For the system shown in Figure 5.6, determine the value of K such that the error due to a ramp input of slope 20 will result in an error of 0.01 in the steady state.

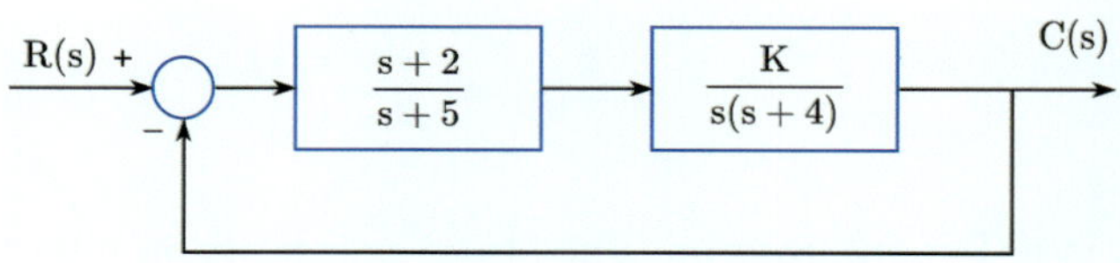

Figure 5.6: Unity feedback control system

SOLUTION:

Now the open-loop transfer function of the system

$$G(s) = \frac{K(s+2)}{s(s+4)(s+5)}$$

This is a Type-1 system and it exhibits finite error due to ramp input. The static velocity error constant is

$$K_v = \lim_{s\to 0} sG(s) = \lim_{s\to 0} s\frac{K(s+2)}{s(s+4)(s+5)} = \frac{2K}{20} = \frac{K}{10}$$

Given that the error due to input 20tu(t) equals to 0.01. That is

$$e_{ss} = \frac{20}{K_v} = \frac{20}{\frac{K}{10}} = \frac{200}{K} = 0.01$$

This gives

$$K = \frac{200}{0.01} = 20,000$$

EXAMPLE 5.6

Problem

Consider the unity feedback system shown in Figure 5.1 where

$$G(s) = \frac{K(s+1)}{s^2(s+3)}$$

Determine the value of K such that the error due to an input $15t^2u(t)$ would give an error of 0.05 in the steady state.

SOLUTION:

Now the open-loop transfer function of the system

$$G(s) = \frac{K(s+1)}{s^2(s+3)}$$

This is a Type-2 system. The static acceleration constant

$$K_a = \lim_{s\to 0} s^2 G(s) = \lim_{s\to 0} s^2 \frac{K(s+1)}{s^2(s+3)} = \frac{K}{3}$$

Hence, the steady state error due to input $15t^2u(t)$ equals to

$$e_{ss} = \frac{2\times 15}{K_a} = \frac{30\times 3}{K} = \frac{90}{K}$$

For desired error $e_{ss} = 0.05$,

$$K = \frac{90}{0.05} = 1800$$

CONCEPT REINFORCEMENT PROBLEM 5.3

Problem

Consider the unity feedback system shown in Figure 5.1 where

$$G(s) = \frac{K(s+10)}{s(s+2)(s+6)(s+15)}$$

a. Determine the static velocity error constant K_v of the system.

b. Determine the value of "K" which would yield a steady state error in position of 0.018 for a ramp input of slope 0.1.

c. What is the minimum possible steady-state position error for the ramp input of slope 0.1?

EXAMPLE 5.7

Problem

For the system shown in Figure 5.1,

a. Express the steady state error for a unit step input in terms of K_1, K_2, where $E(s) = R(s) - C(s)$.

b. Find the value of K_1 which would result in the steady state error to a unit step input zero.

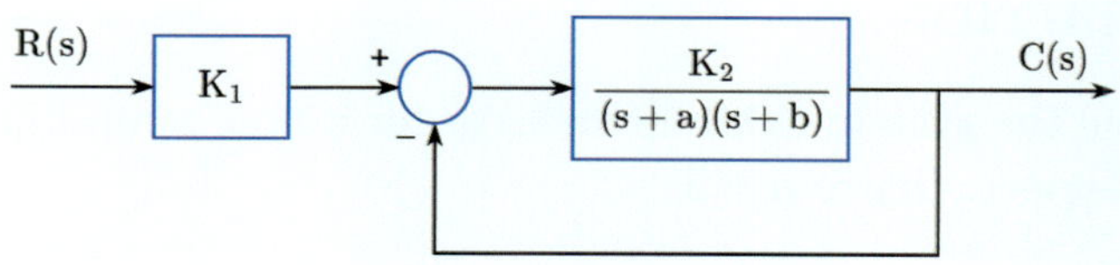

Figure 5.7: Unity Feedback Control System

SOLUTION:

a. Now from the block diagram

$$C(s) = K_1R(s)\frac{K_2}{(s+a)(s+b)+K_2} = \frac{K_1K_2}{(s+a)(s+b)+K_2}R(s)$$

The error $E(s) = R(s) - C(s)$ can be expressed as:

$$\begin{aligned} E(s) &= R(s)\left[1 - \frac{K_1K_2}{(s+a)(s+b)+K_2}\right] \\ &= R(s)\left[\frac{(s+a)(s+b)+K_2(1-K_1)}{(s+a)(s+b)+K_2}\right] \end{aligned}$$

b. The steady state value of error for unit step input $R(s) = \frac{1}{s}$ is

$$e_{ss} = \lim_{s\to 0} sE(s) = \frac{ab + K_2(1-K_1)}{ab + K_2}$$

For zero steady state error

$$\frac{ab + K_2(1-K_1)}{ab + K_2} = 0 \implies ab + K_2(1-K_1) = 0$$

or,

$$K_1 = 1 + \frac{ab}{K_2}$$

EXAMPLE 5.8

Problem

For the system shown in Figure 5.1, where

$$G(s) = \frac{K}{s^n(s + a)}$$

Determine the values of "n," "K," and "a" which would result in 5% overshoot and a steady state error of 0.001 for a unit ramp input.

SOLUTION:

Since the system exhibits an error for a unit ramp input, the system is of Type-1. Hence n = 1.

Now the closed-loop transfer function of the system is

$$T(s) = \frac{K}{s^2 + as + K}$$

Comparing this with the closed-loop transfer function of a prototype second-order system, we get

$$\omega_n^2 = K, \quad \Longrightarrow \quad \omega_n = \sqrt{K}, \quad \text{and} \quad 2\xi\omega_n = a$$

Now for a 5% percent overshoot i.e., $\%M_p = 5$,

$$\begin{aligned} \xi &= \frac{\gamma}{\sqrt{\pi^2 + \gamma^2}}, \quad \text{where} \quad \gamma = -\ln(\%M_p/100) \\ &= 0.6901 \end{aligned}$$

Hence

$$a = 2\xi\omega_n = 2 \times 0.6901\sqrt{K} = 1.3802\sqrt{K}$$

The static velocity error constant of the system

$$K_v = \lim_{s \to 0} sG(s) = \frac{K}{a}$$

Since $e_{ss} = 0.001$, $K_v = 1000$. Hence

$$\frac{K}{a} = 1000$$

or

$$\frac{K}{1.3802\sqrt{K}} = 1000 \quad \Longrightarrow \quad K = 1.9050e + 06, \quad \text{and} \quad a = 1.905e3$$

CONCEPT REINFORCEMENT PROBLEM 5.4

Problem

The forward path transfer function of a unity feedback system shown in Figure 5.1, where

$$G(s) = \frac{K(s+a)}{s(s+b)}$$

Determine the values of K, a, and b which would result in 5% steady state error for a unit ramp input and locate the closed-loop poles at $-2 \pm j3$.

Understanding Zero Steady-State Error: A Common Misconception

One of the most common misconceptions among students learning feedback control is the belief that when the error signal becomes zero, the controller output must also be zero. This leads to the puzzling question: "If the controller output is zero, how can the plant output equal the reference input rather than zero?"

This confusion stems from incorrectly applying the principle that "zero input to the controller means zero output from the controller" to steady-state control situations. However, this reasoning overlooks the fundamental difference between transient and steady-state behavior in control systems. **Even when the input to the controller (error) is zero, its output, which becomes the input to the plant via the actuator, is not zero.**

The Reality: Steady-State vs. Transient Behavior

Transient Phase During the transient response phase:

- Error $\neq 0$ (output has not yet reached the reference)
- Controller is actively responding to the error
- Control signal is changing to drive the error toward zero
- System output is evolving toward the desired value

At steady-state equilibrium:

- Error $= 0$ (output equals reference)
- Controller output has settled to a constant value
- **This constant value is precisely what the plant needs to maintain the desired output**
- The system has found its equilibrium point

Consider a PI controller with the transfer function:

$$u(t) = K_p \cdot e(t) + K_i \int_0^t e(\tau)d\tau$$

When the system reaches steady-state where $e(t) = 0$:

- Proportional term: $K_p \times 0 = 0$
- Integral term: $K_i \int_0^t e(\tau)d\tau = \text{constant value} \neq 0$

The integral term "remembers" all the accumulated error from the past and maintains this memory as a constant offset. This accumulated value represents the control effort required to sustain the desired output.

Understanding Zero Steady State Error

1. **Cruise Control System**

   When your car reaches the desired speed (error = 0), the engine does not shut off. Instead, the throttle remains at whatever position is necessary to maintain that speed against air resistance, road grade, and other disturbances. The "error" between desired and actual speed is zero, but the throttle (controller output) maintains a specific non-zero position.

2. **Room Temperature Control**

   When a room reaches the desired temperature (error = 0), the heating system doesn't turn completely off. Instead, it provides just enough heat to balance the heat loss through walls, windows, and ventilation. The thermostat maintains whatever heating level is needed to sustain the setpoint temperature.

3. **Water Tank Level Control**

   Consider a water tank with an inlet valve controlled to maintain a specific water level. When the water level equals the desired level (error = 0), the inlet valve doesn't close completely. It remains partially open to exactly balance the water flowing out of the tank.

   The Physical Reality

   The key insight is understanding what equilibrium means in a physical system: The plant requires a specific input to produce any given output. When output equals reference, the plant still needs that specific input to maintain this output. Thus, when the input to the controller (error) is zero, its output must equal the steady-state value required by the plant to maintain the desired output. This is where steady-state error analysis becomes crucial for system design.

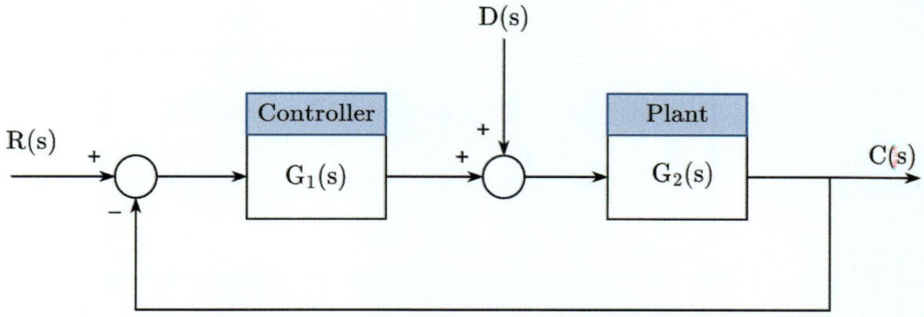

Figure 5.8: Feedback control system with disturbance

5.3 Steady State Error for Disturbances

As discussed before, we can design a feedback system to minimize the effects of disturbance. Consider a closed-loop unity feedback control system shown in Figure 5.8 where a disturbance D(s) is entering between the plant and the controller.

The objective is to derive the expression of the steady state error in the presence of a disturbance. Now

$$C(s) = [G_1(s)E(s) + D(s)]\,G_2(s) = G_1(s)G_2(s)E(s) + G_2(s)D(s) \tag{5.23}$$

But

$$E(s) = R(s) - C(s) \quad \Longrightarrow \quad C(s) = R(s) - E(s) \tag{5.24}$$

Substituting Eqn(5.24) in Eqn(5.23) and eliminating C(s) gives

$$E(s) = \frac{1}{1 + G_1(s)G_2(s)}R(s) - \frac{G_2(s)}{1 + G_1(s)G_2(s)}D(s) \tag{5.25}$$

Eqn(5.25) consists of two parts, where the first part relates the error to the reference input and the second part relates the error to the disturbance signal.

Let us apply the final value theorem to Eqn(5.25) to get the steady state error as:

$$\begin{aligned} e(\infty) &= \lim_{s\to 0} sE(s) = \lim_{s\to 0}\frac{s}{1 + G_1(s)G_2(s)}R(s) - \lim_{s\to 0}\frac{sG_2(s)}{1 + G_1(s)G_2(s)}D(s) \\ &= e_R(\infty) + e_D(\infty) \end{aligned} \tag{5.26}$$

where

$$e_R(\infty) = \lim_{s\to 0} \frac{s}{1 + G_1(s)G_2(s)} R(s) \tag{5.27}$$

and

$$e_D(\infty) = -\lim_{s\to 0} \frac{sG_2(s)}{1 + G_1(s)G_2(s)} D(s) \tag{5.28}$$

If we assume that the disturbance is a step signal, i.e., $D(s) = 1/s$, then

$$\begin{aligned} e_D(\infty) &= -\lim_{s\to 0} \frac{sG_2(s)}{1 + G_1(s)G_2(s)} \frac{1}{s} = -\lim_{s\to 0} \frac{G_2(s)}{1 + G_1(s)G_2(s)} \\ &= -\lim_{s\to 0} \frac{1}{\frac{1}{G_2} + G_1(s)} = -\frac{1}{\frac{1}{\lim\limits_{s\to 0} G_2} + \lim\limits_{s\to 0} G_1(s)} \end{aligned} \tag{5.29}$$

The above equation shows that the error due to disturbance can be reduced by increasing the DC gain of $G_1(s)$ or decreasing the dc gain of $G_2(s)$.

EXAMPLE 5.9

Problem

Determine the steady state error due to a unit step input and the unit step disturbance, and the total steady state error for the system shown in Figure 5.9

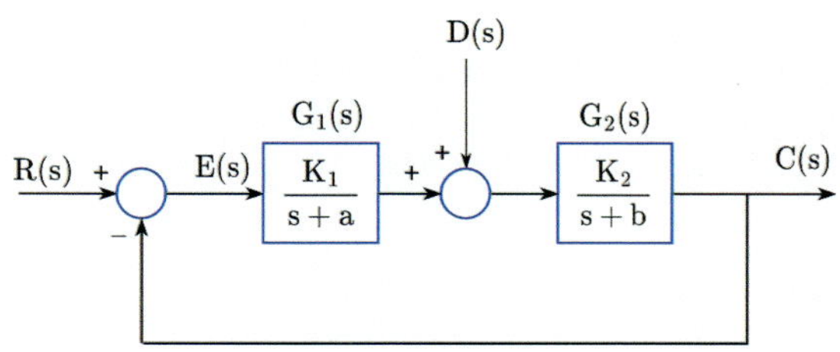

Figure 5.9: Unity feedback control system with disturbance

SOLUTION:

Given

$$G_1(s) = \frac{K_1}{s+a} \quad \& \quad G_2(s) = \frac{K_2}{s+b}$$

For step input $R(s) = 1/s$. Hence, the steady state error due to the step input is computed as:

$$e_R(\infty) = \lim_{s\to 0} \frac{s}{1 + G_1(s)G_2(s)} \frac{1}{s} = \lim_{s\to 0} \frac{1}{1 + G_1(s)G_2(s)}$$

Let us fist compute $1 + G_1(s)G_2(s)$ which is given by

$$1 + G_1(s)G_2(s) = 1 + \frac{K_1}{(s+a)} \frac{K_2}{(s+b)} = \frac{(s+a)(s+b) + K_1K_2}{(s+a)(s+b)}$$

Hence

$$\lim_{s\to 0} 1 + G_1(s)G_2(s) = \lim_{s\to 0} \frac{(s+a)(s+b) + K_1K_2}{(s+a)(s+b)} = \frac{K_1K_2 + ab}{ab}$$

Hence

$$e_R(\infty) = \frac{ab}{K_1K_2 + ab}$$

To compute $e_D(\infty)$ let us compute the following:

$$\lim_{s\to 0} \frac{1}{G_2} = \lim_{s\to 0} \frac{1}{\dfrac{K_2}{s+b}} = \frac{b}{K_2}$$

and

$$\lim_{s\to 0} G_1(s) = \lim_{s\to 0} \frac{K_1}{s+a} = \frac{K_1}{a}$$

Thus

$$e_D(\infty) = -\lim_{s\to 0} \frac{1}{\dfrac{1}{G_2} + G_1(s)} = \frac{-1}{\left[\dfrac{b}{K_2} + \dfrac{K_1}{a}\right]} = -\frac{K_2a}{K_1K_2 + ab}$$

Hence, the total error becomes

$$e(\infty) = e_R(\infty) + e_D(\infty) = \frac{ab}{K_1K_2 + ab} - \frac{K_2a}{K_1K_2 + ab} = \frac{a(b - K_2)}{K_1K_2 + ab}$$

CONCEPT REINFORCEMENT PROBLEM 5.5

Problem

Determine the steady state error due to a unit step input and the unit step disturbance, and the total steady state error for the system shown in Figure 5.10.

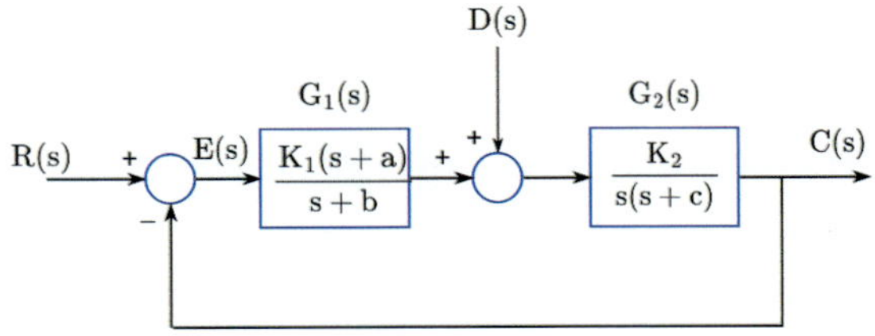

Figure 5.10: Unity feedback control system with disturbance

EXAMPLE 5.10

Problem

For the system shown in Figure 5.11, determine the values of K_1 and K_2 such that the steady state error due to a unit step input is zero and the steady state error due to a unit step disturbance equals -0.0001.

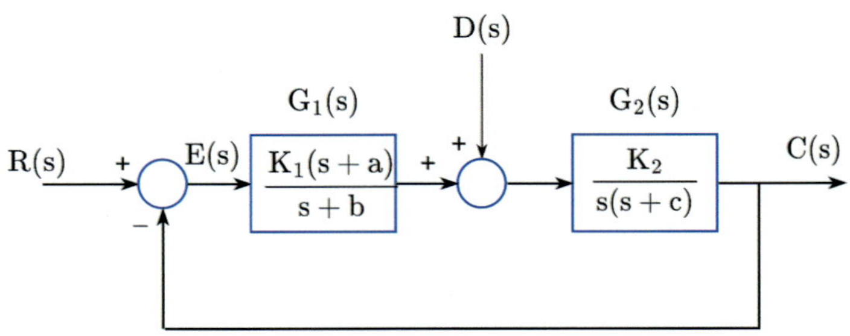

Figure 5.11: Unity feedback control system with disturbance

SOLUTION:

Given

$$G_1(s) = \frac{K_1(s+a)}{s+b} \quad \& \quad G_2(s) = \frac{K_2}{s(s+c)}$$

For step input $R(s) = 1/s$. Hence, the steady state error due to the step input is computed as:

$$\begin{aligned} e_R(\infty) &= \lim_{s\to 0} \frac{s}{1 + G_1(s)G_2(s)} \frac{1}{s} \\ &= \lim_{s\to 0} \frac{1}{1 + G_1(s)G_2(s)} \end{aligned}$$

Let us first compute $1 + G_1(s)G_2(s)$ which is given by

$$1 + G_1(s)G_2(s) = 1 + \frac{K_1(s+a)}{s+b}\frac{K_2}{s(s+c)}$$
$$= \frac{s(s+b)(s+c) + K_1K_2(s+a)}{s(s+b)(s+c)}$$

Hence

$$\lim_{s\to 0} 1 + G_1(s)G_2(s) = \lim_{s\to 0} \frac{s(s+b)(s+c) + K_1K_2(s+a)}{s(s+b)(s+c)} = \infty$$

Hence

$$e_R(\infty) = \frac{1}{\infty} = 0$$

Thus, the steady state error due to a unit step input is zero irrespective of any value of K_1 and K_2.

To compute $e_D(\infty)$, let us compute the following:

$$\lim_{s\to 0} \frac{1}{G_2} = \lim_{s\to 0} \frac{1}{\dfrac{K_2}{s(s+c)}} = \lim_{s\to 0} \frac{s(s+c)}{K_2} = 0$$

and

$$\lim_{s\to 0} G_1(s) = \lim_{s\to 0} \frac{K_1(s+a)}{s+b} = \frac{K_1 a}{b}$$

Thus

$$e_D(\infty) = -\lim_{s\to 0} \frac{1}{\dfrac{1}{G_2} + G_1(s)} = -\frac{b}{K_1 a}$$

If $e_D(\infty) = -0.0001$, the value of K_1 is obtained as:

$$\frac{-b}{K_1 a} = -0.0001 \quad \Longrightarrow \quad K_1 = \frac{10,000b}{a}$$

CONCEPT REINFORCEMENT PROBLEM 5.6

Problem

Determine the steady state error due to a unit ramp input and a unit step disturbance, and the total steady state error for the system shown

in Figure 5.12

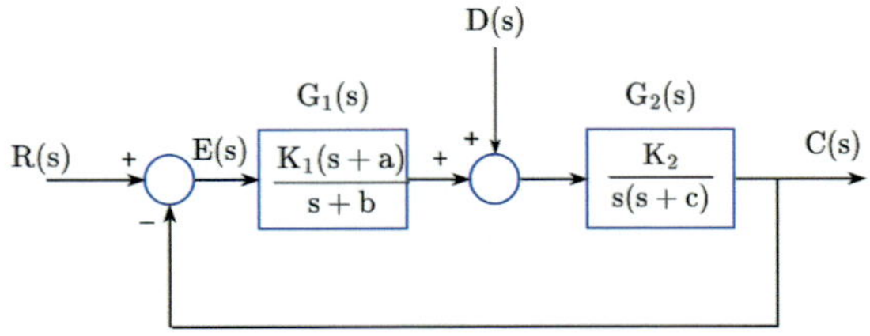

Figure 5.12: Unity feedback control system with disturbance

5.4 Steady State Error from Closed-Loop Transfer Function

We have discussed the methods to compute the steady state error of a closed-loop unity feedback system for different inputs from the knowledge of the forward path transfer function. In this section, we discuss how this can be computed from the closed-loop transfer function.

Denote the closed-loop transfer function as:

$$T(s) = \frac{C(s)}{R(s)} \tag{5.30}$$

The Laplace transform of the error is computed as

$$E(s) = R(s) - C(s) = R(s)\left[1 - T(s)\right] \tag{5.31}$$

Then the steady state error, in terms of the closed-loop transfer function, becomes

$$e_{ss} = \lim_{s \to 0} \left[sE(s)\right] = \lim_{s \to 0} \left\{sR(s)\left[1 - T(s)\right]\right\} \tag{5.32}$$

5.4.1 Steady State Error for Different Test Inputs from Closed-Loop Transfer Function

Step Input: For this case $R(s) = 1/s$ and $sR(s) = 1$. Hence, the steady state error is given by

$$e_{ss} = \lim_{s \to 0} \left[1 - T(s)\right] = 1 - T(0) \tag{5.33}$$

Thus, if $T(0) = 1$, then the steady state error $e_{ss} = 0$. Note that $T(0)$ is called the DC gain of the closed-loop system. Hence

Key Point

If the DC gain of the closed-loop transfer function equals one, i.e., $T(0) = 1$, the steady-state error due to a step input becomes zero.

Ramp Input: Note that the steady state error of a system to a ramp input is finite provided its steady state error due to a step input is zero. Thus, unless $T(0) = 1$, the error due to the ramp input won't be finite. With the assumption that $T(0) = 1$, let us compute the steady state error due to the ramp input, i.e., when $r(t) = t$. Now

$$\begin{aligned} e_{ss} &= \lim_{s \to 0} sR(s)\,[1 - T(s)] \\ &= \lim_{s \to 0} s\frac{1}{s^2}\,[1 - T(s)] = \lim_{s \to 0} \left[\frac{1 - T(s)}{s}\right] \end{aligned} \tag{5.34}$$

This will be finite provided $T(0) = 1$. However when $T(0) = 1$, this takes the form of 0/0. Therefore, we apply L'Hospital's rule to compute e_{ss} as follows:

$$\begin{aligned} e_{ss} &= \lim_{s \to 0} \left[-\frac{dT(s)}{ds}\right] = \lim_{s \to 0} \left[-\frac{1}{T(s)}\frac{dT(s)}{ds}\right] \\ &= \lim_{s \to 0} \left[-\frac{d\ln T(s)}{ds}\right], \quad (\because\ T(0) = 1) \end{aligned} \tag{5.35}$$

Let

$$T(s) = \frac{K(s + z_1)(s + z_2)\ldots\ldots(s + z_m)}{(s + p_1)(s + p_2)\ldots\ldots(s + p_n)} \tag{5.36}$$

Then

$$\ln T(s) = \ln K + \sum_{i=1}^{m} \ln\,(s + z_i) - \sum_{i=1}^{n} \ln\,(s + p_i) \tag{5.37}$$

and

$$\begin{aligned} e_{ss} &= \lim_{s \to 0} \left[-\frac{d\ln T(s)}{ds}\right] = \lim_{s \to 0} \left[-\sum_{i=1}^{m}\frac{1}{s + z_i} + \sum_{i=1}^{n}\frac{1}{s + p_i}\right] \\ &= \sum_{i=1}^{n}\frac{1}{p_i} - \sum_{i=1}^{m}\frac{1}{z_i} = \frac{1}{K_v} \end{aligned} \tag{5.38}$$

Thus, for ramp input

$$\boxed{e_{ss} = \sum_{i=1}^{n}\frac{1}{p_i} - \sum_{i=1}^{m}\frac{1}{z_i} = \frac{1}{K_v}} \tag{5.39}$$

Parabolic Input: For parabolic input $r(t) = \frac{1}{2}t^2$, $R(s) = 1/s^3$. The steady state error is given by

$$e_{ss} = \lim_{s \to 0} \left[\frac{1 - T(s)}{s^2} \right] \tag{5.40}$$

Note that the steady state error of a system to a parabolic input is finite provided its steady state error due to a step input, as well as a ramp input, is zero. Thus, unless

$$T(0) = 1 \tag{5.41}$$

and

$$\lim_{s \to 0} \frac{dT}{ds} = \lim_{s \to 0} \left[\frac{d \ln T(s)}{ds} \right] = 0 \tag{5.42}$$

The steady state error will be infinite. The latter condition is satisfied if

$$\sum_{i=1}^{m} \frac{1}{z_i} = \sum_{i=1}^{n} \frac{1}{p_i} \tag{5.43}$$

Applying L'Hospital's rule twice gives

$$e_{ss} = \lim_{s \to 0} \left[-\frac{1}{2} \frac{d^2 \ln T(s)}{ds^2} \right] \tag{5.44}$$

where

$$\ln T(s) = \ln K + \sum_{i=1}^{m} \ln (s + z_i) - \sum_{i=1}^{n} \ln (s + p_i) \tag{5.45}$$

By differentiating this twice gives

$$\frac{d^2 \ln T(s)}{ds^2} = -\sum_{i=1}^{m} \frac{1}{(s + z_i)^2} + \sum_{i=1}^{n} \frac{1}{(s + p_i)^2} \tag{5.46}$$

and

$$\boxed{e_{ss} = \frac{1}{K_a} = \frac{1}{2} \left[\sum_{i=1}^{m} \frac{1}{z_i^2} - \sum_{i=1}^{n} \frac{1}{p_i^2} \right]} \tag{5.47}$$

EXAMPLE 5.11

Problem

The closed-loop transfer function of a system is given by

$$T(s) = \frac{K(s+a)}{(s+b)(s+c)}$$

i. Compute the steady state error due to a step input.
ii. Determine the value of K which ensures that the steady state error due to the step input is zero.
iii. For this value of K, compute the static velocity error constant and the steady state error due to a unit ramp input in terms of the parameters "a," "b," and "c".
iv. Compute K_v and e_{ss} when the system is excited by a unit ramp input, considering a = 1, b = 4, and c = 5.

SOLUTION:

i. The steady state error to step input is $1 - T(0)$. Thus

$$e_{ss} = 1 - \frac{Ka}{bc}$$

ii. The steady state error to step input will be zero if $T(0) = 1$. Now

$$T(0) = \frac{Ka}{bc}$$

Hence, for zero steady state error

$$\frac{Ka}{bc} = 1 \quad \Longrightarrow \quad K = \frac{bc}{a}$$

iii. The velocity error constant K_v is computed from

$$\sum_{i=1}^{n} \frac{1}{p_i} - \sum_{i=1}^{m} \frac{1}{z_i} = \frac{1}{K_v} \quad \Longrightarrow \quad \frac{1}{b} + \frac{1}{c} - \frac{1}{a} = \frac{1}{K_v}$$

Thus

$$K_v = \frac{abc}{a(b+c) - bc} \quad \& \quad e_{ss} = \frac{a(b+c) - bc}{abc}$$

iv. For a = 1, b = 4 and c = 5, $K_v = -\frac{20}{11}$ and $e_{ss} = -\frac{11}{20}$. The negative value of the steady state error means the output is greater than the input.

EXAMPLE 5.12

The closed-loop transfer function of a system is given by

$$T(s) = \frac{a^2 + b^2}{(s + a + jb)(s + a - jb)} = \frac{a^2 + b^2}{s^2 + 2as + (a^2 + b^2)}$$

i. Compute the steady state error due to a step input.

ii. Compute the static velocity error constant and the steady state error due to a unit ramp input in terms of the parameters a and b.

iv. Determine the steady state error e_{ss} when the system is excited by a unit ramp input, considering $a = 3$, and $b = 4$.

SOLUTION:

i. The steady state error to step input is T(0). Thus

$$1 - T(0) = 1 - \frac{a^2 + b^2}{a^2 + b^2} = 0$$

ii. The velocity error constant K_v is computed from

$$\sum_{i=1}^{n} \frac{1}{p_i} - \sum_{i=1}^{m} \frac{1}{z_i} = \frac{1}{K_v} \implies \frac{1}{a + jb} + \frac{1}{a - jb} - 0 = \frac{1}{K_v}$$

$$\implies \frac{2a}{a^2 + b^2} = \frac{1}{K_v}$$

Thus

$$K_v = \frac{a^2 + b^2}{2a} \quad \& \quad e_{ss} = \frac{2a}{a^2 + b^2}$$

iv. For $a = 3$, and $b = 4$, $K_v = \frac{25}{6}$ and $e_{ss} = \frac{6}{25}$.

CONCEPT REINFORCEMENT PROBLEM 5.7

The closed-loop transfer function of a system with proportional and derivative feedback is given by

$$T(s) = \frac{K(s + z)}{s^2 + 2as + (a^2 + b^2)}$$

i. Determine the values of K and z such that the steady state error due to ramp input is zero. Compute the steady state error due to a step input.

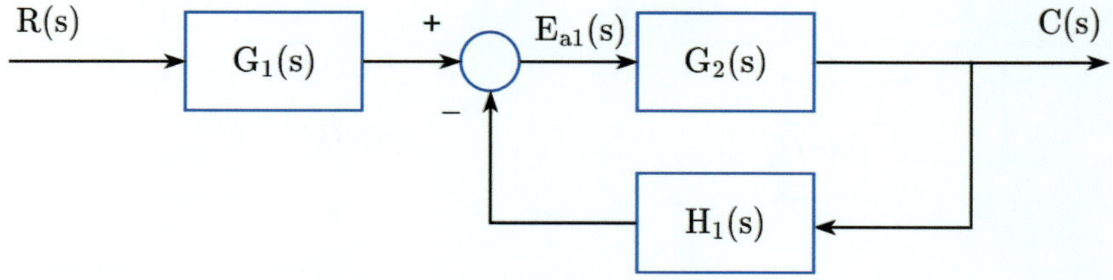

Figure 5.13: Closed-loop control system with non-unity feedback

ii. Compute the static acceleration error constant and the steady state error due to the unit parabolic input $r(t) = 1/2t^2$.

iv. Determine the steady state error e_{ss} when the system is excited by a unit parabolic input, considering $a = 2$, and $b = 5$.

5.5 Steady State Error for Non-unity Feedback Systems

Most of the practical control systems are not unity feedback systems. This is because real-world systems involve components like sensors, actuators, and filters that inherently modify the feedback path. For instance, sensors often introduce scaling factors or dynamics, while actuators may exhibit delays or nonlinear behaviors. Additionally, filters are commonly added to suppress noise, and compensators are frequently used to enhance stability and performance. In the following, we discuss how to compute the steady state error for such systems.

The schematic of a general feedback system is shown in Figure 5.13 where $G_1(s)$ may represent the transfer function of the input transducer, $G_2(s)$ and $H(s)$ represent the transfer function of the plant and the controller, respectively.

Unlike the unity feedback system, the error is no longer the difference between the input and the output and is called the *actuating signal*, $E_a(s)$. Suppose we want to use the same approach to compute the static error coefficients as used for a unity feedback system. In that case, the first step is to convert the non-unity feedback system into a unity feedback system. This is carried out by following the procedure:

i. Push the transducer block $G_1(s)$ in Figure 5.13 to the right of the summing junction, which gives the conventional non-unity feedback system

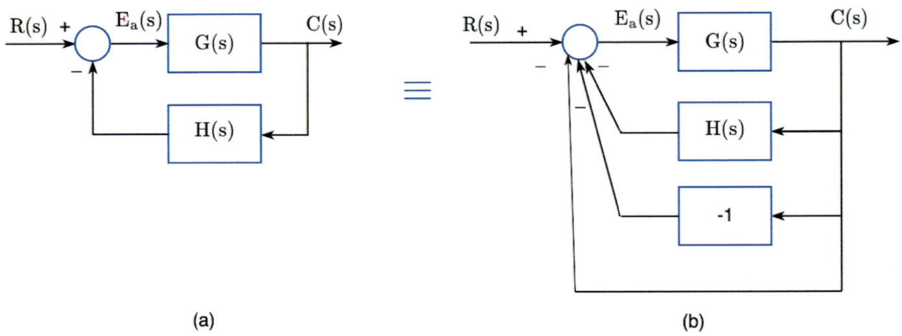

Figure 5.14: Closed-loop control system with non-unity feedback

shown in Figure 5.14a where

$$G(s) = G_1(s)G_2(s) \quad \& \quad H(s) = H_1(s)/G_1(s) \tag{5.48}$$

ii. Add and subtract unity feedback paths to the system of Figure 5.14a, which yields the system shown in Figure 5.14b.

iii. Next, combine H(s) with a negative unity feedback path to give the system shown in Figure 5.15a.

iv. At the final step, combine the system of Figure 5.15a, which consists of G(s) and H(s) − 1. This gives the unity feedback system with equivalent forward path transfer function $G_e(s)$, shown in Figure 5.15b, where

$$G_e(s) = \frac{G(s)}{1 + G(s)H(s) - G(s)} \tag{5.49}$$

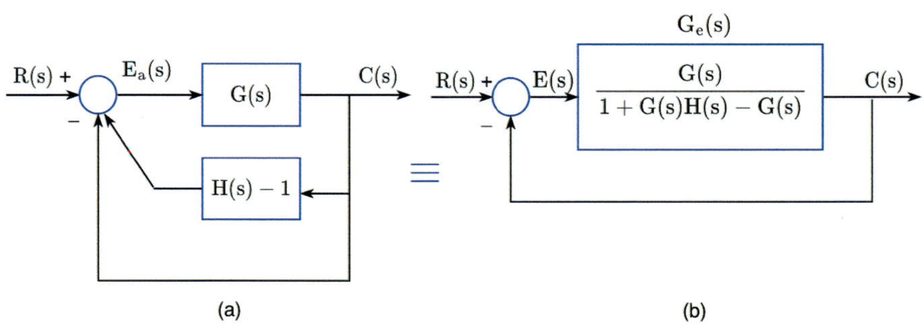

Figure 5.15: Equivalent unity feedback system of the non-unity feedback system

EXAMPLE 5.13

Problem

For the system shown in Figure 5.16, find the following

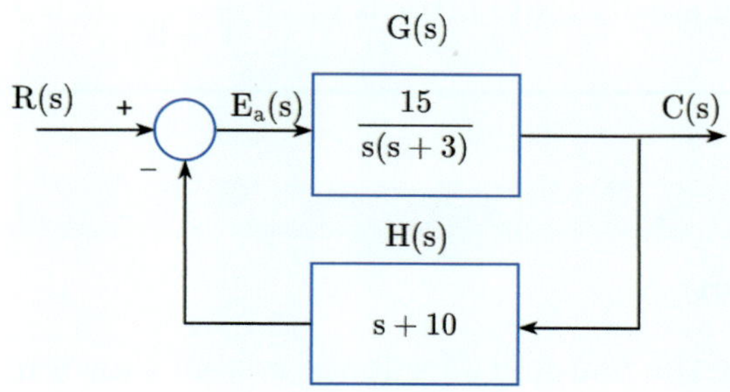

Figure 5.16: Closed-loop system with non-unity feedback

i. The system type
ii. The appropriate static error constant

SOLUTION:

Method-1:

i. Now, the closed-loop transfer of the system is given by

$$T(s) = \frac{G(s)}{1 + G(s)H(s)} = \frac{15}{s^2 + 18s + 150}$$

The type of the system is 0.

ii. The appropriate error constant is the position error constant K_p. The error is $1 - T(0)$ where

$$T(0) = \frac{15}{150} = \frac{1}{10}$$

Thus, the error due to position input is $1 - T(0) = 0.9$.

Method-2:

i. The forward path transfer function of the equivalent unity feedback system is given by

$$G_e(s) = \frac{G(s)}{1 + G(s)H(s) - G(s)} = \frac{15}{s^2 + 18s + 135}$$

The type of the system is 0

ii. The appropriate static position error constant K_p is given by

$$K_p = \lim_{s\to 0} G_e(s) = \lim_{s\to 0} \frac{15}{s^2 + 18s + 135} = \frac{1}{9}$$

The steady state error is $e_{ss} = \dfrac{1}{1 + K_p} = 0.9$

CONCEPT REINFORCEMENT PROBLEM 5.8

Problem

Consider the non-unity feedback system shown in Figure 5.17

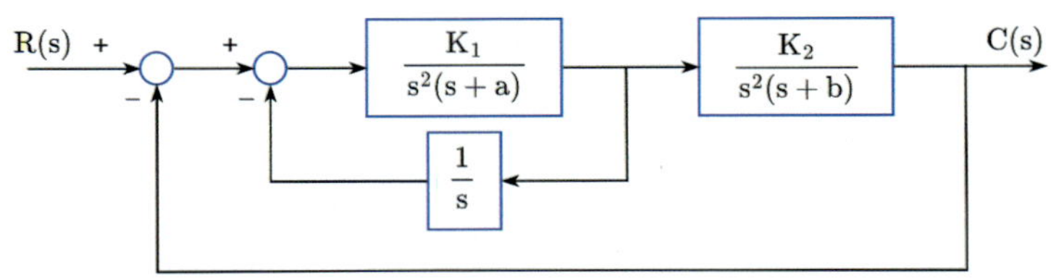

Figure 5.17: Non-unity feedback control System

Find the system type, the appropriate error constant, and the steady state error for a step input 10u(t) and ramp input 10tu(t).

5.5.1 Steady State Error for Non-unity Feedback Systems with Disturbance

Consider a more general system shown in Figure 5.18

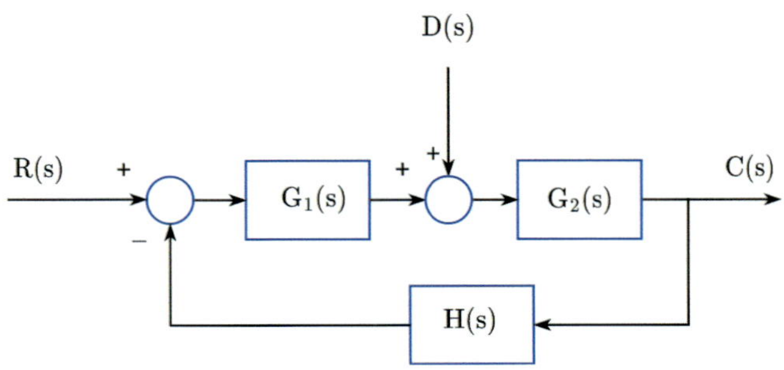

Figure 5.18: Non-unity feedback system with disturbance

The steady-state error for this system, $r(\infty) - c(\infty)$, is

$$
\begin{aligned}
e(\infty) = \lim_{s\to 0} sE(s) &= \lim_{s\to 0} s\left[1 - \frac{G_1(s)G_2(s)}{1+G_1(s)G_2(s)H(s)}\right]R(s) \\
&- \lim_{s\to 0} s\left[\frac{G_2(s)}{1+G_1(s)G_2(s)H(s)}D(s)\right]
\end{aligned} \tag{5.50}
$$

Assuming the system is acted upon by step inputs and step disturbances, where $R(s) = D(s) = 1/s$, the steady state error is

$$
\begin{aligned}
e(\infty) = \lim_{s\to 0} sE(s) &= \left[1 - \frac{\lim_{s\to 0}[G_1(s)G_2(s)]}{\lim_{s\to 0}[1+G_1(s)G_2(s)H(s)]}\right] \\
&- \left[\frac{\lim_{s\to 0} G_2(s)}{\lim_{s\to 0}[1+G_1(s)G_2(s)H(s)]}\right]
\end{aligned} \tag{5.51}
$$

For the steady state error to be zero, the following conditions must be satisfied.

$$
\frac{\lim_{s\to 0}[G_1(s)G_2(s)]}{\lim_{s\to 0}[1+G_1(s)G_2(s)H(s)]} = 1 \quad \text{and} \quad \frac{\lim_{s\to 0} G_2(s)}{\lim_{s\to 0}[1+G_1(s)G_2(s)H(s)]} = 0 \tag{5.52}
$$

Note that Eqn(5.52) can always be satisfied if $G_1(s)$ is a Type 1 system, $G_2(s)$ is a Type 0 system, and $H(s)$ is a Type 0 system with a DC gain of unity.

Now consider the system shown in Figure 5.13. The actuating signal, $E_{a1}(s)$ is given by

$$
\begin{aligned}
E_{a1}(s) &= G_1(s)R(s) - C(s)H(s) \\
&= G_1(s)R(s) - G_2(s)E_{a1}(s)H(s) \quad (\because C(s) = G_2(s)E_{a1}(s))
\end{aligned} \tag{5.53}
$$

Hence, the steady state value of the actuating signal is

$$
e_{a1}(\infty) = \lim_{s\to 0} \frac{sR(s)G_1(s)}{1+G_2(s)H_1(s)} \tag{5.54}
$$

5.6 Sensitivity of Steady State Error to Parameter Variations

Practical systems often face changes in parameters due to ageing, wear and tear, or environmental factors. The sensitivity analysis, as discussed in Chapter 1, evaluates how variations or uncertainties in the system parameters or external disturbances affect system performance. Ideally, a control system should exhibit **zero sensitivity to uncertainties** and **unity sensitivity to the reference input**. However, this ideal scenario is impossible to achieve. For instance, reducing sensitivity to parameter variations may increase sensitivity to noise. Further, high feedback gain improves tracking (achieving unity sensitivity to reference input) but can amplify noise or compromise stability margins.

In practice, the design goal is to achieve **low sensitivity to uncertainties** (rather than zero) while maintaining **unity sensitivity to the reference input** within operational bandwidth. In this section, we illustrate, through examples, how to compute the sensitivity of steady state errors due to parameter variations.

EXAMPLE 5.14

Problem

For the unity feedback system shown in Figure 5.19, determine the sensitivity of the steady state error due to ramp input to changes in parameter a, b, and gain K

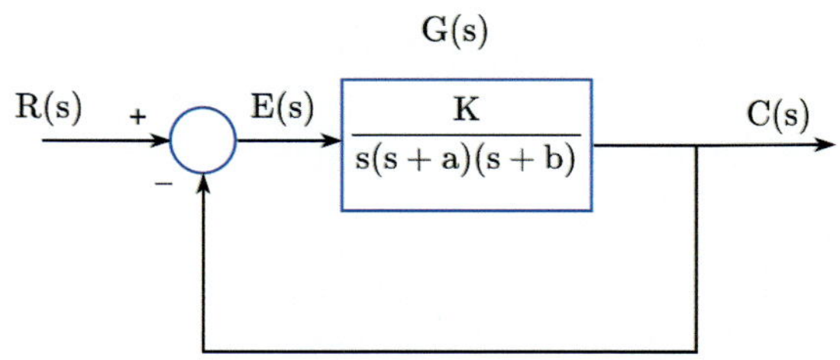

Figure 5.19: Unity feedback control System

SOLUTION:

Now

$$G(s) = \frac{K}{s(s+a)(s+b)}$$

The static velocity error constant

$$K_v = \lim_{s \to 0} sG(s) = \frac{K}{ab}$$

Hence, the steady state error due to the ramp input is

$$e(\infty) = \frac{1}{K_v} = \frac{ab}{K}$$

The sensitivity of $e(\infty)$ to changes in parameter "a" is

$$S_e^a = \frac{a}{e}\frac{\delta e}{\delta a} = \frac{aK}{ab}\left[\frac{b}{K}\right] = 1$$

The sensitivity of $e(\infty)$ to changes in parameter "b" is

$$S_e^b = \frac{b}{e}\frac{\delta e}{\delta b} = \frac{bK}{ab}\left[\frac{a}{K}\right] = 1$$

The sensitivity of $e(\infty)$ to changes in the gain "K" is

$$S_e^K = \frac{K}{e}\frac{\delta e}{\delta K} = \frac{K}{\frac{ab}{K}} ab\frac{\delta e}{\delta K} = \frac{K^2}{ab}\left[-\frac{ab}{K^2}\right] = -1$$

EXAMPLE 5.15

Problem

For the system shown in Figure 5.20, determine the sensitivity of the steady state error due to step input to changes in parameter a.

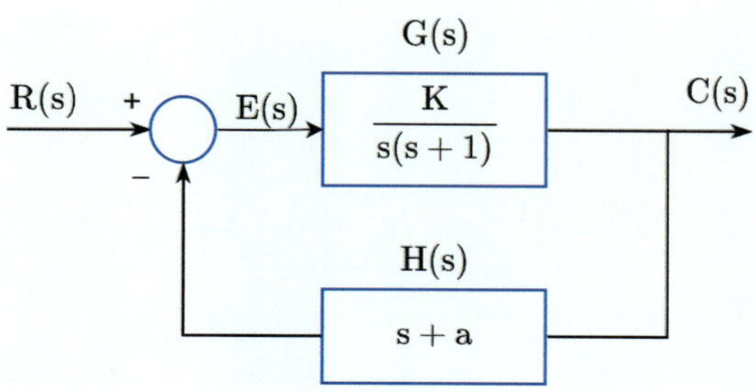

Figure 5.20: Unity feedback control System

SOLUTION:

Given

$$G(s) = \frac{K}{s(s+1)}, \quad H(s) = s + a$$

The forward path transfer function of the equivalent unity feedback system is given by

$$G_e(s) = \frac{G(s)}{1 + G(s)H(s) - G(s)} = \frac{K}{s^2 + (K+1)s + K(a-1)}$$

The static position error constant

$$K_p = \lim_{s \to 0} G(s) = \frac{1}{a-1}$$

Hence, the steady state error due to the step input is

$$e(\infty) = \frac{1}{1+K_p} = \frac{a-1}{a}$$

The sensitivity of $e(\infty)$ to changes in parameter "a" is

$$S_e^a = \frac{a}{e}\frac{\delta e}{\delta a} = \frac{a}{\left(\frac{a-1}{a}\right)}\left[\frac{a-a+1}{a^2}\right] = \frac{1}{a-1} \tag{5.55}$$

CONCEPT REINFORCEMENT PROBLEM 5.9

Problem

For the system shown in Figure 5.21, determine the sensitivity of the steady state error due to step input to changes in parameter K_1 and K_2. Assume $b = 1$.

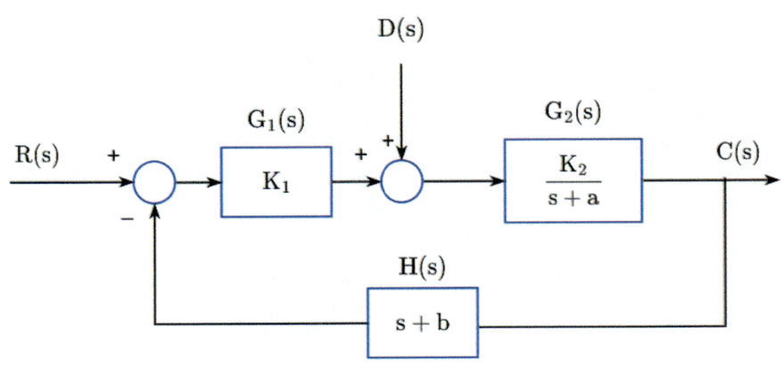

Figure 5.21: Non-unity feedback control System

5.7 Multiple Choice Skills Test Questions

M 5.1. If the static position error constant of a system is finite and non-zero, then the system is of

A. Type-2 **B.** Type-1 **C.** Type-0 **D.** Type-0 and unstable

M 5.2. If the static acceleration error constant of a system is finite and non-zero, then the system is of

A. Type-2 **B.** Type-1 **C.** Type-0 **D.** Type-0 and unstable

M 5.3. The open-loop transfer function of a unity feedback system is given by

$$G(s) = \frac{K(s+5)}{(s+1)(s+10)(s+15)}$$

The steady state error of the system for a unit step input is

A. $\frac{30}{K}$ **B.** $\frac{K}{30}$ **C.** $\frac{K+30}{K}$ **D.** $\frac{30}{K+30}$

M 5.4. Consider the system shown in Figure 5.22

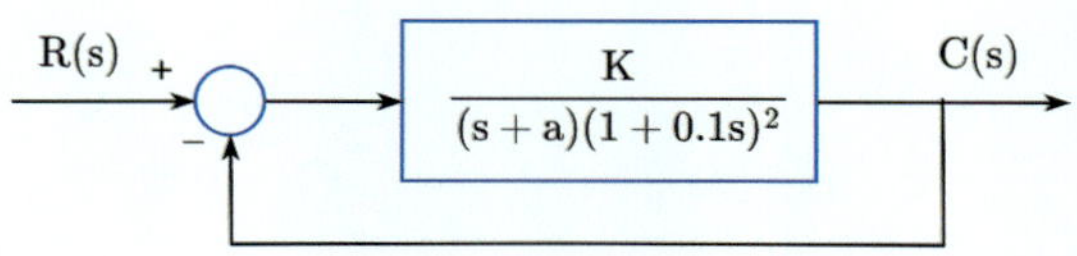

Figure 5.22: Closed-loop control system

The value of K for a steady state error of 10% to a step input is

A. $K = 10a$ **B.** $K = 9a$ **C.** $K = 0.9a$ **D.** $K = 18a$

M 5.5. The open-loop transfer function of a unity feedback system is $\frac{20}{s+5}$. The steady state error to a step input will be

A. $\frac{1}{4}$ **B.** ∞ **C.** $\frac{1}{5}$ **D.** 0

M 5.6. Consider the system shown in Figure 5.23.

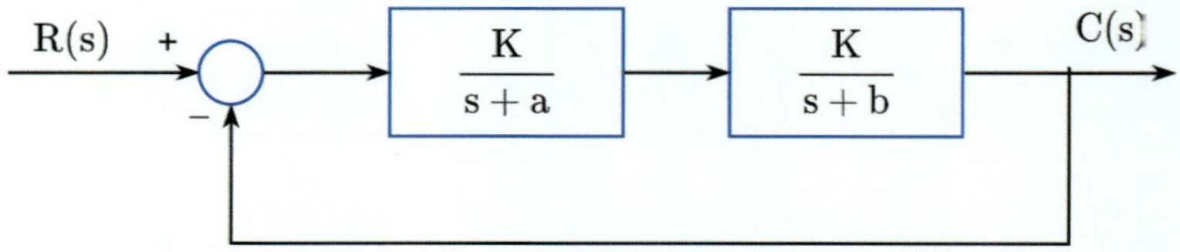

Figure 5.23: Unity Feedback Control System

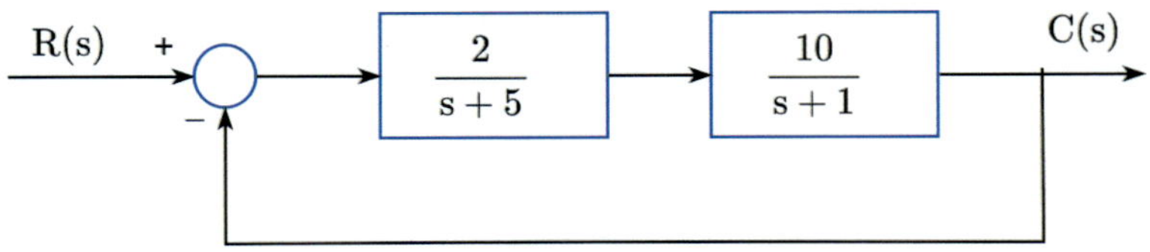

Figure 5.24: Unity Feedback Control System

The value of K which exhibits 10% steady state error to a unit step input is

A. $3\sqrt{ab}$ **B.** 3ab **C.** $\sqrt{3ab}$ **D.** $\frac{3}{\sqrt{ab}}$

M 5.7. The block diagram shown in Figure 5.24 shows a unity feedback closed-loop control system. The steady state error in the output due to a unit step input is

A. 25% **B.** 20% **C.** 10% **D.** 5%

M 5.8. If the static acceleration error constant K_a of a system is 100, then, for this system

A. $K_p = \infty$ and $K_v = 0$
B. $K_p = \infty$ and $K_v = \infty$
C. $K_p = 0$ and $K_v = \infty$
D. $K_p = 0$ and $K_v = 0$

M 5.9. If the static acceleration error constant K_a of a system is 1000, then the steady state error due to an input $50t^2u(t)$ is

A. 0.001 **B.** 0.01 **C.** 0.05 **D.** 0.1

M 5.10. Consider the system shown in Figure 5.25.

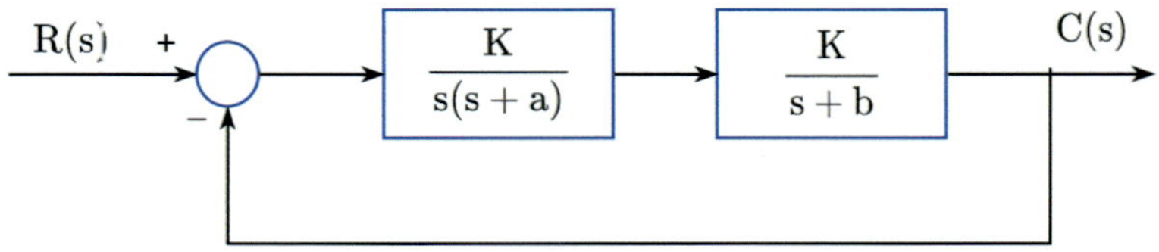

Figure 5.25: Unity Feedback Control System

The value of K which exhibits 10% steady state error to a unit ramp input is

A. $3.333\sqrt{ab}$ **B.** 10ab **C.** $\sqrt{10ab}$ **D.** $10\sqrt{ab}$

M 5.11. The steady state error for a unity feedback system for the system with forward path transfer function $G(s) = \dfrac{10(s+3)(s+4)(s+8)}{s^2(s+2)(s+15)}$ is 2. The input to the system

A. $\frac{1}{2}t^2u(t)$ **B.** $32t^2u(t)$ **C.** $16t^2u(t)$ **D.** $4t^2u(t)$

M 5.12. The steady state error for a unity feedback system for the system with forward path transfer function $G(s) = \dfrac{10}{s(s+2)(s^2+2s+2)}$ is 4. The input to the system is a ramp input of slope

A. 10 **B.** 20 **C.** 5 **D.** one

M 5.13. The steady state error for a unity feedback system for the system with forward path transfer function $G(s) = \dfrac{100(s+2)}{s(s+b)}$ due to a ramp input tu(t)is 0.04. The value of b equals to

A. $b = 3$ **B.** $b = 16$ **C.** $b = 8$ **D.** $b = 4$

M 5.14. The closed-loop transfer function T(s) of a unity feedback system is

$$T(s) = \frac{a_2 s + a_3}{s^3 + a_1 s^2 + a_2 s + a_3}$$

The steady state error due to a unit ramp input is

A. 0 **B.** $\frac{a_2}{a_1}$ **C.** $\frac{a_1}{a_3}$ **D.** $\frac{a_2}{a_3}$

M 5.15. The closed-loop transfer function T(s) of a unity feedback system is

$$T(s) = \frac{a_4}{s^4 + a_1 s^3 + a_2 s^2 + a_3 s + a_4}$$

The steady state error due to a unit ramp input is

A. $\frac{1}{a_3}$ **B.** $\frac{a_3}{a_4}$ **C.** $\frac{a_4}{a_3}$ **D.** $\frac{a_3}{a_1}$

M 5.16. If the closed-loop transfer function T(s) of a unity feedback system is given by

$$T(s) = \frac{a_{n-1}s + a_n}{s^n + a_1 s^{n-1} + \cdots + a_{n-1}s + a_n}$$

then the steady state error for a unit ramp input is

A. $\dfrac{a_n}{a_{n-1}}$ **B.** $\dfrac{a_{n-1}}{a_n}$ **C.** 0 **D.** $\dfrac{a_{n-2}}{a_{n-1}}$

M 5.17. The closed-loop transfer function T(s) of a unity feedback system is

$$T(s) = \frac{K(s+4)}{(s+1)(s+2)(s+3)}$$

The steady state error due to a unit step input is zero if

A. $K = \dfrac{4}{6}$ **B.** $K = \dfrac{3}{6}$ **C.** $K = \dfrac{6}{4}$ **D.** $\dfrac{1}{3}$

Answers

M 5.1. C **M 5.2**. A **M 5.3**. D **M 5.4**. B **M 5.5**. C **M 5.6**. A **M 5.7**. B **M 5.8**. B **M 5.9**. D **M 5.10**. D **M 5.11**. B **M 5.12**. A **M 5.13**. C **M 5.14**. A **M 5.15**. B **M 5.16**. C **M 5.17**. C

5.8 Knowledge Assessment Problems

KaP 5.1 Determine the type of the following unity feedback systems whose forward path transfer function G(s) is given.

a. **S1:** $G(s) = \dfrac{K(s+1)}{(s+2)^2(s+8)}$

b. **S2:** $\dfrac{K}{s(s+5)(s^2+2s+6)}$

c. **S3:** $\dfrac{Ke^{-0.5s}(1+2s)}{(1+0.5s)(1+5s)(1+10s)}$

d. **S4:** $\dfrac{K(1+0.25s)}{s^2(1+0.2s)}$

e. **S5:** $\dfrac{10e^{-2s}}{s^3(1+0.1s)(1+4s)}$

f. **S6:** $\dfrac{5(s^2+s+1)}{s^3(1+10s)}$

KaP 5.2 Find the system type for the system shown in Figure 5.26, and determine the appropriate static error constant, which is non-zero and finite.

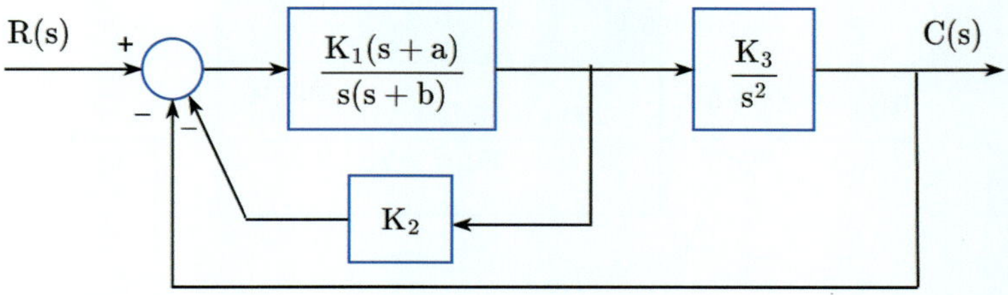

Figure 5.26: Unity Feedback Control System

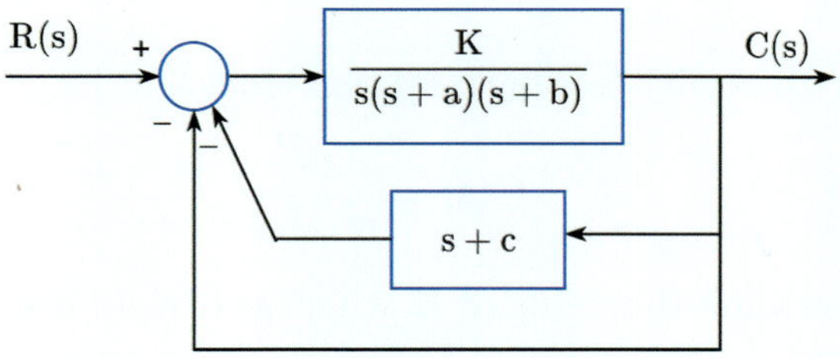

Figure 5.27: Unity Feedback Control System

KaP 5.3 Find the system type for the system shown in Figure 5.27, and determine the appropriate static error constant, which is non-zero and finite.

KaP 5.4 Find the system type for the system shown in Figure 5.28, and determine the appropriate static error constant, which is non-zero and finite.

KaP 5.5 Consider the unity feedback system shown in Figure 5.1 where

$$G(s) = \frac{K(s+10)}{(s+2)(s^2+8s+15)}$$

a. Determine the value of K which will yield a static position error constant of 500.

b. For the value of K found in Part (a), compute the steady state error for inputs $20u(t)$,$10tu(t)$, and $15t^2u(t)$.

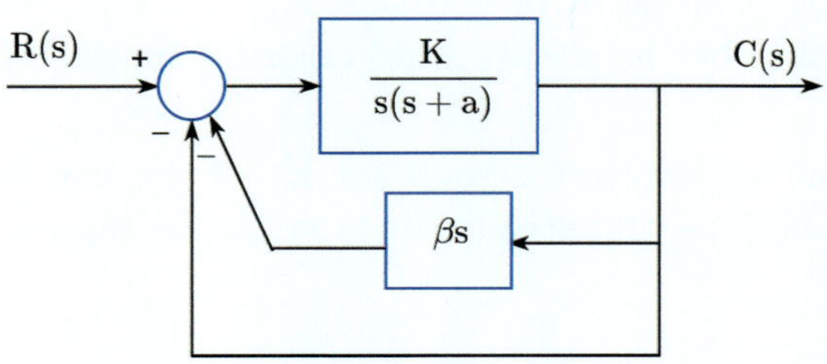

Figure 5.28: Unity Feedback Control System

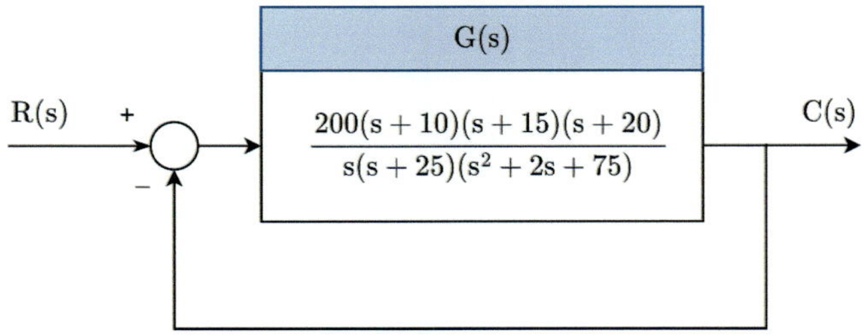

Figure 5.29: Unity Feedback Control System

KaP 5.6 Consider the unity feedback system shown in Figure 5.1 where

$$G(s) = \frac{K(s+1)(s+5)}{s^2(s^2+6s+25)}$$

a. Determine the value of K which will yield a static error constant of 1000.

b. For the value of K found in Part (a), compute the steady state error for inputs 10u(t),30tu(t), and $50t^2u(t)$.

KaP 5.7 Consider the unity feedback system shown in Figure 5.1 where

$$G(s) = \frac{K}{s(s+1)(s+3)(s+5)}$$

a. Determine the static velocity error constant K_v of the system.

b. What is the minimum possible steady-state position error for a unity ramp input?

KaP 5.8 Consider the system shown in Figure 5.29

Find the steady state error of this system when the input

i. $r(t) = t$

ii. $r(t) = 10t$

KaP 5.9 The static position error constant K_p of a unity feedback system is 10. What would be the steady state error for inputs 15u(t) and 15tu(t)?

KaP 5.10 The static velocity error constant K_v of a unity feedback system is 50. What would be the steady state error for inputs 10u(t), 20tu(t), and $15t^2u(t)$?

KaP 5.11 The static acceleration error constant K_a of a unity feedback system is 20. What would be the steady state error for inputs 5u(t), 10tu(t), and $20t^2u(t)$?

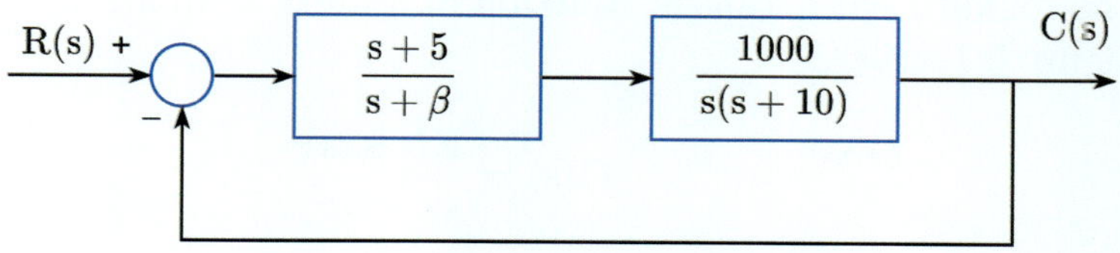

Figure 5.30: Unity Feedback Control System

KaP 5.12 For the unity feedback system shown in Figure 5.1

$$G(s) = \frac{K(s+b)}{s^2(s+a)}, \quad K, a, b > 0$$

Determine the steady state error to an input $20t^2u(t)$ if

i. $a > b$

ii. $a < b$

KaP 5.13 For the system shown in Figure 5.30, determine the value of β such that the error due to a ramp input of slope 10 will result in an error of 0.05 in the steady state.

KaP 5.14 For the system shown in Figure 5.31,

a. Express the steady state error for a unit step input in terms of K_1, K_2, where $E(s) = R(s) - C(s)$.

b. Find the value of K_1 which would result in the steady state error to a unit step input zero.

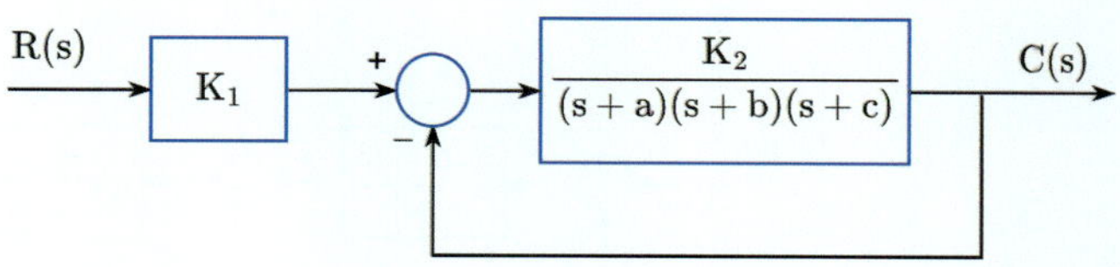

Figure 5.31: Unity Feedback Control System

KaP 5.15 For the system shown in Figure 5.1, where

$$G(s) = \frac{K}{s(s+a)}$$

Determine the values of K, and a which would result in 15% overshoot and steady state error of 0.01 for a unit ramp input.

KaP 5.16 The forward path transfer function of a unity feedback system shown in Figure 5.1, where

$$G(s) = \frac{K(s+a)}{(s+b)^2}$$

Determine the values of K, a, and b which would result 5% steady state error for a unit step input, damping ratio 0.5 and natural frequency $\omega_n = 5\text{rad/sec}$.

KaP 5.17 Determine the steady state error due to a unit step input and the unit step disturbance, and the total steady state error for the system shown in Figure 5.32

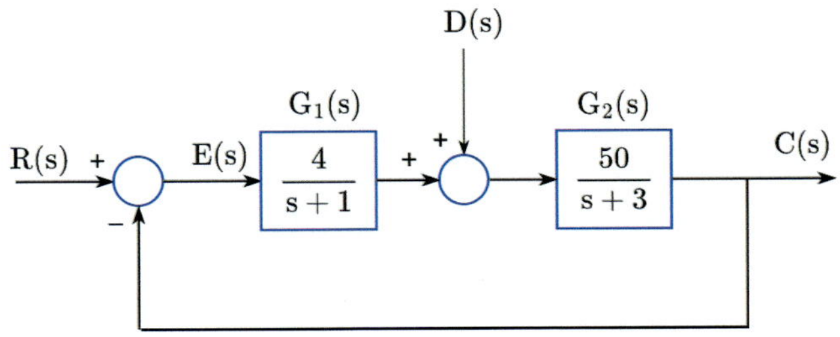

Figure 5.32: Unity Feedback Control System with Disturbance

KaP 5.18 Determine the gains K_1 and K_2 of the system shown in Figure 5.33 such that steady state error due to unit ramp input is 0.002 and that due to unit step disturbance equals −0.0001.

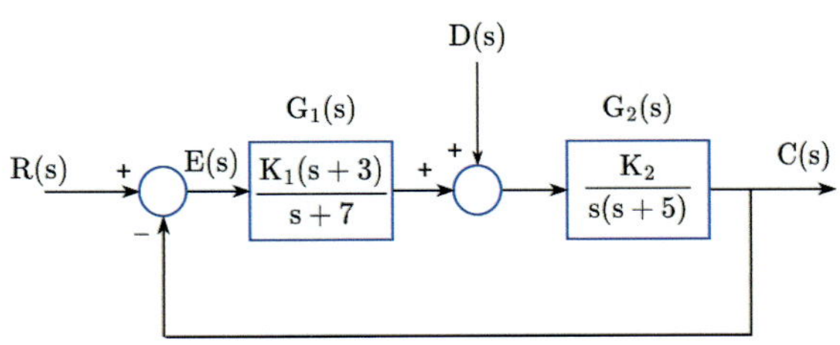

Figure 5.33: Unity Feedback Control System with Disturbance

KaP 5.19 Consider the non-unity feedback system shown in Figure 5.34

Find the system type, the appropriate error constant, and the steady state error for a unit step input.

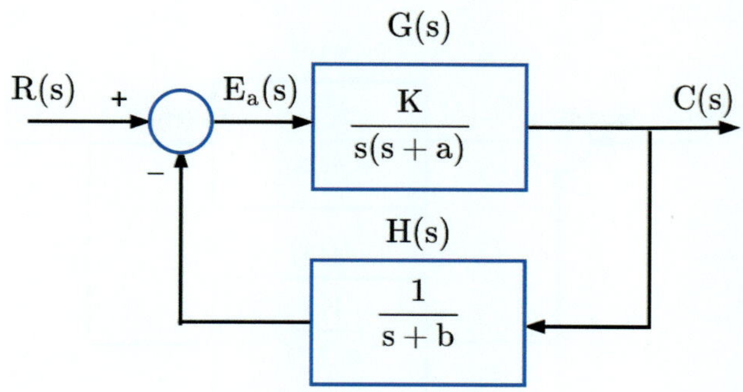

Figure 5.34: Non-unity feedback control System

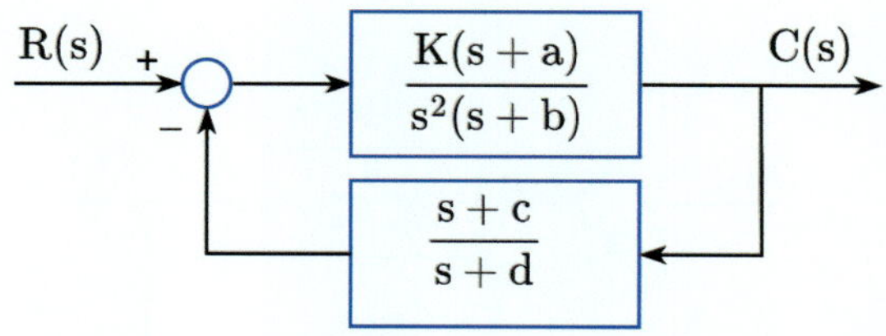

Figure 5.35: Non-unity Feedback Control System

KaP 5.20 Consider the non-unity feedback system shown in Figure 5.35

Find the system type, the appropriate error constant, and the steady state error for a unit step input.

KaP 5.21 Consider the non-unity feedback system shown in Figure 5.36

Find the system type, the appropriate error constant, and the steady state error for a unit step input.

KaP 5.22 Consider the non-unity feedback system shown in Figure 5.37

Find the system type, and determine the value of α which gives zero steady state error to a unit step input.

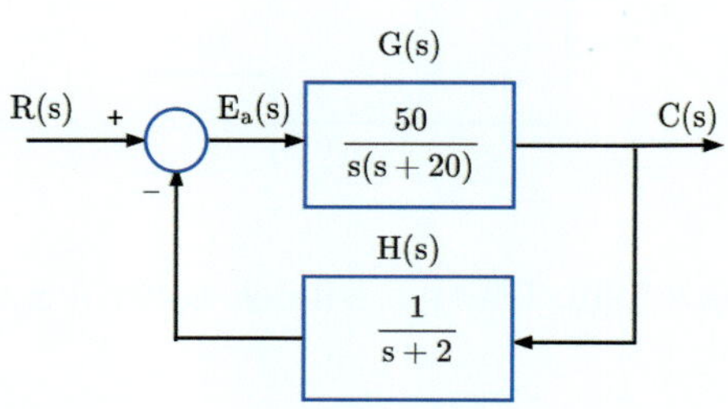

Figure 5.36: Non-unity Feedback Control System

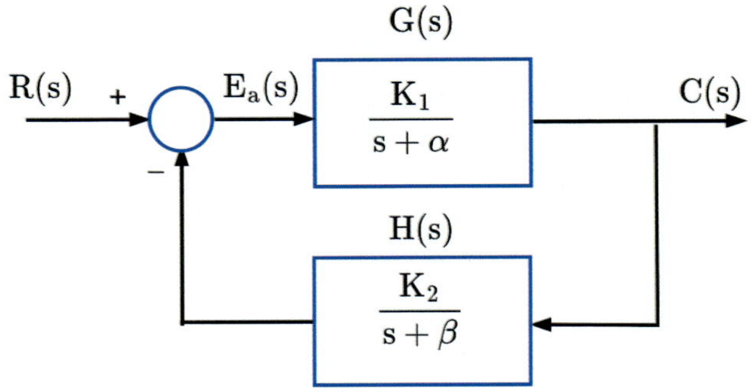

Figure 5.37: Non-unity Feedback Control System

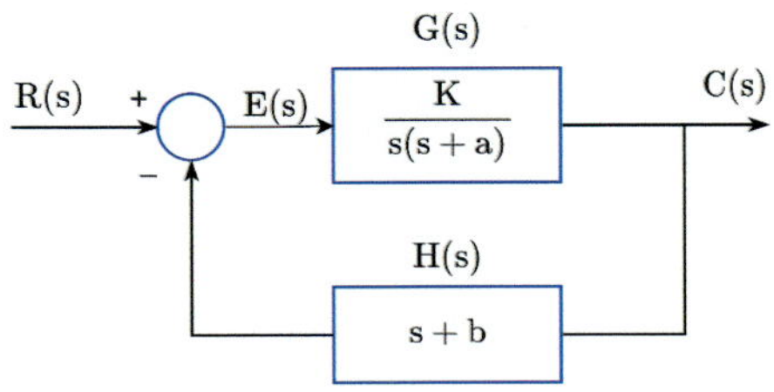

Figure 5.38: Non-unity Feedback Control System

KaP 5.23 For the system shown in Figure 5.38, determine the sensitivity of the steady state error due to step input to changes in parameter b.

KaP 5.24 For the system shown in Figure 5.39, determine the sensitivity of the steady state error due to step input to changes in K_1 and K_2, when $K_1 = 200$, $K_2 = 0.5$, a = 5 and b = 1.

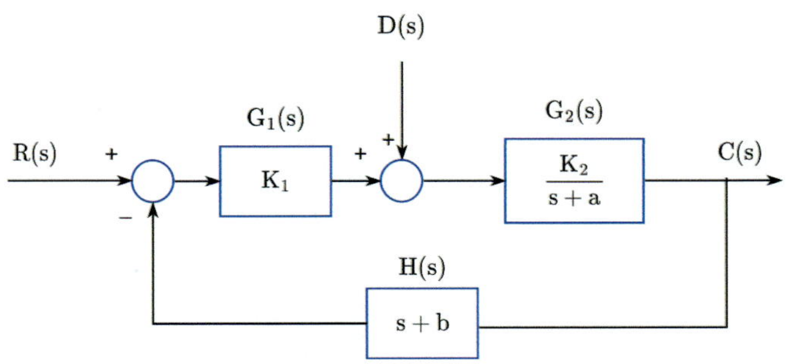

Figure 5.39: Unity feedback control System

KaP 5.25 For the unity feedback system shown in Figure 5.40, determine the sensitivity of the steady state error due to step input to changes in parameter a, b,c, and gain K

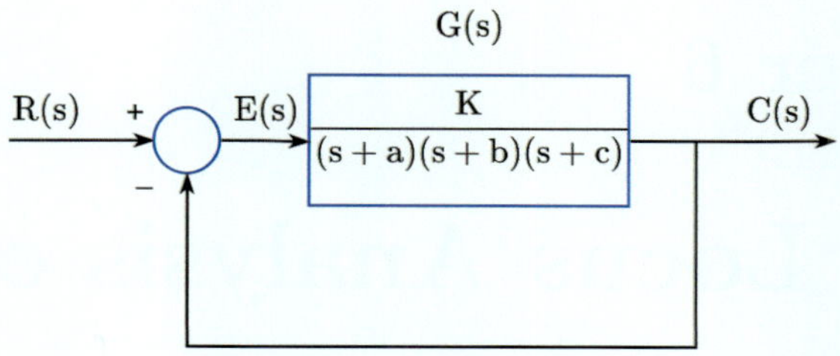

Figure 5.40: Unity feedback control System

Chapter 6

Root Locus Analysis of Linear Systems

Root locus: where poles go on romantic journeys toward their soulmate zeros, but like any good soap opera, expect dramatic breakups, surprise reunions, and some awkward exits to infinity.

DESIRED LEARNING OUTCOMES

Upon completion of this chapter, readers should be able to:

- Explain the concept of the root locus and its importance in control system design
- Sketch the root locus quickly without relying on computer tools
- Analyze the effects of gain on transient response
- Construct the root locus for a positive-feedback system
- Evaluate the impact of time delay on the root locus
- Assess how the addition of poles and zeros affects transient response and system stability
- Improve transient response by adjusting system gain
- Investigate the sensitivity of the root locus to variations in system parameters

DOI: 10.1201/9781003571308-6

6.1 Introduction

In previous chapters, we have seen how the dynamic performance of a linear system changes with the location of poles and zeros of the closed-loop transfer function. The closed-loop poles, which are the roots of the characteristic equation, provide information about the absolute and relative stability of the system, whereas the transient response is determined by both the location of poles and zeros of the closed-loop transfer function. In stability studies presented in Chapter 4, we saw that it is possible to adjust the location of the poles (roots of the characteristic equation) by varying the loop gain. Therefore, for control engineers, it is of interest to know how the roots of the characteristic equation change in the s-plane as the gain is varied.

In 1948, W.R. Evans [21] proposed a graphical procedure, called the root locus method, for sketching the movement of poles of the closed-loop transfer function in the s-plane as one of the parameters is varied from zero to infinity. Originally, this method was confined to the study of control systems where the open-loop gain was taken as the variable parameter. However, it can easily be generalized to study the behavior of roots of any algebraic equation with one or more variable parameters. Some of the advantages of this method are:

1. It can qualitatively describe the performance of a system, such as percent overshoot, settling time, and peak time, as various system parameters are changed.
2. It can be used for designing controllers for higher-order systems to yield the desired overshoot and settling time.
3. The information about absolute stability, relative stability, and marginal stability can be graphically represented.

6.2 Concept of Root Locus

Before discussing the general rules for plotting root locus, let us illustrate the concept of root locus considering a simple second-order position control system with velocity feedback, shown in Figure 6.1.

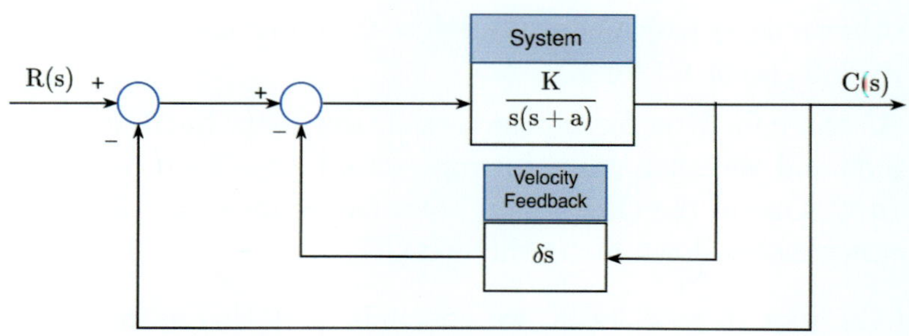

Figure 6.1: Second order position control system with velocity feedback

The transfer function of the closed-loop system is given by (using Mason's rule)

$$\frac{C(s)}{R(s)} = \frac{K}{s^2 + (K\delta + a)s + K} \tag{6.1}$$

The closed-loop characteristic polynomial is

$$\Delta(s) = s^2 + (K\delta + a)s + K \tag{6.2}$$

Consider two cases to investigate how the closed-loop poles move in the s-plane as one of the parameters, such as gain K or velocity feedback parameter δ, changes from zero to infinity.

Case I: Assume no velocity feedback, that is, $\delta = 0$:

In the absence of velocity feedback, the closed-loop characteristic equation is

$$\Delta(s) = s^2 + as + K = 0 \tag{6.3}$$

The roots of this equation are the poles of the closed-loop system and are given by

$$s_1 = \frac{-a + \sqrt{a^2 - 4K}}{2}, \quad \text{and} \quad s_2 = \frac{-a - \sqrt{a^2 - 4K}}{2}$$

Table 6.1 shows the location of poles as the gain K is varied assuming $a = 4$ and it is graphically displayed in Figure 6.2.

As seen from Figure 6.2, the root locus consists of two straight lines AB and DE, intersecting at C. The observations are summarized as:

i. When $K = 0$, the roots of the characteristic polynomial are at $s_1 = -a = -4$ (point A) and $s_2 = 0$ (point B), which are also the poles of the open-loop transfer function with $\delta = 0$.

ii. For the example considered, as K is increased from 0 to 4, we have real roots. The pole s_1 which is at -4 for $K = 0$, moves toward right, and the other pole s_2 which is at 0 for $K = 0$, moves toward the left. They meet at -2.0 (point C) for $K = 4$.

iii. As the gain K is further increased, the roots break away from the real axis and we get a pair of complex conjugate roots with real part equal to -2. One of the closed-loop poles moves up along the line CD, and the other moves down along the line CE.

From the plot of root locus, we conclude that this system is stable for all values of gain. The transient behavior of the system can be inferred from the root locus, which is summarized below.

Table 6.1: Pole location as gain changes for system of Figure 6.1 for a=4, and $\delta = 0$

K	Pole 1	Pole 2
0	-4	0
1	-3.7321	-0.2679
2	-3.4142	-0.5858
3	-3.0	-1.0
4	-2.0	-2.0
5	$-2 - j1$	$-2 + j1$
6	$-2 - j1.4142$	$-2 + j1.4142$
7	$-2 - j1.732$	$-2 + j1.732$
8	$-2 - j2$	$-2 + j2$
9	$-2 - j2.236$	$-2 + j2.236$
10	$-2 - j2.4495$	$-2 + j2.4495$

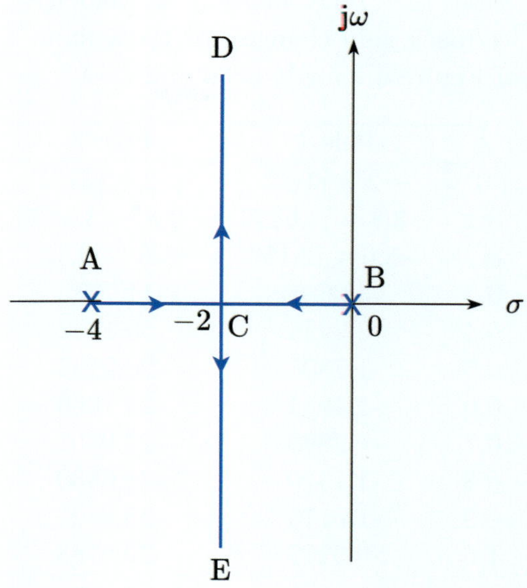

Figure 6.2: Root locus for Eqn(6.3) for a=4

a. All the poles are real if the gain $K < 4$ and the system is overdamped.
b. At $K = 4$, the poles are real and equal, and hence, at this value of K, the system is critically damped.
c. If the gain $K > 4$, the poles are complex conjugate and the system becomes underdamped.
d. Looking at the underdamped region of the root locus, the real part of the roots is the same. Since the settling time is inversely proportional to the real part of complex poles, it remains the same irrespective of the value of the gain.
e. However, as the gain is increased, the system becomes very lightly damped, with the damping ratio $\xi = \frac{a}{2\sqrt{K}} = \frac{2}{\sqrt{K}}$.

Case II: Assume velocity feedback, that is, $\delta \neq 0$:

Consider the case when the open-loop gain K is fixed. The closed-loop characteristic equation for $a = 4$ and $K = 36$ is

$$\Delta(s) = s^2 + (K\delta + a)s + K = (s^2 + 4s + 36) + 36\delta s = 0 \tag{6.4}$$

This can be expressed as:

$$1 + \frac{36\delta s}{s^2 + 4s + 36} = 0 \tag{6.5}$$

Table 6.2: Pole location as velocity feedback gain changes for the system of Figure 6.1 for K = 36 and a=4.

δ	Pole 1	Pole 2
0	$-2+j4\sqrt{2}$	$-2-j4\sqrt{2}$
0.1	$-3.8+j4.6433$	$-3.8-j4.6433$
0.2	$-5.6+j2.154$	$-5.6-j2.154$
0.3	-3.0687	-11.7313
0.4	-2.2258	-16.1742
0.5	-1.7805	-20.2195
0.6	-1.4934	-24.1066
0.7	-1.2899	-27.9101
0.8	-1.1370	-31.6630
0.9	-1.0175	-35.3825
1.0	-0.9212	-39.0788

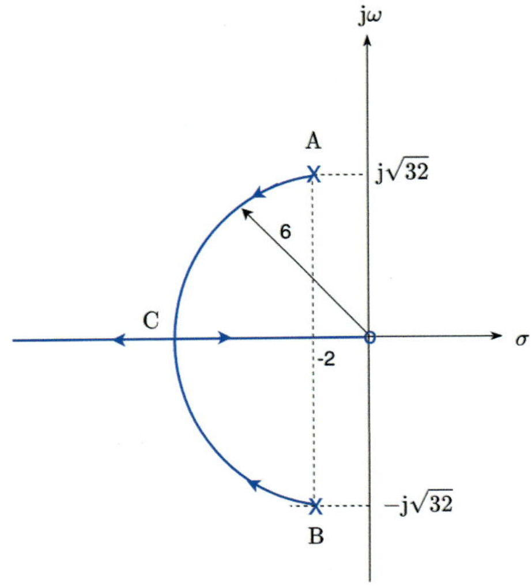

Figure 6.3: Root locus for Eqn(6.5)

Eqn(6.5) can be considered as the characteristic equation of a unity feedback system with the forward path transfer function

$$G'(s) = \frac{36\delta s}{s^2+4s+36} = \frac{K's}{s^2+4s+36}, \quad \text{where } K' = 36\delta \tag{6.6}$$

This system has a zero at the origin and a pair of complex poles at $s = -2 \pm j4\sqrt{2}$.

Table 6.2 shows the location of roots of Eqn(6.4), that is, the closed-loop poles as the velocity feedback gain δ is changed. The root locus is plotted in Figure 6.3.

We summarize our observations as:

i. When $\delta = 0$, the roots of $\Delta(s)$ are located at $s = -2 \pm j\sqrt{32}$, labeled A and B in Figure 6.3.

ii. As δ is increased from zero to 0.2, the roots are complex conjugate and they move along a circular arc of radius 6 with center at the origin.

iii. With further increase of δ, we get real roots, with one moving to the left and the other to the right.

Remark

The velocity feedback essentially increases the damping ratio of the system without altering the undamped natural frequency.

6.3 General Root Locus Problem

In the previous section, we sketched the root locus of a second-order system by computing analytically the roots. However, for higher-order systems, the problem of sketching the root loci is much more involved. In this section, the basic principles of constructing root locus are discussed.

Consider the following algebraic equation of a complex variable s:

$$P(s) + KQ(s) = 0 \quad \text{where} \tag{6.7}$$

$$P(s) = s^n + a_1 s^{n-1} + a_2 s^{n-2} + \cdots + a_{n-1}s + a_n \tag{6.8}$$

and

$$Q(s) = s^m + b_1 s^{m-1} + b_2 s^{m-2} + \cdots + b_{m-1}s + b_m \tag{6.9}$$

The coefficients $a_1, a_2, \ldots, a_n$ and $b_1, b_2, \ldots, b_m$ are assumed to be real and constant, and K is a real constant which can vary from $-\infty$ to $+\infty$. There is no restriction on the relative magnitudes of n and m. The objective is to determine the trajectories of the roots of Eqn(6.7) as the parameter K is varied from $-\infty$ to $+\infty$.

Since our interest lies in control systems, we will discuss the root locus technique for positive values of K, that is, for $0 < K < \infty$, considering a single-loop control system shown in Figure 6.4.

The closed-loop transfer function is given by

$$\frac{C(s)}{R(s)} = \frac{G(s)}{1 + G(s)H(s)} \tag{6.10}$$

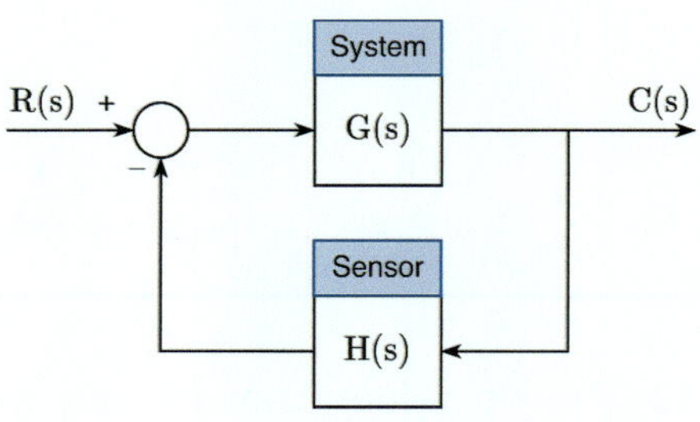

Figure 6.4: Closed-loop system

The characteristic equation of the closed-loop system satisfies

$$1 + G(s)H(s) = 0 \tag{6.11}$$

Assume that the open-loop transfer function G(s)H(s) contains a real parameter K as a multiplying factor, such that

$$G(s)H(s) = K\frac{Q(s)}{P(s)} = KG_1(s)H_1(s) \tag{6.12}$$

where P(s) and Q(s) are polynomials defined in Eqns(6.8) and (6.9), respectively. The closed-loop characteristic equation is expressed as

$$1 + G(s)H(s) = 1 + K\frac{Q(s)}{P(s)} = \frac{P(s) + KQ(s)}{P(s)} = 0 \tag{6.13}$$

Comparing Eqn(6.7) with the numerator of Eqn(6.13), we find that they are identical. Thus, when the open-loop transfer function is expressed in the form of Eqn(6.12), the root locus problem associated with a control system fits into the general root locus problem in Eqn(6.7).

Note that any characteristic equation containing the variable parameter K can be expressed in the form of Eqn(6.7) or Eqn(6.12). Before we proceed further, let us illustrate this with an example.

EXAMPLE 6.1

Suppose the characteristic equation of a system is

$$s^4 + 5s^3 + 8s^2 + K(s^2 + 4s - 5) = 0 \tag{6.14}$$

In order to express Eqn(6.14) in the form of Eqn(6.13), we divide both sides of Eqn(6.14) by the terms which do not contain K. In Eqn(6.14), the terms $s^4 + 5s^3 + 8s^2$ do not contain the variable K. Dividing by this term on both sides gives

$$1 + \frac{K(s^2 + 4s - 5)}{s^4 + 5s^3 + 8s^2} = 1 + K\frac{Q(s)}{P(s)} = 0 \tag{6.15}$$

where

$$\frac{Q(s)}{P(s)} = \frac{s^2 + 4s - 5}{s^4 + 5s^3 + 8s^2} \tag{6.16}$$

Now, from Eqn(6.12), the open-loop transfer function is expressed as:

$$G(s)H(s) = KG_1(s)H_1(s) \tag{6.17}$$

where $G_1(s)H_1(s)$ does not contain the variable parameter K. Hence, Eqn(6.11) can be written as:

$$G_1(s)H_1(s) = -\frac{1}{K} \tag{6.18}$$

Since $G_1(s)H_1(s)$ is a complex quantity, satisfying Eqn(6.18) is equivalent to satisfying the following two basic conditions for all points on the root locus.

Angle Condition

$$\begin{aligned} \angle G_1(s)H_1(s) &= (2q+1)\pi, \quad K \geq 0 \\ &= \text{odd multiples of } \pi \text{ radians or } 180^0 \end{aligned} \tag{6.19}$$

$$\begin{aligned} \angle G_1(s)H_1(s) &= 2q\pi, \quad K \leq 0 \\ &= \text{even multiples of } \pi \text{ radians or } 180^0 \end{aligned} \tag{6.20}$$

where $q = 0, \pm 1, \pm 2, \ldots\ldots$

Magnitude Condition

$$\left|G_1(s)H_1(s)\right| = \frac{1}{|K|} \tag{6.21}$$

Note: Among the above two conditions, the angle condition is more significant than the magnitude condition, because some values of s that satisfy Eqn(6.21) may not satisfy Eqn(6.19) and Eqn(6.20), while every s which satisfies Eqn(6.19) and Eqn(6.20) can be used to find a value of K using Eqn(6.21). Therefore, in practice:

1. The angle conditions in Eqn(6.19) or Eqn(6.20) are used to find trajectories of root loci in the s-plane.
2. After the root loci are drawn, the values of K on the loci are found using the magnitude condition in Eqn(6.21).

Graphical Evaluation of Angle and Magnitude of $G(s)H(s)$

To construct the root locus, we must know the location of poles and zeros of $G(s)H(s)$. Expressing this in pole-zero form gives

$$G(s)H(s) = KG_1(s)H_1(s) = \frac{K(s+z_1)(s+z_2)\ldots(s+z_m)}{(s+p_1)(s+p_2)\ldots\ldots(s+p_n)} = \frac{K\prod_{i=1}^{m}(s+z_i)}{\prod_{j=1}^{n}(s+p_j)} \tag{6.22}$$

where the zeros and poles of $G(s)H(s)$ are real or complex-conjugate pairs.

From the magnitude condition in Eqn(6.21), we have

$$\left|G_1(s)H_1(s)\right| = \frac{\prod_{i=1}^{m} |s+z_i|}{\prod_{j=1}^{n} |s+p_j|} = \frac{1}{|K|}, \quad -\infty < K < \infty \tag{6.23}$$

From the angle conditions in Eqns(6.19) and (6.20), we have

$$\begin{aligned} \angle G_1(s)H_1(s) &= \sum_{k=1}^{m} \angle(s+z_k) - \sum_{j=1}^{n} \angle(s+p_j) = (2q+1) \times 180^\circ, \text{ for } 0 \le K < \infty \\ &= \sum_{k=1}^{m} \angle(s+z_k) - \sum_{j=1}^{n} \angle(s+p_j) = (2q) \times 180^\circ, \text{ for } -\infty < K \le 0 \end{aligned} \tag{6.24}$$

where $q = 0, \pm1, \pm2 \ldots\ldots$ From these, we conclude the following:

1. *For positive values of* K*, the difference between the sums of the angles of the vectors drawn from the zeros and those from the poles of* G(s)H(s) *to any point* $\mathbf{s_1}$ *on the root locus is an odd multiple of* 180°.

2. *For negative values of* K*, the difference between the sums of the angles of the vectors drawn from the zeros and those from the poles of* G(s)H(s) *to any point* $\mathbf{s_1}$ *on the root locus is an even multiple of* 180°*, including zero degrees.*

After the construction of root loci, the value of gain K at any point $\mathbf{s_1}$ on the root loci is determined using the magnitude condition of Eqn(6.21) as:

$$|K| = \frac{\prod_{j=1}^{n} |s_1+p_j|}{\prod_{i=1}^{m} |s_1+z_i|} \tag{6.25}$$

For example, consider a system with an open-loop transfer function

$$G(s)H(s) = \frac{K(s+z_1)}{(s+p_1)(s+p_2)(s+p_3)} \tag{6.26}$$

where p_2 and p_3 are complex conjugate poles. The locations of poles and zeros of the system are assigned arbitrarily and is shown in Figure 6.5.

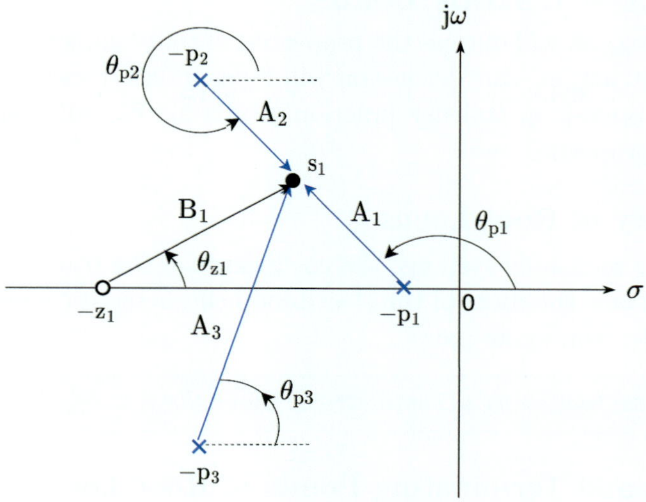

Figure 6.5: Diagram showing angle measurements from open-loop poles and open-loop zero to test point s_1

Let us select an arbitrary test point at $s = s_1$ in the s-plane and draw vectors from poles and zeros to this test point. For the test point s_1 to be on the root locus for **positive values of** K, it must satisfy the angle condition of Eqn(6.24); that is, the angles of the vectors shown in Figure 6.5 must satisfy

$$\angle(s_1 + z_1) - \angle(s_1 + p_1) - \angle(s_1 + p_2) - \angle(s_1 + p_3)$$
$$= \theta_{z1} - \theta_{p1} - \theta_{p2} - \theta_{p3} = (2q + 1) \times 180^\circ \tag{6.27}$$

where $q = 0, \pm 1, \pm 2, \ldots$. Likewise, if s_1 is a point on the root locus for **negative values of** K, then it must satisfy

$$\angle(s_1 + z_1) - \angle(s_1 + p_1) - \angle(s_1 + p_2) - \angle(s_1 + p_3)$$
$$= \theta_{z1} - \theta_{p1} - \theta_{p2} - \theta_{p3} = (2q \times 180^\circ) \tag{6.28}$$

Note that the angles of the vectors are measured in **counter clockwise direction**, that is, the positive real axis is taken as the reference.

If the test point s_1 satisfies the angle condition, then the magnitude of the open-loop gain K at s_1 is given by:

$$\left|K\right| = \frac{|s_1 + p_1||s_1 + p_2||s_1 + p_3|}{|s_1 + z_1|} = \frac{A_1 A_2 A_3}{B_1} \tag{6.29}$$

The sign of K is positive if s_1 satisfies the angle condition in Eqn(6.27) and is negative if it satisfies the condition in Eqn(6.28).

6.4 Properties of Root Locus

In the following, we will discuss the properties of root loci, which will be useful for their construction. Our discussion will focus on a typical feedback control system with open-loop transfer function G(s)H(s). We will not provide any proof of the properties here.

6.4.1 Symmetry of Root Locus

For physically realizable systems, the coefficients of the transfer functions are always real. Thus, the roots of the closed-loop characteristic equation must be real or complex conjugate pairs.

Rule-1: *The root locus is symmetrical about the real axis.*

6.4.2 Starting and Terminating Points of Root Locus

The root locus begins when gain is zero ($K = 0$) and terminates when gain is infinity ($K = \infty$). This is due to the root locus conditions of Eqn(6.18).

$$G_1(s)H_1(s) = -\frac{1}{K} \tag{6.30}$$

Thus, at $K = 0$, $G_1(s)H_1(s) = \infty$ which corresponds to the poles of open-loop transfer function. As K approaches infinity ($K \to \infty$), the branches of root locus approach zeros of $G_1(s)H_1(s)$. If the number of open-loop poles is greater than the number of open-loop zeros, that is $n > m$, then $(n - m)$ branches will terminate at zeros at infinity.[1]

Rule-2: *The root locus starts from finite and infinite poles of the open-loop transfer function and terminates at the finite and infinite zeros of the open-loop transfer function.*

EXAMPLE 6.2

Problem

Determine the starting and terminating points of the root locus for the

[1]Zeros at infinity reflect the natural low-pass filtering behavior of most physical systems, which cannot respond effectively at arbitrarily high frequencies. Like your ears at a rock concert. After a certain pitch, they just shut down. At infinite frequency, the system basically shrugs and walks away. This behavior prevents the system from amplifying very fast signal variations or noise at infinite frequency, providing inherent protection against high-frequency disturbances that could destabilize operation. Zeros at infinity ensure transfer functions remain proper (physically realizable), meaning the system has finite energy and cannot provide unlimited amplification.

system with open-loop transfer function

$$G(s)H(s) = \frac{K(s+1)(s+5)}{s(s+2)(s+6)}$$

SOLUTION:

Starting Points (K=0): The poles of the open-loop transfer function are at $s = 0$, $s = -2$ and at $s = -6$. Hence, the root locus starts from these three poles when $K = 0$.

Terminating Points ($K = \infty$): The zeros of open-loop transfer function are at $s = -1$, and $s = -5$. Since there are fewer zeros (2) than poles (3), the root locus behavior is:

- Two root locus branches terminate at the two finite zeros (**s=-1 and s=-5**)
- The remaining root locus branch terminates at a zero at infinity

Summary:(see Figure 6.6).

- **3 poles → 3 starting points**
- **2 finite zeros + 1 zero at infinity → 3 terminating points**

This accounts for all root locus branches since the number of branches equals the number of poles.

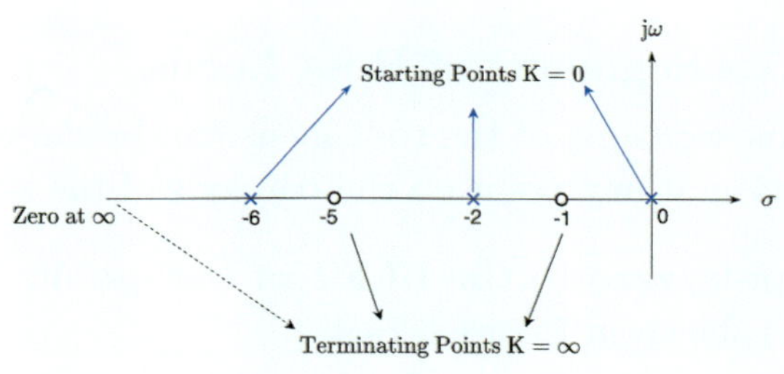

Figure 6.6: Starting and terminating points of root locus

6.4.3 Total Number of Branches of Root Locus

Definition 6.1. A *branch* of the root locus is defined as the path traversed by one root when the gain K varies between 0 and ∞.

Therefore, there will be one branch for each root of the closed-loop characteristic polynomial (see Eqns(6.8),(6.9)). Thus, the number of branches equals the order of this polynomial.

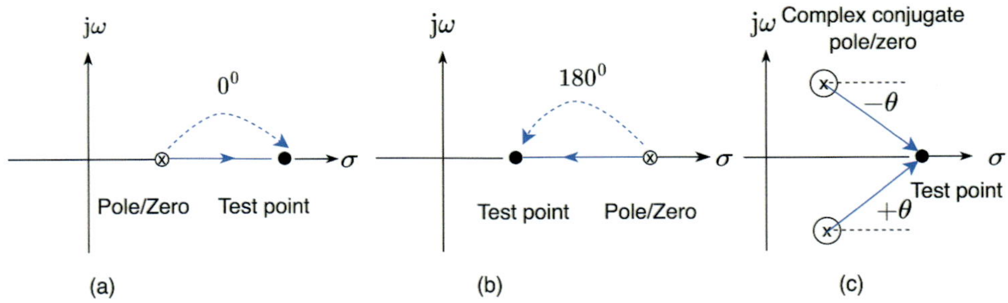

Figure 6.7: Angle condition on real axis

Number of Branches: If **n** denotes the number of finite open-loop poles and **m** is the number of finite open-loop zeros, then the number of root locus branches **N** is:

$$\mathrm{N} = \max(\mathrm{n}, \mathrm{m})$$

or equivalently [2]

$$\mathrm{N} = \begin{cases} \mathrm{n} & \text{if} \quad \mathrm{n} \geq \mathrm{m} \\ \mathrm{m} & \text{if} \quad \mathrm{n} < \mathrm{m} \end{cases}$$

Rule-3: *The number of branches of the root locus is equal to either the number of poles or the number of zeros, whichever is greater.*

6.4.4 Real Axis Segments of Root Locus

The real axis segments of the root locus can be determined by applying the angle condition at any point on the real axis of the s-plane.

1. **Real pole/zero to the LEFT of test point:** The angle contribution is 0°(as shown in Figure 6.7a).
2. **Real pole/zero to the RIGHT of test point:** The angle contribution is 180°(as shown in Figure 6.7b).
3. **Complex conjugate poles/zeros:** The angle contribution at any point on the real axis is 0°(see Figure 6.7c). This is because complex poles/zeros come in conjugate pairs, and their individual angle contributions cancel each other out.

[2] In practice, most physical systems have $\mathrm{n} \geq \mathrm{m}$ (more poles than zeros), so the number of branches typically equals the number of poles. The case where $\mathrm{n} < \mathrm{m}$ is less common and represents idealized systems that require additional poles for physical realizability.

Rule-4: *Any point on the real axis is a part of the root locus if and only if the number of poles and zeros to its right is odd.*

EXAMPLE 6.3

Problem

Sketch the real-axis segment of the root locus for the system whose open-loop transfer function is

$$G(s)H(s) = \frac{K(s+2)(s+4)}{(s+3)(s+5)(s+6)(s+8)(s^2+4s+13)}$$

SOLUTION:

The poles of the open-loop transfer function are located at $s = -3$, $s = -5$, $s = -6$, $s = -8$, $s = -2 + j3$, and at $s = -2 - j3$ while the zeros are located at $s = -2$, and at $s = -4$. The pole-zero plot is shown in Figure 6.8.

Since the complex conjugate poles at $s = -2 \pm j3$ contribute zero angle to any point on the real axis, their locations do not influence the determination of real axis segments of the root locus.

To determine the real axis segments, we apply Rule 4 by examining the number of real poles and zeros to the right of various test points on the real axis. The real poles and zeros, arranged from left to right, are: **-8,-6,-5,-4,-3,-2**.

Moving from left to right along the real axis, for each segment on the real axis

- **Section $-\infty - F$ $(-\infty, -8)$:** 6 poles/zeros to the right → **even** → not on root locus
- **Section E-F (-8, -6):** 5 poles/zeros to the right → **odd** → on root locus
- **Section D-E (-6, -5):** 4 poles/zeros to the right → **even** → not on root locus
- **Section C-D (-5, -4):** 3 poles/zeros to the right → **odd** → on root locus
- **Section B-C (-4, -3):** 2 poles/zeros to the right → **even** → not on root locus
- **Section A-B (-3, -2):** 1 pole/zero to the right → **odd** → on root locus

- **Section** $(-2, +\infty)$**:** 0 poles/zeros to the right → **even** → not on root locus

The root locus exists on real axis segments: **A-B**, **C-D**, and **E-F** as shown in Figure 6.8.

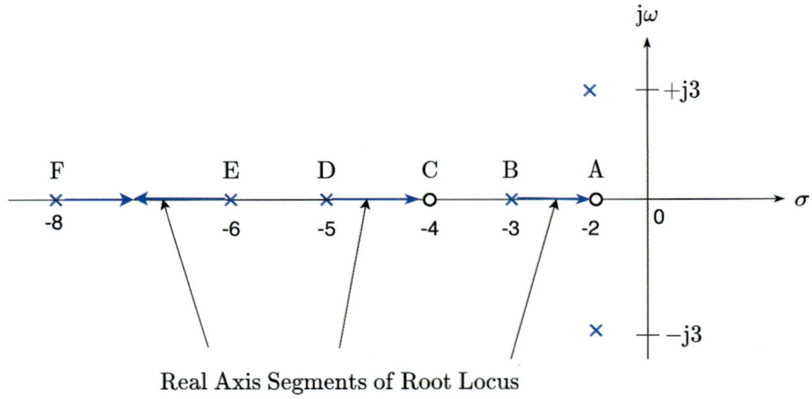

Figure 6.8: Real axis segments of root locus

6.4.5 Behavior of Root Locus at $|s| = \infty$

When the order of the polynomial P(s) (Eqn(6.8)) is not equal to the order of polynomial Q(s) (Eqn(6.9)), that is $n \neq m$, then **(n-m)** branches of root locus must terminate at zeros at infinity. This raises the obvious question: where are those infinite zeros located, and what do those root loci look like as they approach infinity? The answer is provided in Rule 5.

Rule-5: *If the number of finite zeros* **m** *is less than the number of finite poles* **n**, *then as the gain* $K \to \infty$, **(n-m)** *branches will become asymptotic to straight lines that intersect the real axis at a point* σ *(called the* **centroid**, *and are inclined to the real axis at angles* θ_k, *called the angle of asymptotes where*

Centroid:

$$\sigma = \frac{\sum_{i=1}^{n} p_i - \sum_{j=1}^{m} z_i}{n - m} = \frac{\text{Sum of Open Loop Poles} - \text{Sum of Open Loop Zeros}}{\text{Number of Open Loop Poles} - \text{Number of Open Loop Zeros}} \tag{6.31}$$

Angles of Asymptotes:

$$\theta_a = \frac{(2k+1)\pi}{n-m} \text{ [rad]} = \frac{(2k+1)\times 180^0}{n-m} \text{ [deg]} \quad (6.32)$$

where **k=0,1,2,...,(n-m-1)**

These asymptotes represent the directions in which the root locus branches travel as they head toward zeros at infinity. They provide a systematic way to sketch the high-gain behavior of the system without detailed calculations.

EXAMPLE 6.4

Problem

The open-loop transfer function of a system is

$$G(s)H(s) = \frac{K(s+4)}{s(s+1)(s+2)(s+3)}$$

For $0 < K < \infty$, determine

a. Number of finite poles, zeros, and number of zeros at infinity.

b. Number of separate root loci

c. The angle of asymptotes and the centroid

SOLUTION: Given $G(s)H(s) = \dfrac{K(s+4)}{s(s+1)(s+2)(s+3)}$

a. There are 4-poles located at $s = 0$, $s = -1$, $s = -2$ and at $s = -3$. Hence $n = 4$.

There is one finite zero at $s = -4$ in the open-loop transfer function. Hence **m=1**, and the number of zeros at infinity equals **n-m=4-1=3**.

b. The total number of root locus branches equals the number of poles, which is **n=4**. Since there are fewer zeros than poles ($m < n$), **(n-m)=3** branches will approach asymptotes as $K \to \infty$.

c. The pole-zero plot is shown in Figure 6.9. The angles of the asymptotes are computed from:

$$\theta_a = \frac{(2k+1)\pi}{(n-m)} = \frac{(2k+1)}{(n-m)} \times 180^0; \quad k = 0, 1, \text{ and } 2$$

- $\theta_a = \dfrac{\pi}{3} = 60^\circ$ for k=0
- $\theta_a = \dfrac{3\pi}{3} = 180^\circ$ for k=1

- $\theta_a = \dfrac{5\pi}{3} = 300°$ for k=2

The centroid (real axis intercept) is:

$$\sigma = \frac{\sum_{i=1}^{n} p_i - \sum_{j=1}^{m} z_j}{n - m} = \frac{[0 - 1 - 2 - 3] - [-4]}{3} = -\frac{2}{3} = -0.67$$

Figure 6.9 shows the asymptotes with pole-zero plot.

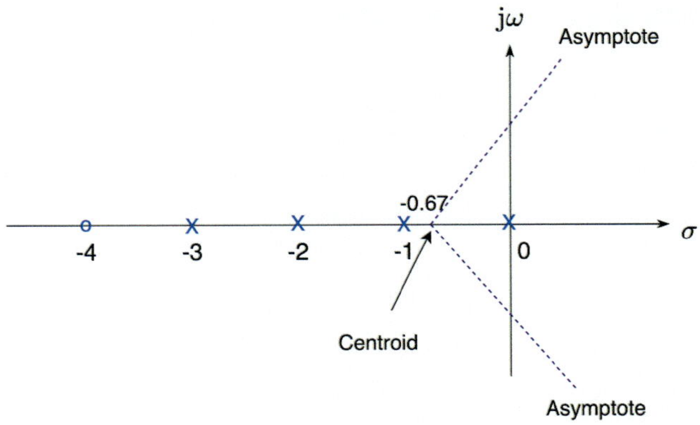

Figure 6.9: Centroid and asymptotes

CONCEPT REINFORCEMENT PROBLEM 6.1

Problem

The open-loop transfer function of a system is

$$G(s)H(s) = \frac{K}{s(s+6)(s^2+2s+17)}$$

For $0 < K < \infty$, determine

a. Number of finite poles, zeros, and number of zeros at infinity.

b. Number of separate root loci

c. The angle of asymptotes and the centroid

6.4.6 Breakaway and Re-entry Points

Definition 6.2. The points on the root locus where two or more branches separate and move in different directions are called *breakaway points*. Similarly, the points on the root loci where two or more branches converge and meet are called *re-entry or break-in points.*[a]

[a]While breakaway and break-in points commonly occur on the real axis, they can also occur at complex locations (off the real axis).

Figure 6.10 shows examples of **real-axis** breakaway and break-in points.

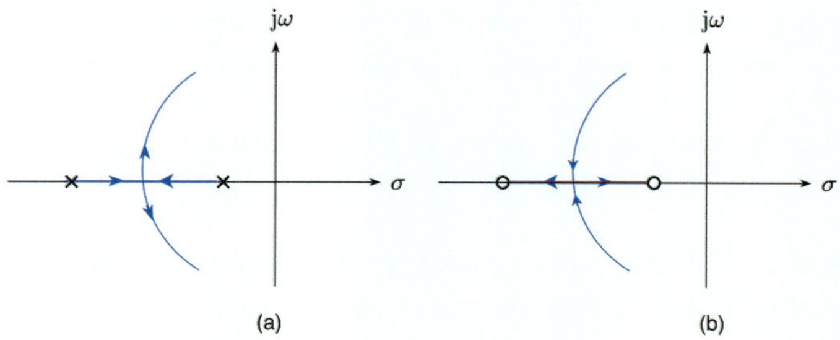

Figure 6.10: (a) Breakaway point and (b) Break-in (re-entry) point

There may be points on the real axis where more than two branches of the root locus meet. The angles at which the branches leave the real axis are given by[3]

$$\phi = \pm\frac{180^0}{\mathrm{r}} \tag{6.33}$$

where r denotes the multiplicity of the roots. Figure 6.11 shows two examples where more than two branches meet and depart.

Some of the properties of these points are:

Properties of Breakaway Point

1. At breakaway or reentry points, the characteristic equation has multiple roots.
2. If a root locus lies between two adjacent open-loop poles on the real axis, then there exists at least one breakaway point between the two poles.

[3]Eqn(6.33) applies specifically to breakaway/break-in points on the real axis. For complex breakaway/break-in points, the analysis requires applying the full angle condition at those points and solving $\frac{\mathrm{dK}}{\mathrm{ds}} = 0$ in the complex plane.

3. Similarly, if a root locus lies between two adjacent open-loop zeros on the real axis (one zero may be at infinity), then there exists at least one reentry (break-in) point between the two zeros.

4. If the root locus lies between an open-loop pole and a zero (either finite or infinite) on the real axis, breakaway or reentry points may or may not exist. When they do exist, both types of points can occur on the same segment.

5. Breakaway and re-entry points are not confined to the real axis. If the characteristic equation has multiple complex roots, then the breakaway points become complex, and these points occur in conjugate pairs due to the inherent symmetry of the root locus about the real axis (since the system has real coefficients).

6. A breakaway point represents the maximum value of K for which the roots can be real.

7. Similarly, a break-in (re-entry) point represents the minimum value of K for which the roots can be real.

   Note: Properties 6 and 7 apply specifically to real breakaway/break-in points. At these points, $dK/ds = 0$, making them critical points where the gain K reaches local maxima (breakaway) or minima (break-in) along the real axis.

6.4.7 Computation of Breakaway/Break-in (Re-entry) Points

Several methods can determine the breakaway and break-in points. The simplest of these is based on the fact that at the breakaway/break-in point, the characteristic equation has multiple roots.

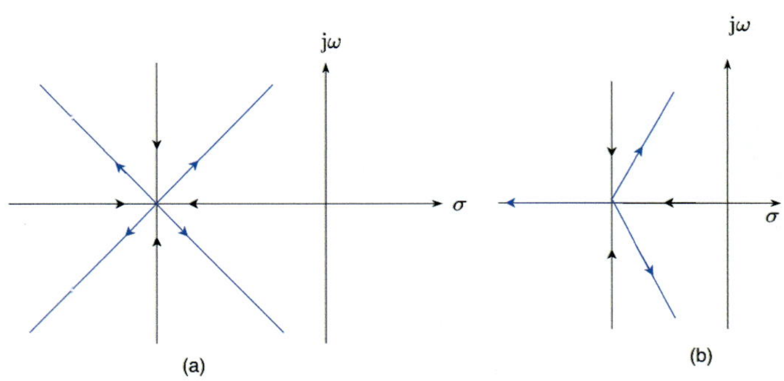

Figure 6.11: Higher order breakaway points: (a) Fourth-order breakaway point (b) Third-order breakaway point

Consider the characteristic equation of the system

$$F(s) = P(s) + KQ(s) = 0 \tag{6.34}$$

where P(s) and Q(s) do not contain K. Note that Eqn(6.34) has multiple roots at points where

$$\frac{dF(s)}{ds} = 0 \tag{6.35}$$

Differentiating Eqn(6.34) with respect to s gives

$$\frac{dF(s)}{ds} = P'(s) + KQ'(s) = 0, \text{ where } P'(s) = \frac{dP(s)}{ds}, \text{and}$$

$$Q'(s) = \frac{dQ(s)}{ds} \tag{6.36}$$

$$\text{Hence, for } \frac{dF(s)}{ds} = 0, \quad K = -\frac{P'(s)}{Q'(s)} \tag{6.37}$$

Substituting the value of K from Eqn(6.37) in Eqn(6.34), we get

$$F(s) = P(s) - \frac{P'(s)}{Q'(s)}Q(s) = 0$$

or,

$$P(s)Q'(s) - P'(s)Q(s) = 0 \tag{6.38}$$

Thus, Eqn(6.38) is satisfied at breakaway/break-in points. Now from Eqn(6.37), $K = -\frac{P(s)}{Q(s)}$. Hence

$$\frac{dK}{ds} = \frac{-[P'(s)Q(s) - Q'(s)P(s)]}{Q^2(s)} = \frac{P(s)Q'(s) - P'(s)Q(s)}{Q^2(s)} \tag{6.39}$$

Since from Eqn(6.38) we find $P(s)Q'(s) - P'(s)Q(s) = 0$, therefore, the breakaway/break-in points are the roots of

$$\boxed{\frac{dK}{ds} = 0} \tag{6.40}$$

Key Takeaways

i. All the roots of Eqn(6.40) are not breakaway or break-in points. Only those points that satisfy the angle condition are valid breakaway/break-in points.

ii. Suppose, all the roots of $\frac{dK}{ds} = 0$ are real. If a real root of Eqn(6.40) lies on the root locus segment of the real axis, then it is either an actual breakaway or break-in point.

iii. If the roots of Eqn(6.40) are complex, then check the value of K corresponding to the complex root of $\frac{dK}{ds} = 0$. If K is positive, then the corresponding complex root is either a breakaway or break-in point; else if K is negative, then the point is neither a breakaway nor a break-in point.

Rule-6: *The breakaway and break-in (re-entry) points on the root locus are determined from the roots of* $\frac{dK}{ds} = 0$. *However, not all roots of this equation correspond to actual breakaway/break-in points—they must be verified to lie on the root locus and satisfy the angle condition.*
For breakaway/break-in points on the real axis where **r** *branches meet, they separate at angles of* $\pm 180°/r$ *from the real axis.*

The following MATLAB function computes the candidates of breakaway and break-in points

```
function candidates = find_breakaway_candidates(num, den)
    % Find candidate breakaway/break-in points by solving dK/ds = 0
    [num_dk, den_dk] = polyder(den, num);
    candidates = roots(num_dk);

    % Keep only real candidates
    candidates = candidates(abs(imag(candidates)) < 1e-6);
    candidates = real(candidates);
end
```

After obtaining the candidate breakaway and break-in points, they can be classified using the MATLAB code in file `classify_breakaway_simple.m`(in supplementary material)

The following two examples demonstrate the procedure of manually finding the breakaway and break-in points.

EXAMPLE 6.5

Problem

Determine the breakaway/break-in points of the root locus for the system with open-loop transfer function

$$G(s)H(s) = \frac{K}{(s+a)(s+b)}$$

Compute the gain at the breakaway point.

SOLUTION:

The closed-loop characteristic equation is given by

$$1 + G(s)H(s) = 1 + \frac{K}{(s+a)(s+b)} = 0$$

This gives

$$\frac{(s+a)(s+b)+K}{(s+a)(s+b)} = 0$$
$$\Longrightarrow K = -(s+a)(s+b) = -\left[s^2 + (a+b)s + ab\right]$$

The breakaway/break-in points are obtained from the solution of $\frac{dK}{ds} = 0$. Now

$$\frac{dK}{ds} = -2s - (a+b) = 0$$

This gives the breakaway point at $s = -\frac{(a+b)}{2}$

Substituting this into expression for K gives $K = \frac{(a-b)^2}{4}$

CONCEPT REINFORCEMENT PROBLEM 6.2

Problem

Determine the breakaway/break-in points of the root locus for the system with open-loop transfer function

$$G(s)H(s) = \frac{K}{s(s+2)(s+4)}$$

EXAMPLE 6.6

Problem

Determine the breakaway/break-in points of the root locus for the system with open-loop transfer function

$$G(s)H(s) = \frac{K(s+c)}{(s+a)(s+b)}, \quad \text{assume } a < b < c$$

SOLUTION:

The closed-loop characteristic equation is given by

$$1 + G(s)H(s) = 1 + \frac{K(s+c)}{(s+a)(s+b)} = 0$$

This gives

$$\frac{s^2 + (a+b)s + ab + K(s+c)}{(s+a)(s+b)} = 0$$

or

$$K = -\frac{s^2 + (a+b)s + ab}{s+c}$$

or

$$\frac{dK}{ds} = -\frac{[2s + (a+b)]\,[(s+c)] - \left[s^2 + (a+b)s + ab\right]}{(s+c)^2}$$

Making $\frac{dK}{ds} = 0$ gives $s^2 + 2cs + ac + bc - ab = 0$

The solution of the above equation gives the breakaway point at

$$s_1, s_2 = -c \pm \sqrt{c^2 - (ac + bc - ab)}$$

One of the solution s_1 which lies between the poles at $s = -a$ and $s = -b$ is a breakaway point and other solution s_2 which lies between the finite zero at $s = -c$ and zero at infinity is a break-in point. For example, if $a = 2$, $b = 8$, and $c = 10$, we get two solutions $s_1 = -6$ and $s_2 = -14$. The former is a breakaway point and the latter is a break-in point as shown in Figure 6.12.

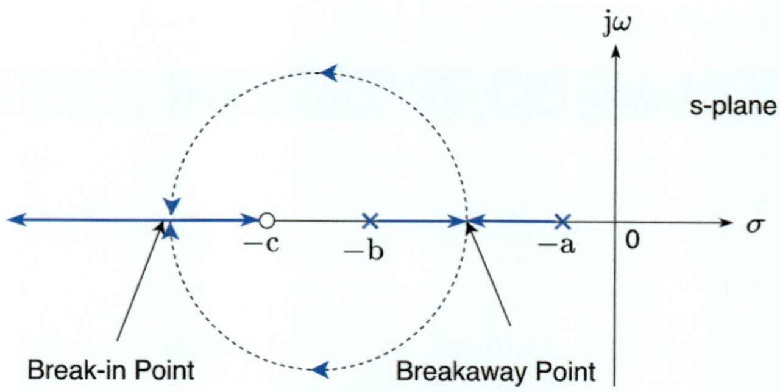

Figure 6.12: Breakaway and break-in points

CONCEPT REINFORCEMENT PROBLEM 6.3

Problem

For the system with an open-loop transfer function

$$G(s)H(s) = \frac{K}{(s^2 + a)(s^2 - b)}, \quad a, b > 0$$

determine the conditions when

i. All the breakaway/break-in points are real

ii. Some of the breakaway/break-in points are imaginary

CONCEPT REINFORCEMENT PROBLEM 6.4

Problem

The open-loop transfer function of a unity feedback system is given by

$$G(s)H(s) = \frac{K}{s(s^2 + bs + c)}$$

The parameters $b, c > 0$ and are selected such that $G(s)H(s)$ has a pair of complex conjugate poles.

a. Determine the conditions, in terms of the parameters b and c, which would result in a distinct breakaway/break-in point on the real axis.

b. Determine the condition such that the root loci will break in and break away from the same point on the real axis (breakaway/break-in points are non-distinct).

The following examples compute the breakaway and break-in points of several systems using MATLAB

EXAMPLE 6.7

In this example, we compute the breakaway and break-in points using MATLAB.

a. **System-1:** $G(s)H(s) = \dfrac{1}{(s+2)(s+4)(s+8)(s+30)}$. The root locus has two breakaway points at $s = -23.753$ and $s = -2.885$. The MATLAB code is found in `Example6_6a.m`.

b. **System-2:** $G(s)H(s) = \dfrac{s}{s^2+9}$. The root locus has one break-in point at $s = -3$. The MATLAB code is found in `Example6_6b.m`.

c. **System-3:** $G(s)H(s) = \dfrac{s+15}{(s+2)(s+5)}$. The root locus has one breakaway point at $s = -3.59$ and one break-in point at $s = -26.402$. The MATLAB code is found in `Example6_6c.m`.

6.4.8 Computation of Angle of Arrival and Departure

For simple poles on the real axis, the root locus departs along the real axis. However, if the pole is complex, it can depart at any angle. To find the departure angle from a complex pole, we apply the angle condition of Eqn(6.19) to a point infinitesimally close (ϵ-close) to the complex pole.

Consider a test point s very close to pole p_j. We assume that all angles from other poles and zeros are drawn directly to pole p_j, since the test point is infinitesimally close to it.

Determining Angle of Departure

Let us rewrite the angle criterion by isolating the angle between the locus and the selected pole p_j for $0 \leq K < \infty$ as:

$$\underbrace{\sum_{i=1}^{m} \angle(s - z_i)}_{\Theta_{zd}} - \underbrace{\sum_{\substack{i=1 \\ i\neq j}}^{n} \angle(s + p_i)}_{\Theta_{pd}} - \underbrace{\angle(s + p_j)}_{\theta_d} = (2q+1) \times 180^0, \quad q = 0, 1, 2 \ldots \tag{6.41}$$

where

Θ_{zd}=sum of angles from all zeros to pole p_j

Θ_{pd}=sum of angles from all other poles to pole p_j

θ_d=angle of departure from pole p_j

Hence, the angle of departure at pole p_j is computed from

$$\theta_d = \pm 180^\circ(2q + 1) + (\Theta_{zd} - \Theta_{pd}) \tag{6.42}$$

Without loss of generality, choose $q = 0$, since the solutions repeat for all integer values of q. Hence

$$\boxed{\theta_d = 180^0 + (\Theta_{zd} - \Theta_{pd})} \tag{6.43}$$

The angle of departure represents the initial direction in which the root locus branch leaves the complex pole as K increases from zero. The locations of all other poles and zeros in the system uniquely determine this angle.

Consider an example of a system whose open-loop pole-zero configuration is shown in Figure 6.13.

Following the procedure, we find

$$\Theta_{pd} = \theta_1 + \theta_3 + \theta_5$$
$$\text{and } \Theta_{zd} = \theta_2 + \theta_4$$

Hence

$$\theta_d = 180^0 + \Theta_{zd} - \Theta_{pd}$$

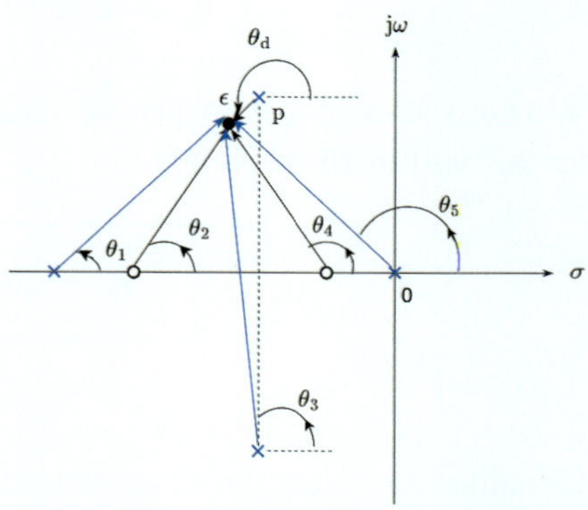

Figure 6.13: Pole-zero configuration of open-loop transfer function and determination of angle of departure

The following MATLAB function computes the angle of departure of the root locus from a specified complex pole.

```
function theta = departure_angle(num, den, pole)
    p = roots(den); z = roots(num);
    other_p = p(abs(p - pole) > 1e-6);
    theta = (pi + sum(angle(pole - z)) - ...
             sum(angle(pole - other_p))) * 180/pi;
end
```

Determining Angle of Arrival

Let us rewrite the angle criterion by isolating the angle between the locus and the selected zero z_j for $0 \le K < \infty$ as:

$$\underbrace{\sum_{\substack{i=1\\ i\neq j}}^{m} \angle(s+z_i)}_{\Theta_{za}} + \underbrace{\angle(s+z_j)}_{\theta_a} - \underbrace{\sum_{i=1}^{n} \angle(s+p_i)}_{\Theta_{pa}} = (2q+1)\times 180^0, \quad q = 0,1,2\ldots \tag{6.44}$$

where

Θ_{za}=sum of angles from all other zeros to zero z_j

Θ_{pa}=sum of angles from all poles to zero z_j

θ_a=angle of arrival at zero z_j

Hence, the angle of arrival at zero z_j is computed from

$$\theta_a = \pm 180^0(2q+1) + (\Theta_{pa} - \Theta_{za}) \tag{6.45}$$

Without loss of generality we can assume $q = 0$, since the solutions are same for all values of q. Hence

$$\boxed{\theta_a = \pm 180^0 + (\Theta_{pa} - \Theta_{za})} \tag{6.46}$$

Consider an example of a system whose open-loop pole-zero configuration is shown in Figure 6.14

Following the procedure, we find

$$\Theta_{pa} = \theta_2 + \theta_3 + \theta_5 \quad \text{and}$$
$$\Theta_{za} = \theta_1 + \theta_4$$

Hence

$$\theta_a = 180^0 + \Theta_{pa} - \Theta_{za}$$

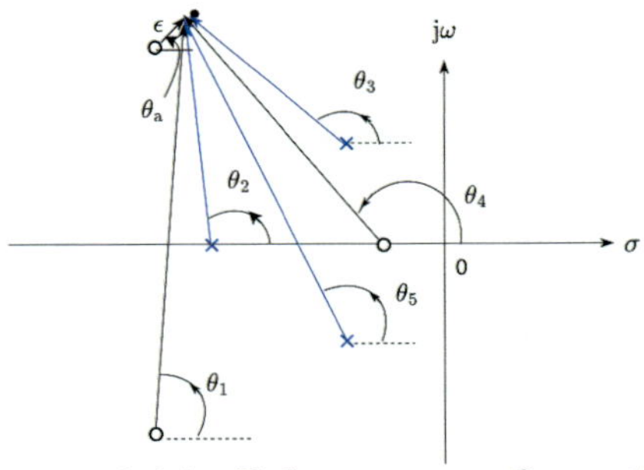

Figure 6.14: Pole-zero configuration of open-loop transfer function and determination of angle of arrival

The MATLAB function to compute the angle of arrival at a specified complex zero is given below.

```
function theta = arrival_angle(num, den, zero)
    p = roots(den); z = roots(num);
    other_z = z(abs(z - zero) > 1e-6);
    theta = (pi + sum(angle(zero - p)) - ...
              sum(angle(zero - other_z))) * 180/pi;
end
```

Steps to calculate the angle of departure at a complex pole p

Computing Angle of Departure

1. Select a test point on the root locus branch starting from p which is very close (to be called ϵ-close) to the complex pole p.
2. Compute the sum of the angles from all **remaining poles** (excluding pole p) to the selected test point, and denote this sum as Θ_{pd}.
3. Compute the sum of the angles from **all zeros** to the selected test point, and denote it as Θ_{zd}.
4. Compute the net angle contribution from all poles and zeros (except from pole p)

$$\phi = \Theta_{\mathrm{zd}} - \Theta_{\mathrm{pd}}$$

5. Let θ_{d} denote the angle contribution of the pole p at the selected point.
6. Apply the angle criterion to compute θ_{d}

$$\phi - \theta_{\mathrm{d}} = \pm 180^0(2\mathrm{q}+1); \quad \mathrm{q} = 0, 1, 2, \ldots$$

   or

$$\theta_{\mathrm{d}} = \pm 180^0(2\mathrm{q}+1) + \phi; \quad \mathrm{q} = 0, 1, 2, \ldots$$

7. The steps for calculating the angle of arrival at complex zeros are identical, except that the roles of poles and zeros are interchanged in the angle criterion. This will be illustrated through examples.

EXAMPLE 6.8

Problem

Determine the angle of departure at $s = +j2$ for $K > 0$ for the system with loop transfer function

$$\frac{Ks}{(s+2)(s^2+4)}$$

SOLUTION:

The system has a zero at $s = 0$ and poles at $s = -2$ and $s = \pm j2$.

Draw vectors from all other poles and zeros to the complex pole at $s = \pm j2$, as shown in Figure 6.15

Compute the total angle contribution from all other poles, denoted as Θ_{pd}, as:

$$\Theta_{pd} = \theta_{p1} + \theta_{p2} = \tan^{-1}\left(\frac{2}{2}\right) + 90^0 = 135^0$$

Compute the total angle contribution from all the zeros, denoted as Θ_{zd}, given by

$$\Theta_{zd} = \theta_{z1} = +90^0$$

The angle of departure is

$$\theta_d = 180^0 + \Theta_{zd} - \Theta_{pd} = 180^0 + 90^0 - 135^0 = 135^0$$

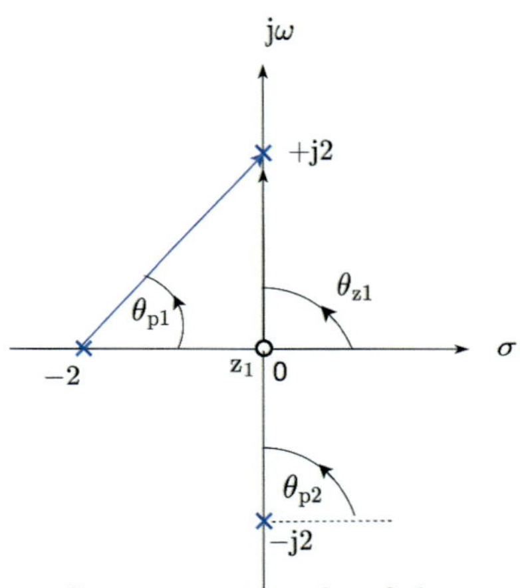

Figure 6.15: Computing angle of departure for system

CONCEPT REINFORCEMENT PROBLEM 6.5

Problem

For the loop transfer of the following systems, determine the angle of departure at the designated pole for K > 0.

S1: $\dfrac{K}{s(s+3)(s^2+6s+45)}$ at $s = -3 + j6$

S2: $\dfrac{K(s^2-4s+8)}{s^2+2s+2}$ at $s = -1 + j1$

S3: $\dfrac{K}{s^2(s^2+4s+20)}$ at $s = -2 + j4$

EXAMPLE 6.9

Problem

Determine the angle of arrival at $s = 2 + j2$ for $K > 0$ for the system with loop transfer function

$$\frac{K(s^2-4s+8)}{s^2+2s+2}$$

SOLUTION:

The system has complex conjugate zeros at $s = +2 \pm j2$ and complex conjugate poles at $s = -1 \pm j1$.

Draw vectors from all poles and other zeros to the designated complex zero at $s = 2 + j2$ as shown in Figure 6.16

Compute the total angle contribution from all poles, denoted as Θ_{pa}, given by

$$\Theta_{pa} = \theta_{p1} + \theta_{p2} = \tan^{-1}\left(\frac{1}{3}\right) + \tan^{-1}\left(\frac{3}{3}\right) = 63.43^0$$

Compute the total angle contribution from other zeros, denoted as Θ_{zd}, given by

$$\Theta_{za} = \theta_{z2} = +90^0$$

The angle of arrival is $\theta_a = 180^0 + \Theta_{pa} - \Theta_{za} = 153.43^0$

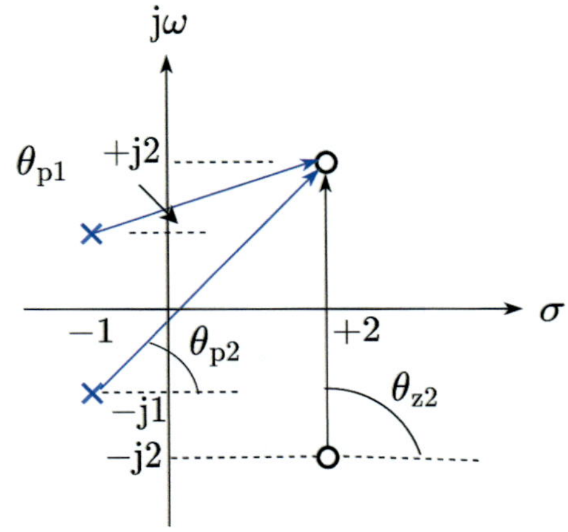

Figure 6.16: Computing angle of arrival for system

CONCEPT REINFORCEMENT PROBLEM 6.6

Problem

For the loop transfer of the following systems, determine the angle of arrival at the designated pole for $K > 0$.

S1: $\dfrac{K(s^2 + 6s + 18)}{s(s^2 + 10s + 29)}$ at $s = -3 + j3$

S2: $\dfrac{K(s^2 + 4s + 8)}{s^2(s + 3)(s + 5)}$ at $s = -2 + j2$

S3: $\dfrac{K(s^2 + 6s + 10)}{s(s^2 + 9)}$ at $s = -3 + j1$

Rule-7:

Note: All angles are measured from the positive real axis in the counterclockwise direction.

a. The **angle of departure** θ_d of the root locus from a complex pole is computed as:

$$\theta_d = 180^0 - (\Theta_{pd} - \Theta_{zd}) \tag{6.47}$$

where Θ_{pd} = sum of angles from all other poles to the complex pole under consideration, and

Θ_{zd} = sum of angles from all zeros to the complex pole under consideration.

b. The **angle of arrival** θ_a of the root locus at a complex zero is computed as:

$$\theta_a = 180^0 - (\Theta_{za} - \Theta_{pa}) \tag{6.48}$$

where Θ_{pa} = sum of angles from all poles to the complex zero under consideration, and

Θ_{za} = sum of angles from all other zeros to the complex zero under consideration.

c. For higher-order poles and zeros:

The angles of departure θ_{dm} from a complex pole of multiplicity m are:

$$\theta_{dm} = \frac{[(2q+1)180^0 - (\Theta_{pd} - \Theta_{zd})]}{m}, \quad q = 0, 1, 2, \ldots, (m-1) \tag{6.49}$$

The angles of arrival θ_{am} at a complex zero of multiplicity m are:

$$\theta_{am} = \frac{[(2q+1)180^0 + (\Theta_{pa} - \Theta_{za})]}{m}, \quad q = 0, 1, 2, \ldots, (m-1) \tag{6.50}$$

where m is the multiplicity of the pole/zero, and q generates all m departure/arrival angles.

Root Locus Construction Steps

1. Determine the open-loop poles, zeros, and number of branches from the given G(s)H(s).
2. Draw the pole-zero plot and determine the segments of the real axis on which the root locus exists using the odd-even rule.
3. Calculate the angles of asymptotes and determine the centroid for branches going to infinity.
4. Calculate the breakaway and break-in points (if any) by solving $dK/ds = 0$ and verify they lie on the root locus.
5. Determine the intersection points of the root locus with the imaginary axis (if any) and find the corresponding gain values.
6. Calculate the angles of departure from complex poles and angles of arrival at complex zeros (if any).
7. Sketch additional points on the root locus for intermediate gain values if needed for accuracy.
8. From the above information, draw the complete root locus plot.

9. Analyze the stability and predict the transient performance of the closed-loop system using the root locus.

These concepts are demonstrated through the following examples.

EXAMPLE 6.10

Problem: Consider a unity feedback system with forward path transfer function

$$G(s) = \frac{K}{s(s+4)(s+6)} \tag{6.51}$$

We need to sketch the root locus showing how the closed-loop poles move as the gain K varies from zero to infinity.

SOLUTION:

First, we identify the basic system characteristics:

- Number of finite poles: $n = 3$ (located at $s = 0, -4, -6$)
- Number of finite zeros: $m = 0$
- Number of branches: $n = 3$ (each starting from a pole)
- Asymptotic behavior: All branches terminate at infinity since $n > m$

Real Axis Segments

The root locus exists on the real axis in regions where the total number of poles and zeros to the right is odd. Therefore, the locus exists on:

- Segment: $0 > s > -4$
- Segment: $s < -6$

Asymptotic Analysis

Since $n - m = 3$, we have three asymptotes approaching infinity.

The centroid is located at:

$$\sigma_a = \frac{\sum \text{poles} - \sum \text{zeros}}{n-m} = \frac{0 + (-4) + (-6) - 0}{3} = -\frac{10}{3} \approx -3.33$$

The asymptote angles are:

$$\theta_k = \frac{(2k+1)\pi}{n-m} = \frac{(2k+1)\cdot 180^0}{3}, \quad k = 0, 1, 2$$

This gives us angles of 60°, 180°, and 300° (or equivalently -60°).

Breakaway Point Analysis

The closed-loop characteristic equation is

$$1 + G(s)H(s) = 1 + \frac{K}{s(s+4)(s+6)} = 0$$

Rearranging to solve for K:

$$\frac{s^3 + 10s^2 + 24s + K}{s(s+4)(s+6)} = 0 \implies K = -s^3 - 10s^2 - 24s$$

The breakaway points occur where $\frac{dK}{ds} = 0$.

$$\frac{dK}{ds} = -3s^2 - 20s - 24 = 0$$

Solving this quadratic equation gives

- $s_1 = -1.57$ (valid breakaway point on the segment $0 > s > -4$)
- $s_2 = -5.09$ (invalid, not on a root locus segment)

v. **Imaginary Axis Crossing:**

To find where the locus crosses the imaginary axis, we use the Routh-Hurwitz criterion on the characteristic equation:

$$1 + G(s) = 1 + \frac{K}{s(s+4)(s+6)} = 0 \implies s^3 + 10s^2 + 24s + K = 0$$

s^3	1	24
s^2	10	K
s^1	$\frac{240-K}{10}$	0
s^0	K	

For marginal stability, the s^1 row element must be zero:

$$240 - K = 0 \implies K = 240$$

At this critical gain, the auxiliary equation from the s^2 row is:

$$10s^2 + 240 = 0 \text{ or, } s^2 = -24 \implies s = \pm j2\sqrt{6}$$

The system crosses the imaginary axis at $s = \pm j4.90$ when $K = 240$.

Root Locus Summary

- Three branches starting from poles at $s = 0, -4, -6$
- One breakaway point at $s = -1.57$
- Imaginary axis crossing at $s = \pm j2\sqrt{6}$ for $K = 240$
- Three asymptotes at angles 60°, 180°, and 300° meeting at $\sigma_a = -10/3$
- System is stable for $0 < K < 240$

A sketch of the complete root locus is shown in Figure 6.17.

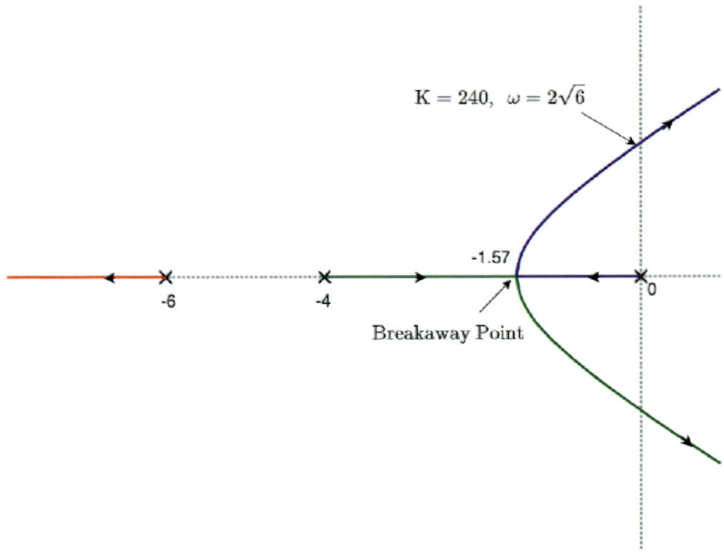

Figure 6.17: Root locus for system of Eqn(6.51)

EXAMPLE 6.11

Problem: Consider a unity feedback system with forward path transfer function

$$G(s) = \frac{K(s+6)}{s(s+4)} \tag{6.52}$$

It is desired to sketch the locus showing how the closed-loop poles move as the gain K varies from zero to infinity.

SOLUTION:

System Parameters

First, we identify the basic system characteristics:

- Number of finite poles: $n = 2$ (located at $s = 0, -4$)
- Number of finite zeros: $m = 1$ (located at $s = -6$)
- Number of branches: $n = 2$ (each starting from a pole)
- Termination: One branch terminates at the finite zero $s = -6$, and one branch terminates at infinity

Real Axis Segments

The root locus exists on the real axis in regions where the total number of poles and zeros to the right is odd. Analyzing each segment:

- Segment $s > 0$: No poles/zeros to the right $\rightarrow$ even count $\rightarrow$ no locus
- Segment $0 > s > -4$: One pole at $s = 0$ to the right $\rightarrow$ odd count $\rightarrow$ locus exists
- Segment $-4 > s > -6$: Two poles at $s = 0, -4$ to the right $\rightarrow$ even count $\rightarrow$ no locus
- Segment $s < -6$: Two poles and one zero to the right $\rightarrow$ odd count $\rightarrow$ locus exists

Therefore, the locus exists on segments: $0 > s > -4$ and $s < -6$.

Asymptotic Analysis

Since $n - m = 2 - 1 = 1$, we have one asymptote approaching infinity.

The centroid is located at:

$$\sigma_a = \frac{\sum \text{poles} - \sum \text{zeros}}{n - m} = \frac{0 + (-4) - (-6)}{1} = \frac{2}{1} = 2$$

The asymptote angle is:

$$\theta_0 = \frac{(2 \cdot 0 + 1)\pi}{1} = \pi = 180^\circ$$

This gives us a single asymptote along the negative real axis from the centroid at $s = 2$.

Breakaway and Break-in Point Analysis

The closed-loop characteristic equation is

$$1 + G(s)H(s) = 1 + \frac{K(s+6)}{s(s+4)} = 0$$

Rearranging to solve for K

$$K = -\frac{s(s+4)}{s+6} = -\frac{s^2+4s}{s+6}$$

The breakaway/break-in points occur where $\frac{dK}{ds} = 0$:

$$\begin{aligned}\frac{dK}{ds} &= -\frac{\left[(s+6)(2s+4) - (s^2+4s)\right]}{(s+6)^2} \\ &= -\frac{2s^2+12s+4s+24-s^2-4s}{(s+6)^2} = -\frac{s^2+12s+24}{(s+6)^2}\end{aligned}$$

Setting $\frac{dK}{ds} = 0$ gives

$$s^2 + 12s + 24 = 0$$

The solution of this gives

- $s_1 = -2.54$ and $s_2 = -9.46$. Since s_1 lies between the poles at $s = 0$ and $s = -4$, it is a *breakaway point.*
- The other solution s_2 lies between the finite zero at $s = -6$ and zero at infinity. Hence, it is a *break-in point.*

Imaginary Axis Crossing Analysis:

Let us construct the Routh table from the characteristic equation

$$1 + G(s) = 1 + \frac{K(s+6)}{s(s+4)} = 0 \quad \Longrightarrow \quad s^2 + (K+4)s + 6K = 0$$

The Routh table is

s^2	1	6K
s^1	K + 4	0
s^0	6K	

For the system to be on the verge of instability, the s^1 coefficient must be zero:

$$K + 4 = 0 \quad \Longrightarrow \quad K = -4$$

Since this critical value is negative, and we are considering positive values of K (as is typical in root locus analysis), the system will not cross the imaginary axis for any positive gain values. This means the system remains stable for all $K > 0$.

A sketch of the complete root locus is shown in Figure 6.18.

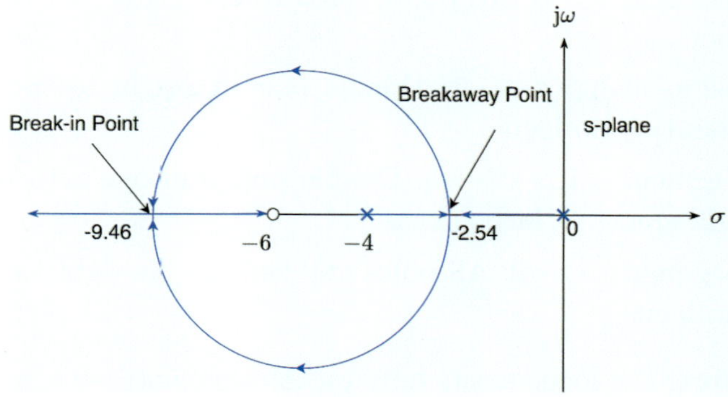

Figure 6.18: Root locus for system of Eqn(6.52)

EXAMPLE 6.12

Problem: Sketch the root locus of the system with open-loop transfer function

$$G(s)H(s) = \frac{K(s+1)}{s^2(s+9)} \tag{6.53}$$

as the gain K varies from zero to infinity. Additionally, determine:

a. The roots when all poles are equal and real
b. The corresponding gain value for condition (a).

SOLUTION:

System Parameters

The open-loop transfer function reveals:

- Number of finite poles: $n = 3$ (double pole at $s = 0$, simple pole at $s = -9$)
- Number of finite zeros: $m = 1$ (located at $s = -1$)
- Number of branches: $n = 3$ (starting from each pole)
- Termination: One branch terminates at $s = -1$, two branches terminate at infinity

Real Axis Segments

The root locus exists on the real axis where the total number of open-loop poles and zeros to the right is odd:

- Segment $s > 0$: No poles/zeros to the right → even count → no locus
- Segment $0 > s > -1$: Double pole at origin to the right → even count → no locus
- Segment $-1 > s > -9$: Double pole and one zero to the right → odd count → locus exists
- Segment $s < -9$: All poles and zero to the right → even count → no locus

Therefore, the locus exists only on the segment: $-9 < s < -1$.

Asymptotic Analysis

The centroid is located at:

$$\sigma_a = \frac{\sum \text{poles} - \sum \text{zeros}}{n - m} = \frac{0 + 0 + (-9) - (-1)}{2} = \frac{-8}{2} = -4$$

The asymptote angles are:

$$\theta_k = \frac{(2k+1)\pi}{n-m} = \frac{(2k+1)\pi}{2}, \quad k = 0, 1$$

This gives us angles of 90°and 270°(or equivalently ±90°).

Breakaway Point Analysis

The closed-loop characteristic equation is

$$1 + G(s)H(s) = 1 + \frac{K(s+1)}{s^2(s+9)} = 0$$

Rearranging to solve for K:

$$K = -\frac{s^2(s+9)}{s+1} = -\frac{s^3 + 9s^2}{s+1}$$

The breakaway points occur where $\frac{dK}{ds} = 0$:

$$\begin{aligned}
\frac{dK}{ds} &= -\frac{(3s^2 + 18s)(s+1) - (s^3 + 9s^2)(1)}{(s+1)^2} \\
&= -\frac{3s^3 + 3s^2 + 18s^2 + 18s - s^3 - 9s^2}{(s+1)^2} \\
&= -\frac{2s^3 + 12s^2 + 18s}{(s+1)^2} = -\frac{2s(s^2 + 6s + 9)}{(s+1)^2} \\
&= -\frac{2s(s+3)^2}{(s+1)^2}
\end{aligned}$$

Setting $\frac{dK}{ds} = 0$:

$$2s(s+3)^2 = 0$$

This yields:

- $s_1 = 0$ (double pole location - breakaway point)
- $s_2 = s_3 = -3$ (repeated root - breakaway point on real axis segment)

a. Now we need all roots to be equal and real. This occurs at the breakaway point $s = -3$, where the characteristic equation becomes:

$$(s+3)^3 = s^3 + 9s^2 + 27s + 27 = 0$$

Comparing with the characteristic equation

$$s^3 + 9s^2 + Ks + K = 0$$

gives $K = 27$. Therefore, all roots are equal and real at $s = -3$ and the corresponding gain is $K = 27$

Determination of Imaginary Axis Crossover:

The characteristic equation in standard form is:

$$1 + G(s) = 1 + \frac{K(s+1)}{s^2(s+9)} = 0 \quad \Longrightarrow \quad s^3 + 9s^2 + Ks + K = 0$$

The Routh table is:

s^3	1	K
s^2	9	K
s^1	$\frac{8K}{9}$	0
s^0	K	

For marginal stability, the s^1 coefficient must be zero:

$$\frac{8K}{9} = 0 \implies K = 0$$

When $K = 0$, the auxiliary equation from the s^2 row gives $s = 0$, indicating the system is tangent to the imaginary axis at the origin but does not cross it. For all $K > 0$, the system remains stable with all poles in the left half-plane.

The presence of the double pole at the origin creates a unique breakaway behavior, while the finite zero at $s = -1$ ensures one branch remains on the real axis.

Root Locus Summary

- Three branches starting from poles at $s = 0$ (double) and $s = -9$
- Breakaway point at $s = 0$ where two branches separate from the double pole
- Critical breakaway point at $s = -3$ where all three branches converge (gain $K = 27$)
- One branch terminates at the finite zero $s = -1$
- Two asymptotes at $\pm 90°$ with centroid at $s = -4$
- System is tangent to the imaginary axis at the origin but never crosses it
- System is stable for all $K > 0$
- All roots are equal and real ($s = -3$) when $K = 27$

A sketch of the complete root locus is shown in Figure 6.19.

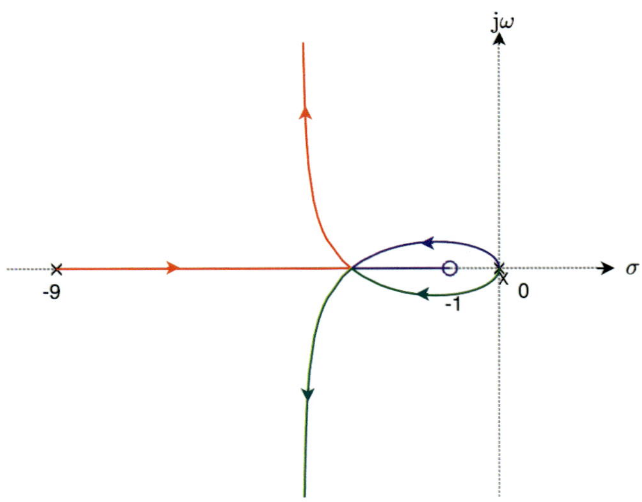

Figure 6.19: Root locus of system of Eqn(6.53)

CONCEPT REINFORCEMENT PROBLEM 6.7

Problem: Sketch the root locus of the system with the open-loop transfer function

$$G(s)H(s) = \frac{K(s+2)}{s^2(s+6)}$$

as the gain K varies from zero to infinity.

CONCEPT REINFORCEMENT PROBLEM 6.8

Problem: Consider a unity feedback system with forward path transfer function

$$G(s) = \frac{K}{s(s^2 + bs + c)}$$

Sketch the locus of the closed-loop poles as the gain K varies from zero to infinity for the following values of the parameters b and c.

i. $b = 6$ and $c = 25$
ii. $b = 5$ and $c = 8$
iii. $b = 3$ and $c = 3$

EXAMPLE 6.13

Problem: Consider a unity feedback system with forward path transfer function

$$G(s) = \frac{K(s^2 + 4s + 8)}{s^2(s + 5)} \tag{6.54}$$

Sketch the locus of the closed-loop poles as the gain K varies from zero to infinity.

SOLUTION:

First, we analyze the zeros by factoring the numerator quadratic:

$$s^2 + 4s + 8 = 0 \implies s = \frac{-4 \pm \sqrt{16 - 32}}{2} = \frac{-4 \pm j4}{2} = -2 \pm j2$$

The system characteristics are:

- Number of finite poles: $n = 3$ (double pole at $s = 0$, simple pole at $s = -5$)
- Number of finite zeros: $m = 2$ (complex conjugate pair at $s = -2 \pm j2$)
- Number of branches: $n = 3$ (starting from each pole)
- Termination: Two branches terminate at finite zeros $s = -2 \pm j2$, one branch terminates at infinity

Asymptotic Behavior Analysis

For the single asymptote (since $n - m = 1$), we calculate:

Centroid location:

Centroid location:

$$\sigma_a = \frac{\sum \text{poles} - \sum \text{zeros}}{n - m} \tag{6.55}$$

$$= \frac{(0 + 0 - 5) - (-2 - 2)}{3 - 2} \tag{6.56}$$

$$= \frac{-5 - (-4)}{1} = -1 \tag{6.57}$$

Asymptote direction:

$$\theta_a = \frac{(2k + 1)\pi}{n - m} = \frac{(2k + 1) \times 180^0}{1} \tag{6.58}$$

$$= 180^0 \quad \text{for } k = 0 \tag{6.59}$$

Breakaway Point Identification

Due to the second-order pole at the origin, the root locus exhibits a breakaway point at $s = 0$.

Real axis segments on root loci:

The root locus exists between $s = -5$ and an infinite zero on the real axis.

Determination of Imaginary Axis Crossover:

The closed-loop characteristic equation becomes:

$$1 + G(s) = 1 + \frac{K(s^2 + 4s + 8)}{s^2(s + 5)} = 0 \implies s^3 + (K + 5)s^2 + 4Ks + 8K = 0$$

Constructing the Routh-Hurwitz array

s^3	1	$4K$
s^2	$K + 5$	$8K$
s^1	$\dfrac{4K(K + 5) - 8K}{K + 5}$	0
s^0	$8K$	

Let us solve for K from s^1 row to determine the stability condition:

$$4K(K + 5) - 8K = 0 \implies K(K + 3) = 0 \implies K = 0, \text{ and } K = -3$$

Since we consider $K > 0$, no imaginary axis crossing occurs for positive gains.

v. **Determination of Angle of Arrival:**

Let us compute the angle of arrival at the upper complex zero. Consider the Figure 6.20

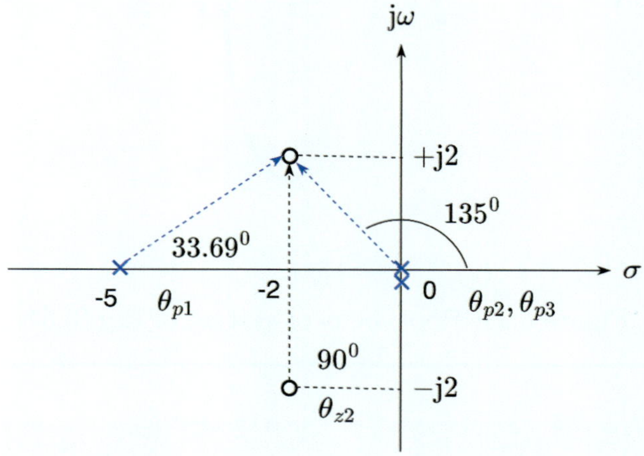

Figure 6.20: Angle of arrival computation for system of Eqn(6.54)

The contribution of angle from poles at $s = -5$ and pole at $s = 0, 0$ are

$$\theta_{p1} = \tan^{-1}\frac{2}{3} = 33.69^0, \quad \theta_{p2} = 135^0, \text{and } \theta_{p3} = 135^0, \quad \theta_{z2} = 90^0$$

Thus, the angle contribution of all the poles at the upper complex zero is

$$\Theta_p = \theta_{p1} + \theta_{p2} + \theta_{p3} = 33.69^0 + 135^0 + 135^0 = 303.69^0, \text{ and } \Theta_z = 90^\circ$$

Similarly, the total angle contribution of other zeros at the selected complex zero is

$$\Theta_z = \theta_{z2} = 90^\circ$$

Hence, the angle of arrival at the upper complex zero

$$\theta_{a1} = 180^0 - \Theta_z + \Theta_p = 180^\circ + 303.69^\circ - 90^\circ = 33.69^\circ$$

By symmetry, the arrival angle at the lower complex zero is -33.69°.

A sketch of the complete root locus is shown in Figure 6.21.

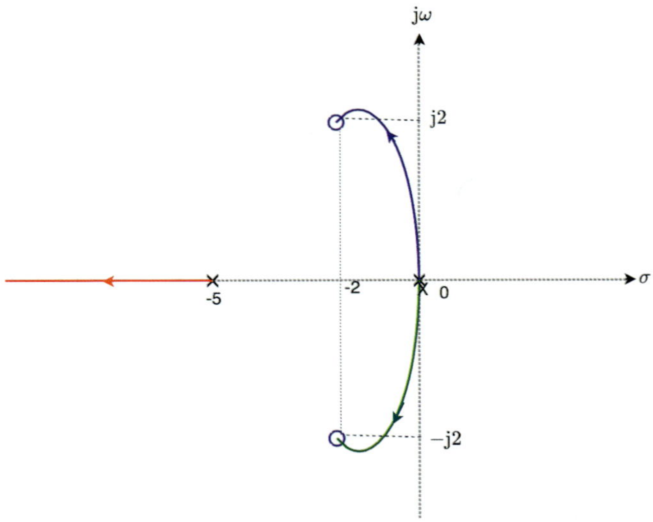

Figure 6.21: Root locus of system of Eqn(6.54)

CONCEPT REINFORCEMENT PROBLEM 6.9

Problem: Consider a unity feedback system with forward path transfer function

$$G(s) = \frac{K(s+3)}{s^2 + 2s + 5}$$

Sketch the locus of the closed-loop poles as the gain K varies from zero to infinity.

CONCEPT REINFORCEMENT PROBLEM 6.10

Problem: Consider a unity feedback system with forward path transfer function

$$G(s) = \frac{K}{(s^2 + a)(s^2 - b)}, \quad a, b > 0$$

Sketch the locus of the closed-loop poles as the gain K varies from zero to infinity when

i. $a = 1, \quad b = 4$

ii. $a = 4, \quad b = 1$

EXAMPLE 6.14

Problem: Analyze a unity feedback control system with the forward path transfer function

$$G(s) = \frac{K(s+2)}{(s+1)(s+3)(s+5)(s+8)} \tag{6.60}$$

Complete the following analysis:

(a) Construct the root locus diagram for $K > 0$, identifying all significant features including breakaway points and critical gain values.

(b) Calculate the gain value that places a closed-loop pole at $s = -1.5$ on the real axis.

(c) Determine the gain K that yields a damping ratio of $\xi = 0.5$ for the dominant complex pole pair.

(d) Evaluate the accuracy of the second-order approximation at the computed gain.

SOLUTION:

a. **Root Locus Construction**

System Configuration Analysis

- **Open-loop poles ($n = 4$):** Located at $s = -1, -3, -5, -8$
- **Open-loop zeros ($m = 1$):** Single zero at $s = -2$
- **Root locus branches:** Four branches originate from the poles
- **Branch termination:** One branch terminates at the finite zero; three branches extend to infinity

Real Axis Segments

Root locus segments exist on the real axis between:

- $s = -2$ and $s = -1$
- $s = -5$ and $s = -3$
- $s = -\infty$ and $s = -8$

Asymptotic Properties

For the three asymptotes ($n - m = 3$):

Centroid calculation:

$$\begin{aligned}\sigma_a &= \frac{\sum \text{poles} - \sum \text{zeros}}{n - m} \\ &= \frac{(-1-3-5-8)-(-2)}{4-1} \\ &= \frac{-17+2}{3} = -5.0\end{aligned}$$

Asymptote angles:

$$\theta_a = \frac{(2k+1)\pi}{n-m} \times \frac{180^0}{\pi}, \quad k = 0, 1, 2$$
$$= 60°, 180°, 300° \quad \text{for } k = 0, 1, 2$$

Stability Analysis via Routh-Hurwitz

The characteristic equation is

$$1 + G(s) = 0 \implies s^4 + 17s^3 + 95s^2 + (K + 199)s + (2K + 120) = 0$$

Constructing the Routh array:

s^4	1	95	$120 + 2K$
s^3	17	$K + 199$	
s^2	$\frac{1416 - K}{17}$	$2K + 120$	
s^1	$\frac{(-K^2 + 639K + 247104)}{1416 - K}$	0	
s^0	$2K + 120$		

Critical gain determination:

Setting the s^1 row numerator to zero:

$$K^2 - 639K - 247104 = 0 \implies (K - 910)(K + 271.4181) = 0$$

Since $K > 0$, the critical gain $K = 910$. Substituting this value of K in s^2 row gives ultimate frequency of oscillation

$$29.7647s^2 = -1940 \quad \text{or}, s = \pm j8.0733$$

Breakaway Point Analysis:

Using the breakaway condition $\frac{dK}{ds} = 0$, the breakaway point on the real axis segment between $s = -5$ and $s = -3$ is calculated to be at $s = -3.79$.

A sketch of the complete root locus is shown in Figure 6.22.

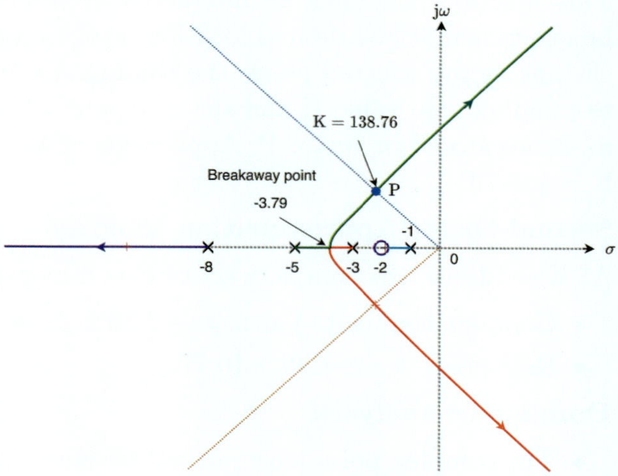

Figure 6.22: Root locus for system of Eqn(6.60)

b. The gain at the real pole -1.5 is

$$K = \frac{0.5 \times 1.5 \times 3.5 \times 6.5}{0.5} = 34.125$$

This can alternately be computed as

$$K = \left| \frac{(s+1)(s+3)(s+5)(s+8)}{s+2} \right|_{s=-1.5} = 34.125$$

c. **Determination of Gain for $\xi = 0.5$:**

To achieve a damping ratio of $\xi = 0.5$, we must locate closed-loop poles that lie along constant damping ratio lines. These lines extend from the origin at an angle of:

$$\theta = \cos^{-1}(0.5) = 60^\circ$$

measured from the negative real axis.

The intersection points where these constant-ξ contours meet the root locus branches are found at approximately $s = -2.18 \pm j3.88$. Applying the magnitude condition at these locations produces:

$$K = 138.76$$

This gain value can be determined through analytical calculation or by employing MATLAB's interactive root locus tool. To find the gain corresponding to any specific point on the root locus, utilize the command:

```
[K P]=rlocfind(sys)
```

This function generates an interactive crosshair cursor that can be positioned at any desired location on the root locus plot. Upon clicking at the selected point, the command returns both the corresponding gain value K and the complete set of closed-loop pole locations stored in vector P. Application of this method confirms K = 138.76.

d. **Second-Order Approximation Validity**

At K = 138.76, the complete set of closed-loop poles is:

- Complex conjugate pair: $s = -2.18 \pm j3.88$
- Real poles: $s = -1.86, -10.77$

Dominance analysis:

- The complex poles have natural frequency $\omega_n = 4.46$ rad/s
- The real pole at $s = -1.86$ is relatively close to the complex pair
- The real pole at $s = -10.77$ is sufficiently distant to be non-dominant

Approximation assessment:

The second-order approximation has limited validity due to the proximity of the real pole at $s = -1.86$ to the dominant complex pair. This nearby pole will contribute significantly to the transient response, causing deviations from the expected second-order behavior. For improved accuracy, a third-order analysis including this real pole would be more appropriate.

The pole at $s = -10.77$ is sufficiently far from the dominant poles (approximately 2.4 times farther) and can be reasonably neglected in the approximation.

6.5 Generalized Root Locus

The generalized root locus technique represents an extension of conventional root locus methodology, providing the capability to investigate how closed-loop pole locations vary with respect to any system parameter of interest, not merely the forward path gain. This enhanced approach empowers control engineers to examine system stability and transient characteristics as functions of diverse parameters, including time constants, controller coefficients, or individual feedback elements that may significantly impact overall system behavior. The implementation procedure for generalized root locus involves the following:

Generalized Root Locus

a. Establish the closed-loop transfer function T(s) for the complete system.

b. Reformulate T(s) such that the parameter under investigation (designated as p) appears explicitly as the varying parameter.

c. Restructure the denominator of T(s) into the standard form:

$$1 + p\frac{N(s)}{D(s)} = 1 + KG(s)H(s)$$

through algebraic manipulation that segregates terms containing and excluding parameter p, thereby facilitating the construction of root trajectories as p undergoes variation.

This methodology facilitates comprehensive system analysis across a wider spectrum of design scenarios, particularly valuable when examining non-gain parameters that influence pole dynamics. Such versatility proves essential in adaptive control applications and systems exhibiting sensitivity to component tolerances or environmental variations. The subsequent example demonstrates this analytical process.

EXAMPLE 6.15

Problem: Consider the unity feedback system shown in Figure 6.23

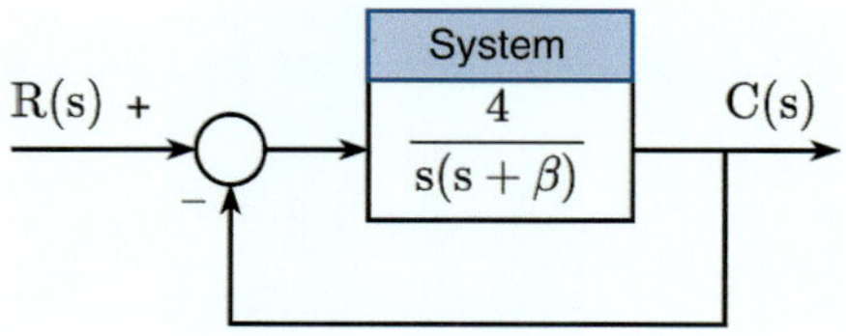

Figure 6.23: Original system

Construct the root locus diagram of this system as parameter β is varied from zero to infinity.

SOLUTION:

Based on Figure 6.23, the closed-loop transfer function can be expressed as

$$T(s) = \frac{4}{s^2 + \beta s + 4} = \frac{4}{(s^2 + 4) + \beta s} = \frac{\dfrac{4}{s^2 + 4}}{1 + \dfrac{\beta s}{s^2 + 4}}$$

The denominator of T(s), representing the closed-loop characteristic polynomial, is algebraically rearranged into the standard form $1 + KG(s)H(s) = 1 + \beta \frac{s}{s^2 + 4}$.

We can now construct the root locus for the equivalent system having open-loop transfer function $\beta \frac{s}{s^2 + 4}$ with β ranging from zero to ∞, employing standard root locus construction techniques.

The system contains two poles positioned on the $j\omega$ axis at $\pm j2$ and a single zero located at the s-plane origin. Consequently, the root locus branches will originate from $s = +j2$ and $s = -j2$. One branch will converge to the finite zero at $s = 0$, while the remaining branch will approach a zero at infinity.

Break-in Point Analysis:

Starting with the closed-loop characteristic equation:

$$s^2 + \beta s + 4 = 0 \implies \beta = -\frac{(s^2 + 4)}{s}$$

Setting $\frac{d\beta}{ds} = 0$ yields $s^2 - 4 = 0 \implies s = \pm 2$

Since $s = -2$ lies between the two zeros (i.e., the finite zero at 0 and zero at infinity), it constitutes a valid *break-in point*. The other solution $s = +2$ does not represent a legitimate *breakaway/break-in* location, as root locus segments do not exist in that portion of the real axis.

A sketch of the complete root locus is shown in Figure 6.24.

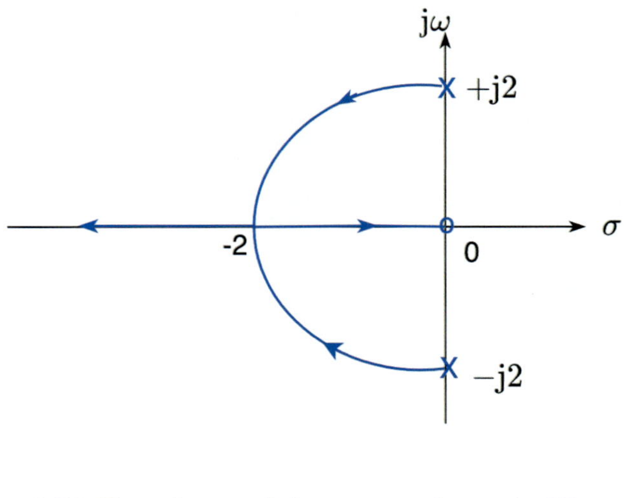

Figure 6.24: Root locus of the system shown in Figure 6.23

CONCEPT REINFORCEMENT PROBLEM 6.11

Problem:

The open-loop transfer function of a position control system is given by

$$G(s)H(s) = \frac{20}{s(s+2)(s+p)}$$

a. Sketch the root locus, considering p as the variable parameter

b. Determine p such that the damping ratio of the closed-loop system is $1/\sqrt{2}$.

6.6 Root Locus Analysis for Positive Feedback Systems

Our previous discussion focused on root locus characteristics for negative feedback configurations, where the feedback signal opposes the reference input. However, when analyzing positive feedback systems, where the feedback signal reinforces the reference input, the root locus properties undergo substantial modifications. The architecture of such a positive feedback arrangement is depicted in Figure 6.25.

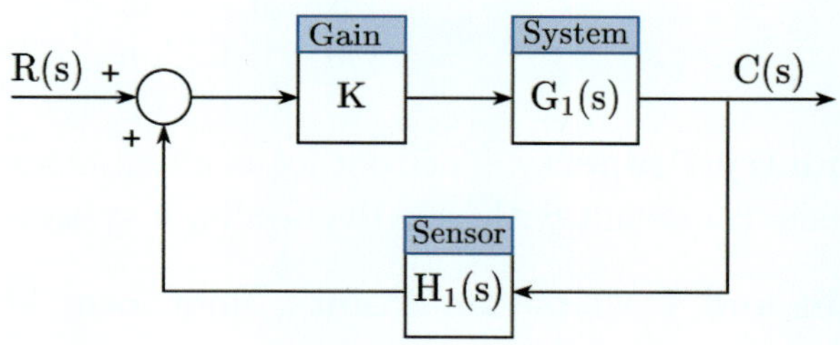

Figure 6.25: System with positive feedback

The closed-loop transfer function becomes:

$$T(s) = \frac{KG_1(s)}{1 - KG_1(s)H_1(s)}$$

with the corresponding characteristic equation:

$$1 - KG_1(s)H_1(s) = 0$$

or equivalently,

$$KG_1(s)H_1(s) = 1$$

Consequently, the magnitude and phase conditions for positive feedback systems are:

$$\begin{aligned} &\left|KG_1(s)H_1(s)\right| = 1 \\ \text{and } &\angle KG_1(s)H_1(s) = 2\pi q, \quad q = 0, \pm1, \pm2, \pm3, \ldots \end{aligned}$$

Observe the parallels and distinctions between these criteria and those governing negative feedback systems. While the magnitude condition remains identical, the phase criterion undergoes a fundamental change. For positive feedback systems, the phase contribution from G(s)H(s) at any root locus point s equals 0°, contrasting with the 180° requirement for negative feedback configurations. We shall now outline the construction principles for positive feedback root loci and examine how these deviate from the established rules for negative feedback systems presented in the previous section. We consider an open-loop transfer function containing n poles and m zeros.

Root Locus Construction for Positive Feedback Systems

1. **Branch Count N.** This criterion remains identical to negative feedback systems **(unchanged)**:

$$N = \begin{cases} n & \text{if } n > m \\ m & \text{if } m > n \end{cases}$$

2. **Symmetry Property.** The root locus maintains symmetry about the real axis, consistent with negative feedback systems **(unchanged)**.

3. **Origin and Destination Points.** Root locus branches for positive feedback systems initiate at the finite and infinite poles of G(s)H(s) and conclude at the finite and infinite zeros of G(s)H(s). This behavior mirrors that of negative feedback systems **(unchanged)**.

4. **Real-Axis Segments.** For positive feedback configurations, a real-axis point belongs to the root locus if and only if the total count of poles and zeros positioned to its right is **even**. This represents a **fundamental change** from negative feedback systems, where real-axis segments exist when the number of poles and zeros to the right is odd. Consequently, *positive feedback root loci occupy real-axis regions where negative feedback loci are absent.*

5. **Asymptotic Behavior: Centroid and Angular Orientation.** The centroid location for asymptotes remains unaltered. However, the asymptotic angles undergo **modification**. When $n > m$, exactly $n - m$ branches approach straight-line asymptotes. For positive feedback sys-

tems, the real-axis intercept σ_a and asymptotic angle θ_a are given by:

$$\sigma_a = \frac{\sum \text{finite poles} - \sum \text{finite zeros}}{n - m} \tag{6.61}$$

and

$$\theta_a = \frac{2q\pi}{n - m}, \quad q = 0, \pm 1, \pm 2, \pm 3, \ldots \tag{6.62}$$

Angular measurements are expressed in radians relative to the positive real axis. Note the critical distinction: Equation(6.62) employs $2q\pi$ rather than $(2q + 1)\pi$.

6. **Departure and Arrival Angles.** This criterion experiences **significant alteration**. The departure angle θ_d for a branch leaving a complex pole becomes:

$$\theta_d = 0^\circ + \Theta_{zd} - \Theta_{pd} \tag{6.63}$$

while the arrival angle θ_a for a branch approaching a complex zero is:

$$\theta_a = 0^0 + \Theta_{pa} - \Theta_{za} \tag{6.64}$$

where Θ_{pd}, Θ_{zd}, Θ_{za}, and Θ_{pa} represent the angular contributions defined in the previous section (6.4.8).

Important Note: Additional construction procedures concerning imaginary-axis crossings, break-in points, and breakaway locations follow identical methodologies as those employed for negative feedback systems.

The following example demonstrates these principles in practice.

EXAMPLE 6.16

Problem: Construct the root locus diagram for a positive feedback system having the forward path transfer function:

$$G(s)H(s) = \frac{K(s + 4)}{s(s - 5)(s^2 + 4s + 8)} \tag{6.65}$$

SOLUTION:

i. **System Configuration and Branch Characteristics**
Open-loop pole and zero identification:

The system contains n = 4 poles (located at $s = 0, +5, -2 \pm j2$) and m = 1 zero (positioned at $s = -4$)

Total number of root locus branches equals n = 4

Branch initiation and termination behavior: The root locus branches originate from $s = 0$, $s = +5$, and the complex conjugate pole pair at $s = -2 \pm j2$. Given the presence of a single finite zero (m = 1), one branch will converge to the finite zero at $s = -4$, while the remaining three branches will extend toward zeros at infinity.

ii. **Real-axis locus segments**: For this positive feedback configuration, root locus segments exist between $s = 0$ and $s = -4$, and from $s = 5$ extending toward positive infinity on the real axis.

iii. **Asymptotic properties: Centroid and angular directions**: With n = 4 and m = 1, the asymptotic parameters are:

$$\begin{aligned}\sigma_a &= \frac{\sum \text{finite poles} - \sum \text{finite zeros}}{n - m} \\ &= \frac{(0 + 5 - 2 - 2) - (-4)}{4 - 1} = 1.67 \\ \theta_a &= \frac{2k\pi}{n - m} = \frac{2k}{n - m} \times 180^0; \quad k = 0, 1, 2 \\ &= 0^\circ, 120^\circ, 240^\circ\end{aligned}$$

Analysis reveals that the root locus branches do not intersect the imaginary axis.

Departure angle computation:

Referring to Figure 6.26 for geometric analysis:

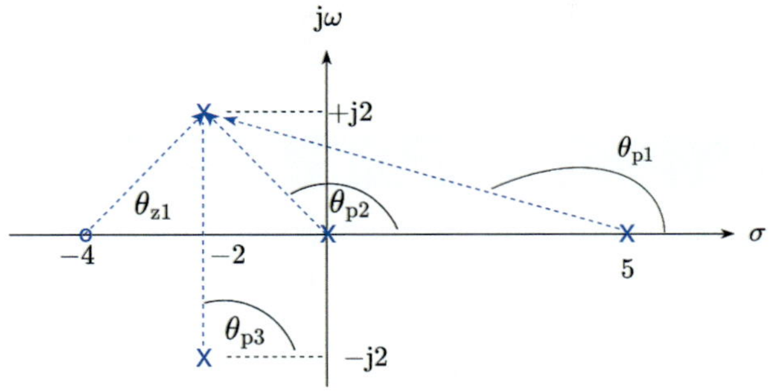

Figure 6.26: Departure angle analysis

The angular contributions from poles at $s = +5$, $s = 0$, and $s = -2 - j2$ to the upper complex pole are:

$$\theta_{p1} = 90^\circ + \tan^{-1}\left(\frac{7}{2}\right) = 164.05^\circ, \quad \theta_{p2} = 135^\circ, \quad \theta_{p3} = 90^\circ$$

The total angular contribution from all poles at the upper complex pole becomes:

$$\Theta_p = \theta_{p1} + \theta_{p2} + \theta_{p3} = 164.05^\circ + 135^\circ + 90^\circ = 389.05^\circ$$

The angular contribution from the zero at $s = -4$ to the upper complex pole is:

$$\Theta_z = \theta_{z1} = 45^\circ$$

Therefore, the departure angle from the upper complex pole is calculated as:

$$\theta_{d1} = 0^\circ + \Theta_z - \Theta_p = 45^\circ - 389.05^\circ = -344.05^\circ = 15.95^\circ$$

By symmetry, the departure angle from the lower complex pole is -15.95°.

The complete root locus configuration is illustrated in Figure 6.27.

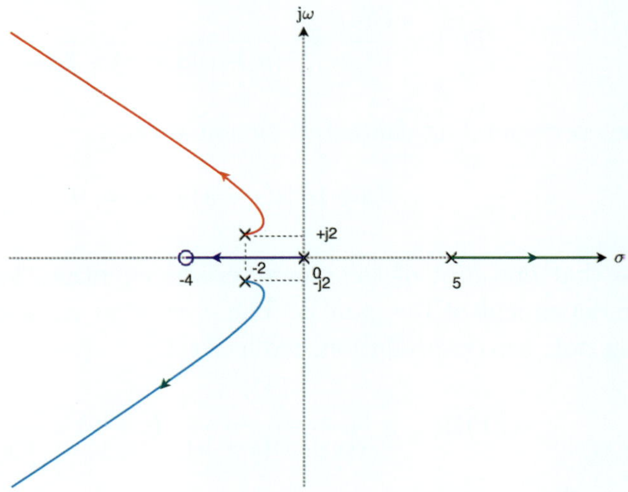

Figure 6.27: Root locus for positive feedback system of Equation (6.65)

CONCEPT REINFORCEMENT PROBLEM 6.12

Problem: Sketch the root locus for a positive feedback system with forward path transfer function

$$G(s)H(s) = \frac{K(s+4)}{s(s+1)(s+2))}$$

6.7 Root Locus Analysis for Systems with Pole-Zero Cancellation

Consider the system configuration shown in Figure 6.28, where a common factor exists between the denominator of the forward path transfer function G(s) and the numerator of the feedback path transfer function H(s).

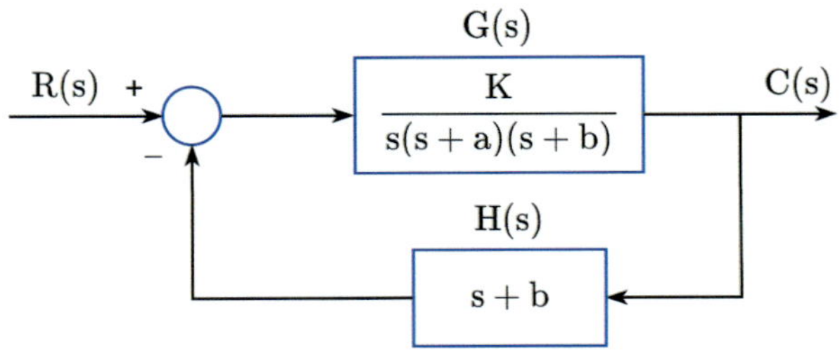

Figure 6.28: System with common factor in G(s)H(s)

The closed-loop transfer function becomes:

$$T(s) = \frac{C(s)}{R(s)} = \frac{K}{s(s+a)(s+b) + K(s+b)}$$

with the corresponding characteristic equation:

$$(s+b)\,[s(s+a) + K] = 0$$

Observe that one root of the characteristic equation, located at $s = -b$, remains independent of the gain K. The open-loop transfer function G(s)H(s), following pole-zero cancellation, reduces to:

$$G(s)H(s) = \frac{K}{s(s+a)(s+b)} \cdot (s+b) = \frac{K}{s(s+a)}$$

The simplified characteristic equation becomes:

$$1 + G(s)H(s) = 0 \implies s^2 + as + K = 0$$

Consequently, the root locus for $G(s)H(s) = \dfrac{K}{s(s+a)}$ demonstrates the variation of two roots as gain K changes. The complete root set for any specified value of K comprises the two roots derived from the root locus analysis plus the cancelled pole at $s = -b$, which remains fixed regardless of gain variations.

6.8 Effects of Delay on Root Locus

Time-delay systems are commonly encountered across numerous engineering applications, including chemical and biological processes, hydraulic systems, manufacturing operations, and transportation networks. These delays manifest in various system components such as actuators, sensors, control algorithms, and communication channels. The presence of time delays in dynamical systems has been identified as a primary source of performance degradation and can potentially destabilize an otherwise stable closed-loop system.

A representative control system incorporating time delay is illustrated in Figure 6.29, where the delay element $G_d(s) = e^{-Ts}$ is located within the forward path.

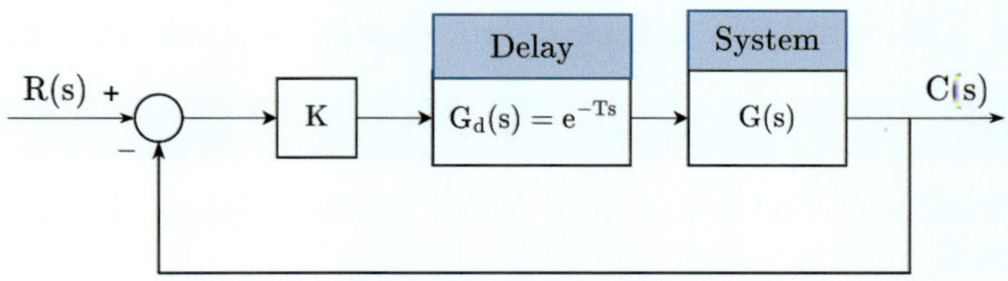

Figure 6.29: Control system with delay in forward path

The transfer function of a time delay element is inherently irrational in nature. Since root locus construction principles require the loop transfer function to possess rational characteristics (expressed as polynomial ratios), conventional root locus techniques cannot be directly applied to systems containing pure time delays.

Nevertheless, root locus methodology can be employed for analyzing delay systems when the delay term is approximated using a rational function. One of the most widely adopted delay approximation techniques is the **Padé approximation**, which represents the delay as:

$$G_d(s) = e^{-Ts} = \frac{e^{-Ts/2}}{e^{Ts/2}} \approx \frac{1 - \dfrac{Ts}{2}}{1 + \dfrac{Ts}{2}} = \frac{-(s - 2/T)}{s + 2/T}$$

Through the application of Padé approximation, the time-delay system depicted in Figure 6.29 becomes equivalent to the positive feedback configuration shown in Figure 6.30.

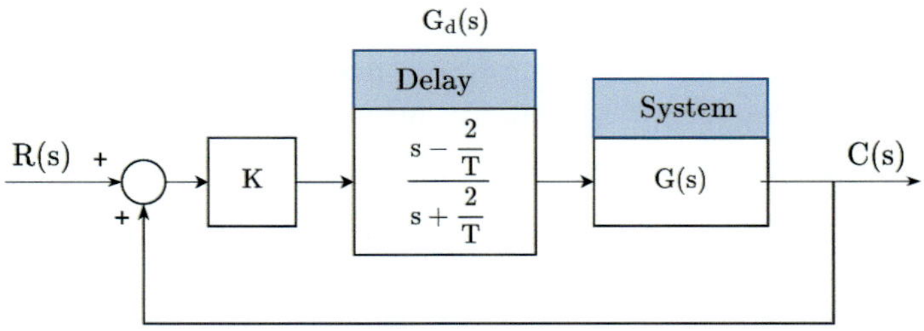

Figure 6.30: Control system with time delay approximated using Padé approximation

Consequently, **the methodologies developed for positive feedback systems can be applied to construct the root locus of time-delay systems**. This approach will be demonstrated through an example.

EXAMPLE 6.17

Problem: Sketch the root locus for the system shown in Figure 6.29 where

$$G(s) = \frac{1}{(s+4)(s+5)}, \quad \text{and} \quad G_d(s) = e^{-Ts} \tag{6.66}$$

For $T = 1$, determine the maximum gain K which ensures stability.

SOLUTION:

The open-loop transfer function for $T = 1$ is given by

$$KG_d(s)G(s) = \frac{-K(s-2)}{(s+2)(s+4)(s+5)}$$

The closed-loop characteristic equation is given by

$$1 - \frac{K(s-2)}{(s+2)(s+4)(s+5)} = 0$$

or

$$s^3 + 11s^2 + (38 - K)s + (2K + 40) = 0$$

From Routh's stability test, we find that the system is stable if $K < 29.0769$. Note that, without the delay, the system is stable for all values of gain K.

A sketch of the complete root locus is shown in Figure 6.31.

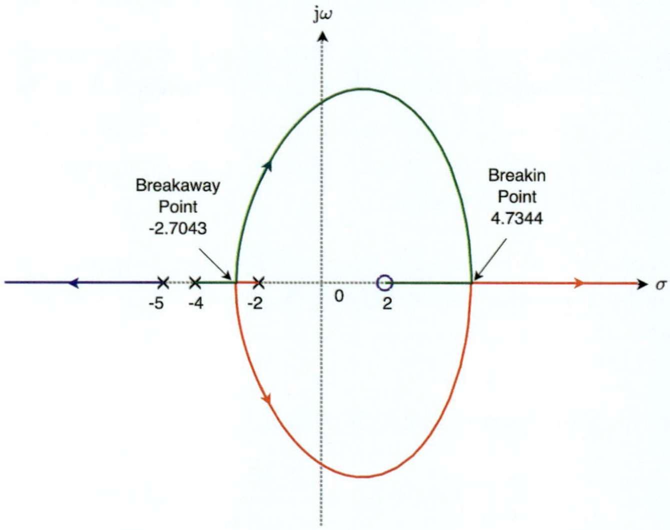

Figure 6.31: Root locus for time delay system of Eqn(6.66)

6.9 Design Concepts of Root Locus: Effects of Addition of Poles and Zeros

The root locus technique can be applied for both *analysis* and *design* of control systems. In analysis, the poles and zeros of the loop transfer function are known. From these, we can determine the closed-loop poles for any value of the gain K and hence evaluate the transient performance of the system.

In design, the root locus is used to select suitable parameter values that meet performance objectives. If the poles and zeros of the loop transfer function are fixed and only the gain K is adjustable, the designer has the freedom to choose an appropriate value of K to satisfy the design criteria. When a single parameter (such as K) is insufficient, additional parameters are introduced. From another perspective, it is important to understand how the root locus shape changes when poles or zeros are added to the open-loop transfer function. Such poles and zeros are typically introduced through compensators or controllers such as PI, PID, phase-lead, phase-lag, and lag-lead controllers (to be discussed in Chapter 8).

6.9.1 Effects of Adding Poles to Loop Transfer Function G(s)H(s)

Effects of Adding Poles on Root Locus

1. Adding a pole to the open-loop transfer function generally makes the closed-loop system less stable, as the root locus tends to shift toward the right half of the s-plane.
2. However, the addition of a pole to G(s)H(s) improves steady-state accuracy by reducing the steady-state error.

Consider an open-loop transfer function

$$G(s)H(s) = \frac{K}{s(s+a)}, \quad a > 0 \tag{6.67}$$

The centroid σ_a and asymptote angles θ_a are

$$\sigma_a = -\frac{a}{2}, \quad \theta_a = 90^\circ, 270^\circ\ (-90^\circ) \tag{6.68}$$

The root locus of Eqn(6.67), shown in Figure 6.32(i), indicates that the system remains stable for $0 < K < \infty$.

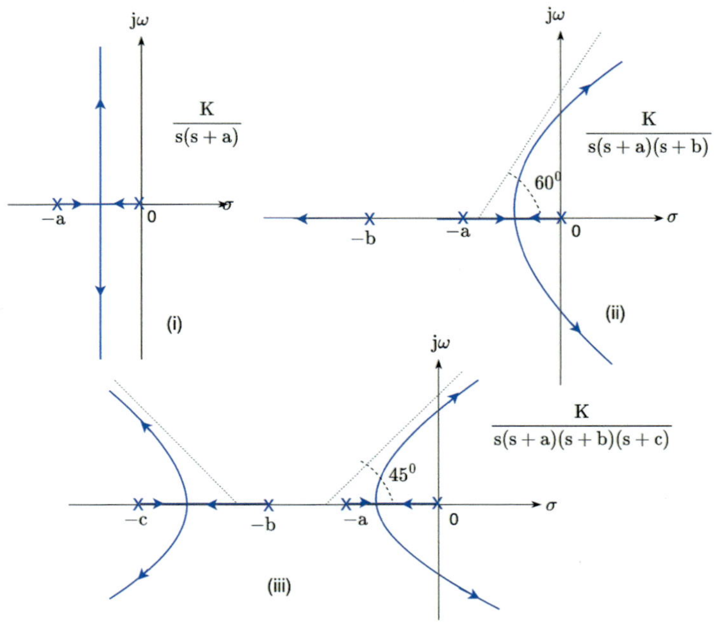

Figure 6.32: Effect of adding poles on the root locus

Now, add a pole at $s = -b$, with $b > a$. The open-loop transfer function becomes

$$G(s)H(s) = \frac{K}{s(s+a)(s+b)} \tag{6.69}$$

The centroid and asymptote angles are

$$\sigma_a = -\frac{(a+b)}{3}, \quad \theta_a = 60°, 180°, 300° \ (-60°) \tag{6.70}$$

Comparing Eqns(6.70) and Eqn(6.68), the centroid shifts from $-\frac{a}{2}$ to $-\frac{a+b}{3}$, while the asymptote angles change from $\pm 90°$ to $\pm 60°$. The root locus of Eqn(6.69), shown in Figure 6.32(ii), reveals that the loci bend toward the right half-plane. Furthermore, the system becomes unstable if $K > ab(a+b)$, whereas the system in Eqn(6.67) is stable for all $K > 0$.

Adding another pole at $s = -c$, with $c > b$, gives

$$G(s)H(s) = \frac{K}{s(s+a)(s+b)(s+c)} \tag{6.71}$$

The centroid and asymptote angles are

$$\sigma_a = -\frac{(a+b+c)}{4}, \quad \theta_a = \pm 45°, \ \pm 135° \tag{6.72}$$

The root locus of Eqn(6.71), shown in Figure 6.32(iii), illustrates that the complex branches bend further into the right half-plane. Thus, as more poles are added, the system stability deteriorates progressively. In a later chapter, we will show that adding a pole to the loop transfer function has an effect similar to introducing **integral control** into the system.

6.9.2 Effects of Adding Zeros to Loop Transfer Function $G(s)H(s)$

Effects of Adding Zeros on Root Locus

Adding a zero in the left half-plane to the open-loop transfer function generally enhances system stability. The root locus shifts toward the left half of the s-plane, moving the closed-loop poles further away from the imaginary axis.

Consider the open-loop transfer function

$$G(s)H(s) = \frac{K}{s(s+a)}, \quad a > 0 \tag{6.73}$$

The corresponding root locus is shown in Figure 6.32(i), with a breakaway point at $s = -\frac{a}{2}$.

Now, let us add a zero at $s = -b$ with $b > a$. The open-loop transfer function becomes

$$G(s)H(s) = \frac{K(s+b)}{s(s+a)} \tag{6.74}$$

With the addition of a real zero in the left half-plane, the complex-conjugate branches of the root locus bend further left, forming a circular arc, as shown in Figure 6.33(i).

If instead we add a pair of complex-conjugate zeros at $s = -\alpha \pm j\beta$, the loop transfer function becomes

$$G(s)H(s) = \frac{K(s^2+cs+d)}{s(s+a)} = \frac{K(s+\alpha+j\beta)(s+\alpha-j\beta)}{s(s+a)} \tag{6.75}$$

In this case, the root locus branches are attracted toward the locations of the complex zeros and bend leftward, as illustrated in Figure 6.33(ii).

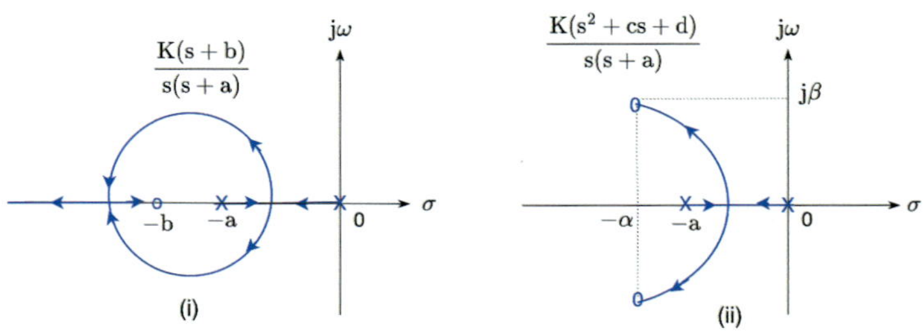

Figure 6.33: Effects of adding zeros on the root locus

Key Takeaways

- Adding a real zero shifts the root locus further left along the real axis, while adding complex-conjugate zeros redirects the branches along curved paths toward the zero locations. In both cases, the overall effect is to improve stability by pulling the closed-loop poles deeper into the left half-plane.
- We may therefore conclude that as more zeros are added, the stability of the system improves progressively. In a later chapter, we will show that the effect of adding a zero to the loop transfer function is closely related to introducing **derivative control** into the system.

6.10 Improving Transient Response via Gain Adjustment

The root locus method is a powerful design tool because it illustrates how the closed-loop poles move as the gain K varies. Since the transient response of a system is determined by the locations of its poles, adjusting K allows the designer to achieve desired performance specifications such as settling time, rise time, overshoot, and damping ratio.

This approach is effective provided that the desired pole locations lie on the root locus. If the locus does not pass through those locations, additional compensators must be introduced to reshape it, as discussed in Chapter 8.

A step-by-step procedure for designing a control system using root locus and gain adjustment is as follows:

Step 1: Plot the Root Locus. Sketch the root locus of the open-loop transfer function G(s)H(s).

> *Hint:* Poles closer to the imaginary axis yield slower responses, while poles further to the left provide faster settling and improved damping.

Step 2: Determine Desired Pole Locations. Translate transient response specifications (*settling time, overshoot, damping ratio*, etc.) into the corresponding desired pole locations in the s-plane.

Step 3: Select the Gain. Choose a value of K that positions the closed-loop poles at or near the desired locations identified in Step 2.

EXAMPLE 6.18

Problem: The block diagram of a position control system with velocity feedback is shown in Figure 6.34. Using the root locus method, determine the value of the open-loop gain K such that the closed-loop damping ratio is 0.5. Assume a = 2, b = 6, and $\delta = 0.1$.

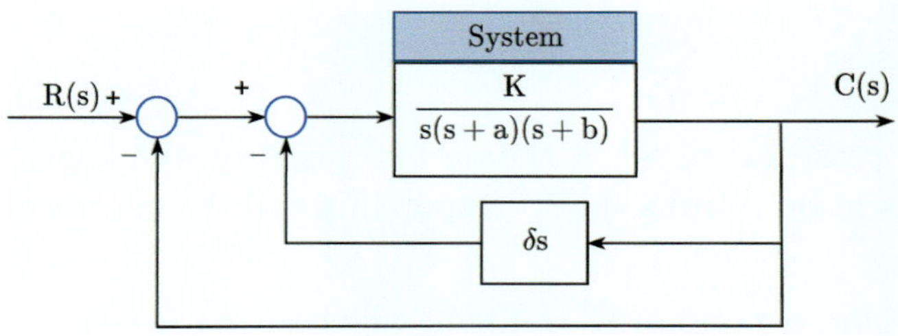

Figure 6.34: A third-order position control system with velocity feedback

Solution: By applying standard block diagram reduction, the closed-loop transfer function is

$$\frac{C(s)}{R(s)} = \frac{K}{s(s+a)(s+b) + K\delta s + K} \tag{6.76}$$

and the characteristic equation becomes

$$\Delta(s) = s(s+a)(s+b) + K\delta s + K = 0.$$

This can be rewritten as

$$1 + G'(s) = 1 + \frac{K'\left(s + \frac{1}{\delta}\right)}{s(s+a)(s+b)} = 0, \quad \text{where} \quad K' = K\delta.$$

Thus, for $a = 2$, $b = 6$, and $\delta = 0.1$, we analyze the auxiliary system

$$G'(s) = \frac{K'(s+10)}{s(s+2)(s+6)}.$$

The root locus can now be sketched using the procedure illustrated earlier. The real-axis segments of the root locus lie between $s = 0$ and $s = -2$, and between $s = -6$ and $s = -10$. Two branches go to infinity along asymptotes at $\pm 90°$, intersecting the real axis at $+1$.

Using the Routh–Hurwitz criterion, the imaginary-axis crossing occurs at $s = \pm j\sqrt{60} = \pm j7.74$ when $K = 48$. The breakaway point is located at $s = -0.9562$ (obtainable using MATLAB or the analytical procedure described earlier).

To satisfy $\xi = 0.5$, draw a line at $\cos^{-1}(\xi) = 60°$ with respect to the negative real axis. The intersection point P of this line with the root locus gives the desired pole location. The corresponding gain K can be determined manually or using MATLAB.

In MATLAB, the command

```
[K, P] = rlocfind(sys)
```

displays a crosshair to select the desired point on the locus. Clicking at the required location returns the corresponding gain K and the closed-loop poles in P.

From the plot, we obtain $K' = 1.866$, which gives

$$K = \frac{K'}{\delta} = \frac{1.866}{0.1} = 18.66.$$

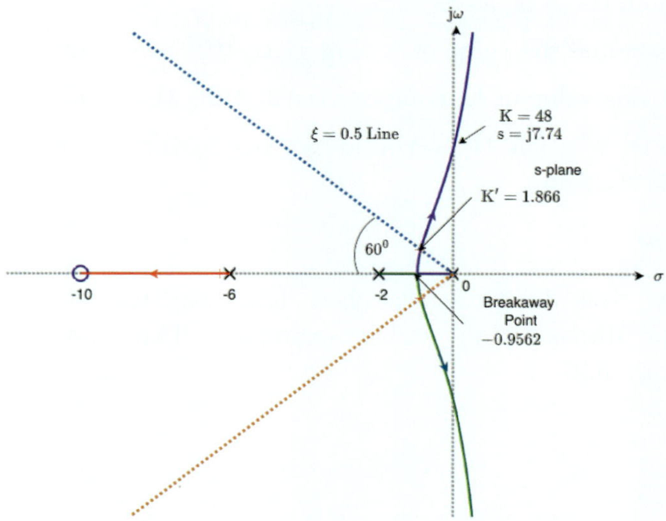

Figure 6.35: Root locus of the system in Eqn(6.76)

CONCEPT REINFORCEMENT PROBLEM 6.13

Problem:

The block diagram of a position control system with velocity feedback is shown in Figure 6.36. Using the root locus method, determine the value of α so that the damping ratio of the closed-loop poles may be 0.6.

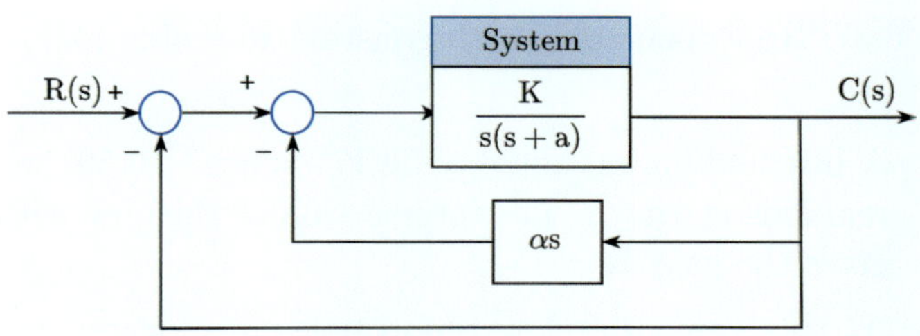

Figure 6.36: A second order position control system with velocity feedback

EXAMPLE 6.19

Problem:

The open-loop transfer function of a unity feedback system is

$$G(s) = \frac{K}{(s+2)(s+6)(s+10)}$$

a. Sketch the root locus.

b. Determine the value of K that gives 10% overshoot.

c. For this value of K, compute the settling time and peak time.

d. Assess whether the second-order approximation is valid for the selected gain.

Solution:

a. The root locus is sketched following the standard procedure discussed in earlier examples. The plot is shown in Figure 6.37.

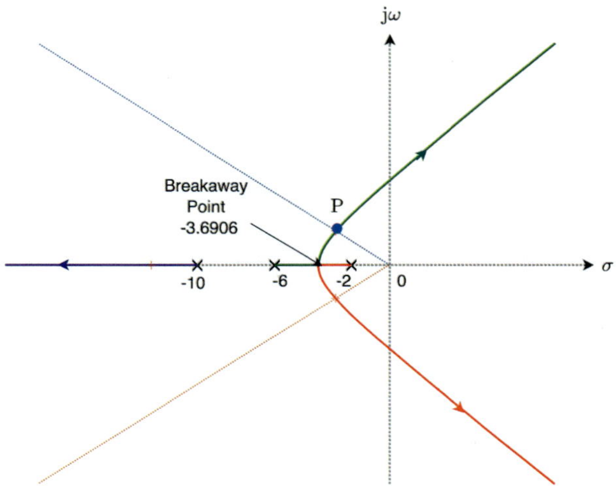

Figure 6.37: Root locus plot with constant damping ratio $\xi = 0.59$ line

b. A line making an angle of $53.17° = \cos^{-1}(0.59)$ with the negative real axis is drawn. The intersection of this line with the root locus gives the gain K = 154.

c. At this gain, the closed-loop poles are located at -12.2861, and $-2.8569 \pm j3.6928 = -\sigma \pm j\omega_d$. The settling time and peak time are

$$T_s = \frac{4}{\sigma} = \frac{4}{2.8569} = 1.40 \text{ s}, \qquad T_p = \frac{\pi}{\omega_d} = \frac{\pi}{3.6928} = 0.857 \text{ s}$$

d. The second-order approximation is valid since the non-dominant pole is more than five times farther from the dominant poles.

CONCEPT REINFORCEMENT PROBLEM 6.14

Problem

Consider the unity feedback system shown in Figure 6.38

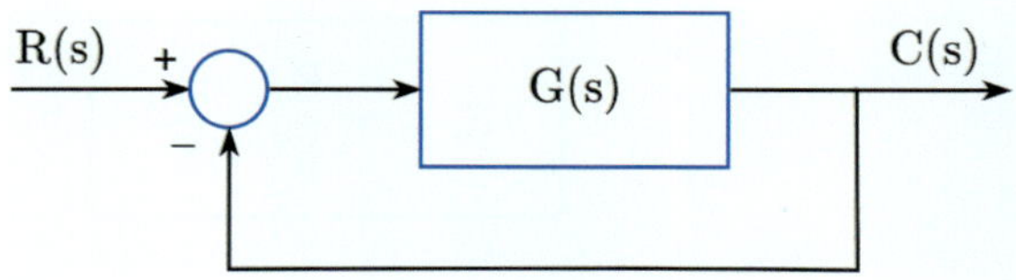

Figure 6.38: Unity feedback control system

a. Sketch the root locus of this system when

$$G(s) = \frac{K}{(s+1)(s+2)(s+4)(s+5)}$$

as the gain K varies from zero to infinity.

b. Find the breakaway points

c. Find the range of gain K which ensures stability

d. Find the gain K which gives a damping ratio of 0.7 for the dominant second-order pair.

To improve the stability, consider the modified open-loop transfer function with an added zero:

$$G(s) = \frac{K(s+z)}{(s+1)(s+2)(s+4)(s+5)}$$

It is desired that the root locus to cross the $j\omega$-axis at j6.0 rad/sec.

e. Determine the value of z and sketch the new root locus.

f. Repeat part(c) for the new locus.

g. Compare the results of Part (c) and Part (f) and comment on your observations about the transient response.

6.11 Sensitivity of the Root Locus

The sensitivity of the root locus refers to how the locations of the poles of a closed-loop system change in response to variations in system parameters, such as gain or the locations of open-loop poles and zeros. This sensitivity analysis is crucial for control system design, as it helps to understand how stable the system will remain under parameter changes and disturbances. As discussed before in Chapter 1, the sensitivity is the ratio of the fractional change in a function to the fractional change in a parameter as the change in the parameter approaches zero. Therefore, the sensitivity of **closed-loop poles to the system gain** K can be defined as:

$$S_{s,K} = \frac{K}{s} \cdot \frac{ds}{dK} \tag{6.77}$$

where $S_{s,K}$ is the sensitivity of the closed-loop pole s with respect to the gain K. The actual change in the closed-loop poles can be approximated as

$$\Delta s = s\left(S_{s,K}\right) \cdot \frac{\Delta K}{K} \tag{6.78}$$

where Δs is the change in pole location, and $\Delta K/K$ is the fractional change in the gain K.

Similarly, the sensitivity of **closed-loop pole** s **to a zero location** z is defined as:

$$S_{s,z} = \frac{z}{s} \cdot \frac{ds}{dz} \tag{6.79}$$

where $S_{s,z}$ is the sensitivity of the closed-loop pole s with respect to a zero at z and $\frac{ds}{dz} \approx \frac{\Delta s}{\Delta z}$ quantifies how much the pole s moves in response to a small change in the zero.

The following example demonstrates the computation of sensitivity in practice.

EXAMPLE 6.20

Problem: Compute the pole sensitivity due to a change in gain for a unity feedback system with forward path transfer function

$$G(s) = \frac{K}{s(s+a)}$$

Determine the change in pole locations for a 10% increase in K, assuming K = 2 and a = 4.

Solution:

The closed-loop characteristic equation is

$$1 + G(s) = 0 \quad \Rightarrow \quad s^2 + as + K = 0$$

The sensitivity of the closed-loop pole to gain K is

$$\begin{aligned} S_{s,K} &= \frac{ds}{dK} \cdot \frac{K}{s} \\ &= \frac{-1}{2s+a} \cdot \frac{K}{s} = \frac{-K}{s(2s+a)} \qquad \left(\because \ \tfrac{ds}{dK} = -\tfrac{1}{2s+a}\right) \end{aligned}$$

The poles are

$$s_{1,2} = \frac{-a \pm \sqrt{a^2 - 4K}}{2} \tag{6.80}$$

For Pole s_1:

$$2s_1 + a = \sqrt{a^2 - 4K}$$

$$S_{s_1,K} = \frac{-2K}{\left(-a + \sqrt{a^2 - 4K}\right)\left(\sqrt{a^2 - 4K}\right)} \tag{6.81}$$

For Pole s_2:

$$S_{s_2,K} = \frac{2K}{\left(-a - \sqrt{a^2 - 4K}\right)\left(\sqrt{a^2 - 4K}\right)} \tag{6.82}$$

Change in the poles: For a 10% change in K, the pole shifts are

$$\Delta s_1 \approx 0.10 \cdot S_{s_1,K} \cdot s_1, \qquad \Delta s_2 \approx 0.10 \cdot S_{s_2,K} \cdot s_2 \tag{6.83}$$

Substituting $K = 2$ and $a = 4$:

$$s_1 = -0.586, \quad s_2 = -3.414, \quad S_{s_1,K} \approx 1.215, \quad S_{s_2,K} \approx -0.207$$

Therefore, for a 10% increase in K:

$$\Delta s_1 \approx -0.071, \qquad \Delta s_2 \approx +0.071$$

Thus, the poles shift as

$$s_1\!: -0.586 \ \rightarrow \ -0.657, \qquad s_2\!: -3.414 \ \rightarrow \ -3.343$$

Conclusion: The first pole s_1 is more sensitive to changes in K compared to s_2. As the gain increases by 10%, s_1 moves further left (more stable), while s_2 moves slightly right (less stable).

CONCEPT REINFORCEMENT PROBLEM 6.15

Problem:

Consider the unity feedback system shown in Figure 6.38 where

$$G(s) = \frac{K}{s(s+4)(s+6)}$$

Evaluate the pole sensitivity of the closed-loop system when the second-order under-damped closed-loop poles are set for

a. $\xi = 0.591$

b. $\xi = 0.3$

c. Which of the above two is the least sensitive?

6.12 Multiple Choice Skills Test Questions

M 6.1. If $KG(s)H(s) = 10\angle 180^0$, then s is a point on the root locus provided

A. $K = 10$ **B.** $K = 20$ **C.** $K = \frac{1}{10}$ **D.** $K = \frac{1}{5}$

M 6.2. If the gain of the system is reduced to a zero value, the roots of the system in the s-plane

A. coincides with zeros

B. coincides with poles

C. move away from zeros

D. has no relation with poles or zeros

M 6.3. The characteristic equation of a feedback control system is given by $s^3 + 6s^2 + (K+8)s + 3K = 0$ where $K > 0$. The centroid of the root locus in the s-plane is located at

A. (-3,0) **B.** (+3,0) **C.** (-1.5,0) **D.** (-2,0)

M 6.4. The open-loop transfer function of a control system is given by

$$G(s)H(s) = \frac{K}{s(s+1)(s+3)(s+5)}$$

Which of the following statements related to root loci conditions is/are correct?

1. There will be 3 asymptotes

2. There will be four separate root loci
3. Asymptotes intersect the real axis at $\sigma = -1/3$
4. Angle of asymptotes are $45^0, 135^0, 224^0$ and 315^0

A. 2 and 4 **B.** 4 alone **C.** 1 and 3 **D.** 1 and 2

M 6.5. Which of the following is NOT on the root locus of a system with an open-loop transfer function

$$G(s)H(s) = \frac{K}{s(s+2)(s+5)}$$

A. -1.5 **B.** -5 **C.** $-\infty$ **D.** -3

M 6.6. The open-loop transfer function of a unity feedback system is

$$G(s) = \frac{K}{s(s+1)(s+2)}$$

The gain K for which $s = -0.5$ is

A. 0 **B.** 1.875 **C.** 0.375 **D.** 24

M 6.7. The breakaway point of the root locus of the system with open-loop transfer function

$$G(s)H(s) = \frac{K}{(s+2)(s+6)} \text{ is}$$

A. -1 **B.** -4 **C.** -3 **D.** -2

M 6.8. Consider the points $s_1 = -3 + j4$ and $s_1 = -3 - j2$ in the s-plane. Then, for a system with an open-loop transfer function

$$G(s)H(s) = \frac{K}{(s+1)^4}$$

A. s_1 is on the root locus, but not s_2

B. s_2 is on the root locus, but not s_1

C. Both s_1 and s_2 are on the root locus

D. Neither s_1 and s_2 are on the root locus

M 6.9. Which of the following sketches of root locus shown in Figure 6.39 is a valid root locus?

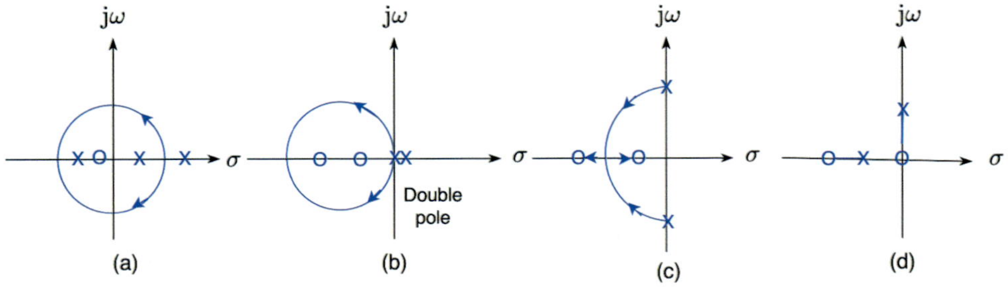

Figure 6.39

A. a **B.** b **C.** c **D.** d

M 6.10. The root locus plot of a system is shown in Figure 6.40. The open-loop transfer function of the system is

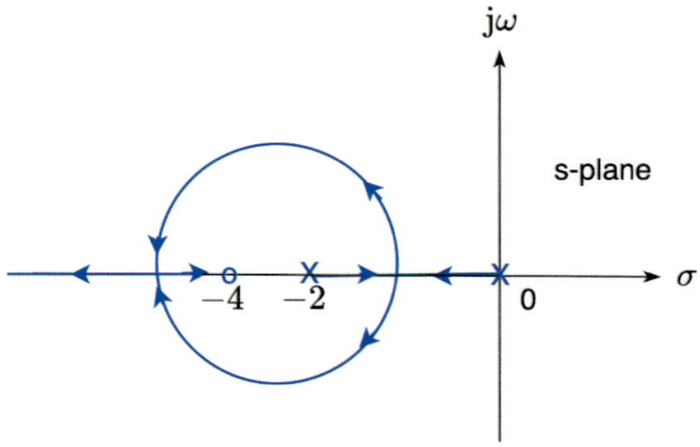

Figure 6.40

A. $\frac{K}{s(s+2)(s+4)}$ **B.** $\frac{K(s+4)}{s(s+2)}$ **C.** $\frac{K(s+2)}{s(s+4)}$ **D.** $\frac{K}{s(s+4)}$

M 6.11. The root locus plot of a system is shown in Figure 6.41. The system is underdamped only if

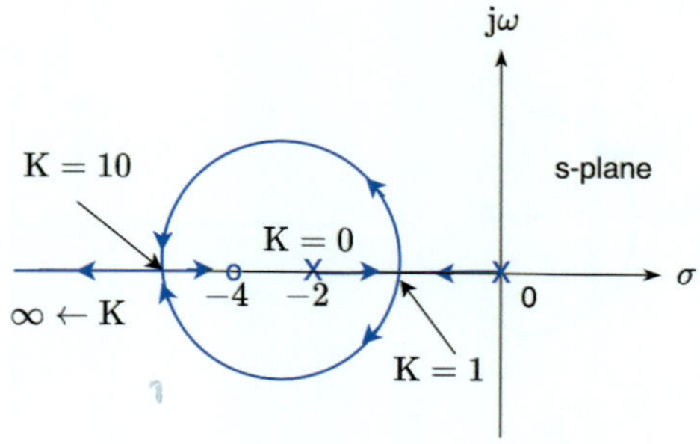

Figure 6.41

A. $1 < K < 10$ **C.** $K > 10$

B. $0 \leq K < 1$ and $K < 10$ **D.** $K = 0.5$

M 6.12. Consider the following four systems with open-loop transfer functions

$$\textbf{S1:}\ \frac{K(s-4)}{(s+1)(s+2)},\quad \textbf{S2:}\ \frac{K}{(s+1)(s^2+4)},\quad \textbf{S3:}\ \frac{K}{(s^2+16)},\quad \textbf{S4:}\ \frac{K(s+2)(s+4)}{s^2+9}$$

The correct correspondence between the root locus paths in Figure 6.42 and the systems is

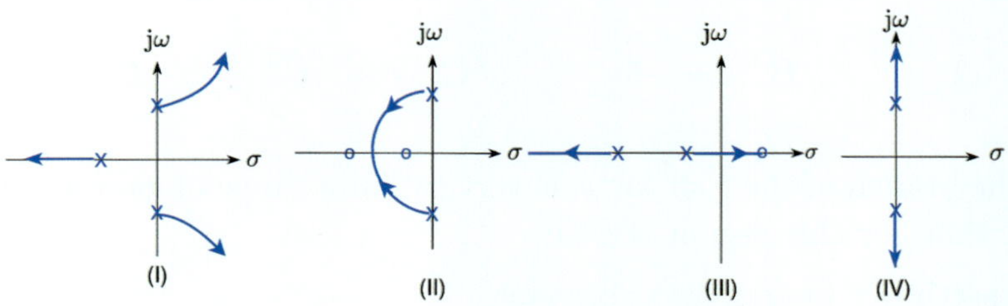

Figure 6.42

A. S1 – III, S2 – I, S3 – IV and S4 – II

B. S1 – II, S2 – IV, S3 – I and S4 – III

C. S1 – I, S2 – III, S3 – IV and S4 – II

D. S1 – III, S2 – IV, S3 – II and S4 – I

M 6.13. The open-loop transfer function of a system is

$$G(s)H(s) = \frac{K(s+3)}{(s+1)^2}$$

The correct root locus of the system is (see Figure 6.43)

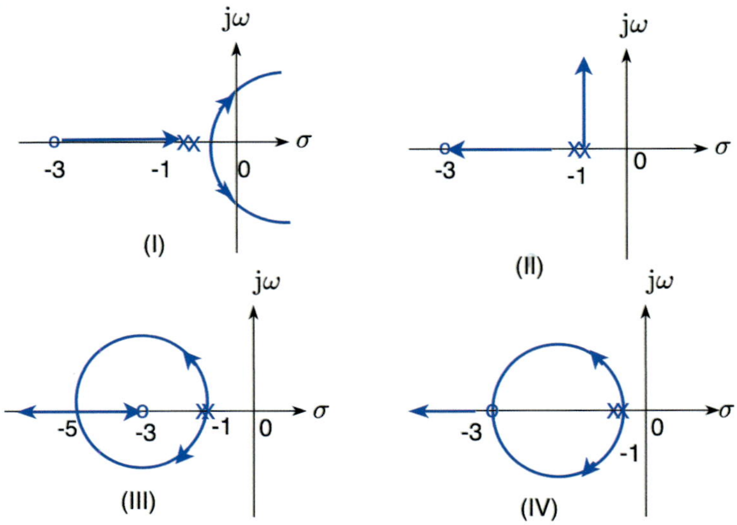

Figure 6.43

A. II **B.** I **C.** III **D.** IV

M 6.14. The closed-loop characteristic equation of a system is $s^2 + 6Ks + 2s + 5 = 0$ where $K > 0$. The break-in point of the root locus on the real axis is

A. $s = -1$ **B.** $s = -5$ **C.** $s = -\sqrt{5}$ **D.** $s = \sqrt{5}$

M 6.15. If the branch of the root locus is vertical for a range of gain on the vertical section, then for this region of gain

A. the settling time remains constant

B. the peak time remains constant

C. the steady state error remains constant

D. the peak overshoot remains constant

M 6.16. Addition of poles closer to the imaginary axis to the open-loop transfer function makes the closed-loop system

A. more stable and damped

B. less stable and increased damping

C. less stable, increased oscillatory response with increased settling time

D. over damped

Answers

M 6.1. C **M 6.2**. B **M 6.3**. C **M 6.4**. A **M 6.5**. D **M 6.6**. C **M 6.7**. B **M 6.8**. B **M 6.9**. C **M 6.10**. B **M 6.11**. A **M 6.12**. A **M 6.13**. C **M 6.14**. C **M 6.15**. A **M 6.16**. C

6.13 Knowledge Assessment Problems

KaP 6.1 Compute the angles of asymptotes and centroid of the asymptotes of the root loci of the linear system with the following open-loop transfer functions when K varies from 0 to ∞.

a. $G(s)H(s) = \dfrac{K}{s(s+a)(s+b)}$

b. $G(s)H(s) = \dfrac{K}{s(s+4)(s^2+6s+25)}$

c. $G(s)H(s) = \dfrac{K(s^2+2s+2)}{(s-2)(s+0.5)(s+1.5)(s+4)}$

d. $G(s)H(s) = \dfrac{K(s+5)}{(s^2+2s+3)(s^2+4s+20)}$

KaP 6.2 For the loop transfer of the following systems, determine the angle of departure at the designated pole for $K > 0$.

S1: $\dfrac{Ks(s+2)}{s^2+6s+10}$ at $s = -3 + j1$

S2: $\dfrac{K(s^2+4s+13)}{s(s^2+1)}$ at $s = +j1$

S3: $\dfrac{K(s^2+16)(s+1)}{(s^2+4)(s+10)}$ at $s = +j2$

S4: $\dfrac{K(s^2+2s+2)}{(s+2)(s^2+8s+52)}$ at $s = -4 + j6$

KaP 6.3 For the loop transfer of the following systems, determine the angle of arrival at the designated zero for $K > 0$.

S1: $\dfrac{K(s^2-6s+10)}{(s+1)(s+2)}$ at $s = 3 + j1$

S2: $\dfrac{K(s^2+4s+8)}{s(s^2+8s+52)}$ at $s = -2 + j2$

S3: $\dfrac{K(s^2+4s+5)}{(s^2+2s+5)(s+3)(s+4)}$ at $s = -2 + j1$

S4: $\dfrac{K(s^2-4s+8)}{((s+2)(s+4)(s+5)(s+6)}$ at $s = 2 + j2$

KaP 6.4 Sketch the root loci for $K \geq 0$ of the systems with the following characteristic equations. Do not show calculations

a. $s^2 + (K + 1)s + 6K = 0$
b. $s^3 + 4s^2 + 8s + K = 0$
c. $(s^2 - 1)(s + 3) + K(s^3 + 2s^2 + 2s) = 0$
d. $s^3 + K(s^2 + 7s + 10) = 0$

KaP 6.5 Sketch the root locus for the unity feedback system with the following forward path transfer functions:

a. $G(s) = \dfrac{K}{s(s+1)(s+2)(s+4)}$

b. $G(s) = \dfrac{K(s+2)(s+5)}{s(s+1)}$

c. $G(s) = \dfrac{K(s^2+4)}{s^2}$

d. $G(s) = \dfrac{K(s-2)}{(s+1)(s+5)(s+20)}$

KaP 6.6 A unity feedback system has the loop transfer function

$$G(s) = \frac{K(s+2)}{s(s+1)}$$

a Determine the breakaway and break-in points of the root locus and sketch the root locus for $K > 0$.

b. Determine the gain K when the two characteristic roots have a ξ of $1/\sqrt{2}$.

KaP 6.7 Sketch the root locus for the unity feedback system with the forward path transfer function

$$G(s) = \frac{K(s^2 - a)}{s^2 + b}, \quad a, b > 0$$

and determine the value of gain K such that the system is stable.

KaP 6.8 Consider the unity feedback system shown in Figure 6.38 where

$$G(s) = \frac{K(s^2 + a)}{(s+b)(s+c)}, \quad a, b, c > 0$$

i. Sketch the root locus of the system
ii. Compute the breakaway/break in point for $a = 4$, $b = 1$, and $c = 2$,
iii. Compute the gain K when the roots are equal.

KaP 6.9 For the unity feedback system with forward path transfer function

$$G(s) = \frac{K(s-2)(s-4)}{s(s+3)} \tag{6.84}$$

sketch the root locus and find the following:

a. The breakaway and break-in points
b. The value of gain at the $j\omega$-axis crossing
c. The range of gain that ensures system stability.

KaP 6.10 The open-loop transfer function of a unity feedback system is given by

$$G(s)H(s) = \frac{K}{(s+a)^{\alpha}}, \quad a > 0 \tag{6.85}$$

Construct the root loci of the characteristic equation of the closed-loop system for $K > 0$ with

i $\alpha = 1$ ii $\alpha = 2$ iii $\alpha = 3$ iv $\alpha = 4$ v $\alpha = 5$

For each case, determine the angle of asymptotes, the ultimate gain and ultimate frequency (if any), breakaway points (if any).

KaP 6.11 A unity feedback system has the loop transfer function

$$G(s)H(s) = \frac{(1+p)s - p}{s^2 + 4s + 29}$$

Sketch the root locus as p varies $0 < p < \infty$

KaP 6.12 Sketch the root locus of a system with open-loop transfer function

$$G(s)H(s) = \frac{Ks(s+4)}{s^2 + 2s + 5}$$

a. Sketch the root locus
b. Find the gain when the roots are both equal

KaP 6.13 A unity feedback system has the loop transfer function

$$G(s) = \frac{K(s+4)}{s(s+2)} \tag{6.86}$$

a Determine the breakaway and break-in points of the root locus and sketch the root locus for $K > 0$.

b. Determine the gain K when the two characteristic roots have a ξ of $1/\sqrt{2}$.

KaP 6.14 The loop transfer function of a unity feedback system

$$L(s) = G(s)H(s) = \frac{Ks}{(s+5)(s^2 - 2s + 5)}$$

Sketch the root locus and determine the maximum damping ratio ξ of the stable complex roots and the corresponding gain K

KaP 6.15 Sketch the root locus of a unity feedback system with forward path transfer function

$$G(s) = \frac{K(s-1)(s-3)}{s^2+4s+13} \tag{6.87}$$

as the gain K varies from zero to infinity and find the following:

a. The break-in point
b. The gain at which the root locus crosses the imaginary axis and the corresponding frequency.
c. The angle of departure from the complex poles
d. The gain K where the root locus crosses the 0.5 damping ratio line.

KaP 6.16 The loop transfer function of a unity feedback system. Sketch the root locus of a unity feedback system with a forward path

$$L(s) = G(s) = \frac{K(s+0.5)}{s(s-0.5)(s+3)} \tag{6.88}$$

a. Determine the range of gain K for stability.
b. Sketch the root locus as the gain K varies from zero to infinity.
c. Determine the maximum value of damping ratio ξ of the stable complex roots.

KaP 6.17 Sketch the root locus for a system whose open-loop transfer function is given as:

$$G(s)H(s) = \frac{K(s+2)}{s(s-5)} \tag{6.89}$$

KaP 6.18 For the feedback system shown in Figure 6.38,

$$G(s) = \frac{K(s+1)}{s^2}$$

Sketch the root locus and determine the value of K which results in the damping ratio $\xi = 0.707$.

KaP 6.19 For the unity feedback system shown in Figure 6.38 where

$$G(s) = \frac{K}{s(s+4)(s+6)(s+10)}$$

do the following:

a. Sketch the root locus

b. Find the range of K that will ensure closed-loop stability from the root locus. Do not use Routh's test.

c. Find the value of K that will give 10% overshoot

d. Locate all the poles and investigate if the second-order approximation is valid.

KaP 6.20 Compute the pole sensitivity due to a change in the zero z for a unity feedback system with forward path transfer function

$$G(s) = \frac{K(s+z)}{s(s+a)}$$

Compute the sensitivity for $K = 10$, $a = 5$ and $z = 2$. Comment on your observations.

Chapter 7

Frequency Domain Analysis of Linear Systems

Switching domains is like changing languages-same story, completely different accent, but now the resonance finally makes sense.

DESIRED LEARNING OUTCOMES

Upon completion of this chapter, readers should be able to:

- Understand the necessity of linear system analysis in the frequency domain
- Compute and interpret commonly used frequency domain specifications
- Establish the correlation between time domain and frequency domain specifications
- Analyze the effects of pole and zero addition on frequency domain specifications and system performance
- Construct and interpret system behavior using Bode plots, polar plots, and Nyquist plots
- Investigate the relative stability of linear systems using gain and phase margin concepts
- Apply constant M-circles (magnitude circles) and N-circles (phase circles) for closed-loop frequency response analysis

DOI: 10.1201/9781003571308-7

7.1 Introduction

We have studied the performance of control systems in the time domain due to certain test signals and have analytically derived various performance measures such as *maximum overshoot, peak time, rise time, settling time*, etc., considering a prototype second-order system. We have investigated the absolute stability of a system using Routh-Hurwitz stability criteria without computing the exact roots of the closed-loop characteristic equation. Furthermore, we have analyzed how closed-loop poles move in the complex s-plane with parameter changes using the Root Locus method. These methods provide sufficient insight into the transient behavior of closed-loop systems, which guides controller design. However, time domain methods have certain limitations, which are summarized as follows:

1. It requires an accurate mathematical model of the system.
2. For higher-order systems, time domain analysis becomes cumbersome. The analytical expressions for various performance measures, such as *maximum overshoot, peak time, rise time*, etc., which are derived for second-order systems, will no longer be accurate for higher-order systems.
3. It does not apply to systems whose open-loop transfer function is not a rational function of the complex variable s. Systems with pure delay belong to this category because for these systems, the transfer function is the ratio of two polynomials multiplied by the factor $e^{-T_d s}$, where T_d is the transport lag in the system. For such cases, we cannot apply the Routh-Hurwitz criterion to investigate its stability. The root locus method can be applied only after replacing the irrational factor $e^{-T_d s}$ by an approximate rational function (say, Pade approximation).
4. It is not convenient to analyze the effects of noise and disturbance on the system performance.
5. Analysis is restricted to specific inputs such as step, ramp, and parabolic.

An alternate approach of analysis and design of control systems is based on the frequency domain, where we excite the system by a sinusoidal input of varying frequency and measure the steady state response. Some of the motivating factors in favor of frequency domain analysis are:

Motivation for Frequency Domain Analysis

Frequency Domain Analysis:Motivation

1. It does not require any mathematical model of the system. The frequency response of a stable system can be precisely measured experimentally by using readily available sinusoidal signal generators and high-precision gain-phase meters.

2. If there is a need to get a transfer function model of the system, it can be obtained from the experimentally collected frequency response data. Determination of the transfer function model from the step response data is difficult and often inaccurate.

3. This approach can easily be applied to systems with time delay whose transfer function model is not rational.

4. There exist correlating relations between the frequency domain and time domain performances in linear systems. Therefore, it is possible to predict the time domain properties of the system from frequency domain characteristics.

5. The availability of several graphical tools makes the controller design easy in the frequency domain, both for lower and higher-order systems.

6. The effects of noise and parameter variations can be easily assessed in the frequency domain rather than the time domain.

7. The relative stability of a system can easily be investigated in the frequency domain, in contrast to the Routh-Hurwitz criteria, which provide information about the absolute stability of a system in the time domain.

Despite the above motivating factors, frequency domain methods have certain limitations.

Limitations of Frequency Domain Analysis

1. It is not preferred for systems with large time constants. This is because the frequency response of a system is its steady-state response to a sinusoidal input. If the system time constant is large, it takes a long time to reach the steady state. Hence, for slow systems, the time-domain approach may be preferred.

2. This approach is not suitable for uninterruptible systems; to collect frequency response data, the system has to be isolated, and a sinusoidal

signal has to be applied to the system. For such systems, time domain analysis using either step or impulse input is preferred.

We must emphasize that the frequency domain and time domain approaches should not be considered as rivals; instead, they are complementary to each other. The various analysis and design methods, such as Routh-Hurwitz stability criterion, Root-locus method (in time domain), Nyquist criterion, Bode and Polar plot (in frequency domain), are not mutually exclusive. Each of these methods in different domains has its strengths and weaknesses. It is the designer who has to judiciously exploit the advantages of each method and approach.

7.2 The Steady State Frequency Response

The frequency response of a stable system is the steady-state response to a sinusoidal input. In this method, we vary the frequency of the input signal over a certain range and obtain the resulting steady-state response. Consider a stable, linear, time-invariant system shown in Figure 7.1 with input u(t) and output y(t).

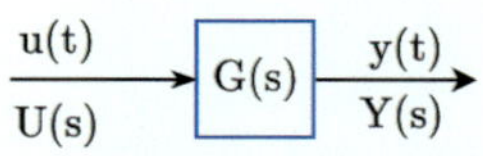

Figure 7.1: A linear time-invariant system

When it is excited by a sinusoidal input with amplitude A and frequency ω

$$\mathrm{u(t) = A}\sin\omega\mathrm{t} \tag{7.1}$$

the steady state output of the system y(t) will be a sinusoid of the same frequency ω and is expressed as:

$$\mathrm{y(t) = Y}\sin(\omega\mathrm{t} + \phi) \tag{7.2}$$

where Y and ϕ denote respectively the amplitude and phase of the output.

The output signal of a typical stable linear time-invariant system to a sinusoidal input is shown in Figure 7.2.

The amplitude and phase can be computed from the transfer function model of the system. Let G(s) is the transfer function of the system, then

$$\mathrm{Y(s) = G(s)U(s)} \quad \Longrightarrow \quad \mathrm{Y(j\omega) = G(j\omega)U(j\omega)} \tag{7.3}$$

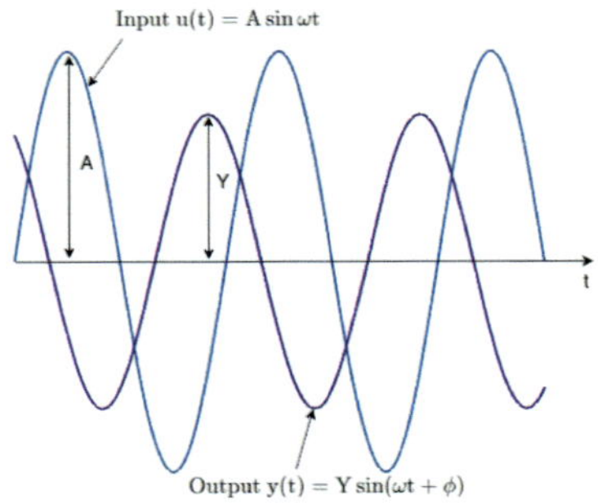

Figure 7.2: Input and output signals of a typical linear time-invariant system

From Eqn(7.3), the magnitude relation between the input and output:

$$\left|\mathrm{Y}(\mathrm{j}\omega)\right| = \left|\mathrm{G}(\mathrm{j}\omega)\right|\left|\mathrm{U}(\mathrm{j}\omega)\right| \tag{7.4}$$

and the phase relation

$$\angle\mathrm{Y}(\mathrm{j}\omega) = \angle\mathrm{G}(\mathrm{j}\omega) + \angle\mathrm{U}(\mathrm{j}\omega) \tag{7.5}$$

If the input is of the form of Eqn(7.1) and the output is of the form Eqn(7.2), then

$$\mathrm{Y} = \mathrm{A}\left|\mathrm{G}(\mathrm{j}\omega)\right| \quad \text{and } \phi = \angle\mathrm{G}(\mathrm{j}\omega) \tag{7.6}$$

Key Points

a. If a stable, linear, time-invariant system is excited by a sinusoidal input, then at steady state, the output will be sinusoidal.

b. The frequency of the output will be equal to that of the input. That is, in a linear time-invariant system, the energy transfer from the input to the output occurs at the same frequency. However, when the system is nonlinear, or linear time-varying, this may not be the case [a].

c. The amplitude and phase of the output will, in general, be different from those of the input.

d. The sinusoidal transfer function of any linear time-invariant system can be obtained by replacing jω with s in the transfer function of the system.

[a]This is discussed in the companion volume on digital and nonlinear systems, which is under development

EXAMPLE 7.1

Problem Determine the steady state output equation for the system shown in Figure 7.3 when the system is excited by an input $u(t) = A \sin \omega t$.

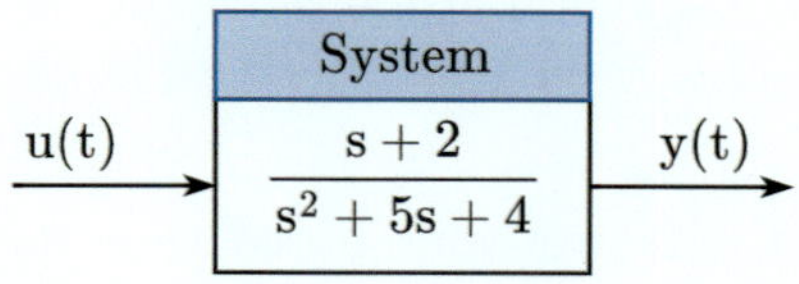

Figure 7.3: Second order control system

SOLUTION:

First let us compute $G(j\omega)$. Given

$$G(s) = \frac{s+2}{s^2+5s+4} = \frac{s+2}{(s+1)(s+4)} \implies G(j\omega) = \frac{j\omega+2}{(j\omega+1)(j\omega+4)}$$

Its magnitude $\left|G(j\omega)\right| = M(\omega) = \dfrac{\sqrt{\omega^2+2^2}}{\sqrt{(\omega^2+1^2)(\omega^2+4^2)}}$

and phase $\angle G(j\omega) = \phi(\omega) = \tan^{-1}\dfrac{\omega}{2} - \tan^{-1}\dfrac{\omega}{1} - \tan^{-1}\dfrac{\omega}{4}$

Thus, for the input $u(t) = A \sin \omega t$, the steady state value of the output is

$$y_{ss}(t) = AM(\omega)\sin(\omega t + \phi(\omega))$$

CONCEPT REINFORCEMENT PROBLEM 7.1

Consider a linear time-invariant (LTI) system with transfer function $G(s)$ which is excited by an input $u(t) = A \sin \omega t$.

a. Show that the steady-state output response of the system is given by:

$$y_{ss}(t) = \left|G(j\omega)\right| A \sin(\omega t + \phi) \text{ where } \phi = \angle G(j\omega)$$

b. Explain why replacing s with $j\omega$ in the transfer function yields the frequency response of the system.

c. Determine the steady state sinusoidal response of a system with $G(s) = \dfrac{s}{s+5}$ to an input $u(t) = 10\sin(4t - 25^0)$

7.2.1 Frequency Response Evaluation via Experiment

If the model of the system is not known, we can determine the magnitude $M(\omega)$ and phase $\phi(\omega)$ experimentally (provided the system is stable) as follows:

1. Excite the system with a sinusoid varying in frequency.
2. Choose a suitable amplitude of the input sinusoid; it should neither be too large to overload the system nor too small to get masked by noise.

3 Wait until the transients die and then measure the amplitude ratio (magnification factor) from

$$M(\omega) = |G(s)|_{s=j\omega} = \frac{\text{Amplitude of sinusoidal output}}{\text{Amplitude of sinusoidal input}} \tag{7.7}$$

4. Obtain the phase angle $\phi(\omega)$ by measuring the differences in phase between input and output signals (may use a spectral analyzer).

7.3 Frequency Response of Closed-Loop Systems

Consider the closed-loop system shown in Figure 7.4.

The closed-loop transfer function is given by

$$T(s) = \frac{C(s)}{R(s)} = \frac{G(s)}{1 + G(s)H(s)} \tag{7.8}$$

The sinusoidal steady state closed-loop transfer function is obtained by replacing $s = j\omega$ in Eqn(7.8), which gives

$$T(j\omega) = \frac{C(j\omega)}{R(j\omega)} = \frac{G(j\omega)}{1 + G(j\omega)H(j\omega)} \tag{7.9}$$

The magnitude of $T(j\omega)$ is

$$\left|T(j\omega)\right| = M(\omega) = \left|\frac{G(j\omega)}{1 + G(j\omega)H(j\omega)}\right| \tag{7.10}$$

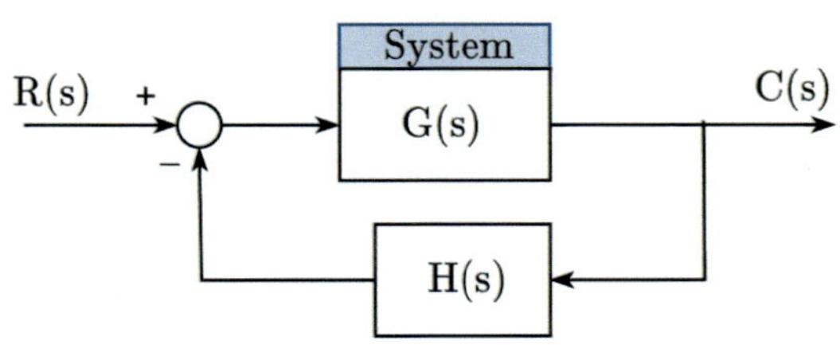

Figure 7.4: Closed-loop control system

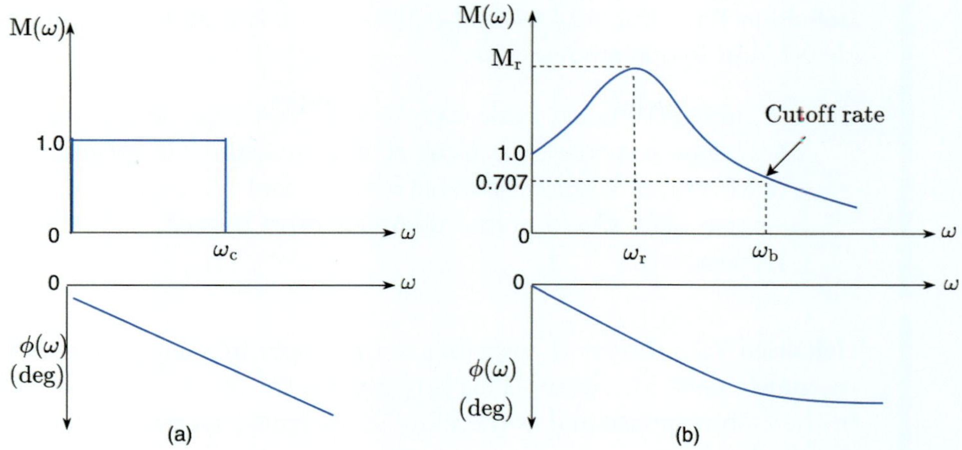

Figure 7.5: Gain and phase characteristics: (a) An ideal low-pass filter and (b) typical control system

and the phase of $T(j\omega)$ is

$$\angle T(j\omega) = \phi(\omega) = \angle G(j\omega) - \angle 1 + G(j\omega)H(j\omega) \qquad (7.11)$$

Note that there exist significant similarities between the design of a control system and filter design. Therefore, let us examine the gain and phase characteristics of an ideal low-pass filter, shown in Figure 7.5(a), which has a sharp cutoff frequency at $\omega = \omega_c$.

If a control system would be designed such that its characteristics are similar to that of an ideal low pass filter, i.e., $M(\omega) = 1$ at all frequencies, then its output $C(j\omega)$ would be identical to the input $R(j\omega)$. However, from Eqn(7.10), we see that, for $M(\omega)$ to be unity at all frequencies, the magnitude of $G(j\omega)$ must be infinite. The possibility of $G(j\omega)$ magnitude being infinity is ruled out in control systems; even a very high value of $|G(j\omega)|$ often makes the closed-loop system unstable. Moreover, one of the objectives of a control system is that it should be able to reject and suppress noise and unwanted signals. This is possible provided the frequency response of the system has a finite cutoff frequency ω_c.

The gain and phase characteristics of a typical control system is shown in Figure 7.5b.

7.4 Frequency Domain Specifications

The time domain specifications, such as the *maximum overshoot, peak time, rise time, settling time*, etc., cannot be directly used in frequency domain analysis and design. The following are some of the commonly used frequency domain specifications used in practice:

Definition 7.1. *Resonant peak* M_r: It is the maximum magnitude of the closed-loop frequency response.

A large resonance peak corresponds to a large overshoot in the transient response. For most control systems, an optimum value of M_r would be somewhere between 1.1 and 1.5. For a second-order system, this would give a damping ratio between 0.54 and 0.36, respectively.

Definition 7.2. *Resonant frequency* ω_r: It is the frequency at which the resonance peak M_r occurs. This is related to the frequency of oscillation in the step response and is therefore an alternate measure of the speed of transient response.

Definition 7.3. *Bandwidth* ω_b: The bandwidth ω_b is the frequency at which $\left|M(j\omega)\right|$ drops to 70.7% of, or 3dB down from, its zero frequency value.

a. It measures the ability of a system to reproduce the input signal. For example, for a system to accurately follow arbitrary inputs, it must have a large bandwidth. However, systems with large bandwidths have poor noise rejection characteristics.

b. It gives a measure of the transient response characteristics. A large bandwidth corresponds to a faster rise time because higher-frequency signals are more easily passed to the output. Conversely, if the bandwidth is small, only signals of relatively low frequencies are passed to the output. This results in slow and sluggish time response.

c. A system with a large bandwidth requires high-performance components; thus, the cost of components usually increases with the increase bandwidth.

Definition 7.4. *Cut-off rate*: The cut-off rate is the slope of the magnitude near the bandwidth (cut-off frequency).

This indicates the ability of the system to distinguish between the signal and the noise. A large high-frequency cut-off rate (steep roll-off at high frequencies) means that the system has a good signal-to-noise ratio. However, a high cut-off rate is often accompanied by a large M_r which, in turn, corresponds to a system with low damping with relatively small stability margin.

The above frequency domain performance criteria are shown in Figure 7.5b. Other frequency domain criteria such as *gain margin* and *phase margin* will be defined in later sections of this chapter.

7.5 Computation of Frequency Domain Specifications for the Prototype Second-Order System

Consider a prototype second-order system shown in Figure 7.6.

Now, the closed-loop transfer function

$$\mathrm{T(s)} = \frac{\mathrm{C(s)}}{\mathrm{R(s)}} = \frac{\omega_{\mathrm{n}}^2}{\mathrm{s}^2 + 2\xi\omega_{\mathrm{n}}\mathrm{s} + \omega_{\mathrm{n}}^2} \tag{7.12}$$

Replacing s with $\mathrm{j}\omega$ in Eqn(7.12) gives

$$\mathrm{T(j}\omega) = \frac{\mathrm{C(j}\omega)}{\mathrm{R(j}\omega)} = \frac{\omega_{\mathrm{n}}^2}{(\mathrm{j}\omega)^2 + 2\xi\omega_{\mathrm{n}}(\mathrm{j}\omega) + \omega_{\mathrm{n}}^2}$$

$$= \frac{\omega_{\mathrm{n}}^2}{-\omega^2 + \mathrm{j}2\xi\omega_{\mathrm{n}}\omega + \omega_{\mathrm{n}}^2} = \frac{1}{1 - \dfrac{\omega^2}{\omega_{\mathrm{n}}^2} + \mathrm{j}2\xi\dfrac{\omega}{\omega_{\mathrm{n}}}} \tag{7.13}$$

If we define $\mathrm{u} = \dfrac{\omega}{\omega_{\mathrm{n}}}$, called the normalized frequency, then Eqn(7.12) can be expressed as:

$$\mathrm{T(j}\omega) = \frac{1}{(1-\mathrm{u}^2) + \mathrm{j}2\xi\mathrm{u}} \tag{7.14}$$

From Eqn(7.14), we find

$$\left|\mathrm{T(j}\omega)\right| = \mathrm{M(u)} = \frac{1}{\sqrt{(1-\mathrm{u}^2)^2 + (2\xi\mathrm{u})^2}} \tag{7.15}$$

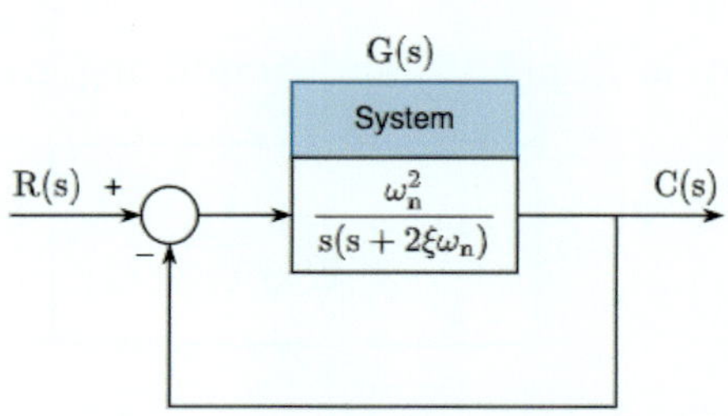

Figure 7.6: Prototype second-order system

and

$$\angle T(j\omega) = \phi(u) = -\tan^{-1}\frac{2\xi u}{1-u^2} \tag{7.16}$$

Thus, the output of the system for an input $r(t) = \sin\omega t$ is given by (from Eqns(7.15) and (7.16)

$$\begin{aligned} c(t) &= M\sin(\omega t + \phi) \\ &= \frac{1}{\sqrt{(1-u^2)^2 + (2\xi u)^2}}\sin\left[\omega t - \tan^{-1}\frac{2\xi u}{1-u^2}\right] \end{aligned} \tag{7.17}$$

7.5.1 Computation of Resonant Frequency ω_r and Resonant Peak M_r

The resonant frequency is determined by setting the derivative of M(u) with respect to u to zero. Thus

$$\frac{dM(u)}{du} = -\frac{1}{2}\left[(1-u^2)^2 + (2\xi u)^2\right]^{-3/2}\left[4u^3 - 4u + 8u\xi^2\right] = 0$$

from which we get

$$4u^3 - 4u + 8u\xi^2 = 4u\left(u^2 - 1 + 2\xi^2\right) = 0$$

This gives $u = 0$, and $u = \sqrt{1-2\xi^2}$

The first solution $u = 0$ indicates that the slope of M versus ω curve is zero at $\omega = 0$. The resonant frequency is obtained from the second solution as:

$$\boxed{u_r = \sqrt{1-2\xi^2} \implies \omega_r = \omega_n\sqrt{1-2\xi^2}} \tag{7.18}$$

Since frequency is a real quantity, Eqn(7.18) is valid provided $2\xi^2 < 1$, or $\xi \le \frac{1}{\sqrt{2}}$. This means, if $\xi > 0.707$, the resonant frequency $\omega_r = 0$ and $M_r = 1$.

Substituting $u = u_r$ in Eqn(7.15) and simplifying gives

$$\boxed{M_r = \frac{1}{2\xi\sqrt{1-\xi^2}}} \tag{7.19}$$

From Eqn(7.18) and (7.19), we find that, for the prototype second-order system

Note

1. The resonant peak M_r is a function of the damping ratio ξ only, similar to overshoot M_p (which is also a function of ξ only).

2. However, the resonant frequency ω_r is a function of both ξ and ω_n

The variations of resonant peak M_r and normalized resonant frequency $u_r = \dfrac{\omega_r}{\omega_n}$ with damping factor ξ are shown respectively in Figures 7.7 and 7.8.

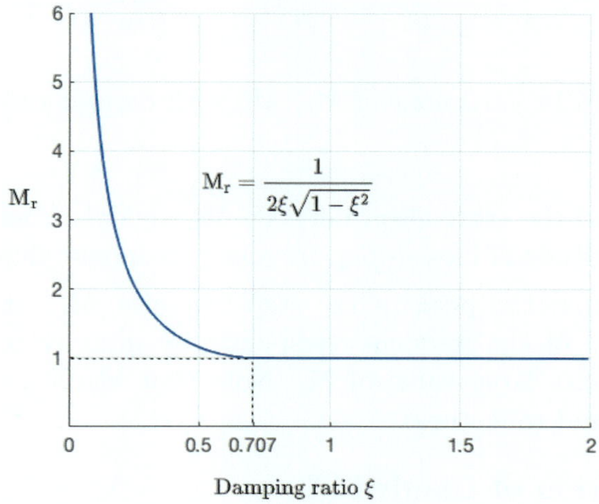

Figure 7.7: Variation of M_r with damping ratio ξ for prototype second order system

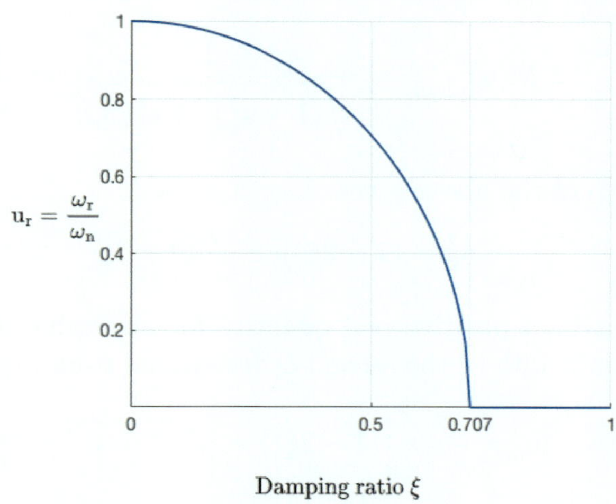

Figure 7.8: Variation of normalized resonant frequency $u_r = \sqrt{1-2\xi^2}$ with damping ratio ξ for prototype second order system

The similarity between M_r and M_p can be seen from Figure 7.9.

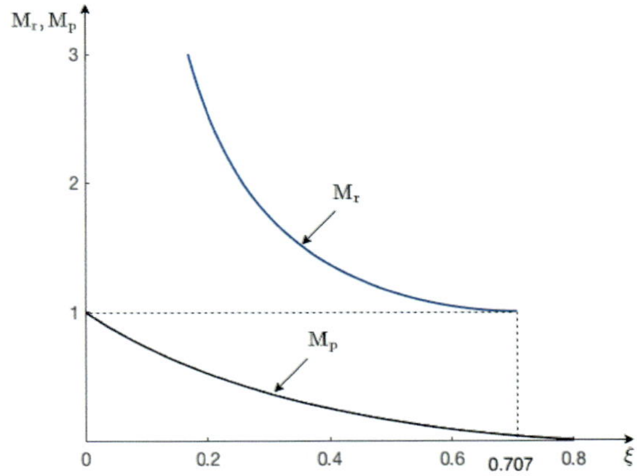

Figure 7.9: Variations of M_r, M_p with damping ratio ξ

Let us find the correlation between M_p and M_r from the step response and magnitude of closed-loop frequency response shown in Figure 7.10

We find that the peak in the step response, M_p, and the peak in the magnitude of the frequency response are directly related. A large M_p value implies large value of M_r. Note that M_r is a measure of relative stability and robustness.

7.5.2 Computation of Bandwidth ω_b

The bandwidth of a closed-loop system is defined as the frequency at which $M = |T(j\omega)|$ drops to $\frac{1}{\sqrt{2}}$ of its value at $\omega = 0$. Thus

$$M(u)\big|_{u=u_b} = \frac{1}{\sqrt{\left(1-u_b^2\right)^2 + \left(2\xi u_b\right)^2}} = \frac{1}{\sqrt{2}}$$

Simplification of the above gives

$$u_b^2 = \left(1-2\xi^2\right) \pm \sqrt{4\xi^4 - 4\xi^2 + 2} \tag{7.20}$$

Since u must be a positive real quantity for any value of ξ, we choose plus sign. The bandwidth of the second-order system from Eqn(7.20) is

$$u_b = \frac{\omega_b}{\omega_n} = \left[\left(1-2\xi^2\right) + \sqrt{4\xi^4 - 4\xi^2 + 2}\right]^{1/2}$$

or

$$\boxed{\omega_b = \omega_n \left[\left(1-2\xi^2\right) + \sqrt{4\xi^4 - 4\xi^2 + 2}\right]^{1/2}} \tag{7.21}$$

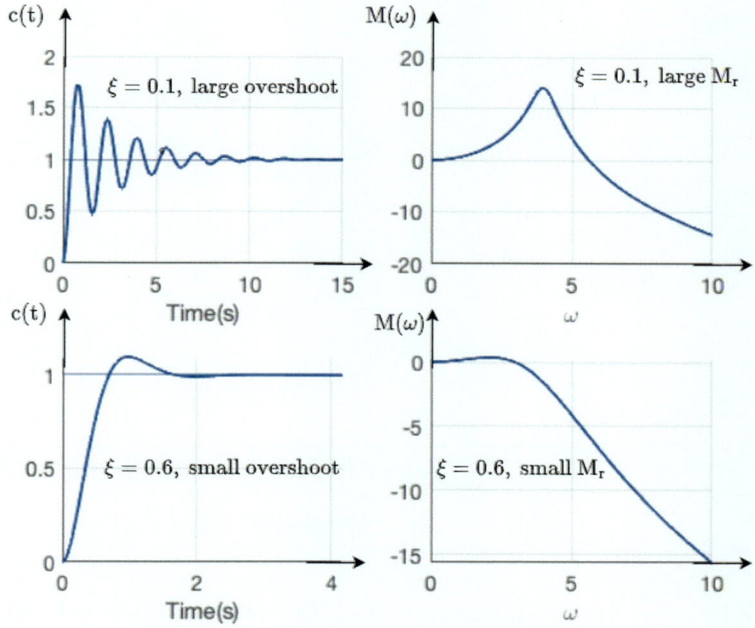

Figure 7.10: Step response and closed-loop frequency response (magnitude) for $\xi = 0.1$ and $\xi = 0.6$

In terms of settling time $T_s = \dfrac{4}{\xi \omega_n}$, the bandwidth is expressed as:

$$\omega_b = \frac{4}{\xi T_s}\left[\left(1 - 2\xi^2\right) + \sqrt{4\xi^4 - 4\xi^2 + 2}\right]^{1/2} \tag{7.22}$$

Similarly, in terms of peak time T_p,

$$\omega_b = \frac{\pi}{T_p\sqrt{1 - \xi^2}}\left[\left(1 - 2\xi^2\right) + \sqrt{4\xi^4 - 4\xi^2 + 2}\right]^{1/2} \tag{7.23}$$

The variation of bandwidth $BW/\omega_n = \omega_b/\omega_n$ with the damping ratio ξ is shown in Figure 7.11.

From Figure 7.11, we see that BW/ω_n monotonically decreases with increasing value of ξ. Notice from Eqn(7.21) that BW is directly proportional to the natural frequency ω_n.

Let us find if there exists a correlation between the BW and the rise time t_r.

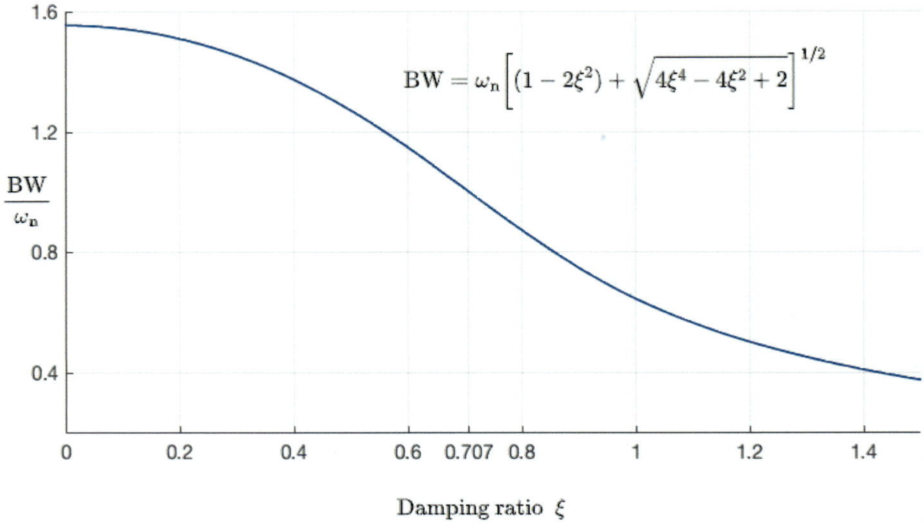

Figure 7.11: $\frac{\mathrm{BW}}{\omega_{\mathrm{n}}}$ versus damping ratio ξ for prototype second order system

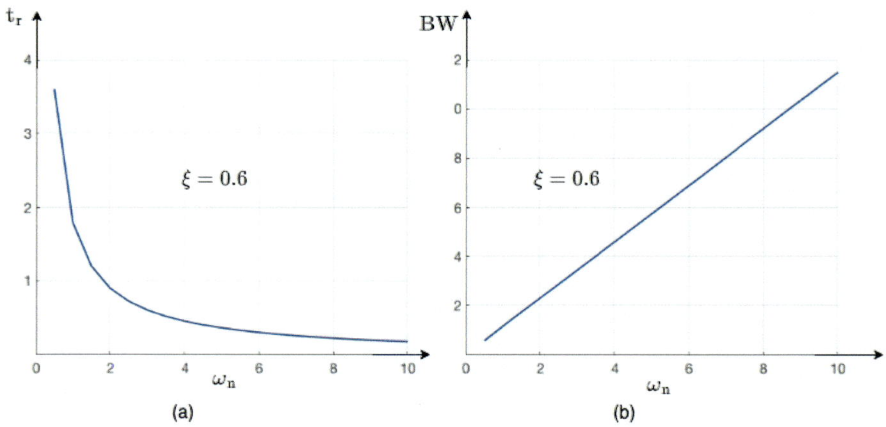

Figure 7.12: (a) $\mathrm{t_r}$ versus ω_n (b) BW versus ω_n for prototype second order system when $\xi = 0.6$

We know that the rise time $\mathrm{t_r}$ can be expressed as (see Eqn(3.58))

$$\mathrm{t_r} = \frac{1 - 0.4167\xi + 2.917\xi^2}{\omega_{\mathrm{n}}} \tag{7.24}$$

From Eqn(7.24), we could see that as $\mathrm{t_r}$ is inversely related to ω_{n} (see Figure 7.12(a)); on the contrary, BW is directly related with ω_{n} (see Figure 7.12(b)).

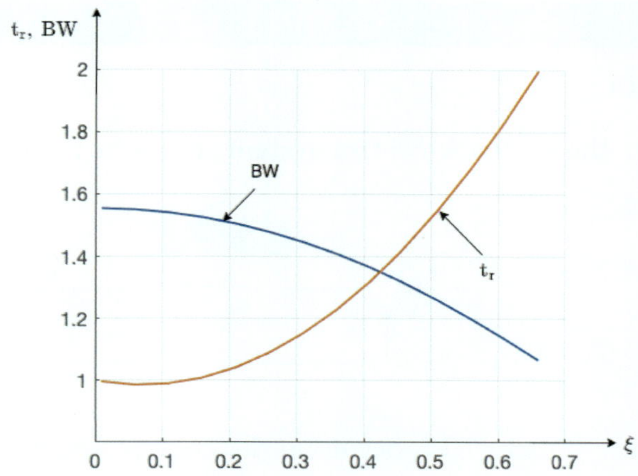

Figure 7.13: Correlation between the rise time t_r and bandwidth BW for prototype second order system when $\omega_n = 1.0$

Thus, BW and rise time are inversely proportional as shown in Figure 7.13.

From our discussion about the frequency domain specifications and some time domain performance measures discussed in Chapter 3, we summarize the relationships between these two domains.

Correlation Between Frequency and Time Domain Specifications

1. Both M_p and M_r are purely a function of the damping ratio ξ. There exists a direct correlation between them; a large value of M_p implies a large value of M_r.

2. The resonant frequency ω_r is directly related to the natural frequency ω_n. Since the settling time t_s is inversely proportional to the product $\xi\omega_n$, that is, $t_s \approx \dfrac{4}{\xi\omega_n}$, for a given ξ, a higher value of ω_r results faster settling time.

3. The bandwidth (BW) and rise time are inversely proportional. If the bandwidth of the system increases, the system exhibits a faster response (due to a decrease in rise time t_r).

4. The BW increases (t_r decreases) with increase of the natural frequency ω_n. Therefore, the system response becomes faster with an increase of ω_n.

5. Increase of damping ratio ξ decreases the BW and increases t_r, resulting in slower system response.

EXAMPLE 7.2

Problem

Consider the feedback control system shown in Figure 7.14.

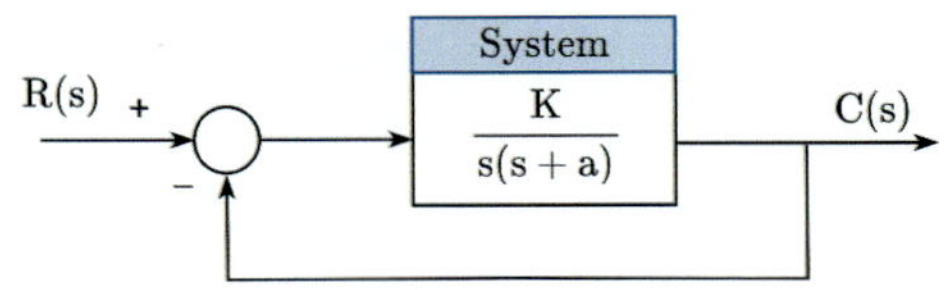

Figure 7.14: Second order control system

i. Determine the value of K and a such that $M_r = 1.2$ and $\omega_r = 10$ rad/sec.
ii. For the values of K and a determined in part(i), calculate the peak overshoot to a unit step input and the bandwidth of the system.

SOLUTION:

i. Since M_r is a function of only ξ, let us determine ξ from given M_r as follows:

$$M_r = \frac{1}{2\xi\sqrt{1-\xi^2}} \quad \Longrightarrow \quad M_r^2 = \frac{1}{4\xi^2(1-\xi^2)}$$

This gives

$$4\xi^2 M_r^2 - 4\xi^4 M_r^2 = 1 \quad \Longrightarrow \quad 4\xi^4 M_r^2 - 4\xi^2 M_r^2 + 1 = 0$$

For $M_r = 1.2$, the solution of

$$4\xi^4 M_r^2 - 4\xi^2 M_r^2 + 1 = 5.76\xi^4 - 5.76\xi^2 + 1 = 0$$

gives $\xi = 0.8811$ and $\xi = 0.4729$. The only admissible solution is $\xi = 0.4729$. ($\because \quad \xi \ngtr 0.707$).

The closed-loop transfer function of the system is

$$\frac{C(s)}{R(s)} = \frac{K}{s^2 + as + K} = \frac{\omega_n^2}{s^2 + 2\xi\omega_n s + \omega_n^2}$$

Hence, $\omega_n = \sqrt{K}$ and $a = 2\xi\sqrt{K}$. From given $\omega_r = 10$ rad/sec,

$$\omega_r = \omega_n\sqrt{1-2\xi^2} \quad \Longrightarrow \quad \omega_n = \frac{\omega_r}{\sqrt{1-2\xi^2}}$$

$$\Longrightarrow \quad \omega_n = 13.4502 \text{ rad /sec}$$

This gives $K = \omega_n^2 = 180.9068$ and $a = 2\xi\sqrt{K} = 12.7206$.

ii. For $\xi = 0.4729$, the percent peak overshoot is

$$\%M_p = \exp\left(-\pi\xi/\sqrt{1-\xi^2}\right) \times 100 = 18.53\%$$

and the bandwidth is

$$BW = \omega_n \left[(1-2\xi^2) + \sqrt{4\xi^4 - 4\xi^2 + 2}\right]^{1/2} = 17.513 \text{ rad/sec}$$

EXAMPLE 7.3

Problem

The closed-loop transfer function of a standard second-order system is

$$T(s) = \frac{C(s)}{R(s)} = \frac{\omega_n^2}{s^2 + 2\xi\omega_n s + \omega_n^2} \tag{7.25}$$

It is desired that the maximum overshoot must not exceed 10% and the settling time must be less than 1sec. Find the corresponding limiting values of natural frequency ω_n, M_r and bandwidth (BW).

SOLUTION:

Given maximum overshoot $M_p = 0.1$. Now

$$\xi = \frac{\gamma}{\sqrt{\pi^2 + \gamma^2}}, \quad \text{where} \quad \gamma = -(\ln\%M_p/100)$$

This gives $\xi = 0.5912$. Now given

$$T_s = \frac{4}{\xi\omega_n} = 1 \quad \Longrightarrow \quad \omega_n = \frac{4}{\xi T_s} = \frac{4}{0.5912} = 6.7658 \text{ rad/sec}$$

The minimum value of ω_n should be 6.7658 rad/sec.

The limiting values of M_r and bandwidth are

$$M_r = \frac{1}{2\xi\sqrt{1-\xi^2}} = 1.0487$$

and

$$BW = \omega_n \left[(1-2\xi^2) + \sqrt{4\xi^4 - 4\xi^2 + 2}\right]^{1/2} = 7.8485 \text{ rad/sec}$$

CONCEPT REINFORCEMENT PROBLEM 7.2

Problem

Determine the closed-loop bandwidth BW and resonance peak M_r of a prototype second-order system with the following performance characteristics.

a. $T_s = 5$ seconds, $T_p = 2.5$seconds

b. $T_r = 3$seconds, $\xi = 0.4$

7.6 Effects of Adding a Zero to the Forward Path Transfer Function

In Chapter 3, we derived analytical expressions for computing various time domain performance measures considering a prototype second-order system. Further, we investigated the effects of adding a pole or zero on the time response. In this chapter, we have derived analytical expressions for various frequency domain performance measures that are valid for the prototype second-order system. When the system order deviates from the prototype, the analysis becomes more involved. However, it is of interest to know how the frequency response will change with the addition of a pole or zero. In the following, we study the effects of adding zero to the forward path transfer function on the characteristics of the closed-loop frequency response.

Let us add a zero at $s = -1/T_z$ to the forward path transfer function of G(s) of the prototype second order system. The resulting closed-loop system is shown in Figure 7.15.

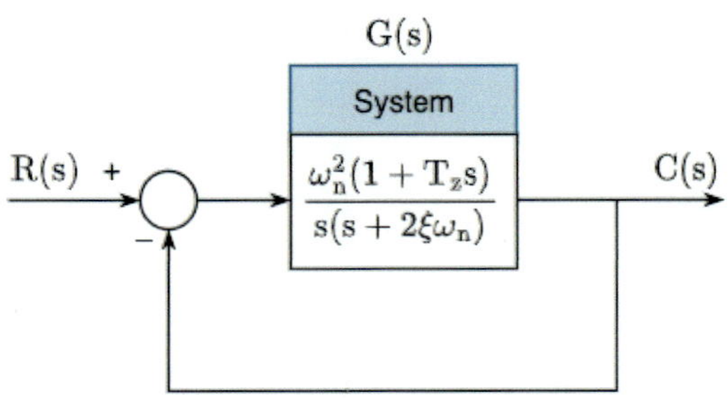

Figure 7.15: Second order system with addition of a zero

The closed-loop transfer function becomes

$$T(s) = \frac{\omega_n^2(1+T_z s)}{s^2 + (2\xi\omega_n + \omega_n^2 T_z)\, s + \omega_n^2} \tag{7.26}$$

Although, with the addition of a zero, the system is still second order, due to the presence of three parameters ξ, ω_n, and T_z, it becomes difficult (though not impossible) to obtain analytical expressions for M_r, ω_r, and BW. The bandwidth of the system can be derived as:

$$BW = \left(-\beta + \frac{1}{2}\sqrt{\beta^2 + 4\omega_n^4}\right)^{1/2} \tag{7.27}$$

where $\beta = 4\xi^2\omega_n^2 + 4\xi\omega_n^3 T_z - 2\omega_n^2 - \omega_n^4 T_z^2$. It is often difficult to see how each individual parameter in Eqn(7.27) affects the bandwidth. However, for simplicity of analysis, we analyze how BW is affected as the location of the added zero is changed. Figure 7.16 shows the variation of BW with T_z for $\xi = 0.6$ and $\omega_n = 1$.

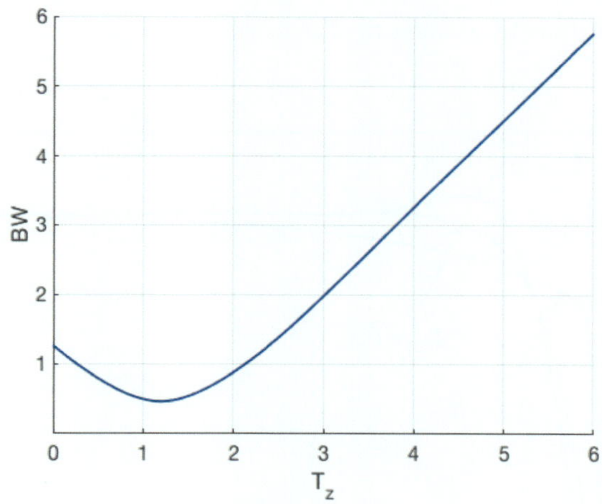

Figure 7.16: Bandwidth of a second order system with forward path transfer function $G(s) = \dfrac{(1 + T_z s)}{s(s + 1.2)}$

From Figure 7.16, we find that, over a range of small values of T_z, the bandwidth decreases. However, after that small range, the bandwidth increases monotonically with an increase in T_z, that is, the bandwidth increases as the zero becomes closer to the $j\omega$-axis. This is further established from the plot of $M(\omega) = |T(j\omega)|$ of the closed-loop system of Figure 7.15 shown in Figure 7.17 as T_z takes on various values.

Since bandwidth and rise time are inversely related, an increase in bandwidth (due to zero addition) will result in faster rise time. However, as the zero becomes very close to the origin, the effective time constant of the closed-loop system will be large. This will increase the settling time. This is illustrated in Figure 7.18 considering $\omega_n = 1.0, \xi = 0.707$

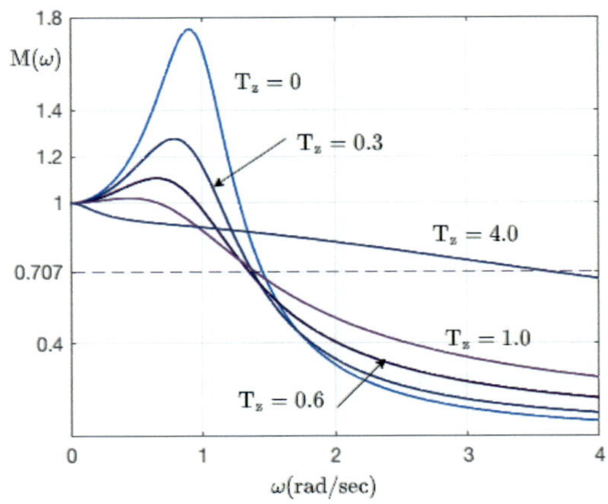

Figure 7.17: Magnification curve of a second-order system of Figure 7.15. $\omega_n = 1,\ \xi = 0.3$ for different values of T_z

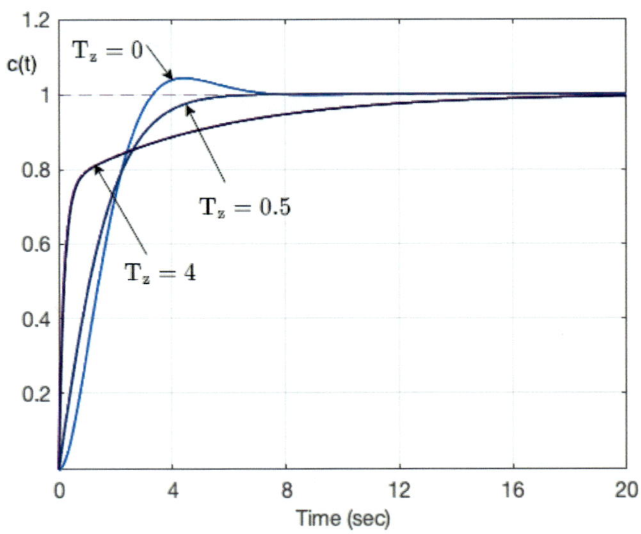

Figure 7.18: Output response of a second order system of Figure 7.15. $\omega_n = 1,\ \xi = 0.3$

From these observations, we conclude the following:

Key Takeaways

1. Adding a zero to the forward-path transfer function increases the bandwidth of the closed-loop system.
2. The addition of a zero in the forward path reduces the rise time (faster rise time).

3. However, if the zero is very close to the origin, the effective time constant of the closed-loop system increases, which makes the settling time longer.

7.7 Effects of Adding a Pole to the Forward Path Transfer Function

If a pole at $s = -1/T_p$ is added to the forward path, the order of the resulting closed-loop system will be three (see Figure 7.19).

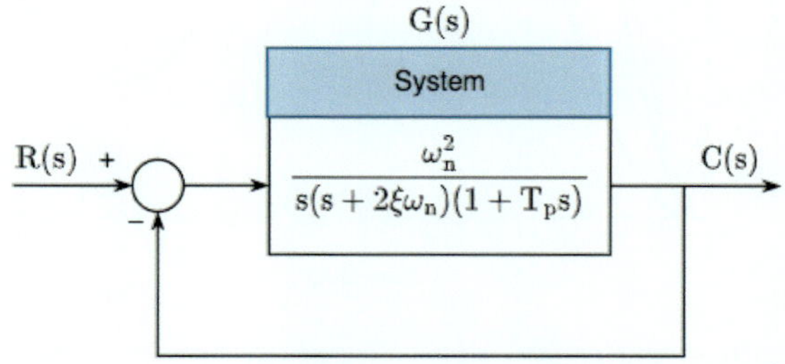

Figure 7.19: Second order system with addition of a pole

The mathematical derivation of the BW of the closed-loop system is quite tedious. Therefore, we will qualitatively investigate the effects of adding a pole on the bandwidth and rise time of the system. Figure 7.20 shows the plot of $M(\omega)$ versus ω for $\omega_n = 1.0, \xi = 0.6$, and various values of T_p. The choice of these parameters ensures that the closed-loop system is stable.

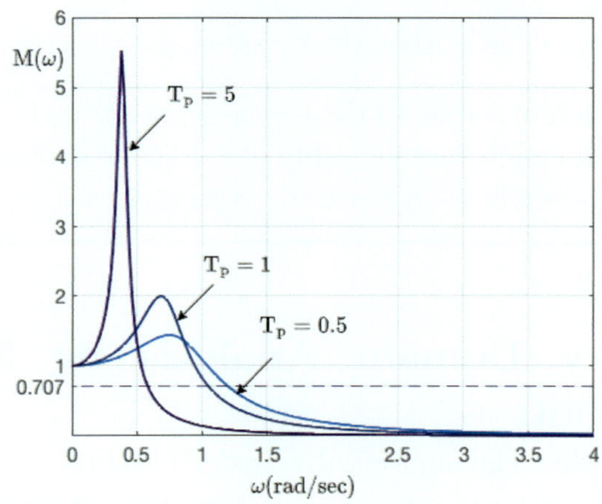

Figure 7.20: Magnification curve of a second-order system of Figure 7.19. $\omega_n = 1,\ \xi = 0.6$

From Figure 7.20, we find that when T_p is large, that is, the pole added to G(s) is closer to the jω-axis, the bandwidth decreases. However, the peak magnitude M_r increases. The time response to the unit step input shown in Figure 7.21 conforms to the inverse relation between BW and the rise time, that is, the rise time increases with the closeness of the added pole to the $j\omega$ -axis.

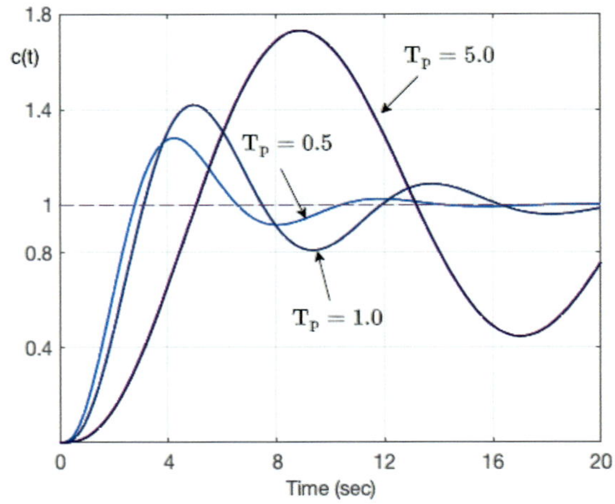

Figure 7.21: Output response of a second-order system of Figure 7.19. $\omega_n = 1,\ \xi = 0.6$

From these observations, we conclude the following:

Key Takeaways

1. Adding a pole to the forward-path transfer function decreases the bandwidth of the closed-loop system, resulting in a lesser stability margin.
2. The rise time of the system increases with the addition of a pole.
3. Addition of a pole to the forward-path transfer function increases M_r, which alternately implies that the closed-loop system exhibits larger overshoot in the unit-step response.

7.8 Frequency Domain Analysis by Frequency Response Plots

We have seen that the sinusoidal transfer function is a complex function of the frequency ω. A lot of information about the dynamic response of a system can be obtained from knowledge of the magnitude $M(\omega)$ and phase $\phi(\omega)$ of its transfer function. The frequency response, thus, consists of three variables-frequency ω, magnitude $M(\omega)$ and phase $\phi(\omega)$. This part intends to analyze the frequency domain performance of a system, considering three commonly used frequency domain representations:

1. Bode plot or logarithmic plot
2. Polar plot or Nyquist plot and
3. Log-magnitude-versus phase plot (Nichols plots)

7.8.1 Bode Plot

The bode plot of the complex function $G(j\omega)H(j\omega)$ is composed of two plots;

1. A plot of magnitude $|G(j\omega)H(j\omega)|$ in decibels (dB) versus $\log_{10}\omega$ and
2. The phase $\phi(\omega) = \angle G(j\omega)H(j\omega)$ in degrees versus $\log_{10}\omega$.

Without loss of generality, we assume $H(j\omega) = 1$. The Bode plot is also called a **corner plot** or an **asymptotic plot**; because this plot can be sketched by using straight-line approximations which are asymptotic to the actual plot. Frequency domain analysis using Bode plot offers many advantages listed below

Advantages of Bode Plot

1. Since the magnitude of the open-loop transfer function $G(j\omega)H(j\omega)$ is expressed in decibels (dB), multiplicative and divisive components in the transfer function are transformed into simple additions and subtractions, respectively. This logarithmic scaling greatly simplifies analysis and interpretation.
2. Similarly, the overall phase angle is obtained by algebraically adding or subtracting the individual phase contributions from each factor. This modular approach enables intuitive understanding of how each component affects the total system phase.
3. Another key advantage is that the magnitude plot can be approximated using straight-line asymptotes (also known as "breakpoint approximations") with minimal loss in accuracy. These approximations allow for quick and reasonably accurate hand-sketching of the Bode plot without the need for extensive calculations.

As a result of these features, the Bode plot becomes a powerful graphical tool for understanding system behavior, designing controllers, and assessing stability margins.

Consider the function

$$G(s) = \frac{K_1(s+z_1)(s+z_2)\ldots\ldots(s+z_m)}{s^N(s+p_1)(s+p_2)\ldots\ldots(s+p_n)}e^{-T_d s} \tag{7.28}$$

where K_1 and T_d are real constants, and the z's and the p's may be real or complex (in conjugate pairs) numbers. This form of G(s), called as pole-zero form, is preferred for root-locus construction. However, for manual construction of

Bode plot, it is preferred to represent Eqn(7.28) in time-constant form as:

$$G(s) = \frac{K(1+T_1 s)(1+T_2 s)\ldots\ldots(1+T_m s)}{s^N(1+T_a s)(1+T_b s)\ldots\ldots(1+T_n s)} e^{-T_d s} \tag{7.29}$$

where K is a real constant, the T′s may be real or complex (in conjugate pairs) numbers, and T_d is the real time delay. In general, the open-loop transfer function G(s) consists of five types of factors, which are discussed below.

7.8.2 Basic Factors of $G(j\omega)$

1. Constant factor: Gain K
2. Poles or zeros of order N at the origin: $(j\omega)^{\pm N}$
3. Real poles and zeros not at the origin
4. Complex poles and zeros
5. Ideal time delays

Constant factor: K

Here the gain $G(j\omega) = K$. Let M denotes the magnitude of the function $G(j\omega)$ i.e., $M = |G(j\omega)|$. Then the log-magnitude of this constant factor is

$$|G(j\omega)|_{dB} = 20\log_{10} M = 20\log_{10}|K| \tag{7.30}$$

and the phase angle $\phi(\omega)$ is

$$\phi(\omega) = \begin{cases} 0^0 & K > 0 \\ 180^\circ & K < 0 \end{cases} \tag{7.31}$$

Both the gain and phase curves are horizontal lines, as shown in the Bode diagram of Figure 7.22.

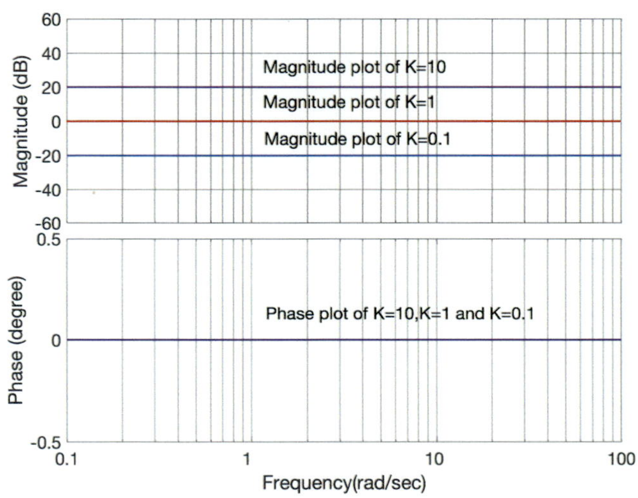

Figure 7.22: Bode magnitude and phase plot for constant gain

Poles (or Zeros) at the Origin

When there is a simple pole at the origin, then $G(j\omega) = \frac{1}{j\omega}$. Hence

$$|G(j\omega)|_{dB} = 20\log_{10}\left|\frac{1}{j\omega}\right| = -20\log_{10}\omega \ \ dB \tag{7.32}$$

for $\omega \geq 0$ and the phase angle $\phi(\omega)$ is

$$\phi(\omega) = -90^\circ \tag{7.33}$$

Eqn(7.32) can be expressed as:

$$20 \ \log_{10} M = -20\log_{10}\omega \ \ dB \tag{7.34}$$

If we plot dB, $20\log_{10} M$, versus $\log_{10}\omega$, Eqn(7.34) becomes a straight line:

$$y = -20x \tag{7.35}$$

where $y = 20 \ \log_{10} M$, and $x = \log_{10}\omega$. Note that this is a straight line passing through 0 dB at $\omega = 1$ with a slope of -20 dB per unit change in $\log_{10}\omega$. From Eqn(7.35), we see that if x changes by one unit, then y changes by -20. But a unit change in $x = \log_{10}\omega$ in rectangular coordinates is equivalent to one **decade** of variation in ω. This is because

$$\log_{10}\left(\frac{\omega_2}{\omega_1}\right) = 1, \quad \Longrightarrow \quad \omega_2 = 10\omega_1$$

The frequency interval $\omega_2 = 10\omega_1$ is called a decade, that is, the frequency intervals from 1 to 10, 10 to 100, and so on in the semilog coordinates. Thus, the slope of the straight line described by Eqn(7.34) and Eqn(7.35) is -20 dB/decade.

We can use **octaves**, instead of **decades**, to represent the separation of two frequencies. If the frequency interval between two frequencies is 2, that is, $\omega_2/\omega_1 = 2$, then it is called an **octave**.

Relation Between Decade and Octave

The number of decades between any two frequencies ω_1 and ω_2 is given by

$$\text{Number of decades} = \frac{\log_{10}(\omega_2/\omega_1)}{\log_{10} 10} = \log_{10}\left(\frac{\omega_2}{\omega_1}\right) \tag{7.36}$$

Similarly, the number of octaves between any two frequencies ω_1 and ω_2 is given by

$$\text{Number of octaves} = \frac{\log_{10}(\omega_2/\omega_1)}{\log_{10} 2} = \frac{1}{0.301}\log_{10}\left(\frac{\omega_2}{\omega_1}\right) \tag{7.37}$$

Hence, the relation between octaves and decades is

$$\text{Number of octaves} = \frac{1}{0.301}\ \text{decades} = 3.32\ \text{decades} \tag{7.38}$$

Thus,

$$20\ \text{dB/decade} = 20 \times 0.301 \approx 6\ \text{dB/Octave} \tag{7.39}$$

For multiple poles at the origin, i.e., a pole of multiplicity N, $G(j\omega) = \frac{1}{(j\omega)^N}$. The log-magnitude of this factor is

$$|G(j\omega)|_{dB} = 20\log_{10}\left|\frac{1}{(j\omega)^N}\right| = -20N\log_{10}\omega \quad dB \tag{7.40}$$

for $\omega \geq 0$ and the phase angle $\phi(\omega)$ is

$$\phi(\omega) = -90^0 N \tag{7.41}$$

For a **zero** at the origin, $G(j\omega) = j\omega$. The log-magnitude of this factor is

$$|G(j\omega)|_{dB} = +20\log_{10}\omega \quad dB \tag{7.42}$$

for $\omega \geq 0$ and the phase angle $\phi(\omega)$ is

$$\phi(\omega) = +90^\circ \tag{7.43}$$

If there are zeros of multiplicity N at the origin, then $G(j\omega) = (j\omega)^N$. The log-magnitude of this factor is

$$|G(j\omega)|_{dB} = 20\log_{10}\left|(j\omega)^N\right| = +20N\log_{10}\omega \quad dB \tag{7.44}$$

for $\omega \geq 0$ and the phase angle $\phi(\omega)$ is

$$\phi(\omega) = +90^0 N \tag{7.45}$$

- The slope of the magnitude plot due to the pole of multiplicity N at the origin is −20N dB/decade and the phase angle $\phi(\omega) = -90^0$N.
- Similarly, the slope of the magnitude plot due to zero of multiplicity N at the origin is +20N dB/decade and the phase angle $\phi(\omega) = +90^0$N.

The Bode plot for systems with poles and zeros at the origin is shown in Figures 7.23 and 7.24, respectively.

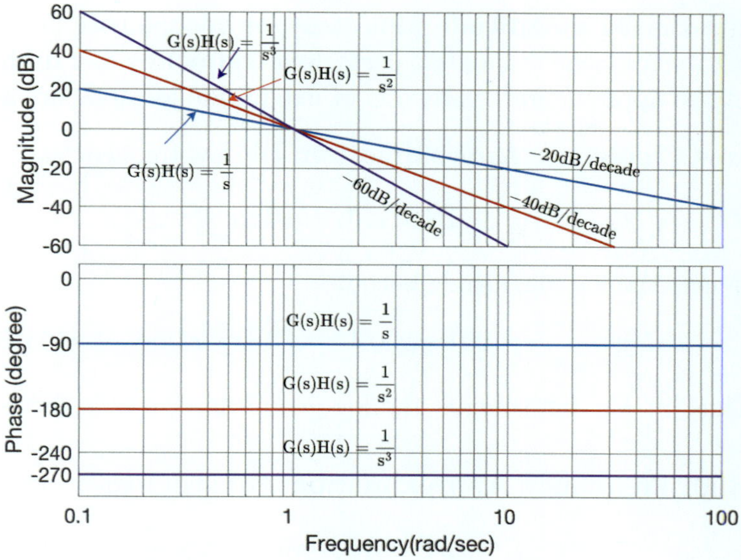

Figure 7.23: Bode magnitude and phase plot for systems with poles at the origin

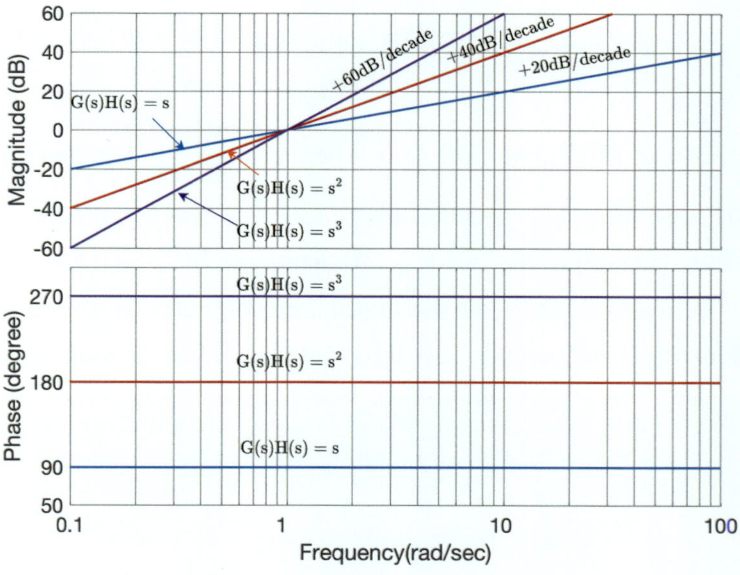

Figure 7.24: Bode magnitude plot for systems with zeros at the origin

Simple Pole on Real Axis

For a simple pole on the real axis, we have $G(j\omega) = \dfrac{1}{1 + j\omega T}$, where T is a positive real constant. The log-magnitude of this factor is

$$\left|G(j\omega)\right|_{dB} = 20\log_{10}\left|\frac{1}{1 + j\omega T}\right| = -20\log_{10}\sqrt{1 + \omega^2 T^2}\ \ dB \tag{7.46}$$

To sketch the **log magnitude plot** easily, we approximate Eqn(7.46) by first splitting the frequencies into two parts: **low frequency** and **high frequency**, and get the approximate expression of magnitude in each part as follows:

Low Frequency Approximation: For low frequencies

$$\omega \mathrm{T} << 1, \quad \text{or,} \quad \omega << \frac{1}{\mathrm{T}}$$

Hence, the log-magnitude becomes

$$\left|\mathrm{G}(\mathrm{j}\omega)\right|_{\mathrm{dB}} = -20\log_{10}\sqrt{1+\omega^2\mathrm{T}^2} \approx 20\log_{10}1 \approx 0 \text{ dB} \tag{7.47}$$

Thus, at low frequencies, the log-magnitude curve is the constant 0-dB line.

High Frequency Approximation: For high frequencies

$$\omega \mathrm{T} >> 1, \quad \text{or,} \quad \omega >> \frac{1}{\mathrm{T}}$$

Hence, the log-magnitude becomes

$$\begin{aligned}\left|\mathrm{G}(\mathrm{j}\omega)\right|_{\mathrm{dB}} &= -20\log_{10}\sqrt{1+\omega^2\mathrm{T}^2} \\ &\approx -20\log_{10}\sqrt{\omega^2\mathrm{T}^2} = -20\log_{10}\omega\mathrm{T} \\ &= -20\log_{10}\omega - 20\log_{10}\mathrm{T}\end{aligned} \tag{7.48}$$

From Eqn(7.48), we find that when $\omega = \frac{1}{\mathrm{T}}$, the log magnitude equals 0 dB, and after one decade, where $\omega = \frac{10}{\mathrm{T}}$, the log magnitude is -20 dB. Thus, the value of $-20\log_{10}\omega\mathrm{T}$ decreases by 20 dB for every decade of ω. Hence, for $\omega >> \frac{1}{\mathrm{T}}$, the log-magnitude curve is a straight line with a slope of -20dB/decade (or -6dB/octave). Thus, in summary

a. The log-magnitude of the factor corresponding to a simple pole on the real axis can be approximated by two straight-line asymptotes.
   1. The **low frequency asymptote** is a straight-line of 0 dB for the frequency range $0 < \omega < \frac{1}{\mathrm{T}}$.
   2. The **high frequency asymptote** is a straight line with a slope of -20dB/decade (-6 dB/octave) for the frequency range $\frac{1}{\mathrm{T}} < \omega < \infty$.

b. The frequency at which these two asymptotes meet is called the *corner frequency* or *break frequency*. For the factor $\frac{1}{1+\mathrm{j}\omega\mathrm{T}}$, the corner frequency occurs at $\omega = \frac{1}{\mathrm{T}}$. It divides the frequency-response curve into two regions: one curve for the low-frequency region and the second curve for the high-frequency region.

Error in Asymptotic Magnitude Plot[1]

There exists some error in the asymptotic plot, which is the difference between the actual value and the approximated value. These are calculated as follows:

i. The error at the corner frequency $\omega = \frac{1}{T}$ is

$$-20\log_{10}\sqrt{1+1} + 20\log_{10} 1 = -10\log_{10} 2 = -3.03\text{dB} \cong -3.0 \text{ dB}$$

ii. The error at the frequency below the corner frequency, specifically at $\omega = \frac{\omega_c}{\beta}$ where $\beta > 1$, is given by

$$-20\log_{10}\sqrt{1+\frac{1}{\beta^2}} \text{ dB}$$

iii. The error at the frequency above the corner frequency, that is, at $\omega = \alpha\omega_c$ where $\alpha > 1$

$$-20\log_{10}\sqrt{1+\alpha^2} + 20\log_{10}\alpha \text{ dB}$$

The errors in asymptotic approximations for a real pole of order r with corner frequency ω_c are given in Table 7.1.

ω	$\omega_c/10$	$\omega_c/5$	$\omega_c/2$	ω_c	$2\omega_c$	$5\omega_c$	$10\omega_c$
Error(dB)	−0.043r	−0.17r	−0.96r	−3r	−0.96r	−0.17r	−0.043r

Table 7.1: Approximation errors in the asymptotic magnitude plot at various frequencies

Although the exact log magnitude plot can be obtained by applying the correction for the errors, with the widespread availability of MATLAB and similar computational tools, it is no longer necessary to compute these approximation errors manually, as the exact Bode plots can be generated with high precision.

Phase Angle versus $\log\omega$

The exact phase angle of the factor $\frac{1}{1+j\omega T}$ is

$$\phi = -\tan^{-1}\omega T \tag{7.49}$$

[1]While asymptotic Bode plots are approximations, they are generally adequate for nearly all practical control system design purposes.

Similar to a log-magnitude plot, the phase angle versus $\log\omega$ plot can be approximated by three-segment straight line plot. Let us compute the phase angle at different frequency regions. For example

i. When $\omega = 0$, the phase angle $\phi = 0°$. Thus, at frequencies, where $\omega << \frac{1}{\mathrm{T}}$, the phase angle $\phi \approx 0^0$.

ii. At the corner frequency $\omega = \frac{1}{\mathrm{T}}$, the phase angle $\phi = -\tan^{-1}\frac{\mathrm{T}}{\mathrm{T}} = -45°$

iii. When $\omega = \infty$, the phase angle $\phi = -90°$. Hence at frequencies, where $\omega >> \frac{1}{\mathrm{T}}$, the phase angle $\phi \approx -90^0$.

To accurately sketch the phase plot, it is typically necessary to compute the phase angle at several representative frequencies for each term and then add them together to obtain the phase angle. For example, the phase angle of the transfer function $\frac{1}{1 + j\omega\mathrm{T}}$ at some selected frequency points is given in Table 7.2.

ω	$\omega_c/10$	$\omega_c/2$	ω_c	$2\omega_c$	$10\omega_c$
$\phi(\omega)$	-5.7^0	-26.6^0	-45^0	-63.4^0	-84.3^0

Table 7.2: Phase angle at various frequencies

The complete Bode plot for a simple pole is shown in Figure 7.25.

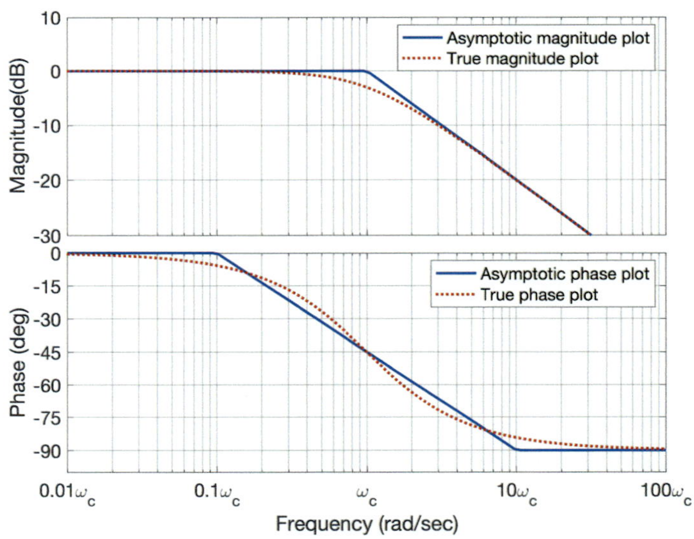

Figure 7.25: Bode plot of simple pole; upper(magnitude) and lower (phase)

Simple Zero on Real Axis

The Bode plot of simple zero on the real axis, where $G(j\omega) = 1 + j\omega\mathrm{T}$ follows similar procedure as described for simple pole with a difference that the plot has a slope of +20dB/decade, instead of -20dB/decade and the phase angle is $+\tan^{-1}\omega T$.

> Similar to a real pole, the log-magnitude on the real axis can be approximated by two straight line asymptotes.
>
> 1. The **low frequency asymptote** is a straight-line of 0 dB for the frequency range $0 < \omega < \dfrac{1}{T}$.
> 2. The **high frequency asymptote** is a straight line with a slope of +20dB/decade (+6 dB/octave) for the frequency range $\dfrac{1}{T} < \omega < \infty$.

The errors in asymptotic approximations for real zero of order r with corner frequency ω_c is obtained from Table 7.1 by replacing +r with −r. The complete Bode plot is shown in Figure 7.26.

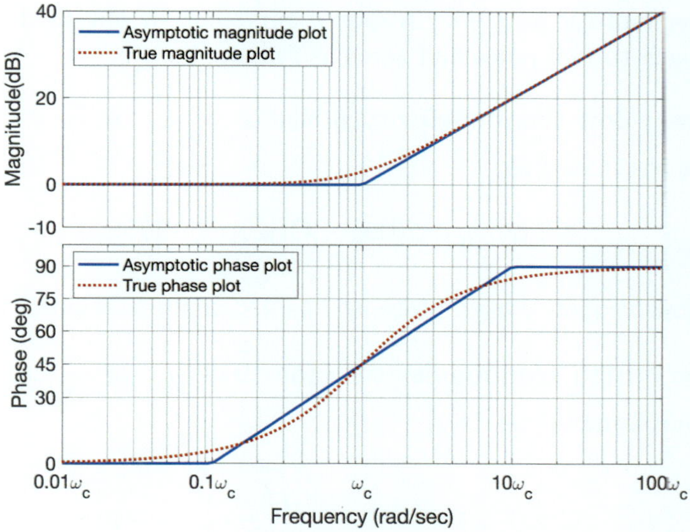

Figure 7.26: Bode plot of simple zero; upper(magnitude), lower(phase)

Quadratic Factors with Complex Conjugate Poles and Zeros

Consider a system that has complex poles with a transfer function

$$G(s) = \frac{\omega_n^2}{s^2 + 2\xi\omega_n s + \omega_n^2} = \frac{1}{1 + \left(\dfrac{2\xi}{\omega_n}\right)s + \left(\dfrac{1}{\omega_n^2}\right)s^2} \tag{7.50}$$

Note that our interest is for the cases when $0 < \xi < 1$; because if $\xi > 1$, the roots will be real and the quadratic factor can be expressed as the product of two first-order factors with simple poles.

Substituting $s = j\omega$ in Eqn(7.50) gives

$$G(j\omega) = \frac{1}{1 + 2\xi\left(j\frac{\omega}{\omega_n}\right) - \left(\frac{\omega^2}{\omega_n^2}\right)} = \frac{1}{\left[1 - \frac{\omega^2}{\omega_n^2}\right] + j2\xi\left(\frac{\omega}{\omega_n}\right)} \tag{7.51}$$

The magnitude of $G(j\omega)$ in dB is

$$|G(j\omega)|_{dB} = 20\log_{10}|G(j\omega)| = -20\log_{10}\sqrt{\left[1 - \frac{\omega^2}{\omega_n^2}\right]^2 + 4\xi^2\left(\frac{\omega}{\omega_n}\right)^2} \tag{7.52}$$

Sketching Magnitude (dB) Plot

The asymptotic frequency response curve can be obtained following a similar approach to real poles and zeros as follows:

For **very low frequencies** such that $\omega << \omega_n$ or $\frac{\omega}{\omega_n} << 1$, Eqn(7.52) can be approximated as:

$$|G(j\omega)|_{dB} = 20\log_{10}|G(j\omega)| \cong -20\log_{10} 1 = 0 \text{ dB} \tag{7.53}$$

The low frequency asymptote is thus a straight horizontal line at 0 dB.

At **very high frequencies** such that $\omega >> \omega_n$ or $\frac{\omega}{\omega_n} >> 1$, Eqn(7.52) can be approximated as:

$$|G(j\omega)|_{dB} \cong -20\log_{10}\sqrt{\left(\frac{\omega}{\omega_n}\right)^4} = -40\log_{10}\frac{\omega}{\omega_n} \text{ dB} \tag{7.54}$$

The high frequency asymptote of Eqn(7.54) is thus a straight horizontal line having a slope of −40 dB/decade, because

$$-40\log_{10}\frac{10\omega}{\omega_n} = -40 - 40\log_{10}\frac{\omega}{\omega_n}$$

The frequency at the intersection of low frequency and high frequency asymptotes is the *corner frequency*. It is obtained by equating Eqn(7.54) to Eqn(7.53) which gives $\omega = \omega_n$. Thus, for the quadratic factors, the *corner frequency* equals the natural frequency ω_n.

Error in Asymptotic Magnitude Plot

Note that while deriving the low and high frequency asymptotes, we have not considered the effect of the damping factor ξ. But the magnitude and phase curve of second order G(jω) depends both on the corner frequency ω_n as well as on the damping ratio ξ. For a second-order system, at $\omega = \omega_n$, we expect a resonant peak whose magnitude is dependent on the value of ξ. Therefore, unlike the case of real pole and zero, in this case, the actual magnitude curve of G(jω) will differ very much from the asymptotic curve. The errors due to asymptotic approximation at two octaves below and above the corner frequency are evaluated as follows:

i. The error at the corner frequency $\omega = \omega_n$ is

$$-10\log_{10}(2\xi)^2 = -20\log_{10}2\xi \quad \text{(from Eqn(7.52))}$$

ii. The error at the frequency one octave below the corner frequency, that is, at $\omega = \frac{\omega_n}{2}$, is

$$-10\log_{10}\left(\frac{9}{16}+\xi^2\right) \quad \text{(from Eqn(7.52))}$$

iii. The error at the frequency one octave above the corner frequency, that is, at $\omega = 2\omega_n$, is

$$-10\log_{10}\left(9+16\xi^2\right)+40\log_{10}(2) \quad \text{(from Eqn(7.52) and (7.54))}$$

Figure 7.27 shows the exact log-magnitude curves together with the straight line asymptotes for several values of ξ. Note that, at the corner frequency $\omega = \omega_n$, the error is zero for $\xi = 0.5$. When $0.5 < \xi < 1$, the error is negative and the true log-magnitude plot lies below the asymptotic plot. For $0 < \xi < 0.5$, the error is positive and the true log-magnitude plot lies above the asymptotic plot. Since for $\xi \geq 1$, the poles are real, we can use the methods described for a simple real pole before.

Sketching Phase Plot

The phase angle of the quadratic factor with complex poles is

$$\phi = -\tan^{-1}\left[\frac{2\xi\left(\dfrac{\omega}{\omega_n}\right)}{1-\left(\dfrac{\omega^2}{\omega_n^2}\right)}\right] \tag{7.55}$$

Similar to the magnitude, the phase angle is a function of both ω and ξ.

i. At $\omega = 0$, $\phi = -\tan^{-1}0 = 0^0$

ii. At $\omega = \omega_n$,

$$\phi = -\tan^{-1}\left(\frac{2\xi}{0}\right)0 = -\tan^{-1}(\infty) = -90^0.$$

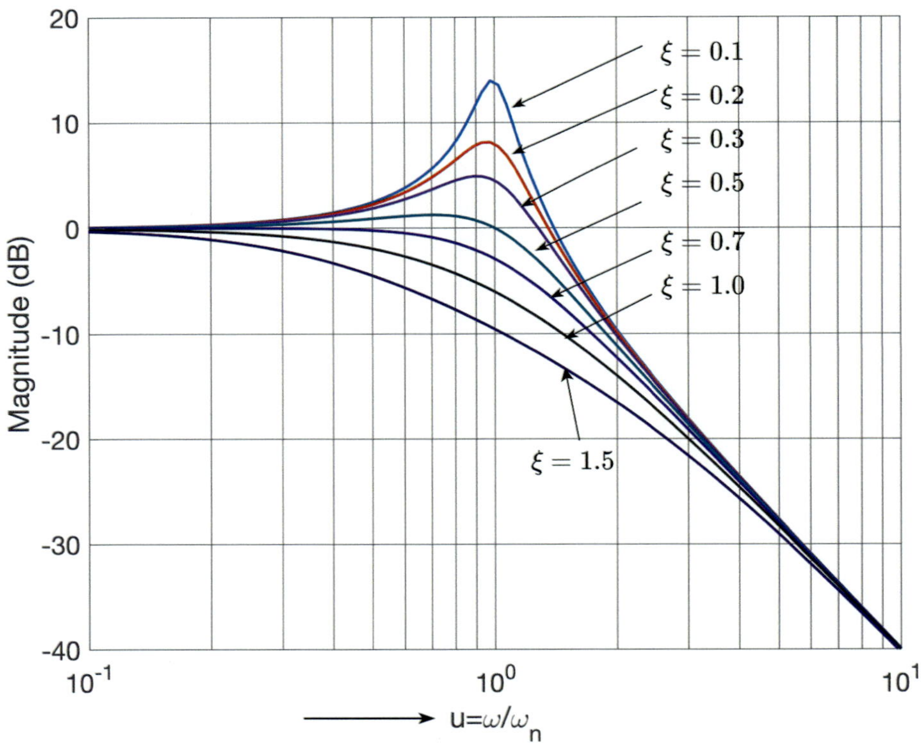

Figure 7.27: Magnitude Bode plot for $1/(s^2 + 2\xi\omega_n s + \omega_n^2)$

iii. At $\omega = \infty$, $\phi = -\tan^{-1}\infty = -180^0$.

The approximate phase angle plot is 0^0 until one tenth of the corner frequency, then -90^0/decade slope, passing through $-90°$ at the corner frequency and continues until 10 times the corner frequency. Beyond 10-times the corner frequency, the approximate phase angle is -180^0. The phase plots for several values of ξ are shown in Figure 7.28.

The analysis carried out for the Bode plot of the quadratic factor of complex poles can be applied to the second-order transfer function with complex zeros. For

$$G(j\omega) = 1 + 2\xi\left(j\frac{\omega}{\omega_n}\right) + \left(j\frac{\omega}{\omega_n}\right)^2 \tag{7.56}$$

We can obtain the magnitude and phase curves by merely inverting the log magnitude and phase angle of the factor

$$G(j\omega) = \frac{1}{1 + 2\xi\left(j\frac{\omega}{\omega_n}\right) + \left(j\frac{\omega}{\omega_n}\right)^2} \tag{7.57}$$

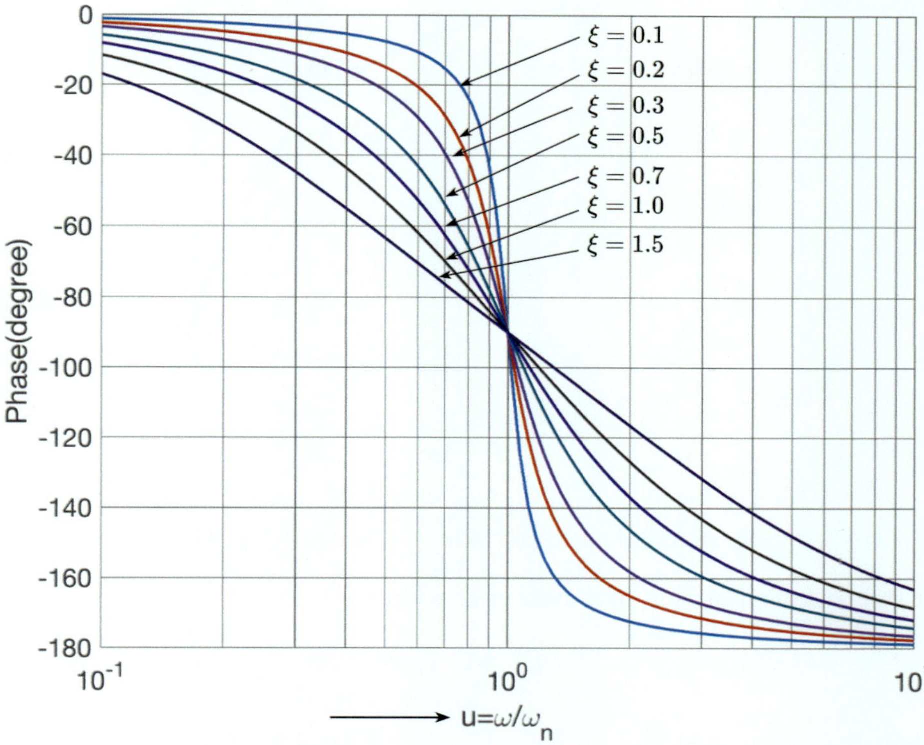

Figure 7.28: Phase Bode plot for $1/(s^2 + 2\xi\omega_n s + \omega_n^2)$

Pure Time Delay $e^{-j\omega T_d}$

Consider a system with pure time delay or transportation lag

$$G(j\omega) = e^{-j\omega T_d} \tag{7.58}$$

The magnitude of Eqn(7.58) is always unity, because

$$|G(j\omega)| = |\cos\omega T_d - j\sin\omega T_d| = 1$$

Hence, the log-magnitude of pure time delay (transportation lag) $e^{-j\omega T_d}$ is equal to 0 dB. The phase angle of the time delay is

$$\angle G(j\omega) = \phi(\omega) = \begin{cases} -\omega T_d & \text{(radians)} \\ -57.3\ \omega T_d & \text{(degrees)} \end{cases} \tag{7.59}$$

The phase angle varies linearly with ω. Its characteristic is shown in Figure 7.29.

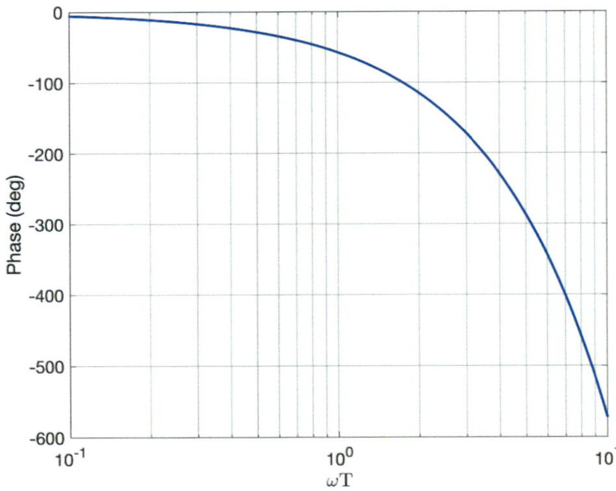

Figure 7.29: Phase angle characteristic of transport lag: $G(j\omega) = e^{-j\omega T}$

If the transfer function of a system is given by

$$G(j\omega) = G_1(j\omega)e^{-j\omega T_d} \tag{7.60}$$

then the magnitude plot of $|G(j\omega)|_{dB}$ is identical to $|G_1(j\omega)|_{dB}$. The phase plot of $\angle G(j\omega)$ is obtained by subtracting ωT_d radians from the phase plot of $G_1(j\omega)$ at various ω.

7.9 General Procedure for Constructing Bode Plots

While MATLAB provides the easiest method for plotting Bode diagrams, designers often need to sketch Bode plots manually to gain quick, first-hand insight into system behavior. The following systematic procedure outlines the steps for manually constructing the Bode plot of $G(j\omega)H(j\omega)$.

Procedure to Construct Bode Plot

1. **Express** $G(j\omega)H(j\omega)$ **in time constant form**

   - Convert $G(j\omega)H(j\omega)$ into time constant form as given in Eqn (7.29). **This step is critical**; failure to perform this conversion before sketching is a common mistake that leads to incorrect gain calculations.

2. **Identify the corner frequencies**

   - Determine the corner frequencies for each factor of $G(j\omega)H(j\omega)$
   - Arrange these corner frequencies in ascending order

- Note whether each corner frequency corresponds to a pole or to a zero

3. **Construct the Low-Frequency Segment**

   The asymptotic magnitude plot consists of straight line segments. The initial segment is determined as follows:

   a. **No poles or zeros at origin:**
      - Draw a horizontal line at magnitude $20\ \log_{10}K$ dB

   b. **n poles at origin (term** $K/(j\omega)^n$**):**
      - Draw a line with slope $-20n$ dB/decade
      - Line passes through the intersection of $\omega = 1$ and $20\ \log_{10}K$ dB

   c. **n zeros at origin (term** $K(j\omega)^n$**):**
      - Draw a line with slope $+20n$ dB/decade
      - Line passes through the intersection of $\omega = 1$ and $20\ \log_{10}K$ dB

4. **Extend to First Corner Frequency**

   Draw the first segment until the first corner frequency (due to a pole or zero) is encountered.

   a. If the first corner frequency corresponds to a zero of order (multiplicity) n, change the slope of the plot by $+20n$ dB/decade at this corner frequency.
   b. If the first corner frequency corresponds to a pole of order (multiplicity) n, change the slope of the plot by $-20n$ dB/decade at this corner frequency.

5. **Continue Through All Corner Frequencies**

   - Draw each new segment with the updated slope until the next corner frequency
   - Repeat the slope adjustment process for each subsequent corner frequency
   - The final (asymptotic) plot consists of straight-line segments with slope changes at each corner frequency

6. **Handle Complex Conjugate Pairs**

   - Complex conjugate poles of multiplicity n:
     Slope changes by $-40n$ dB/decade

- Complex conjugate zeros of multiplicity n:
  Slope changes by +40n dB/decade

7. **Apply Corner Frequency Corrections**

   - Use standard correction factors to adjust the asymptotic plot at each corner frequency for greater accuracy (see Table 7.1).

8. **Draw the Smooth Magnitude Curve**

   - Connect the corrected points with a smooth curve that remains asymptotic to the straight-line segments.

     Note: The common practice is to select the minimum frequency around one tenth of the lowest break frequency, and the maximum frequency is no lower than 10 times the highest break frequency.

9. **Calculate Phase Angles**

   - Compute phase angles at strategically selected frequencies and create a reference table.
   - A convenient approach is to calculate the phase contribution from each factor separately, then add them algebraically. For example, for a factor $(1+j\omega T)^{\pm 1}$ with corner frequency $\omega_c = \frac{1}{T}$, select the frequencies as:

$$\omega = \begin{bmatrix} \frac{\omega_c}{10} & \frac{\omega_c}{2} & \omega_c & 2\omega_c & 10\omega_c \end{bmatrix} = \begin{bmatrix} \frac{1}{10T} & \frac{1}{2T} & \frac{1}{T} & \frac{2}{T} & \frac{10}{T} \end{bmatrix}$$

     The phase angles at these frequencies are

$$\phi = \begin{bmatrix} \pm 5.7^0 & \pm 26.6^0 & \pm 45^0 & \pm 63.4^0 & \pm 84.3^0 \end{bmatrix}$$

     where the plus sign corresponds to a zero and the negative sign corresponds to a pole. For a factor $K(j\omega)^{\pm r}$, the phase angle equals $\pm 90^0 r$ at all the frequencies. Plot these points and sketch a smooth curve, which is the Bode phase plot.

10. **Complete the Phase Plot**

EXAMPLE 7.4

Problem:

Sketch a Bode plot for a feedback system with the open-loop transfer function

$$G(s)H(s) = \frac{10(s+5)}{(s+1)(s+10)}$$

SOLUTION:

Step-1: Convert to Time Constant Form

The open-loop transfer function must first be expressed in time constant form:

$$\begin{aligned} G(s)H(s) &= \frac{10(s+5)}{(s+1)(s+10)} \\ &= \frac{5\left(1+\frac{s}{5}\right)}{(1+s)\left(1+\frac{s}{10}\right)} \end{aligned}$$

Step-2:Identify System Components(Factors)

From the time constant form, we can identify:

a. **Constant gain factor** $K = 5$

b. **Corner frequency 1:** $\omega_{c1} = 1$ rad/sec (real pole)

c. **Corner frequency 2:** $\omega_{c2} = 5$ rad/sec (real zero) and

d. **Corner frequency 3:** $\omega_{c3} = 10$ rad/sec (real pole)

Note that corner frequencies are arranged in ascending order: 1, 5, 10 rad/sec.

Step-3: Construct the Magnitude Plot

The asymptotic magnitude plot consists of four distinct segments:

Segment 1: Low Frequency $\omega < 1$ rad/sec

Since $G(j\omega)H(j\omega)$ contains no poles or zeros at the origin, the initial segment is:

- Horizontal line at magnitude $= 20\log_{10}(5) \approx 14$dB
- Slope: 0 dB/decade
- Duration: From low frequency until $\omega_{c1} = 1$ rad/sec

Segment 2: First Transition ($1 < \omega < 5$ rad/sec

At $\omega_{c1} = 1$ rad/sec, the simple pole contributes -20 dB/decade:

- New slope: 0+(-20)=-20 dB/decade
- Duration: From $\omega_{c1} = 1$ rad/sec until $\omega_{c2} = 5$ rad/sec

Segment 3: Second Transition ($5 < \omega < 10$ rad/sec)

At $\omega_{c2} = 5$ rad/sec, the simple zero contributes +20 dB/decade:

- New slope: -20+(+20)=0 dB/decade
- Horizontal line parallel to frequency axis
- Duration: From $\omega_{c2} = 5$ rad/sec until $\omega_{c3} = 10$ rad/sec

Segment 4: Final Segment ($\omega > 10$ rad/sec)

At $\omega_{c3} = 10$ rad/sec, the simple pole contributes −20 dB/decade:

- Final slope: 0+(-20)=-20 dB/decade
- Duration: From $\omega_{c3} = 10$ rad/sec to high frequency

Step-4: Apply Corner Frequency Corrections

For greater accuracy, apply standard correction factors at each corner frequency using error correction tables/graph

Step-5: Construct the Phase Plot

Convert to frequency domain representation:

$$G(j\omega)H(j\omega) = \frac{5(1 + j0.2\omega)}{(1 + j\omega)(1 + j0.1\omega)}$$

Individual Phase Contributions:

- Constant gain: $\angle 5 = 0°$
- Zero term: $\angle(1 + j0.2\omega) = \tan^{-1}(0.2\omega)$
- Pole term 1: $\angle(1 + j\omega)^{-1} = -\tan^{-1}(\omega)$
- Pole term 2: $\angle(1 + j0.1\omega)^{-1} = -\tan^{-1}(0.1\omega)$

(a) **Total Phase Angle:**

$$\phi = \angle G(j\omega)H(j\omega) = -\tan^{-1}\omega - \tan^{-1}0.1\omega + \tan^{-1}0.2\omega$$

A table of phase angle for a few arbitrarily selected values of ω is given in Table 7.3.

ω	$1 + j0.2\omega$	$(1 + j\omega)^{-1}$	$(1 + j0.1\omega)^{-1}$	$\phi(\omega)$
0.5	5.76^0	-26.5^0	-2.86^0	-23.71^0
2.5	26.5^0	-68.19^0	-14^0	-55.67^0
5	45^0	-78.69^0	-26.5^0	-60.25^0
10	63.4^0	-84.3^0	-45^0	-65.85^0
100	87.14^0	-89.4^0	-84.3^0	-86.57^0

Table 7.3: Phase angle at selected frequencies

The Bode magnitude and phase plot is shown in Figure 7.30.

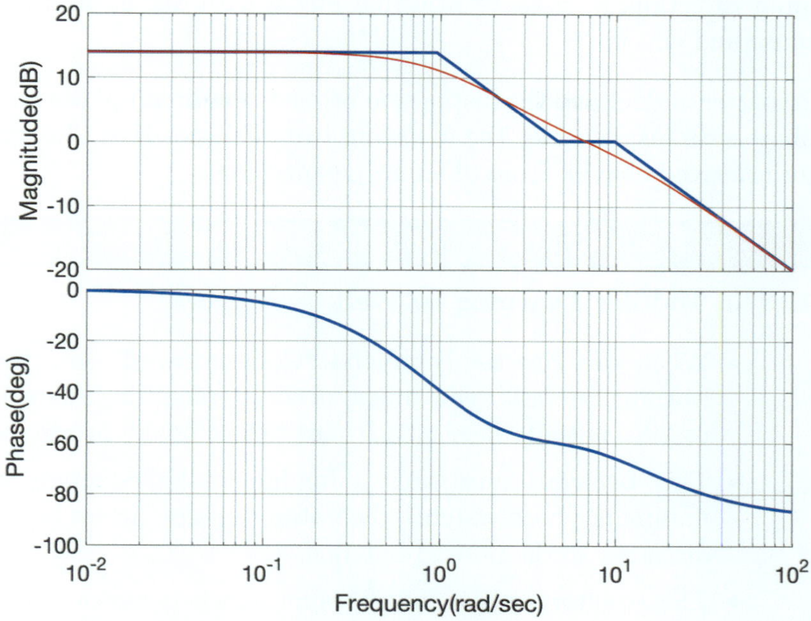

Figure 7.30: Bode plot of system of example 7.4

The complete Bode plot shows both the asymptotic magnitude plot with its four distinct segments and the smooth phase plot calculated from the individual contributions of each system component. The magnitude plot demonstrates how poles and zeros systematically alter the frequency response slope at their respective corner frequencies.

CONCEPT REINFORCEMENT PROBLEM 7.3

Problem Sketch the Bode plot for the following feedback systems with open-loop transfer functions

a. **S1:** $\mathrm{G(s)H(s)} = \dfrac{160(\mathrm{s}+0.5)(\mathrm{s}+10)}{\mathrm{s}(\mathrm{s}+2)(\mathrm{s}+40)}$

b. **S2:** $\mathrm{G(s)H(s)} = \dfrac{6400(\mathrm{s}+10)}{\mathrm{s}(\mathrm{s}+2)(\mathrm{s}^2+8\mathrm{s}+1600)}$

7.10 Nonminimum-Phase Systems

Until now, our exploration of frequency response analysis has implicitly assumed that systems exhibit minimum-phase behavior. From an intuitive standpoint, minimum-phase systems behave in a predictable and straightforward manner: when the input increases, the output increases correspondingly, and when the input decreases, the output decreases correspondingly. In essence, the output response moves in harmonious synchronization with the input direction; a characteristic that aligns with our natural expectations of cause and effect.

To better understand the distinction between minimum-phase and nonminimum-phase systems, consider the following two analogies that illustrate this counterintuitive behavior from different perspectives.

Understanding Nonminimum Phase System

1. **The Toddler Learning to Walk**

   - When you give the command "Come forward" to a toddler just learning to walk, you might observe something unexpected: the toddler initially moves backward before finally moving forward.
   - This behavior illustrates a fundamental characteristic of non-minimum-phase systems; the initial response moves in the opposite direction to the desired outcome.
   - This counterintuitive initial response, where the output moves opposite to the expected direction before eventually responding correctly, is the hallmark of non-minimum-phase behavior. Just as the toddler takes a step back to gain balance before moving forward, nonminimum-phase systems exhibit an initial "wrong-way" response that can be surprising and challenging for control system designers.
   - In contrast, a minimum-phase system would be like an experienced walker who immediately steps forward when asked, with no initial backward movement. The response is immediate, predictable, and in the same direction as the command.

2. **The Hesitant Parent**

   - A nonminimum-phase system can be compared to the behavior of some parents when their child asks for something. The child's request is like the input to the system.
   - The immediate reaction of the parents is "No": an initial response in the wrong direction. However, after some time, they change their mind and give the child what was asked for, producing the final output in the desired direction.
   - This initial opposite reaction, followed by eventual agreement, captures the essence of non-minimum phase behavior. The more emphatic the initial "No," the longer it typically takes to reach the eventual "Yes," similar to how non-minimum phase systems with right-half plane zeros farther from the origin exhibit more severe inverse responses and longer settling times.

- A minimum-phase system, in contrast, would be like parents who immediately respond appropriately to the request, with no initial rejection or hesitation.

Both analogies highlight the same fundamental characteristic: the counterintuitive inverse response that distinguishes non-minimum-phase systems from their minimum-phase counterparts. Whether viewed through the lens of physical balance (the toddler) or decision-making processes (the parent), the essence remains the same: an initial movement contrary to the ultimate destination.

Having developed an intuitive understanding through these analogies, let us now examine the frequency domain characteristics that formally distinguish minimum-phase from non-minimum-phase systems.

Among all systems with the same magnitude characteristics, the minimum phase system has the least amount of phase (delay). For example, if G(s) is the transfer function model of a minimum phase system, then there exists no other transfer function $\overline{\mathrm{G(s)}}$ which has the same magnitude $\left|\mathrm{G(j\omega)}\right| = \left|\overline{\mathrm{G(j\omega)}}\right|, \ \forall\ \omega$ but has a smaller lag.

Transfer Function Characteristics

When nonminimum-phase systems are represented by transfer function models, they exhibit two key mathematical properties:

1. **Right Half-Plane Zeros**

   Some zeros of the transfer function lie in the right half of the s plane. Specifically, linear systems with at least one zero in the open right-half plane are classified as non-minimum phase systems.

2. **Non-Unique Magnitude-Phase Relationship**

   For non-minimum phase systems, the magnitude and phase-angle characteristics are not uniquely related. This is a critical distinction from minimum phase systems, where knowing the magnitude response completely determines the phase response and vice versa.

For example, given the transfer functions

$$\mathrm{G_1(j\omega)} = \frac{1}{1+\mathrm{j\omega T}}, \text{ and } \mathrm{G_2(j\omega)} = \frac{1}{1-\mathrm{j\omega T}}$$

While the transfer functions $\mathrm{G_1(j\omega)}$ and $\mathrm{G_2(j\omega)}$ exhibit identical magnitude response, they possess distinctly different phase characteristics. For any given magnitude response, there exists exactly one minimum phase system that produces that particular frequency response characteristic. In contrast, infinitely many non-minimum phase systems can generate the same magnitude response while exhibiting different phase behaviors, as can be seen from the following example.

Consider two systems with transfer functions

$$G_1(j\omega) = \frac{1 + j\omega T}{1 + j\omega T_1}, \quad G_2(j\omega) = \frac{1 - j\omega T}{1 + j\omega T_1}, \quad 0 < T < T_1$$

The magnitude of the transfer functions of these two systems is equal, that is,

$$\left|G_1(j\omega)\right| = \left|G_2(j\omega)\right| = \frac{\sqrt{1 + \omega^2 T^2}}{\sqrt{1 + \omega^2 T_1^2}}$$

But their phase angle characteristics are different as shown in Figure 7.31.

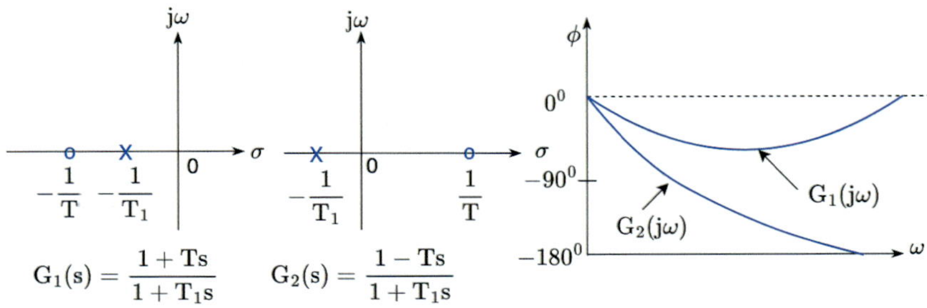

Figure 7.31: Pole zero plot of a minimum phase system $G_1(s)$ and nonminimum-phase system $G_2(s)$ and their phase angle characteristics

These two systems differ from each other by the factor $G(j\omega)$, which equals to

$$G(j\omega) = \frac{G_2(j\omega)}{G_1(j\omega)} = \frac{1 - j\omega T}{1 + j\omega T}$$

This factor always has a magnitude of unity; however, its phase angle equals to $-2\tan^{-1}\omega T$ which varies from 0^0 to -180^0 as ω is increased from zero to infinity. From this relation we find that for a given $G_1(j\omega)$, we can get infinite number of $G_2(j\omega)$ where $\left|G_2(j\omega)\right| = \left|G_1(j\omega)\right|$.

At high frequencies ($\omega = \infty$), the phase angle of a minimum phase system approaches $-90^0(n - m)$ where m and n are the degrees of the numerator and denominator polynomials of the transfer function, respectively.

In contrast,for **non-minimum phase system**, the phase at $\omega = \infty$ differs from $-90^0(n - m)$. In either case, the slope of the log-magnitude curve at high frequencies is equal to $-20(n - m)$dB/decade. This relationship can be used to distinguish between minimum-phase and non-minimum-phase systems.

Detection of Minimum/Non-minimum Phase System:

To determine whether a system is minimum phase or non-minimum phase from its frequency response characteristics, follow this systematic two-step verification process:

1. **Examine High-Frequency Asymptote Slope**
   - For a transfer function with n poles and m zeros, the expected high-frequency slope should be $-20(n - m)$ dB/decade for both minimum and non-minimum phase systems. This slope reflects the pole-zero excess and is not affected by whether zeros are in the left or right half-plane.
2. **Verify Phase ϕ at Infinite Frequency**
   - Minimum phase: $\phi(\infty) = -90°(n - m)$
   - Non-minimum phase $\phi(\infty) \neq -90°(n - m)$

Non-minimum phase systems can arise in various engineering applications, such as in the control of electronic circuits, mechanical systems, or chemical processes. Understanding and dealing with non-minimum phase behavior is important in control system design to ensure stable and robust performance. The following problem gives examples of several such systems.

CONCEPT REINFORCEMENT PROBLEM 7.4

Problem

Explain why the following systems are non-minimum phase.

i. Heating of mercury in a glass test tube
ii. Water level control in the boiler drum system
iii. Digestive system
iv. Balancing a pole on your hand
v. Boost Converter in Voltage Mode Control
vi. Hydro-turbine

7.11 Relationship Between System Type and Bode Magnitude (dB) Curve

In Chapter 5, we derived the expressions of static position, velocity, and acceleration constants for a unity feedback system, which are reproduced here as:

$$K_p = \lim_{s\to 0} G(s)$$
$$K_v = \lim_{s\to 0} sG(s)$$
$$K_a = \lim_{s\to 0} s^2G(s)$$

Recall that, for a given system, only one of the static error constants is finite and significant. For example, K_p, K_v, and K_a are finite for Type-0, Type-1, and Type-2 systems, respectively. These constants describe the low-frequency behavior of Type-0, Type-1, and Type-2 systems. Higher values of the finite error constants imply higher loop gain at lower frequencies.

The type of a system determines the slope of the Bode magnitude (dB) plot at low frequencies. Thus, by examining the low-frequency region of the magnitude plot, we can determine the type and magnitude of static error constants. This will be demonstrated in the following.

Determination of static position error constant

Consider a unity-feedback control system whose open-loop transfer function, expressed in time-constant form, is given by

$$G(s) = \frac{K(1+T_1s)(1+T_2s)\ldots\ldots(1+T_ms)}{s^N(1+T_as)(1+T_bs)\ldots\ldots(1+T_ns)}$$

For a Type 0 system: $G(j\omega) = \dfrac{K(1+j\omega T_1)(1+j\omega T_2)\ldots\ldots(1+j\omega T_m)}{(1+j\omega T_a)(1+j\omega T_b)\ldots\ldots(1+j\omega T_n)}$

Since this system does not have any pole or zero at the origin, the magnitude of $G(j\omega)$ at low frequencies equals K. This is equal to the static position error constant. That is

$$K_p = \lim_{s\to 0} G(s) = \lim_{\omega\to 0} G(j\omega) = K$$

The low frequency asymptote of the Type-0 system is flat and is coincident with a horizontal line at $20\log_{10} K_p$ dB, as shown in Figure 7.32.

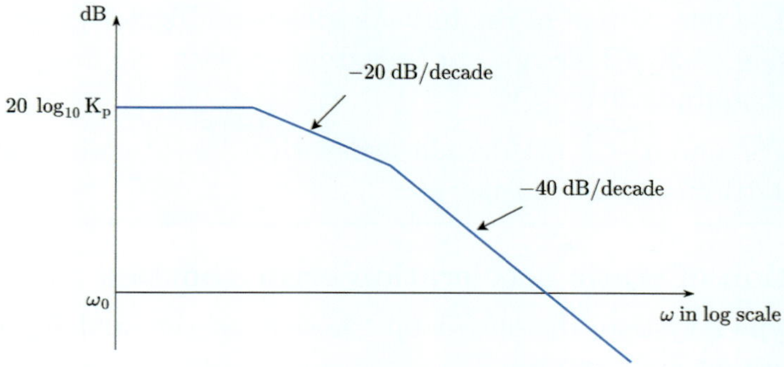

Figure 7.32: Evaluation of K_p: Log magnitude curve of a Type 0 system

Determination of static velocity error constant

For a Type 1 system, the open-loop transfer function and static velocity error constant are given by

$$G(j\omega) = \frac{K(1 + j\omega T_1)(1 + j\omega T_2)\ldots\ldots(1 + j\omega T_m)}{j\omega(1 + j\omega T_a)(1 + j\omega T_b)\ldots\ldots(1 + j\omega T_n)} \text{ and } K_v = \lim_{s\to 0} sG(s) = K$$

Since this has a single pole at the origin, the magnitude (dB) curve of this system at low frequencies will be asymptotic to a straight line with a slope of −20 dB/decade as shown in Figure 7.33.

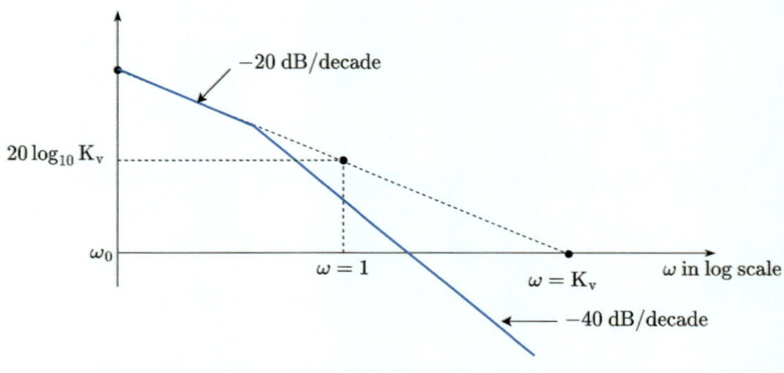

Figure 7.33: Evaluation of K_v: Log magnitude curve of a Type 1 system

Now $G(j\omega) \cong \dfrac{K}{j\omega} = \dfrac{K_v}{j\omega}$, for $\omega << 1$. Let us compute the magnitude (dB) of the initial segment of the plot at $\omega = 1$ and $\omega = K_v$. This gives

$$20\log_{10}\left|\frac{K_v}{j\omega}\right|_{\omega=1} = 20\log_{10} K_v \quad \text{and} \quad 20\log_{10}\left|\frac{K_v}{j\omega}\right|_{\omega=K_v} = 0$$

From these we conclude the following

1. The intersection of the initial segment of dB plot (a straight line of slope −20 dB/decade) or its extension with the line $\omega = 1$ has the magnitude $20\log_{10}\mathrm{K_v}$.
2. The initial −20 dB/decade segment (or its extension) intersects the 0-dB line at a frequency $\omega = \mathrm{K_v}$.

Determination of static acceleration error constant

For a Type 2 system, the open-loop transfer function and static acceleration error constant are given by

$$\mathrm{G}(j\omega) = \frac{\mathrm{K}(1+j\omega\mathrm{T_1})(1+j\omega\mathrm{T_2})\ldots\ldots(1+j\omega\mathrm{T_m})}{(j\omega)^2(1+j\omega\mathrm{T_a})(1+j\omega\mathrm{T_b})\ldots\ldots(1+j\omega\mathrm{T_n})} \quad \text{and } \mathrm{K_a} = \lim_{s\to 0} s^2\mathrm{G}(s) = \mathrm{K}$$

Since this has a double pole at the origin, the magnitude (dB) curve of this system at low frequencies will be asymptotic to a straight line with a slope of −40 dB/decade as shown in Figure 7.34.

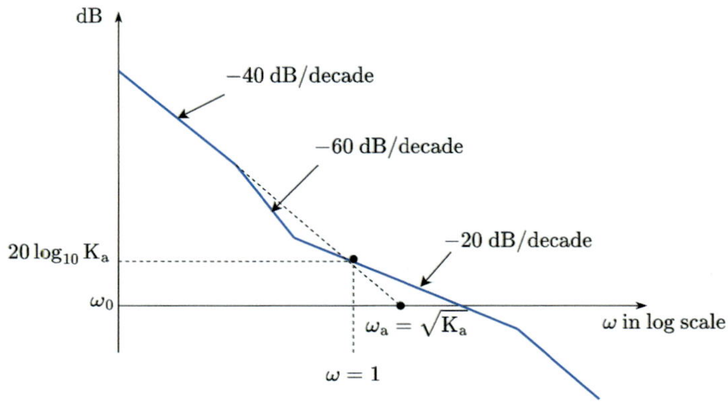

Figure 7.34: Evaluation of $\mathrm{K_a}$: Log magnitude curve of a Type 2 system

Now $\mathrm{G}(j\omega) \cong \dfrac{\mathrm{K}}{(j\omega)^2} = \dfrac{\mathrm{K_a}}{(j\omega)^2}$, for $\omega << 1$

The magnitude (dB) of the initial segment of the plot at $\omega = 1$ is

$$20\log_{10}\left|\frac{\mathrm{K_a}}{(j\omega)^2}\right|_{\omega=1} = 20\log_{10}\mathrm{K_a}$$

Suppose at the frequency $\omega = \omega_a$, the initial −40 dB/decade segment (or its extension) intersects the 0-dB line. Then

$$20\log_{10}\left|\frac{\mathrm{K_a}}{\omega^2}\right|_{\omega=\omega_a} = 20\log_{10}1$$

This gives $\omega_a = \sqrt{\mathrm{K_a}}$. From these, we conclude the following:

1. The intersection of the initial segment of dB plot (a straight line of slope −40 dB/decade) or its extension with the line $\omega = 1$ has the magnitude $20\log_{10}\mathrm{K_a}$.
2. The initial −40 dB/decade segment (or its extension) intersects the 0-dB line at a frequency $\omega = \sqrt{\mathrm{K_a}}$.

7.12 Experimental Determination of Transfer Functions

Transfer function models provide valuable insight into system behavior and enable comprehensive analysis. However, analytical determination of transfer functions for many practical systems is often challenging or impossible. In such cases, experimental frequency response data can be collected over the desired frequency range and used to determine the system's transfer function using the systematic procedure outlined below.

Step-by-Step Procedure

Transfer Function fronm Experimental Data

1. **Data Plotting:** Plot the dB gain and phase angle versus frequency on semilog graph paper to visualize the frequency response characteristics.
2. **Asymptote Construction:** Draw asymptotes on the log-magnitude curve, ensuring that:
   - Slopes must be integral multiples of ±20 dB/decade (±6 dB/octave)
   - Asymptotes follow the standard Bode plot construction rules
3. **Corner Frequency Adjustment:** Adjust corner frequencies on the asymptotic plot so that the dB value at each corner frequency differs from the actual log-magnitude plot by an amount that closely matches the standard dB correction for the identified factor type.
4. **Initial Segment Analysis:** The initial segment of the asymptotic plot reveals crucial information about the system type and gain:

   a. **Flat Initial Segment (Type-0 System):** If the initial segment is horizontal at β dB:
      - **Indicates:** No poles/zeros at the origin (Type-0 system)
      - **Gain calculation:**

$$20\log_{10}\mathrm{K} = \beta \quad \text{or} \quad \mathrm{K} = 10^{(\beta/20)} = \text{antilog}\left(\frac{\beta}{20}\right)$$

b. **Sloped Initial Segment:**

- **Slope = −20n dB/decade:** Indicates n poles at the origin
- **Slope = +20n dB/decade:** Indicates n zeros at the origin

Example calculation: For an initial slope of −40 dB/decade with magnitude β at frequency ω_i:

$$20\log_{10}\frac{K}{\omega^2}\bigg|_{\omega=\omega_i} = \beta$$

$$K = \omega_i^2 \cdot 10^{(\beta/20)} = \omega_i^2\left[\text{antilog}\left(\frac{\beta}{20}\right)\right]$$

5. **Factor Identification from Corner Frequencies:** Beyond the gain and origin poles/zeros, other factors are determined by analyzing slope changes at corner frequencies:

a. **Negative Slope Change:** If the slope changes by −20n dB/decade at corner frequency ω_1:

- **Transfer function factor:** $\dfrac{1}{\left(1+\dfrac{j\omega}{\omega_1}\right)^n}$
- **Physical meaning:** n poles at $s = -\omega_1$

b. **Positive Slope Change:** If the slope changes by +20n dB/decade at corner frequency ω_2:

- **Transfer function factor:** $\left(1+\dfrac{j\omega}{\omega_2}\right)^n$
- **Physical meaning:** n zeros at $s = -\omega_2$

6. **Resolving Complex Pole Ambiguity:** When the slope changes by −40 dB/decade at frequency ω_3, two possibilities exist:

a. **Double Real Pole:**

- **Criterion:** Error between asymptotic and actual curves ≈ 6 dB
- **Transfer function factor:** $\dfrac{1}{\left(1+\dfrac{j\omega}{\omega_3}\right)^2}$

b. **Complex Conjugate Pole Pair:**

- **Criterion:** Error is significantly different from 6 dB (usually larger)
- **Transfer function factor:** $\dfrac{1}{1+\left(\dfrac{j\omega}{\omega_3}\right)^2+j\left(2\xi\dfrac{\omega}{\omega_3}\right)}$

- **Damping factor ξ:** Determined from standard error correction charts

7. **Minimum vs. Non-Minimum Phase Classification:** This critical step distinguishes between minimum and non-minimum phase systems:

Method:

- Construct a theoretical phase curve from the determined log-magnitude curve
- Compare with experimental phase data

Classification Criteria:

a. **Minimum Phase System:**
   - No appreciable difference between theoretical and experimental phase curves
   - Both phase curves approach $-90°(n - m)$ at high frequencies, where n = denominator degree, m = numerator degree

b. **Non-Minimum Phase System:**
   - Computed phase angle is 180° less negative than experimental data
   - Indicates that at least one zero lies in the right half-plane
   - High-frequency phase may not equal $-90°(n - m)$

This experimental approach provides a powerful tool for system identification when theoretical modeling is impractical or unavailable.

In the following, we present a few illustrative examples

EXAMPLE 7.5

Problem

Estimate the transfer function of the system whose asymptotic Bode magnitude plot is shown in Figure 7.35

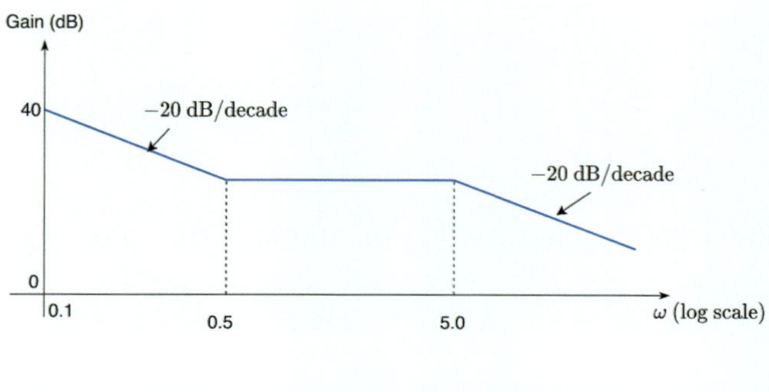

Figure 7.35: Asymptotic Log Magnitude (dB) plot of example

SOLUTION:

Since the slope of the first segment is −20 dB/decade, the system is a type-1 system with a term $\frac{K}{j\omega}$. From the plot, we observe that at $\omega = 0.1$, the gain is +40dB. Thus at $\omega = 0.1$

$$20\log_{10}\left(\frac{K}{0.1}\right) = 40 \quad \Longrightarrow \quad K = 0.1 \times 10^{\left(\frac{40}{20}\right)} = 10$$

There are two corner frequencies; $\omega_{c1} = 0.5$ rad/sec and $\omega_{c2} = 5$ rad/sec. Since at ω_{c1}, the slope changes from −20 dB/decade to 0 dB/decade, this corresponds to a simple zero.

At the second corner frequency $\omega_{c2} = 5$ rad/sec, the slope changes from 0 dB/decade to −20 dB/decade. Hence, this is due to a simple pole.

The transfer function is therefore given by

$$G(s)H(s) = \frac{10\left(1 + \frac{1}{\omega_{c1}}s\right)}{s\left(1 + \frac{1}{\omega_{c2}}s\right)} = \frac{10(1+2s)}{s(1+0.2s)} = \frac{100(s+0.5)}{s(s+5)}$$

CONCEPT REINFORCEMENT PROBLEM 7.5

Problem

Estimate the transfer function of the systems whose asymptotic Bode magnitude plots are shown in Figures 7.36 and 7.37. Assume the system is minimum-phase.

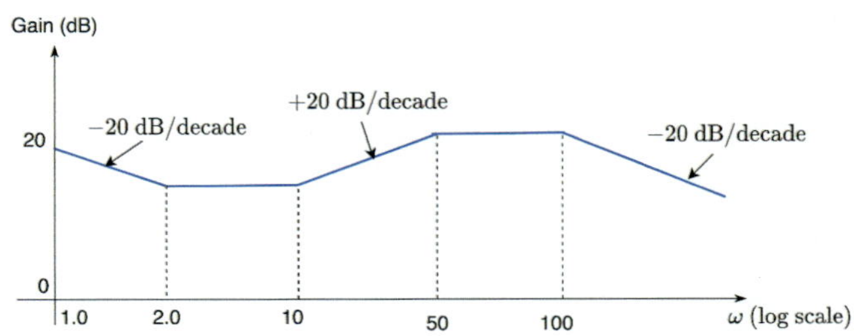

Figure 7.36: Asymptotic log magnitude (dB) plot of example

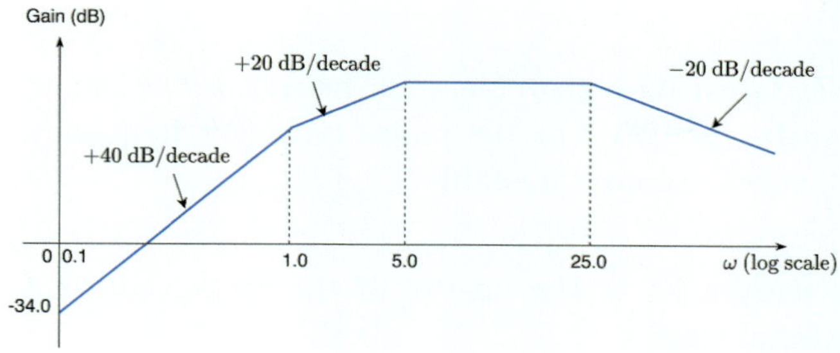

Figure 7.37: Asymptotic log magnitude (dB) plot of example

7.13 Relative Stability Analysis Via Bode Plot

In Chapter 4, we investigated system stability using the Routh-Hurwitz criterion, which provides information about *absolute stability*. This criterion gives a binary answer to stability: either *yes* or *no*. However, in most practical applications, this binary assessment is insufficient. We need to understand the *degree of stability*: if the system is stable, how close is it to instability? Can this stability be maintained under varying operating conditions, modeling uncertainties, and other practical constraints? If the system is unstable, what specific measures are required to achieve stability?

Relative stability analysis answers these critical questions. Information about relative stability can be readily obtained from the open-loop frequency response. Unless otherwise stated, the present analysis assumes that the open-loop transfer function is stable and minimum phase.

Key Stability Measures

Two commonly used measures of relative stability are the *gain margin* and the *phase margin*. Before discussing these measures, we must first establish the following fundamental definitions:

Definition 7.5. Gain Crossover Frequency: The gain crossover frequency ω_{gc} is the frequency at which the magnitude of the open-loop transfer function $L(j\omega)$ equals unity (0 dB) in a Bode plot.

Definition 7.6. Phase Crossover Frequency: The phase crossover frequency ω_{pc} is the frequency at which the open-loop transfer function $L(j\omega)$ has a phase angle of $-180°$, that is, where $\angle L(j\omega) = -180°$.

Gain Margin

Definition 7.7. Gain Margin: The gain margin is the amount by which the loop gain can be increased at the phase crossover frequency ω_{pc} before the closed-loop system becomes unstable.

The gain margin K_g is the inverse of the magnitude of $L(j\omega)$ at the phase crossover frequency:

$$K_g = \frac{1}{|L(j\omega_{pc})|}, \quad K_g(dB) = 20\log_{10} K_g = -20\log_{10}|L(j\omega_{pc})| \tag{7.61}$$

Interpretation:

- **Positive gain margin (dB):** $K_g > 1 \Rightarrow$ System is stable
- **Negative gain margin (dB):** $K_g < 1 \Rightarrow$ System is unstable

For a stable minimum phase system, the gain margin indicates how much the gain can be increased before instability occurs. For an unstable system, it shows how much the gain must be reduced to achieve closed-loop stability.

Phase Margin

Definition 7.8. Phase Margin: The phase margin is the amount of additional phase lag at the gain crossover frequency ω_{gc} required to bring the system to the verge of instability.

The phase margin γ is 180° plus the phase angle ϕ_{gc} of the open-loop transfer function at the gain crossover frequency:

$$\gamma = 180^\circ + \phi_{gc}, \quad \text{where } \phi_{gc} = \angle L(j\omega_{gc}) \tag{7.62}$$

Stability Criterion

For system stability, both the gain margin and phase margin must be positive.

This dual requirement ensures adequate stability margins against variations in both system gain and phase characteristics, providing robustness against modeling uncertainties and operational variations.

7.13.1 Procedure to Determine Gain and Phase Margins

Determination of Gain Margin

1. **Plot Construction:** Draw the Bode magnitude and phase plots with aligned frequency axes, placing the phase plot directly below the magnitude plot for easy comparison.

2. **Locate Phase Crossover Point:** Identify the intersection of the $-180°$ horizontal line with the phase plot. Note the corresponding *phase crossover frequency* ω_{pc}.

3. **Find Corresponding Magnitude:** Draw a vertical line from $\omega = \omega_{\mathrm{pc}}$ upward to intersect the magnitude plot. Record the magnitude value $|\mathrm{L}(\mathrm{j}\omega_{\mathrm{pc}})|$ (in dB) at this intersection point.

4. **Calculate Gain Margin:** The gain margin is computed as:

$$\mathrm{GM} = \mathrm{K_g} = -|\mathrm{L}(\mathrm{j}\omega_{\mathrm{pc}})| \text{ (dB)}$$

5. **Stability Assessment:**

   - **Positive GM:** Intersection on magnitude plot is *below* 0 dB line $\Rightarrow$ **Stable system**
   - **Negative GM:** Intersection on magnitude plot is *above* 0 dB line $\Rightarrow$ **Unstable system**

Determination of Phase Margin

a. **Locate Gain Crossover Point:** Identify the intersection of the 0 dB horizontal line with the magnitude plot. Note the corresponding *gain crossover frequency* ω_{gc}.

b. **Find Corresponding Phase:** Draw a vertical line from $\omega = \omega_{\mathrm{gc}}$ downward to intersect the phase plot. Record the phase angle $\angle \mathrm{L}(\mathrm{j}\omega_{\mathrm{gc}})$ at this intersection point.

c. **Calculate Phase Margin:** The phase margin is computed as:

$$\mathrm{PM} = \gamma = 180° + \angle \mathrm{L}(\mathrm{j}\omega_{\mathrm{gc}})$$

d. **Stability Assessment:**

   - **Positive PM:** Phase plot intersection is *above* $-180°$ line $\Rightarrow$ **Stable system**
   - **Negative PM:** Phase plot intersection is *below* $-180°$ line $\Rightarrow$ **Unstable system**

Key Observations

- **Dual Requirement:** For overall system stability, *both* gain margin and phase margin must be positive.
- **Graphical Interpretation:** The margins represent the "safety distance" from the critical stability boundaries (0 dB and $-180°$).
- **Design Guidelines:** Typical acceptable margins are:
  - Gain Margin: GM $\geq$ 6 dB
  - Phase Margin: PM $\geq 30°$ to $60°$
- **Multiple Crossovers:** If multiple phase or gain crossovers exist, analyze each crossover point to determine the most restrictive margins.

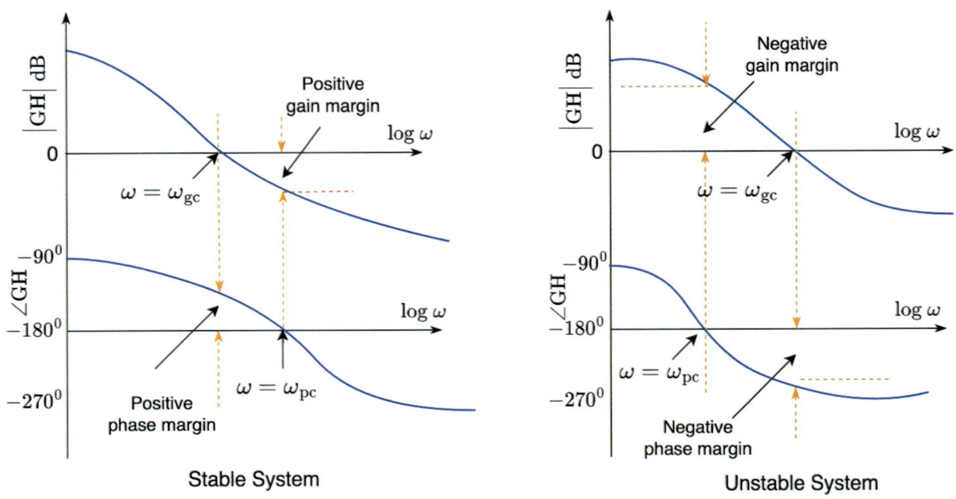

Figure 7.38: Gain and phase margin from Bode plot

Summary: Relative Stability Analysis Via Bode Plot

a

i. **Stable System:** For system stability, *both* GM and PM must be positive.

- **Frequency relationship:** $\omega_{gc} < \omega_{pc}$
- **Physical meaning:** The system has adequate margins against both gain and phase variations
- **Graphical indicator:** Magnitude plot crosses 0 dB before phase plot reaches $-180°$

ii. **Unstable System (Case 1):** If *both* GM and PM are negative, the system is unstable.

- **Frequency relationship:** $\omega_{gc} > \omega_{pc}$
- **Physical meaning:** The system lacks sufficient margins for stability
- **Corrective action:** Reduce overall loop gain or add phase lead compensation

iii. **Unstable System (Case 2):** If *either* GM *or* PM is negative, the system is unstable.

- **Partial instability:** One margin criterion fails while the other passes
- **Design implication:** Both margins must be simultaneously satisfied
- **Common scenario:** Systems with insufficient phase lead or excessive gain

iv. **Marginally Stable System:** If both GM and PM equal zero, the system is marginally stable.

- **Frequency relationship:** $\omega_{gc} = \omega_{pc}$
- **Critical condition:** System operates exactly at the stability boundary
- **Practical concern:** Any small perturbation can cause instability
- **Design recommendation:** Avoid this operating condition in practice

[a] This analysis is valid for minimum phase transfer functions

Design Guidelines

- **Recommended Margins:**
  - Gain Margin: GM $\geq$ 6 dB (factor of 2)
  - Phase Margin: PM $\geq$ 30°to 60°
- **Frequency Relationship Rule:** For stable minimum-phase systems, the gain crossover frequency should always be less than the phase crossover frequency ($\omega_{gc} < \omega_{pc}$).
- **Practical Verification:** Always check both margins; system stability requires satisfaction of both criteria simultaneously.

CONCEPT REINFORCEMENT PROBLEM 7.6

Find the relationship between the phase margin ϕ_{PM} and the damping ratio ξ for a prototype second-order system. Show that $\xi \approx 0.01\phi_{PM}$

EXAMPLE 7.6

Problem:

Determine the phase and gain margins of the system shown in Figure 7.39 for K = 10 and K = 250

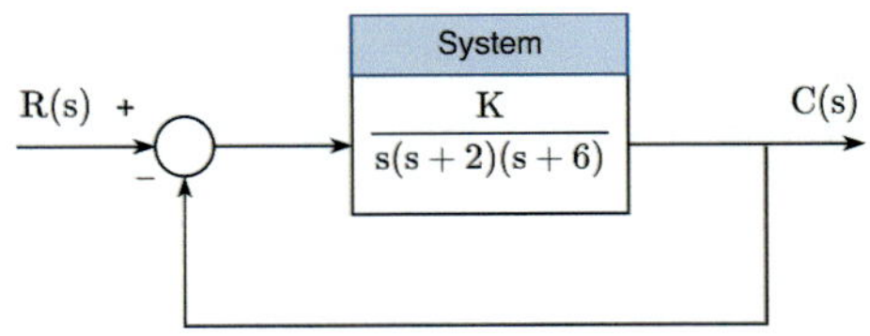

Figure 7.39: Unity feedback system

SOLUTION:

The Bode diagram of the open-loop transfer function of the system for K = 10 is shown in Figure 7.40

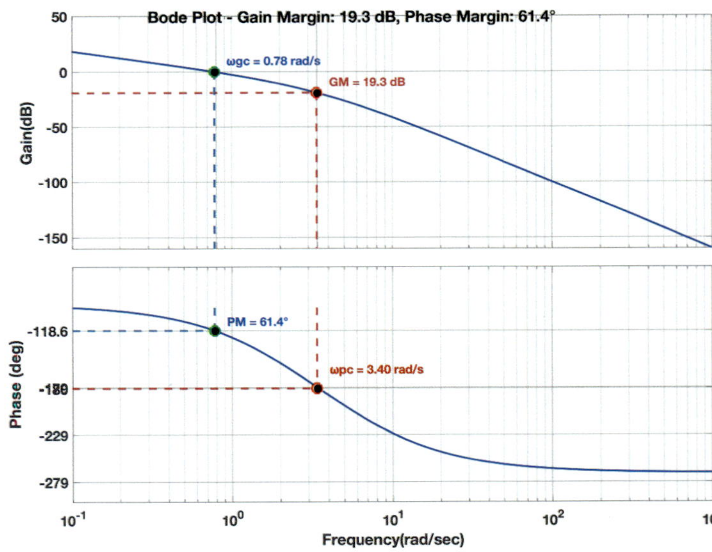

Figure 7.40: Bode diagram of system shown in Figure 7.39 for K = 10

From the plot, the gain at the phase crossover frequency $\omega_{pc} = 3.39$ rad/sec equals -19.3 dB and the phase angle at the gain crossover frequency $\omega_{gc} = 0.76$ rad/sec equals -118.5^0. Hence

$$\text{Gain margin} = 19.3 \text{ dB} \quad \text{and} \quad \text{Phase margin} = 61.5^\circ$$

The Bode diagram for K = 250 is shown Figure 7.41

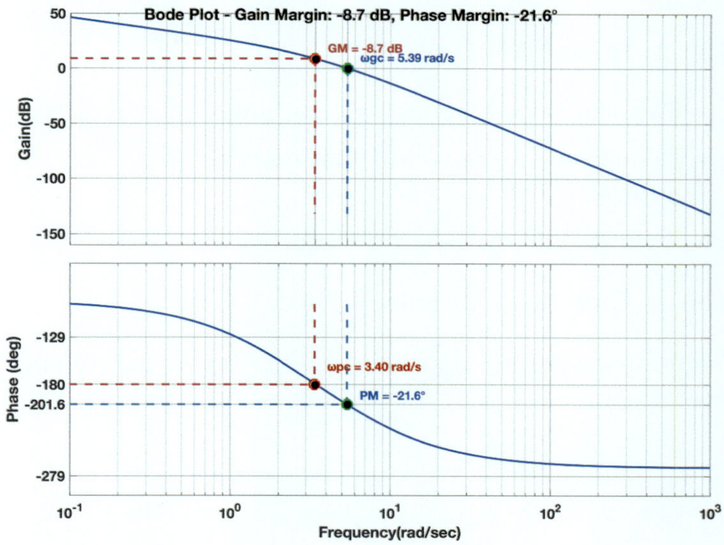

Figure 7.41: Bode diagram of system shown in Figure 7.39 for K = 250

From the plot, the gain at the phase crossover frequency $\omega_{\text{pc}} = 3.39$ rad/sec equals 8.7 dB and the phase angle at the gain crossover frequency $\omega_{\text{gc}} = 5.38$ rad/sec equals 201.5^0. Hence

$$\text{Gain margin} = -8.7 \text{ dB} \quad \text{and} \quad \text{Phase margin} = -21.5^\circ$$

Thus, it is found that increasing the gain from K = 10 to K = 250 increases the gain crossover frequency and makes both the gain and phase margin negative. The system is unstable at this gain value.

You can sketch this plot using MATLAB file `Example7_6.m`.

7.13.2 Procedure for Adjusting Gain for Desired Gain Margin

In practical control system design, the initial gain selection may not provide adequate stability margins. When the existing gain margin is insufficient for reliable operation or when design specifications require a specific gain margin value, the system gain must be systematically adjusted. Since changing the loop gain shifts the magnitude plot vertically while leaving the phase plot unchanged, we can calculate the exact gain adjustment needed to achieve any desired gain margin.

Objective: Determine the loop gain K_d required to achieve a specified gain margin GM_d.

Step 1: Initial Analysis

- Plot the Bode diagram with an initial gain $K = K_1$
- Determine the resulting gain margin GM_1 using the standard procedure
- Note the phase crossover frequency ω_{pc}

Step 2: Calculate Required Gain

The desired gain K_d that provides the target gain margin GM_d is computed from:

$$20 \log_{10} K_d - 20 \log_{10} K_1 = GM_1 - GM_d \tag{7.63}$$

Solving for K_d:

$$K_d = K_1 \times 10^{\left[\frac{GM_1 - GM_d}{20}\right]} \tag{7.64}$$

Physical Interpretation:

- If $GM_d > GM_1$: Need to *decrease* gain ($K_d < K_1$)
- If $GM_d < GM_1$: Need to *increase* gain ($K_d > K_1$)

Procedure for Adjusting Gain for Desired Phase Margin

While gain margin adjustment focuses on the phase crossover frequency, achieving a desired phase margin requires a different approach. The phase margin is measured at the gain crossover frequency, so we must determine the frequency where the phase angle provides the desired margin, then adjust the gain to make this frequency the new gain crossover point. This procedure is particularly useful when transient response specifications dictate a minimum phase margin requirement for adequate damping and stability robustness.

Objective: Determine the loop gain K_d required to achieve a specified phase margin PM_d.

Step 1: Initial Bode Plot

- Plot the Bode diagram with an initial gain $K = K_1$

- This establishes the baseline magnitude and phase characteristics

Step 2: Identify Target Frequency

Determine the frequency ω_d at which the phase angle equals:

$$\phi = -180^\circ + \phi_d \tag{7.65}$$

where ϕ_d is the desired phase margin.

Step 3: Read Magnitude at Target Frequency

- Obtain the magnitude G_d (in dB) at frequency ω_d
- This represents how much the gain must be adjusted to make ω_d the new gain crossover frequency

Step 4: Calculate Required Gain

The desired gain K_d that provides the target phase margin ϕ_d is computed from:

$$20\log_{10} K_1 - 20\log_{10} K_d = G_d \tag{7.66}$$

Solving for K_d:

$$K_d = K_1 \times 10^{\left(-\frac{G_d}{20}\right)} \tag{7.67}$$

Physical Interpretation:

- The magnitude plot is shifted vertically by $-G_d$ dB
- The new gain crossover occurs at ω_d, where the phase margin is exactly ϕ_d

Key Takeaways

- **Gain Scaling Effect:** Changing the loop gain K shifts the entire magnitude plot vertically without affecting the phase plot.
- **Verification:** After calculating the new gain, verify the results by plotting the adjusted Bode diagram and confirming the achieved margins.
- **Design Trade-offs:**
  - Adjusting gain for desired GM may change the PM
  - Adjusting gain for desired PM may change the GM
  - Both margins should be checked after any gain adjustment

- **Practical Considerations:** These procedures assume the system remains minimum-phase and that no additional crossover frequencies are introduced by the gain change.

EXAMPLE 7.7

Problem:

The open-loop transfer function of a unity feedback control system is given by

$$G(s) = \frac{K}{s(s+4)(s+8)(s+10)(s+15)}$$

Determine the value of K such that the system may have

a. a gain margin of 20 dB
b. a phase margin of 45^0

SOLUTION:

a.

Step-1: The Bode diagram of the open-loop transfer function of the system for K = 100 is shown in Figure 7.42

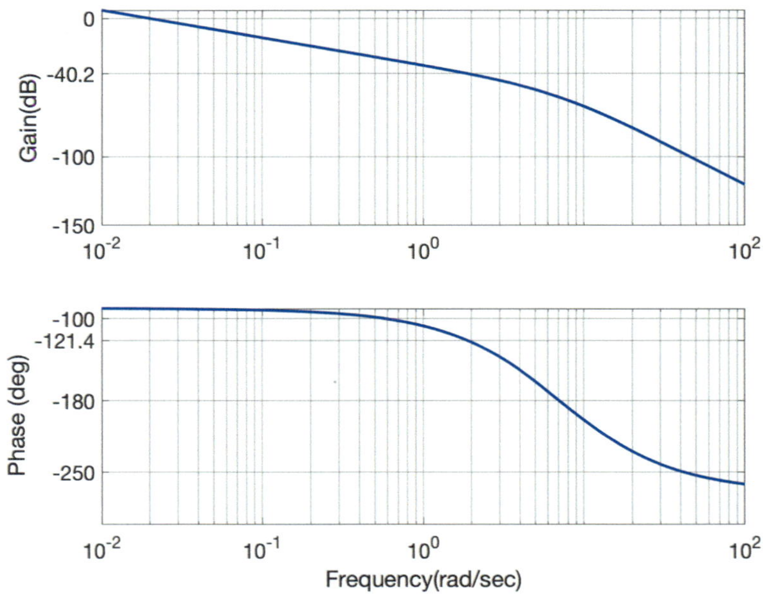

Figure 7.42: Bode diagram of system for K = 100

From the plot, the gain of the system at the phase crossover frequency is found to be -47dB. Hence the gain margin of the system for K = 100 equals to 47 dB; that is $G_{m1} = 47$dB.

The desired gain margin $G_{md} = 20$ dB. The gain has to be adjusted as:

$$K_d = K_1 \times 10^{\left[\frac{G_{m1} - G_{md}}{20}\right]} = 100 \times 10^{1.35} = 2238.7$$

b.

The desired phase margin is 45^0. From the Bode diagram of Figure 7.42, we determine the following.

$\omega_d = 1.48$ rad/sec=the frequency where the phase angle is -135^0.

The gain G_d at this frequency equals to -38dB. Thus, the gain that gives this phase margin is obtained from

$$K_d = K_1 \times 10^{\left(-\frac{G_d}{20}\right)} = 100 \times 10^{1.9} = 7943.3$$

Note: Due to the graphical nature of the calculations, there may be some discrepancies in the answers.

CONCEPT REINFORCEMENT PROBLEM 7.7

Problem:

For the system shown in Figure 7.43, determine the value of K which gives a step response with a 10% overshoot. Plot the step response using MATLAB.

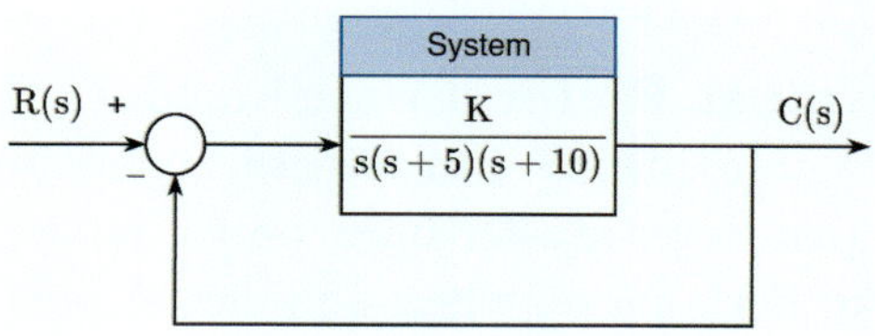

Figure 7.43: Unity feedback system

7.14 Polar Plots

The polar plot of a function of complex variable s, G(s), is a plot of the magnitude $|G(j\omega)|$ versus the phase angle of $\angle G(j\omega)$ on polar coordinates as ω is varied from zero to infinity. It can mathematically be viewed as the mapping of the positive half of the imaginary axis of the s-plane onto the G(jω) plane. (More on complex mapping to be discussed later). An example of a polar plot in the G(jω) plane is shown in Figure 7.44

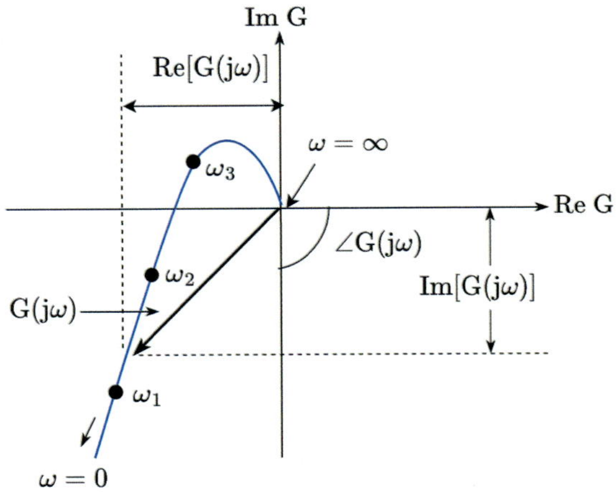

Figure 7.44: Polar plot

Let us first outline a few key features of polar plots before delving into examples of some familiar functions.

Important Features of Polar Plot

1. **Unified Representation:** The polar plot represents the frequency response of the system over the entire frequency range in a single plot, unlike the Bode plot, which depicts the frequency response characteristics in two separate plots (magnitude and phase plots). This unified representation provides a complete picture of the system's frequency domain behavior in one comprehensive diagram.

2. **Individual Factor Contributions:** However, the polar plot does not explicitly indicate the contributions of each factor of the open-loop transfer function to the overall frequency response. Unlike Bode plots, where individual pole and zero effects can be traced, the polar plot shows only the net combined effect of all system components.

3. **Polar Coordinate System:** The plot is constructed on polar graph paper featuring:

   - **Concentric circles:** Represent magnitude values
   - **Radial lines:** Represent phase angle values
   - **Point representation:** Each point is uniquely defined by its

magnitude (distance from origin) and phase angle (angular position)

4. **Phase Angle Convention:** The phase angle measurement follows standard mathematical convention:
   - **Positive angles:** Measured counter clockwise from the positive real axis (reference axis at 0°)
   - **Negative angles:** Measured clockwise from the positive real axis
   - **Reference direction:** The positive real axis serves as the 0° reference line
5. **Frequency Parameter:** Although not explicitly shown on the plot axes, frequency ω serves as the parameter that traces the curve. Each point on the polar plot corresponds to the system's response at a specific frequency, with the curve typically traced from $\omega = 0$ to $\omega = \infty$.

General Procedure for Plotting Polar Plot

The systematic steps for constructing the polar plot of G(s)H(s) are outlined below:

Procedure for Plotting Polar Pot

1. **Frequency Domain Substitution:** Replace s with $j\omega$ in the open-loop transfer function G(s)H(s) to obtain $G(j\omega)H(j\omega)$.
2. **Extract Magnitude and Phase Expressions:** Derive analytical expressions for:
   - Magnitude: $|G(j\omega)H(j\omega)|$
   - Phase angle: $\angle G(j\omega)H(j\omega)$
3. **Frequency Response Calculation:** Compute the magnitude and phase of $G(j\omega)H(j\omega)$ for various frequency values and create a comprehensive table.

   Key Boundary Points:

   a. **Starting Point ($\omega = 0$):** The magnitude $|G(j0)H(j0)|$ and phase angle $\angle G(j0)H(j0)$ define the beginning point of the polar plot.

   b. **Ending Point ($\omega = \infty$):** The magnitude and phase angle of $G(j\omega)H(j\omega)$ as $\omega \to \infty$ determine the terminating point of the polar plot.

4. **Alternative Method - Rectangular Coordinates:** If $G(j\omega)H(j\omega)$ can be expressed in rectangular form as:

$$G(j\omega)H(j\omega) = G_R(\omega) + jG_I(\omega)$$

where:

- $G_R(\omega)$ = Real part of $G(j\omega)H(j\omega)$
- $G_I(\omega)$ = Imaginary part of $G(j\omega)H(j\omega)$

Then, the polar plot can be constructed on ordinary Cartesian graph paper by plotting $G_I(\omega)$ versus $G_R(\omega)$ as ω varies from 0 to ∞.

5. **Plot Construction:**

- **Polar method:** Plot points using magnitude and phase coordinates on polar graph paper
- **Rectangular method:** Plot points using real and imaginary coordinates on Cartesian graph paper
- **Curve tracing:** Connect the plotted points with a smooth curve, indicating the direction of increasing frequency with arrows

Important Notes

- **Frequency Direction:** Always indicate the direction of increasing frequency on the polar plot using arrows or frequency labels.
- **Critical Frequencies:** Pay special attention to frequencies where the phase changes rapidly or the magnitude has local extrema.
- **Symmetry Properties:** For systems with real coefficients, the polar plot for negative frequencies is the complex conjugate of the positive frequency response.
- **Method Selection:** Choose between polar coordinates and rectangular coordinates based on which form makes the calculations more straightforward for the given transfer function.

We now present several functions that frequently appear in polar plot analysis.

Integral Factor:

Consider the transfer function of an integral, which is given by

$$G(s) = \frac{1}{s}$$

Substituting s = jω in the above equation gives $G(j\omega) = \frac{1}{j\omega} = \frac{1}{\omega}\angle{-90^\circ}$

Thus,

$$\lim_{\omega\to 0} G(j\omega) = \lim_{\omega\to 0} \frac{1}{\omega}\angle{-90^\circ} = \infty\angle{-90^\circ} \quad \text{and}$$
$$\lim_{\omega\to \infty} G(j\omega) = \lim_{\omega\to \infty} \frac{1}{\omega}\angle{-90^\circ} = 0\angle{-90^\circ}$$

The polar plot of the integral factor is the negative imaginary axis as shown in Figure 7.45a.

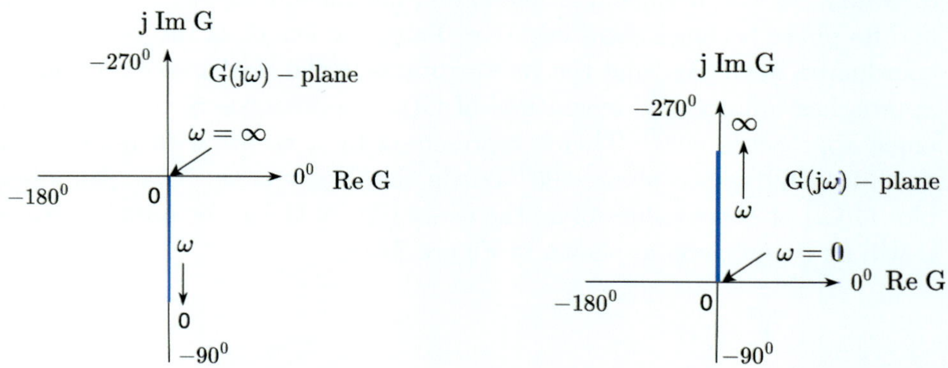

(a) Polar plot of pure integral factor (b) Polar plot of pure derivative factor

Figure 7.45: Polar plots of pure integral and derivative factors

Derivative Factor

Consider the transfer function of a derivative factor

$$G(s) = s \quad \text{or,} \quad G(j\omega) = j\omega = \omega\angle 90^\circ$$

Thus

$$\lim_{\omega\to 0} G(j\omega) = \lim_{\omega\to 0} \omega\angle 90^\circ = 0\angle 90^\circ \text{ and}$$
$$\lim_{\omega\to \infty} G(j\omega) = \lim_{\omega\to \infty} \omega\angle 90^\circ = \infty\angle 90^\circ$$

Hence, the polar plot of the derivative factor is the positive imaginary axis as shown in Figure 7.45b.

First Order Factor Due to Real Pole

For this factor, we have

$$G(s) = \frac{1}{1 + Ts}$$

Substitute s = jω in the above gives

$$G(j\omega) = \frac{1}{1 + j\omega T} = \frac{1}{\sqrt{1 + \omega^2 T^2}} \angle{-\tan^{-1}\omega T}$$

Hence

$$\lim_{\omega \to 0} G(j\omega) = \lim_{\omega \to 0} (1)\angle 0^\circ = 1\angle 0^\circ \text{ and}$$

$$\lim_{\omega \to \infty} G(j\omega) = \lim_{\omega \to \infty} \left(\frac{1}{\omega T}\right) \angle{-\tan^{-1}\omega T} = 0\angle{-90^\circ}$$

Thus, we find that at $\omega = 0$, the magnitude of $G(j\omega)$ equals 1 and its phase angle is 0^0. It is represented by a vector of unit length directed in the 0^0 direction. As the frequency ω increases, the magnitude of $G(j\omega)$ decreases, and its phase becomes more negative. Thus, the length of the vector in polar coordinates decreases, and the vector rotates in the clockwise direction. As ω approaches infinity, the magnitude of $G(j\omega)$ approaches zero and the phase angle approaches -90^0. This is represented by a vector with infinitesimally small length directed along -90^0-axis in the $G(j\omega)$-plane. If we compute and plot $G(j\omega)$ at other values of ω, the exact plot of $G(j\omega)$ for positive values of ω will be a semicircle as shown in Figure 7.46a.

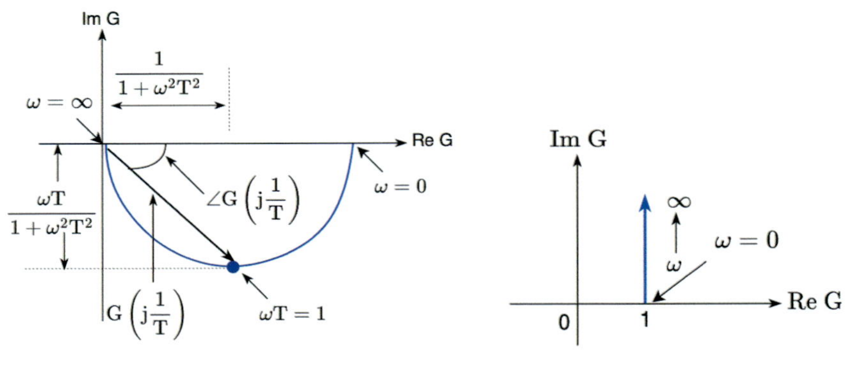

(a) Polar plot of simple pole (b) Polar plot of simple zero

Figure 7.46: Polar plots of simple pole and zero

The formal proof of this is left as an exercise.

First Order Factor Due to Real Zero

For this factor, we have

$$G(s) = 1 + Ts \implies G(j\omega) = 1 + j\omega T = \sqrt{1 + \omega^2 T^2}\, \angle \tan^{-1}\omega T$$

Hence

$$\lim_{\omega \to 0} G(j\omega) = \lim_{\omega \to 0} (1)\angle 0^\circ = 1\angle 0^\circ \text{ and}$$

$$\lim_{\omega \to \infty} G(j\omega) = \lim_{\omega \to \infty} (\omega T)\angle \tan^{-1}\omega T = \infty\angle 90^\circ$$

The polar plot of $1 + j\omega T$ is the upper half of the straight line passing through the point $(1, 0)$ in the complex $G(j\omega)$-plane and is parallel to the $j\omega$-axis as shown in Figure 7.46b. Comparing this with Figure 7.46a, we find that the appearance of the polar plot of $1 + j\omega T$ is completely different from that of $\dfrac{1}{1 + j\omega T}$.

Quadratic factor due to complex conjugate poles

For this factor, we have

$$G(s) = \frac{\omega_n^2}{s^2 + 2\xi\omega_n s + \omega_n^2} \implies G(j\omega) = \frac{\omega_n^2}{(j\omega)^2 + 2\xi\omega_n(j\omega) + \omega_n^2}$$

Hence

$$\lim_{\omega\to 0} G(j\omega) = \lim_{\omega\to 0}\left(\frac{\omega_n^2}{\omega_n^2}\right) = 1\angle 0^\circ$$

and

$$\lim_{\omega\to\infty} G(j\omega) = \lim_{\omega\to\infty}\frac{\omega_n^2}{(j\omega)^2} = 0\angle{-180^\circ}$$

Thus, the polar plot of the factor due to complex conjugate poles starts from $1\angle 0^\circ$ and ends at $0\angle{-180^\circ}$ as the frequency ω increases from zero to infinity. The high frequency part of $G(j\omega)$-phasor is tangent to the negative real axis.

The polar plots of this factor are shown in Figure 7.47a for small and large values of ξ. Although the exact shape of the polar plot is dependent on the value of the damping ratio ξ, the general shape of the plot is the same, both for the underdamped $(1 > \xi < 0)$ and the overdamped case $(\xi > 1)$.

Let us evaluate the frequency at which the $G(j\omega)$ locus intersects the imaginary axis. Now

$$G(j\omega) = \frac{\omega_n^2}{(j\omega)^2 + 2\xi\omega_n(j\omega) + \omega_n^2} = \frac{1}{\left(1 - \dfrac{\omega^2}{\omega_n^2}\right) + j2\xi\left(\dfrac{\omega}{\omega_n}\right)}$$

At the imaginary axis intersection, the real part of $G(j\omega)$ will be zero. From the above we find that at $\omega = \omega_n$, $\text{Re}[G(j\omega)] = 0$. At this frequency

$$\text{Im}[G(j\omega)] = -\frac{1}{2\xi} \quad \text{and} \quad \angle G(j\omega)\big|_{\omega=\omega_n} = -90^\circ$$

Some of the important points of this analysis are:

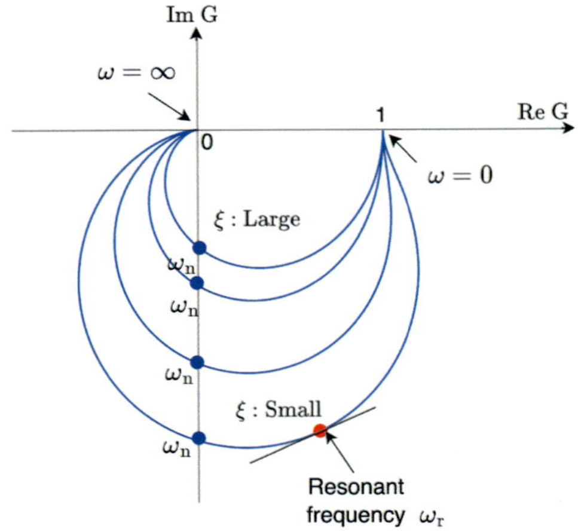

(a) Polar plot of a quadratic factor corresponding to complex conjugate poles

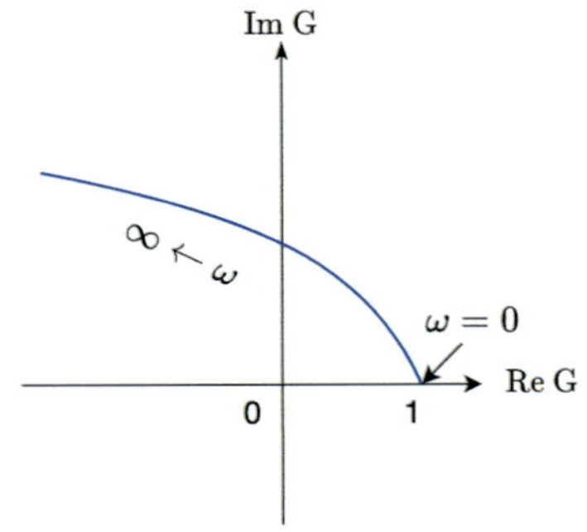

(b) Polar plot of a quadratic factor corresponding to complex conjugate zeros

Figure 7.47: Polar plots of quadratic factors

a. The $G(j\omega)$ locus intersects the imaginary axis at the undamped natural frequency ω_n.

b. The frequency point whose distance from the origin is maximum is the resonant frequency ω_r; as shown in Figure 7.47a.

c. The resonant peak M_r is obtained by computing the ratio of magnitude of $G(j\omega)$ at resonant frequency ω_r to the magnitude of $G(j\omega)$ at $\omega = 0$. That is

$$M_r = \frac{|G(j\omega_r)|}{|G(j0)|}$$

d. If the damping ratio ξ increases well beyond unity, the $G(j\omega)$ locus approaches a semicircle (similar to the polar plot of Figure 7.47a). This is because, for the system which is heavily damped, both the characteristic roots become real, with one being much smaller than the other. The effects of the larger root on the system response become very small, and the smaller root essentially determines the system dynamics. The second-order system behaves as a first-order system. This fact is reflected in the polar plot.

Quadratic Factor Due to Complex Conjugate Zeros

For this factor, we have

$$G(s) = \frac{s^2 + 2\xi\omega_n s + \omega_n^2}{\omega_n^2}$$

This implies

$$G(j\omega) = \frac{(j\omega)^2 + j\omega 2\xi\omega_n + \omega_n^2}{\omega_n^2}$$

$$= 1 + 2\xi\left(j\frac{\omega}{\omega_n}\right) + \left(j\frac{\omega}{\omega_n}\right)^2$$

$$= \left(1 - \frac{\omega^2}{\omega_n^2}\right) + j\left(\frac{2\xi\omega}{\omega_n}\right)$$

Hence

$$\lim_{\omega\to 0} G(j\omega) = \lim_{\omega\to 0}(1) = 1\angle 0^\circ \text{ and}$$

$$\lim_{\omega\to\infty} G(j\omega) = \lim_{\omega\to\infty}\left(\frac{j\omega}{\omega_n}\right)^2 = \infty\angle -180^\circ$$

The real and imaginary parts of $G(j\omega)$ are

$$\text{Re}[G(j\omega)] = 1 - \frac{\omega^2}{\omega_n^2} \text{ and } \text{Im}[G(j\omega)] = \frac{2\xi\omega}{\omega_n}$$

From the above, we find that for $\omega > 0$, the imaginary part of $G(j\omega)$ is positive and is monotonically increasing, whereas the real part of $G(j\omega)$ is monotonically decreasing from unity. The locus intersects the imaginary axis at $\omega = \omega_n$ where $\text{Im}[G(j\omega) = 2\xi$. The general shape of the polar plot is shown in Figure 7.47b.

EXAMPLE 7.8

Problem:

Sketch the polar plot of the system with the transfer function

$$G(s) = \frac{K}{s(1 + Ts)}$$

SOLUTION:

Now the sinusoidal transfer function is expressed as:

$$G(j\omega) = \frac{K}{j\omega(1 + j\omega T)}$$

The magnitude and phase of $G(j\omega)$ at $\omega = 0$ and $\omega = \infty$ are computed as:

$$\lim_{\omega\to 0} G(j\omega) = \lim_{\omega\to 0}\frac{K}{j\omega} = \infty\angle -90^\circ$$

and

$$\lim_{\omega\to\infty} = \lim_{\omega\to\infty} \frac{K}{(j\omega \cdot jT\omega)} = \lim_{\omega\to\infty} \frac{K}{-\omega^2} = 0\angle{-180^\circ}$$

After determining the values of $G(j\omega)$ locus at $\omega = 0$ and $\omega = \infty$, let us determine the intersection of $G(j\omega)$ locus on real and imaginary axis (if any). Express $G(j\omega)$ as the sum of real and imaginary parts as:

$$G(j\omega) = \frac{K(-j\omega)(1 - j\omega T)}{j\omega(1 + j\omega T)(-j\omega)(1 - j\omega T)}$$

$$= \frac{-KT}{1+\omega^2T^2} - j\frac{K}{\omega(1+\omega^2T^2)}$$

$$= \mathrm{Re}[G(j\omega)] + j\mathrm{Im}[G(j\omega)]$$

The intersection point of $G(j\omega)$ locus with the real axis is obtained by making

$$\mathrm{Im}[G(j\omega)] = \frac{K}{\omega(1+\omega^2T^2)} = 0$$

This gives $\omega = \infty$. Since at this frequency $G(j\omega) = 0\angle{-90^\circ}$, the locus intersects the real axis at the origin.

Similarly, the intersection point of $G(j\omega)$ locus with the imaginary axis is obtained by making

$$\mathrm{Re}[G(j\omega)] = \frac{-KT}{1+\omega^2T^2} = 0$$

Again, the only real solution of the above occurs at $\omega = \infty$, which is the origin of the $G(j\omega)$-plane. Hence, we conclude that the polar plot of $G(j\omega) = \dfrac{K}{j\omega(1 + j\omega T)}$ does not intersect any one of the axes at any finite non-zero frequency.

Further, at $\omega = 0$, $\mathrm{Im}[G(j\omega)] = \infty$ and $\mathrm{Re}[G(j\omega)] = -KT$. The general shape of the polar plot of this function is shown in Figure 7.48, which shows that the plot is asymptotic to the vertical line passing through the point $(-KT, 0)$.

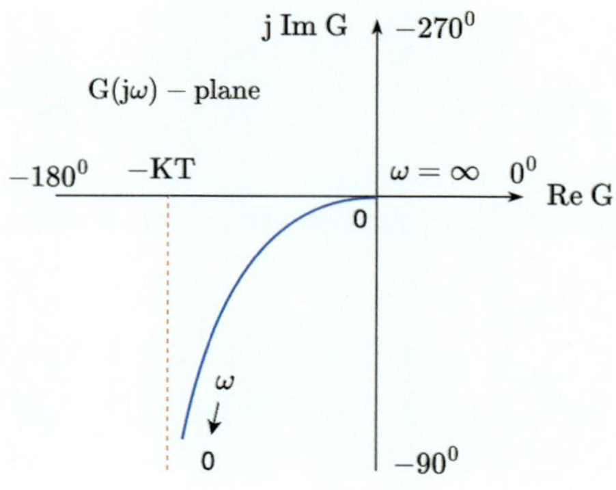

Figure 7.48: Polar plot of $\dfrac{K}{s(1+Ts)}$

CONCEPT REINFORCEMENT PROBLEM 7.8

Problem

Sketch the polar plot of the following systems with loop transfer functions

a. $G(s) = \dfrac{K}{(1+T_1s)(1+T_2s)}$

b. $G(s) = \dfrac{K}{s(1+T_1s)(1+T_2s)}$

EXAMPLE 7.9

Problem:

Sketch the polar plot of the system with the transfer function

$$G(s) = \frac{K}{(1+T_1s)(1+T_2s)(1+T_3s)}$$

SOLUTION:

Now the sinusoidal transfer function is expressed as:

$$G(j\omega) = \frac{K}{(1+j\omega T_1)(1+j\omega T_2)(1+j\omega T_3)}$$

The magnitude and phase of $G(j\omega)$ at $\omega = 0$ and $\omega = \infty$ are computed as:

$$\lim_{\omega\to 0} G(j\omega) = \lim_{\omega\to 0}(K) = K\angle 0^\circ$$

and

$$\lim_{\omega\to\infty} G(j\omega) = \lim_{\omega\to\infty} \frac{K}{j^3 T_1 T_2 T_3 \omega^3} = 0\angle -270^\circ$$

Let us rationalize $G(j\omega)$ and express it as the sum of real and imaginary parts as:

$$G(j\omega = \frac{K(1-j\omega T_1)(1-j\omega T_2)(1-j\omega T_3)}{(1+j\omega T_1)(1-j\omega T_1)(1+j\omega T_2)(1-j\omega T_2)(1+j\omega T_3)(1-j\omega T_3)}$$

$$= \frac{K[1-\omega^2(T_1T_2+T_2T_3+T_3T_1)]}{(1+\omega^2T_1^2)(1+\omega^2T_2^2)(1+\omega^2T_3^2)}$$
$$-j\frac{K[\omega(T_1+T_2+T_3)-\omega^3T_1T_2T_3]}{(1+\omega^2T_1^2)(1+\omega^2T_2^2)(1+\omega^2T_3^2)}$$

$$= \mathrm{Re}[G(j\omega)] + j\mathrm{Im}[G(j\omega)]$$

The intersection point of $G(j\omega)$ locus with the imaginary axis is obtained by making

$$\mathrm{Re}[G(j\omega)] = \frac{K[(1-\omega^2(T_1T_2+T_2T_3+T_3T_1)]}{(1+\omega^2T_1^2)(1+\omega^2T_2^2)(1+\omega^2T_3^2)} = 0$$

This gives

$$\omega = \frac{1}{\sqrt{T_1T_2+T_2T_3+T_3T_1}}$$

Let us determine the frequency where the imaginary part of $G(j\omega) = 0$. Now

$$\mathrm{Im}[G(j\omega)] = \frac{K[\omega(T_1+T_2+T_3)-\omega^3T_1T_2T_3]}{(1+\omega^2T_1^2)(1+\omega^2T_2^2)(1+\omega^2T_3^2)} = 0$$

This satisfied for

$$\omega = 0, \quad \omega = \infty, \text{ and } \omega = \sqrt{\frac{T_1+T_2+T_3}{T_1T_2T_3}}$$

The general shape of the polar plot of this function is shown in Figure 7.49.

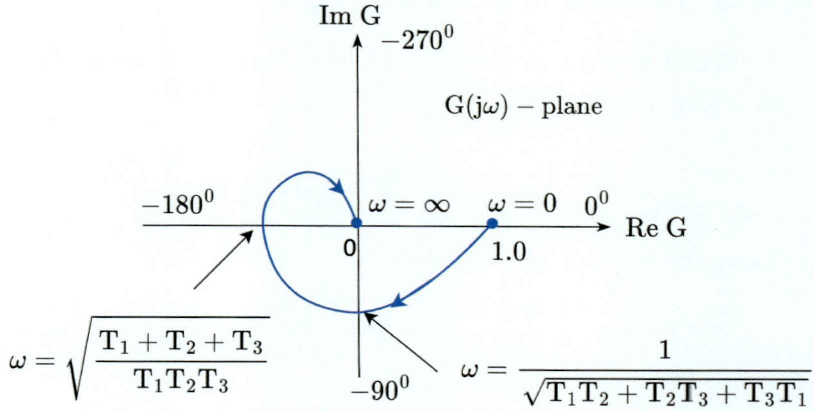

Figure 7.49: Polar plot of $\dfrac{K}{(1 + T_1s)(1 + T_2s)(1 + T_3s)}$

7.15 Effects of Addition of Poles and Zeros to G(s) on Shape of Polar Plots

We know that the performance of a feedback control system is affected by adding poles and zeros to the open-loop transfer function G(s). In Chapter 6, we observed how the shape of the root locus changes with the addition of poles and zeros to the open-loop transfer function. It is of interest to know how the shape of the polar plot changes due to the addition of poles/zeros.

Consider a simple Type 0, first-order system with transfer function

$$G(s) = \frac{1}{1 + Ts} \tag{7.68}$$

Its polar plot is shown in Figure 7.50.

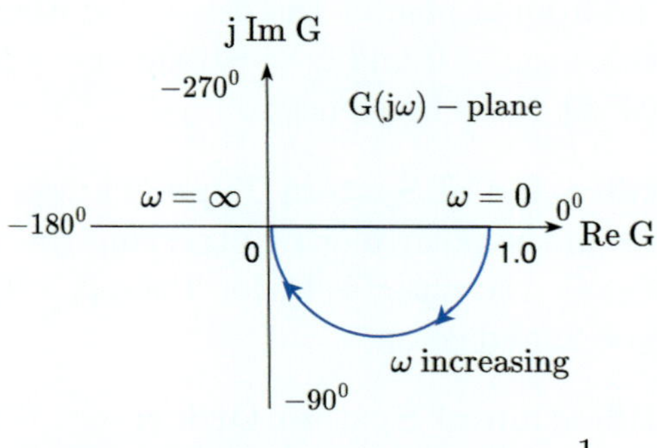

Figure 7.50: Polar plot of $G(s) = \dfrac{1}{1 + Ts}$

Effects of Addition of Poles at Origin

If we add n poles at the origin to the transfer function $G(s) = \dfrac{1}{1+Ts}$, it is modified to

$$G(s) = \frac{1}{s^n(1+Ts)}$$

Thus, for $n = 1, 2$ and 3, the transfer functions are:

$$G_1(s) = \frac{1}{s(1+Ts)},\quad G_2(s) = \frac{1}{s^2(1+Ts)},\quad \text{and } G_3(s) = \frac{1}{s^3(1+Ts)}$$

The corresponding polar plots are shown in Figure 7.51.

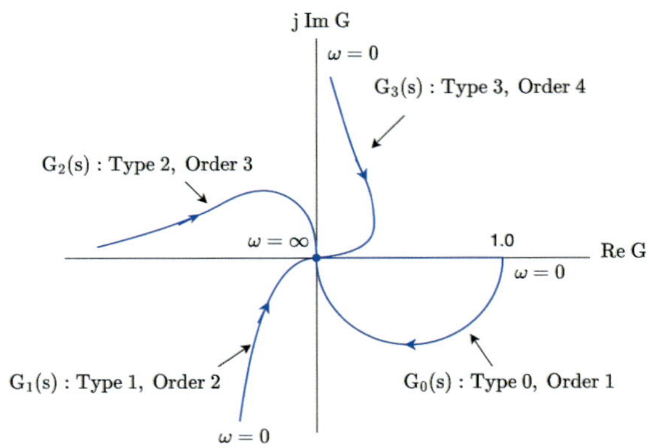

Figure 7.51: Polar plots of Type 0, Type 1, Type 2 and Type 3 systems

From these plots, we observe the following:

Key Observations

i. Each additional pole at the origin decreases the phase angle by 90° both at $\omega = 0$ and $\omega = \infty$, thereby rotating the entire plot by $-90°$ at these frequencies.

ii. **Identification of System Type:** The system type can be identified from the polar plot by observing the phase angle at $\omega = 0$. Specifically, the phase is 0° for Type 0, $-90°$ for Type 1, $-180°$ for Type 2, and so on.

iii. **Identification of System Order:**

Similarly, the order of the system can be determined by examining the phase at $\omega = \infty$. Although the magnitude approaches zero for all systems, the phase angle equals $-90°$ for first-order, $-180°$ for second-order, $-270°$ for third-order, and so forth.

iv. The addition of poles at the origin reduces the stability margin of the system. For instance, a system with two or more poles at the origin (Type 2 or higher) is often unstable.

Effects of Addition of Non-Origin Poles

Consider a simple Type 0 first-order system given by

$$G_0(s) = \frac{1}{1 + T_1 s}$$

By successively adding simple poles to $G_0(s)$, we obtain

$$G_1(s) = \frac{1}{(1 + T_1 s)(1 + T_2 s)}$$
$$G_2(s) = \frac{1}{(1 + T_1 s)(1 + T_2 s)(1 + T_3 s)}$$
$$G_3(s) = \frac{1}{(1 + T_1 s)(1 + T_2 s)(1 + T_3 s)(1 + T_4 s)}$$

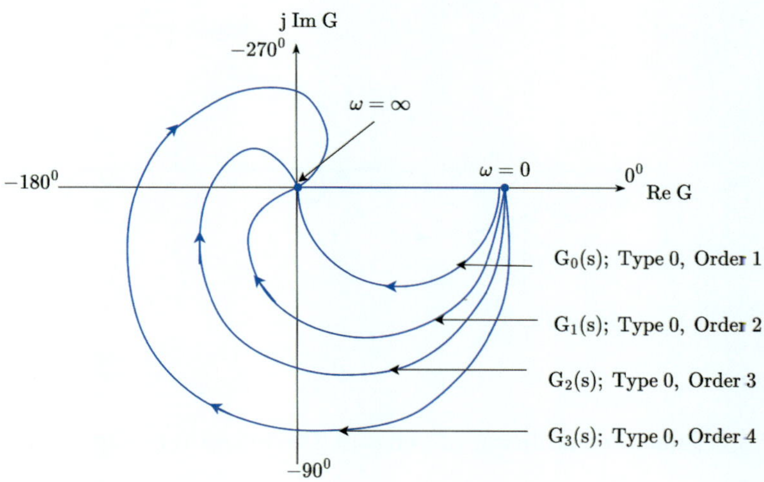

Figure 7.52: Effects of adding non-origin poles on shape of polar plots

The polar plots of these functions are shown in Figure 7.52. From these plots, we observe the following:

Key Points

i. Each additional non-origin simple pole rotates the polar plot by $-90°$ at $\omega = \infty$.

ii. Adding poles to the open-loop transfer function generally reduces the stability margin of the closed-loop system (to be discussed later).

7.15.1 Effects of Addition of Zeros at Origin

By successively adding zeros at the origin to G(s) of Eqn(7.68), we obtain

$$G_1(s) = \frac{s}{1+Ts}, \quad G_2(s) = \frac{s^2}{1+Ts}, \text{ and } G_3(s) = \frac{s^3}{1+Ts}$$

The polar plots of these functions are shown in Figure 7.53. From these plots, we notice that the addition of a zero at the origin causes the polar plot to rotate counter clockwise by 90° both at $\omega = 0$ and at $\omega = \infty$.

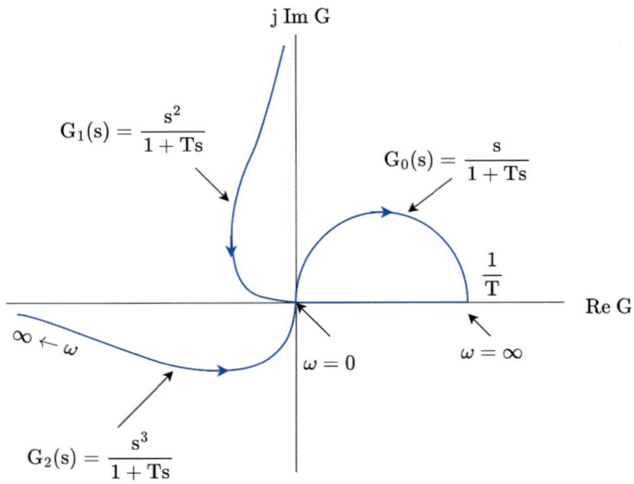

Figure 7.53: Effects of addition of zeros at the origin on shape of polar plots

From these plots, we observe that the addition of each zero at the origin rotates the polar plot counter clockwise by 90° at both $\omega = 0$ and $\omega = \infty$.

7.16 Relative Stability Analysis Using Polar Plots

In Section 7.13, we discussed how to compute the gain and phase margins from the Bode plot. These margins can also be obtained from the polar plot and from the Nyquist plot (to be discussed in Section 7.18). Consider the polar plot of a typical minimum-phase open-loop system shown in Figure 7.54. The plot intersects:

- the negative real axis at frequency $\omega = \omega_{pc}$, with an intercept **a**, and
- the unit circle (centered at the origin) at frequency $\omega = \omega_{gc}$.

The line joining the origin to the point corresponding to $\omega = \omega_{gc}$ makes an angle γ with the negative real axis.

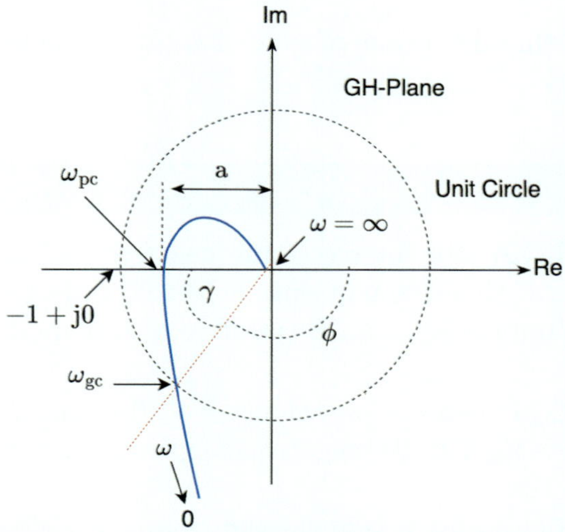

Figure 7.54: Relative stability analysis by polar plot

Before discussing how these intercepts and phase angles influence the stability of the closed-loop system, let us first define the following terms from the perspective of polar plots in the $G(j\omega)H(j\omega) = L(j\omega)$ plane.

Definition 7.9. Phase Crossover Point The phase crossover point in the $L(j\omega)$ plane is the point at which the polar plot of $L(j\omega)$ intersects the negative real axis i.e., when the phase of $L(j\omega)$ is $-180°$ (or an odd multiple thereof)

Definition 7.10. Phase Crossover Frequency The phase crossover frequency ω_{pc} is the frequency at which the polar plot of $L(j\omega)$ intersects the negative real axis. Equivalently, it is defined as the frequency at which

$$\angle L(j\omega_{pc}) = 180°(\text{or an odd multiple thereof}).$$

Gain Margin

As discussed earlier in Section 7.13, the gain margin, K_g is defined as the reciprocal of the magnitude of $L(j\omega)$ at the phase crossover frequency. From the polar plot in Figure 7.54, we observe

$$\left| L(j\omega_{pc}) \right| = a$$

Hence, the gain margin is

$$K_g = \frac{1}{\left|L(j\omega_{pc})\right|} = \frac{1}{a}, \quad K_g(dB) = 20\log_{10}K_g = -20\log_{10}a \tag{7.69}$$

Based on this definition of gain margin, we note the following important points:

Gain Margin: A Few Key Points

a. If the $L(j\omega)$ plot intersects the negative real axis between 0 and -1, i.e., $a < 1$, then the plot does not enclose the $(-1 + j0)$ point, and the gain margin is $K_g > 0$ dB (positive gain margin).

b. If the $L(j\omega)$ plot encloses the $(-1 + j0)$ point, i.e., $a > 1$, then the gain margin is $K_g < 0$ dB (negative gain margin), and the system is unstable.

c. If the $L(j\omega)$ plot passes through the $(-1 + j0)$ point, i.e., $a = 1$, then the gain margin is $K_g = 0$ dB. The system is marginally stable.

d. If the $L(j\omega)$ plot does not intersect the negative real axis, then $a = 0$ and the gain margin is $K_g = \infty$.

e. **Gain margin of non-minimum-phase systems:** Points (a)–(d) apply only to **minimum-phase systems**. For **non-minimum-phase systems**, stability in general cannot be determined solely from the sign of the gain margin. Such systems may remain stable even when the phase crossover point lies to the left of $(-1 + j0)$, i.e., with a negative gain margin. The stability of these systems depends on the number of encirclements of the $(-1 + j0)$ point by the $L(j\omega)$ plot, as determined by the Nyquist stability criterion (to be discussed in Section 7.18). Nevertheless, the closeness of the phase crossover to the $(-1 + j0)$ point still indicates relative stability.

Definition 7.11. Gain Crossover Point: The gain crossover point is the point on the polar plot of $L(j\omega)$ where the plot intersects the unit circle centered at the origin.

Definition 7.12. Gain Crossover Frequency The gain crossover frequency,

ω_{gc}, is the frequency at which the magnitude of $L(j\omega)$ equals unity, that is,

$$\left|L(j\omega_{gc})\right| = 1$$

Phase Margin

As discussed before in Section 7.13, the phase margin is the amount of phase lag which can be added at the gain crossover frequency to bring the closed-loop system to the verge of instability. The phase margin is shown by an angle γ in Figure 7.54 where we see that

$$\angle L(j\omega)\Big|_{\omega=\omega_{gc}} = -180^0 + \gamma$$

Thus, the phase margin is given by

$$\boxed{\gamma = 180^0 + \phi, \quad \text{where} \;\; \phi = \angle L(j\omega)\Big|_{\omega=\omega_{gc}}} \tag{7.70}$$

Note that the angle of $L(j\omega)\Big|_{\omega=\omega_{gc}}$ in the above expression is measured negatively, that is, in the clockwise direction from the positive real axis. The phase margin γ is measured positively in the counter clockwise direction from the negative real axis.

If we introduce additional phase lag γ at $\omega = \omega_{gc}$, then

$$\angle L(j\omega)\Big|_{\omega=\omega_{gc}} = -180^\circ$$

while $\left|L(j\omega_{gc}\right| = 1$. The polar plot will therefore pass through $-1 + j0$ point, driving the system to the verge of instability.

Additional Remarks on Gain and Phase Margins

a. The concepts of gain margin (GM) and phase margin (PM) for relative stability analysis are applicable only to open-loop stable, minimum-phase systems.

b. For minimum-phase systems to be closed-loop stable, both GM and PM must be positive (see Figure 7.55).

c. A system can have a satisfactory GM but an unsatisfactory PM, or vice versa, as illustrated in Figure 7.56.

d. Neither GM nor PM alone is sufficient to assess the relative stability of a system. Both must be considered together. For example, systems with the same GM may have different PM values (Figure 7.57a). The system S_1 with a higher PM is relatively more stable than S_2. In many cases, the phase margin provides a better (more informative) measure of relative stability.

e. The polar plot of higher-order systems with complex dynamics may have multiple gain crossover frequencies (Figure 7.57b). For stable systems with multiple gain crossovers, the PM should be evaluated at the highest gain crossover frequency ω_{gc1}. The corresponding phase margin is ϕ, as shown in Figure 7.57b.

f. A large GM or PM indicates a highly stable system; however, such systems are often sluggish in response.

g. A GM close to unity or a PM close to zero corresponds to a highly oscillatory system.

h. Although the desired GM and PM values depend on specific design requirements (e.g., response speed, disturbance rejection), a simple guideline is that a GM of about 6 dB or a PM of 35°–45° ensures a reasonable degree of relative stability.

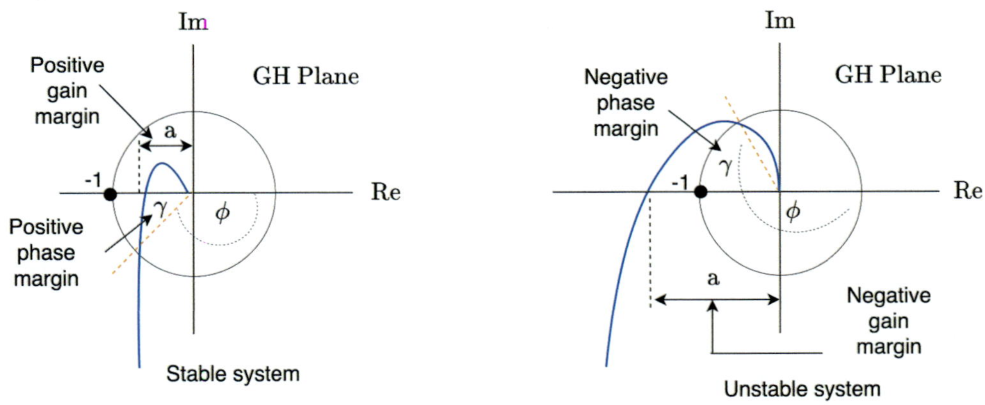

Figure 7.55: Gain and phase margin from polar plot

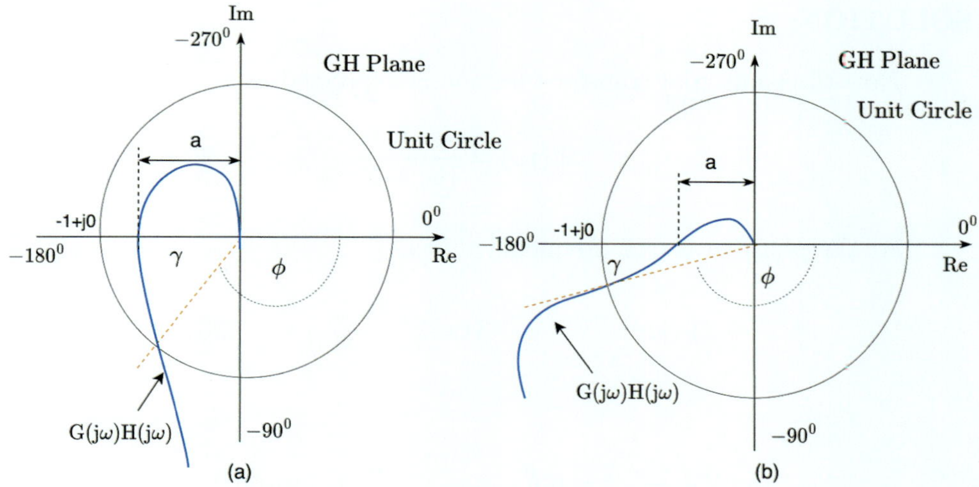

Figure 7.56: (a) Polar plot of a system with good phase margin but poor gain margin (b) Polar plot of a system with good gain margin but poor phase margin

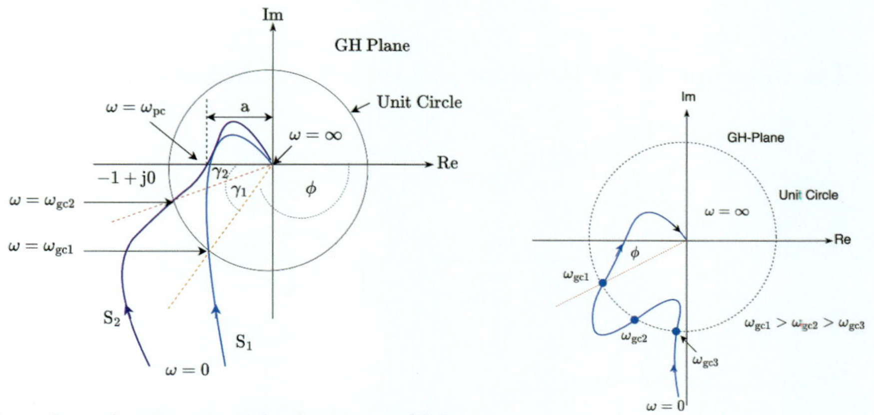

(a) Polar plot of systems with the same GM but different degrees of relative stability due to different phase margins

(b) Polar plot of systems with multiple gain crossover frequencies

Figure 7.57

EXAMPLE 7.10

Problem:

Determine the phase crossover frequency, gain margin, and stability condition of a system with an open-loop transfer function

$$L(s) = G(s)H(s) = \frac{K}{(s+\beta)^3}, \quad K, \beta > 0$$

SOLUTION:

Now the sinusoidal transfer function is expressed as:

$$\mathrm{L}(j\omega) = \frac{\mathrm{K}}{(j\omega + \beta)^3}$$

At the phase crossover frequency $\omega = \omega_{pc}$

$$\angle \mathrm{L}(j\omega)\Big|_{\omega=\omega_{pc}} = -3\tan^{-1}\left(\frac{\omega_{pc}}{\beta}\right) = -180^\circ$$

or,

$$\tan^{-1}\left(\frac{\omega_{pc}}{\beta}\right) = 60^0 \quad \Longrightarrow \quad \frac{\omega_{pc}}{\beta} = \tan(60^0) = \sqrt{3}$$

This gives the phase crossover frequency

$$\omega_{pc} = \sqrt{3}\beta$$

The intercept "a" of the polar plot on the real axis is determined as:

$$\mathrm{a} = \Big|\mathrm{L}(j\omega\Big|_{\omega=\omega_{pc}} = \frac{\mathrm{K}}{\left[\sqrt{\beta^2 + \omega_{pc}^2}\right]^3}$$

$$= \frac{\mathrm{K}}{\left[\sqrt{\beta^2 + 3\beta^2}\right]^3} = \frac{\mathrm{K}}{\left[\sqrt{4\beta^2}\right]^3} = \frac{\mathrm{K}}{8\beta^3}$$

Hence the gain margin $\mathrm{K_g}$

$$\mathrm{K_g} = \frac{1}{\mathrm{a}} = \frac{8\beta^3}{\mathrm{K}}$$

and

$$\mathrm{K_g}\ \mathrm{dB} = 20\ \log_{10}(8\beta^3) - 20\ \log_{10}\mathrm{K}$$

The system is stable if

$$0 < \mathrm{K} < 8\beta^3$$

You can verify this result using the Routh-Hurwitz criterion.

CONCEPT REINFORCEMENT PROBLEM 7.9

Problem:

The open-loop transfer function of a unity feedback system is

$$G(s) = \frac{1 + as}{s^2}$$

Determine the value of "a" such that the phase margin of the closed-loop system is

a. 30^0 b. 45^0 c. 60^0

EXAMPLE 7.11

Problem:

The block diagram of a space vehicle control system with a derivative controller is shown in Figure 7.58. Show that the gain K which would result in a phase margin of θ equals $a \sin\theta \tan\theta$.

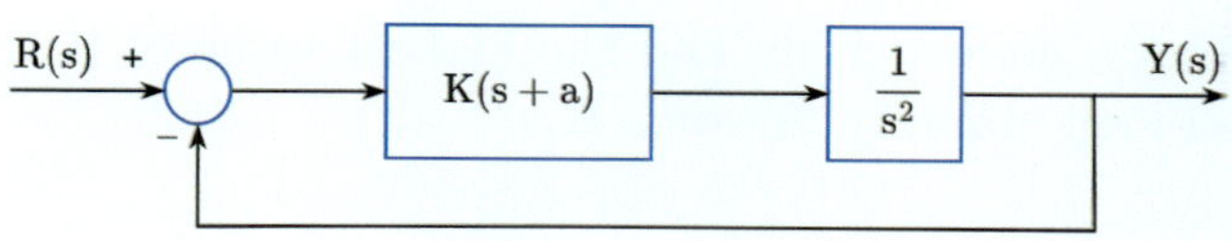

Figure 7.58: Space vehicle control system

SOLUTION:

Now

$$L(s) = G(s)H(s) = \frac{K(s + a)}{s^2}$$

Hence

$$L(j\omega) = \frac{K(j\omega + a)}{(j\omega)^2}, \quad \text{and} \quad \angle L(j\omega) = -180^0 + \tan^{-1}\left(\frac{\omega}{a}\right)$$

Since the desired phase margin is θ, then at the gain crossover frequency ω_{gc}, $\angle L(j\omega_{gc}) = -180^0 + \theta$. That is

$$-180^0 + \tan^{-1}\left(\frac{\omega_{gc}}{a}\right) = -180 + \theta \quad \Longrightarrow \quad \tan^{-1}\left(\frac{\omega_{gc}}{a}\right) = \theta$$

This gives the gain crossover frequency

$$\omega_{gc} = a \tan\theta$$

Note that the magnitude of $L(j\omega)$ must be equal to 1 (0 dB) at ω_{gc}. That is

$$\left|\frac{K(j\omega + a)}{(j\omega)^2}\right|_{\omega = a\tan\theta} = 1 \implies \left.\frac{K(\sqrt{a^2 + \omega^2})}{\omega^2}\right|_{\omega = a\tan\theta} = 1$$

This gives

$$K = \frac{a^2 \tan^2\theta}{\sqrt{a^2 + a^2\tan^2\theta}} = \frac{a^2\tan^2\theta}{a\sec\theta} = a\sin\theta\tan\theta$$

For example, if $a = 2$, then to achieve a phase margin of 45°, the gain K should be $\sqrt{2}$.

CONCEPT REINFORCEMENT PROBLEM 7.10

Problem:

For the system shown in Figure 7.59, determine the phase crossover frequency, gain margin, and the highest value of K that would ensure closed-loop stability. Assume $K, n > 0$.

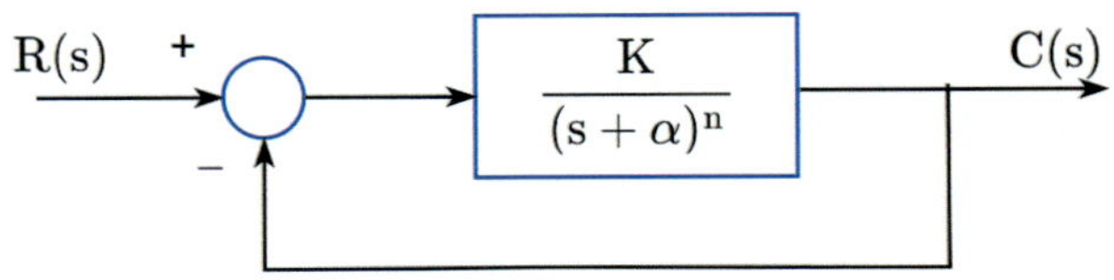

Figure 7.59: Closed-loop system

EXAMPLE 7.12

Problem:

The open-loop transfer function of a unity feedback system is given by

$$G(s) = \frac{Ke^{-Ts}}{s}$$

Find the maximum value of K which will ensure stability of the system.

SOLUTION:

Now

$$L(s) = \frac{Ke^{-Ts}}{s}$$

Hence

$$L(j\omega) = \frac{Ke^{-j\omega T}}{j\omega}, \quad \text{and} \quad \angle L(j\omega) = -90^0 - \omega T$$

We are interested in finding the gain that ensures stability. Therefore, the phase crossover frequency where the phase angle is -180^0 is obtained as:

$$-90^0 - \omega T = -180, \quad \Longrightarrow \quad \omega = \frac{\pi}{2T}$$

At this frequency, the magnitude of $L(j\omega)$ should be less than 1.

That is

$$\left|L(j\omega)\right| = \frac{K}{\omega} = \frac{K}{(\pi/2T)} < 1 \quad \Longrightarrow \quad K < \frac{\pi}{2T}$$

CONCEPT REINFORCEMENT PROBLEM 7.11

Problem:

Determine the frequency where the Polar plot intersects the real axis and find the range of K for closed-loop stability of a unity feedback system with open-loop transfer function

$$L(s) = G(s) = \frac{Ke^{-Ts}}{1 + as}$$

7.17 Stability Analysis in Frequency Domain using Nyquist Stability Criterion

Stability is the fundamental requirement for any practical control system. It cannot be compromised regardless of other performance considerations. In Chapters 4 and 6, we examined stability from a time-domain perspective using the Routh-Hurwitz criterion and root-locus method. Both approaches are based on the principle that a linear system is stable if and only if all roots of its characteristic equation lie in the left half of the complex s-plane (LHP).

The Routh-Hurwitz criterion determines absolute stability by analyzing the signs of characteristic polynomial coefficients, while the root-locus method provides insights into relative stability. However, both methods have significant limitations: they require the open-loop transfer function to be a rational function of the complex variable s. When the system contains pure time delays of the form e^{-Ts}, these time-domain methods become inapplicable.

This section presents the **Nyquist Stability Criterion**, developed by H. Nyquist in 1932, which analyzes both absolute and relative stability from a frequency-domain perspective. This powerful method offers several advantages over time-domain approaches:

- Works directly with frequency response data (analytical or experimental)
- Does not require explicit transfer function models
- Handles systems with pure delays e^{-Ts} effectively
- Provides both stability determination and relative stability margins

The Nyquist criterion is founded on **Cauchy's Principle of Argument** from complex variable theory, which relates the behavior of analytic functions to contour mappings in the complex plane. Before formally presenting the Nyquist criterion, we will first review the essential concepts of mapping analytic functions of complex variables.

7.17.1 Mapping Contours in Complex s-Plane and the Principle of Argument

Let F(s) be a function of the complex variable $s = \sigma + j\omega$. Since s is a complex variable, the function F(s) will be complex and can be expressed as:

$$F(s) = u(\sigma, \omega) + jv(\sigma, \omega) \tag{7.71}$$

This can be represented on a complex F(s)-plane with coordinates u and v. Eqn(7.71) suggests that, any point in the complex s-plane, where F(s) is analytic, can uniquely be mapped into F(s)-plane by computing values of u and v for given value of s. Let us first illustrate mapping a point in the s-plane unto F(s)-plane, followed by mapping of a closed curve (contour) in the s-plane into F(s)-plane.

EXAMPLE 7.13

Consider the function

$$F(s) = \frac{3s+2}{s+4}$$

Suppose we want to map the point $s = -2 + j3$ in the s-plane into the F(s)-plane. Now at $s = -2 + j3$,

$$F(-2 + j3) = \frac{3(-2 + j3) + 2}{-2 + j3 + 4} = \frac{-4 + j9}{2 + j3} = 1.4615 + j2.3077$$

Figure 7.60 shows the mapping of the point in s-plane into F(s) plane.

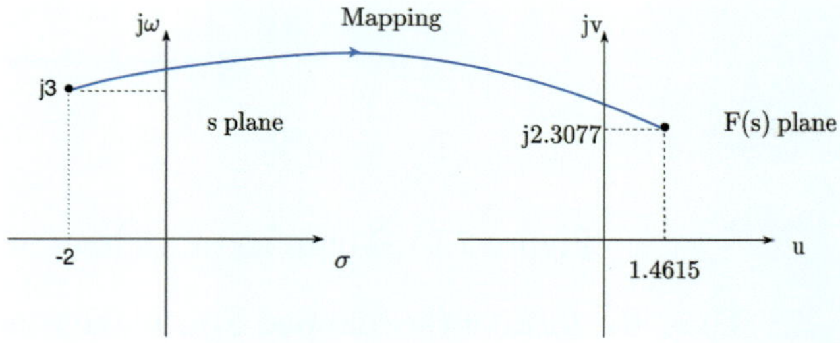

Figure 7.60: Mapping of a function F(s)

EXAMPLE 7.14

Consider the function

$$F(s) = 3s + 1$$

and a unit contour in the s-plane as shown in Figure 7.61a. Suppose we want to map the contour in the s-plane into the F(s)-plane. Then

$$u + jv = F(s) = 3s + 1 = 3(\sigma + j\omega) + 1$$

This gives

$$u = 3\sigma + 1, \quad \text{and} \quad v = 3\omega$$

Thus

$$\begin{aligned} F(s)\big|_{s=s_A} &= F(s)\big|_{s=1+j1} = 4 + j3 \\ F(s)\big|_{s=s_B} &= F(s)\big|_{s=1-j1} = 4 - j3 \\ F(s)\big|_{s=s_C} &= F(s)\big|_{s=-1-j1} = -2 - j3 \\ F(s)\big|_{s=s_D} &= F(s)\big|_{s=-1+j1} = -2 + j3 \end{aligned}$$

Figure 7.61 shows the mapping of the unit contour in s-plane into F(s) plane.

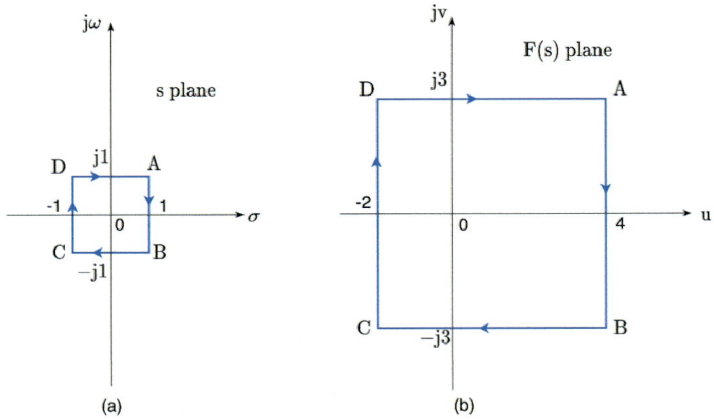

Figure 7.61: Mapping of a Function F(s)

Thus, the form of the mapped F(s)-plane contour is identical with the s-plane contour with the center shifted and the magnitude scaled. This type of mapping, which preserves the angles of the s-plane contour on the F(s) plane, is called a **conformal mapping**.

Before proceeding with the Nyquist criterion, we establish the fundamental concepts of complex function mapping:

Mapping of Complex Functions

1. **Analytic Functions:** A function F(s) is analytic in a region of the s-plane if both the function and all its derivatives exist and are continuous throughout that region. Points where F(s) or its derivatives fail to exist are called **singular points** (e.g., poles, branch points).
2. **Uniqueness of Forward Mapping:** For analytic functions, each point in the s-plane maps to exactly one point in the F(s)-plane, ensuring a unique forward transformation. However, the inverse mapping from F(s)-plane to s-plane is generally not unique; multiple s-plane points may map to the same F(s)-plane point.
3. **Curve Mapping:** The mapping concept extends naturally from points to curves. A smooth curve in the s-plane transforms into a smooth curve in the F(s)-plane, provided F(s) is analytic at every point along the original curve. For instance, the polar plot of a transfer function G(s) (studied in Section 7.14) represents the mapping of the positive jω-axis from the s-plane to the G(s)-plane.
4. **Contour Mapping:** This principle extends to closed contours: any closed curve in the s-plane that avoids all singular points of F(s) maps to a unique closed contour in the F(s)-plane. This property forms the theoretical foundation for the Nyquist stability criterion.

Cauchy's Principle of Argument

Definition 7.13. Let F(s) be a function that is analytic within and on a closed contour Γ_s in the complex s-plane, except for a finite number of poles. Suppose that:

- Γ_s encloses Z zeros and P poles of F(s) (counting multiplicities)
- Γ_s does not pass through any zeros or poles of F(s)
- Γ_s is traversed in the clockwise direction

Then the corresponding mapped contour Γ_F in the F(s)-plane will encircle the origin N times in the clockwise direction, where:

$$\boxed{N = Z - P}$$

Here, N represents the net number of clockwise encirclements of the origin in the F(s)-plane.

Notes:

- If $N > 0$: Net clockwise encirclements (more zeros than poles inside Γ_s)
- If $N < 0$: Net counterclockwise encirclements (more poles than zeros inside Γ_s)
- If $N = 0$: No net encirclements (equal number of poles and zeros, or contour passes through origin)

The proof of this theorem is omitted here. The interested reader may refer to [10]. The basic idea will be illustrated in the following examples.

EXAMPLE 7.15

Consider the function

$$F(s) = s + a, \quad a > 0 \tag{7.72}$$

which has a zero at $s = -a$.

Let us consider two contours Γ_1 and Γ_2, shown in Figure 7.62a. The contour Γ_1 is a circle of radius less than a and the contour Γ_2 is a circle of radius greater than a.

These contours are mapped into F(s)-plane using the relation in Eqn(7.72) and they are labeled also as Γ_1 and Γ_2.

In the s-plane, the contour Γ_1 does not enclose any pole or zero of F(s), that is Z = 0 and P = 0. Hence

$$N = Z - P = 0 - 0 = 0$$

and the map of Γ_1 in the F-plane does not encircle the origin (N = 0).

However, in the s-plane, the contour Γ_2 encloses one zero F(s), that is Z = 1 and P = 0. Hence

$$N = Z - P = 1 - 0 = 1$$

and the map of Γ_2 in the F-plane encircles the origin once in the clockwise direction (N = +1).

Figure 7.62(b) shows the mapped contours into F(s) plane.

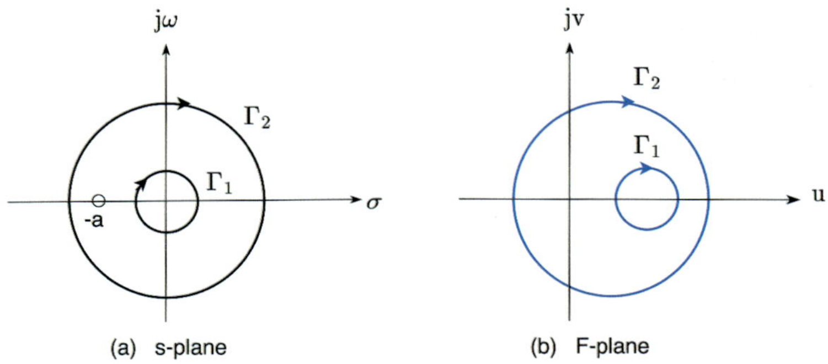

Figure 7.62: Mapping of a Function F(s)

Observe that these mappings satisfy the principle of argument.

EXAMPLE 7.16

Consider a system whose pole-zero pattern is shown in Figure 7.63a

The contour Γ_s encloses 2-poles and 4-zeros of a function F(s), that is P = 2 and Z = 4. Therefore

$$N = Z - P = 4 - 2 = 2$$

Hence, according to principle of argument, the Γ_F contour will encircle the origin of the F-plane twice in the clockwise direction, that is the direction of encirclement of Γ_F contour is the same as the direction of Γ_s contour.

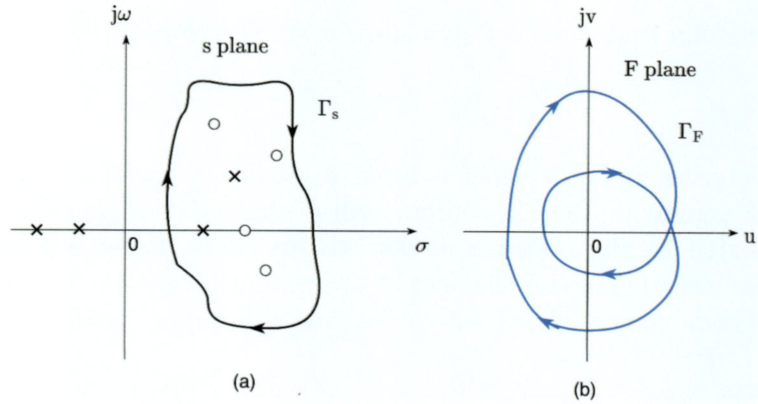

Figure 7.63: Mapping of a Contour from s-plane to F(s) plane

Next, consider a system whose pole-zero pattern is shown in Figure 7.64a

Figure 7.64(b) shows the mapped contours into F(s) plane.

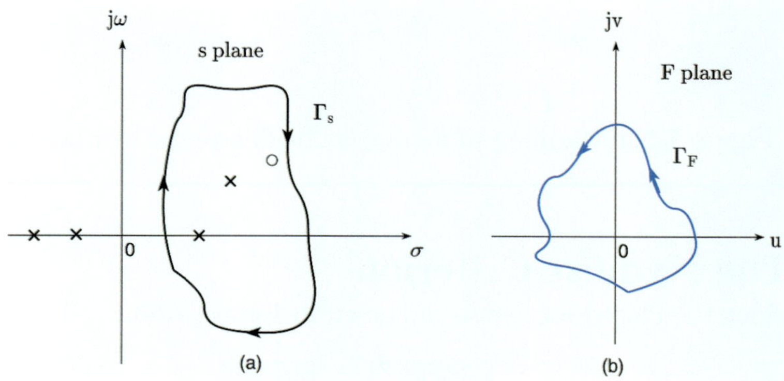

Figure 7.64: Mapping of a contour from s-plane to F(s) plane

The contour Γ_s encloses 2-poles and 1-zero of a function F(s), that is P = 2 and Z = 1. Therefore

$$N = Z - P = 1 - 2 = -1$$

Note the sign of N which is negative. Hence, according to the principle of argument, the Γ_F contour will encircle the origin of the F-plane once in the counterclockwise direction. Because N is negative, the direction of encirclement of Γ_F contour is opposite to the direction of encirclement of Γ_s contour, as shown in Figure 7.64b.

The Γ_s contour shown in Figure 7.65a encloses 5-poles and 2-zeros of a function F(s), that is P = 5 and Z = 2. Therefore

$$N = Z - P = 2 - 5 = -3$$

Note the sign of N which is negative. Hence, according to the principle of argument, the Γ_F contour will encircle the origin of the F-plane **thrice in the counterclockwise direction**. Because N is negative, the direction of encirclement of Γ_F contour is opposite to the direction of encirclement of Γ_s contour, as shown in Figure 7.65b.

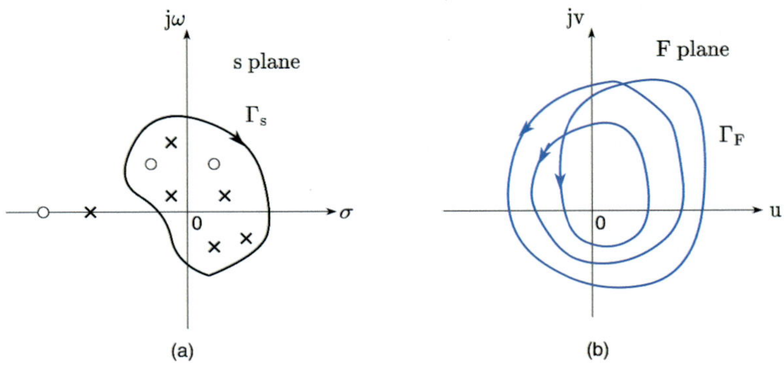

Figure 7.65: Mapping of a contour from s-plane to F(s) plane

7.18 The Nyquist Criterion

Consider the closed-loop system shown in Figure 7.66.

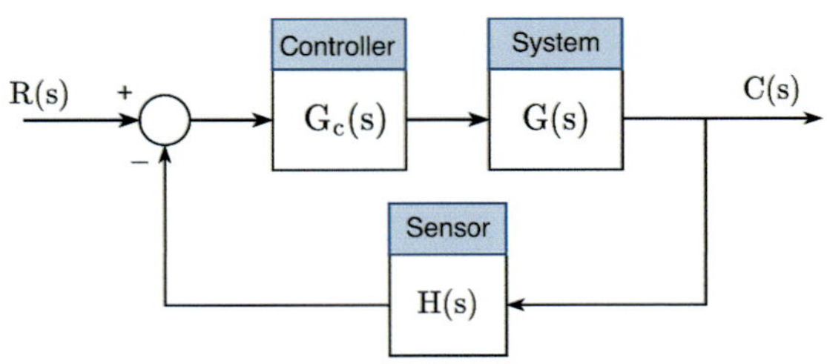

Figure 7.66: Closed-loop control system

The closed-loop transfer function of this system is given by

$$M(s) = \frac{G_c(s)G(s)}{1 + G_c(s)G(s)H(s)} \tag{7.73}$$

where $G_c(s)G(s)H(s)$ is assumed to have the standard pole-zero form:

$$G_c(s)G(s)H(s) = \frac{K(s+z_1)(s+z_2)\dots(s+z_m)}{s^j(s+p_1)(s+p_2)\dots(s+p_n)}e^{-T_d s}; \quad m < n \tag{7.74}$$

where the z_is and p_is are real or complex-conjugate, j is the system *Type* and T_d is the real time delay.

To determine the stability of a closed-loop system, we must determine the roots of the closed-loop characteristic equation

$$F(s) = \Delta(s) = 1 + L(s) = 0 \tag{7.75}$$

For the single-loop control system of Figure 7.66,

$$L(s) = G_c(s)G(s)H(s)$$

For a multi-loop system, discussed in Chapter 2, the characteristic equation is

$$F(s) = \Delta(s) = 1 - \Sigma L_n + \Sigma L_m L_q \cdots = 0$$

where $\Delta(s)$ is the graph determinant. In subsequent analysis, the closed-loop characteristic equation of single-loop or multi-loop systems will be represented by Eq.(7.75) where L(s) is the loop transfer function.

Before stating the Nyquist criterion, the pole-zero relationships of various system transfer functions are summarized below.

i. Zeros of loop transfer function=zeros of L(s)
ii. Poles of loop transfer function= poles of L(s)
iii. Poles of closed-loop transfer function=zeros of $(1 + L(s))$=roots of the characteristic equation $(1 + L(s)) = 0$=poles of L(s)

The Nyquist Contour

For closed-loop stability, all zeros of the characteristic function

$$F(s) = 1 + L(s) = 0$$

must lie in the left-half plane (LHP) of the complex s-plane. Any zero in the right-half plane (RHP) renders the system unstable. Nyquist's ingenious approach applies Cauchy's Principle of Argument to determine the RHP zero count without explicitly solving the characteristic equation.

Nyquist Stability Analysis Procedure

Step 1: Construct the Nyquist Contour (Γ_s)

The Nyquist contour is a closed path in the s-plane that encloses the entire RHP. The specific construction depends on the pole-zero locations of F(s):

Case(a): If F(s) has no poles or zeros on the jω-axis (including the origin), then Γ_s consists of:

- The entire jω-axis from $\omega = -\infty$ to $\omega = +\infty$
- A semicircular arc of infinite radius in the RHP

This standard contour is illustrated in Figure 7.67a.

Case (b): If F(s) has poles or zeros on the jω-axis, modify Γ_s to avoid these singularities by introducing infinitesimal semicircular detours around each singular point, as shown in Figure 7.67b.

Step 2: Map to the F(s)**-Plane**

Transform the Nyquist contour Γ_s into the corresponding contour Γ_F in the F(s)-plane and count the net number of clockwise encirclements N of the origin.

Step 3: Apply Cauchy's Principle

Using the relationship $N = Z - P$, determine the number of RHP zeros:

- $Z =$ number of F(s) zeros in RHP (unknown)
- $P =$ number of F(s) poles in RHP (known from open-loop system)
- $N =$ net clockwise encirclements of origin (measured from mapping)

For stability: $Z = N + P = 0$, which requires $N = -P$.

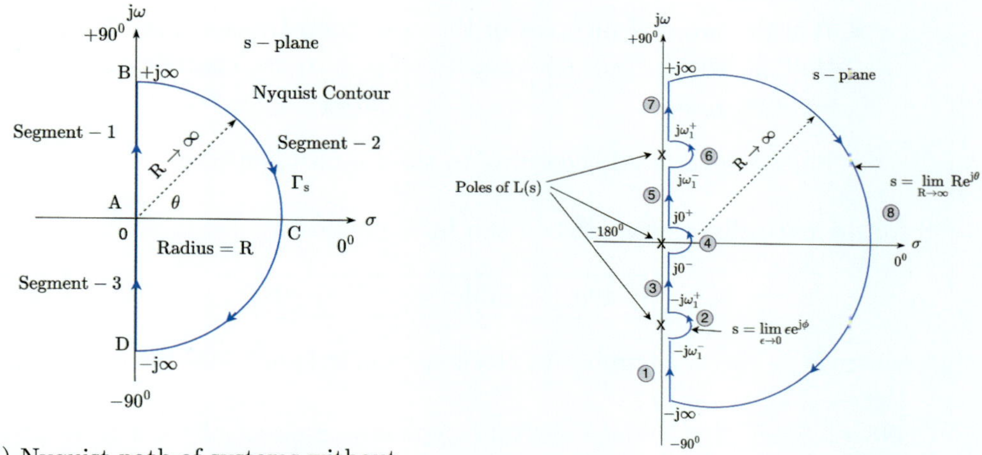

(a) Nyquist path of systems without any singularities on jω-axis

(b) Polar plot of pure derivative factor

Figure 7.67: Nyquist paths

Nyquist Stability Criterion

Let F(s) has P poles and Z zeros within the Nyquist contour. Note that the poles of F(s) are also the poles of L(s). However, the zeros of F(s) are different from those of L(s). Further, the zeros of F(s) are the closed-loop poles or roots of the characteristic equation

$$F(s) = 1 + L(s) = 0 \tag{7.76}$$

For the system to be stable, we must have $Z = 0$; that is the characteristic equation must not have any root within the Nyquist contour. From the principle of argument, a map of Nyquist contour in the F-plane will encircle the origin of the F-plane N-times in the clockwise direction (direction of traversal of Nyquist path in s plane), where

$$\boxed{N = Z - P} \tag{7.77}$$

Thus, the system is stable if and only if $N = Z - P = 0 - P = -P$. The **critical point**, in this case, is the origin of F(s)-plane. The Nyquist criterion may be stated as follows:

Theorem 7.1. A closed-loop feedback system is stable if and only if the mapped contour Γ_F in the F(s)-plane encircles the origin exactly P times in the counterclockwise direction, where:

- Γ_F is the image of the Nyquist contour Γ_s under the mapping $F(s) = 1 + L(s)$

- P is the number of poles of the open-loop transfer function L(s) that lie within (are enclosed by) the Nyquist contour Γ_s in the s-plane
- Poles are counted according to their multiplicities

Mathematically, this condition can be expressed as:

$$\boxed{\text{System is stable} \iff N = -P} \quad (7.78)$$

where N is the net number of clockwise encirclements of the origin by Γ_F.
Note: Since the Nyquist contour typically encloses the entire right-half plane, P represents the number of unstable (RHP) poles in the open-loop system L(s).

The Nyquist criterion, as originally formulated, involves mapping the characteristic function

$$F(s) = 1 + L(s)$$

and analyzing encirclements of the origin in the F(s)-plane. However, since the open-loop transfer function L(s) is typically available in factored form while $1 + L(s)$ is not, a more practical approach involves working directly with L(s).

Consider the shifted function:

$$F'(s) = F(s) - 1 = L(s)$$

This transformation maps the Nyquist contour Γ_s from the s-plane into the L(s)-plane, creating the contour Γ_L. The key insight is that the origin of the F(s)-plane corresponds to the critical point $(-1 + j0)$ in the L(s)-plane.

This coordinate transformation allows us to restate the stability analysis in terms of the more convenient L(s)-plane:

Theorem 7.2. A closed-loop feedback system is stable if and only if the mapped Nyquist contour Γ_L in the L(s)-plane encircles the critical point $(-1 + j0)$ exactly P times in the counterclockwise direction, where P is the number of poles of L(s) that lie within the Nyquist contour Γ_s in the s-plane.

Practical Advantage: This formulation is particularly useful because:

- The function L(s) is directly available from the system description

- Frequency response data can be used without requiring analytical expressions
- The critical point $(-1 + j0)$ is easily identified on standard Nyquist plots

Polar Plot vs. Complete Nyquist Plot: When to Use Each

Polar Plot vs Nyquist Plot

a. Polar Plot Sufficiency: When the open-loop transfer function L(s) has no poles or zeros in the right-half plane (RHP), the polar plot of $L(j\omega)$ for $\omega \in [0, \infty)$ contains sufficient information for closed-loop stability analysis.

b. Why Polar Plots Work for Stable Open-Loop Systems: The polar plot represents only the positive frequency portion of the complete Nyquist contour. For minimum-phase systems (no RHP poles or zeros), the remaining portions of the Nyquist contour contribute predictably:

- Negative frequency axis: Mirror image of positive frequencies
- Infinite semicircle: Maps to origin, contributing no encirclements

c. Complete Nyquist Plot Necessity: The full Nyquist plot becomes essential when L(s) contains RHP poles or zeros because:

- The principle of argument requires knowledge of P (RHP poles)
- Proper encirclement counting demands the complete closed contour
- Partial information from polar plots can lead to incorrect stability conclusions

d. Practical Design Guideline: For most common control systems with stable open-loop dynamics, polar plot analysis provides an efficient and sufficient stability assessment. Reserve complete Nyquist analysis for systems with open-loop instabilities or when detailed encirclement counting is required.

Key Takeaway: Polar plots offer computational simplicity for stable open-loop systems, while complete Nyquist plots provide the mathematical rigor needed for unstable open-loop systems.

The stability investigation using the Nyquist criterion will be illustrated by a number of examples.

EXAMPLE 7.17

Problem

Sketch Nyquist plot and interpret the stability of a system with open-loop transfer function

$$L(s) = G(s)H(s) = \frac{K}{(1 + T_1 s)(1 + T_2 s)}$$

SOLUTION:

Step-1: Express Magnitude and Phase in Frequency Domain

Now given system $G(s)H(s) = \dfrac{K}{(1 + T_1 s)(1 + T_2 s)}$

Hence $G(j\omega)H(j\omega) = \dfrac{K}{(1 + j\omega T_1)(1 + j\omega T_2)}$

- **Magnitude:** $\left|G(j\omega)H(j\omega)\right| = \dfrac{K}{(\sqrt{1 + \omega^2 T_1^2})(\sqrt{1 + \omega^2 T_2^2}}$
- **Phase:** $\angle G(j\omega)H(j\omega) = -\tan^{-1}\omega T_1 - \tan^{-1}\omega T_2$

Step 2: Define the Nyquist Contour (Γ_s)

Since the system has no imaginary axis poles, use the standard Nyquist contour shown in Figure 7.68:

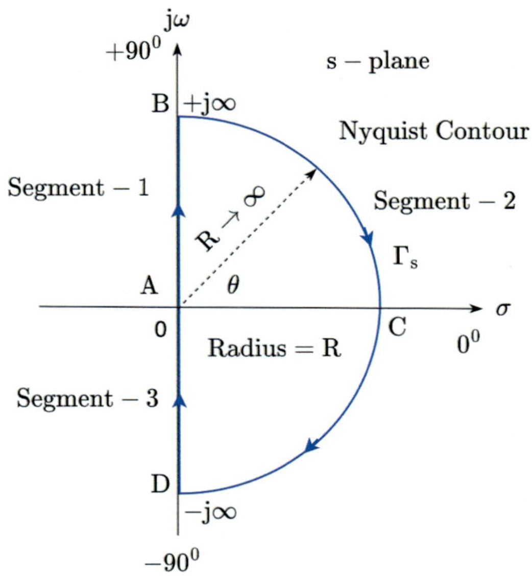

Figure 7.68: Nyquist contour of Example 7.17

- **Segment 1 (AB)**: Positive imaginary axis from $\omega = 0$ to $\omega = +\infty$
- **Segment 2 (BCD)**: Semicircle of infinite radius (clockwise from $+90^\circ$ to -90°)
- **Segment 3 (DA)**: Negative imaginary axis from $\omega = -\infty$ to $\omega = 0$

Step 3: Map Each Segment to the L(s) Plane

Segment 1: Positive Imaginary Axis ($\omega : 0 \to +\infty$)

- **At** $\omega = 0$: $|L(j0)| = K$, $\angle L(j0) = 0^\circ$. Because

$$\lim_{\omega \to 0} L(j\omega) = \lim_{\omega \to 0} \frac{K}{(1 + j\omega T_1)(1 + j\omega T_2)} = K\angle 0^\circ \text{ (map of point A)}$$

- **At** $\omega = \infty$: $|L(j\infty)| = 0$, $\angle L(j\infty) = -180^\circ$. Because

$$\begin{aligned}\lim_{\omega \to \infty} L(j\omega) &= \lim_{\omega \to \infty} \frac{K}{(1 + j\omega T_1)(1 + j\omega T_2)} \\ &= \lim_{\omega \to \infty} \frac{K}{j^2 T_1 T_2 \omega^2} = \lim_{\omega \to \infty} \frac{-K}{\omega^2 T_1 T_2} \\ &= 0\angle -180^\circ \text{ (map of point B)}\end{aligned}$$

- **Behavior**: Starts at $(K, 0^\circ)$ and spirals inward to origin at -180°

Segment-1: Positive Imaginary Axis Mapping

At $\omega = 0$	$\lvert L(j\omega)\rvert$	$= K$	$\angle L(j\omega) = 0^\circ$
At $\omega = \infty$	$\lvert L(j\omega)\rvert$	$= 0$	$\angle L(j\omega) = -180^\circ$

Segment 2: Infinite Semicircle ($\theta : +90^\circ \to -90^\circ$)

For points on the infinite semicircle: $s = \lim_{R \to \infty} Re^{j\theta}$

$$\lim_{R \to \infty} L(s)\Big|_{s=Re^{j\theta}} = \lim_{R \to \infty} \left|\frac{K}{R^2 T_1 T_2}\right| e^{-2j\theta} \tag{7.79}$$

- **Magnitude**: Always zero (since $|L(s)| \propto 1/R^2$ as $R \to \infty$)
- **Phase**: Changes from -180° to $+180^\circ$
- **Result**: Point at origin in L(s) plane

Segment-2:Infinite Semicircle Mapping

At $\omega = +\infty$, $\theta = +90^0$	$\lvert L(j\omega)\rvert$	$= 0$	$\angle L(j\omega) = -180^\circ$
At $\omega = -\infty$, $\theta = -90^0$	$\lvert L(j\omega)\rvert$	$= 0$	$\angle L(j\omega) = +180^\circ$

Segment 3: Negative Imaginary Axis ($\omega : -\infty \to 0$)

- **Property**: $L(-j\omega)$ is complex conjugate of $L(j\omega)$. Hence, it is not necessary to evaluate the mapping of the lower half of the imaginary axis of Figure 7.68
- **Result**: The Nyquist plot is the mirror image of Segment 1 about the real axis
- **At** $\omega = -\infty$: $|L(-j\infty)| = 0$, $\angle L(-j\infty) = +180^\circ$
- **At** $\omega = 0$: $|L(j0)| = K$, $\angle L(j0) = 0^\circ$

Segment-3: Negative Imaginary Axis Mapping

At $\omega = \infty^-$	$\lvert L(j\omega)\rvert = 0$	$\angle L(j\omega) = 180^\circ$
At $\omega = 0$	$\lvert L(j\omega)\rvert = K$	$\angle L(j\omega) = 0^\circ$

Step 4: Complete Nyquist Plot Construction

Combine all three segments to form the complete Nyquist plot:

(a) **Lower curve (Segment 1)**: Polar plot from $K\angle 0^\circ$ to $0\angle -180^\circ$

(b) **Point at origin (Segment 2)**: Infinite semicircle maps to origin

(c) **Upper curve (Segment 3)**: Mirror image of lower curve

The Nyquist plot is sketched in Figure 7.69.

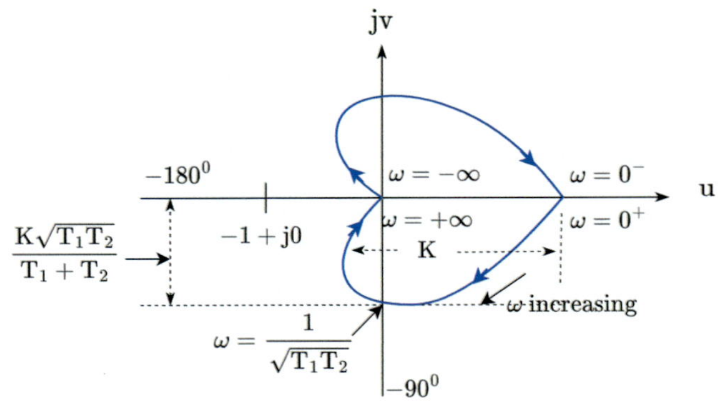

Figure 7.69: Nyquist plot for $L(s) = \dfrac{K}{(1 + T_1s)(1 + T_2s)}$

EXAMPLE 7.18

Problem

Sketch the Nyquist plot and interpret the stability of a system with an open-loop transfer function

$$\mathrm{L(s) = G(s)H(s) = \frac{K}{s(1+Ts)}}$$

SOLUTION:

Step 1: Nyquist contour in the s-plane. Since L(s) has a pole at the origin, the Nyquist contour must include a small semicircle of radius $\epsilon \to 0$. The contour Γ_s, shown in Figure 7.70a, consists of four segments.

Step 2: Mapping Γ_s into the L(s)-plane.

Segment 1 (positive imaginary axis, $\omega = 0^+ \to +\infty$) On this segment,

$$\begin{aligned} \mathrm{L(j\omega)} &= \frac{\mathrm{K}}{\mathrm{j\omega(1+j\omega T)}} \\ \lim_{\omega\to 0^+} \mathrm{L(j\omega)} &= \infty\angle-90^\circ \\ \lim_{\omega\to +\infty} \mathrm{L(j\omega)} &= \lim_{\omega\to+\infty} \frac{\mathrm{K}}{+\mathrm{j\omega(1+j\omega T)}} \\ &= \lim_{\omega\to+\infty}\frac{\mathrm{K}}{\mathrm{j^2T\omega^2}} = \lim_{\omega\to+\infty}\frac{-\mathrm{K}}{\omega^2\mathrm{T}} \\ &= 0\angle{-180^\circ} \end{aligned}$$

Thus, Segment 1 maps to the polar plot from $\infty\angle-90^\circ$ to $0\angle-180^\circ$.

At $\omega = 0^+$	$\lvert \mathrm{L(j\omega)}\rvert = \infty$	$\angle \mathrm{L(j\omega)} = -90^\circ$
At $\omega \to +\infty$	$\lvert \mathrm{L(j\omega)}\rvert = 0$	$\angle \mathrm{L(j\omega)} = -180^\circ$

Segment 2 (semicircle of infinite radius) For this segment,

$$\mathrm{s} = \lim_{\mathrm{R}\to\infty} \mathrm{Re^{j\theta}}, \qquad \theta : +90^\circ \to -90^\circ$$

Substituting into L(s):

$$\lim_{\mathrm{R}\to\infty}\mathrm{L(s)} = \lim_{\mathrm{R}\to\infty}\frac{\mathrm{K}}{\mathrm{R^2T}}\mathrm{e}^{-2\mathrm{j}\theta} = 0\,\mathrm{e}^{-2\mathrm{j}\theta}$$

Hence, as θ changes from $+90^\circ$ at $\omega = +\infty$ to -90° at $\omega = -\infty$, the Nyquist plot sweeps from -180° to $+180^\circ$. The magnitude of L(s) when R is infinite is always zero.

At $\theta = +90^\circ$ ($\omega = +\infty$)	$\lvert L\rvert = 0$	$\angle L = -180^\circ$
At $\theta = -90^\circ$ ($\omega = -\infty$)	$\lvert L\rvert = 0$	$\angle L = +180^\circ$

Segment 3 (negative imaginary axis, $\omega = -\infty \to 0^-$)

By symmetry,

$$\mathrm{L}(-\mathrm{j}\omega) = \overline{\mathrm{L}(\mathrm{j}\omega)}, \qquad \omega > 0$$

so this segment is the mirror image of Segment 1 with respect to the real axis.

At $\omega = -\infty$	$\lvert \mathrm{L}(\mathrm{j}\omega)\rvert = 0$	$\angle \mathrm{L}(\mathrm{j}\omega) = -180^\circ$
At $\omega = 0^-$	$\lvert \mathrm{L}(\mathrm{j}\omega)\rvert = \infty$	$\angle \mathrm{L}(\mathrm{j}\omega) = +90^\circ$

Segment 4 (small semicircle around origin) For this segment,

$$\mathrm{s} = \lim_{\epsilon \to 0} \epsilon e^{\mathrm{j}\theta}, \qquad \theta : -90^\circ \to +90^\circ$$

Substituting:

$$\lim_{\epsilon \to 0} \mathrm{L(s)} = \lim_{\epsilon \to 0} \frac{\mathrm{K}}{\epsilon} \mathrm{e}^{-\mathrm{j}\theta}$$

Thus, the magnitude tends to ∞, and the angle sweeps from $+90^\circ$ (at $\omega = 0^-$) to -90° (at $\omega = 0^+$).

At $\omega = 0^-$ ($\theta = -90^\circ$)	$\lvert \mathrm{L}(\mathrm{j}\omega)\rvert = \infty$	$\angle L = +90^\circ$
At $\omega = 0^+$ ($\theta = +90^\circ$)	$\lvert \mathrm{L}(\mathrm{j}\omega)\rvert = \infty$	$\angle L = -90^\circ$

Step 3: Final Nyquist plot. Combining all segments, the complete Nyquist plot is obtained as shown in Figure 7.70b.

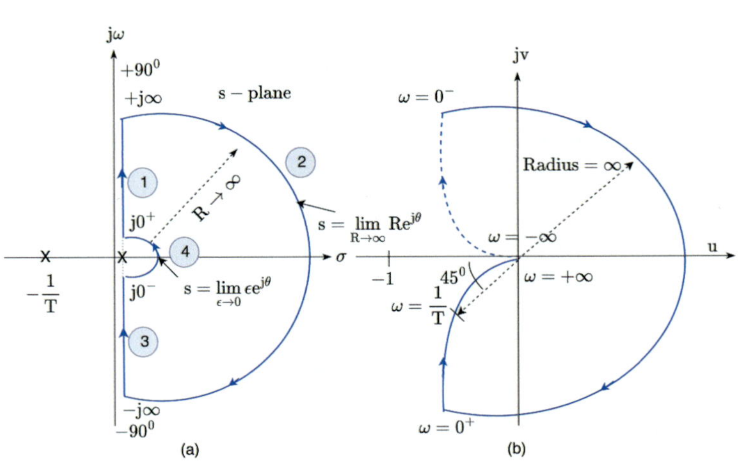

Figure 7.70: Nyquist plot of system with one pole at the origin

EXAMPLE 7.19

Problem

Sketch the Nyquist plot and interpret the stability of a system with open-loop transfer function

$$L(s) = G(s)H(s) = \frac{K}{s(1 + T_1 s)(1 + T_2 s)}$$

SOLUTION:

Step-1: Nyquist Contour

Since L(s) has a pole at the origin, the Nyquist path must include a small semicircle of radius ϵ, where $\epsilon \to 0$. A sketch of the Nyquist path Γ_s is shown in Figure 7.70a, which has four segments.

Let us map each segment of Γ_s into the L(s) = G(s)H(s) plane.

Step-2: Mapping Nyquist Contour into GH Plane

Step-2.1: Mapping Segment-1

Segment-1 corresponds to $\omega = 0^+$ to $\omega = +\infty$. For this part,

$$L(s)\Big|_{s=j\omega} = L(j\omega) = \frac{K}{j\omega(1 + j\omega T_1)(1 + j\omega T_2)}$$

Hence

$$\lim_{\omega\to 0} L(j\omega) = \lim_{\omega\to 0} \frac{K}{j\omega(1 + j\omega T_1)(1 + j\omega T_2)} = \infty\angle{-90^\circ}$$

$$\lim_{\omega\to \infty} L(j\omega) = \lim_{\omega\to \infty} \frac{K}{+j\omega(1 + j\omega T_1)(1 + j\omega T_2)}$$

$$= \lim_{\omega\to \infty} \frac{K}{j^3 T_1 T_2 \omega^3} = 0\angle{-270^\circ}$$

Segment-1: Polar Plot

At $\omega = 0^+$	$\lvert L(j\omega)\rvert = \infty$	$\angle L(j\omega) = -90^\circ$
At $\omega = \infty$	$\lvert L(j\omega)\rvert = 0$	$\angle L(j\omega) = -270^\circ$

Step-2.2: Mapping Segment-3

Segment-3 corresponds to $\omega = -\infty$ to $\omega = 0^-$. Since $L(s)\big|_{s=-j\omega} = L(-j\omega)$, this segment is the mirror image of Segment-1 with respect to the real axis.

Segment-3: Mirror Image of Polar Plot

At $\omega = -\infty$	$\lvert L(j\omega)\rvert = 0$	$\angle L(j\omega) = +270^\circ$
At $\omega = 0^-$	$\lvert L(j\omega)\rvert = \infty$	$\angle L(j\omega) = +90^\circ$

Step-2.3: Mapping Segment-2 (Infinite Semicircle)

Segment-2 corresponds to the infinite-radius semicircle,

$$s = \lim_{R\to\infty} Re^{j\theta}, \quad \theta : +90^\circ \to -90^\circ$$

Substituting gives

$$\lim_{R\to\infty} L(s)\bigg|_{s=Re^{j\theta}} = \lim_{R\to\infty} \frac{K}{R^3 T_1 T_2} e^{-3j\theta} = 0 \cdot e^{-3j\theta}$$

Thus, the mapping shrinks to the origin, sweeping from -270° to $+270^\circ$.

Segment-2: Infinite Semicircle

At $\theta = +90^\circ$	$\lvert L\rvert = 0$	$\angle L = -270^\circ$
At $\theta = -90^\circ$	$\lvert L\rvert = 0$	$\angle L = +270^\circ$

Step-2.4: Mapping Segment-4 (Small Semicircle)

Segment-4 corresponds to the small semicircle around the origin,

$$s = \lim_{\epsilon\to 0} \epsilon e^{j\theta}, \quad \theta : -90^\circ \to +90^\circ$$

Substituting gives

$$\lim_{\epsilon\to 0} L(s)\bigg|_{s=\epsilon e^{j\theta}} = \lim_{\epsilon\to 0} \frac{K}{\epsilon} e^{-j\theta} = \infty \cdot e^{-j\theta}$$

Thus, the Nyquist plot goes to infinity in the L(s)-plane, sweeping from $+90^\circ$ to -90°.

Segment-4: Small Semicircle

At $\omega = 0^-$, $\theta = -90^\circ$	$\lvert L\rvert = \infty$	$\angle L = +90^\circ$
At $\omega = 0^+$, $\theta = +90^\circ$	$\lvert L\rvert = \infty$	$\angle L = -90^\circ$

The complete Nyquist plot is shown in Figure 7.71.

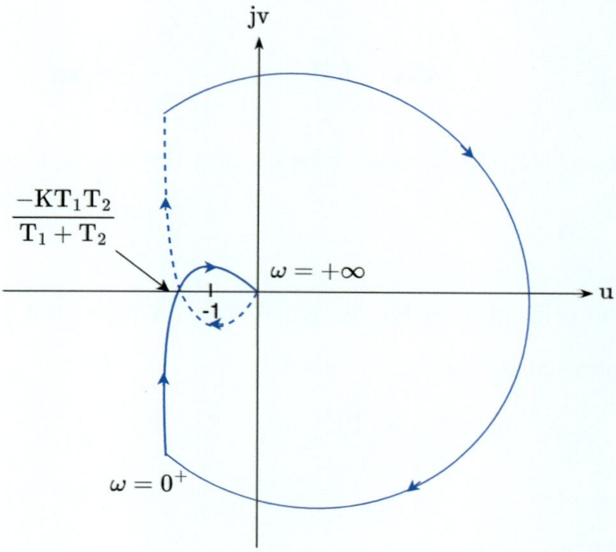

Figure 7.71: Nyquist plot of L(s) with three poles, including one at the origin

Step-3: Stability Interpretation

The open-loop transfer function has no right-half-plane poles (the poles are at $s = 0$ and $s = -1/T_1$, $-1/T_2$ with $T_1, T_2 > 0$). Hence $P = 0$ and, by the Nyquist criterion, the number of closed-loop RHP poles Z equals the number of clockwise encirclements N of $(-1 + j0)$ by the Nyquist plot:

$$Z = P + N = N$$

To locate where the Nyquist plot crosses the real axis, set $\mathrm{Im}(L(j\omega)) = 0$. Writing

$$L(j\omega) = \frac{K}{j\omega(1 + j\omega T_1)(1 + j\omega T_2)}$$

and letting

$$A(\omega) = 1 - T_1 T_2 \omega^2, \qquad B(\omega) = \omega(T_1 + T_2),$$

one finds (after algebra) that $\mathrm{Im}(L(j\omega)) = 0$ iff $A(\omega) = 0$. Thus, the crossing frequency is

$$\omega_0 = \frac{1}{\sqrt{T_1 T_2}}.$$

Evaluating the real part at ω_0 gives the real-axis intercept

$$\Re\{L(j\omega_0)\} = -\frac{K T_1 T_2}{T_1 + T_2}.$$

Therefore, the Nyquist plot crosses the real axis at

$$x_c = -\frac{K T_1 T_2}{T_1 + T_2}.$$

The critical gain K_{cr} at which the Nyquist plot passes through -1 is obtained by setting $x_c = -1$:

$$-\frac{K_{cr} T_1 T_2}{T_1 + T_2} = -1 \qquad \Rightarrow \qquad K_{cr} = \frac{T_1 + T_2}{T_1 T_2}$$

Hence:

- If $(0 < K < K_{cr}$, then $x_c > -1$. The Nyquist plot does not encircle $(-1, 0)$, so $N = 0$ and $Z = 0$; the closed-loop system is stable.
- If $K = K_{cr}$, the Nyquist plot passes through $(-1, 0)$ and the closed-loop system is marginally stable: a complex conjugate pair lies on the imaginary axis at $\pm j\omega_0$.
- If $K > K_{cr}$, the Nyquist plot crosses to the left of -1; the Nyquist plot then produces two clockwise encirclements of $(-1, 0)$ (so $N = 2$), and hence $Z = 2$. The closed-loop system is unstable with two RHP poles (a complex conjugate pair in the RHP).

(Remark: the marginal frequency is $\omega_0 = 1/\sqrt{T_1 T_2}$ and the stability boundary occurs at $K = K_{cr} = (T_1 + T_2)/(T_1 T_2)$.)

EXAMPLE 7.20

Problem

Sketch the Nyquist plot and interpret the stability of a system with an open-loop transfer function

$$L(s) = G(s)H(s) = \frac{K}{s^2(1 + Ts)}$$

SOLUTION:

Since $L(s)$ has a double pole at the origin, the Nyquist contour includes a small semicircle of radius $\epsilon \to 0$ around the origin (similar to Figure 7.70a).

Replace s by jω to obtain the real-frequency polar plot:

$$\begin{aligned} L(j\omega) &= \frac{K}{-\omega^2(1+j\omega T)} \\ &= \frac{K}{\omega^2\sqrt{1+(\omega T)^2}} \angle(-180^\circ - \tan^{-1}(\omega T)). \end{aligned}$$

Thus the phase varies from -180° at $\omega \to 0^+$ to -270° as $\omega \to \infty$. (Equivalently, interpreted modulo 360°, the angle moves from $+180^\circ$ toward $+90^\circ$.) Evaluating real and imaginary parts shows the imaginary part of L(jω) is positive for $\omega > 0$, so the polar locus lies in the upper half of the complex plane (positive imaginary part).

$$\lim_{\omega \to 0^+} L(j\omega) = \lim_{\omega \to 0^+} \frac{K}{\omega^2} \angle -180^\circ = \infty \angle -180^\circ,$$

$$\lim_{\omega \to +\infty} L(j\omega) = \lim_{\omega \to +\infty} \frac{K}{\omega^3} \angle -270^\circ = 0 \angle -270^\circ.$$

For the small semicircle near the origin, $s = \epsilon e^{j\theta}$ with θ from -90° to $+90^\circ$, and

$$\lim_{\epsilon \to 0} L(s)\Big|_{s=\epsilon e^{j\theta}} = \lim_{\epsilon \to 0} \frac{K}{\epsilon^2 T} e^{-2j\theta} = \infty \cdot e^{-2j\theta}.$$

Hence the mapping of the small semicircle goes to infinite radius in the L-plane; its angle sweeps from $+180^\circ$ (at $\omega = 0^-$) to -180° (at $\omega = 0^+$), i.e., it passes once around a full 360° as the small semicircle is traversed.

The complete Nyquist plot is sketched in Figure 7.72.

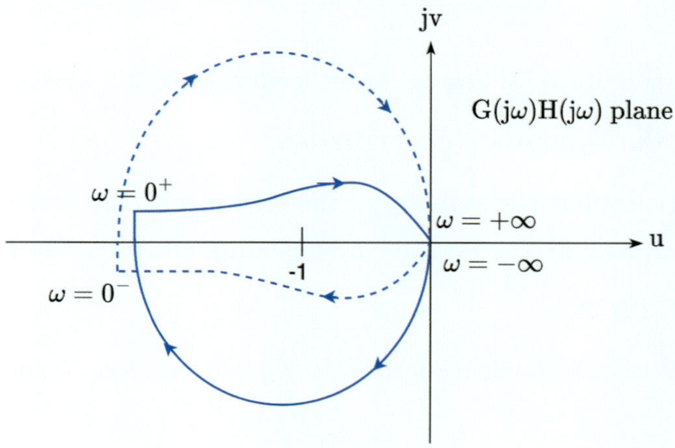

Figure 7.72: Nyquist plot of system with two poles at the origin

Since the contour encircles the $-1 + j0$ point twice, there are two roots of the closed-loop system in the right-hand plane, and irrespective of the value of gain K, the system is unstable. Verify this using Routh's criterion.

CONCEPT REINFORCEMENT PROBLEM 7.12

Problem

The open-loop transfer function of a system is

$$L(s) = G(s)H(s) = \frac{s+a}{(s+b)(s-c)}, \quad a, b, c > 0$$

Using the Nyquist stability criterion, investigate the stability of the closed-loop system.

EXAMPLE 7.21

Problem

Consider the closed-loop system with derivative feedback shown in Figure 7.73

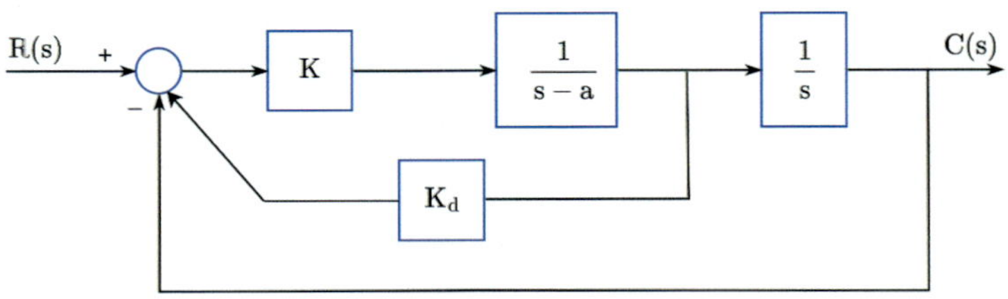

Figure 7.73: Second order feedback control system

Using the Nyquist stability criterion,

a. Determine the stability of the system without derivative feedback
b. Investigate the stability by including derivative feedback

SOLUTION:

a. Without derivative feedback, $K_d = 0$, the loop transfer function

$$L(s) = G(s)H(s) = \frac{K}{s(s-a)}, \quad a > 0$$

The open-loop transfer function has one open-loop pole in the right half of the s-plane and therefore $P = 1$. Thus, for this system to be stable, we require $N = -P = -1$, that is, the Nyquist contour must encircle the -1 point once in the counter clockwise direction.

Let us obtain the Nyquist contour of this system. Since this system has one pole at the origin, the Nyquist path of this system is the same as shown in Figure 7.70a.

Let us determine the mapping of the small semicircle around the origin represented by the phasor

$$s = \lim_{\epsilon \to 0} \epsilon e^{j\theta}$$

where θ changes from -90^0 at $\omega = 0^-$ to $+90^0$ at $\omega = 0^+$.

Substituting the above value of s into $L(s) = G(s)H(s)$ gives

$$\lim_{\epsilon \to 0} L(s)\bigg|_{s=\epsilon e^{j\theta}} = \lim_{\epsilon \to 0} \frac{K}{-a\epsilon e^{j\theta}} = \lim_{\epsilon \to 0} \left|\frac{K}{a\epsilon}\right| \angle -180^0 - \theta$$

Its mapping is summarized as:

Segment-4:	**Mapping**	**Small Semicircle**
At $\omega = 0^-$, $\theta = -90^\circ$	$\lvert L(j\omega)\rvert = \infty$	$\angle L(j\omega) = -90^\circ$
At $\omega = 0^+$, $\theta = +90^\circ$	$\lvert L(j\omega)\rvert = \infty$	$\angle L(j\omega) = -270^\circ$

Therefore, this portion of the Nyquist contour is a semicircle of infinite magnitude in the left-half of $L(s) = G(s)H(s)$ plane, as shown in Figure 7.74.

Let us map segment 2; a semicircle of infinite radius, where the points on this semicircle are represented by the phasor

$$s = \lim_{R \to \infty} Re^{j\theta}$$

where θ changes from $+90^0$ to -90^0.

Substituting the above value of s into $L(s) = G(s)H(s)$ gives

$$\lim_{R \to \infty} L(s)\bigg|_{s=Re^{j\theta}} = \lim_{R \to \infty} \left|\frac{K}{R^2}\right| e^{-2j\theta}$$

As θ changes from $+90^0$ at $\omega = +\infty$ to -90^0 at $\omega = -\infty$, the Nyquist plot moves from an angle of -180^0 at $\omega = +\infty$ to an angle of $+180^0$ at $\omega = -\infty$. The magnitude of $L(s)$ contour when R is infinite is always zero.

Segment-2: Mapping Infinite Semicircle

At $\omega = +\infty$, $\theta = +90^0$	$\lvert L(j\omega)\rvert = 0$	$\angle L(j\omega) = -180°$
At $\omega = -\infty$, $\theta = -90^0$	$\lvert L(j\omega)\rvert = 0$	$\angle L(j\omega) = +180°$

Now

$$L(j\omega) = \frac{K}{j\omega(j\omega - a)} = \frac{K(-j\omega - a)(-j\omega)}{\omega^2(a^2 + \omega^2)}$$
$$= \frac{-K\omega^2}{\omega^2(a^2 + \omega^2)} + j\frac{Ka\omega}{\omega^2(a^2 + \omega^2)}$$

It is obvious that

1. Both Real(L(jω)) and Im(L(jω)) are zero for $\omega = \infty$.
2. Real $L(j\omega)\Big|_{\omega=0} = -\frac{K}{a^2}$

When $s = j\omega$

$$L(j\omega) = \frac{K}{j\omega(j\omega - a)} = \frac{K}{\sqrt{\omega^4 + a^2\omega^2}}\angle -90^0 - \tan^{-1}(-\omega/a)$$
$$= \frac{K}{\sqrt{\omega^4 + a^2\omega^2}}\angle 90^0 + \tan^{-1}(\omega/a)$$

By computing L(jω) at several values of ω, the Nyquist plot is sketched as shown in 7.74.

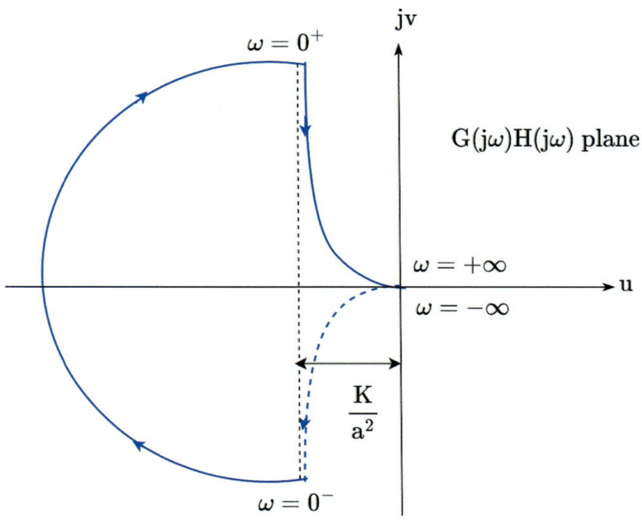

Figure 7.74: Nyquist plot for $L(s) = \frac{K}{s(s - a)}$

From Figure 7.74, we find that the Nyquist plot encircles the -1 point once in a clockwise direction. Thus, $N = +1$ and since there is one pole at $s = +a$ in the right-hand plane, $P = 1$. Hence

$$Z = N + P = 2$$

and the system is unstable, irrespective of the value of the gain K.

b. With the inclusion of derivative feedback with $K_d > 0$, the loop transfer function is

$$L(s) = G(s)H(s) = \frac{K(1 + K_d s)}{s(s - a)}$$

The Nyquist plot due to mapping the small semicircle when $s = \epsilon e^{j\theta}$ is the same as the system without derivative feedback.

Mapping the semicircle of infinite radius gives

$$\lim_{R\to\infty} L(s)\bigg|_{s=Re^{j\theta}} = \lim_{R\to\infty} \left|\frac{KK_d}{R}\right| e^{-j\theta}$$

As θ changes from $+90^0$ at $\omega = +\infty$ to -90^0 at $\omega = -\infty$, the Nyquist plot moves from an angle of -90^0 at $\omega = +\infty$ to an angle of $+90^0$ at $\omega = -\infty$.

Let us determine when the locus $L(j\omega)$ crosses the real axis (u-axis) by considering

$$\begin{aligned} L(j\omega) &= \frac{K(1 + jK_d\omega)}{j\omega(j\omega - a)} = \frac{K(1 + jK_d\omega)(-\omega^2 + ja\omega)}{\omega^4 + a^2\omega^2} \\ &= \frac{-K(\omega^2 + a\omega^2 K_d)}{\omega^4 + a^2\omega^2} + j\frac{K(a\omega - K_d\omega^3)}{\omega^4 + a^2\omega^2} \end{aligned}$$

The $L(j\omega)$-locus intersects the real axis at the point where the imaginary part of $L(j\omega)$ is zero. Therefore, at the real axis intersection

$$a\omega - K_d\omega^3 = 0 \quad \text{or,} \quad \omega(a - K_d\omega^2) = 0$$

At this point

$$\omega^2 = \frac{a}{K_d}$$

The real part of $L(j\omega)$ at the intersection is

$$\frac{-K(\omega^2 + a\omega^2 K_d)}{\omega^4 + a^2\omega^2}\bigg|_{\omega^2 = a/K_d} = -\frac{KK_d}{a}$$

Therefore, when $-\frac{KK_d}{a} < -1$ or $KK_d > a$, the Nyquist map will encircle -1 point once in the counter clockwise direction, and therefore $N = -1$. Then, the number of zeros of the system in the right-hand plane is

$$Z = N + P = -1 + 1 = 0$$

Thus, the system is stable when $KK_d > a$. The complete Nyquist plot is shown in Figure 7.75.

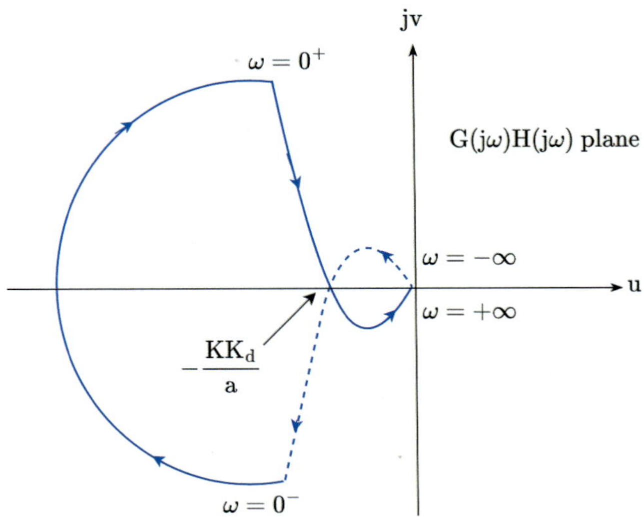

Figure 7.75: Nyquist plot for $L(s) = \frac{K(1 + K_d s)}{s(s - a)}$

EXAMPLE 7.22

Problem

Using the Nyquist stability criterion, investigate the stability of the system with the loop transfer function

$$L(s) = G(s)H(s) = \frac{K(1 + 5s)}{s^2(1 + s)(1 + 2s)} \tag{7.80}$$

SOLUTION:

1. **Nyquist contour and the small semicircle around the origin.**

   Since L(s) has a double pole at the origin, the Nyquist contour must go around the origin with a small semicircle $s = \epsilon e^{j\theta}$, θ varies from $-90°$ (from 0^-) to $+90°$ (from 0^+) (same as Figure 7.70a).

The mapping of this small semicircle around the origin is computed by substituting $s = \epsilon e^{j\theta}$ into $L(s) = G(s)H(s)$ which gives

$$\lim_{\epsilon \to 0} L(s)\Big|_{s=\epsilon e^{j\theta}} = \lim_{\epsilon \to 0}\left[\frac{K(1+5\epsilon e^{j\theta})}{\epsilon^2 e^{j2\theta}(1+\epsilon e^{j\theta})(1+2\epsilon e^{j\theta})}\right]$$

$$= \lim_{\epsilon \to 0}\left[\frac{K}{\epsilon^2 e^{j2\theta}}\right] = \infty e^{-j2\theta}$$

Its mapping is summarized as:

Segment-4: Mapping Small Semicircle

At $\omega = 0^-$, $\theta = -90^0$	$\lvert L(j\omega)\rvert = \infty$	$\angle L(j\omega) = +180°$
At $\omega = 0^+$, $\theta = +90^0$	$\lvert L(j\omega)\rvert = \infty$	$\angle L(j\omega) = -180°$

Thus, the small semicircle maps to an arc of very large radius with angle -2θ. At $\theta = -90°$, the angle is $+180°$; at $\theta = +90°$, the angle is $-180°$. This is shown in Figure 7.76b.

2. **Mapping Segment-2; Infinite Radius Semicircle**

The semicircle of infinite radius where the points on this semicircle are represented by the phasor

$$s = \lim_{R \to \infty} Re^{j\theta}$$

where θ changes from $+90^0$ to -90^0.

Substituting the above value of s into $L(s) = G(s)H(s)$ gives

$$\lim_{R \to \infty} L(s)\Big|_{s=Re^{j\theta}} = \lim_{R \to \infty} \frac{K(1+5Re^{j\theta})}{R^2 e^{-2j\theta}(1+Re^{j\theta})(1+2Re^{j\theta})} = 0e^{-j3\theta}$$

As θ changes from $+90^0$ at $\omega = +\infty$ to -90^0 at $\omega = -\infty$, the Nyquist plot moves from an angle of -270^0 at $\omega = +\infty$ to an angle of $+270^0$ at $\omega = -\infty$. The magnitude of L(s) contour when R is infinite is always zero.

Segment-2: Mapping Infinite Semicircle

At $\omega = +\infty$, $\theta = +90^0$	$\lvert L(j\omega)\rvert = 0$	$\angle L(j\omega) = -270°$
At $\omega = -\infty$, $\theta = -90^0$	$\lvert L(j\omega)\rvert = 0$	$\angle L(j\omega) = +270°$

3. **Mapping along Imaginary Axis** $j\omega$

Here

$$L(j\omega) = G(j\omega)H(j\omega) = \frac{K(1 + j5\omega)}{(j\omega)^2(1 + j\omega)(1 + j2\omega)}$$

Hence

$$\left|GH\right| = \frac{K\sqrt{1 + 25\omega^2}}{\omega^2\sqrt{1 + \omega^2}\sqrt{1 + 4\omega^2}}$$
$$\angle GH = -180^0 + \tan^{-1} 5\omega - \tan^{-1} \omega - \tan^{-1} 2\omega$$

4. **Computation of Real axis cross over**

At the real axis cross-over point, we have

$$-180^0 + \tan^{-1} 5\omega - \tan^{-1} \omega - \tan^{-1} 2\omega = -180^\circ$$

This implies

$$\tan^{-1} 5\omega - \tan^{-1} \frac{\omega + 2\omega}{1 - 2\omega^2} = 0$$

or,

$$\tan^{-1} \frac{5\omega - \dfrac{3\omega}{1 - 2\omega^2}}{1 + 5\omega\dfrac{3\omega}{1 - 2\omega^2}} = 0 \quad \Longrightarrow \quad \tan^{-1} \frac{5\omega - 10\omega^3 - 3\omega}{1 - 2\omega^2 + 15\omega^2} = 0$$

This implies

$$2\omega(1 - 5\omega^2) = 0 \quad \Longrightarrow \quad \omega = 0 \text{ or } \omega^2 = 0.2$$
$$\Longrightarrow \quad \omega = \sqrt{0.2} = 0.4472 \text{ rad/sec}$$

5. **Magnitude at that crossing.**

Substituting $\omega^2 = 0.2$:

$$\left|G(j\omega)H(j\omega)\right|_{\omega=0.4472} = \frac{K\sqrt{1 + 25(0.2)}}{(0.2)\sqrt{1 + 0.2}\sqrt{1 + 4(0.2)}}$$
$$= \frac{K\sqrt{6}}{0.2\sqrt{1.2}\sqrt{1.8}} = 8.333K$$

So the Nyquist plot will cross the negative real axis at $-8.333K$.

The complete Nyquist plot is shown in Figure 7.76.

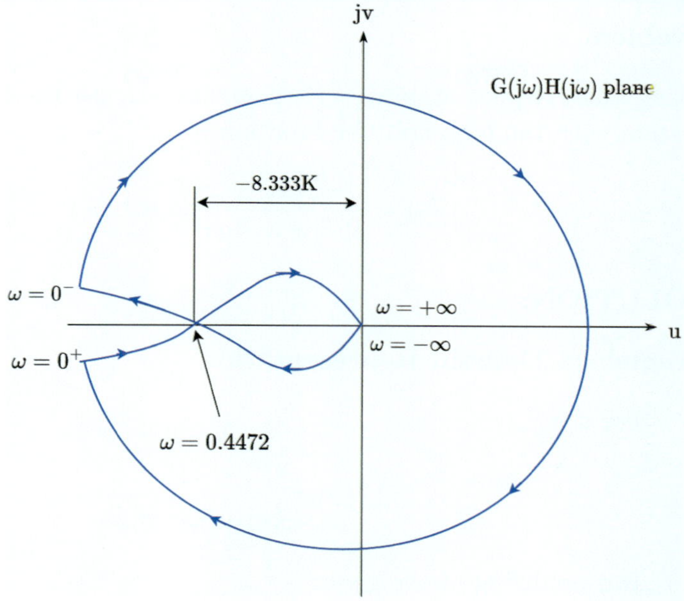

Figure 7.76: Nyquist plot for $L(s) = \dfrac{K(1+5s)}{s^2(1+s)(1+2s)}$

6. **Stability Condition**

The open-loop L(s) has no RHP poles, so $P = 0$. For the closed-loop to have no RHP poles, we need the Nyquist plot to encircle -1 zero times ($N = 0$, so $Z = P + N = 0$). Because the Nyquist plot scales with K, the negative-real-axis crossing is at $-8.333K$. If that crossing lies to the right of -1 (i.e., $-8.333K > -1$, the plot does not encircle $-1 + j0$, so the closed-loop is stable.

If the critical point $-1 + j0$ lies to the left of this crossing point, the closed-loop system will be stable because, in that case, the Nyquist plot does not encircle the critical point. Thus, the stable range of K is

$$-8.333K > -1 \implies 8.333K < 1 \implies K < 0.12$$

CONCEPT REINFORCEMENT PROBLEM 7.13

Problem

Using the Nyquist stability criterion, investigate the stability of the system with the loop transfer function

$$L(s) = G(s)H(s) = \frac{K(s-b)}{(s+a)^2}, \quad a, b > 0 \tag{7.81}$$

EXAMPLE 7.23

Problem

Using the Nyquist stability criterion, investigate the stability of the system with the loop transfer function

$$L(s) = \frac{K}{s(s^2 + as + b)}, \quad a, b > 0 \tag{7.82}$$

SOLUTION:

1. **Frequency Domain Representation**

For $s = j\omega$

$$L(j\omega) = \frac{K}{j\omega(b - \omega^2 + ja\omega)}$$

Rationalizing above gives

$$\begin{aligned} L(j\omega) &= \frac{K(-j\omega)(b - \omega^2 - ja\omega)}{\omega^2\left[(b - \omega^2)^2 + a^2\omega^2\right]} \\ &= \frac{-Ka\omega^2}{\omega^2\left[(b - \omega^2)^2 + a^2\omega^2\right]} - j\frac{K\omega(b - \omega)}{\omega^2\left[(b - \omega^2)^2 + a^2\omega^2\right]} \end{aligned}$$

2. **Real-axis Crossing**

At the real axis cross-over point, the imaginary part is zero, i.e.,

$$b - \omega^2 = 0 \quad \Longrightarrow \quad \omega^2 = b \;\; \text{or,} \;\; \omega = \pm\sqrt{b}$$

3. **Magnitude at Real-axis Crossing**

At this value of ω,

$$\left|L(j\omega)\right|_{\omega^2=b} = \frac{-K}{\omega\sqrt{[(b - \omega^2)^2 + a^2\omega^2]}} = -\frac{K}{ab}$$

So the Nyquist plot crosses the real axis at $\omega = \pm\sqrt{b}$ with an intercept of $(-K/ab)$. The plot is sketched in Figure 7.77.

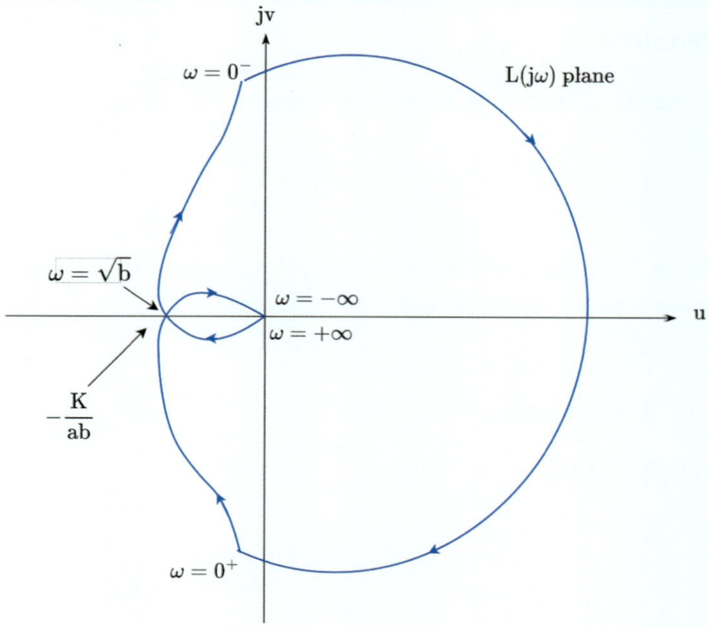

Figure 7.77: Nyquist plot for $L(s) = \dfrac{K}{s(s^2 + as + b)}$

4. **Stability Analysis**

The open-loop L(s) has no RHP poles because $a, b > 0$. So $P = 0$.

When $K > ab$, the Nyquist contour encircles -1 point twice (two clockwise encirclements). Using the convention "clockwise encirclements counted positive," $N = +2$. Therefore

$$Z = N + P = +2$$

So two closed-loop poles lie in the RHP and the closed-loop system is unstable. If $K < ab$, the system is stable

CONCEPT REINFORCEMENT PROBLEM 7.14

Problem

Consider a system with a loop transfer function

$$L(s) = \frac{K(1 + 0.4s)(s + 1)}{(1 + 5s)(s - 1)} \tag{7.83}$$

Using Nyquist stability criterion, determine range of K for closed-loop system to be stable.

EXAMPLE 7.24

Problem

Consider a system with a loop transfer function

$$L(s) = \frac{Ke^{-Ts}}{s+a}$$

Using Nyquist stability criterion, determine range of K for closed-loop system to be stable when

i. T = 0.2 and a = 2.

ii. T = 0.6 and a = 2.

SOLUTION:

Now

$$L(j\omega) = \frac{Ke^{-j\omega T}}{a+j\omega}$$

Hence

$$\lim_{\omega \to 0} L(j\omega) = \frac{K}{a}\angle 0^\circ \quad \text{and}$$

$$\lim_{\omega \to \infty} L(j\omega) = 0$$

Note that since at $\omega = \infty$, the magnitude $|L(j\omega)| = 0$, the angle information is irrelevant

To determine the real axis cross-over point, let us express $L(j\omega)$ as follows.

$$L(j\omega) = \frac{K(a-j\omega)(\cos\omega T - j\sin\omega T)}{(a+j\omega)(a-j\omega)}$$

or,

$$\frac{L(j\omega)}{K} = \frac{a\cos\omega T - \omega\sin\omega T}{a^2+\omega^2} - j\frac{\omega\cos\omega T + a\sin\omega T}{a^2+\omega^2}$$

Equating $\text{Im}\left[\frac{L(j\omega)}{K}\right]$ to zero gives

$$\omega\cos\omega T + a\sin\omega T = 0 \quad \text{or,} \quad \tan\omega T = -\frac{\omega}{a} \implies \omega T = \tan^{-1}\left(\frac{-\omega}{a}\right)$$

Expanding $\tan^{-1}\left(\frac{-\omega}{\text{a}}\right)$ and retaining first two terms of the series, the above reduces to

$$\omega\text{T}= -\frac{\omega}{\text{a}} - \frac{1}{3}\left(-\frac{\omega}{\text{a}}\right)^3 \implies \frac{\omega^3}{3\text{a}^3} - \frac{\omega}{\text{a}} - \omega\text{T} = 0$$
$$\implies \omega\left[\omega^2 - 3\text{a}^2(1+\text{T})\right] = 0$$

This gives

$$\omega = 0, \quad \omega \approx \sqrt{3\text{a}^2(1+\text{T})}$$

i. When $\text{T} = 0.2$ and $\text{a} = 2$, we have $\omega = \sqrt{3\text{a}^2(1+\text{T})} = 3.7947$. Now

$$\text{Re}\left[\frac{\text{L}(\text{j}\omega)}{\text{K}}\right]\Bigg|_{\omega=3.7947} = \frac{\text{a}\cos\omega\text{T} - \omega\sin\omega\text{T}}{\text{a}^2+\omega^2}\Bigg|_{\omega=3.7947} = -0.0631$$

Hence

$$\text{Re}\left[\text{L}(\text{j}\omega)\right]\Big|_{\omega=3.7947} = -0.0631\text{K}$$

Since $\tan^{-1}\left(-\frac{\omega}{\text{a}}\right)$ is an infinite series, there will be infinite points of intersection on the real axis. For the smallest value of ω, just greater than zero, the real axis intersection is -0.0631K. For higher values of ω, the polar plot spirals around the origin as shown in Figure 7.78a. The closed-loop stability is ensured for $0.0631\text{K} < 1$ or $\text{K} < 15.858$

ii. When $\text{T} = 0.6$ and $\text{a} = 2$, we have $\omega \approx \sqrt{3\text{a}^2(1+\text{T})} = 4.3818$. Now

$$\text{Re}\left[\frac{\text{L}(\text{j}\omega)}{\text{K}}\right]\Bigg|_{\omega=4.3818} = \frac{\text{a}\cos\omega\text{T} - \omega\sin\omega\text{T}}{\text{a}^2+\omega^2}\Bigg|_{\omega=4.3818} = -0.1677$$

Hence

$$\text{Re}\left[\text{L}(\text{j}\omega)\right]\Big|_{\omega=4.3818} = -0.1677\text{K}$$

The closed-loop stability is ensured for $0.1677\text{K} < 1$ or $\text{K} < 5.9613$

The plot is sketched in Figure 7.78b.

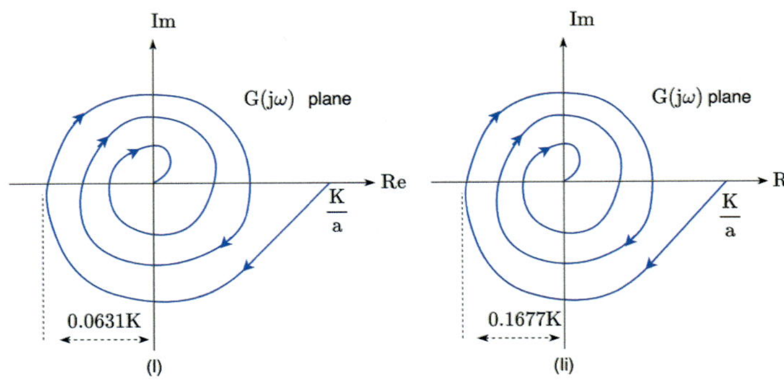

Figure 7.78: Nyquist plot for $L(s) = \frac{Ke^{-Ts}}{s+a}$; (i) for T=0.2,a=2; (ii) T=0.6, a=2

7.19 Closed-Loop Frequency Response of Unity Feedback Systems

We have discussed in Section 7.5 the correlations between frequency response and transient response, considering a prototype second-order system. While designing controllers using frequency domain compensation methods, first, the time domain specifications are translated into frequency domain specifications, and after the completion of the design, we translate the frequency response back to give an approximate time response.

Amongst all the frequency response specifications, the resonance peak M_r, resonant frequency ω_r and bandwidth ω_b are three key performance metrics of a system. Before the widespread adoption of computers, determining these parameters analytically, particularly for higher-order systems, was challenging. To address this, alternative graphical methods such as constant M and N circles and the Nichols chart were developed. Although their usage has diminished in recent years, we present these methods here due to their historical significance.

First, we discuss about **constant magnitude loci (M circles)** and **constant phase angle loci (N circles)** which can be used to determine closed-loop frequency response from open-loop frequency response (polar plot). Our analysis will focus on a unity feedback system, though this method can be applied to non-unity feedback systems with some modifications.

7.19.1 Constant M Circles

Before computers became widespread, analyzing closed-loop frequency response characteristics; particularly resonance peak M_r, resonant frequency (ω_r), and bandwidth (ω_b) was challenging for higher-order systems. Constant M-circles provided a graphical solution to determine these parameters directly from the open-loop frequency response without computing the complete closed-loop response.

Consider a unity feedback system with forward path transfer function G(s). The closed-loop transfer is

$$T(s) = \frac{C(s)}{R(s)} = \frac{G(s)}{1 + G(s)} \tag{7.84}$$

Replacing s by $j\omega$ for sinusoidal steady state response, we have

$$T(j\omega) = \frac{C(j\omega)}{R(j\omega)} = \frac{G(j\omega)}{1 + G(j\omega)} \tag{7.85}$$

Let us define the magnitude and phase angle of the closed-loop frequency response as:

$$\left|T(j\omega)\right| = M \quad \text{and} \quad \angle T(j\omega) = \alpha. \text{ Hence}$$

$$T(j\omega) = \frac{C(j\omega)}{R(j\omega)} = Me^{j\alpha} \tag{7.86}$$

Since $G(j\omega)$ is a complex number, it can be expressed in rectangular form as follows:

$$G(j\omega) = \text{Re } G(j\omega) + \text{Im } G(j\omega) = x + jy \tag{7.87}$$

Then, the magnitude of the closed-loop transfer function is expressed as:

$$M = \frac{\left|x + jy\right|}{\left|1 + x + jy\right|} = \frac{\sqrt{x^2 + y^2}}{\sqrt{(1 + x)^2 + y^2}} \tag{7.88}$$

or,

$$M^2 = \frac{x^2 + y^2}{(1 + x)^2 + y^2} \tag{7.89}$$

Rearranging Eqn(7.89) gives

$$\left(1 - M^2\right)x^2 + \left(1 - M^2\right)y^2 - 2M^2x = M^2 \tag{7.90}$$

If $M = 1$, then from Eqn(7.90), we obtain $x = \frac{-1}{2}$. This is an equation of a straight line parallel to the y-axis and passing through the point $\left(-\frac{1}{2}, 0\right)$.

If $M \neq 1$, Eqn(7.90) can be expressed as:

$$x^2 + \frac{2M^2}{M^2-1}x + \frac{M^2}{M^2-1} + y^2 = 0 \quad (7.91)$$

By adding $\left[\frac{M^2}{M^2-1}\right]^2$ on both sides of Eqn(7.91) gives

$$x^2 + \frac{2M^2}{M^2-1}x + \left[\frac{M^2}{M^2-1}\right]^2 + \frac{M^2}{M^2-1} + y^2 = \left[\frac{M^2}{M^2-1}\right]^2 \quad (7.92)$$

This gives

$$\boxed{\left[x + \frac{M^2}{M^2-1}\right]^2 + y^2 = \left[\frac{M}{M^2-1}\right]^2, \quad M \neq 1} \quad (7.93)$$

Eqn(7.93) is the equation of a circle with center at

$$x = \text{Re } G(j\omega) = -\frac{M^2}{M^2-1}, \quad y = 0 \quad (7.94)$$

The radius of this circle is

$$r = \left|\frac{M}{M^2-1}\right| \quad (7.95)$$

For different values of M, Eqn(7.93) describes a family of circles in the $G(j\omega)$-plane. These are called the **constant-M loci** or the **constant-M circles.** Figure 7.79 shows a typical set of constant-M circles in the $G(j\omega)$-plane. These circles give an integrated view of constant M circles for a few values of M greater than 1, less than 1, including for $M = 1$.

Some of the features of constant-M circles are:

a. The constant-M circles are symmetrical to the $M = 1$ line and the real axis.

b. The circles to the left of the $M = 1$ locus correspond to values of M greater than 1, and those to the right of the $M = 1$ locus correspond to values of M less than 1.

c. As M becomes larger compared to 1, the M circles become smaller and their center converges to $-1 + j0$ for $M = \infty$. Similarly, as M becomes smaller, compared to 1, M circles become smaller and their center moves toward the origin. Note that the center of M circles lies to the right of the origin for $0 < M < 1$.

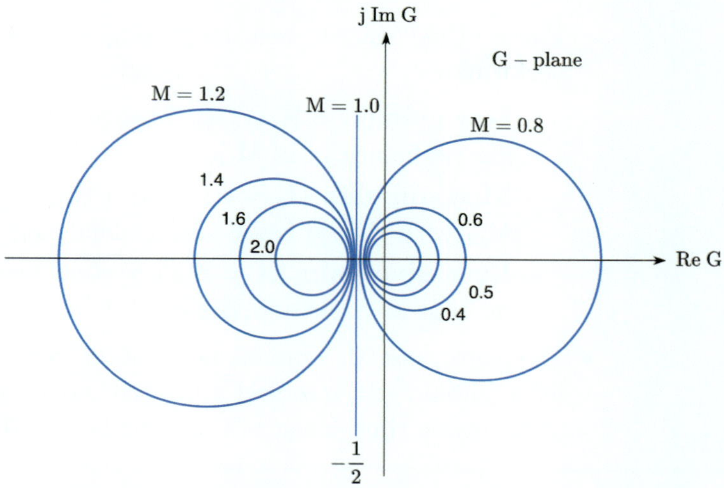

Figure 7.79: Family of constant M-Circles

7.19.2 Determining Closed-Loop Magnitude Response Using Constant M Circles

Constant M circles enable us to extract closed-loop magnitude characteristics directly from open-loop frequency response plots (polar plots) without computing closed-loop transfer function magnitudes at individual frequencies.

Graphical Analysis Procedure

Step-1: Construct a family of constant M circles on the complex plane.

Step-2: Overlay the open-loop frequency response G(jω) polar plot onto the family of M circles.

Key Analysis Points

- **Magnitude Reading:** The intersection of G(jω) locus and the constant-M circles directly provide closed-loop magnitude values M at corresponding frequencies marked on the G(jω) plot as shown in Figure 7.80.

  Resonance Peak Identification: The smallest-radius M-circle that is tangent to the G(jω) locus indicates:

  - **Resonance peak magnitude (M_r):** The M-value of the tangent circle
  - **Resonance frequency (ω_r):** The frequency at the tangency point

  Bandwith Determination: The intersection of G(jω) with the M = 0.707 circle identifies the system bandwidth ω_b.

Gain Effect Analysis:

The method reveals how increasing system gain K affects performance:

- **Low gain** ($K = K_1$**:**) $G(j\omega)$ tangent to M-circle M_1, giving resonance peak M_{r1}
- **Medium gain** ($K = K_2 > K_1$**:**) Higher resonance peak M_{r2} with shifted resonant frequency ω_{r2}
- **Excessive gain** ($K = K_3$**:**) System becomes unstable, making M-circle analysis invalid

- To ensure that M_r remains below a desired threshold, $G(j\omega)$ locus should not intersect the corresponding M-circle and, at the same time, must not enclose $(-1 + j0)$ point.

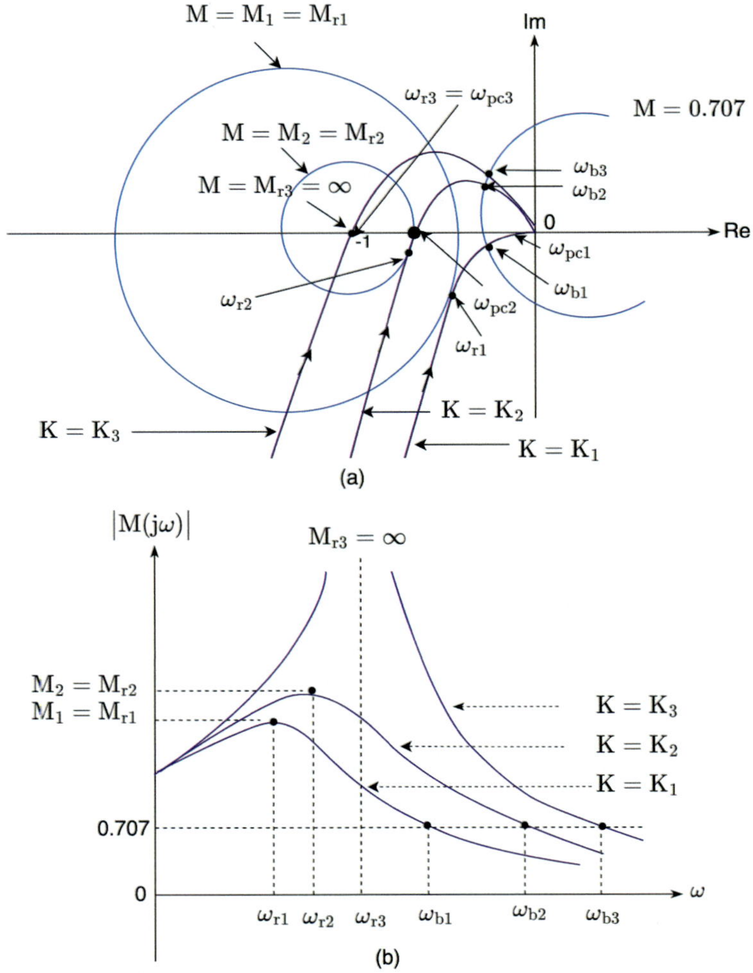

Figure 7.80: (a) Constant M-Circles together with polar plot; (b) Corresponding magnitude response

7.19.3 Constant N Circles (Constant Phase Loci)

From the perspective of controller design, the phase response is not as significant as the magnitude response, because the information about M_r, ω_r, and bandwidth can be obtained from the magnitude response. However, the concept of constant M-circles can be extended to analyze phase response, resulting in constant N-circles or constant phase loci.

From Eqn(7.85), the phase angle of $T(j\omega)$ is given by

$$\angle T(j\omega) = \phi = \angle \frac{G(j\omega)}{1 + G(j\omega)} = \angle \frac{x + jy}{1 + x + jy} \tag{7.96}$$

Then

$$\phi = \tan^{-1}\left(\frac{y}{x}\right) - \tan^{-1}\left(\frac{y}{1+x}\right) \tag{7.97}$$

Let us define $N = \tan\phi$ and use the identity

$$\tan(a - b) = \frac{\tan a - \tan b}{1 + \tan a \cdot \tan b}$$

Then, Eqn(7.97) becomes

$$N = \tan\phi = \frac{\dfrac{y}{x} - \dfrac{y}{1+x}}{1 + \dfrac{y}{x}\left(\dfrac{y}{1+x}\right)} = \frac{y}{x^2 + x + y^2} \tag{7.98}$$

or

$$x^2 + x + y^2 - \frac{y}{N} = 0 \tag{7.99}$$

Adding $\left[\dfrac{1}{4} + \dfrac{1}{(2N)^2}\right]$ on both sides of Eqn(7.99) gives

$$x^2 + x + \frac{1}{4} + y^2 - \frac{y}{N} + \frac{1}{(2N)^2} = \frac{1}{4} + \frac{1}{(2N)^2} \tag{7.100}$$

or

$$\boxed{\left(x + \frac{1}{2}\right)^2 + \left(y - \frac{1}{2N}\right)^2 = \frac{1}{4} + \frac{1}{(2N)^2}} \tag{7.101}$$

Eqn(7.101) is an equation of a circle with center at $x = -\dfrac{1}{2}$, $y = \dfrac{1}{2}N$ and radius $\left(\dfrac{1}{4} + \dfrac{1}{(2N)^2}\right)^{\frac{1}{2}}$. For a given value of N, we get a family of constant N circles as shown in Figure 7.81.

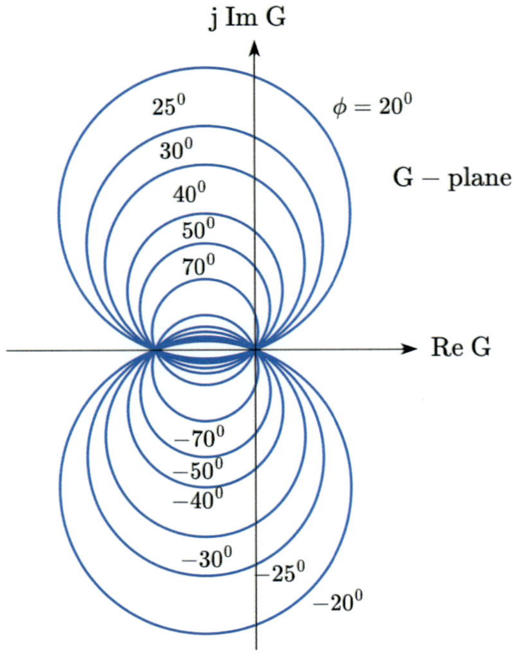

Figure 7.81: Constant N-Circles

Some of the notable points about N circles are:

1. **Common Intersection Points:**

   All N-circles, regardless of their N value, pass through two fixed points in the G(jω) plane:

   - **Origin:** (0,0)
   - **Critical point:** (-1,0)

   This occurs because both points satisfy the N-circle equation for any value of N.

2. **Multi-Valued Nature and Phase Ambiguity:**

   N-circles exhibit a fundamental ambiguity: circles corresponding to phase angles ϕ_1 and $\phi_1 \pm 180°$n, (where n = 1, 2, 3...) are identical. This creates potential confusion when interpreting phase values from the graphical construction.

   Recommended Practice: To avoid phase interpretation errors:

   - Begin analysis at zero frequency where $\phi = 0°$

- Track the phase progression continuously as frequency increases
- Ensure phase angle curves remain continuous (no sudden jumps)

7.20 Nichols Chart

While **M** and **N** circles provided valuable tools for analyzing closed-loop performance, they had a significant limitation: changes in open-loop gain cannot be easily visualized since these circles exist in linear (not logarithmic)(dB) coordinates. Therefore, changes in gain require scaling every point of G(jω) locus, making gain analysis cumbersome.

The Nichols chart, developed by N.B. Nichols, addresses this limitation through a polar-to-rectangular coordinate transformation. It provides a more intuitive and streamlined approach to control system analysis and design. In Nichols' chart, the magnitude of each point on the constant **M** circles is converted to decibels (dB) and represented on the chart using a polar-to-Cartesian transformation. The magnitude in dB is plotted on the vertical axis (ordinate), while the phase angle (in degrees) is plotted on the horizontal axis (abscissa). Similarly, the **N** circles, which represent constant phase, are transformed and overlaid onto the Nichols chart to facilitate visual analysis of closed-loop performance. The Nichols chart for the system with

$$G(s) = \frac{5}{s(s+1)(s+4)}$$

is shown in Figure 7.82.

Key Features and Applications of the Nichols Chart

Geometric Properties:

- Constant M-circles from polar coordinates transform into elliptical contours in the logarithmic gain-phase plane
- Chart exhibits symmetry about the $-180°$ vertical axis
- The critical stability point $-1 + j0$ from polar plots maps directly to $(0\text{dB}, -180°)$
- M and N contours display periodic behavior, repeating every 360° with symmetrical patterns at 180° intervals, all centered around the critical point

Closed-Loop Analysis Procedure:

Step-1: Plot the open-loop frequency response G(jω) on the Nichols grid

Step-2: Extract closed-loop characteristics where $G(j\omega)$ intersects the contours:

- **Magnitude and Phase:** Direct reading from M and N contour intersections
- **Resonance Peak** (M_r)**:** Identified as the smallest $M > 1$ contour that is tangent to the $G(j\omega)$ curve
- **Resonant Frequency** (ω_r)**:** Frequency corresponding to the tangency point
- **Bandwidth:** Frequency where $G(j\omega)$ intersects the $M = -3$ dB contour

Stability Margin Determination:

Gain Margin (GM): Vertical separation in dB between:

- Phase crossover point (where $G(j\omega)$ crosses $-180°$ line) (see Figure 7.82 where the gain margin is 12dB)
- Critical point (0 dB, $-180°$)

Phase Margin (PM): Horizontal separation in degrees between:

- Gain crossover point (where $G(j\omega)$ crosses 0 dB axis) (see Figure 7.82 where the phase margin is 45°)
- Critical point (0 dB, $-180°$)

Primary Advantage—Gain Sensitivity:

Open-loop gain changes (factor K) affect only the vertical position of the $G(j\omega)$ locus without altering its shape:

- **Increased gain:** Shifts entire curve upward
- **Decreased gain:** Shifts entire curve downward

This unique property makes the Nichols chart exceptionally powerful for gain-based controller design and optimization, where it allows us to visualize immediately how gain adjustments impact both performance and stability margins.

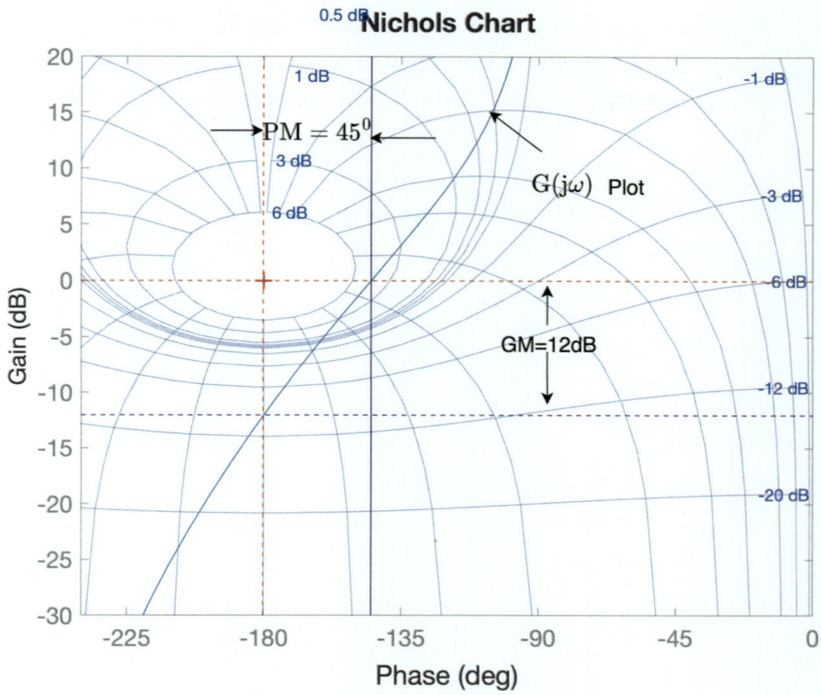

Figure 7.82: Nichols chart of $G(s) = \dfrac{5}{s(s+1)(s+4)}$

EXAMPLE 7.25

Problem

Consider the system with the open-loop transfer function

$$G(s) = \frac{K}{s(s+2)(s+3)}$$

a. Determine the gain and phase margins of this system using the Nichols chart for $K = 30$.

b. Find these margins if the gain is decreased to 5 and increased to 100. Observe how the plot changes in the Nichols chart.

SOLUTION:

a. The Nichols chart displaying open-loop frequency response for $K = 30$ is shown in Figure 7.83. At this gain, the gain margin is 0dB and the phase margin is 0^0, indicating that the system is marginally stable.

b. Next, when $K = 5$, the $G(j\omega)$ locus shifts downward, resulting in a gain margin of 15.5dB and a phase margin of 55^0. The system is therefore stable for this gain value.

However, when the gain is increased from K = 30 to K = 100, the locus shifts upward, reducing the gain margin to -10dB and phase margin to -28^0. Under this condition, the system becomes unstable.

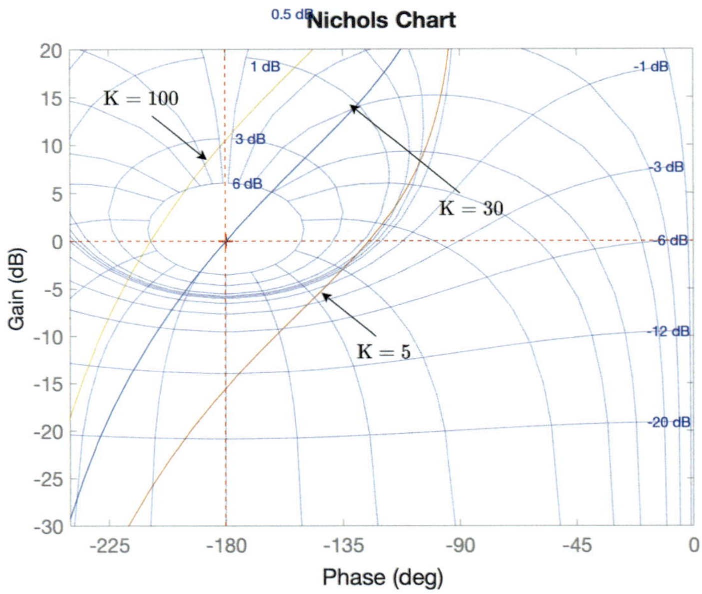

Figure 7.83: Nichols chart of $G(s) = \dfrac{K}{s(s+2)(s+3)}$

7.21 Multiple Choice Skills Test Questions

M 7.1. Frequency response test is not recommended for systems with

A. small time constants **C.** any time constant

B. large time constants **D.** very low time constants

M 7.2. Consider a linear time-invariant system with transfer function

$$G(s) = \frac{1}{s+1}$$

If the input is $\sin t$ and the steady state output is $A\sin(t+\phi)$, then the value of A equals to

A. $\sqrt{2}$ **B.** $\frac{1}{2}$ **C.** $\frac{1}{\sqrt{2}}$ **D.** 1

M 7.3. The forced sinusoidal response to the input $u(t) = 9\cos(3t - 30^0)$ for the system with transmittance $F(s) = \dfrac{s}{s+3}$ is

A. $\dfrac{9}{\sqrt{2}}\cos(3t + 15^0)$

B. $\dfrac{9}{\sqrt{2}}\cos(3t - 15^0)$

C. $\dfrac{9}{\sqrt{2}}\cos(3t - 55^0)$

D. $9\cos(3t + 15^0)$

M 7.4. The amplitude ratio $A(\omega)$ and phase shift function $\phi(\omega)$ for irrational function $G(s) = \dfrac{e^{-2s}}{s}$ are

A. $A(\omega) = -2\omega$ and $\phi(\omega) = -\dfrac{\pi}{2}$

B. $A(\omega) = \dfrac{-2}{\omega}$ and $\phi(\omega) = -2\omega$

C. $A(\omega) = \dfrac{1}{\omega}$ and $\phi(\omega) = -2\omega - \dfrac{\pi}{2}$

D. $A(\omega) = \dfrac{2}{\omega}$ and $\phi(\omega) = -2\omega$

M 7.5. For loop transmittance $G(s) = \dfrac{\omega_n^2}{s^2 + 2\xi\omega_n s + \omega_n^2}$, the value of $\left|G(j\omega_n\right|$ is

A. $\dfrac{1}{4\xi}$ **B.** $\dfrac{\xi}{2}$ **C.** $\dfrac{1}{2\xi}$ **D.** $\dfrac{2}{\xi}$

M 7.6. The resonance peak M_r does not exit for systems with

A. $\xi = 0$ **B.** $\xi < 0.707$ **C.** $\xi > 0.707$ **D.** all values of ξ

M 7.7. The resonance peak M_r of a second order system is

A. $M_r = \dfrac{1}{2\xi\sqrt{1-\xi^2}}$

B. $M_r = \xi\sqrt{1-\xi^2}$

C. $M_r = \dfrac{1}{\xi\sqrt{1-\xi^2}}$

D. $M_r = \dfrac{1}{\xi\sqrt{1-2\xi^2}}$

M 7.8. The resonance frequency ω_r of a second order system is

A. $\omega_r = \omega_n\sqrt{1-\xi^2}$

B. $\omega_r = \dfrac{\omega_n}{\sqrt{1-\xi^2}}$

C. $\omega_r = \dfrac{\omega_n}{\sqrt{1-2\xi^2}}$

D. $\omega_r = \omega_n\sqrt{1-2\xi^2}$

M 7.9. Consider the following statements related to resonance peak M_r and resonant frequency ω_r of a control system.

A. M_r is a function of damping ratio ξ and ω_r is independent of ξ.

B. M_r is a function of damping ratio ξ and ω_r is a function of both ξ and natural frequency ω_n.

C. Both M_r and ω_r are functions of ξ only.

D. Both M_r and ω_r are functions of ξ and ω_n

Which of the above is correct?

M 7.10. Let M_p and M_r denote respectively the peak overshoot and resonance peak of a control system, and ξ denotes the damping ratio. Consider the following statements:

I Both M_p and M_r are functions of ξ.

II M_p is a function of damping ratio ξ and M_r is independent of ξ.

III Neither M_p nor M_r are functions of ξ.

IV A large of M_p implies a large value of M_r

Which of the following is correct?

A. Statement I is only correct

B. Statements II and III are correct

C. Statements II and IV are correct

D. Statements I and IV are correct

M 7.11. For a given value of damping ratio, higher values of ω_r yield

A. Larger settling time (System settles slowly)

B. Constant settling time

C. Smaller settling time (System settles faster)

D. Infinite settling time

M 7.12. Consider the following statements

I Bandwidth (BW) and rise time t_r are directly proportional

II Increase of ξ decreases BW and decreases t_r

III Bandwidth (BW) and rise time t_r are inversely proportional

IV BW decreases with increase in natural frequency ω_n

V Rise time t_r decreases with increase of natural frequency ω_n

Which of the following is correct?

A. Statement I, II, and IV are correct

B. Statements III and V are correct

C. Statements I and II are correct

D. Statements-I is only correct

M 7.13. Consider the following statements

I Addition of a zero (closer to the j $j\omega$-axis) to the forward-path transfer function results in sluggish system response.

II Adding a zero very close to the origin increases the settling time.

III Addition of a zero to the forward path does not significantly affect the transient response

IV Addition of a zero to the forward-path transfer function increases the bandwidth.

Which of the following is correct?

A. Statement-I is only correct

B. Statements-II and IV are correct

C. Statements-IV is only correct

D. Statements-III is only correct

M 7.14. Addition of a pole to the forward path transfer function

A. Increases M_r and causes large overshoot of closed-loop system

B. Decreases M_r and causes large overshoot of closed-loop system

C. Decreases M_r and causes smaller overshoot of closed-loop system

D. Increases M_r and causes smaller overshoot of closed-loop system

M 7.15. The frequency where the low frequency and high frequency asymptotes intersect is called

A. Resonant frequency

B. Cutoff frequency

C. Damped frequency

D. Corner frequency

M 7.16. The initial slope of the Bode magnitude plot of a system with a simple open-loop pole at the origin is

A. +20 dB/decade

B. -40 dB/decade

C. -20 dB/decade

D. 0 dB/decade

M 7.17. If the initial slope of the Bode magnitude plot is -40 dB/decade, this implies that

A. There is a double pole at the origin

B. There is a quadratic factor in the open-loop transfer function

C. There is a factor with repeated pole of order 2 of corner frequency 1

D. A single pole at the origin

M 7.18. The slope of high frequency asymptotes of the Bode magnitude plot for a system with 10 poles and 2 zeros is

A. -240 dB/decade

B. +40 dB/decade

C. -200 dB/decade

D. -160 dB/decade

M 7.19. Consider a stable system with transfer function

$$G(s) = \frac{s^m + b_1 s^{m-1} + b_2 s^{m-2} + \cdots + b_0}{s^n + a_1 s^{n-1} + a_2 s^{n-2} + \cdots + a_n}$$

where $b_1, \ldots, b_m$ and $a_1, \ldots, a_n$ are real valued constants. The slope of the Bode log magnitude curve of G(s) converges to -60 dB/decade as $\omega \to \infty$. A possible pair of values of m and n are

A. $m = 2$ and $n = 3$

B. $m = 0$ and $n = 3$

C. $m = 1$ and $n = 1$

D. $m = 3$ and $n = 5$

M 7.20. The open-loop transfer function of a system is given by

$$G(s)H(s) = \frac{Ks^2}{(1 + 0.5s)(1 + 2s)(1 + 5s)}$$

The initial slope of the Bode magnitude plot would be

A. +20dB/decade

B. -20dB/decade

C. -40dB/decade

D. +40dB/decade

M 7.21. The Bode magnitude plot of a feedback control system is shown in Figure 7.84. The value of ω_1 and ω_2 are respectively equal to

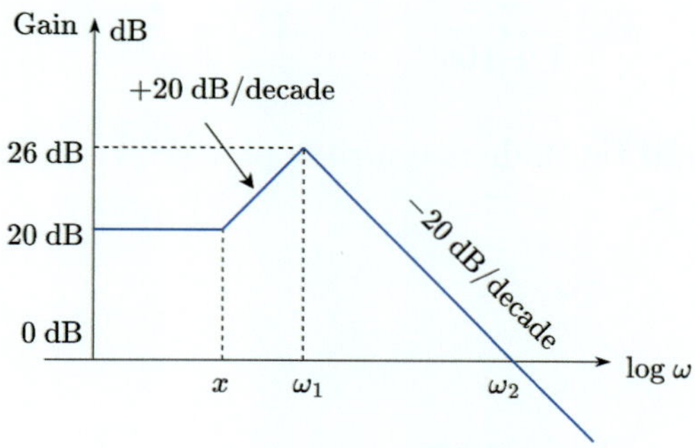

Figure 7.84

A. x and $40x$

B. $2x$ and $40x$

C. $10x$ and $40x$

D. $2x$ and $4x$

M 7.22. For the asymptotic Bode magnitude plot shown in Figure 7.85, the system transfer function is

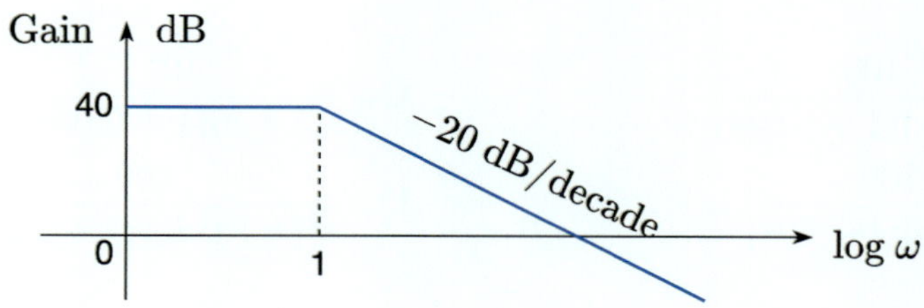

Figure 7.85

A. $\frac{10}{1+s}$ **B.** $\frac{1000}{1+s}$ **C.** $\frac{100}{1+s}$ **D.** $\frac{100}{1+0.1s}$

M 7.23. For the asymptotic Bode magnitude plot shown in Figure 7.86, the system transfer function is

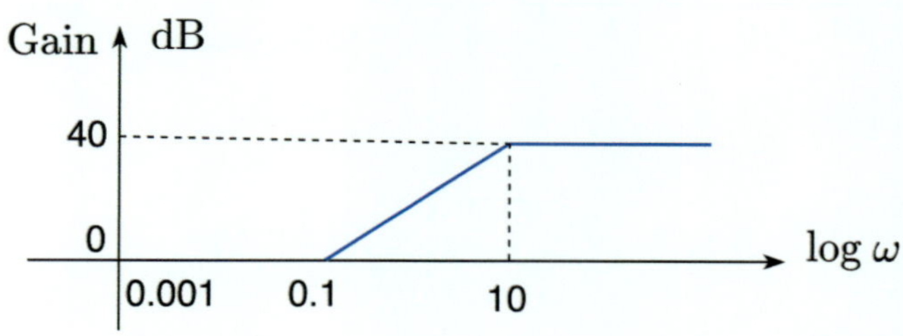

Figure 7.86

A. $\dfrac{1+0.1s}{1+10s}$ **B.** $\dfrac{1+s}{1+10s}$ **C.** $\dfrac{1}{1+10s}$ **D.** $\dfrac{1+10s}{1+0.1s}$

M 7.24. For the asymptotic Bode magnitude plot shown in Figure 7.87, the system transfer function is

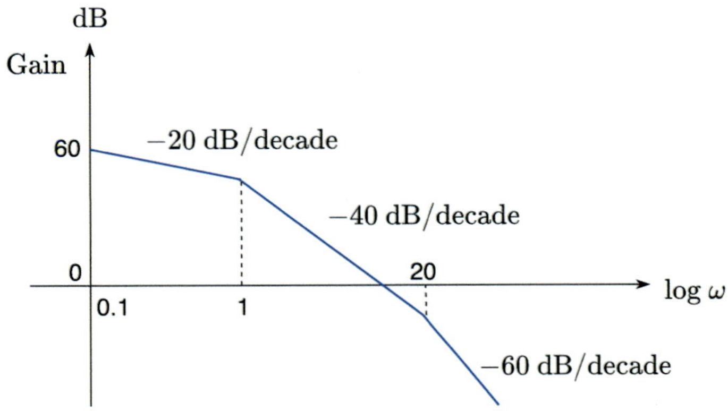

Figure 7.87

A. $\dfrac{100}{s(1+s)(1+0.05s)}$ **C.** $\dfrac{10}{s(1+s)(1+0.05s)}$

B. $\dfrac{100}{s(1+s)(1+20s)}$ **D.** $\dfrac{100}{s^2(1+0.1s)(1+0.05s)}$

M 7.25. For the asymptotic Bode magnitude plot shown in Figure 7.88, the system transfer function is

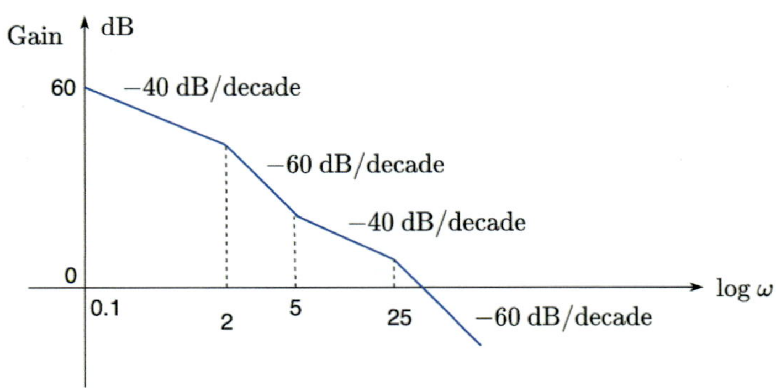

Figure 7.88

A. $\dfrac{100(s+2)}{s^2(s+5)(s+25)}$ **C.** $\dfrac{10(s+5)}{s^2(s+2)(s+25)}$

B. $\dfrac{100(s+5)}{s(s+2)(s+25)}$ **D.** $\dfrac{100(s+5)}{s^2(s+2)(s+25)}$

M 7.26. For a minimum phase system to be stable

A. Both gain and phase margin must be negative

B. Gain margin should be positive; but the phase margin is negative

C. Both gain and phase margin must be positive

D. Gain margin should be negative, and phase margin is positive

M 7.27. The gain cross over frequency ω_{gc} and phase cross over frequency ω_{pc} of a system are 6rad/sec and 10 rad/sec, respectively. This implies that

A. the system is stable with positive phase margin

B. the system is unstable with positive phase margin

C. the system is stable with negative phase margin

D. the system is unstable with a negative phase margin

M 7.28. If for a system, the gain cross over frequency $\omega_{\mathrm{gc}} = 12\mathrm{rad/sec}$ and the phase cross over frequency $\omega_{\mathrm{pc}} = 8\mathrm{rad/sec}$, then

A. the system is stable with positive phase margin

B. the system is unstable with negative phase margin

C. the system is stable with negative phase margin

D. the system is unstable with positive phase margin

M 7.29. If $\omega_{\mathrm{gc}} = \omega_{\mathrm{pc}}$, then

A. the system is stable with negative phase margin

B. the system is unstable with negative gain margin

C. the system is marginally stable with positive phase margin

D. the system is marginally stable with zero phase margin

M 7.30. For a system, if the magnitude $\left|\mathrm{G(j\omega)H(j\omega)}\right| = -15$ dB at the phase cross over frequency, then the system is

A. marginally stable and gain margin (GM) equals to 0dB

B. stable and GM=+15dB

C. stable and GM=-15dB

D. unstable and GM=+15dB

M 7.31. If the phase angle $\angle\mathrm{G(j\omega)H(j\omega)}$ at the gain cross over frequency equals -190^0, then the phase margin is

A. $+10^0$ **B.** $+190^0$ **C.** -10^0 **D.** -190^0

M 7.32. The magnitude of open-loop transfer function $\left|G(j\omega)H(j\omega)\right|$ at the phase crossover frequency equals to -20dB and the phase angle $\angle G(j\omega)H(j\omega)$ at the gain cross over frequency is -150^0. This implies

A. GM $= -20$dB, PM $= +30°$, system stable

B. GM $= +20$dB, PM $= -30°$, system unstable

C. GM $= -20$dB, PM $= -30°$, system unstable

D. GM $= +20$dB, PM $= +30°$, system stable

M 7.33. The magnitude of open-loop transfer function $\left|G(j\omega)H(j\omega)\right|$ at the phase crossover frequency equals to +10dB and the phase angle $\angle G(j\omega)H(j\omega)$ at the gain cross over frequency is -150^0. This implies

A. GM $= -10$dB, PM $= +30°$, system stable

B. GM $= -10$dB, PM $= +30°$, system unstable

C. GM $= +10$dB, PM $= -30°$, system stable

D. GM $= -10$dB, PM $= +150°$, system stable

M 7.34. The magnitude of open-loop transfer function $\left|G(j\omega)H(j\omega)\right|$ at the phase crossover frequency equals to -25dB and the phase angle $\angle G(j\omega)H(j\omega)$ at the gain cross over frequency is -200^0. This implies

A. GM $= +25$dB, PM $= +20°$, system stable

B. GM $= +25$dB, PM $= +20°$, system stable

C. GM $= +25$dB, PM $= -20°$, system unstable

D. GM $= -25$dB, PM $= -20°$, system unstable

M 7.35. The gain margin of a system is 0dB, and the phase margin is 0^0. It represents

A. a stable system

B. a marginally stable system

C. a conditionally stable system

D. unstable system

M 7.36. The gain margin (in dB) of a system having a loop transfer function

$$G(s)H(s) = \frac{5}{s(s+2)} \quad \text{is}$$

A. 0 **B.** 5 **C.** 2.5 **D.** ∞

M 7.37. If the Γ_s contour (clockwise direction) encloses 4 zeros and 3 poles of F(s), the corresponding contour Γ_F in the F(s) plane will encircle the origin of F(s)-plane

A. once in the clockwise direction

B. thrice in clockwise direction

C. once in the counterclockwise direction

D. twice in counterclockwise direction

M 7.38. If the Γ_s contour (clockwise direction) encloses 4 poles and 2 zeros of F(s), the corresponding contour Γ_F in the F(s) plane will encircle the origin of F(s)-plane

A. once (clockwise)

B. thrice (counterclockwise)

C. twice (clockwise)

D. twice (counterclockwise)

M 7.39. If the Γ_s contour (clockwise direction) encloses 3 poles and 3 zeros of F(s), the corresponding contour Γ_F in the F(s) plane will encircle the origin of F(s)-plane

A. zero times

B. thrice in counterclockwise direction

C. twice in the clockwise direction

D. thrice in clockwise direction

M 7.40. The phase cross over frequency ω_{pc} of the system with loop transfer function $L(s) = \dfrac{e^{-5s}}{s}$ is

A. 3.1831 **B.** 0.6283 **C.** 0.314 **D.** 5.0

M 7.41. For the system with loop transfer function $L(s) = \dfrac{e^{-Ts}}{s}$, the value of T which results in a gain margin of 20dB is

A. $\dfrac{\pi}{2}$ **B.** $\dfrac{2}{\pi}$ **C.** π **D.** $\dfrac{\pi}{5}$

M 7.42. If the gain of the system is doubled, the gain margin of the system

A. becomes one-fourth of its original value

B. is doubled

C. is halved

D. remains unaltered

M 7.43. The gain margin of a feedback system with a loop transfer function

$$G(s)H(s) = \frac{20}{(s+4)^3}$$

is

A. 12.08 dB **B.** 28.16 dB **C.** 25.6 dB **D.** 64.85 dB

M 7.44. The damping ratio of a system with a phase margin of 35^0 is approximately equal to

A. 0.035 **B.** 0.07 **C.** 0.35 **D.** 0

M 7.45. If the polar plot of a system intersects the real axis at -0.01, the gain margin in dB is

A. 20 dB **B.** 40 dB **C.** 60 dB **D.** 100 dB

M 7.46. The damping ratio of a second-order system is 0.4. Its phase margin is approximately equal to

A. $+40^0$ **B.** -40^0 **C.** $+60^0$ **D.** $+20^0$

M 7.47. If the polar plot touches the real axis only at the origin, then the gain margin is

A. 0 **B.** negative **C.** finite positive **D.** ∞

M 7.48. Addition of a pole to the loop transfer function

A. makes the closed-loop system more stable and decreases the bandwidth

B. makes the closed-loop system less stable and increases the bandwidth

C. makes the closed-loop system less stable and decreases the bandwidth

D. makes the closed-loop system more stable and increases the bandwidth

M 7.49. Addition of a zero to the forward path transfer function

A. increases the bandwidth of the closed-loop system

B. decreases the bandwidth of the closed-loop system

C. does not change the bandwidth of the closed-loop system

D. None of the above

Answers

M 7.1. B **M 7.2.** C **M 7.3.** A **M 7.4.** C **M 7.5.** D **M 7.6.** C **M 7.7.** A **M 7.8.** D **M 7.9.** B **M 7.10.** D **M 7.11.** C **M 7.12.** B **M 7.13.** B **M 7.14.** A **M 7.15.** D **M 7.16.** C **M 7.17.** A **M 7.18.** D **M 7.19.** B **M 7.20.** D **M 7.21.** B **M 7.22.** C **M 7.23.** D **M 7.24.** A **M 7.25.** D **M 7.26.** C **M 7.27.** A **M 7.28.** B **M 7.29.** D **M 7.30.** B **M 7.31.** C **M 7.32.** D **M 7.33.** B **M 7.34.** C **M 7.35.** B **M 7.36.** D **M 7.37.** A **M 7.38.** D **M 7.39.** A **M 7.40.** C **M 7.41.** A **M 7.42.** C **M 7.43.** B **M 7.44.** C **M 7.45.** B **M 7.46.** A **M 7.47.** D **M 7.48.** C **M 7.49.** A

7.22 Knowledge Assessment Problems

KaP 7.1 Find the steady state output c(t) for each of the following systems with transfer function $G(s) = \dfrac{C(s)}{R(s)}$ for the given input excitation.

i. $G(s) = \dfrac{K(1 + T_1 s)}{1 + T_2 s}, \quad r(t) = A \sin \omega t$

ii. $G(s) = \dfrac{s}{s + 10}, \quad r(t) = 10 \sin(5t - 30^0)$

iii. $G(s) = \dfrac{10(s + 1)}{s(s + 2)(s + 10)}, \quad r(t) = 5 \cos(t + 45^0)$

iv. $G(s) = \dfrac{80e^{-0.1s}}{s(s + 4)(s + 10)}, \quad r(t) = 12 \sin(2t + 60^0)$

KaP 7.2 Consider the feedback control system shown in Figure 7.89.

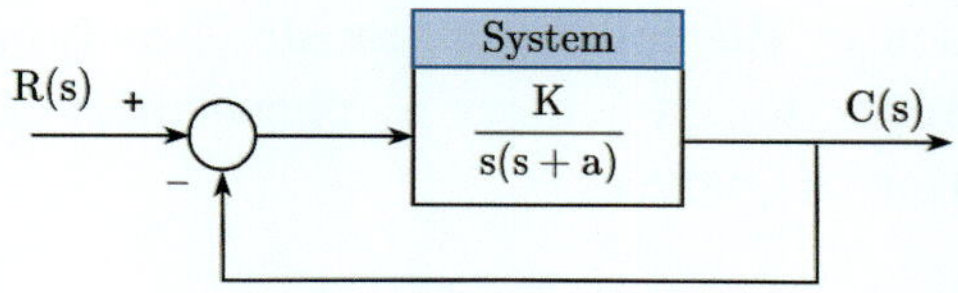

Figure 7.89: Second order control system

Analytically determine resonance peak M_r, resonant frequency ω_r, bandwidth ω_b and percentage peak overshoot for following values of K and a.

i. K = 25 and a = 2

ii. K = 81 and a = 9

iii. K = 16 and a = 3.2

KaP 7.3 Compute the resonant peak M_r, resonant frequency ω_r and closed-loop bandwidth ω_b of a prototype second order system with the following performance characteristics:

a. $M_p = 15\%$, $T_p = 1.5$ seconds
b. $T_s = 7$ seconds, $T_p = 3.5$ seconds
c. $\xi = 0.3$, $T_s = 2$ seconds
d. $\xi = 0.6$, $T_p = 2$ seconds

KaP 7.4 Consider the feedback control system shown in Figure 7.89.

i. Determine the value of K and a such that $M_r = 1.5$ and $\omega_r = 15$ rad/sec.
ii. For the values of K and a determined in part(i), calculate the percentage peak overshoot to unit step input, settling time, and bandwidth of the system.

KaP 7.5 Use MATLAB to verify your answer in **KaP 7.4**.

KaP 7.6 Sketch the asymptotic gain plots for the following open-loop transfer functions

a. $L(s) = \dfrac{40s}{(s+1)(s+10)}$

b. $L(s) = \dfrac{100(s+2)}{s(s+1)(s+5)}$

c. $L(s) = \dfrac{(s+2)(s+4)}{s(s+1)(s+5)(s+10)}$

d. $L(s) = \dfrac{s(s+10)(s+20)}{(s+5)(s+15)(s+25)(s+50)}$

KaP 7.7 For each of the following transfer functions, sketch the Bode plot and determine the gain margin, phase margin, gain cross-over, and phase crossover frequencies

a. $L(s) = \dfrac{1000}{(s+10)(s+500)}$

b. $L(s) = \dfrac{20(s+100)}{(s+2)(s+50)}$

c. $L(s) = \dfrac{500}{s(s+5)(s+10)}$

d. $L(s) = \dfrac{100}{(s+1)(s^2+10s+2)}$

KaP 7.8 The forward-path transfer function of a unity feedback control system is

$$G(s) = \frac{K(s+1)}{s(s+2)(s+5)(s+10)}$$

Find the value of gain K such that the system may have a

a. Gain margin of 20dB

b. Phase margin of 45^0.

KaP 7.9 A system has a loop transfer function

$$L(s) = \frac{K(s+100)}{s(s+10)(s+40)}$$

a. Sketch the Bode plot for $K = 1000$ and determine the gain and phase margin. Is the system stable?

b. Find the gain K which ensures a phase margin of 40^0.

KaP 7.10 The loop transfer function of a unity feedback system is

$$L(s) = \frac{K}{(s+2)^2}$$

a. Sketch the Bode diagram for $K = 100$ and determine the phase margin.

b. Find the gain K such that the phase margin is at least 50^0.

KaP 7.11 Consider a unity feedback system with a loop transfer function is

$$L(s) = \frac{K(s+4)}{s^2(s+20)^2}$$

Determine the value of K that will yield a phase margin $\geq 20^0$.

KaP 7.12 A closed-loop system with unity feedback has a loop transfer function

$$L(s) = \frac{K(s+5)}{s^2}$$

a. Determine the gain K so that the phase margin is 45^0.

b. For the gain K selected in part (a), determine the gain margin.

c. Predict the bandwidth of the closed-loop system.

KaP 7.13 The forward path transfer function of a unity feedback system is

$$G(s) = \frac{Ke^{-Ts}}{s(s^2+10s+100)}$$

Using Bode plot, find the maximum value of K for stability. Assume $T = 1\text{sec}$.

KaP 7.14 Consider the system shown in Figure 7.90

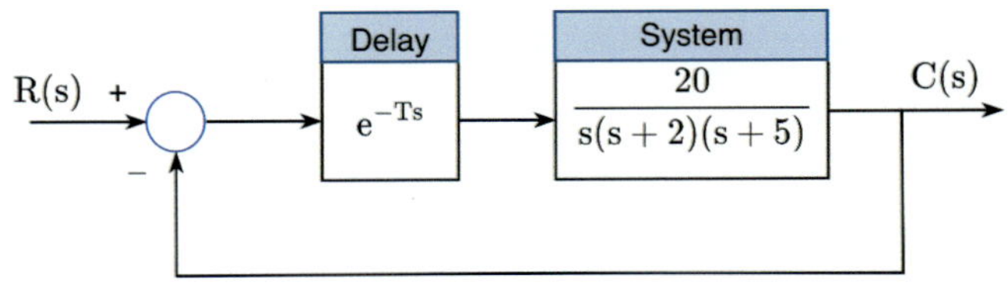

Figure 7.90: System with time delay

Determine the gain and phase margin for time delays $T = 0, 0.1, 0.2, 0.5$ and 1. Comment on the effects of delay T on the relative stability of the system.

KaP 7.15 Determine the phase crossover frequency, gain margin, and stability condition of a system with an open-loop transfer function

$$L(s) = G(s)H(s) = \frac{K}{s(s+\alpha)^2}, \quad K, \alpha > 0$$

KaP 7.16 A unity feedback system has a loop transfer function

$$L(s) = \frac{K}{s-a}, \quad a > 0$$

Plot the Nyquist plot for $K < a$, $K = a$ and $K > a$ and comment on stability of the closed-loop system.

KaP 7.17 Consider the system shown in Figure 7.91.

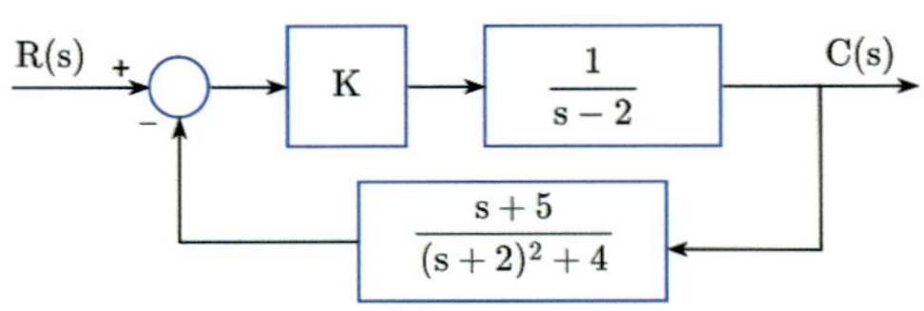

Figure 7.91: Feedback control system

a. Use MATLAB and draw the Bode plot for $K = 1$ and estimate the range of K for which the system is stable.

b. Verify the stable range of K using `margin` function in MATLAB

c. Sketch the Nyquist plot of the system, and use it to verify the number of unstable roots for the unstable ranges of K

KaP 7.18 Consider the system shown in Figure 7.92.

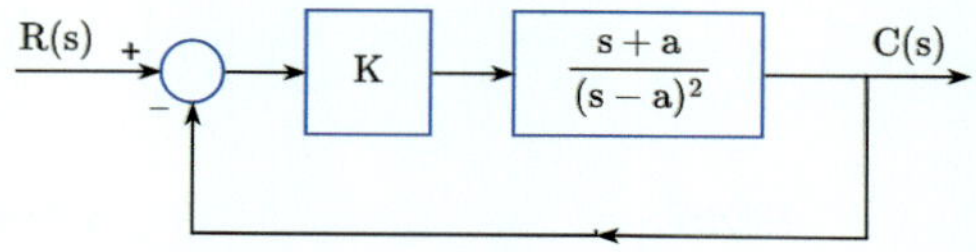

Figure 7.92: Feedback control system

Investigate the stability of the system from the Nyquist plot.

KaP 7.19 Consider the system shown in Figure 7.93.

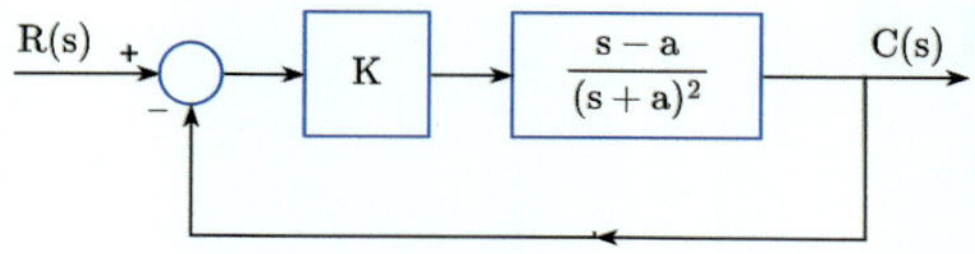

Figure 7.93: Feedback control system

Investigate the stability of the system from the Nyquist plot.

KaP 7.20 Draw the Nyquist plot for the system with open-loop transfer function

$$L(s) = \frac{K(s-1)(s-2)}{(s+1)(s+2)}$$

and investigate its stability.

KaP 7.21 Investigate the stability of the systems using the Nyquist criterion, whose loop transfer functions are

i. $L(s) = \dfrac{K}{(s+a)(s-b)}, \quad a, b > 0$

ii. $L(s) = \dfrac{K}{(s-a)(s+a)^2}, \quad a > 0$

iii. $L(s) = \dfrac{K(s^2+a)}{(s-a)^3}, \quad a > 0$

KaP 7.22 Using Nyquist plot, investigate the stability of the following systems with open-loop transfer functions

a. $L(s) = \dfrac{K(s+2)(s^2+5)}{5(s-1)^3}$ b. $L(s) = \dfrac{K(s+1)(s^2-4s+1)}{(s-1)^3}$

Chapter 8

Controller Design of Linear Systems

Controllers are like personalities: proportional ones react instantly, integral ones hold grudges, and derivative ones panic at the slightest change

DESIRED LEARNING OUTCOMES

Upon completion of this chapter, readers should be able to:

- Understand different compensation strategies and controller configurations
- Design of P, PI, and PID, lead, lag, and lag-lead compensators in the time domain using Root locus
- Analytically design lag, lead, and lag-lead compensators
- Design various compensators in the frequency domain using the Bode plot
- Design rate feedback compensators
- Design PID controller

8.1 Introduction

In the preceding chapters, we have explored how to analyze and quantify the behavior of control systems using various performance measures in both time and frequency domains- such as *peak overshoot, rise time, settling time, steady state error, bandwidth, gain, and phase margins.* The primary purpose of this analysis is to derive insights and guidelines that aid in the design of controllers.

In control engineering, "*Analysis often precedes Synthesis (Design).*" While

DOI: 10.1201/9781003571308-8

analysis answers the question "How does the system behave?," synthesis is concerned with "How can the system be made to behave in a desired manner?."

Given a controlled process, the controller design process typically consists of three essential steps:

Steps of Controller Design

1. **Define the desired performance characteristics** including criteria such as accuracy, stability, and speed of response. These must be explicitly stated before beginning the design and are referred to as the performance specifications.
2. **Determine the type and configuration of the controller or compensator** that will help meet these specifications.
3. **Calculate the parameters of the controller** to achieve the desired performance objectives.

These steps are elaborated in the sections that follow.

8.1.1 Performance Specifications

Performance specifications play a central role in shaping the controller design. These specifications, which outline the desired behavior of the system, may be expressed either in the time domain or in the frequency domain. Common time-domain specifications include *maximum overshoot, rise time*, and *settling time*; each formally defined in the context of second-order systems (see Chapter 3). On the other hand, frequency-domain specifications include *gain margin, phase margin, bandwidth*, and *resonant peak* M_r (see Chapter 7).

For a prototype second-order system, there are well-established analytical relationships connecting time-domain and frequency-domain performance indices. Although such direct correlations are harder to establish in higher-order systems, the intuition derived from second-order behavior often guides the design of practical controllers.

Controller design can be carried out either in the time domain or in the frequency domain. While time-domain specifications offer intuitive insight (such as how much the output overshoots a set point), frequency-domain specifications provide a deeper understanding of stability margins and robustness. A designer may choose one approach over the other depending on the application at hand.

Historically, time-domain methods were favored, especially for systems that could be well approximated as second-order. However, for more complex or higher-order systems, frequency-domain methods have proven more versatile and insightful, particularly when used with tools like Bode plots, Nyquist diagrams, and Nichols charts.

The rise of user-friendly simulation and computational tools—such as MATLAB, Simulink, and LabVIEW—has greatly expanded the accessibility and applicability of both time- and frequency-domain methods in modern controller design.

8.1.2 Approaches to Controller Design and System Compensation

The most straightforward way to improve the performance of a control system is to modify the process itself, if feasible. For instance, a servomechanism's transient response may be improved by using a more efficient motor; structural vibrations can be mitigated using additional damping devices; an aircraft's aerodynamic design might be modified to improve dynamic response. However, in most practical cases, such process modifications are either infeasible or insufficient. Therefore, we must resort to incorporating corrective elements, known as compensators, into the control system.

For simple control systems, it may be possible to get a satisfactory response by adjusting the gain. But, in most practical cases, the gain adjustment alone may not provide the desired performance. For instance, increasing the gain may reduce the steady state error, but it can simultaneously worsen transient behavior or even destabilize the system. To address this trade-off, compensators are introduced. These are auxiliary subsystems that modify the closed-loop system dynamics to achieve desired specifications. They may be electrical, mechanical, hydraulic, or pneumatic.

> A compensator is a device that can be inserted within the structure of the feedback system to equalize or compensate for performance deficiency.

8.2 Controller Configurations

Series compensation and feedback (or parallel) compensation

After identifying the performance objectives, the next step in the controller design process is to determine the appropriate location for the compensator within the system's structure. Following this, the parameters of the compensator are selected to ensure that the desired design objectives are met.

Figures 8.1, 8.2, and 8.3 illustrate several commonly used configurations of compensators and controllers for single-input single-output (SISO) systems. These configurations vary in structure and functionality, and are briefly described in the following.

Series (Cascade) Compensation: Figure 8.1a shows the most commonly employed system configuration, in which the compensator $G_c(s)$

is inserted in series with the plant. This configuration is known as **series or cascade compensation**. It directly modifies the forward path dynamics of the control system and is widely used due to its straightforward implementation and effectiveness in shaping the overall system response.

Feedback (Parallel) Compensation: In the configuration shown in Figure 8.1b, the compensator is located in the feedback path of the system. This arrangement is referred to as feedback or parallel compensation. It is particularly useful for modifying the characteristics of the feedback signal, such as improving disturbance rejection or stability, without affecting the main control path directly.

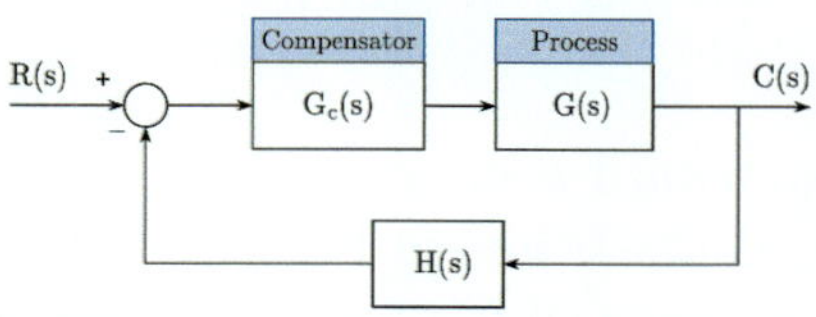

(a) Series or cascade compensation

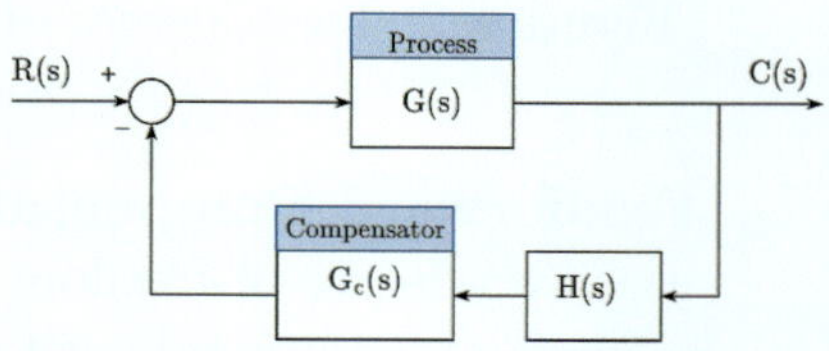

(b) Feedback or parallel compensation

Figure 8.1: Single degree of freedom controller configuration

Two Degrees of Freedom Controller Configurations

Series (cascade) and feedback (parallel) compensation, as previously discussed, are both single-degree-of-freedom systems, meaning each uses a single controller. While such a controller may contain multiple tunable parameters, the number of performance objectives that can be achieved simultaneously is fundamentally limited.

For instance, a single-degree-of-freedom controller may be designed to achieve satisfactory stability margins, but the system might still exhibit poor robustness to parameter variations. Alternatively, a controller might be tuned to yield desired damping, yet the system could display significant overshoot in its step response due to the presence of closed-loop zeros.

These limitations can be mitigated by adopting two-degree-of-freedom (2-DOF) control structures, which provide greater design flexibility and the ability to independently shape reference tracking and disturbance rejection characteristics. The controller configurations, shown in Figure 8.2 and Figure 8.3, have two degrees of freedom and can overcome some of the limitations of a single-degree-of-freedom controller.

Figure 8.2a and 8.2b illustrate two common two-degree-of-freedom configurations

Series Feedback Compensation: This configuration, shown in Figure 8.2a and 8.2b, employs two compensators: one in the forward path and the other in the feedback path. This setup provides the designer with additional degrees of freedom to satisfy multiple performance requirements concurrently, such as minimizing steady-state error while simultaneously improving disturbance rejection and transient behavior.

(a) Series-Feedback compensation

(b) Series-Feedback compensation

Figure 8.2: Two degrees of freedom controller configuration

Feedforward Compensation: Feedforward control is another widely used two-degree-of-freedom structure, as illustrated in Figure 8.3. In this scheme, an additional compensator is introduced to process the reference input signal directly, bypassing the feedback loop. In Figure 8.3a, the feedforward compensator is placed in series with the closed-loop system, and in Figure 8.3b, the feedforward controller is placed in parallel with the forward path. Feedforward compensation allows independent shaping of the system's reference tracking behavior, often enabling improved performance in the presence of known disturbances or plant dynamics.

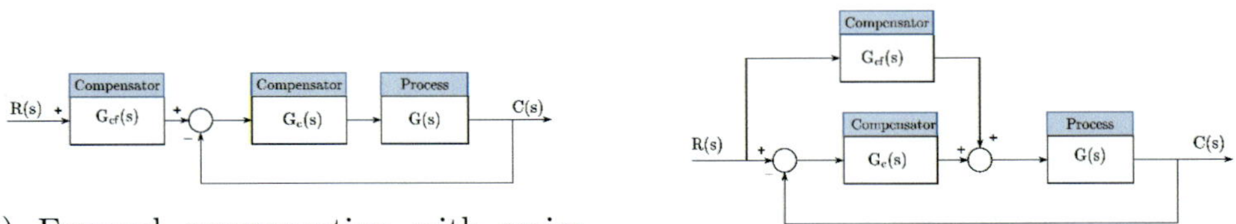

(a) Forward compensation with series compensation

(b) Feedforward compensation

Figure 8.3: Two degrees of freedom controller configuration

8.3 Commonly Used Compensators and Their Structures

In practical control system design, several common dynamic controllers are most widely employed due to their simplicity, analytical tractability, and effectiveness in shaping both transient and steady-state performance. These are listed below.

Commonly Used Compensators

1. Proportional and Derivative (PD) compensator

2. Proportional and Integral (PI) compensator
3. Proportional, Integral and Derivative (PID) compensator
4. Lead Compensator
5. Lag Compensator
6. Lag-Lead Compensator

Each of these compensators is characterized by a distinct pole-zero configuration, which dictates its influence on system behavior and can be realized using different active and passive circuit components.

8.3.1 PD Controller

The structure of a control system with a PD controller in series with the process is shown in Figure 8.4

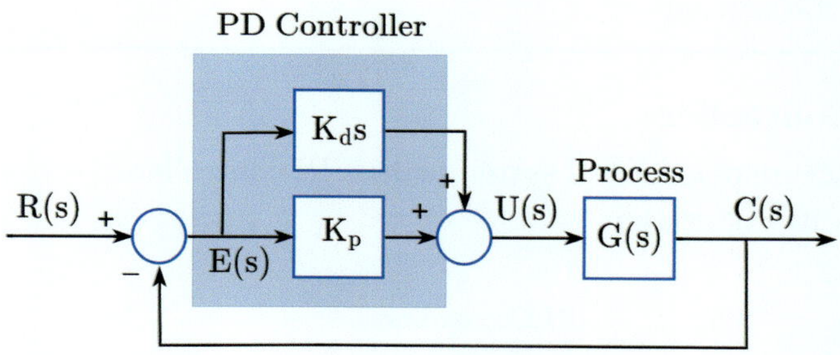

Figure 8.4: Feedback control system with PD controller

The transfer function of this controller is

$$G_c(s) = K_p + K_d s \tag{8.1}$$

The output of this controller is

$$u(t) = K_p e(t) + K_d \frac{de(t)}{dt} \tag{8.2}$$

For simplicity, assume that the transfer function G(s) in Figure 8.4 is a prototype second order system where

$$G(s) = \frac{\omega_n^2}{s(s + 2\xi\omega_n)}$$

Hence, the forward path transfer function of the compensated system is

$$G_c(s)G(s) = \frac{\omega_n^2(K_p + K_d s)}{s(s + 2\xi\omega_n)} \tag{8.3}$$

Eqn(8.3) shows that the PD control is equivalent to adding a simple zero at $s = -K_p/K_d$ to the forward-path transfer function. Due to the addition of zero, this controller exhibits various characteristics, which are summarized below.

Key Features of PD Controller

1. Improves the response speed of the system by anticipating error trends.
2. Reduces overshoot and oscillations compared to a pure proportional controller.
3. Adds damping to the system, enhancing stability.
4. However, unlike PI or PID, it does not eliminate steady-state error in systems without integral action.
5. The derivative action amplifies high-frequency noise; filtering may be needed.

8.3.2 PI Controller

The structure of a control system with a PI controller in series with the process is shown in Figure 8.5

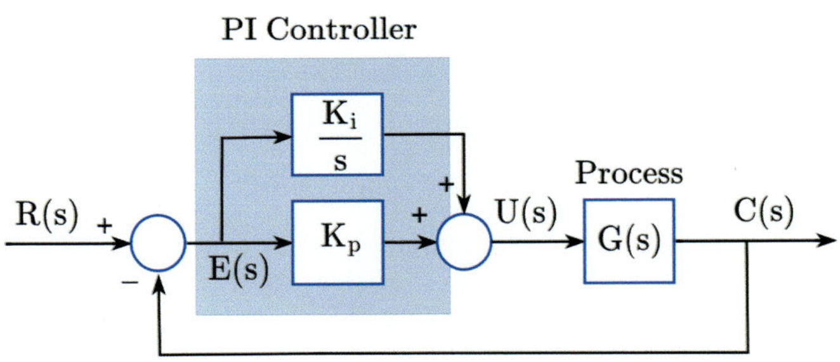

Figure 8.5: Feedback control system with PI controller

The transfer function of this controller is

$$G_c(s) = K_p + \frac{K_i}{s} \tag{8.4}$$

The output of this controller is

$$u(t) = K_p e(t) + K_i \int e(t)dt \tag{8.5}$$

For a standard second-order system, the forward path transfer function of the compensated system is

$$G_c(s)G(s) = \frac{\omega_n^2(K_p s + K_i)}{s^2(s + 2\xi\omega_n)} \tag{8.6}$$

Eqn(8.6) shows that the PI controller adds both a zero at $s = -K_i/K_p$ and a pole at $s = 0$ to the forward path transfer function. The addition of a pole at the origin increases the system type by 1, thereby improving the steady-state error of the original system by one order. However, the compensated system may be less stable or even unstable if the controller parameters K_p and K_i are not properly tuned. The following summarizes some of the key characteristics of this controller.

Key Features of PI Controller

1. Reduces the steady state error due to the step input to zero since it increases the system Type.
2. May reduce the speed of response and cause oscillations if not tuned properly. Reduces overshoot and oscillations compared to a pure proportional controller.
3. May reduce phase margin and affect the stability of the system.
4. Makes it less sensitive to high-frequency noise compared to the PD controller.

8.3.3 PID Controller

The PID controller, shown in Figure 8.6, is a combination of PD and PI controller.

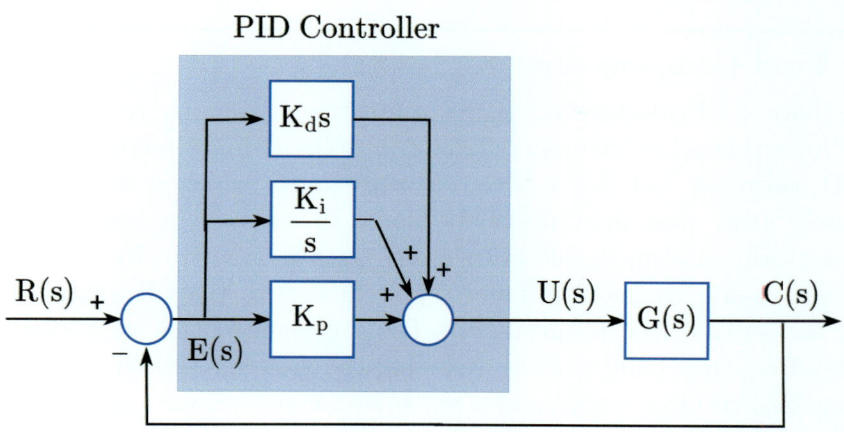

Figure 8.6: Feedback control system with PID controller

The transfer function of this controller is

$$G_c(s) = K_p + \frac{K_i}{s} + K_d s = \frac{K_p s + K_i + K_d s^2}{s} \tag{8.7}$$

The output of this controller is

$$u(t) = K_p e(t) + K_i \int e(t)dt + K_d \frac{de(t)}{dt} \tag{8.8}$$

When the PID controller is in the forward path, the open-loop transfer function becomes:

$$G_c(s)G(s) = \frac{\omega_n^2(K_d s^2 + K_p s + K_i)}{s^2(s + 2\xi\omega_n)} \tag{8.9}$$

The system becomes more flexible with 3 parameters to control the performance. The integral term ensures zero steady-state error for step input; the derivative term helps improve damping and response speed, and the proportional term adjusts overall responsiveness and gain. Some of the characteristics of this controller are summarized below:

Key Features of PID Controller

1. Combines proportional (P), integral (I), and derivative (D) actions to address transient and steady-state performance simultaneously.
2. The proportional (P) action improves rise time and reduces steady-state error to some extent.
3. The integral action eliminates the steady-state error to step inputs and may affect the stability by reducing the phase margin.
4. The derivative action improves damping, reducing overshoot, and makes the system response fast by decreasing rise time.

8.3.4 Lead Compensator

A phase lead compensator serves a functionally similar role to that of a PD (Proportional-Derivative) controller, but the two are not identical. An ideal PD controller includes a pure derivative term, which is not physically realizable using passive components alone. In contrast, a lead compensator is practically implementable using either passive or active circuits. It introduces a zero nearer to the origin and a pole farther to the left along the real axis of the s-plane (see Figure 8.9b). The inclusion of this pole helps limit the high-frequency gain, thereby reducing the system's sensitivity to noise. The standard transfer function of a phase lead compensator is given by:

$$G_c(s) = G_{lead}(s) = \frac{s + z_c}{s + p_c}, \quad \text{with } 0 < z_c < p_c \tag{8.10}$$

Key Features of Lead Compensator

1. Improves transient response by increasing the damping ratio, which helps reduce overshoot and oscillations.
2. Speeds up system response by decreasing rise time (increased bandwidth) and settling time.
3. Introduces a positive phase shift (phase lead) within a specified frequency range, which helps to improve phase margin and thus enhances system stability.
4. Increases the gain crossover frequency, allowing faster system dynamics without compromising stability.

8.3.5 Lag Compensator

A phase lag compensator is functionally similar to a PI (Proportional-Integral) controller, but they are not the same. A PI controller contains an ideal integrator, which introduces a pole at the origin, a condition that is difficult to implement with passive components in analog systems. In contrast, a lag compensator can be readily realized using either passive or active networks. It introduces a zero and a pole, with the pole placed closer to the origin and the zero farther to the left on the real axis of the s-plane (see Figure 8.10b). This configuration increases low-frequency gain, improving steady-state accuracy, while minimizing its effect on transient performance. The pole also ensures that the high-frequency gain remains limited, which helps maintain robustness against noise. The standard transfer function of a lag compensator is of the form

$$\mathrm{G_c(s)} = \mathrm{G_{lag}(s)} = \frac{\mathrm{s} + \mathrm{z_c}}{\mathrm{s} + \mathrm{p_c}}, \quad \text{with } 0 < \mathrm{p_c} < \mathrm{z_c} \tag{8.11}$$

Key Features of Lag Compensator

1. Improves steady-state accuracy by increasing low-frequency gain, which helps reduce steady-state error in systems like Type 0 and Type 1.
2. Has minimal effect on the transient response. It maintains the existing damping ratio and rise time when properly designed.
3. Introduces a negative phase shift (phase lag) over a limited frequency range, which may slightly reduce phase margin if not carefully tuned.
4. Shifts the gain crossover frequency to a lower value, which can result in slower response but greater noise attenuation and improved robustness in certain applications.

8.3.6 Lag-Lead Compensator

A lag-lead compensator combines the characteristics of both lag and lead compensators in a single configuration. It is typically used when both steady-state performance and transient response need improvement. While a lag compensator enhances steady-state accuracy by increasing low-frequency gain, the lead portion contributes additional phase margin, thereby improving stability and transient response.

The lag-lead compensator introduces two zeros and two poles in its transfer function: the lag component places its pole near the origin and its zero further left on the real axis, whereas the lead component does the opposite - placing the zero closer to the origin and the pole further left. This results in a compensator that can be carefully tuned to meet both steady-state and dynamic specifications without significantly degrading either.

The pole-zero representation of this compensator is shown in Figure 8.11b. The transfer function of this compensator is of the form

$$G_c(s) = \underbrace{\left(\frac{s + z_{c1}}{s + p_{c1}}\right)}_{\text{Lead}} \underbrace{\left(\frac{s + z_{c2}}{s + p_{c2}}\right)}_{\text{Lag}}, \quad \text{with } z_{c1} < p_{c1}, \text{ and } z_{c2} > p_{c2} \tag{8.12}$$

Key Features of Lag-Lead Compensator

1. Enhances both steady-state performance (via lag section) and transient response (via lead section) within a single compensator structure.

2. The lag section increases low-frequency gain, improving steady-state accuracy by reducing steady-state error, and has minimal effect on transient dynamics when properly tuned.

3. The lead section introduces a positive phase shift, which improves phase margin, damping, and stability, and speeds up the system response by reducing rise time and settling time, while increasing the bandwidth.

8.4 Realization of Basic Compensators

Compensators can be physically implemented through various engineering approaches, including electrical circuits, mechanical systems, and pneumatic devices, each offering distinct advantages depending on the specific application requirements, operating environment, and system constraints. However, electrical implementation using electronic components remains the most widely adopted approach due to its versatility, accuracy, and ease of integration with modern control systems.

A critical distinction exists between different types of compensators regarding their implementation requirements. The classical PD, PI, and PID controllers require active components, particularly operational amplifiers, for accurate, scalable, and practical realization. This is because passive components alone cannot fully achieve their intended functionality; pure integration and differentiation operations necessitate active circuitry. In contrast, the lead, lag, and lag-lead compensators, which respectively approximate the behavior of PD, PI, and PID controllers, offer the significant advantage of being implementable using either active or passive elements.

The following sections present the active circuit realizations of these compensators, followed by realizations of lag, lead, and lag-lead compensators using passive components.

8.4.1 Active-Circuit Realization of Compensators

The transfer function of an inverting operational amplifier, shown in Figure 8.7 as

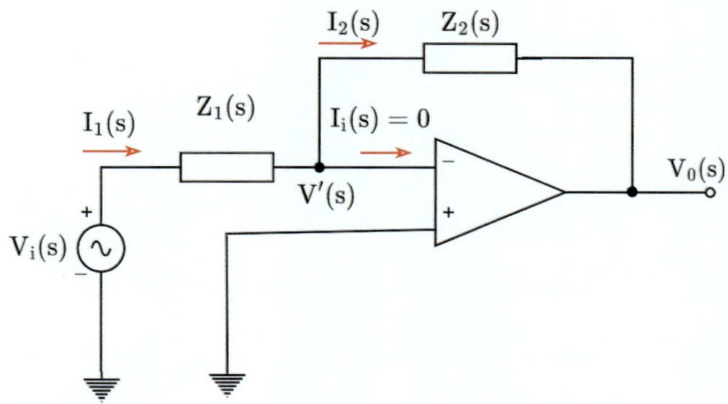

Figure 8.7: Inverting operational amplifier

$$\boxed{\frac{V_o(s)}{V_i(s)} = -\frac{Z_2(s)}{Z_1(s)}} \tag{8.13}$$

The compensators and controllers, discussed earlier, can be implemented using the circuit shown in Figure 8.7 by appropriately selecting $Z_1(s)$ and $Z_2(s)$. The accompanying Table 8.1 summarizes the realization of various control actions, including gain, pure integrator, pure differentiator, PD, PI, PID controllers, as well as lag, lead compensators. Furthermore, by cascading these basic compensators, more complex configurations, such as the lag-lead compensator (illustrated in Figure 8.8), can be constructed.

The lag-lead compensator can be implemented by cascading the lag compensator with the lead compensator as shown in Figure 8.8.

Table 8.1: Analog realizations of various controllers using operational amplifiers

Function	$Z_1(s)$	$Z_2(s)$	$G_c(s) = -\frac{Z_2(s)}{Z_1(s)}$
Gain	R_1	R_2	$-\frac{R_2}{R_1}$
Integration	R	C	$-\frac{1}{RCs}$
Differentiation	C	R	$-RCs$
PI Controller	R_1	R_2 C (series)	$-\frac{R_2}{R_1}\frac{s+\frac{1}{R_2C}}{s}$
PD Controller	C R_1 (parallel)	R_2	$-R_2C\left(s+\frac{1}{R_1C}\right)$
PID Controller	C_1 R_1 (parallel)	R_2 C_2 (series)	$-\left[\left(\frac{R_2}{R_1}+\frac{C_1}{C_2}\right)+R_2C_1s+\frac{\frac{1}{R_1C_2}}{s}\right]$
Lag Compensator	C_1 R_1 (parallel)	C_2 R_2 (parallel)	$-\frac{C_1}{C_2}\frac{s+\frac{1}{R_1C_1}}{s+\frac{1}{R_2C_2}}$, $R_2C_2 > R_1C_1$
Lead Compensator	C_1 R_1 (parallel)	C_2 R_2 (parallel)	$-\frac{C_1}{C_2}\frac{s+\frac{1}{R_1C_1}}{s+\frac{1}{R_2C_2}}$, $R_1C_1 > R_2C_2$

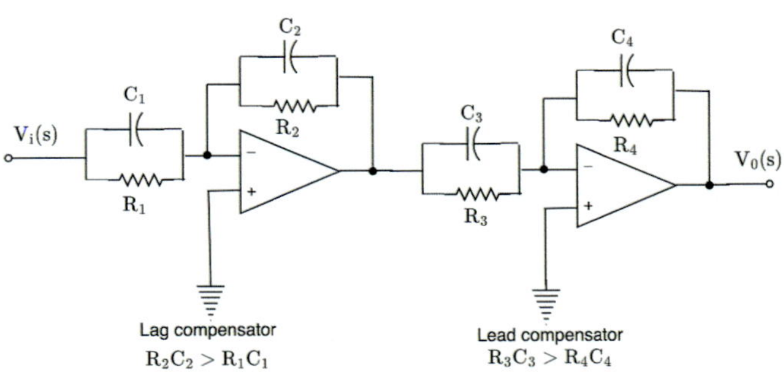

Figure 8.8: Lag-lead compensator implemented with an inverting operational amplifier

8.4.2 Passive Circuit Realization of Compensators

Passive compensators utilize only resistors and capacitors to achieve the desired transfer function characteristics, offering simplicity, reliability, and cost-effectiveness. These RC networks can effectively realize lead, lag, and lag-lead compensators without requiring external power sources or active components.

The design of passive compensators involves strategically placing poles and zeros through appropriate selection of resistor and capacitor values, with each compensator type requiring a specific RC network configuration to achieve its characteristic frequency response, which will be discussed below.

Realization of Lead Compensator

A simple and widely used analog realization of a lead compensator employs a first-order RC network, such as the one shown in Figure 8.9a.

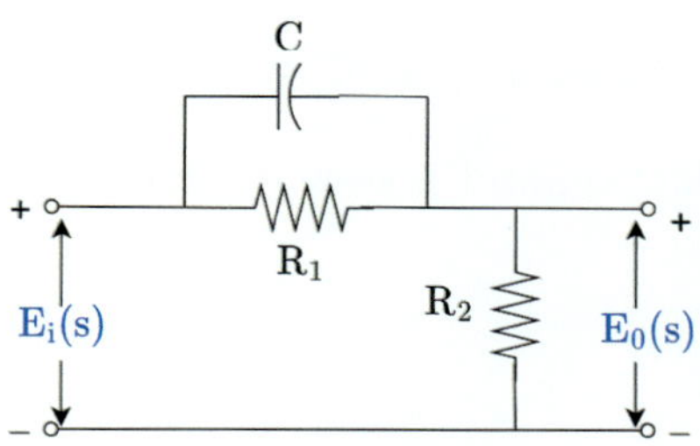

(a) A phase lead network

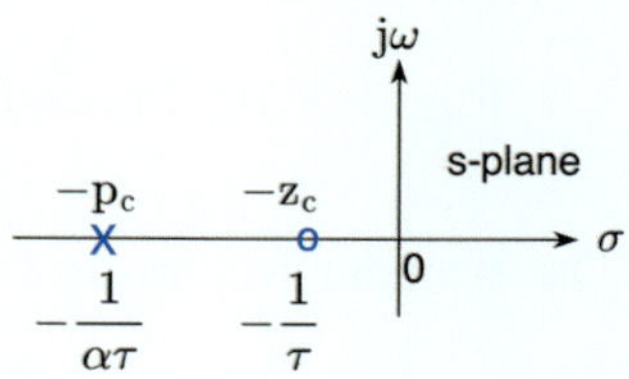

(b) Pole-zero plot of phase-lead network

Figure 8.9: Phase-lead network: realization and pole-zero plot

Assuming the source impedance to be zero and the output load impedance to be infinite, the transfer function of the lead network is

$$\frac{E_0(s)}{E_i(s)} = \frac{R_2}{R_2 + R_1 \parallel \dfrac{1}{C}} = \frac{R_2}{R_2 + \dfrac{R_1/Cs}{R_1 + (1/Cs)}}$$
$$= \frac{R_2(1 + R_1Cs)}{(R_2 + R_1)\left[1 + \left(\dfrac{R_1R_2}{R_1 + R_2}\right)Cs\right]} \tag{8.14}$$

If we define $\tau = R_1C$ and $\alpha = \dfrac{R_2}{R_1 + R_2} < 1$, then the transfer function of Eqn(8.14) can be expressed as:

$$\frac{E_0(s)}{E_i(s)} = \alpha\frac{1 + \tau s}{1 + \alpha\tau s} = \frac{s + \dfrac{1}{\tau}}{s + \dfrac{1}{\alpha\tau}} \tag{8.15}$$

Comparing this with Eqn(8.10) gives $z_c = 1/\tau$ and $p_c = 1/(\alpha\tau)$. The pole-zero plot of this compensator is shown in Figure 8.9b, which shows that the zero is closer to the origin than the pole.

Realization of Lag Compensator

Similar to the lead compensator, a straightforward and widely adopted analog realization of a lag compensator uses a first-order RC network, as depicted in Figure 8.10a.

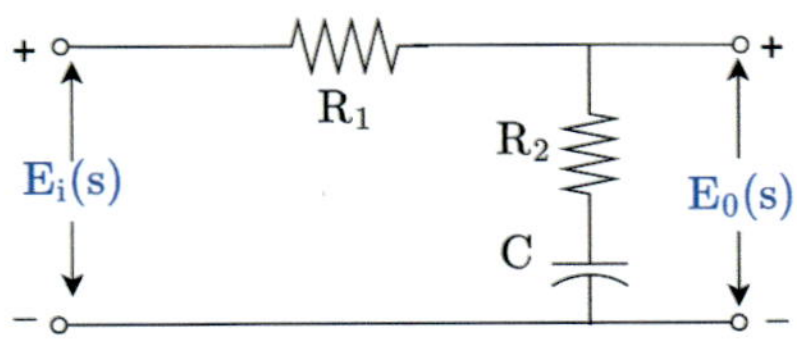

(a) A phase-lag network

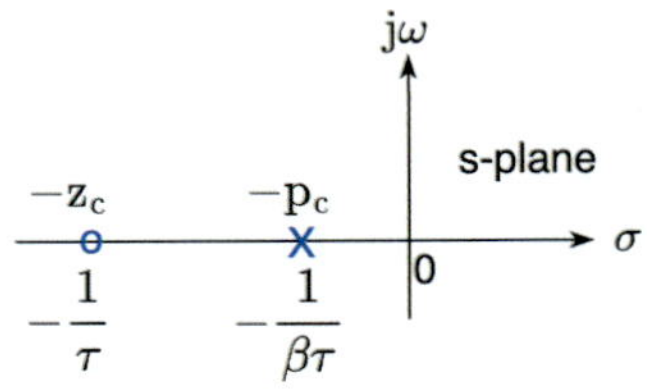

(b) Pole-zero plot of phase-lag network

Figure 8.10: Phase-lag network: realization and pole-zero plot

The transfer function of the lag network is

$$\frac{E_0(s)}{E_i(s)} = \frac{R_2 + \dfrac{1}{Cs}}{R_1 + R_2 + \dfrac{1}{Cs}} = \frac{R_2Cs + 1}{(R_1 + R_2)\,Cs + 1} \tag{8.16}$$

If we define $\tau = R_2C$ and $\beta = \dfrac{R_1 + R_2}{R_2} > 1$, then the transfer function of Eqn(8.16) can be expressed as:

$$\frac{E_0(s)}{E_i(s)} = \frac{1 + \tau s}{1 + \beta\tau s} = \frac{1}{\beta}\left(\frac{s + \dfrac{1}{\tau}}{s + \dfrac{1}{\beta\tau}}\right) = \frac{1}{\beta}\left(\frac{s + z_c}{s + p_c}\right); \quad \beta = \frac{z_c}{p_c} > 1 \tag{8.17}$$

Comparing this with standard form in Eqn(8.11) gives $z_c = 1/\tau$ and $p_c = 1/(\beta\tau)$. This realization includes a multiplication factor of $1/\beta$, introducing an additional degree of freedom. As in the case of the lead compensator, this extra flexibility can be exploited for purposes such as impedance matching or gain adjustment in practical implementations. The pole-zero plot of this compensator is shown in Figure 8.10b, which shows that the pole is closer to the origin than the zero.

Realization of Lag-Lead Compensator

As stated before, a lag-lead compensator combines the effects of both lag and lead networks to improve steady-state accuracy and transient response simultaneously. It is typically realized by cascading a lag network with a lead network, each implemented using first-order RC circuits as shown in Figure 8.11a.

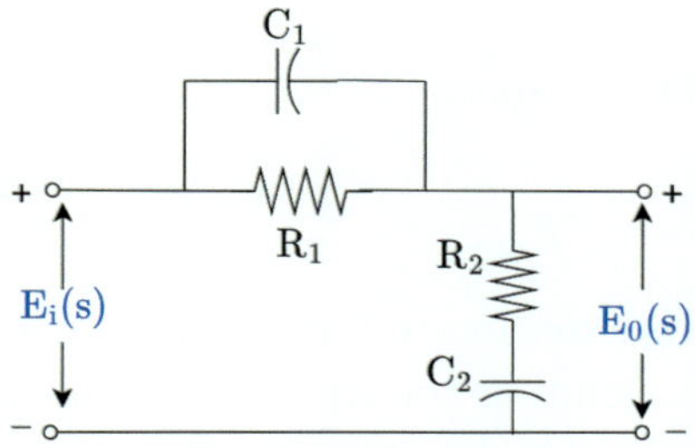

(a) A phase lag-lead network

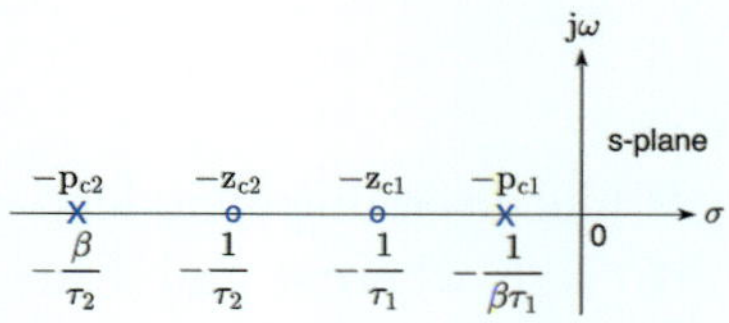

(b) Pole-zero plot of phase lag-lead network

Figure 8.11: Phase lag-lead network: realization and pole-zero plot

The general form of this compensator is

$$G_c(s) = \underbrace{\left(\frac{s+z_{c1}}{s+p_{c1}}\right)}_{\text{Lag-Section}} \underbrace{\left(\frac{s+z_{c2}}{s+p_{c2}}\right)}_{\text{Lead-Section}} = \underbrace{\left(\frac{s+\frac{1}{\tau_1}}{s+\frac{1}{\beta\tau_1}}\right)}_{\text{Lag-Section}} \underbrace{\left(\frac{s+\frac{1}{\tau_2}}{s+\frac{1}{\alpha\tau_2}}\right)}_{\text{Lead-Section}} ; \beta > 1, \quad \alpha < 1$$

$$= \frac{\left(s+\frac{1}{\tau_1}\right)\left(s+\frac{1}{\tau_2}\right)}{s^2 + \left(\frac{1}{\beta\tau_1} + \frac{1}{\alpha\tau_2}\right)s + \frac{1}{\alpha\beta\tau_1\tau_2}} \tag{8.18}$$

Eqn(8.18) can be realized by a single network shown in Figure 8.11a. From this figure, the transfer function of the lag-lead network is

$$\frac{E_0(s)}{E_i(s)} = \frac{R_2 + 1/(C_2 s)}{R_2 + \frac{1}{C_2 s} + \frac{R_1/(C_1 s)}{R_1 + 1/(C_1 s)}}$$

$$= \frac{\left(s+\frac{1}{R_1C_1}\right)\left(s+\frac{1}{R_2C_2}\right)}{\left[s^2 + \left(\frac{1}{R_1C_1} + \frac{1}{R_2C_2} + \frac{1}{R_2C_1}\right)s + \frac{1}{R_1R_2C_1C_2}\right]} \tag{8.19}$$

Comparing Eqns(8.18) with (8.19), gives

$$R_1C_1 = \tau_1 \tag{8.20}$$

$$R_2C_2 = \tau_2 \tag{8.21}$$

$$R_1R_2C_1C_2 = \alpha\beta\tau_1\tau_2 \tag{8.22}$$

$$\frac{1}{R_1C_1} + \frac{1}{R_2C_1} + \frac{1}{R_2C_2} = \frac{1}{\beta\tau_1} + \frac{1}{\alpha\tau_2} \tag{8.23}$$

From Eqns(8.18), (8.19), (8.20), and (8.21), we have

$$\alpha\beta = 1 \tag{8.24}$$

While designing a lag-lead compensator, it is common to impose the constraint $\alpha\beta = 1$, where α and β are the parameters defining the lead and lag sections, respectively. This constraint simplifies the design by ensuring that the compensator has unity gain at very low and very high frequencies, making it easier to meet gain-related specifications. However, it is important to emphasize that this constraint is not mandatory. Designers are free to choose $\alpha\beta \neq 1$ if the application requires more flexibility in shaping the gain or phase characteristics. Allowing $\alpha\beta \neq 1$ gives more degrees of freedom, enabling the compensator to simultaneously satisfy a wider range of performance specifications—especially in systems with tight or conflicting design requirements.

Nevertheless, adopting $\alpha\beta = 1$ can significantly streamline the design process and is particularly useful in classical control design where analytical solutions and Bode plot insights are preferred. With this constraint, Eqn(8.18) is expressed as:

$$\begin{aligned} G_c(s) &= \underbrace{\left(\frac{s + \frac{1}{\tau_1}}{s + \frac{1}{\beta\tau_1}}\right)}_{\text{Lag-Section}} \underbrace{\left(\frac{s + \frac{1}{\tau_2}}{s + \frac{\beta}{\tau_2}}\right)}_{\text{Lead-Section}}; \\ &= \underbrace{\left(\frac{s + z_{c1}}{s + p_{c1}}\right)}_{\text{Lag-Section}} \underbrace{\left(\frac{s + z_{c2}}{s + p_{c2}}\right)}_{\text{Lead-Section}}; \beta = \frac{z_{c1}}{p_{c1}} = \frac{p_{c2}}{z_{c2}} > 1 \end{aligned} \tag{8.25}$$

8.5 Cascade Compensation in Time Domain: Root Locus-Based Design

In Chapter 6, we explored how to sketch the root locus of a system and extract useful information regarding its transient behavior and stability for a given gain value. The root locus is especially valuable because it shows how the transient response evolves as the system gain K varies. For instance, assuming the system remains stable, increasing K typically reduces the steady-state error but increases the maximum percent overshoot. Thus, when the gain K is a free design parameter, the root locus method serves as a valuable tool to adjust K such that the closed-loop poles align with desired performance criteria; if those criteria can be satisfied on the original root locus. However, in most practical situations, the desired transient performance does not correspond to any point on the existing root locus. This limits our flexibility in controller design.

To overcome this, we can introduce compensators that reshape the root locus, offering greater design flexibility. Consider, for example, the root locus of a typical third-order system, shown in Figure 8.12a.

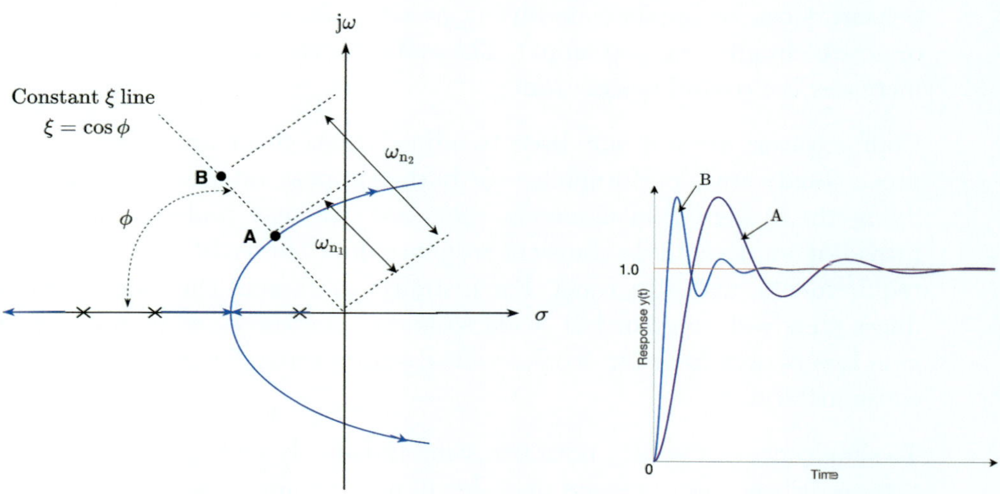

(a) Concept of gain adjustment design with root locus

(b) Typical unit step responses for dominant roots at A and B

Figure 8.12: Concept of root locus-based compensator design

Suppose the desired performance specifications—such as percent overshoot and settling time are translated into a pair of complex conjugate dominant poles, one of which lies at point B. Notice that point B does not lie on the current root locus.

The settling time of the uncompensated system, for a given overshoot, is determined by the real part of the root at point A, that is,

$$T_{s_1} = \frac{4}{\xi \omega_{n_1}}$$

To achieve faster response (i.e., lower settling time) while maintaining the same overshoot, we need to increase the natural frequency from ω_{n1} to ω_{n2} as illustrated in Figure 8.12a. The corresponding time responses at points A and B are shown in Figure 8.12b. Since point B does not lie on the root locus of the uncompensated system, this desired behavior cannot be achieved through gain adjustment alone.

One possible, but typically impractical solution would be to replace the existing plant with a new one whose root locus naturally passes through the target point B. However, this is typically expensive and infeasible. A more practical and widely used approach is to introduce a compensator that strategically adds poles and zeros so that the modified root locus of the compensated system passes through point B. This approach offers design flexibility without requiring alterations to the physical plant.

By placing the compensator at the low-power end, before the plant, we avoid imposing additional load or complicating system design. Moreover, such compensators can be implemented using passive components (e.g., RC networks) or active circuits (e.g., op-amps). The main drawback is that the compensator increases the overall system order.

Compensators are not only used to adjust transient behavior but also to improve steady-state performance—or both. Suppose we are restricted to modifying the system gain alone. In that case, we may find that a gain value providing an acceptable transient response does not satisfy steady-state error requirements, and vice versa. For instance, increasing the gain typically reduces steady-state error but leads to higher overshoot, while decreasing the gain lowers overshoot but worsens steady-state error. Excessive gain can even cause instability.

To simultaneously satisfy both transient and steady-state performance specifications, it becomes necessary to design dynamic compensators (controllers)—a topic that will be explored in the sections that follow. Our focus will be on designing:

1. PI and phase-lag compensators for improving steady-state accuracy,
2. PD and phase-lead compensators for enhancing transient response, and
3. PID and phase lag-lead compensators for achieving improvements in both steady-state and transient performance.

8.6 Improving Steady State Error Via Cascade Compensation

In Chapter 5, we have seen that the steady state error can be improved by placing an open-loop pole at the origin, which increases the system *Type*. For example, a Type-0 system, which gives finite steady state error due to a step input, results in zero steady state error if we increase the system Type to 1 by adding a pole at the origin. Let us assume that a system exhibits a satisfactory transient response. Our objective is to improve the steady state error of the system without affecting the transient response. This can be achieved in two ways.

1. Using a proportional plus integral (PI) compensator with transfer function

$$G_{PI}(s) = K_p + \frac{K_i}{s} = \frac{K_p s + K_i}{s} = K_p \frac{s + \frac{K_i}{K_p}}{s}$$
$$= \frac{K(s+a)}{s}, \quad \text{where} \quad K = K_p, \quad \text{and} \quad a = \frac{K_i}{K_p}$$

This adds a pole at $s = 0$ and a zero at $s = -\frac{K_i}{K_p} = -a$. The added pole increases the system type by 1, and therefore, for a stable system, the steady state error reduces to zero.

2. Using lag-compensator with transfer function

$$G_{lag}(s) = \frac{K(s+a)}{s+b}, \quad \text{where} \quad a > b$$

This compensator does not introduce a pole at the origin (an integrator) and therefore cannot reduce the steady state error to zero. However, by proper placement of its poles and zeros, we can significantly reduce the steady-state error.

Key Points

a. While the PI compensator eliminates steady-state error, it requires an active network implementation using components like an operational amplifier.

b. The lag compensator reduces steady-state error to practically acceptable levels, though not to zero. In contrast, it offers the advantage of passive network realization without requiring an external power supply.

Cascade PI Compensator Design

Consider the system shown in Figure 8.13.

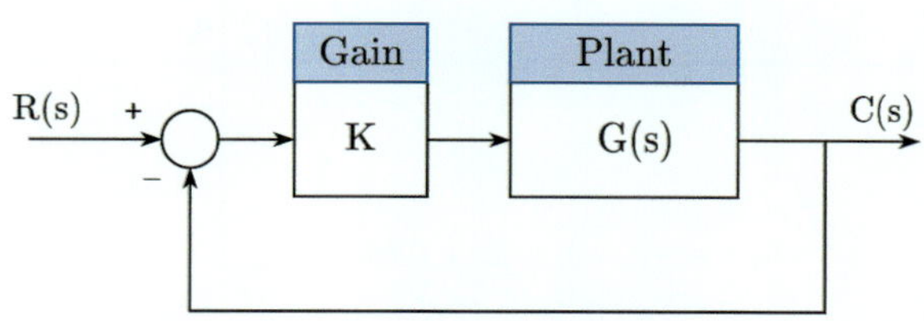

Figure 8.13: Closed-loop system

The root locus of this uncompensated system is shown in Figure 8.14. It is assumed that the desired transient response is generated by the dominant closed-loop poles at **A** for $K = K_A$ as shown in Figure 8.14.

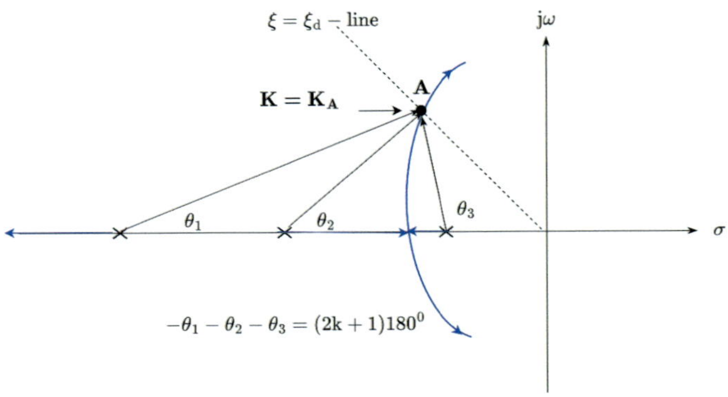

Figure 8.14: Operating point is on root locus

The objective is to reduce the steady state error to zero without affecting the transient response. As a first step, let us add a pole at the origin ($s = 0$) which increases the system type as shown in Figure 8.15.

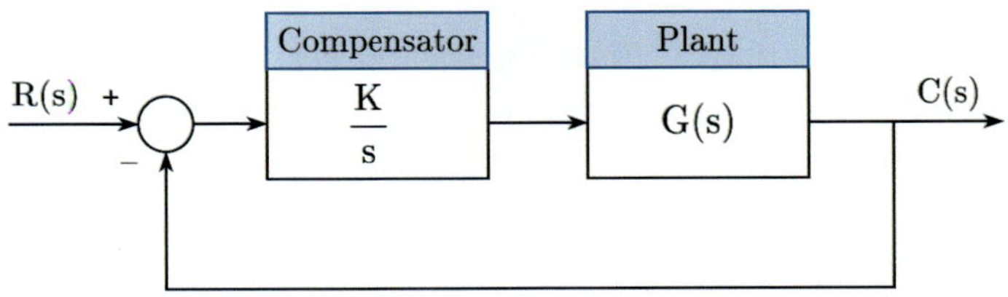

Figure 8.15: Closed-loop system with an ideal integrator

The angular contribution of the open-loop poles of the modified system at point **A** is no longer 180^0 and therefore its root locus does not pass through the point **A** as seen from Figure 8.16.

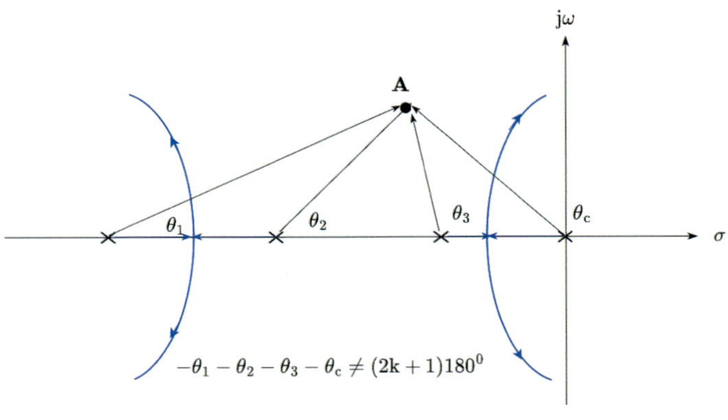

Figure 8.16: Operating point is not on root locus

Let us add a zero close to the origin at $s = -a$ in addition to the pole at $s = 0$ as shown in Figure 8.17.

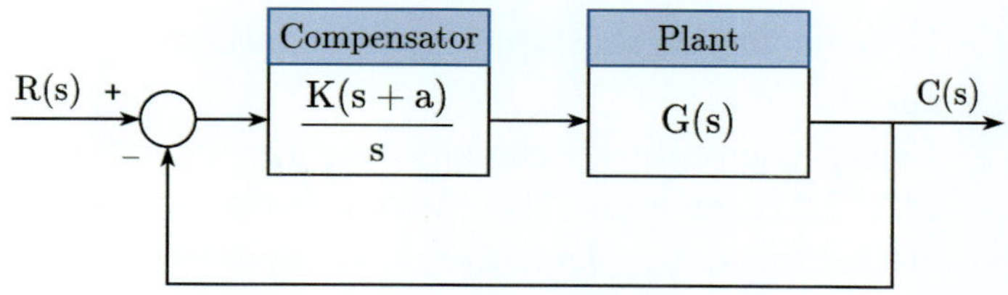

Figure 8.17: Closed-loop system

For small values of **a**, the angular contribution of the compensator zero and pole cancel out ($\theta_{pc} \approx \theta_{zc}$). The desired operating point **A** remains on the root locus; the system type is increased (see Figure 8.18). The gain at point **A**, $K = K_A$, remains approximately the same as before compensation due to the ratio of lengths from compensator pole and compensator zero being approximately unity.

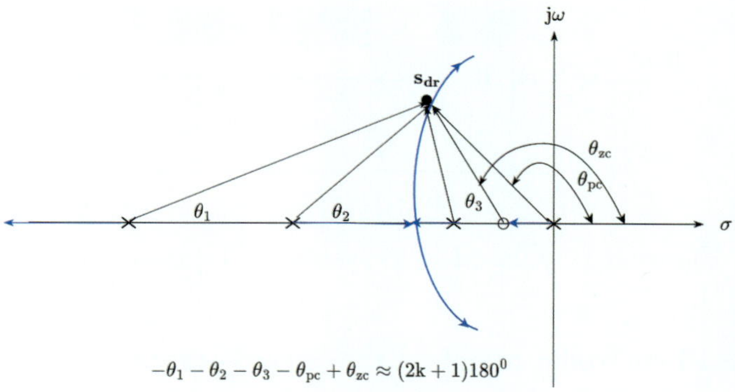

Figure 8.18: Operating point s_{dr} is on root locus due to addition of zero

Steps of Designing Cascade PI-Compensator

Assumption: The desired transient response specification can be met through simple gain adjustment of the uncompensated system.

Step-1: Sketch the root locus of the uncompensated system.

Step-2: Translate the transient response requirements into a pair of complex conjugate dominant poles s_{dr} and mark their location on the uncompensated system root locus plot.

Step-3: Determine the gain K that places the dominant root at s_{dr}.

- Draw a line of constant damping ratio that passes through s_{dr}.
- Either compute it manually, use the `rlocfind` function in MATLAB, or apply the custom routine `RL_with_zeta.m` included in the supplementary code material.

Step-4: Place the PI compensator pole at the origin and position the zero very close to the origin, say at $\mathbf{a = 0.1}$ or $\mathbf{a = 0.01}$.

EXAMPLE 8.1

Problem

Consider a system shown in Figure 8.19 which operates with a damping ratio of 0.2.

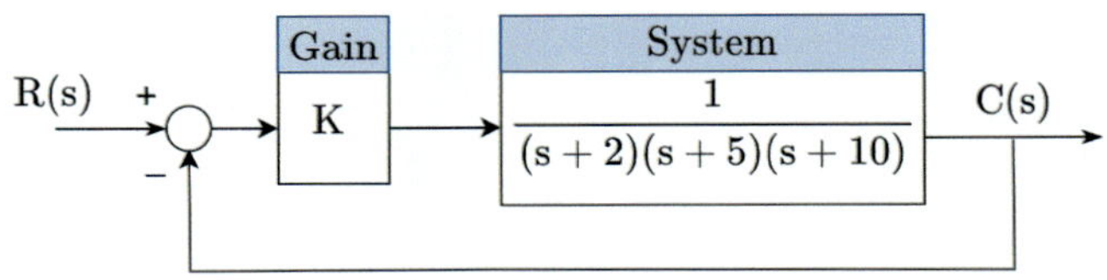

Figure 8.19: Closed-loop system of Example-8.1

Design a PI controller to obtain zero steady-state error for step inputs without appreciably changing the transient response.

SOLUTION:

Step-1: The root locus of the uncompensated system with open-loop transfer function

$$G(s) = \frac{K}{(s+2)(s+5)(s+10)}$$

is shown in Figure 8.20.

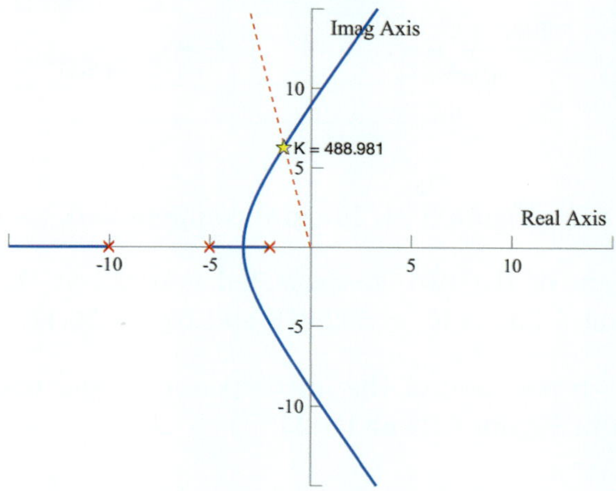

Figure 8.20: Root locus for uncompensated system of Figure 8.19

Step-2: Let us draw a line which makes an $\theta = \cos^{-1}(0.2) = 78.46^0$ with the negative real axis and find the gain at the point of intersection of this line with the root locus.

Step-3: As can be seen from Figure 8.20, this line intersects the locus at the point $s = s_d = -1.3570 + j6.2759$. The gain of the uncompensated system with the desired damping ratio is $K = 488.91$. This can be computed by applying the magnitude criterion at this point or using the `rlocfind` function in MATLAB. Here we used the supplementary code `RL_with_zeta.m`.

Note that, at this gain, the system static position error coefficient K_p is

$$K_p = \frac{488.981}{100} = 4.8898$$

The steady state position error is

$$e_{ss} = \frac{1}{1 + K_p} = \frac{1}{5.8898} = 0.1698$$

Step-4:

The block diagram of the compensated system is shown in Figure 8.21. Let us design two PI compensators as:

$$G_{c1}(s) = \frac{s + 0.1}{s}, \quad G_{c2}(s) = \frac{s + 1}{s}$$

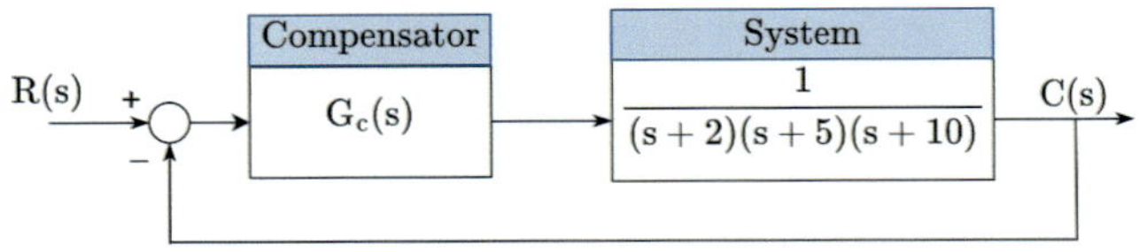

Figure 8.21: Integral compensated system

The gain of the two compensated systems at the intersection of the constant-ξ line is $K_1 = 502.278$ and $K_2 = 450.12$.

The step response of the uncompensated and compensated system is shown in Figure 8.22 and 8.23

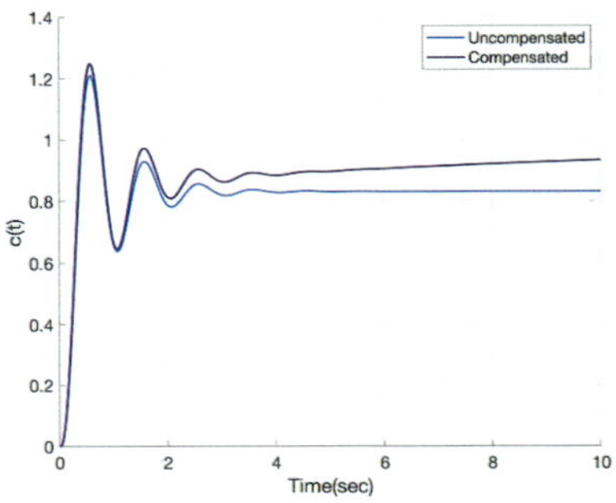

Figure 8.22: Response when $G_{c1}(s) = \dfrac{s + 0.1}{s}$

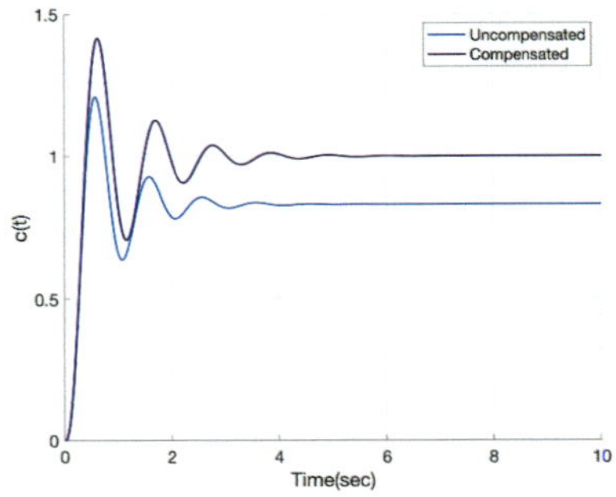

Figure 8.23: Response when $G_{c2}(s) = \dfrac{s + 1}{s}$

The compensator can eliminate the steady-state error without significantly altering the transient response. However, observe that when the zero of the compensator G_{c1} is closer to the $j\omega$-axis, the system response becomes more sluggish, with a longer settling time and narrower bandwidth. In contrast, when the zero is placed farther from the $j\omega$-axis, as in the case of G_{c2}, the response is faster, as illustrated in Figure 8.23.

The MATLAB code for this example can be found in the supplementary file `Example8_1.m`.

Cascade Lag Compensator Design

We know that a PI compensator could improve the steady state performance of a system by placing a pole at the origin and locating its zero arbitrarily close to the pole. This ensures that the transient response of the uncompensated system is preserved. However, this compensator design requires an active integrator. Instead of placing the compensator pole at the origin, let us shift it to the left, while keeping it close to the origin and placing the compensator zero arbitrarily close to the left of the pole as shown in Figure 8.24c. Such a compensator is called a lag compensator, and some of its important features are:

a. It can be realized by using a passive network and therefore does not require active amplifiers and possible additional power supplies.

b. Since it is implemented as a dipole, the transient response of the uncompensated system is not significantly altered by the compensator.

c. Although this compensator does not increase the system type, it can significantly reduce the steady-state error, as will be demonstrated below.

Consider an uncompensated type-N system shown in Figure 8.24a. The appropriate static error constant for this system is

$$K_u = \lim_{s \to 0} s^N KG(s) = K\frac{z_1 z_2 \ldots z_m}{p_1 p_2 \ldots, p_n} = K\frac{\prod_{i=1}^{m} z_i}{\prod_{j=1}^{n} p_j} \tag{8.26}$$

Let us consider a lag compensator of the form

$$G_c(s) = \frac{s + z_c}{z + p_c}, \quad z_c > p_c > 0 \text{ i.e. } \frac{z_c}{p_c} > 1$$

If we insert this compensator in series with the system, as shown in Figure 8.24b, the new static error constant is

$$\begin{aligned} K_c &= \lim_{s \to 0} s^N KG_c(s)G(s) = K\frac{(z_1 z_2 \ldots z_m)(z_c)}{(p_1 p_2 \ldots, p_n)(p_c)} \\ &= K\left(\frac{z_c}{p_c}\right)\frac{\prod_{i=1}^{m} z_i}{\prod_{j=1}^{n} p_j} \end{aligned} \tag{8.27}$$

From Eqn(8.26) and Eqn(8.27) we get

$$K_c = \frac{z_c}{p_c} K_u = \beta K_u; \quad \text{where } \beta = \frac{z_c}{p_c} > 1 \tag{8.28}$$

The error constant is increased by a factor of β. To reduce the steady state error, the ratio $\beta = \frac{z_c}{p_c}$ should be as large as possible. For example, if $z_c = 0.1$ and $p_c = 0.01$, the error constant is increased by a factor of $\beta = 10$. Note that if β approaches infinity, the cascade lag compensator becomes a cascade PI compensator. However, the factor β does have an upper limit, typically about 100. Because physical realization (fabrication) of the compensator with high β requires large values of resistors and capacitors. If β is of the order of 1000, it is preferable to fabricate it with pneumatic components.

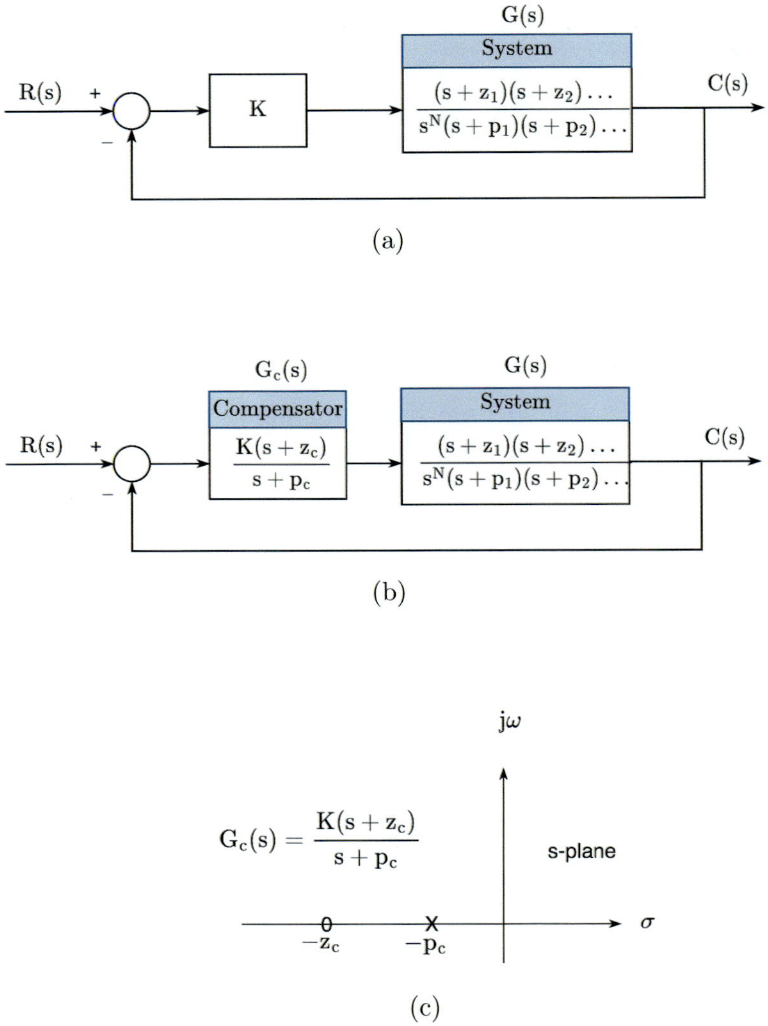

Figure 8.24: (a) Type-N uncompensated system (b) Type-N compensated system (c) pole-zero plot of lag compensator

Let us find how the root locus of the uncompensated system changes with the insertion of the lag compensator. Figure 8.25a shows the root locus of the uncompensated system where the dominant pole is located at point $\mathbf{s_{dr}}$. If the compensator pole and zero are placed relatively close to each other, their angular contribution at the point $\mathbf{s_{dr}}$ is approximately zero degrees. Therefore, the root locus of the compensated system is altered very slightly, as shown in Figure 8.25b, and the gain at the desired dominant pole is virtually the same for both the uncompensated and compensated systems.

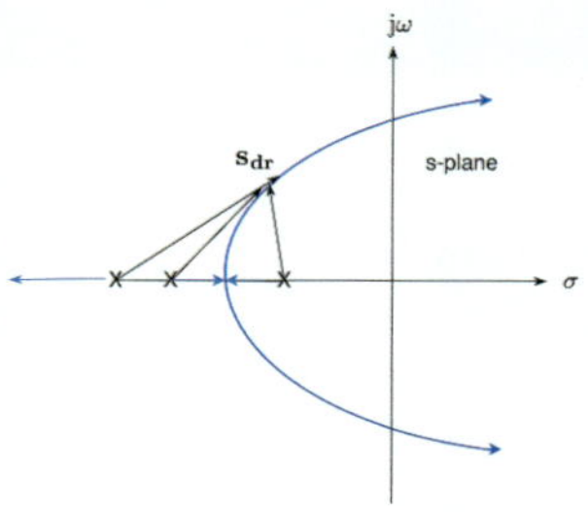

(a) Before lag compensation

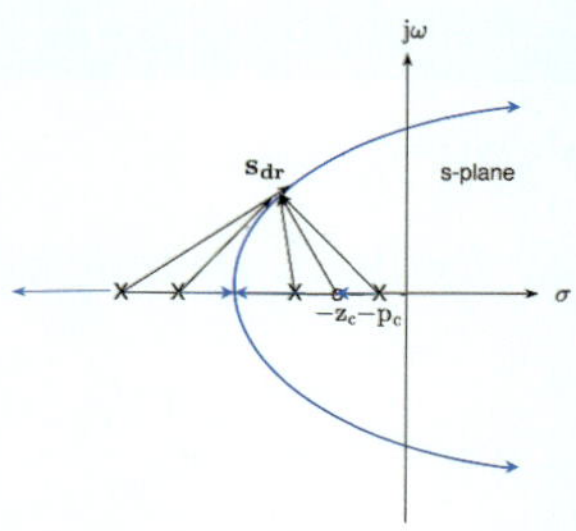

(b) After lag compensation

Figure 8.25: Root locus: (a) before lag compensation, (b) after lag compensation

Lag Compensator Design Procedure

Assumption: The transient response specification can be achieved by simple gain adjustment of the uncompensated system.

Step-1: Sketch the root locus of the uncompensated system.

Step-2: Translate the transient response specifications into a pair of complex dominant roots $\mathbf{s_{dr}}$ and locate these roots on the uncompensated root locus plot.

Step-3: Find the gain K at the dominant root $\mathbf{s_{dr}}$ and compute the corresponding static error constant specified in the problem.

Step-4: Determine the factor by which the error constant of the uncompensated system should be increased to satisfy the specifications. Select the β parameter of the lag compensator to be slightly higher than this factor.

Step-5: Choose the zero of the compensator sufficiently close to the origin of the s-plane in comparison to the natural frequency ω_n. As a guide rule

- Construct a line making an angle of 10^0 (or less) with the desired ξ-line from $\mathbf{s_{dr}}$.
- The intersection of this line with the real axis at $s = -z_c$ gives the location of compensator zero.

Step-6: Locate the compensator pole at $s = -p_c = -\dfrac{z_c}{\beta}$.

Note: The pole-zero pair of the compensator should contribute an angle

$$\left|\theta_{z_{lag}} - \theta_{p_{lag}}\right| < 5^0$$

which causes a negligible change in the root locus around the dominant root $\mathbf{s_{dr}}$ and thereby ensures that the transient behavior of the system is preserved.

EXAMPLE 8.2

Problem:

A unity feedback system with forward transfer function

$$G(s) = \frac{K}{(s+2)(s+5)(s+10)}$$

is operating with a damping ratio of 0.2. Design a phase lag controller to improve the steady-state error by 10-fold.

SOLUTION:

The system operates at a damping ratio of 0.2. The root locus of the uncompensated system is shown in Figure 8.26. The value of gain K at the intersection of the constant damping line and uncompensated root locus is found using the routine `RL_with_zeta.m`.

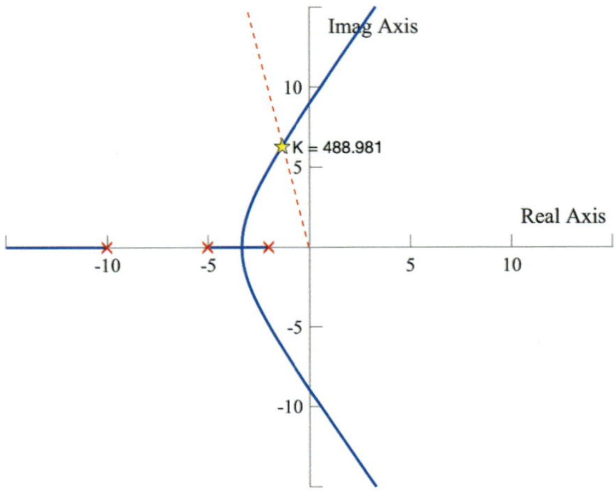

Figure 8.26: Root locus of uncompensated system of Example 8.2

This gives

$$K = 488.98$$

The static position error constant at this gain $K_{p0} = 4.8898$ and the steady state error

$$e_{ss} = \frac{1}{1 + K_{p0}} = 0.1698$$

It is desired to improve the steady state error by 10 fold, that is $e_{ss} = 0.01698$. The required K_{p1} to achieve this is

$$K_{p1} = \frac{1 - e_{ss}}{e_{ss}} = 57.89$$

The required ratio of the compensator zero to pole is

$$\frac{z_c}{p_c} = \frac{K_{p1}}{K_{po}} = \frac{57.898}{4.88} \approx 12$$

Let us select (arbitrarily) $p_c = 0.01$. This gives $z_c = 0.12$. The compensator transfer function is

$$G_c(s) = \frac{s + 0.12}{s + 0.01}$$

The compensated system is shown in Figure 8.27.

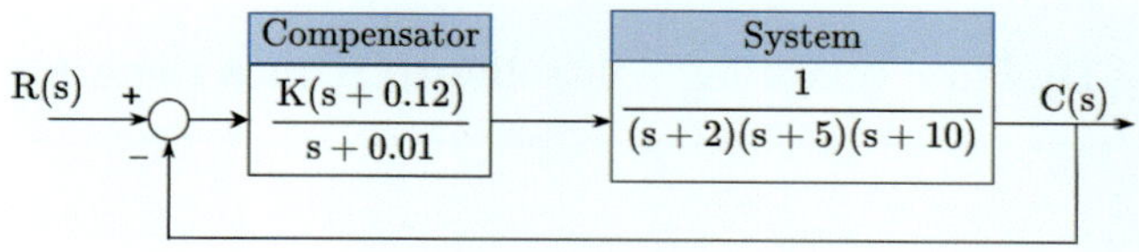

Figure 8.27: Lag compensated system

The gain of the compensated system at the desired damping ratio is $K = 494.6421$. The step response of the uncompensated and compensated system is shown in Figure 8.28

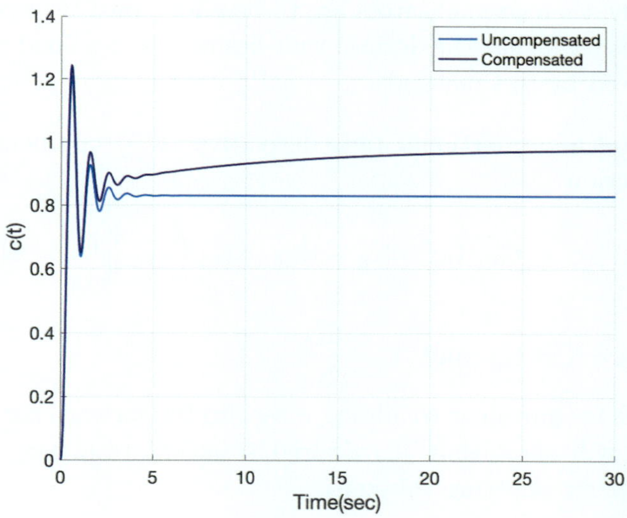

Figure 8.28: System response for uncompensated and PI compensated system

The compensator could reduce the steady state error without appreciably changing the transient response.

The MATLAB code for this example can be found in the supplementary file `Example8_2.m`.

CONCEPT REINFORCEMENT PROBLEM 8.1

Problem

Consider a unity feedback system with forward path transfer function

$$G(s) = \frac{K(s+10)}{s(s+20)(s^2+6s+18)}$$

The system is desired to operate at a damping ratio $\xi = 0.4$

a. Design a PI controller such that the error due to ramp input is zero

b. Design a phase lag controller to reduce the error by 10-fold.

8.7 Improving Transient Response Via Cascade Compensation

It is well known that the transient response specifications of a system can be related to a corresponding dominant root $\mathbf{s_d}$ in the s-plane. If this point $\mathbf{s_d}$ lies on the root locus of the uncompensated system, then we can achieve this response by simple gain adjustment. However, if this point does not lie on the root locus, then compensators are to be added into the system to reshape the root locus such that the desired root lies on the modified root locus. This can be achieved by two methods.

1. Using a proportional plus derivative (PD) compensator with transfer function

$$G_{PD}(s) = K_p + K_d s = K_d\left(s + \frac{K_p}{K_d}\right) = K(s + z_c)$$

where $K = K_d$ and $z_c = \frac{K_p}{K_d}$

This is equivalent to adding a zero to the forward path transfer function. We will show that the desired transient response can be obtained by properly selecting this zero.

2. Using lead-compensator with transfer function

$$G_{lead}(s) = \frac{K(s+a)}{s+b}, \quad \text{where } b > a$$

This is equivalent to adding a zero and a pole on the negative real axis with the zero to the right of the pole.

Design of Cascade PD Compensator

Consider a system shown in Figure 8.29 with a PD compensator $G_c(s)$ where

$$G_c(s) = s + z_c$$

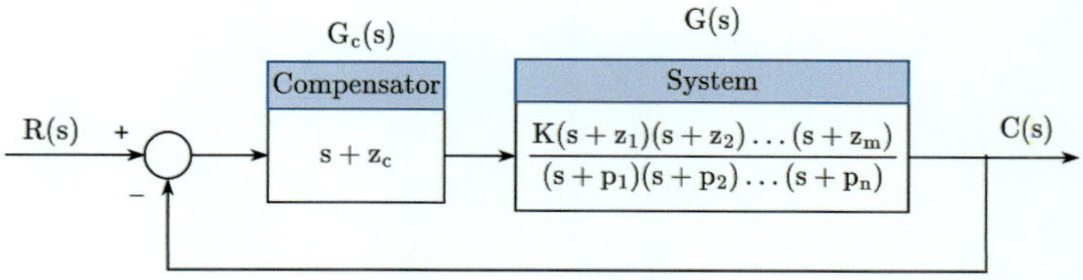

Figure 8.29: Type-0 PD compensated system

Without loss of generality, assume that the transfer function of the system is of the form

$$G(s) = \frac{K(s+z_1)(s+z_2)\dots(s+z_m)}{(s+p_1)(s+p_2)\dots(s+p_n)}$$

The objective is to place the zero (z_c) of the PD compensator at a suitable location such that the root locus of the compensated system goes through the dominant closed-loop pole location ($\mathbf{s_{dp}}$) as dictated by desired transient response specifications. This location is found by applying the angle condition of the root locus, which states that for a selected point ($\mathbf{s_{dp}}$), to be on the root locus, the difference sum of the angles between the point ($\mathbf{s_{dp}}$) and the open-loop zeros, and the sum of angles between the point ($\mathbf{s_{dp}}$) and the open-loop poles must be 180^0. Let

$\theta_c = \angle s_{dp} + z_c$ is the angle contributed by the zero at desired point $\mathbf{s_{dp}}$.

$\Theta_{pdp} = \sum_{j=1}^{n} \angle s_{dp} + p_j$ is the sum of angles from open-loop poles of the uncompensated system at the desired point ($\mathbf{s_{dp}}$) and

$\Theta_{zdp} = \sum_{i=1}^{m} \angle s_{dp} + z_i$ is the sum of angles from open-loop zeros of the uncompensated system at the desired point ($\mathbf{s_{dp}}$).

Now applying the angle rule of root locus to the compensated system, we get

$$\angle G_c(s_{dp})G(s_{dp}) = \theta_c + \Theta_{zdp} - \Theta_{pdp} = 180^0 \tag{8.29}$$

or,

$$\boxed{\theta_c = 180^0 - \Theta_{zdp} + \Theta_{pdp}} \tag{8.30}$$

The zero of the compensator can be calculated from applying a simple trigonometric rule as demonstrated in Figure 8.30.

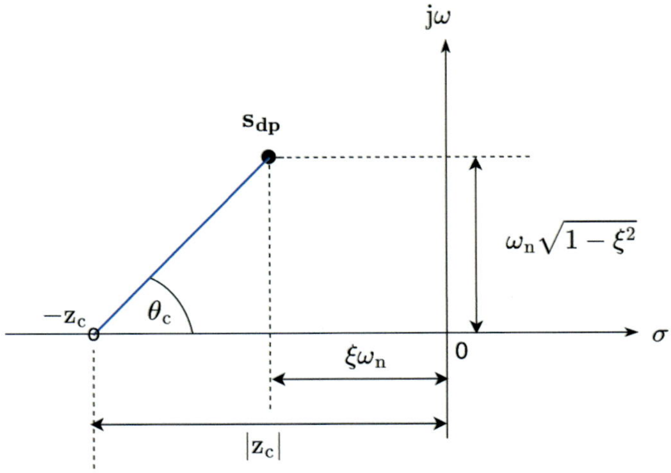

Figure 8.30: Determination of zero location of PD compensator

Now from Figure 8.30,

$$\tan(\theta_c) = \frac{\omega_n\sqrt{1-\xi^2}}{|z_c| - \xi\omega_n} \implies |z_c| = \frac{\omega_n}{\tan\theta_c}\left[\xi\tan\theta_c + \sqrt{1-\xi^2}\right] \tag{8.31}$$

The steps for designing the PD compensator are detailed below.

PD Compensator Design Procedure

Step-1: Translate the desired transient response specifications (damping ratio and natural frequency) into a pair of complex dominant poles s_{dp}.

Step-2: Sketch the root locus of the uncompensated system.

Step-3: Determine the required angle contribution by the PD compensator using the formula in Eqn(8.30).

Step-4: Compute the absolute value of PD compensator zero using the formula in Eqn(8.31).

EXAMPLE 8.3

Problem: Consider a unity feedback system with forward path transfer function

$$G(s) = \frac{K(s+15)}{(s+1)(s+2)(s+18)}$$

Design a PD controller such that the percent overshoot is less than 20% and the 5% settling time is 2.0 sec.

SOLUTION:

Step-1: Let us translate the desired specifications into a pair of complex conjugate poles. Now

$$M_p = 20\% \implies \xi = 0.4559$$

Given

$$T_s = \frac{4}{\xi\omega_n} = 2 \implies \omega_n = \frac{4}{\xi T_s} = 4.3864$$

This gives the pair of desired complex conjugate poles at $s_d = -2 \pm j3.904$.

The root locus of the uncompensated system with the constant ξ line is shown in Figure 8.31.

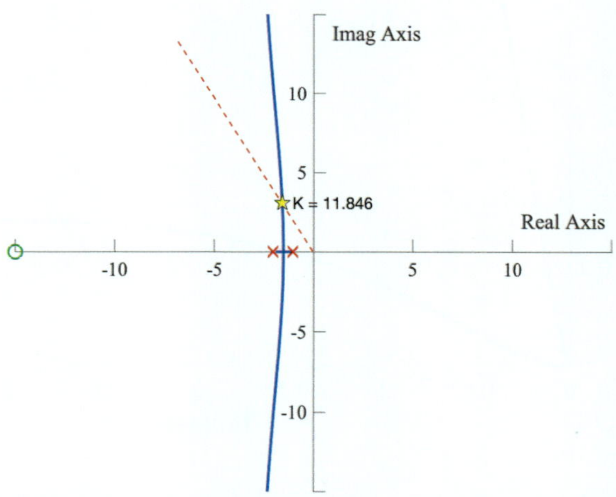

Figure 8.31: Root locus of uncompensated system with constant ξ line

It is found that the desired dominant poles do not lie on the uncompensated system root locus. Therefore, let us compute the phase contribution of the PD controller using Eqn(8.30), which gives

$$\theta_c = 180^0 - \Theta_{zdp} + \Theta_{pdp} = 11.36^0$$

The PD controller zero is calculated from Eqn(8.31) which gives

$$z_c = 21.42$$

The gain of the compensated system at the desired pole equals 0.9635. Note that, at this gain, the system static position error coefficient K_p is

$$K_p = \frac{15 \times 21.42 \times 0.9635}{36} = \frac{309.57}{36} \approx 8.6$$

The steady state position error is

$$e_{ss} = \frac{1}{1 + K_p} = \frac{1}{9.6} = 0.1042$$

The steady state error due to the step input becomes 0.1007. The response of the compensated system is compared with the uncompensated system and is shown in Figure 8.32.

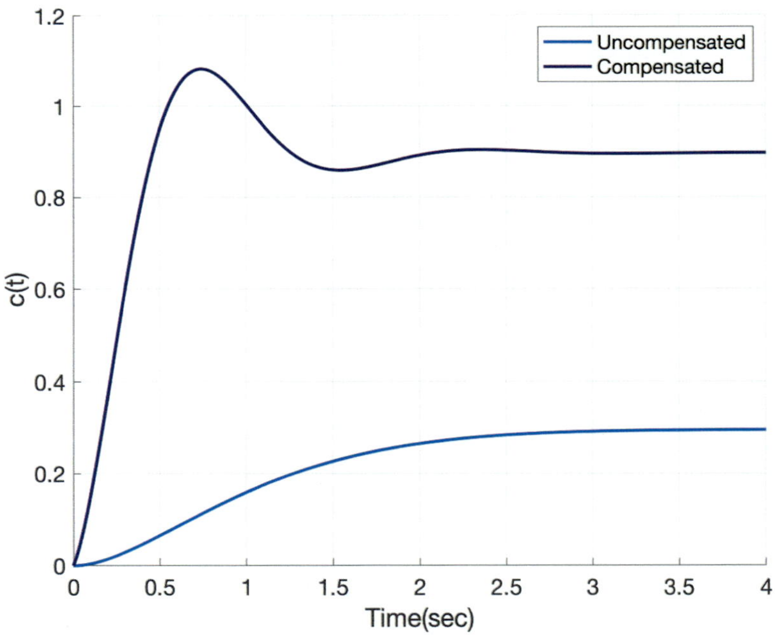

Figure 8.32: System response for uncompensated and PD compensated system

The MATLAB code for this example can be found in the supplementary file `Example8_3.m`.

Design of Lead Compensator

As an active PI compensator (ideal integral compensator) can be approximated by a passive lag network, similarly, an active PD compensator (ideal derivative compensator) can be approximated by a passive lead network (**phase lead controller**). Unlike the active PD compensator, the phase lead controller cannot generate a single zero due to the use of passive elements. Instead, the compensator has a pole-zero pair. However, if we place the compensator pole farther from the origin than the zero, then the net angle contribution of the compensator remains positive, and its behavior resembles that of a PD compensator. Some of the merits and demerits of a passive lead network over an active PD controller are:

a. This does not require additional power supplies.

b. Due to the absence of pure differentiation, the noise due to differentiation is reduced.

c. This does not reduce the number of root locus branches that may cross the imaginary axis and move to the right half plane. Since the PD compensator adds a single zero and no pole, the number of root locus branches that tend to move to the RHP reduces.

The design principle of the phase lead compensator is the same as that of the PD compensator. It is based on the angle condition of root locus, which states that the phase lead added to the system should be such that the desired dominant poles (having desired transient response characteristics) must lie on the root locus of the compensated system.

The transfer function of the phase-lead network is

$$G_c(s) = \frac{s + \dfrac{1}{T}}{s + \dfrac{1}{\alpha T}} = \frac{s + z_c}{s + p_c}, \quad \text{where } \alpha < 1 \text{ and } p_c > z_c > 0 \tag{8.32}$$

Let us find the point $\mathbf{s_{dp}}$ in the s-plane which satisfies the desired transient response specification. Then the design of the phase-lead compensator can be carried out following a procedure similar to PD compensator design. To ensure that the point $\mathbf{s_{dp}}$ lies on the root locus, the angle contribution of the compensator at $\mathbf{s_{dp}}$ should satisfy the following:

$$\angle G_c(s_{dp}) = 180^0 - \angle G(s_{dp}) \tag{8.33}$$

that is

$$\theta_c = \angle(s_{dp} + z_c) - \angle(s_{dp} + p_c) = 180^0 - \sum_{i=1}^{m} \angle(s_{dp} + z_i) - \sum_{i=1}^{n} \angle(s_{dp} + p_i) \tag{8.34}$$

Next, we would find the location of the pole-zero pair of the lead compensator. Note that there is no unique method to find such a pair, as demonstrated in Figure 8.32.

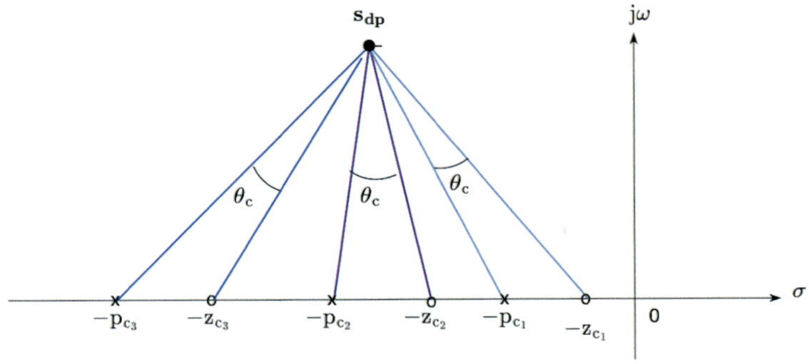

Figure 8.33: Possible locations of poles and zeros of lead compensators which have the same angular contribution

From Figure 8.33, it is important to recognize that an infinite number of lead compensators can be designed to satisfy a given transient response specification. These designs vary in the exact location of the compensator zero and pole, and yet all guide the root locus to intersect the desired closed-loop pole location. However, to make the design process more structured, one can fix the location of the compensator zero based on design preferences or constraints (e.g., phase boost requirements, implementation limits). Then, using a simple trigonometric relationship derived from the angle condition (as illustrated in Figure 8.34), the location of the required compensator pole can be calculated precisely. This approach reduces trial-and-error and results in a more systematic lead compensator design method.

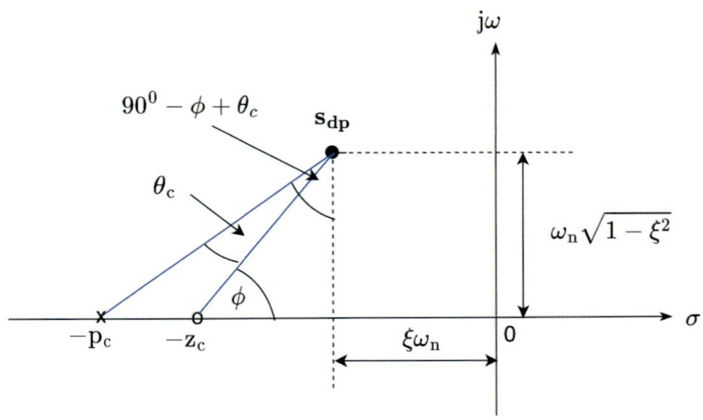

Figure 8.34: Determination of pole location of lead compensator

Suppose the zero of the compensator is selected to be at $-z_c$, then using Figure 8.34, the pole of the compensator is obtained as:

$$p_c = \xi\omega_n + \omega_n\sqrt{1-\xi^2}\tan\left(90^0 - \phi + \theta_c\right), \quad \phi = \angle s_{dp} + z_c \tag{8.35}$$

Note that $\phi > \theta_c$. While different compensators may be designed to yield the same transient response by guiding the root locus through the desired pole location, their impact on steady-state error can vary significantly. This is because steady-state error is governed by the system's type and static error constants, which are affected by the location of compensator poles and zeros. In particular, for a phase-lead compensator, where the zero z_c is closer to the origin than the pole p_c (i.e., $\frac{z_c}{p_c} < 1$), the compensator reduces the static error constant (e.g., K_p,K_v or K_a). As a result, it typically increases the steady-state error, unless additional corrective measures (like a PI or lag compensator) are introduced.

The compensator can be designed by following these sequential steps:

Phase-Lead Compensator Design Procedure

Step-1: Translate the desired transient response specifications (damping ratio and natural frequency) into a pair of complex conjugate dominant poles $-\xi\omega_n \pm j\omega_n\sqrt{1-\xi^2}$.

Step-2: Sketch the root locus of uncompensated system.

a. If the dominant pole lies on the root locus, then we can achieve the desired transient response by adjustment of gain alone.
b. However, if the dominant pole does not lie on the root locus, then compute the phase angle contribution of the lead compensator using Eqn(8.34). If this angle is very large, it is advisable to use two or more lead compensators instead of a single one. In such cases, the compensator will be of the form

$$G^n_{lead} = \left(\frac{s+z_c}{s+p_c}\right)^n, \quad p_c, z_c > 0 \tag{8.36}$$

so that each single-phase compensator contributes a phase lead of $\frac{\theta_c}{n}$, where n-is the number of stages of the compensator.

Step-3: Identify suitable locations for the pole-zero pair of the lead compensator. While their placement is not unique, it is essential to ensure that the chosen locations do not compromise the dominance of the desired complex conjugate poles. The following guidelines can assist in making this selection:

i. Place the compensator zero directly beneath the desired closed-loop-pole on the real axis, that is, at $-\xi\omega_n$. Then, determine the location of the compensator pole using the relation:

$$p_c = \xi\omega_n + (\omega_n\sqrt{1-\xi^2})\tan\theta_c \tag{8.37}$$

ii. If there are no specific constraints, then we may arbitrarily choose either the compensator pole or zero and then compute the other using Eqn(8.35).

iii. When static error constants are not specified, it is advantageous to maximize the ratio $\alpha = \frac{z_c}{p_c}$. A larger value of α increases the static error constants, thereby reducing the steady state error.

Step-4: Determine the gain of the complete lead compensated system at the desired pole and check whether the design objectives are achieved.

Step-5: Repeat the above steps if the desired specifications are not met.

EXAMPLE 8.4

Problem: Consider the system with the open-loop transfer system

$$G(s) = \frac{K(s+5)}{(s+10)(s^2+4s+8)}$$

Design a lead compensator that will reduce the settling time by a factor of 2 while maintaining 20% overshoot. Compare the step response of the compensated system with the uncompensated one.

SOLUTION:

Let us determine the location of dominant poles of the uncompensated system which operates at 20% overshoot. Note that

$$M_p = 20\% \quad \Longrightarrow \quad \xi = 0.4559$$

The constant ξ line corresponding to this ξ intersects the uncompensated system root locus at $-3.955 + j7.65$. The gain of the uncompensated system at this point is $K = 74.4$. The root locus of the uncompensated system is shown in Figure 8.35.

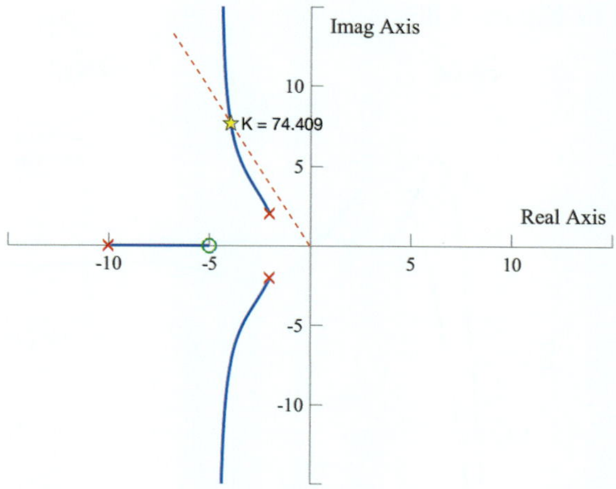

Figure 8.35: Root locus of uncompensated system operating at 20% overshoot

The settling time of the uncompensated system

$$T_{s1} = \frac{4}{3.955} = 1.0144 \text{ sec}$$

Determination of Desired Dominant Poles:

To achieve a twofold reduction in settling time, the real part of the desired pole must be doubled, shifting it to $2 \times (-3.955) = -7.91$. Since the overshoot of the compensated system remains unchanged, the imaginary part is determined by $\omega_d = -7.91 \ \tan\theta$, where $\theta = \cos^{-1}\xi$. Consequently, the desired dominant poles are located at

$$s = s_d = -7.91 \pm j15.44$$

Selection and Computation of Compensator Zero and Pole:

We now compute the compensator's phase angle contribution at the desired pole location using Eqn(8.34). This gives $\theta_c = 24.0^0$. We begin by arbitrarily selecting the compensator zero as $z_c = -5$ and compute the corresponding pole using Eqn(8.35), which yields $p_c = -11.59$. The resulting lead compensator transfer function is

$$G_c(s) = \frac{s+5}{s+11.59}$$

The gain of the compensated system at the desired pole location is $K = 278$. The static position error of the uncompensated system at $K = 74.4$ equals 17.7% and of the compensated system at $K = 278$ is 11.76%.

The step response of the uncompensated and compensated system is shown in Figure 8.36

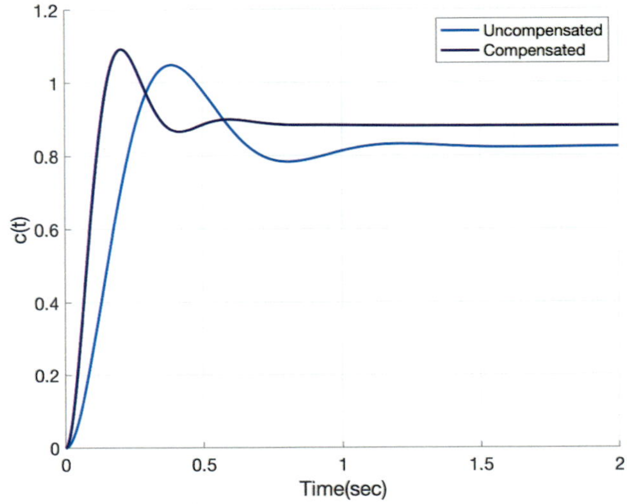

Figure 8.36: System response for uncompensated and lead compensated system

It is observed that while the compensated system meets the desired transient response specifications, it does not eliminate the steady-state error, although a noticeable reduction is achieved compared to the uncompensated system.

The MATLAB code for this example can be found in the supplementary file `Example8_4.m`.

8.8 Improving Both Steady-State Error and Transient Response

We studied the design principles of two groups of compensators, which are used for different purposes. The first group, the PD (phase-lead) compensators, improves the transient response, whereas the other group, the PI (phase-lag) compensators, improves the steady state performance of a system. The design procedure of these two groups does not interfere with each other. If our desire is to improve both the transient and steady state performance independently, then we must combine these two controllers. This combination can be carried out by two approaches:

Approach-1: First improve the transient performance by designing either PD or phase-lead compensators (discussed in Section 8.7), followed by designing PI or phase-lag compensators to improve steady state performance.

Disadvantage: There will be a slight decrease in the speed of response (transient response) due to improvement in steady state error.

Approach-2: First improve the steady state performance by designing either PI or phase-lag compensators (discussed in Section 8.6), followed by designing PD or phase-lead compensators to improve transient performance.

Disadvantages: Depending on the nature of the system, improving the transient response in the second stage could further deteriorate or improve the steady state error.

In this chapter, we will first design controllers to improve transient response, followed by designing controllers to improve steady state performance. We will focus on two controllers PID controller and the lag-lead controller.

Design of PID Controller

The structure of a PID controller is shown in Figure 8.6.

The transfer function of this controller is expressed as:

$$\begin{aligned} G_c(s) &= K_p + \frac{K_i}{s} + K_d s = \frac{K_p s + K_i + K_d s^2}{s} \\ &= K_d \frac{s^2 + \frac{K_p}{K_d} s + \frac{K_i}{K_d}}{s} = \frac{K_d(s + z_{c1})(s + z_{c2})}{s} \\ &= K_d\,(s + z_{c1}) \times \frac{(s + z_{c2})}{s} \\ &= G_{PD}(s) \times G_{PI}(s) \end{aligned} \tag{8.38}$$

This shows that the transfer function of a PID controller is the product of the transfer functions of PD and PI compensators. As seen before, the design procedure of PD and PI compensators does not conflict with each other. Hence, the design of the PID controller is the combination of the design methods of PD and PI controllers. The complete procedure is summarized below.

PID Controller Design Procedure

Step-1: Evaluate the transient and steady state characteristics of the uncompensated system.

Step-2: Design PD controller to meet the transient response specifications. The various steps for this design are given in PD compensator design

Step-3: Check via simulation that all transient response specifications are met.

Step-4: If the transient response is not according to the desired, redesign the PD compensator.

Step-5: Design PI controller to meet the steady state error requirements. The various steps for this design are given in PI compensator design

Step-6: Determine the gains K_p, K_i, and K_d of the PID controller shown in Figure 8.6

Step-7: Simulate to ensure that all design specifications are met.

Step-8: Redesign if all the requirements have not been met.

EXAMPLE 8.5

Problem: Consider the system with the open-loop transfer system

$$G(s) = \frac{K(s+10)}{(s+2)(s+4)(s+15)}$$

Design a PID controller that will reduce the peak time by a factor of 1.5 while maintaining 15% overshoot. Compare the step response of the compensated system with the uncompensated one.

SOLUTION:

Let us determine the location of dominant poles of the uncompensated system, which operates at 15% overshoot. Note that

$$M_p = 15\% \implies \xi = 0.5169$$

The constant ξ line equivalent to this ξ intersects the uncompensated system root locus at $-4.0 + j6.7$. The gain of the uncompensated system at this point is $K = 66.86$. The root locus of the uncompensated system is shown in Figure 8.37.

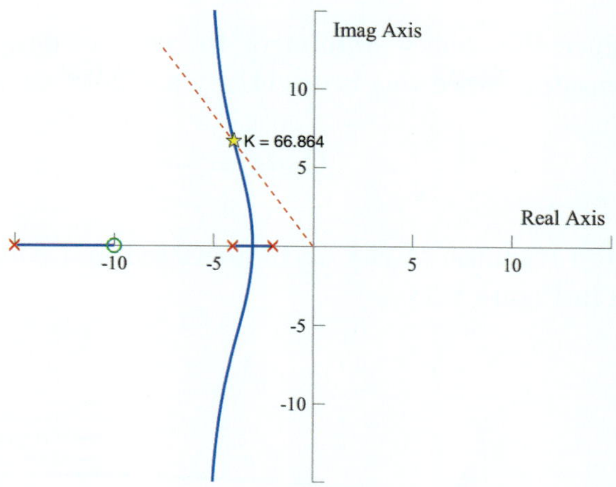

Figure 8.37: Root locus of uncompensated system operating at 20% overshoot

The peak time of the uncompensated system

$$T_{p1} = \frac{\pi}{6.7} = 0.47 \text{ sec}$$

Determination of Desired Dominant Poles:

To achieve a 1.5 reduction in peak time, the imaginary part of the desired pole must be shifted to $1.5 \times (-6.7) = -10.05$. Since the overshoot of the compensated system remains unchanged, the real part is determined by $\sigma = \dfrac{-10.05}{\tan\theta} = -6.0$, where $\theta = \cos^{-1}\xi$. Consequently, the desired dominant poles are located at

$$s = s_d = -6.06 \pm j10.05$$

Design of PD Compensator:

We now compute the phase angle contribution of the PD controller using Eqn(8.30). This gives $\theta_c = 13.4^0$. The zero of the PD controller is calculated using Eqn(8.31), which gives $z_c = 48.22$. The resulting PD compensator transfer function is

$$G_{PD}(s) = s + 48.22$$

The gain of the compensated system at the desired pole location is $K = 3.1$. We simulated the PD-compensated system and observed (see Figure 8.38) that the peak time is reduced with improvement in steady state error compared to the uncompensated system.

Design of PI Compensator

To reduce the steady state error to zero, we design an ideal integral compensator. Since this is essentially a dipole, we consider

$$G_{PI}(s) = \frac{s + 0.5}{s}$$

The step response of the uncompensated and compensated system is shown in Figure 8.38

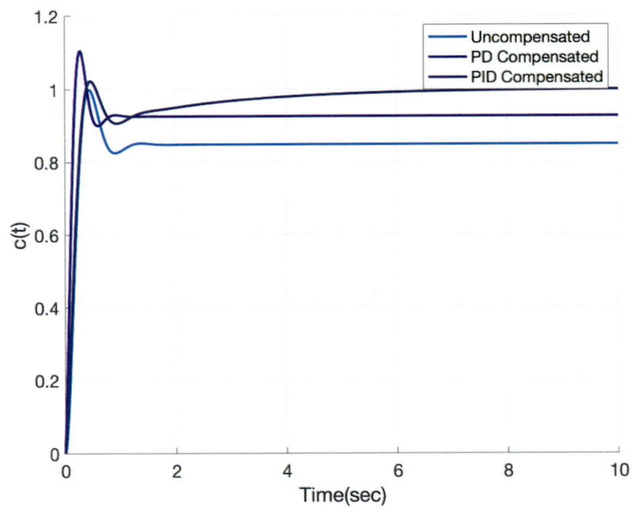

Figure 8.38: Step responses for uncompensated, PD-compensated and PID-compensated system

It is observed that the PD compensated system satisfies the desired transient response specifications but fails to eliminate the steady-state error, though it achieves a significant reduction compared to the uncompensated system. In contrast, the PID compensated system successfully eliminates the steady-state error; however, it does not fully meet the transient response criteria.

The MATLAB code for this example can be found in the supplementary file `Example8_5.m`.

Design of Lag-Lead Compensator

From the preceding discussions, we have seen that the lead compensator could improve the transient response and increase the stability margin of the system. Its ability to improve steady state behavior is very limited and minimal. On the contrary, a lag compensator can improve the steady-state behavior. If we desire to simultaneously improve both the transient and steady state behavior, then the best attributes of each of the lag and lead compensators can be utilized to achieve this. Its design principle is similar to that of a PID controller, discussed before, where the ideal PD and PI controllers are serially combined. Note that the lag-lead compensator is essentially the passive counterpart of the active PID controller.

The transfer function of the lag-lead compensator is of the form

$$\begin{aligned} G_c(s) &= \underbrace{\frac{(s+z_{c_1})}{(s+p_{c_1})}}_{\text{Lag section}} \underbrace{\frac{(s+z_{c_2})}{(s+p_{c_2})}}_{\text{Lead section}}; \quad \beta = \frac{z_{c1}}{p_{c_1}} = \frac{p_{c2}}{z_{c_2}} > 1 \\ &= G_{lag}(s)G_{lead}(s) \end{aligned} \tag{8.39}$$

where $p_{c_2} > z_{c_2} > z_{c_1} > p_{c_1} > 0$. We have assumed that $z_{c_1} z_{c_2} = p_{c_1} p_{c_2}$.

Similar to the design of a PID controller, we first design a lead compensator to improve the transient response, followed by the design of a lag compensator to achieve the steady state error requirement. The procedure for designing this compensator is summarized below.

Lag-Lead Controller Design Procedure

Step-1: Plot the root locus of the uncompensated system and evaluate the transient and steady state characteristics of the uncompensated system.

Step-2: Design a lead compensator to meet the transient response specifications. The various steps for this design are given in the lead compensator design.

Step-3: Check via simulation that all transient response specifications are met.

Step-4: If the transient response is not according to the desired, redesign the lead compensator.

Step-5: Evaluate the steady state error performance of the lead compensated system and determine the deficiency in the steady state error requirement.

Step-6: Design the lag compensator to meet the steady state error requirements. The various steps for this design are given in the lag compensator design.

Step-7: Simulate to ensure that all design specifications are met.

Step-8: Redesign if all the requirements have not been met.

EXAMPLE 8.6

Problem: Consider the system with the open-loop transfer system

$$G(s) = \frac{K}{s(s+5)(s+10)}$$

Design a lag-lead compensator for this system so that it operates at 25% overshoot and a two-fold reduction in settling time. The compensator should improve the steady state error due to the ramp input by 10-fold.

SOLUTION:

Step 1 The damping ratio corresponding to 25% overshoot is

$$M_p = 25\% \quad \Longrightarrow \quad \xi = 0.4037$$

The root locus of the uncompensated system is shown in Figure 8.39.

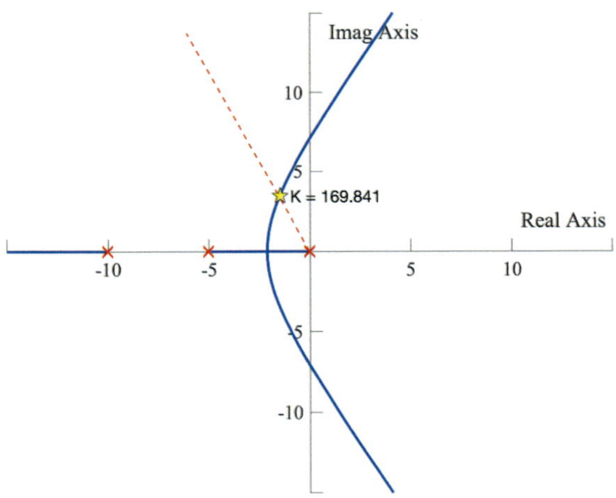

Figure 8.39: Root locus of uncompensated system operating at 25% overshoot

The constant ξ line corresponding to this ξ intersects the uncompensated system root locus at $-1.5 + j3.45$. The gain of the uncompensated system at this point is $K = 169.84$. Therefore, the open-loop transfer function of the uncompensated system is

$$G_{UC}(s) = \frac{169.84}{s(s+5)(s+10)} \tag{8.40}$$

The static error constant K_v of the uncompensated system is 3.396.

Step 2: Design of Lead Compensator:

We have to select the location of dominant poles such that the settling time is reduced by 2. Since the settling time is inversely proportional to the real part of the dominant pole, to achieve a 2 times reduction in settling time, the real part of the desired pole must be doubled, which gives

$$-\xi\omega_{\mathrm{n}} = -\sigma = -2(1.5) = -3$$

The imaginary part of the desired pole is

$$\omega_{\mathrm{d}} = \sigma\tan(\cos^{-1}\xi) = \sigma\tan(66.18^0) = 6.7985$$

Thus, the desired dominant poles are located at

$$\mathrm{s_d} = -3 + \mathrm{j}6.7985$$

In this example, we select the zero of the compensator to be coincident with the open-loop pole at -5. This will cancel one of the poles. We now compute the phase angle contribution of the lead controller using Eqn(8.30). This gives $\theta_{\mathrm{c}} = 52^0$. The pole of the lead controller is calculated using Eqn(8.35) (see Figure 8.34), which gives $\mathrm{p_c} = 19.53$. The resulting lead compensator transfer function is

$$\mathrm{G_{lead}(s)} = \frac{\mathrm{s}+5}{\mathrm{s}+19.53}$$

The gain of the lead compensated system at the desired pole location is $\mathrm{K} = 1247.8$. The open-loop transfer function of the lead-compensated system is

$$\mathrm{G_{LC}(s)} = \frac{1247.8}{\mathrm{s(s+10)(s+19.53)}} \tag{8.41}$$

The static error constant $\mathrm{K_v}$ equals to 6.3911. We simulated the lead compensated system and observed that the settling time is reduced with improvement in steady state error compared to the uncompensated system.

Step-6: Design of Lag Compensator

To reduce the steady state error, we design a lag compensator. Now the static velocity error constant of uncompensated system $\mathrm{K_{vuc}}$ and lead compensated system $\mathrm{K_{vlead}}$ are

$$\mathrm{K_{vuc}} = \lim_{\mathrm{s}\to 0}\mathrm{s}\frac{169.84}{\mathrm{s(s+5)(s+10)}} = 3.3968 \text{ and}$$

$$\mathrm{K_{vlead}} = \lim_{\mathrm{s}\to 0}\mathrm{s}\frac{1280.8}{\mathrm{s(s+10)(s+19.53)}} = 6.3911$$

Note that the static velocity error has been reduced by 1.8805 times. Since the desired improvement is 20-fold, the lag compensator must improve the steady-state error by a factor $20/1.8805 = 10.6$ over the lead compensated system.

Let us arbitrarily choose the pole of the lag compensator at 0.01, which then places the zero of the compensator at 0.105. Thus

$$G_{lag}(s) = \frac{s + 0.1064}{s + 0.01}$$

The open-loop transfer function of lag-lead compensated system is

$$G_{LLC}(s) = \frac{K(s + 0.1064)}{s(s + 10)(s + 19.53)(s + 0.01)}$$

The gain of the compensated system at the desired location is $K = 1400$.

The step response of the uncompensated and compensated system is shown in Figure 8.40

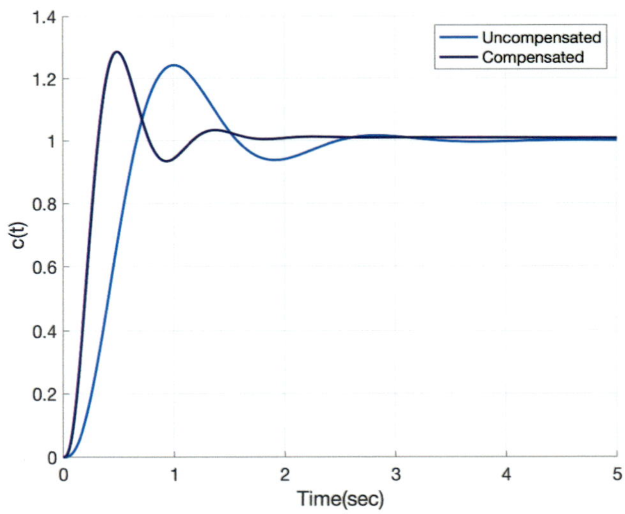

Figure 8.40: Step responses for uncompensated and lag-lead-compensated system

It is observed that the lead-lag compensated system satisfies the desired transient response specifications.

The MATLAB code for this example can be found in the supplementary file `Example8_6.m`.

8.9 Frequency Domain Based Compensator Design

In preceding sections, we have extensively utilized the root locus of a system for achieving desired transient and steady state responses. For example, it was observed that if the desired transient response performance measures such as *peak overshoot, settling time, steady state error*, can be mapped into a point on the root locus, then this can be achieved by simple adjustment of the gain. However, if these specifications are mapped to a point that is not on the root locus, we must add different compensators to alter and reshape the root locus so that the desired closed-loop roots lie on the root locus.

An alternative approach to compensator design is through frequency domain methods, which offer a different perspective on achieving system performance specifications. It is challenging to make a general statement about the superiority of one approach over another. Both these approaches have some merits and demerits. Several decades ago, the frequency domain approach was given preference because its implementation did not require a computer. But, in the present day, with the availability of abundant computational resources, the superiority of a method should not be judged from a computational perspective. It should be noted that frequency domain and time domain methods provide complementary design perspectives, each offering distinct advantages for different types of analysis and applications.

In this section, our design will focus to satisfy various frequency domain performance measures such as *phase margin* (ϕ_{pm}), *resonant peak* ($\mathrm{M_r}$), *bandwidth* (ω_{b}), *resonant frequency*(ω_{r}). This design process will be parallel to the root locus-based design in the time domain. If time-domain specifications are given, they should first be translated into the frequency domain to carry out frequency-domain design. The correlation between the specifications in the two domains is explicitly expressed in Chapter 7 for a second-order system. These correlations can be considered valid for higher-order systems that are dominated by a pair of complex-conjugate poles. For a second-order system, the correlations between the time and frequency domain specifications are reproduced here for quick reference.

$$\text{(i) Phase margin } \phi_{\mathrm{m}} = \tan^{-1}\left[\frac{2\xi}{\left(\sqrt{(4\xi^4+1)}-2\xi^2\right)^{1/2}}\right] \approx 100\xi \tag{8.42}$$

$$\text{(ii) Gain crossover frequency } \omega_{\mathrm{gc}} = \omega_{\mathrm{n}}\sqrt{\sqrt{(4\xi^4+1)}-2\xi^2} \tag{8.43}$$

$$\text{(iii) Resonant peak } \mathrm{M_r} = \frac{1}{2\xi\sqrt{1-\xi^2}} \tag{8.44}$$

iv) Resonant frequency $\omega_r = \omega_n\sqrt{1-2\xi^2}$ (8.45)

$$\text{(v) Bandwidth } \omega_b = \omega_n\left[\left(1-2\xi^2\right)+\sqrt{4\xi^4-4\xi^2+2}\right]^{1/2} \tag{8.46}$$

$$= \frac{4}{t_s\xi}\left[\left(1-2\xi^2\right)+\sqrt{4\xi^4-4\xi^2+2}\right]^{1/2};\quad t_s = \text{Settling time} \tag{8.47}$$

$$= \frac{\pi}{t_p\sqrt{1-\xi^2}}\left[\left(1-2\xi^2\right)+\sqrt{4\xi^4-4\xi^2+2}\right]^{1/2};\quad t_p = \text{Peak time} \tag{8.48}$$

The design of the compensator, represented by $G_c(s)$, can be developed using either a Nyquist plot, Nichols chart, or Bode diagram. If the given specifications are in terms of M_r and ω_r, it is convenient to use Nyquist plots. If the compensated system should have a specified phase margin (ϕ_{pm}), then it is easier to use a Bode plot. In this chapter, our design will be based on Bode plots, because they are easier to draw (by hand) and modify. Further, the effect of a cascade transfer function can easily be accounted for on a Bode plot by the simple addition of the frequency response of the network.

8.9.1 Frequency Domain Based Design of PD Compensator

The transfer function of the PD controller is

$$G_c(s) = K_p + K_d s = K_d\left(s+\frac{K_p}{K_d}\right) = K_d\,(s+z) \tag{8.49}$$

where $z = \dfrac{K_p}{K_d}$ is to be designed to satisfy frequency domain specifications. The Bode plot of Eqn(8.49) with $K_p = K_d = 1$ is shown in Figure 8.41.

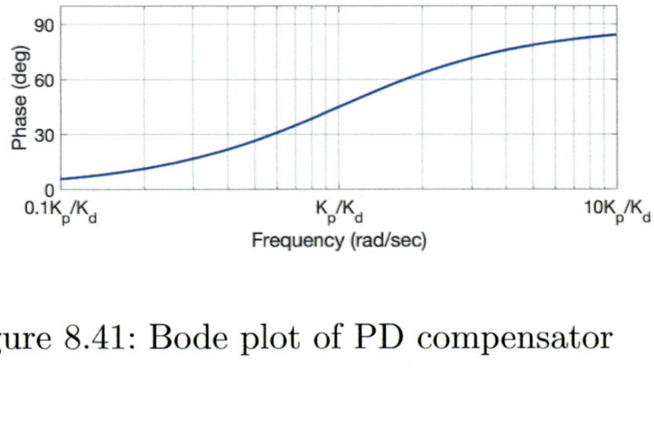

Figure 8.41: Bode plot of PD compensator

Typically, the desired specifications are

a. Desired phase margin ϕ_d (based on percentage overshoot or desired damping ratio)
b. The desired gain crossover frequency ω_{gc} (often selected to match the bandwidth), and it is assumed that the steady state specifications are satisfied by the plant.

The procedure for designing PD compensator in frequency domain is detailed below.

PD compensator Design Procedure in Frequency Domain

Step-1: Analyze the uncompensated system

Evaluate the open-loop frequency response to determine the gain crossover frequency and the existing phase margin ω_{gc}.

Step-2: Determine required phase lead

Calculate the necessary phase boost to be provided by the PD compensator using the formula:

$$\phi_{PD} = \phi_d - \phi_{uc} + \phi_{safety\ margin}$$

Step-3: Locate the zero of the PD compensator

Assume the PD compensator is given by $G_c(s) = K_d(s + z)$. To achieve the required phase lead ϕ_{PD} at ω_{gc}, place the zero such that

$$\phi_{PD} = \tan^{-1}\left(\frac{\omega_{gc}}{z}\right) \implies z = \frac{\omega_{gc}}{\tan(\phi_{PD})}$$

Step-4: Compute the compensator gain K_d

At ω_{gc}, the total loop gain should be 1 (0 dB):

$$\left|G_c(j\omega_{gc}G(j\omega_{gc}\right| = 1 \implies K_d = \frac{1}{\left|j\omega_{gc} + z\right| \cdot \left|G(j\omega_{gc}\right|}$$

EXAMPLE 8.7

Problem

Consider a unity feedback system with the plant:

$$G(s) = \frac{60}{s(s+2)(s+6)}$$

Design a PD compensator of the form

$$G_c(s) = K_p + K_d s = K_d\left(s + \frac{K_p}{K_d}\right) = K_d(s + z)$$

such that the open-loop gain crossover frequency ω_{gc} of $G_c(s)G(s)$ is 3 rad/sec and the desired phase margin is 45^0.

SOLUTION:

Step-1: From the Bode plot of the uncompensated system, shown in Figure 8.42, we found the phase angle at the desired $\omega_{gc} = 3\text{rad/sec} = -173^0$. Thus, the phase margin at this frequency is

$$\phi_{uc} = 7^0$$

Step-2: The desired phase margin $\phi_d = 45^0$. Hence, the required phase boost from the PD compensator is

$$\phi_{PD} = \phi_d - \phi_{uc} + \phi_{safety} = 45^0 - 7^0 + 5^0 = 43^0$$

Step-3: The zero of the PD compensator is obtained from

$$z = \frac{\omega_{gc}}{\tan(\phi_{PD})} = \frac{3}{\tan(43^0)} = 3.2171$$

Step-4: Let us calculate K_d from

$$K_d = \frac{1}{|j\omega_{gc} + z| \cdot |G(j\omega_{gc})|} = 0.2749$$

This gives $K_p = K_d \cdot z = 0.8845$ and the PD compensator

$$G_c(s) = 0.8845 + 0.2749s = 0.2749(s + 3.2171)$$

The open-loop transfer function of the compensated system is

$$L(s) = \frac{16.5(s + 3.2171)}{s(s+2)(s+6)}$$

The Bode plot of both compensated and uncompensated systems is shown in Figure 8.42.

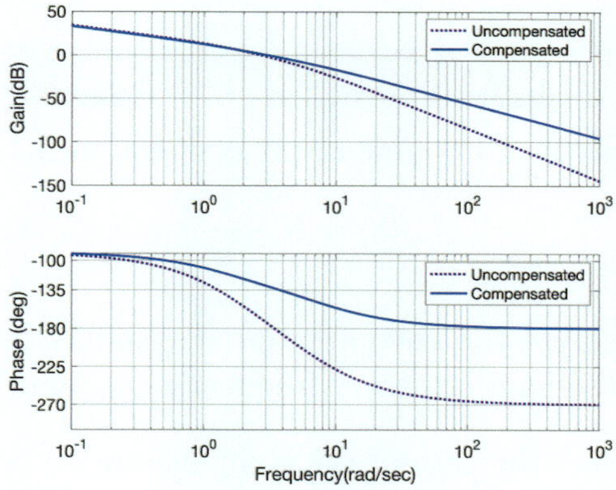

Figure 8.42: Bode plot of the system with and without PD compensator

The phase margin of the compensated system is 50^0 and the gain crossover frequency is 3 rad/sec.

The MATLAB implementation can be found in the file `Example8_7.m` within the supplementary materials.

8.9.2 Frequency Domain Based Design of PI Compensator

The transfer function of the PI controller is expressed as:

$$\mathrm{G_c(s)} = \mathrm{K_p} + \frac{\mathrm{K_i}}{\mathrm{s}} = \frac{\mathrm{K_i}\left(1 + \dfrac{\mathrm{K_p}}{\mathrm{K_i}}\mathrm{s}\right)}{\mathrm{s}} \tag{8.50}$$

The Bode plot of $\mathrm{G_c(j\omega)}$ is shown in Figure 8.43

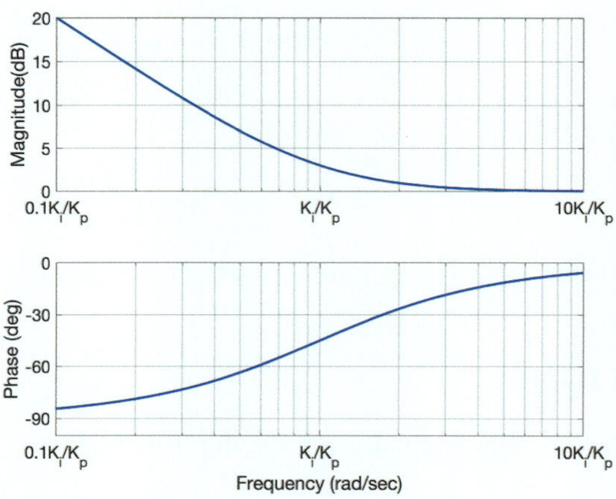

Figure 8.43: Bode plot of the PI compensator

The procedure for designing PI compensator in frequency domain is detailed below.

PI compensator Design Procedure in Frequency Domain

Assume that the PI compensator is of the form

$$G_c(s) = K_c\frac{s+z}{s} = K_cG'_c(s)$$

Step-1: Analyze the uncompensated system G(s) using Bode plot and determine the gain crossover frequency ω_{gc}, and phase margin.

Step-2: Select the zero of the compensator to be

$$\frac{\omega_{gc}}{10} < z < \frac{\omega_{gc}}{3}$$

This ensures that the zero contributes positive phase around ω_{gc} and improves phase margin without overshooting it.

Step-3: Let $L(s) = G_c(s)G(s) = K_cG'_c(s)G(s)$. To achieve unit gain (0 dB) at the desired gain crossover frequency ω_{gc}, enforce:

$$\left|L(j\omega_{gc}\right| = 1 \implies K_c = \frac{1}{\left|L(j\omega_{gc})\right|} = \frac{1}{\left|G'_c(j\omega_{gc})G(j\omega_{gc})\right|}$$

This ensures that the magnitude condition is satisfied, and the gain crossover frequency remains at the desired point.

EXAMPLE 8.8

Problem

Consider a unity feedback system with the plant:

$$G(s) = \frac{10}{(s+4)^2}$$

Design a PI controller of the form

$$G_c(s) = K_c\frac{s+z}{s}$$

such that the open-loop gain crossover frequency ω_{gc} of $G_c(s)G(s)$ is 3 rad/sec.

SOLUTION:

Let us denote the open-loop transfer function:

$$L(s) = G_c(s)G(s) = K_c \cdot \frac{s+z}{s} \cdot \frac{10}{(s+4)^2}$$

The objective is to find z and K_c such that:

$$|L(j\omega_{gc})| = 1 \quad \text{at} \quad \omega_{gc} = 3$$

The selection of zero is flexible, where it should be selected to lie between $\frac{\omega_{gc}}{10}$ and $\frac{\omega_{gc}}{3}$. Let us select

$$z = \frac{\omega_{gc}}{4} = \frac{3}{4} = 0.75$$

This gives the controller

$$G_c(s) = K_c \frac{s + 0.75}{s}$$

where the gain K_c is computed by solving $|L(j\omega_{gc})| = 1$ where

$$L(j\omega) = G_c(j\omega)G(j\omega) = K_c \left(\frac{j\omega + 0.75}{j\omega}\right) \cdot \left(\frac{10}{(j\omega + 4)^2}\right)$$

This gives $K_c = 2.4245$. Thus, the open-loop transfer function of the compensated system is

$$L(s) = 24.25 \left(\frac{s + 0.75}{s(s + 4)^2}\right)$$

The Bode plot of both compensated and uncompensated systems is shown in Figure 8.44.

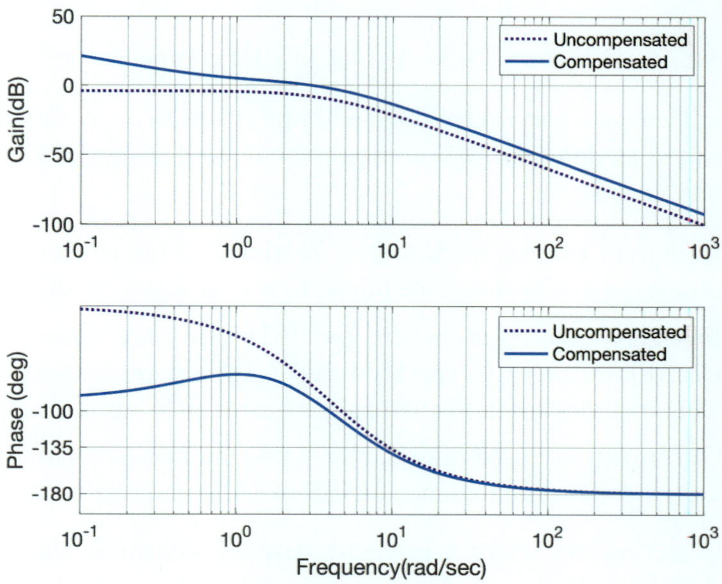

Figure 8.44: Bode plot of the system with and without PI compensator

Note that we intentionally design the controller gain K_c and zero location z so that the open-loop gain $|L(j\omega) = G_c(j\omega)G(j\omega)| = 1$ at a desired higher frequency ω_{gc}.

A complete MATLAB implementation is available in the supplementary file `Example8_8.m`.

8.9.3 Frequency Domain Based Design of PID Compensator

The Bode plot of the PID controller is shown in Figure 8.45.

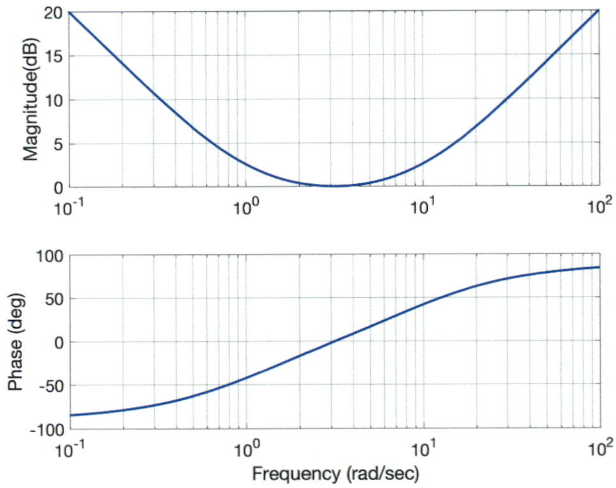

Figure 8.45: Bode plot of PID compensator

Note: There are several established approaches to designing PID controllers, each suited to different system requirements and design preferences. Common methods include:

- Empirical tuning, such as the Ziegler–Nichols method, which relies on observing the system's response to a step input or sustained oscillation.
- Pole-zero placement, where the PID gains are chosen to position the closed-loop poles to meet desired transient performance specifications. This includes methods like root locus or pole cancellation.
- Optimization-based design, which uses numerical techniques to minimize a performance index (such as ISE or IAE), subject to system constraints.

In this section, we adopt a modular strategy known as the "PD followed by PI" method, which provides a systematic and practical approach for PID controller design, particularly in the frequency domain using Bode plots. This two-step method involves:

1. Designing a PD controller to shape the transient response by improving overshoot, phase margin, and speed. It provides phase lead near the desired gain crossover frequency, helping to achieve adequate stability and responsiveness.
2. Adding a PI controller to eliminate steady-state error by introducing an integrator (a pole at the origin). It also includes a low-frequency zero to offset the phase lag introduced by the integrator, while minimally affecting the high-frequency phase margin when properly placed.

This approach provides a structured framework that balances performance tuning and robustness. The procedure to design such a controller is summarized below:

PID Compensator Design Procedure in Frequency Domain

Step-1: Analyze the Uncompensated System

- Find the phase margin ϕ_{uc} and gain crossover frequency $\omega_{\mathrm{gc,uc}}$ of uncompensated system.
- Define performance specifications such as desired phase margin, ϕ_{d}, desired bandwidth, etc.

Step-2: Choose Controller Form

Let the PID controller be of the form:

$$G_c(s) = K\frac{(s+z_d)(s+z_i)}{s}$$

Step-3: Design the PD Compensator

- Compute the required phase lead that needs to be provided by the PD part of the compensator

$$\phi_{PD} = \phi_d - \phi_{uc} + \phi_\epsilon$$

where $\phi_\epsilon \approx 5^0 - 12^0$.

- Choose crossover frequency

$$\omega_{gc} \approx \omega_{gc,uc} \quad \text{(or slightly higher for faster response)}$$

- Compute zero location of PD part:

$$z_d = \frac{\omega_{gc}}{\tan(\phi_{PD})}$$

- Compute Gain K such that

$$\left|G_{PD}(j\omega_{gc})G(j\omega_{gc})\right| = 1 \implies K = \frac{1}{\left|(j\omega_{gc} + z_d)G(j\omega_{gc})\right|}$$

- The PD part of compensator $G_{PD} = K(s + z_d)$

Step-4: Verify PD Compensated System

- Plot the Bode plot of $G_{PD}(s()G(s)$
- Check that: the Phase margin $\approx$ desired value and the crossover frequency are acceptable.
- If not, iterate on z_d or ω_{gc}

Step-5: Add the PI Compensator

Choose zero of PI:

- Select it at low frequency, such as:

$$z_i << \omega_{gc}, \text{ e.g. } z_i = \omega_{gc}/8$$

- This ensures minimal effect on phase margin.

Step-6: Define PI compensator:

$$G_{PI} = \frac{s + z_i}{s}$$

Step-7: Combine Full PID Compensator

$$G_{PID}(s) = K \cdot \frac{(s + z_d)(s + z_i)}{s}$$

Step-8: Verify Final Design

EXAMPLE 8.9

Problem

Consider a unity feedback system with the plant:

$$G(s) = \frac{100}{(s + 1)(s + 2)(s + 5)}$$

Design a PID compensator of the form

$$G_{PID}(s) = K\frac{(s + z_d)(s + z_i)}{s}$$

such that

a. Phase margin $\approx 50^0$
b. Gain crossover frequency = 4 rad/sec
c. Steady state error to step input is zero

SOLUTION:

Step-1: From the Bode plot of the uncompensated system, shown in Figure 8.46, we found the phase angle at the desired $\omega_{gc} = 4$ rad/sec $= -178^0$. Thus, the phase margin at this frequency is

$$\phi_{uc} = 2^0$$

Step-2: The desired phase margin $\phi_d = 45^0$. Hence, the required phase boost from the PD part of the compensator is

$$\phi_{PD} = \phi_d - \phi_{uc} + \phi_{safety} = 50^0 - 2^0 + 5^0 = 53^0$$

Step-3: The zero of the PD compensator is obtained from

$$z = \frac{\omega_{gc}}{\tan(\phi_{PD})} = \frac{4}{\tan(53^0)} \approx 4$$

Step-4: Let us calculate $K = K_d$ from

$$K_d = \frac{1}{|j\omega_{gc} + z| \cdot |G(j\omega_{gc})|} = 0.2357$$

This gives the PD part of the compensator

$$G_{PD}(s) = K_d(s + z_d) = 0.2357(s + 4)$$

Design of PI part of compensator:

$$G_{PI}(s) = \frac{s + z_i}{s}$$

Let us choose $z_i = \frac{\omega_{gc}}{8} = \frac{4}{8} = 0.5$. Hence, the PID compensator is

$$G_{PID} = G_{PD} \times G_{PI} = 0.2357\frac{(s + 4)(s + 0.5)}{s}$$

This gives the compensated system with an open-loop transfer function

$$G_{comp}(s) = G \times G_{PID} = 23.6\frac{(s + 4)(s + 0.5)}{s(s + 1)(s + 2)(s + 5)}$$

The Bode plot of both compensated and uncompensated systems is shown in Figure 8.46.

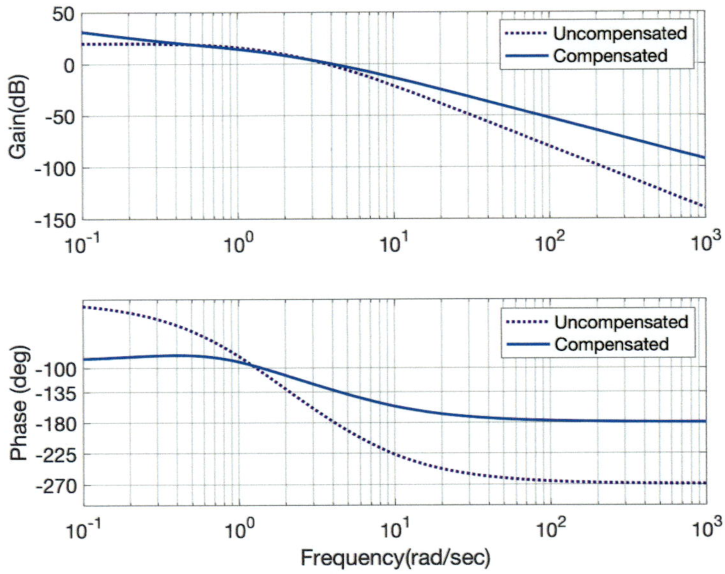

Figure 8.46: Bode plot of the system with and without PID compensator

The phase margin of the compensated system is 49^0 and the gain crossover frequency is 4 rad/sec. The closed-loop step response is shown in Figure 8.47

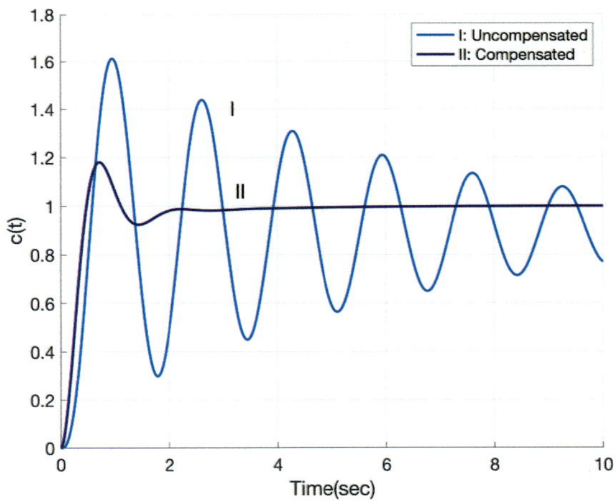

Figure 8.47: Step response of uncompensated and PID compensated system

The full MATLAB source code is contained in `Example8_9.m` (supplementary material).

8.9.4 Frequency Domain Based Design of Lead Compensator

Now, the transfer function of a lead compensator is given by

$$G_c(s) = \frac{s + z_c}{s + p_c} = \frac{s + \frac{1}{\tau}}{s + \dfrac{1}{\alpha\tau}} = \alpha\left(\frac{1 + \tau s}{1 + \alpha\tau s}\right); \quad \alpha = \frac{z_c}{p_c} < 1, \quad \tau > 0 \tag{8.51}$$

where α is the attenuation factor of the lead compensator. It has a zero at $s = -1/\tau$ and a pole at $s = -1/(\alpha\tau)$.

The sinusoidal transfer function of the lead compensator is given by

$$G_c(j\omega) = \alpha\left[\frac{1 + j\omega\tau}{1 + j\omega\alpha\tau}\right]; \quad \alpha < 1$$

At zero frequency, the network has a gain of $\alpha < 1$ or an attenuation of $1/\alpha$. Often, in the frequency domain compensation technique, it is convenient to cancel the d.c. attenuation of the network with an amplification $A = 1/\alpha$. The lead compensator is then visualized as a combination of a network and an amplifier, as shown in Figure 8.48

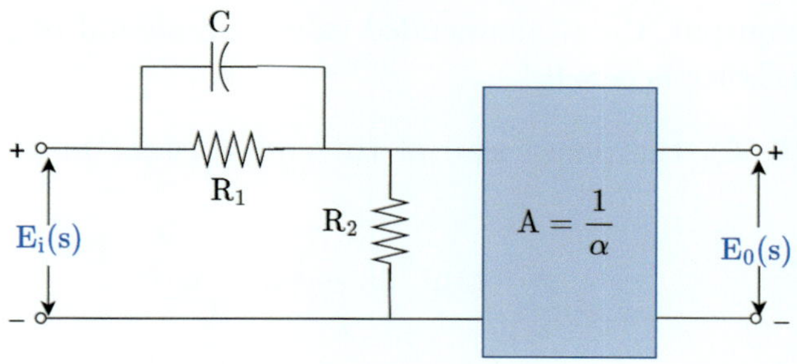

Figure 8.48: Phase-lead network with an amplifier

The sinusoidal transfer function of the lead compensator is then given by

$$G_c(j\omega) = \frac{1 + j\omega\tau}{1 + j\omega\alpha\tau}; \quad \alpha < 1 \tag{8.52}$$

Since $\alpha < 1$, the network output of the compensator leads the sinusoidal input under steady state and hence the name lead compensator. The two corner frequencies are at $\dfrac{1}{\tau}$ and at $\dfrac{1}{\alpha\tau}$. The Bode diagram of the lead compensator is shown in Figure 8.49.

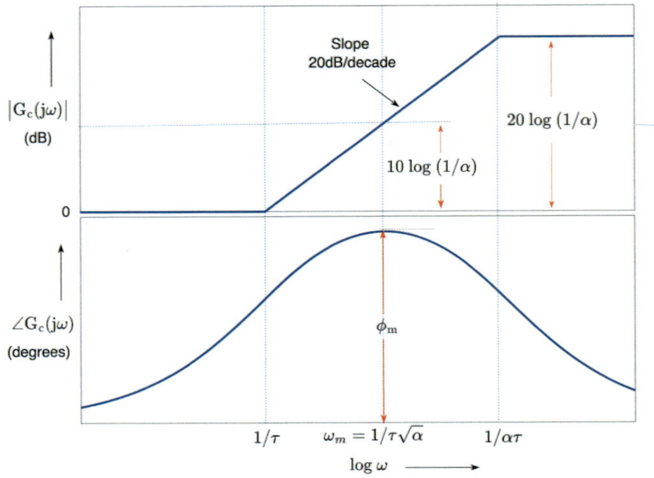

Figure 8.49: Bode plot of phase lead network

From Figure 8.49, it is seen that the lead compensator essentially behaves as a high-pass filter. The high-frequency noise signals are amplified by a factor $1/\alpha > 1$, while the low-frequency (control) signals undergo unit amplification. Therefore, the signal-to-noise ratio at the output of the lead compensator is poorer than at its input. To prevent excessive deterioration of signal-to-noise ratio at the output, the recommended value of α should be greater than 0.07. A common choice is $\alpha = 0.1$.

From Eqn(8.52), the phase lead of the compensator at any frequency ω is given by

$$\phi = \tan^{-1}\omega\tau - \tan^{-1}\alpha\omega\tau$$

or

$$\tan\phi = \frac{-\omega\tau}{1 + \alpha\omega^2\tau^2} \tag{8.53}$$

The frequency ω_{m} at which the compensator provides maximum phase lead is computed by solving $\dfrac{\mathrm{d}\phi}{\mathrm{d}\omega} = 0$, which gives

$$\boxed{\omega_{\mathrm{m}} = 1/(\tau\sqrt{\alpha}) = \sqrt{(1/\tau)(1/\alpha\tau)}} \tag{8.54}$$

From Eqn(8.9.4), we see that ω_{m} is the geometric mean of the two corner frequencies. At $\omega = \omega_m$, the maximum phase lead ϕ_m is given by

$$\tan\phi_{\mathrm{m}} = \frac{(1-\alpha)}{(2\sqrt{\alpha})}, \quad \text{or,} \quad \sin\phi_{\mathrm{m}} = (1-\alpha)/(1+\alpha)$$

In terms of ϕ_{m}, α is expressed as:

$$\boxed{\alpha = \frac{1 - \sin\phi_m}{1 + \sin\phi_m}} \tag{8.55}$$

For a required maximum phase lead ϕ_{m}, Eqn(8.55) will be used to determine the α parameter of the network.

The magnitude of $\mathrm{G_c(j\omega)}$ at $\omega = \omega_m$ is given by

$$\left|\mathrm{G_c(j\omega_m)}\right| = \left|\frac{1 + \mathrm{j}\omega_{\mathrm{m}}\tau}{1 + \mathrm{j}\alpha\omega_{\mathrm{m}}\tau}\right| = \frac{1}{\sqrt{\alpha}} \tag{8.56}$$

or, the dB gain of the lead compensator is

$$\boxed{\left|\mathrm{G_c(j\omega_m)}\right|_{dB} = 20\log_{10}\frac{1}{\sqrt{\alpha}} = 10\log\left(\frac{1}{\alpha}\right)} \tag{8.57}$$

The variation of ϕ_{m} versus $1/\alpha$ is shown in the Figure 8.50.

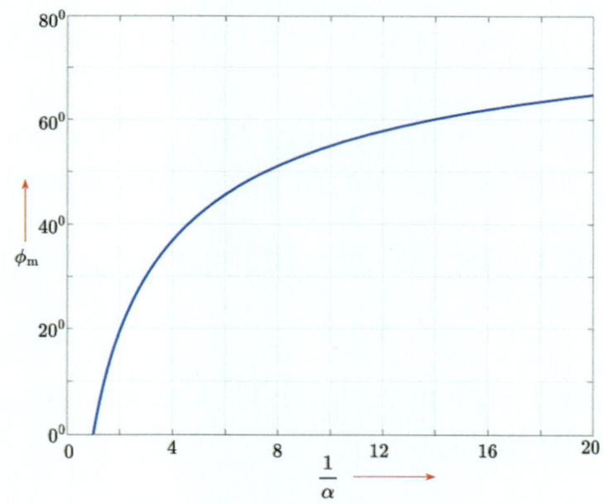

Figure 8.50: Variation of ϕ_{m} with α of a lead network

From Figure 8.50 it is observed that for phase leads greater than 60^0, the α-parameter of the network becomes very low, that is, the attenuation due to the network becomes very high. Therefore, for phase leads greater than 60^0, it is preferred to use two cascaded lead networks with moderate values of α instead of a single lead network with a minimal value of α.

The systematic steps for designing a lead compensator are summarized below.

Lead Compensator Design Procedure in Frequency Domain

Step-1: Map the desired specifications to desired phase margin using the relationships given in Eqns(8.42)–(8.48).

Step-2: Determine loop gain, K, to satisfy specified error constant.

Step-3: Using this value of K, determine the phase margin ϕ_{uc} of the uncompensated system.

Step-4: Determine the additional phase-lead required using the relation

$$\phi_{\mathrm{ld}} = \phi_{\mathrm{s}} - \phi_{\mathrm{uc}} + \epsilon$$

where $\phi_{\mathrm{s}} =$ is the specified phase margin; ϕ_{uc} is the phase margin of the uncompensated system; and $\epsilon =$ is a margin of safety required due to increase in the cross over frequency because of compensation (usually 10^0). Note that if the given specifications are in the time domain, such as peak overshoot or damping ratio, these should be translated into phase margin.

Step-5: Determine the α parameter of the network from

$$\alpha = \frac{1 - \sin\,\phi_{\mathrm{ld}}}{1 + \sin\,\phi_{\mathrm{ld}}}$$

If the required ϕ_{m} is more than 60^0, it is recommended to use two identical networks, each contributing a maximum lead of $\phi_{\mathrm{ld}}/2$.

Step-6: Calculate the dB gain 10 $\log(1/\alpha)$ provided by the network at ω_{m}.

a. Locate the frequency at which the uncompensated system has a gain of $-10\,\log(1/\alpha)$.

b. This is the new cross-over frequency $\omega_{\mathrm{c2}} = \omega_{\mathrm{m}}$ of the compensated system.

Step-7: Compute two corner frequencies of the network as:

$$\omega_1 = 1/\tau = \omega_{\mathrm{m}}\sqrt{\alpha}, \quad \text{and} \quad \omega_2 = 1/\alpha\tau = \omega_{\mathrm{m}}/\sqrt{\alpha}$$

The phase lead compensator becomes

$$\mathrm{G_{lead}(s)} = \frac{\mathrm{s} + \dfrac{1}{\tau}}{\mathrm{s} + \dfrac{1}{\alpha\tau}} = \alpha\frac{1 + \tau\mathrm{s}}{1 + \alpha\tau\mathrm{s}}$$

Step-8: Insert an amplifier with gain $\frac{1}{\alpha}$ to neutralize the attenuation of the lead compensator.

Step-9: Draw the magnitude and phase plots of the compensated system and check the resulting phase margin.

Step-10: If the desired specifications are not met, try with another choice of cross over frequency ω_{c_2} till the specifications are met.

EXAMPLE 8.10

Problem

The open-loop transfer function of pitch dynamics of a F-16 fighter aircraft is given by

$$G(s) = \frac{K}{(1 + T_1 s)(1 + T_2 s)}$$

where $T_1 = 0.3$sec and $T_2 = 2$sec. Design a lead compensator to meet the following specifications.

a. $K_p = 100$
b. Phase margin $= 45^0$

SOLUTION:

Step-1: The gain K which provides static position error constant $K_p = 100$ is obtained from

$$K_p = \lim_{s \to 0} G(s) = \lim_{s \to 0} \frac{K}{(1 + 0.3s)(1 + 2s)} = K = 100$$

Step-2: From the Bode diagram of the system with $K = 100$, shown in Figure 8.51, we determine the phase margin $\phi_{uc} = 17^0$ and the gain cross-over frequency $\omega_{c1} = 13$ rad/sec.

Step-3: The additional phase lead required to have the desired phase margin of 45^0 of the compensated system is

$$\phi_{ld} = \phi_s - \phi_{uc} + \epsilon = 45^0 - 17^0 + 5^0 = 33^0$$

Note that we have taken the tolerance $\epsilon = 5^0$. It is justified because the magnitude plot does not roll off rapidly.

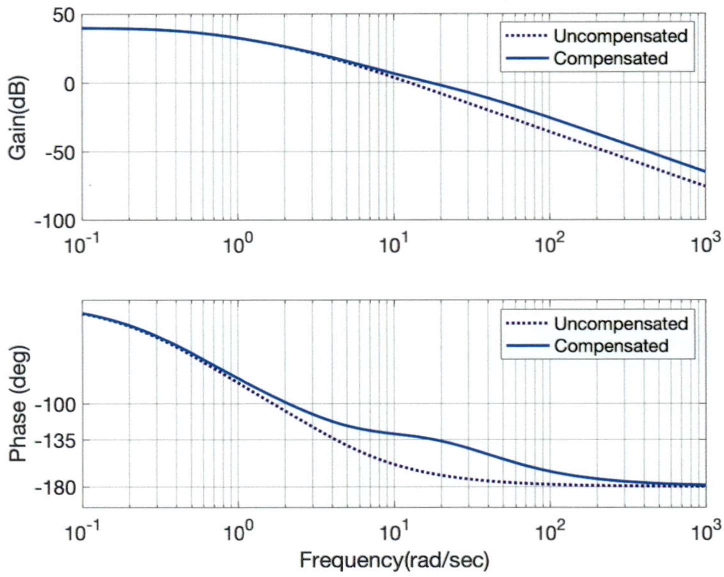

Figure 8.51: Bode plot of pitch dynamics of F-16 fighter aircraft with and without lead compensator plane

Step-4: Since the phase lead required is not too large, that is, 33^0, we use a single lead network. The attenuation factor of the lead network is

$$\alpha = \frac{1 - \sin 33^0}{1 + \sin 33^0} = 0.2948 \approx 0.3$$

Step-5: Let us find the frequency $\omega_{\rm m}$ where the uncompensated system has a gain of $-10\log(1/\alpha) = -5.3$ dB ≈ -6 dB. From the Bode plot, this is found to be 18 rad/sec. This frequency is the gain crossover frequency of the compensated system, that is, $\omega_{\rm c2} = \omega_{\rm m} = 18$ rad/sec.

Step-6: The two corner frequencies of the lead network are

$$\omega_1 = \frac{1}{\tau} = \omega_{\rm m}\sqrt{\alpha} = 9.77 \quad \text{and} \quad \omega_2 = \frac{1}{\alpha\tau} = \omega_{\rm m}/\sqrt{\alpha} = 33.15$$

Step-7: The transfer function of the lead compensator is

$$\mathrm{G_c(s)} = \frac{\mathrm{s} + 9.77}{\mathrm{s} + 33.17} = 0.3\frac{1 + 0.1\mathrm{s}}{1 + 0.03\mathrm{s}}$$

Step-8: If we insert an amplifier of gain $1/\alpha$, the open-loop transfer function of the compensated system is

$$\mathrm{G(s)} = \frac{100(1 + 0.1\mathrm{s})}{(1 + 0.3\mathrm{s})(1 + 2\mathrm{s})(1 + 0.03\mathrm{s}}$$

The Bode plot for G(s) is shown in Figure 8.51 (solid line), where the phase margin is found to be 45^0. Note that the gain crossover frequency has increased from 13 rad/sec to 18 rad/sec.

The MATLAB implementation can be found in the file `Example8_10.m` within the supplementary materials.

CONCEPT REINFORCEMENT PROBLEM 8.2

Problem:

The open-loop transfer function of a unity feedback system is given by

$$G_p(s) = \frac{K(s+0.5)}{(s+0.4)(s+2.5)(s+60)}$$

Design a lead compensator to meet the following specifications.

a. The steady state error due to a unit ramp input should be less than 0.02, that is $e_{ss} < 0.02$ and
b. Phase margin $= 40^0$

EXAMPLE 8.11

Problem

The open-loop transfer function of a unity feedback system is given by

$$G_f(s) = \frac{K}{s^2(s+4)}$$

Design a lead compensator to meet the following specifications.

a. $K_a = 20\ \text{sec}^{-2}$
b. Phase margin $= 35^0$

SOLUTION:

Step-1: The gain K which results static acceleration error constant $K_a = 20\ \text{sec}^{-2}$ is

$$K_a = \lim_{s\to 0} s^2 G(s) = \lim_{s\to 0} s^2 \frac{K}{s^2(s+4)} = \frac{K}{4} = 20 \quad \Longrightarrow \quad K = 80$$

Step-2: The phase margin ϕ_{uc} and gain crossover frequency ω_{c1} of the uncompensated system are determined from its Bode plot (with $K = 80$) shown in Figure 8.52. This gives $\phi_{uc} = -43^0$ and $\omega_{c1} = 3.8$ rad/sec

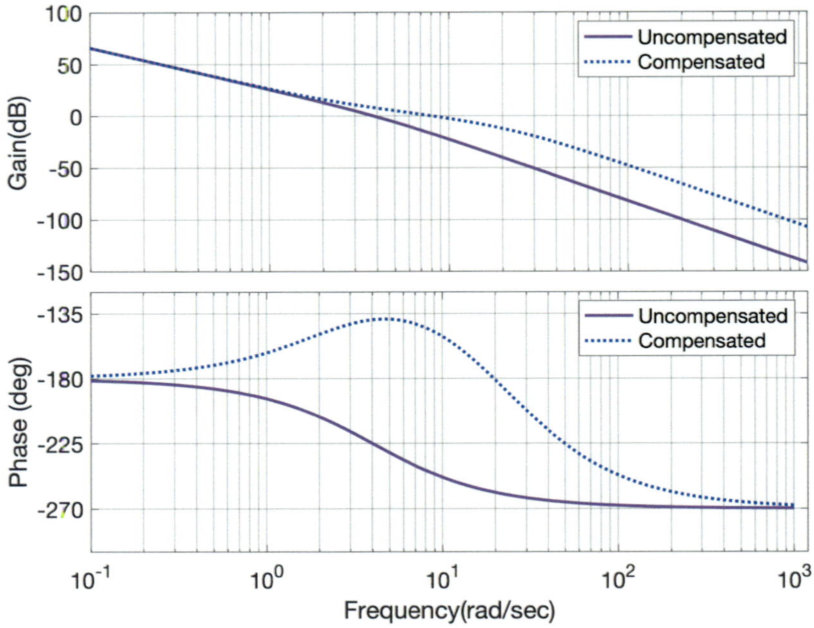

Figure 8.52: Bode plot for Type-2 system with and without lead compensator

This shows that the uncompensated system is absolutely unstable.

Step-3: The additional phase lead required to have the desired phase margin of 35^0 of the compensated system is

$$\phi_{ld} = \phi_s - \phi_{uc} + \epsilon = 35^0 - (-43^0) + 20^0 = 98^0$$

Note that we have taken the tolerance $\epsilon = 20^0$ which normally should be around 10^0. This is because, for this system, the final slope of the Bode magnitude plot is -60dB/decade, and the angle $\angle G_f(j\omega)$ therefore decreases at a faster pace.

Step-4: Since the phase lead required is too large, that is, 98^0, it is not advisable to use a single lead network. Instead, we use a double lead network such that each one will provide a phase lead of $98/2 = 49^0$. Thus, the attenuation factor of each lead network is

$$\alpha = \frac{1 - \sin 49^0}{1 + \sin 49^0} = 0.1377$$

Step-5: Since we use a double lead network, let us find the frequency at which the uncompensated system has a gain of

$2 \times -10\log(1/\alpha) = -17.2$ dB. This frequency should be the gain crossover frequency of the compensated system $\omega_{c2} = \omega_m$. From the Bode plot, this is found to be 8 rad/sec, that is $\omega_{c2} = \omega_m = 8$ rad/sec.

Step-6: The two corner frequencies of the lead network are

$$\omega_1 = \frac{1}{\tau} = \omega_m\sqrt{\alpha} = 2.9843 \quad \text{and} \quad \omega_2 = \frac{1}{\alpha\tau} = \omega_m/\sqrt{\alpha} = 21.67$$

Step-7: The transfer function of each section of the lead compensator is

$$G_c^1(s) = \frac{s + 2.995}{s + 21.75} = 0.1377\frac{1 + 0.331s}{1 + 0.0461s}$$

Step-8: The open-loop transfer function of the compensated system is

$$G(s) = \frac{80\,(1 + 0.331s)^2}{s^2(s + 4)\,(1 + 0.0461s)^2}$$

The Bode plot for G(s) is shown in Figure 8.52 (dotted line), where the phase margin is found to be 35^0. Note that the gain crossover frequency has increased from 3.80 rad/sec to 8.04 rad/sec.

Refer to supplementary file `Example8_11.m` for the MATLAB implementation.

CONCEPT REINFORCEMENT PROBLEM 8.3

Problem:

The linearized model for automatic gauge control in a hot-strip finishing mill is shown in Figure 8.53

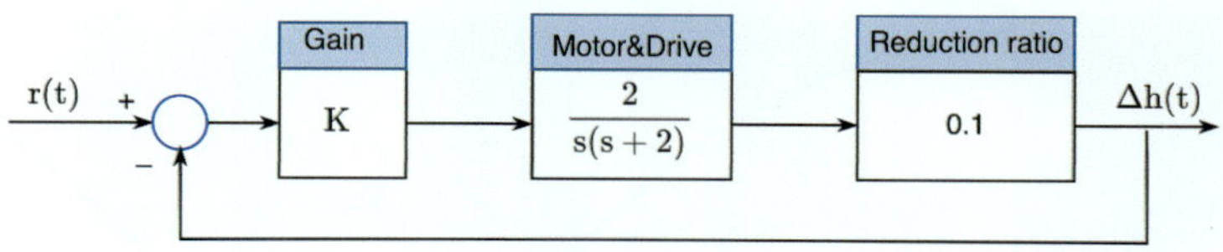

Figure 8.53: Gauge control system for a hot-strip finishing mill

Design a lead compensator to satisfy the following specifications

a. The steady state error to a unit ramp input should not exceed 0.05.

b. The phase margin should be at least 45^0.

8.9.5 Frequency Domain Based Design of Lag Compensator

In Section 8.4, we derived the transfer function of a lag compensator, which is given by

$$\begin{aligned} G_c(s) &= \frac{1}{\beta}\left(\frac{s+\frac{1}{\tau}}{s+\dfrac{1}{\beta\tau}}\right) = \frac{1}{\beta}\left(\frac{s+z_c}{s+p_c}\right); \quad \beta = \frac{z_c}{p_c} > 1, \quad \tau > 0 \\ &= \frac{1+\tau s}{1+\beta\tau s} \end{aligned} \tag{8.58}$$

Since $\beta > 1$, its zero at $s = -1/\tau$ is always located to the left of the pole at $s = -1/(\beta\tau)$.

The sinusoidal transfer function of the lag compensator is given by

$$G_c(j\omega) = \left[\frac{1+j\omega\tau}{1+j\beta\omega\tau}\right]; \quad \beta > 1$$

Since $\beta > 1$, the steady state output has a lagging phase angle concerning sinusoidal input and hence the name lag network. The Bode diagram of the lag compensator is shown in Figure 8.54.

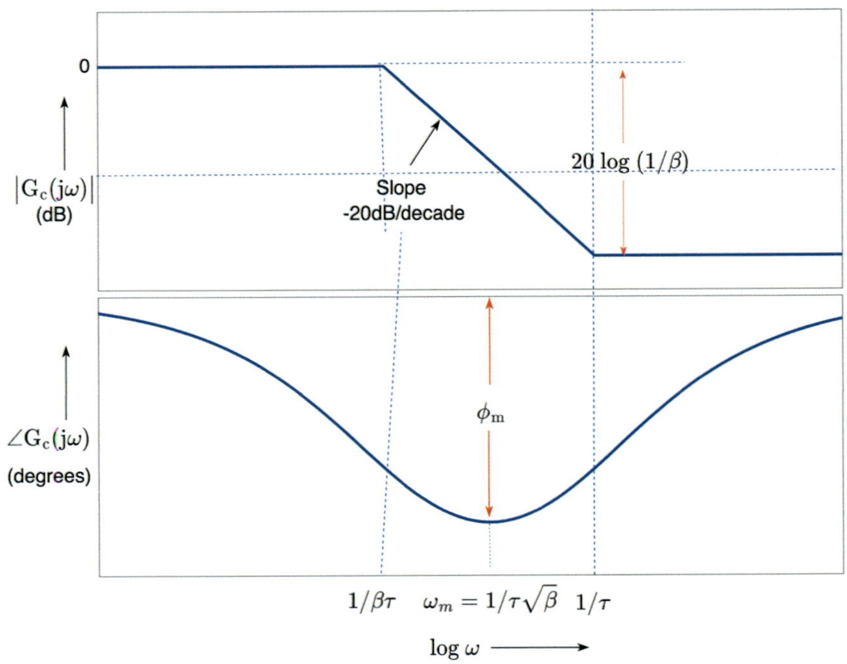

Figure 8.54: Bode plot for lag compensator

The lower and upper corner frequencies of the compensator are at $\frac{1}{\beta\tau}$ and $\frac{1}{\tau}$, respectively. From Figure 8.54, it is seen that it has a d.c. gain of unity (0 dB) and a high-frequency gain of $1/\beta$ ($-20\log\beta$ dB). The gain decreases from 0-dB at $\omega = \frac{1}{\beta\tau}$ to $-20\log\beta$ dB with a slope of -20db/decade and remains constant thereafter. This means the lag compensator attenuates the high-frequency noise (essentially behaves as a low-pass filter) and thereby improves the signal-to-noise ratio.

Now the phase lag of the compensator is given by

$$\phi = \tan^{-1}\omega\tau - \tan^{-1}\beta\omega\tau, \quad \Longrightarrow \quad \tan\phi = \frac{-\omega\tau}{1+\beta\omega^2\tau^2}$$

The frequency ω_m at which the compensator provides maximum phase lag is computed by solving $\frac{d\phi}{d\omega} = 0$, which gives

$$\omega_{\rm m} = \frac{1}{(\tau\sqrt{\beta})} = \sqrt{(1/\tau)(1/\beta\tau)}$$

Similar to the lead compensator, $\omega_{\rm m}$ is the geometric mean of the two corner frequencies of the compensator. At $\omega = \omega_{\rm m}$, the maximum phase lag $\phi_{\rm m}$ is given by

$$\tan\phi_{\rm m} = \frac{(1-\beta)}{(2\sqrt{\beta})}, \quad \text{or,} \quad \sin\phi_{\rm m} = \frac{\beta-1}{\beta+1}$$

The design procedure of the lag compensator is summarized below.

Lag Compensator Design Procedure in Frequency Domain

Step-1: Map the desired transient response (time domain) specifications into frequency domain specifications.

Step-2: Determine the loop gain, K, which satisfies the specified error constant.

Step-3: Using this value of K, sketch the Bode plot and determine the gain cross-over frequency $\omega_{\rm g}$ and the phase margin of the uncompensated system.

Step-4: Determine the the frequency $\omega'_{\rm g}$ where the phase angle of the uncompensated system is

$$-180^0 + \phi_d + \epsilon$$

where ϕ_d = desired phase margin and ϵ is the tolerance.

a. Select this frequency as the new gain crossover frequency ω_g'.

b. Select ϵ between 5^0 15^0, which would compensate for the phase lag of the network at the new crossover frequency ω_g'.

Step-5: Determine β parameter of the network as follows:

a. Find the gain of the uncompensated system at ω_g'.

b. Equate this gain to the required high-frequency attenuation ($20\ log\ \beta$) and compute the β parameter.

Step-6: Choose the upper corner frequency $\omega_2 = 1/\tau$ of the network one octave to one decade below ω_g', i.e.,

$$\omega_2 = 1/\tau = \frac{\omega_g'}{2} \text{ to } \frac{\omega_g'}{10}$$

a. Note that a large value of τ calculated from the above guideline is undesirable. Because, for large τ, the circuit realization requires excessively large capacitors.

b. Compute the lower corner frequency $\omega_1 = 1/(\beta\tau)$; as β is known.

Step-7: Draw the magnitude and phase plots of the compensated system and check if the desired specifications are satisfied. If not, redesign the compensator with a different value of ϵ beginning from step-3 above.

EXAMPLE 8.12

Problem

The open-loop transfer function of a unity feedback system is given by

$$G(s) = \frac{K}{s(s+5)(s+10)}$$

Design a lag compensator to meet the following specifications.

a. $K_v = 20\ \text{sec}^{-1}$

b. Phase margin $= 45^0$

SOLUTION

Step-1: The gain K which provides static velocity error constant $K_v = 10\ sec^{-1}$ is obtained from

$$K_v = \lim_{s\to 0} sG(s) = \lim_{s\to 0} s\frac{K}{s(s+5)(s+10)} = \frac{K}{20} = 10 \quad \Longrightarrow \quad K = 500$$

Step-2: The phase margin ϕ_{uc} and gain cross over frequency ω_g of the uncompensated system are obtained from its Bode diagram (shown in Figure 8.55). It gives $\phi_{uc} = 12^0$ and $\omega_g = 6$rad/sec

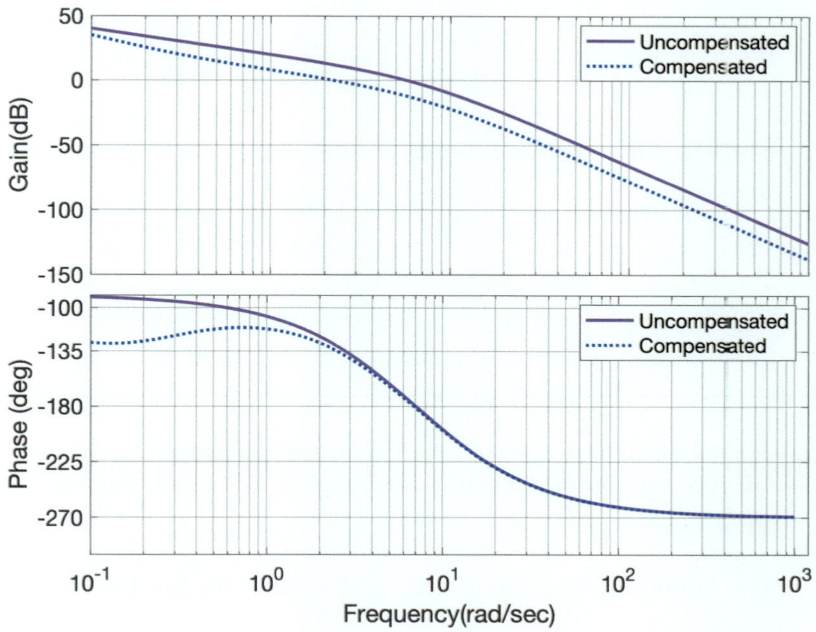

Figure 8.55: Bode plot of the system with and without lag compensator

Step-3: Determine the new gain crossover frequency. This is the frequency where the uncompensated system has a phase angle of $-180^0 + 45^0 + 5^0 = -135^0$. The specified phase margin is increased by 5^0 to compensate for the small phase angle contribution of the lag compensator at the new gain crossover frequency. From the Bode plot, we found $\omega'_g = 2.5$rad

Step-4: Determine the β parameter. Now, the required reduction in gain (dB) from the relation

$$20\log\beta = 20\log\left|G_{uc}\right|_{\omega=\omega'g} = 11.2\text{dB} \quad \Longrightarrow \quad \beta = 4$$

Step-5: Choose the corner frequencies of the compensator. Let us place the upper corner frequency

$$\omega_2 = \frac{\omega_g'}{10} = 0.25 \text{ rad/sec}$$

The lower corner frequency is then

$$\omega_1 = \frac{\omega_2}{\beta} = 0.0625 \text{ rad/sec}$$

This gives the transfer function of the lag compensator

$$G_{lag}(s) = \frac{1}{\beta}\left(\frac{s+\omega_2}{s+\omega_1}\right) = \frac{1}{4}\left(\frac{s+0.25}{s+0.0625}\right) = \frac{1+4s}{1+16s}$$

Step-6: The open-loop transfer function of the compensated system is

$$G_{comp}(s) = \frac{500(1+4s)}{s(s+5)(s+10)(1+16s)}$$

The Bode plot of the compensated system is shown in Figure 8.55 (dotted line), where the phase margin is found to be 48^0.

CONCEPT REINFORCEMENT PROBLEM 8.4

Problem

The open-loop transfer function of a unity feedback system is given by

$$G(s) = \frac{K}{(s+3)(s+7)(s+8)}$$

It is operating at 16.3% overshoot. Design a lag compensator that would reduce the steady state error by 10-fold.

8.9.6 Frequency Domain Based Design of Lag-Lead Compensator

The transfer function of Eqn(8.23) can be expressed as

$$G_c(s) = \frac{(1+\tau_1 s)(1+\tau_2 s)}{(1+\beta\tau_1 s)(1+(\tau_2/\beta)s} \tag{8.59}$$

The sinusoidal transfer function is therefore given by

$$G_c(j\omega) = \frac{(1 + j\omega\tau_1)\,(1 + j\omega\tau_2)}{(1 + j\omega\beta\tau_1)\,(1 + j\omega(\tau_2/\beta)} \tag{8.60}$$

The Bode plot is shown in Figure 8.56

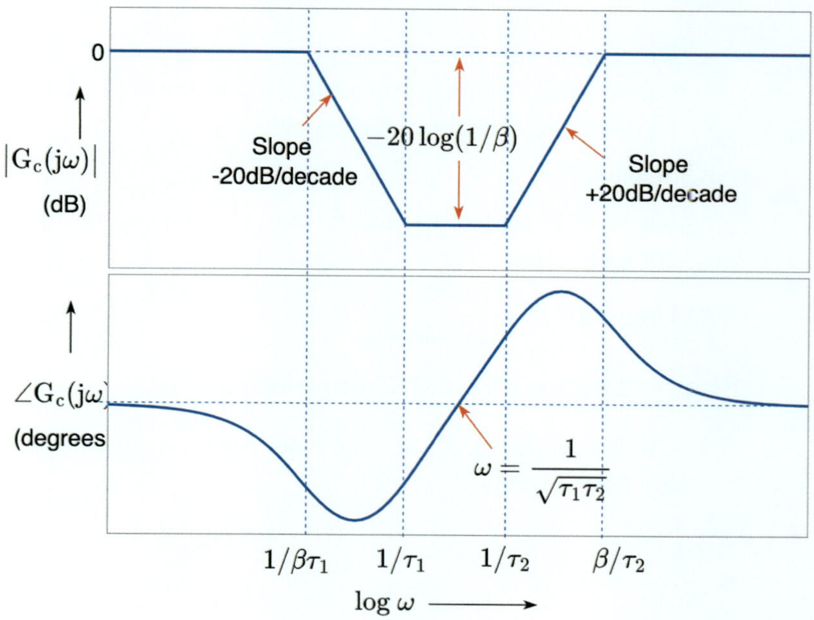

Figure 8.56: Bode plot for lag-lead compensator

The design procedure of the lag-lead compensator is summarized below.

Lag-Lead Compensator Design Procedure in Frequency Domain

Step-1: Using the formulae in Eqns(8.42)–(8.48), derived for the second order system,

i. Map the desired damping ratio and peak overshoot into phase margin and

ii. Map the desired rise time, peak time, or settling time into closed-loop bandwidth ω_{BW}.

Note: Such translations are not needed if the desired phase margin and bandwidth are directly specified.

Step-2: Determine loop gain K, to satisfy the specified error constant.

Step-3: Using this value of K, plot the Bode diagram.

Step-4: Select a new gain crossover frequency ω_{gcn} tentatively near the specified ω_{BW}. The phase crossover frequency ω_{pc} of the uncompensated system may also be a choice for the new gain crossover frequency, ω_{gcn}.

Step-5: Determination of β-parameter: The β parameter will be determined from the phase lead contributed by the lead compensator.

a. Determine the phase margin ϕ_{pmn} at the new gain crossover frequency and compute the additional amount of phase lead required to meet the desired phase margin ϕ_d.

b. Select $\phi_{ld} = \phi_d - \phi_{pmn} + \epsilon$

   where ϵ is added to nullify the angular contribution of the lag compensator.

c. Compute $\beta = \dfrac{1 + \sin\phi_{ld}}{1 - \sin\phi_{ld}}$

Step-6: Design of Lead Compensator

i. Compute the zero of the lead compensator (lower corner frequency) from

$$\omega_{L1} = \frac{1}{\tau_1} = \frac{\omega_{gcn}}{\sqrt{\beta}}$$

ii. Compute the pole of the lead compensator (higher corner frequency) from

$$\omega_{H1} = \beta\omega_{L1} = \frac{\beta}{\tau_1}$$

The transfer function of lead compensator

$$G_{lead}(s) = \frac{s + \omega_{L1}}{s + \omega_{H1}}$$

Step-7: Design of Lag-Compensator:

i. Select the zero of the lag compensator (the higher corner frequency) to be located one decade below the new gain crossover frequency, that is

$$\omega_{H2} = \frac{1}{\tau_2} = \frac{\omega_{gcn}}{10}$$

ii. Select the pole of the lag compensator (the lower corner frequency)

$$\omega_{L2} = \frac{1}{\beta\tau_2} = \frac{\omega_{H2}}{\beta}$$

The transfer function of the lag compensator

$$G_{lag}(s) = \frac{s + \omega_{H2}}{s + \omega_{L2}}$$

Let us illustrate the design procedure through an example.

EXAMPLE 8.13

Problem

The open-loop transfer function of a unity feedback system is given by

$$G(s) = \frac{K}{s(s+8)(s+30)}$$

Design a lag-lead compensator to meet the following specifications.

a. $K_v = 10 \text{ sec}^{-1}$

b. Phase margin $= 55^0$

SOLUTION

Step-1: The bandwidth ω_b of the system with the desired settling time $t_s = 1$ sec and damping ratio $\xi = 0.5$ is 10.2rad/sec.

Step-2: The gain K which provides static velocity error constant $K_v = 20 \text{ sec}^{-1}$ is obtained from

$$K_v = \lim_{s\to 0} sG(s) = \lim_{s\to 0} s\frac{K}{s(s+8)(s+30)} = \frac{K}{240} = 10 \implies K = 2400$$

Step-3: The Bode plot of the uncompensated system for $K = 2400$ shown in Figure 8.57

Step-4: Let us select the new gain crossover frequency ω_{gcn} near the desired closed-loop bandwidth of the system. In this example select $\omega_{gcn} = 8$ rad/sec.

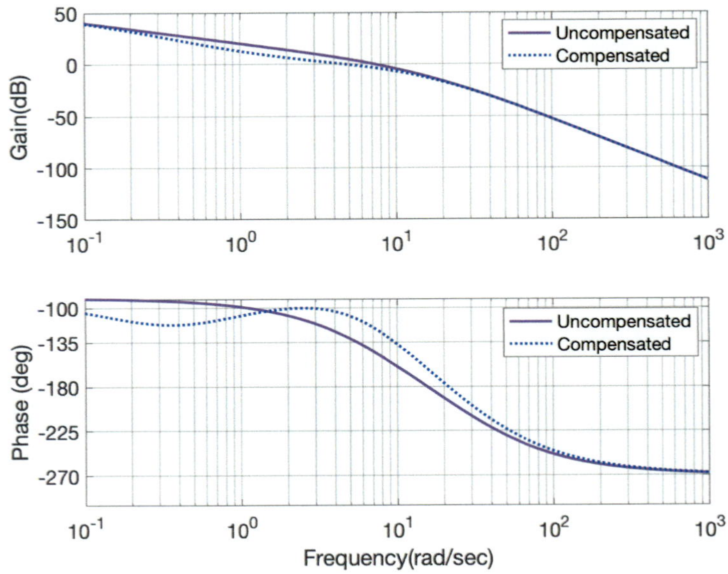

Figure 8.57: Bode plot of the system of Example 8.13 with and without the compensator

Step-5: Select the zero of the lag compensator (the higher corner frequency) to be one decade below the new gain crossover frequency. Thus

$$\omega_{\mathrm{H}} = \frac{1}{\tau_2} = \frac{\omega_{\mathrm{gcn}}}{10} = 0.8 \text{ rad/sec}$$

Step-6: We will determine β parameter as follows:

Find the phase margin at the new gain crossover frequency ϕ_{pmn} from Bode plot. This is equal to 30^0. Let us add 5^0 tolerance to the desired phase margin to compensate for the phase contributed by the lag compensator at $\omega = \omega_{\mathrm{gcn}}$. Thus, the additional phase lead required to meet the desired phase margin is

$$\phi_{\mathrm{ld}} = \phi_{\mathrm{d}} - \phi_{\mathrm{pmn}} + \epsilon = 55^0 - 30^0 + 5^0 = 30^0$$

Hence

$$\beta = \frac{1 + \sin\ \phi_{\mathrm{ld}}}{1 - \sin\ \phi_{\mathrm{ld}}} = 3$$

Step-7: Design of Lead Compensator:

The lower corner frequency of the lead compensator is computed from the relation $\omega_{\text{gcn}} = \dfrac{\sqrt{\beta}}{\tau_1}$ which gives

$$\omega_{\text{L}} = \frac{1}{\tau_1} = \frac{\omega_{\text{gcn}}}{\sqrt{\beta}} = 4.618$$

and the upper corner frequency of lead compensator $\omega_{\text{H}} = \dfrac{\beta}{\tau_1} = 13.85.$

Hence, the transmittance (transfer function) of the lead compensator is

$$\text{G}_{\text{lead}}(\text{s}) = \frac{\text{s} + 4.618}{\text{s} + 13.86}$$

Design of Lag compensator:

The upper corner frequency of lag compensator is

$$\omega_{\text{H}} = \frac{1}{\tau_2} = \frac{\omega_{\text{gcn}}}{10} = 0.8$$

The lower corner frequency of lag compensator is

$$\omega_{\text{L}} = \frac{1}{\beta\tau_2} = \frac{0.8}{3} = 0.2667$$

Hence, the transmittance (transfer function) of the lag compensator is

$$\text{G}_{\text{lag}}(\text{s}) = \frac{\text{s} + 0.8}{\text{s} + 0.2667}$$

Thus, the transfer function of the compensated system is

$$\text{G}_{\text{comp}}(\text{s}) = \left(\frac{\text{s} + 4.618}{\text{s} + 13.86}\right) \cdot \left(\frac{\text{s} + 0.8}{\text{s} + 0.2667}\right) \cdot \left(\frac{2400}{\text{s}(\text{s} + 8)(\text{s} + 30)}\right)$$

CONCEPT REINFORCEMENT PROBLEM 8.5

Problem

The open-loop transfer function of a unity feedback system is given by

$$\text{G}(\text{s}) = \frac{\text{K}(\text{s} + 6)}{\text{s}(\text{s} + 4)(\text{s} + 12)}$$

Design a lag-lead compensator to meet the following specifications.

a. $K_v \geq 12.5\ \text{sec}^{-1}$

b. Percentage overshoot=10% and

c. Settling time=0.5 second

8.10 Design of Compensators using Analytical Methods

In earlier sections, we discussed methods for designing lead, lag, and lag-lead compensators. Although computer programs were developed to assist in these designs, the approaches were primarily based on trial and error. In this section, we present a more rigorous and systematic analytical method for designing these compensators.

To avoid confusion with the notations used in the trial-and-error design methods in previous sections, we adopt a different set of notations throughout this section. These notations will streamline the development and analysis of the compensators, making it easier to follow the derivations and calculations

8.10.1 Analytical Design of Lead Compensator

Let the transfer function of the lead compensator be of the form

$$G_c(s) = \frac{1+\alpha_1 \tau s}{1+\tau s} = \frac{\alpha_1(s+a)}{s+\alpha_1 a}, \quad \text{where } a = \frac{1}{\alpha_1 \tau}, \quad \text{and } \alpha_1 > 1 \tag{8.61}$$

then it is possible to show the following: (left as an exercise)

i. The maximum phase lead provided by the compensator is

$$\phi_m = \sin^{-1}\frac{\alpha_1 - 1}{\alpha_1 + 1} \tag{8.62}$$

ii. For a specified phase lead ϕ_c and decibel gain M_c at a particular frequency ω_c, the parameters of the lead compensator are

$$a = \frac{\omega_c}{\alpha_1}\left(\frac{\alpha_1^2 - c}{c-1}\right)^{\frac{1}{2}} \quad \text{where } c = 10^{M_c/10} \tag{8.63}$$

and α_1 is the positive real solution of the quadratic equation

$$\left(p^2 - c + 1\right)\alpha_1^2 + 2p^2 c\alpha_1 + \left(p^2 c + c - 1\right)c = 0 \tag{8.64}$$

where $p = \tan\phi_c$. Note that for the design of the lead compensator, the following conditions must be satisfied.

i. $p > 0$ which $\Longrightarrow$ $\phi_c > 0$
ii. $c > 1$ which $\Longrightarrow$ $M_c > 0$ and
iii. $c > p^2 + 1$

The design steps for a lead compensator are:

Analytical Design Procedure for Lead Compensator

Step-1: Select the desired gain crossover frequency of compensated system ω_c.

Step-2: From the desired phase margin ϕ_c, compute $p = \tan\phi_c$.

Step-3: Verify that $\phi_c > 0$ and gain $M_c > 0$, where M_c is the dB gain at ω_c.

Step-4: Determine whether a single stage will be sufficient by testing $c > p^2 + 1$.

Step-5: Determine α_1 from Eqn(8.64) and "a" from Eqn(8.63).

EXAMPLE 8.14

Problem

Determine the transfer function of a lead compensator that will provide a phase lead of 35^0 and gain 7dB at 3 rad/sec.

SOLUTION:

Given $\phi_c = 35^0$, $\omega_c = 3$ rad/sec and $M_c = 7$ dB. Hence we have

$$p = \tan(35^0) = 0.7002; \quad c = 10^{7/10} = 5.0119$$

The positive real solution of α_1 is computed from solving the equation

$$\begin{aligned}&\left(p^2 - c + 1\right)\alpha_1^2 + 2p^2c\alpha_1 + \left(p^2c + c - 1\right)c\\ &= -3.5216\alpha_1^2 + 4.915\alpha_1 + 32.4255 = 0\end{aligned}$$

This gives $\alpha_1 = 3.8112$. Hence

$$a = \frac{\omega_c}{\alpha_1}\left(\frac{\alpha_1^2 - c}{c - 1}\right)^{\frac{1}{2}} = 1.2121$$

Hence, the transfer function of the lead compensator is given by

$$G_c(s) = \frac{\alpha_1(s + a)}{s + \alpha_1 a} = \frac{3.8112(s + 1.2121)}{s + 4.62}$$

Analytical Design of Lag Compensator

Consider the lag compensator of the form

$$G_c(s) = \frac{1}{\alpha_1}\frac{(s+\alpha_1 a)}{s+a} \quad \alpha_1 > 1 \tag{8.65}$$

$$= \frac{\beta_1(s+b)}{s+\beta_1 b}, \quad \beta < 1 \tag{8.66}$$

where $\beta_1 = 1/\alpha_1$, $b = \alpha_1 a$ and $\beta_1 b = a = b/\alpha_1$.

Suppose the specified phase shift is ϕ_c and the decibel gain at the specified frequency ω_c is M_c, then it can be shown that (left as an exercise)

$$b = \frac{\omega_c}{\beta_1}\left(\frac{\beta_1^2 - c}{c-1}\right)^{\frac{1}{2}} \quad \text{where} \quad c = 10^{M_c/10} \tag{8.67}$$

and β_1 is the positive real solution of the quadratic equation

$$\left(p^2 - c + 1\right)\beta_1^2 + 2p^2 c\beta_1 + \left(p^2 c + c - 1\right)c = 0 \tag{8.68}$$

where $p = -\tan\phi_c$. Note that to design a lag compensator, the following conditions must be met.

i. $p < 0$ which $\implies$ $\phi_c < 0$(phase lag)
ii. $0 < c < 1$ which $\implies$ $M_c < 0$ (attenuation) and
iii. $p^2c + c - 1 < 0 \implies c < \frac{1}{1+p^2}$

The design steps for a lag compensator are:

[Analytical Design Procedure for Lag Compensator

Step-1: Select the desired gain crossover frequency of compensated system ω_c.

Step-2: From the desired phase margin ϕ_c, compute $p = -\tan\phi_c$.

Step-3: Verify that the phase lag is applicable: $p = -\tan\phi_c < 0$ and gain $M_c < 0$, where M_c is the dB gain at ω_c.

Step-4: Determine whether a single stage will be sufficient by testing $c < \frac{1}{p^2+1}$.

Step-5: Determine β_1 from Eqn(8.68) and "b" from Eqn(8.67).

EXAMPLE 8.15

Problem

Determine the transfer function of a lag compensator that will provide a phase lag of 45^0 and an attenuation of 18dB at $\omega = 5$ rad/sec.

SOLUTION:

Given $\phi_c = 45^0$, $\omega_c = 5$ rad/sec and $M_c = 18$ dB. Hence we have

$$p = -\tan(45^0) = -1; \quad c = 10^{-18/10} = 0.0158$$

The positive real solution of β_1 is computed from solving the equation

$$\begin{aligned}&\left(p^2 - c + 1\right)\beta_1^2 + 2p^2 c\beta_1 + \left(p^2 c + c - 1\right)c \\ &= 1.9842\beta_1^2 + 0.0317\beta_1 - 0.0153 = 0\end{aligned}$$

This gives $\beta_1 = 0.0803$. Hence

$$b = \frac{\omega_c}{\beta_1}\left(\frac{\beta_1^2 - c}{c - 1}\right)^{\frac{1}{2}} = 6.083$$

Hence, the transfer function of the lag compensator is given by

$$G_c(s) = \frac{\beta_1(s + b)}{s + \beta_1 b} = \frac{0.0803(s + 6.083)}{s + 0.4886}$$

Analytical Design of Lag-Lead Compensator

Consider the lag-lead compensator of the form

$$G_c(s) = \frac{\alpha_1(s + a)}{s + \alpha_1 a} \cdot \frac{\beta_1(s + b)}{s + \beta_1 b} \quad \text{where } \alpha_1 = \frac{1}{\beta_1} \text{ and } \alpha_1 > 1 \tag{8.69}$$

$$= \frac{(s + a)(s + b)}{(s + \alpha_1 a)\left(s + \dfrac{b}{\alpha_1}\right)} \tag{8.70}$$

Suppose the specified phase lead is ϕ_c and the gain (attenuation) at the specified frequency ω_c is M_c. Assume that the lag portion of the compensator results in a small phase lag ϕ_2 at ω_c. Then it can be shown that

$$\boxed{a = \omega_c\left(\frac{1 - c}{\alpha_1^2 c - 1}\right)^{\frac{1}{2}} \quad \text{where } c = 10^{M_c/10}} \tag{8.71}$$

where α_1 is the positive real solution of the quadratic equation

$$\left(\mathrm{p}^2\mathrm{c}+\mathrm{c}-1\right)\mathrm{c}\alpha_1^2+2\mathrm{p}^2\mathrm{c}\alpha_1+\left(\mathrm{p}^2-\mathrm{c}+1\right)\mathrm{c}=0 \tag{8.72}$$

where $\mathrm{p}=\tan(\phi_c+\phi_2)$.

The parameter b is the smaller solution of the quadratic equation

$$\left(\frac{\mathrm{b}}{\omega_c}\right)^2+\frac{\alpha_1-1}{\tan(-\phi_2)}\left(\frac{\mathrm{b}}{\omega_c}\right)+\alpha_1=0 \tag{8.73}$$

Note that for the design of a lag-lead compensator, the following conditions must be met.

i. $\mathrm{p}>0$

ii. $0<\mathrm{c}<1$ which $\implies$ $M_c<0$ (attenuation) and

iii. $(\mathrm{p}^2\mathrm{c}+\mathrm{c}-1)\mathrm{c}<0 \implies \mathrm{c}<\dfrac{1}{1+\mathrm{p}^2}$

The design steps for a lag-lead compensator are:

Analytical Design Procedure for Lag-Lead Compensator

Step-1: Select the desired gain crossover frequency of compensated system ω_c.

Step-2: Let ϕ_2 is the phase lag contribution of lag part ω_c. Usually very small (say 2^0).

Step-3: From the desired phase margin ϕ_c, and assumed value of ϕ_2, compute $\mathrm{p}=\tan(\phi_c+\phi_2)$.

Step-4: Determine α_1 from Eqn(8.72).

Step-5: Compute parameter "a" from Eqn(8.71) and parameter "b" from Eqn(8.73).

EXAMPLE 8.16

Problem

Determine the transfer function of a lag-lead compensator that will provide a phase lag of 50^0 and an attenuation of 20dB at $\omega=8$ rad/sec.

SOLUTION:

Given $\phi_c = 50^0$, $\omega_c = 8$ rad/sec and $M_c = 20$ dB. Let the contribution of the lag part at $\omega_c = 8$ be given by $\phi_2 = 2^0$. Hence we have

$$p = \tan(50^0 + 2^0) = 1.28; \quad c = 10^{-20/10} = 0.01$$

The positive real solution of α_1 is computed from solving the equation

$$\begin{aligned}&\left(p^2 - c + 1\right)\alpha_1^2 + 2p^2c\alpha_1 + \left(p^2c + c - 1\right)c\\ &= -0.0097\alpha_1^2 + 0.0328\alpha_1 + 2.6283 = 0\end{aligned}$$

This gives $\alpha_1 = 18.1986$. Hence

$$a = \omega_c\left(\frac{1-c}{\alpha_1^2c - 1}\right)^{\frac{1}{2}} = 5.2351 \quad \text{and} \quad \frac{\alpha_1 - 1}{\tan(-\phi_2)} = -492.50$$

This gives the quadratic equation

$$\left(\frac{b}{\omega_c}\right)^2 - 492.5\left(\frac{b}{\omega_c}\right) + 18.1986 = 0$$

Solving the above and using the smaller root gives

$$\frac{b}{\omega_c} = 0.0370 \quad \Longrightarrow \quad b = 0.2956$$

Hence, the transfer function of the lag-lead compensator is given by

$$\begin{aligned}G_c(s) &= \frac{(s+a)(s+b)}{(s+\alpha_1 a)(s+\beta_1 b)} = \frac{(s+5.2351)(s+0.2956)}{(s+95.27)(s+0.0162)}\\ &= \frac{s^2 + 5.531s + 1.548}{s^2 + 95.29s + 1.548}\end{aligned}$$

8.11 Feedback Compensation

So far, we have focused on the design of cascade compensators, where the compensator is placed in series with the plant to reshape the root locus and achieve desired performance objectives. An alternative approach involves placing the compensator in the feedback path to modify the root locus. Although the design of feedback compensators may initially appear more complex, it can often be simplified by reformulating the problem as an equivalent series compensation task. In the following section, we demonstrate how a feedback compensation problem can be transformed into an equivalent series compensation problem, simplifying the design process.

Principle for Designing Feedback Compensated System

The generic structures of series and feedback compensator are shown in Figures 8.58a and 8.58b, respectively.

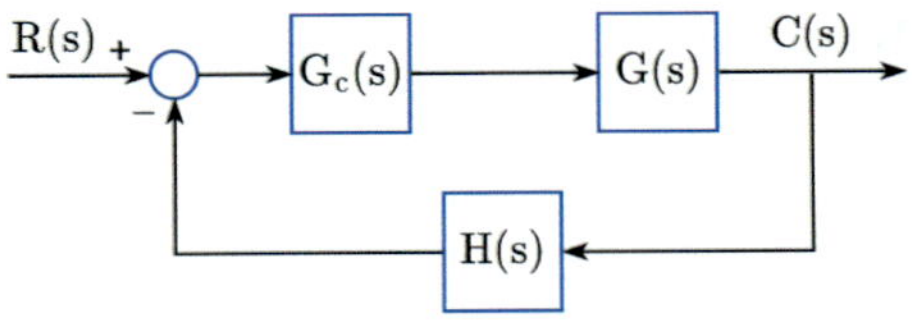

(a) Series Compensator

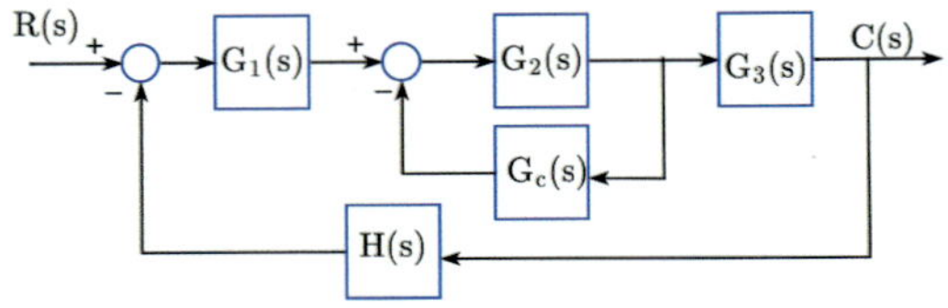

(b) Feedback Compensator

Figure 8.58: Generic structures of series and feedback compensators.

The transfer function for the series compensated system is

$$\frac{C(s)}{R(s)} = \frac{G_c(s)G(s)}{1 + G_c(s)G(s)H(s)}$$

The characteristic equation is

$$1 + G_c(s)G(s)H(s) = 0$$

From the given G(s) and H(s), the series compensator $G_c(s)$ is designed to meet the given specifications.

Now consider the feedback compensated system. The closed-loop transfer is

$$\frac{C(s)}{R(s)} = \frac{G_1G_2G_3}{1 + G_2G_c + G_1G_2G_3GH}$$

The characteristic equation is

$$1 + G_2G_c + G_1G_2G_3H = 0$$

Divide this equation by the sum of the terms that do not contain G_c. This gives

$$1 + \frac{G_2G_c}{1 + G_1G_2G_3H} = 0 \tag{8.74}$$

If we define

$$G_f = \frac{G_2}{1 + G_1G_2G_3H}$$

then Eqn(8.74) becomes

$$1 + G_cG_f = 0$$

Since G_f is a fixed transfer function, the design of G_c can be carried out following the design approach of series compensation.

8.11.1 Methods of Feedback Compensation

The feedback compensation can be carried out following two approaches.

Approach-1: Equivalent Series Compensation:

Consider a typical feedback system, where G(s) represents the forward path transfer function and H(s) represents the feedback path transfer function.

i. Plot the root locus of open-loop transfer function G(s)H(s)
ii. Identify the location of dominant poles that correspond to the desired transient performance specifications.
iii. Design a compensator using the same methodology applied in cascade compensation.

It is important to note that, in this case, the compensator's poles and zeros are introduced through the feedback path H(s), in contrast to cascade compensation where they are added to the forward path G(s).

Compensator Design for Rate Feedback System

Most commonly found example of a feedback compensated system is the velocity feedback (rate feedback system). The structure of such a system is shown in Figure 8.59a. It consists of two loops. The outer loop (also called as major loop) is unity feedback, while the inner loop (also called the minor loop) provides the rate of change of output. A popular feedback compensator is a rate sensor that acts as a differentiator.

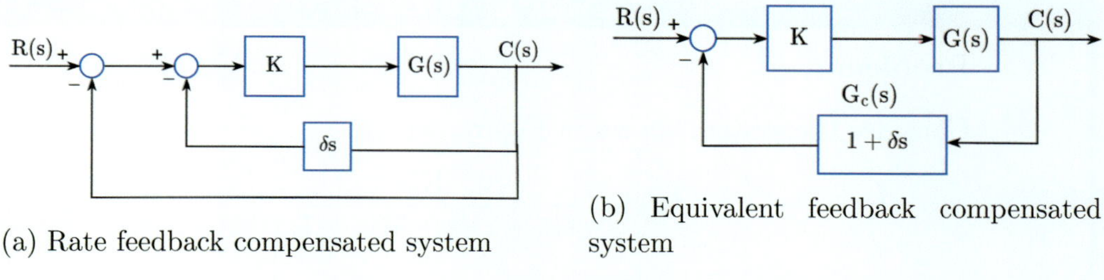

(a) Rate feedback compensated system

(b) Equivalent feedback compensated system

Figure 8.59: Structure of rate feedback system

The objective is to design a feedback compensator by determining the value of K and δ to meet the design goals.

The system of Figure 8.59a can be represented by Figure 8.59b, which combines the minor and major feedback loops into a single feedback path. The loop gain from Figure 8.59b is

$$\mathrm{KG(s)G_c(s)} = \mathrm{KG(s)}\,[1+\delta \mathrm{s}] = \mathrm{K}\delta\left(\mathrm{s}+\frac{1}{\delta}\right)\mathrm{G(s)} \tag{8.75}$$

Hence, placing the rate sensor with transmittance δs in the minor loop of Figure 8.59a, is equivalent to placing a compensator with transmittance

$$G_c(s) = (1 + \delta s) = \delta \left(s + \frac{1}{\delta} \right) \tag{8.76}$$

From Eqn(8.76), it is seen that the output from the feedback compensator $G_c(s)$ consists of two components: one is proportional to its input and the other is the derivative of its input. This type of compensator, as has been discussed before, is essentially a PD compensator. Its design here involves adding a zero at $s = -1/\delta$ to the existing open-loop poles and zeros of the uncompensated system G(s). The parameter δ is to be selected such that the root locus is reshaped to pass through the desired dominant roots. The choice of both δ as well as the gain K yields the desired response. The characteristics of the rate compensator are summarized as follows:

1. The addition of one open-loop zero decreases the number of asymptotes by 1. The angles of the asymptotes increase, thereby improving the relative stability.
2. The steady state error may improve depending on the value of the gain K.

The procedure of designing a rate feedback compensator will be illustrated through an example.

EXAMPLE 8.17

Problem

Consider the system shown in Figure, where

$$G(s) = \frac{K}{s(s+2)(s+5)(s+7)}$$

Design rate feedback compensation, as shown in Figure 8.59, to ensure that the settling time is 4 seconds while it operates at 20% overshoot.

SOLUTION:

The desired peak overshoot of 20% translates to $\xi = 0.4559$. Let us first design PD compensator.

Design of PD Compensator

The root locus of the uncompensated system is shown in Figure 8.60. The dominant poles are found at $-0.5471 + j1.2123$ at $K = 76.26$ from the intersection of constant $\xi \approx 0.46$ line with root locus, as shown in Figure 8.60

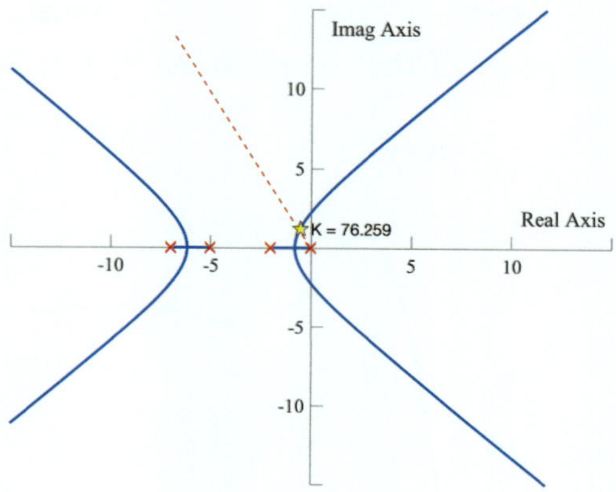

Figure 8.60: Root locus of uncompensated system

Dominant Poles of Compensated System

Since the desired settling time Ts = 4 seconds, it translates to the desired pole at

$$s_d = -1 + j1.95$$

The PD compensator zero is calculated from Eqn(8.31) which gives $z_c = 3.019$. Thus the PD compensator

$$G_c(s) = s + 3.019$$

- The root locus of the equivalent compensated system is shown in Figure 8.61

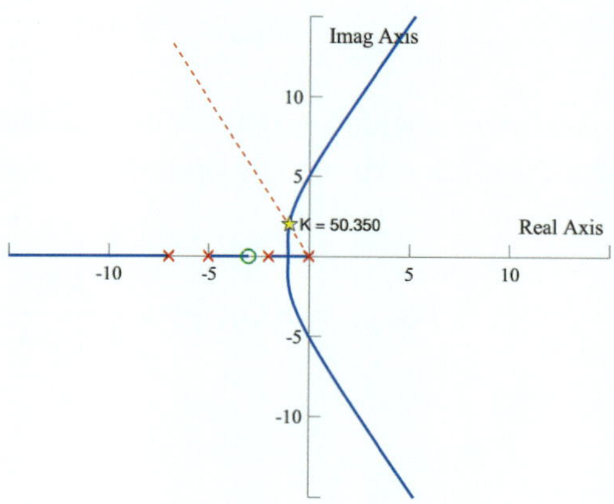

Figure 8.61: Root locus of the compensated system

The gain at the design point, which is $K\delta$, from Figure 8.61, is found to be 50.35. Since δ is reciprocal of compensator zero z_c, this gives $K = 152$.

The step response of the compensated system is shown in Figure 8.62.

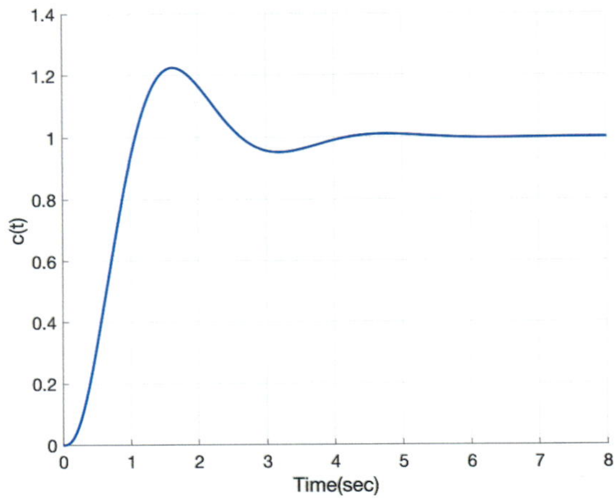

Figure 8.62: Step response of compensated system

The MATLAB code is in the file `Example8_17.m`

Minor Loop Feedback Compensation

When the rate feedback compensation system includes both minor and major feedback loops, as illustrated in the Figure 8.63, we adopt a different strategy known as minor loop feedback compensation. Consider the minor loop of the system with the open-loop transfer function

$$L_{minor} = \delta sG(s)$$

Identify the parameter δ which satisfies the transient response specifications of the minor loop. Next, determine the open-loop transfer function of the major loop as:

$$L_{major} = KG_m(s) = \frac{KG(s)}{1 + \delta sG(s)}$$

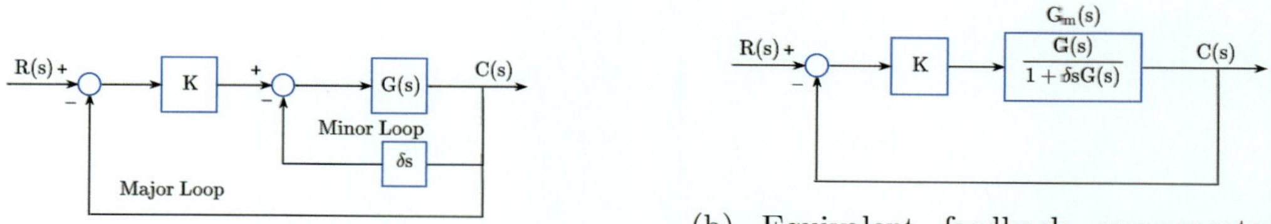

(a) Rate feedback compensated system

(b) Equivalent feedback compensated system

Figure 8.63: Structure of rate feedback system

Approach-2: Minor Loop

i. Plot the root locus of the minor loop, treating δ as the variable parameter.

ii. Determine the value of δ that meets the desired transient response specification for the minor loop.

iii. Next, plot the root locus of the **major loop** using the loop transfer function $KG_m(s)$ with K as the variable parameter.

iv. Identify the gain K that satisfies the transient response specification of the major loop.

EXAMPLE 8.18

Problem

Consider the system shown in Figure 8.63, where

$$G(s) = \frac{1}{s(s+5)(s+7)}$$

Design minor loop rate feedback compensation to yield a damping of 0.707 for the minor loop's dominant poles and a damping ratio of 0.5 for the closed-loop system's dominant poles.

SOLUTION:

Step-1: Minor Loop Compensation

Let us consider the minor loop. The open-loop transfer function of the minor loop is

$$G(s)\delta s = \frac{\delta s}{s(s+5)(s+7)}$$

The root locus of the minor loop with δ as the variable parameter is shown in Figure 8.64.

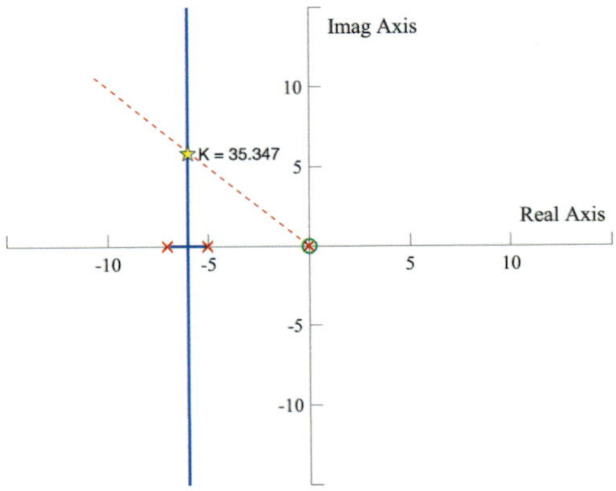

Figure 8.64: Root locus for minor loop

It is found that the dominant root, when searched on $\xi = 0.707$ line, is at $-6 + j5.86$ with $\delta = 35.34$. This places the minor loop poles at $-6 + j5.86$ and meets the minor loop design specification ($\xi = 0.707$).

Step-2: Major Loop Compensation

The open-loop transfer function of the major loop is

$$G_{\text{major}}(s) = \frac{KG(s)}{1 + G(s)\delta s} = \frac{K}{s\left[s^2 + 12s + (35 + \delta s)\right]}$$
$$= \frac{K}{s(s^2 + 12s + 66)}$$

The root locus of the major loop is shown in Figure 8.65. Searching the dominant root along $\xi = 0.5$ dominant poles, the complex conjugate roots (closed-loop poles) are found at $-2.71 + j5.22$ at $K = 227.85$.

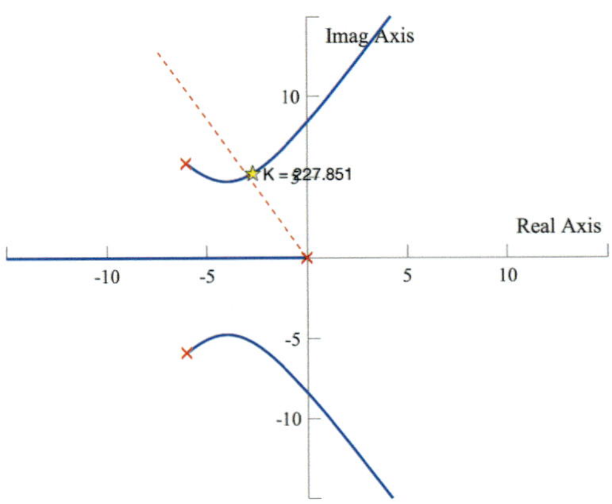

Figure 8.65: Root locus for closed-loop system

The step response is shown in Figure 8.66.

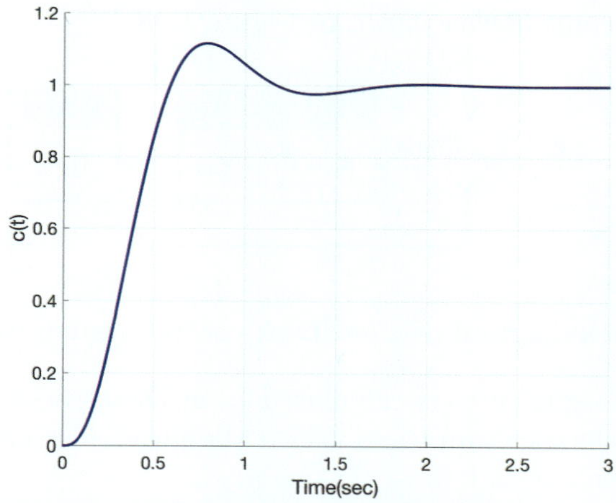

Figure 8.66: Step response of the system

The MATLAB code can be found in `Example8_18.m`.

CONCEPT REINFORCEMENT PROBLEM 8.6

Problem

For the feedback compensation system shown in Figure 8.67 do the following:

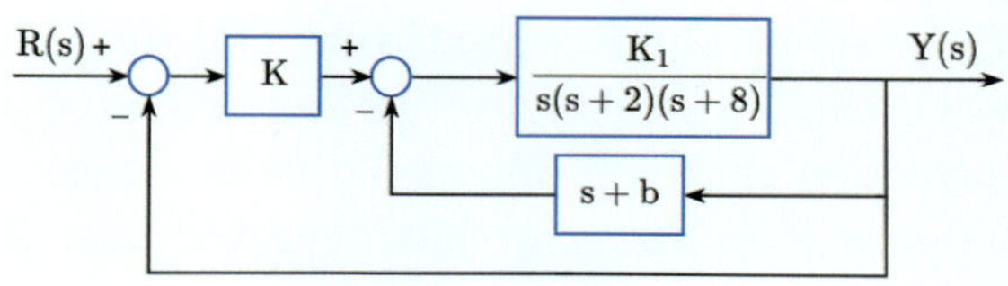

Figure 8.67: Rate feedback system

a. Find the value of K_1 and b in the minor feedback loop such that the settling time is 1.5 seconds with 10% peak overshoot in the step response.
b. Find the value of K so that the response of the major loop exhibits an overshoot of 16.5%

8.12 Design of PID Controller: Basics and Tuning Rules

A Proportional–Integral–Derivative (PID) controller is one of the most commonly used controllers in industry due to its simplicity, robustness, and effectiveness across a wide range of systems. The controller attempts to minimize the error between a desired setpoint and the actual output of a system by adjusting the control input. The block diagram of a feedback control system with a PID controller is shown in Figure 8.68.

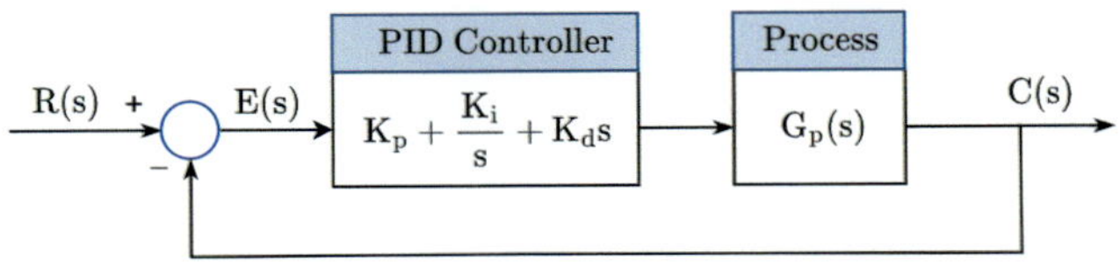

Figure 8.68: Block diagram of a feedback control system using a PID controller

The control signal u(t) is calculated from the error $e(t) = r(t) - c(t)$, where r(t) is the reference input and c(t) is the measured output. This is expressed as:

$$u(t) = K_p e(t) + K_i \int_0^t e(\tau)d\tau + K_d \frac{de(t)}{dt}$$

where K_p is the proportional gain, K_i is the integral gain, and K_d is the derivative gain. Before we explain the design procedure of a PID controller, let us first understand its operation intuitively.

A PID controller determines the control input by evaluating three aspects of the error signal: the difference between the desired output and the actual output of the system. The **Proportional (P)** term considers how much error exists between the desired and actual outputs at the current moment, producing a correction that is proportional to this error. The **Integral (I)** term accounts for how long the error has persisted over time, by integrating the error and applying a correction to eliminate steady-state error. The **Derivative (D)** term anticipates future behavior by considering how the error is changing by measuring its rate of change, thus helping to reduce overshoot and improve system stability by adding a damping effect. Together, these three actions form a control law that responds not only to the present error but also to its history and predicted future behavior, resulting in improved performance for a wide range of dynamic systems. The operation of a PID controller can be explained in an intuitive manner and is summarized below.

8.12.1 Intuitive Understanding of PID Controller

1. **Proportional(P) Control**- "How big is the error?"
   - The **P** term responds to the **present** error.
   - If the error is large, the control effort is large.

- Control signal:

$$u_P(t) = K_p \cdot e(t)$$

- Effect: Reduces rise time, but alone cannot eliminate steady-state error.

2. **Integral(I) Control**-"How long has the error persisted?"

   - The **I** term responds to the accumulated error over time.
   - It integrates the error, so even small errors that persist will cause a growing control signal.
   - Control signal

$$u_I(t) = K_i \int_0^t e(\tau)d\tau$$

   - Effect: Eliminates steady-state error but can cause overshoot and slower response if overused.

3. **Derivative (D) Control**-"How fast is the error changing?"

   - The **D** term predicts future behavior of the error based on its rate of change.
   - It dampens the system response.
   - Control signal:

$$u_D(t) = K_d \cdot \frac{de(t)}{dt}$$

   - Effect: Improves stability and reduces overshoot, especially helpful for systems with inertia.

To reinforce the understanding, let us consider the following real-life analogy.

Understanding PID Controller

Analogy: Think of a driver trying to keep a car in the center of the lane:

- **P:** Reacts to how far off-center the car is.
- **I:** Remembers how long the car has been drifting off course and accumulates corrections.

- **D:** Anticipates where the car is heading by observing how fast the drift is changing, and applies brakes if needed.

8.12.2 Tuning a PID Controller: Common PID Tuning Rules

Before delving into PID controller tuning, it's essential to understand why tuning is both necessary and valuable. While the basic structure of a PID controller is simple and intuitive, with proportional, integral, and derivative terms working together to reduce error and improve response, its performance heavily depends on the appropriate choice of its three parameters: K_p, K_i, and K_d. Poorly tuned controllers can lead to oscillations, instability, slow responses, or excessive overshoot, making the system inefficient or even unsafe. Since real-world systems vary widely in their dynamics and disturbances, tuning allows us to tailor the controller to achieve desired performance in terms of speed, accuracy, and robustness. Various methods exist for this purpose, ranging from empirical techniques like the Ziegler–Nichols rules to automated tools like MATLAB's PID Tuner, each offering different levels of precision, convenience, and adaptability depending on the application.

There are several heuristic and analytical approaches to tune PID parameters (K_p, K_i, K_d). Each method provides starting values, which are then fine-tuned based on system performance (rise time, overshoot, steady-state error). In the following, we summarize the tuning procedure for three different methods:

1. **Ziegler–Nichols Tuning Based on the S-Shaped Step Response (Open-Loop Method)**

This method is based on the open-loop response of the system to a step input, particularly when the system exhibits an S-shaped curve shown in Figure 8.69

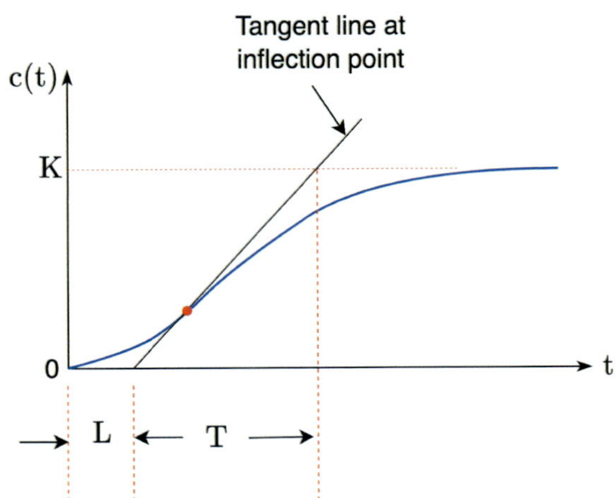

Figure 8.69: S-shaped response curve

Zeigler-Nichols Tuning of PID Tuning (Open-Loop) Method

- Apply a step input to the process and record the open-loop response.
- Approximate the response curve with a tangent at the inflection point to identify: (see Figure 8.69).
  a. L: Delay time (time between step input and the tangent line's intersection with the time axis;
  b. T: Time constant (time between the tangent line's intersection with the time axis and the point where it meets the final value).
- Use standard Ziegler–Nichols tuning formulas, shown in Table 8.2, to compute PID parameters:

Table 8.2: Ziegler-Nichols PID tuning using reaction curve (based on step-response of plant)

Controller Type	K_p	K_i	K_d
P: K_p	$\frac{T}{L}$	–	–
PI: $K_p + \frac{K_i}{s}$	$0.9\frac{T}{L}$	$0.27\frac{T}{L^2}$	–
PID: $K_p + \frac{K_i}{s} + K_d s$	$1.2\frac{T}{L}$	$0.6\frac{T}{L^2}$	0.5L

2. **Ziegler–Nichols Closed-Loop Tuning**

This method is used when the process model is unknown or difficult to obtain. It provides initial tuning values based on system response at the edge of stability.

Zeigler-Nichols Tuning of PID Tuning(Closed-Loop)Method

Step-1: Set the controller to P-only mode

- Turn off integral **(I)** and derivative **(D)** actions:

$$u(t) = K_p e(t)$$

Step-2: Increase the proportional gain K_p

- Gradually increase K_p until the output of the closed-loop system starts to oscillate with constant amplitude.
- This value is known as the ultimate gain K_u

Step-3: Measure the ultimate period T_u

- Once the output enters sustained oscillations (neither growing nor decaying), measure the time period T_u of one full cycle of the output oscillation.

Step-4: Compute the PID gains using the Ziegler–Nichols formulas, shown in Table 8.3

Note: If the model of the system is known, then the ultimate gain K_u and ultimate period T_u can be determined by various methods, such as using Routh's table and Bode plot. The following MATLAB commands extract the critical parameters directly from the frequency response analysis:

```
[Gm, Pm, Wcg, Wcp] = margin(G_openloop);
K_ultimate = Gm;          % Ultimate gain
omega_ultimate = Wcg;     % Ultimate frequency
Tu=2*pi/omega_ultimate;   % Ultimate period
```

Table 8.3: Ziegler-Nichols PID tuning using ultimate gain K_u and ultimate period T_u

Controller Type	K_p	K_i	K_d
P: K_p	$0.5K_u$	–	–
PI: $K_p + \frac{K_i}{s}$	$0.45K_u$	$1.2K_p/T_u = \frac{0.54K_u}{T_u}$	–
PID: $K_p + \frac{K_i}{s} + K_d s$	$0.6K_u$	$2K_p/T_u = \frac{1.2K_u}{T_u}$	$K_pT_u/8 = 0.6\frac{K_uT_u}{8}$

3. **MATLAB-Based PID Tuning**

Modern control system design has been significantly enhanced by computational tools, with MATLAB providing comprehensive functions for automated PID controller tuning. These tools combine classical control theory with optimization algorithms to achieve superior performance compared to traditional manual tuning methods.

Automated PID Design Functions

MATLAB offers two primary approaches for PID controller design: the `pidtune()` for algorithmic tuning and the `pidTuner` app for interactive design. The `pidtune()` function employs loop-shaping techniques to achieve specified performance objectives, while maintaining adequate stability margins. The basic syntax requires only the plant transfer function and controller type.

Performance-Based Tuning Options

The strength of MATLAB's approach lies in its ability to tune controllers based on specific performance criteria rather than empirical rules. The `pidtuneOptions()` function allows engineers to specify design requirements such as phase margin, crossover frequency, and bandwidth:

This approach enables three distinct tuning philosophies: **robust tuning** (high phase margins, conservative response), **balanced tuning** (moderate performance with good stability), and **aggressive tuning** (fast response with reduced margins). The selection depends on system requirements and operating conditions. Here are the specific MATLAB commands for each tuning type:

1. **Balanced (Default):**

```
C = pidtune(plant, 'PID');
```

2. **Aggressive:**

```
opt_agg = pidtuneOptions('PhaseMargin', 45);
C_aggressive = pidtune(G_openloop, 'PID', opt_agg);
```

3. **Robust:**

```
opt_rob = pidtuneOptions('PhaseMargin', 75,'CrossoverFrequency',5);
C_robust = pidtune(G_openloop, 'PID', opt_rob);
```

Two key parameters govern the tuning behavior: phase margin and crossover frequency. The phase margin setting creates a trade-off between aggressiveness and robustness, with lower values (45°) favoring speed over stability margins and higher values (75°) prioritizing robust operation. The crossover frequency specification similarly balances response speed against conservative control, where higher frequencies accelerate system response while lower frequencies promote stability and reduce control effort.

Note

- `pidtune()` assumes unity feedback.
- It designs a PID controller to go in series with the plant. Hence, the open-loop transfer function in the above code is actually the forward path transfer function.

EXAMPLE 8.19

Problem:

The block diagram of an AVR system, controlled by a PID controller, is shown in Figure 8.70.

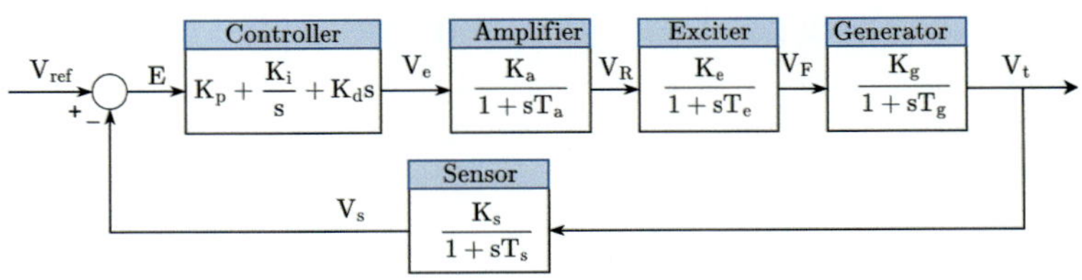

Figure 8.70: Block diagram of AVR with PID controller

The various parameters are: $K_a = K_e = K_g = K_s = 1$. The time constants of amplifier, exciter, generator, and sensor are: $T_a = 0.1$, $T_e = 0.4$, $T_g = 1.0$, and $T_s = 0.05$.

Compute the critical gain and critical period and tune the PID gains using Zeigler-Nichols tuning rules given in Table 8.3.

Use MATLAB's `pidtune()` function to design PID controllers with three different tuning approaches: the phase margin in aggressive tuning is 45^0 and in robust tuning is 75^0. The crossover frequency is 10rad/sec in aggressive and 2 rad/sec in robust tuning.

SOLUTION:

The ultimate gain and the corresponding oscillation frequency are determined from constructing Routh's table for the characteristic polynomial

Table 8.4: Completed Routh Table

s^4	1	307.5	$500 + 500K_a$
s^3	33.5	775	0
s^2	284.36	$500 + 500K_a$	
s^1	$58.9K_a - 716.1$	0	0
s^0	$500 + 500K_a$		

For marginal stability, the s^1 row entry must equal zero: This gives $K_a = 716.15/58.91 = 12.16$. This is the critical gain. For the system to be stable

$$K_{cu} = K_a < 12.16$$

To determine the critical frequency,the auxiliary equation from s^2 row is

$$284.365s^2 + 6,580 = 0 \implies s^2 = -23.15$$

This gives $\omega_c = \sqrt{23.15} = 4.81$ rad/sec

The ultimate period $T_u = \omega_c/2\pi = 1.305$.

The gains of PID controller are

$$K_p = 0.6K_u = 7.3 \quad K_i = \frac{1.2K_u}{T_u} = 11.18 \text{ and } K_d = \frac{0.6K_uT_u}{8} = 1.19$$

Here's a table with various tuning parameters for the AVR PID controller:

Tuning Method	K_p	K_i	K_d	**Characteristics**
Ziegler-Nichols	7.29	11.16	1.191	–
Balanced	2.175	2.08	0.551	Default performance
Aggressive	14.0	8.5	5.77	Fast response, less robust
Robust	2.88	2.319	0.793	High stability, slower response

The comparative performance of various PID controllers to a step reference is shown in Figure 8.71.

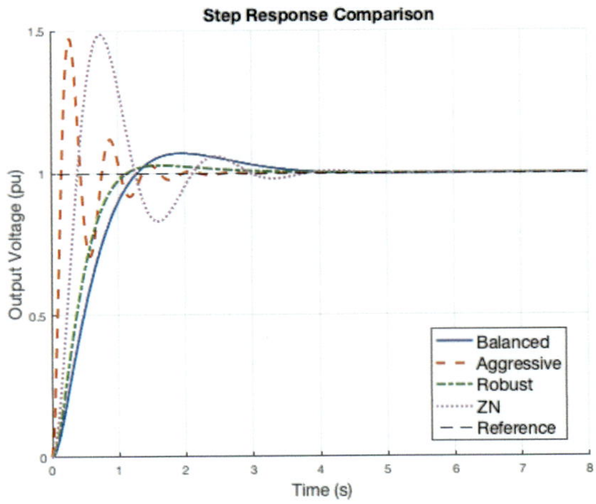

Figure 8.71: Step response comparison with different PID controller

The MATLAB code for implementing this is in `Example8_19.m`.

8.13 Multiple Choice Skills Test Questions

M 8.1. The transfer of a passive RC circuit, acting as a controller, is:

$$G_c(s) = \frac{s + z_1}{s + p_1}$$

This circuit will behave as a phase lead controller if

A. $p_1 < z_1$ **B.** $p_1 = 0$ **C.** $p_1 > z_1$ **D.** $p_1 = z_1$

M 8.2. The gain crossover frequency and bandwidth of a control system are ω_{gcuc} and ω_{bwuc}, respectively. The system is compensated by a cascade phase lead network. The gain crossover frequency and bandwidth of compensated system are ω_{gcc} and ω_{bwc}, respectively, then

A. $\omega_{gcc} < \omega_{gcuc}; \omega_{bwc} < \omega_{bwuc}$

B. $\omega_{gcc} > \omega_{gcuc}; \omega_{bwc} > \omega_{bwuc}$

C. $\omega_{gcc} > \omega_{gcuc}; \omega_{bwc} < \omega_{bwuc}$

D. $\omega_{gcc} < \omega_{gcuc}; \omega_{bwc} > \omega_{bwuc}$

M 8.3. The maximum phase shift from a lead compensator with transfer function

$$G_c(s) = \frac{1 + 0.12s}{1 + 0.04s} \text{ is equal to}$$

A. 60^0 **B.** 45^0 **C.** 15^0 **D.** 30^0

M 8.4. A system is compensated by a PD controller. Compared to the uncompensated system, the compensated system has

A. higher noise amplification

B. a higher type number

C. reduced damping

D. larger overshoot

M 8.5. A phase lead compensator is designed to improve the phase margin of a system. At what frequency does it contribute its maximum phase lead?

A. At its zero frequency

B. At its pole frequency

C. At the geometric mean of its zero and pole frequencies

D. At the gain crossover frequency

M 8.6. When a lag compensator is added to a stable system, which of the following typically occurs?

A. The gain crossover frequency shifts to a higher frequency

B. The system becomes unstable

C. The gain crossover frequency shifts to a lower frequency

D. The phase margin increases significantly

M 8.7. In the design of a PID controller $\mathrm{G_c(s) = K_p + \frac{K_i}{s} + K_d s}$ Which term is mainly responsible for reducing the steady-state error to zero for a step input?

A. Proportional gain $\mathrm{K_p}$

B. Integral gain $\mathrm{K_i}$

C. Derivative gain $\mathrm{K_d}$

D. All three terms equally affect steady-state error

M 8.8. A lead compensator typically results in which combination of effects on a system?

A. Higher bandwidth and improved phase margin

B. Lower bandwidth and improved steady-state error

C. Lower bandwidth and improved phase margin

D. Higher bandwidth and worse phase margin

M 8.9. In frequency domain design, the zero of a PI compensator is often placed:

A. At the origin

B. At the gain crossover frequency

C. A decade below the gain crossover frequency

D. At a frequency higher than the dominant poles

M 8.10. Which of the following best describes the role of the derivative action in a PD controller?

A. It increases the steady-state error

B. It amplifies low-frequency signals

C. It adds phase lag near crossover frequency

D. It anticipates the error and adds damping to the response

M 8.11. In frequency domain compensator design, once the compensator shape is fixed (e.g., location of pole/zero), the gain K_c is typically chosen to:

A. Minimize phase margin

B. Cancel the system poles

C. Satisfy the magnitude condition at the desired crossover frequency

D. Maximize bandwidth

M 8.12. The transfer function of a compensator is given as

$$G_c(s) = K\frac{s+a}{s+b}$$

$G_c(s)$ is a lead compensator if

A. $a = 2,\ b = 4$

B. $a = 3,\ b = 2$

C. $a = -2,\ b = -1$

D. $a = 4,\ b = 2$

M 8.13. The phase of the above compensator is maximum at

A. $\sqrt{2}$ rad/s **B.** 2 rad/s **C.** 8 rad/s **D.** $2\sqrt{2}$ rad/s

M 8.14. What is the typical phase contribution of a lag compensator near the gain crossover frequency?

A. Large positive phase lead

B. No phase effect

C. Small phase lag

D. Sudden phase shift

M 8.15. A lag compensator is introduced in a control system primarily to:

A. Improve transient response

B. Increase phase margin

C. Reduce steady-state error without significantly affecting phase margin

D. Increase the natural frequency of oscillation

M 8.16. To improve phase margin without significantly affecting low-frequency gain, the preferred compensator is:

A. Lag compensator

B. PI controller

C. Lead compensator

D. Derivative-only controller

M 8.17. When a PI controller is applied to a Type-0 system (no pole at origin), the system becomes:

A. Type-0 system with finite position error

B. Type-1 system with zero position error for step input

C. Unstable due to integrator pole

D. Type-1 system with infinite ramp error

Answers

M 8.1. C **M 8.2**. B **M 8.3**. D **M 8.4**. A **M 8.5**. C **M 8.6**. C **M 8.7**. B **M 8.8**. A **M 8.9**. C **M 8.10**. D **M 8.11**. C **M 8.12**. A **M 8.13**. D **M 8.14**. C **M 8.15**. C **M 8.16**. C **M 8.17**. B

8.14 Knowledge Assessment Problems

KaP 8.1 Consider the unity feedback system shown in Figure 8.72.

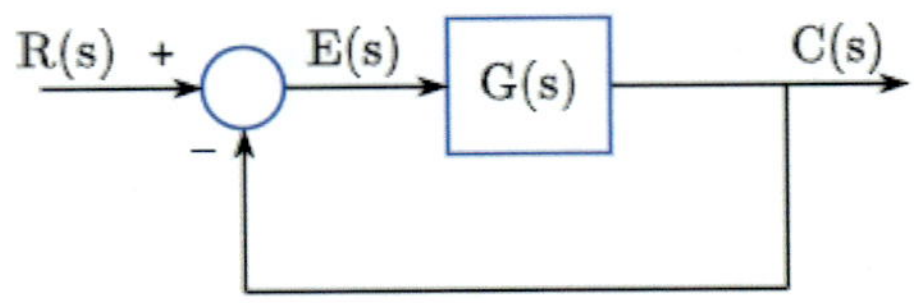

Figure 8.72: Unity feedback system

Design a PI controller for this system, which would drive the steady state error to the step input to zero with a damping ratio of 0.6. Given

$$G(s) = \frac{1}{(s+2)(s+4)(s+12)}$$

KaP 8.2 Consider the unity feedback system shown in Figure 8.72, where

$$G(s) = \frac{K}{s(s+2)(s+6)}$$

Design a PI controller such that the ramp response error will go to zero. The maximum overshoot is limited to 25%.

KaP 8.3 A unity feedback system with forward transfer function

$$G(s) = \frac{K}{(s+1)(s+4)(s+6)}$$

is operating with an overshoot of 15%. Design a phase lag controller to reduce the steady-state error by a factor of 10.

KaP 8.4 Repeat Problem **KaP 8.3** for

$$G(s) = \frac{K}{s(s+4)(s+6)}$$

KaP 8.5 Design a PD controller for the system shown in Figure 8.72 with

$$G(s) = \frac{K(s+8)}{(s+1)(s+2)(s+4)}$$

such that the system operates with a dominant pole having a damping ratio of 0.5 and the settling time is reduced by a factor of 2.

KaP 8.6 Design a PD controller for the system shown in Figure 8.72 with

$$G(s) = \frac{K}{s(s+3)(s+12)}$$

such that the system operates with an overshoot of 10% and the settling time is reduced by a factor of 3.

KaP 8.7 Consider the satellite attitude control system with J = 1 shown in Figure 8.73

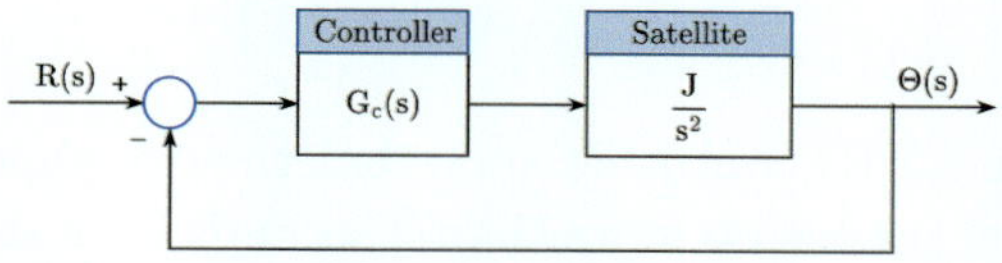

Figure 8.73: Satellite attitude control system

Design a lead compensator to give a settling time of 1.5 secs with a damping ratio of 0.6. Place the compensator zero at -1.0.

KaP 8.8 Consider the unity feedback control system shown in Figure 8.72 with

$$G(s) = \frac{K}{s(s+15)(s+30)}$$

The system is operating at 15% overshoot. Design a compensator to decrease the settling time by a factor of 2 without affecting the percent overshoot.

KaP 8.9 Consider a unity feedback system with forward path transfer function

$$G(s) = \frac{K}{s(s+1)(s+5)}$$

Design a compensator to yield the following specifications: settling time=2 seconds; damping ratio=0.707; the steady state error is to be improved by a factor of 2.

KaP 8.10 A unity feedback system has the forward transfer function

$$G(s) = \frac{K}{s^2(s+5)(s+15)}$$

Design a lead compensator to yield a closed-loop step response with 12% overshoot and a settling time of 2.5 seconds.

KaP 8.11 For the unity feedback system shown in Figure 8.72, with

$$G(s) = \frac{K}{(s+1)(s+5)}$$

design a PID controller which would result in a peak time of 1 second and a damping ratio of 0.8, with zero error for a step input.

KaP 8.12 Consider a unity feedback system with forward transfer function

$$G(s) = \frac{K}{(s+5)(s+8)(s+15)}$$

Design a PID controller such that the overshoot is limited to 15% and the settling time is no more than 2 seconds for a step input. The compensated system should have zero steady-state error for both step and ramp inputs.

KaP 8.13 For the unity feedback system of Figure 8.72 with

$$G(s)(= \frac{K}{(s+1)(s+3)(s+5)(s+10)}$$

Design a lag-lead compensator that will reduce the settling time of the uncompensated system by 1.5 times with a damping ratio of 0.6, and improve the steady state error by a factor of 20. Place zero of the lead compensator at -3 and the pole of the lag compensator at -0.01.

KaP 8.14 The forward path transfer function of a unity feedback system is

$$G(s) = \frac{K}{s(s+2)(s+30)}$$

Design a lead compensator to meet the following specifications.

a. $K_v \geq 20\ \text{sec}^{-1}$

b. Phase margin $= 45^0$

KaP 8.15 The forward path transfer function of a unity feedback system is

$$G(s) = \frac{K}{s(s+10)(s+50)}$$

Design a lead compensator to meet the following specifications.

a. $K_v \geq 20\ \text{sec}^{-1}$

b. Peak overshoot 16.3%

KaP 8.16 The forward path transfer function of a unity feedback system

$$G(s) = \frac{K}{(s+2)(s+20)(s+25)}$$

The system is operating at 18% overshoot. Design a compensator which would result in zero steady state error to step input and the static velocity error constant $K_v \geq 5\%$.

KaP 8.17 Design both a lag and a lead compensator for the unity feedback system shown in Figure 8.72 with

$$G(s) = \frac{K(s+6)}{(s+4)(s+8)(s+10)}$$

to yield a static position error constant $K_p = 60$ and a phase margin of 45^0.

KaP 8.18 Consider a unity feedback system with forward path transfer function

$$G(s) = \frac{K}{(s+5)^2}$$

Design a lag network such that the steady-state error to a step input is approximately 10% and the phase margin of the system to be approximately 60^0.

KaP 8.19 Consider a unity feedback system with forward path transfer function

$$G(s) = \frac{K}{s(s+10)(s+15)}$$

Design a lag compensator to meet the following specifications

a. $K_v \geq 20\%$

b. Peak overshoot=4.33%

KaP 8.20 The transfer of a DC servo motor is given by

$$G(s) = \frac{K}{s(s+4)}$$

Design a lag compensator to meet the following specifications

a. $K_v \geq 20\%$

b. Phase margin=45^0

KaP 8.21 The open-loop transfer function of a unity feedback system is given by

$$G(s) = \frac{K}{s(s+5)(s+10)}$$

Design a lag-lead compensator to meet the following specifications.

a. $K_v \geq 2 \text{ sec}^{-1}$

b. Phase margin=45^0

KaP 8.22 The open-loop transfer function of a unity feedback system is given by

$$G(s) = \frac{K}{s(s+2)(s+5)}$$

Design a lag-lead compensator to meet the following specifications.

a. $K_v \geq 20 \text{ sec}^{-1}$

b. Percentage overshoot=15%

KaP 8.23 Consider a unity feedback system with the plant:

$$G(s) = \frac{100}{(s+1)(s+2)}$$

Design a PD compensator of the form

$$G_c(s) = K_p + K_d s = K_d \left(s + \frac{K_p}{K_d} \right) = K_d(s+z)$$

such that the open-loop gain crossover frequency ω_{gc} of $G_c(s)G(s)$ is the same as that of the uncompensated system and the desired phase margin is 45^0.

KaP 8.24 Consider a unity feedback system with the plant:

$$G(s) = \frac{15(s+9)}{s^2+4s+40}$$

Design a PI controller of the form

$$G_c(s) = K_c \frac{s+z}{s}$$

such that the open-loop gain crossover frequency ω_{gc} of $G_c(s)G(s)$ is 10 rad/sec.

KaP 8.25 The block diagram of a feedback compensator is shown in Figure 8.74. Design the parameters K and δ to achieve the following specification:

i. Peak overshoot $\leq$ 15%

ii. Settling time $\leq$ 2 seconds

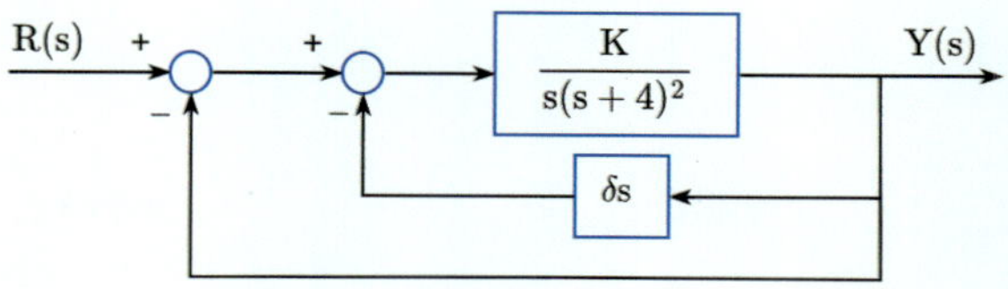

Figure 8.74: Rate feedback system

KaP 8.26 The schematic of a feedback compensator is shown in Figure 8.75. Design the parameters K and δ to achieve the following specification:

i. Peak overshoot $\leq 10\%$

ii. Settling time ≤ 3 seconds

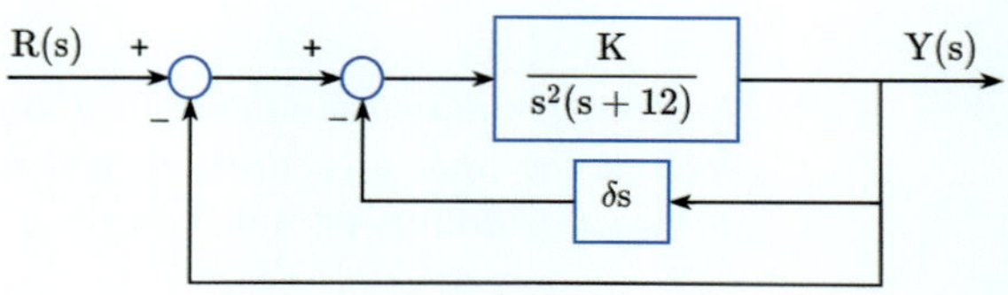

Figure 8.75: Rate feedback system

Chapter 9

State Variable Modeling of Dynamic Systems

State-space is like managing a group project—you can't just look at the final presentation, you need to know what each member is doing (or not doing)

DESIRED LEARNING OUTCOMES

Upon completion of this chapter, readers should be able to:

- Understand the philosophy of state space modeling
- Represent various systems in state space form
- Linearize nonlinear systems and represent them in generic state space form
- Develop state space models for practical systems

9.1 Introduction

Up to this point in this book, most of our analysis and design of control systems are based on transfer function models of the physical systems. These transfer functions are derived from constant coefficient linear differential equation models by applying the Laplace transform. Although the transfer functions have been proven to be a powerful tool for classical control system analysis and design using root locus and various frequency response methods such as Bode plot, polar, and Nyquist plots, it has several limitations and drawbacks as follows:

DOI: 10.1201/9781003571308-9

Limitations of Transfer Function Models

1. Defined only for linear time-invariant systems; (not defined for nonlinear systems ?)[a].
2. It becomes highly cumbersome for use with Multiple Input Multiple Output (MIMO) systems.
3. For a given input, it reveals information about the system's output only and provides no information regarding the system's internal state. This is a serious limitation; as we may have a situation where the output is stable, however, some other elements of the system exceed their specified ratings.
4. It may be necessary and advantageous to provide feedback proportional to some of the internal variables of the system, rather than the output alone, to stabilize and improve the performance of the system.
5. Independent of the input of the system
6. It is a function of the complex variable "s"; not a function of the real variable, time, or any other variable that is used as the independent variable.

[a] In linear systems, the frequency response function (FRF) fully characterizes the input–output behavior in the frequency domain. For nonlinear systems, an analogous concept exists, called the *Generalized Frequency Response Function (GFRF)*. GFRFs extend the idea of FRFs to capture nonlinear dynamics, enabling analysis of nonlinear systems in the frequency domain [4].

To overcome the limitations of transfer function models, a particularly useful time domain model, called the *state space model*, was introduced to the engineering community of the United States during the late 1950s and early 1960s by a small number of mathematically oriented engineers and applied mathematicians. This model has been the basis of *modern control theory*. More details about this can be found in [22].

To understand the fundamental difference between transfer function and state-space approaches, let us first explore what this new perspective offers through practical analogies.

Understanding Different Modeling Approaches

The Car Analogy:Transfer Function vs. State-Space

- **Transfer Function View (black-box test):** Suppose you want to evaluate a car. You don't look under the hood or check its components. Instead, you take it for a test drive: press the accelerator, brake, and steering wheel, and observe how it responds. From this **input–output behavior**, you can judge the car's performance (speed, handling, stopping distance) without knowing anything about the engine, transmission, or suspension.

- **State-Space View (under-the-hood look):** Now imagine opening the hood and examining the engine, gearbox, suspension, and electronics. You see how torque is generated, how gears shift, and how different subsystems interact internally. This gives you the **state variables**: the inner workings that together determine the car's observable performance.

The Family Analogy:Transfer Function vs. State-Space Models

- **Transfer Function View (black-box view):** Imagine meeting a celebrity family on the red carpet. You interact only with the star parent, giving interviews. From their responses, you can infer the family's overall image and reputation. But you don't know the kids' personalities, sibling rivalries, or what goes on behind closed doors. You see only the polished **input–output behavior**, not the inner workings.

- **State-Space View (inside view):** Now imagine getting a backstage pass or being invited inside their mansion. Suddenly, you see how the kids joke, tease, and support each other, how each family member contributes, and how the daily chaos and routines shape the polished public image you saw on the red carpet. These hidden dynamics are like **state variables**: the internal details that drive the system's outward behavior.

In short:

- Transfer function = *test-driving the car* or *meeting the family on the red carpet*

- State-space = *lifting the hood* or *hanging out backstage to see how everything really works*

Philosophy of State Variable Modeling

Having understood the conceptual difference between black-box and internal-state perspectives, we can now examine the mathematical philosophy behind state variable modeling. The key principles are:

Key Concept

1. A system having **n-number of independent energy storing elements** can be modeled by **a single** n-th order differential equation.
2. The state variable model essentially **splits the single (one) differential equation of order "n" into n-number of first-order differential equations**.
3. Thus, a single n-th order differential equation becomes n 1st order differential equations.

This mathematical transformation is more than just a computational convenience. It provides direct access to the internal dynamics of the system. Before we formally discuss the concept of state variable representation, let us illustrate this concept with a few simple, well-known examples.

EXAMPLE 9.1

Consider the simple mechanical system that consists of a mass that is acted upon by the forces of friction and a spring. This system has two independent energy-storing elements: mass and the spring, and one energy dissipating element: the damper.

The dynamics of this system due to application of a force u(t) is given by a **single second order differential equation**

$$m\frac{d^2x}{dt^2} + f\frac{dx}{dt} + kx = u(t) \tag{9.1}$$

where m is the mass, f is the damping coefficient, and k is the spring constant.

Our objective is to split this second-order differential equation into two first-order differential equations. There are many possible ways to achieve this (to be discussed later). In this example, let us introduce two variables $x_1 = x$ and $x_2 = \dot{x}$ and express Eqn(9.1) in terms of these variables as:

$$\begin{aligned}
\dot{x}_1 &= \dot{x} = x_2 \\
\dot{x}_2 &= \ddot{x} = -\frac{k}{m}x - \frac{f}{m}\dot{x} + \frac{1}{m}u(t) \\
&= -\frac{k}{m}x_1 - \frac{f}{m}x_2 + \frac{1}{m}u(t)
\end{aligned} \tag{9.2}$$

In matrix form, Eqn(9.2) can be represented as

$$\begin{bmatrix} \dot{x}_1 \\ \dot{x}_2 \end{bmatrix} = \begin{bmatrix} 0 & 1 \\ -\dfrac{k}{m} & -\dfrac{f}{m} \end{bmatrix} \begin{bmatrix} x_1 \\ x_2 \end{bmatrix} + \begin{bmatrix} 0 \\ \dfrac{1}{m} \end{bmatrix} u \tag{9.3}$$

If we consider the displacement x as the output y, then

$$y = \begin{bmatrix} 1 & 0 \end{bmatrix} \begin{bmatrix} x_1 \\ x_2 \end{bmatrix} \tag{9.4}$$

EXAMPLE 9.2

Problem

Figure 9.1 shows a simply supported beam with a concentrated load acting on the midspan. Assuming that the mass of the beam is negligible compared to the load, write the equation of motion.

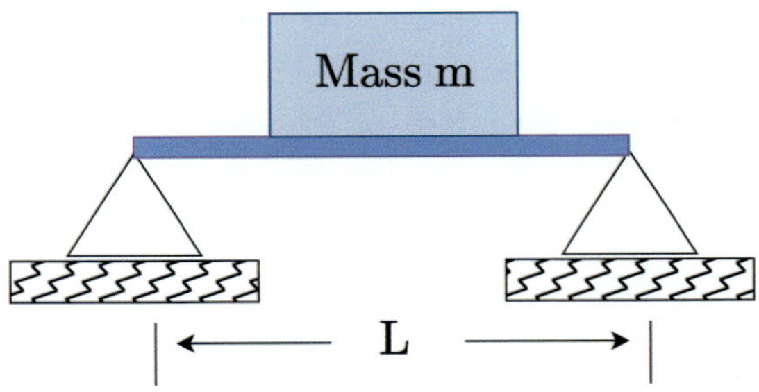

Figure 9.1: Simple supported beam loaded at the center

SOLUTION:

From the theory of strength of materials, the deflection at the midspan of a simply supported beam due to a concentrated load P at its center is

$$\delta = \frac{PL^3}{48EI}$$

where δ is the deflection, L is the length between supports, I is the moment of inertia, and E is modulus of elasticity.

For small deflections, the beam acts like a linear spring, that is $k = \dfrac{P}{\delta} = \dfrac{48EI}{L^3}$.

The dynamics is

$$m\ddot{x} + \frac{48EI}{L^3}x = f(t) \tag{9.5}$$

Let us introduce 2-state variables $x_1 = x$ and $x_2 = \dot{x}$. Let output $y = x$ and input $u = f(t)$. Hence Eqn(9.5) is expressed as:

$$\begin{aligned} \dot{x}_1 &= x_2 \\ \dot{x}_2 &= -\frac{k}{m}x_1 + \frac{1}{m}u \end{aligned} \tag{9.6}$$

In matrix form, it is expressed as:

$$\begin{bmatrix} \dot{x}_1 \\ \dot{x}_2 \end{bmatrix} = \begin{bmatrix} 0 & 1 \\ -\dfrac{k}{m} & 0 \end{bmatrix} \begin{bmatrix} x_1 \\ x_2 \end{bmatrix} + \begin{bmatrix} 0 \\ \dfrac{1}{m} \end{bmatrix} u, \quad \text{and} \quad y = \begin{bmatrix} 1 & 0 \end{bmatrix} \begin{bmatrix} x_1 \\ x_2 \end{bmatrix}$$

EXAMPLE 9.3

Problem

A wooden cylinder of radius r is partially immersed in a bath of distilled water as shown in Figure 9.2. Let ρ_1 equal the density of the water and ρ_2 that of the wood. Write the homogeneous equation of motion.

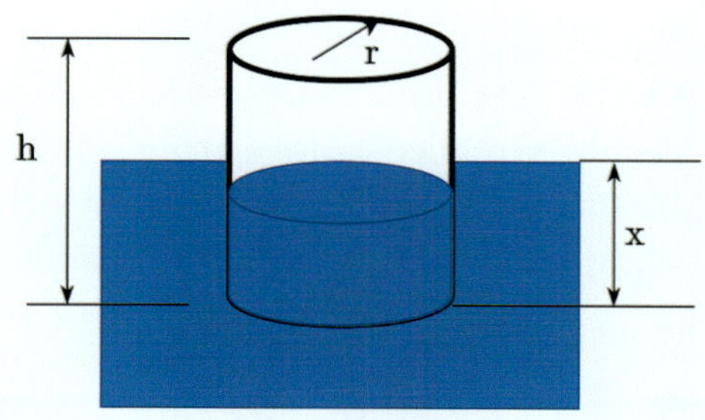

Figure 9.2: Wooden cylinder bobbing in water

SOLUTION:

The cylinder is originally in equilibrium. There are then no unbalanced forces. If the cylinder is depressed momentarily, the upward force experienced by the cylinder is due to the displaced volume of water and equals $g\rho_1 x\pi r^2$.

The mass of the cylinder is $\rho_2\pi r^2 h$.

But $\rho_2 = \gamma\rho_1$, where γ is the specific gravity of the wood. We have the equation of motion for the cylinder.

$$\gamma\rho_1\pi r^2 h\ddot{x} = -\rho_1\pi x r^2 g \quad\Longrightarrow\quad \ddot{x} + \frac{g}{\gamma h}x = 0 \tag{9.7}$$

Select $x_1 = x$ and $x_2 = \dot{x}$ and the output $y = x$. Hence Eqn(9.7) is expressed as:

$$\begin{aligned}\dot{x}_1 &= x_2\\ \dot{x}_2 &= -\frac{g}{\gamma h}x_1\end{aligned} \tag{9.8}$$

In matrix form, it is expressed as:

$$\begin{bmatrix}\dot{x}_1\\ \dot{x}_2\end{bmatrix} = \begin{bmatrix}0 & 1\\ -\dfrac{g}{\gamma h} & 0\end{bmatrix}\begin{bmatrix}x_1\\ x_2\end{bmatrix}, \quad \text{and} \quad y = \begin{bmatrix}1 & 0\end{bmatrix}\begin{bmatrix}x_1\\ x_2\end{bmatrix}$$

9.2 Modeling in State Space

Before we formally discuss the concept of state variable representation of a system, let us familiarize ourselves with some common terms associated with state variable analysis and design.

System State:

Definition 9.1. *System State:* The state of a dynamic system is the smallest set of variables (called *state variables*) such that the knowledge of these variables at time $t = t_0$, together with the knowledge of the input for $t \geq t_0$, completely determines the evolution of the system for any time $t \geq t_0$.

Key Insight

i. The state is a compact representation of the history of the system, which can be utilized to predict the future behavior of a system due to any external inputs.

ii. The concept of "*state*" applies to almost all types of systems, including electrical, mechanical, chemical, biological, economic, social systems, and so on.

State Variables:

During the state space approach of system analysis and design, we assume that the behavior of dynamic systems is represented by ordinary differential equations. We can predict the behavior of the physical systems by solving the differential equation models of the systems. Since the complete solution of a differential equation of order n requires precisely n-number of initial conditions, it follows that the state of such a system will be specified by the values of n quantities, called *state variables.*

As an example, in an electrical network, if we know the voltage across each capacitor and current through each inductor at any given instant of time, then its future response for given inputs can be determined.

Definition 9.2. *State Variables:* The state variables of a dynamic system are the smallest set of variables whose values at $t = t_0$, together with the input for $t \geq t_0$, completely determine the future state of the system.

Essential Point

i. The state variables need not be physically meaningful and measurable quantities; although it is preferred to choose measurable quantities as state variables, such as voltage, current, displacement, velocity, temperature, pressure, etc.

ii. Although the number of state variables in the model of a physical process is *unique*, the choice of state variables is not unique (to be proved later).

State Vector:

If the behavior of a given system can be completely described by n-number of state variables (say $x_1, x_2, \ldots, x_n$), then these n-variables can be represented as n-components of a vector $\mathbf{x}$, called as *state vector.*

Definition 9.3. *State Vector:* A state vector is a vector whose value at $t = t_0$, that is $\mathbf{x}(\mathbf{t_0})$ together with the input $u(t)$, $t \geq t_0$ determines uniquely the system state $\mathbf{x}(t)$ for any time $t \geq t_0$.

State Space:

Definition 9.4. *State Space:* State space is the set of all possible states of a dynamical system; each state of the system corresponds to a unique point in the state space. For an n-th order system, the coordinates of the n-dimensional space consist of the x_1 axis, x_2 axis, $\ldots$, x_n axis.

Let us illustrate this modeling method by considering a few simple, foundational examples that demonstrate the key principles of state-space representation.

9.2.1 State-Space Modeling of Electrical Networks

In this subsection, we consider modeling simple electrical networks in state space. For electrical circuits, the state variables are typically the voltages across capacitors and currents through inductors, since these quantities cannot change instantaneously and represent the energy storage elements in the circuit.

EXAMPLE 9.4

Problem:

Consider the simple R-L-C circuit shown in Figure 9.3

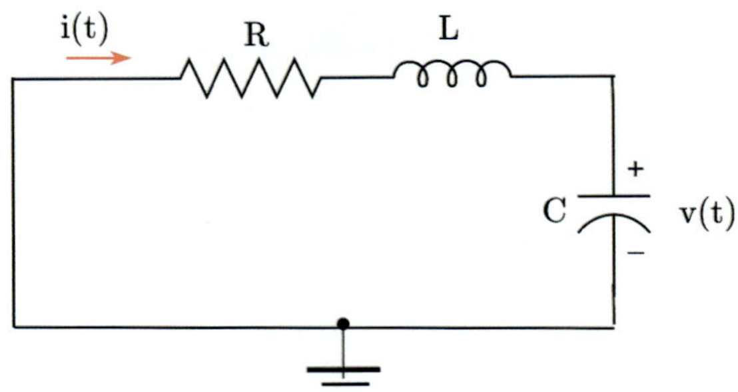

Figure 9.3: R-L-C Circuit

Represent this system in state space, considering the voltage across the capacitor and current through the inductor as state variables. Assume the voltage across the capacitor is the output and the applied voltage is the input.

SOLUTION:

The dynamics of the circuit are described by

$$\mathrm{Ri(t)} + \mathrm{L}\frac{\mathrm{di(t)}}{\mathrm{dt}} + \frac{1}{\mathrm{C}}\int \mathrm{i(t)dt} = \mathrm{v(t)} \tag{9.9}$$

$$\frac{1}{\mathrm{C}}\int \mathrm{i(t)dt} = \mathrm{v_c(t)} = \mathrm{v_0(t)} \tag{9.10}$$

If we select two state variables $x_1 = v_c(t)$ and $x_2 = i(t)$, $u(t) = v(t)$ and $y(t) = v_c(t)$, then, Eqn(9.9) can be expressed as:

$$Rx_2(t) + L\dot{x}_2(t) + x_1(t) = u(t)$$

or

$$\dot{x}_2 = -\frac{1}{L}x_1 - \frac{R}{L}x_2 + \frac{1}{L}u \tag{9.11}$$

Similarly, Eqn(9.10) can be expressed as:

$$C\frac{dv_c(t)}{dt} = i(t) \quad \Longrightarrow \quad \dot{x}_1 = \frac{1}{C}x_2 \tag{9.12}$$

The dynamics of the circuit from Eqns(9.9) and (9.10) can be expressed in matrix form as:

$$\begin{bmatrix}\dot{x}_1\\ \dot{x}_2\end{bmatrix} = \begin{bmatrix}0 & \frac{1}{C}\\ -\frac{1}{L} & -\frac{R}{L}\end{bmatrix}\begin{bmatrix}x_1\\ x_2\end{bmatrix} + \begin{bmatrix}0\\ \frac{1}{L}\end{bmatrix}u \tag{9.13}$$

Since voltage across the capacitor is taken as the output, then

$$y = \begin{bmatrix}1 & 0\end{bmatrix}\begin{bmatrix}x_1\\ x_2\end{bmatrix} \tag{9.14}$$

EXAMPLE 9.5

Problem

Obtain the state variable model of the system shown in Figure 9.4, taking the current through the inductors and voltage across the capacitor as state variables. The voltage across the capacitor is the output.

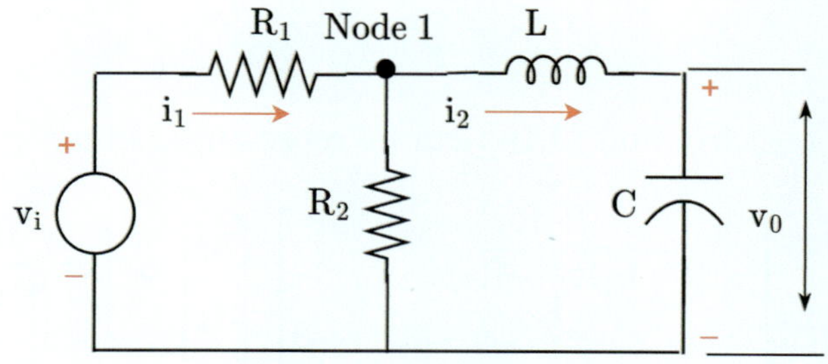

Figure 9.4: Two Loop Electrical Network

SOLUTION:

Let the current through inductor L and voltage across the capacitor C is selected as state variables. Thus, $x_1 = i_2$ and $x_2 = v_C$

Consider node-1 with a voltage denoted as v_1. Then

$$L\frac{di_2}{dt} + v_C = v_1 \quad \text{or} \quad L\dot{x}_1 = v_1 - v_C \tag{9.15}$$

Applying KCL in node-1 gives

$$\frac{u - v_1}{R_1} = i_2 + \frac{v_1}{R_2} \quad \text{or,} \quad \frac{v_1}{R_1} + \frac{v_1}{R_2} = \frac{u}{R_1} - x_1$$

$$\text{or,} \quad v_1 \left[\frac{R_1 + R_2}{R_1 R_2}\right] = \frac{u - R_1 x_1}{R_1}$$

$$\text{or,} \quad v_1 = \frac{(u - R_1 x_1) R_2}{R_1 + R_2} \tag{9.16}$$

Substituting the value of v_1 from Eqn(9.16) into Eqn(9.15) gives

$$L\dot{x}_1 = v_1 - v_C$$

$$\text{or,} \quad L\dot{x}_1 = \frac{(u - R_1 x_1) R_2}{R_1 + R_2} - v_C$$

$$\text{or,} \quad L\dot{x}_1 = -\frac{R_1 R_2}{R_1 + R_2} x_1 - x_2 + \frac{R_2}{R_1 + R_2} u$$

$$\text{or,} \; \dot{x}_1 = -\frac{R_1 R_2}{L(R_1 + R_2)} x_1 - \frac{1}{L} x_2 + \frac{R_2}{L(R_1 + R_2)} u \tag{9.17}$$

Further, the current through the capacitor is expressed as:

$$C\frac{dV_C}{dt} = i_2 \quad \text{or,} \quad C\dot{x}_2 = x_1$$

$$\text{or,} \quad \dot{x}_2 = \frac{1}{C} x_1 \tag{9.18}$$

The output is the voltage across the capacitor. Hence

$$y = v_C = x_2 \tag{9.19}$$

Eqns(9.17), (9.18), and (9.19) can be represented in matrix form as:

$$\begin{bmatrix} \dot{x}_1 \\ \dot{x}_2 \end{bmatrix} = \begin{bmatrix} -\dfrac{R_1 R_2}{L(R_1 + R_2)} & \dfrac{-1}{L} \\ \dfrac{1}{C} & 0 \end{bmatrix} \begin{bmatrix} x_1 \\ x_2 \end{bmatrix} + \begin{bmatrix} \dfrac{R_2}{L(R_1 + R_2)} \\ 0 \end{bmatrix} u$$

$$y = \begin{bmatrix} 0 & 1 \end{bmatrix} \begin{bmatrix} x_1 \\ x_2 \end{bmatrix} \tag{9.20}$$

CONCEPT REINFORCEMENT PROBLEM 9.1

Problem

Obtain the state variable model of the system shown in Figure 9.5, taking the current through the inductors and voltage across the capacitor as state variables. The voltage across the resistor is the output.

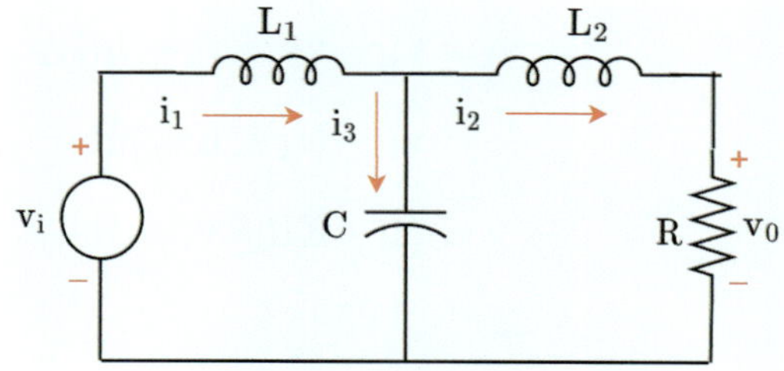

Figure 9.5: Multi Loop Electrical Network

EXAMPLE 9.6

Problem

Represent the electrical network shown in Figure 9.6 in state space, considering the voltage across the capacitor and current through the inductor as state variables. The voltage across the resistor R_2 is the output.

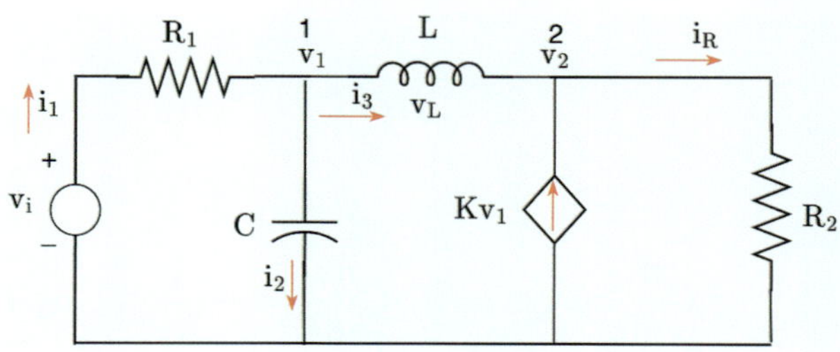

Figure 9.6: Electrical network with dependent source

SOLUTION:

Denote the state variables as $x_1 = v_1$ and $x_2 = i_3$. Now

$$C\frac{dv_1}{dt} = i_2 \quad \text{and} \quad L\frac{di_3}{dt} = v_L$$

The voltage across the inductor

$$\begin{aligned} v_L &= v_1 - v_2 = v_1 - i_R R_2 \\ &= v_1 - (i_3 + Kv_1) R_2 \\ &= (1 - KR_2) v_1 - R_2 i_3 \end{aligned}$$

Hence

$$L\frac{di_3}{dt} = (1 - KR_2) v_1 - R_2 i_3$$

or

$$L\dot{x}_2 = (1 - KR_2) x_1 - R_2 x_2$$

This gives

$$\dot{x}_2 = \frac{(1 - KR_2)}{L} x_1 - \frac{R_2}{L} x_2 \tag{9.21}$$

The current through the capacitor

$$\begin{aligned} i_2 &= i_1 - i_3 = \frac{v_i - v_1}{R_1} - i_3 \\ &= -\frac{1}{R_1} v_1 - i_3 + \frac{v_i}{R_1} \end{aligned}$$

Hence

$$C\frac{dv_1}{dt} = i_2 \implies C\frac{dv_1}{dt} = -\frac{1}{R_1} v_1 - i_3 + \frac{v_i}{R_1}$$

or

$$C\dot{x}_1 = -\frac{1}{R_1} x_1 - x_2 + \frac{u}{R_1}$$

This gives

$$\dot{x}_1 = -\frac{1}{R_1 C} x_1 - \frac{1}{C} x_2 + \frac{u}{R_1 C} \tag{9.22}$$

The output

$$y = i_R R_2 = (i_3 + K v_1) R_2 = K R_2 x_1 + R_2 x_2 \tag{9.23}$$

Eqns(9.21), (9.22), and (9.23) can be expressed in matrix form as:

$$\begin{bmatrix} \dot{x}_1 \\ \dot{x}_2 \end{bmatrix} = \begin{bmatrix} -\dfrac{1}{R_1 C} & -\dfrac{1}{C} \\ \dfrac{(1 - KR_2)}{L} & -\dfrac{R_2}{L} \end{bmatrix} + \begin{bmatrix} \dfrac{1}{R_1 C} \\ 0 \end{bmatrix}$$

$$y = \begin{bmatrix} KR_2 & R_2 \end{bmatrix} \begin{bmatrix} x_1 \\ x_2 \end{bmatrix} \tag{9.24}$$

CONCEPT REINFORCEMENT PROBLEM 9.2

Problem

Determine the state variable model of the circuit shown in Figure 9.7, considering the current through the inductor and voltage across the capacitor as state variables. The circuit is excited by the current source i and voltage source v_0. Consider v_{R_1}, i_L, and i_{R_2} as outputs.

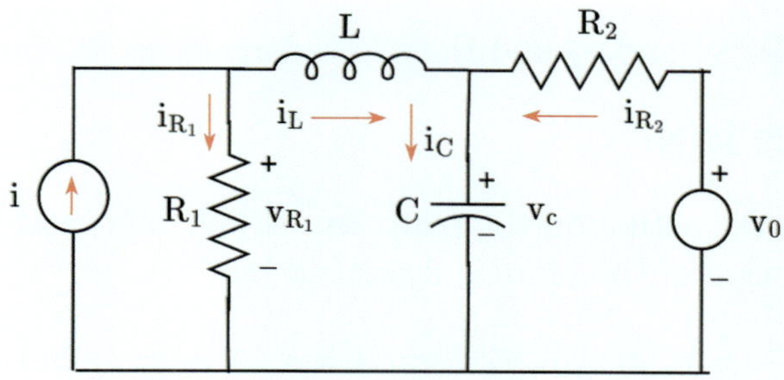

Figure 9.7: Three loop network

9.2.2 State-Space Modeling of Mechanical Systems

In this subsection, we consider modeling simple mechanical systems in state space. For mechanical systems, the state variables are typically the positions and velocities of masses, as these represent the system's kinetic and potential energy storage.

EXAMPLE 9.7

Problem

Obtain the state variable model of the system shown in Figure 9.8, where the plate of mass m_1 is connected to the mass m_2.

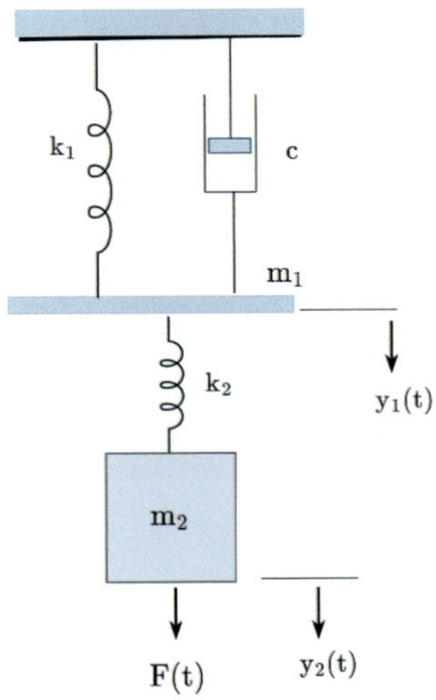

Figure 9.8: In-line mass spring system

Consider y_1 and y_2 and their first derivatives as state variables.

SOLUTION:

The forces acting on the plate are $-k_1y_1$,$-c\dot{y}_1$, and $k_2(y_2 - y_1)$. Thus, the motion of the plate is described by

$$m_1\ddot{y}_1 + c\dot{y}_1 + k_1y_1 - k_2(y_2 - y_1) = 0$$

$$\text{or,}\ m_1\ddot{y}_1 + c\dot{y}_1 + (k_1 + k_2)y_1 - k_2y_2 = 0 \tag{9.25}$$

The forces acting on the mass m_2 are $-k_2(y_2 - y_1)$ and the applied force F. Thus, its motion is described by

$$m_2\ddot{y}_2 + k_2(y_2 - y_1) = F \tag{9.26}$$

Let us select the state variables as:

$$x_1 = y_1,\quad x_2 = \dot{y}_1,\quad x_3 = y_2,\quad x_4 = \dot{y}_2 \ \text{ and }\ u = F$$

Then the dynamics in Eqn(9.25) and (9.26) can be expressed as:

$$\begin{aligned}
\dot{x}_1 &= x_2 \\
\dot{x}_2 &= -\frac{(k_1 + k_2)}{m_1}x_1 - \frac{c}{m_1}x_2 + \frac{k_2}{m_1}x_3 \\
\dot{x}_3 &= x_4 \\
\dot{x}_4 &= \frac{k_2}{m_2}x_1 - \frac{k_2}{m_2}x_3 + \frac{1}{m_2}u
\end{aligned} \tag{9.27}$$

Eqn(9.27) can be expressed in matrix form as:

$$\begin{bmatrix} \dot{x}_1 \\ \dot{x}_2 \\ \dot{x}_3 \\ \dot{x}_4 \end{bmatrix} = \begin{bmatrix} 0 & 1 & 0 & 0 \\ -\dfrac{(k_1 + k_2)}{m_1} & -\dfrac{c}{m_1} & \dfrac{k_2}{m_1} & 0 \\ 0 & 0 & 0 & 1 \\ \dfrac{k_2}{m_2} & 0 & -\dfrac{k_2}{m_2} & 0 \end{bmatrix} \begin{bmatrix} x_1 \\ x_2 \\ x_3 \\ x_4 \end{bmatrix} + \begin{bmatrix} 0 \\ 0 \\ 0 \\ \dfrac{1}{m_2} \end{bmatrix} u \tag{9.28}$$

EXAMPLE 9.8

Problem

Obtain the state variable model of the system shown in Figure 9.9 considering y_1 and y_2 and their first derivatives as state variables, where the plate of mass m_1 is connected to the mass m_2.

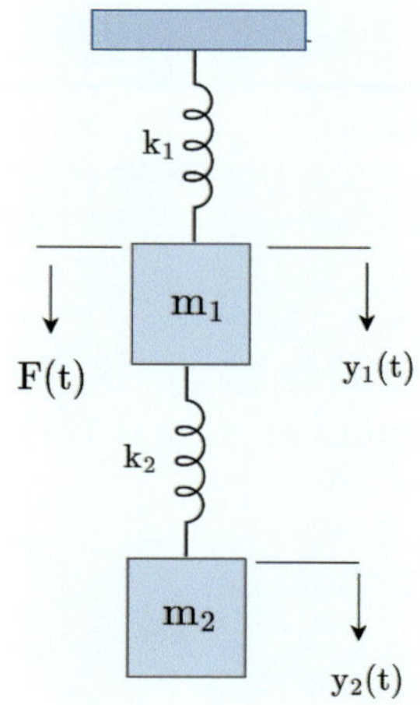

Figure 9.9: Two mass system

SOLUTION:

The dynamics associated with mass m_1 and m_2 are respectively given by

$$m_1\ddot{y}_1 + k_1 y_1 - k_2\left(y_2 - y_1\right) = F \tag{9.29}$$

$$m_2\ddot{y}_2 + k_2\left(y_2 - y_1\right) = 0 \tag{9.30}$$

Let us select the state variables as:

$$x_1 = y_1, \quad x_2 = \dot{y}_1, \quad x_3 = y_2, \quad x_4 = \dot{y}_2 \text{ and } u = F$$

Then the dynamics in Eqn(9.29) and (9.30) can be expressed as:

$$\begin{aligned}
\dot{x}_1 &= x_2 \\
\dot{x}_2 &= -\frac{(k_1 + k_2)}{m_1}x_1 + \frac{k_2}{m_1}x_3 + \frac{1}{m_1}u \\
\dot{x}_3 &= x_4 \\
\dot{x}_4 &= \frac{k_2}{m_2}x_1 - \frac{k_2}{m_2}x_3
\end{aligned} \tag{9.31}$$

Eqn(9.31) can be expressed in matrix form as:

$$\begin{bmatrix}\dot{x}_1\\ \dot{x}_2\\ \dot{x}_3\\ \dot{x}_4\end{bmatrix} = \begin{bmatrix} 0 & 1 & 0 & 0\\ -\dfrac{(k_1+k_2)}{m_1} & 0 & \dfrac{k_2}{m_1} & 0\\ 0 & 0 & 0 & 1\\ \dfrac{k_2}{m_2} & 0 & -\dfrac{k_2}{m_2} & 0\end{bmatrix}\begin{bmatrix}x_1\\ x_2\\ x_3\\ x_4\end{bmatrix} + \begin{bmatrix}0\\ \dfrac{1}{m_1}\\ 0\\ 0\end{bmatrix}u \tag{9.32}$$

CONCEPT REINFORCEMENT PROBLEM 9.3

Problem

Obtain the state variable model of the system shown in Figure 9.10, considering the outputs x, y, and their first derivatives as state variables.

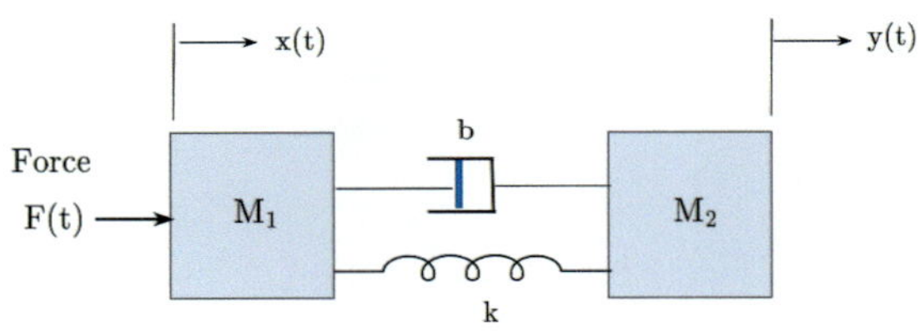

Figure 9.10: Coupled mass spring system

EXAMPLE 9.9

Problem

Obtain the state variable model of the system shown in Figure 9.11, considering y_1, y_2, y_3, and their first derivatives as state variables.

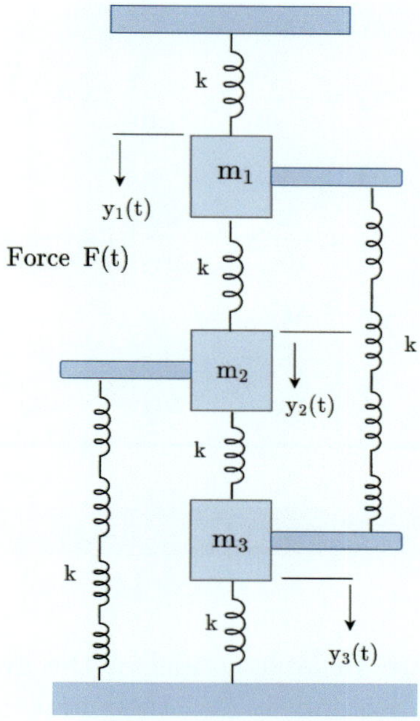

Figure 9.11: Three mass three spring system

SOLUTION:

The equation of motion for m_1 is given by

$$m_1\ddot{y}_1 + ky_1 + k(y_1 - y_2) + k(y_1 - y_3) = F$$
$$\text{or, } m_1\ddot{y}_1 + 3ky_1 - ky_2 - ky_3 = F \tag{9.33}$$

The dynamics of the mass m_2 and m_3 is described by

$$m_2\ddot{y}_2 + k(y_2 - y_1) + k(y_2 - y_3) + ky_2 = 0$$
$$m_3\ddot{y}_3 + ky_3 + k(y_3 - y_1) + k(y_3 - y_2) = 0 \tag{9.34}$$

Rearranging Eqns(9.33) and (9.34) gives

$$m_1\ddot{y}_1 + 3ky_1 - ky_2 - ky_3 = F$$
$$m_2\ddot{y}_2 + 3ky_2 - ky_1 - ky_3 = 0$$
$$m_3\ddot{y}_3 + 3ky_3 - ky_1 - ky_2 = 0 \tag{9.35}$$

Let us select the state variables as

$$x_1 = y_1, \quad x_2 = \dot{y}_1, \quad x_3 = y_2, \quad x_4 = \dot{y}_2, x_5 = y_3, \quad x_6 = \dot{y}_3, \quad \& \quad u = F$$

Then, the dynamics in Eqn(9.35) can be expressed as:

$$\begin{aligned}
\dot{x}_1 &= x_2 \\
\dot{x}_2 &= -\frac{3k}{m_1}x_1 + \frac{k}{m_1}x_3 + \frac{k}{m_1}x_5 + \frac{1}{m_1}u \\
\dot{x}_3 &= x_4 \\
\dot{x}_4 &= \frac{k}{m_2}x_1 - \frac{3k}{m_2}x_3 + \frac{k}{m_2}x_5 \\
\dot{x}_5 &= x_6 \\
\dot{x}_6 &= \frac{k}{m_3}x_1 + \frac{k}{m_3}x_3 - \frac{3k}{m_3}x_5
\end{aligned} \tag{9.36}$$

CONCEPT REINFORCEMENT PROBLEM 9.4

Problem

Obtain the state variable model of the system shown in Figure 9.12 considering y_1, y_2, and y_3 and their first derivatives as state variables where y_1, y_2, and y_3 represent respectively the displacement of masses m_1, m_2, and m_3.

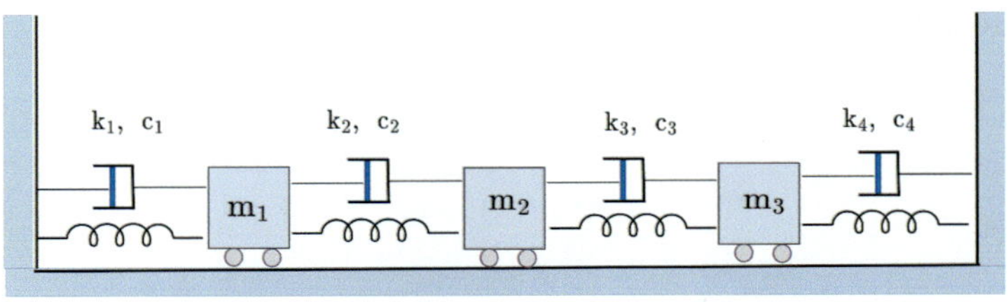

Figure 9.12: Horizontal three degree freedom system

CONCEPT REINFORCEMENT PROBLEM 9.5

Problem

Two identical pendulums, each with mass m are coupled by a spring with stiffness k at a distance d from the fixed end, as shown in the Figure 9.13. The rod which connects the mass is massless with length L.

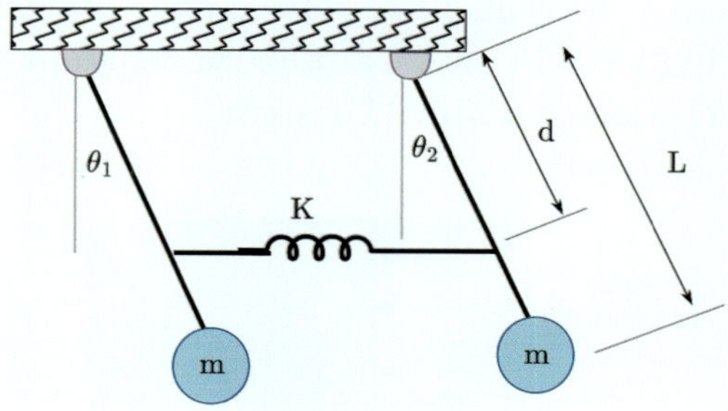

Figure 9.13: Two mass coupled pendulum system

Determine the state equations of the system considering θ_1, $\dot{\theta}_1$, θ_2, and $\dot{\theta}_2$ as state variables.

9.3 State-space Equations of Dynamic Systems

These examples are typical of the general form of the dynamic equations of a physical process. The state variable model of an n-th order multiple-input-multiple-output system with m-number of inputs and p-number of outputs is described as:

$$\begin{aligned} \dot{x}_1 &= f_1(x_1, x_2, \ldots, x_n, u_1, u_2, \ldots, u_m, t) \\ \dot{x}_2 &= f_2(x_1, x_2, \ldots, x_n, u_1, u_2, \ldots, u_m, t) \\ &\vdots \\ \dot{x}_n &= f_n(x_1, x_2, \ldots, x_n, u_1, u_2, \ldots, u_m, t) \end{aligned} \tag{9.37}$$

where x_i is a state variable, u_i is an input variable and t is the time. The notation $\dot{x}_i$ is used to represent $\frac{dx_i}{dt}$. The outputs y_1, y_2, …, y_p, are described by a set of algebraic equations relating states and inputs to output variables as:

$$\begin{aligned} y_1 &= g_1(x_1, x_2, \ldots, x_n, u_1, u_2, \ldots, u_m, t) \\ y_2 &= g_2(x_1, x_2, \ldots, x_n, u_1, u_2, \ldots, u_m, t) \\ &\vdots \\ y_p &= g_p(x_1, x_2, \ldots, x_n, u_1, u_2, \ldots, u_m, t) \end{aligned} \tag{9.38}$$

To simplify the notation, the state variables $x_1, x_2, \ldots, x_n$, control variables $u_1, u_2, \ldots, u_m$, and output variables $y_1, y_2, \ldots, y_p$ are collected in vectors

$$\mathbf{x} = \begin{bmatrix} x_1 \\ x_2 \\ \vdots \\ x_n \end{bmatrix}, \quad \mathbf{u} = \begin{bmatrix} u_1 \\ u_2 \\ \vdots \\ u_m \end{bmatrix}, \quad \text{and} \quad \mathbf{y} = \begin{bmatrix} y_1 \\ y_2 \\ \vdots \\ y_p \end{bmatrix}, \tag{9.39}$$

where **x**, **u**, and **y** are called the *state vector*, *input vector*, and *output vector*, respectively. Figure 9.14 shows the block diagram visualization of the state space representation of a MIMO system.

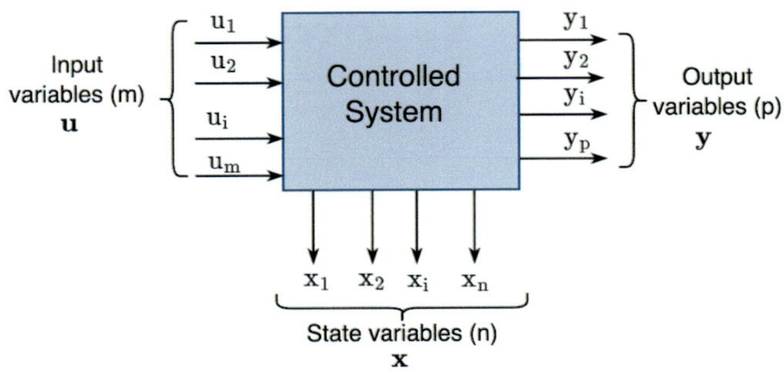

Figure 9.14: Structure of a state space-based control system

Define

$$\mathbf{f}(\mathbf{x}, \mathbf{u}, t) = \begin{bmatrix} f_1(x_1, \ldots, x_n, u_1, \ldots, u_m; t \\ f_2(x_1, \ldots, x_n, u_1, \ldots, u_m; t \\ \vdots \\ f_n(x_1, \ldots, x_n, u_1, \ldots, u_m; t \end{bmatrix} \quad \text{and}$$

$$\mathbf{g}(\mathbf{x}, \mathbf{u}, t) = \begin{bmatrix} g_1(x_1, \ldots, x_n, u_1, \ldots, u_m; t \\ g_2(x_1, \ldots, x_n, u_1, \ldots, u_m; t \\ \vdots \\ g_p(x_1, \ldots, x_n, u_1, \ldots, u_m; t \end{bmatrix}$$

Eqns(9.37) and (9.38) can, therefore, be expressed as:

$$\dot{\mathbf{x}}(t) = \mathbf{f}(\mathbf{x}, \mathbf{u}; t) \tag{9.40}$$

$$\mathbf{y}(t) = \mathbf{g}(\mathbf{x}, \mathbf{u}; t) \tag{9.41}$$

If the vector functions **f** and **g** involve time t explicitly, then the system is referred to as *time varying* system.

9.3.1 Linearization of Nonlinear State Equations

It is important to recognize that most physical systems inherently exhibit nonlinear characteristics. Therefore, the functions **f** and **g** in Eqns(9.40) and (9.41) are generally nonlinear in nature. However, it is often possible and practical to obtain a linearized approximation of these nonlinear equations, which remains valid in a small neighborhood around a chosen operating point, typically an equilibrium point. This approximation enables the application of powerful linear control techniques to analyze and design controllers for nonlinear systems.

The procedure of linearizing the nonlinear system described by Eqns(9.40) and (9.41) is discussed below. It is assumed that the system is time-invariant, that is, the functions (vector fields) **f** and **g** depend only on the state vector **x** and control input **u**, and do not involve time t explicitly. This simplification allows the use of constant Jacobian matrices evaluated at an equilibrium point for linear approximation.

Given Nonlinear System

Let the nonlinear system be described by

$$\dot{\mathbf{x}} = \mathbf{f}(\mathbf{x}, \mathbf{u}) \tag{9.42}$$

$$\mathbf{y} = \mathbf{g}(\mathbf{x}, \mathbf{u}) \tag{9.43}$$

where $\mathbf{x} \in \mathbb{R}^n$ is the state vector, $\mathbf{u} \in \mathbb{R}^m$ is the input vector, $\mathbf{y} \in \mathbb{R}^p$ is the output vector, $f : \mathbb{R}^n \times \mathbb{R}^m \to \mathbb{R}^n$ and $g : \mathbb{R}^n \times \mathbb{R}^m \to \mathbb{R}^p$ are differentiable vector functions.

Linearization Procedure

Step-1: Choose an Operating Point (Equilibrium)

Let (x_0, u_0) be the equilibrium point around which the system will be linearized, such that

$$f(x_0, u_0) = 0, \quad \text{and} \quad y = g(x_0, u_0) \tag{9.44}$$

Step-2: Define small perturbations

$$\delta x = x - x_0, \ \ \delta u = u - u_0, \ \text{ and } \delta y = y - y_0$$

Assuming small deviations, we express the dynamics in terms of δx, δu, and δy

Ste-3: Linearize Using First-Order Taylor Expansion:

$$\begin{aligned} \delta\dot{x} &= A\delta x + B\delta u \\ \delta y &= C\delta x + D\delta u \end{aligned} \tag{9.45}$$

where the system matrices are given by Jacobian evaluated at the operating point (x_0, u_0):

$$A = \left.\frac{\partial f}{\partial x}\right|_{(x_0,u_0)} \in \mathbb{R}^{n\times n}, \ B = \left.\frac{\partial f}{\partial u}\right|_{(x_0,u_0)} \in \mathbb{R}^{n\times m},$$

$$C = \left.\frac{\partial g}{\partial x}\right|_{(x_0,u_0)} \in \mathbb{R}^{p\times n}, \text{ and } D = \left.\frac{\partial g}{\partial u}\right|_{(x_0,u_0)} \in \mathbb{R}^{p\times m}$$

Step-4: Final Linearized State-Space Model

The linearized, time-invariant state-space model around the equilibrium point is:

$$\delta\dot{x} = A\delta x + B\delta u \tag{9.46}$$

$$\delta y = C\delta x + D\delta u \tag{9.47}$$

Note

It's common to **drop the delta (δ) notation** to keep expressions in Eqn(9.46) and (9.47) clean and readable.

With the implicit understanding that x, u and $\dot{x}$ are deviations from the equilibrium point, Eqns(9.46) and (9.47) are expressed as:

$$\dot{\mathbf{x}}(t) = \mathbf{A}\mathbf{x}(t) + \mathbf{B}\mathbf{u}(t) \tag{9.48}$$

$$\mathbf{y}(t) = \mathbf{C}\mathbf{x}(t) + \mathbf{D}\mathbf{u}(t) \tag{9.49}$$

where $\mathbf{x}(t)$ is $n \times 1$ state vector, $\mathbf{u}(t)$ is $m \times 1$ input vector, $\mathbf{A}$ is an $(n \times n)$ **system matrix** and $\mathbf{B}$ is an $(n \times m)$ **input matrix** defined as:

$$\mathbf{A} = \begin{bmatrix} a_{11} & a_{12} & \dots & a_{1n} \\ a_{21} & a_{22} & \dots & a_{2n} \\ \vdots & \vdots & & \vdots \\ a_{n1} & a_{n2} & \dots & a_{nn} \end{bmatrix}, \text{ and } \mathbf{B} = \begin{bmatrix} b_{11} & b_{12} & \dots & b_{1m} \\ b_{21} & b_{22} & \dots & b_{2m} \\ \vdots & \vdots & & \vdots \\ b_{n1} & b_{n2} & \dots & b_{nm} \end{bmatrix}$$

The output variables $\mathbf{y}(t)$ is a $p \times 1$ output vector, $\mathbf{C}$ is a $(p \times n)$ **output matrix** and $\mathbf{D}$ is a $(p \times m)$ **direct transmission matrix** defined as:

$$\mathbf{C} = \begin{bmatrix} c_{11} & c_{12} & \dots & c_{1n} \\ c_{21} & c_{22} & \dots & c_{2n} \\ \vdots & \vdots & & \vdots \\ c_{p1} & c_{p2} & \dots & c_{pn} \end{bmatrix}, \text{ and } \mathbf{D} = \begin{bmatrix} d_{11} & d_{12} & \dots & d_{1m} \\ d_{21} & d_{22} & \dots & d_{2m} \\ \vdots & \vdots & & \vdots \\ d_{p1} & d_{p2} & \dots & d_{pm} \end{bmatrix}$$

Figure 9.15 shows the block diagram representation of the state space model described in Eqns(9.48) and (9.49).

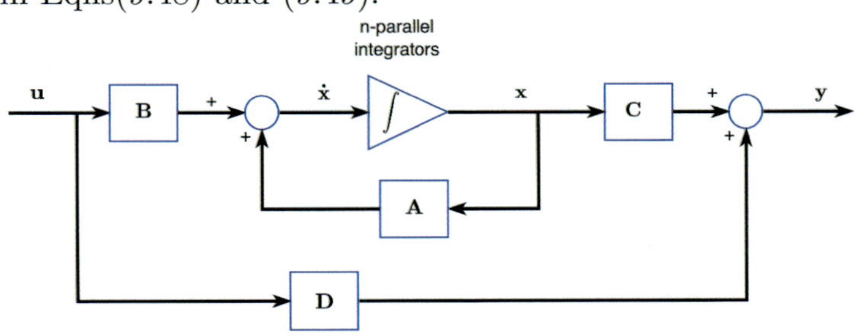

Figure 9.15: Structure of a state space based control system

EXAMPLE 9.10

Problem

Consider a simple second-order nonlinear system

$$\dot{x}_1 = x_2$$
$$\dot{x}_2 = -x_1 - 0.5x_1^2 - 0.2x_2 + u + 0.1u^2$$

a. Find the equilibrium points
b. Obtain the linearized model around the equilibrium points

SOLUTION:

The system dynamics can be expressed as:

$$\dot{\mathbf{x}} = f(\mathbf{x}, u) = \begin{bmatrix} x_2 & -x_1 - 0.5x_1^2 - 0.2x_2 + u + 0.1u^2 \end{bmatrix}^T$$

At equilibrium, $\dot{\mathbf{x}} = 0$. Hence

$$0 = x_{2e}$$
$$0 = -x_{1e} - 0.5x_{1e}^2 - 0.2x_{2e} + u_e + 0.1u_e^2$$

From the first equation: $x_{2e} = 0$. Substituting this into the second equation:

$$0 = -x_{1e} - 0.5x_{1e}^2 + u_e + 0.1u_e^2$$

Rearranging:

$$u_e + 0.1u_e^2 = x_{1e} + 0.5x_{1e}^2$$

Before we proceed further, consider the following:

The Relationship Between State and Input at Equilibrium

When we have a nonlinear system, the equilibrium condition $\dot{\mathbf{x}} = 0$ creates **a constraint equation** that relates the equilibrium state and input. This means:

1. We can choose either $\mathbf{x_e}$ or u_e (but typically not both independently)
2. The other is determined by the equilibrium constraint

In this example:

- Equilibrium constraint: $u_e + 0.1u_e^2 = x_{1e} + 0.5x_{1e}^2$

- This is one equation with two unknowns (x_{1e} and u_e)

Three Common Approaches:

Approach-1: Choose x_{1e}, solve for u_e

Example: Pick $x_{1e} = 2$, solve $u_e + 0.1u_e^2 = x_{1e} + 0.5x_{1e}^2 = 4$. Taking a positive value gives $u_e = 3.062$.

Approach-2: Choose u_e, Solve for x_{1e}

Example: Pick $u_e = 1$, Solve: $x_{1e} + 0.5x_{1e}^2 = u_e + 0.1u_e^2 = 1.1$. Taking positive root gives $x_{1e} = 0.788$.

Approach-3: Choose based on physical/practical constraints

- Maybe we want the system to operate at a specific force level (u)
- Or maybe we want a particular position (x_1)
- Or maybe there are bounds: $|x_1| \leq 5, \ |u| \leq 10$

In practice, we often start by choosing where we want the system to operate (the equilibrium state), then find what constant input is needed to maintain that condition.

Let us follow (Approach-1) and choose $x_{1e} = 2$. This gives $u_e = 3.062$. So our equilibrium point is: $(\mathbf{x}_e, u_e) = ([2, 0]^T, 3.062)$.

Linearization Process

Compute the partial derivatives at $(\mathbf{x_e}, u_e)$.

$$\mathbf{A} = \left.\frac{\partial f}{\partial x}\right|_{(x_e,u_e)} = \begin{bmatrix} \frac{\partial f_1}{\partial x_1} & \frac{\partial f_1}{\partial x_2} \\ \frac{\partial f_2}{\partial x_1} & \frac{\partial f_2}{\partial x_2} \end{bmatrix}_{(x_e,u_e)}$$

$$= \begin{bmatrix} 0 & 1 \\ -1 - x_1 & -0.2 \end{bmatrix}_{(x_e,u_e)} = \begin{bmatrix} 0 & 1 \\ -3 & -0.2 \end{bmatrix} \text{ and}$$

$$\mathbf{B} = \begin{bmatrix} \frac{\partial f_1}{\partial u} \\ \frac{\partial f_2}{\partial u} \end{bmatrix}_{(x_e,u_e)} = \begin{bmatrix} 0 \\ 1 + 0.2u_e \end{bmatrix} = \begin{bmatrix} 0 \\ 1.612 \end{bmatrix}$$

Hence the linearized system

$$\begin{bmatrix} \delta\dot{x}_1 \\ \delta\dot{x}_2 \end{bmatrix} = \begin{bmatrix} 0 & 1 \\ -3 & -0.2 \end{bmatrix} \begin{bmatrix} \delta x_1 \\ \delta x_2 \end{bmatrix} + \begin{bmatrix} 0 \\ 1.612 \end{bmatrix} \delta u$$

CONCEPT REINFORCEMENT PROBLEM 9.6

Problem

Consider the one-axis model of a synchronous machine with the following nonlinear state equations:

$$\dot{\delta} = \omega - \omega_s$$
$$\dot{\omega} = \frac{1}{M}(T_m - T_e)$$
$$\dot{E}_q = \frac{1}{T'_d}\left(E_f - E'_q - (X_d - X'_d)\, I_d\right)$$

where δ=rotor angle, ω= rotor speed, E'_q=internal EMF, T_m=mechanical torque(input), T_e= electrical torque=$\frac{E'_q V}{X'_d}\sin(\delta)$, $I_d \approx \frac{V - E'_q \cos(\delta)}{X'_d}$, $M, T'_d, X_d, X'_d, E_f, V, \omega_s$ are constants. Assume $M = 10, T'_d = 6$, $X_d = 1.8, X'_d = 0.3$, $E_f = 1.0$ and $V = 1.0$.

The operating point: $\delta_0 = 30^0$, $\omega_0 = \omega_s = 1$, and $E'_{q_0} = 1.0$.

Linearize the model around the operating/equilibrium points.

9.4 Converting Block Diagrams to State Variable Models

Block diagrams are commonly used to represent dynamic systems through interconnected functional blocks, each describing a specific mathematical operation or relationship. In Chapter 2, we discussed methods to simplify these block diagrams. However, for more advanced analysis, such as stability, controllability, and observability studies, these diagrams must be expressed in state-variable form. This process involves defining state variables, input-output relationships, and constructing the system, input, output, and transmission matrices (**A**, **B**, **C**, and **D**) which collectively describe the system's dynamics in a compact matrix representation.

In the following examples, we will demonstrate how to derive state variable models from block diagrams. Note that, for simplicity, in block diagrams, a variable of the form $X_1(s)$ is represented as x_1. This simplification makes the diagrams less cluttered and easier to interpret while maintaining the essential mathematical relationships. The conversion between the s-domain to the time domain, along with the corresponding variables, will be clear from the context.

EXAMPLE 9.11

Problem

Obtain the state variable model of the system shown in Figure 9.16, considering x_1, x_2 as states and the reference signal as input.

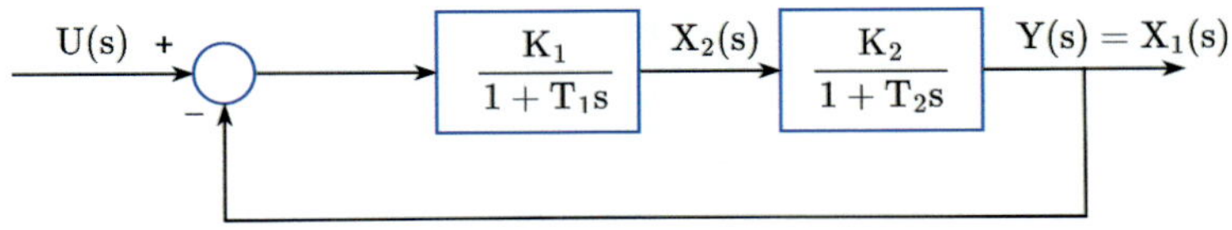

Figure 9.16: Closed-loop control system

SOLUTION:

From the schematic, consider the first block from the right and derive the state equations as follows:

$$X_2(s)\left[\frac{K_2}{1+T_2s}\right] = X_1(s) \implies X_1(s)\left[1+T_2s\right] = K_2X_2(s)$$

$$\text{or,}\quad X_1(s) + T_2sX_1(s) = K_2X_2(s)$$

$$\implies sX_1(s) = -\frac{1}{T_2}X_1(s) + \frac{K_2}{T_2}X_2(s) \tag{9.50}$$

By taking the inverse Laplace transform of Eqn(9.50) gives

$$\boxed{\dot{x}_1 = -\frac{1}{T_2}x_1 + \frac{K_2}{T_2}x_2} \tag{9.51}$$

Next, we consider the second block from the right

$$\left[U(s) - X_1(s)\right]\left[\frac{K_1}{1+T_1s}\right] = X_2(s)$$

$$\implies X_2(s)\left[1+T_1s\right] = K_1U(s) - K_1X_1(s)$$

Upon simplification, this gives

$$sX_2(s) = -\frac{K_1}{T_1}X_1(s) - \frac{1}{T_2}X_2(s) + \frac{K_1}{T_1}U(s) \tag{9.52}$$

By taking the inverse Laplace transform of Eqn(9.52) gives

$$\boxed{\dot{x}_2 = -\frac{K_1}{T_1}x_1 - \frac{1}{T_1}x_2 + \frac{K_1}{T_1}u} \tag{9.53}$$

The output $Y(s) = X_1(s)$. Taking its inverse Laplace transform gives

$$y = x_1 \tag{9.54}$$

In matrix form Eqns (9.51),(9.53), and (9.54) can be expressed as:

$$\begin{bmatrix} \dot{x}_1 \\ \dot{x}_2 \end{bmatrix} = \begin{bmatrix} -\frac{1}{T_2} & \frac{K_2}{T_2} \\ -\frac{K_1}{T_1} & -\frac{1}{T_1} \end{bmatrix} \begin{bmatrix} x_1 \\ x_2 \end{bmatrix} + \begin{bmatrix} 0 \\ \frac{K_1}{T_1} \end{bmatrix} u, \quad y = \begin{bmatrix} 1 & 0 \end{bmatrix} \begin{bmatrix} x_1 \\ x_2 \end{bmatrix} \tag{9.55}$$

EXAMPLE 9.12

Problem

Write the state equations for the system shown in Figure 9.17 using the indicated state variables.

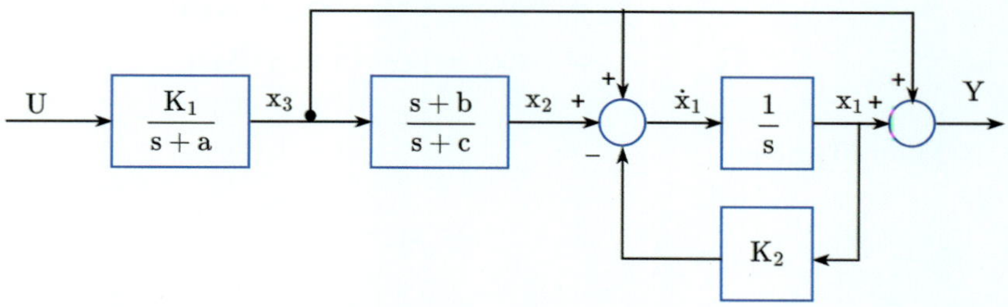

Figure 9.17: Block diagram of Example 9.12

SOLUTION:

From the feedback block on the right of Figure 9.17, we find the following relations:

$$sX_1(s) = X_2(s) + X_3(s) - K_2X_1(s)$$

The inverse Laplace transform of the preceding equation gives

$$\dot{x}_1 = -K_2x_1 + x_2 + x_3 \tag{9.56}$$

Next, consider the first block from the left of the Figure 9.17. This gives

$$\frac{K_1}{s+a}U(s) = X_3(s) \quad \text{or,} \quad sX_3(s) = -aX_3(s) + K_1U(s)$$

This gives

$$\dot{x}_3 = -ax_3 + K_1u \tag{9.57}$$

Finally, consider the second block from the right. This gives

$$X_3(s)\left[\frac{s+b}{s+c}\right] = X_2(s)$$

or,

$$sX_2(s) = -cX_2(s) + bX_3(s) + sX_3(s) \tag{9.58}$$

This gives

$$\dot{x}_2 = -cx_2 + bx_3 + \dot{x}_3 \tag{9.59}$$

Substituting value of $\dot{x}_3$ from Eqn(9.57) in Eqn(9.59) gives

$$\begin{aligned}\dot{x}_2 &= -cx_2 + bx_3 - ax_3 + K_1u \\ &= -cx_2 - [a-b]\,x_3 + K_1u\end{aligned} \tag{9.60}$$

The output is

$$y = x_1 + x_3 \tag{9.61}$$

In matrix form, these can be expressed as:

$$\begin{aligned}\begin{bmatrix}\dot{x}_1\\ \dot{x}_2\\ \dot{x}_3\end{bmatrix} &= \begin{bmatrix}-K_2 & 1 & 1\\ 0 & -c & -(a-b)\\ 0 & 0 & -a\end{bmatrix}\begin{bmatrix}x_1\\ x_2\\ x_3\end{bmatrix} + \begin{bmatrix}0\\ K_1\\ K_1\end{bmatrix}u \\ y &= \begin{bmatrix}1 & 0 & 1\end{bmatrix}\begin{bmatrix}x_1\\ x_2\\ x_3\end{bmatrix}\end{aligned} \tag{9.62}$$

CONCEPT REINFORCEMENT PROBLEM 9.7

Problem

Write the state equations for the system shown in Figure 9.18 using the indicated state variables.

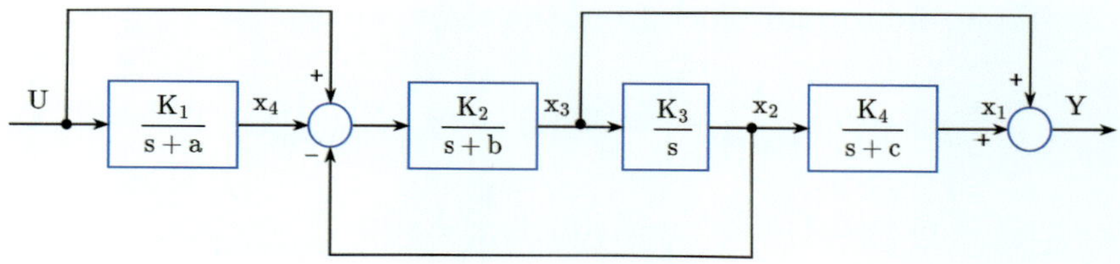

Figure 9.18: Block diagram of CRP 9.7

EXAMPLE 9.13

Problem

Obtain the state variable model of the system shown in Figure 9.19, taking the state variables as: $\mathbf{x} = \begin{bmatrix} \Delta f & \Delta P_g & \Delta X_g \end{bmatrix} = \begin{bmatrix} x_1 & x_2 & x_3 \end{bmatrix}$. Consider the incremental change in the speed changer positicn as input u and the disturbance signal ΔP_d.

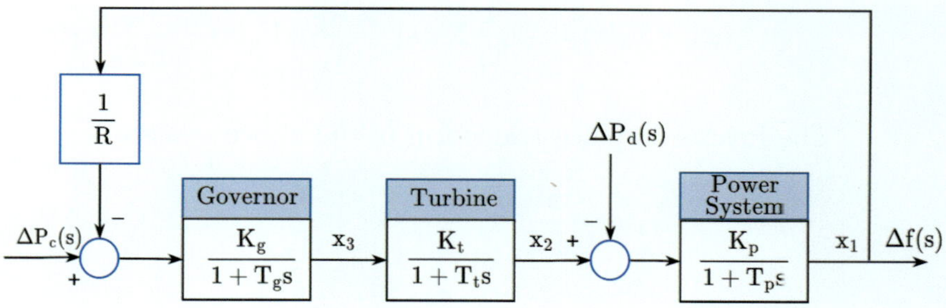

Figure 9.19: Single area thermal power system

SOLUTION:

Now from the rightmost block in Figure 9.19, we find the following relations:

$$X_1(s)\,[1+T_ps] = K_p[X_2(s) - \Delta P_d(s)]$$
$$\Longrightarrow \quad X_1(s) + T_psX_1(s) = K_pX_2(s) - K_p\Delta P_d(s)$$

Simplifying the above gives

$$sX_1(s) = -\frac{1}{T_p}X_1(s) + \frac{K_p}{T_p}X_2(s) - \frac{K_p}{T_p}\Delta P_d(s) \tag{9.63}$$

The inverse Laplace transform of the preceding equation gives

$$\dot{x}_1 = -\frac{1}{T_p}x_1 + \frac{K_p}{T_p}x_2 - \frac{K_p}{T_p}\Delta P_d \tag{9.64}$$

From the second block from the right, we get

$$X_3(s)\left[\frac{K_t}{1+T_t s}\right] = X_2(s) \implies X_2(s)\left[1+T_t s\right] = K_t X_3(s)$$

$$\text{or,}\ sX_2(s) = -\frac{1}{T_t}X_2(s) + \frac{K_t}{T_t}X_3(s) \tag{9.65}$$

Taking the inverse Laplace transform of the above gives

$$\dot{x}_2 = -\frac{1}{T_t}x_2 + \frac{K_t}{T_t}x_3 \tag{9.66}$$

From the first block from left

$$\left[\Delta P_c(s) - \frac{1}{R}X_1(s)\right]\frac{K_g}{1+T_g s} = X_3(s)$$

$$\text{or,}\quad X_3(s)\left[1+T_g s\right] = K_g \Delta P_c(s) - \frac{K_g}{R}X_1(s)$$

$$\text{or,}\quad X_3(s) + T_g s X_3(s) = K_g \Delta P_c(s) - \frac{K_g}{R}X_1(s)$$

$$\text{or,}\quad sX_3(s) = -\frac{1}{T_g}X_3(s) + \frac{K_g}{T_g}\Delta P_c(s) - \frac{K_g}{RT_g}X_1(s) \tag{9.67}$$

The inverse Laplace transform of the above results

$$\dot{x}_3 = -\frac{K_g}{RT_g}x_1 - \frac{1}{T_g}x_3 + \frac{K_g}{T_g}u, \quad (\because\ \ u = \Delta P_c) \tag{9.68}$$

Eqns (9.64), (9.66), and (9.68) can be expressed in matrix form as:

$$\begin{bmatrix}\dot{x}_1\\ \dot{x}_2\\ \dot{x}_3\end{bmatrix} = \begin{bmatrix}-\frac{1}{T_p} & \frac{K_p}{T_p} & 0\\ 0 & -\frac{1}{T_t} & \frac{K_t}{T_t}\\ -\frac{K_g}{RT_g} & 0 & -\frac{1}{T_g}\end{bmatrix}\begin{bmatrix}x_1\\ x_2\\ x_3\end{bmatrix} + \begin{bmatrix}0\\ 0\\ \frac{K_g}{T_g}\end{bmatrix}u + \begin{bmatrix}-\frac{K_p}{T_p}\\ 0\\ 0\end{bmatrix}\Delta P_d$$

$$y = \begin{bmatrix}1 & 0 & 0\end{bmatrix}\begin{bmatrix}x_1\\ x_2\\ x_3\end{bmatrix} \tag{9.69}$$

CONCEPT REINFORCEMENT PROBLEM 9.8

Problem

Write the state equations for the system shown in Figure 9.20 using the indicated state variables.

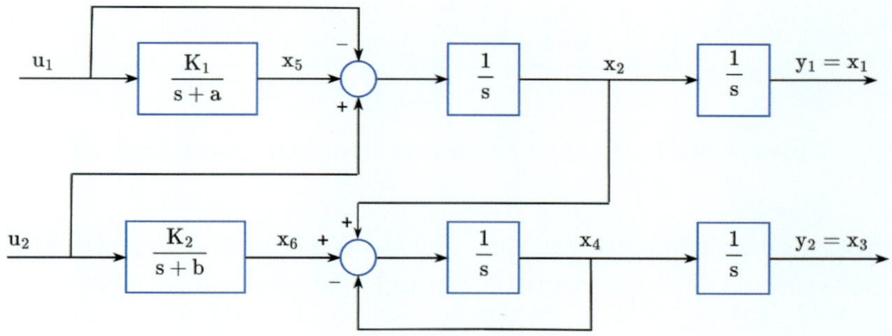

Figure 9.20: MIMO Feedback Control System

9.5 State Space Modeling of Practical Systems

Having established the fundamental principles through simple examples, we now apply state-space modeling to more complex, practical engineering systems that arise frequently in real-world applications.

State-space modeling provides a unified and compact framework for representing a wide range of physical systems across multiple engineering domains, whether mechanical, electrical, thermal, hydraulic, or electromechanical. By expressing complex system dynamics in terms of first-order differential (or difference) equations, this approach enables the application of modern control analysis and design techniques, including controllability and observability analysis, pole placement, observer design, and optimal control methods (to be discussed in later chapters).

This systematic approach not only simplifies the mathematical representation of multivariable systems but also provides deep insights into system behavior and facilitates the design of sophisticated control strategies. In this section, we present representative state-space models of various practical systems, demonstrating the versatility and power of this modeling approach across different engineering disciplines.

9.5.1 Armature Controlled DC Servo Motor

An armature-controlled DC servo motor is a widely used actuator in control systems where the field current is kept constant and control is achieved by

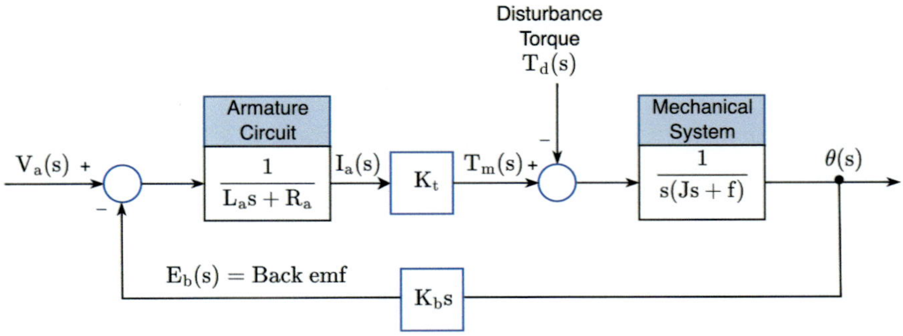

Figure 9.21: Block diagram of armature controlled DC motor

varying the armature voltage. The system consists of electrical and mechanical subsystems that are coupled through the electromagnetic torque.

The transfer function model of this system has been derived in Chapter 2 (see section 2.10.1. The block diagram of an armature-controlled DC motor is shown in Figure 9.21

Let us derive the state variable model of DC servo motor considering $x_1 = \theta$, $x_2 = \dot{\theta} = \omega$, and $x_3 = i_a$ as state variables. The input $u = v_a$.

Now the dynamics of the system is

$$J\frac{d^2\theta}{dt^2} + f\frac{d\theta}{dt} = K_t i_a - T_d \quad \text{and}$$

$$L_a\frac{di_a}{dt} + R_a i_a = v_a - K_b\frac{d\theta}{dt}$$

In terms of the state variables, it is expressed as:

$$J\dot{x}_2 + fx_2 = K_t x_3 - T_d \implies \dot{x}_2 = -\frac{f}{J}x_2 + \frac{K_t}{J}x_3 - \frac{1}{J}T_d \text{ and}$$

$$L_a\dot{x}_3 + R_a x_3 + K_b x_2 = u \implies \dot{x}_3 = -\frac{K_b}{L_a}x_2 - \frac{R_a}{L_a}x_3 + \frac{1}{L_a}u$$

Thus, the state equations can be written as

$$\begin{bmatrix}\dot{x}_1\\ \dot{x}_2\\ \dot{x}_3\end{bmatrix} = \begin{bmatrix}0 & 1 & 0\\ 0 & -\frac{f}{J} & \frac{K_t}{J}\\ 0 & -\frac{K_b}{L_a} & -\frac{R_a}{L_a}\end{bmatrix}\begin{bmatrix}x_1\\ x_2\\ x_3\end{bmatrix} + \begin{bmatrix}0\\ 0\\ \frac{1}{L_a}\end{bmatrix}u + \begin{bmatrix}0\\ -\frac{1}{J}\\ 0\end{bmatrix}T_d$$

$$y = \begin{bmatrix}1 & 0 & 0\end{bmatrix}\begin{bmatrix}x_1\\ x_2\\ x_3\end{bmatrix}$$

If the inductance of the armature is neglected, then

$$i_a = \frac{v_a - E_b}{R_a} = \frac{u - K_b x_2}{R_a}$$

The dynamics becomes

$$J\frac{d^2\theta}{dt^2} + f\frac{d\theta}{dt} = K_t i_a - T_d$$

The state equations are

$$\begin{aligned} \dot{x}_1 &= x_2 \\ \dot{x}_2 &= -\frac{f}{J}x_2 - \frac{K_t K_b}{JR_a}x_2 + \frac{K_t}{JR_a}u - \frac{T_d}{J} \end{aligned}$$

or

$$\begin{bmatrix} \dot{x}_1 \\ \dot{x}_2 \end{bmatrix} = \begin{bmatrix} 0 & 1 \\ 0 & -\dfrac{fR_a + K_t K_b}{JR_a} \end{bmatrix} \begin{bmatrix} x_1 \\ x_2 \end{bmatrix} + \begin{bmatrix} 0 \\ \dfrac{K_t}{JR_a} \end{bmatrix} u - \begin{bmatrix} 0 \\ \dfrac{1}{J} \end{bmatrix} T_d$$

CONCEPT REINFORCEMENT PROBLEM 9.9

Problem

Represent the dynamics of an armature-controlled DC motor, shown in Figure 9.21 in state space, considering the angular position, velocity, and acceleration as state variables. Assume $T_d = 0$.

9.5.2 Simple Pendulum System

Consider the schematic of an idealized pendulum with a "massless" string or rod of length **L** and a bob of mass **m** shown in Figure 9.22(a). The purpose of studying this is that the dynamics of a majority of multi-link robotics manipulators are simply the dynamics of a large number of coupled pendula. Also, the dynamics of a single pendulum are rich enough to introduce most of the concepts from nonlinear dynamics that we will use in this text, but tractable enough for us to (mostly) understand in the next few pages.

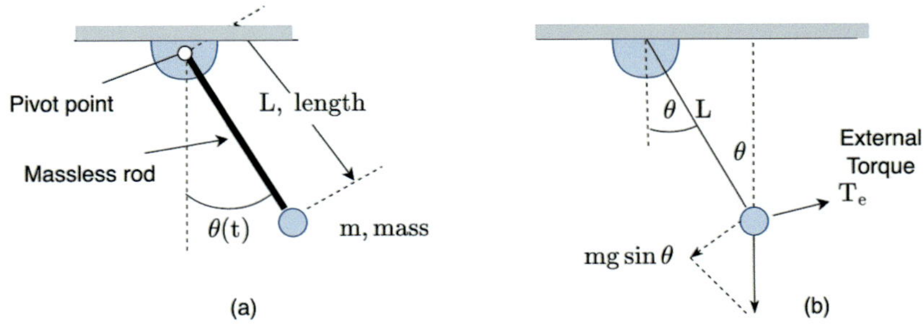

Figure 9.22: (a) Schematic of a simple pendulum (b) Simplified free body diagram

The detailed diagram is shown in Figure 9.22(b).

The dotted vertical line shows the rest position of the bob. When the bob is moved from its rest position and let go, it swings back and forth. The time it takes the pendulum to swing from its farthest right position to its farthest left position and back to its next farthest right position is the period of the pendulum.

Let us apply the rotational version of Newton's second law, which states that the total torque equals moment of inertia times angular acceleration. Thus, the pendulum torque plus the external torque T_e must be equal to the product of inertia with acceleration. Thus,

$$T_e + T_{pendulum} = J\ddot{\theta} \tag{9.70}$$

where J is the moment of inertia of the pendulum around the point of rotation. It is equal to mL^2. The pendulum torque is given by

$$T_{pendulum} = -mg\sin\theta \times L \tag{9.71}$$

This is the torque due to gravity trying to restore the pendulum to its equilibrium position. Hence, the dynamics of the system are given by

$$J\frac{d^2\theta}{dt^2} + mgL\sin\theta = T_e \tag{9.72}$$

We will construct a model to describe how the angle θ of the pendulum varies as a function of time "t". If $x_1 = \theta$, $x_2 = \dot{\theta}$, and $u = T_e$, then the state space model of this system is expressed as:

$$\begin{aligned}
\dot{x}_1 &= x_2 = f_1(x_1, x_2, u) \\
\dot{x}_2 &= -\frac{mgL}{J}\sin x_1 + \frac{1}{J}u \\
&= -\frac{g}{L}\sin x_1 + \frac{1}{mL^2}u = f_2(x_1, x_2, u) \quad (\because\ J = mL^2)
\end{aligned} \tag{9.73}$$

Assume that the output is the horizontal displacement of the pendulum bob, i.e.,

$$y = L\sin\theta = L\sin x_1 = h(x, u)$$

Let us linearize the state space model of the system around the equilibrium point $\theta_0 = x_{10} = 0$ (pendulum hanging down), $\dot{\theta}_0 = x_{20} = 0$ and input $u_0 = 0$, that is, there is no external torque such that the pendulum is at equilibrium.

For small values of x_1, $\sin x_1 \approx x_1$. The linearized state model is given by

$$\begin{bmatrix} \dot{x}_1 \\ \dot{x}_2 \end{bmatrix} = \begin{bmatrix} 0 & 1 \\ -\dfrac{g}{L} & 0 \end{bmatrix} \begin{bmatrix} x_1 \\ x_2 \end{bmatrix} + \begin{bmatrix} 0 \\ \dfrac{1}{mL^2} \end{bmatrix} u$$

$$y = \begin{bmatrix} L & 0 \end{bmatrix} \begin{bmatrix} x_1 \\ x_2 \end{bmatrix} \tag{9.74}$$

Alternatively, the Jacobian matrix evaluated at the chosen operating point is

$$A = \left.\frac{\partial f}{\partial x}\right|_{(x_0,u_0)} = \begin{bmatrix} \dfrac{\partial f_1}{\partial x_1} & \dfrac{\partial f_1}{\partial x_2} \\ \dfrac{\partial f_2}{\partial x_1} & \dfrac{\partial f_2}{\partial x_2} \end{bmatrix}_{(x_0,u_0)} = \begin{bmatrix} 0 & 1 \\ -\dfrac{g}{L}\cos x_1 & 0 \end{bmatrix}_{(0,0)} = \begin{bmatrix} 0 & 1 \\ -\dfrac{g}{L} & 0 \end{bmatrix}$$

$$B = \begin{bmatrix} \dfrac{\partial f_1}{\partial u} \\ \dfrac{\partial f_2}{\partial u} \end{bmatrix}_{(x_0,u_0)} = \begin{bmatrix} 0 \\ \dfrac{1}{mL^2} \end{bmatrix} \text{ and } \mathbf{C} = \left.\frac{\partial h}{\partial x}\right|_{((x_0,u_0)} = \begin{bmatrix} L\cos(x_1) & 0 \end{bmatrix} = \begin{bmatrix} L & 0 \end{bmatrix}$$

So the linearized system is same as Eqn(9.74)

9.5.3 Buck Converter Circuit

Consider the Buck converter shown in Figure 9.23. This is a DC-DC converter that steps down the input voltage to a lower output voltage by switching a transistor on and off rapidly, using an inductor and a capacitor to filter and smooth the output. The output voltage v_0 is directly proportional to the duty cycle D. Let us derive the state space model of the Buck converter shown in considering the inductor current and the capacitor voltage as state variables, that is, $x_1 = i_L$ and $x_2 = v_c$. The voltage across the capacitor/resistor is the output.

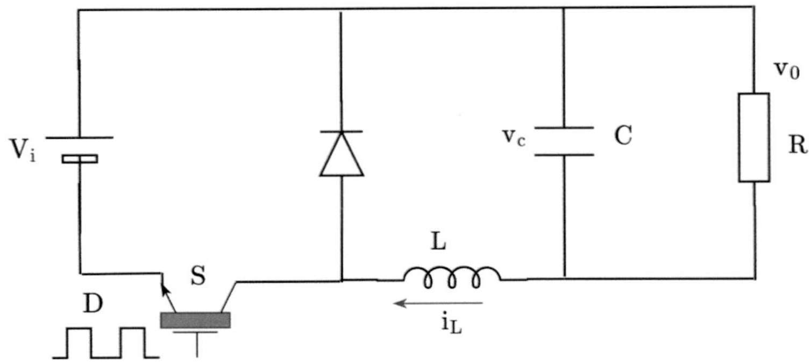

Figure 9.23: Buck converter

Let us write the dynamics of the system when the switch is ON and OFF.

SWITCH ON:

$$L\frac{di_L(t)}{dt} = V_{in} - v_c \implies \dot{x}_1 = -\frac{1}{L}x_2 + \frac{1}{L}u$$
$$C\frac{dv_c(t)}{dt} = i_L - \frac{v_c}{R} \implies \dot{x}_2 = \frac{1}{C}x_1 - \frac{1}{RC}x_2 \quad (9.75)$$

When Eqn(9.75) is represented in the form

$$\dot{\mathbf{x}} = A_1\mathbf{x} + B_1 u, \quad y = Cx \quad \text{gives} \quad (9.76)$$

$$A_1 = \begin{bmatrix} 0 & -\frac{1}{L} \\ \frac{1}{C} & -\frac{1}{RC} \end{bmatrix}, \quad B_1 = \begin{bmatrix} \frac{1}{L} \\ 0 \end{bmatrix}, \quad C = \begin{bmatrix} 0 & 1 \end{bmatrix}$$

SWITCH OFF:

$$L\frac{di_L(t)}{dt} = -v_c \implies \dot{x}_1 = -\frac{1}{L}x_2$$
$$(9.77)$$
$$C\frac{dv_c(t)}{dt} = i_L - \frac{v_c}{R} \implies \dot{x}_2 = \frac{1}{C}x_1 - \frac{1}{RC}x_2 \quad (9.78)$$

When Eqn(9.79) is represented in the form

$$\dot{\mathbf{x}} = A_2\mathbf{x} + B_2 u, \quad y = Cx \quad \text{gives} \quad (9.79)$$

$$A_2 = \begin{bmatrix} 0 & -\frac{1}{L} \\ \frac{1}{C} & -\frac{1}{RC} \end{bmatrix}, \quad B_2 = \begin{bmatrix} 0 \\ 0 \end{bmatrix}, \quad C = \begin{bmatrix} 0 & 1 \end{bmatrix}$$

Let D is the duty cycle. Then, the dynamics of the converter can be expressed as:

$$\dot{\mathbf{x}} = \mathrm{A}\mathbf{x} + \mathrm{Bu}, \quad \mathrm{y} = \mathrm{Cx} \quad \text{where} \tag{9.80}$$

$$\mathrm{A} = \mathrm{DA}_1 + (1 - \mathrm{D})\,\mathrm{A}_2, \text{ and } \mathrm{B} = \mathrm{DB}_1 + (1 - \mathrm{D})\,\mathrm{B}_2$$

CONCEPT REINFORCEMENT PROBLEM 9.10

Problem

Derive the state space model of the Boost converter shown in Figure 9.24 considering the inductor current and the capacitor voltage as state variables, that is, $x_1 = i_L$ and $x_2 = v_c$. The voltage across the capacitor is the output.

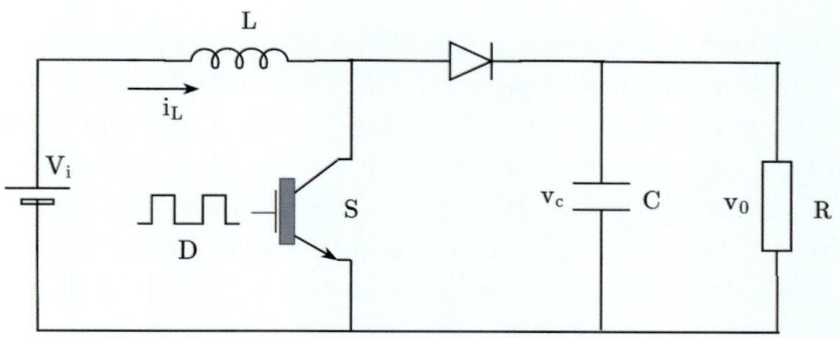

Figure 9.24: Boost converter

9.5.4 Inverted Pendulum

The inverted pendulum represents one of the most fundamental and challenging problems in control systems engineering. This seemingly simple system (consisting of a pendulum balanced upside-down on a movable cart) serves as a cornerstone for understanding advanced control concepts and has profound implications for numerous real-world applications. The principles governing inverted pendulum control are directly applicable to many modern technologies, including humanoid robots and Segways. More broadly, the inverted pendulum serves as a benchmark system for testing and validating new control algorithms before implementing them in more complex, safety-critical applications such as aircraft autopilot systems, robotic manipulators, and industrial automation.

Despite its apparent simplicity, the inverted pendulum presents significant modeling and control challenges that make it an ideal educational and research platform. The system exhibits inherent instability. Any small deviation from the upright position leads to exponential growth unless actively controlled. This creates a non-minimum phase behavior that complicates controller design. The system demonstrates strong nonlinear coupling between the cart's horizontal motion and the pendulum's rotational dynamics. The mathematical model involves coupled differential equations with trigonometric nonlinearities that cannot be easily separated or linearized without approximations. Additionally, the system is underactuated, meaning it has fewer control inputs (one horizontal force) than degrees of freedom (two), requiring sophisticated control strategies to achieve simultaneous position and angle regulation. These characteristics make the inverted pendulum an excellent testbed for exploring advanced control techniques while providing insights into the fundamental principles that govern dynamic system behavior. The following derivation presents the complete mathematical model that captures these complex dynamics.

The free body diagram of an inverted pendulum mounted on a motor-driven cart is shown in Figure 9.25. It is assumed that the pendulum rod is massless, and the hinge is frictionless. Therefore, the whole pendulum mass is concentrated in the center of gravity (COG) located at the center of the pendulum ball. The mass of the cart and the ball at the upper end of the inverted pendulum are denoted respectively by M and m. The external force u(t) is acting on the cart in the x-direction, and a gravity force acts on the point mass at all times. The coordinate system considered is shown in Figure 9.25, where x(t) represents the cart position, and θ(t) is the tilt angle referenced to the vertically upward direction.

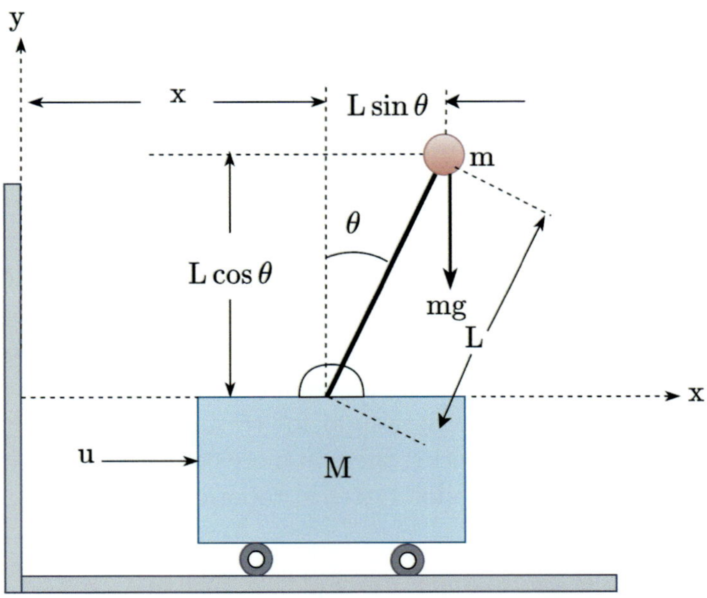

Figure 9.25: Schematic of Inverted Pendulum

Let θ denote the angle of the rod from the vertical line. Let (x_g, y_g) denote the (x, y) coordinates of the center of gravity of the rod. Then

$$\begin{aligned} x_g &= x + L\sin\theta \\ y_g &= L\cos\theta \end{aligned} \tag{9.81}$$

A force balance on the system in the x-direction can be written as

$$M\frac{d^2x}{dt^2} + m\frac{d^2x_g}{dt^2} = u \tag{9.82}$$

where L is the length of the pendulum. Now

$$\begin{aligned} \frac{d^2}{dt^2}[x + L\sin\theta] &= \ddot{x} + L\frac{d^2}{dt^2}[\sin\theta] = \ddot{x} + L\frac{d}{dt}\left[\cos\theta \times \dot{\theta}\right] \\ &= \ddot{x} + L\left[\ddot{\theta}\cos\theta - \sin\theta\,\dot{\theta}^2\right] \end{aligned} \tag{9.83}$$

Hence Eqn(9.82) can be expressed as:

$$\boxed{(M + m)\,\ddot{x} - mL\sin\theta\,\dot{\theta}^2 + mL\cos\theta\,\ddot{\theta} = u} \tag{9.84}$$

Similar to the force-balance equation, the torque-balance equation can be written as:

$$(F_x\cos\theta)\,L - (F_x\sin\theta)\,L = (mg\sin\theta)\,L \tag{9.85}$$

where $F_x = m\frac{d^2}{dt^2}x_g$ and $F_y = m\frac{d^2}{dt^2}y_g$ are the components of the force in x and y directions, respectively.

Manipulating Eqn(9.85) gives

$$\boxed{m\ddot{x}\cos\theta + mL\ddot{\theta} = mg\sin\theta \implies \ddot{x}\cos\theta + L\ddot{\theta} = g\sin\theta} \tag{9.86}$$

A linearized model of the inverted pendulum can be obtained from Eqns(9.84) and (9.86) by assuming that during normal operating position (pendulum is in upright position), the pendulum displacement θ is very small so that $\sin\theta \approx \theta$, $\cos\theta \approx 1$, $\theta\dot{\theta}^2 \approx 0$.

Under these assumptions, we can obtain the linearized equations of the inverted pendulum from Eqns(9.84) and (9.86) as:

$$(M + m)\,\ddot{x} + mL\ddot{\theta} = u \tag{9.87}$$

$$\ddot{x} + L\ddot{\theta} = g\theta \tag{9.88}$$

Now $\ddot{x} = g\theta - L\ddot{\theta}$. Substituting this into Eqn(9.87) gives

$$(M+m)\left[g\theta - L\ddot{\theta}\right] + mL\ddot{\theta} = u \implies ML\ddot{\theta} = (M+m)g\theta - u \tag{9.89}$$

Thus

$$\boxed{\ddot{\theta} = \frac{(M+m)g}{ML}\theta - \frac{1}{ML}u} \tag{9.90}$$

Substituting the value of $\ddot{\theta}$ from Eqn(9.90) into Eqn(9.88) gives

$$\boxed{\ddot{x} = g\theta - L\ddot{\theta} = -\frac{mg}{M}\theta + \frac{1}{M}u} \tag{9.91}$$

Define the state variables as: $x_1 = \theta$, $x_2 = \dot{\theta}$, $x_3 = x$, and $x_4 = \dot{x}$

The state equations are written as:

$$\begin{aligned}
\dot{x}_1 &= x_2 \\
\dot{x}_2 &= \frac{(M+m)g}{ML}x_1 - \frac{1}{ML}u \\
\dot{x}_3 &= x_4 \\
\dot{x}_4 &= -\frac{m}{M}gx_1 + \frac{1}{M}u
\end{aligned} \tag{9.92}$$

If the cart position and angle of the pendulum are considered as output, then the outputs of the system are expressed as:

$$\mathbf{y} = \begin{bmatrix} y_1 \\ y_2 \end{bmatrix} = \begin{bmatrix} \theta \\ x \end{bmatrix} \tag{9.93}$$

In matrix form, it is represented by

$$\begin{bmatrix} \dot{x}_1 \\ \dot{x}_2 \\ \dot{x}_3 \\ \dot{x}_4 \end{bmatrix} = \begin{bmatrix} 0 & 1 & 0 & 0 \\ \frac{(M+m)g}{ML} & 0 & 0 & 0 \\ 0 & 0 & 0 & 1 \\ -\frac{mg}{M} & 0 & 0 & 0 \end{bmatrix} \begin{bmatrix} x_1 \\ x_2 \\ x_3 \\ x_4 \end{bmatrix} + \begin{bmatrix} 0 \\ -\frac{1}{ML} \\ 0 \\ \frac{1}{M} \end{bmatrix} u$$

$$\mathbf{y} = \begin{bmatrix} 1 & 0 & 0 & 0 \\ 0 & 0 & 1 & 0 \end{bmatrix} \begin{bmatrix} x_1 \\ x_2 \\ x_3 \\ x_4 \end{bmatrix} \tag{9.94}$$

9.5.5 Unidirectional Inductive Power Transfer System

Unidirectional inductive power transfer (IPT) systems enable wireless transmission of electrical energy from a stationary primary source to a mobile or isolated secondary load through electromagnetic induction, eliminating the need for physical electrical connections. These systems consist of a primary coil energized by a high-frequency AC source and a secondary coil positioned within the magnetic field, where power flows in only one direction from transmitter to receiver. The complex electromagnetic coupling, sensitivity to geometric parameters, and dynamic operating conditions necessitate accurate mathematical modeling for systematic design, control optimization, and reliable operation of these wireless power transfer systems.

Consider the IPT system shown in Figure 9.26.

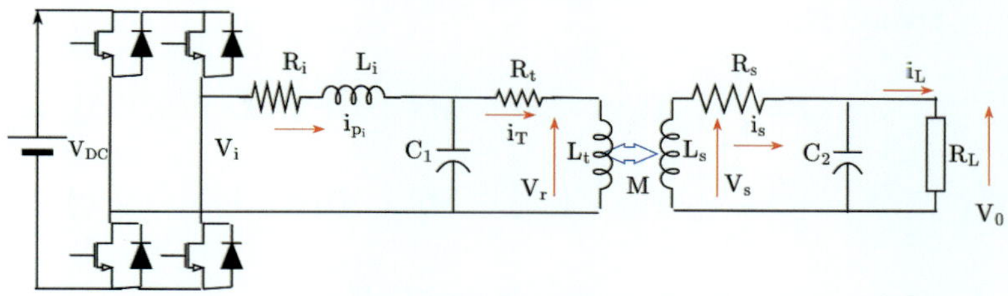

Figure 9.26: Unidirectional IPT system

The dynamic model of the system is developed by selecting the state variables as:

$$\mathbf{x} = \begin{bmatrix} x_1 & x_2 & x_3 & x_4 & x_5 \end{bmatrix}^T = \begin{bmatrix} i_{p_i} & v_{C_1} & i_T & i_s & v_{C_2} \end{bmatrix}^T$$

where i_{p_i} is the current through the primary input inductor L_{pi}, v_{C_1} is the voltage across primary capacitor C_1, i_T is the current through the primary track inductor L_t, i_s is the current through the pick-up inductor L_s, and v_{C_2} is the voltage across secondary capacitor C_2. Let the input u of the system be equal to v_{DC}, and the voltage across the secondary capacitor is the output.

The instantaneous value of induced voltage $v_s(t)$ of the pick-up coil L_s due to track current i_T is

$$v_s(t) = M\frac{di_T(t)}{dt}$$

where M is the mutual inductance between the track inductance L_t and pick-up coil inductance L_s. Further, the instantaneous value of the voltage $v_r(t)$ reflected back into the track due to current $i_s(t)$ in the pick-up coil is

$$v_r(t) = -M\frac{di_s(t)}{dt}$$

The dynamics of the circuit using principles of circuit theory is

$$R_i i_{p_i} + L_i \frac{di_{p_i}}{dt} + v_{C_1} = v_i \implies R_i x_1 + L_i \dot{x}_1 + x_2 = u \tag{9.95a}$$

$$C_1 \frac{dv_{C_1}}{dt} + i_T = i_{pi} \implies C_1 \dot{x}_2 + x_3 = x_1 \tag{9.95b}$$

$$R_t i_T + L_t \frac{di_T(t)}{dt} + v_r - v_{C_1} = 0 \implies R_t x_3 + L_t \dot{x}_3 - M\dot{x}_4 - x_2 = 0 \tag{9.95c}$$

$$R_s i_s + L_s \frac{di_s(t)}{dt} + v_{C_2} = v_s(t) \implies R_s x_4 + L_s \dot{x}_4 + x_5 = M\dot{x}_3 \tag{9.95d}$$

$$C_2 \frac{dv_{C_2}}{dt} = i_s(t) - i_L(t) \implies C_2 \dot{x}_5 = x_4 - \frac{1}{R_L} x_5 \tag{9.95e}$$

We need to eliminate $\dot{x}_4$ and $\dot{x}_3$ from Eqn(9.95c) and Eqn(9.95d), respectively. Now

$$\dot{x}_4 = \frac{1}{L_s}\left[-R_s x_4 - x_5 + M\dot{x}_3\right] \text{ (from Eqn(9.95d)) and} \tag{9.96}$$

$$\dot{x}_3 = \frac{1}{L_t}\left[-R_t x_3 + x_2 + M\dot{x}_4\right] \text{ (from Eqn(9.95c))} \tag{9.97}$$

After some minor manipulation (left as an exercise), the dynamics of the IPT system are expressed as:

$$\begin{aligned}
\dot{x}_1 &= -\frac{R_i}{L_i} x_1 - \frac{1}{L_i} x_2 + \frac{1}{L_i} u \\
\dot{x}_2 &= \frac{1}{C_1} x_1 - \frac{1}{C_1} x_3 \\
\dot{x}_3 &= \frac{\gamma}{L_t} x_2 - \frac{\gamma R_t}{L_t} x_3 - \gamma\beta R_s x_4 - \gamma\beta x_5 \\
\dot{x}_4 &= \gamma\beta x_2 - \gamma\beta R_t x_3 - \frac{\gamma R_s}{L_s} x_4 - \frac{\gamma}{L_s} x_5 \\
\dot{x}_5 &= \frac{1}{C_2} x_4 - \frac{1}{R_L C_2} x_5
\end{aligned} \tag{9.98}$$

where $\beta = \dfrac{M}{L_s L_t}$ and $\gamma = \dfrac{1}{1 - M\beta} = \dfrac{L_s L_t}{L_s L_t - M^2}$

9.5.6 Chemostat: Nutrient–Microbe Dynamics

A chemostat can be represented as a nonlinear dynamic system where nutrient concentration and microbial biomass are the key state variables. By accounting for nutrient inflow, outflow, and consumption-driven biomass growth, the process can be formulated in a state-space form suitable for control analysis and design.

Consider a continuous stirred-tank bioreactor (chemostat) where nutrient and microorganism concentrations are denoted by a(t), and b(t), respectively. The system operates under the following assumptions: (i) nutrient is supplied at concentration a_0 with dilution rate D, giving supply rate Da_0 (ii) both nutrient and microorganism are removed at dilution rate D; (iii) microorganisms consume nutrient according to consumption function C[a(t)], with total consumption rate kC[a(t)]b(t); and (iv) nutrient consumption directly translates to microorganism growth.

Applying mass balance principles, the governing equations are:

$$\begin{aligned}\dot{a}(t) &= Da_0 - Da(t) - kC\left[a(t)\right]b(t)\\ \dot{b}(t) &= kC[a(t)]b(t) - Db(t)\end{aligned} \tag{9.99}$$

Defining state variables $x_1 = a$, $x_2 = b$, and input $u = a_0$, the state-space representation becomes:

$$\begin{aligned}\dot{x}_1 &= -Dx_1 - kC(x_1)x_2 + Du\\ \dot{x}_2 &= -Dx_2 + kC(x_1)x_2\end{aligned} \tag{9.100}$$

The input can alternately be defined as $u = Da_0$, and the output may be chosen as $y = x_1$ (nutrient), $y = x_2$(microorganism), or $y = [x_1\ \ x_2]^T$(both concentrations).

This nonlinear system model is fundamental in bioprocess control, with typical parameters: $D = 0.1 - 2.0h^{-1}$, $k = 0.1 - 0.8g/g$, and the function $C(x_1) = x_1$ for Lotka-Volterra kinetics; $C(x_1) = \dfrac{x_1}{m + x_1}L$ for Michaelis-Menten kinetics.

9.5.7 Magnetic Levitation System

Consider the magnetic levitation system shown in Figure 9.27

The current through the coil induces a magnetic force that can balance the force of gravity and cause the ball to be suspended in mid-air.

Let h=the vertical position of the ball, i = the current through the electromagnet, v =applied voltage, M=mass of the ball, g = gravity L = coil inductance and R = resistance of the coil, K= = coefficient that determines the magnetic force exerted on the ball.

The magnetic force is proportional to the square of the current but inversely proportional to the square of the gap distance because the magnetic field strength weakens with distance according to the inverse square law. The dynamics of the magnetic levitation system are given by

$$\begin{aligned}M\frac{d^2h(t)}{dt^2} &= Mg - \frac{Ki^2}{h^2}\\ Ri + L\frac{di}{dt} &= v(t)\end{aligned} \tag{9.101}$$

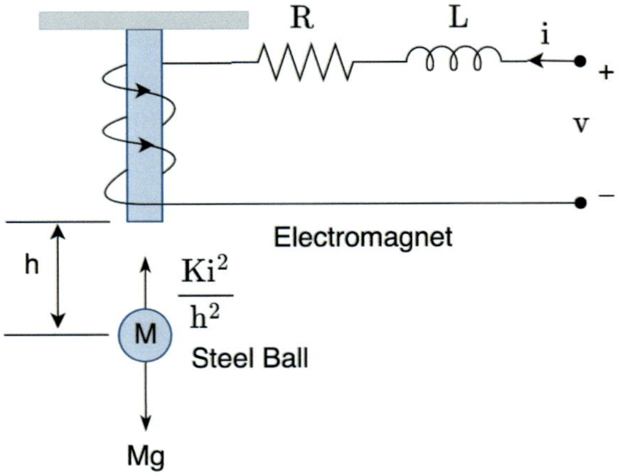

Figure 9.27: Magnetic levitation system

Select the vertical position (displacement), vertical velocity, and current as state variables, the position being the output, and the applied voltage being the input. That is, $x_1 = h$, $x_2 = \dot{h}$, $x_3 = i$, $y = h$ and $u = v$. The dynamics of the system in terms of these variables is given by

$$\begin{aligned} \dot{x}_1 &= x_2 = f_1(x_1, x_2, x_3) \\ \dot{x}_2 &= g - \frac{Kx_3^2}{Mx_1^2} = f_2(x_1, x_2, x_3) \\ \dot{x}_3 &= -\frac{R}{L}x_3 + \frac{1}{L}u = f_3(x_1, x_2, x_3, u) \end{aligned} \tag{9.102}$$

$$y = x_1 \tag{9.103}$$

The Jacobian matrix of the system is

$$J = \begin{bmatrix} \frac{\partial f_1(.)}{x_1} & \frac{\partial f_1(.)}{x_2} & \frac{\partial f_1(.)}{x_3} \\ \frac{\partial f_2(.)}{x_1} & \frac{\partial f_2(.)}{x_2} & \frac{\partial f_2(.)}{x_3} \\ \frac{\partial f_3(.)}{x_1} & \frac{\partial f_3(.)}{x_2} & \frac{\partial f_3(.)}{x_3} \end{bmatrix} = \begin{bmatrix} 0 & 1 & 0 \\ \frac{2Kx_3^2}{Mx_1^3} & 0 & -\frac{2Kx_3}{Mx_1^2} \\ 0 & 0 & -\frac{R}{L} \end{bmatrix} \tag{9.104}$$

At equilibrium point $(h = h_0,\ i = i_0)$

$$Mg = \frac{Ki_0^2}{h_0^2} \implies i_0 = \sqrt{\frac{Mgh_0^2}{K}}$$

Let us linearize the system about the equilibrium point $h_0(t) = x_{1s} =$ constant. Then

$$x_{2s} = \frac{dx_{1s}}{dt} = 0$$

Thus, the linearized system around the equilibrium point, when expressed in standard state space form, gives

$$\mathbf{A} = \begin{bmatrix} 0 & 1 & 0 \\ \frac{2Kx_{03}^2}{Mx_{01}^3} & 0 & -\frac{2Kx_{03}}{Mx_{01}^2} \\ 0 & 0 & -\frac{R}{L} \end{bmatrix} \text{ and } \mathbf{B} = \begin{bmatrix} 0 \\ 0 \\ \frac{1}{L} \end{bmatrix}$$

The parameters of the system considered are M = 0.05kg, g = 9.81 m/s^2, L = 0.01 H, R = 1Ω, K = 0.0001 Nm^2/A^2. Let the equilibrium position $h_0 = 0.02$ m. This gives $i_0 = \sqrt{Mgh_0^2/K} = 1.4$A. Therefore, the system and input matrices become

$$\mathbf{A} = \begin{bmatrix} 0 & 1 & 0 \\ 980 & 0 & -14 \\ 0 & 0 & -100 \end{bmatrix}, \quad \mathbf{B} = \begin{bmatrix} 0 \\ 0 \\ 100 \end{bmatrix}$$

9.5.8 Rumor Spreading Dynamics

Consider a fixed population T and a rumor spreading through it. Individuals are in one of three classes: Ignorants I (have not heard the rumor), Spreaders S (actively transmit the rumor), and Stiflers R (have heard the rumor but no longer spread it). Thus

$$T = I + S + R$$

Assume homogeneous mixing with average contact rate k. An Ignorant becomes a Spreader on contact with a Spreader with probability β. A Spreader becomes a Stifler when meeting another Spreader or a Stifler, with probability α. Normalizing by T such that $I + S + R = 1$ yields the deterministic mean-field ISR model below.

$$\begin{aligned} \dot{I} &= -\beta kIS \\ \dot{S} &= \beta kIS - \alpha kS(S+R) \quad (I+S+R=1) \\ \dot{R} &= \alpha kS(S+R) \end{aligned} \tag{9.105}$$

9.5.9 Liquid Level Control System

The three-level tank system, shown in Figure 9.28, consists of three tanks connected in series, where liquid flows from the top tank to the bottom tank through orifices with flow resistances.

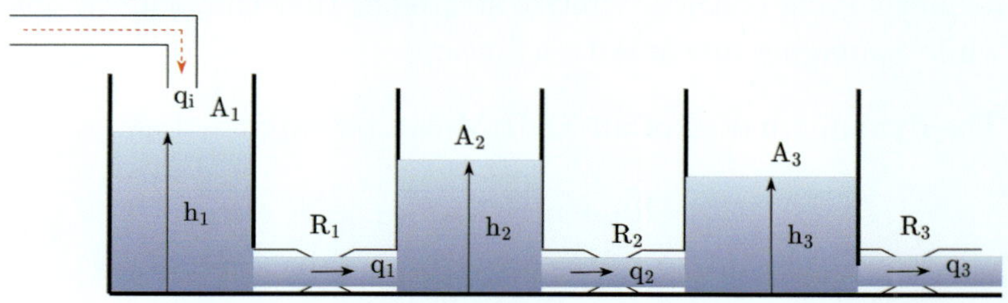

Figure 9.28: Liquid level system

Let A_1, A_2, and A_3 denote the cross-sectional areas of tanks 1, 2, and 3, respectively. h_1, h_2, and h_3 are liquid levels in tanks 1, 2, and 3, respectively. R_1, R_2, and R_3 denote flow resistances at orifices 1, 2, and 3, respectively. q_i is the Input flow rate to tank 1 (control input), and q_1,q_2, and q_3 are output flow rates from tanks 1, 2, and 3, respectively.

For each tank, the rate of change of liquid volume equals the difference between input and output flow rates. The basic equations describing the flow dynanmics are:

$$q_i - q_1 = q_i - (h_1 - h_2)/R_1 = A_1\dot{h}_1 \tag{9.106}$$

$$q_1 - q_2 = (h_1 - h_2)/R_1 - (h_2 - h_3)/R_2 = A_2\dot{h}_2 \tag{9.107}$$

$$q_2 - q_3 = (h_2 - h_3)/R_2 - h_3/R_3 = A_3\dot{h}_3 \tag{9.108}$$

Defining $\mathbf{x} = \begin{bmatrix} x_1 & x_2 & x_3 \end{bmatrix}^T = \begin{bmatrix} h_1 & h_2 & h_3 \end{bmatrix}^T$ as the state vector, the state equations of the system will be

$$\dot{\mathbf{x}} = \begin{bmatrix} -\dfrac{1}{A_1R_1} & \dfrac{1}{A_1R_1} & 0 \\ \dfrac{1}{A_2R_1} & -\dfrac{1}{A_2R_1} - \dfrac{1}{A_2R_2} & \dfrac{1}{A_2R_2} \\ 0 & \dfrac{1}{A_3R_2} & -\dfrac{1}{A_3R_2} - \dfrac{1}{A_3R_3} \end{bmatrix} \mathbf{x} + \begin{bmatrix} \dfrac{1}{A_1} \\ 0 \\ 0 \end{bmatrix} q_i$$

Output definition depends on the application. If we want to control the liquid level in all tanks, then the output vector is defined as:

$$y = \begin{bmatrix} y_1 & y_2 & y_3 \end{bmatrix}^T$$

9.5.10 Single Link Flexible Joint Robotic Manipulator

Modern robotic manipulators increasingly incorporate flexible joints to reduce weight, improve energy efficiency, and enhance safety in human-robot interaction. However, joint flexibility introduces additional complexity in the form of elastic deformation between the actuator and the link, creating a two-degree-of-freedom system where the motor position and link position can differ significantly. Accurate modeling of these flexible joint dynamics is essential for developing effective control strategies that can achieve precise positioning while managing unwanted oscillations.

The dynamic model of an SFJRM can be expressed as:

$$\begin{aligned} I\ddot{\theta}_1 + mgL\sin\theta_1 + k(\theta_1 - \theta_2) &= 0 \\ J\ddot{\theta}_2 - k(\theta_1 - \theta_2) &= T \end{aligned} \tag{9.109}$$

where θ_1, θ_2 are angular positions of the link and motor respectively, I, J are moments of inertia, m and L are mass and length of the link respectively, and k is the spring constant representing link flexibility, and T is the applied torque.Define the state variables as:

$$\mathrm{x_1} = \theta_1, \quad \mathrm{x_2} = \dot{\theta}_1, \quad \mathrm{x_3} = \theta_2 \quad \text{and} \quad \mathrm{x_4} = \dot{\theta}_2 \text{ and } \mathrm{u} = \mathrm{T}$$

The nonlinear model of the manipulator is expressed as:

$$\begin{aligned}
\dot{\mathrm{x}}_1 &= \mathrm{x_2} \\
\dot{\mathrm{x}}_2 &= -\frac{\mathrm{mgL}}{\mathrm{I}}\sin \mathrm{x_1} - \frac{\mathrm{k}}{\mathrm{I}}(\mathrm{x_1} - \mathrm{x_3}) \\
\dot{\mathrm{x}}_3 &= \mathrm{x_4} \\
\dot{\mathrm{x}}_4 &= \frac{\mathrm{k}}{\mathrm{J}}(\mathrm{x_1} - \mathrm{x_3}) + \frac{1}{\mathrm{J}}\mathrm{u}
\end{aligned} \tag{9.110}$$

The linearized model around is obtained by taking partial derivatives of the nonlinear functions around the equilibrium point ($\mathrm{x_{1n}, x_{2n}, x_{3n}, x_{4n}}$ and $\mathrm{u_n}$).

$$\dot{\mathbf{x}} = \mathbf{A}\mathbf{x} + \mathbf{B}u$$

where

$$\mathbf{A} = \begin{bmatrix} 0 & 1 & 0 & 0 \\ -\dfrac{\mathrm{mgL}\cos \mathrm{x_{1n}} + \mathrm{k}}{\mathrm{I}} & 0 & \dfrac{\mathrm{k}}{\mathrm{I}} & 0 \\ 0 & 0 & 0 & 1 \\ \dfrac{\mathrm{k}}{\mathrm{J}} & 0 & -\dfrac{\mathrm{k}}{\mathrm{J}} & 0 \end{bmatrix}, \quad \mathbf{B} = \begin{bmatrix} 0 \\ 0 \\ 0 \\ \dfrac{1}{\mathrm{J}} \end{bmatrix}$$

Assuming that the output variable is the angular position of the link, that is $\mathrm{y} = \mathrm{x_1}$, then, the output matrix

$$\mathbf{C} = \begin{bmatrix} 1 & 0 & 0 & 0 \end{bmatrix}$$

The Generic State-Space Model

We have discussed the modeling methodology for representing several practical systems in state-space form, demonstrating the versatility and unifying power of this mathematical framework across diverse engineering domains such as electrical circuits, power systems, mechanical structures, biological processes, and social dynamics. These examples illustrate how complex physical phenomena, regardless of their underlying nature, can be systematically captured using the state-space approach.

In the subsequent chapters, our focus will shift from system-specific models to the generic linear state-space representation:

$$\dot{\mathbf{x}} = \mathbf{A}\mathbf{x} + \mathbf{B}\mathbf{u}$$
$$\mathrm{y} = \mathbf{C}\mathbf{x} + \mathbf{D}\mathbf{u}$$

The theory and design methodologies developed in subsequent chapters apply universally to all systems expressible in this form. This encompasses both inherently linear systems and linearized models of nonlinear systems around equilibrium points. Consequently, the mathematical tools and control strategies presented have broad applicability across diverse engineering applications.

9.6 Multiple Choice Skills Test Questions

M 9.1. In the state variable representation, a system described by an n-th order differential equation is decomposed into

A. 2n-number of first order differential equations

B. $\frac{n}{2}$ number of first order differential equations

C. n-number of first order differential equations

D. n-number of second order differential equations

M 9.2. Consider the circuit shown in Figure 9.29.

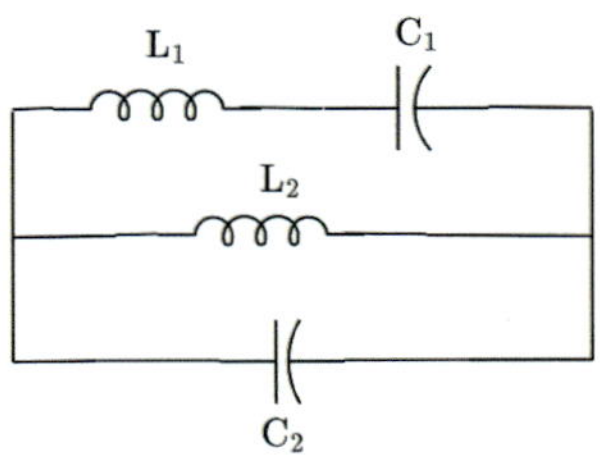

Figure 9.29: Linear Circuit

The number of state variables needed to represent this system in state space is

A. 3 **B.** 4 **C.** 2 **D.** 5

M 9.3. Consider the circuit shown in Figure 9.30.

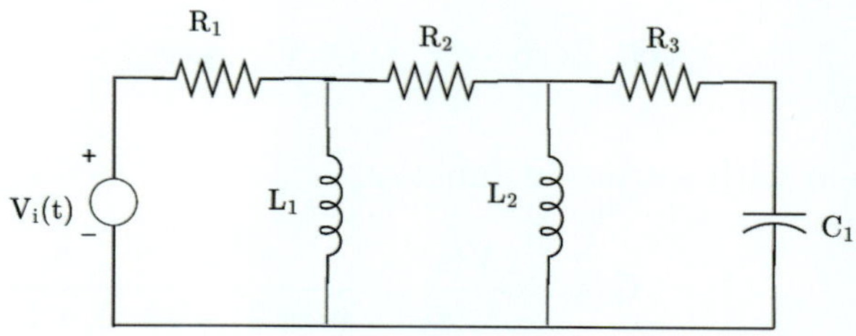

Figure 9.30: Linear Circuit

The number of state variables needed to represent this system in state space is

A. 3 **B.** 4 **C.** 2 **D.** 5

M 9.4. Consider the following statements related to the state variable model of a linear time-invariant system, which is described by

$$\dot{\mathbf{x}}(t) = \mathbf{A}\mathbf{x}(t) + \mathbf{B}u(t)$$
$$y(t) = \mathbf{C}\mathbf{x}(t) + Du(t) \tag{9.111}$$

I **A**: transmission matrix, **B**: input matrix

II **A**: system matrix, **B**: input matrix

III **C**: transmission matrix, **D**: output matrix

IV **C**: output matrix, **D**: transmission matrix

A. Statements I and IV are true
B. Statements I, III, and IV are true
C. Statements II and IV are true
D. Statements II and III are true

M 9.5. The transmission matrix D of systems with proper transfer functions is

A. 1 **B.** Zero **C.** ∞ **D.** Non-zero

M 9.6. The transmission matrix D of systems with strictly proper transfer functions is

A. 1 **B.** Zero **C.** ∞ **D.** Non-zero

M 9.7. A system with a transfer function

$$G(s) = \frac{Y(s)}{U(s)} = \frac{s+2}{(s+2)(s^2+4s+3)}$$

is represented by a state space model. The dimension of the system matrix equals

A. 3 **B.** 2 **C.** 4 **D.** 1

M 9.8. Consider the system shown in Figure 9.31.

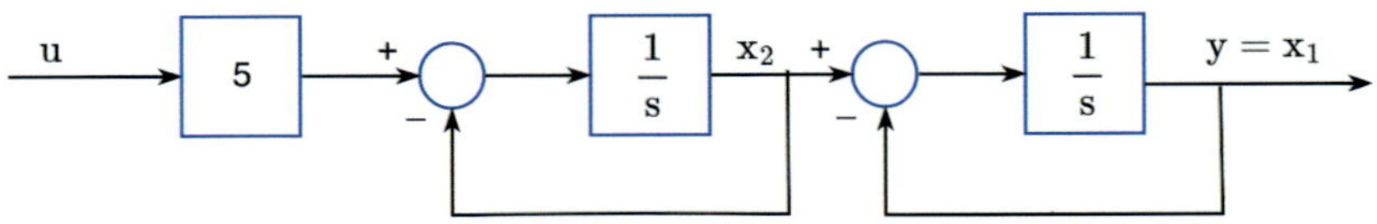

Figure 9.31: Linear System

When this system is represented in state space model of Eqn(9.111), then various matrices becomes

A. $\mathbf{A} = \begin{bmatrix} 1 & -1 \\ 0 & -1 \end{bmatrix}$, $\mathbf{B} = \begin{bmatrix} 0 \\ 1 \end{bmatrix}$, $\mathbf{C} = \begin{bmatrix} 1 & 0 \end{bmatrix}$, $D = 0$

B. $\mathbf{A} = \begin{bmatrix} -1 & 1 \\ 0 & -1 \end{bmatrix}$, $\mathbf{B} = \begin{bmatrix} 0 \\ 5 \end{bmatrix}$, $\mathbf{C} = \begin{bmatrix} 1 & 0 \end{bmatrix}$, $D = 0$

C. $\mathbf{A} = \begin{bmatrix} -1 & 1 \\ 0 & -1 \end{bmatrix}$, $\mathbf{B} = \begin{bmatrix} 5 \\ 0 \end{bmatrix}$, $\mathbf{C} = \begin{bmatrix} 1 & 0 \end{bmatrix}$, $D = 0$

D. $\mathbf{A} = \begin{bmatrix} -1 & 1 \\ 0 & -1 \end{bmatrix}$, $\mathbf{B} = \begin{bmatrix} 0 \\ 5 \end{bmatrix}$, $\mathbf{C} = \begin{bmatrix} 0 & 1 \end{bmatrix}$, $D = 0$

M 9.9. Consider the system shown in Figure 9.32.

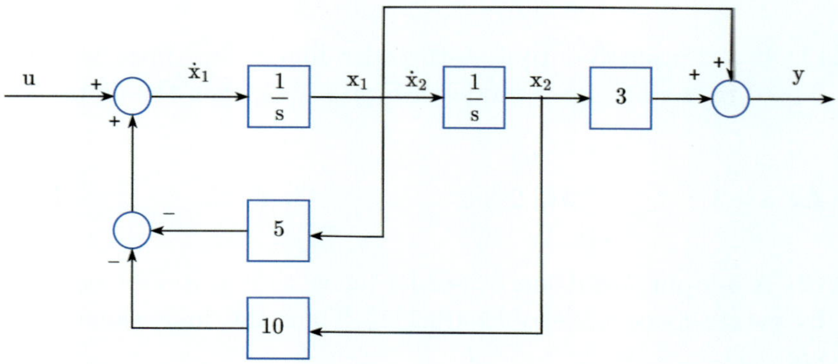

Figure 9.32: Linear System

When this system is represented in state space model of Eqn(9.111), then various matrices becomes

A. $\mathbf{A} = \begin{bmatrix} -5 & -10 \\ 1 & 0 \end{bmatrix}$, $\mathbf{B} = \begin{bmatrix} 1 \\ 0 \end{bmatrix}$, $\mathbf{C} = \begin{bmatrix} 1 & 3 \end{bmatrix}$, $D = 0$

B. $\mathbf{A} = \begin{bmatrix} 1 & 0 \\ -5 & -10 \end{bmatrix}$, $\mathbf{B} = \begin{bmatrix} 0 \\ 1 \end{bmatrix}$, $\mathbf{C} = \begin{bmatrix} 1 & 3 \end{bmatrix}$, $D = 0$

C. $\mathbf{A} = \begin{bmatrix} -5 & 1 \\ -10 & 0 \end{bmatrix}$, $\mathbf{B} = \begin{bmatrix} 0 \\ 1 \end{bmatrix}$, $\mathbf{C} = \begin{bmatrix} 1 & 3 \end{bmatrix}$, $D = 0$

D. $\mathbf{A} = \begin{bmatrix} -10 & -5 \\ 0 & 1 \end{bmatrix}$, $\mathbf{B} = \begin{bmatrix} 1 \\ 0 \end{bmatrix}$, $\mathbf{C} = \begin{bmatrix} 3 & 1 \end{bmatrix}$, $D = 0$

M 9.10. The dynamics of a system is described by state equations

$$\dot{x}_1 = 3x_1 + u_1 + u_2$$

$$\dot{x}_2 = 2x_2 + u_2, \quad y = x_1, u = \begin{bmatrix} u_1 \\ u_2 \end{bmatrix}$$

When this system is represented in state space model of Eqn(9.111), then various matrices become

A. $\mathbf{A} = \begin{bmatrix} 2 & 0 \\ 0 & 3 \end{bmatrix}$, $\mathbf{B} = \begin{bmatrix} 0 & 1 \\ 1 & 1 \end{bmatrix}$, $\mathbf{C} = \begin{bmatrix} 1 & 0 \end{bmatrix}$, $D = 0$

B. $\mathbf{A} = \begin{bmatrix} 3 & 0 \\ 0 & 2 \end{bmatrix}$, $\mathbf{B} = \begin{bmatrix} 1 & 0 \\ 0 & 1 \end{bmatrix}$, $\mathbf{C} = \begin{bmatrix} 1 & 0 \end{bmatrix}$, $D = 0$

C. $\mathbf{A} = \begin{bmatrix} 3 & 0 \\ 0 & 2 \end{bmatrix}$, $\mathbf{B} = \begin{bmatrix} 1 & 1 \\ 0 & 1 \end{bmatrix}$, $\mathbf{C} = \begin{bmatrix} 1 & 0 \end{bmatrix}$, $D = 0$

D. $\mathbf{A} = \begin{bmatrix} 0 & 3 \\ 2 & 0 \end{bmatrix}$, $\mathbf{B} = \begin{bmatrix} 1 & 1 \\ 0 & 1 \end{bmatrix}$, $\mathbf{C} = \begin{bmatrix} 1 & 0 \end{bmatrix}$, $D = 0$

M 9.11. If a 2-input 3-output, 5-th order linear time-invariant system is represented by a state space model of Eqn(9.111), then the dimension of matrix **C** is

A. 5×3 **B.** 2×3 **C.** 2×5 **D.** 3×5

M 9.12. A 3-input, 2-output, 6th-order linear time-invariant system is represented by a state space model of Eqn(9.111). Then, the dimensions of various matrices are

A. $\mathbf{A} = 6 \times 6$, $\mathbf{B} = 6 \times 3$, $\mathbf{C} = 2 \times 6$, and $\mathbf{D} = 2 \times 3$

B. $\mathbf{A} = 6 \times 6$, $\mathbf{B} = 3 \times 6$, $\mathbf{C} = 6 \times 2$, and $\mathbf{D} = 2 \times 3$

C. $\mathbf{A} = 6 \times 6$, $\mathbf{B} = 3 \times 6$, $\mathbf{C} = 2 \times 6$, and $\mathbf{D} = 3 \times 2$

D. $\mathbf{A} = 6 \times 6$, $\mathbf{B} = 6 \times 3$, $\mathbf{C} = 2 \times 2$, and $\mathbf{D} = 3 \times 3$

Answers

M 9.1. C **M 9.2**. B **M 9.3**. A **M 9.4**. C **M 9.5**. B **M 9.6**. B **M 9.7**. A **M 9.8**. B **M 9.9**. A **M 9.10**. C **M 9.11**. D **M 9.12**. A

9.7 Knowledge Assessment Problems

KaP 9.1 Consider the simple R-L-C circuit shown in Figure 9.33

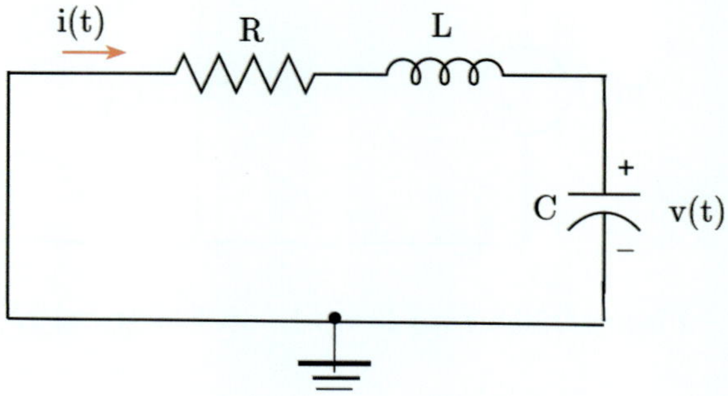

Figure 9.33: R-L-C Circuit

Represent this system in state space, considering the voltage across the capacitor and its derivative as state variables.

KaP 9.2 Obtain the state variable model of the system shown in Figure 9.34, taking the current through the inductors and voltage across the capacitor as state variables. The voltage across the resistor R_2 is the output.

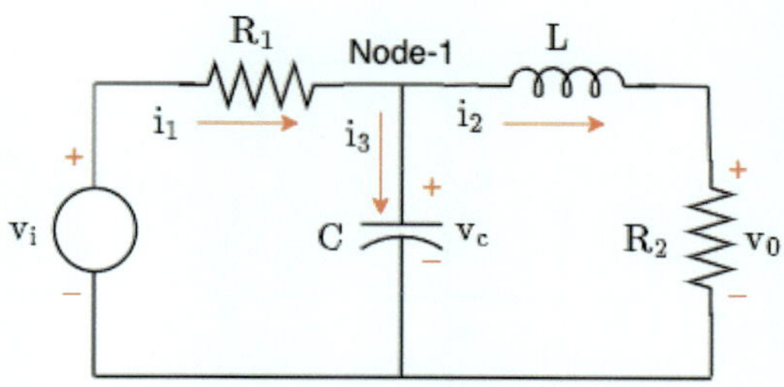

Figure 9.34: Two Loop Loop Electrical Network

KaP 9.3 Obtain the state variable model of the system shown in Figure 9.35, taking the current through the inductors and voltage across the capacitor as state variables. The current through the inductor is the output.

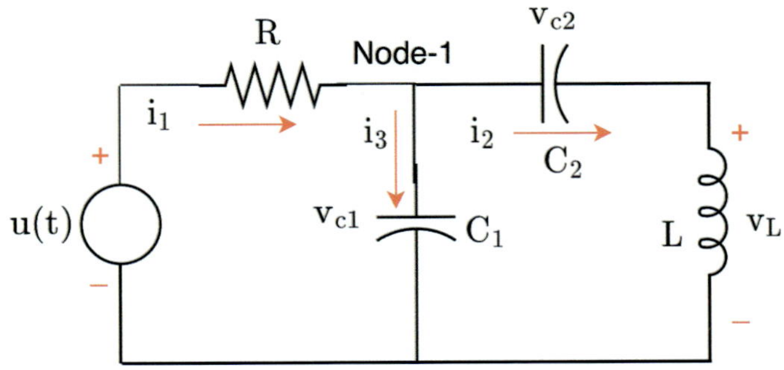

Figure 9.35: Two Loop Loop Electrical Network

KaP 9.4 Obtain state variable models for each network shown in Figure 9.36, taking the current through the inductors and voltage across the capacitor as state variables. The applied voltage is the input.

a. For the network in Figure 9.52a, determine the transfer function between the voltage across the inductor L_2 and the input for $R_1 = R_2 = 1\Omega$ and $L_1 = L_2 = 1H$.

b. Using MATLAB, compute the transfer function of the network in Figure 9.52b for $L_1 = 2H, L_2 = 3H$, $R = 1\Omega$ and $C = 0.5F$ considering the voltage across the capacitor of the network in Figure 9.52b as output.

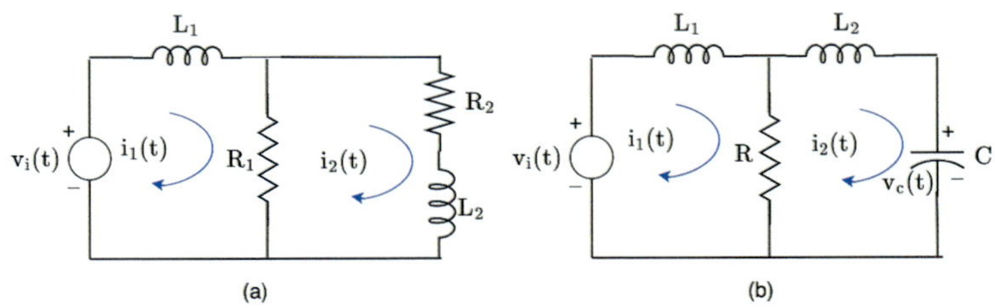

Figure 9.36: Multi Loop Loop Electrical Network

KaP 9.5 Obtain the state variable model of the system shown in Figure 9.37, taking the current through the inductor and voltage across the capacitor as state variables. The voltage across the capacitance is the output.

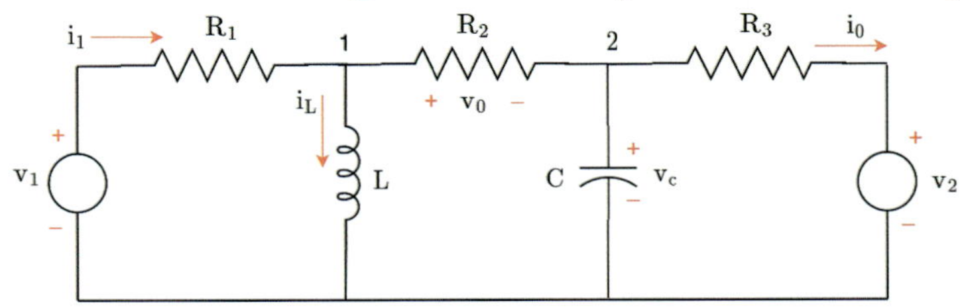

Figure 9.37: Multi loop electrical network with two inputs

KaP 9.6 Consider the circuit shown in Figure 9.38 where v_i is the input voltage to the circuit and v_0 denotes the output.

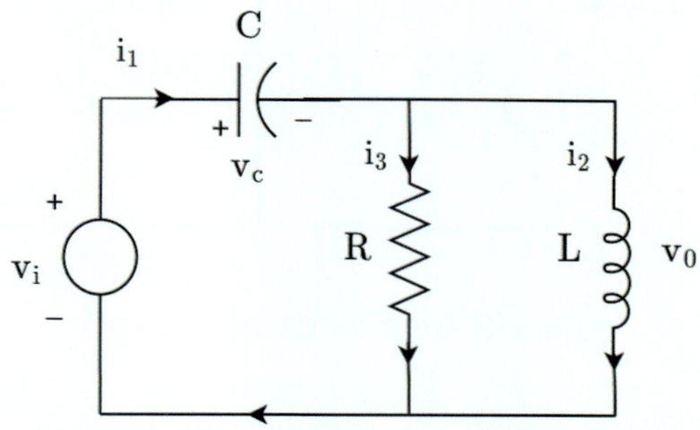

Figure 9.38: Two Loop RLC Network

a. Represent this circuit in state space, considering the voltage across the capacitor and current through the inductor as state variables. The input $u = v_i$ and the voltage across the inductor is the outputs.

b. Compute the transfer function of the system assuming $R = 2\Omega$, $C = 0.5F$ and $L = 1H$.

KaP 9.7 Obtain the state variable model of the system shown in Figure 9.39, taking the current through the inductor and voltage across the capacitor as state variables. The voltage across the inductor is the output, and the current i and the voltage $v_o(t)$ are the inputs.

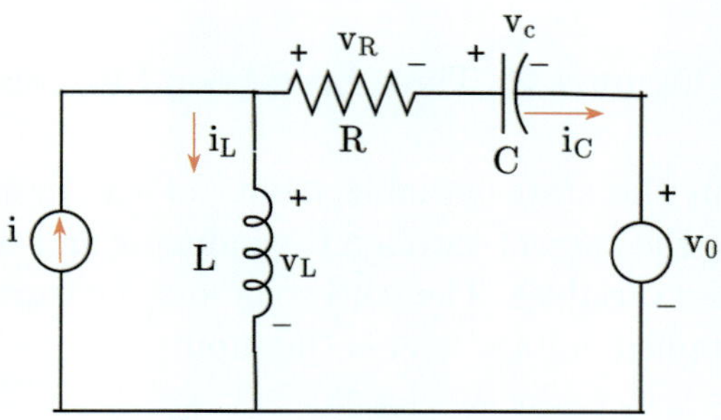

Figure 9.39: Electrical network

KaP 9.8 Obtain the state variable model of the system shown in Figure 9.40, taking the current through the inductor and voltage across the capacitor as state variables. The voltage across the inductor is the output, and the current i and the voltage $v_o(t)$ are the inputs.

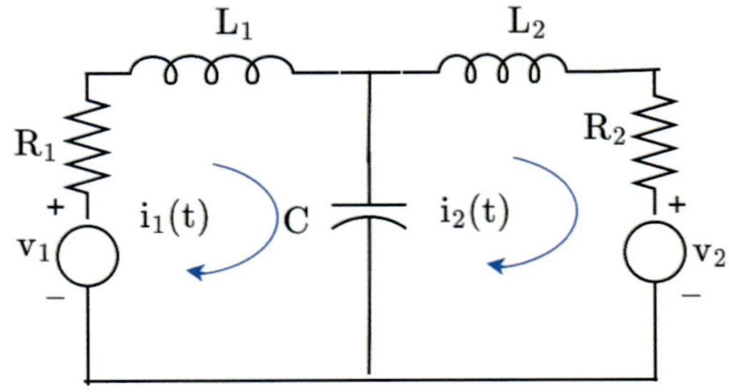

Figure 9.40: Electrical network

KaP 9.9 Obtain the state variable model of the system shown in Figure 9.41, taking the current through the inductors and voltage across the capacitor as state variables. The voltage across the capacitor is the output, and the applied voltage $v_i(t)$ is the input.

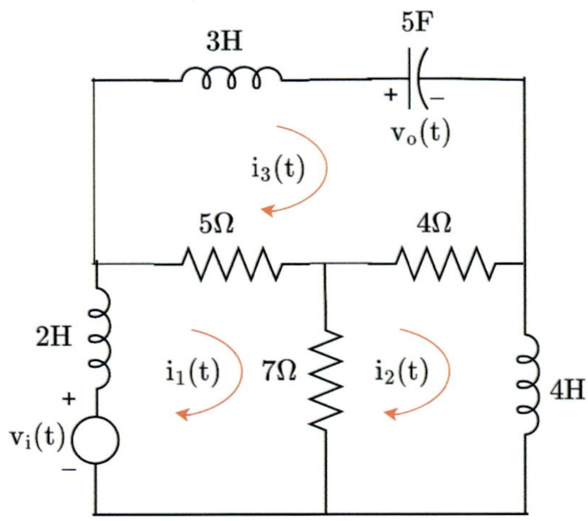

Figure 9.41: Three Loop Loop Electrical Network

KaP 9.10 Obtain the state variable model of the system shown in Figure 9.42, taking the current through the inductor and voltage across the capacitors as state variables. The voltage across the capacitor C_2 is the output, and the applied voltage v(t) is the input.

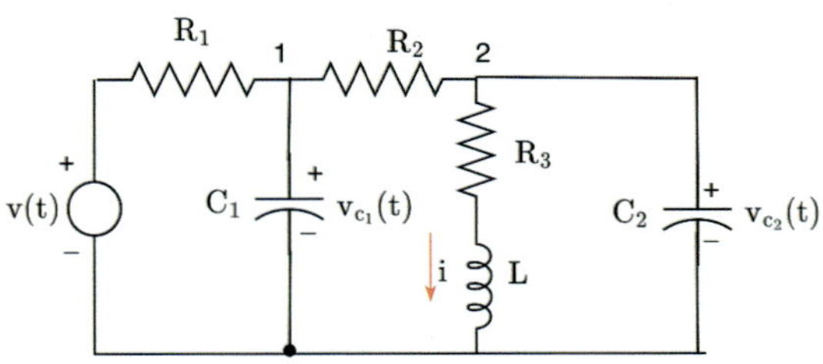

Figure 9.42: Three Loop Loop Electrical Network

KaP 9.11 Obtain the state variable model of the system shown in Figure 9.43, taking the current through the inductor and voltage across the capacitors as state variables. The voltage across the capacitor C_2 is the output, and the applied voltage v(t) is the input.

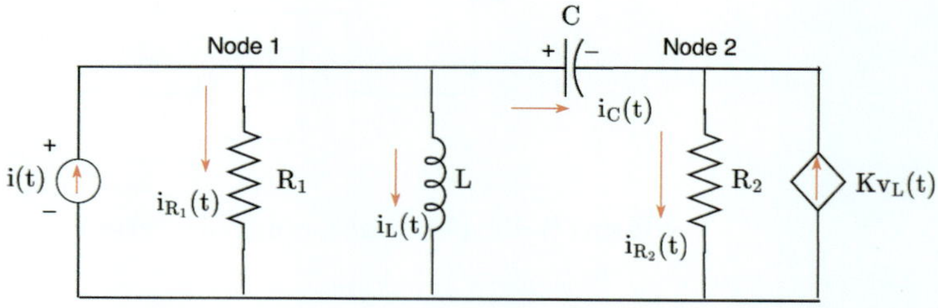

Figure 9.43: Electrical network with dependent source

KaP 9.12 Obtain the state variable model of the system shown in Figure 9.44, considering the outputs y_1, y_2 and their first derivatives as state variables.

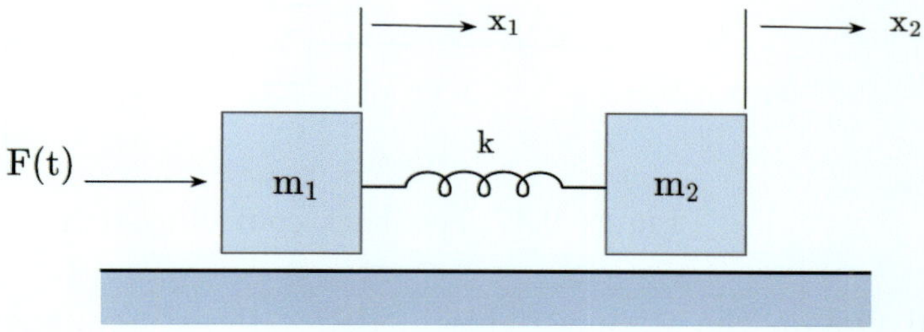

Figure 9.44: Two masses in frictionless surface

KaP 9.13 Obtain the state variable model of the system shown in Figure 9.45, considering the outputs y_1, y_2, and their first derivatives as state variables.

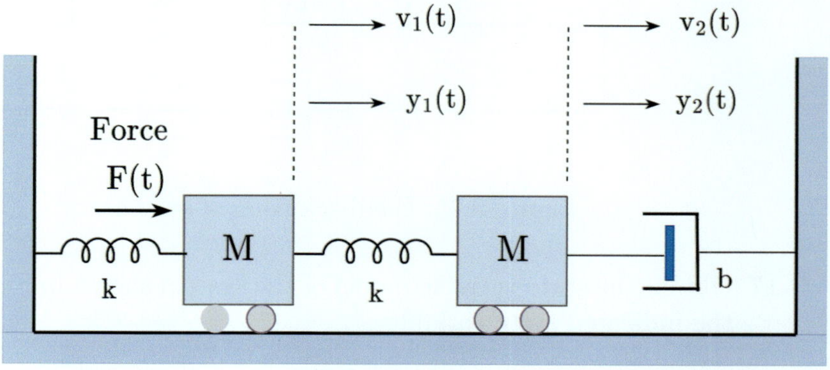

Figure 9.45: Coupled mass spring system

KaP 9.14 Obtain the state variable model of the system shown in Figure 9.46, taking the output as the state variable and the reference as the input.

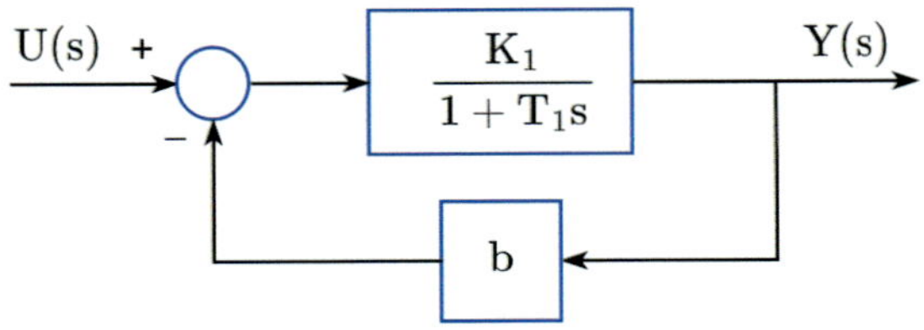

Figure 9.46: Feedback control system

KaP 9.15 Obtain the state variable model of the system shown in Figure 9.47, taking the output as the state variable and the reference as the input.

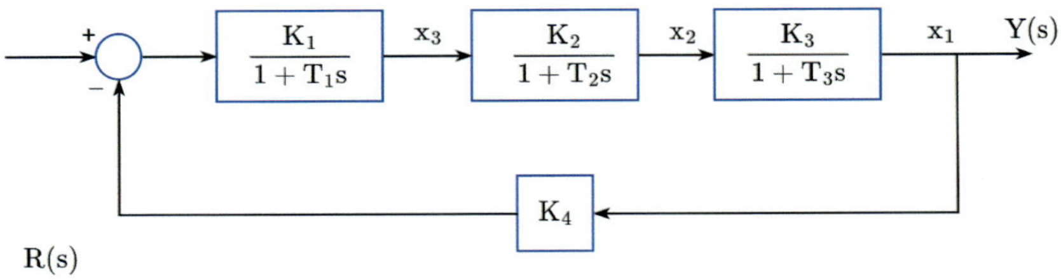

Figure 9.47: Feedback control system

KaP 9.16 Obtain the state variable model of the system shown in Figure 9.48, taking the output as the state variable and the reference as the input.

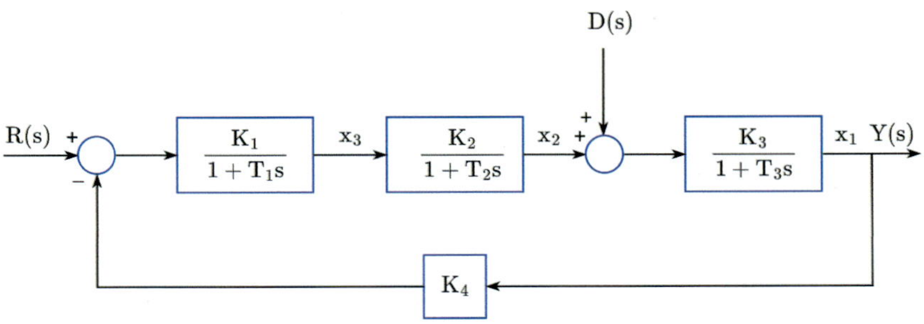

Figure 9.48: Feedback control system

KaP 9.17 Obtain the state variable model of the system shown in Figure 9.49 using the indicated state variables.

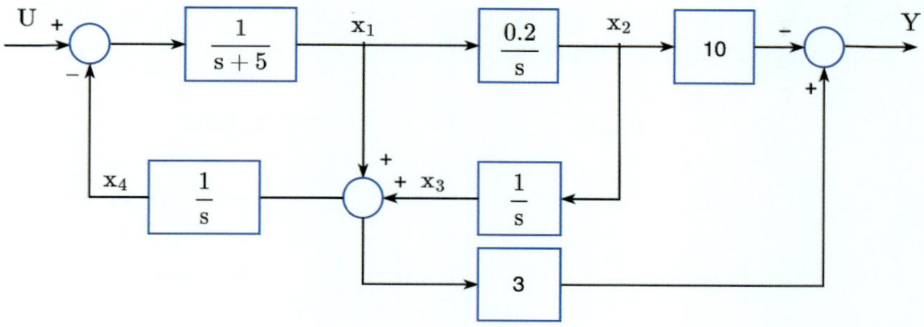

Figure 9.49: Feedback control system

KaP 9.18 The block diagram representation of a system is shown in Figure 9.50. Represent this system in state space

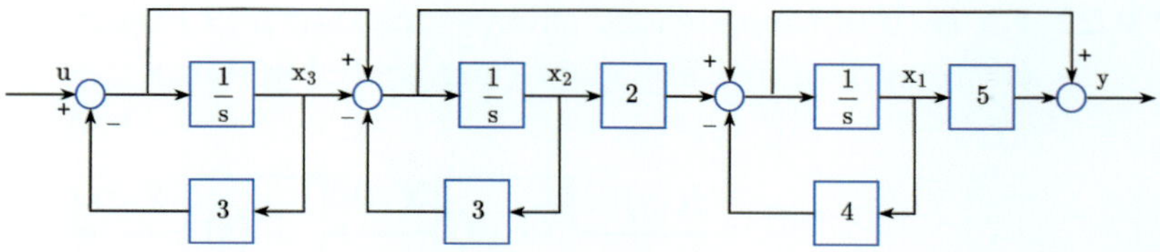

Figure 9.50: Linear System

KaP 9.19 The block diagram representation of a system is shown in Figure 9.51. Represent this system in state space

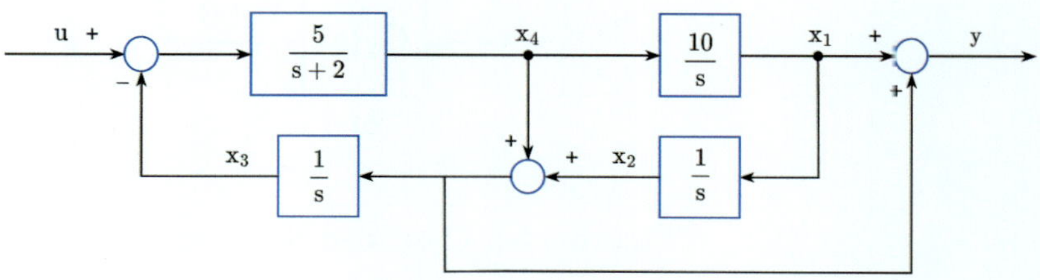

Figure 9.51: Linear System

KaP 9.20 The block diagram representation of a system is shown in Figure 9.52. Represent this system in state space

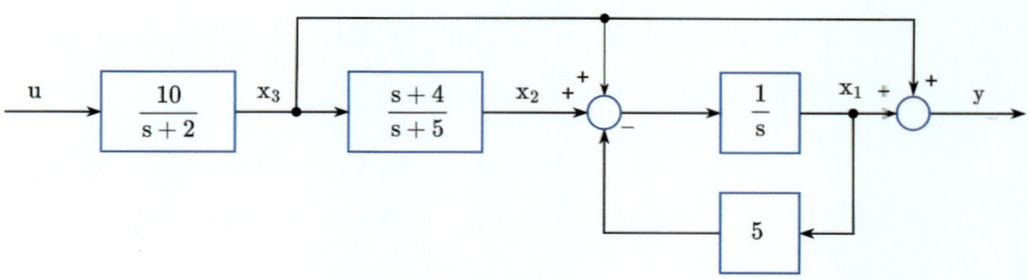

Figure 9.52: Linear System

KaP 9.21 The block diagram representation of a system is shown in Figure 9.53. Represent this system in state space

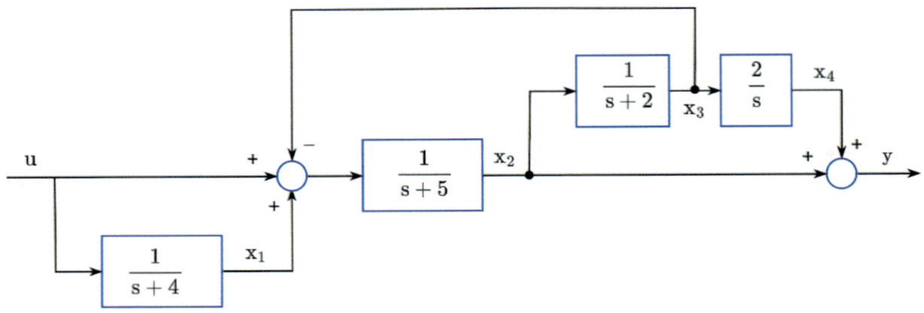

Figure 9.53: Linear System

KaP 9.22 For each of the cascaded subsystems shown in Figure 9.54, represent the entire system in state space (augmented system).

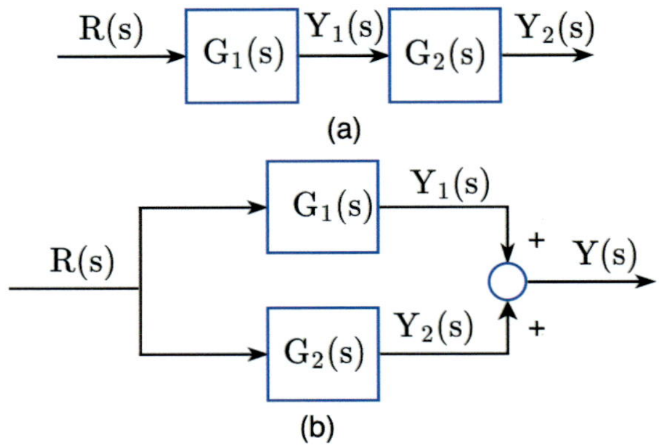

Figure 9.54: Cascaded Systems

a. For the system in Figure 9.54a, $G_1(s)$ and $G_2(s)$ are represented in state space as

$$G_1(s): \quad \dot{x}_1 = \mathbf{A_1 x_1} + \mathbf{B_1} r, \quad y_1 = \mathbf{C_1 x_1} \quad \text{and}$$
$$G_2(s): \quad \dot{x}_2 = \mathbf{A_2 x_2} + \mathbf{B_2} y_1, \quad y_2 = \mathbf{C_2 x_2}$$

b. For the system in Figure 9.54b, $G_1(s)$ and $G_2(s)$ are represented in state space as

$$G_1(s): \quad \dot{\mathbf{x}} = \mathbf{A_1 x_1} + \mathbf{B_1} r, \quad y_1 = \mathbf{C_1 x_1} \quad \text{and}$$
$$G_2(s): \quad \dot{x}_2 = \mathbf{A_2 x_2} + \mathbf{B_2} r, \quad y_2 = \mathbf{C_2 x_2}$$

KaP 9.23 Consider the system shown in Figure 9.55, which consists of two subsystems connected in closed-loop.

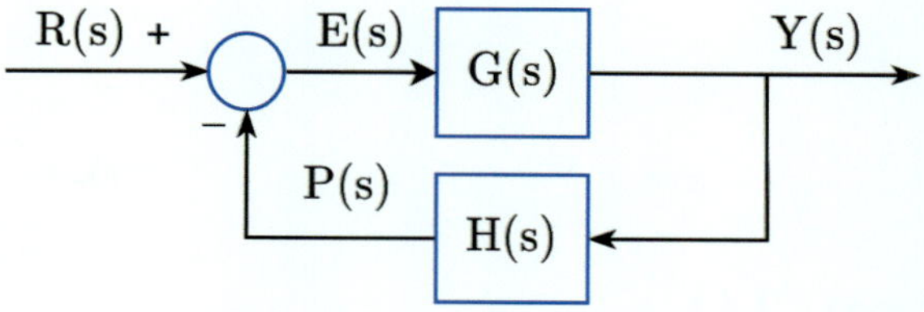

Figure 9.55: Cascaded system

The subsystems G(s) and H(s) are represented in state space as

$$\begin{aligned} G(s): \quad & \dot{x}_1 = \mathbf{A_1 x_1} + \mathbf{B_1} e, \quad y_1 = \mathbf{C_1 x_1} \quad \text{and} \\ H(s): \quad & \dot{x}_2 = \mathbf{A_2 x_2} + \mathbf{B_2} y, \quad p = \mathbf{C_2 x_2} \end{aligned}$$

Represent the closed-loop system in state space.

KaP 9.24 The block diagram representation of a two-area power system is shown in Figure 9.56. Represent this system in state space, considering the states as:

$$\mathbf{x} = \begin{bmatrix} \Delta f_1 & \text{Delta}P_{m1} & \Delta X_{E1} & \Delta P_{tie1} & \Delta f_2 & \Delta P_{m2} & \Delta X_{E2} \end{bmatrix}$$

Consider $u_1 = \Delta P_{c1}$,$u_2 = \Delta P_{d1}$ $u_3 = \Delta P_{c2}$, and $u_4 = \Delta P_{d2}$

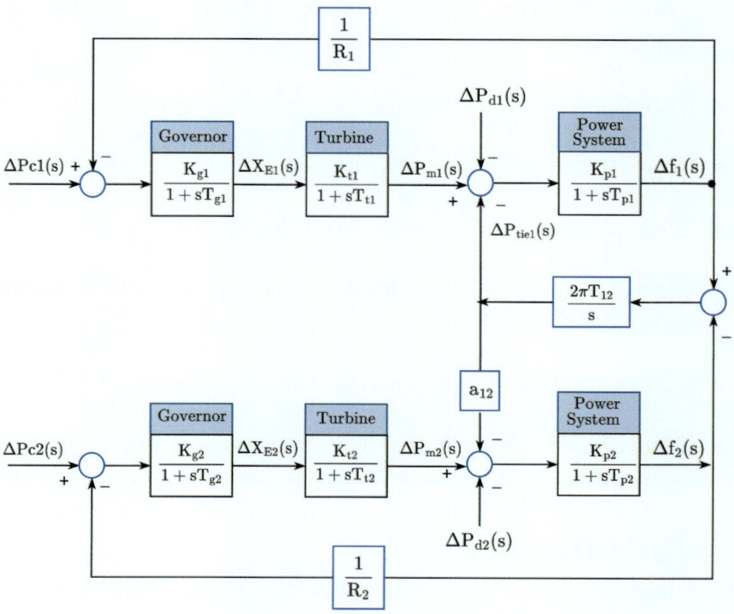

Figure 9.56: Two area power system

Chapter 10

State Variable Analysis of Dynamic Systems

Transfer function analysis watches the waiter bring food. State variable analysis storms into the kitchen to see who's actually cooking and whether they washed their hands.

DESIRED LEARNING OUTCOMES

Upon completion of this chapter, readers should be able to:

- Understand the relationship between state variable and transfer function models
- Explain why state variable models are not unique
- Represent a linear system in various canonical forms
- Diagonalize a system with complex poles
- Apply similarity transformations to convert systems into different canonical forms
- Compute the solution of the state equations
- Determine steady-state error from state equations

10.1 Introduction

In Chapter 9, we have discussed the conceptual basis for using state variables to represent system dynamics. The versatility and applicability of state-variable formulations across various types of systems, including electrical, mechanical, thermal, and hydraulic systems, have been demonstrated. One

DOI: 10.1201/9781003571308-10

of the key advantages of state variable formulation is its ability to handle a wide range of system types, including linear, nonlinear, time-invariant, time-varying, single-input-single-output (SISO), and multi-input-multi-output (MIMO) systems, in a unified manner. Thus, it offers a comprehensive framework to represent diverse physical systems in contrast to transfer functions, which are applicable only to linear time-invariant systems. Further, most of the modern and advanced control methods, such as optimal and robust control, are based on state space models. But to effectively design state-space model-based controllers, it's crucial to analyze the behavior of the systems thoroughly. The objective of this chapter is, therefore, to equip readers with essential skills in state variable analysis and state equations, which would lay the groundwork for further exploration in modern and optimal control design methodologies.

10.2 Relationship Between State Equations and Transfer Functions

Classical control theory primarily employs transfer functions to represent and analyze systems. In contrast, modern control analysis is centered around state-space models defined in terms of state vectors. Nevertheless, the transfer function approach offers distinct advantages that continue to be valuable. For this reason, it is useful to understand how transfer functions can be derived directly from the state-space representation of linear systems.

Consider a linear time-invariant system described by

$$\dot{\mathbf{x}}(t) = \mathbf{A}\mathbf{x}(t) + \mathbf{B}\mathbf{u}(t) \tag{10.1}$$

$$\mathbf{y}(t) = \mathbf{C}\mathbf{x}(t) + \mathbf{D}\mathbf{u}(t) \tag{10.2}$$

where $\mathbf{x}(t) \in \mathbb{R}^{n}$, $\mathbf{u}(t) \in \mathbb{R}^{m}$, and $\mathbf{y}(t) \in \mathbb{R}^{p}$. The matrices $\mathbf{A} \in \mathbb{R}^{n\times n}$, $\mathbf{B} \in \mathbb{R}^{n\times m}$, $\mathbf{C} \in \mathbb{R}^{p\times n}$, and $\mathbf{D} \in \mathbb{R}^{p\times m}$.

Taking the Laplace transform of both sides of Eqn(10.1) gives

$$s\mathbf{X}(s) - x(0) = \mathbf{A}\mathbf{X}(s) + \mathbf{B}\mathbf{U}(s) \implies (s\mathbf{I} - \mathbf{A})\,\mathbf{X}(s) = \mathbf{x}(0) + \mathbf{B}\mathbf{U}(s)$$

Solving for $\mathbf{X}(s)$, we get

$$\mathbf{X}(s) = (s\mathbf{I} - \mathbf{A})^{-1}\left[\mathbf{x}(\mathbf{0}) + \mathbf{B}\mathbf{U}(s)\right] \tag{10.3}$$

Taking the Laplace transform of both sides of Eqn(10.2) gives

$$\mathbf{Y}(s) = \mathbf{C}\mathbf{X}(s) + \mathbf{D}\mathbf{U}(s) \tag{10.4}$$

Substituting the value of X(s) from Eqn(10.3) into Eqn(10.4), and assuming zero the initial conditions ($\mathbf{x}(0) = 0$), we obtain

$$\mathbf{Y}(s) = \left[\mathbf{C}\,(s\mathbf{I} - \mathbf{A})^{-1}\,\mathbf{B} + \mathbf{D}\right]\mathbf{U}(s) \tag{10.5}$$

Thus, the transfer function matrix of the system is given by

$$\mathbf{G}(s) = \frac{\mathbf{Y}(s)}{\mathbf{U}(s)} = \left[\mathbf{C}\left(s\mathbf{I} - \mathbf{A}\right)^{-1}\mathbf{B} + \mathbf{D}\right] \tag{10.6}$$

EXAMPLE 10.1

Problem

Determine the transfer functions of the R-L-C circuit from its state variable model described in Example 9.4, considering

a. Voltage across the capacitor as the output
b. Current through the inductor as output

SOLUTION:

The system and input matrices of R-L-C circuit in Example 9.4 are given by

$$\mathbf{A} = \begin{bmatrix} 0 & \frac{1}{C} \\ -\frac{1}{L} & -\frac{R}{L} \end{bmatrix}, \text{ and } \mathbf{B} = \begin{bmatrix} 0 \\ \frac{1}{L} \end{bmatrix}$$

Let us first compute $(s\mathbf{I} - \mathbf{A})^{-1}$ as follows:

$$(s\mathbf{I} - \mathbf{A})^{-1} = \left\{\begin{bmatrix} s & 0 \\ 0 & s \end{bmatrix} - \begin{bmatrix} 0 & \frac{1}{C} \\ -\frac{1}{L} & -\frac{R}{L} \end{bmatrix}\right\}^{-1} = \left\{\begin{bmatrix} s & -\frac{1}{C} \\ \frac{1}{L} & s + \frac{R}{L} \end{bmatrix}\right\}^{-1}$$

$$= \frac{1}{\Delta}\begin{bmatrix} s + \frac{R}{L} & \frac{1}{C} \\ -\frac{1}{L} & s \end{bmatrix} = \frac{1}{\Delta}\begin{bmatrix} s_{11} & s_{12} \\ s_{21} & s_{22} \end{bmatrix}$$

where $\Delta = s^2 + \frac{R}{L}s + \frac{1}{LC}$.

a. If the voltage across the capacitor is taken as the output, the output matrix becomes

$$\mathbf{C} = \begin{bmatrix} 1 & 0 \end{bmatrix}$$

The transfer function is given by

$$G(s) = \mathbf{C}\,(\mathbf{sI} - \mathbf{A})^{-1}\,\mathbf{B}$$

$$= \begin{bmatrix} 1 & 0 \end{bmatrix} \frac{1}{\Delta} \begin{bmatrix} s_{11} & s_{12} \\ s_{21} & s_{22} \end{bmatrix} \begin{bmatrix} 0 \\ \frac{1}{L} \end{bmatrix} = \frac{1}{\Delta} \begin{bmatrix} 1 & 0 \end{bmatrix} \begin{bmatrix} \frac{s_{12}}{L} \\ \frac{s_{22}}{L} \end{bmatrix} = \frac{s_{12}}{L\Delta}$$

$$= \frac{1/LC}{s^2 + \frac{R}{L}s + \frac{1}{LC}}$$

b. If the current in the inductor is taken as the output, the output matrix becomes

$$\mathbf{C} = \begin{bmatrix} 0 & 1 \end{bmatrix}$$

The transfer function is given by

$$G(s) = \mathbf{C}\,(\mathbf{sI} - \mathbf{A})^{-1}\,\mathbf{B}$$

$$= \begin{bmatrix} 0 & 1 \end{bmatrix} \frac{1}{\Delta} \begin{bmatrix} s_{11} & s_{12} \\ s_{21} & s_{22} \end{bmatrix} \begin{bmatrix} 0 \\ \frac{1}{L} \end{bmatrix} = \frac{1}{\Delta} \begin{bmatrix} 0 & 1 \end{bmatrix} \begin{bmatrix} \frac{s_{12}}{L} \\ \frac{s_{22}}{L} \end{bmatrix} = \frac{s_{22}}{L\Delta}$$

$$= \frac{s}{L\left(s^2 + \frac{R}{L}s + \frac{1}{LC}\right)}$$

CONCEPT REINFORCEMENT PROBLEM 10.1

Problem

Consider the circuit shown in Figure 10.1.

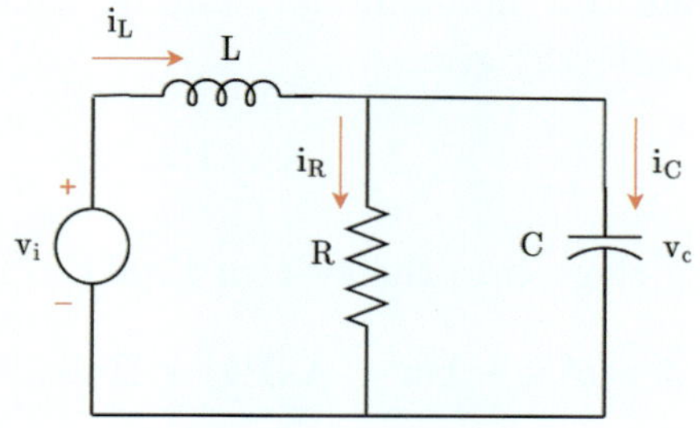

Figure 10.1: Two loop electrical network

a. Obtain the state variable model of the system, taking the current through the inductor and voltage across the capacitor as state variables. The voltage across the capacitance is the output.

b. Determine the transfer function from the state variable model assuming $R = 1\Omega, C = 0.25F$ and $L = 0.5H$.

10.3 Nonuniqueness of State Variable Representation

State-variable models are not unique; they can be expressed in different forms depending on the choice of state variables. Different selections of state variables lead to different state-space representations of the same system. While some choices may simplify analysis or controller design, others may provide deeper insight into the system's internal dynamics.

In this section, we establish the non-uniqueness of state-variable representations using the theory of **similarity transformation**. This fundamental concept in linear algebra and system theory enables one representation of a system to be transformed into another while preserving essential properties. In the context of state-variable models, similarity transformations are often employed to obtain equivalent state-space forms to achieve simplification, facilitate analysis, or aid in control design (topics to be discussed later).

Theorem 10.1. The state variable representation of a system is not unique

Proof:

Consider a system described by

$$\dot{\mathbf{x}} = \mathbf{A}\mathbf{x} + \mathbf{B}\mathrm{u}, \quad \mathrm{y} = \mathbf{C}\mathbf{x} + \mathbf{D}\mathrm{u} \tag{10.7}$$

Define another set of state variables $\mathbf{z}$ which is related to $\mathbf{x}$ by the relation

$$\mathbf{x} = \mathbf{P}\mathbf{z}; \quad \text{Note: } \implies \mathbf{z} = \mathbf{P}^{-1}\mathbf{x} \tag{10.8}$$

where $\mathbf{P}$ is a constant non-singular matrix. Since $\mathbf{P}$ is a constant matrix, differentiating Eqn(10.8) gives

$$\dot{\mathbf{x}} = \mathbf{P}\dot{\mathbf{z}} \tag{10.9}$$

Now substituting these into the system Eqn(10.7) gives

$$\mathbf{P}\dot{\mathbf{z}} = \dot{\mathbf{x}} = \mathbf{A}\mathbf{x} + \mathbf{B}\mathrm{u} = \mathbf{A}[\mathbf{P}\mathbf{z}] + \mathbf{B}\mathrm{u} = \mathbf{A}\mathbf{P}\mathbf{z} + \mathbf{B}\mathrm{u} \tag{10.10}$$

Pre multiplying $\mathbf{P}^{-1}$ in (10.10) gives

$$\begin{gathered}\dot{\mathbf{z}} = \mathbf{P}^{-1}\mathbf{A}\mathbf{P}\mathbf{z} + \mathbf{P}^{-1}\mathbf{B}\mathrm{u} = \mathbf{A}_{\mathbf{z}}\mathbf{z} + \mathbf{B}_{\mathbf{z}}\mathrm{u} \text{ where} \\ \mathbf{A}_{\mathbf{z}} = \mathbf{P}^{-1}\mathbf{A}\mathbf{P}, \mathbf{B}_{\mathbf{z}} = \mathbf{P}^{-1}\mathbf{B}\end{gathered} \tag{10.11}$$

The output equation modifies to

$$y = \mathbf{Cx} + \mathbf{D}u = \mathbf{CPz} + \mathbf{D}u = \mathbf{C_z z} + \mathbf{D_z}u; \ \mathbf{C_z} = \mathbf{CP}, \ \mathbf{D_z} = \mathbf{D} \tag{10.12}$$

Invariance Properties of Similarity Transformation

Theorem 10.2. The characteristic equation, eigenvalues, and transfer function are invariant under similarity transformation. Eigenvectors are not invariant but transform according to the similarity map (they change coordinates)

Proof: Invariance of Characteristic Equation

The characteristic equation of the transformed system is given by $|\lambda\mathbf{I} - \mathbf{A_z}| = 0$. Now

$$\begin{aligned} |\lambda\mathbf{I} - \mathbf{A_z}| &= |\lambda\mathbf{I} - \mathbf{P^{-1}AP}| = |\lambda\mathbf{P^{-1}P} - \mathbf{P^{-1}AP}| \\ &= |\mathbf{P^{-1}}(\lambda\mathbf{I} - \mathbf{A})\mathbf{P}| = |\mathbf{P^{-1}}|\,|\lambda\mathbf{I} - \mathbf{A}|\,|\mathbf{P}| \\ &= |\mathbf{P^{-1}}|\,|\mathbf{P}|\,|\lambda\mathbf{I} - \mathbf{A}| = |\lambda\mathbf{I} - \mathbf{A}| \end{aligned} \tag{10.13}$$

Thus, the characteristic equation does not change, which further implies that the eigenvalues and eigenvectors of $\mathbf{A}$ remain unchanged under similarity transformation.

Proof: Invariance of Transfer Function

We will use the identity

$$(s\mathbf{I} - \mathbf{P^{-1}AP})^{-1} = \mathbf{P^{-1}}(s\mathbf{I} - \mathbf{A})^{-1}\mathbf{P} \tag{10.14}$$

The transfer function of the transformed system is given by

$$\begin{aligned} \mathbf{G_z}(s) &= \mathbf{C_z}(s\mathbf{I} - \mathbf{A_z})^{-1}\mathbf{B_z} + \mathbf{D} \\ &= \mathbf{CP}(s\mathbf{I} - \mathbf{P^{-1}AP})^{-1}\mathbf{P^{-1}B} + \mathbf{D} \\ &= \mathbf{CP}\cdot\mathbf{P^{-1}}(s\mathbf{I} - \mathbf{A})^{-1}\mathbf{P}\cdot\mathbf{P^{-1}B} + \mathbf{D} \\ &= \mathbf{C}(\mathbf{PP^{-1}})(s\mathbf{I} - \mathbf{A})^{-1}(\mathbf{PP^{-1}})\mathbf{B} + \mathbf{D} \\ &= \mathbf{C}\cdot\mathbf{I}\cdot(s\mathbf{I} - \mathbf{A})^{-1}\cdot\mathbf{I}\cdot\mathbf{B} + \mathbf{D} \\ &= \mathbf{C}(s\mathbf{I} - \mathbf{A})^{-1}\mathbf{B} + \mathbf{D} = \mathbf{G}(s) \end{aligned} \tag{10.15}$$

EXAMPLE 10.2

Problem:

Consider a system whose dynamics are described by

$$\dot{\mathbf{x}} = \begin{bmatrix} 1 & 2 \\ -4 & -3 \end{bmatrix} \mathbf{x} + \begin{bmatrix} 1 \\ 2 \end{bmatrix} \mathbf{u}, \quad \mathbf{y} = \begin{bmatrix} 1 & 1 \end{bmatrix} \mathbf{x}$$

Transform this system to another system with a new state vector

$$\mathbf{z} = \begin{bmatrix} 2 & 1 \\ 1 & 3 \end{bmatrix} \mathbf{x}$$

Compute the transfer function from both systems and compare.

SOLUTION:

When the system is represented in the standard form

$$\dot{\mathbf{x}} = \mathbf{A}\mathbf{x} + \mathbf{B}\mathbf{u}, \quad \mathbf{y} = \mathbf{C}\mathbf{x} + \mathbf{D}\mathbf{u}$$

gives

$$\mathbf{A} = \begin{bmatrix} 1 & 2 \\ -4 & -3 \end{bmatrix}, \quad \mathbf{B} = \begin{bmatrix} 1 \\ 2 \end{bmatrix}, \quad \mathbf{C} = \begin{bmatrix} 1 & 1 \end{bmatrix}, \quad \mathbf{D} = 0$$

The system is to be represented by a new state variable $\mathbf{z}$ where $\mathbf{x} = \mathbf{P}\mathbf{z} \quad \Longrightarrow \quad \mathbf{z} = \mathbf{P}^{-1}\mathbf{x}$. Comparing this with the given data, we have

$$\mathbf{P}^{-1} = \begin{bmatrix} 2 & 1 \\ 1 & 3 \end{bmatrix}$$

The system in terms of the new state vector is given by

$$\begin{aligned} \dot{\mathbf{z}} &= \mathbf{P}^{-1}\mathbf{A}\mathbf{P}\mathbf{z} + \mathbf{P}^{-1}\mathbf{B}\mathbf{z} = \mathbf{A}_{\mathbf{z}}\mathbf{z} + \mathbf{B}_{\mathbf{z}}\mathrm{u} \\ \mathrm{y} &= \mathbf{C}\mathbf{P}\mathbf{z} + \mathbf{D}\mathrm{u} = \mathbf{C}_{\mathbf{z}}\mathbf{z} + \mathbf{D}\mathrm{u} \end{aligned}$$

Let us first compute $\mathbf{P}$ as follows:

$$\mathbf{P} = \left[\mathbf{P}^{-1}\right]^{-1} = \begin{bmatrix} 2 & 1 \\ 1 & 3 \end{bmatrix}^{-1} = \frac{1}{5}\begin{bmatrix} 3 & -1 \\ -1 & 2 \end{bmatrix} = \begin{bmatrix} 0.6 & -0.2 \\ -0.2 & 0.4 \end{bmatrix}$$

The various matrices in the transformed variables are

$$\mathbf{A_z} = \mathbf{P}^{-1}\mathbf{A}\mathbf{P} = \begin{bmatrix} 2 & 1 \\ 1 & 3 \end{bmatrix} \begin{bmatrix} 1 & 2 \\ -4 & -3 \end{bmatrix} \begin{bmatrix} 0.6 & -0.2 \\ -0.2 & 0.4 \end{bmatrix}$$

$$= \begin{bmatrix} -1.4 & 0.8 \\ -5.2 & -0.6 \end{bmatrix}$$

$$\mathbf{B_z} = \mathbf{P}^{-1}\mathbf{B} = \begin{bmatrix} 2 & 1 \\ 1 & 3 \end{bmatrix} \begin{bmatrix} 1 \\ 2 \end{bmatrix} = \begin{bmatrix} 4 \\ 7 \end{bmatrix}$$

$$\mathbf{C_z} = \mathbf{C}\mathbf{P} = \begin{bmatrix} 1 & 1 \end{bmatrix} \begin{bmatrix} 0.6 & -0.2 \\ -0.2 & 0.4 \end{bmatrix} = \begin{bmatrix} 0.4 & 0.2 \end{bmatrix}$$

Thus, the state equations in transformed states are given by

$$\dot{\mathbf{z}} = \begin{bmatrix} -1.4 & 0.8 \\ -5.2 & -0.6 \end{bmatrix} \mathbf{z} + \begin{bmatrix} 4 \\ 7 \end{bmatrix} \mathbf{u}, \qquad \mathbf{y} = \begin{bmatrix} 0.4 & 0.2 \end{bmatrix} \mathbf{z}$$

Let us now compare the transfer functions. From the original state equation, the transfer function is computed as:

$$(\mathbf{sI} - \mathbf{A})^{-1} = \left[\begin{bmatrix} s & 0 \\ 0 & s \end{bmatrix} - \begin{bmatrix} 1 & 2 \\ -4 & -3 \end{bmatrix} \right]^{-1} = \begin{bmatrix} s-1 & -2 \\ 4 & s+3 \end{bmatrix}^{-1}$$

$$= \frac{1}{s^2 + 2s + 5} \begin{bmatrix} s+3 & 2 \\ -4 & s-1 \end{bmatrix}$$

The closed-loop transfer function is computed as:

$$\mathbf{T(s)} = \mathbf{C}(\mathbf{sI} - \mathbf{A})^{-1}\mathbf{B} + \mathbf{D}$$

$$= \begin{bmatrix} 1 & 1 \end{bmatrix} \frac{1}{s^2 + 2s + 5} \begin{bmatrix} s+3 & 2 \\ -4 & s-1 \end{bmatrix} \begin{bmatrix} 1 \\ 2 \end{bmatrix} = \frac{3s + 1}{s^2 + 2s + 5}$$

The transfer function of the transformed system is computed as:

$$(\mathbf{sI} - \mathbf{A_z})^{-1} = \left[\begin{bmatrix} s & 0 \\ 0 & s \end{bmatrix} - \begin{bmatrix} -1.4 & 0.8 \\ -5.2 & -0.6 \end{bmatrix} \right]^{-1} = \begin{bmatrix} s+1.4 & -0.8 \\ 5.2 & s+0.6 \end{bmatrix}^{-1}$$

$$= \frac{1}{s^2 + 2s + 5} \begin{bmatrix} s+0.6 & 0.8 \\ -5.2 & s+1.4 \end{bmatrix}$$

The closed-loop transfer function is

$$\mathbf{T(s)} = \mathbf{C_z}(\mathbf{sI} - \mathbf{A_z})^{-1}\mathbf{B_z} + \mathbf{D}$$

$$= \begin{bmatrix} 0.4 & 0.2 \end{bmatrix} \frac{1}{s^2 + 2s + 5} \begin{bmatrix} s+0.6 & 0.8 \\ -5.2 & s+1.4 \end{bmatrix} \begin{bmatrix} 4 \\ 7 \end{bmatrix} = \frac{3s + 1}{s^2 + 2s + 5}$$

Thus, it is found that the transfer function is invariant with this similarity transformation.

10.4 Representation of Linear Systems in Canonical Forms

We have established that state-space representations of a system are not unique; a single system can be described through infinitely many equivalent state-space realizations. Despite this non-uniqueness, it is often beneficial to convert transfer function models into specific standardized forms known as canonical representations.

Canonical forms are structured state-space realizations that emphasize particular system characteristics, making them valuable tools for analysis and design in control theory. The most commonly used canonical forms include the controllable canonical form, observable canonical form, and diagonal canonical form. These standardized representations streamline mathematical manipulation and reveal fundamental system properties that might be obscured in other realizations.

It is important to emphasize that regardless of which canonical form is chosen, the underlying dynamic behavior of the system remains invariant; all representations describe the same input-output relationship and system response. This raises a natural question: **why represent a system in different canonical forms if they all describe the same dynamics?**

The answer lies in the distinct advantages each form provides. Different canonical forms highlight different aspects of system behavior and facilitate specific types of analysis or design tasks, as summarized below.

Why Different Canonical Forms?

i. The controllable canonical form substantially simplifies the mathematical computations required for designing state feedback controllers using pole placement techniques. This structured representation streamlines the controller design process by organizing the system matrices in a way that directly supports feedback gain calculations.

ii. The observable canonical form facilitates the systematic design of state observers, which are essential for estimating unmeasured system states from available output measurements. This representation optimally arranges the system structure to support observability analysis and observer synthesis.

iii. The diagonal canonical form provides superior physical insight into system behavior by decoupling the system dynamics through matrix diagonalization. This representation reveals the individual modes of the system, making it easier to understand how each eigenvalue contributes to the overall system response and stability characteristics.

The choice of canonical form can significantly streamline calculations, provide clearer insights into system behavior, and facilitate targeted design methodologies. In the following sections, we will examine the characteristics and applications of five fundamental canonical forms:

1. **Phase Variable Canonical Form:**Applied when the transfer function contains no finite zeros, resulting in a simplified structure that directly relates to the system's characteristic equation.
2. **Controllable Canonical Form:**Used when the transfer function contains finite zeros, providing optimal structure for state feedback controller design and pole placement techniques.
3. **Observable Canonical Form:**The mathematical dual of the controllable form, specifically designed to facilitate state observer design and system identification from output measurements.
4. **Diagonal Canonical Form:** Applicable when the system has distinct eigenvalues, achieving complete diagonalization that decouples the system dynamics into independent modal components.
5. **Jordan Canonical Form:**Required when the system has repeated eigenvalues (poles with multiplicity greater than one), resulting in partial diagonalization with Jordan blocks that capture the system's modal structure.

10.4.1 Phase Variable Canonical Form: All Pole Transfer Function

This canonical representation is specifically designed for systems whose transfer functions contain only poles and no finite zeros (i.e., all-pole systems). It transforms the system into a form where the poles dominate the dynamics, offering insights into the system's stability and transient response. We will begin with a specific example before presenting the general formulation.

EXAMPLE 10.3

Consider a system represented by the transfer function

$$\frac{Y(s)}{U(s)} = \frac{10}{s^2 + 5s + 6} \tag{10.16}$$

Step-1: Convert the transfer function model to differential equation model by replacing terms of the form $s^2Y(s)$ by $\ddot{y}(t)$, and $sY(s)$ by $\dot{y}(t)$. This gives

$$s^2Y(s) + 5sY(s) + 6Y(s) = 10U(s)$$
$$\text{or,} \quad \ddot{y} + 5\dot{y} + 6y = 10u(t) \tag{10.17}$$

Step-2: Decide the number of state variables to represent the system.

Since the order of the system is two, we require 2 state variables to represent this system.

Step-3: Selection of the State Variables

In the phase variable canonical form, the state variables are systematically defined as the output and its successive derivatives. This direct relationship creates an intuitive state-space structure. Let us define:

$$\begin{aligned} x_1 &= y(t) \\ x_2 &= \dot{y}(t) \end{aligned} \tag{10.18}$$

Step-4: Write down the state equations in terms of state variables.

From Eqns(10.17) and (10.18) we get

$$\begin{aligned} \dot{x}_1 &= \dot{y} = x_2 \\ \dot{x}_2 &= \ddot{y} = -6y - 5\dot{y} + 10u \\ &= -6x_1 - 5x_2 + 10u \end{aligned} \tag{10.19}$$

In matrix form, this can be expressed as:

$$\begin{aligned} \begin{bmatrix} \dot{x}_1 \\ \dot{x}_2 \end{bmatrix} &= \begin{bmatrix} 0 & 1 \\ -6 & -5 \end{bmatrix} \begin{bmatrix} x_1 \\ x_2 \end{bmatrix} + \begin{bmatrix} 0 \\ 10 \end{bmatrix} u \\ y &= \begin{bmatrix} 1 & 0 \end{bmatrix} \begin{bmatrix} x_1 \\ x_2 \end{bmatrix} \end{aligned} \tag{10.20}$$

CONCEPT REINFORCEMENT PROBLEM 10.2

Problem:

Represent the following systems in phase variable canonical form.

a. $\dfrac{Y(s)}{U(s)} = \dfrac{25}{s^4 + 10s^3 + 35s^2 + 50s + 24}$

b. $\dfrac{Y(s)}{U(s)} = \dfrac{10}{s^4 + 3s^2 + 7s + 9}$

Phase Variable Canonical Form (General Case)

Consider an n-th order system with no zeros whose transfer function is given by

$$\frac{Y(s)}{U(s)} = \frac{b_0}{s^n + a_1 s^{n-1} + a_2 s^{n-2} + \cdots + a_{n-1}s + a_n} \tag{10.21}$$

The differential equation corresponding to this transfer function representation is given by

$$\frac{d^n y}{dt^n} + a_1 \frac{d^{n-1}y}{dt^{n-1}} + a_2 \frac{d^{n-2}y}{dt^{n-2}} + \cdots + a_{n-1}\frac{dy}{dt} + a_n y = b_0 u \tag{10.22}$$

Since the order of the system equals n, the system can be represented by n-number of state variables. Define the state variables as:

$$x_1 = y, \quad x_2 = \frac{dy}{dt}, \quad x_3 = \frac{d^2y}{dt^2}, \ldots, \quad x_{n-1} = \frac{d^{n-2}y}{dt^{n-2}} \tag{10.23}$$

Hence,

$$\begin{aligned}
\dot{x}_1 &= x_2 \\
\dot{x}_2 &= x_3 \\
&\vdots \\
\dot{x}_{n-1} &= x_n \\
\dot{x}_n &= -a_n x_1 - a_{n-1}x_2 - \cdots - a_2 x_{n-1} - a_1 x_n + b_0 u
\end{aligned} \tag{10.24}$$

When this is represented in the standard form of a state space model

$$\dot{\mathbf{x}} = \mathbf{A}\mathbf{x} + \mathbf{B}u, \quad y = \mathbf{C}\mathbf{x} + \mathbf{D}u \tag{10.25}$$

the system matrix **A**, input matrix **B**, output matrix **C**, and the transmission matrix **D** are:

$$\mathbf{A} = \begin{bmatrix} 0 & 1 & 0 & \cdots & 0 & 0 \\ 0 & 0 & 1 & \cdots & 0 & 0 \\ \vdots & \vdots & \vdots & \cdots & \vdots & \vdots \\ \vdots & \vdots & \vdots & \cdots & \vdots & \vdots \\ 0 & 0 & 0 & \cdots & 0 & 1 \\ -a_n & -a_{n-1} & -a_{n-2} & \cdots & -a_2 & -a_1 \end{bmatrix}, \quad \mathbf{B} = \begin{bmatrix} 0 \\ 0 \\ \vdots \\ \vdots \\ 0 \\ 1 \end{bmatrix},$$

$$\mathbf{C} = \begin{bmatrix} 1 & 0 & \cdots & \cdots & 0 \end{bmatrix}, \text{ and } D = 0$$

Summary:Phase Variable Canonical Form

1. The elements of the last row of matrix **A** (from column-1 to column-n) consist of the negative coefficients of the denominator polynomial arranged in ascending powers of "s" (from right to left in the polynomial).
2. All elements of matrix **B** are zero except the last element, which equals 1.
3. All elements of matrix **C** are zero except the first element, which equals 1.

10.4.2 Controllable Canonical Form

This canonical representation applies to systems whose transfer functions contain both poles and zeros. The structure is specifically designed to emphasize the system's **controllability** properties (discussed in Chapter 11). It is particularly valuable for control design methodologies, including pole placement, optimal control, and state feedback control synthesis.

We will examine two distinct cases based on the nature of the transfer function:

Case-1: Systems with Strictly Proper Transfer Function (degree of numerator < degree of denominator)

Case-2: Systems with Proper Transfer Function (degree of numerator ≤ degree of denominator)

Systems with Strictly Proper Transfer Function

To illustrate the derivation of the controllable canonical form for an nth-order system, we begin with a third-order example having the strictly proper transfer function:

$$\frac{Y(s)}{U(s)} = \frac{b_1 s^2 + b_2 s + b_3}{s^3 + a_1 s^2 + a_2 s + a_3} \tag{10.26}$$

The differential equation corresponding to this transfer function is

$$\dddot{y} + a_1 \ddot{y} + a_2 \dot{y} + a_3 y = b_1 \ddot{u} + b_2 \dot{u} + b_3 u \tag{10.27}$$

Challenge with Direct State Variable Selection:

Unlike the phase variable form, we cannot directly choose the output and its derivatives as state variables:

$$x_1 = y, \quad x_2 = \dot{y}, \quad x_3 = \ddot{y}$$

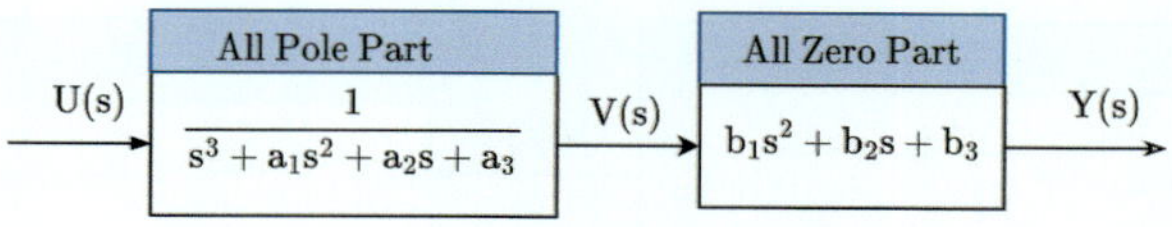

Figure 10.2: Decomposition of transfer function into all pole and all zero parts

This choice would yield:

$$\begin{aligned}\dot{x}_1 &= x_2\\ \dot{x}_2 &= x_3\\ \dot{x}_3 &= -a_3x_1 - a_2x_2 - a_1x_3 + b_1\ddot{u} + b_2\dot{u} + b_3u\end{aligned} \tag{10.28}$$

This representation violates the fundamental requirement of state-space models: **state equations must not contain derivatives of the input**. The presence of $\dot{u}$ and $\ddot{u}$ terms makes this an invalid state-space realization.

Solution Approach:

To eliminate input derivatives, we must define an alternative set of state variables using a systematic transformation procedure, which we will develop next.

1. Split the transfer function into two parts; one with **all pole parts** and other with **all zero parts**, as shown in Figure 10.2

   The all-pole part is written as:

$$\frac{V(s)}{U(s)} = \frac{1}{s^3 + a_1s^2 + a_2s + a_3} \tag{10.29}$$

   The all-zero part is written as:

$$Y(s) = \left(b_1s^2 + b_2s + b_3\right)V(s) = b_1\ddot{v} + b_2\dot{v} + b_3v \tag{10.30}$$

2. Consider the all-pole part of the system and define the state variables as:$x_1 = v$, $x_2 = \dot{v}$, and $x_3 = \ddot{v}$. This gives the following representation, called, controllable canonical form

$$\begin{bmatrix}\dot{x}_1\\ \dot{x}_2\\ \dot{x}_3\end{bmatrix} = \begin{bmatrix}0 & 1 & 0\\ 0 & 0 & 1\\ -a_3 & -a_2 & -a_1\end{bmatrix}\begin{bmatrix}x_1\\ x_2\\ x_3\end{bmatrix} + \begin{bmatrix}0\\ 0\\ 1\end{bmatrix}u, \tag{10.31}$$

   The output equation can be written as

$$y = \begin{bmatrix}b_3 & b_2 & b_1\end{bmatrix}\begin{bmatrix}x_1\\ x_2\\ x_3\end{bmatrix} \tag{10.32}$$

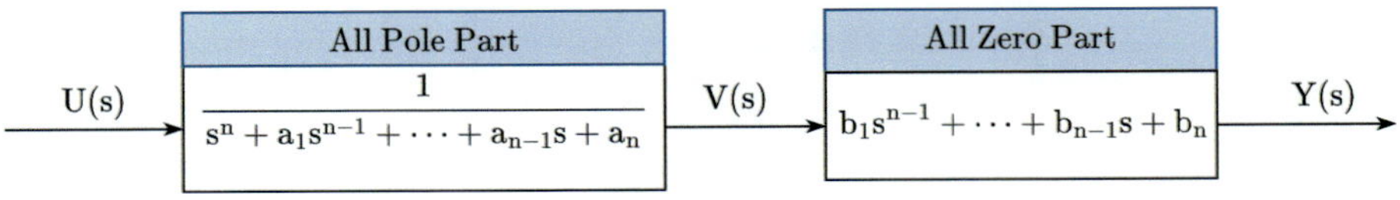

Figure 10.3: Decomposition of transfer function into all pole and all zero parts

Let us now extend this approach to the general case. Consider an n-th order system with a strictly proper transfer function:

$$\frac{Y(s)}{U(s)} = \frac{b_1 s^{n-1} + \cdots + b_{n-1}s + b_n}{s^n + a_1 s^{n-1} + a_2 s^{n-2} + \cdots + a_{n-1}s + a_n} \tag{10.33}$$

Split this transfer function into two parts as shown in Figure 10.3

Select the state variables as the output of the all pole part and its subsequent derivatives i.e., let $x_1 = v$, $x_2 = \dot{v}, \ldots$, $x_n = v^{(n)}$. Then, the controllable canonical form of this system is given as:

$$\dot{\mathbf{x}} = \mathbf{A}\mathbf{x} + \mathbf{B}u, \quad y = \mathbf{C}\mathbf{x} + \mathbf{D}u$$

where

$$\mathbf{A} = \begin{bmatrix} 0 & 1 & 0 & \cdots & 0 & 0 \\ 0 & 0 & 1 & \cdots & 0 & 0 \\ \vdots & \cdots & \cdots & \cdots & \vdots & \vdots \\ \vdots & \cdots & \cdots & \cdots & \vdots & \vdots \\ 0 & 0 & 0 & \cdots & 0 & 1 \\ -a_n & -a_{n-1} & -a_{n-2} & \cdots & -a_2 & -a_1 \end{bmatrix}, \mathbf{B} = \begin{bmatrix} 0 \\ 0 \\ \vdots \\ \vdots \\ 0 \\ 1 \end{bmatrix},$$

$$\mathbf{C} = \begin{bmatrix} b_n & b_{n-1} & \cdots & \cdots & b_2 & b_1 \end{bmatrix}, \text{ and } \mathbf{D} = 0$$

Systems with Proper Transfer Function

We will first consider an example of a third-order system and then extend this to the n-th order system. Consider a third-order system with a proper transfer function given by

$$\frac{Y(s)}{U(s)} = \frac{b_0 s^3 + b_1 s^2 + b_2 s + b_3}{s^3 + a_1 s^2 + a_2 s + a_3} \tag{10.34}$$

By direct division, we can represent this system by

$$\begin{aligned} \frac{Y(s)}{U(s)} &= b_0 + \frac{(b_1 - a_1 b_0)s^2 + (b_2 - a_2 b_0)s + (b_3 - a_3 b_0)}{s^3 + a_1 s^2 + a_2 s + a_3} \\ &= \beta_0 + \frac{\beta_1 s^2 + \beta_2 s + \beta_3}{s^3 + a_1 s^2 + a_2 s + a_3} \quad \text{where} \\ \beta_0 &= b_0, \quad \beta_1 = b_1 - a_1 b_0, \quad \beta_2 = b_2 - a_2 b_0, \quad \beta_3 = b_3 - a_3 b_0 \end{aligned} \tag{10.35}$$

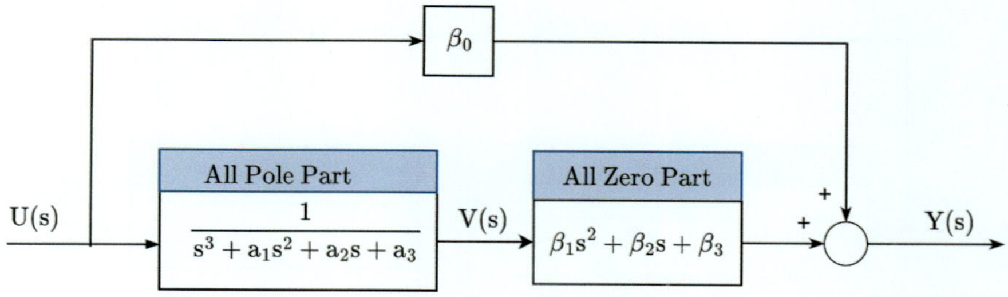

Figure 10.4: Decomposition of transfer function into all pole and all zero parts

Following a similar approach as with a strictly proper transfer function, this system can be represented as:

Define the state variables as: $x_1 = v$, $x_2 = \dot{v}$ and $x_3 = \ddot{v}$. This gives the following representation called as **controllable canonical form**.

$$\begin{bmatrix} \dot{x}_1 \\ \dot{x}_2 \\ \dot{x}_3 \end{bmatrix} = \begin{bmatrix} 0 & 1 & 0 \\ 0 & 0 & 1 \\ -a_3 & -a_2 & -a_1 \end{bmatrix} \begin{bmatrix} x_1 \\ x_2 \\ x_3 \end{bmatrix} + \begin{bmatrix} 0 \\ 0 \\ 1 \end{bmatrix} u, \tag{10.36}$$

The output equation can be written as:

$$y = \begin{bmatrix} \beta_3 & \beta_2 & \beta_1 \end{bmatrix} \begin{bmatrix} x_1 \\ x_2 \\ x_3 \end{bmatrix} + \beta_0 u \tag{10.37}$$

This can be extended to the general case of an n-th order system with a proper transfer function:

$$\frac{Y(s)}{U(s)} = \frac{b_0 s^n + b_1 s^{n-1} + \cdots + b_{n-1} s + b_n}{s^n + a_1 s^{n-1} + a_2 s^{n-2} + \cdots + a_{n-1} s + a_n} \tag{10.38}$$

By direct division, we can represent this system by

$$\begin{aligned} \frac{Y(s)}{U(s)} &= b_0 + \frac{(b_1 - a_1 b_0)s^{n-1} + (b_2 - a_2 b_0)s^{n-2} + \cdots + (b_n - a_n b_0)}{s^n + a_1 s^{n-1} + a_2 s^{n-2} + \cdots + +a_{n-1} s + a_n} \\ &= \beta_0 + \frac{\beta_1 s^{n-1} + \beta_2 s^{n-2} + \cdots + \beta_{n-1} s + \beta_n}{s^n + a_1 s^{n-1} + a_2 s^{n-2} + \cdots + a_{n-1} s + a_n} \quad \text{where} \\ \beta_0 &= b_0, \quad \beta_1 = b_1 - a_1 b_0, \ldots, \beta_{n-1} = b_{n-1} - a_{n-1} b_0, \beta_n = b_n - a_n b_0 \end{aligned}$$

This system can be decomposed into all pole part and all zero part as shown in Figure 10.5.

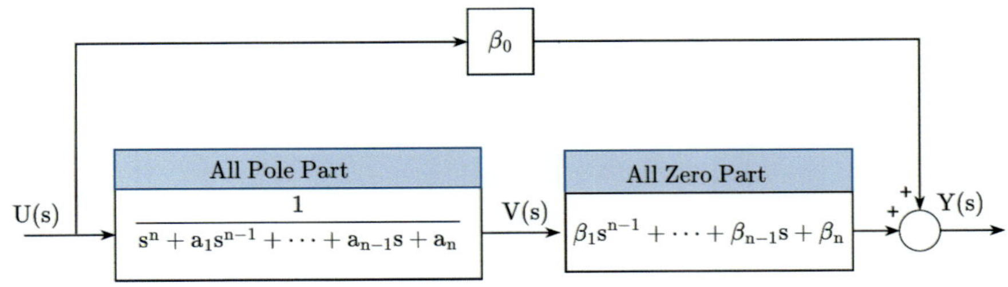

Figure 10.5: Decomposition of transfer function into all pole and all zero parts

Following a similar approach to that used for the strictly proper transfer function, this system can be represented in controllable canonical form as:

$$\dot{\mathbf{x}} = \mathbf{A}\mathbf{x} + \mathbf{B}\mathrm{u}, \quad \mathrm{y} = \mathbf{C}\mathbf{x} + \mathbf{D}\mathrm{u} \text{ where} \tag{10.39}$$

$$\mathbf{A} = \begin{bmatrix} 0 & 1 & 0 & \dots & 0 & 0 \\ 0 & 0 & 1 & \dots & 0 & 0 \\ \vdots & \dots & \dots & \dots & \vdots & \vdots \\ \vdots & \dots & \dots & \dots & \vdots & \vdots \\ 0 & 0 & 0 & \dots & 0 & 1 \\ -\mathrm{a}_n & -\mathrm{a}_{n-1} & -\mathrm{a}_{n-2} & \dots & -\mathrm{a}_2 & -\mathrm{a}_1 \end{bmatrix}, \mathbf{B} = \begin{bmatrix} 0 \\ 0 \\ \vdots \\ \vdots \\ 0 \\ 1 \end{bmatrix}$$

$$\mathbf{C} = \begin{bmatrix} \beta_n & \beta_{n-1} & \dots & \dots & \beta_2 & \beta_1 \end{bmatrix}, \mathbf{D} = \beta_0$$

10.4.3 Observable Canonical Form

Unlike the controllable canonical form, the observable canonical form emphasizes the system's observability properties (to be discussed in Chapter 11) rather than controllability. This representation is specifically designed to facilitate the design of state observers for estimating unmeasured states from available output measurements. It proves particularly valuable for observer-based control strategies and state estimation algorithms.

The observable canonical form is the mathematical dual of the controllable canonical form and can be systematically derived from it using the duality principle. Let the controllable canonical representation of a single input single output system with transfer function G(s) be given by

$$\begin{aligned} \dot{\mathbf{x}} &= \mathbf{A}\mathbf{x} + \mathbf{B}\mathrm{u} \\ \mathrm{y} &= \mathbf{C}\mathbf{x} + \mathrm{D}\mathrm{u} \end{aligned} \tag{10.40}$$

We will demonstrate that this system can alternatively be expressed in observable canonical form as:

$$\begin{aligned} \dot{\mathbf{x}} &= \mathbf{A}^{\mathbf{T}}\mathbf{x} + \mathbf{C}^{\mathbf{T}}\mathrm{u} \\ \mathrm{y} &= \mathbf{B}^{\mathbf{T}}\mathbf{x} + \mathrm{D}\mathrm{u} \end{aligned} \tag{10.41}$$

Proof:

The transfer function of the controllable canonical system is:

$$G(s) = \mathbf{C}(s\mathbf{I} - \mathbf{A})^{-1}\mathbf{B} + D \tag{10.42}$$

Taking the transpose of equation (10.42):

$$\begin{aligned} G^T(s) &= \left[\mathbf{C}(s\mathbf{I} - \mathbf{A})^{-1}\mathbf{B} + D\right]^T \\ &= \mathbf{B}^T(s\mathbf{I} - \mathbf{A}^T)^{-1}\mathbf{C}^T + D^T \end{aligned} \tag{10.43}$$

Note that Eqn(10.42) represents the transfer function of the controllable canonical form (10.40), while Eqn(10.43) represents the transfer function of the observable canonical form (10.41). Since $G^T(s) = G(s)$ for scalar transfer functions and $D^T = D$, for scalar feedthrough terms, these two representations are equivalent.

Therefore, the system matrices of the observable and controllable canonical forms are related by the duality transformation:

$$\mathbf{A}\mathrm{obs} = \mathbf{A}^T\mathrm{cont}, \quad \mathbf{B}\mathrm{obs} = \mathbf{C}^T\mathrm{cont}, \quad \mathbf{C}\mathrm{obs} = \mathbf{B}^T\mathrm{cont}$$

where the subscripts "obs" and "cont" refer to observable and controllable forms, respectively.

For a general nth-order system, the observable canonical form corresponding to Eqn(10.39) can be written as:

$$\mathbf{A} = \begin{bmatrix} 0 & 0 & \dots & \dots & 0 & -a_n \\ 1 & 0 & \dots & \dots & 0 & -a_{n-1} \\ \vdots & \dots & \dots & \dots & \vdots & \vdots \\ \vdots & \dots & \dots & \dots & \vdots & \vdots \\ 0 & 0 & \dots & \dots & 0 & -a_2 \\ 0 & 0 & \dots & \dots & 1 & -a_1 \end{bmatrix}, \mathbf{B} = \begin{bmatrix} \beta_n \\ \beta_{n-1} \\ \vdots \\ \vdots \\ \beta_2 \\ \beta_1 \end{bmatrix},$$
$$\mathbf{C} = \begin{bmatrix} 0 & 0 & \dots & \dots & 0 & 1 \end{bmatrix}, \mathbf{D} = \beta_0 \tag{10.44}$$

Compare Eqn(10.44) with Eqn(10.39).

10.4.4 Diagonal Canonical Form

This canonical representation is particularly advantageous for systems with distinct eigenvalues (poles). The diagonal form decouples the system dynamics, offering better physical insight into the system behavior by clearly revealing the independent contribution of each modal component to the overall response. Before generalizing the approach, we will first examine a second-order system described by:

$$\frac{Y(s)}{U(s)} = \frac{b_0 s^2 + b_1 s + b_2}{s^2 + a_1 s + a_2} \tag{10.45}$$

Let us assume that the two poles of the system p_1 and p_2 are distinct (i.e., $p_1 \neq p_2$). Using partial fraction expansion, this transfer function can be decomposed as:

$$\frac{Y(s)}{U(s)} = \frac{b_0 s^2 + b_1 s + b_2}{s^2 + a_1 s + a_2} = b_0 + \frac{c_1}{s + p_1} + \frac{c_2}{s + p_2} \tag{10.46}$$

This gives the output as:

$$\boxed{Y(s) = b_0 U(s) + \frac{c_1}{s + p_1} U(s) + \frac{c_2}{s + p_2} U(s)} \tag{10.47}$$

Define the state variables x_1 and x_2 where

$$X_1(s) = \frac{1}{s + p_1} U(s), \quad \text{and } X_2(s) = \frac{1}{s + p_2} U(s) \tag{10.48}$$

Now

$$\begin{aligned} X_1(s) &= \frac{1}{s + p_1} U(s) \quad \Longrightarrow \quad sX_1(s) = -p_1 X_1(s) + U(s) \\ X_2(s) &= \frac{1}{s + p_2} U(s) \quad \Longrightarrow \quad sX_2(s) = -p_2 X(s) + U(s) \end{aligned} \tag{10.49}$$

The output equation is given by

$$Y(s) = c_1 X_1(s) + c_2 X_2(s) + b_0 U(s) \tag{10.50}$$

Taking the inverse Laplace transform of Eqns(10.49) and (10.50) these equations give

$$\begin{aligned} \dot{x}_1 &= -p_1 x_1 + u \\ \dot{x}_2 &= -p_2 x_2 + u \\ y &= c_1 x_1 + c_2 x_2 + b_0 u \end{aligned} \tag{10.51}$$

In matrix form, it is expressed as:

$$\begin{bmatrix}\dot{x}_1\\ \dot{x}_2\end{bmatrix} = \begin{bmatrix}-p_1 & 0\\ 0 & -p_2\end{bmatrix}\begin{bmatrix}x_1\\ x_2\end{bmatrix} + \begin{bmatrix}1\\ 1\end{bmatrix} u$$

$$y = \begin{bmatrix}c_1 & c_2\end{bmatrix}\begin{bmatrix}x_1\\ x_2\end{bmatrix} + b_0 u \tag{10.52}$$

Let us extend this to an n-th order system described by

$$\frac{Y(s)}{U(s)} = \frac{b_0 s^n + b_1 s^{n-1} + \cdots + b_{n-1}s + b_n}{s^n + a_1 s^{n-1} + \cdots + a_{n-1}s + a_n}$$
$$= b_0 + \frac{c_1}{s+p_1} + \frac{c_2}{s+p_2} + \cdots + \frac{c_n}{s+p_n}, \quad p_i \neq p_j, \forall\, i,j = 1,\ldots n \tag{10.53}$$

The diagonal canonical model of the system is given by

$$\dot{\mathbf{x}} = \begin{bmatrix}-p_1 & 0 & 0 & \ldots & 0\\ 0 & -p_2 & 0 & \ldots & 0\\ \vdots & \ldots & & \ldots & \vdots\\ 0 & 0 & \ldots & -p_{n-1} & 0\\ 0 & 0 & \ldots & 0 & -p_n\end{bmatrix}\mathbf{x} + \begin{bmatrix}1\\ 1\\ \vdots\\ 1\\ 1\end{bmatrix} u$$

$$y = \begin{bmatrix}c_1 & c_2 & \ldots & c_{n-1} & c_n\end{bmatrix}\mathbf{x} + b_0 u \tag{10.54}$$

10.4.5 Jordan Canonical Forms (Non distinct Roots)

This canonical representation is a generalization of the diagonal canonical form, specifically designed for systems with poles of multiplicity greater than one (repeated poles). For such systems, complete diagonalization is not possible, resulting in a partially diagonalized system matrix with a block-diagonal structure containing Jordan blocks along the diagonal. This structure facilitates the analysis and design of control strategies for systems with complex pole patterns.

Consider an n-th order system described by

$$\frac{Y(s)}{U(s)} = \frac{b_0 s^n + b_1 s^{n-1} + \cdots + b_{n-1}s + b_n}{s^n + a_1 s^{n-1} + \cdots + a_{n-1}s + a_n} \tag{10.55}$$

Assume that this system contains non-distinct roots. As an example assume that $p_1 = p_2 = p_3$ and $p_4 = p_5$, $p_6, \ldots, p_n$ are distinct.The partial fraction expansion of Eqn(10.55) gives

$$\frac{Y(s)}{U(s)} = \frac{b_0 s^n + b_1 s^{n-1} + \cdots + b_{n-1}s + b_n}{s^n + a_1 s^{n-1} + \cdots + a_{n-1}s + a_n}$$
$$= b_0 + \frac{c_1}{(s+p_1)^3} + \frac{c_2}{(s+p_1)^2} + \frac{c_3}{(s+p_1)}$$
$$+ \frac{c_4}{(s+p_4)^2} + \frac{c_5}{s+p_4} + \frac{c_6}{s+p_6} + \frac{c_7}{s+p_7} \cdots + \frac{c_n}{s+p_n} \tag{10.56}$$

or,

$$\begin{aligned}Y(s) &= b_0U(s) + \frac{c_1}{(s+p_1)^3}U(s) + \frac{c_2}{(s+p_1)^2}U(s) + \frac{c_3}{(s+p_1)}U(s) \\ &+ \frac{c_4}{(s+p_4)^2}U(s) + \frac{c_5}{s+p_4}U(s) + \frac{c_6}{s+p_6}U(s) + \frac{c_7}{s+p_7}U(s) + \cdots \\ &+ \frac{c_n}{s+p_n}U(s)\end{aligned} \tag{10.57}$$

Define

$$\begin{aligned}X_1(s) &= \frac{1}{(s+p_1)^3}U(s),\ X_2(s) = \frac{1}{(s+p_1)^2}U(s),\ X_3(s) = \frac{1}{s+p_1}U(s) \\ X_4(s) &= \frac{1}{(s+p_4)^2}U(s),\ X_5(s) = \frac{1}{s+p_4}U(s) \\ X_6(s) &= \frac{1}{s+p_6}U(s)\ldots\ X_n(s) = \frac{1}{s+p_n}U(s)\end{aligned} \tag{10.58}$$

From Eqn(10.58), we can find the following relations among various state variables

$$\begin{aligned}
\frac{X_1(s)}{X_2(s)} &= \frac{1}{s+p_1} &\implies sX_1(s) = -p_1X_1(s) + X_2(s) &\implies \dot{x}_1 = -p_1x_1 + x_2 \\
\frac{X_2(s)}{X_3(s)} &= \frac{1}{s+p_1} &\implies sX_2(s) = -p_1X_2(s) + X_3(s) &\implies \dot{x}_2 = -p_1x_2 + x_3 \\
\frac{X_3(s)}{U(s)} &= \frac{1}{s+p_1} &\implies sX_3(s) = -p_1X_3(s) + U(s) &\implies \dot{x}_3 = -p_1x_3 + u \\
\frac{X_4(s)}{X_5(s)} &= \frac{1}{s+p_4} &\implies sX_4(s) = -p_4X_4(s) + X_5(s) &\implies \dot{x}_4 = -p_4x_4 + x_5 \\
\frac{X_5(s)}{U(s)} &= \frac{1}{s+p_4} &\implies sX_5(s) = -p_4X_5(s) + U(s) &\implies \dot{x}_5 = -p_4x_5 + u \\
\frac{X_6(s)}{U(s)} &= \frac{1}{s+p_6} &\implies sX_6(s) = -p_6X_6(s) + U(s) &\implies \dot{x}_6 = -p_6x_6 + u \\
&\vdots \\
\frac{X_n(s)}{U(s)} &= \frac{1}{s+p_n} &\implies sX_n(s) = -p_nX_n(s) + U(s) &\implies \dot{x}_n = -p_nx_n + u
\end{aligned} \tag{10.59}$$

The output equation can be expressed as

$$Y(s) = b_0U(s) + c_1X_1(s) + c_2X_2(s) + c_3X_3(s) + c_4X_4(s) + \cdots + c_nX_n(s) \tag{10.60}$$

Further, if the state variable model of a system having both real and complex characteristic roots, is

$$\begin{bmatrix}\dot{x}_1\\\dot{x}_2\\\dot{x}_3\\\dot{x}_4\\\dot{x}_5\\\dot{x}_6\end{bmatrix}=\begin{bmatrix}-p_1&0&0&0&0&0\\0&-p_2&0&0&0&0\\0&0&0&1&0&0\\0&0&-b&-a&0&0\\0&0&0&0&0&1\\0&0&0&0&-d&-c\end{bmatrix}\begin{bmatrix}x_1\\x_2\\x_3\\x_4\\x_5\\x_6\end{bmatrix}+\begin{bmatrix}1\\1\\0\\1\\0\\1\end{bmatrix}u$$

$$y=\begin{bmatrix}k_1&k_2&k_4&k_3&k_6&k_5\end{bmatrix}\begin{bmatrix}x_1\\x_2\\x_3\\x_4\\x_5\\x_6\end{bmatrix} \tag{10.70}$$

then, its transfer function can be obtained by inspection and is expressed as:

$$T(s)=\frac{Y(s)}{U(s)}=\frac{k_1}{s+p_1}+\frac{k_2}{s+p_2}+\frac{k_3s+k_4}{s^2+as+b}+\frac{k_5s+k_6}{s^2+cs+d} \tag{10.71}$$

CONCEPT REINFORCEMENT PROBLEM 10.5

Represent the system with following transfer function in state space.

$$T(s)=\frac{Y(s)}{U(s)}=\frac{2s^2+5s+12}{(s+2)(s^2+2s+10)}$$

10.5 Transformation to Different Canonical Forms

We have established that state-space representations are not unique and explored the invariance properties of similarity transformations. Having discussed the characteristics of various canonical forms, a natural question arises:

Key Question

Given any arbitrary state-space representation, is it possible to transform it into:

1. Controllable Canonical Form
2. Observable Canonical Form
3. Diagonal Canonical Form

Answer: Yes, it is generally possible to transform an arbitrary state-space representation into these canonical forms through appropriate similarity transformations. However, each transformation requires specific system properties to be satisfied.

The transformation process involves systematically manipulating the system matrices to achieve the desired structural properties while preserving the system's input-output behavior and dynamic characteristics. The key insight is that by selecting an appropriate transformation matrix, we can restructure the internal representation without altering the fundamental system dynamics.

Transformation Principle

Through suitable selection of the transformation matrix $\mathbf{P}$ in the similarity transformation:

$$\mathbf{A}_{\text{new}} = \mathbf{P}^{-1}\mathbf{A}\mathbf{P}, \qquad \mathbf{B}_{\text{new}} = \mathbf{P}^{-1}\mathbf{B}, \qquad \mathbf{C}_{\text{new}} = \mathbf{C}\mathbf{P}$$

and ensuring satisfaction of specific system conditions (controllability, observability, or diagonalizability), any arbitrary state-space model can be converted into the desired canonical form.

Important Conditions:

- Controllable Canonical Form: System must be controllable
- Observable Canonical Form: System must be observable
- Diagonal Canonical Form: System matrix must have distinct eigenvalues (diagonalizable)

In the following sections, we will examine the specific procedures and conditions required for each transformation.

10.5.1 Transformation to Controllable Canonical Form

Consider a single-input single-output system described by

$$\begin{aligned} \dot{\mathbf{x}} &= \mathbf{A}\mathbf{x} + \mathbf{B}\mathrm{u} \\ \mathrm{y} &= \mathbf{C}\mathbf{x} + \mathbf{D}\mathrm{u} \end{aligned} \tag{10.72}$$

where $\mathbf{x}$ is the $\mathrm{n} \times 1$ state vector, and y and u are respectively the scalar output and input.

Let the characteristic equation of this system is

$$\left|\mathrm{s}\mathbf{I} - \mathbf{A}\right| = \mathrm{s}^{\mathrm{n}} + \mathrm{a}_1\mathrm{s}^{\mathrm{n}-1} + \cdots + \mathrm{a}_{\mathrm{n}-1}\mathrm{s} + \mathrm{a}_{\mathrm{n}} = 0 \tag{10.73}$$

Theorem 10.3. If the system of Eqn(10.72) is completely state controllable, then it can be transformed into controllable canonical form using the transformation matrix

$$\mathbf{P} = \mathbf{Q_C W} \quad \text{where}$$

$$\mathbf{Q_C} = \begin{bmatrix} \mathbf{B} & \mathbf{AB} & \mathbf{A^2B} & \dots & \dots & \mathbf{A^{n-1}B} \end{bmatrix} \quad \text{and}$$

$$\mathbf{W} = \begin{bmatrix} a_{n-1} & a_{n-2} & \dots & \dots & a_1 & 1 \\ a_{n-2} & a_{n-3} & \dots & \dots & 1 & 0 \\ \vdots & \vdots & \vdots & \vdots & \vdots & \vdots \\ \vdots & \vdots & \vdots & \vdots & \vdots & \vdots \\ a_1 & 1 & \dots & \dots & 0 & 0 \\ 1 & 0 & \dots & \dots & 0 & 0 \end{bmatrix}$$

where $a_1, a_2, \dots, a_{n-1}$ are the coefficients of the characteristic polynomial of $\mathbf{A}$.

The dynamics of the transformed system are given by

$$\begin{aligned} \dot{\mathbf{z}} &= \mathbf{P^{-1}APz} + \mathbf{P^{-1}Bu} \\ y &= \mathbf{CPz} + \mathbf{D}u \end{aligned} \tag{10.74}$$

where $\mathbf{x} = \mathbf{Pz}$ and

$$\mathbf{P^{-1}AP} = \begin{bmatrix} 0 & 1 & 0 & \dots & 0 & 0 \\ 0 & 0 & 1 & \dots & 0 & 0 \\ \vdots & \dots & \dots & \dots & \vdots & \vdots \\ \vdots & \dots & \dots & \dots & \vdots & \vdots \\ 0 & 0 & 0 & \dots & 0 & 1 \\ -a_n & -a_{n-1} & -a_{n-2} & \dots & -a_2 & -a_1 \end{bmatrix}, \quad \mathbf{P^{-1}B} = \begin{bmatrix} 0 \\ 0 \\ \vdots \\ \vdots \\ 0 \\ 1 \end{bmatrix} \tag{10.75}$$

This transformation is valid provided $\mathbf{P^{-1}}$ exists. Since:

$$\mathbf{P^{-1}} = (\mathbf{Q_C W})^{-1} = \mathbf{W^{-1} Q_C^{-1}} \tag{10.76}$$

It can be shown that the determinant of $\mathbf{W} = (-1)^{n-1}$, ensuring that $\mathbf{W}^{-1}$ always exists. Therefore, $\mathbf{P^{-1}}$ exists if and only if $\mathbf{Q_C^{-1}}$ exists, which occurs when the controllability matrix has full rank (i.e., the system is completely controllable).

Note: The output matrix **CP** in the transformed system does not follow a specific canonical pattern, as the transformation is designed to achieve the controllable canonical structure for the **A** and **B** matrices.

EXAMPLE 10.7

Problem:

Consider a system represented as:

$$\dot{\mathbf{x}} = \mathbf{Ax} + \mathbf{Bu}, \quad y = \mathbf{Cx}, \quad \text{where}$$
$$\mathbf{A} = \begin{bmatrix} -2 & 3 \\ 3 & -2 \end{bmatrix}, \quad \mathbf{B} = \begin{bmatrix} 2 \\ 0 \end{bmatrix}, \quad \mathbf{C} = \begin{bmatrix} 0 & 1 \end{bmatrix}$$

Transform this system into controllable canonical form.

SOLUTION

Given $\mathbf{A} = \begin{bmatrix} -2 & 3 \\ 3 & -2 \end{bmatrix}, \quad \mathbf{B} = \begin{bmatrix} 2 \\ 0 \end{bmatrix}, \quad \mathbf{C} = \begin{bmatrix} 0 & 1 \end{bmatrix}.$

The characteristic equation of the system is given by

$$|s\mathbf{I} - \mathbf{A}| = \left| \begin{bmatrix} s & 0 \\ 0 & s \end{bmatrix} - \begin{bmatrix} -2 & 3 \\ 3 & -2 \end{bmatrix} \right| = \left| \begin{bmatrix} s+2 & -3 \\ -3 & s+2 \end{bmatrix} \right| = s^2 + 4s - 5 = 0$$

Comparing this with the general form of characteristic equation as:

$$s^2 + a_1 s + a_2 = s^2 + 4s - 5$$

gives $a_1 = 4$ and $a_2 = -5$.

The transformation matrix **P**, which transform the system into controllable form, is

$$\mathbf{P} = \mathbf{Q_C W}, \quad \text{where } \mathbf{W} = \begin{bmatrix} a_1 & 1 \\ 1 & 0 \end{bmatrix}, \quad \text{and} \quad \mathbf{Q_C} = \begin{bmatrix} \mathbf{B} & \mathbf{AB} \end{bmatrix}$$

Substituting the values of the matrices and the parameters gives

$$\mathbf{W} = \begin{bmatrix} 4 & 1 \\ 1 & 0 \end{bmatrix}, \quad \mathbf{Q_C} = \begin{bmatrix} 2 & -4 \\ 0 & 6 \end{bmatrix}, \quad \text{and} \quad \mathbf{P} = \mathbf{Q_C W} = \begin{bmatrix} 4 & 2 \\ 6 & 0 \end{bmatrix}$$

The transformed system is given by

$$\dot{\mathbf{z}} = \begin{bmatrix} 0 & 1 \\ 5 & -4 \end{bmatrix} \mathbf{z} + \begin{bmatrix} 0 \\ 1 \end{bmatrix} u$$
$$y = \begin{bmatrix} 6 & 0 \end{bmatrix} \mathbf{z}$$

10.5.2 Geometric Interpretation of Controllable Canonical Form

We have shown that any controllable system can be transformed into controllable canonical form using the transformation matrix $\mathbf{P}$ constructed from the controllability matrix. While this algebraic procedure is systematic, one might wonder: what does this transformation actually accomplish? We already obtained controllable canonical form earlier through direct realization from the transfer function (Section 10.4.2). What additional insight does the controllability-matrix approach provide?

The answer is that this transformation reveals how to **reorganize any existing state-space description** regardless of how messy or unstructured into the cleanest possible form for control design. To understand what this reorganization achieves, consider the following analogy.

Controllable Canonical Form: Organizing a Chaotic Kitchen

- Imagine the kitchen of a teenager, say **Aashay**, which is in complete chaos: spices scattered across five cabinets, knives in the bathroom, a pan under the bed, and somehow the rice cooker ended up in the garage.
- You can cook anything **(the kitchen is "controllable")**, but every recipe is a treasure hunt—"Where did I put the salt? Was it next to the shampoo?" This represents a system with redundant or poorly organized state variables: all the necessary components exist, but the description is so chaotic that practical control becomes nearly impossible.
- So you reorganize everything into a logical assembly line: ingredients → prep station → stove → plating area. Now when you reach for something, you know exactly where it is and what comes next. The flow is crystal clear.
- That's **controllable canonical form:** you haven't changed what **Aashay** can cook **(same dynamics)**, you've just rearranged the **"state variables" (kitchen items)** so that your **"input" (cooking actions)** flows through a clean, predictable chain. Perfect for designing your control strategy—now you can actually follow a recipe without losing your mind!
- **Mathematically:** The chaotic original system $\dot{\mathbf{x}} = \mathbf{A}\mathbf{x} + \mathbf{B}\mathbf{u}$ becomes $\dot{\mathbf{z}} = \mathbf{A}_z\mathbf{z} + \mathbf{B}_z u$, where $\mathbf{A}_z$ is in companion form and $\mathbf{B}_z = [0, 0, \ldots, 0, 1]^T$ (see Eqn(10.11). All the control effort flows directly into the last state, which cascades cleanly through the others. It's the kitchen assembly line in matrix form.

Just as reorganizing the kitchen doesn't change what can be cooked, the transformation **P** doesn't change the system's controllability or transfer function—it merely rearranges the state variables to expose the most transparent control structure. This is why the transformation approach is invaluable when working with physical models that don't naturally arise in canonical form.

10.5.3 Transformation to Observable Canonical Form

Consider a single-input single-output system described by Eqn(10.72) with the characteristic equation given in Eqn(10.73).

Theorem 10.4. If the system of Eqn(10.72) is completely observable, then it can be transformed into observable canonical form using the transformation matrix

$$\mathbf{P} = (\mathbf{W}\mathbf{Q_O})^{-1} \quad \textbf{where}$$

$$\mathbf{Q_O} = \begin{bmatrix} \mathbf{C} \\ \mathbf{CA} \\ \mathbf{CA^2} \\ \vdots \\ \vdots \\ \mathbf{CA^{n-1}} \end{bmatrix} \quad \text{and } \mathbf{W} = \begin{bmatrix} a_{n-1} & a_{n-2} & \dots & \dots & a_1 & 1 \\ a_{n-2} & a_{n-3} & \dots & \dots & 1 & 0 \\ \vdots & \vdots & \vdots & \vdots & \vdots & \vdots \\ \vdots & \vdots & \vdots & \vdots & \vdots & \vdots \\ a_1 & 1 & \dots & \dots & 0 & 0 \\ 1 & 0 & \dots & \dots & 0 & 0 \end{bmatrix}$$

The dynamics of the transformed system are given by

$$\begin{aligned} \dot{\mathbf{z}} &= \mathbf{P}^{-1}\mathbf{APz} + \mathbf{P}^{-1}\mathbf{B}u \\ y &= \mathbf{CPz} + \mathbf{D}u \end{aligned} \tag{10.77}$$

where $\mathbf{x} = \mathbf{Pz}$ and

$$\mathbf{P^{-1}AP} = \begin{bmatrix} 0 & 0 & 0 & \dots & 0 & -a_n \\ 1 & 0 & 0 & \dots & 0 & -a_{n-1} \\ 0 & 1 & \dots & \dots & 0 & -a_{n-2} \\ \vdots & \dots & \dots & \dots & \vdots & \vdots \\ \vdots & \dots & \dots & \dots & \vdots & \vdots \\ 0 & 0 & 0 & \dots & 0 & 1 \\ 0 & 0 & \dots & \dots & 1 & -a_1 \end{bmatrix}, \quad \mathbf{CP} = \begin{bmatrix} 0 & 0 & \dots & \dots & 0 & 1 \end{bmatrix} \tag{10.78}$$

This transformation is valid provided $\mathbf{P}^{-1}$ exists. Since

$$\mathbf{P}^{-1} = (\mathbf{W}\mathbf{Q_O})^{-1} = \mathbf{Q_O^{-1}}\mathbf{W}^{-1} \tag{10.79}$$

Since the determinant of $\mathbf{W} = (-1)^{n-1} \neq 0$, the matrix $\mathbf{W}^{-1}$ always exists. Therefore, $\mathbf{P}^{-1}$ exists if and only if $\mathbf{Q_O}^{-1}$ exists, which occurs when the observability matrix has full rank (i.e., the system is completely observable).

Note: The input matrix $\mathbf{P}^{-1}\mathbf{B}$ in the transformed system does not follow a specific canonical pattern, as the transformation is designed to achieve the observable canonical structure for the **A** and **C** matrices.

EXAMPLE 10.8

Problem:

Consider a system represented as:

$$\dot{\mathbf{x}} = \mathbf{A}\mathbf{x} + \mathbf{B}\mathrm{u}, \quad \mathrm{y} = \mathbf{C}\mathbf{x}, \quad \text{where}$$
$$\mathbf{A} = \begin{bmatrix} -3 & 1 \\ 1 & -3 \end{bmatrix}, \quad \mathbf{B} = \begin{bmatrix} 1 \\ 2 \end{bmatrix}, \quad \mathbf{C} = \begin{bmatrix} 2 & 3 \end{bmatrix}$$

Transform this system into observable canonical form.

SOLUTION:

Given

$$\mathbf{A} = \begin{bmatrix} -3 & 1 \\ 1 & -3 \end{bmatrix}, \quad \mathbf{B} = \begin{bmatrix} 1 \\ 2 \end{bmatrix}, \quad \mathbf{C} = \begin{bmatrix} 2 & 3 \end{bmatrix}$$

The characteristic equation of the system is given by

$$|\mathrm{s}\mathbf{I} - \mathbf{A}| = \left| \begin{bmatrix} \mathrm{s} & 0 \\ 0 & \mathrm{s} \end{bmatrix} - \begin{bmatrix} -3 & 1 \\ 1 & -3 \end{bmatrix} \right| = \left| \begin{bmatrix} \mathrm{s}+3 & -1 \\ -1 & \mathrm{s}+3 \end{bmatrix} \right| = \mathrm{s}^2 + 6\mathrm{s} + 8 = 0$$

Comparing this with the general form of characteristic equation as:

$$\mathrm{s}^2 + \mathrm{a}_1\mathrm{s} + \mathrm{a}_2 = \mathrm{s}^2 + 6\mathrm{s} + 8$$

gives $\mathrm{a}_1 = 6$ and $\mathrm{a}_2 = 8$.

The transformation matrix **P**, which transforms the system into observable form, is

$$\mathbf{P} = (\mathbf{W}\mathbf{Q_O})^{-1}, \quad \text{where } \mathbf{W} = \begin{bmatrix} \mathrm{a}_1 & 1 \\ 1 & 0 \end{bmatrix}, \quad \text{and} \quad \mathbf{Q_O} = \begin{bmatrix} \mathbf{C} \\ \mathbf{CA} \end{bmatrix}$$

Substituting the values of the matrices and the parameters gives

$$\mathbf{W} = \begin{bmatrix} 6 & 1 \\ 1 & 0 \end{bmatrix}, \quad \mathbf{Q_O} = \begin{bmatrix} 2 & 3 \\ -3 & -7 \end{bmatrix},$$
$$\mathbf{P} = (\mathbf{W}\mathbf{Q_O})^{-1} = \begin{bmatrix} 0.6 & -2.2 \\ -0.4 & 1.8 \end{bmatrix}$$

The transformed system is given by

$$\dot{\mathbf{z}} = \begin{bmatrix} 0 & -8 \\ 1 & -6 \end{bmatrix} \mathbf{z} + \begin{bmatrix} 31 \\ 8 \end{bmatrix} \mathrm{u}, \qquad \mathrm{y} = \begin{bmatrix} 0 & 1 \end{bmatrix} \mathbf{z}$$

10.5.4 Geometric Interpretation of Observable Canonical Form

We have shown that any observable system can be transformed into observable canonical form through coordinate transformation using the observability matrix. Any observable system can be transformed via $\mathbf{z} = \mathbf{P}^{-1}\mathbf{x}$ where $\mathbf{P}$ is constructed from $\mathcal{Q}_\mathcal{O}$. While the algebraic procedure is systematic, the geometric meaning deserves attention. What does this transformation accomplish? How does it differ from the controllable canonical form we derived earlier?

The key insight is that observable canonical form reorganizes the state variables to reveal the cleanest possible path from internal states to measured output. Whereas controllable canonical form structures **how inputs enter the system**, observable canonical form structures bf how information leaves the system through measurements. To understand what this means intuitively, consider the following analogy.

Observable Canonical Form—"The Gossip Chain in Your Apartment Building"

- In your apartment building, every juicy piece of news eventually reaches you, but **the route is unpredictable:** sometimes the security guard tells you, sometimes the neighbor's kid, sometimes that auntie on the second floor who knows everyone's business.
- To make sense of things, you **reorganize the gossip network into a clean chain:** the auntie tells the security guard, the guard tells your neighbor, and your neighbor tells you.
- Now **every hidden detail flows through a structured path**, and from your neighbor's final summary you can reconstruct the entire story.
- That is **observable canonical form:** you restructure the system so the flow from hidden states to the measured output is orderly, making it easy to infer what's happening inside.

This reorganization is precisely what the transformation matrix $\mathbf{P}$ accomplishes: it reorders the state variables so that the output y directly observes only the first state ($\mathbf{C}_\mathrm{z} = [1, 0, \ldots, 0]$) (see Eqn(10.12), and information about the other states flows through the companion matrix structure $\mathbf{A_z}$ in a predictable cascade). The "tangled web" of the original $(\mathbf{A}, \mathbf{C})$ pair is transformed into a transparent information chain—making observer design as systematic as following a well-organized gossip network.

Just as the gossip chain doesn't change what information exists (only how it reaches you), the transformation doesn't change the system's observability or transfer function—it merely reveals the most transparent structure for reconstructing internal states from measurements.

10.5.5 Transformation to Diagonal Canonical Form

Consider a single-input single-output system described by Eqn(10.72). We present two different methods for obtaining the diagonal canonical form:

Theorem 10.5.

General Diagonalization Method: If the system matrix $\mathbf{A}$ in Eqn(10.72) has "n" distinct eigenvalues, then it can be transformed into diagonal canonical form using the transformation matrix $\mathbf{P}$ whose columns are the eigenvectors of matrix $\mathbf{A}$:

$$\mathbf{P} = \begin{bmatrix} \mathbf{v_1} & \mathbf{v_2} & \mathbf{v_3} & \dots & \dots & \mathbf{v_n} \end{bmatrix} \tag{10.80}$$

where $\mathbf{v_i}, i = 1, 2, \dots, \mathrm{n}$ denotes the eigenvector associated with the eigenvalue λ_i.

Theorem 10.6.

Vandermonde Matrix Method If the system matrix $\mathbf{A}$ in Eqn(10.72) is already in controllable canonical form and has "n" distinct eigenvalues, then it can be transformed into diagonal canonical form using the transformation matrix $\mathbf{P}$:

$$\mathbf{P} = \begin{bmatrix} 1 & 1 & \dots & \dots & 1 \\ \lambda_1 & \lambda_2 & \dots & \dots & \lambda_n \\ \lambda_1^2 & \lambda_2^2 & \dots & \dots & \lambda_n^2 \\ \vdots & \vdots & \vdots & \vdots & \vdots \\ \lambda_1^{n-1} & \lambda_2^{n-1} & \dots & \dots & \lambda_n^{n-1} \end{bmatrix} \tag{10.81}$$

where λ_1, λ_2,..., λ_n are n-distinct eigenvalues of $\mathbf{A}$. This matrix $\mathbf{P}$ is called the Vandermonde matrix.

For both methods, the dynamics of the transformed system are given by:

$$\begin{aligned} \dot{\mathbf{z}} &= \mathbf{P^{-1}APz} + \mathbf{P^{-1}Bu} \\ \mathrm{y} &= \mathbf{CPz} + \mathbf{Du} \end{aligned} \tag{10.82}$$

$$\text{where} \quad \mathbf{P}^{-1}\mathbf{A}\mathbf{P} = \begin{bmatrix} \lambda_1 & 0 & 0 & \dots & 0 \\ 0 & \lambda_2 & 0 & \dots & 0 \\ 0 & 0 & \lambda_3 & \dots & 0 \\ \vdots & \dots & \dots & \dots & \vdots \\ 0 & 0 & \dots & \dots & \lambda_n \end{bmatrix}$$

Note that the other matrices $\mathbf{P}^{-1}\mathbf{B}$ and $\mathbf{CP}$ do not follow specific canonical patterns. The Vandermonde method is particularly useful when starting from the controllable canonical form, and both methods require the system matrix to have distinct eigenvalues for successful diagonalization. We demonstrate this with an example.

EXAMPLE 10.9

Problem: Consider a system

$$\begin{bmatrix} \dot{x}_1 \\ \dot{x}_2 \end{bmatrix} = \begin{bmatrix} -5 & 3 \\ 3 & -5 \end{bmatrix} \begin{bmatrix} x_1 \\ x_2 \end{bmatrix} + \begin{bmatrix} 1 \\ 2 \end{bmatrix} u = \mathbf{Ax} + \mathbf{B}u$$

$$y = \begin{bmatrix} 1 & 1 \end{bmatrix} \begin{bmatrix} x_1 \\ x_2 \end{bmatrix} = \mathbf{Cx}$$

Represent this system in the form

$$\dot{\mathbf{z}} = \mathbf{P}^{-1}\mathbf{APz} + \mathbf{P}^{-1}\mathbf{B}u$$

such that $\mathbf{P}^{-1}\mathbf{AP}$ is diagonal.

SOLUTION:

Note that if the columns of $\mathbf{P}$ are the eigenvectors of $\mathbf{A}$, then $\mathbf{P}^{-1}\mathbf{AP}$ becomes diagonal. Hence the problem of diagonalization reduces to finding the eigenvectors of the system matrix $\mathbf{A}$.

Step-1: Find the eigenvalues of system matrix $\mathbf{A}$. This is obtained by solving $|\lambda\mathbf{I} - \mathbf{A}| = 0$. Now

$$|\lambda\mathbf{I} - \mathbf{A}| = \left| \begin{bmatrix} \lambda & 0 \\ 0 & \lambda \end{bmatrix} - \begin{bmatrix} -5 & 3 \\ 3 & -5 \end{bmatrix} \right| = \left| \begin{bmatrix} \lambda+5 & -3 \\ -3 & \lambda+5 \end{bmatrix} \right|$$
$$= \lambda^2 + 10\lambda + 16 = 0$$

This gives the eigenvalues at $\lambda_1 = -8$ and $\lambda_2 = -2$.

Step-2: Determine the eigenvectors corresponding to each of the distinct eigenvalues. This is obtained by solving

$$(\lambda_i\mathbf{I} - \mathbf{A})\,\mathbf{v}_i = 0$$

where $\mathbf{v}_i = \begin{bmatrix} v_{1,i} & v_{2,i} & \dots & v_{n,i} \end{bmatrix}^T$ is the eigenvector

The eigenvectors corresponding to eigenvalue $\lambda_1 = -8$ are obtained from

$$\mathbf{Av_1} = \lambda_1\mathbf{v_1} \implies \begin{bmatrix} -5 & 3 \\ 3 & -5 \end{bmatrix}\begin{bmatrix} \mathrm{v}_{11} \\ \mathrm{v}_{21} \end{bmatrix} = -8\begin{bmatrix} \mathrm{v}_{11} \\ \mathrm{v}_{21} \end{bmatrix}$$

This results in 2-number of simultaneous equations:

$$\begin{aligned} -5\mathrm{v}_{11} + 3\mathrm{v}_{21} &= -8\mathrm{v}_{11} \\ 3\mathrm{v}_{11} - 5\mathrm{v}_{21} &= -8\mathrm{v}_{21} \end{aligned}$$

Solving above gives $\mathrm{v}_{11} = -\mathrm{v}_{21}$. We may select $\mathbf{v_1} = \begin{bmatrix} \mathrm{c} \\ -\mathrm{c} \end{bmatrix}$

Step-2: Similarly,the eigenvectors corresponding to eigenvalue $\lambda_2 = -2$ is obtained from

$$\mathbf{Av_2} = \lambda_2\mathbf{v_2} \implies \begin{bmatrix} -5 & 3 \\ 3 & -5 \end{bmatrix}\begin{bmatrix} \mathrm{v}_{12} \\ \mathrm{v}_{22} \end{bmatrix} = -2\begin{bmatrix} \mathrm{v}_{12} \\ \mathrm{v}_{22} \end{bmatrix}$$

This gives

$$\begin{aligned} -5\mathrm{v}_{12} + 3\mathrm{v}_{22} &= -2\mathrm{v}_{12} \\ 3\mathrm{v}_{12} - 5\mathrm{v}_{22} &= -2\mathrm{v}_{22} \end{aligned}$$

Its solution is $\mathrm{v}_{12} = \mathrm{v}_{22}$. We may select $\mathbf{v_2} = \begin{bmatrix} \mathrm{c} \\ \mathrm{c} \end{bmatrix}$

Step-3:

Considering one possible choice of $\mathbf{P}$ as:

$$\mathbf{P} = \begin{bmatrix} 1 & 1 \\ -1 & 1 \end{bmatrix}$$

First compute $\mathbf{P}^{-1}$, then compute $\mathbf{P}^{-1}\mathbf{AP}$, $\mathbf{P}^{-1}\mathbf{B}$ and $\mathbf{CP}$.

$$\mathbf{P}^{-1} = \begin{bmatrix} 0.5 & -0.5 \\ 0.5 & 0.5 \end{bmatrix}$$

Thus,

$$\mathbf{P}^{-1}\mathbf{AP} = \begin{bmatrix} 0.5 & -0.5 \\ 0.5 & 0.5 \end{bmatrix}\begin{bmatrix} -5 & 3 \\ 3 & -5 \end{bmatrix}\begin{bmatrix} 1 & 1 \\ -1 & 1 \end{bmatrix} = \begin{bmatrix} -8 & 0 \\ 0 & -2 \end{bmatrix}$$

$$\mathbf{P}^{-1}\mathbf{B} = \begin{bmatrix} 0.5 & -0.5 \\ 0.5 & 0.5 \end{bmatrix}\begin{bmatrix} 1 \\ 2 \end{bmatrix} = \begin{bmatrix} -0.5 \\ 1.5 \end{bmatrix}$$

$$\mathbf{CP} = \begin{bmatrix} 1 & 1 \end{bmatrix}\begin{bmatrix} 1 & 1 \\ -1 & 1 \end{bmatrix} = \begin{bmatrix} 0 & 2 \end{bmatrix}$$

Thus, the diagonal system becomes

$$\begin{bmatrix} \dot{z}_1 \\ \dot{z}_2 \end{bmatrix} = \begin{bmatrix} -8 & 0 \\ 0 & -2 \end{bmatrix} \begin{bmatrix} z_1 \\ z_2 \end{bmatrix} + \begin{bmatrix} -0.5 \\ 1.5 \end{bmatrix} u$$

$$y = \begin{bmatrix} 0 & 2 \end{bmatrix} \begin{bmatrix} z_1 \\ z_2 \end{bmatrix}$$

CONCEPT REINFORCEMENT PROBLEM 10.6

Problem:

Consider the following systems

$$\textbf{S1}: \quad \begin{bmatrix} \dot{x}_1 \\ \dot{x}_2 \end{bmatrix} = \begin{bmatrix} -3 & -1 \\ 6 & 2 \end{bmatrix} \begin{bmatrix} x_1 \\ x_2 \end{bmatrix} + \begin{bmatrix} 1 \\ 0 \end{bmatrix} u = \mathbf{Ax} + \mathbf{B}u$$

$$y = \begin{bmatrix} 1 & 2 \end{bmatrix} \begin{bmatrix} x_1 \\ x_2 \end{bmatrix} = \mathbf{Cx}$$

$$\textbf{S2}: \quad \begin{bmatrix} \dot{x}_1 \\ \dot{x}_2 \end{bmatrix} = \begin{bmatrix} -5 & -1 \\ 3 & -1 \end{bmatrix} \begin{bmatrix} x_1 \\ x_2 \end{bmatrix} + \begin{bmatrix} 2 \\ 5 \end{bmatrix} u = \mathbf{Ax} + \mathbf{B}u$$

$$y = \begin{bmatrix} 1 & 2 \end{bmatrix} \begin{bmatrix} x_1 \\ x_2 \end{bmatrix} = \mathbf{Cx}$$

$$\textbf{S3}: \quad \begin{bmatrix} \dot{x}_1 \\ \dot{x}_2 \end{bmatrix} = \begin{bmatrix} -3 & 2 \\ -3 & 4 \end{bmatrix} \begin{bmatrix} x_1 \\ x_2 \end{bmatrix} + \begin{bmatrix} 1 \\ -1 \end{bmatrix} u = \mathbf{Ax} + \mathbf{B}u$$

$$y = \begin{bmatrix} 1 & 0 \end{bmatrix} \begin{bmatrix} x_1 \\ x_2 \end{bmatrix} = \mathbf{Cx}$$

Represent these systems in the form

$$\dot{\mathbf{z}} = \mathbf{P}^{-1}\mathbf{APz} + \mathbf{P}^{-1}\mathbf{B}u$$

such that $\mathbf{P}^{-1}\mathbf{AP}$ is diagonal.

10.5.6 Geometric Interpretation of Diagonal Canonical Form

We have shown that any system with distinct eigenvalues can be transformed into diagonal canonical form through coordinate transformation using the modal matrix. The system is transformed via $\mathbf{z} = \mathbf{P}^{-1}\mathbf{x}$ where $\mathbf{P}$ is constructed from the eigenvectors. While the algebraic procedure is systematic, the geometric meaning deserves attention. What does this transformation accomplish? How does it differ from the controllable and observable canonical forms we derived earlier?

The key insight is that diagonal canonical form decouples the state variables to reveal the system's independent modes of evolution. Whereas controllable canonical form structures **how inputs enter the system** and observable canonical form structures **how information leaves the system**, diagonal canonical form structures **how the system naturally evolves on its own**. Each diagonal entry represents an independent exponential mode, evolving without interfering with the others. To understand what this means intuitively, consider the following analogy.

Diagonal Canonical Form—"Your Group of Friends at a Cafe"

- When your group of friends meets at a café, chaos reigns: one person vents about work while another debates cricket, someone's ordering food as another launches into politics—everyone talking over each other in coupled, intertwined chaos.
- Now imagine if everyone magically **agreed to speak one at a time**, each sticking to their own topic in perfectly separate turns. Suddenly you can follow each friend's story independently, with no cross-talk to untangle.
- That's **diagonal canonical form:** you **decompose a complicated, coupled system into independent "modes,"** each evolving on its own timeline, making the math beautifully simple. If only real friends were as cooperative as eigenvectors!

This decoupling is precisely why diagonal form is so powerful: the system matrix $\mathbf{A}$ becomes a diagonal matrix of eigenvalues, the state transition matrix becomes $e^{\mathbf{\Lambda}t}$ (exponentially simple), and we can analyze each mode's stability and time constant independently. When the conversation is structured this way, everything becomes transparent.

10.5.7 Geometric Interpretation of Jordan Canonical Form

In Section 10.4.4, we derived the diagonal canonical form for systems with distinct eigenvalues, where the system completely decouples into independent modes. But what happens when eigenvalues repeat? When a system has repeated eigenvalues with insufficient eigenvectors (defective matrices), complete decoupling becomes impossible. The best we can achieve is Jordan canonical form — a near-diagonal structure where the system breaks into independent clusters, each cluster containing tightly coupled variables that cannot be separated.

The key insight is that Jordan canonical form represents the closest we can get to diagonal form when complete decoupling is geometrically impossible. Each

Jordan block corresponds to a cluster of states that must evolve together, linked by their shared eigenvalue and geometric deficiency. While different blocks remain independent of each other, the states within each block are inseparably coupled. To understand what this means intuitively, consider the following analogy.

Jordan Canonical Form — "The Group Project with Clusters of Friends Who Can't Work Independently"

- Imagine you're coordinating a big group project with several friends. Ideally, each friend would take one task and work independently — that would be like diagonal form: clean, simple, no interference.
- But life is never that easy. Some friends insist on working in pairs or trios, because one knows the topic, another knows the formatting, and the third just shows up for the snacks.
- These **mini-groups can't be split** — they depend on each other — **but each group is still independent of the others**. So the entire project becomes a set of clusters: Group A (3 people who refuse to separate), Group B (2 people joined at the hip), Group C (1 person working happily alone).
- That's **Jordan canonical form:** when a system can't be fully diagonalized, you **break it into independent clusters (Jordan blocks)**, each block containing variables that are inseparable, internally linked, and must move together — **but different blocks don't interfere with each other**. It's the mathematical version of friend circles in group projects: some groups act like one giant person.

This block-diagonal structure is why Jordan form matters: when full diagonalization fails, we still achieve partial decoupling. The system matrix becomes block-diagonal with Jordan blocks on the diagonal, each block capturing the coupled dynamics of its defective eigenvalue. Different blocks evolve independently, but within each block, the states cascade through each other — mathematically elegant, if slightly messier than the ideal diagonal case.

10.6 Solution of State Equations

The solution of state equations involves determining the temporal evolution of a system's state variables given specific initial conditions and input signals. This fundamental analysis is essential for understanding dynamic system behavior and serves as the foundation for effective controller design strategies. The solution can be obtained either through numerical simulations or by various analytical procedures such as matrix exponential methods, Laplace transform techniques.

In this section, we will derive the general analytical solution for linear time-invariant (LTI) systems using systematic mathematical procedures. Our approach will follow a structured progression:

1. **Homogeneous case:** Systems with no external forcing inputs ($u(t) = 0$
2. **Non-homogeneous case:** Systems with external inputs present

10.6.1 Solution of Homogeneous State Equations: The State Transition Matrix

Consider a system described by the scalar differential equation

$$\dot{x}(t) = ax(t) \tag{10.83}$$

Since this is a first-order scalar differential equation, the solution of this can be obtained using the classical methods of solving differential equations. It is given by

$$x(t) = e^{at}x(0) \tag{10.84}$$

where $x(0)$ is the initial condition.

Let us consider a system described by a vector differential equation

$$\dot{\mathbf{x}}(t) = \mathbf{A}\mathbf{x}(t) \tag{10.85}$$

Assume that the solution of Eqn(10.85) is of the form

$$\mathbf{x}(t) = \mathbf{a_0} + \mathbf{a_1}t + \mathbf{a_2}t^2 + \cdots + \mathbf{a_k}t^k + \cdots \tag{10.86}$$

Differentiating Eqn(10.86) gives

$$\dot{\mathbf{x}}(t) = \mathbf{a_1} + 2\mathbf{a_2}t + 3\mathbf{a_3}t^2 \cdots + k\mathbf{a_k}t^{k-1} + \cdots \tag{10.87}$$

Substituting the value of $\mathbf{x}(t)$ and $\dot{\mathbf{x}}(t)$ into Eqn(10.85) gives

$$\mathbf{a_1}+2\mathbf{a_2}t+3\mathbf{a_3}t^2\cdots+k\mathbf{a_k}t^{k-1}+\cdots = \mathbf{A}\left(\mathbf{a_0} + \mathbf{a_1}t + \mathbf{a_2}t^2 + \cdots + \mathbf{a_k}t^k + \cdots\right) \tag{10.88}$$

Equating the coefficients of equal powers of "t" on both sides of Eqn(10.88) gives

$$\begin{aligned}
\mathbf{a_1} &= \mathbf{A}\mathbf{a_0} \\
\mathbf{a_2} &= \frac{1}{2}\mathbf{A}\mathbf{a_1} = \frac{1}{2}\mathbf{A}^2\mathbf{a_0} \\
\mathbf{a_3} &= \frac{1}{3}\mathbf{A}\mathbf{a_2} = \frac{1}{6}\mathbf{A}^3\mathbf{a_0} \\
&\vdots \quad \vdots \\
\mathbf{a_k} &= \frac{1}{k!}\mathbf{A}^k\mathbf{a_0}
\end{aligned}$$

The initial value of $\mathbf{x}(t)=\mathbf{x}(0)$ is obtained by substituting $t = 0$ in Eqn(10.86) which gives

$$\mathbf{x}(\mathbf{0}) = \mathbf{a_0}$$

The solution of $\mathbf{x}(t)$ is therefore becomes

$$\mathbf{x}(t) = \left(\mathbf{I} + \mathbf{A}t + \frac{1}{2!}\mathbf{A^2}t^2 + \cdots + \frac{1}{k!}\mathbf{A^k}t^k + \cdots\right)\mathbf{x}(0) \tag{10.89}$$

Note that each term in the parentheses on the right side of Eqn(10.89) is an $n \times n$ matrix. This is an infinite power series and resembles a scalar exponential, and is called the matrix exponential. Thus

$$e^{\mathbf{A}t} = \mathbf{I} + \mathbf{A}t + \frac{1}{2!}\mathbf{A^2}t^2 + \cdots + \frac{1}{k!}\mathbf{A^k}t^k + \cdots \tag{10.90}$$

The solution to the state equation (10.83) can be written as

$$\boxed{\mathbf{x}(t) = e^{\mathbf{A}t}\mathbf{x}(0)} \tag{10.91}$$

Eqn(10.91) shows that the transition of the states from their initial values $\mathbf{x}(0)$ to a state $\mathbf{x}(t)$ takes place via the matrix exponential $e^{\mathbf{A}t}$. Therefore, e^{At} is called as *state transition matrix* and is denoted by $\mathbf{\Phi(t)}$. Before we discuss the computational procedure of the state transition matrix and its properties, some of the important characteristics of this are:

Charactersitics of State Transition Matrix

1. **Free Response Characterization:** The state transition matrix governs the system's free response, representing the evolution of states due to initial conditions alone. These initial conditions physically correspond to the energy stored in the system's energy-storage elements (capacitors, inductors, springs, masses, etc.) at the initial time.

2. **System Matrix Dependence:** The state transition matrix depends solely on the system matrix **A** and is independent of the input matrix **B** or output matrix **C**. This makes it an intrinsic property of the system dynamics.

3. **Complete State Evolution:** In the absence of external inputs ($u(t) = 0$) the state transition matrix completely determines how the system states transition from any initial time t_0 to any future time t. It provides a complete mapping of the unforced system dynamics.

4. **Physical Interpretation:** The state transition matrix essentially captures the system's "natural" behavior; how it evolves when left to its own devices without external driving forces. This is analogous to studying the free vibration of a mechanical system or the natural response of an electrical circuit after the sources are removed.

10.6.2 Computation of State Transition Matrix by Laplace Transform Approach

Let us first consider a scalar case. A system described by a scalar differential equation is expressed as:

$$\dot{\mathrm{x}} = \mathrm{ax} \tag{10.92}$$

Taking the Laplace transform of Eqn(10.92) gives

$$\mathrm{sX(s) - x(0) = aX(s)} \tag{10.93}$$

where X(s) is the Laplace transform of x(t). Solving Eqn(10.93) for X(s) gives

$$\mathrm{X(s) = \frac{x(0)}{s - a} = (s - a)^{-1}x(0)}$$

Taking the inverse Laplace transform of X(s) gives

$$\mathrm{x(t) = e^{at}x(0)} \tag{10.94}$$

Let us extend this to a more general case where we consider the vector differential equation

$$\dot{\mathbf{x}} = \mathbf{A}\mathbf{x} \tag{10.95}$$

Taking the Laplace transform of Eqn(10.95) gives

$$\mathrm{s}\mathbf{X}(\mathbf{s}) - \mathbf{x}(\mathbf{0}) = \mathbf{A}\mathbf{X}(\mathbf{s}) \text{ or, } \mathrm{s}\mathbf{X}(\mathbf{s}) - \mathbf{A}\mathbf{X}(\mathbf{s}) = \mathrm{x}(0) \implies (\mathrm{s}\mathbf{I} - \mathbf{A})\,\mathbf{X}(\mathbf{s}) = \mathrm{x}(0) \tag{10.96}$$

Pre-multiplying Eqn(10.96) by $(s\mathbf{I} - \mathbf{A})^{-1}$ gives

$$\mathbf{X}(\mathrm{s}) = (\mathrm{s}\mathbf{I} - \mathbf{A})^{-1}\mathrm{x}(0) \tag{10.97}$$

Taking the inverse Laplace transform of Eqn(10.97) gives the solution $\mathbf{x}(\mathrm{t})$ as:

$$\boxed{\mathbf{x}(\mathbf{t}) = \mathcal{L}^{-1}\left[(\mathbf{sI} - \mathbf{A})^{-1}\right]\mathbf{x}(\mathbf{0})} \tag{10.98}$$

$$\text{But } (\mathrm{s}\mathbf{I} - \mathbf{A})^{-1} = \frac{\mathbf{I}}{\mathrm{s}} + \frac{\mathbf{A}}{\mathrm{s}^2} + \frac{\mathbf{A}^2}{\mathrm{s}^3} + \cdots \tag{10.99}$$

Thus, the inverse Laplace transform of $(s\mathbf{I} - \mathbf{A})^{-1}$ becomes:

$$\mathcal{L}^{-1}\left[(s\mathbf{I} - \mathbf{A})^{-1}\right] = \mathbf{I} + \mathbf{A}t + \frac{\mathbf{A}^2t^2}{2!} + \frac{\mathbf{A}^3t^3}{3!} + \cdots = e^{\mathbf{A}t} \tag{10.100}$$

From Eqns(10.98) and (10.100), the solution of Eqn(10.95) is given by

$$\mathbf{x}(t) = e^{\mathbf{A}t}\mathbf{x}(0) \tag{10.101}$$

Eqn(10.101) shows that the transition of the states from their initial values $\mathbf{x}(0)$ to a state $\mathbf{x}(t)$ takes place via the matrix exponential $e^{\mathbf{A}t}$. Therefore, $e^{\mathbf{A}t}$ is called as *state transition matrix* and is denoted by $\mathbf{\Phi}(t)$. Thus

$$\boxed{\mathbf{\Phi}(t) = \mathcal{L}^{-1}\left[(s\mathbf{I} - \mathbf{A})^{-1}\right]} \tag{10.102}$$

The solution of homogeneous state equation (10.95) can be expressed as:

$$\boxed{\mathbf{x}(t) = \mathbf{\Phi}(\mathbf{t})\mathbf{x}(0) = e^{\mathbf{A}t}\mathbf{x}(0)} \tag{10.103}$$

10.6.3 Properties of State Transition Matrix

We will examine the fundamental properties of the state transition matrix $\Phi(t)$, each of which reveals important insights about system behavior.

Property-1: Identity Property (Initial Condition Property)

$$\boxed{\mathbf{\Phi}(0) = \mathbf{I}}$$

Proof:

$$\mathbf{\Phi}(t) = e^{\mathbf{A}t} \implies \mathbf{\Phi}(\mathbf{0}) = \mathbf{e}^{\mathbf{A}.\mathbf{0}} = \mathbf{I}$$

Physical Interpretation: At $t = 0$, no time has elapsed, so the states remain unchanged from their initial values. The identity matrix ensures that

$$\mathbf{x}(0) = \mathbf{\Phi}(\mathbf{0})\mathbf{x}(\mathbf{0}) = \mathbf{I}\mathbf{x}(\mathbf{0}) = \mathbf{x}(\mathbf{0})$$

Property-2: Invertibility Property (Time Reversal Property)

$$\boxed{\mathbf{\Phi}^{-1}(t) = \mathbf{\Phi}(-t)}$$

Proof:

$$\mathbf{\Phi}(t) = e^{\mathbf{A}t} = \left[e^{-\mathbf{A}t}\right]^{-1} = [\mathbf{\Phi}(-t)]^{-1}$$

Taking the inverse gives $\mathbf{\Phi}^{-1}(t) = \mathbf{\Phi}(-t)$.

Physical Interpretation: This property enables time reversal in state evolution. If we know the current state and want to determine the initial state, we can "run time backward." This means that the state transition is bilateral in time, where the time transition can take place in either direction. For example, from the state equation: $\mathbf{x}(t) = \mathbf{\Phi}(t)\mathbf{x}(0)$, we can solve for initial condition:

$$\mathbf{x}(\mathbf{0}) = \mathbf{\Phi}^{-1}(t)\mathbf{x}(t) = \mathbf{\Phi}(-t)\mathbf{x}(t)$$

This relationship is particularly useful in system identification, where we can determine initial conditions from measurements; fault diagnosis, where we can trace system states back to fault inception; and optimal control, where we perform backward integration in two-point boundary value problems.

Property-3: Commutativity Property

$$\boxed{\mathbf{\Phi}(t_1)\mathbf{\Phi}(t_2) = \mathbf{\Phi}(t_2)\mathbf{\Phi}(t_1)}$$

Proof:

$$\mathbf{\Phi}(t_1 + t_2) = e^{\mathbf{A}(t_1+t_2)} = e^{\mathbf{A}t_1}e^{\mathbf{A}t_2} = \mathbf{\Phi}(t_1)\mathbf{\Phi}(t_2) = \mathbf{\Phi}(t_2)\mathbf{\Phi}(t_1)$$

Physical Interpretation: The order of applying time transitions doesn't matter. This reflects the time-invariant nature of the system.

Property-4: Power Property (Repeated Transition Property)

$$\boxed{[\mathbf{\Phi}(t)]^n = \mathbf{\Phi}(nt)}$$

Proof:

$$[\mathbf{\Phi}(t)]^n = \left[e^{\mathbf{A}t}\right]^n = e^{\mathbf{A}nt} = \mathbf{\Phi}(nt)$$

Physical Interpretation: Applying the same time transition n times is equivalent to a single transition over time nt. This is useful for discrete-time analysis and periodic systems.

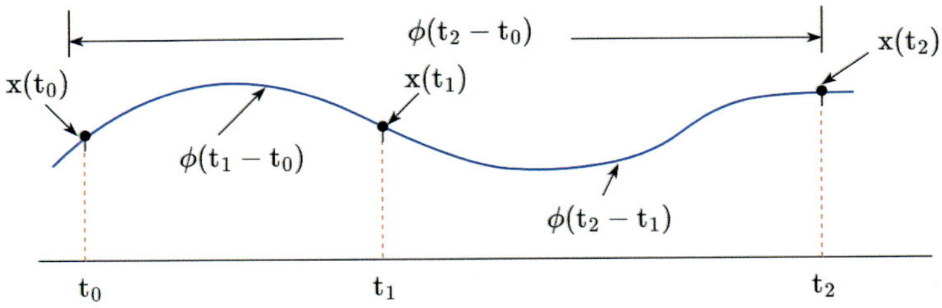

Figure 10.6: Property of the state transition matrix

Property-5: Semigroup Property (Composition Property)

$$\boldsymbol{\Phi}(t_2 - t_1)\boldsymbol{\Phi}(t_1 - t_0) = \boldsymbol{\Phi}(t_2 - t_0)$$

Proof:

$$\boldsymbol{\Phi}(t_2 - t_1)\boldsymbol{\Phi}(t_1 - t_0) = e^{\mathbf{A}(t_2 - t_1)}e^{\mathbf{A}(t_1 - t_0)} = e^{\mathbf{A}(t_2 - t_0)} = \boldsymbol{\Phi}(t_2 - t_0)$$

Physical Interpretation: It demonstrates that complex transitions can be decomposed into sequential intermediate transitions, enabling modular analysis of system evolution over extended time periods. Thus, the state evolution from time t_0 to t_2 can be achieved through any intermediate time t_1, as illustrated in Figure 10.6. The system state first transitions from t_0 to t_1 and then from t_1 to t_2, with the overall effect being identical to a direct transition from t_0 to t_2.

Property-6: Derivative Property

$$\frac{d\boldsymbol{\Phi}(t)}{dt} = \mathbf{A}\boldsymbol{\Phi}(t) = \boldsymbol{\Phi}(t)\mathbf{A}$$

Proof:

$$\frac{d\Phi(t)}{dt} = \frac{d}{dt}(e^{\mathbf{A}t}) = \mathbf{A}e^{\mathbf{A}t} = e^{\mathbf{A}t}\mathbf{A} = \mathbf{A}\Phi(t) = \Phi(t)\mathbf{A}$$

Physical Interpretation: The rate of change of the state transition matrix is directly proportional to the system matrix $\mathbf{A}$. This property confirms that the state transition matrix satisfies the same matrix differential equation that governs the system dynamics. It also shows that $\mathbf{A}$ commutes with $\boldsymbol{\Phi}(\mathbf{t})$, which is a fundamental requirement for the matrix exponential solution.

These six fundamental properties of the state transition matrix form the mathematical foundation for analyzing linear time-invariant systems. Together, they establish that the transition matrix provides a complete characterization of system dynamics; the state evolution is deterministic, reversible, and decomposable, and the mathematical structure is consistent with the underlying differential equation.

With this theoretical foundation established, we can now apply these concepts to concrete examples that demonstrate how to compute and utilize the state transition matrix in practical engineering scenarios. The following examples will illustrate both analytical and computational approaches to finding $\mathbf{\Phi(t)}$ for various system configurations.

EXAMPLE 10.10

Problem

Compute the state transition matrix of the system

$$\dot{\mathbf{x}} = \mathbf{A}\mathbf{x} \quad \text{where} \quad \mathbf{A} = \begin{bmatrix} 0 & 1 \\ -4 & -5 \end{bmatrix}$$

SOLUTION:

The state transition matrix $\mathbf{\Phi}(\mathrm{t}) = \mathcal{L}^{-1}\left[(\mathrm{s}\mathbf{I} - \mathbf{A})^{-1}\right]$. Now

$$(\mathrm{s}\mathbf{I} - \mathbf{A}) = \begin{bmatrix} \mathrm{s} & 0 \\ 0 & \mathrm{s} \end{bmatrix} - \begin{bmatrix} 0 & 1 \\ -4 & -5 \end{bmatrix} = \begin{bmatrix} \mathrm{s} & -1 \\ 4 & \mathrm{s}+5 \end{bmatrix}$$

The inverse of $(\mathrm{s}\mathbf{I} - \mathbf{A})$ is given by

$$\begin{aligned}
(\mathrm{s}\mathbf{I} - \mathbf{A})^{-1} &= \begin{bmatrix} \mathrm{s} & -1 \\ 4 & \mathrm{s}+5 \end{bmatrix}^{-1} = \frac{1}{\mathrm{s}(\mathrm{s}+5)+4} \begin{bmatrix} \mathrm{s}+5 & 1 \\ -4 & \mathrm{s} \end{bmatrix} \\
&= \begin{bmatrix} \dfrac{\mathrm{s}+5}{(\mathrm{s}+1)(\mathrm{s}+4)} & \dfrac{1}{(\mathrm{s}+1)(\mathrm{s}+4)} \\ \dfrac{-4}{(\mathrm{s}+1)(\mathrm{s}+4)} & \dfrac{\mathrm{s}}{(\mathrm{s}+1)(\mathrm{s}+4)} \end{bmatrix} \\
&= \begin{bmatrix} \dfrac{4/3}{s+1} - \dfrac{1/3}{s+4} & \dfrac{-1/3}{s+4} + \dfrac{1/3}{s+1} \\ \dfrac{4/3}{s+4} + \dfrac{-4/3}{s+1} & \dfrac{-1/3}{s+1} + \dfrac{4/3}{s+4} \end{bmatrix}
\end{aligned}$$

Taking Laplace inverse of $(\mathrm{s}\mathbf{I} - \mathbf{A})^{-1}$ gives the state transition matrix as follows:

$$\mathbf{\Phi}(t) = \mathcal{L}^{-1}\left[(s\mathbf{I} - \mathbf{A})^{-1}\right] = \mathcal{L}^{-1}\begin{bmatrix} \frac{4/3}{s+1} - \frac{1/3}{s+4} & \frac{-1/3}{s+4} + \frac{1/3}{s+1} \\ \frac{4/3}{s+4} + \frac{-4/3}{s+1} & \frac{-1/3}{s+1} + \frac{4/3}{s+4} \end{bmatrix}$$

$$= \begin{bmatrix} \frac{4}{3}e^{-t} - \frac{1}{3}e^{-4t} & \frac{1}{3}e^{-t} - \frac{1}{3}e^{-4t} \\ \frac{4}{3}e^{-4t} - \frac{4}{3}e^{-t} & \frac{4}{3}e^{-4t} - \frac{1}{3}e^{-t} \end{bmatrix}$$

CONCEPT REINFORCEMENT PROBLEM 10.7

Problem:

Compute the state transition matrix of the system

$$\dot{\mathbf{x}} = \begin{bmatrix} 0 & 1 \\ -3 & -2 \end{bmatrix}\mathbf{x}$$

EXAMPLE 10.11

Problem

Consider the system

$$\dot{\mathbf{x}} = \mathbf{A}\mathbf{x} + \mathbf{B}u, \text{ where } \mathbf{A} = \begin{bmatrix} 0 & a \\ -a & 0 \end{bmatrix} \text{ and } \mathbf{B} = \begin{bmatrix} 0 \\ 1 \end{bmatrix}$$

Compute the state transition matrix of this system and find its inverse.

SOLUTION:

The state transition matrix $\mathbf{\Phi}(t) = \mathcal{L}^{-1}\left[(s\mathbf{I} - \mathbf{A})^{-1}\right]$. Now

$$(s\mathbf{I} - \mathbf{A}) = \begin{bmatrix} s & 0 \\ 0 & s \end{bmatrix} - \begin{bmatrix} 0 & a \\ -a & 0 \end{bmatrix} = \begin{bmatrix} s & -a \\ a & s \end{bmatrix}$$

The inverse of $(s\mathbf{I} - \mathbf{A})$ is given by

$$(s\mathbf{I} - \mathbf{A})^{-1} = \begin{bmatrix} s & -a \\ a & s \end{bmatrix}^{-1} = \frac{1}{s^2 + a^2}\begin{bmatrix} s & a \\ -a & s \end{bmatrix} = \begin{bmatrix} \frac{s}{s^2 + a^2} & \frac{a}{s^2 + a^2} \\ \frac{-a}{s^2 + a^2} & \frac{s}{s^2 + a^2} \end{bmatrix}$$

Taking Laplace inverse of $(\mathbf{sI} - \mathbf{A})$ gives

$$\mathbf{\Phi}(t) = \mathcal{L}^{-1}\left[(s\mathbf{I} - \mathbf{A})^{-1}\right] = \mathcal{L}^{-1}\begin{bmatrix} \frac{s}{s^2+a^2} & \frac{a}{s^2+a^2} \\ \frac{-a}{s^2+a^2} & \frac{s}{s^2+a^2} \end{bmatrix}$$

$$= \begin{bmatrix} \cos(at) & \sin(at) \\ -\sin(at) & \cos(at) \end{bmatrix}$$

The inverse of $\mathbf{\Phi}(t)$ is computed from the relation $\mathbf{\Phi}^{-1}(t) = \mathbf{\Phi}(-t)$. This gives

$$\mathbf{\Phi}^{-1}(t) = \begin{bmatrix} \cos(a(-t)) & \sin(a(-t)) \\ -\sin(a(-t)) & \cos(a(-t)) \end{bmatrix} = \begin{bmatrix} \cos(at) & -\sin(at) \\ \sin(at) & \cos(at) \end{bmatrix}$$

EXAMPLE 10.12

Problem:

Which of the following matrices can be the state transition matrices?

a. $\mathbf{\Phi}(t) = \begin{bmatrix} -e^{-5t} & 0 \\ 0 & 2-e^{-4t} \end{bmatrix}$ **b.** $\mathbf{\Phi}(t) = \begin{bmatrix} 1 & 0 \\ 2(1-e^{-3t}) & e^{-3t} \end{bmatrix}$

c. $\mathbf{\Phi}(t) = \begin{bmatrix} 1-e^{-2t} & 0 \\ 1 & e^{-2t} \end{bmatrix}$ **d.** $\mathbf{\Phi}(t) = \begin{bmatrix} \cos t & \sin t \\ -\sin t & \cos t \end{bmatrix}$

SOLUTION:

Let us compute $\mathbf{\Phi}(0)$ and $\mathbf{\Phi}^{-1}(t)$ and check if $\mathbf{\Phi}(0) = \mathbf{I}$ and $\mathbf{\Phi}^{-1}(t) = \mathbf{\Phi}(-t)$.

a. $\mathbf{\Phi}(t) = \begin{bmatrix} -e^{-5t} & 0 \\ 0 & 2-e^{-4t} \end{bmatrix} \implies \mathbf{\Phi}(\mathbf{0}) = \begin{bmatrix} -1 & 0 \\ 0 & 1 \end{bmatrix} \neq \mathbf{I}$

Hence, this matrix cannot be a state transition matrix

b. $\mathbf{\Phi}(t) = \begin{bmatrix} 1 & 0 \\ 2(1-e^{-3t}) & e^{-3t} \end{bmatrix}, \implies \mathbf{\Phi}(0) = \begin{bmatrix} 1 & 0 \\ 0 & 1 \end{bmatrix} = \mathbf{I}$

Now $\mathbf{\Phi}^{-1}(t) = \begin{bmatrix} 1 & 0 \\ 2(1-e^{-3t}) & e^{-3t} \end{bmatrix}^{-1} = \begin{bmatrix} 1 & 0 \\ 2(1-e^{3t}) & e^{3t} \end{bmatrix} = \mathbf{\Phi}(-t)$

Hence, this matrix can be a state transition matrix.

c. $\mathbf{\Phi}(t) = \begin{bmatrix} 1-e^{-2t} & 0 \\ 1 & e^{-2t} \end{bmatrix}, \implies \mathbf{\Phi}(0) = \begin{bmatrix} 0 & 0 \\ 1 & 1 \end{bmatrix} \neq \mathbf{I}$

Hence, this matrix cannot be a state transition matrix.

d. $\mathbf{\Phi}(t) = \begin{bmatrix} \cos t & \sin t \\ -\sin t & \cos t \end{bmatrix}, \implies \mathbf{\Phi}(0) = \begin{bmatrix} 1 & 0 \\ 0 & 1 \end{bmatrix} = \mathbf{I}$

Now $\mathbf{\Phi}^{-1}(t) = \begin{bmatrix} \cos t & -\sin t \\ \sin t & \cos t \end{bmatrix} = \mathbf{\Phi}(-t)$

Hence, this matrix can be a state transition matrix.

EXAMPLE 10.13

Problem:

For the system represented by the state equation

$$\dot{\mathbf{x}} = \mathbf{A}\mathbf{x}(t)$$

the state response $\mathbf{x}(t)$ is given by

$$\mathbf{x}(t) = \begin{bmatrix} e^{at} \\ e^{-at} \end{bmatrix}$$

Determine the system matrix $\mathbf{A}$.

SOLUTION

Let

$$\mathbf{A} = \begin{bmatrix} a_{11} & a_{12} \\ a_{21} & a_{22} \end{bmatrix}$$

Now, from the solution of the state equation, we have

$$\mathbf{x}(t) = \begin{bmatrix} e^{at} \\ e^{-at} \end{bmatrix} \implies \dot{\mathbf{x}}(t) = \begin{bmatrix} ae^{at} \\ -ae^{-at} \end{bmatrix}$$

In terms of the elements of $\mathbf{A}$ matrix

$$\dot{\mathbf{x}}(t) = \begin{bmatrix} a_{11} & a_{12} \\ a_{21} & a_{22} \end{bmatrix} \begin{bmatrix} e^{at} \\ e^{-at} \end{bmatrix} = \begin{bmatrix} ae^{at} \\ -ae^{-at} \end{bmatrix}$$

This gives

$$a_{11}e^{at} + a_{12}e^{-at} = ae^{at}$$
$$a_{21}e^{at} + a_{22}e^{-at} = -ae^{-at}$$

Equating like powers of an exponential gives

$$a_{11} = a, \quad a_{12} = 0, \quad a_{21} = 0 \quad \text{and} \quad a_{22} = -a$$

Hence the system matrix

$$\mathbf{A} = \begin{bmatrix} a & 0 \\ 0 & -a \end{bmatrix}$$

CONCEPT REINFORCEMENT PROBLEM 10.8

Problem:

Consider the system represented by the state equation

$$\dot{\mathbf{x}} = \mathbf{A}\mathbf{x}(t)$$

Determine the system matrix if

i. $\mathbf{x}(\mathrm{t}) = \begin{bmatrix} \mathrm{e}^{-2\mathrm{t}} \\ -2\mathrm{e}^{-2\mathrm{t}} \end{bmatrix}$, for $\mathbf{x}(0) = \begin{bmatrix} 1 \\ -2 \end{bmatrix}$ and

$\mathbf{x}(\mathrm{t}) = \begin{bmatrix} \mathrm{e}^{-\mathrm{t}} \\ -\mathrm{e}^{-\mathrm{t}} \end{bmatrix}$, for $\mathbf{x}(0) = \begin{bmatrix} 1 \\ -1 \end{bmatrix}$

ii. $\mathbf{x}(\mathrm{t}) = \begin{bmatrix} \cos \mathrm{at} \\ -\sin \mathrm{at} \end{bmatrix}$, for $\mathbf{x}(0) = \begin{bmatrix} 1 \\ 0 \end{bmatrix}$ and

$\mathbf{x}(\mathrm{t}) = \begin{bmatrix} \sin \mathrm{at} \\ \cos \mathrm{at} \end{bmatrix}$, when $\mathbf{x}(0) = \begin{bmatrix} 0 \\ 1 \end{bmatrix}$

10.7 Solution of Non-homogeneous State Equations

Consider a non-homogeneous system described by a scalar differential equation

$$\dot{\mathrm{x}} = \mathrm{ax} + \mathrm{bu} \tag{10.104}$$

Eqn(10.104) can be rewritten as:

$$\dot{\mathrm{x}} - \mathrm{ax} = \mathrm{bu} \tag{10.105}$$

Let us multiply both sides of Eqn(10.105) by $\mathrm{e}^{-\mathrm{at}}$. This gives

$$\mathrm{e}^{-\mathrm{at}}\left[\dot{\mathrm{x}} - \mathrm{ax}\right] = \mathrm{e}^{-\mathrm{at}}\mathrm{bu} \tag{10.106}$$

But $\mathrm{e}^{-\mathrm{at}}\left[\dot{\mathrm{x}} - \mathrm{ax}\right] = \dfrac{\mathrm{d}}{\mathrm{dt}}\left[\mathrm{e}^{-\mathrm{at}}\mathrm{x}(\mathrm{t})\right]$. Hence

$$\frac{\mathrm{d}}{\mathrm{dt}}\left[\mathrm{e}^{-\mathrm{at}}\mathrm{x}(\mathrm{t})\right] = \mathrm{e}^{-\mathrm{at}}\mathrm{bu} \tag{10.107}$$

Integrating Eqn(10.107) between 0 to t gives

$$\mathrm{e}^{-\mathrm{at}}\mathrm{x}(\mathrm{t}) = \mathrm{x}(0) + \int_0^{\mathrm{t}} \mathrm{e}^{-\mathrm{a}\tau}\mathrm{bu}(\tau)\,\mathrm{d}\tau. \text{ Hence} \tag{10.108}$$

$$\mathrm{x}(\mathrm{t}) = \mathrm{e}^{\mathrm{at}}\mathrm{x}(0) + \int_0^{\mathrm{t}} \mathrm{e}^{\mathrm{a}(\mathrm{t}-\tau)}\mathrm{bu}(\tau)\,\mathrm{d}\tau \tag{10.109}$$

The first term on the right side of Eqn(10.109) is the response due to initial conditions and the second term is the response due to the input.

Let us extend it to the more general case of a vector differential equation

$$\dot{\mathbf{x}} = \mathbf{Ax} + \mathbf{Bu} \tag{10.110}$$

Eqn(10.110) can be rewritten as:

$$\dot{\mathbf{x}} - \mathbf{Ax} = \mathbf{Bu} \tag{10.111}$$

Let us multiply both sides of Eq.(10.111) by $\mathrm{e}^{-\mathbf{A}\mathrm{t}}$. This gives

$$\mathrm{e}^{-\mathbf{A}\mathrm{t}}\left[\dot{\mathbf{x}} - \mathbf{Ax}\right] = \mathrm{e}^{-\mathbf{A}\mathrm{t}}\mathbf{Bu} \tag{10.112}$$

But $\mathrm{e}^{-\mathbf{A}\mathrm{t}}\left[\dot{\mathbf{x}} - \mathbf{Ax}\right] = \dfrac{\mathrm{d}}{\mathrm{dt}}\left[\mathrm{e}^{-\mathbf{A}\mathrm{t}}\mathbf{x}(\mathrm{t})\right]$. Hence

$$\frac{\mathrm{d}}{\mathrm{dt}}\left[\mathrm{e}^{-\mathbf{A}\mathrm{t}}\mathrm{x}(\mathrm{t})\right] = \mathrm{e}^{-\mathbf{A}\mathrm{t}}\mathbf{Bu} \tag{10.113}$$

Integrating Eq.(10.113) between 0 to t gives

$$\mathrm{e}^{-\mathbf{A}\mathrm{t}}\mathbf{x}(\mathrm{t}) = \mathbf{x}(0) + \int_0^{\mathrm{t}} \mathrm{e}^{-\mathbf{A}\tau}\mathbf{Bu}(\tau)\,\mathrm{d}\tau \tag{10.114}$$

Hence

$$\begin{aligned}\mathbf{x}(\mathrm{t}) &= \mathrm{e}^{\mathbf{A}\mathrm{t}}\mathrm{x}(0) + \int_0^{\mathrm{t}} \mathrm{e}^{\mathbf{A}(\mathrm{t}-\tau)}\mathbf{Bu}(\tau)\,\mathrm{d}\tau \\ &= \mathbf{\Phi}(\mathrm{t})\mathbf{x}(0) + \int_0^{\mathrm{t}} \mathbf{\Phi}(\mathrm{t}-\tau)\mathbf{Bu}(\tau)\,\mathrm{d}\tau\end{aligned} \tag{10.115}$$

10.7.1 Solution of Nonhomogeneous State Equations by Laplace Transform Approach

Consider the system described by the state equation

$$\dot{\mathbf{x}} = \mathbf{Ax} + \mathbf{Bu} \tag{10.116}$$

and the output equation

$$\mathbf{y} = \mathbf{Cx} + \mathbf{Du} \tag{10.117}$$

Taking the Laplace transform of both sides of state equation Eqn10.116) yields

$$\mathrm{s}\mathbf{X}(\mathrm{s}) - \mathrm{x}(0) = \mathbf{AX}(\mathrm{s}) + \mathbf{BU}(\mathrm{s}) \implies (\mathrm{s}\mathbf{I} - \mathbf{A})\,\mathbf{X}(\mathrm{s}) = \mathrm{x}(0) + \mathbf{BU}(\mathrm{s}) \tag{10.118}$$

where $\mathbf{I}$ is an $(n \times n)$ identity matrix where n is the order of the system.

By premultiplying $(s\mathbf{I} - \mathbf{A})^{-1}$ in Eqn(10.118), gives the solution as:

$$\mathbf{X}(s) = (s\mathbf{I} - \mathbf{A})^{-1} x(0) + (s\mathbf{I} - \mathbf{A})^{-1} \mathbf{BU}(s) \tag{10.119}$$

Taking inverse Laplace transform of Eqn(10.119) gives the state equations as in Eqn(10.115).

Remark

Although we can compute the state and output response using Eqn(10.115), it is less cumbersome to compute these responses using the Laplace transform approach.

Computation Procedure

Given the state and output equations as:

$$\dot{\mathbf{x}} = \mathbf{Ax} + \mathbf{Bu}$$
$$\mathbf{y} = \mathbf{Cx} + \mathbf{Du}$$

Step-1: Compute $(s\mathbf{I} - \mathbf{A})^{-1}$

Step-2: Compute the Laplace transform of states as:

$$\mathbf{X}(s) = (s\mathbf{I} - \mathbf{A})^{-1} [x(0) + \mathbf{BU}(s)]$$

Step-3: Compute the transform of output y(t) from

$$\mathbf{Y}(s) = \mathbf{CX}(s) + \mathbf{DU}(s)$$

Step-4: Compute the inverse Laplace transform of $\mathbf{X}(s)$ and $\mathbf{Y}(s)$ to get $\mathbf{x}(t)$ and $\mathbf{y}(t)$

EXAMPLE 10.14

Problem:

Consider the system

$$\dot{\mathbf{x}} = \begin{bmatrix} 0 & 4 \\ -4 & 0 \end{bmatrix} \mathbf{x} + \begin{bmatrix} 0 \\ 1 \end{bmatrix} e^{-t}, \quad y = \begin{bmatrix} 0 & 1 \end{bmatrix} \mathbf{x}, \mathbf{x}(0) = \begin{bmatrix} 0 \\ 0 \end{bmatrix}$$

Obtain both the state and output response.

SOLUTION:

From the system equation, we have

$$\mathbf{A} = \begin{bmatrix} 0 & 4 \\ -4 & 0 \end{bmatrix}, \mathbf{B} = \begin{bmatrix} 0 \\ 1 \end{bmatrix}$$

Let us first compute $(\mathrm{s}\mathbf{I} - \mathbf{A})^{-1}$ Now

$$\mathrm{s}\mathbf{I} - \mathbf{A} = \begin{bmatrix} \mathrm{s} & 0 \\ 0 & \mathrm{s} \end{bmatrix} - \begin{bmatrix} 0 & 4 \\ -4 & 0 \end{bmatrix} = \begin{bmatrix} \mathrm{s} & -4 \\ 4 & \mathrm{s} \end{bmatrix}$$

$$(\mathrm{s}\mathbf{I} - \mathbf{A})^{-1} = \begin{bmatrix} \mathrm{s} & -4 \\ 4 & \mathrm{s} \end{bmatrix}^{-1} = \frac{1}{\mathrm{s}^2 + 16} \begin{bmatrix} \mathrm{s} & 4 \\ -4 & \mathrm{s} \end{bmatrix}$$

$$= \begin{bmatrix} \dfrac{\mathrm{s}}{\mathrm{s}^2 + 16} & \dfrac{4}{\mathrm{s}^2 + 16} \\ \dfrac{-4}{\mathrm{s}^2 + 16} & \dfrac{\mathrm{s}}{\mathrm{s}^2 + 16} \end{bmatrix} = \begin{bmatrix} \mathbf{s_{11}} & \mathbf{s_{12}} \\ \mathbf{s_{21}} & \mathbf{s_{22}} \end{bmatrix}$$

Next compute $\mathrm{x}(0) + \mathbf{B}\mathrm{U}(\mathrm{s})$ which for the given data becomes

$$\mathrm{x}(0) + \mathbf{B}\mathrm{U}(\mathrm{s}) = \begin{bmatrix} 0 \\ 0 \end{bmatrix} + \begin{bmatrix} 0 \\ 1 \end{bmatrix} \frac{1}{\mathrm{s}+1} = \begin{bmatrix} 0 \\ \dfrac{1}{\mathrm{s}+1} \end{bmatrix}$$

Compute $\mathbf{X}(\mathrm{s}) = (\mathrm{s}\mathbf{I} - \mathbf{A})^{-1}[\mathrm{x}(0) + \mathbf{B}\mathrm{U}(\mathrm{s})]$. This gives

$$\mathbf{X}(\mathrm{s}) = (\mathrm{s}\mathbf{I} - \mathbf{A})^{-1}[\mathrm{x}(0) + \mathbf{B}\mathrm{U}(\mathrm{s})] = \begin{bmatrix} \mathbf{s_{11}} & \mathbf{s_{12}} \\ \mathbf{s_{21}} & \mathbf{s_{22}} \end{bmatrix} \begin{bmatrix} 0 \\ \dfrac{1}{\mathrm{s}+1} \end{bmatrix} = \begin{bmatrix} \dfrac{\mathbf{s_{12}}}{\mathrm{s}+1} \\ \dfrac{\mathbf{s_{22}}}{\mathrm{s}+1} \end{bmatrix}$$

$$= \begin{bmatrix} \dfrac{4}{(\mathrm{s}+1)(\mathrm{s}^2+16)} \\ \dfrac{\mathrm{s}}{(\mathrm{s}+1)(\mathrm{s}^2+16)} \end{bmatrix} = \begin{bmatrix} \dfrac{4}{17}\dfrac{1}{\mathrm{s}+1} - \dfrac{4}{17}\dfrac{\mathrm{s}}{\mathrm{s}^2+16} + \dfrac{4}{17}\dfrac{1}{\mathrm{s}^2+16} \\ -\dfrac{1}{17}\dfrac{1}{\mathrm{s}+1} + \dfrac{1}{17}\dfrac{\mathrm{s}}{\mathrm{s}^2+16} + \dfrac{16}{17}\dfrac{1}{\mathrm{s}^2+16} \end{bmatrix}$$

Hence

$$\mathbf{x}(t) = \begin{bmatrix} \dfrac{4}{17}\mathrm{e}^{-\mathrm{t}} - \dfrac{4}{17}\cos 4\mathrm{t} + \dfrac{1}{17}\sin 4\mathrm{t} \\ \dfrac{1}{17}\cos 4\mathrm{t} - \dfrac{1}{17}\mathrm{e}^{-\mathrm{t}} + \dfrac{4}{17}\sin 4\mathrm{t} \end{bmatrix}$$

$$\mathrm{y}(\mathrm{t}) = \begin{bmatrix} 0 & 1 \end{bmatrix} \mathrm{x}(\mathrm{t}) = \frac{1}{17}\cos 4\mathrm{t} - \frac{1}{17}\mathrm{e}^{-\mathrm{t}} + \frac{4}{17}\sin 4\mathrm{t}$$

CONCEPT REINFORCEMENT PROBLEM 10.9

Problem:

Consider the system

$$\dot{\mathbf{x}} = \begin{bmatrix} -1 & 0 \\ -2 & -4 \end{bmatrix} \mathbf{x} + \begin{bmatrix} 1 \\ 0 \end{bmatrix} u, y = \begin{bmatrix} 0 & 1 \end{bmatrix} \mathbf{x}, \mathbf{x}(0) = \begin{bmatrix} 0 \\ 0 \end{bmatrix}$$

Obtain both the state and output response for a unity step input.

10.8 Relationship Between Eigenvalues, Poles, BIBO Stability, Asymptotic Stability, and Internal Stability

10.8.1 Asymptotic Stability

Consider a system described in state space:

$$\dot{\mathbf{x}}(t) = \mathbf{A}\mathbf{x}, \quad \mathbf{x}(0) = \mathbf{x}_0 \tag{10.120}$$

Definition 10.1. The system of Eqn(10.120) is said to be asymptotically stable if all the states approach zero as the time approaches infinity, that is,

$$\mathbf{x}(t) \to 0 \quad \text{as } t \to \infty$$

Without loss of generality, assume that the eigenvalues of $\mathbf{A}$ are distinct. Then, with appropriate transformation matrix $\mathbf{P}$, $\mathbf{x}(t) = \mathbf{P}\mathbf{z}(t)$, the system of Eqn(10.120) can be represented as:

$$\dot{\mathbf{z}}(t) = \mathbf{\Lambda}\mathbf{z}(t), \quad \text{where } \Lambda = \begin{bmatrix} \lambda_1 & 0 & \dots & 0 \\ 0 & \lambda_2 & \dots & \dots \\ \dots & 0 & \dots & \dots \\ \vdots & \vdots & \vdots & \vdots \\ 0 & 0 & \dots & \lambda_n \end{bmatrix}$$

The dynamics of individual subsystems are

$$\dot{z}_i(t) = \lambda_i z_i(t) \quad \text{whose solution is given by}$$

$$z_i(t) = e^{\lambda_i t} z_i(0) \tag{10.121}$$

From Eqn(10.121), it is obvious that if the eigenvalues of the system have negative real parts, the individual states go to zero asymptotically with time, that is, $z_i(t) \to 0$ as $t \to \infty$.

Theorem 10.7. The system is asymptotically stable if all the eigenvalues of $\mathbf{A}$ have negative real parts (i.e., lie in the left half plane).

10.8.2 Input-Output Stability of LTI Systems

Another important notion of stability, beyond asymptotic stability, is *Bounded Input-Bounded Output (BIBO) stability* which ensures that bounded inputs produce bounded outputs, regardless of the initial conditions. While *asymptotic stability* focuses on the natural response due to initial conditions, BIBO stability concerns forced response behavior under zero initial conditions.

As studied in Chapter 4, a system will be BIBO stable provided all the poles lie in the LHP. Let us investigate this from the state space model of the system.

BIBO Stability from State Variable Models

Consider a LTI system represented by state equations

$$\begin{aligned} \dot{\mathbf{x}} &= \mathbf{Ax} + \mathbf{Bu} \\ \mathbf{y} &= \mathbf{Cx} + \mathbf{Du} \end{aligned} \tag{10.122}$$

The transfer function of the system is given by

$$\mathbf{G}(\mathrm{s}) = \frac{\mathrm{N(s)}}{\mathrm{D(s)}} = \mathbf{C}\left(\mathrm{s}\mathbf{I} - \mathbf{A}\right)^{-1}\mathbf{B} + \mathbf{D} = \frac{\mathbf{C}\ \mathrm{adj}\left(\mathrm{s}\mathbf{I} - \mathbf{A}\right)\mathbf{B}}{\det\left(\mathrm{s}\mathbf{I} - \mathbf{A}\right)} + \mathbf{D} \tag{10.123}$$

The denominator of the transfer function is $\det(\mathrm{s}\mathbf{I} - \mathbf{A})$. The eigenvalues of the system are determined from the equation

$$\det(\mathrm{s}\mathbf{I} - \mathbf{A}) = 0$$

Key Relationships

i. Without pole-zero cancellations, system poles coincide with eigenvalues.

ii. The system is BIBO stable if all eigenvalues lie in the left half plane.

iii. When conditions (i) and (ii) hold, **BIBO stability** and **asymptotic stability** are equivalent.

10.8.3 Internal Stability

Internal stability is the strongest stability concept, requiring all internal signals to remain bounded for bounded inputs.

Definition 10.2. A system is *internally stable* if all signals within the system remain bounded for all bounded inputs.

Internal stability is stronger than BIBO stability (external stability)[1] because it requires every internal signal to be bounded, not just the measured output. A system can be BIBO stable (bounded output for bounded input) while having unbounded internal states, a potentially dangerous situation hidden from external observation.

To understand this critical distinction between what appears stable externally versus what's actually happening internally, consider a situation that perfectly captures this phenomenon.

Understanding Internal vs External Stability

The Unexpected Guest Analogy:

Internal stability versus BIBO stability is like hosting an unexpected dinner guest:

The Scenario: You and your spouse had carefully planned a quiet romantic dinner at home: just the two of you, special recipes prepared, candles ready. Suddenly, your old university friend shows up unannounced at your door, expecting to stay for dinner. Due to courtesy and friendship, you warmly welcome him in.

External Stability—BIBO Stability (What the Guest Observes):

Your spouse greets the guest with a warm smile, serves tea, and engages in pleasant conversation. Dinner is served beautifully, everyone laughs and chats, and the guest leaves thinking, "What wonderful hospitality! Everything was perfect!" **From the guest's perspective (input $\rightarrow$ output), the system appears completely stable and well-behaved.**

Internal Instability (What's Really Happening Inside):

Your spouse is seething inside: the ruined plans, the wasted preparation, the lost romantic evening. Every smile is forced, every polite word costs emotional energy, and the frustration is growing exponentially with each passing minute: $e^{\alpha t}$ where $\alpha > 0$. She's giving you "the look" that

[1] While BIBO stability and external stability are not technically identical concepts, throughout this discussion, we use the terms interchangeably. BIBO stability is the most common form of external stability, and for the purposes of understanding the contrast with internal stability, this simplification is pedagogically useful without loss of conceptual clarity.

you know means serious trouble later. **The internal state variables (emotional state, stress level, frustration) are unbounded and unstable, but completely hidden from the output (observable behavior).**

The Explosion (When the Facade Breaks):

The moment the guest closes the door and leaves: "How could you?! We had PLANS! You didn't even ask me! That was supposed to be OUR evening!" The internal instability that was hidden suddenly becomes catastrophically observable.

The Control Systems Parallel:

The guest (measuring the output) sees only pleasant hospitality: bounded input produces bounded output (BIBO stability). Inside the system, however, hidden state variables (such as the spouse's frustration, stress, and anger) are growing without bound. These internal signals are unstable even though the output looks fine. Pole-zero cancellation or unobservable modes can conceal unstable internal dynamics from the input-output relationship; the transfer function appears stable, but internal states are diverging.

The lesson: A system that is BIBO stable is NOT necessarily internally stable. Just because your measurements look good doesn't mean disaster isn't brewing beneath the surface.

Why This Matters in Control Design: Never design controllers or estimators that create pole-zero cancellations with unstable poles. You might hide an unstable mode from the output. Still, it remains, growing exponentially, and will eventually destroy your system, just like that conversation waiting for you after the guest leaves.

Real-world example: In a motor speed control system, the speed (output) may remain perfectly stable and well-regulated, tracking the reference beautifully. However, due to pole-zero cancellation hiding an unstable current mode, the motor current (internal state) grows without bound: $\mathrm{i(t) = i_0 e^{\alpha t}}$. The speed appears to be fine in your measurements, but the current continues to increase exponentially. Eventually, the windings overheat, insulation breaks down, or protection circuits trip. The system fails catastrophically even though the "output" (speed) appeared stable throughout.

Mathematical Connection: A linear system described by:

$$\dot{\mathbf{x}}(t) = \mathbf{A}\mathbf{x}(\mathrm{t}) + \mathbf{B}\mathbf{u}(\mathrm{t})$$
$$\mathrm{y(t)} = \mathbf{C}\mathbf{x}(\mathrm{t}) + \mathbf{D}\mathbf{u}(\mathrm{t})$$

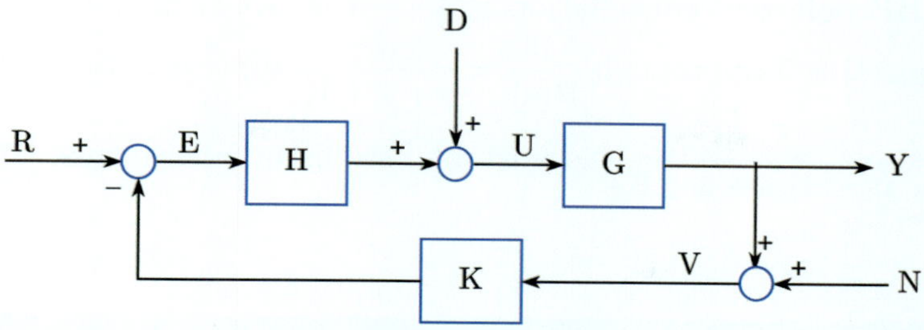

Figure 10.7: Block diagram of closed-loop feedback system showing disturbance and noise inputs

is internally stable if and only if all eigenvalues of $\mathbf{A}$ have strictly negative real parts: $\mathrm{Re}(\lambda_i) < 0, \quad \forall i$. This ensures that all internal states (not just the output) remain bounded for bounded inputs.

Key Distinction: *BIBO Stability* requires only that poles of the transfer function G(s) be in the left half-plane, while *internal stability* requires that all eigenvalues of $\mathbf{A}$ be in the left half-plane. Due to pole-zero cancellation, some eigenvalues of $\mathbf{A}$ may not appear as poles in G(s). If those canceled eigenvalues are unstable, we have BIBO stability without internal stability.

Consider a system represented by the general block diagram in Figure 10.7.

The system will be internally stable if all possible transfer functions between all inputs and outputs are stable. In this case, there are 3 inputs $(\mathrm{R}, \mathrm{D}, \mathrm{N})$ and there are 3 outputs $(\mathrm{E},\ \mathrm{U},\ \mathrm{V})$ taken at the output of the summing junction. This gives a total of 9 transfer functions, which are to be tested for internal stability (a daunting task indeed). However, the following theorem can be used to investigate the internal stability of a system.

Theorem 10.8. A feedback system is internally stable if and only if:

i. The characteristic polynomial $1 + \mathrm{KGH}$ has no zeros in the right half-plane (including $\mathrm{j}\omega$ axis)

ii. The open-loop transfer function KGH has no pole-zero cancellations in the right-half plane

Consider the case where

$$\mathrm{G(s)} = \frac{1}{\mathrm{s-a}}, \quad \mathrm{H(s)} = \frac{\mathrm{s-a}}{\mathrm{s+a}} \quad \text{and} \quad \mathrm{K, a} > 0$$

The RHP pole zero cancellation at $s = +a$ in KGH makes the system internally unstable, even though $\frac{Y(s)}{R(s)} = \frac{1}{s + a + K}$ appears BIBO stable. However, $\frac{Y(s)}{D(s)} = \frac{s + a}{(s - a)(s + a + K)}$ is unstable, so disturbances cause unbounded responses.

EXAMPLE 10.15

Let us compute the transfer function, eigenvalues, and poles for each of the following systems and investigate the stability properties.

System-1 $\mathbf{A} = \begin{bmatrix} 0 & 1 \\ -6 & -5 \end{bmatrix}, \quad \mathbf{B} = \begin{bmatrix} 0 \\ 1 \end{bmatrix}, \quad \mathbf{C} = \begin{bmatrix} 1 & 1 \end{bmatrix}, \quad \mathbf{D} = 0$

We can compute the transfer function of the system following the procedure discussed before. It is given by

$$G(s) = \frac{s + 1}{s^2 + 5s + 6} = \frac{s + 1}{(s + 2)(s + 3)}$$

The eigenvalues of the system are computed for λ by solving $|\lambda\mathbf{I} - \mathbf{A}| = 0$ which gives $\lambda_1 = -2$ and $\lambda_2 = -3$. Thus, the eigenvalues are the same as the poles. Further, since all the eigenvalues are in the left half plane, the system is asymptotically stable. Further, since there is no pole-zero cancellation, the external and internal stability are the same.

System-2 $\mathbf{A} = \begin{bmatrix} 0 & 1 \\ 4 & -3 \end{bmatrix}, \quad \mathbf{B} = \begin{bmatrix} 0 \\ 1 \end{bmatrix}, \quad \mathbf{C} = \begin{bmatrix} -1 & 1 \end{bmatrix}, \quad \mathbf{D} = 0$

The transfer function of the system is

$$G(s) = \frac{s - 1}{s^2 + 3s - 4} = \frac{s - 1}{(s - 1)(s + 4)} = \frac{1}{s + 4}$$

The eigenvalues of the system are $\lambda_1 = -4$ and $\lambda_2 = +1$ and the pole is at -4. Thus, the eigenvalues are not the same as the poles. Further, since not all the eigenvalues are in the left half plane, the system is not asymptotically stable. Also, the system is not internally stable because of the cancellation of the pole and zero in the RHP.

System-3 $\mathbf{A} = \begin{bmatrix} 0 & 1 \\ 8 & -2 \end{bmatrix}, \quad \mathbf{B} = \begin{bmatrix} 0 \\ 1 \end{bmatrix}, \quad \mathbf{C} = \begin{bmatrix} 4 & 1 \end{bmatrix}, \quad \mathbf{D} = 0$

The transfer function of the system is computed as:

$$G(s) = \frac{s + 4}{s^2 + 2s - 8} = \frac{s + 4}{(s - 2)(s + 4)} = \frac{1}{s - 2}$$

The eigenvalues of the system are $\lambda_1 = -4$ and $\lambda_2 = +2$. The pole is at +2. Thus, the eigenvalues are not the same as the poles. Further, since not all the eigenvalues are in the left half plane, the system is not asymptotically as well as internally stable.

CONCEPT REINFORCEMENT PROBLEM 10.10

Problem

Consider the multi-loop feedback system shown in Figure 10.8

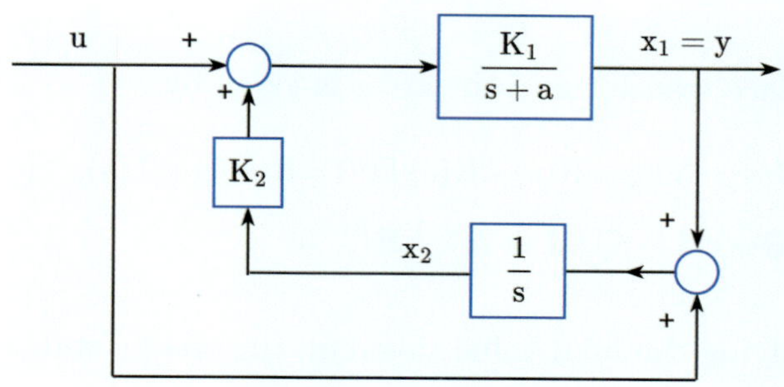

Figure 10.8: Multi-loop feedback control system

a. Obtain the state variable model of the system, taking the state variables as indicated in the figure.

b. Compute the transfer function $G(s) = \dfrac{Y(s)}{U(s)}$ of the system

c. Determine the eigenvalues and pole/poles of the system for

   i. $a = 3$, $K_1 = 1$, and $K_2 = 10$

   ii. $a = 3$, $K_1 = 1$, and $K_2 = 4$

   For each case, comment on the stability properties.

10.9 Computation of Steady State Errors from State Variable Models

In Chapter 4, we have derived the expression for steady state errors for systems which are represented by transfer function models. Here we will compute the steady state errors for systems modeled by state variable models using two different methods

1. Analysis via the final value theorem
2. Analysis via input substitution

Analysis via Final Value Theorem

Consider a closed-loop represented by the state space model

$$\begin{aligned}\dot{\mathbf{x}} &= \mathbf{A}\mathbf{x} + \mathbf{B}\mathrm{r} \\ \mathrm{y} &= \mathbf{C}\mathbf{x}\end{aligned} \tag{10.124}$$

The closed-loop transfer function between the output "y" and and reference input "r" is

$$\mathrm{T(s)} = \frac{\mathrm{Y(s)}}{\mathrm{R(s)}} = \mathbf{C}\left(\mathrm{s}\mathbf{I} - \mathbf{A}\right)^{-1}\mathbf{B}$$

The Laplace transform of the error is given by

$$\begin{aligned}\mathrm{E(s)} &= \mathrm{R(s)} - \mathrm{Y(s)} = \mathrm{R(s)} - \mathrm{R(s)T(s)} = \mathrm{R(s)}\left[1 - \mathrm{T(s)}\right] \quad (\because \quad \mathrm{Y(s)} = \mathrm{R(s)T(s)}) \\ &= \mathrm{R(s)}\left[1 - \mathbf{C}\left(\mathrm{s}\mathbf{I} - \mathbf{A}\right)^{-1}\mathbf{B}\right]\end{aligned}$$

Now applying the final value theorem, the steady state error is given by

$$\boxed{\mathrm{e_{ss}} = \lim_{\mathrm{s}\to 0} \mathrm{sE(s)} = \lim_{\mathrm{s}\to 0} \mathrm{sR(s)}\left[1 - \mathbf{C}\left(\mathrm{s}\mathbf{I} - \mathbf{A}\right)^{-1}\mathbf{B}\right]} \tag{10.125}$$

The following example illustrates its application.

EXAMPLE 10.16

Problem:

Determine the steady state error due to a step input for the system represented in controllable canonical form as

$$\dot{\mathbf{x}} = \mathbf{A}\mathbf{x} + \mathbf{B}\mathrm{r}, \quad \mathrm{y} = \mathbf{C}\mathbf{x}$$

where $\mathbf{A} = \begin{bmatrix} 0 & 1 \\ -\mathrm{a}_2 & -\mathrm{a}_1 \end{bmatrix}$, $\mathbf{B} = \begin{bmatrix} 0 \\ 1 \end{bmatrix}$, $\mathbf{C} = \begin{bmatrix} \mathrm{b}_1 & \mathrm{b}_0 \end{bmatrix}$

SOLUTION:

Let us compute the transfer function $\mathbf{T(s)} = \mathbf{C}\left(\mathbf{sI} - \mathbf{A}\right)^{-1}\mathbf{B}$. Since the system is in controllable canonical form, the transfer function can be determined by inspection. Thus

$$\mathbf{T(s)} = \mathbf{C}\left(\mathbf{sI} - \mathbf{A}\right)^{-1}\mathbf{B} = \frac{\mathrm{b}_0\mathrm{s} + \mathrm{b}_1}{\mathrm{s}^2 + \mathrm{a}_1\mathrm{s} + \mathrm{a}_2}$$

Now the steady state error due to the input $r(t) = 1$ is given by

$$\begin{aligned} e_{ss} &= \lim_{s\to 0} sR(s)\left(1 - \mathbf{C}\left(s\mathbf{I} - \mathbf{A}\right)^{-1}\mathbf{B}\right) \\ &= \lim_{s\to 0} s\frac{1}{s}\left(1 - \frac{b_0 s + b_1}{s^2 + a_1 s + a_2}\right) \\ &= \lim_{s\to 0} \frac{s^2 + (a_1 - b_0)s + (a_2 - b_1)}{s^2 + a_1 s + a_2} \\ &= \frac{a_2 - b_1}{a_2} \end{aligned}$$

Analysis via Input Substitution

This method of steady state error analysis does not require inversion of the matrix $(s\mathbf{I} - \mathbf{A})$ and can be applied for multiple-input-multiple-output systems. In the following, we derive the steady state error expressions for unit step and unit ramp inputs.

Unit Step Input: Consider the system of Eqn(10.124). Let $\mathbf{x_{ss}}$ denote the steady state value of the states, which is represented as:

$$\mathbf{x_{ss}} = \begin{bmatrix} V_1 \\ V_2 \\ \vdots \\ V_n \end{bmatrix} = \mathbf{V} \quad \text{where } V_i, i = 1, 2, \ldots, n \text{ are constants}$$

At steady state, the dynamics of the system cease to zero, that is, $\dot{\mathbf{x}}_{\mathbf{ss}} = 0$. When the input is a unit step, its steady state value (say $r_{ss} = 1$). Therefore, at steady state the system of Eqn(10.124) becomes

$$\begin{aligned} \mathbf{0} &= \mathbf{A}\mathbf{x_{ss}} + \mathbf{B}r_{ss} = \mathbf{AV} + \mathbf{B} \quad \Longrightarrow \quad \mathbf{V} = -\mathbf{A}^{-1}\mathbf{B} \\ y_{ss} &= \mathbf{C}\mathbf{x_{ss}} = \mathbf{CV} \end{aligned}$$

where y_{ss} is the steady-state output. Therefore, the steady state error, which is the difference between the steady-state input and steady-state output, is given by

$$\boxed{e_{ss} = r_{ss} - y_{ss} = 1 - \mathbf{CV} = 1 + \mathbf{CA}^{-1}\mathbf{B}} \tag{10.126}$$

Ramp Input: When the input $r = t$, the steady state value of the states is of the form

$$\mathbf{x_{ss}} = \begin{bmatrix} V_1 t + W_1 \\ V_2 t + W_2 \\ \vdots \\ V_n t + W_n \end{bmatrix} = \mathbf{Vt} + \mathbf{W} \text{ where } V_i, W_i, i = 1, 2, \ldots, n \text{ are constants.}$$

Therefore

$$\dot{\mathbf{x}}_{ss} = \begin{bmatrix} V_1 \\ V_2 \\ \vdots \\ V_n \end{bmatrix} = \mathbf{V}$$

Therefore, for ramp input, the steady state of the system of Eqn(10.124) becomes

$$\dot{\mathbf{x}}_{ss} = \mathbf{A}\mathbf{x}_{ss} + \mathbf{B}r_{ss} \implies \mathbf{V} = \mathbf{A}\left(\mathbf{V}t + \mathbf{W}\right) + \mathbf{B}t \quad (10.127a)$$

$$y_{ss} = \mathbf{C}\mathbf{x}_{ss} = \mathbf{C}\left(\mathbf{V}t + \mathbf{W}\right) \quad (10.127b)$$

Let us balance Eqn(10.127a) by equating the matrix coefficients of t which gives

$$\mathbf{AV} + \mathbf{B} = \mathbf{0}, \ \text{or}, \ \mathbf{AV} = -\mathbf{B}, \ \text{or}, \ \mathbf{V} = -\mathbf{A}^{-1}\mathbf{B} \quad (10.128)$$

Similarly, equating the coefficients of constant terms in Eqn(10.127a) gives

$$\mathbf{AW} = \mathbf{V}, \ \text{or}, \ \mathbf{W} = \mathbf{A}^{-1}\mathbf{V} \quad (10.129)$$

Hence, $\mathbf{x}_{ss} = \mathbf{V}t + \mathbf{W} = -\mathbf{A}^{-1}\mathbf{B}t + \mathbf{A}^{-1}\mathbf{V} = -\mathbf{A}^{-1}\mathbf{B}t - \left(\mathbf{A}^{-1}\right)^2\mathbf{B}$

The steady state output is given by

$$y_{ss} = \mathbf{C}\mathbf{x}_{ss} = -\mathbf{C}\left[\mathbf{A}^{-1}\mathbf{B}t + \left(\mathbf{A}^{-1}\right)^2\mathbf{B}\right] \quad (10.130)$$

Therefore, the steady state error becomes

$$e_{ss} = \lim_{t\to\infty}\left(t - y_{ss}\right) = \lim_{t\to\infty}\left[\left(1 + \mathbf{C}\mathbf{A}^{-1}\mathbf{B}\right)t + \mathbf{C}\left(\mathbf{A}^{-1}\right)^2\mathbf{B}\right] \quad (10.131)$$

EXAMPLE 10.17

Problem:

Determine the steady state error due to step and ramp input using both the, final value theorem approach and input substitution approach, for the system represented in controllable canonical form as

$$\dot{\mathbf{x}} = \mathbf{A}\mathbf{x} + \mathbf{B}r, \quad y = \mathbf{C}\mathbf{x} \text{ where}$$

$$\mathbf{A} = \begin{bmatrix} 0 & 1 \\ -\mathrm{K} & -\mathrm{a} \end{bmatrix}, \quad \mathbf{B} = \begin{bmatrix} 0 \\ \mathrm{K} \end{bmatrix}, \quad \mathbf{C} = \begin{bmatrix} 1 & 0 \end{bmatrix}$$

SOLUTION:

Final Value Theorem Method:

Let us compute $\mathbf{C}(\mathbf{sI} - \mathbf{A})^{-1}\mathbf{B}$. Note that the transfer function of the system can be found by inspection and is

$$\mathbf{C}(\mathbf{sI} - \mathbf{A})^{-1}\mathbf{B} = \frac{\mathrm{K}}{\mathrm{s}^2 + \mathrm{as} + \mathrm{K}}$$

Now the steady state error due to the step input $\mathrm{r(t)} = 1$ is given by

$$\begin{aligned} \mathrm{e_{ss}} &= \lim_{\mathrm{s}\to 0} \mathrm{sR(s)}\left(1 - \mathbf{C}(\mathbf{sI} - \mathbf{A})^{-1}\mathbf{B}\right) \\ &= \lim_{\mathrm{s}\to 0} \mathrm{s}\frac{1}{\mathrm{s}}\left(1 - \frac{\mathrm{K}}{\mathrm{s}^2 + \mathrm{as} + \mathrm{K}}\right) \\ &= \lim_{\mathrm{s}\to 0} \frac{\mathrm{s}^2 + \mathrm{as}}{\mathrm{s}^2 + \mathrm{as} + \mathrm{K}} \\ &= 0 \end{aligned}$$

The steady state error due to the ramp input $\mathrm{r(t)} = \mathrm{t}$ is given by

$$\begin{aligned} \mathrm{e_{ss}} &= \lim_{\mathrm{s}\to 0} \mathrm{sR(s)}\left(1 - \mathbf{C}(\mathbf{sI} - \mathbf{A})^{-1}\mathbf{B}\right) \\ &= \lim_{\mathrm{s}\to 0} \mathrm{s}\frac{1}{\mathrm{s}^2}\left(1 - \frac{\mathrm{K}}{\mathrm{s}^2 + \mathrm{as} + \mathrm{K}}\right) \\ &= \lim_{\mathrm{s}\to 0} \frac{1}{\mathrm{s}}\frac{\mathrm{s}^2 + \mathrm{as}}{\mathrm{s}^2 + \mathrm{as} + \mathrm{K}} \\ &= \lim_{\mathrm{s}\to 0} \frac{\mathrm{s} + \mathrm{a}}{\mathrm{s}^2 + \mathrm{as} + \mathrm{K}} \\ &= \frac{\mathrm{a}}{\mathrm{K}} \end{aligned}$$

Input Substitution Method:

For a step input, the steady state error is given by

$$\begin{aligned} \mathrm{e_{ss}} &= 1 + \mathbf{CA}^{-1}\mathbf{B} \\ &= 1 + \begin{bmatrix} 1 & 0 \end{bmatrix} \begin{bmatrix} 0 & 1 \\ -\mathrm{K} & -\mathrm{a} \end{bmatrix}^{-1} \begin{bmatrix} 0 \\ \mathrm{K} \end{bmatrix} \\ &= 1 + \begin{bmatrix} 1 & 0 \end{bmatrix} \begin{bmatrix} -\frac{\mathrm{a}}{\mathrm{K}} & -\frac{1}{\mathrm{K}} \\ 1 & 0 \end{bmatrix} \begin{bmatrix} 0 \\ \mathrm{K} \end{bmatrix} \\ &= 1 + \begin{bmatrix} 1 & 0 \end{bmatrix} \begin{bmatrix} -1 \\ 0 \end{bmatrix} = 1 - 1 = 0 \end{aligned}$$

For a ramp input, the steady state error is given by

$$e_{ss} = \lim_{t\to\infty}\left[\left(1 + \mathbf{CA^{-1}B}\right)t + \mathbf{C}\left(\mathbf{A^{-1}}\right)^{\mathbf{2}}\mathbf{B}\right]$$

We found that the term $\left(1 + \mathbf{CA^{-1}B}\right)$ equals to zero. Let us compute the second term

$$\begin{aligned}
e_{ss} &= \lim_{t\to\infty}\left[\left(1 + \mathbf{CA^{-1}B}\right)t + \mathbf{C}\left(\mathbf{A^{-1}}\right)^{\mathbf{2}}\mathbf{B}\right] \\
&= \lim_{t\to\infty}\left[\mathbf{C}\left(\mathbf{A^{-1}}\right)^{\mathbf{2}}\mathbf{B}\right] = \lim_{t\to\infty}\begin{bmatrix}1 & 0\end{bmatrix}\begin{bmatrix}-\frac{a}{K} & -\frac{1}{K} \\ 1 & 0\end{bmatrix}^2\begin{bmatrix}0 \\ K\end{bmatrix} \\
&= \lim_{t\to\infty}\begin{bmatrix}1 & 0\end{bmatrix}\begin{bmatrix}\frac{a^2}{K^2} - \frac{1}{K} & \frac{a}{K^2} \\ -\frac{a}{K} & -\frac{1}{K}\end{bmatrix}\begin{bmatrix}0 \\ K\end{bmatrix} \\
&= \lim_{t\to\infty}\begin{bmatrix}1 & 0\end{bmatrix}\begin{bmatrix}\frac{a}{K} \\ -1\end{bmatrix} \\
&= \frac{a}{K}
\end{aligned}$$

10.10 Multiple Choice Skills Test Questions

M 10.1. The transfer function of the system with following the state equations, is

$$\dot{x}(t) = -\alpha x(t) + \beta u(t), \quad y(t) = \gamma x(t)$$

A. $\dfrac{\alpha\gamma}{s+\beta}$ **B.** $\dfrac{\gamma}{s+\alpha}$ **C.** $\dfrac{\alpha\beta}{s+\gamma}$ **D.** $\dfrac{\gamma\beta}{s+\alpha}$

M 10.2. The state variable model of a system is given by

$$\begin{bmatrix}\dot{x}_1 \\ \dot{x}_2\end{bmatrix} = \begin{bmatrix}0 & 1 \\ -\alpha & -2\beta\end{bmatrix}\begin{bmatrix}x_1 \\ x_2\end{bmatrix} + \begin{bmatrix}0 \\ \alpha\end{bmatrix}u$$

$$y = \begin{bmatrix}1 & 0\end{bmatrix}\begin{bmatrix}x_1 \\ x_2\end{bmatrix}$$

where y is the output and u is the input. The damping ratio ξ and undamped natural frequency ω_n rad/sec of the system are given by

A. $\xi = \sqrt{\alpha};\ \omega_n = \dfrac{\beta}{\sqrt{\alpha}}$

B. $\xi = \dfrac{\beta}{\sqrt{\alpha}};\ \omega_n\sqrt{\alpha}$

C. $\xi = \dfrac{\sqrt{\alpha}}{\beta};\ \omega_n\sqrt{\beta}$

D. $\xi = \sqrt{\beta};\ \omega_n = \sqrt{\alpha}$

M 10.3. Which of the following statements is true?

I. The eigenvalues as well as the poles of a system are always the same.
II. The eigenvalues and poles of a system are the same, provided there is no pole-zero cancellation in the transfer function.
III. Poles are subsets of eigenvalues; but not vice versa.
IV. Eigenvalues are subsets of poles, but not vice versa.
V. Eigenvalues are natural modes of the system

A. Statements II, III, and V are true.
B. Statements I and IV are true.
C. Statements I and IV are true.
D. Statements IV is true.

M 10.4. The system with the transfer function $G(s) = \dfrac{Y(s)}{U(s)} = \dfrac{s-1}{s^2+s-2}$ is

A. not asymptotically stable; but BIBO unstable

B. not asymptotically stable; but BIBO stable

C. both asymptotically and BIBO unstable

D. both asymptotically and BIBO stable

M 10.5. Consider the system

$$\dot{\mathbf{x}}(t) = \mathbf{A}\mathbf{x}(t) + \mathbf{B}u(t)$$
$$y(t) = \mathbf{C}\mathbf{x}(t) + Du(t)$$

If it is represented by a new set of state variables $\mathbf{z} = \mathbf{T}\mathbf{x}$ where $\mathbf{T}$ is a constant nonsingular matrix, then the state equations in terms of $\mathbf{z}$ variables are given by

A. $\dot{\mathbf{z}} = \mathbf{T}^{-1}\mathbf{A}\mathbf{T}\mathbf{z} + \mathbf{T}^{-1}\mathbf{B}u,\ y = \mathbf{C}\mathbf{T}^{-1}\mathbf{z} + Du$
B. $\dot{\mathbf{z}} = \mathbf{T}^{-1}\mathbf{A}\mathbf{T}\mathbf{z} + \mathbf{T}^{-1}\mathbf{B}u,\ y = \mathbf{C}\mathbf{T}\mathbf{z} + Du$
C. $\dot{\mathbf{z}} = \mathbf{T}\mathbf{A}\mathbf{T}^{-1}\mathbf{z} + \mathbf{T}\mathbf{B}u,\ y = \mathbf{C}\mathbf{T}^{-1}\mathbf{z} + Du$
D. $\dot{\mathbf{z}} = \mathbf{T}\mathbf{A}\mathbf{T}^{-1}\mathbf{z} + \mathbf{T}\mathbf{B}u,\ y = \mathbf{C}\mathbf{T}\mathbf{z} + Du$

M 10.6. The state space model of a system is given by

$$\dot{\mathbf{x}} = \mathbf{A}\mathbf{x} + \mathbf{B}u, \quad y = \mathbf{C}\mathbf{x} + Du$$

where $\mathbf{A} = \begin{bmatrix} 0 & 1 & 0 \\ 0 & 0 & 1 \\ -1 & -2 & -3 \end{bmatrix}$, $\mathbf{B} = \begin{bmatrix} 0 \\ 0 \\ 10 \end{bmatrix}$, $\mathbf{C} = \begin{bmatrix} 2 & 3 & 1 \end{bmatrix}$

The transfer function model of the system is

A. $\dfrac{10(s^2+3s+2)}{s^3+3s^2+2s+1}$ **C.** $\dfrac{10(2s^2+3s+1)}{s^3+s^2+2s+3}$

B. $\dfrac{10(2s^2+3s+1)}{s^3+3s^2+2s+1}$ **D.** $\dfrac{s^2+3s+1)}{s^3+s^2+2s+3}$

M 10.7. Consider a system with transfer function

$$G(s) = \frac{Y(s)}{U(s)} = \frac{5s+2}{s^2+8s+15}$$

If the phase variable canonical representation of this system is given by

$$\dot{\mathbf{x}} = \begin{bmatrix} 0 & 1 \\ -a & -b \end{bmatrix}\mathbf{x} + \begin{bmatrix} 0 \\ 1 \end{bmatrix}u, \quad y = \begin{bmatrix} c & d \end{bmatrix}\mathbf{x}$$

then, the values of the parameters a, b, c and d are

A. $a = 15$, $b = 8$, $c = 2$ and $d = 5$ **C.** $a = 15$, $b = 8$, $c = 5$ and $d = 2$

B. $a = 8$, $b = 15$, $c = 2$ and $d = 5$ **D.** $a = -8$, $b = -15$, $c = 2$ and $d = 5$

M 10.8. A system has the state space representation

$$\dot{\mathbf{x}} = \begin{bmatrix} -1 & 0 & 0 \\ 0 & -3 & 0 \\ 0 & 0 & -5 \end{bmatrix}\mathbf{x} + \begin{bmatrix} 1 \\ 1 \\ 1 \end{bmatrix}u, \quad y = \begin{bmatrix} 1 & 2 & -1 \end{bmatrix}\mathbf{x}$$

If this system is represented as:

$$\dot{\mathbf{x}} = \begin{bmatrix} 0 & 1 & 0 \\ 0 & 0 & 1 \\ -a & -b & -c \end{bmatrix}\mathbf{x} + \begin{bmatrix} 0 \\ 0 \\ 1 \end{bmatrix}u, \quad y = \begin{bmatrix} 12 & 16 & 22 \end{bmatrix}\mathbf{x}$$

then the parameters a, b, and c are

A. $a = 9$, $b = 23$, and $c = 15$ **C.** $a = 15$, $b = 23$, and $c = 9$

B. $a = -15$, $b = -23$, and $c = -9$ **D.** $a = -9$, $b = -23$, and $c = -15$

M 10.9. A system is represented by the state equation

$$\begin{bmatrix} \dot{x}_1 \\ \dot{x}_2 \end{bmatrix} = \begin{bmatrix} -a & 0 \\ 0 & b \end{bmatrix}\begin{bmatrix} x_1 \\ x_2 \end{bmatrix} + \begin{bmatrix} 1 \\ 1 \end{bmatrix}u$$

The state transition matrix of the system is

A. $\begin{bmatrix} e^{at} & 0 \\ 0 & e^{-bt} \end{bmatrix}$ **B.** $\begin{bmatrix} e^{at} & 0 \\ 0 & e^{bt} \end{bmatrix}$ **C.** $\begin{bmatrix} e^{-at} & 0 \\ 0 & e^{-bt} \end{bmatrix}$ **D.** $\begin{bmatrix} e^{-at} & 0 \\ 0 & e^{bt} \end{bmatrix}$

M 10.10. The state transition matrix of a system $\phi(t) = \begin{bmatrix} \cos(2t) & \sin(2t) \\ -\sin(2t) & \cos(2t) \end{bmatrix}$

The inverse of this matrix $\phi^{-1}(t)$ is

A. $\begin{bmatrix} \cos(2t) & \sin(2t) \\ -\sin(2t) & \cos(2t) \end{bmatrix}$

B. $\begin{bmatrix} -\cos(2t) & \sin(2t) \\ -\sin(2t) & -\cos(2t) \end{bmatrix}$

C. $\begin{bmatrix} \cos(2t) & -\sin(2t) \\ \sin(2t) & \cos(2t) \end{bmatrix}$

D. $\begin{bmatrix} -\cos(2t) & \sin(2t) \\ \sin(2t) & \cos(2t) \end{bmatrix}$

M 10.11. Consider the following state space representation of a system

$$\begin{bmatrix} \dot{x}_1 \\ \dot{x}_2 \end{bmatrix} = \begin{bmatrix} \alpha & 0 \\ 0 & \beta \end{bmatrix} \begin{bmatrix} x_1 \\ x_2 \end{bmatrix}, \quad y = \begin{bmatrix} 1 & 1 \end{bmatrix} \begin{bmatrix} x_1 \\ x_2 \end{bmatrix}, \quad \mathbf{x}(0) = \begin{bmatrix} 1 \\ 1 \end{bmatrix}$$

The output $y(t)$ at $t = \log_e(2)$ is

A. $\alpha + \beta$ **B.** $2^\alpha + 2^\beta$ **C.** $\alpha^2 + \beta^2$ **D.** $\alpha^2 - \beta^2$

M 10.12. Consider a double integrator whose dynamics is described by

$$\dot{\mathbf{x}} = \begin{bmatrix} 0 & 1 \\ 0 & 0 \end{bmatrix} \mathbf{x}$$

The state response of this system to initial condition $\mathbf{x}(0)$ is given by

A. $\mathbf{x}(t) = \begin{bmatrix} 1 & t \\ 0 & 1 \end{bmatrix} \mathbf{x}(0)$

B. $\mathbf{x}(t) = \begin{bmatrix} -1 & t \\ 0 & -1 \end{bmatrix} \mathbf{x}(0)$

C. $\mathbf{x}(t) = \begin{bmatrix} 1 & t^2 \\ 0 & 1 \end{bmatrix} \mathbf{x}(0)$

D. $\mathbf{x}(t) = \begin{bmatrix} t & t^2 \\ 0 & 1 \end{bmatrix} \mathbf{x}(0)$

M 10.13. The state and output equation of a linear time-invariant system is given by

$$\begin{bmatrix} \dot{x}_1 \\ \dot{x}_2 \end{bmatrix} = \begin{bmatrix} 10 & -5 \\ 0 & 5 \end{bmatrix} \begin{bmatrix} x_1 \\ x_2 \end{bmatrix} + \begin{bmatrix} 0 \\ 1 \end{bmatrix} u, \quad y = \begin{bmatrix} 1 & 1 \end{bmatrix} \begin{bmatrix} x_1 \\ x_2 \end{bmatrix}$$

If $x_1(0) = 0.1$, $x_2(0) = -1$, $u(0) = 0$, then $\left.\dfrac{dy}{dt}\right|_{t=0}$ equals to

A. -1 **B.** 1 **C.** 2 **D.** 0

M 10.14. For the system represented by the state equation

$$\dot{\mathbf{x}} = \mathbf{A}\mathbf{x}(t)$$

the state response $\mathbf{x}(t)$ is given by

$$\mathbf{x}(t) = \begin{bmatrix} e^{2t} \\ e^{-2t} \end{bmatrix} \quad \text{when} \quad \mathbf{x}(0) = \begin{bmatrix} 1 \\ 0 \end{bmatrix}$$

The system matrix $\mathbf{A}$ is given by

A. $\mathbf{A} = \begin{bmatrix} -2 & 0 \\ 0 & 2 \end{bmatrix}$ **B.** $\mathbf{A} = \begin{bmatrix} 0 & 2 \\ -2 & 0 \end{bmatrix}$ **C.** $\mathbf{A} = \begin{bmatrix} 2 & 0 \\ 0 & -2 \end{bmatrix}$ **D.** $\mathbf{A} = \begin{bmatrix} 0 & -2 \\ 2 & 0 \end{bmatrix}$

M 10.15. Given a system

$$\begin{bmatrix} \dot{x}_1 \\ \dot{x}_2 \end{bmatrix} = \begin{bmatrix} -5 & 1 \\ -4 & -1 \end{bmatrix} \begin{bmatrix} x_1 \\ x_2 \end{bmatrix} + \begin{bmatrix} 0 \\ 1 \end{bmatrix} u$$

The steady state value of states $\mathbf{x}_{ss}$ for a step input u with zero initial conditions is

A. $\mathbf{x}_{ss} = \begin{bmatrix} 0.1111 \\ 0.5556 \end{bmatrix}$ **C.** $\mathbf{x}_{ss} = \begin{bmatrix} 0.5556 \\ 0 \end{bmatrix}$

B. $\mathbf{x}_{ss} = \begin{bmatrix} 0.5556 \\ 0.1111 \end{bmatrix}$ **D.** $\mathbf{x}_{ss} = \begin{bmatrix} -0.5556 \\ -0.1111 \end{bmatrix}$

M 10.16. Given a system

$$\begin{bmatrix} \dot{x}_1 \\ \dot{x}_2 \end{bmatrix} = \begin{bmatrix} 0 & 1 \\ -4 & -8 \end{bmatrix} \begin{bmatrix} x_1 \\ x_2 \end{bmatrix} + \begin{bmatrix} 0 \\ 1 \end{bmatrix} u, \quad y = \begin{bmatrix} 1 & 1 \end{bmatrix} \begin{bmatrix} x_1 \\ x_2 \end{bmatrix}$$

The steady state error for a step input is

A. 0.25 **B.** 0.75 **C.** 1.25 **D.** -0.75

Answers

M 10.1. D **M 10.2**. B **M 10.3**. A **M 10.4**. B **M 10.5**. C **M 10.6**. A **M 10.7**. A **M 10.8**. C **M 10.9**. D **M 10.10**. B **M 10.11**. B **M 10.12**. A **M 10.13**. B **M 10.14**. C **M 10.15**. A **M 10.16**. B

10.11 Knowledge Assessment Problems

KaP 10.1 The differential equation models of several linear time-invariant systems are given below. Represent these systems in state-space.

a. $\ddot{y}(t) + 3y(t) = 20u(t)$

b. $\ddot{y}(t) + 5\dot{y}(t) + y(t) = 10u(t)$

c. $\dddot{y}(t) + 2\ddot{y}(t) + 5\dot{y}(t) + y(t) + \int_0^t y(\tau)d\tau = u(t)$

d. $\ddddot{y}(t) + 2.5\dddot{y}(t) + 3.5\dot{y}(t) + y(t) = 8u(t)$

KaP 10.2 Determine the transfer function $G(s) = \dfrac{Y(s)}{U(s)}$ of the multi-loop feedback system shown in Figure 10.9 from its state space model.

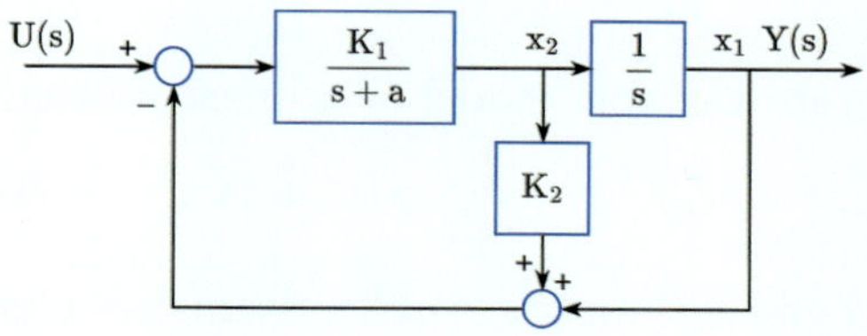

Figure 10.9: Multi-loop system

KaP 10.3 Find the transfer function matrices of the following systems represented in standard state space form with various system matrix **A**, input matrix **B**, output matrix **C**, and transmission matrix **D**,

a. $\mathbf{A} = \begin{bmatrix} -3 & 4 \\ -2 & 0 \end{bmatrix}$, $\mathbf{B} = \begin{bmatrix} 1 & -1 \\ -6 & 0 \end{bmatrix}$, $\mathbf{C} = \begin{bmatrix} 1 & 2 \end{bmatrix}$, & $D = \begin{bmatrix} 0 & 0 \\ 0 & 0 \end{bmatrix}$

b. $\mathbf{A} = \begin{bmatrix} -2 & 4 \\ -1 & 1 \end{bmatrix}$, $\mathbf{B} = \begin{bmatrix} 1 \\ 2 \end{bmatrix}$, $\mathbf{C} = \begin{bmatrix} -3 & 5 \\ 2 & -1 \end{bmatrix}$, & $D = 0$

c. $\mathbf{A} = \begin{bmatrix} -1 & 2 \\ -3 & -4 \end{bmatrix}$, $\mathbf{B} = \begin{bmatrix} 1 & 0 \\ -2 & 5 \end{bmatrix}$, $\mathbf{C} = \begin{bmatrix} -1 & 0 \\ 4 & -1 \end{bmatrix}$, & $D = \begin{bmatrix} 0 & 0 \\ 0 & 0 \end{bmatrix}$

KaP 10.4 Verify the results of the above problem in MATLAB.

KaP 10.5 Represent the following systems in controllable, observable, and diagonal canonical forms.

i. $\dfrac{Y(s)}{U(s)} = \dfrac{5s + 3}{s^2 + 7s + 10}$

ii. $\dfrac{Y(s)}{U(s)} = \dfrac{3s^2 + 23s + 38}{s^2 + 8s + 12}$

iii. $\dfrac{Y(s)}{U(s)} = \dfrac{4s^2 + 47s + 79}{(s + 7)(s^2 + 5s + 4)}$

iv. $\dfrac{Y(s)}{U(s)} = \dfrac{s^3 + 33s^2 + 148s + 184}{(s^2 + 5s + 6)(s^2 + 12s + 32)}$

KaP 10.6 Consider a system represented as:

$$\dot{\mathbf{x}} = \mathbf{A}\mathbf{x} + \mathbf{B}\mathrm{u}, \quad \mathrm{y} = \mathbf{C}\mathbf{x}, \quad \text{where}$$

$$\mathbf{A} = \begin{bmatrix} -3 & 1 \\ 1 & -3 \end{bmatrix}, \quad \mathbf{B} = \begin{bmatrix} 1 \\ 2 \end{bmatrix}, \quad \mathbf{C} = \begin{bmatrix} 2 & 3 \end{bmatrix}$$

Transform this system into controllable canonical form.

KaP 10.7 Consider a system represented as:

$$\dot{\mathbf{x}} = \mathbf{A}\mathbf{x} + \mathbf{B}\mathrm{u}, \quad \mathrm{y} = \mathbf{C}\mathbf{x}, \quad \text{where}$$

$$\mathbf{A} = \begin{bmatrix} -2 & 3 \\ 3 & -2 \end{bmatrix}, \quad \mathbf{B} = \begin{bmatrix} 2 \\ 0 \end{bmatrix}, \quad \mathbf{C} = \begin{bmatrix} 0 & 1 \end{bmatrix}$$

Transform this system into observable canonical form.

KaP 10.8 The state variable model of a linear system is of the form

$$\dot{\mathbf{x}}(\mathrm{t}) = \mathbf{A}\mathbf{x}(\mathrm{t}) + \mathbf{B}\mathrm{u}(\mathrm{t})$$

The coefficient matrices of 2 systems are given as follows. Represent these two systems in the form

$$\dot{\mathbf{z}} = \mathbf{P}^{-1}\mathbf{A}\mathbf{P}\mathbf{z} + \mathbf{P}^{-1}\mathbf{B}u$$

such that $\mathbf{P}^{-1}\mathbf{A}\mathbf{P}$ is diagonal.

a. $\mathbf{A} = \begin{bmatrix} -6 & 2 \\ 2 & -6 \end{bmatrix}, \quad \mathbf{B} = \begin{bmatrix} 0 \\ 1 \end{bmatrix}$, and $\mathbf{C} = \begin{bmatrix} 1 & 2 \end{bmatrix}$

b. $\mathbf{A} = \begin{bmatrix} -3 & 1 \\ 1 & -3 \end{bmatrix}, \quad \mathbf{B} = \begin{bmatrix} 1 \\ 2 \end{bmatrix}$, and $\mathbf{C} = \begin{bmatrix} 2 & 3 \end{bmatrix}$

KaP 10.9 The state variable model of a system in phase variable form is

$$\dot{\mathbf{x}} = \begin{bmatrix} 0 & 1 \\ -\mathrm{a} & -\mathrm{b} \end{bmatrix}\mathbf{x} + \begin{bmatrix} 0 \\ \mathrm{d} \end{bmatrix}\mathrm{u}(\mathrm{t}), \quad \mathrm{y} = \begin{bmatrix} 1 & 0 \end{bmatrix}\mathbf{x}$$

If the diagonal canonical form of the system is

$$\dot{\mathbf{z}} = \begin{bmatrix} -4 & 0 \\ 0 & -2 \end{bmatrix}\mathbf{x} + \begin{bmatrix} 1 \\ 1 \end{bmatrix}\mathrm{u}(\mathrm{t}), \quad \mathrm{y} = \begin{bmatrix} -5 & 5 \end{bmatrix}\mathbf{x}$$

determine the parameters a, b and d.

KaP 10.10 The state variable model of a linear system is described by

$$\dot{\mathbf{x}}(\mathrm{t}) = \mathbf{A}\mathbf{x}(\mathrm{t}) + \mathbf{B}\mathrm{u}(\mathrm{t}), \quad \mathrm{y} = \mathbf{C}\mathbf{x}(\mathrm{t}) \tag{10.132}$$

The coefficient matrices are given as follows. Explain why the state equations cannot be transformed into the controllability canonical form.

a. $\mathbf{A} = \begin{bmatrix} \alpha_1 & 0 \\ 0 & \alpha_2 \end{bmatrix}$, $\mathbf{B} = \begin{bmatrix} 0 \\ 1 \end{bmatrix}$

b. $\mathbf{A} = \begin{bmatrix} -3 & 1 \\ -2 & 1.5 \end{bmatrix}$, $\mathbf{B} = \begin{bmatrix} 1 \\ 4 \end{bmatrix}$

c. $\mathbf{A} = \begin{bmatrix} 1 & 2 & 0 \\ 0 & 1 & 0 \\ 0 & 0 & 3 \end{bmatrix}$, $\mathbf{B} = \begin{bmatrix} 1 \\ 0 \\ 0 \end{bmatrix}$

KaP 10.11 The coefficient matrices of system of Eqn(10.132) are given as follows. Explain why the state equations cannot be transformed into the observable canonical form.

a. $\mathbf{A} = \begin{bmatrix} 0 & 1 \\ -2 & -3 \end{bmatrix}$, $\mathbf{C} = \begin{bmatrix} 1 & 1 \end{bmatrix}$

b. $\mathbf{A} = \begin{bmatrix} 0 & 1 \\ -0.4 & -1.3 \end{bmatrix}$, $\mathbf{C} = \begin{bmatrix} 0.8 & 1 \end{bmatrix}$

c. $\mathbf{A} = \begin{bmatrix} 1 & 1 & 0 \\ 0 & 1 & 0 \\ 0 & 0 & 2 \end{bmatrix}$, $\mathbf{C} = \begin{bmatrix} 1 & 0 & 0 \end{bmatrix}$

KaP 10.12 Compute the state transition matrices for the following systems.

i. $\dot{\mathbf{x}} = \begin{bmatrix} 0 & 1 \\ -2 & -3 \end{bmatrix} \mathbf{x}$

ii. $\dot{\mathbf{x}} = \begin{bmatrix} a & 0 \\ 0 & -a \end{bmatrix} \mathbf{x}$

iii. $\dot{\mathbf{x}} = \begin{bmatrix} 0 & a \\ 0 & 0 \end{bmatrix} \mathbf{x}$

iv. $\dot{\mathbf{x}} = \begin{bmatrix} -a & a \\ 0 & 0 \end{bmatrix} \mathbf{x}$

KaP 10.13 Show that the state transition matrix for a diagonal system is diagonal, with the system modes along the diagonal.

KaP 10.14 Which of the following matrices can be the state transition matrices?

a. $\mathbf{\Phi}(t) = \begin{bmatrix} 1 - e^{-t} & 0 \\ 0 & e^{-2t} \end{bmatrix}$

b. $\mathbf{\Phi}(t) = \begin{bmatrix} e^{-2t} & -te^{-2t} \\ 0 & e^{-2t} \end{bmatrix}$

c. $\mathbf{\Phi}(t) = \begin{bmatrix} -e^{at} & 0 \\ 0 & e^{at} \end{bmatrix}$

d. $\mathbf{\Phi}(t) = \begin{bmatrix} 1 & 5t \\ 0 & 1 \end{bmatrix}$

KaP 10.15 The dynamics of a second-order system are described by the following differential equation

$$\ddot{y}(t) + a_1\dot{y}(t) + a_2 y(t) = u(t)$$

a. Represent this system in state space considering the state variables as: $x_1 = y$ and $x_2 = \dot{y}$. Find the state transition matrix for $a_1 = 2$ and $a_2 = 1$.

b. Represent the system by a new set of state variables as: $x_1 = y$ and $x_2 = y + \dot{y}$. Compute the state transition matrix for $a_1 = 2$ and $a_2 = 1$.

c. Show that the characteristic equations, $\left|s\mathbf{I} - \mathbf{A}\right| = 0$, for parts (a) and (b) are identical.

KaP 10.16 Obtain both the state and output response for a unity step input for following systems.

a. $\dot{\mathbf{x}} = \begin{bmatrix} 0 & 1 \\ -2 & -3 \end{bmatrix}\mathbf{x} + \begin{bmatrix} 0 \\ 1 \end{bmatrix}u, y = \begin{bmatrix} 1 & 0 \end{bmatrix}\mathbf{x}, \mathbf{x}(0) = \begin{bmatrix} 0 \\ 0 \end{bmatrix}$

b. $\dot{\mathbf{x}} = \begin{bmatrix} 0 & 3 \\ -3 & 0 \end{bmatrix}\mathbf{x} + \begin{bmatrix} 0 \\ 1 \end{bmatrix}u(t), y = \begin{bmatrix} 0 & 1 \end{bmatrix}\mathbf{x}, \mathbf{x}(0) = \begin{bmatrix} 0 \\ 0 \end{bmatrix}$

KaP 10.17 Given the system

$$\dot{\mathbf{x}} = \begin{bmatrix} 0 & 1 \\ 7 & -4 \end{bmatrix}\mathbf{x} + \begin{bmatrix} 1 \\ 2 \end{bmatrix}\mathrm{u}$$
$$\mathrm{y} = \begin{bmatrix} 1 & 3 \end{bmatrix}\mathbf{x}$$

with zero initial conditions. Determine the steady state values of states and output for unit step input u.

KaP 10.18 Consider the system

$$\dot{\mathbf{x}} = \mathbf{A}\mathbf{x} + \mathbf{B}\mathrm{u}$$
$$\mathrm{y} = \mathbf{C}\mathbf{x} + \mathrm{D}\mathrm{u} \quad \text{where}$$

$$\mathbf{A} = \begin{bmatrix} 3 & 2 \\ 3 & 4 \end{bmatrix}, \ \mathbf{B} = \begin{bmatrix} 1 \\ -1 \end{bmatrix}, \ \mathbf{C} = \begin{bmatrix} 1 & 0 \end{bmatrix}, \text{ and } \mathrm{D} = 0$$

Compute the steady state error for a step input using both the final-value theorem and the input substitution methods.

Chapter 11

Controller Design via State Space

State feedback is like conducting an orchestra with invisible musicians: you build observers to spy on them, then pretend you knew what they were doing all along.

DESIRED LEARNING OUTCOMES

Upon completion of this chapter, readers should be able to:

- Assess system controllability and observability
- Apply both time-domain and frequency-domain tests for controllability and observability
- Identify uncontrollable and unobservable modes from diagonal system forms
- Determine structural causes of uncontrollability and unobservability
- Apply stabilizability and detectability criteria to partially controllable/observable systems
- Design state feedback controllers using pole placement
- Incorporate feedforward gains into state feedback designs for reference tracking
- Design observer-based state feedback controllers with integral action for disturbance rejection
- Construct reduced-order observers to estimate unmeasured states

DOI: 10.1201/9781003571308-11

- Design optimal controllers using Linear Quadratic Regulator (LQR) and Linear Quadratic Gaussian (LQG) methods

11.1 Introduction

Our previous discussions have primarily emphasized modeling and analysis of linear time-invariant systems, with Chapter 8 introducing basic controller (compensator) design using root locus and frequency domain techniques. Chapter 2 established how to derive transfer function models from physical systems and differential equations, which then enabled a comprehensive analysis of LTI system behavior across both time and frequency domains to determine key performance specifications.

Chapter 9 introduced state space models; a more versatile representation capable of handling diverse system types, including linear, nonlinear, time-invariant, and time-varying systems. This modeling approach transcends the limitations of classical transfer function methods and establishes the groundwork for modern control theory.

Is Controller Design Necessary?

Before embarking on controller design, we must address a fundamental question: Is a controller actually necessary for the given system? This decision forms the cornerstone of rational control system engineering, as unnecessary control adds complexity, cost, and potential failure modes without corresponding benefits.

The necessity of controller design depends primarily on whether the open-loop system can satisfy the specified performance requirements. When a system's inherent dynamics already meet all desired specifications, additional control intervention becomes superfluous. However, when open-loop performance falls short of requirements, a controller becomes essential to bridge the gap between actual and desired system behavior.

Controller Objectives and Priorities

The objectives of any controller depend fundamentally on the stability characteristics of the open-loop system. For unstable systems, stabilization takes absolute priority, as no meaningful performance discussion can occur without basic stability. Once stability is assured, secondary objectives such as response time optimization, disturbance rejection enhancement, and stability margin improvement can be pursued.

Stable systems present a different scenario entirely. Here, the controller's role shifts from the critical task of stabilization to the refinement of performance characteristics. This may involve reducing steady-state errors, shaping transient response, improving robustness to parameter variations, or enhancing disturbance rejection capabilities.

To illustrate these controller design principles in a more intuitive context, consider the following everyday analogy that parallels the decision-making process in control system engineering.

Understanding Necessity of Controller Design:The Student-Tutor Decision

Consider the decision of whether to engage a private tutor **(analogously a controller)** for a student preparing for examinations. This everyday scenario mirrors the controller design decision process with remarkable clarity.

Initially, we observe the student's performance in class tests without any external tutoring assistance. This baseline performance reveals the student's inherent abilities, study habits, and natural aptitude across different subjects. Much like analyzing open-loop system response, this initial assessment provides crucial insights into strengths, weaknesses, and areas requiring attention.

The decision to hire a tutor emerges from comparing actual performance against desired grade percentages. If the student **(analogously the system)** consistently achieves satisfactory results that meet or exceed expectations, additional tutoring may be unnecessary. The student's existing approach proves adequate for the desired outcomes.

However, when performance falls below acceptable thresholds, the tutoring option becomes relevant. Yet, before committing to this intervention, we must assess a critical factor: the student's willingness to listen and cooperate with tutorial guidance. This represents the **controllability** assessment in our analogy.

A receptive student who follows tutorial advice and implements suggested study strategies represents a controllable system where intervention can produce meaningful improvement. Conversely, a student who remains unresponsive to guidance, regardless of the tutor's expertise, represents an uncontrollable system where even excellent tutorial intervention may yield limited results.

Interestingly, some initially uncontrollable situations can be transformed through strategic modifications. Adjusting teaching methods, communication styles, or providing appropriate incentives might make an unresponsive student more receptive to tutorial guidance. This parallels efforts in control engineering to modify system structure or input configurations to achieve controllability.

Systematic Design Methodology

The controller design process follows a logical sequence that prevents unnecessary complexity while ensuring adequate performance.

Controller Design Steps

Step 1: Define System Specifications

- Establish stability requirements
- Specify transient response characteristics (rise time, settling time, overshoot)
- Define steady-state accuracy requirements
- Determine disturbance rejection capabilities needed

Step 2: Analyze Open-Loop Performance

- Evaluate system behavior without control intervention
- Compare actual performance against desired specifications
- If specifications are met → No controller needed (design process complete)
- If specifications are not met → Proceed to Step 3

Step 3: Assess System Controllability

- Determine whether control inputs can meaningfully influence system states
- Verify that desired performance is theoretically achievable through control
- If uncontrollable → Modify system structure or abandon control approach
- If controllable → Proceed to Step 4

Step 4: Design and Select Appropriate Controller

a. Choose controller type based on system dynamics and objectives: For example
    - PID controllers for simple systems
    - State feedback for modern control approaches
    - Robust controllers for uncertain systems

b. Consider practical implementation constraints

c. Validate design through simulation and testing

The complete systematic procedure is illustrated in the flowchart of Figure 11.1.

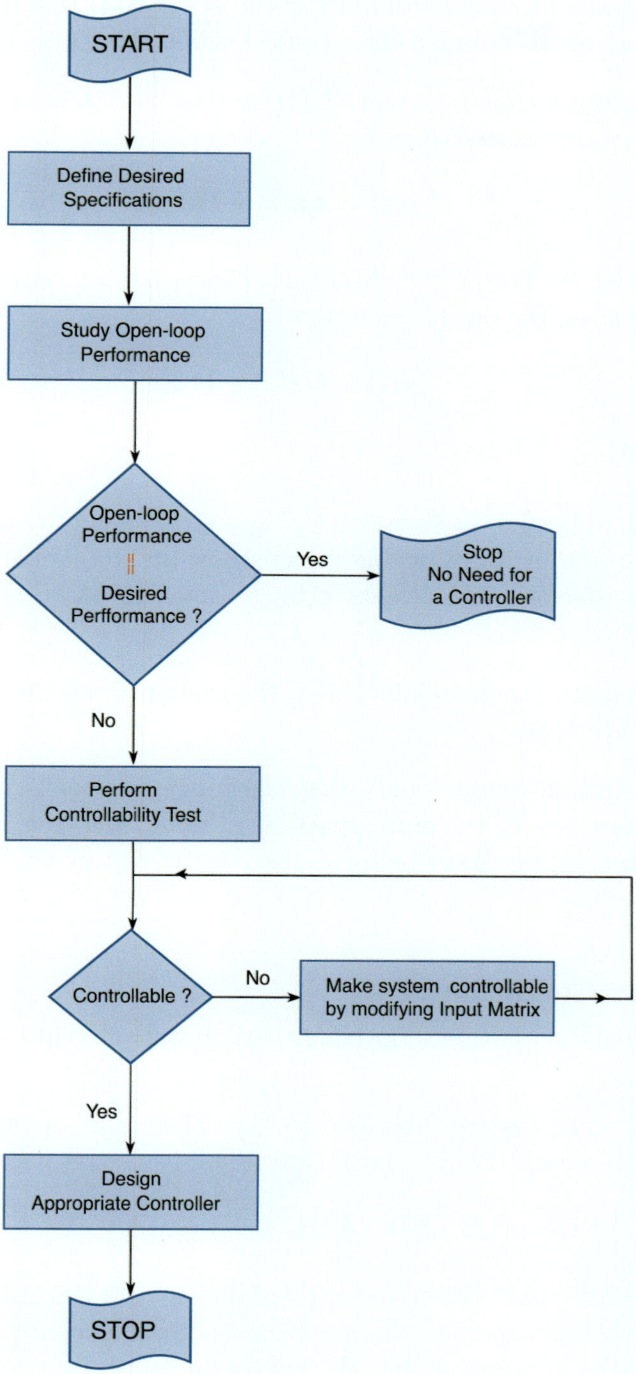

Figure 11.1: Systematic procedure for controller design

11.2 Controllability of Control Systems

Roughly speaking, a system is controllable if every state variable of the system can be controlled to move from any initial state $x(t_0)$ to any final state $x(t_f)$ in finite time by an *unconstrained* control signal.

Consider a linear time-invariant (LTI) system with **n**-number of states and **m**-number of inputs described by

$$\dot{\mathbf{x}}(t) = \mathbf{A}\mathbf{x}(t) + \mathbf{B}\mathbf{u}(t), \tag{11.1}$$

where $\mathbf{A} \in \mathbb{R}^{\mathbf{n}\times\mathbf{n}}$, $\mathbf{B} \in \mathbb{R}^{\mathbf{n}\times\mathbf{m}}$. Note that controllability only relates to inputs and states; thus, the output equation

$$\mathbf{y}(t) = \mathbf{C}\mathbf{x}(t) + \mathbf{D}\mathbf{u}(t)$$

is irrelevant.

Definition 11.1. The system described by Eqn(11.1) is said to be state controllable at time $t = t_0$ if it is possible to construct an admissible control input which will transfer any initial state $x(t_0)$ to any final state $x(t_1)$ in a finite time interval $t_0 \leq t \leq t_1$.

a. If every state is controllable, then the system is said to be *completely state controllable.*

b. This definition requires only that the input be capable of driving the state anywhere in the state space in a finite time; the specific trajectory taken by the state vector is not constrained by the controllability requirement.

Controllability Conditions:

The following theorem gives the conditions of controllability for linear systems.

Theorem 11.1. The system described in Eqn(11.1) is completely state controllable if and only if the $(n \times nm)$ controllability matrix

$$\mathbf{Q_C} = \begin{bmatrix}\mathbf{B} & \mathbf{AB} & \mathbf{A^2B} & \dots & \dots & \mathbf{A^{n-1}B}\end{bmatrix} \tag{11.2}$$

is of rank n. This means that the controllability matrix has full column rank; it has n linearly independent columns. Because the controllability matrix $\mathbf{Q_C}$ involves only the matrices $\mathbf{A}$ and $\mathbf{B}$, this condition is also referred to as the pair $[\mathbf{A}, \mathbf{B}]$ is controllable.

Remark

If the matrix $\mathbf{Q_C}$ is non-square, then the system of Eqn(11.1) is controllable provided the n × n matrix $\mathbf{Q_C}^T\mathbf{Q_C}$ is nonsingular.

EXAMPLE 11.1

Problem:

Investigate the controllability of the system $\dot{\mathbf{x}} = \mathbf{A}\mathbf{x} + \mathbf{B}\mathbf{u}$ for

i. $\mathbf{A} = \begin{bmatrix} a & b \\ 0 & c \end{bmatrix}, \mathbf{B} = \begin{bmatrix} 0 \\ 1 \end{bmatrix}$ ii. $\mathbf{A} = \begin{bmatrix} a & b \\ 0 & c \end{bmatrix}, \mathbf{B} = \begin{bmatrix} 1 \\ 0 \end{bmatrix}$

SOLUTION:

i. Now the controllability matrix $\mathbf{Q_C}$ is

$$\mathbf{Q_C} = \begin{bmatrix} \mathbf{B} & \mathbf{AB} \end{bmatrix} = \begin{bmatrix} 0 & b \\ 1 & c \end{bmatrix}$$

The determinant of $\mathbf{Q_C}$ is b. Thus, the system is controllable.

ii. With change in the input matrix $\mathbf{B}$, the controllability matrix $\mathbf{Q_C}$ becomes

$$\mathbf{Q_C} = \begin{bmatrix} \mathbf{B} & \mathbf{AB} \end{bmatrix} = \begin{bmatrix} 1 & a \\ 0 & 0 \end{bmatrix}$$

The determinant of $\mathbf{Q_C}$ is 0. Thus, the system is uncontrollable.

Remark

Changing the input matrix can change the controllability property of the system.

11.2.1 Geometric Interpretation of Controllability

In the preceding sections, we have established the mathematical criterion for controllability: a linear system $\dot{\mathbf{x}} = \mathbf{A}\mathbf{x} + \mathbf{B}\mathbf{u}$ is controllable if and only if the controllability matrix

$$\mathbf{Q_C} = [\mathbf{B} \mid \mathbf{AB} \mid \mathbf{A}^2\mathbf{B} \mid \cdots \mid \mathbf{A}^{n-1}\mathbf{B}] \tag{11.3}$$

has full rank. While this condition is straightforward to verify computationally, as demonstrated in Example 11.1, the underlying geometric and physical significance may not be immediately apparent.

A natural question arises: why does the rank of this particular matrix determine our ability to steer the system? What does it mean, intuitively, for a system to be controllable by one set of inputs but uncontrollable by another? More importantly, can an uncontrollable system become controllable simply by changing where or how we apply inputs—that is, by changing the **B** matrix? To develop insight into these questions, we now present an analogy that illuminates the fundamental nature of controllability.

Controllability — "Who Does He Listen To?"

Imagine you ask your friend Robin to bring you a glass of water. The question isn't "Can Robin bring a glass of water?" (obviously yes), but "Can you make him bring it?"

- If Robin immediately gets up and brings it, from your perspective, Robin is **controllable.**
- If Robin refuses because he's lazy, unmotivated, or "in the middle of something," he appears **uncontrollable.**
- But wait — if Robin's **crush** walks in and makes the same request, and he sprints to the kitchen like an Olympic athlete, then clearly Robin **can** be controlled — just not by you.

The system (Robin) hasn't changed. His internal dynamics (his ability to walk, fetch water, his natural laziness) remain identical. What changed is **who provides the input**. This is the essence of controllability: it depends critically on the pairing between the system and the actuator.

The key observations:

- Controllability is **not** an intrinsic property of the system dynamics (matrix **A**) alone. Rather, it is a property of the **system-input pair** $(\mathbf{A}, \mathbf{B})$.
- A given system may be perfectly controllable under one choice of input matrix **B**, yet completely uncontrollable under a different choice. The **B** matrix encodes both **where** actuators are placed in the system and **how effectively** they couple to different state variables.
- The rank condition on $\mathbf{Q_C}$ mathematically captures whether your particular actuator configuration can "reach" all the system's state dimensions through repeated application of the dynamics.
- In the Robin analogy: you represent one input configuration $\mathbf{B}_1$ (low rank $\mathbf{Q_C}$, uncontrollable), while his crush represents another $\mathbf{B}_2$ (full rank $\mathbf{Q_C}$, controllable). Same system **A**, different actuators, completely different controllability.

The Profound Lesson:
An uncontrollable system can often be **made** controllable by changing the actuator configuration—relocating sensors, adding new input channels, or selecting different points of influence. Before asking "Is this system controllable?," ask "Controllable by **whom** (or **what**)?" Controllability is a property of the **(system, actuator) pair**, not the system alone. Just as Robin's behavior depends on who's asking, a system's controllability depends entirely on which inputs you have access to and where they're applied.

11.3 Observability of Control Systems

It is a property of a dynamic system that determines whether its internal states can be fully inferred from its outputs over a finite time interval. In simpler terms, it indicates the extent to which the internal states of the system can be determined or "observed" based on its outputs. A system is said to be completely observable if every state variable of the system influences or affects at least one of the system's outputs. In other words, if all the internal states contribute information to the system's outputs, the system is considered completely observable.

Consider a linear time-invariant (LTI) system with **n**-number of states, **m**-number of inputs and **p**-number of outputs described by

$$\begin{aligned} \dot{\mathbf{x}}(\mathrm{t}) &= \mathbf{A}\mathbf{x}(\mathrm{t}) + \mathbf{B}\mathbf{u}(\mathrm{t}) \\ \mathbf{y}(\mathrm{t}) &= \mathbf{C}\mathbf{x}(\mathrm{t}) + \mathbf{D}\mathbf{u}(\mathrm{t}) \end{aligned} \tag{11.4}$$

where $\mathbf{A} \in \mathbb{R}^{\mathbf{n}\times\mathbf{n}}$, $\mathbf{B} \in \mathbb{R}^{\mathbf{n}\times\mathbf{m}}$,$\mathbf{C} \in \mathbb{R}^{\mathbf{p}\times\mathbf{n}}$, and $\mathbf{D} \in \mathbb{R}^{\mathbf{p}\times\mathbf{m}}$.

Definition 11.2. The system described in Eqn(11.4) is said to be completely observable if every initial state $\mathbf{x}(\mathrm{t}_0)$ can be determined from the measurements of $\mathbf{y}(\mathrm{t})$ over a finite time interval $\mathrm{t}_0 \leq \mathrm{t} \leq \mathrm{t}_\mathrm{f}$.

In other words, the system is observable if, given any input $\mathbf{u}(\mathrm{t})$, there exists a finite time $\mathrm{t}_\mathrm{f} \geq \mathrm{t}_0$ such that the knowledge of $\mathbf{u}(\mathrm{t})$ for $\mathrm{t}_0 \leq \mathrm{t} \leq \mathrm{t}_\mathrm{f}$, matrices **A**, **B**, **C**, and **D**; and the output $\mathbf{y}(\mathrm{t})$ for $\mathrm{t}_0 \leq \mathrm{t} \leq \mathrm{t}_\mathrm{f}$ are sufficient to determine $\mathbf{x}(\mathrm{t}_0)$

a. If every state of the system is observable, then the system is said to be *completely observable*.

b. This definition requires that all internal states contribute distinguishable information to the outputs over the finite time interval.

Observability Conditions:

The following theorem gives the conditions of observability for linear systems.

Theorem 11.2. The system described in Eqn(11.4) is completely state observable if and only if the (n × np) **observability matrix**

$$\mathbf{Q_O} = \begin{bmatrix} \mathbf{C} \\ \mathbf{CA} \\ \mathbf{CA^2} \\ \vdots \\ \vdots \\ \mathbf{CA^{n-1}} \end{bmatrix} \tag{11.5}$$

is of rank n. Because the observability matrix $\mathbf{Q_O}$ involves only the matrices **A** and **C**, this condition is also referred to as the pair $[\mathbf{A}, \mathbf{C}]$ is observable.

Remark

If the matrix $\mathbf{Q_O}$ is non-square, then the system of Eqn(11.4) is observable provided the n × n matrix $\mathbf{Q_O}^T\mathbf{Q_O}$ is nonsingular.

EXAMPLE 11.2

Problem:

Investigate the observability of the system $\dot{\mathbf{x}} = \mathbf{Ax}$ and $y = \mathbf{Cx}$ for

i. $\mathbf{A} = \begin{bmatrix} a & b \\ 0 & c \end{bmatrix}, \mathbf{C} = \begin{bmatrix} 1 & 0 \end{bmatrix}$ ii. $\mathbf{A} = \begin{bmatrix} a & b \\ 0 & c \end{bmatrix}, \mathbf{C} = \begin{bmatrix} 0 & 1 \end{bmatrix}$

SOLUTION:

i. Now the observability matrix $\mathbf{Q_O}$ is

$$\mathbf{Q_O} = \begin{bmatrix} \mathbf{C} \\ \mathbf{CA} \end{bmatrix} = \begin{bmatrix} 1 & 0 \\ a & b \end{bmatrix}$$

The determinant of $\mathbf{Q_O}$ is b. Thus, the system is observable.

ii. With change in the output matrix **C**, the observability matrix $\mathbf{Q_O}$ becomes

$$\mathbf{Q_O} = \begin{bmatrix} \mathbf{C} \\ \mathbf{CA} \end{bmatrix} = \begin{bmatrix} 0 & 1 \\ 0 & c \end{bmatrix}$$

The determinant of $\mathbf{Q_O}$ is 0. Thus, the system is unobservable.

Key Insight

Consider the second case of the above example where the state equations are:

$$\dot{x}_1 = ax_1 + bx_2$$
$$\dot{x}_2 = cx_2$$
$$y = x_2$$

While dynamic coupling exists between the states, the system remains unobservable because **the coupling direction prevents unmeasured states from influencing the measured output.** The first state x_1 is influenced by x_2 through the term bx_2, but x_2 evolves independently according to $\dot{x}_2 = cx_2$.

This creates the observability problem: the information flow is from the measured state x_2 to the unmeasured state (x_1, rather than from unmeasured states to measured outputs).

- We can observe $x_2(t)$ directly from the output $y(t)$
- We know how x_2 evolves: $x_2(t) = x_2(0)e^{ct}$
- However, knowledge of $x_2(t)$ and its evolution provides no information about the initial condition $x_1(0)$

The measurable state x_2 carries no information about x_1's initial value because x_2's behavior is completely independent of x_1. Although x_1 responds to x_2, this influence flows in only one direction, making x_1 fundamentally unobservable from output measurements.

Remark

i. The choice of output matrix **C** critically determines observability by selecting which state combinations are measured.

ii. For complete observability, unmeasured states must be able to influence measured outputs through the system dynamics. Unidirectional coupling, where measured states influence unmeasured states (but not vice versa), leads to unobservability.

iii. Dynamic coupling alone is insufficient for observability. The coupling must create pathways through which all states can affect the measured outputs.

11.3.1 Understanding Observability: Beyond the Rank Condition

At this point, we can compute observability matrices and apply the rank test. We have seen that the condition

$$\text{rank}\,\mathbf{Q_O} = \text{rank}\begin{bmatrix}\mathbf{C}\\ \mathbf{CA}\\ \vdots \\ \mathbf{CA}^{n-1}\end{bmatrix} = \text{n}$$

determines whether a system is observable. But a deeper question remains: **What does this condition really tell us?** Why should the rank of this specific matrix determine our ability to infer the internal state?

The answer lies in recognizing that observability is fundamentally about the **relationship** between system dynamics and available measurements. It is not whether internal states exist, but whether **we can figure out what those states are** from what we can measure. The same system may be observable or unobservable depending entirely on which outputs we can access.

Observability: What Can You Really Tell About Someone?

Observability isn't about whether a system has internal states—it's about whether **you**, with **your specific measurements**, can determine what those states are.

The Setup: Aashay is an expressive teenager experiencing some internal emotional state. The question isn't "Does Aashay have emotions?" but "Can **you** determine what he's feeling?"

Rich Sensor Suite—Face-to-Face (Observable): You observe facial expressions, body language, voice tone, and eye contact. Even when Aashay says "I'm fine," you deduce he's stressed. With this rich sensor configuration, the internal state is **observable**—you can reconstruct it from available measurements.

Minimal Sensors—Text Only (Unobservable): You text: "How are you?" He replies: "I'm fine." With only text content as your sensor, you cannot determine if he's genuinely fine, stressed, sad, or excited. Different internal states produce identical outputs. The state is **unobservable**.

Enhanced Sensors—Text with Metadata (Observable Again): Now suppose you upgrade your sensor suite to capture not just text content, but also response time, typing speed, and punctuation patterns. The same "I'm fine" message now reveals: instant reply (unusual), terse period (normally casual), rapid typing (high arousal). With these additional sensor channels, you can reconstruct his stressed state.

The **system (Aashay's emotional dynamics) hasn't changed**—same **A** matrix. What changed is your **C** matrix: you added sensor channels that capture additional state information. Observability is restored simply by enriching the measurement configuration.

The Key Insight: Observability is a property of the **(system, sensor) pair**. The same system can be observable with one sensor configuration but unobservable with another. **Just as an uncontrollable system can be made controllable by relocating actuators, an unobservable system can be made observable by adding sensors, repositioning them, or measuring different aspects of the output.**

Mathematical Translation: For $\dot{\mathbf{x}} = \mathbf{A}\mathbf{x} + \mathbf{B}\mathbf{u}$, $\mathbf{y} = \mathbf{C}\mathbf{x}$:

- $\mathbf{x}$: Internal state (Aashay's emotions: stress, happiness, energy)
- $\mathbf{y}$: Measurements (words, facial cues, physiology, metadata)
- $\mathbf{C}$: Which state aspects "leak out" into your sensor suite

The observability matrix $\mathcal{O}$ tests whether information leaking through $\mathbf{C}$, combined with dynamics $\mathbf{A}$, suffices to reconstruct all state dimensions. Adding sensors increases the row dimension of $\mathbf{C}$, potentially restoring full rank to $\mathcal{O}$.

The Lesson: Before asking "Is this observable?," ask "Observable through **which sensor configuration**?" Observability depends on the **(system, sensor) pair**, not the system alone.

CONCEPT REINFORCEMENT PROBLEM 11.1

Problem:

Consider the system given by

$$\begin{aligned}\dot{x}_1 &= ax_1 + bu\\ \dot{x}_2 &= -bx_1 + bx_2 - bu\\ y &= b(x_1 + x_2)\end{aligned}$$

Determine the conditions of controllability and observability.

EXAMPLE 11.3

Problem:

Consider the system given by

$$\dot{\mathbf{x}} = \begin{bmatrix} a & -a \\ -a & a \end{bmatrix} \mathbf{x} + \begin{bmatrix} k_1 \\ k_2 \end{bmatrix} u, \qquad y = \begin{bmatrix} \beta_1 & \beta_2 \end{bmatrix} \mathbf{x}$$

Determine the conditions of controllability and observability.

SOLUTION:

Now from the state equations, the system matrix **A**, input matrix **B** and the output matrix **C** is

$$\mathbf{A} = \begin{bmatrix} a & -a \\ -a & a \end{bmatrix}, \quad \mathbf{B} = \begin{bmatrix} k_1 \\ k_2 \end{bmatrix} \quad C = \begin{bmatrix} \beta_1 & \beta_2 \end{bmatrix}$$

The controllability matrix $\mathbf{Q_C}$ is computed as:

$$\mathbf{Q_C} = \begin{bmatrix} \mathbf{B} & \mathbf{AB} \end{bmatrix} = \begin{bmatrix} k_1 & a(k_1 - k_2) \\ k_2 & a(k_2 - k_1) \end{bmatrix}$$

The determinant of $\mathbf{Q_C}$ is

$$\begin{aligned} \det(\mathbf{Q_C}) &= ak_1(k_2 - k_1) - ak_2(k_1 - k_2) \\ &= a(k_2^2 - k_1^2) \end{aligned}$$

Thus, the system is controllable, provided $k_1^2 \neq k_2^2$.

The observability matrix $\mathbf{Q_O}$ is computed as:

$$\mathbf{Q_O} = \begin{bmatrix} \mathbf{C} \\ \mathbf{CA} \end{bmatrix} = \begin{bmatrix} \beta_1 & \beta_2 \\ a(\beta_1 - \beta_2) & a(\beta_2 - \beta_1) \end{bmatrix}$$

The determinant of $\mathbf{Q_O}$ is

$$\begin{aligned} \det(\mathbf{Q_O}) &= a\beta_1(\beta_2 - \beta_1) - a\beta_2(\beta_1 - \beta_2) \\ &= a(\beta_2^2 - \beta_1^2) \end{aligned}$$

Thus, the system is observable, provided $\beta_1^2 \neq \beta_2^2$

CONCEPT REINFORCEMENT PROBLEM 11.2

Problem:

For the system shown in Figure 11.2, determine the values of parameter "a" which will make the system uncontrollable. Assume $K_1, K_2 > 0$.

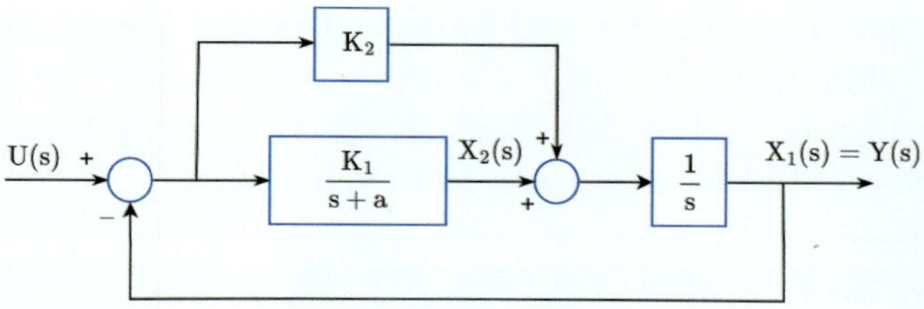

Figure 11.2: Feedback control system

11.3.2 The Independence of Controllability and Observability

At this point, we have established separate tests for controllability and observability. A natural question arises: are these properties related? If a system is controllable, must it also be observable? The answer is definitively no. Controllability and observability are independent properties—a system can possess one without the other, both, or neither.

This independence stems from a fundamental asymmetry: controllability depends on the $(\mathbf{A}, \mathbf{B})$ pair (how inputs couple to state dynamics), while observability depends on the $(\mathbf{A}, \mathbf{C})$ pair (how states map to measurements). Since $\mathbf{B}$ and $\mathbf{C}$ are structurally independent matrices, there is no reason to expect the two properties to coincide.

While this can be verified algebraically through rank tests, the intuition behind this independence is best understood through concrete examples. We present two analogies that illustrate the four possible combinations, focusing on the two surprising cases where the properties diverge.

Observable but Not Controllable: Aashay as a Teenager

Consider Aashay, who is now a teenager: you can clearly read his internal state through facial expressions, body language, and tone of voice. When he says "I'm fine," you know immediately he's upset. Your intuition provides rich sensors—the output matrix $\mathbf{C}$ has full rank, and his emotions "leak out" through multiple observable channels. The system is **observable**.

But can you control him? You say "Go to bed!" and he keeps playing. You say "Eat your vegetables!" and he refuses. You say "Stop being upset!" and he withdraws further. Your commands (input matrix $\mathbf{B}$) have weak coupling to his behavioral states. Certain modes—stubbornness, curiosity, emotional responses—evolve according to their own internal dynamics (**A**), largely immune to your inputs. The system is **not controllable**.

Why this happens: Aashay has autonomous dynamics driven by his own will, preferences, and biological needs. Your actuator (**B**) is limited—like trying to steer a spacecraft with only one thruster, you can influence some directions but not others. Many internal states lie in the **uncontrollable subspace** from your inputs.

Controllable but Not Observable: Your Laptop's Operating System

Consider your laptop's operating system: you have extensive control through keyboard and mouse inputs. You can launch programs, modify files, adjust settings, shut down processes. Your input matrix **B** has full rank—you can steer the system to any desired state through appropriate command sequences. The system is **controllable**.

But can you observe what's happening internally? The CPU is executing millions of instructions per second, memory is being allocated and freed, background processes are running, thermal management is adjusting clock speeds. Yet your only outputs are what appears on screen and perhaps some simple indicators (battery level, WiFi status). The output matrix **C** is severely rank-deficient—vast portions of the internal state remain hidden. The system is **not observable**.

Why this happens: The designers chose to expose only a minimal set of outputs to the user. While the system's internal state space is enormous (billions of possible states across registers, memory, disk I/O), the measurement vector $\mathbf{y} = \mathbf{C}\mathbf{x}$ projects this onto a tiny subspace—essentially just pixel values and a few status indicators. Most internal states lie in the **unobservable subspace**.

The engineering insight: This is often intentional. Full observability would overwhelm users with information and expose security vulnerabilities. Designers provide controllability (you can make the system do what you need) while hiding complexity through limited observability.

These analogies reveal why the independence of controllability and observability matters in practice. A system can be:

- Both controllable and observable (ideal for feedback control design)
- Observable but not controllable (you can diagnose problems but cannot fix them—monitoring systems, medical diagnostics)

- Controllable but not observable (you can command the system but cannot verify the result—many embedded systems, security-critical applications)
- Neither controllable nor observable (typically indicates poor actuator/sensor placement or fundamental structural limitations)

Practical Implication

- When designing a control system, actuator placement (determining **B**) and sensor placement (determining **C**) are independent design problems that must both be solved.
- Controllability ensures you can reach desired states; observability ensures you can measure whether you got there. Both are necessary for effective feedback control.

11.4 Frequency Domain Criteria for Testing Controllability and Observability

We studied the conditions of controllability and observability in the time domain using the state space models of the systems. The question arises whether there exists any frequency domain criteria to test the controllability and observability of a linear time-invariant system. Can the transfer function of a system provide information about these properties? The answer is given in the following theorem.

Theorem 11.3. If there are no pole-zero cancellations in the input-output transfer function of a single-input single-output linear time-invariant system, the system is both controllable and observable i.e., it is always possible to represent this system as a completely controllable and observable system. If the transfer function has pole-zero cancellation, then the system is either uncontrollable or unobservable, or both, depending on how the state variables are defined.

The physical interpretation is that pole-zero cancellations represent dynamics that are either not controllable from the input, or not observable from the output, or both. These "hidden modes" exist in the system but don't appear in the input-output transfer function relationship.

We will illustrate this concept through examples without the proof of this theorem.

EXAMPLE 11.4

Problem:

Consider a system

$$\dot{\mathbf{x}} = \mathbf{A}\mathbf{x} + \mathbf{B}\mathrm{u}, \quad \text{and} \quad \mathrm{y} = \mathbf{C}\mathbf{x} \quad \text{where}$$

$$\mathbf{A} = \begin{bmatrix} -2 & -7 \\ 0 & -3 \end{bmatrix}, \quad \mathbf{B} = \begin{bmatrix} 10 \\ 5 \end{bmatrix} \quad \mathbf{C} = \begin{bmatrix} 0 & 4 \end{bmatrix}$$

Investigate the controllability and observability property of the system. Compute the transfer function of the system and comment on your observation.

SOLUTION:

The controllability matrix $\mathbf{Q_C} = \begin{bmatrix} \mathbf{B} & \mathbf{AB} \end{bmatrix} = \begin{bmatrix} 10 & -55 \\ 5 & -15 \end{bmatrix}$.

The determinant of $\mathbf{Q_C}$ is 125. Thus, the system is **controllable.**

The observability matrix $\mathbf{Q_O} = \begin{bmatrix} \mathbf{C} \\ \mathbf{CA} \end{bmatrix} = \begin{bmatrix} 0 & 4 \\ 0 & -12 \end{bmatrix}$

The determinant of $\mathbf{Q_O}$ is 0. Thus, the system is **unobservable.**

Computation of Transfer Function:

The transfer function of the system is computed as:

$$\begin{aligned}
\mathrm{T(s)} &= \mathbf{C}(\mathrm{s}\mathbf{I} - \mathbf{A})^{-1}\mathbf{B} + \mathrm{D} \\
&= \begin{bmatrix} 0 & 4 \end{bmatrix} \left(\begin{bmatrix} \mathrm{s} & 0 \\ 0 & \mathrm{s} \end{bmatrix} - \begin{bmatrix} -2 & -7 \\ 0 & -3 \end{bmatrix} \right)^{-1} \begin{bmatrix} 10 \\ 5 \end{bmatrix} \\
&= \begin{bmatrix} 0 & 4 \end{bmatrix} \begin{bmatrix} \mathrm{s}+2 & 7 \\ 0 & \mathrm{s}+3 \end{bmatrix}^{-1} \begin{bmatrix} 10 \\ 5 \end{bmatrix} \\
&= \begin{bmatrix} 0 & 4 \end{bmatrix} \frac{1}{(\mathrm{s}+2)(\mathrm{s}+3)} \begin{bmatrix} \mathrm{s}+3 & -7 \\ 0 & \mathrm{s}+2 \end{bmatrix} \begin{bmatrix} 10 \\ 5 \end{bmatrix} \\
&= \frac{20\mathrm{s}+40}{(\mathrm{s}+2)(\mathrm{s}+3)} = \frac{20(\mathrm{s}+2)}{(\mathrm{s}+2)(\mathrm{s}+3)}
\end{aligned}$$

Observe that there is a pole-zero cancellation which confirms Theorem 11.3.

CONCEPT REINFORCEMENT PROBLEM 11.3

Problem:

Consider the system given by

$$\dot{\mathbf{x}} = \begin{bmatrix} 2 & 3 \\ 6 & -1 \end{bmatrix} \mathbf{x} + \begin{bmatrix} 1 \\ -2 \end{bmatrix} \mathrm{u}, \quad \mathrm{y} = \begin{bmatrix} 2 & 0 \end{bmatrix} \mathbf{x}$$

Investigate the controllability and observability property of the system. Compute the transfer function of the system and comment on your observation.

EXAMPLE 11.5

Problem

Consider the circuit shown in Figure 11.3.

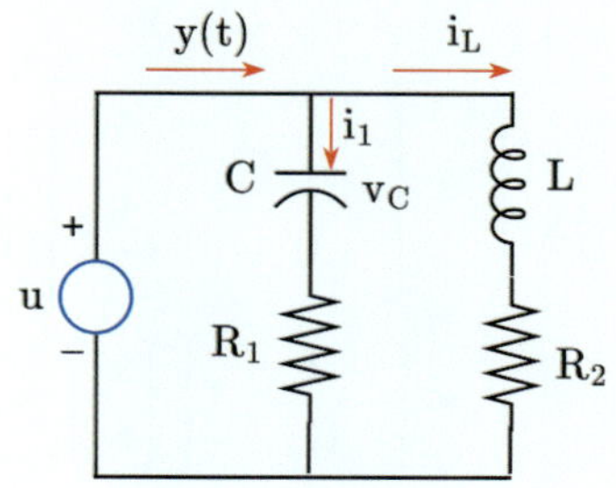

Figure 11.3: Two loop electrical network

a. Obtain the state variable model of the system, taking the current through the inductor and voltage across the capacitor as state variables. The current from the source voltage is considered as output.
b. Find the conditions on R_1, R_2, C, and L which causes the system to be uncontrollable. Further, find similar conditions for the system to be unobservable.
c. Explain the physical significance of losing either controllability or observability.
d. Using the condition of uncontrollability, compute the transfer function of the system and show that this conforms with the frequency domain conditions of controllability and observability.

SOLUTION:

a. Given $x_1 = v_C$ and $x_2 = i_L$. Applying KVL to the first loop gives

$$R_1C\dot{x}_1 + x_1 = u \implies \dot{x}_1 = -\frac{1}{R_1C}x_1 + \frac{1}{R_1C}u \tag{11.6}$$

Similarly, applying KVL to the second loop, we get

$$L\dot{x}_2 + R_2x_2 = u \implies \dot{x}_2 = -\frac{R_2}{L}x_2 + \frac{1}{L}u \tag{11.7}$$

The output is given by

$$y = i_1 + i_L = C\dot{x}_1 + x_2 \tag{11.8}$$

Substituting the value of $\dot{x}_1$ from Eqn(11.6) into Eqn(11.8) gives

$$y = -\frac{1}{R_1}x_1 + x_2 + \frac{1}{R_1}u \tag{11.9}$$

In matrix form Eqns(11.6), (11.7), and (11.8) can be expressed as:

$$\begin{bmatrix}\dot{x}_1\\ \dot{x}_2\end{bmatrix} = \begin{bmatrix}-\dfrac{1}{R_1C} & 0\\ 0 & -\dfrac{R_2}{L}\end{bmatrix}\begin{bmatrix}x_1\\ x_2\end{bmatrix} + \begin{bmatrix}\dfrac{1}{R_1C}\\ \dfrac{1}{L}\end{bmatrix}u$$

$$y = \begin{bmatrix}-\dfrac{1}{R_1} & 1\end{bmatrix}\begin{bmatrix}x_1\\ x_2\end{bmatrix} + \frac{1}{R_1}u \tag{11.10}$$

b. Let us form the controllability matrix

$$Q_C = \begin{bmatrix}B & AB\end{bmatrix} = \begin{bmatrix}\dfrac{1}{R_1C} & -\dfrac{1}{(R_1C)^2}\\ \dfrac{1}{L} & -\dfrac{R_2}{L^2}\end{bmatrix}$$

$\det(Q_C) = -\dfrac{R_2}{R_1CL^2} + \dfrac{1}{L(R_1C)^2}$. For the system to be uncontrollable

$$\det(Q_C) = 0 \implies R_1R_2C = L \text{ or } \frac{1}{R_1C} = \frac{R_2}{L}$$

Next, we form the observability matrix

$$Q_O = \begin{bmatrix}C\\ CA\end{bmatrix} = \begin{bmatrix}-\dfrac{1}{R_1} & 1\\ \dfrac{1}{R_1^2C} & -\dfrac{R_2}{L}\end{bmatrix}$$

For the system to be unobservable $\det(Q_O) = 0$. This gives

$$\frac{R_2}{R_1 L} - \frac{1}{R_1^2 C} = 0 \implies R_1 R_2 C = L \text{ or } \frac{1}{R_1 C} = \frac{R_2}{L}$$

c. We found, when the system is uncontrollable/unobservable $\frac{1}{R_1 C} = \frac{R_2}{L}$. The system dynamics become

$$\begin{bmatrix} \dot{x}_1 \\ \dot{x}_2 \end{bmatrix} = \begin{bmatrix} -\frac{1}{R_1 C} & 0 \\ 0 & -\frac{1}{R_1 C} \end{bmatrix} \begin{bmatrix} x_1 \\ x_2 \end{bmatrix} + \begin{bmatrix} \frac{1}{R_1 C} \\ \frac{1}{L} \end{bmatrix} u$$

$$y = \begin{bmatrix} -\frac{1}{R_1} & 1 \end{bmatrix} \begin{bmatrix} x_1 \\ x_2 \end{bmatrix} + \frac{1}{R_1} u \tag{11.11}$$

The system matrix is diagonal with two modes with the same time constant, hence cannot be changed independently using only one control u, i.e., it is not controllable. The system output is a linear combination of modes having the same frequency. Therefore, we cannot determine the composition of the combination, i.e., it is not observable.

d. With the conditions of uncontrollability and unobservability, we can compute the transfer function (left as an exercise), which gives

$$\begin{aligned} G(s) &= \frac{(s + R_2/L)(s + R_1/L)}{R_1(s + 1/R_1 C)(s + R_2/L)} \\ &= \frac{(s + R_1/L)}{R_1(s + 1/R_1 C)} \end{aligned}$$

Note the pole-zero cancellation in the transfer function. It therefore conforms to the frequency domain condition of controllability and observability.

CONCEPT REINFORCEMENT PROBLEM 11.4

Problem

Consider the electric circuit shown in Figure 11.4

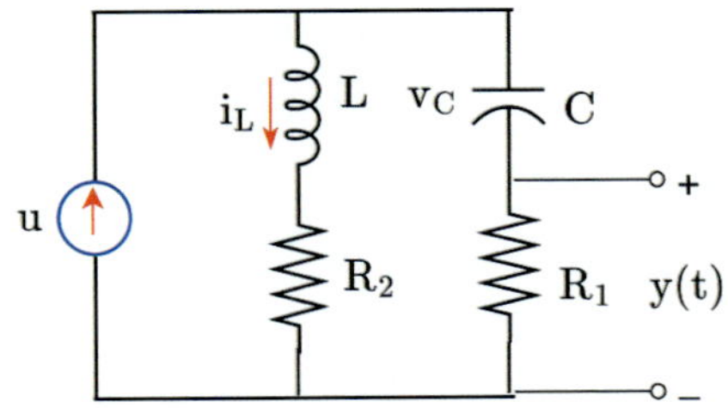

Figure 11.4: Two loop electrical network

a. Obtain the state variable model of the system, taking the current through the inductor and voltage across the capacitor as state variables. The current through the inductor is considered the output, and the input is the current.

b. Find the condition on R, L, and C which ensures controllability of the system.

c. What conditions on R, L, and C will guarantee observability of the system?

11.5 Output Controllability

In practice, we often desire to control the output instead of the state of the system. Note that complete state controllability is neither necessary nor sufficient for controlling the output. Therefore, we should investigate the output controllability property of a system.

11.5.1 Definition of Output Controllability

Consider a linear time-invariant (LTI) system with $\mathbf{n}$-number of states and $\mathbf{m}$-number of inputs and $\mathbf{p}$ number of outputs described by

$$\begin{aligned}\dot{\mathbf{x}}(t) &= \mathbf{A}\mathbf{x}(t) + \mathbf{B}\mathbf{u}(t) \\ \mathbf{y}(t) &= \mathbf{C}\mathbf{x}(t) + \mathbf{D}\mathbf{u}(t) \qquad (11.12)\end{aligned}$$

where the state vector $\mathbf{x} \in \mathbb{R}^{\mathbf{n}}$, the input vector $\mathbf{u} \in \mathbb{R}^{\mathbf{m}}$ and the output vector $\mathbf{y} \in \mathbb{R}^{\mathbf{p}}$. The system matrix $\mathbf{A} \in \mathbb{R}^{\mathbf{n}\times\mathbf{n}}$, the input matrix $\mathbf{B} \in \mathbb{R}^{\mathbf{n}\times\mathbf{m}}$, the output matrix $\mathbf{C} \in \mathbb{R}^{\mathbf{p}\times\mathbf{n}}$, and the transmission matrix $\mathbf{D} \in \mathbb{R}^{\mathbf{p}\times\mathbf{m}}$.

Definition 11.3. The system described in Eqn(11.12) is said to be completely output controllable if it is possible to construct an admissible control input $\mathbf{u}(t)$ which will transfer any given initial output $\mathbf{y}(t_0)$ to any final output $\mathbf{y}(t_1)$ in a finite time interval $t_0 \leq t \leq t_1$.

Output Controllability Condition:

The following theorem provides the conditions for output controllability of linear systems.

Theorem 11.4. The system described in Eqn(11.12) is completely output controllable if and only if the $(p \times (nm + m))$ matrix

$$\begin{bmatrix} \mathbf{CB} & \mathbf{CAB} & \mathbf{CA}^2\mathbf{B} & \ldots & \mathbf{CA}^{n-1}\mathbf{B} & \mathbf{D} \end{bmatrix} \tag{11.13}$$

has rank p.

Key Point

The system of Eqn(11.12) is output controllable if the transmission matrix **D** has full rank (rank(D)=p). For a single-output system, it should be non-zero.

Full-rank D guarantees output controllability.

But systems can still be output controllable even with rank-deficient or zero D (through the state dynamics terms in the controllability matrix)

EXAMPLE 11.6

Consider the system

$$\dot{\mathbf{x}} = \mathbf{A}\mathbf{x} + \mathbf{B}\mathrm{u} \quad \text{and } \mathrm{y} = \mathbf{C}\mathbf{x} + \mathrm{Du} \text{ where}$$

$$\mathbf{A} = \begin{bmatrix} a & b \\ 0 & c \end{bmatrix}, \quad \text{and } \mathbf{B} = \begin{bmatrix} 1 \\ 0 \end{bmatrix}$$

Investigate the state and output controllability of the system when

a. $\mathbf{C} = \begin{bmatrix} \alpha & 0 \end{bmatrix}$, $D = 0$
b. $\mathbf{C} = \begin{bmatrix} 0 & \beta \end{bmatrix}$, $D = 0$
c. $\mathbf{C} = \begin{bmatrix} 0 & \beta \end{bmatrix}$ and $D = \gamma$

SOLUTION:

The controllability matrix $\mathbf{Q_C}$ is computed as:

$$\mathbf{Q_C} = \begin{bmatrix} \mathbf{B} & \mathbf{AB} \end{bmatrix} = \begin{bmatrix} 1 & a \\ 0 & 0 \end{bmatrix}$$

The rank of $\mathbf{Q_C}$ is 1. Therefore, the system is **not completely state controllable**.

Let us compute the output controllability matrix $\mathbf{Q_{OC}}$ as follows:

a. For this case, the output matrix $\mathbf{C} = \begin{bmatrix} \alpha & 0 \end{bmatrix}$. Hence

$$\mathbf{CB} = \begin{bmatrix} \alpha & 0 \end{bmatrix} \begin{bmatrix} 1 \\ 0 \end{bmatrix} = \alpha$$

and

$$\mathbf{CAB} = \begin{bmatrix} \alpha & 0 \end{bmatrix} \begin{bmatrix} \mathrm{a} & \mathrm{b} \\ 0 & \mathrm{c} \end{bmatrix} \begin{bmatrix} 1 \\ 0 \end{bmatrix} = \begin{bmatrix} \alpha & 0 \end{bmatrix} \begin{bmatrix} \mathrm{a} \\ 0 \end{bmatrix} = \mathrm{a}\alpha$$

$$\mathbf{Q_{OC}} = \begin{bmatrix} \mathbf{CB} & \mathbf{CAB} \end{bmatrix} = \begin{bmatrix} \alpha & \mathrm{a}\alpha \end{bmatrix}$$

The rank of the $\mathbf{Q_{OC}}$ equals to 1 i.e., it is equal to the number of outputs. Hence, the system is **output controllable.**

b. In this case, the output matrix is changed as $\mathbf{C} = \begin{bmatrix} 0 & \beta \end{bmatrix}$. Hence

$$\mathbf{CB} = \begin{bmatrix} 0 & \beta \end{bmatrix} \begin{bmatrix} 1 \\ 0 \end{bmatrix} = 0$$

and

$$\mathbf{CAB} = \begin{bmatrix} 0 & \beta \end{bmatrix} \begin{bmatrix} \mathrm{a} & \mathrm{b} \\ 0 & \mathrm{c} \end{bmatrix} \begin{bmatrix} 1 \\ 0 \end{bmatrix} = \begin{bmatrix} 0 & \beta \end{bmatrix} \begin{bmatrix} \mathrm{a} \\ 0 \end{bmatrix} = 0$$

The output controllability matrix $\mathbf{Q_{OC}}$ therefore becomes

$$\mathbf{Q_{OC}} = \begin{bmatrix} \mathbf{CB} & \mathbf{CAB} \end{bmatrix} = \begin{bmatrix} 0 & 0 \end{bmatrix}$$

The rank of the $\mathbf{Q_{OC}}$ equals to 0. Hence the system is **not output controllable.**

c. For this case, the transmission matrix $\mathbf{D} = \gamma$. Thus, the output controllability matrix becomes

$$\mathbf{Q_{OC}} = \begin{bmatrix} \mathbf{CB} & \mathbf{CAB} & \mathrm{D} \end{bmatrix} = \begin{bmatrix} 0 & 0 & \gamma \end{bmatrix}$$

The rank of the $\mathbf{Q_{OC}}$ equals to 1. Hence, the system is **output controllable.**. Note that a non-zero value of the transmission matrix makes the output controllable.

11.6 Alternate Conditions of Controllability

Consider the system defined by

$$\begin{aligned} \dot{\mathbf{x}} &= \mathbf{Ax} + \mathbf{Bu} \\ \mathbf{y} &= \mathbf{Cx} + \mathbf{Du} \end{aligned} \tag{11.14}$$

where the state vector $\mathbf{x} \in \mathbb{R}^n$, input vector $\mathbf{u} \in \mathbb{R}^m$, output vector $\mathbf{y} \in \mathbb{R}^p$, system matrix $\mathbf{A} \in \mathbb{R}^{n\times n}$, input matrix $\mathbf{B} \in \mathbb{R}^{n\times m}$, output matrix $\mathbf{C} \in \mathbb{R}^{p\times n}$, and transmission matrix $\mathbf{D} \in \mathbb{R}^{p\times m}$.

When the system matrix $\mathbf{A}$ has n linearly independent eigenvectors (i.e., $\mathbf{A}$ is diagonalizable), it is possible to find a transformation matrix $\mathbf{P}$ whose columns are the eigenvectors of $\mathbf{A}$. This transformation converts the system into diagonal form:

$$\begin{aligned} \dot{\mathbf{z}} &= \mathbf{\Lambda z} + \overline{\mathbf{B}}\mathbf{u} \\ \mathbf{y} &= \overline{\mathbf{C}}\mathbf{z} + \mathbf{Du} \end{aligned} \tag{11.15}$$

where $\mathbf{x} = \mathbf{Pz}$, and:

- $\mathbf{\Lambda} = \mathbf{P^{-1}AP}$ is the diagonal matrix of eigenvalues
- $\overline{\mathbf{B}} = \mathbf{P^{-1}B}$ is the transformed input matrix
- $\overline{\mathbf{C}} = \mathbf{CP}$ is the transformed output matrix

The matrices have the form:

$$\mathbf{\Lambda} = \mathbf{P^{-1}AP} = \begin{bmatrix} \lambda_1 & 0 & \dots & 0 \\ 0 & \lambda_2 & \dots & 0 \\ \vdots & \vdots & \ddots & \vdots \\ 0 & 0 & \dots & \lambda_n \end{bmatrix}$$

$$\overline{\mathbf{B}} = \mathbf{P^{-1}B} = \begin{bmatrix} \overline{b}_{11} & \overline{b}_{12} & \dots & \overline{b}_{1m} \\ \overline{b}_{21} & \overline{b}_{22} & \dots & \overline{b}_{2m} \\ \vdots & \vdots & \ddots & \vdots \\ \overline{b}_{n1} & \overline{b}_{n2} & \dots & \overline{b}_{nm} \end{bmatrix}$$

where λ_i $(i = 1, 2, \dots, n)$ are the eigenvalues of matrix $\mathbf{A}$.

Important Note:

Diagonalizability requires n linearly independent eigenvectors. This is guaranteed when eigenvalues are distinct, but can also occur with repeated eigenvalues if their geometric multiplicity equals their algebraic multiplicity.

Theorem 11.5. If the system matrix $\mathbf{A}$ is diagonalizable, then the system is completely state controllable if and only if no row of $\mathbf{P}^{-1}\mathbf{B}$ consists entirely of zeros.
Equivalent Statement: For each $\mathrm{i} = 1, 2, \ldots, \mathrm{n}$, at least one element $\bar{\mathrm{b}}_{\mathrm{ij}}$ $(\mathrm{j} = 1, 2, \ldots, \mathrm{m})$ must be non-zero.

Physical Interpretation:

- If all elements in row i of $\mathbf{P}^{-1}\mathbf{B}$ are zero ($\bar{\mathrm{b}}_{\mathrm{ij}} = 0$ for all j), then the i-th diagonal state $\mathrm{z_i}$ is uncontrollable.
- This corresponds to an eigenmode with eigenvalue λ_i that cannot be influenced by any control input.
- The original system becomes uncontrollable because this eigenmode cannot be controlled.

Note: This test provides an intuitive geometric interpretation: each eigenmode must be reachable by at least one control input for complete controllability.

EXAMPLE 11.7

Consider the system with diagonal canonical representation as:

$$\begin{bmatrix} \dot{\mathrm{x}}_1 \\ \dot{\mathrm{x}}_2 \end{bmatrix} = \begin{bmatrix} -3 & 0 \\ 0 & -5 \end{bmatrix} \begin{bmatrix} \mathrm{x}_1 \\ \mathrm{x}_2 \end{bmatrix} + \begin{bmatrix} 0.5 \\ 1 \end{bmatrix} \mathbf{u} = \mathbf{Ax} + \mathbf{Bu}$$

Since all the elements of the input matrix are non-zero, this system is completely state controllable. Verify this by computing the rank of the controllability matrix (left as an exercise).

Consider the same system with a different input matrix described as:

$$\begin{bmatrix} \dot{\mathrm{x}}_1 \\ \dot{\mathrm{x}}_2 \end{bmatrix} = \begin{bmatrix} -3 & 0 \\ 0 & -5 \end{bmatrix} \begin{bmatrix} \mathrm{x}_1 \\ \mathrm{x}_2 \end{bmatrix} + \begin{bmatrix} 0.5 \\ 0 \end{bmatrix} \mathbf{u} = \mathbf{Ax} + \mathbf{Bu}$$

Since not all the elements of the input matrix are non-zero, this system is not completely state controllable.

The question arises, which of the states is uncontrollable ?. This can be determined by inspecting the input matrix of the diagonal system. Since, in this case, the second element of the input matrix is zero, the state x_2 is uncontrollable.

Let us demonstrate this by considering a few more systems.

$$\mathbf{S_1}: \begin{bmatrix} \dot{x}_1 \\ \dot{x}_2 \\ \dot{x}_3 \end{bmatrix} = \begin{bmatrix} -1 & 0 & 0 \\ 0 & -3 & 0 \\ 0 & 0 & -6 \end{bmatrix} \begin{bmatrix} x_1 \\ x_2 \\ x_3 \end{bmatrix} + \begin{bmatrix} 0.5 \\ -0.1 \\ 0 \end{bmatrix} \mathbf{u}$$

$$\mathbf{S_2}: \begin{bmatrix} \dot{x}_1 \\ \dot{x}_2 \\ \dot{x}_3 \\ \dot{x}_4 \end{bmatrix} = \begin{bmatrix} -2 & 0 & 0 & 0 \\ 0 & -4 & 0 & 0 \\ 0 & 0 & -1 & 0 \\ 0 & 0 & 0 & -9 \end{bmatrix} \begin{bmatrix} x_1 \\ x_2 \\ x_3 \\ x_4 \end{bmatrix} + \begin{bmatrix} 0.0 \\ 1.0 \\ 0.0 \\ 3.0 \end{bmatrix} \mathbf{u}$$

Since all the elements of the input matrices of **S1** and **S2** are not non-zero, both these systems uncontrollable. Following similar analysis as above, in **S1**, the state x_3 is uncontrollable and in **S2**, both x_1 and x_3 are uncontrollable. Verify this in MATLAB.

The next theorem addresses the controllability condition for systems with repeated eigenvalues.

Consider the system of Eqn(11.14) where the system matrix **A has repeated eigenvalues**. In this case, complete diagonalization is not possible; instead, the system can be transformed into Jordan canonical form. Suppose there exists a transformation matrix **S** such that the system of Eqn(11.14) is transformed into Jordan canonical form as:

$$\begin{aligned} \dot{\mathbf{z}} &= \mathbf{S}^{-1}\mathbf{A}\mathbf{S}\mathbf{z} + \mathbf{S}^{-1}\mathbf{B}\mathbf{u} = \mathbf{J}\mathbf{z} + \mathbf{S}^{-1}\mathbf{B}\mathbf{u} \\ \mathbf{y} &= \mathbf{C}\mathbf{S}\mathbf{z} + \mathbf{D}\mathbf{u} \end{aligned} \tag{11.16}$$

where $\mathbf{x} = \mathbf{S}\mathbf{z}$ and $\mathbf{S}^{-1}\mathbf{A}\mathbf{S} = \mathbf{J}$ is the Jordan canonical form.

The Jordan matrix **J** consists of Jordan blocks along the diagonal:

$$\mathbf{J} = \text{diag}(\mathbf{J}_1, \mathbf{J}_2, \ldots, \mathbf{J}_k)$$

where each Jordan block $\mathbf{J}_i$ has the form:

$$\mathbf{J}_i = \begin{bmatrix} \lambda_i & 1 & 0 & \cdots & 0 \\ 0 & \lambda_i & 1 & \cdots & 0 \\ \vdots & \vdots & \ddots & \ddots & \vdots \\ 0 & 0 & 0 & \cdots & 1 \\ 0 & 0 & 0 & \cdots & \lambda_i \end{bmatrix}$$

Theorem 11.6. If the system matrix **A** has repeated eigenvalues and is transformed to Jordan canonical form, then the system of Eqn(11.14) is completely state controllable, if and only if, for each Jordan block $\mathbf{J}_i$ in $\mathbf{J}$, the row of $\mathbf{S}^{-1}\mathbf{B}$ corresponding to the last row of that Jordan block has at least one non-zero element.

Interpretation

- In Jordan form, each Jordan block represents a chain of states. Controllability requires that control inputs can reach the "bottom" state of each Jordan chain. If the last row of a Jordan block in $\mathbf{S}^{-1}\mathbf{B}$ consists entirely of zeros, that entire Jordan chain becomes uncontrollable.
- Multiple Jordan blocks can correspond to the same eigenvalue when the geometric multiplicity of an eigenvalue is less than its algebraic multiplicity. The controllability condition must be satisfied for each individual Jordan block, regardless of whether they share the same eigenvalue.

EXAMPLE 11.8

Consider the system with Jordan canonical representation as:

$$\mathbf{S_1}: \begin{bmatrix} \dot{x}_1 \\ \dot{x}_2 \\ \dot{x}_3 \end{bmatrix} = \begin{bmatrix} -2 & 1 & 0 \\ 0 & -2 & 0 \\ 0 & 0 & -4 \end{bmatrix} \begin{bmatrix} x_1 \\ x_2 \\ x_3 \end{bmatrix} + \begin{bmatrix} 0 \\ 5 \\ 3 \end{bmatrix} \mathbf{u} = \mathbf{Ax} + \mathbf{B}u$$

This system has one Jordan block associated with the eigenvalue -2. Since the last element of the input matrix corresponding to the last row of this Jordan block is equal to 5 (non-zero) and other elements of the input matrix, corresponding to the distinct eigenvalue -4, are 3 (non-zero), this system is completely state controllable.

Let us investigate the controllability of a few more systems by inspection, which are represented by Jordan canonical form:

$$\mathbf{S_2}: \begin{bmatrix} \dot{x}_1 \\ \dot{x}_2 \\ \dot{x}_3 \end{bmatrix} = \begin{bmatrix} -4 & 1 & 0 \\ 0 & -4 & 0 \\ 0 & 0 & -5 \end{bmatrix} \begin{bmatrix} x_1 \\ x_2 \\ x_3 \end{bmatrix} + \begin{bmatrix} 2 \\ 0 \\ 3 \end{bmatrix} \mathbf{u}$$

The system $\mathbf{S_2}$ is uncontrollable; since the element of input matrix corresponding to the last row of the first Jordan block is zero.

$$\mathbf{S_3}: \begin{bmatrix} \dot{x}_1 \\ \dot{x}_2 \\ \dot{x}_3 \\ \dot{x}_4 \\ \dot{x}_5 \end{bmatrix} = \begin{bmatrix} -3 & 1 & 0 & & 0 \\ 0 & -3 & 1 & & \\ 0 & 0 & -3 & & \\ & & & -7 & 1 \\ & & & 0 & -7 \end{bmatrix} \begin{bmatrix} x_1 \\ x_2 \\ x_3 \\ x_4 \\ x_5 \end{bmatrix} + \begin{bmatrix} 0 & 1 \\ 0 & 0 \\ 3 & 0 \\ 0 & 0 \\ 1 & 2 \end{bmatrix} \mathbf{u}$$

System S_3 is controllable because the row corresponding to the last row of the first Jordan block has at least one non-zero element (3), and the row corresponding to the last row of the second Jordan block also has at least one non-zero element (both 1 and 2 are non-zero).

11.7 Alternate Conditions of Observability

The observability of a system depends on the system matrix and the output matrix. The input matrix is irrelevant for observability analysis. Therefore, consider the system defined by

$$\dot{\mathbf{x}} = \mathbf{A}\mathbf{x}, \ \mathbf{y} = \mathbf{C}\mathbf{x} \tag{11.17}$$

where the state vector $\mathbf{x} \in \mathbb{R}^{\mathrm{n}}$, the output $\mathbf{y} \in \mathbb{R}^{\mathrm{p}}$, the system matrix $\mathbf{A} \in \mathbb{R}^{\mathrm{n}\times\mathrm{n}}$ and the output matrix $\mathbf{C} \in \mathbb{R}^{\mathrm{p}\times\mathrm{n}}$.

Assume that the eigenvalues of $\mathbf{A}$ are distinct. Let us transform the system into diagonal form using the transformation matrix $\mathbf{P}$ such that

$$\mathbf{P}^{-1}\mathbf{A}\mathbf{P} = \mathbf{\Lambda} \tag{11.18}$$

where $\mathbf{\Lambda}$ is a diagonal matrix with eigenvalues λ_{i} $(\mathrm{i} = 1, 2, \ldots, \mathrm{n})$ of matrix $\mathbf{A}$ along its diagonal. The transformed system becomes

$$\dot{\mathbf{z}} = \mathbf{\Lambda}\mathbf{z}, \quad \mathbf{y} = \overline{\mathbf{C}}\mathbf{z} \quad \text{where} \quad \mathbf{x} = \mathbf{P}\mathbf{z}, \quad \overline{\mathbf{C}} = \mathbf{C}\mathbf{P} \tag{11.19}$$

The output can be expressed as:

$$\mathbf{y}(\mathrm{t}) = \mathbf{C}\mathbf{P}e^{\mathbf{\Lambda}t}\mathbf{z}(0) = \mathbf{C}\mathbf{P}\begin{bmatrix} \mathrm{e}^{\lambda_1 \mathrm{t}} & 0 & \ldots & 0 \\ 0 & \mathrm{e}^{\lambda_2 \mathrm{t}} & \ldots & 0 \\ \vdots & \vdots & \ddots & \vdots \\ 0 & 0 & \ldots & \mathrm{e}^{\lambda_\mathrm{n} \mathrm{t}} \end{bmatrix}\mathbf{z}(0) \tag{11.20}$$

Theorem 11.7. The system of Eqn(11.17) is completely observable if and only if no column of the $(\mathrm{p} \times \mathrm{n})$ matrix $\mathbf{C}\mathbf{P}$ consists entirely of zeros.

Note

If the i-th column of $\mathbf{CP}$ consists entirely of zeros, then the initial condition $z_{\mathrm{i}}(0)$ corresponding to eigenmode λ_{i} will not appear in any output equation. Therefore, this eigenmode cannot be determined from observations of the output $\mathbf{y}(\mathrm{t})$, making the system unobservable.

EXAMPLE 11.9

Consider the system $\mathbf{S_1}$ with diagonal canonical representation as:

$$\mathbf{S_1} \quad \begin{bmatrix}\dot{x}_1\\ \dot{x}_2\end{bmatrix} = \begin{bmatrix}-2 & 0\\ 0 & -3\end{bmatrix}\begin{bmatrix}x_1\\ x_2\end{bmatrix}, \quad \mathbf{y} = \begin{bmatrix}1 & 5\end{bmatrix}\begin{bmatrix}x_1\\ x_2\end{bmatrix}$$

Since all the elements of the output matrix are non-zero, this system is completely observable.

Consider the same system with a different output matrix described as:

$$\begin{bmatrix}\dot{x}_1\\ \dot{x}_2\end{bmatrix} = \begin{bmatrix}-2 & 0\\ 0 & -3\end{bmatrix} = \begin{bmatrix}x_1\\ x_2\end{bmatrix}, \quad \mathbf{y} = \begin{bmatrix}2 & 0\end{bmatrix} = \begin{bmatrix}x_1\\ x_2\end{bmatrix}$$

Since not all the elements of the output matrix are non-zero, this system is not completely observable.

In order to determine which of the states is unobservable, inspect the columns of the output matrix of the diagonal system. Since, in this case, the second column of the output matrix is zero, the state x_2 is unobservable.

Let us demonstrate this, considering another system whose dynamics are given by

$$\mathbf{S_2}: \quad \begin{bmatrix}\dot{x}_1\\ \dot{x}_2\\ \dot{x}_3\end{bmatrix} = \begin{bmatrix}-1 & 0 & 0\\ 0 & -4 & 0\\ 0 & 0 & -7\end{bmatrix}\begin{bmatrix}x_1\\ x_2\\ x_3\end{bmatrix}, \quad \mathbf{y} = \begin{bmatrix}0 & 1 & 0\end{bmatrix}\begin{bmatrix}x_1\\ x_2\\ x_3\end{bmatrix}$$

Since the first and third columns of the output matrix are zero, the system $\mathbf{S_2}$ is unobservable. Following a similar argument, we conclude that the states x_1 and x_3 are unobservable.

The next theorem addresses the observability condition for systems with repeated eigenvalues. Consider the system of Eqn(11.17) where the system matrix $\mathbf{A}$ has repeated eigenvalues. Let us transform the system matrix $\mathbf{A}$ into Jordan canonical form using the transformation matrix $\mathbf{S}$ such that

$$\mathbf{S}^{-1}\mathbf{AS} = \mathbf{J} \quad \text{and} \quad \mathbf{x} = \mathbf{Sz}$$

The system of Eqn(11.17) can be expressed as:

$$\begin{aligned}\dot{\mathbf{z}} &= \mathbf{Jz}\\ \mathbf{y} &= \mathbf{CSz}\\ \mathbf{y}(t) &= \mathbf{CS}e^{\mathbf{J}t}\mathbf{z}(0)\end{aligned} \tag{11.21}$$

The Jordan matrix $\mathbf{J}$ consists of Jordan blocks along the diagonal:

$$\mathbf{J} = \mathrm{diag}(\mathbf{J}_1, \mathbf{J}_2, \ldots, \mathbf{J}_\mathrm{k})$$

where each Jordan block $\mathbf{J}_i$ has the form:

$$\mathbf{J}_i = \begin{bmatrix} \lambda_\mathrm{i} & 1 & 0 & \cdots & 0 \\ 0 & \lambda_\mathrm{i} & 1 & \cdots & 0 \\ \vdots & \vdots & \ddots & \ddots & \vdots \\ 0 & 0 & 0 & \cdots & 1 \\ 0 & 0 & 0 & \cdots & \lambda_\mathrm{i} \end{bmatrix}$$

Theorem 11.8. If the system matrix $\mathbf{A}$ has repeated eigenvalues and is transformed to Jordan canonical form, then the system of Eqn(11.17) is completely observable if and only if:

> For each Jordan block $\mathbf{J}_\mathrm{i}$ in $\mathbf{J}$, the column of $\mathbf{CS}$ corresponding to the first row of that Jordan block has at least one non-zero element.

Interpretation

- In Jordan form, each Jordan block represents a chain of states. Observability requires that the output can detect the "top" state of each Jordan chain. If the column of $\mathbf{CS}$ corresponding to the first row of a Jordan block consists entirely of zeros, that entire Jordan chain becomes unobservable.
- Multiple Jordan blocks can correspond to the same eigenvalue when the geometric multiplicity is less than the algebraic multiplicity. The observability condition must be satisfied for each individual Jordan block, regardless of whether they share the same eigenvalue. This is the dual of the controllability condition, which requires the last row of each Jordan block to have at least one non-zero element in the transformed input matrix.

EXAMPLE 11.10

Consider the system with Jordan canonical representation as:

$$\mathbf{S_1} : \begin{bmatrix} \dot{\mathrm{x}}_1 \\ \dot{\mathrm{x}}_2 \\ \dot{\mathrm{x}}_3 \end{bmatrix} = \begin{bmatrix} -3 & 1 & 0 \\ 0 & -3 & 0 \\ 0 & 0 & -5 \end{bmatrix} \begin{bmatrix} \mathrm{x}_1 \\ \mathrm{x}_2 \\ \mathrm{x}_3 \end{bmatrix}, \begin{bmatrix} \mathrm{y}_1 \\ \mathrm{y}_2 \end{bmatrix} = \begin{bmatrix} 5 & 0 & 2 \\ 1 & 0 & 6 \end{bmatrix}$$

This system has one Jordan block associated with the eigenvalue -3. Notice that the first column of the output matrix corresponding to the first row of the Jordan block is non-zero. The last column of the output matrix is non-zero, which corresponds to the distinct eigenvalue, and this system is completely observable. However, by changing the output to

$$\begin{bmatrix} y_1 \\ y_2 \end{bmatrix} = \begin{bmatrix} 0 & 5 & 2 \\ 0 & 1 & 6 \end{bmatrix}$$

will make the system unobservable.

Similarly, consider another system

$$\mathbf{S_2}: \begin{bmatrix} \dot{x}_1 \\ \dot{x}_2 \\ \dot{x}_3 \\ \dot{x}_4 \\ \dot{x}_5 \end{bmatrix} = \begin{bmatrix} -2 & 1 & 0 & & \\ 0 & -2 & 1 & & \\ 0 & 0 & -2 & & \\ & & & -4 & 1 \\ 0 & & & 0 & -4 \end{bmatrix} \begin{bmatrix} x_1 \\ x_2 \\ x_3 \\ x_4 \\ x_5 \end{bmatrix}$$

$$\begin{bmatrix} y_1 \\ y_2 \end{bmatrix} = \begin{bmatrix} 6 & 2 & 1 & 0 & 0 \\ 0 & 3 & 1 & 8 & 0 \end{bmatrix} \begin{bmatrix} x_1 \\ x_2 \\ x_3 \\ x_4 \\ x_5 \end{bmatrix}$$

This system has two Jordan blocks associate with the eigenvalues -2 and -4. Notice that the first column of output matrix corresponding to first row of first Jordan block as well as the fourth column of output matrix corresponding to the first row of second Jordan block are non-zero. Hence, this system is completely observable. However, by changing the output to

$$\begin{bmatrix} y_1 \\ y_2 \end{bmatrix} = \begin{bmatrix} 0 & 2 & 1 & 0 & 4 \\ 0 & 3 & 1 & 0 & 6 \end{bmatrix}$$

will make the system unobservable.

EXAMPLE 11.11

This example will further demonstrate how to assess the controllability and observability of a system from its diagonal representation. Consider a system with the following system, input, output, and transmission matrices

$$\mathbf{A} = \begin{bmatrix} -3 & 0 & 0 & 0 & 0 \\ 0 & -5 & 0 & 0 & 0 \\ 0 & 0 & -8 & 0 & 0 \\ 0 & 0 & 0 & -4 & 0 \\ 0 & 0 & 0 & 0 & -1 \end{bmatrix}, \quad \mathbf{B} = \begin{bmatrix} 1 \\ 1 \\ 0 \\ 0 \\ 1 \end{bmatrix},$$
$$\mathbf{C} = \begin{bmatrix} 1 & 0 & 1 & 0 & 1 \end{bmatrix}, \quad \mathbf{D} = 0$$

Since **A** is a diagonal matrix, we can determine the controllability and observability properties by inspection. In this example, these are as follows:

x_1: Controllable and observable

x_2: Controllable, but unobservable

x_3: Uncontrollable, but observable

x_4: Uncontrollable and unobservable

x_5: Controllable and observable

The transfer function of this system is given by

$$\frac{Y(s)}{U(s)} = \mathbf{C}(s\mathbf{I} - \mathbf{A})^{-1}\mathbf{B} + \mathbf{D} = \frac{(s+1)(s+2)(s+5)(s+8)}{(s+1)(s+2)(s+3)(s+5)(s+8)}$$

Notice the pole-zero cancellation in the transfer function.

11.8 Possible Reasons of Uncontrollability and Unobservability

Often, we wonder what causes the system to be either uncontrollable or unobservable or both. There are many possible causes of losing these properties, which may include pole-zero cancellation, symmetry in the system. The most common reason is due to redundant state variables. This is because most of the systems are often complex, and while representing these systems by state variable models, the designer may include extra and unnecessary state variables. This may lead to loss of controllability or observability or both.

Understanding Loss of Controllability and Observability

1. **The Redundant Coordinates Analogy:** Imagine describing the position of a car driving on a straight highway:

   - You could use just the distance traveled (one state variable)

- Or you could redundantly describe both x and y coordinates, even though y remains constant
- The y coordinate is neither controllable (steering doesn't change it on a straight road) nor observable (measuring forward motion tells you nothing about y)
- This redundant description creates uncontrollability and unobservability

2. **The Pole-Zero Cancellation Analogy:** Think of a mechanical system with a mass-spring-damper:

    - You add a vibration absorber that perfectly cancels out a particular resonance frequency
    - Now you cannot excite that frequency (uncontrollable) nor detect it in measurements (unobservable)
    - The cancellation "hides" this mode from both input and output

3. **The Symmetric Door (Symmetry):** Consider pushing a perfectly centered revolving door panel.

    - It doesn't rotate at all.
    - Your force cancels itself out.
    - And if you attach a sensor right in the middle, it can't measure rotation either.
    - Symmetry deletes both control and observation.

The key insight: Uncontrollability and unobservability often arise not from fundamental physical limitations, but from *how we choose to describe the system*—through redundant states, unfortunate cancellations, or failing to account for symmetries that makes certain dynamics **ignore us** (uncontrollable) or **hide from us** (unobservable).

The following examples demonstrate the effects of non-minimal state space representation on controllability and observability of a system.

EXAMPLE 11.12

Consider the simple RLC-circuit shown in Figure 9.3. Let the input $u(t) = v(t)$ and the output $y(t) = v_c(t)$. Defining the state variables as $x_1 = i(t)$ and $x_2 = v_c(t)$, the circuit dynamics in state space can be expressed as:

$$\begin{bmatrix} \dot{x}_1 \\ \dot{x}_2 \end{bmatrix} = \begin{bmatrix} -\frac{R}{L} & -\frac{1}{L} \\ \frac{1}{C} & 0 \end{bmatrix} \begin{bmatrix} x_1 \\ x_2 \end{bmatrix} + \begin{bmatrix} \frac{1}{L} \\ 0 \end{bmatrix} u, \quad y = \begin{bmatrix} 0 & 1 \end{bmatrix} \begin{bmatrix} x_1 \\ x_2 \end{bmatrix}$$

For R = 3, L = 1 and C = 1/2, this becomes

$$\begin{bmatrix} \dot{x}_1 \\ \dot{x}_2 \end{bmatrix} = \begin{bmatrix} -3 & -1 \\ 2 & 0 \end{bmatrix} \begin{bmatrix} x_1 \\ x_2 \end{bmatrix} + \begin{bmatrix} 1 \\ 0 \end{bmatrix} u, \quad y = \begin{bmatrix} 0 & 1 \end{bmatrix} \begin{bmatrix} x_1 \\ x_2 \end{bmatrix}$$

Let us test the controllability and observability of the system. The controllability matrix $\mathbf{Q_C}$ and the observability matrix $\mathbf{Q_O}$

$$\mathbf{Q_C} = \begin{bmatrix} \mathbf{B} & \mathbf{AB} \end{bmatrix} = \begin{bmatrix} 1 & -3 \\ 0 & 2 \end{bmatrix} \text{ and } \mathbf{Q_O} = \begin{bmatrix} \mathbf{C} \\ \mathbf{CA} \end{bmatrix} = \begin{bmatrix} 0 & 1 \\ 2 & 0 \end{bmatrix}$$

The rank of both $\mathbf{Q_C}$ and $\mathbf{Q_O}$ is 2. Hence, the system is completely state controllable and observable.

Let us introduce charge q(t) as an additional state variable and study these properties. Let $x_1 = q(t)$, $x_2 = i(t)$ and $x_3 = v_c(t)$. In terms of these state variables, the circuit dynamics can be expressed in matrix form as:

$$\begin{bmatrix} \dot{x}_1 \\ \dot{x}_2 \\ \dot{x}_3 \end{bmatrix} = \begin{bmatrix} 0 & 1 & 0 \\ 0 & -\frac{R}{L} & -\frac{1}{L} \\ 0 & \frac{1}{C} & 0 \end{bmatrix} \begin{bmatrix} x_1 \\ x_2 \\ x_3 \end{bmatrix} + \begin{bmatrix} 0 \\ \frac{1}{L} \\ 0 \end{bmatrix} u, \quad y = \begin{bmatrix} 0 & 0 & 1 \end{bmatrix} \begin{bmatrix} x_1 \\ x_2 \\ x_3 \end{bmatrix}$$

For R = 3, L = 1 and C = 1/2, this becomes

$$\begin{bmatrix} \dot{x}_1 \\ \dot{x}_2 \\ \dot{x}_3 \end{bmatrix} = \begin{bmatrix} 0 & 1 & 0 \\ 0 & -3 & -1 \\ 0 & 2 & 0 \end{bmatrix} \begin{bmatrix} x_1 \\ x_2 \\ x_3 \end{bmatrix} + \begin{bmatrix} 0 \\ 1 \\ 0 \end{bmatrix} u, \quad y = \begin{bmatrix} 0 & 0 & 1 \end{bmatrix} \begin{bmatrix} x_1 \\ x_2 \\ x_3 \end{bmatrix}$$

Let us test the controllability and observability of the system. The controllability matrix $\mathbf{Q_C}$ and the observability matrix $\mathbf{Q_O}$

$$\mathbf{Q_C} = \begin{bmatrix} \mathbf{B} & \mathbf{AB} & \mathbf{A^2B} \end{bmatrix} = \begin{bmatrix} 0 & 1 & -3 \\ 1 & -3 & 7 \\ 0 & 2 & -6 \end{bmatrix}$$

$$\mathbf{Q_O} = \begin{bmatrix} \mathbf{C} \\ \mathbf{CA} \\ \mathbf{CA^2} \end{bmatrix} = \begin{bmatrix} 0 & 0 & 1 \\ 0 & 2 & 0 \\ 0 & -6 & -2 \end{bmatrix}$$

The rank of both $\mathbf{Q_C}$ and $\mathbf{Q_O}$ is 2. Hence, the system is both **uncontrollable and unobservable.** The transfer function between the output voltage and the input voltage is computed as:

$$T(s) = \frac{Y(s)}{U(s)} = \mathbf{C}(s\mathbf{I} - \mathbf{A})^{-1}\mathbf{B} = \frac{2s}{s(s^2 + 3s + 2)} = \frac{2}{s^2 + 3s + 2}$$

Note that there is a pole-zero cancellation in the transfer function.

Let us change the output to $y = q(t) = x_1(t)$. This gives the output matrix

$$\mathbf{C} = \begin{bmatrix} 1 & 0 & 0 \end{bmatrix}$$

The observability matrix $\mathbf{Q_O} = \begin{bmatrix} \mathbf{C} \\ \mathbf{CA} \\ \mathbf{CA^2} \end{bmatrix} = \begin{bmatrix} 1 & 0 & 0 \\ 0 & 1 & 0 \\ 0 & -3 & -1 \end{bmatrix}$

The rank the observability matrix is 3; hence the system is observable . However, the system is uncontrollable. The transfer function between the charge and the input voltage is computed, and it is given by

$$T(s) = \frac{Y(s)}{U(s)} = \frac{Q(s)}{V(s)} = \frac{s}{s(s^2 + 3s + 2)} = \frac{1}{s^2 + 3s + 2}$$

Note that there is a pole-zero cancellation in the transfer function.

CONCEPT REINFORCEMENT PROBLEM 11.5

Problem

Consider the simple RLC-circuit shown in Figure 11.5 where $R = 3\Omega$, $L = 1H$ and $C = 0.5F$.

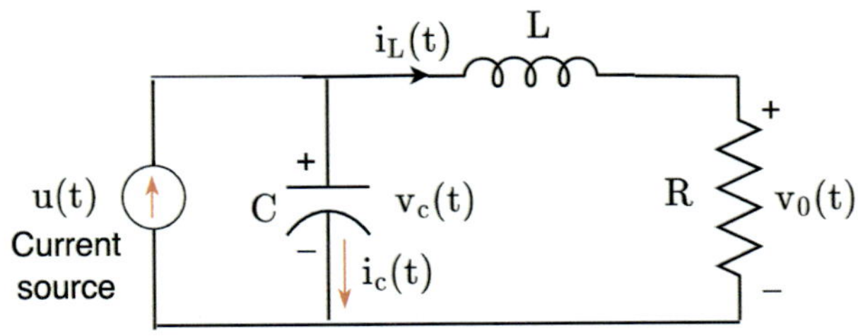

Figure 11.5: An RLC circuit with current source input

Let i_L is the current through the inductor and v_c denotes the voltage across the capacitor and q is charge flowing through inductor. Investigate controllability and observability properties of the circuit for the following choice of state and output variables. The output $y(t)$ is the voltage across the resistor R, and the input $u(t)$ to the circuit is a current source.

a. $x_1 = i_L(t)$, and $x_2 = v_c(t)$

b. $x_1 = q(t)$, $x_2 = i_L(t)$, and $x_3 = v_c(t)$

11.9 Principle of Duality

A close examination of controllability and observability conditions reveals apparent analogies between these two properties. Consider the system $\mathbf{S_1}$ defined by

$$\begin{aligned} \dot{\mathbf{x}} &= \mathbf{Ax} + \mathbf{Bu} \\ \mathbf{y} &= \mathbf{Cx} \end{aligned} \tag{11.22}$$

where $\mathbf{x} \in \mathbb{R}^n$, $\mathbf{u} \in \mathbb{R}^m$, $\mathbf{y} \in \mathbb{R}^p$, $\mathbf{A} \in \mathbb{R}^{n\times n}$, $\mathbf{B} \in \mathbb{R}^{n\times m}$ and $\mathbf{C} \in \mathbb{R}^{p\times n}$.

The dual system $\mathbf{S_2}$ is defined as:

$$\begin{aligned} \dot{\mathbf{z}} &= \mathbf{A^T z} + \mathbf{C^T v} \\ \mathbf{w} &= \mathbf{B^T z} \end{aligned} \tag{11.23}$$

where $\mathbf{z} \in \mathbb{R}^n$ is the state vector, $\mathbf{v} \in \mathbb{R}^p$ is the input vector, $\mathbf{w} \in \mathbb{R}^m$ is the output vector, and the superscript T denotes matrix transpose.

Theorem 11.9. System $\mathbf{S_1}$ is completely state controllable if and only if the dual system $\mathbf{S_2}$ is completely observable.
Similarly, system $\mathbf{S_1}$ is completely observable if and only if the dual system $\mathbf{S_2}$ is completely state controllable.

Practical Significance: This duality principle allows us to convert any controllability problem into an equivalent observability problem and vice versa, providing alternative approaches to system analysis and design.

11.10 Stabilizability and Detectability

We have discussed the controllability properties of a linear system in detail. Often, the controllability property can be considered too strong. In practice, a weaker condition than controllability is sufficient for most situations. This is called the *stabilizability* property. Similarly, the observability property can be considered too strong, and a weaker condition than observability is sufficient for most situations. This is called the *detectability* property and is the dual of the *stabilizability* property.

The following theorems state these properties.

Theorem 11.10. A partially controllable system is *stabilizable* if the uncontrollable modes are stable and the unstable modes are controllable.

Theorem 11.11. A partially observable system is *detectable* if the unobservable modes are stable and the unstable modes are observable.

We discuss two methods to check these properties.

Checking Stabilizability and Detectability:Sensitivity Matrix Method

Step-1: Compute the eigenvalues of the system λ_i, $i = 1, 2, \ldots, n$, where n denotes the system order.

Step-2: Form the sensitivity matrix corresponding to each eigenvalue of the n-th order system:

i. **For stabilizability**, the sensitivity matrix is

$$\mathbf{S}(\lambda_i) = \begin{bmatrix}\mathbf{A} - \lambda_i\mathbf{I} & \mathbf{B}\end{bmatrix}$$

ii. **For detectability**, the sensitivity matrix is

$$\mathbf{S}(\lambda_i) = \begin{bmatrix}\mathbf{A}^T - \lambda_i\mathbf{I} \\ \mathbf{C}^T\end{bmatrix}$$

and compute the rank of $\mathbf{S}(\lambda_i)$.

Step-3: If rank of $\mathbf{S}(\lambda_i) < n$, then λ_i corresponds to an *uncontrollable* (for stabilizability) or *unobservable* (for detectability) mode.

Step-4: The system is:

- **Stabilizable** if all uncontrollable eigenvalues have negative real parts (are stable)
- **Detectable** if all unobservable eigenvalues have negative real parts (are stable)

Checking Stabilizability and Detectability:Diagonalization Method

Step-1: Convert the system into diagonal (modal) form

Step-2: Identify the stable and unstable modes by examining the eigenvalues

Step-3: Determine the controllability and observability of each mode by examining respectively the rows of the input matrix and columns of the output matrix of the diagonalized system.

Step-4: Check conditions:

- **Stabilizable**: All unstable modes must be controllable
- **Detectable**: All unstable modes must be observable

The following example illustrates this procedure.

EXAMPLE 11.13

Problem

Consider system with system matrix **A**, input matrix **B** and output matrix **C** as:

$$\mathbf{A} = \begin{bmatrix} 2 & 1 \\ 0 & -3 \end{bmatrix}, \quad \mathbf{B} = \begin{bmatrix} 1 \\ 0 \end{bmatrix}, \quad \mathbf{C} = \begin{bmatrix} 0 & 1 \end{bmatrix}$$

Investigate the stabilizability and detectability properties of the system.

SOLUTION:

Controllability and Observability Analysis

The controllability matrix $\mathbf{Q_C} = \begin{bmatrix} 1 & 2 \\ 0 & 0 \end{bmatrix}$ is rank deficient; so the system is uncontrollable.

The observability matrix $\mathbf{Q_O} = \begin{bmatrix} 0 & 1 \\ 0 & -3 \end{bmatrix}$ is rank deficient; so the system is uncontrollable.

Method-1: The eigenvalues of the system are at $\lambda_1 = 2$ (unstable mode) and $\lambda_2 = -3$ (stable mode).

Stabilizability Analysis:

The sensitivity matrix corresponding to each of these eigenvalues is

$$\mathrm{S}(2) = \begin{bmatrix} \mathbf{A} - 2\mathbf{I} & \mathbf{B} \end{bmatrix} = \begin{bmatrix} 0 & 1 & 1 \\ 0 & -5 & 0 \end{bmatrix} \implies \mathrm{Rank}\,[\mathrm{S}(2)] = 2$$

$$\mathrm{S}(-3) = \begin{bmatrix} \mathbf{A} + 3\mathbf{I} & \mathbf{B} \end{bmatrix} = \begin{bmatrix} 5 & 1 & 1 \\ 0 & 0 & 0 \end{bmatrix} \implies \mathrm{Rank}\,[\mathrm{S}(-3)] = 1$$

This implies that $\lambda_1 = 2$ is controllable and $\lambda_3 = -3$ is uncontrollable. Since the uncontrollable mode is stable, the system is **stabilizable.**

Detectability Analysis:

The sensitivity matrix corresponding to each of these eigenvalues is

$$S(2) = \begin{bmatrix} \mathbf{A}^T - 2\mathbf{I} & \mathbf{C}^T \end{bmatrix} = \begin{bmatrix} 0 & 0 & 0 \\ 1 & -5 & 1 \end{bmatrix} \implies \text{Rank}\,[S(2)] = 1$$

$$S(-3) = \begin{bmatrix} \mathbf{A}^T + 3\mathbf{I} & \mathbf{C}^T \end{bmatrix} = \begin{bmatrix} 5 & 0 & 0 \\ 1 & 0 & 1 \end{bmatrix} \implies \text{Rank}\,[S(-3)] = 2$$

This implies that $\lambda_1 = 2$ is unobservable and $\lambda_3 = -3$ is observable. Since the unobservable mode is unstable, the system is **not detectable**.

Method-2: When the system is represented in diagonal form (using the diagonalization procedure discussed before), the system equation is

$$\begin{bmatrix} \dot{x}_1 \\ \dot{x}_2 \end{bmatrix} = \begin{bmatrix} 2 & 0 \\ 0 & -3 \end{bmatrix} \begin{bmatrix} x_1 \\ x_2 \end{bmatrix} + \begin{bmatrix} 1 \\ 0 \end{bmatrix} u$$

$$y = \begin{bmatrix} 0 & 0.9806 \end{bmatrix} \begin{bmatrix} x_1 \\ x_2 \end{bmatrix}$$

By inspecting the second row of matrix **B**, which is zero, we conclude that the mode $\lambda_2 = -3$ is uncontrollable. But since it is a stable mode, the system is stabilizable. Similarly, the first column of the output matrix is zero, which means the mode $\lambda_1 = +2$ is unobservable. Since this is an unstable mode, the system is not detectable.

To build intuition for these concepts, consider the following analogies that illustrate why stabilizability and detectability are often more practical requirements than full controllability and observability.

Understanding Stabilizability

1. **The Unruly Classroom Analogy:** Think of a classroom with different groups of students:

   - Some students are naturally **well-behaved (stable modes)** and don't need teacher intervention
   - Some students are **disruptive (unstable modes)** but respond to the teacher's instructions **(controllable)**
   - The classroom is "**stabilizable**" if you can keep order by controlling only the disruptive students, while the well-behaved ones take care of themselves

   The key insight: You don't need to control *everyone* (full controllability), just the troublemakers. The quiet students in the back who don't respond to your instructions aren't a problem: they're already behaving.

2. **The Wildfire Management Analogy:** Imagine managing fires in different regions:

   - Some areas have small, dying embers that will naturally burn out **(stable, uncontrollable modes)**
   - Other areas have growing fires that you can reach with your firefighting equipment **(unstable, controllable modes)**
   - The situation is "stabilizable" if you can extinguish all the dangerous fires with your resources

You don't need firefighting access to every smoldering patch; only to the ones that could grow out of control.

Understanding Detectability

1. **The Home Security System Analogy:** Consider monitoring your home:

   - Some rooms are naturally secure with locked, reinforced doors **(stable, unobservable modes)**—even though you can't see inside, nothing bad can happen there
   - Other rooms are vulnerable but have cameras (unstable, observable modes)—you can detect any problems
   - Your system is "detectable" if you can observe all potential threats, even if you can't see the secure areas

2. **The Patient Monitoring Analogy:** A doctor monitoring a patient:

   - Some body systems are healthy and self-regulating (stable, unobservable modes)—even without constant monitoring, they'll be fine
   - Critical systems with potential issues have sensors attached (unstable, observable modes)—any deterioration is immediately detected
   - The monitoring setup is "detectable" if all potentially dangerous conditions can be observed

The key insight: You don't need to observe *everything* (full observability), just the parts that could go wrong. The healthy systems you can't directly measure aren't a concern—they're stable on their own.

The Duality Connection

These concepts are dual:

- **Stabilizability**: "Can I control everything that might cause problems?"
- **Detectability**: "Can I see everything that might cause problems?"

Both relax the stronger requirements (controllability/observability) by recognizing that **stable dynamics take care of themselves**—they need neither control nor observation.

11.11 Pole Placement via State Feedback: Motivation

We will discuss the design of the pole placement controller by state feedback. Before we present the design of this controller for a generic linear time-invariant system and discuss the necessary and sufficient condition, the concept of pole placement is illustrated considering the following examples.

EXAMPLE 11.14

Consider a simple first-order system described by

$$\begin{aligned} \dot{x}(t) &= 2x(t) + u(t), \quad x(0) = x_0 \\ y(t) &= x(t) \end{aligned} \tag{11.24}$$

where y is the output of the system and u is the input to the system. The transfer function of the open-loop system is given by

$$\frac{Y(s)}{U(s)} = \frac{1}{s-2} \tag{11.25}$$

The pole of the open-loop system is located at +2; thus the open-loop system is unstable. Let us stabilize this system by feeding back the state with a gain k such that

$$u(t) = -kx(t) \tag{11.26}$$

Substituting the control law in Eqn(11.26) into Eqn(11.24) gives the closed-loop system

$$\dot{x}(t) = 2x(t) - kx(t) = (2-k)x(t) \tag{11.27}$$

The pole of the closed-loop system is at $2 - k$ and its solution is given by

$$x(t) = e^{(2-k)t}x(0) \tag{11.28}$$

From Eqn(11.28), it can be seen that if the state feedback gain $k > 2$, then the closed-loop system will be asymptotically stable. Further, since the pole of the closed-loop system is located at $2 - k$, we can place the closed-loop pole anywhere on the real axis by a suitable choice of state feedback gain k.

EXAMPLE 11.15

Consider the system

$$\begin{aligned} \dot{x}_1 &= a_1 x_1 + b_1 u \\ \dot{x}_2 &= a_2 x_2 + b_2 u \end{aligned} \tag{11.29}$$

Using the state feedback control law $u = -k_1 x_1 - k_2 x_2$, the closed-loop system dynamics becomes

$$\begin{aligned} \dot{x}_1 &= a_1 x_1 - b_1 k_1 x_1 - b_1 k_2 x_2 \\ \dot{x}_2 &= a_2 x_2 - b_2 k_1 x_1 - b_2 k_2 x_2 \end{aligned} \tag{11.30}$$

In matrix form, this can be represented as:

$$\begin{bmatrix} \dot{x}_1 \\ \dot{x}_2 \end{bmatrix} = \begin{bmatrix} a_1 - b_1 k_1 & -b_1 k_2 \\ -b_2 k_1 & a_2 - b_2 k_2 \end{bmatrix} \begin{bmatrix} x_1 \\ x_2 \end{bmatrix} \tag{11.31}$$

The characteristic equation of the system is given by

$$s^2 + [b_1 k_1 + b_2 k_2 - (a_1 + a_2)]\, s + a_1 a_2 - a_1 b_2 k_2 - a_2 b_1 k_1 = 0 \tag{11.32}$$

Our objective is to find if it is possible to place the poles anywhere in the complex plane. Therefore, consider placing the poles in arbitrary locations, which results in the desired characteristic polynomial

$$\Delta_d(s) = s^2 + \alpha s + \beta \tag{11.33}$$

The state feedback gains are computed by equating the like powers in Eqn(11.32) and Eqn(11.33). This gives

$$\begin{aligned} b_1 k_1 + b_2 k_2 &= \alpha + (a_1 + a_2) \\ -a_2 b_1 k_1 - a_1 b_2 k_2 &= \beta - a_1 a_2 \end{aligned} \tag{11.34}$$

Eqn(11.34) can be expressed in matrix form as:

$$\begin{bmatrix} b_1 & b_2 \\ -a_2 b_1 & -a_1 b_2 \end{bmatrix} \begin{bmatrix} k_1 \\ k_2 \end{bmatrix} = \begin{bmatrix} \alpha + (a_1 + a_2) \\ \beta - a_1 a_2 \end{bmatrix} \tag{11.35}$$

Eqn(11.35) has a solution for k_1 and k_2 if the determinant of the coefficient matrix

$$b_1 b_2 (a_2 - a_1) \neq 0 \tag{11.36}$$

Thus, for arbitrary pole placement $a_1 \neq a_2$ and both b_1 and b_2 should not be zero.

Let us determine the controllability of the system and investigate if the above conditions can be related to the controllability condition. The controllability matrix of this system is given by

$$\mathbf{Q_C} = \begin{bmatrix} \mathbf{B} & \mathbf{AB} \end{bmatrix} = \begin{bmatrix} b_1 & a_1 b_1 \\ b_2 & a_2 b_2 \end{bmatrix} \tag{11.37}$$

The system is controllable provided $\mathbf{Q_C}$ has full rank i.e if

$$b_1 b_2 (a_2 - a_1) \neq 0 \tag{11.38}$$

This is the same condition as given in Eqn(11.36). This example shows that for arbitrary pole placement, the system must be controllable. In the following, we will discuss the general problem of pole placement by state feedback for a single-input single-output system. The design of pole placement for multi-input multi-output systems is considerably more complex and beyond the scope of this book.

11.12 Design of Pole Placement Controller via State Feedback

Consider a single-input single-output system described by

$$\begin{aligned} \dot{\mathbf{x}}(t) &= \mathbf{Ax}(t) + \mathbf{B}u(t) \\ y(t) &= \mathbf{Cx}(t) \end{aligned} \tag{11.39}$$

where $\mathbf{x} \in \mathbb{R}^n$ is the state vector, u is the scalar control signal and y denotes the scalar output of the system. The matrices $\mathbf{A} \in \mathbb{R}^{n\times n}$, $\mathbf{B} \in \mathbb{R}^{n\times 1}$ and $\mathbf{C} \in \mathbb{R}^{1\times n}$.

The objective of the state-feedback controller using pole placement is to place the eigenvalues of the closed-loop system at arbitrary (desired) locations using the control law

$$\boxed{u(t) = -\mathbf{Kx}(t) + \mathbf{N}r(t)} \tag{11.40}$$

where $\mathbf{r}(t)$ is the reference input, $\mathbf{K} \in \mathbb{R}^{1\times n}$ is the state feedback gain matrix, and $\mathbf{N}$ is called the feed forward gain.

Substituting the control input of Eqn(11.40) into the open-loop state equation (11.39) gives the closed-loop dynamics as:

$$\begin{aligned} \dot{\mathbf{x}}(t) &= (\mathbf{A}-\mathbf{BK})\,\mathbf{x}(t)+\mathbf{BN}r(t) \\ y(t) &= \mathbf{Cx}(t) \end{aligned} \tag{11.41}$$

The closed-loop characteristic equation is given by

$$\left|s\mathbf{I}-(\mathbf{A}-\mathbf{BK})\right| = \left|s\mathbf{I}-\mathbf{A}+\mathbf{BK}\right| = 0$$

If the desired poles are to be located at μ_1, μ_2,..., μ_n, then the state feedback gain matrix $\mathbf{K}$ is determined from

$$\left|s\mathbf{I}-\mathbf{A}+\mathbf{BK}\right| = (s-\mu_1)(s-\mu_2)\dots(s-\mu_n)$$

The following theorem gives the necessary and sufficient condition to place the poles in desired locations.

Theorem 11.12. If the system of Eqn(11.39) is completely state controllable, that is, if the pair $(\mathbf{A},\mathbf{B})$ is controllable, then there exists a feedback gain matrix $\mathbf{K}$ that can place the eigenvalues of $(\mathbf{A}-\mathbf{BK})$ at arbitrary locations.

11.13 Proof of Pole-Placement via Controllable Canonical Form

We provide a simple proof of the existence of a state-feedback gain matrix $\mathbf{K}$ for arbitrary pole placement.

If the pair $(\mathbf{A},\mathbf{B})$ is controllable, there exists a nonsingular transformation $\mathbf{T}$ such that

$$\mathbf{x} = \mathbf{Tz}$$

which transforms the system in Eqn (11.39) into the *controllable canonical form*:

$$\begin{aligned} \dot{\mathbf{z}} &= \mathbf{T}^{-1}\mathbf{ATz}+\mathbf{T}^{-1}\mathbf{B}u \\ &= \underbrace{\begin{bmatrix} 0 & 1 & 0 & \dots & 0 \\ 0 & 0 & 1 & \dots & 0 \\ \vdots & \vdots & \vdots & \ddots & \vdots \\ 0 & 0 & 0 & \dots & 1 \\ -a_n & -a_{n-1} & -a_{n-2} & \dots & -a_1 \end{bmatrix}}_{\mathbf{A}_c}\mathbf{z} + \underbrace{\begin{bmatrix} 0 \\ 0 \\ \vdots \\ 0 \\ 1 \end{bmatrix}}_{\mathbf{B}_c} u \end{aligned} \tag{11.42}$$

Here, the coefficients $\mathrm{a_i}$ are from the characteristic polynomial of $\mathbf{A}$:

$$\det(\lambda\mathbf{I} - \mathbf{A}) = \mathrm{s^n} + \mathrm{a_1 s^{n-1}} + \mathrm{a_2 s^{n-2}} + \cdots + \mathrm{a_{n-1} s} + \mathrm{a_n} \tag{11.43}$$

Under this transformation, the state-feedback control law (11.40) becomes

$$\mathrm{u} = -\mathbf{KTz} + \mathbf{N}\mathrm{r} = -\mathbf{K}_\mathrm{c}\mathbf{z} + \mathbf{N}\mathrm{r}, \quad \text{where } \mathbf{K}_\mathrm{c} = \mathbf{KT} \tag{11.44}$$

The closed-loop dynamics in the transformed space are:

$$\begin{aligned}\dot{\mathbf{z}} &= \mathbf{A}_\mathrm{c}\mathbf{z} + \mathbf{B}_\mathrm{c}\mathrm{u} = \mathbf{A}_\mathrm{c}\mathbf{z} + \mathbf{B}_\mathrm{c}(-\mathbf{K}_\mathrm{c}\mathbf{z} + \mathbf{N}\mathrm{r}) \\ &= (\mathbf{A}_\mathrm{c} - \mathbf{B}_\mathrm{c}\mathbf{K}_\mathrm{c})\mathbf{z} + \mathbf{B}_\mathrm{c}\mathbf{N}\mathrm{r}\end{aligned} \tag{11.45}$$

Since similarity transformations preserve eigenvalues, the eigenvalues of $(\mathbf{A} - \mathbf{BK})$ are the same as those of $(\mathbf{A}_c - \mathbf{B}_c\mathbf{K}_c)$. If we choose

$$\boxed{\mathbf{K}_\mathrm{c} = [\delta_\mathrm{n}, \delta_\mathrm{n-1}, \ldots, \delta_1]} \tag{11.46}$$

then the closed-loop dynamics in controllable canonical form are

$$\dot{\mathbf{z}} = \begin{bmatrix} 0 & 1 & 0 & \cdots & 0 \\ 0 & 0 & 1 & \cdots & 0 \\ \vdots & \vdots & \vdots & \ddots & \vdots \\ 0 & 0 & 0 & \cdots & 1 \\ -(\mathrm{a_n} + \delta_\mathrm{n}) & -(\mathrm{a_{n-1}} + \delta_\mathrm{n-1}) & \cdots & -(\mathrm{a_1} + \delta_1) & \end{bmatrix} \mathbf{z} + \begin{bmatrix} 0 \\ 0 \\ \vdots \\ 0 \\ 1 \end{bmatrix} \mathbf{N}\mathrm{r} \tag{11.47}$$

The characteristic equation can then be determined by inspection:

$$\mathrm{s^n} + (\mathrm{a_1} + \delta_1)\mathrm{s^{n-1}} + (\mathrm{a_2} + \delta_2)\mathrm{s^{n-2}} + \cdots + (\mathrm{a_n} + \delta_\mathrm{n}) = 0 \tag{11.48}$$

If the desired poles are $\mu_1, \mu_2, \ldots, \mu_n$, the desired characteristic equation is

$$(\mathrm{s} - \mu_1)(\mathrm{s} - \mu_2)\ldots(\mathrm{s} - \mu_\mathrm{n}) = \mathrm{s^n} + \alpha_1\mathrm{s^{n-1}} + \cdots + \alpha_\mathrm{n} = 0 \tag{11.49}$$

Equating coefficients of like powers of s in Eqn(11.48) and Eqn(11.49) gives

$$\boxed{\delta_1 = \alpha_1 - \mathrm{a_1}, \quad \delta_2 = \alpha_2 - \mathrm{a_2}, \quad \ldots, \quad \delta_\mathrm{n} = \alpha_\mathrm{n} - \mathrm{a_n}}$$

Finally, the original state-feedback gain matrix is recovered via

$$\boxed{\mathbf{K} = \mathbf{K}_c\mathbf{T}^{-1} = [\alpha_\mathrm{n} - \mathrm{a_n}, \alpha_\mathrm{n-1} - \mathrm{a_{n-1}}, \ldots, \alpha_1 - \mathrm{a_1}]\mathbf{T}^{-1}} \tag{11.50}$$

This method is also known as the Bass-Gura approach. This can be summarized as follows:

Bass-Gura Approach of Eigenvalue Assignment: Summary

System:

$$\dot{\mathbf{x}}(t) = \mathbf{A}\mathbf{x}(t) + \mathbf{B}u(t)$$

Objective: Determine the feedback gain matrix **K** to place the eigenvalues at $\mu_1, \mu_2, \ldots, \mu_n$ using the control law

$$u(t) = -\mathbf{K}\mathbf{x} + Nr$$

Step-1: Check that the system is controllable; proceed only if controllable.

Step-2: Determine the coefficients $a_1, a_2, \ldots, a_n$ of the open-loop characteristic equation

$$\det(s\mathbf{I} - \mathbf{A}) = s^n + a_1 s^{n-1} + a_2 s^{n-2} + \cdots + a_{n-1}s + a_n$$

Step-3 Compute the transformation matrix **T** using Bass-Gura method:

$$\mathbf{T} = \mathbf{Q_C}\mathbf{W}$$

where the controllability matrix $\mathbf{Q_C}$ is

$$\mathbf{Q_C} = \begin{bmatrix}\mathbf{B} & \mathbf{AB} & \ldots & \mathbf{A}^{n-1}\mathbf{B}\end{bmatrix}, \quad \text{and}$$

and the upper-triangular matrix **W** is

$$\mathbf{W} = \begin{bmatrix} a_{n-1} & a_{n-2} & \ldots & a_1 & 1 \\ a_{n-2} & a_{n-3} & \ldots & 1 & 0 \\ \vdots & \vdots & \vdots & \vdots & \vdots \\ a_1 & 1 & \ldots & 0 & 0 \\ 1 & 0 & \ldots & 0 & 0 \end{bmatrix}$$

Note: If the system is in controllable canonical form, with

$$\mathbf{B} = \begin{bmatrix}0 & 0 & \ldots & K_u\end{bmatrix}^T$$

then

$$\mathbf{T} = K_u\mathbf{I}, \quad \mathbf{T}^{-1} = \frac{1}{\mathbf{K_u}}\mathbf{I}$$

Step-4: Compute the desired closed-loop characteristic equation corresponding to the desired eigenvalues $\mu_1, \mu_2, \ldots, \mu_n$:

$$(s-\mu_1)(s-\mu_2)\ldots(s-\mu_n) = s^n + \alpha_1 s^{n-1} + \cdots + \alpha_{n-1}s + \alpha_n$$

Step-5: Compute the state feedback gain matrix **K** from:

$$\mathbf{K} = \begin{bmatrix}\alpha_n - a_n & \alpha_{n-1} - a_{n-1} & \ldots & \alpha_2 - a_2 & \alpha_1 - a_1\end{bmatrix} \mathbf{T}^{-1}$$

Determination of State Feedback Gain K by Direct Substitution

When the order of the system is low (say $n \leq 3$), then it is simpler to compute the gain matrix **K** by direct substitution. The design procedure for a third-order system ($n = 3$) is given below.

Step-1: Define the state feedback gain matrix **K** as:

$$\mathbf{K} = \begin{bmatrix}k_1 & k_2 & k_3\end{bmatrix}$$

Step-2: Compute the desired characteristic equation as:

$$(s-\mu_1)(s-\mu_2)(s-\mu_3) = s^3 + \alpha_1 s^2 + \alpha_2 s + \alpha_3$$

Step-3: Determine the state feedback gains k_1, k_2 and k_3 by equating like powers of s from:

$$\left|s\mathbf{I} - \mathbf{A} + \mathbf{BK}\right| = s^3 + \alpha_1 s^2 + \alpha_2 s + \alpha_3$$

Note: It is assumed that the system is controllable; else matrix **K** cannot be determined.

Determination of State Feedback Gain K using Ackermann's Formula

This formula essentially utilizes the Cayley-Hamilton theorem, which states that every square matrix satisfies its own characteristic equation. The steps of determining the state feedback gain are given here without a full derivation of this formula.

Step-1: Compute the desired characteristic equation

$$(s-\mu_1)(s-\mu_2)\ldots(s-\mu_n) = s^n + \alpha_1 s^{n-1} + \alpha_2 s^{n-2} + \cdots + \alpha_{n-1}s + \alpha_n$$

Step-2: Compute the controllability matrix

$$\mathbf{Q_C} = \begin{bmatrix}\mathbf{B} & \mathbf{AB} & \ldots & \mathbf{A}^{n-1}\mathbf{B}\end{bmatrix}$$

Ensure that $\mathbf{Q_C}$ is invertible (system is controllable)

Step-3: Using the coefficients of the desired characteristic equation and system matrix $\mathbf{A}$, compute

$$\phi(\mathbf{A}) = \mathbf{A}^{n} + \alpha_1\mathbf{A}^{n-1} + \alpha_2\mathbf{A}^{n-2} + \cdots + \alpha_{n-1}\mathbf{A} + \alpha_n\mathbf{I}$$

Step-4: Compute the state feedback gain:

$$\mathbf{K} = \begin{bmatrix}0 & 0 & 0 & \ldots & \ldots & 1\end{bmatrix}\begin{bmatrix}\mathbf{B} & \mathbf{AB} & \mathbf{A}^2\mathbf{B} & \ldots & \mathbf{A}^{n-1}\mathbf{B}\end{bmatrix}^{-1}\phi(\mathbf{A})$$

Let us understand this through the following examples.

EXAMPLE 11.16

Problem

The transfer function model of a rigid satellite system is given by

$$\frac{\theta(s)}{T(s)} = \frac{1}{Js^2}$$

where J is the moment of inertia, T is the torque input and θ represents the angle. Design a state feedback controller which would yield a damping ratio $\xi = 0.707$ and peak time $T_p = 1.5708$ sec. Assume $J = 1\ \text{Kg} - \text{m}^2$.

SOLUTION:

The state variable model of the system with $x_1 = \theta$, $x_2 = \dot{\theta}$ and input $u(t) = T(t)$ is described by

$$\begin{bmatrix}\dot{x}_1\\ \dot{x}_2\end{bmatrix} = \begin{bmatrix}0 & 1\\ 0 & 0\end{bmatrix} + \begin{bmatrix}0\\ 1\end{bmatrix}u, \quad y = \begin{bmatrix}1 & 0\end{bmatrix}\begin{bmatrix}x_1\\ x_2\end{bmatrix}$$

Testing Controllability: From the system dynamics, we have

$$\mathbf{A} = \begin{bmatrix}0 & 1\\ 0 & 0\end{bmatrix}, \quad \mathbf{B} = \begin{bmatrix}0\\ 1\end{bmatrix}, \quad \mathbf{Q_C} = \begin{bmatrix}\mathbf{B} & \mathbf{AB}\end{bmatrix} = \begin{bmatrix}0 & 1\\ 1 & 0\end{bmatrix}$$

We find $|\mathbf{Q_C}| = -1$. Hence, the controllability matrix has full rank and the system is *controllable.*

Desired Characteristic Equation:

Now $T_p = \dfrac{\pi}{\omega_d} \implies \omega_d = 2.$

The natural frequency $\omega_n = \dfrac{\omega_d}{\sqrt{1-\xi^2}} = 2.828$. This gives the desired pole locations at $-\xi\omega_n \pm j\omega_d = -2 \pm j2$.

The closed-loop characteristic equation corresponding to the desired pole location is given by

$$\Delta_d(s) = (s+2+j2)(s+2-j2) = s^2+4s+8 = 0 \quad = s^2+\alpha_1 s+\alpha_2 = 0$$

Comparing this with $s^2+\alpha_1 s+\alpha_2 = 0$ gives $\alpha_1 = 4$ and $\alpha_2 = 8$

Method-1: Bass-Gura Approach:

Step:1.1 Determine the characteristic equation of the system as:

$$|s\mathbf{I}-\mathbf{A}| = \left|\begin{bmatrix} s & 0 \\ 0 & s \end{bmatrix} - \begin{bmatrix} 0 & 1 \\ 0 & 0 \end{bmatrix}\right| = \left|\begin{bmatrix} s & -1 \\ 0 & s \end{bmatrix}\right| = 0 \implies s^2 = 0$$

Comparing this with $s^2+a_1 s+a_2 = 0$ gives $a_1 = 0$ and $a_2 = 0$

Step-1.2 Determine the transformation matrix $\mathbf{T} = \mathbf{Q_C W}$ where

$$\mathbf{W} = \begin{bmatrix} a_1 & 1 \\ 1 & 0 \end{bmatrix} = \begin{bmatrix} 0 & 1 \\ 1 & 0 \end{bmatrix}$$

Thus, the transformation matrix

$$\mathbf{T} = \mathbf{Q_C W} = \begin{bmatrix} 0 & 1 \\ 1 & 0 \end{bmatrix}\begin{bmatrix} 0 & 1 \\ 1 & 0 \end{bmatrix} = \begin{bmatrix} 1 & 0 \\ 0 & 1 \end{bmatrix}$$

The state feedback gain is given as:

$$\begin{aligned}\mathbf{K} &= \begin{bmatrix} \alpha_2 - a_2 & \alpha_1 - a_1 \end{bmatrix}\mathbf{T}^{-1} \\ &= \begin{bmatrix} 8 & 4 \end{bmatrix}\begin{bmatrix} 1 & 0 \\ 0 & 1 \end{bmatrix}^{-1} = \begin{bmatrix} 8 & 4 \end{bmatrix} = \begin{bmatrix} k_1 & k_2 \end{bmatrix}\end{aligned}$$

Hence, the state feedback control law is given as

$$u = -k_1x_1 - k_2x_2 = -8x_1 - 4x_2$$

Method-2: Direct Substitution Approach

Step-2.1: Define $\mathbf{K} = \begin{bmatrix} k_1 & k_2 \end{bmatrix}$

Step 2.2: Express the closed-loop characteristic equation in terms of state feedback gains k_1 and k_2 from $\left|s\mathbf{I}-\mathbf{A}+\mathbf{BK}\right|$. This gives

$$\begin{aligned}|s\mathbf{I}-\mathbf{A}+\mathbf{BK}| &= \left|\begin{bmatrix} s & 0 \\ 0 & s \end{bmatrix} - \begin{bmatrix} 0 & 1 \\ 0 & 0 \end{bmatrix} + \begin{bmatrix} 0 \\ 1 \end{bmatrix}\begin{bmatrix} k_1 & k_2 \end{bmatrix}\right| \\ &= \left|\begin{bmatrix} s & -1 \\ 0 & s \end{bmatrix} + \begin{bmatrix} 0 & 0 \\ k_1 & k_2 \end{bmatrix}\right| = \left|\begin{bmatrix} s & -1 \\ k_1 & s+k_2 \end{bmatrix}\right| \\ &= s^2 + k_2 s + k_1\end{aligned}$$

Step 2.3: Equate $\left|sI - A + BK\right|$ to the desired characteristic equation and determine the feedback gains. Now

$$s^2 + k_2 s + k_1 = s^2 + 4s + 8$$

This gives $\mathbf{K} = \begin{bmatrix} k_1 & k_2 \end{bmatrix} = \begin{bmatrix} 8 & 4 \end{bmatrix}$

Method-3: Ackermann's Method

Step-3.1: Compute $\phi(\mathbf{A}) = \mathbf{A}^2 + \alpha_1 \mathbf{A} + \alpha_2 \mathbf{I}$. This gives

$$\phi(\mathbf{A}) = \begin{bmatrix} 0 & 1 \\ 0 & 0 \end{bmatrix}^2 + 4\begin{bmatrix} 0 & 1 \\ 0 & 0 \end{bmatrix} + 8\begin{bmatrix} 1 & 0 \\ 0 & 1 \end{bmatrix} = \begin{bmatrix} 8 & 4 \\ 0 & 8 \end{bmatrix}$$

Step-3.2: Compute

$$\begin{aligned} \mathbf{K} &= \begin{bmatrix} 0 & 1 \end{bmatrix} \begin{bmatrix} \mathbf{B} & \mathbf{AB} \end{bmatrix}^{-1} \phi(\mathbf{A}) \\ &= \begin{bmatrix} 0 & 1 \end{bmatrix} \begin{bmatrix} 0 & 1 \\ 1 & 0 \end{bmatrix}^{-1} \begin{bmatrix} 8 & 4 \\ 0 & 8 \end{bmatrix} = \begin{bmatrix} 8 & 4 \end{bmatrix} \end{aligned}$$

CONCEPT REINFORCEMENT PROBLEM 11.6

Problem

The linearized model of the glucose level control system is expressed as:

$$\begin{aligned} \dot{x}_1 &= -ax_1 - bx_2 \\ \dot{x}_2 &= -cx_1 - dx_2 + u \end{aligned}$$

where x_1 denotes the deviation of the extracellular insulin level from the mean, x_2 is the deviation of the extracellular glucose level from the mean, and u is the rate of glucose intravenous injection. The objective is to stabilize the glucose level. Design a state feedback controller which will regulate the glucose level with 4.32% overshoot and settle at 4 secs. The parameters of the system are $a = 0.78$, $b = 0.208$, $c = 4.34$ and $d = 2.92$

EXAMPLE 11.17

Problem

The transfer function of a plant is given by

$$G(s) = \frac{Y(s)}{R(s)} = \frac{K}{s(s+5)(s+20)}$$

Design a state feedback controller which will ensure 5% peak overshoot, peak time 0.5 sec and 5% velocity error. Place the third pole 10 times farther from the dominant pole pair.

SOLUTION:

To achieve a velocity error of 5%, the velocity error constant K_v must be 20. This gives $K = 2000$.

Let us translate the desired specifications into a dominant pole pair. Now, given $\%M_p = 5$. The damping ratio corresponding to this overshoot is

$$\xi = \frac{\gamma}{\sqrt{\pi^2 + \gamma^2}} = 0.6901; \quad (\gamma = -\ln\%M_p/100)$$

Given $T_p = \dfrac{\pi}{\xi\omega_d} = 0.5$ sec. Hence $\omega_d = \dfrac{\pi}{\xi T_p} = 6.2832; \quad \omega_n = 8.6819$

The desired poles are located at $-\xi\omega_n \pm j\omega_d = -5.9915 \pm j6.2832$ and the corresponding characteristic equation is

$$s^3 + 71.897s^2 + 793.33s + 4516.1 = s^3 + \alpha_1 s^2 + \alpha_2 s + \alpha_3$$

The characteristic equation of the plant is

$$s^3 + 25s^2 + 100s = s^3 + a_1 s^2 + a_2 s + a_3$$

Hence the state feedback gain matrix **K** is

$$\mathbf{K} = \frac{1}{2000}\begin{bmatrix}\alpha_3 - a_3 & \alpha_2 - a_2 & \alpha_1 - a_1\end{bmatrix} = \begin{bmatrix}2.2581 & 0.3467 & 0.0234\end{bmatrix}$$

We can alternatively verify this result in MATLAB, where

$$\mathbf{A} = \begin{bmatrix}0 & 1 & 0\\ 0 & 0 & 1\\ 0 & -100 & -25\end{bmatrix}, \quad \mathbf{B} = \begin{bmatrix}0\\ 0\\ 2000\end{bmatrix}, \quad \mathbf{C} = \begin{bmatrix}1 & 0\end{bmatrix}$$

$$P = \begin{bmatrix}-5.9915 + j6.2832 & -5.9915 - j6.2832 & -59.915\end{bmatrix},$$

$$K = \text{acker}(A, B, P) = \begin{bmatrix}2.2581 & 0.3467 & 0.0234\end{bmatrix}$$

Design of Pole-Placement Controllers with MATLAB

For systems of order higher than two, it is generally preferable to use MATLAB for controller design. The algebraic computations required for pole placement in higher-order systems are tedious and prone to numerical errors, even though they are theoretically possible by hand. MATLAB provides two commands to compute the state-feedback gain for pole-placement design: `acker` and `place`.

Use acker: Suitable for small systems and textbook-style examples. It is based on Ackermann's formula and can handle repeated poles in controllable systems. However, it is not numerically robust for large or ill-conditioned problems.

Use place: Preferred for practical applications. It employs a numerically stable algorithm, making it more reliable for large-scale systems. Note, however, that **place** does not permit repeated poles. All poles must be distinct.

CONCEPT REINFORCEMENT PROBLEM 11.7

Problem

Consider a system represented by

$$\dot{\mathbf{x}} = \mathbf{A}\mathbf{x} + \mathbf{B}\mathrm{u}, \quad \mathrm{y} = \mathbf{C}\mathbf{x}$$

where $\mathbf{A} = \begin{bmatrix} -1 & 2 & 0 \\ 4 & 0 & 7 \\ -2 & 1 & 5 \end{bmatrix}$, $\mathbf{B} = \begin{bmatrix} 5 \\ 0 \\ 1 \end{bmatrix}$, and $\mathbf{C} = \begin{bmatrix} 1 & 2 & 1 \end{bmatrix}$.

Using MATLAB

a. Design a state feedback regulator $\mathrm{u} = -\mathbf{K}\mathbf{x}$, such that the poles of the closed-loop system are at $-4 \pm \mathrm{j}6$ and at -10.
b. Obtain the response of the system to the following initial condition:

$$\mathbf{x}(0) = \begin{bmatrix} 1 & 0 & 0 \end{bmatrix}^T$$

11.13.1 Selection of Pole Locations: Some Guide Lines

The state feedback control gives ample freedom to the designer to place the closed-loop poles anywhere in the s-plane. Although this is apparently overwhelming, on the contrary, it is a burden on the designer to judiciously choose pole locations. There exists no unique location in the s-plane to place the poles which would yield the desired response for a variety of systems. This depends on various complex and interrelated criteria and still remains a major issue in state feedback control. However, there are a few guidelines that may be followed during pole placement.

Guidelines for Pole Placement

Guidelines for Pole Placement

1. **Avoid Excessive Pole Movement**

   Do not select the closed-loop poles far away from the open-loop poles, because this requires high control effort.

2. **Resist Placing Poles Too Far Left**

   Resist the temptation to place the poles too far into the left half plane (LHP).

   a. Although this will result in faster response, this is possible only at the cost of high-precision sensors and stronger (more powerful) actuators. The control signals may be very large, and the system may become nonlinear.

   b. This will increase the system bandwidth and, therefore, the system will be more sensitive to noise.

3. **Keep Acceptable Poles**

   If any of the poles in the LHP are in an acceptable location, do not shift it to another location.

4. **Handle Unstable/Marginal Poles**

   If there are poles in the right half plane (RHP) or close to the $j\omega$-axis, they must be shifted. There are a few thumb rules. For example

   a. If the open-loop pole is located at "+a", this should be pushed to "−a".

   b. The complex conjugate poles should be placed at locations to meet transient response specifications.

11.13.2 Limitations of State Feedback Control

As we have discussed, the placement of closed-loop poles in the complex plane requires considerable expertise, since one must establish a precise correlation between pole locations and the desired dynamic performance. Some of the major limitations of state feedback control are summarized below:

Limitations of State Feedback Control

1. **Practical implementation issues:**

   a. State feedback often demands unrealistically high actuator bandwidth. Practical devices have finite bandwidth, making such implementations difficult.

   b. In most cases, it is not possible to measure all the states directly; additional state observers or estimators are required.

2. **Steady-state accuracy:** State feedback by itself cannot guarantee zero steady-state error. It is generally not sufficient for reference tracking or disturbance rejection without integral action.

3. **Model dependence:** The design is highly sensitive to the accuracy of the system model. Any parameter uncertainty or unmodeled dynamics can lead to poor performance or even instability.

4. **Robustness:** The controller may not perform well under external disturbances or noise, especially if the pole placement is aggressive (very fast poles).

5. **Control effort:** Assigning poles far into the left-half plane to achieve fast response often results in very large control inputs, which may saturate actuators or damage hardware.

11.14 Regulation and Tracking

Most of the problems in control engineering can generally be classified into two main categories:

1. **Regulator Problem:**

   In a regulation problem, the objective is to drive the system output to zero as time approaches infinity while rejecting disturbances, regardless of the initial state. That is, $\lim_{t \to \infty} y(t) = 0$. This is essentially a **stabilization problem**. A common approach to solving this problem is to treat disturbances as unknown but belonging to specific function classes such as step, ramp, or sinusoidal disturbances.

   Examples of Regulation Problems:

   - **Building Vibration Control:** Minimize building sway due to wind disturbances using active mass dampers.
   - **Inverted Pendulum Stabilization:** Keep a pendulum upright despite external pushes by adjusting cart position.
   - **Aircraft Altitude Hold:** Return aircraft to cruise altitude after turbulence-induced deviations.

- **Active Suspension:** Minimize vehicle body displacement over road irregularities using active dampers.

2. **Tracking Problem:**

In the tracking problem, the goal is to make the output y(t) follow (track) a given reference signal r(t)

a. Suppose the reference signal is a constant, i.e., $r(t) = R \neq 0$. In that case, the tracking problem becomes similar to a regulation problem, but to drive the output to R rather than zero. The mathematical structure is similar, but the reference point differs.

Examples of Constant Reference Tracking:

- **Temperature Control:** Maintain furnace temperature at 850°C despite heat losses.
- **DC Motor Speed Control:** Keep motor speed constant at 1500 RPM under varying loads.
- **Water Level Control:** Maintain tank water level at 2.5 meters with varying outlet demands.

b. However, if the reference signal r(t) varies with time, the problem becomes more challenging and is referred to as a **servomechanism problem**. This requires designing a control law that ensures the system output accurately follows the changing reference signal.

Examples of Servomechanism Problems:

- **CNC Machine Control:** Tool precisely follows programmed 3D machining trajectory.
- **Radar Antenna Tracking:** Antenna tracks moving aircraft with time-varying position commands.
- **Robotic Arm Control:** Robot end-effector follows complex pick-and-place trajectories.
- **Solar Panel Tracking:** Panel orientation follows sun position throughout the day.
- **Autonomous Vehicle:** Car follows lane centerline or GPS-planned route.

11.15 Tracking Problem

Until now, we have focused on designing state feedback controllers for regulation problems, where the reference signal is set to zero. The objective of the controller has been to drive the system states to zero while rejecting disturbances. However, we have not yet discussed whether the state feedback controller can eliminate the steady state error when subjected to standard test signals such as step, ramp, or sinusoidal inputs.

Before proceeding further, let us introduce a feedforward gain and determine its value to eliminate the steady-state error. We will also compute the steady-state error in response to a step input and explore how feedforward compensation can improve tracking performance.

11.15.1 Steady State Error in State Feedback Control System without Feedforward

Consider the closed-loop state feedback control system shown in Figure 11.6

The system dynamics is described by

$$\begin{aligned} \dot{\mathbf{x}}(\mathrm{t}) &= \mathbf{A}\mathbf{x}(\mathrm{t}) + \mathbf{B}\mathrm{u}(\mathrm{t}) \\ \mathrm{y}(\mathrm{t}) &= \mathbf{C}\mathbf{x}(\mathrm{t}) \end{aligned} \tag{11.51}$$

Let us apply the control signal

$$\mathrm{u}(\mathrm{t}) = -\mathbf{K}\mathbf{x}(\mathrm{t}) + \mathrm{r}(\mathrm{t}), \qquad \mathbf{K} \in \mathbb{R}^{1\times n} \tag{11.52}$$

The dynamics of the closed-loop system, therefore, becomes

$$\begin{aligned} \dot{\mathbf{x}}(\mathrm{t}) &= (\mathbf{A} - \mathbf{B}\mathbf{K})\,\mathbf{x}(\mathrm{t}) + \mathbf{B}\mathrm{r}(\mathrm{t}) \\ \mathrm{y}(\mathrm{t}) &= \mathbf{C}\mathbf{x}(\mathrm{t}) \end{aligned} \tag{11.53}$$

At steady state, both the states and output reach a constant value. If $\mathbf{x_{ss}}$ and $\mathrm{y_{ss}}$ denote the steady state values of the states and output, then at steady state

$$\dot{\mathbf{x}}_{\mathrm{ss}} = 0 = (\mathbf{A} - \mathbf{B}\mathbf{K})\mathbf{x}_{\mathbf{ss}}(\mathrm{t}) + \mathbf{B}\mathrm{r}(\mathrm{t}) \tag{11.54}$$

This gives the steady state value of the states and output as:

$$\boxed{\mathbf{x_{ss}} = -\left(\mathbf{A} - \mathbf{B}\mathbf{K}\right)^{-1}\mathbf{B}\mathrm{r}, \quad \text{and} \quad \mathrm{y_{ss}} = -\mathbf{C}\left(\mathbf{A} - \mathbf{B}\mathbf{K}\right)^{-1}\mathbf{B}\mathrm{r}} \tag{11.55}$$

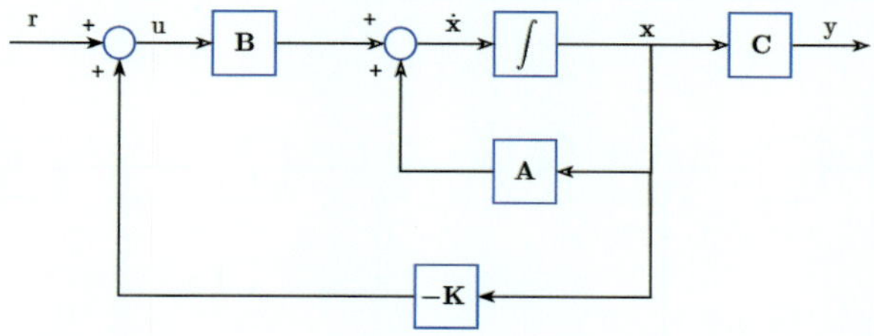

Figure 11.6: Closed-loop state feedback control system without feedforward

The steady state error due to constant reference equals to

$$e_{ss} = r - y_{ss} = r + C\left(\mathbf{A} - \mathbf{BK}\right)^{-1}\mathbf{B}r = \left[1 + \mathbf{C}\left(\mathbf{A} - \mathbf{BK}\right)^{-1}\mathbf{B}\right]r \tag{11.56}$$

It can be seen that the state feedback control will not eliminate the steady state error.

11.15.2 Tracking with Feedforward Gain

Consider the system of Eqn(11.51). The objective is to drive the output y to a given reference value **r** and hold it there. Let us introduce a gain **N** in the forward path, as shown in Figure 11.7.

The control signal of Eqn(11.52) now modifies to

$$u(t) = -\mathbf{Kx}(t) + \mathbf{N}r(t), \quad \mathbf{K} \in \mathbb{R}^{1\times n} \tag{11.57}$$

The closed-loop system, therefore, becomes

$$\begin{aligned} \dot{\mathbf{x}}(t) &= \left(\mathbf{A} - \mathbf{BK}\right)\mathbf{x}(t) + \mathbf{BN}r(t) \\ y(t) &= \mathbf{Cx}(t) \end{aligned} \tag{11.58}$$

The feedback gain matrix **K** is determined using the standard procedure of pole placement. Note that **N** does not affect the stability of the closed-loop system, but it affects the steady-state solution. Following a similar procedure, as in the last subsection, the steady state values of states and output are given as:

$$\mathbf{x}_{ss} = -\left(\mathbf{A} - \mathbf{BK}\right)^{-1}\mathbf{BN}r, \quad y_{ss} = -\mathbf{C}\left(\mathbf{A} - \mathbf{BK}\right)^{-1}\mathbf{BN}r \tag{11.59}$$

The steady state error due to a constant reference equals to

$$e_{ss} = r - y_{ss} = r + C\left(\mathbf{A} - \mathbf{BK}\right)^{-1}\mathbf{BN}r = \left[1 + \mathbf{C}\left(\mathbf{A} - \mathbf{BK}\right)^{-1}\mathbf{BN}\right]r \tag{11.60}$$

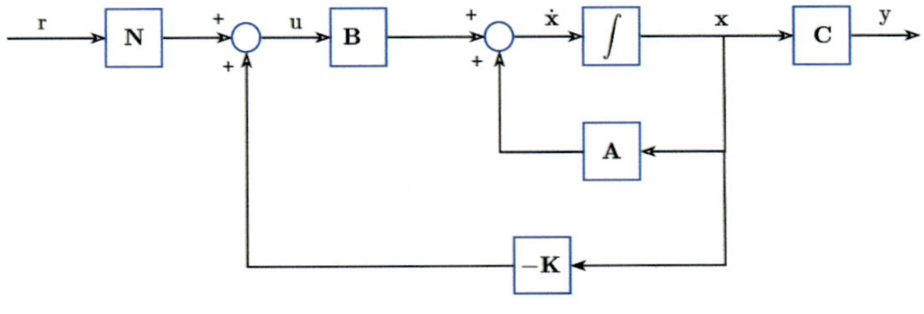

Figure 11.7: Closed-loop state feedback control with feedforward gain

From Eqn(11.60), it is obvious that if we select

$$\mathbf{N} = -\frac{1}{\mathbf{C}\left(\mathbf{A}-\mathbf{BK}\right)^{-1}\mathbf{B}} \tag{11.61}$$

then, the steady state error due to the step input will be zero.

Although the state feedback controller can eliminate the steady state error due to the step command by proper choice of the feedforward gain $\mathbf{N}$, it is not robust against parameter variations. Because its value is dependent on the system parameters (see Eqn(11.61)). Further, as this is not within the feedback loop, the system response will be highly sensitive to variations in feedforward path elements. This will be illustrated in the example to follow.

Problems with Feedforward Gain

1. **Parameter Dependency:** The feedforward gain N must be calculated based on exact system parameters (A, B, C matrices). Any deviation from these nominal values results in steady-state error.
2. **Open-Loop Vulnerability:** Since feedforward operates outside the feedback loop, it has no mechanism to self-correct for parameter variations or disturbances.
3. **High Sensitivity:** Unlike feedback systems that can tolerate some parameter uncertainty within stability margins, feedforward systems are highly sensitive to any parameter changes.

Let us illustrate this with an example.

EXAMPLE 11.18

Problem:

The linearized equations of motion of a simple pendulum are given by

$$\mathrm{mL}^2\frac{\mathrm{d}^2\theta}{\mathrm{dt}^2} + \mathrm{b}\frac{\mathrm{d}\theta}{\mathrm{dt}} + \mathrm{mgL}\theta = \mathrm{T(t)}$$

where m is the mass of the pendulum bob, θ is the angular displacement, L is the length of the pendulum, measured from the pivot point to the center of mass of the bob, $\mathrm{J} = \mathrm{mL}^2$ is the moment of inertia about the pivot point, b and T is the applied torque (see Section 9.5.2).

a. Represent this system in state space, considering the output θ and its derivative $\dot{\theta}$ as state variables and the applied torque as input.

b. Design a state feedback controller $u = -\mathbf{K}\mathbf{x} + r$ which places the closed-loop poles at $-5 \pm j$.
c. Design a state feedback controller with pre-compensator in the forward path of the form $u = -\mathbf{K}\mathbf{x} + Nr$ which will place the poles at the same location as in (b) and eliminate the steady state error due to step input.
d. Investigate the robustness of this controller by varying the mass of the pendulum bob by 20%.
e. Compare the closed-loop output responses for the controllers designed in all the above cases.

The parameters of the system are: $m = 0.5$ kg, $L = 0.5$ m, $b = 0.05$. Consider the reference as 5^0.

SOLUTION:

a. The state space model of the system is

$$\begin{bmatrix}\dot{x}_1\\ \dot{x}_2\end{bmatrix} = \underbrace{\begin{bmatrix}0 & 1.0\\ -\dfrac{g}{L} & -\dfrac{b}{mL^2}\end{bmatrix}}_{\mathbf{A}}\begin{bmatrix}x_1\\ x_2\end{bmatrix} + \underbrace{\begin{bmatrix}0\\ \dfrac{1}{mL^2}\end{bmatrix}}_{\mathbf{B}}u, \quad y = \underbrace{\begin{bmatrix}1 & 0\end{bmatrix}}_{\mathbf{C}}\begin{bmatrix}x_1\\ x_2\end{bmatrix}$$

b. For the given parameters, the system matrix $\mathbf{A}$ and the input matrix $\mathbf{B}$ are

$$\mathbf{A} = \begin{bmatrix}0 & 1\\ -19.62 & -0.4\end{bmatrix}, \quad \text{and} \quad \mathbf{B} = \begin{bmatrix}0\\ 8\end{bmatrix}$$

To design the state feedback controller $u = -\mathbf{K}\mathbf{x} + r$, first test the controllability of the system. The controllability matrix

$$\mathbf{Q_C} = \begin{bmatrix}\mathbf{B} & \mathbf{AB}\end{bmatrix} = \begin{bmatrix}0 & 8\\ 8 & -3.2\end{bmatrix}$$

It is of full rank and the system is therefore controllable.

Design of State feedback Gain K

Let us design the state feedback gain $\mathbf{K} = \begin{bmatrix}k_1 & k_2\end{bmatrix}$ by the direct substitution method as discussed before. Now

$$\begin{aligned}
\left|s\mathbf{I} - \mathbf{A} + \mathbf{BK}\right| &= \left|\begin{bmatrix}s & 0\\ 0 & s\end{bmatrix} - \begin{bmatrix}0 & 1\\ -19.62 & -0.4\end{bmatrix} + \begin{bmatrix}0\\ 8\end{bmatrix}\begin{bmatrix}k_1 & k_2\end{bmatrix}\right|\\
&= \left|\begin{bmatrix}s & -1\\ 19.62 & s+0.4\end{bmatrix} + \begin{bmatrix}0 & 0\\ 8k_1 & 8k_2\end{bmatrix}\right|\\
&= \left|\begin{bmatrix}s & -1\\ 8k_1 + 19.62 & 8k_2 + s + 0.4\end{bmatrix}\right|\\
&= s^2 + (8k_2 + 0.4)s + 8k_1 + 19.62
\end{aligned}$$

The desired characteristic equation is

$$\Delta_d(s) = (s + 5 + j)(s + 5 - j) = s^2 + 10s + 26$$

Step 2.3: Equate $\left|sI - A + BK\right|$ to the desired characteristic equation and determine the feedback gains. Now

$$s^2 + (8k_2 + 0.4)s + 8k_1 + 19.62 = s^2 + 10s + 26$$

This gives

$$\begin{aligned} 8k_1 + 19.62 &= 26 \implies k_1 = 0.7975 \text{ and} \\ 8k_2 + 0.4 &= 10 \implies k_2 = 1.2 \end{aligned}$$

c. The feedforward gain is

$$N = -\frac{1}{C(A - BK)^{-1}B} = 3.25$$

The steady state error due to this pre-compensator is

$$e_{ss} = 1 + C(A - BK)^{-1} * BN = 0$$

d. When the system is perturbed by varying the mass by 20%, the system and input matrix becomes

$$\overline{\mathbf{A}} = \begin{bmatrix} 0 & 1 \\ -19.62 & -0.333 \end{bmatrix}, \quad \overline{\mathbf{B}} = \begin{bmatrix} 0 \\ 6.667 \end{bmatrix}$$

The steady state error due to step-input for this perturbed condition is

$$\bar{e}_{ss} = 1 + C(\overline{A} - \overline{B}K)^{-1} * \overline{B}N = 0.1331$$

e. The response of the closed-loop system is compared for a reference of 5^0 and is shown in Figure 11.8.

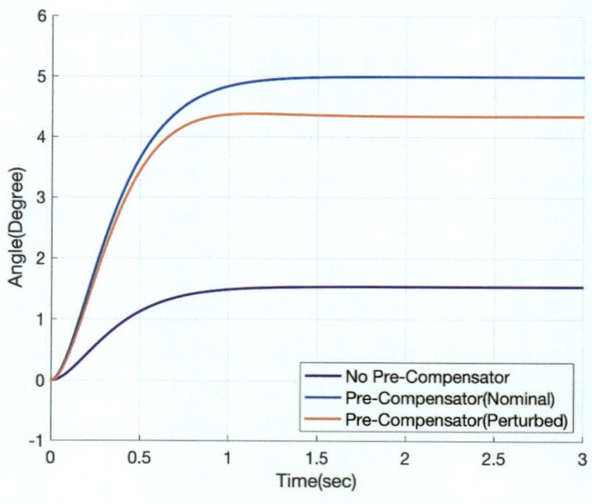

Figure 11.8: Tracking performance comparison with state feedback controller with feed forward gain

The MATLAB code can be found in `Example11_18.m`.

CONCEPT REINFORCEMENT PROBLEM 11.8

Problem

Consider a first-order system with transfer function

$$G(s) = \frac{Y(s}{U(s)} = \frac{3}{s-2}$$

a. Represent this system by a state variable model using output as the state variable.

b. Design a state feedback controller

$$u = -kx + r$$

which places the closed-loop pole at $s = -4$.

c. Compute the steady state error due to a step input.

d. Design a state feedback controller with a pre-compensator in the forward path of the form

$$u = -kx + Nr$$

which will place the pole at the same location as in (b) and eliminate the steady state error due to the step input.

11.15.3 Tracking with Integral Controller

It is found that the feed-forward compensation is not robust against parametric uncertainty. But in practice, it is often not possible to get an accurate model of a system, and we expect some degree of uncertainty in the model whose effects can be minimized by the use of feedback. In fact, this is one of the main objectives of feedback. An alternative to feedforward compensation is to use an integral controller in series with the plant (as shown in Figure 11.9).

In the following, we show that this configuration of state feedback controller with integral action will provide zero steady-state error due to a step command. By introducing integral action, the order of the control system is increased by one due to the addition of one new state variable, denoted as x_i,

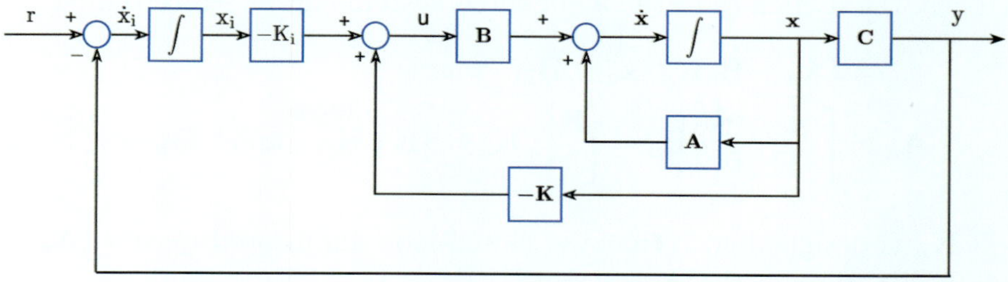

Figure 11.9: State feedback control with integral control

where

$$\dot{x}_i = r - y = r - \mathbf{C}\mathbf{x} = -\mathbf{C}\mathbf{x} + r \tag{11.62}$$

From Figure 11.9, the state equations can be written as:

$$\begin{aligned} \dot{\mathbf{x}} &= \mathbf{A}\mathbf{x} + \mathbf{B}u \\ \dot{x}_i &= -\mathbf{C}\mathbf{x} + r \\ y &= \mathbf{C}\mathbf{x} \end{aligned} \tag{11.63}$$

Equation (11.63) can be written as augmented vectors and matrices as:

$$\begin{aligned} \underbrace{\begin{bmatrix} \dot{\mathbf{x}} \\ \dot{x}_i \end{bmatrix}}_{\dot{\mathbf{x}}_a} &= \underbrace{\begin{bmatrix} \mathbf{A} & \mathbf{0}_{n\times 1} \\ -\mathbf{C} & 0 \end{bmatrix}}_{\mathbf{A_a}} \underbrace{\begin{bmatrix} \mathbf{x} \\ x_i \end{bmatrix}}_{\mathbf{x_a}} + \underbrace{\begin{bmatrix} \mathbf{B} \\ 0 \end{bmatrix}}_{\mathbf{B_a}} u + \underbrace{\begin{bmatrix} \mathbf{0}_{n\times 1} \\ 1 \end{bmatrix}}_{\mathbf{B_1}} r \\ y &= \begin{bmatrix} \mathbf{C} & 0 \end{bmatrix} \begin{bmatrix} \mathbf{x} \\ x_i \end{bmatrix} \end{aligned} \tag{11.64}$$

The state feedback control law with integral action is given by

$$\boxed{u = -\mathbf{K_a}\mathbf{x_a} = -\mathbf{K}\mathbf{x} - K_i x_i} \tag{11.65}$$

where $\mathbf{K} = \mathbf{K_a}(1:n)$ and $K_i = \mathbf{K_a}(n+1)$.

Substituting the control law of Eqn(11.65) into the augmented state equation (11.64) gives

$$\begin{aligned} \underbrace{\begin{bmatrix} \dot{\mathbf{x}} \\ \dot{x}_i \end{bmatrix}}_{\dot{\mathbf{x}}_a} &= \underbrace{\begin{bmatrix} \mathbf{A} & \mathbf{0}_{n\times 1} \\ -\mathbf{C} & 0 \end{bmatrix}}_{\mathbf{A_a}} \underbrace{\begin{bmatrix} \mathbf{x} \\ x_i \end{bmatrix}}_{\mathbf{x_a}} + \underbrace{\begin{bmatrix} \mathbf{B} \\ 0 \end{bmatrix}}_{\mathbf{B_a}} \begin{bmatrix} -\mathbf{K} & -K_i \end{bmatrix} \begin{bmatrix} \mathbf{x} \\ x_i \end{bmatrix} + \underbrace{\begin{bmatrix} \mathbf{0}_{n\times 1} \\ 1 \end{bmatrix}}_{B_1} r \\ &= \underbrace{\begin{bmatrix} \mathbf{A} & \mathbf{0}_{n\times 1} \\ -\mathbf{C} & 0 \end{bmatrix}}_{\mathbf{A_a}} \underbrace{\begin{bmatrix} \mathbf{x} \\ x_i \end{bmatrix}}_{\mathbf{x_a}} - \underbrace{\begin{bmatrix} \mathbf{B} \\ 0 \end{bmatrix}}_{\mathbf{B_a}} \underbrace{\begin{bmatrix} \mathbf{K} & K_i \end{bmatrix}}_{\mathbf{K_a}} \begin{bmatrix} \mathbf{x} \\ x_i \end{bmatrix} + \underbrace{\begin{bmatrix} \mathbf{0}_{n\times 1} \\ 1 \end{bmatrix}}_{\mathbf{B_1}} r \end{aligned} \tag{11.66}$$

Thus, the dynamics of the augmented system can be expressed as:

$$\dot{\mathbf{x}}_a = (\mathbf{A_a} - \mathbf{B_a K_a})\,\mathbf{x_a} + \mathbf{B_1}\mathrm{r} \quad \text{where}$$
$$\mathbf{A_a} = \begin{bmatrix} \mathbf{A} & 0 \\ -\mathbf{C} & 0 \end{bmatrix}, \mathbf{B_a} = \begin{bmatrix} \mathbf{B} \\ 0 \end{bmatrix}, \mathbf{K_a} = \begin{bmatrix} \mathbf{K} & \mathrm{K_i} \end{bmatrix}, \text{ and } \quad \mathbf{B_1} = \begin{bmatrix} 0 \\ 1 \end{bmatrix} \tag{11.67}$$

If $\mathbf{K_a}$ is designed such that the closed-loop augmented matrix $(\mathbf{A_a} - \mathbf{B_a K_a})$ is stable, then in the steady state

$$\lim_{\mathbf{t}\to\infty} \dot{\mathbf{x}}_\mathbf{a} = \mathbf{0} \implies \lim_{\mathbf{t}\to\infty} \mathbf{y(t)} = \mathbf{r}$$

By substituting the control signal $\mathrm{u} = -\mathrm{K}\mathbf{x} - \mathrm{K_i x_i}$ into the open-loop system in Eqn(11.64) results the closed-loop system

$$\underbrace{\begin{bmatrix} \dot{\mathbf{x}} \\ \dot{\mathrm{x}}_\mathrm{i} \end{bmatrix}}_{\dot{\mathbf{x}}_a} = \underbrace{\begin{bmatrix} \mathbf{A} - \mathbf{BK} & -\mathrm{BK_i} \\ -\mathbf{C} & 0 \end{bmatrix}}_{\mathbf{A_{cl}}} \underbrace{\begin{bmatrix} \mathbf{x} \\ \mathrm{x_i} \end{bmatrix}}_{\mathbf{x_a}} + \underbrace{\begin{bmatrix} 0_{\mathrm{n}\times 1} \\ 1 \end{bmatrix}}_{\mathbf{B_1}} \mathrm{r}$$
$$\mathrm{y} = \underbrace{\begin{bmatrix} \mathbf{C} & 0 \end{bmatrix}}_{\mathbf{C_a}} \underbrace{\begin{bmatrix} \mathbf{x} \\ \mathrm{x_i} \end{bmatrix}}_{\mathbf{x_a}} \tag{11.68}$$

EXAMPLE 11.19

Problem

Design a state feedback with an integral controller for the plant

$$\dot{\mathbf{x}} = \begin{bmatrix} -5 & -2 \\ 2 & 0 \end{bmatrix} \mathbf{x} + \begin{bmatrix} 1 \\ 0 \end{bmatrix} \mathrm{u}, \quad \mathrm{y} = \begin{bmatrix} 0 & 1 \end{bmatrix} \mathbf{x}$$

to yield a step response with 5% overshoot, a settling time of 2 seconds, and zero steady state error. Place the third pole at -10.

SOLUTION:

Step-1: Mapping Desired Specifications to Pole Locations:

Given $\%\mathrm{M_p} = 5.0$, settling time $\mathrm{T_s} = 2$ seconds.

Compute $\gamma = -\ln(\%\mathrm{M_p}/100) = 2.9957$

$$\xi = \frac{\gamma}{\sqrt{\gamma^2 + \pi^2}} = 0.6901$$
$$\mathrm{T_s} = \frac{4}{\xi \omega_\mathrm{n}} = 2 \implies \omega_\mathrm{n} = 2.8981 \text{ and } \omega_\mathrm{d} = 2.0974$$

This gives the desired pole locations $-\xi\omega_n \pm j\omega_d = -2 \pm j2.0974$.

Step-2: Determine the augmented matrices as:

$$\mathbf{A_a} = \begin{bmatrix} \mathbf{A} & 0 \\ -\mathbf{C} & 0 \end{bmatrix} = \begin{bmatrix} -5 & -2 & 0 \\ 2 & 0 & 0 \\ 0 & -1 & 0 \end{bmatrix}, \quad \mathbf{B_a} = \begin{bmatrix} \mathbf{B} \\ 0 \end{bmatrix} = \begin{bmatrix} 1 \\ 0 \\ 0 \end{bmatrix}$$

$$\mathbf{C_a} = \begin{bmatrix} \mathbf{C} & 0 \end{bmatrix} = \begin{bmatrix} 0 & 1 & 0 \end{bmatrix}, \quad \text{and} \quad \mathbf{B_1} = \begin{bmatrix} 0 \\ 0 \\ 1 \end{bmatrix}$$

Step-3: Test the controllability of the pair $\begin{bmatrix} \mathbf{B_a} & \mathbf{A_a B_a} & \mathbf{A_a^2 B_a} \end{bmatrix}$.

$$\mathbf{Q_C} = \begin{bmatrix} \mathbf{B_a} & \mathbf{A_a B_a} & \mathbf{A_a^2 B_a} \end{bmatrix} = \begin{bmatrix} 1 & -5 & 21 \\ 0 & 2 & -10 \\ 0 & 0 & -2 \end{bmatrix}$$

Its rank is 3, and hence the system is controllable.

Step-4: Determine the closed-loop characteristic equation corresponding to the desired pole locations. Let us place the poles at $p_1 = -2 + j2.0974$ and $p_2 = -2 - j2.0974$. This gives

$$\mathbf{\Delta_{cl}}(s) = s^3 + 14s^2 + 48.4s + 84$$

Step-5: Determine the state feedback gain matrix $\mathbf{K_a} = \begin{bmatrix} \mathbf{K} & K_i \end{bmatrix}$

using either of the methods discussed before (left as an exercise). This gives

$$K_a = \begin{bmatrix} K & K_i \end{bmatrix} = \begin{bmatrix} 9 & 22.1995 & -41.995 \end{bmatrix}$$

Step-6: Determine the closed-loop system.

Substituting the gains in Eqn(11.68) gives the closed-loop system

$$\dot{\mathbf{x}}_{cl} = \begin{bmatrix} \dot{\mathbf{x}} \\ \dot{x}_i \end{bmatrix} = A_{cl}\mathbf{x}_{cl} + \mathbf{B}_{cl}r, \quad y = \mathbf{C}_{cl}\mathbf{x}_{cl}$$

where

$$\mathbf{A}_{cl} = \begin{bmatrix} -14 & -24.1995 & 41.995 \\ 2 & 0 & 0 \\ 0 & -1 & 0 \end{bmatrix}, \quad \mathbf{B}_{cl} + \begin{bmatrix} 0 \\ 0 \\ 1 \end{bmatrix}, \text{ and } \mathbf{C}_{cl} = \begin{bmatrix} 0 & 1 & 0 \end{bmatrix}$$

Step-6: Compute Steady State Error:

Now the steady state error to the step input is

$$e_{ss} = 1 + C_{cl}A_{cl}^{-1}B_{cl} = 0$$

Alternatively, we can determine the steady state error from the closed-loop transfer function of the system, which is computed as

$$T(s) = \frac{Y(s)}{R(s)} = \frac{83.99}{s^3 + 14s^2 + 48.4s + 83.99}$$

This gives $e_{ss} = 1 - T(0) = 1 - 1 = 0$.

The MATLAB code can be found in file `Example11_19.m`.

CONCEPT REINFORCEMENT PROBLEM 11.9

Problem

For the simple pendulum system considered in Example 11.18

a. Design a state feedback integral controller which would place the closed-loop poles p_1,p_2 at $-5 \pm j$. Place the third pole p_3 at -10. Show that the steady state error due to a step input is zero.

b. Investigate the robustness of this controller considering the perturbed pendulum system where the mass m is changed by 20%.

11.16 Design of Observer

So far, in all our discussions on state feedback controller design, we have explicitly assumed that

1. The plant model is perfectly known and
2. All the states are accessible (measurable) for feedback.

If assumption-1 is not true, that is, the plant model has some degree of uncertainty, then this problem can be dealt with using the **robust control** design framework. However, in this chapter, we will focus on the second case when assumption 2 is not true. This is a very common phenomenon, as often either it may not be possible to measure all the states, or it is too costly to measure them. In these circumstances, we can try either of the following:

1. Design an output feedback controller. This is similar to the proportional feedback, but is often difficult to design in general(will not be covered here).
2. Estimate the states from the input-output measurements by creating a replica of the original dynamic system and replacing the state feedback

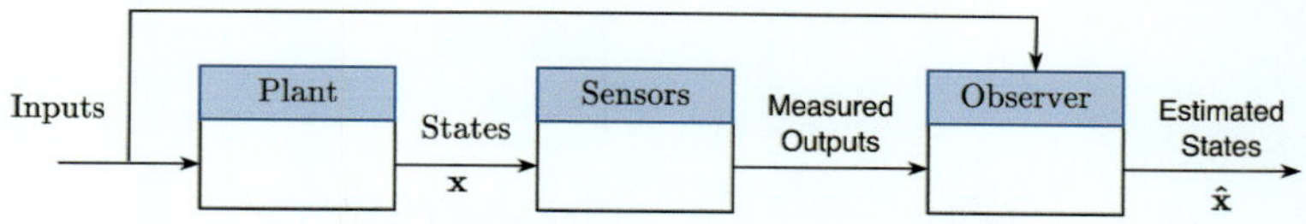

Figure 11.10: Block diagram of system and its observer

control law $\mathbf{u}(t) = -\mathbf{Kx}(t)$ with $\mathbf{u}(t) = -\mathbf{K}\hat{\mathbf{x}}(t)$, where $\hat{\mathbf{x}}(t)$ denotes the estimated states. This can be carried out by an *observer or estimator*

The structure of an observer is shown in Figure 11.10

This is a device that can estimate the states from the measurements of inputs and outputs. It can be realized using either hardware electronic components or software programs. If the observer estimates all the states, it is called a *full-order state observer.* Sometimes, we may not need to estimate all the state variables, as some of the state variables are measurable. An observer that estimates fewer states is called a *reduced-order observer.* In this section, we will discuss full-order observers.

Consider a system defined by

$$\begin{aligned}\dot{\mathbf{x}} &= \mathbf{Ax}(t) + \mathbf{B}u \\ y &= \mathbf{Cx}\end{aligned} \tag{11.69}$$

where matrices **A**, **B**, and **C** are known, the input u is known, the output y is measurable and the initial values of the states $\mathbf{x}(0)$ are unknown

It is further assumed that the system is completely observable. The objective is to create a dynamic system (called *observer*) whose states

$$\hat{\mathbf{x}}(t) = \mathbf{x}(t)$$

We can estimate these states either by using an open-loop observer or a closed-loop observer.

11.16.1 Open-Loop Observer

Consider the system of Eqn(11.69) and create a replica of this system using the known matrices **A**, **B**, and **C** and run the simulation in parallel with the original system as shown in Figure 11.11 which gives

$$\begin{aligned}\dot{\hat{\mathbf{x}}} &= \mathbf{A}\hat{\mathbf{x}}(t) + \mathbf{B}u \\ \hat{y} &= \mathbf{C}\hat{\mathbf{x}}\end{aligned} \tag{11.70}$$

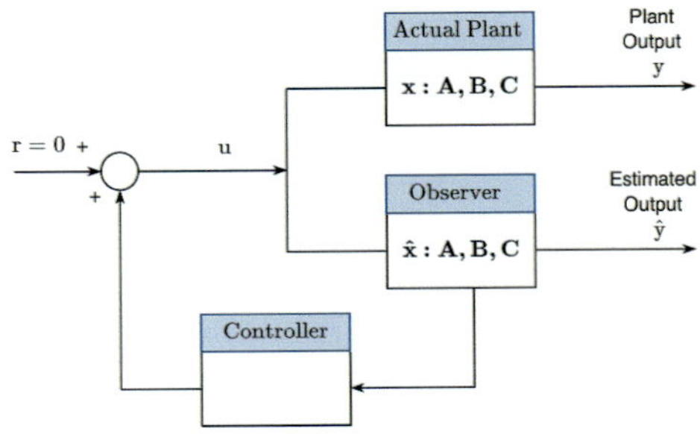

Figure 11.11: Open-loop observer

Subtracting Eqn (11.70) from Eqn(11.69) gives

$$\dot{\mathbf{x}} - \dot{\hat{\mathbf{x}}} = \mathbf{A}\left(\mathbf{x} - \hat{\mathbf{x}}\right)$$
$$\mathrm{y} - \hat{\mathrm{y}} = \mathbf{C}\left(\mathbf{x} - \hat{\mathbf{x}}\right) \tag{11.71}$$

If we define the state estimation error as: $\mathbf{e} = \mathbf{x} - \hat{\mathbf{x}}$, then the error dynamics can be expressed as: (from Eqn(11.71))

$$\dot{\mathbf{e}} = \mathbf{A}\mathbf{e} \tag{11.72}$$

This has a solution

$$\boxed{\mathbf{e}(\mathrm{t}) = \mathrm{e}^{\mathbf{A}\mathrm{t}}\mathbf{e}(0)} \tag{11.73}$$

Notice that the dynamics of the error depend on the system matrix **A**. This raises the question: **does this guarantee that the error will remain zero at all times, or will it merely approach zero asymptotically?** We now have two possible scenarios:

1. If the initial values of the estimated states and original states are equal, i.e., $\hat{\mathbf{x}}(0) = \mathbf{x}(0)$, then $\mathbf{e}(0) = 0$ and the open-loop observer will yield the correct estimates. However, we do not know $\mathbf{x}(0)$. Therefore, the possibility of $\mathbf{e}(0) = 0$ is very rare in practice. Suppose $\mathbf{e}(0) \neq 0$, then

2. If the system matrix **A** is stable, then (see Eqn(11.73))

$$\lim_{\mathrm{t}\to\infty} \mathbf{e}(\mathrm{t}) = 0$$

Since the dynamics of the estimation error are determined by the eigenvalues of the system, the designer has no control over the speed of its convergence. In practice, we want the error to converge to zero as fast as possible. However, there is no obvious way to modify the error dynamics to achieve faster convergence. This, therefore, raises the following question:

Can we devise a method to adjust the estimation error dynamics to accelerate the convergence between the actual and estimated states? This question serves as the motivation for designing a closed-loop observer.

11.16.2 Closed-Loop Observer

So far, we have focused on the true and estimated states. Let us now compare the estimated output with the measured output of the system to assess how closely they match. The measured and estimated outputs are respectively given by:

$$\mathrm{y} = \mathbf{Cx} \quad \text{and} \quad \hat{\mathrm{y}} = \mathbf{C}\hat{\mathbf{x}} \tag{11.74}$$

The difference between y and $\hat{\mathrm{y}}$ is given as:

$$\mathrm{y} - \hat{\mathrm{y}} = \mathbf{C}\left(\mathbf{x} - \hat{\mathbf{x}}\right) \tag{11.75}$$

Let us use this difference as additional information through a gain **L** to improve the state estimates at the desired speed. It gives the closed-loop observer shown in Figure 11.12.

A more detailed diagram of the closed-loop observer is depicted in Figure (11.13).

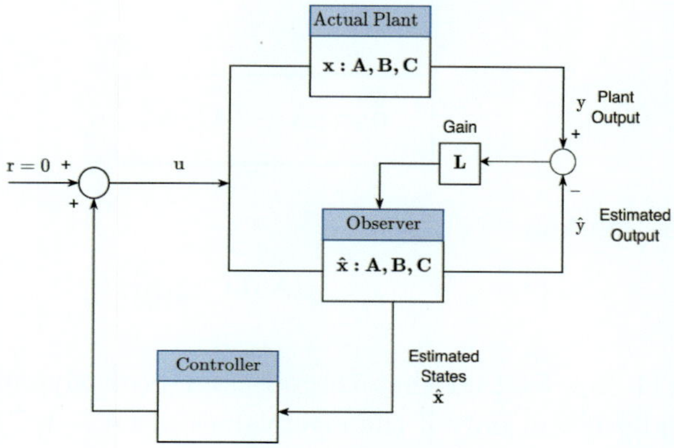

Figure 11.12: Closed-loop observer

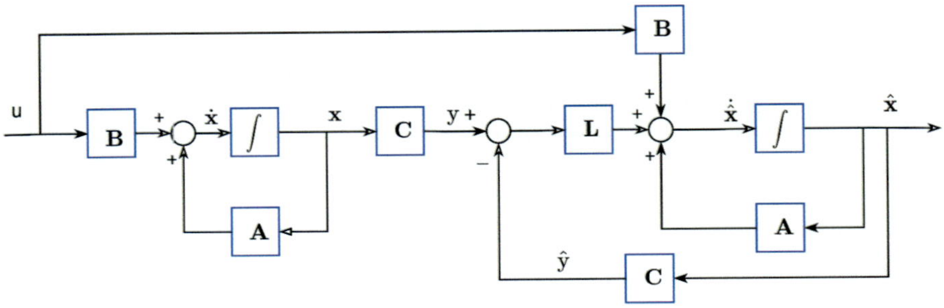

Figure 11.13: Block diagram of system with observer (closed-loop)

Analysis of Closed-Loop Observer

The dynamics of the estimated states are described by

$$\begin{aligned}\dot{\hat{\mathbf{x}}} &= \mathbf{A}\hat{\mathbf{x}} + \mathbf{B}\mathrm{u} + \mathbf{L}\left(\mathrm{y} - \hat{\mathrm{y}}\right)\\ &= \mathbf{A}\hat{\mathbf{x}} + \mathbf{B}\mathrm{u} + \mathbf{L}\left(\mathbf{C}\mathbf{x} - \mathbf{C}\hat{\mathbf{x}}\right)\\ &= \left(\mathbf{A} - \mathbf{L}\mathbf{C}\right)\hat{\mathbf{x}} + \mathbf{B}\mathrm{u} + \mathbf{L}\mathbf{y}\end{aligned} \tag{11.76}$$

Thus, the equation for the full-order state observer is given by

$$\boxed{\dot{\hat{\mathbf{x}}} = \left(\mathbf{A} - \mathbf{L}\mathbf{C}\right)\hat{\mathbf{x}} + \mathbf{B}\mathrm{u} + \mathbf{L}\mathbf{y}} \tag{11.77}$$

Now the dynamics of state estimation error are computed as:

$$\begin{aligned}\dot{\mathbf{e}} &= \dot{\mathbf{x}} - \dot{\hat{\mathbf{x}}}\\ &= \left[\mathbf{A}\mathbf{x} + \mathbf{B}\mathrm{u}\right] - \left[\mathbf{A}\hat{\mathbf{x}} + \mathbf{B}\mathrm{u} + \mathbf{L}\left(\mathrm{y} - \hat{\mathrm{y}}\right)\right]\\ &= \left[\mathbf{A}\mathbf{x} + \mathbf{B}\mathrm{u}\right] - \left[\mathbf{A}\hat{\mathbf{x}} + \mathbf{B}\mathrm{u} + \mathbf{L}\left(\mathbf{C}\mathbf{x} - \mathbf{C}\hat{\mathbf{x}}\right)\right]\\ &= \mathbf{A}\left(\mathbf{x} - \hat{\mathbf{x}}\right) - \mathbf{L}\mathbf{C}\left(\mathbf{x} - \hat{\mathbf{x}}\right)\\ &= \left(\mathbf{A} - \mathbf{L}\mathbf{C}\right)\mathbf{e}\end{aligned} \tag{11.78}$$

Thus

$$\boxed{\dot{\mathbf{e}} = \left(\mathbf{A} - \mathbf{L}\mathbf{C}\right)\mathbf{e}} \tag{11.79}$$

This has the solution

$$\mathbf{e}(t) = \mathrm{e}^{(\mathbf{A}-\mathbf{L}\mathbf{C})\mathrm{t}}\mathbf{e}(0) \tag{11.80}$$

From Eqn(11.80), we find that the observer error dynamics will go to zero asymptotically if and only if the eigenvalues of $(\mathbf{A} - \mathbf{L}\mathbf{C})$ are in the left half plane. Furthermore, since the system matrix $\mathbf{A}$ and the output matrix $\mathbf{C}$ are known, we can design the matrix $\mathbf{L}$ to control the observer error dynamics.

Note that, in the open-loop observer, we do not have any option to modify the error dynamics. Thus, the problem of designing the observer now reduces to the pole assignment problem (discussed earlier) where we determine the observer gain matrix $\mathbf{L}$ to place the eigenvalues of $(\mathbf{A}-\mathbf{LC})$ at desired locations.

The following theorem gives the necessary and sufficient condition to place the observer poles in desired locations.

Theorem 11.13. If the system is completely state observable, i.e., if the pair $[\mathbf{A},\mathbf{C}]$ is observable, then there exists an observer gain matrix $\mathbf{L}$ that can place the eigenvalues of $(\mathbf{A}-\mathbf{LC})$ at any desired locations in the complex plane.

11.16.3 Different Methods of Designing Observer Gain Matrix

The design of the observer follows a similar approach of state feedback controller design for pole assignment. Consider the system described by

$$\begin{aligned}\dot{\mathbf{x}}(t) &= \mathbf{A}\mathbf{x}(t)+\mathbf{B}\mathbf{u}(t)\\ y &= \mathbf{C}\mathbf{x}\end{aligned}$$

The observer error dynamics is defined as: $\dot{\mathbf{e}}=(\mathbf{A}-\mathbf{LC})\,\mathbf{e}$

The objective is to determine the observer gain matrix $\mathbf{L}$ which will place the eigenvalues of $(\mathbf{A}-\mathbf{LC})$ at $\mu_1,\mu_2,\ldots,\mu_n$. The first step in all the methods is to check the observability of the system. If the system is observable, then proceed with the following steps:

Method-1: Transformation Approach

Step-1: Determine the coefficients $a_1,a_2,\ldots,a_n$ of the open-loop characteristic equation

$$\det(s\mathbf{I}-\mathbf{A})=s^n+a_1s^{n-1}+a_2s^{n-2}+\cdots+a_{n-1}s+a_n$$

Step-2 Determine the transformation matrix $\mathbf{T}$ from

$$\mathbf{T}=(\mathbf{W}\mathbf{Q}_O)^{-1}$$

where the $\mathbf{Q_O}$ is the observability matrix i.e.,

$$\mathbf{Q_O} = \begin{bmatrix} \mathbf{C} \\ \mathbf{CA} \\ \vdots \\ \mathbf{CA}^{n-2} \\ \mathbf{CA}^{n-1} \end{bmatrix}, \quad \& \quad \mathbf{W} = \begin{bmatrix} a_{n-1} & a_{n-2} & \dots & a_1 & 1 \\ a_{n-2} & a_{n-3} & \dots & 1 & 0 \\ \vdots & \vdots & \vdots & \vdots & \vdots \\ a_1 & 1 & \dots & 0 & 0 \\ 1 & 0 & \dots & 0 & 0 \end{bmatrix}$$

Note: If the system is in observable canonical form, then $\mathbf{T} = \mathbf{I}$.

Step-3: Compute the characteristic equation corresponding to the desired eigenvalues $\mu_1, \mu_2, \dots, \mu_n$ of the observer:

$$(s - \mu_1)(s - \mu_2)\dots(s - \mu_n) = s^n + \alpha_1 s^{n-1} + \dots + \alpha_{n-1}s + \alpha_n$$

Step-4: Determine the observer gain matrix $\mathbf{L}$ from:

$$\mathbf{L} = \mathbf{T}\begin{bmatrix} \alpha_n - a_n \\ \alpha_{n-1} - a_{n-1} \\ \alpha_{n-2} - a_{n-2} \\ \vdots \\ \alpha_2 - a_2 \\ \alpha_1 - a_1 \end{bmatrix} = (\mathbf{W}\mathbf{Q_O})^{-1}\begin{bmatrix} \alpha_n - a_n \\ \alpha_{n-1} - a_{n-1} \\ \alpha_{n-2} - a_{n-2} \\ \vdots \\ \alpha_2 - a_2 \\ \alpha_1 - a_1 \end{bmatrix}$$

Method-2: Determination Observer Gain L by Direct Substitution

When the order of the system is low (say $n \leq 3$), then it is simpler to compute the gain matrix $\mathbf{L}$ by direct substitution. We illustrate this for a third-order system ($n = 3$).

Step-1: Define the observer gain matrix $\mathbf{L}$ as:

$$\mathbf{L} = \begin{bmatrix} l_1 \\ l_2 \\ l_3 \end{bmatrix}$$

Step-2: Compute the desired characteristic equation of observer dynamics as:

$$(s - \mu_1)(s - \mu_2)(s - \mu_3) = s^3 + \alpha_1 s^2 + \alpha_2 s + \alpha_3$$

Step-3: Determine the observer gains l_1, l_2 and l_3 by equating like powers of s from:

$$\left|s\mathbf{I} - \mathbf{A} + \mathbf{LC}\right| = s^3 + \alpha_1 s^2 + \alpha_2 s + \alpha_3$$

Note: It is assumed that the system is observable; else matrix **L** cannot be determined.

Method-3: Determination of Observer Gain L using Ackermann's Formula

This formula essentially utilizes the Cayley-Hamilton theorem, which states that every square matrix satisfies its characteristic equation. The steps of determining observer gain are given here without a full derivation of this formula.

Step-1: Compute the desired characteristic equation of observer error dynamics

$$(s - \mu_1)(s - \mu_2)\ldots(s - \mu_n) = s^n + \alpha_1 s^{n-1} + \alpha_2 s^{n-2} + \cdots + \alpha_{n-1} s + \alpha_n$$

Step-2: Compute the observability matrix

$$\mathbf{Q_O} = \begin{bmatrix} \mathbf{C} \\ \mathbf{CA} \\ \vdots \\ \mathbf{CA}^{n-1} \end{bmatrix}$$

Step-3: Using the coefficients of the desired characteristic equation and system matrix **A**, compute

$$\phi(\mathbf{A}) = \mathbf{A}^n + \alpha_1 \mathbf{A}^{n-1} + \alpha_2 \mathbf{A}^{n-2} + \cdots + \alpha_{n-1}\mathbf{A} + \alpha_n \mathbf{I}$$

Step-4: Determine the observer gain **L** from

$$\mathbf{L} = \phi(\mathbf{A})\mathbf{Q}_O^{-1} \begin{bmatrix} 0 \\ 0 \\ \vdots \\ 0 \\ 1 \end{bmatrix}$$

Guidelines for Selecting Observer Poles

Selecting Observer Poles

1. Choose the observer poles at least five times faster than the controller poles. This ensures that the observer error converges quickly, allowing the controller to use accurate estimates.
2. Avoid placing observer poles too far in the left half-plane, as this increases system bandwidth and makes the system more susceptible to noise.
3. If the system has oscillatory modes (complex poles), it is often preferable to place observer poles as real to obtain faster and smoother error convergence.
4. In practice, observer poles are usually chosen between 3 to 10 times faster than the dominant controller poles, balancing convergence speed with noise sensitivity.
5. For higher-order systems, distribute observer poles rather than clustering them at a single location, to avoid numerical ill-conditioning.

Let us demonstrate these with an example.

EXAMPLE 11.20

Problem

An extender of a robot arm can be modeled as a linear actuator that extends or contracts the arm. The dynamics of such a system are typically governed by the actuator's motor and mechanical components (mass, damping, stiffness). The transfer function of a typical extender for a robot manipulator is given by

$$\frac{\mathrm{Y(s)}}{\mathrm{U(s)}} = \frac{25}{\mathrm{s}^2 + 6\mathrm{s} + 8}$$

where U(s) is the Laplace transform of the force of the human hand applied to the robot manipulator, and Y(s) is the Laplace transform of the force of the manipulator applied to the load. The state space model of the system is

$$\dot{\mathbf{x}} = \mathbf{A}\mathbf{x} + \mathbf{B}\mathrm{u}$$
$$\mathrm{y} = \mathbf{C}\mathbf{x} + \mathbf{D}\mathrm{u}$$

where $\mathbf{A} = \begin{bmatrix} 0 & 1 \\ -8 & -6 \end{bmatrix}, \quad \mathbf{B} = \begin{bmatrix} 0 \\ 25 \end{bmatrix}, \quad \mathbf{C} = \begin{bmatrix} 1 & 0 \end{bmatrix}, \quad \mathbf{D} = 0$

Design an observer by placing the observer poles at −5 and −5.

SOLUTION:

Step-1: Testing Observability From the system dynamics we have

$$\mathbf{A} = \begin{bmatrix} 0 & 1 \\ -8 & -6 \end{bmatrix} \quad \text{and} \quad \mathbf{C} = \begin{bmatrix} 1 & 0 \end{bmatrix}$$

The observability matrix $\mathbf{Q_O} = \begin{bmatrix} \mathbf{C} \\ \mathbf{CA} \end{bmatrix} = \begin{bmatrix} 1 & 0 \\ 0 & 1 \end{bmatrix}$

We find $\left|\mathbf{Q_O}\right| = 1$. Hence, the observability matrix has full rank and the system is *observable* and we can place the observer poles arbitrarily.

Step-2: Computation of Desired Characteristic Equation $\Delta_d(s)$: Since the desired pole locations are at -5 and -5, we have

$$\Delta_d(s) = (s+5)(s+5) = s^2 + 10s + 25 = 0$$

Equating this to $s^2 + \alpha_1 s + \alpha_2$ gives $\alpha_1 = 10$ & $\alpha_2 = 25$

Method-1: Bass-Gura Approach:

Step:1.1

Determine the characteristic equation of the system as:

$$\left|s\mathbf{I} - \mathbf{A}\right| = \left|\begin{bmatrix} s & 0 \\ 0 & s \end{bmatrix} - \begin{bmatrix} 0 & 1 \\ -8 & -6 \end{bmatrix}\right| = \left|\begin{bmatrix} s & -1 \\ 8 & s+6 \end{bmatrix}\right|$$
$$= s^2 + 6s + 8 = s^2 + a_1 s + a_2 = 0$$

This gives

$$a_1 = 6 \quad \text{and} \quad a_2 = 8$$

Step-1.2 Determine the transformation matrix $\mathbf{T} = (\mathbf{WQ_O})^{-1}$ where

$$\mathbf{W} = \begin{bmatrix} a_1 & 1 \\ 1 & 0 \end{bmatrix} = \begin{bmatrix} 6 & 1 \\ 1 & 0 \end{bmatrix}$$

Thus, the transformation matrix

$$\mathbf{T} = (\mathbf{WQ_O})^{-1} = \left(\begin{bmatrix} 6 & 1 \\ 1 & 0 \end{bmatrix}\begin{bmatrix} 1 & 0 \\ 0 & 1 \end{bmatrix}\right)^{-1}$$
$$= \left(\begin{bmatrix} 6 & 1 \\ 1 & 0 \end{bmatrix}\right)^{-1} = \begin{bmatrix} 0 & 1 \\ 1 & -6 \end{bmatrix}$$

The observer gain is given as:

$$\mathbf{L} = (\mathbf{WQ_O})^{-1}\begin{bmatrix}\alpha_2 - \mathrm{a}_2 \\ \alpha_1 - \mathrm{a}_1\end{bmatrix}$$
$$= \begin{bmatrix}0 & 1 \\ 1 & -6\end{bmatrix}\begin{bmatrix}17 \\ 4\end{bmatrix} = \begin{bmatrix}4 \\ -7\end{bmatrix} = \begin{bmatrix}\mathrm{l}_1 \\ \mathrm{l}_2\end{bmatrix}$$

Hence, the observer equation is expressed as:

$$\dot{\hat{\mathrm{x}}}_1 = \hat{\mathrm{x}}_2 + 4(\mathrm{x}_1 - \hat{\mathrm{x}}_1)$$
$$\dot{\hat{\mathrm{x}}}_2 = -8\hat{\mathrm{x}}_1 - 6\hat{\mathrm{x}}_2 + 25\mathrm{u} - 7(\mathrm{x}_2 - \hat{\mathrm{x}}_2)$$

Method-2: Direct Substitution Approach

Step-2.1: Define $\mathbf{L} = \begin{bmatrix}l_1 \\ l_2\end{bmatrix}$

Step 2.2: Express the closed-loop equation in terms observer gains l_1 and l_2 from $\left|\mathrm{s}\mathbf{I} - \mathbf{A} + \mathbf{LC}\right|$. This gives

$$\left|\mathrm{s}\mathbf{I} - \mathbf{A} + \mathbf{LC}\right| = \left|\begin{bmatrix}\mathrm{s} & 0 \\ 0 & \mathrm{s}\end{bmatrix} - \begin{bmatrix}0 & 1 \\ -8 & -6\end{bmatrix} + \begin{bmatrix}l_1 \\ l_2\end{bmatrix}\begin{bmatrix}1 & 0\end{bmatrix}\right|$$
$$= \left|\begin{bmatrix}\mathrm{s} & -1 \\ 8 & \mathrm{s}+6\end{bmatrix} + \begin{bmatrix}l_1 & 0 \\ l_2 & 0\end{bmatrix}\right| = \left|\begin{bmatrix}\mathrm{s}+\mathrm{l}_1 & -1 \\ 8+l_2 & \mathrm{s}+6\end{bmatrix}\right|$$
$$= \mathrm{s}^2 + (\mathrm{l}_1 + 6)\mathrm{s} + (6\mathrm{l}_1 + \mathrm{l}_2 + 8)$$

Step 2.3: Equate $|\mathrm{s}\mathbf{I} - \mathbf{A} + \mathbf{LC}|$ to the desired characteristic equation and determine the feedback gains. Now

$$\mathrm{s}^2 + (\mathrm{l}_1 + 6)\mathrm{s} - (6\mathrm{l}_1 + \mathrm{l}_2 + 8) = \mathrm{s}^2 + 10\mathrm{s} + 25$$

This gives

$$\mathbf{L} = \begin{bmatrix}l_1 \\ l_2\end{bmatrix} = \begin{bmatrix}4 \\ -7\end{bmatrix}$$

Method-3: Ackermann's Method

Step-3.1: Compute

$$\phi(\mathbf{A}) = \mathbf{A}^2 + \alpha_1\mathbf{A} + \alpha_2\mathbf{I}$$

This gives

$$\phi(\mathbf{A}) = \begin{bmatrix}0 & 1 \\ -8 & -6\end{bmatrix}^2 + 10\begin{bmatrix}0 & 1 \\ -8 & -6\end{bmatrix} + 25\begin{bmatrix}1 & 0 \\ 0 & 1\end{bmatrix} = \begin{bmatrix}17 & 4 \\ -32 & -7\end{bmatrix}$$

Step-3.2: Compute

$$\mathbf{L} = \phi(\mathbf{A}) \begin{bmatrix} \mathbf{C} \\ \mathbf{CA} \end{bmatrix}^{-1} \begin{bmatrix} 0 \\ 1 \end{bmatrix} = \begin{bmatrix} 17 & 4 \\ -32 & -7 \end{bmatrix} \begin{bmatrix} 1 & 0 \\ 0 & 1 \end{bmatrix}^{-1} \begin{bmatrix} 0 \\ 1 \end{bmatrix} = \begin{bmatrix} 4 \\ -7 \end{bmatrix}$$

The estimated states of the system are shown in Figure 11.14.

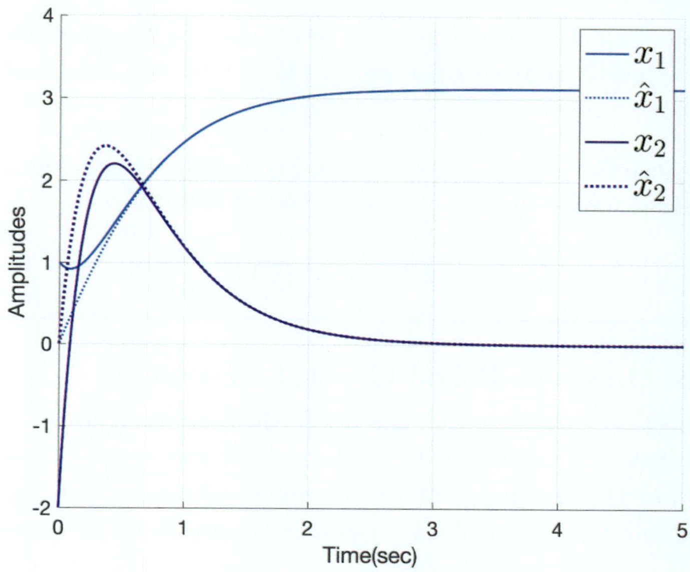

Figure 11.14: State estimation of an extender of a robot arm

The complete MATLAB program is in `Example11_20.m`

11.17 Observer-Based Control System

During the design of a state feedback controller for pole placement, we assumed that all the states are available for feedback. However, if some states are not measurable, an observer can be designed to estimate them, and these estimates replace the true states in the control law.

A common rule of thumb in observer design is to place its poles significantly faster than the closed-loop poles of the plant. This ensures that the estimation error decays quickly, so the state estimates can be treated almost as the true states. Consequently, the observer dynamics can be neglected in the overall analysis.

Hence, the design of an observer-based control system is carried out in two stages as described below.

Stage-1: Determine the state feedback gain matrix $\mathbf{K}$ which places the closed-loop eigenvalues at desired locations

Stage-2: Design the observer gain matrix $\mathbf{L}$ according to the desired observer dynamics.

Note on Observer Initial Conditions

i. Since $(\mathbf{A} - \mathbf{LC})$ is designed to be stable, the state estimation error will converge to zero irrespective of the initial conditions of the observer states.

ii. Because the initial output y(0) is known, the optimal choice (in a least-squares sense) for the observer's initial condition is

$$\hat{\mathbf{x}}(0) = \mathbf{C^T}\left(\mathbf{CC^T}\right)^{-1}\mathrm{y}(0)$$

Stability of Observer-Based Control System

The question arises whether the use of the estimated states $\hat{\mathbf{x}}(\mathrm{t})$, instead of the actual state $\mathbf{x}(\mathrm{t})$, will affect the stability of the system. To study this aspect, we obtain the dynamics of the complete observer-based closed-loop system and examine its eigenvalues. Now the dynamics of the system are

$$\begin{aligned}\dot{\mathbf{x}} &= \mathbf{Ax} + \mathbf{Bu}\\ \mathrm{y} &= \mathbf{Cx}\end{aligned} \tag{11.81}$$

The control input $\mathbf{u}$ in the observer based state feedback control is given by

$$\mathrm{u} = -\mathbf{K}\hat{\mathbf{x}} \tag{11.82}$$

where $\hat{\mathbf{x}}$ is the observed/estimated state. The observer equation is given by

$$\dot{\hat{\mathbf{x}}} = \mathbf{A}\hat{\mathbf{x}} + \mathbf{Bu} + \mathbf{L}\left(\mathbf{y} - \mathbf{C}\hat{\mathbf{x}}\right), \quad \hat{\mathbf{x}} = \hat{\mathbf{x}}_0 \tag{11.83}$$

The closed-loop system, therefore, becomes

$$\dot{\mathbf{x}} = \mathbf{Ax} + \mathbf{B}\left(-\mathbf{K}\hat{\mathbf{x}}\right) = \mathbf{Ax} - \mathbf{BK}\hat{\mathbf{x}} \tag{11.84}$$

$$\dot{\hat{\mathbf{x}}} = \left(\mathbf{A} - \mathbf{LC}\right)\hat{\mathbf{x}} - \mathbf{BK}\hat{\mathbf{x}} + \mathbf{Ly} = \left(\mathbf{A} - \mathbf{LC} - \mathbf{BK}\right)\hat{\mathbf{x}} + \mathbf{LCx} \tag{11.85}$$

This results in an observer-based feedback control system shown in Figure 11.15.

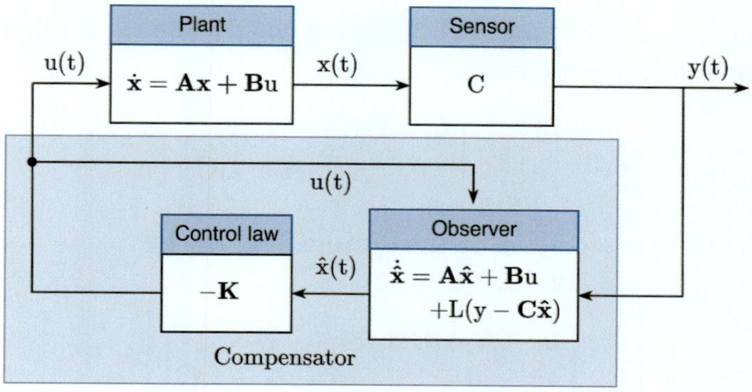

Figure 11.15: Observer-based control system

Combining Eqns(11.84) and (11.85), gives closed-loop composite system as:

$$\begin{bmatrix} \dot{\mathbf{x}} \\ \dot{\hat{\mathbf{x}}} \end{bmatrix} = \begin{bmatrix} \mathbf{A} & -\mathbf{BK} \\ \mathbf{LC} & \mathbf{A} - \mathbf{BK} - \mathbf{LC} \end{bmatrix} \begin{bmatrix} \mathbf{x} \\ \hat{\mathbf{x}} \end{bmatrix} \tag{11.86}$$

The closed-loop system will be stable provided the eigenvalues of the block matrix in Eqn(11.86) are in the left half plane. This will be shown alternately by transforming the closed-loop system matrix into a block triangular one as follows:

Separation Principle

While designing the observer, we have seen that $\hat{\mathbf{x}}(\mathrm{t})$ asymptotically approaches to $\mathbf{x}(\mathrm{t})$ if the eigenvalues of $(\mathbf{A} - \mathbf{LC})$ have negative real parts. Hence, for all practical purposes, $\hat{\mathbf{x}}(t)$ and $\mathbf{x}(t)$ will be identical. In this section, we investigate whether the dynamics of the closed-loop system is affected due to the presence of the observer.

Consider the system described by

$$\begin{aligned} \dot{\mathbf{x}} &= \mathbf{A}\mathbf{x} + \mathbf{B}\mathrm{u} \\ \mathrm{y} &= \mathbf{C}\mathbf{x} \end{aligned}$$

Substituting the control law $\mathrm{u} = -\mathbf{K}\hat{\mathbf{x}}$ into the above gives

$$\begin{aligned} \dot{\mathbf{x}} &= \mathbf{A}\mathbf{x} - \mathbf{BK}\hat{\mathbf{x}} = \mathbf{A}\mathbf{x} + \mathbf{BK}\mathbf{x} - \mathbf{BK}\mathbf{x} - \mathbf{BK}\hat{\mathbf{x}} \\ &= (\mathbf{A} - \mathbf{BK})\,\mathbf{x} + \mathbf{BK}\,(\mathbf{x} - \hat{\mathbf{x}}) \\ &= (\mathbf{A} - \mathbf{BK})\,\mathbf{x} + \mathbf{BK}\mathbf{e} \end{aligned} \tag{11.87}$$

where $\mathbf{e}$ is the difference between the actual state $\mathbf{x}$ and the observed state $\hat{\mathbf{x}}$. The observation error dynamics has been defined in Eqn(11.79) as:

$$\dot{\mathrm{e}} = (\mathbf{A} - \mathbf{LC})\,\mathbf{e} \tag{11.88}$$

Combining Eqns(11.87) and (11.88) gives the dynamics of the observer-based closed-loop control system as:

$$\begin{bmatrix}\dot{\mathrm{x}}\\ \dot{\mathrm{e}}\end{bmatrix} = \begin{bmatrix}\mathbf{A}-\mathbf{BK} & \mathbf{BK}\\ \mathbf{0}_{\mathrm{n}\times\mathrm{n}} & \mathbf{A}-\mathbf{LC}\end{bmatrix}\begin{bmatrix}\mathrm{x}\\ \mathrm{e}\end{bmatrix} \tag{11.89}$$

The characteristic equation of the system is given by

$$\left|\begin{bmatrix}\mathrm{s}\mathbf{I}-\mathbf{A}+\mathbf{BK} & -\mathbf{BK}\\ \mathbf{0} & \mathrm{s}\mathbf{I}-\mathbf{A}+\mathbf{LC}\end{bmatrix}\right| = 0 \tag{11.90}$$

Note that Eqn(11.90) is in block triangular form. We will use the following fact in our subsequent analysis.

Remark

The eigenvalues of a block triangular matrix are equal to the eigenvalues of the matrices along the diagonal blocks.

Using this remark, we can write from Eqn(11.90)

$$\left|\mathrm{s}\mathbf{I}-\mathbf{A}+\mathbf{BK}\right|\cdot\left|\mathrm{s}\mathbf{I}-\mathbf{A}+\mathbf{LC}\right| = 0 \tag{11.91}$$

Thus, the poles of the closed-loop observer-based control system are the union of observer poles and the controller poles. Since controllability of the system allows to place the eigenvalues of $(\mathbf{A}-\mathbf{BK})$ at arbitrary locations and observability of the system allows to place the poles of $(\mathbf{A}-\mathbf{LC})$ arbitrarily, **the design of the controller and the observer can be carried out independently** and then be combined to form the observer based state feedback control system. This property is called the **separation property**. Note that, for an nth order system, with a full-order state observer, the order of the characteristic equation of the entire closed-loop system becomes 2n. However, only n of these poles are related to the plant dynamics, and the other n correspond to the observer dynamics."

11.17.1 Transfer Function of Observer-Based Controller

It is known that the transfer function of a system can provide better physical insight into the behavior of a system. This tempts us to determine the transfer function of the observer-based control system and relate it to classical design.

Consider the plant defined by

$$\dot{\mathrm{x}} = \mathbf{A}\mathrm{x}+\mathbf{B}\mathrm{u}$$
$$\mathrm{y} = \mathbf{C}\mathrm{x}$$

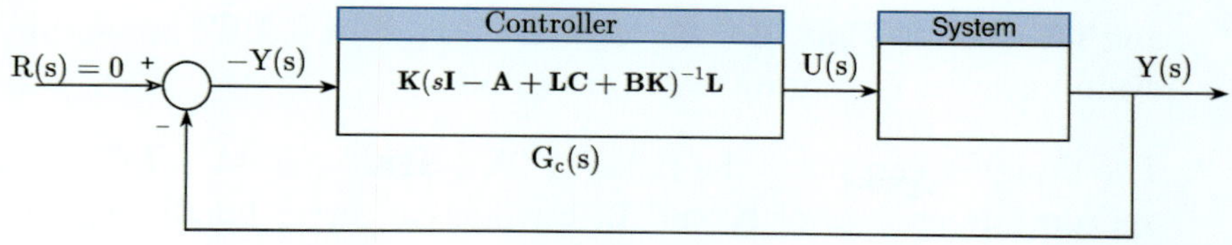

Figure 11.16: Transfer function representation of observer-based state feedback control system

It is assumed that the system is completely observable and is controlled by the observer-based state feedback controller $u = -\mathbf{K}\hat{\mathbf{x}}$. The observer dynamics is given by

$$\dot{\hat{\mathbf{x}}} = \mathbf{A}\hat{\mathbf{x}} + \mathbf{B}u + \mathbf{L}(y - \hat{y}) \quad \text{(see Eqn(11.83)} \qquad (11.92)$$

Substituting $u = -\mathbf{K}\hat{\mathbf{x}}$ into Eqn(11.92) gives the closed-loop dynamics as:

$$\begin{aligned} \dot{\hat{\mathbf{x}}} &= \mathbf{A}\hat{\mathbf{x}} - \mathbf{BK}\hat{\mathbf{x}} - \mathbf{LC}\hat{\mathbf{x}} + \mathbf{L}y \\ &= (\mathbf{A} - \mathbf{LC} - \mathbf{BK})\,\hat{\mathbf{x}} + \mathbf{L}y \end{aligned} \qquad (11.93)$$

Taking Laplace transform of Eqn(11.93) with zero initial conditions give

$$\begin{aligned} s\hat{\mathbf{X}}(s) &= (\mathbf{A} - \mathbf{LC} - \mathbf{BK})\,\hat{\mathbf{X}}(s) + \mathbf{L}Y(s) \\ \Longrightarrow \quad \hat{\mathbf{X}}(s) &= (s\mathbf{I} - \mathbf{A} + \mathbf{LC} + \mathbf{BK})^{-1}\,\mathbf{L}Y(s) \end{aligned} \qquad (11.94)$$

Hence, $U(s) = -\mathbf{K}\hat{\mathbf{X}}(s) = -\mathbf{K}\,(s\mathbf{I} - \mathbf{A} + \mathbf{LC} + \mathbf{BK})^{-1}\,\mathbf{L}Y(s)$

Thus, the transfer function $U(s)/Y(s)$ is expressed as:

$$\boxed{\frac{U(s)}{Y(s)} = -G_c(s) = -\mathbf{K}\,(s\mathbf{I} - \mathbf{A} + \mathbf{LC} + \mathbf{BK})^{-1}\,\mathbf{L}} \qquad (11.95)$$

The block diagram of the closed-loop system is shown in Figure 11.16.

Note that, from the perspective of a classical control system, $G_c(s)$ acts as a dynamic controller for this system. While designing an observer-based control system, there are a few important points to keep in mind.

Key Points

1. The matrix $(\mathbf{A} - \mathbf{LC} - \mathbf{BK})$ appears in the controller transfer function, but it does not represent the actual closed-loop system matrix. The true closed-loop dynamics are governed by a block-triangular system matrix whose eigenvalues are simply the union of the eigenvalues of $(\mathbf{A} - \mathbf{BK})$

and $(\mathbf{A}-\mathbf{LC})$. Thus, if both $(\mathbf{A}-\mathbf{BK})$ and $(\mathbf{A}-\mathbf{LC})$ are stable, the overall system is guaranteed to be stable.

2. The designer can place the poles of $(\mathbf{A}-\mathbf{BK})$ and $(\mathbf{A}-\mathbf{LC})$ through appropriate choices of $\mathbf{K}$ and $\mathbf{L}$, but has no direct handle on classical robustness measures such as gain margin and phase margin. This is a limitation compared with transfer–function–based loop-shaping.

EXAMPLE 11.21

Problem

Consider the system represented in state variable form

$$\dot{\mathbf{x}} = \mathbf{A}\mathbf{x} + \mathbf{B}\mathrm{u} \quad \text{and } \mathrm{y} = \mathbf{C}\mathbf{x} + \mathbf{D}\mathrm{u} \qquad \text{where}$$

$$\mathbf{A} = \begin{bmatrix} 2 & 1 \\ -5 & -10 \end{bmatrix}, \quad \mathbf{B} = \begin{bmatrix} 4 \\ 1 \end{bmatrix}, \quad \mathbf{C} = \begin{bmatrix} 1 & -2 \end{bmatrix}, \text{ and } \mathbf{D} = 0$$

a. Verify that the system is observable and controllable.

b. If so, design a full state-feedback law and an observer by placing the closed-loop system poles at $s_{1,2} = -2 \pm j\sqrt{3}$ and the observer poles at $s_{1,2} = -15$.

SOLUTION:

a. The controllability matrix

$$\mathbf{Q_C} = \begin{bmatrix} 4 & 9 \\ 1 & -30 \end{bmatrix} \text{ and det } (Q_C) = -129 \neq 0$$

The observability matrix

$$\mathbf{Q_O} = \begin{bmatrix} 1 & -2 \\ 12 & 21 \end{bmatrix} \text{ and det } (Q_O) = 45 \neq 0$$

Hence, the system is both controllable and observable.

b. For the design of a state feedback controller, the desired closed-loop characteristic equation is

$$\Delta_d(s) = s^2 + 4s + 7$$

Following the procedure discussed in previous examples, the state feedback gain is computed as:

$$\mathbf{K} = \begin{bmatrix} -0.5116 & -1.9535 \end{bmatrix}$$

For the design of the observer, the desired closed-loop characteristic equation is

$$\Delta_o(s) = s^2 + 30s + 225$$

The observer gain matrix

$$\mathbf{L} = \begin{bmatrix} 13.111 \\ -4.444 \end{bmatrix}$$

To obtain the response to the given initial condition $\mathbf{x}(0)$, first we get the closed-loop system by substituting the control input $u = -\mathbf{K}\hat{\mathbf{x}}$ into the system equation. This gives the closed-loop system

$$\begin{bmatrix} \dot{\mathbf{x}} \\ \dot{\mathbf{e}} \end{bmatrix} = \begin{bmatrix} \mathbf{A} - \mathbf{BK} & \mathbf{BK} \\ \mathbf{0}_{n\times n} & \mathbf{A} - \mathbf{LC} \end{bmatrix} \begin{bmatrix} \mathbf{x} \\ \mathbf{e} \end{bmatrix}$$
$$= \begin{bmatrix} 4.0465 & 8.814 & -2.0465 & -7.8240 \\ -4.4884 & -8.0465 & -0.5116 & -1.9535 \\ 0 & 0 & -11.111 & 27.222 \\ 0 & 0 & -0.5556 & -18.889 \end{bmatrix} \begin{bmatrix} \mathbf{x} \\ \mathbf{e} \end{bmatrix}$$

The state response of both the regulator and observer-based regulator is shown in Figure 11.17.

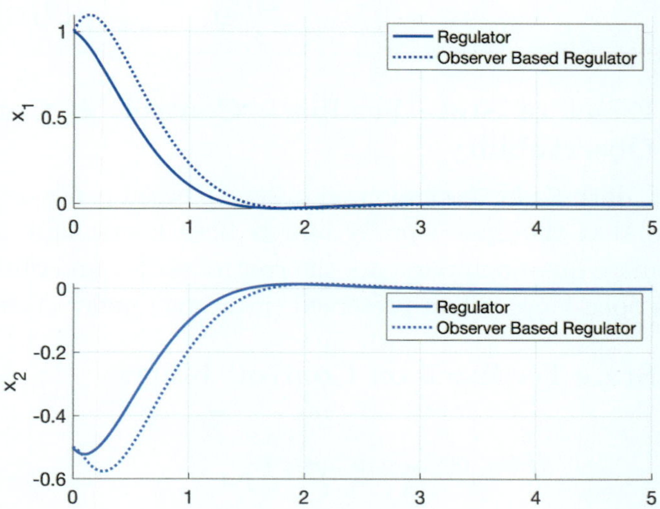

Figure 11.17: States for regulator and observer based state regulator estimation

The MATLAB code is in `Example11_21.m`

CONCEPT REINFORCEMENT PROBLEM 11.10

Problem

Consider a system represented by

$$\dot{\mathbf{x}} = \mathbf{A}\mathbf{x} + \mathbf{B}\mathrm{u}, \quad \mathrm{y} = \mathbf{C}\mathbf{x}$$

where

$$\mathbf{A} = \begin{bmatrix} -1 & 2 & 0 \\ 4 & 0 & 7 \\ -2 & 1 & 5 \end{bmatrix}, \quad \mathbf{B} = \begin{bmatrix} 5 \\ 0 \\ 1 \end{bmatrix}, \quad \text{and } \mathbf{C} = \begin{bmatrix} 1 & 2 & 1 \end{bmatrix}$$

Using MATLAB

a. Design a state feedback regulator $\mathrm{u} = -\mathbf{K}\mathbf{x}$, such that the poles of the closed-loop system are at $-3 \pm \mathrm{j}\sqrt{4}$ and at -5.

b. Design an observer such that its poles are located at -10,-10 and at -20.

c. Obtain the state responses of the regulator and observer-based regulator. Consider the initial condition of the states and estimated states as:

$$\mathbf{x}(0) = \begin{bmatrix} 1 \\ -0.5 \\ -0.5 \end{bmatrix} \quad \hat{\mathbf{x}}(0) = \begin{bmatrix} 0 \\ 0 \\ 0 \end{bmatrix}$$

11.17.2 Effect of State Feedback Control on Controllability and Observability

So far, during our discussion on observer-based control systems, we have assumed that the open-loop system is both controllable and observable. An immediate question arises: are the controllability and observability properties of the open-loop system preserved (invariant) under state feedback?

Effect of State Feedback on Controllability:

Theorem 11.14. *If the open-loop system*

$$\dot{\mathbf{x}} = \mathbf{A}\mathbf{x}(\mathrm{t}) + \mathbf{B}\mathbf{u}(\mathrm{t})$$

is completely state-controllable, then the closed-loop system

$$\dot{\mathbf{x}} = (\mathbf{A} - \mathbf{B}\mathbf{K})\mathbf{x}(\mathrm{t}) + \mathbf{B}\mathbf{r}(\mathrm{t})$$

obtained through state feedback control

$$\mathbf{u}(\mathrm{t}) = -\mathbf{K}\mathbf{x}(\mathrm{t}) + \mathbf{r}(\mathrm{t})$$

is also completely state-controllable.

Alternatively, if the pair $(\mathbf{A}, \mathbf{B})$ *is uncontrollable, then no choice of* $\mathbf{K}$ *can make* $(\mathbf{A} - \mathbf{BK}, \mathbf{B})$ *controllable. Hence, state feedback cannot recover controllability if it is absent in the open-loop system.*

Effect of State Feedback on Observability:

Theorem 11.15. *Even if the open-loop system is controllable and observable, applying state feedback control*

$$\mathbf{u}(t) = -\mathbf{K}\mathbf{x}(t) + \mathbf{r}(t)$$

may render the closed-loop system unobservable.
In other words, unlike controllability, the observability of the closed-loop system is generally **not preserved** *under state feedback.*

Let us illustrate this with an example.

EXAMPLE 11.22

Problem:

Consider the system $\dot{\mathbf{x}} = \mathbf{A}\mathbf{x} + \mathbf{B}\mathrm{u}$, $\mathrm{y} = \mathbf{C}\mathbf{x}$, where

$$\mathbf{A} = \begin{bmatrix} 0 & 1 \\ -3 & -4 \end{bmatrix}, \quad \mathbf{B} = \begin{bmatrix} 1 \\ 1 \end{bmatrix}$$

The system is controlled by a state feedback controller

$$\mathrm{u} = -\mathbf{K}\mathbf{x} + \mathrm{r}, \quad \text{where} \quad \mathbf{K} = \begin{bmatrix} \mathrm{k}_1 & \mathrm{k}_2 \end{bmatrix}$$

Determine the conditions when the closed-loop system with state feedback will be unobservable for the following three cases when

a. $\mathbf{C} = \begin{bmatrix} 1 & 0 \end{bmatrix}$ b. $\mathbf{C} = \begin{bmatrix} 0 & 1 \end{bmatrix}$ c. $\mathbf{C} = \begin{bmatrix} 2 & 1 \end{bmatrix}$

SOLUTION:

The closed-loop system is described by $\dot{\mathbf{x}} = (\mathbf{A} - \mathbf{BK})\,\mathbf{x} + \mathbf{B}\mathrm{r}$. Now

$$\text{Now} \quad (\mathbf{A} - \mathbf{BK}) = \begin{bmatrix} -\mathrm{k}_1 & 1 - \mathrm{k}_2 \\ -3 - \mathrm{k}_1 & -4 - \mathrm{k}_2 \end{bmatrix}$$

a. For $\mathbf{C} = \begin{bmatrix} 1 & 0 \end{bmatrix}$, the observability matrix is given by

$$\mathbf{Q_O} = \begin{bmatrix} \mathbf{C} \\ \mathbf{C}(\mathbf{A} - \mathbf{BK}) \end{bmatrix} = \begin{bmatrix} 1 & 0 \\ -k_1 & 1 - k_2 \end{bmatrix}$$

The closed-loop system is unobservable if $|\mathbf{Q_O}| = \mathbf{0}$. This gives

$$1 - k_2 = 0, \quad \Longrightarrow \quad k_2 = 1$$

b. For $\mathbf{C} = \begin{bmatrix} 0 & 1 \end{bmatrix}$, the observability matrix is given by

$$\mathbf{Q_O} = \begin{bmatrix} \mathbf{C} \\ \mathbf{C}(\mathbf{A} - \mathbf{BK}) \end{bmatrix} = \begin{bmatrix} 0 & 1 \\ -3 - k_1 & -4 - k_2 \end{bmatrix}$$

The closed-loop system is unobservable if $|\mathbf{Q_O}| = \mathbf{0}$. This gives

$$3 + k_1 = 0, \quad \Longrightarrow \quad k_1 = -3$$

c. For $\mathbf{C} = \begin{bmatrix} 2 & 1 \end{bmatrix}$, the observability matrix is given by

$$\mathbf{Q_O} = \begin{bmatrix} \mathbf{C} \\ \mathbf{C}(\mathbf{A} - \mathbf{BK}) \end{bmatrix} = \begin{bmatrix} 2 & 1 \\ -3 - 3k_1 & -3k_2 - 2 \end{bmatrix}$$

The closed-loop system is unobservable if $|\mathbf{Q_O}| = \mathbf{0}$. This gives

$$3k_1 - 6k_2 - 1 = 0$$

11.18 Observer-Based Regulation and Tracking

The regulators and tracking controllers designed in Sections 11.13 and 11.14 are essentially static, because the control signal $\mathbf{u}$ depends on the present values of the states and the reference signal. Furthermore, it requires that all states of the system be measured. However, when all the states of the system cannot be measured, we estimate the unmeasured states using an **observer**, or **state estimator**, which reconstructs the states from the measurements of outputs $\mathbf{y}(\mathbf{t})$ and known inputs $\mathbf{u}(\mathbf{t})$. The combination of state feedback and state estimation gives a **dynamic** feedback controller. The design procedure of the dynamic versions of the regulators and tracking controllers is similar to their static counterparts, except that the estimated states replace the states in the control law. That is, this controller design is carried out in two phases. In the first phase, an estimate of the states is generated by a state observer. Next, these state estimates are treated as if they were equal to the exact state of the system and are used by the static state feedback controller. In the following, we summarize the equations for synthesizing both the static and dynamic controllers.

SOLUTION:

Let us express the system and input matrices as:

$$\mathbf{A} = \left[\begin{array}{c|c} \mathbf{A}_{11} & \mathbf{A}_{12} \\ \hline \mathbf{A}_{21} & \mathbf{A}_{22} \end{array}\right] = \left[\begin{array}{c|c} 0 & 1 \\ \hline -36 & -12 \end{array}\right] \quad \text{and} \quad \mathbf{B} = \left[\begin{array}{c} \mathbf{B}_1 \\ \hline \mathbf{B}_2 \end{array}\right] = \left[\begin{array}{c} 0 \\ \hline 1 \end{array}\right]$$

This gives $\mathbf{A}_{11} = 0$, $\mathbf{A}_{12} = 1$, $\mathbf{A}_{21} = -36$, $\mathbf{A}_{22} = -12$, $\mathbf{B}_1 = 0$ and $\mathbf{B}_2 = 1$.

Now the observer eigenvalue is to be at -15. Let $\mathrm{L} = \mathrm{l}_1$. Thus

$$\lambda - \mathbf{A}_{22} + \mathbf{L}\mathbf{A}_{12} = \lambda + 15 \quad \Longrightarrow \quad 12 + \mathbf{L} = -15 \quad \Longrightarrow \quad \mathbf{L} = 3$$

The observer equation (from Eqn(11.116) is

$$\begin{aligned} \dot{\mathbf{z}} &= \mathbf{Dz} + \mathbf{Fy} + \mathbf{Gu} \\ \hat{\mathbf{x}}_{\mathrm{u}} &= \mathbf{z} + \mathbf{Ly} \end{aligned}$$

where

$$\begin{aligned} \mathbf{D} &= \mathbf{A}_{22} - \mathbf{L}\mathbf{A}_{12} = -12 - 3 = -15 \\ \mathbf{F} &= \mathbf{DL} + \mathbf{A}_{21} - \mathbf{L}\mathbf{A}_{11} = -15 \times 3 - 36 - 0 = -81 \\ \mathbf{G} &= \mathbf{B}_2 - \mathbf{L}\mathbf{B}_1 = 1 \end{aligned}$$

Hence

$$\dot{\mathrm{z}} = -15\mathrm{z} - 81\mathrm{y} + \mathrm{u}, \quad \text{and} \quad \hat{\mathrm{x}}_2 = \mathrm{z} + 3\mathrm{y}$$

Notice that, in this example, we have 2 states and one measurement. Therefore, the observer gain **L** was computed by solving a first-order algebraic equation, which gives a unique solution of **L**. However, for an n-th order system with "m" measurements, there are (m-1) degrees of freedom to solve for **L** and the solution is, therefore, not unique.

Let us investigate the observer performance using MATLAB simulation. For this purpose, the augmented system with the reduced order observer is written as (see Eqn(11.119):

$$\begin{bmatrix} \dot{\mathrm{x}}_1 \\ \dot{\mathrm{x}}_2 \\ \dot{\mathrm{z}} \end{bmatrix} = \begin{bmatrix} 0 & 1 & 0 \\ -36 & -12 & 0 \\ -81 & 0 & -15 \end{bmatrix} \begin{bmatrix} \mathrm{x}_1 \\ \mathrm{x}_2 \\ \mathrm{z} \end{bmatrix} + \begin{bmatrix} 0 \\ 1 \\ 1 \end{bmatrix} \mathrm{u}$$

The estimated state is given by

$$\hat{\mathrm{x}}_2 = \mathrm{z} + \mathrm{Ly} = \mathrm{z} + 3\mathrm{x}_1$$

The zero input response of the system with reduced-order observer with initial condition $x_1(0) = 0.1$, $x_2(0) = 0.3$, and $z(0) = 0.1$, is computed and the estimated state $\hat{x}_2$ is shown in Figure 11.20

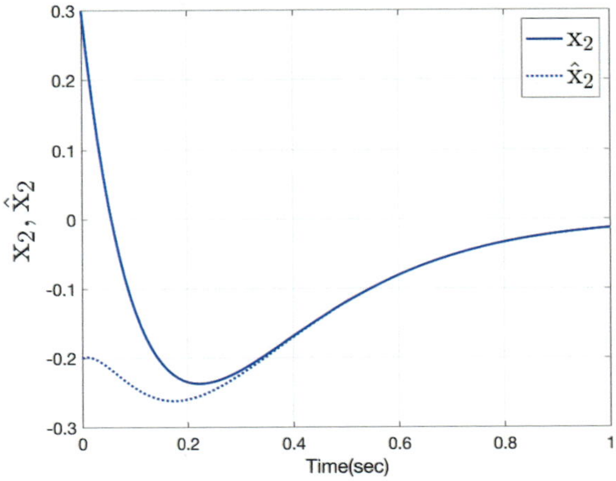

Figure 11.20: Estimated state $\hat{x}_2$ with reduced order observer

The MATLAB code for getting this response is given in `Example11_24.m`

CONCEPT REINFORCEMENT PROBLEM 11.12

Problem

The longitudinal dynamics of a helicopter near hover are:

Pitch rate dynamics: $\dot{q} = M_q q + M_\theta \theta + M_u u + M_{\delta_t} \delta_t$

Pitch angle dynamics: $\dot{\theta} = q$

Horizontal dynamics: $\dot{u} = X_q q + X_\theta \theta + X_u u + X_{\delta_t} \delta_t$

where M_q, M_θ, M_u, and M_{δ_t} denote stability derivatives for pitch rate, X_q, X_θ, X_u, and X_{δ_t} denote stability derivatives for horizontal velocity and δ_t is the rotor tilt angle (the control input).

Using MATLAB, do the following:

a. Express the system dynamics in state space form, considering the state variables $\mathbf{x} = \begin{bmatrix} u & q & \theta \end{bmatrix}^T$. With the given parameters, find the system, input, and output matrices.
b. If the system is controllable, find the state feedback gain that places the poles of the system at $s_{1,2} = -2 \pm j4$ and $s_3 = -5$.

c. Design a full order estimator that places the observer poles at -8 and $-5 \pm j5$.

d. Compute the transfer function using the control gain and full-order state observer.

e. Design a reduced-order estimator with both poles at -10.

f. Compute the transfer function using the control gain and reduced-order state observer.

Assume the parameters as: $M_q = -0.5, M_\theta = 0.1$, $M_u = 0.05$, $M_{\delta_t} = 0.2$

$X_q = 0.1, X_\theta = -0.05$, $X_u = -0.02$, $X_{\delta_t} = 0.1$.

EXAMPLE 11.25

Problem:

The transfer function of a third-order system is given by

$$G(s) = \frac{10}{s(s+2)(s+4)}$$

This is represented in state space form as:

$$\dot{\mathbf{x}} = \mathbf{A}\mathbf{x} + \mathbf{B}u, \qquad \text{where}$$

$$\mathbf{x} = \begin{bmatrix} x_1 \\ x_2 \\ x_3 \end{bmatrix}, \quad \mathbf{A} = \begin{bmatrix} 0 & 1 & 0 \\ 0 & 0 & 1 \\ 0 & -8 & -6 \end{bmatrix}, \quad \text{and} \quad \mathbf{B} = \begin{bmatrix} 0 \\ 0 \\ 10 \end{bmatrix}$$

Suppose the states x_1 and x_2 are measured.

Design a reduced order observer to estimate only x_3 by placing the observer eigenvalue at -25. Compute the observer gain. Write the observer equation and verify the result using MATLAB.

SOLUTION:

Since two states are measurable and we need to reconstruct one of the states, we partition the system and input matrices as:

$$\mathbf{A} = \left[\begin{array}{c|c} \mathbf{A}_{11} & \mathbf{A}_{12} \\ \hline \mathbf{A}_{21} & \mathbf{A}_{22} \end{array}\right], \quad \mathbf{B} = \left[\begin{array}{c} \mathbf{B}_1 \\ \hline \mathbf{B}_2 \end{array}\right], \quad \text{where}$$

$$\mathbf{A}_{11} = \begin{bmatrix} 0 & 1 \\ 0 & 0 \end{bmatrix}, \mathbf{A}_{12} = \begin{bmatrix} 0 \\ 1 \end{bmatrix}, \mathbf{A}_{21} = \begin{bmatrix} 0 & -8 \end{bmatrix}, \quad \mathbf{A}_{22} = -6$$

$$\mathbf{B}_1 = \begin{bmatrix} 0 \\ 0 \end{bmatrix} \quad \text{and} \quad \mathbf{B}_2 = 10$$

Now the observer eigenvalue is to be at -25. Notice that, in this example, there are 3 states and two measurements. So the solution of $\mathbf{L}$ is not unique. We have to fix one element of $\mathbf{L}$ to get a unique solution. Without loss of generality, we assume the structure of $\mathbf{L} = \begin{bmatrix} 0 & l_2 \end{bmatrix}$ and solve for l_2.

The observer eigenvalue is at -25. Hence

$$\lambda\,\mathbf{I} - \mathbf{A}_{22} + \mathbf{L}\mathbf{A}_{12} = \lambda + 25 \quad \Longrightarrow \quad 6 + l_2 = 25 \quad \Longrightarrow \quad l_2 = 19$$

Hence $\mathbf{L} = \begin{bmatrix} 0 & 19 \end{bmatrix}$.

The observer equation (from Eqn(11.116) is

$$\begin{aligned} \dot{\mathbf{z}} &= \mathbf{Dz} + \mathbf{Fy} + \mathbf{Gu} \\ \hat{\mathbf{x}}_{\mathrm{u}} &= \mathbf{z} + \mathbf{Ly} \quad \text{where} \end{aligned}$$

$$\begin{aligned} \mathbf{D} &= \mathbf{A}_{22} - \mathbf{L}\mathbf{A}_{12} = -25 \\ \mathbf{F} &= \mathbf{DL} + \mathbf{A}_{21} - \mathbf{L}\mathbf{A}_{11} = \begin{bmatrix} 0 & -483 \end{bmatrix} \\ \mathbf{G} &= \mathbf{B}_2 - \mathbf{L}\mathbf{B}_1 = 10 \end{aligned}$$

Hence

$$\dot{\mathrm{z}} = -25\mathrm{z} + \begin{bmatrix} 0 & -483 \end{bmatrix} \begin{bmatrix} \mathrm{x}_1 \\ \mathrm{x}_2 \end{bmatrix} + 10\mathrm{u} = -25\mathrm{z} - 483\mathrm{x}_2 + 10\mathrm{u}, \quad \text{and}$$

$$\hat{\mathrm{x}}_3 = \mathrm{z} + \begin{bmatrix} 0 & 19 \end{bmatrix} \begin{bmatrix} \mathrm{x}_1 \\ \mathrm{x}_2 \end{bmatrix} = \mathrm{z} + 19\mathrm{x}_2$$

Let us investigate the observer performance using MATLAB simulation. For this purpose, the augmented system with the reduced order observer is written as (see Eqn(11.119):

$$\begin{bmatrix} \dot{\mathrm{x}}_1 \\ \dot{\mathrm{x}}_2 \\ \dot{\mathrm{x}}_3 \\ \dot{\mathrm{z}} \end{bmatrix} = \begin{bmatrix} 0 & 1 & 0 & 0 \\ 0 & 0 & 1 & 0 \\ 0 & -8 & -6 & 0 \\ 0 & -483 & 0 & -25 \end{bmatrix} \begin{bmatrix} \mathrm{x}_1 \\ \mathrm{x}_2 \\ \mathrm{x}_3 \\ \mathrm{z} \end{bmatrix} + \begin{bmatrix} 0 \\ 0 \\ 10 \\ 10 \end{bmatrix} \mathrm{u}$$

The zero input response of the system with reduced-order observer with initial condition $\mathrm{x}_1(0) = 1$, $\mathrm{x}_2(0) = -0.9$, $\mathrm{x}_3(0) = 0.8$ and $\mathrm{z}(0) = -0.5$, is computed and the estimated state $\hat{\mathrm{x}}_2$ is shown in Figure 11.21

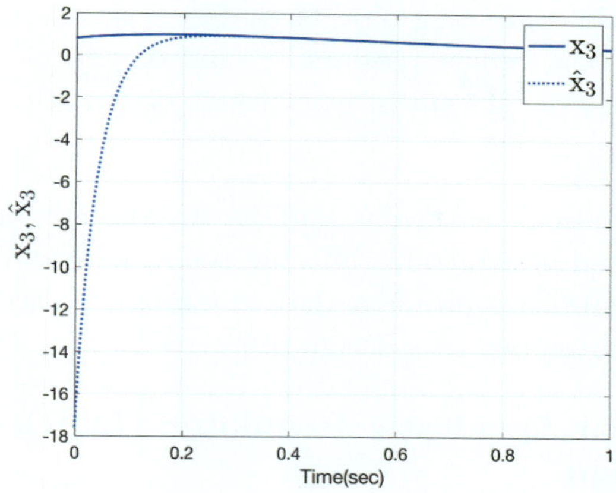

Figure 11.21: Estimated state $\hat{x}_3$ with reduced order observer

The MATLAB code `Example11_25.m` implements this observer.

11.20 Advanced Control Strategies: Moving Beyond Classical State-Space Design

Throughout this chapter, we have explored state-space controller design starting with pole placement techniques, which allow us to arbitrarily assign closed-loop eigenvalues to achieve desired stability and transient response characteristics. While pole placement provides a direct and intuitive approach to controller design, practical implementation reveals several fundamental limitations that motivate more sophisticated control strategies.

Limitations of Pole Placement: The Need for Optimal Control

Pole placement, despite its conceptual elegance, suffers from several critical shortcomings in real-world applications:

- **Design Arbitrariness:** The selection of pole locations is often based on rules of thumb or trial-and-error, with no systematic method to determine truly optimal locations. We may choose poles to meet settling time or damping requirements, but this approach lacks a principled framework for balancing competing performance objectives.
- **No Performance Optimization:** Pole placement does not inherently minimize any meaningful performance criterion. While we can achieve desired closed-loop dynamics, we cannot guarantee that the resulting controller optimizes important practical considerations such as energy consumption, tracking accuracy, or control effort.
- **Sensitivity and Robustness Issues:** Arbitrarily chosen poles may require excessively high controller gains, leading to poor robustness margins and sensitivity to model uncertainties, noise, and parameter variations.

- **Multi-Objective Design Challenges:** In practice, we must balance multiple competing objectives: fast response versus low overshoot, tight regulation versus reasonable control effort, robustness versus performance. Pole placement provides no systematic framework for these trade-offs.

These limitations naturally lead us to consider optimal control strategies, where instead of arbitrarily placing poles, we formulate the controller design as an optimization problem that minimizes a carefully chosen performance index capturing our true design objectives.

11.20.1 Linear Quadratic Regulator (LQR): Systematic Optimal Design

The Linear Quadratic Regulator (LQR) represents one of the most important and widely-used optimal control strategies, specifically designed for linear time-invariant systems. LQR addresses the pole placement limitations through several key advantages:

Advantages of LQR

(a) **Systematic performance optimization:** LQR minimizes a quadratic cost function that explicitly balances state regulation performance against control effort, eliminating the guesswork in controller design.

(b) **Guaranteed stability and robustness:** The resulting controller always provides a stable closed-loop system for controllable systems, with built-in robustness margins (gain margin ≥ 6 dB, phase margin $\geq 60°$).

(c) **Intuitive design parameters:** The weighting matrices $\mathbf{Q}$ and $\mathbf{R}$ provide transparent tuning parameters that directly relate to physical performance objectives.

(d) **Analytical solution:** Unlike many optimal control problems, LQR has a closed-form solution through the Algebraic Riccati Equation, making it computationally tractable.

(e) **Optimal trade-off mechanism:** The $\mathbf{Q}/\mathbf{R}$ ratio provides a systematic way to balance competing objectives like fast response versus low control effort.

Mathematical Framework and Problem Formulation

Step-1: The LQR problem seeks to minimize a quadratic cost function over an infinite time horizon. The cost function J is typically defined as:

$$\boxed{J = \int_0^{\infty} (x^T(t)Qx(t) + u^T(t)Ru(t))\, dt} \tag{11.136}$$

This formulation transforms the subjective art of pole placement into an objective optimization problem where:

- Q and R are symmetric, positive semi-definite and positive definite weighting matrices, respectively.
- Q penalizes the deviation of the state from the desired trajectory (usually the origin) and allows us to specify which states need tight regulation and
- R matrix penalizes the magnitude of the control input, ensuring that control effort is minimized.
- The optimal solution balances these competing objectives mathematically

Step-2: The design of an LQR controller requires that the system be represented in state-space form as:

$$\dot{\mathbf{x}} = \mathbf{Ax} + \mathbf{Bu} \tag{11.137}$$

where $\mathbf{x}(t)$ is the state vector, $\mathbf{u}(t)$ is the control input, $\mathbf{A}$ and $\mathbf{B}$ are the system matrices that define the dynamics.

Step-3: To minimize the cost function, the optimal control law u(t) must be determined using various methods such as Pontryagin's Maximum Principle or Dynamic Programming. The optimal control law for a linear system can be derived as:

$$\boxed{u(t) = -Kx(t)} \tag{11.138}$$

where K is the optimal feedback gain matrix computed by solving the Algebraic Riccati Equation (ARE) given by

$$\boxed{A^TP + PA - PBR^{-1}B^TP + Q = 0} \tag{11.139}$$

where P is positive definite matrix. Once P is found, the optimal feedback gain K is computed as:

$$\boxed{K = R^{-1}B^TP} \tag{11.140}$$

The Riccati equation is typically solved numerically using iterative methods.

Closed-Loop System

With optimal control law applied, the closed-loop dynamics become

$$\dot{\mathbf{x}}(t) = (\mathbf{A} - \mathbf{BK})\mathbf{x}(t), \tag{11.141}$$

and the eigenvalues of $(\mathbf{A} - \mathbf{BK})$ determine system stability.

Properties of LQR

- **Guaranteed Stability:** LQR guarantees a unique, globally optimal stabilizing controller if $(\mathbf{A}, \mathbf{B})$ is controllable and $(\mathbf{Q^{1/2}}, \mathbf{A})$ is observable.
- **Systematic Trade-off:** The weighting matrices **Q** and **R** act as design knobs: higher **Q** enforces smaller state deviations, higher **R** reduces control effort. Thus, the Q/R ratio provides a principled mechanism for balancing performance versus control effort.
- **Built-in Robustness:** The solution is robust in the sense that small plant perturbations do not destabilize the system. Inherent stability margins ($\geq$ 6 dB gain margin, $\geq$ 60° phase margin)

LQR Tuning Guidelines

Bryson's Rule for initial weighting matrix selection:

$$Q = \begin{bmatrix} \frac{\alpha_1^2}{(x_1)^2_{max}} & & & \\ & \frac{\alpha_2^2}{(x_2)^2_{max}} & & \\ & & \ddots & \\ & & & \frac{\alpha_n^2}{(x_n)^2_{max}} \end{bmatrix}$$

and

$$R = \rho \begin{bmatrix} \frac{\beta_1^2}{(u_1)^2_{max}} & & & \\ & \frac{\beta_2^2}{(u_2)^2_{max}} & & \\ & & \ddots & \\ & & & \frac{\beta_m^2}{(u_m)^2_{max}} \end{bmatrix}$$

where

- α_i are relative weights for state components (typically $\alpha_i \geq 0$)
- β_j are relative weights for control components (typically $\beta_j \geq 0$)
- $\rho > 0$ is the overall state-to-control penalty ratio

and $(x_i)_{max}$ represents the maximum acceptable deviation for ith state x_i, and $(u_j)_{max}$ represents the maximum available control effort for jth actuator.

Tuning Procedure

1. Start with basic Bryson's rule ($\alpha_i = \beta_j = 1, \rho = 1$)
2. Simulate closed-loop response
3. Adjust relative weights:
   - If state x_i oscillates too much → increase α_i
   - If control u_j saturates → increase β_j or increase ρ
   - If response is too sluggish → decrease ρ
4. Iterate until satisfactory performance

Remark

Unlike pole-placement methods, which require desired closed-loop pole locations to be pre-specified, LQR provides an automatic trade-off between performance and control effort through the selection of Q and R matrices.

LQR Design Procedure

LQR Design Procedure

1. Define system matrices $\mathbf{A}$, $\mathbf{B}$
2. Choose weighting matrices $\mathbf{Q}$, $\mathbf{R}$
   - $\mathbf{Q}$: Penalize important state deviations ($\mathbf{Q} \geq 0$)
   - $\mathbf{R}$: Penalize control effort ($\mathbf{R} > 0$)
3. Solve the ARE to find $\mathbf{P}$
4. Compute optimal gain: $\mathbf{K} = \mathbf{R}^{-1}\mathbf{B}^{\mathbf{T}}\mathbf{P}$
5. Implement control law: $\mathbf{u} = -\mathbf{Kx}$

EXAMPLE 11.26

Problem

Consider the dynamical system described by

$$\dot{x}(t) = ax(t) + bu(t)$$

Find the optimal control law that minimizes

$$J = \int_0^\infty \left[x^2(t) + ru^2(t)\right] dt, \quad r > 0$$

SOLUTION:

From the system equation, we have

$$A = a, \quad B = b, \quad Q = 1 \quad \text{and } R = r$$

The CARE for this scalar system is

$$A^TP + PA - PBR^{-1}B^TP + Q = 0 \quad \Longrightarrow \quad 2aP - \frac{b^2}{r}P^2 + 1 = 0$$

Rearrange to a quadratic in P

$$\frac{b^2}{r}P^2 - 2aP - 1 = 0$$

Solving this gives

$$P = \frac{2a \pm \sqrt{4a^2 + 4\dfrac{b^2}{r}}}{2\dfrac{b^2}{r}} = \frac{r}{b^2}\left(a \pm \sqrt{a^2 + \frac{b^2}{r}}\right)$$

There are two algebraic roots; the stabilizing (physically meaningful) solution is the one that makes the closed-loop pole negative:

$$P = \frac{r}{b^2}\left(a + \sqrt{a^2 + \frac{b^2}{r}}\right)$$

The optimal state-feedback gain is

$$K = \frac{b}{r}P = \frac{b}{r}\cdot\frac{r}{b^2}\frac{a + \sqrt{a^2 + \dfrac{b^2}{r}}}{b} = \frac{a + \sqrt{a^2 + \dfrac{b^2}{r}}}{b}$$

and the closed-loop pole

$$\lambda_{cl} = a - \frac{b^2}{r}P = a - \left(a + \sqrt{a^2 + \frac{b^2}{r}}\right) = -\sqrt{a^2 + \frac{b^2}{r}} < 0$$

so the closed-loop is stable.

For $a = 2$, $b = 1$ and $r = 1$, we have

$$P = \frac{1}{1}\left(2 + \sqrt{4+1}\right) = 2 + \sqrt{5} \approx 4.236068, \quad K = \frac{1}{1}P \approx 4.236068$$

and the closed-loop pole

$$\lambda_{cl} = 2 - P \approx -2.236068 \quad \text{(stable)}$$

Note that for the LQR to be applicable, $b \neq 0$; because for $b = 0$, the system is uncontrollable.

EXAMPLE 11.27

Problem

Consider a dynamical system described by

$$\dot{x} = au_1 + bu_2, \quad x(0) = 3$$

Find the optimal control law that minimizes

$$J = \int_0^\infty \left(x^2 + r_1u_1^2 + r_2u_2^2\right) dt$$

SOLUTION:

From the system equation, we have

$$A = 0, B = \begin{bmatrix} a & b \end{bmatrix}, Q = 1, R = \begin{bmatrix} r_1 & 0 \\ 0 & r_2 \end{bmatrix}$$

The ARE for this problem

$$A^TP + PA - PBR^{-1}B^TP + Q = 0$$

or

$$-P\begin{bmatrix} a & b \end{bmatrix}\begin{bmatrix} \frac{1}{r_1} & 0 \\ 0 & \frac{1}{r_2} \end{bmatrix}\begin{bmatrix} a \\ b \end{bmatrix}P + 1 = 0$$

This gives

$$1 - P^2\left[\frac{r_2a^2 + r_1b^2}{r_1r_2}\right] = 0 \quad \text{or,} \quad P = \sqrt{\frac{r_1r_2}{r_2a^2 + r_1b^2}}$$

The optimal controller is

$$u = -R^{-1}B^{T}Px = -\begin{bmatrix} \frac{1}{r_1} & 0 \\ 0 & \frac{1}{r_2} \end{bmatrix} \begin{bmatrix} a \\ b \end{bmatrix} \sqrt{\frac{r_1 r_2}{r_2 a^2 + r_1 b^2}} x$$

$$= -\begin{bmatrix} \frac{a}{r_1} \\ \frac{b}{r_2} \end{bmatrix} \sqrt{\frac{r_1 r_2}{r_2 a^2 + r_1 b^2}} x$$

The optimal value of J is

$$J = x^T(0)Px(0)$$

For a=1; b=3; $r_1 = 1$, $r_2 = 2$ and $Q = 1$, we have

$$P = 0.4264, \quad K = \begin{bmatrix} 0.4264 \\ 0.6396 \end{bmatrix}, \quad J = x^T(0)Px(0) = 3 \cdot 0.4264 \cdot 3 = 3.8376$$

Solving LQR Problems Using MATLAB

As illustrated in the previous examples, the Linear Quadratic Regulator (LQR) design involves solving the algebraic Riccati equation. For low-order systems, it is feasible to carry out the solution manually and determine the optimal state feedback gain matrix **K** step by step. However, as the order of the system increases, solving the Riccati equation by hand becomes increasingly tedious and error-prone.

In such cases, it is preferable to use MATLAB, which provides efficient and reliable computational routines for solving Riccati equations and obtaining the optimal gain matrix. MATLAB has built-in functions such as `care` (for solving the continuous-time algebraic Riccati equation) and `lqr` (for directly computing the optimal feedback gain).

EXAMPLE 11.28

Problem: Consider the linear system

$$\dot{\mathbf{x}}(t) = \begin{bmatrix} 0 & 1 \\ -2 & -3 \end{bmatrix} \mathbf{x}(t) + \begin{bmatrix} 0 \\ 1 \end{bmatrix} u(t),$$

with the performance index

$$J = \int_0^\infty \left(\mathbf{x}^T Q \mathbf{x} + u^T R u\right) dt,$$

where $Q = \begin{bmatrix} 1 & 0 \\ 0 & 1 \end{bmatrix}$, $\quad R = 1.$

MATLAB Solution: The optimal gain matrix **K** can be obtained in MATLAB using the following commands:

```
A = [0 1; -2 -3];
B = [0; 1];
Q = eye(2);
R = 1;
K = lqr(A,B,Q,R)
```

Output:

```
K =
    1.0000    2.0000
```

Interpretation: The resulting optimal feedback control law is

$$u(t) = -\mathbf{K}\mathbf{x}(t) = -[\,1 \;\; 2\,]\mathbf{x}(t)$$

This ensures minimization of the quadratic cost function and guarantees closed-loop stability of the system.

EXAMPLE 11.29

Problem

LQR Design for an Inverted Pendulum on a Cart

Consider an inverted pendulum mounted on a cart of mass M with a pendulum of mass m and length L shown in Figure 9.25. The horizontal force u(t) acts on the cart as the control input. The state variable model of this system is described in Eqn(9.94), where the state vector is defined as:

$$\mathrm{x}(t) = \begin{bmatrix}\theta & \dot{\theta} & \mathrm{x} & \dot{\mathrm{x}}\end{bmatrix}^{\mathrm{T}},$$

where x is the cart displacement and θ is the pendulum angle (measured from the vertical).

Design an LQR to stabilize this system, considering $\mathrm{Q} = \mathrm{diag}(100, 1, 10, 1)$ and $\mathrm{R} = 0.01$. Assume $\mathrm{M} = 0.5$ kg, $\mathrm{m} = 0.2$ kg, $\mathrm{L} = 0.3$ m and $\mathrm{g} = 9.81$ m/s^2.

SOLUTION:

For the given parameters, the system and input matrices of the linearized system about the upright equilibrium ($\theta = 0$) are:

$$\dot{\mathbf{x}}(t) = \mathbf{A}\mathbf{x}(t) + \mathbf{B}u(t),$$

with

$$\mathbf{A} = \begin{bmatrix} 0 & 1 & 0 & 0 \\ \dfrac{(M+m)g}{ML} & 0 & 0 & 0 \\ 0 & 0 & 0 & 1 \\ -\dfrac{mg}{M} & 0 & 0 & 0 \end{bmatrix} = \begin{bmatrix} 0 & 1 & 0 & 0 \\ 45.78 & 0 & 0 & 0 \\ 0 & 0 & 0 & 1 \\ -3.924 & 0 & 0 & 0 \end{bmatrix}$$

$$\mathbf{B} = \begin{bmatrix} 0 \\ -\dfrac{1}{ML} \\ 0 \\ \dfrac{1}{M} \end{bmatrix} = \begin{bmatrix} 0 \\ -6.667 \\ 0 \\ 2 \end{bmatrix}$$

Given $Q = \text{diag}(100, 1, 10, 1)$ and $R = 0.01$. Note that the choice of Q and R reflects the desired performance:

- large weight on θ to keep the pendulum upright,
- moderate weight on x to prevent excessive cart displacement,
- small weight on $\dot{x}, \dot{\theta}$ to regulate velocities,
- positive scalar R to penalize large control input.

The optimal state feedback gain is obtained from `lqr` function of MATLAB and is

$$K = \begin{bmatrix} -152.0462 & -21.9754 & -31.6228 & -32.1866 \end{bmatrix}$$

The eigenvalues of the closed-loop system are found to be

$$-69.5521, \quad -9.8561 \text{ and } \quad -1.3606 \pm j1.0796$$

all lying in the left-half plane. Hence, the closed-loop system is asymptotically stable, and the pendulum is successfully balanced.

The state response to initial conditions is shown in Figure 11.22.

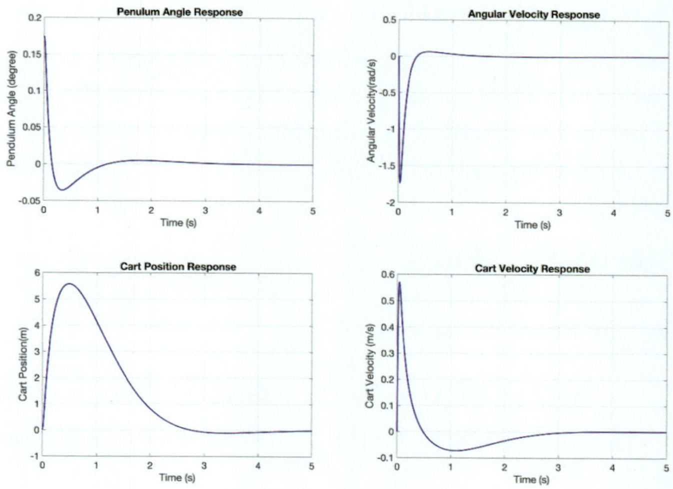

Figure 11.22: Stabilized states with LQR

The MATLAB implementation for solving this problem is in `Example11_29.m`.

11.21 The State Measurement Challenge: From Observer to Kalman Filter

LQR assumes full state feedback, requiring measurement or estimation of all system states. In our earlier observer design, we addressed unmeasurable states by constructing a deterministic state estimator under the assumption of noise-free measurements and perfect model knowledge.

However, real systems operate in noisy environments where:

- Process disturbances affect system dynamics unpredictably
- Measurement noise corrupts sensor readings
- Model uncertainties introduce estimation errors

The classical observer design, while mathematically elegant, cannot optimally handle these stochastic disturbances. This limitation motivates the development of the Kalman filter, an optimal state estimator that minimizes the mean-square estimation error in the presence of known statistical properties of process and measurement noise.

11.21.1 Linear Quadratic Gaussian (LQG): Optimal Control with Optimal Estimation

When we combine the LQR optimal controller with the Kalman optimal estimator, we obtain the Linear Quadratic Gaussian (LQG) controller. The beauty of LQG lies in the separation principle: we can design the LQR controller and Kalman filter independently, then combine them to create an optimal closed-loop system.

The Stochastic Model

Consider the noisy linear system

$$\begin{aligned}\dot{\mathbf{x}}(\mathrm{t}) &= \mathbf{A}\mathbf{x}(\mathrm{t}) + \mathbf{B}\mathbf{u}(\mathrm{t}) + \mathbf{G}\mathbf{w}(\mathrm{t}) \quad \text{(proces equation)} \\ \mathrm{y}(\mathrm{t}) &= \mathbf{C}\mathbf{x}(\mathrm{t}) + \mathrm{v}(\mathrm{t}) \quad \text{(measurement equation)}\end{aligned} \tag{11.142}$$

where

$\mathrm{x}(\mathrm{t})$ = state vector $(\mathrm{n} \times 1)$

$\mathrm{u}(\mathrm{t})$ = input vector $(\mathrm{m} \times 1)$

$\mathrm{w}(\mathrm{t})$ = process noise vector $(\mathrm{r} \times 1)$(white, zero-mean, covariance W)

$\mathrm{v}(\mathrm{t})$ = measurement noise (white, zero-mean, covariance V)

$\mathbf{G}$ = process noise distribution matrix $(\mathrm{n} \times \mathrm{r})$

The process noise $\mathrm{w}(\mathrm{t})$ is modeled as zero-mean white noise with covariance:

$$\mathrm{E}\left[\mathrm{w}(\mathrm{t})\mathrm{w}^{\mathrm{T}}(\tau)\right] = \mathrm{W}\delta(\mathrm{t}-\tau), \text{ and } \mathrm{E}\left[\mathrm{v}(\mathrm{t})\mathrm{v}^{\mathrm{T}}(\tau)\right] = \mathrm{V}\delta(\mathrm{t}-\tau)$$

where W is an $\mathrm{r} \times \mathrm{r}$ positive semi-definite covariance matrix.

Two Modeling Approaches

Case-A: Scalar noise input

When we have a single noise source affecting specific states, design $\mathbf{G}$ as a column vector that maps where the scalar noise $\mathrm{w}(\mathrm{t})$ enters the system.

- If $\mathrm{w}(\mathrm{t})$ is scalar, then $\mathrm{w} \in \mathbb{R}$
- Choose $\mathbf{G}$ as a column vector $(\mathrm{n} \times 1)$ that specifies which states are directly excited by the noise. For example

$$\mathbf{G} = \begin{bmatrix} 0 \\ 1 \end{bmatrix}, \quad \mathrm{w}(\mathrm{t}) \sim N(0, \sigma_{\mathrm{w}}^2)$$

This means noise only affects the second state.

Case-B: Direct covariance injection

When we want to directly specify how noise affects all states through the covariance matrix, then we work with the relationship:

$$\mathrm{E}\left[(\mathrm{Gw})(\mathrm{Gw})^{\mathrm{T}}\right] = \mathrm{GWG}^{\mathrm{T}} = \mathrm{Q}$$

By choosing $\mathbf{G} = \mathbf{I}$ and setting $\mathbf{W}$ equal to the desired noise covariance matrix $\mathbf{Q}$ allows direct control over the noise structure.

For example, if

$$\mathrm{W} = \begin{bmatrix} 0.01 & 0 \\ 0 & 0.1 \end{bmatrix}$$

this directly specifies $\mathrm{Q} = \mathrm{GWG}^{\mathrm{T}}$.

What MATLAB expects (lqe)

In MATLAB's

```
[L, P, E]=lqe(A, G, C, W, V)
```

The function uses $\mathrm{GWG}^{\mathrm{T}}$ internally for the estimator design. Choose your parameters based on your modeling approach: G is the noise distribution matrix, W is the covariance of w, and the estimator design uses $\mathrm{GWG}^{\mathrm{T}}$ internally. In summary

- **For scalar noise:** pick G as a column vector, $\mathrm{W} = \sigma_{\mathrm{w}}^2$
- **For direct covariance::** pick $\mathbf{G} = \mathbf{I}$ and specify W as our desired covariance matrix (a preferred approach).

We can use either of the codes given below.

```
  % Case A: Single noise source affecting second state
A = [0 1; -2 -3];
G = [0; 1];              % Column vector
C = [1 0];
W = 0.01;                % Scalar variance
V = 0.05;
[L, P, E] = lqe(A, G, C, W, V);
```

```
    % Case B: Direct covariance specification
A = [0 1; -2 -3];
G = eye(2);              % Identity matrix
C = [1 0];
W = [0 0; 0 0.01];  % 2x2 covariance matrix
V = 0.05;
[L, P, E] = lqe(A, G, C, W, V);
```

Objective

Design an output-feedback controller that minimizes the same quadratic cost

$$J = \int_0^\infty \left(x^TQx + u^TRu\right) dt \tag{11.143}$$

given that $\mathbf{x}(t)$ is not directly available.

Solution: Separation Principle

The LQG controller is obtained by combining:

1. An LQR regulator that computes the optimal feedback law $u(t) = -K\hat{\mathbf{x}}(t)$.
2. A Kalman filter that estimates $\hat{\mathbf{x}}(t)$ from noisy measurements $\mathbf{y}(t)$.

Step-1: Design LQR Controller

Solve the control ARE for the noise-free system:

$$\boxed{A^TP_c + P_cA - P_cBR^{-1}B^TP_c + Q = 0}$$

Step-2: Design Kalman Filter

The state estimate evolves as

$$\dot{\hat{\mathbf{x}}}(t) = A\hat{\mathbf{x}}(t) + Bu(t) + L\,[]\mathbf{y}(t) - C\hat{\mathbf{x}}(t)], \tag{11.144}$$

where L is the Kalman gain, obtained by solving the dual Riccati equation

$$\boxed{AP_f + P_fA^\top - P_fC^\top V^{-1}CP_f + GWG^T = 0} \tag{11.145}$$

with W and V being the covariance matrices of process and measurement noise, respectively. The filter gain

$$\boxed{L = P_fC^TV^{-1}}$$

Final LQG Controller: Combine via Separation Principle

The resulting dynamic output feedback controller is

$$\dot{\hat{\mathbf{x}}}(t) = \mathbf{A}\hat{\mathbf{x}} + \mathbf{B}\mathbf{u} + \mathrm{L}\,(\mathrm{y} - \mathrm{C}\hat{\mathbf{x}}) \quad \text{(Kalman filter)} \tag{11.146}$$

$$\mathrm{u}(t) = -\mathbf{K}\hat{\mathbf{x}}(t) \tag{11.147}$$

where K is from LQR and $\hat{\mathbf{x}}(t)$ is from the Kalman filter.

Properties of LQG

- By the separation principle, the LQR and Kalman filter designs can be carried out independently.
- LQG is optimal for systems with Gaussian noise and quadratic cost.
- However, unlike pure LQR, LQG does not guarantee robustness margins automatically; this leads to the development of Loop Transfer Recovery (LTR) techniques.

Remark

The LQG framework unifies control (LQR) and estimation (Kalman filtering) into a single optimal output-feedback controller.

LQG Design Procedure

LQG Design Procedure

1. Identify system matrices **A**, **B**, **C**, and **G**
2. Choose LQR weighting matrices **Q**, **R**
3. Estimate noise covariances **W** and **V**
4. Solve control ARE to find **K**
5. Solve filter ARE to find **L**
6. Implement control law

EXAMPLE 11.30

Problem

Consider the following plant

$$\dot{x}(t) = 2x(t) + u(t) + w(t)$$
$$y(t) = x(t) + v(t)$$

where x(t) is the state variable, y(t) is the output, w(t) and v(t) are zero-mean stochastic Gaussian processes uncorrelated in time and with each other. Assume that the plant has the measurement noise covariance, $V = \beta > 0$, and the process noise covariance, $W = \alpha > 0$. Design a Kalman filter to estimate the state x(t) by $\hat{x}(t)$ such that the estimate error covariance is minimized.

SOLUTION:

Comparing with the standard system equation, we have

$$A = 2; B = 1; C = 1, W = \alpha \text{ and } V = \beta$$

Let us solve the CARE

$$AP_e + P_eA^T - P_eC^TV^{-1}CP_e + W = 0$$

or

$$2P_e + 2P_e - P_e \cdot 1 \cdot \frac{1}{\beta} \cdot 1 \cdot P_e + \alpha = 0$$

or

$$4P_e - \frac{P_e^2}{\beta} + \alpha = 0 \implies P_e^2 - 4P_e\beta - \alpha\beta = 0$$

This gives

$$P_e = 2\beta \pm \sqrt{\beta(\alpha + 4\beta)} = \beta\left(2 \pm \sqrt{4 + \frac{\alpha}{\beta}}\right)$$

The estimator covariance must be nonnegative. Note $\sqrt{4 + \alpha/\beta} > 2$ (since $\alpha/\beta > 0$), therefore the stabilizing root is

$$P_e = \beta\left(2 + \sqrt{4 + \frac{\alpha}{\beta}}\right)$$

Now, the Kalman gain is

$$L = P_eC^TV^{-1} = \frac{P_e}{\beta} = 2 + \sqrt{4 + \frac{\alpha}{\beta}}$$

Estimator dynamics

$$\dot{\hat{x}} = 2\hat{x} + u + L(y - \hat{x}) = (2 - L)\hat{x} + u + Ly$$

The eigenvalue of estimation error is eigenvalue of $\mathbf{A} - \mathbf{LC}$ is

$$2 - \mathrm{L} = -\sqrt{4 + \frac{\alpha}{\beta}} < 0$$

So the estimator is asymptotically stable.

For example, if $\alpha = 0.1$ and $\beta = 1$:

$$\mathrm{P_e} = 1 \cdot 2 + \sqrt{4 + 0.1} = 2 + \sqrt{4.1} \approx 4.0248, \quad \mathrm{L} \approx 4.0248$$

and eigenvalue of estimation error dynamics is $2 - \mathrm{L} = 2 - 4.0248 \approx -2.0248$. Thus, it is stable.

EXAMPLE 11.31

Problem:

Consider a system

$$\begin{aligned} \dot{\mathbf{x}}(\mathrm{t}) &= \mathrm{A}\mathbf{x}(\mathrm{t}) + \mathrm{Bu}(\mathrm{t}) + \mathrm{w}(\mathrm{t}) \\ \mathrm{y}(\mathrm{t}) &= \mathrm{C}\mathbf{x}(\mathrm{t}) + \mathrm{v}(\mathrm{t}), \end{aligned}$$

with

$$\mathrm{A} = \begin{bmatrix} 0 & 1 \\ -2 & -3 \end{bmatrix}, \quad \mathrm{B} = \begin{bmatrix} 0 \\ 1 \end{bmatrix}, \quad \mathrm{C} = \begin{bmatrix} 1 & 0 \end{bmatrix}.$$

The process noise $\mathrm{w}(\mathrm{t})$ and measurement noise $\mathrm{v}(\mathrm{t})$ are zero-mean white Gaussian with covariances

$$\mathbb{E}\{\mathrm{w}(\mathrm{t})\mathrm{w}(\tau)^{\mathrm{T}}\} = \mathrm{W}\,\delta(\mathrm{t} - \tau), \qquad \mathbb{E}\{\mathrm{v}(\mathrm{t})\mathrm{v}(\tau)^{\mathrm{T}}\} = \mathrm{V}\,\delta(\mathrm{t} - \tau).$$

Take the numerical values $\mathrm{W} = 0.1\,\mathrm{I}_2, \qquad \mathrm{V} = 0.5$

(a) Find the steady-state estimation error covariance $\mathrm{P_e}$ satisfying the algebraic Riccati equation

$$\mathrm{AP_e} + \mathrm{P_e A^T} + \mathrm{W} - \mathrm{P_e C^T V^{-1} C P_e} = 0.$$

(b) Compute the steady-state Kalman gain

$$\mathbf{L} = \mathbf{P_e C^T V^{-1}}.$$

(c) Verify that the estimator error dynamics matrix $(\mathbf{A} - \mathbf{LC})$ is stable (all eigenvalues in the left half plane).

SOLUTION: For the numerical data above, the solution is

$$P_e = \begin{bmatrix} 0.0295 & -0.0065 \\ -0.0065 & 0.0203 \end{bmatrix}, \qquad L = \begin{bmatrix} 2.9495 \\ -0.6503 \end{bmatrix}.$$

The estimator error dynamics matrix (A − LC) has eigenvalues

$$\lambda = -2.9747 \pm j1.1615$$

which lie in the left-half complex plane; hence, the estimator is asymptotically stable.

Estimator equation (with the computed gain):

$$\dot{\hat{x}} = A\hat{x} + Bu + L(y - C\hat{x})$$

where L is given above.

The complete LQG controller with $Q = I$ and $R = 1$ are implemented MATLAB and the code is in `Example11_31.m`

11.21.2 Features and Characteristics of LQR/LQG Controllers

This section summarizes the key advantages and limitations of Linear Quadratic Regulator (LQR) and Linear Quadratic Gaussian (LQG) controllers, providing insight into their practical applicability and design considerations.

LQR Controller

Advantages

- **Guaranteed Stability Margins:** LQR controllers possess excellent robustness properties, including infinite gain margin for gain increases, at least 6 dB gain margin for gain decreases, and a minimum of 60° phase margin.

- **Systematic Design Approach:** The design process reduces to selecting weighting matrices **Q** and **R**, eliminating the trial-and-error nature of traditional pole placement methods.

- **Optimal Performance:** The controller minimizes a quadratic cost function that systematically balances state regulation against control effort.

- **Computational Efficiency:** The solution involves solving the algebraic Riccati equation, which is mathematically well-conditioned and numerically stable.

Limitations

- **Full State Requirement:** LQR assumes all system states are measurable, which is rarely satisfied in practical applications.
- **Model Dependency:** Controller performance degrades significantly in the presence of plant uncertainties and modeling errors.
- **Weighting Matrix Selection:** Choosing appropriate **Q** and **R** matrices requires engineering experience and insight, as they may not relate directly to conventional design specifications.

LQG Controller

Advantages

- **Separation Principle:** The controller and state estimator can be designed independently, significantly simplifying the overall design process.
- **Handles Noisy Measurements:** The Kalman filter component optimally estimates system states from noisy sensor measurements.
- **Systematic Framework:** LQG provides a unified approach that combines optimal control (LQR) with optimal state estimation (Kalman filter).
- **Stochastic Optimality:** The controller minimizes the expected value of the cost function in the presence of both process and measurement noise.

Limitations

- **Loss of Robustness:** Unlike LQR, LQG controllers do not guarantee stability margins and can exhibit arbitrarily small gain and phase margins.
- **Model Sensitivity:** Performance depends heavily on accurate noise statistical models and precise plant parameters.
- **Estimation Errors:** Inaccurate state estimates directly compromise control performance, potentially leading to instability.

Addressing LQG Limitations

The robustness deficiencies inherent in LQG design have motivated the development of advanced control methodologies. Several approaches have been developed to overcome these limitations:

- **Loop Transfer Recovery (LTR):** A design technique that systematically recovers the desirable robustness properties of LQR controllers within the LQG framework by appropriately shaping the Kalman filter.

- $\mathcal{H}_\infty$ **Control:** Modern robust control methods that explicitly account for model uncertainties and bounded disturbances while providing guaranteed robustness margins.

- **Other Advanced Robust Control Techniques:** Including μ-synthesis for structured uncertainty, sliding mode control for nonlinear systems, and adaptive control methods for time-varying parameters.

These advanced topics extend beyond the scope of this text but represent important extensions of classical optimal control theory. They address the critical need for robustness to uncertainties in real-world control applications where perfect models are unattainable.

11.22 Multiple Choice Skills Test Questions

M 11.1. For the system given below, which of the following statements is true?

$$\dot{\mathbf{x}} = \begin{bmatrix} 1 & 0 \\ 0 & 1 \end{bmatrix} \mathbf{x} + \begin{bmatrix} \alpha \\ \beta \end{bmatrix} \mathrm{u}$$

A. The system is controllable for all non-zero values of α and β

B. The system is controllable if $\alpha = 0$ and $\beta \neq 0$

C. The system is uncontrollable for all values of α and β

D. The system controllability is not dependent on α and β

M 11.2. Consider the following systems

S1: $\dot{\mathrm{x}}_1 = -2\mathrm{x}_1 - 3\mathrm{x}_2 + \mathrm{u}, \quad \dot{\mathrm{x}}_2 = -4\mathrm{x}_2$

S2: $\dot{\mathrm{x}}_1 = -2\mathrm{x}_1 - 3\mathrm{x}_2 + \mathrm{u}, \quad \dot{\mathrm{x}}_2 = -4\mathrm{x}_2 - 6\mathrm{x}_1$

S3: $\dot{\mathrm{x}}_1 = -2\mathrm{x}_1, \quad \dot{\mathrm{x}}_2 = -3\mathrm{x}_1 - 4\mathrm{x}_2 + \mathrm{u}$

S4: $\dot{\mathrm{x}}_1 = -5\mathrm{x}_1 + \mathrm{u}, \quad \dot{\mathrm{x}}_2 = -7\mathrm{x}_2 + \mathrm{u}$

By inspection, determine which of the following is true?

A. S1 and S4 are controllable

B. S2 and S3 are controllable

C. S2 and S4 are uncontrollable

D. S1 and S3 are uncontrollable

M 11.3. Which is true for the following system:

$$\dot{\mathbf{x}} = \begin{bmatrix} -4 & 0 \\ 0 & -3 \end{bmatrix} \mathbf{x} + \begin{bmatrix} 0 \\ 1 \end{bmatrix} \mathrm{u}, \quad \mathrm{y} = \begin{bmatrix} 1 & 0 \end{bmatrix} \mathbf{x}$$

A. uncontrollable and unobservable

B. controllable and observable

C. uncontrollable and observable

D. controllable and unobservable

M 11.4. The transfer function of a system is

$$\mathrm{G(s)} = \frac{\mathrm{Y(s)}}{\mathrm{U(s)}} = \frac{\mathrm{s} + \alpha}{(\mathrm{s}^2 + 12\mathrm{s} + 32)(\mathrm{s}^2 + 2\mathrm{s} + 2)}$$

The real value of α which can make the system either uncontrollable or unobservable or both are

A. $\alpha = 6$ or $\alpha = 5$

B. $\alpha = 12$ or $\alpha = 0$

C. $\alpha = 3$ or $\alpha = 9$

D. $\alpha = 4$ or $\alpha = 8$

M 11.5. For the system represented by a state space model

$$\begin{bmatrix} \dot{\mathrm{x}}_1 \\ \dot{\mathrm{x}}_2 \end{bmatrix} = \begin{bmatrix} -5 & 0 \\ 0 & -7 \end{bmatrix} \begin{bmatrix} \mathrm{x}_1 \\ \mathrm{x}_2 \end{bmatrix} + \begin{bmatrix} 0 \\ 2 \end{bmatrix} \mathrm{u}, \quad \mathrm{y} = \begin{bmatrix} 1 & 0 \end{bmatrix} \begin{bmatrix} \mathrm{x}_1 \\ \mathrm{x}_2 \end{bmatrix}$$

A. state x_1 is controllable and state x_2 is observable

B. state x_1 is both uncontrollable and unobservable

C. state x_1 is controllable and state x_2 is unobservable

D. state x_1 is uncontrollable and state x_2 is unobservable

M 11.6. Which is true for the system represented by state space model

$$\begin{bmatrix} \dot{\mathrm{x}}_1 \\ \dot{\mathrm{x}}_2 \\ \dot{\mathrm{x}}_3 \\ \dot{\mathrm{x}}_4 \end{bmatrix} = \begin{bmatrix} -2 & 0 & 0 & 0 \\ 0 & -4 & 0 & 0 \\ 0 & 0 & -6 & 0 \\ 0 & 0 & 0 & -8 \end{bmatrix} \begin{bmatrix} \mathrm{x}_1 \\ \mathrm{x}_2 \\ \mathrm{x}_3 \\ \mathrm{x}_4 \end{bmatrix} + \begin{bmatrix} 1 \\ 0 \\ 3 \\ 0 \end{bmatrix} \mathrm{u}, \quad \mathrm{y} = \begin{bmatrix} 0 & 1 & 0 & 1 \end{bmatrix} \begin{bmatrix} \mathrm{x}_1 \\ \mathrm{x}_2 \\ \mathrm{x}_3 \\ \mathrm{x}_4 \end{bmatrix}$$

A. states x_1 and x_3 are controllable; but unobservable

B. states x_1 and x_3 are uncontrollable; but observable

C. states x_2 and x_4 are controllable; but unobservable

D. states x_2 and x_4 are uncontrollable; but observable

M 11.7. The following system holds which of the properties?

$$\begin{bmatrix} \dot{x}_1 \\ \dot{x}_2 \end{bmatrix} = \begin{bmatrix} -3 & 0 \\ 0 & 6 \end{bmatrix} \begin{bmatrix} x_1 \\ x_2 \end{bmatrix} + \begin{bmatrix} 0 \\ 1 \end{bmatrix} u, \quad y = \begin{bmatrix} 1 & 0 \end{bmatrix} \begin{bmatrix} x_1 \\ x_2 \end{bmatrix}$$

A. completely controllable
B. all states are uncontrollable
C. stabilizable
D. not stabilizable

M 11.8. Consider two systems S1 and S2 modeled as

$$\textbf{S1: } \dot{\mathbf{x}} = \begin{bmatrix} 4 & 0 \\ 0 & -2 \end{bmatrix} \mathbf{x} + \begin{bmatrix} 1 \\ 0 \end{bmatrix} u \quad \text{and} \quad \textbf{S2: } \dot{\mathbf{x}} = \begin{bmatrix} 4 & 0 \\ 0 & 2 \end{bmatrix} \mathbf{x} + \begin{bmatrix} 1 \\ 0 \end{bmatrix} u$$

A. both **S1** and **S2** are stabilizable
B. only **S2** is stabilizable
C. neither **S1** and **S2** is stabilizable
D. only **S1** is stabilizable

M 11.9. Consider two systems S1 and S2 modeled as

$$\textbf{S1: } \dot{\mathbf{x}} = \begin{bmatrix} 4 & 0 \\ 0 & -2 \end{bmatrix} \mathbf{x} + \begin{bmatrix} 1 \\ 0 \end{bmatrix} u, \quad y = \begin{bmatrix} 1 & 0 \end{bmatrix} \mathbf{x} \quad \text{and}$$

$$\textbf{S2: } \dot{\mathbf{x}} = \begin{bmatrix} 4 & 0 \\ 0 & 2 \end{bmatrix} \mathbf{x} + \begin{bmatrix} 1 \\ 0 \end{bmatrix} u, \quad y = \begin{bmatrix} 1 & 0 \end{bmatrix} \mathbf{x}$$

A. both **S1** and **S2** are detectable
B. only **S1** is detectable
C. only **S2** is detectable
D. neither **S1** and **S2** is detectable

M 11.10. Consider a system

$$\dot{x}(t) = 4x(t) + u(t), \quad x(0) = x_0$$
$$y(t) = x(t)$$

where y is the output of the system and u is the input to the system. It is desired that the state response to initial conditions to decay at a rate e^{-10t}. The state feedback control law which will achieve this is

A. $u(t) = -14x(t)$
B. $u(t) = -6x(t)$
C. $u(t) = 14x(t)$
D. $u(t) = 6x(t)$

M 11.11. Consider the following systems with state space models

$$\textbf{S1: } \dot{\mathbf{x}} = \begin{bmatrix} 4 & 0 \\ 0 & -2 \end{bmatrix} \mathbf{x} + \begin{bmatrix} 1 \\ 0 \end{bmatrix} \mathrm{u}, \qquad \textbf{S2: } \dot{\mathbf{x}} = \begin{bmatrix} -3 & 0 \\ 0 & -2 \end{bmatrix} \mathbf{x} + \begin{bmatrix} 1 \\ 2 \end{bmatrix} \mathrm{u}$$

$$\textbf{S3: } \dot{\mathbf{x}} = \begin{bmatrix} -1 & 0 & 0 \\ 0 & -2 & 0 \\ 0 & 0 & -3 \end{bmatrix} \mathbf{x} + \begin{bmatrix} 1 \\ 0 \\ 4 \end{bmatrix} \mathrm{u}, \quad \textbf{S4: } \dot{\mathbf{x}} = \begin{bmatrix} -1 & 0 & 0 & 0 \\ 0 & -5 & 0 & 0 \\ 0 & 0 & -7 & 0 \\ 0 & 0 & 0 & -4 \end{bmatrix} \mathbf{x} + \begin{bmatrix} 1 \\ 2 \\ 4 \\ 3 \end{bmatrix} \mathrm{u}$$

The system/systems that allow pole placement by state feedback are

A. **S1** and **S3** **B.** **S1** and **S4** **C.** **S2** and **S4** **D.** **S2** and **S3**

M 11.12. Consider a system

$$\dot{\mathrm{x}}(\mathrm{t}) = -2\mathrm{x}(\mathrm{t}) + \mathrm{u}(\mathrm{t}), \quad \mathrm{x}(0) = \mathrm{x}_0$$
$$\mathrm{y}(\mathrm{t}) = \mathrm{x}(\mathrm{t})$$

where y is the output of the system and u is the input to the system. The system is to be controlled by a state feedback control law with feedforward action given by $\mathrm{u} = -\mathrm{Kx} + \mathrm{Nr}$, which ensures that the output tracks the reference signal r with zero steady state error and the closed-loop poles are located at -10. The values of state feedback gain K and the feedforward gain N to achieve this are

A. $\mathrm{K} = 8$ and $\mathrm{N} = 10$

B. $\mathrm{K} = 20$ and $\mathrm{N} = 10$

C. $\mathrm{K} = -8$ and $\mathrm{N} = 10$

D. $\mathrm{K} = 4$ and $\mathrm{N} = 20$

M 11.13. Consider a system

$$\dot{\mathrm{x}}(\mathrm{t}) = -3\mathrm{x}(\mathrm{t}) + \mathrm{u}(\mathrm{t})$$
$$\mathrm{y}(\mathrm{t}) = 2\mathrm{x}(\mathrm{t})$$

where y is the output of the system and u is the input to the system. If the observer state estimation error is to be decayed at a rate $\mathrm{e}^{-10\mathrm{t}}$, the required observer gain L equals to

A. $\mathrm{L} = 3.5$ **B.** $\mathrm{L} = 6.5$ **C.** $\mathrm{L} = -3.5$ **D.** $\mathrm{L} = 7$

M 11.14. If the open-loop system is completely controllable, then the closed-loop system with state feedback controller

A. is completely controllable
B. may be uncontrollable
C. always uncontrollable
D. is always stable

M 11.15. If the open-loop system is controllable and observable, then with the integration of the state feedback controller

A. the controllability and observability properties of the closed-loop system are preserved

B. the observability property of a closed-loop system is not guaranteed to be preserved

C. the observability properties of a closed-loop system are always preserved, but not its controllability property

D. both controllability and observability of the closed-loop system are destroyed

Answers

M 11.1. C **M 11.2**. D **M 11.3**. A **M 11.4**. D **M 11.5**. D **M 11.6**. A **M 11.7**. C **M 11.8**. D **M 11.9**. B **M 11.10**. A **M 11.11**. C **M 11.12**. A **M 11.13**. A **M 11.14**. A **M 11.15**. B

11.23 Knowledge Assessment Problems

KaP 11.1 Investigate the controllability of the following systems given the system matrix **A** and input matrix **B**.

$$\textbf{S1:}\ \ \mathbf{A} = \begin{bmatrix} 0 & 2 \\ 1 & -1 \end{bmatrix},\ \mathbf{B} = \begin{bmatrix} 1 \\ 0 \end{bmatrix}$$

$$\textbf{S2:}\ \ \mathbf{A} = \begin{bmatrix} -3 & 1 \\ -2 & 1.5 \end{bmatrix},\ \mathbf{B} = \begin{bmatrix} 1 \\ 4 \end{bmatrix}$$

$$\textbf{S3:}\ \ \mathbf{A} = \begin{bmatrix} 2 & 1 \\ 0 & 3 \end{bmatrix},\ \mathbf{B} = \begin{bmatrix} 1 \\ 0 \end{bmatrix}$$

$$\textbf{S4:}\ \ \mathbf{A} = \begin{bmatrix} 1 & 2 \\ 1 & 1 \end{bmatrix},\ \mathbf{B} = \begin{bmatrix} 2 \\ \sqrt{2} \end{bmatrix}$$

$$\textbf{S5:}\ \ \mathbf{A} = \begin{bmatrix} 1 & 2 & 3 \\ 0 & 1 & 5 \\ 2 & 1 & 0 \end{bmatrix},\ \mathbf{B} = \begin{bmatrix} 2 \\ 3 \\ 1 \end{bmatrix}$$

$$\textbf{S6:}\ \ \mathbf{A} = \begin{bmatrix} 1 & -2 & 0 \\ 0 & -3 & 0 \\ 0 & 0 & -5 \end{bmatrix},\ \mathbf{B} = \begin{bmatrix} 1 \\ 0 \\ 0 \end{bmatrix}$$

KaP 11.2 Show that a system represented in phase variable canonical form is always controllable. Prove this by considering an example of a third-order system.

KaP 11.3 Investigate the observability of the following systems given the system matrix **A** and output matrix **C**

$$\textbf{S1:}\ \mathbf{A} = \begin{bmatrix} 0 & 2 \\ 1 & -1 \end{bmatrix},\mathbf{C} = \begin{bmatrix} 1 & 0 \end{bmatrix}$$

$$\textbf{S2:}\ \mathbf{A} = \begin{bmatrix} -1 & 0 \\ -3 & -5 \end{bmatrix},\ \mathbf{C} = \begin{bmatrix} 2 & 0 \end{bmatrix}$$

$$\textbf{S3:}\ \mathbf{A} = \begin{bmatrix} 1 & 2 \\ 0 & 3 \end{bmatrix},\ \mathbf{C} = \begin{bmatrix} 0 & 1 \end{bmatrix}$$

$$\textbf{S4:}\ \mathbf{A} = \begin{bmatrix} 1 & 2 & 3 \\ 0 & 1 & 5 \\ 2 & 1 & 0 \end{bmatrix},\ \mathbf{C} = \begin{bmatrix} 1 & 0 & 0 \end{bmatrix}$$

$$\textbf{S5:}\ \mathbf{A} = \begin{bmatrix} 0 & 1 & 0 \\ 0 & 0 & 1 \\ 0 & -2 & -3 \end{bmatrix},\ \mathbf{C} = \begin{bmatrix} 1 & 4 & 3 \end{bmatrix}$$

$$\textbf{S6:}\mathbf{A} = \begin{bmatrix} -1 & 2 & 0 \\ 0 & -3 & 1 \\ 0 & 0 & -5 \end{bmatrix},\mathbf{C} = \begin{bmatrix} 0 & 1 & 0 \end{bmatrix}$$

$$\textbf{S7: } \mathbf{A} = \begin{bmatrix} 1 & 2 & 0 \\ 0 & 3 & 4 \\ 0 & 0 & -2 \end{bmatrix}, \mathbf{C} = \begin{bmatrix} 0 & 0 & 1 \end{bmatrix}$$

KaP 11.4 Show that an nth-order system with n outputs is completely observable if its n × n-output coupling matrix is nonsingular.

KaP 11.5 Consider the system described by

$$\dot{\mathbf{x}} = \begin{bmatrix} a & a \\ 0 & a \end{bmatrix} \mathbf{x} + \begin{bmatrix} b_1 \\ b_2 \end{bmatrix} u$$

$$y = \begin{bmatrix} d_1 & d_2 \end{bmatrix} \mathbf{x}$$

where a, b_1, b_2, d_1 and d_2 are constants. In terms of these constants, determine the conditions of controllability and observability.

KaP 11.6 Investigate the controllability and observability of the following systems and show that the results conform to frequency domain criteria.

a. $\dot{\mathbf{x}} = \begin{bmatrix} 5 & -5 \\ -5 & 5 \end{bmatrix} \mathbf{x} + \begin{bmatrix} 1 \\ -1 \end{bmatrix} u, \quad y = \begin{bmatrix} 2 & -2 \end{bmatrix} \mathbf{x}$

b. $\dot{\mathbf{x}} = \begin{bmatrix} 2 & -2 \\ -2 & 2 \end{bmatrix} \mathbf{x} + \begin{bmatrix} 1 \\ 3 \end{bmatrix} u, \quad y = \begin{bmatrix} 1 & 1 \end{bmatrix} \mathbf{x}$

c. $\dot{\mathbf{x}} = \begin{bmatrix} 1 & -1 \\ -1 & 1 \end{bmatrix} \mathbf{x} + \begin{bmatrix} 3 \\ 3 \end{bmatrix} u, \quad y = \begin{bmatrix} 2 & 5 \end{bmatrix} \mathbf{x}$

d. $\dot{\mathbf{x}} = \begin{bmatrix} 3 & -3 \\ -3 & 3 \end{bmatrix} \mathbf{x} + \begin{bmatrix} 2 \\ -2 \end{bmatrix} u, \quad y = \begin{bmatrix} 1 & -1 \end{bmatrix} \mathbf{x}$

KaP 11.7 Investigate the controllability property of the following circuit

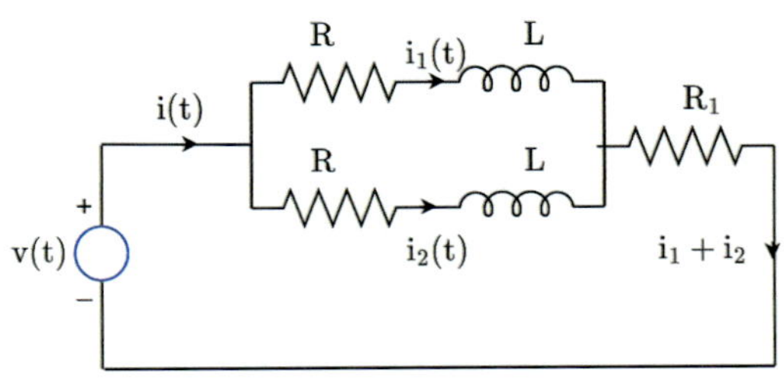

Figure 11.23: Symmetrical circuit

KaP 11.8 Write state equations for systems-each with modes e^{-5t},e^{3t},e^{-8t},$e^{(-2+j4)t}$, and $e^{(-2-j4)t}$- which have the following properties

a. The mode e^{-5t} is uncontrollable.

b. The mode e^{-8t} is unobservable.

c. The mode e^{3t} is both uncontrollable and unobservable.

d. The modes $e^{(-2+j4)t}$ and $e^{(-2-j4)t}$ are uncontrollable.

e. The mode e^{-5t} is both uncontrollable and unobservable.

f. The modes e^{-8t} is uncontrollable and the modes $e^{(-2+j4)t}$ and $e^{(-2-j4)t}$ are unobservable.

KaP 11.9 Consider the system given by

$$\dot{\mathbf{x}} = \begin{bmatrix} a & b \\ 0 & c \end{bmatrix} \mathbf{x} + \begin{bmatrix} 0 \\ 1 \end{bmatrix} u$$

Investigate the state and output controllability of the system when

a. $y = \begin{bmatrix} \alpha & 0 \end{bmatrix} \mathbf{x}$ b $y = \begin{bmatrix} 0 & \beta \end{bmatrix} \mathbf{x}$

KaP 11.10 By inspection, investigate the state controllability of the following systems and determine the states which are uncontrollable (if any)

$$\mathbf{S_1}: \begin{bmatrix} \dot{x}_1 \\ \dot{x}_2 \end{bmatrix} = \begin{bmatrix} -1 & 0 \\ 0 & -2 \end{bmatrix} \begin{bmatrix} x_1 \\ x_2 \end{bmatrix} + \begin{bmatrix} 3 \\ 4 \end{bmatrix} \mathbf{u},$$

$$\mathbf{S_2}: \begin{bmatrix} \dot{x}_1 \\ \dot{x}_2 \end{bmatrix} = \begin{bmatrix} -2 & 0 \\ 0 & -4 \end{bmatrix} \begin{bmatrix} x_1 \\ x_2 \end{bmatrix} + \begin{bmatrix} 0 \\ 1 \end{bmatrix} \mathbf{u}$$

$$\mathbf{S_3}: \begin{bmatrix} \dot{x}_1 \\ \dot{x}_2 \\ \dot{x}_3 \end{bmatrix} = \begin{bmatrix} -1 & 0 & 0 \\ 0 & -7 & 0 \\ 0 & 0 & -10 \end{bmatrix} \begin{bmatrix} x_1 \\ x_2 \\ x_3 \end{bmatrix} + \begin{bmatrix} 0.0 \\ 1.0 \\ 0 \end{bmatrix} \mathbf{u},$$

$$\mathbf{S_4}: \begin{bmatrix} \dot{x}_1 \\ \dot{x}_2 \\ \dot{x}_3 \\ \dot{x}_4 \end{bmatrix} = \begin{bmatrix} -2 & 0 & 0 & 0 \\ 0 & -3 & 0 & 0 \\ 0 & 0 & -5 & 0 \\ 0 & 0 & 0 & -1 \end{bmatrix} \begin{bmatrix} x_1 \\ x_2 \\ x_3 \\ x_4 \end{bmatrix} + \begin{bmatrix} 1.0 \\ 4.0 \\ 0 \\ 7.0 \end{bmatrix} \mathbf{u}$$

KaP 11.11 By inspection, investigate the controllability of the following systems and determine the states which are uncontrollable (if any)

$$\mathbf{S_1}: \begin{bmatrix}\dot{x}_1\\ \dot{x}_2\\ \dot{x}_3\end{bmatrix} = \begin{bmatrix}-4 & 1 & 0\\ 0 & -4 & 0\\ 0 & 0 & -3\end{bmatrix}\begin{bmatrix}x_1\\ x_2\\ x_3\end{bmatrix} + \begin{bmatrix}0\\ 1\\ 2\end{bmatrix}\mathbf{u}$$

$$\mathbf{S_2}: \begin{bmatrix}\dot{x}_1\\ \dot{x}_2\\ \dot{x}_3\end{bmatrix} = \begin{bmatrix}-5 & 1 & 0\\ 0 & -5 & 0\\ 0 & 0 & -2\end{bmatrix}\begin{bmatrix}x_1\\ x_2\\ x_3\end{bmatrix} + \begin{bmatrix}0\\ 2\\ 0\end{bmatrix}\mathbf{u}$$

$$\mathbf{S_3}: \begin{bmatrix}\dot{x}_1\\ \dot{x}_2\\ \dot{x}_3\end{bmatrix} = \begin{bmatrix}-3 & 1 & 0\\ 0 & -3 & 0\\ 0 & 0 & -6\end{bmatrix}\begin{bmatrix}x_1\\ x_2\\ x_3\end{bmatrix} + \begin{bmatrix}2 & 5\\ 0 & 0\\ 1 & 4\end{bmatrix}\begin{bmatrix}\mathbf{u}_1\\ \mathbf{u}_2\end{bmatrix}$$

$$\mathbf{S_4}: \begin{bmatrix}\dot{x}_1\\ \dot{x}_2\\ \dot{x}_3\\ \dot{x}_4\end{bmatrix} = \begin{bmatrix}-2 & 1 & 0 & 0 & 0\\ 0 & -2 & 0 & 0 & 0\\ 0 & 0 & -6 & 1 & 0\\ 0 & 0 & 0 & -6 & 0\\ 0 & 0 & 0 & 0 & -1\end{bmatrix}\begin{bmatrix}x_1\\ x_2\\ x_3\\ x_4\\ x_5\end{bmatrix} + \begin{bmatrix}0\\ -5\\ 0\\ 3\\ 8\end{bmatrix}\mathbf{u}$$

KaP 11.12 Investigate the observability of the following systems

$$\mathbf{S_1} \quad \begin{bmatrix}\dot{x}_1\\ \dot{x}_2\end{bmatrix} = \begin{bmatrix}-1 & 0\\ 0 & -5\end{bmatrix}\begin{bmatrix}x_1\\ x_2\end{bmatrix}, \quad \mathbf{y} = \begin{bmatrix}3 & 1\end{bmatrix}\begin{bmatrix}x_1\\ x_2\end{bmatrix}$$

$$\mathbf{S_2} \quad \begin{bmatrix}\dot{x}_1\\ \dot{x}_2\end{bmatrix} = \begin{bmatrix}-3 & 0\\ 0 & -4\end{bmatrix}\begin{bmatrix}x_1\\ x_2\end{bmatrix}, \quad \mathbf{y} = \begin{bmatrix}0 & 2\end{bmatrix}\begin{bmatrix}x_1\\ x_2\end{bmatrix}$$

$$\mathbf{S_3} \quad \begin{bmatrix}\dot{x}_1\\ \dot{x}_2\\ \dot{x}_3\\ \dot{x}_4\end{bmatrix} = \begin{bmatrix}-1 & 0 & 0 & 0\\ 0 & -3 & 0 & 0\\ 0 & 0 & -5 & 0\\ 0 & 0 & 0 & -7\end{bmatrix}\begin{bmatrix}x_1\\ x_2\\ x_3\\ x_4\end{bmatrix}, \quad \mathbf{y} = \begin{bmatrix}2 & 0 & 1 & 0\end{bmatrix}\begin{bmatrix}x_1\\ x_2\\ x_3\\ x_4\end{bmatrix}$$

$$\mathbf{S_4} \quad \begin{bmatrix}\dot{x}_1\\ \dot{x}_2\\ \dot{x}_3\\ \dot{x}_4\end{bmatrix} = \begin{bmatrix}-2 & 0 & 0 & 0 & 0\\ 0 & -3 & 0 & 0 & 0\\ 0 & 0 & -5 & 0 & 0\\ 0 & 0 & 0 & -10 & 0\\ 0 & 0 & 0 & 0 & -1\end{bmatrix}\begin{bmatrix}x_1\\ x_2\\ x_3\\ x_4\\ x_5\end{bmatrix}, \mathbf{y} = \begin{bmatrix}0 & 7 & 0 & 4 & 0\end{bmatrix}\begin{bmatrix}x_1\\ x_2\\ x_3\\ x_4\end{bmatrix}$$

KaP 11.13 Analyze the stabilizability of systems of knowledge assessment problem Kap11.1. Confirm your results by diagonalization.

KaP 11.14 Analyze the detectability of systems of knowledge assessment problem Kap11.2. Confirm your results by diagonalization.

KaP 11.15 The open-loop transfer function of a plant is given by

$$G(s) = \frac{5}{(s+1)(s+2)(s+3)}$$

Design a state feedback controller which will place the dominant poles of the closed-loop system in the s-plane to yield 10% overshoot and 1sec settling time (with 2% criterion). Place the third pole 10 times farther from the dominant pole pair.

KaP 11.16 The open-loop transfer function of a plant is given by

$$G(s) = \frac{20(s+5)(s+10)}{(s+2)(s+6)(s+8)}$$

Design a state feedback controller which will place the dominant poles of the closed-loop system in the s-plane to yield 5% overshoot and peak time of 0.75 sec. Place the third pole 10 times farther from the dominant pole pair.

KaP 11.17 The transfer function of a plant is given by

$$G(s) = \frac{Y(s)}{R(s)} = \frac{K}{s(s+20)}$$

Design a state feedback controller which will ensure 5% peak overshoot, settling time 2 sec (2% criterion) and the velocity error 10%.

KaP 11.18 Design a full state observer for the following systems, given the system matrix **A**, the output matrix **C**, and the location of observer poles. You can use either of the approaches.

a. $\mathbf{A} = \begin{bmatrix} -5 & 0 \\ 1 & -1 \end{bmatrix}$, $\mathbf{C} = \begin{bmatrix} 0 & 1 \end{bmatrix}$, Observer poles at -10 and -10

b. $\mathbf{A} = \begin{bmatrix} 1 & 2 \\ -3 & 5 \end{bmatrix}$, $\mathbf{C} = \begin{bmatrix} 1 & -2 \end{bmatrix}$, Observer poles at -5 and -5

c. $\mathbf{A} = \begin{bmatrix} -2 & 1 \\ 1 & -1 \end{bmatrix}$, $\mathbf{C} = \begin{bmatrix} 1 & 0 \end{bmatrix}$, Observer poles at -5 and -6

KaP 11.19 Consider a first-order system with transfer function

$$G(s) = \frac{Y(s)}{U(s)} = \frac{2}{s-3}$$

Represent this system by a state variable model using output as the state variable.

a. Design a state feedback controller

$$u = -kx + r$$

which places the closed-loop pole at $s = -5$.

b. Compute the steady state error due to a step input.

c. Design a state feedback controller with a pre-compensator in the forward path of the form

$$u = -kx + Nr$$

which will place the pole at the same location as in (a) and eliminate the steady state error due to the step input.

KaP 11.20 A simple model for the read/write head of a hard disk is described by

$$J\ddot{\theta} + c\dot{\theta} + k\theta = K_T i$$

where J represents the inertia of the head assembly; c denotes the viscous damping coefficient of the bearings; k is the return spring constant; K_T is the motor torque constant; θ, $\dot{\theta}$ and $\ddot{\theta}$ denote angular position, velocity and acceleration, respectively; and i is the input current.

i. Represent this system in state space, considering the output θ and its derivative $\dot{\theta}$ as state variables and the applied current as input.

ii. Design a state feedback controller $u = -\mathbf{Kx} + r$ which places the closed-loop poles at $-3 \pm j4$.

iii. Design a state feedback controller with pre-compensator in the forward path of the form $u = -\mathbf{Kx} + Nr$ which will place the poles at the same location as in (ii) and eliminate the steady state error due to step input.

The parameters of the system are: $J = 0.0008\ kg.m^2$, $c = 0.001\ N - m - s/rad$, $k = 0.005\ N.m/rad$ and $K_T = 0.05\ N.m/A$. Consider the reference as 1^0

KaP 11.21 i. Design a state feedback controller with an integral controller that will place the closed-loop poles at $p_1 = -2$ and $p_2 = -4$ for the system

$$\dot{x} = 5x + 10u, \quad y = x$$

ii. Design the same controller for the system

$$G(s) = \frac{Y(s)}{U(s)} = \frac{3}{s-2}$$

Place the poles at -4 and at -6.

KaP 11.22 Design a state feedback controller with an integral controller for the inverted pendulum such that the cart position tracks a step reference signal. Place the poles of the closed-loop system at $-1 \pm j\sqrt{2}$, -4, -4, and -4

KaP 11.23 The model of an antenna positioner is given by

$$\dot{\mathbf{x}} = \mathbf{A}\mathbf{x} + \mathbf{B}u, \quad y = \mathbf{C}\mathbf{x}$$

where

$$\mathbf{x} = \begin{bmatrix} x_1 \\ x_2 \\ x_3 \end{bmatrix} = \begin{bmatrix} \text{angular position(rad)} \\ \text{velocity(rad/sec)} \\ \text{torque(N} - \text{m)} \end{bmatrix}$$

$$\mathbf{A} = \begin{bmatrix} 0 & 1 & 0 \\ 0 & -1 & 1 \\ 0 & 0 & -5 \end{bmatrix}, \quad \mathbf{B} = \begin{bmatrix} 0 \\ 0 \\ 5 \end{bmatrix}, \quad \text{and} \quad \mathbf{C} = \begin{bmatrix} 1 & 0 & 0 \end{bmatrix}$$

Using MATLAB

a. Design an observer-based state feedback controller, such that the angular position tracks a step reference. Place the controller poles at $-1 \pm j2$, -4 and at -5. The observer poles are to be placed at -10, -15 and at -25

b. Obtain the output response of the state feedback and observer-based state feedback with an integral controller. Consider the initial condition of the states and estimated states as:

$$\mathbf{x}(0) = \begin{bmatrix} 1 \\ 0.5 \\ -0.5 \end{bmatrix} \quad \hat{\mathbf{x}}(0) = \begin{bmatrix} 0 \\ 0 \\ 0 \end{bmatrix}$$

KaP 11.24 The state space model of a dynamic system is expressed as

$$\dot{\mathbf{x}}(t) = \mathbf{A}\mathbf{x}(t) + \mathbf{B}\mathbf{u}(t), \quad \text{and} \quad \mathbf{y} = \mathbf{C}\mathbf{x}(t)$$

a. Design a reduced-order observer for the following systems.

a. $\mathbf{A} \begin{bmatrix} 0 & 1 & 0 \\ 0 & 0 & 1 \\ -3 & -2 & -5 \end{bmatrix}, \quad \mathbf{B} = \begin{bmatrix} 0 \\ 0 \\ 1 \end{bmatrix}, \text{ and } \mathbf{C} = \begin{bmatrix} 1 & 0 & 0 \end{bmatrix}$

The desired observer poles are at -10 and -15.

b. $\mathbf{A} \begin{bmatrix} 0 & 1 & 0 \\ 0 & 0 & 1 \\ -8 & -14 & -7 \end{bmatrix}, \quad \mathbf{B} = \begin{bmatrix} 0 \\ 0 \\ 1 \end{bmatrix}, \text{ and } \mathbf{C} = \begin{bmatrix} 1 & 0 & 0 \\ 0 & 1 & 0 \end{bmatrix}$

The desired observer pole is at -12.

KaP 11.25 The dynamics of a system is described by the vector differential equation

$$\begin{bmatrix} \dot{x}_1 \\ \dot{x}_2 \end{bmatrix} = \begin{bmatrix} 0 & 1 \\ 0 & 0 \end{bmatrix} \begin{bmatrix} x_1 \\ x_2 \end{bmatrix} + \begin{bmatrix} 0 \\ 1 \end{bmatrix} u$$

Design an LQR that would optimize the performance index

$$J = \int_0^\infty (x^T x + u^T u)\, dt$$

KaP 11.26 The dynamics of a mass–spring–damper system with first-order actuator dynamics are described by the continuous-time state-space model

$$\begin{aligned} \dot{x}(t) &= Ax(t) + Bu(t) + w(t) \\ y(t) &= Cx(t) + v(t) \end{aligned}$$

with

$$\mathbf{A} = \begin{bmatrix} 0 & 1 & 0 \\ -4 & -1 & 1 \\ 0 & 0 & -10 \end{bmatrix}, \quad \mathbf{B} = \begin{bmatrix} 0 \\ 0 \\ 10 \end{bmatrix}, \quad \mathbf{C} = \begin{bmatrix} 1 & 0 & 0 \end{bmatrix},$$

where $x = \begin{bmatrix} \text{position} & \text{velocity} & \text{actuator state} \end{bmatrix}^T$, u(t) is the control input, and y(t) is the measured output (position).

Process disturbance w(t) and measurement noise v(t) are modeled as zero-mean Gaussian white noise with covariances

$$\mathbb{E}\left[w(t)w(\tau)^\top\right] = W\,\delta(t-\tau), \qquad \mathbb{E}\left[v(t)v(\tau)\right] = V\,\delta(t-\tau),$$

where $W \geq 0$ and $V > 0$ are given.

Design an LQG controller (dynamic output-feedback controller) for this plant such that:

(a) The closed-loop system is asymptotically stable.

(b) The controller minimizes the infinite-horizon quadratic cost

$$J = \mathbb{E}\left[\int_0^\infty \left(x(t)^\top Q x(t) + u(t)^\top R u(t)\right) dt\right],$$

for given weighting matrices $Q = Q^\top \succeq 0$ and $R = R^\top \succ 0$.

(c) Only the measured output $y(t)$ is available for feedback (full state $x(t)$ is not measured).

(d) The state estimator (Kalman filter) provides the minimum mean-square error state estimate under the assumed noise statistics.

Bibliography

[1] Karl Johan Åström and Tore Hägglund. PID Controllers: Theory, Design, and Tuning. 2nd. 343 pages. Research Triangle Park, NC: ISA - The Instrumentation, Systems and Automation Society, 1995. ISBN: 9781556175169.

[2] Karl Johan Astrom and Richard M. Murray. Feedback Systems: An Introduction for Scientists and Engineers. Princeton University Press, 2008.

[3] Yazdan Bavafa-Toosi. Introduction to Linear Control Systems. Comprehensive textbook for undergraduate courses in linear control systems. London, UK: Academic Press, 2017. ISBN: 9780128127483.

[4] S. A. Billings. Nonlinear System Identification: NARMAX Methods in the Time, Frequency, and Spatio-Temporal Domains. Wiley, 2013.

[5] S. A. Billings and J. C. Peyton-Jones. Mapping nonlinear integro-differential equations into the frequency domain. In: International Journal of Control 52.4 (1990), pp. 863–879.

[6] Robert H. Bishop, ed. The Mechatronics Handbook. 2nd. Two-volume set. CRC Press, 2007. ISBN: 9780849392573.

[7] Hendrik W. Bode. Network Analysis and Feedback Amplifier Design. New York: Van Nostrand, 1945. ISBN: 9780486492811.

[8] W. Bolton. Control Engineering. 2nd. Prentice Hall, 1992. ISBN: 9780582327733.

[9] William L. Brogan. Modern Control Theory. 3rd. Englewood Cliffs, NJ: Prentice-Hall, 1991. ISBN: 9780135897638.

[10] James Ward Brown and Ruel V. Churchill. Complex Variables and Applications. 9th ed. New York: McGraw-Hill Education, 2013. ISBN: 9780073383170.

[11] Chi-Tsong Chen. Analog and Digital Control System Design: Transfer-Function, State-Space, Algebraic Methods. Oxford University Press, 1993.

[12] Chi-Tsong Chen. Linear System Theory and Design. 3rd. Oxford University Press, 1999. ISBN: 0195117778.

[13] Chi-Tsong Chen. Signals and Systems. Oxford University Press, 2004.

[14] Ira Cochin and William Cadwallender. Analysis and Design of Dynamic Systems. 3rd Revised. New York: Prentice Hall, 1997. ISBN: 9780673982582.

[15] John J. D'Azzo and Constantine H. Houpis. Linear Control System Analysis and Design: Conventional and Modern. 4th. New York: McGraw-Hill, 1995. ISBN: 9780071132811.

[16] E. van Dijk, J. N. Spruijt, D. M. O. Kashani, and J. B. Klaassens. PWM-switch modeling of DC–DC converters. In: IEEE Transactions on Power Electronics 10.6 (1995), pp. 659–665.

[17] Richard C. Dorf and Robert H. Bishop. Modern Control Systems. 13th. Pearson, 2016. ISBN: 9780134407623.

[18] John C. Doyle, Bruce A. Francis, and Allen R. Tannenbaum. Feedback Control Theory. 1st. Macmillan Publishing Co., 1990, p. 224. ISBN: 9780023300110.

[19] Ken Dutton, Bill Barraclough, and Steve Thompson. The Art of Control Engineering. 1st. Harlow; Reading, MA: Addison-Wesley/Prentice Hall, 1997, p. 813. ISBN: 9780201175455.

[20] Ramin S. Esfandiari and Bei Lu. Modeling and Analysis of Dynamic Systems. 3rd. Boca Raton, FL: CRC Press, 2018. ISBN: 9781138726420.

[21] W. R. Evans. Control Systems Synthesis by Root Locus Method. In: Transactions of the American Institute of Electrical Engineers 69.1 (Jan. 1950), pp. 66–69. ISSN: 0096-386. DOI: 10.1109/T-AIEE.1950.5060121.

[22] Bernard Friedland. Control System Design: An Introduction to State-Space Methods. Dover Books on Electrical Engineering. Mineola, NY: Dover Publications, 2005. ISBN: 9780486442785.

[23] Z. Gajic and M. Lelic. Modern Control Systems Engineering. Pearson Education, 1996. ISBN: 0131341162.

[24] Murray F. Gardner and John L. Barnes. Transients in Linear Systems Studied by the Laplace Transformation: Volume I – Lumped-Constant Systems. New York: John Wiley & Sons, 1950. ISBN: 9780471291725.

[25] Gene H. Golub and Charles F. Van Loan. Matrix Computations. 4th. Johns Hopkins University Press, 2013. ISBN: 9781421407944. URL: https://epubs.siam.org/doi/book/10.1137/1.9781421407944.

[26] Graham C. Goodwin, Stefan F. Graebe, and Mario E. Salgado. Control System Design. 1st. Upper Saddle River, NJ, USA: Prentice Hall, 2001. ISBN: 9780130957026.

[27] M. Gopal. Digital Control Engineering. New Delhi, India: New Age International, 1988, p. 535.

[28] M. Gopal. Modern Control System Theory. Reprint. New Delhi, India: New Age International, 1993, p. 688. ISBN: 9788122405033.

[29] Donald T. Greenwood. Principles of Dynamics. English. 2nd. Prentice-Hall, 1987, p. 560. ISBN: 9780137099818.

[30] Simon Haykin and Barry Van Veen. Signals and Systems. 2nd. New York, NY, USA: Wiley, 1999. ISBN: 9780471355479.

[31] Roger A. Horn and Charles R. Johnson. Matrix Analysis. 2nd. Cambridge, UK: Cambridge University Press, 2013. ISBN: 9780521839402.

[32] Gene H. Hostetter, Bahram Shahian, Raymond T. Stefani, and Clement J. Savant Jr. Design of Feedback Control Systems. 4th. Oxford University Press, 2002. ISBN: 9780195142495.

[33] Adolf Hurwitz. Ueber die Bedingungen, unter welchen eine Gleichung nur Wurzeln mit negativen reellen Theilen besitzt. In: Mathematische Annalen 46.2 (1895), pp. 273–284. DOI: 10.1007/BF01446812.

[34] Thomas Kailath. Linear Systems. Illustrated. Prentice-Hall, 1980, p. 682. ISBN: 9780135369616.

[35] R. E. Kalman. Canonical Structure of Linear Dynamical Systems. In: Transactions of the ASME, Journal of Basic Engineering 82.Series D (1960), pp. 35–45.

[36] Edward W. Kamen. Linear Systems. English. Illustrated. Prentice-Hall, 1980, p. 682. ISBN: 9780135369616.

[37] Edward W. Kamen and Bonnie S. Heck. Fundamentals of Signals and Systems Using MATLAB. English. 3rd. Prentice Hall, 2000, p. 688. ISBN: 9780131687370.

[38] Vojislav Kecman. State-Space Models of Lumped and Distributed Systems. English. Vol. 112. Lecture Notes in Control and Information Sciences. Springer-Verlag, 1988, p. 283. ISBN: 9783540500827.

[39] Branislav Kisacanin and Gyan C. Agarwal. Linear Control Systems: With Solved Problems and MATLAB Examples. 1st. University Series in Mathematics. Springer, 2001. ISBN: 9780306467431. DOI: 10.1007/978-1-4615-0553-2.

[40] Jerzy Klamka. Controllability of Dynamical Systems. Vol. 48. Mathematics and Its Applications. Springer Netherlands, 1991. ISBN: 9780792308225.

[41] Alfred Korzybski. Science and Sanity: An Introduction to Non-Aristotelian Systems and General Semantics. Lancaster, Pa.; New York City: The International Non-Aristotelian Library Publishing Company, 1933.

[42] Bohdan T. Kulakowski, John F. Gardner, and J. Lowen Shearer. Dynamic Modeling and Control of Engineering Systems. 3rd. Cambridge, UK: Cambridge University Press, 2007. ISBN: 9780521864350.

[43] Benjamin C. Kuo. Digital Control Systems. 2nd. Oxford University Press, 1995. ISBN: 9780195120646.

[44] Benjamin C. Kuo and Farid Golnaraghi. Automatic Control Systems. 8th. John Wiley & Sons, 2002.

[45] T.R. Kurfess. Foundations of Classical Control Theory with Reference to Gain Plots. In: Control Engineering Practice 1.1 (1993), pp. 1–10. DOI: 10.1016/0967-0661(93)90001-B.

[46] Huibert Kwakernaak and Raphael Sivan. Linear Optimal Control Systems. English. Wiley Interscience, 1972, p. 498. ISBN: 9780471565441.

[47] B. P. Lathi. Linear Systems and Signals. 3rd. New York, NY, USA: Oxford University Press, 2017, p. 1430. ISBN: 9780190200176.

[48] B. P. Lathi. Principles of Linear Systems and Signals. 2nd. New York, NY, USA: Oxford University Press, 2009, p. 768. ISBN: 9780195392579.

[49] B. P. Lathi. Signal Processing and Linear Systems. New York, NY, USA: Oxford University Press, 1998, p. 648. ISBN: 9780195219173.

[50] Antonio M. Lepschy, Mario Lorenzini, and Umberto Viaro. Feedback control in ancient water and mechanical clocks. In: IEEE Transactions on Education 35.1 (1992), pp. 3–10.

[51] William S. Levine, ed. The Control Handbook (Three Volume Set). 2nd. Also available as an eBook. Boca Raton, FL: CRC Press, 2011. ISBN: 9781420073669. DOI: 10.1201/9781315218694.

[52] David G. Luenberger. An Introduction to Observers. In: IEEE Transactions on Automatic Control 16.6 (1971), pp. 596–602. DOI: 10.1109/TAC.1971.1099602. URL: https://ieeexplore.ieee.org/document/1099602.

[53] David G. Luenberger. Introduction to Dynamic Systems: Theory, Models, and Applications. John Wiley & Sons, 1979. ISBN: 9780471025948. URL: https://www.wiley.com/en-us/Introduction%2Bto%2BDynamic%2BSystems%3A%2BTheory%2C%2BModels%2Band%2BApplications-p-9780471025948.

[54] Samuel J. Mason. Feedback Theory: Further Properties of Signal-Flow Graphs. In: Proceedings of the IRE 44.7 (1956), pp. 920–926. DOI: 10.1109/JRPROC.1956.275147.

[55] Samuel J. Mason. Feedback Theory: Some Properties of Signal-Flow Graphs. In: Proceedings of the IRE 41 (1953), pp. 1144–1156.

[56] MathWorks, Inc. MATLAB User's Guide. Accessed August 2025. MathWorks, Inc. 2025. URL: https://www.mathworks.com.

[57] Otto Mayr. The Origins of Feedback Control. Originally published in German in 1969. Cambridge, MA: MIT Press, 1970. ISBN: 9780262130677.

[58] Gjerrit Meinsma. Elementary Proof of the Routh-Hurwitz Test. In: Systems & Control Letters 25.4 (1995), pp. 237–242. DOI: 10.1016/0167-6911(94)00089-E.

[59] I. J. Nagrath and M. Gopal. Control Systems Engineering. 5th. New Delhi, India: New Age International, 2015. ISBN: 9788122420081.

[60] Norman S. Nise. Control Systems Engineering. 8th. John Wiley & Sons, 2019. ISBN: 9781119590170.

[61] Harry Nyquist. Regeneration Theory. In: Bell System Technical Journal 11.1 (1932), pp. 126–147. DOI: 10.1002/j.1538-7305.1932.tb02344.x. URL: https://onlinelibrary.wiley.com/doi/abs/10.1002/j.1538-7305.1932.tb02344.x.

[62] Katsuhiko Ogata. Discrete-Time Control Systems. 2nd. Englewood Cliffs, NJ: Prentice Hall, 1995. ISBN: 0130342815.

[63] Katsuhiko Ogata. Modern Control Engineering. 5th. Boston, MA: Prentice Hall, 2010. ISBN: 9780136156734.

[64] Katsuhiko Ogata. Solving Control Engineering Problems with MATLAB. 3rd. MATLAB Curriculum Series. Prentice-Hall, 1994. ISBN: 9780130263537.

[65] P. N. Paraskevopoulos. Modern Control Engineering. English. 1st. CRC Press, 2001, p. 752. ISBN: 9780824789817.

[66] J. C. Peyton-Jones and S. A. Billings. Recursive algorithm for non-linear system identification using the NARMAX model. In: International Journal of Control 49.6 (1989), pp. 1851–1869. DOI: 10.1080/00207178908559767.

[67] Charles L. Phillips and John M. Parr. Feedback Control Systems. 5th. Includes MATLAB examples and applications. Upper Saddle River, NJ, USA: Pearson, 2011. ISBN: 9780131866140.

[68] Charles L. Phillips, John M. Parr, and Eve Ann Riskin. Signals, Systems, and Transforms. 5th. Includes MATLAB examples and applications. Upper Saddle River, NJ: Pearson, 2014. ISBN: 9781292015286.

[69] V. C. Prasad. Compact Form of Mason's Gain Formula for Signal Flow Graphs. In: International Journal of Electrical Engineering Education 47.4 (2010). DOI: 10.7227/IJEEE.47.4.4.

[70] K. Deergha Rao. Signals and Systems. Includes MATLAB code examples. Cham, Switzerland: Springer, 2018. ISBN: 9783319686745.

[71] Francis H. Raven. Automatic Control Engineering. 5th. New York: McGraw-Hill, 1995. ISBN: 9780070513419.

[72] José A. Romagnoli and Ahmet Palazoglu. Introduction to Process Control. 3rd. Boca Raton, FL: CRC Press, 2020. ISBN: 9780367367787. DOI: 10.1201/9780429351396.

[73] Edward John Routh. A Treatise on the Dynamics of a System of Rigid Bodies. 5th revised and enlarged. London: Macmillan and Co., 1891.

[74] Wilson J. Rugh. Linear System Theory. 2nd. Prentice-Hall Information and System Sciences Series. Widely used in graduate-level courses on linear systems. Prentice Hall, 1996. ISBN: 9780134412054. URL: https://www.amazon.com/Linear-System-Theory-Wilson-Rugh/dp/0134412052.

[75] Lorenzo Sciavicco and Bruno Siciliano. Modelling and Control of Robot Manipulators. 2nd. Springer, 2000. ISBN: 9781852332212. DOI: 10.1007/978-1-4471-0449-0. URL: https://link.springer.com/book/10.1007/978-1-4471-0449-0.

[76] Clarence W. de Silva. Modeling and Control of Engineering Systems. Boca Raton, FL: CRC Press, 2009. ISBN: 9781420076868.

[77] Clarence W. de Silva. Modeling of Dynamic Systems with Engineering Applications. 2nd. CRC Press, 2022, p. 420. ISBN: 9780367644215. DOI: 10.1201/9780367644284.

[78] Naresh K. Sinha. Control Systems. 3rd. New Age International, 2008, p. 496. ISBN: 9788122411683.

[79] Sigurd Skogestad and Ian Postlethwaite. Multivariable Feedback Control: Analysis and Design. English. 2nd. Wiley, 2005, p. 592. ISBN: 9780470011676.

[80] Frederick O. Smetana. Introduction to the Control of Dynamic Systems. AIAA Education Series. American Institute of Aeronautics and Astronautics, 1994. ISBN: 9781563470837.

[81] Mark W. Spong, Seth Hutchinson, and M. Vidyasagar. Robot Modeling and Control. Hoboken, NJ: John Wiley & Sons, 2004. ISBN: 9780471649907.

[82] Gilbert Strang. Introduction to Linear Algebra. 5th. Wellesley, MA: Wellesley-Cambridge Press, 2016. ISBN: 9780980232776.

[83] A. K. Swain and S. A. Billings. Generalized Frequency Response Function Matrix for MIMO Nonlinear Systems. In: International Journal of Control 74.8 (2001), pp. 829–844.

[84] A. K. Swain, S. Devarakonda, and U. K. Madawala. Modeling, Sensitivity Analysis, and Controller Synthesis of Multipickup Bidirectional Inductive Power Transfer Systems. In: IEEE Transactions on Industrial Informatics 10.2 (2014), pp. 1372–1380. DOI: 10.1109/TII.2013.2287880. URL: https://ieeexplore.ieee.org/document/6522910.

[85] A. K. Swain, E. M. A. M. Mendes, and S. K. Nguang. Analysis of the Effects of Time Delay in Nonlinear Systems Using Generalised Frequency Response Functions. In: Journal of Sound and Vibration 294.1–2 (2006), pp. 341–354. DOI: 10.1016/j.jsv.2005.11.013.

[86] A. K. Swain, M. J. Neath, U. K. Madawala, and D. J. Thrimawithana. A Dynamic Multivariable State-Space Model for Bidirectional Inductive Power Transfer Systems. In: IEEE Transactions on Power Electronics 27.11 (2012), pp. 4772–4780. DOI: 10.1109/TPEL.2012.2185712. URL: https://ieeexplore.ieee.org/document/6186584.

[87] Louis K. Timothy. State Space Analysis. New York: McGraw-Hill, 1972.

[88] M. E. Van Valkenburg. Network Analysis. 2nd. Englewood Cliffs, NJ: Prentice-Hall, 1964.

[89] David D. Walden, Garry J. Roedler, Kevin J. Forsberg, R. Douglas Hamelin, and Thomas M. Shortell, eds. Systems Engineering Handbook: A Guide for System Life Cycle Processes and Activities. 4th. Hoboken, NJ, USA: John Wiley & Sons, Inc., 2015. ISBN: 9781118999400.

[90] Stanislaw H. Zak. Systems and Control. 1st. The Oxford Series in Electrical and Computer Engineering. New York: Oxford University Press, 2002. ISBN: 9780195150117.

[91] John G. Ziegler and Nathaniel B. Nichols. Optimum Settings for Automatic Controllers. In: ASME Transactions 64 (1942), pp. 759–768.

Index

For Product Safety Concerns and Information please contact our EU representative GPSR@taylorandfrancis.com Taylor & Francis Verlag GmbH, Kaufingerstraße 24, 80331 München, Germany

Batch number: 10774844

Printed by Printforce, the Netherlands